Probability and Statistics
for Computer Science

Probability and Statistics for Computer Science

JAMES L. JOHNSON
Western Washington University

WILEY-
INTERSCIENCE

A JOHN WILEY & SONS, INC., PUBLICATION

Published by John Wiley & Sons, Inc., Hoboken, New Jersey.
Published simultaneously in Canada.

For general information on our other products and services or for technical support, please contact our Customer Care Department within the United States at (800) 762-2974, outside the United States at (317) 572-3993 or fax (317) 572-4002.

Wiley also publishes its books in a variety of electronic formats. Some content that appears in print may not be available in electronic format. For information about Wiley products, visit our web site at www.wiley.com.

Library of Congress Cataloging-in-Publication Data is available.

ISBN 978-0-470-38342-1

10 9 8 7 6 5 4 3 2 1

To my muses
Shelley, Jessica, and Jimmy

Contents

Preface

This text develops introductory topics in probability and statistics with particular emphasis on concepts that arise in computer science. It starts with the basic definitions of probability distributions and random variables and elaborates their properties and applications. Statistics is the major application domain for probability theory and consequently merits mention in the title. Unsurprisingly, then, the text treats the most common discrete and continuous distributions and shows how they find use in decision and estimation problems. It also constructs computer algorithms for generating observations from the various distributions.

However, the text has a major subtheme. It develops in a thorough and rigorous fashion all the necessary supporting mathematics. This approach contrasts with that adopted by most probability and statistics texts, which for economy of space or for fear of mixing presentations of different mathematical sophistication, simply cite supporting results that cannot be proved in the context of the moment. With careful organization, however, it is possible to develop all the needed mathematics beyond differential and integral calculus and introductory matrix algebra, and this text purports to do just that.

Of course, as the book lengthens to accommodate the supporting mathematics, some material from the typical introduction to probability theory must be omitted. I feel the omissions are minor and that all major introductory topics receive adequate attention. Moreover, engagement with the underlying mathematics provides an opportunity to understand probability and statistics at a much deeper level than that afforded by mechanical application of unproved theorems.

Although the presentation is as rigorous as a pure mathematics text, computer science students comprise the book's primary audience. Certain aspects of most computer science curriculums involve probabilistic reasoning, such as algorithm analysis and performance modeling, and frequently students are not sufficiently prepared for these courses. While it is true that most computer science curriculums do require a course in probability and statistics, these courses often fail to provide the necessary depth. This text certainly does not fail in presenting a thorough grounding in elementary probability and statistics. Moreover, it seizes the opportunity to extend the student's command of mathematical analysis. This approach is different than that taken by other probability and statistics texts currently aimed at computer science curriculums. The more rigorous approach does require more work, both from the student and from the instructor, but the rewards are commensurate.

The engineering sciences, like computer science, also tend to use texts that place more emphasis on mechanical application of results than on the mathematical derivation of such results. Consequently, engineering science students will also benefit from the deeper presentation afforded by this text. Nevertheless, the primary audience remains computer science students because many of the illustrative examples are computer science applications. Therefore, from this point forward, I assume that I am addressing a computer science student or instructor.

Computer science students typically follow a traditional curriculum that includes one or two terms of probability and statistics, which follow prerequisite courses in differential and integral calculus and linear algebra. Although these prerequisite courses do introduce limit processes and matrix transformations, they typically emphasize formulas that isolate applications from the underlying theory. For example, if we drain a swimming pool with a sinusoidal cross-section, we can calculate how fast the water level falls without invoking limit operations. We simply set up a standard differential ratio and equate it to the drain flow rate. Why this works is buried in the theory and receives less and less emphasis once a satisfactory collection of calculation templates is available. This text provides an opportunity to reconnect with the theoretical concepts of these prerequisite courses. As it probes deeper into the properties of probability distributions, the text puts these concepts to fruitful use in constructing rigorous proofs.

The book's ambient prose deals with the principal themes and applications of probability, and a sequence of mathematical support modules interrupts this prose at strategic junctures. With some exceptions, these modules appear as needed by the probability concepts under discussion. A reader can omit the modules and still obtain a good grounding in elementary probability and statistics, including philosophical interpretations of probability and ample exercise in the associated numerical techniques. Reading the support modules will, however, strengthen this understanding and will also arouse an appreciation for the mathematics itself.

The encapsulation is as follows. An appendix gathers selected topics from set theory, limit processes, the structure of the real numbers, Riemann-Stieltjes integrals, matrix transformations, and determinants. The treatment first reviews the material at an introductory level. The prepared reader will be familiar with these concepts from previous courses, but the results are nevertheless proved in detail. The less prepared reader will certainly find frequent recourse to the appendix, and the text provides pointers to the appropriate sections. However, even the prepared reader will benefit from the introductory presentations, which serve both as a review of proof technique and as an introduction to the argument style pursued in the main text. Upon completing an introductory review, the appendix then extends the topics as necessary to support the arguments that appear in the main body of the text. Therefore, all chapters depend on the appendix for completeness. Even a reader well grounded in the aforementioned prerequisites can expect to spend some time mastering the specialized tools developed in the appendix.

The appendix, with its eclectic collection of review topics and specialized extensions, provides general mathematical background. There is need, however, for more specific supporting mathematics in connection with particular probabilistic and statistical concepts. Until perhaps halfway through the text, this supporting mathematics appears in mathematical interludes, which occur in each chapter. These interludes introduce particular results that are needed for the first time in that chapter. The first interlude deals with summation techniques, which are useful tools for the combinatoric problems associated with probability over equally likely outcomes. Others treat convergence issues in power series, stability features of Markov matrices, and sufficient statistics. Before taking up continuous distributions, however, it is appropriate to devote a full chapter to the mathematical issues that arise when one attempts to generalize discrete probability to uncountable sets and to the real line in particular. This chapter is actually a brief introduction to measure theory, and its logical place is just prior to the discussion of the common distributions on the real line. Two further interludes follow in subsequent chapters. They deal with limit theorems for continuous random variables and with decompositions of the sample variance. In short, the text exploits opportunities to introduce the mathematical analysis necessary to establish the basic results of probability theory. Moreover, the presentation clearly considers the mathematical analysis and the probability theory to be of equal importance.

The following sketch shows the dependencies among the chapters, with the understanding that portions of the appendix are prerequisite for any given path. The dashed boxes note the mathematical interludes within the chapters.

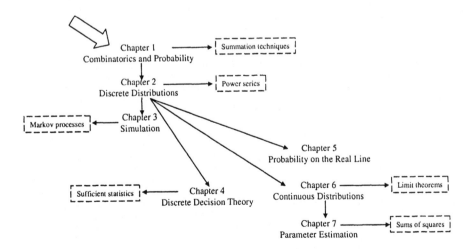

The reader can study the appendix in detail to ensure familiarity with all the background mathematics needed in the text, or can start immediately with the probability discussions of Chapter 1 and refer to the appendix as needed. Because of its breadth, the appendix is more difficult to master in its entirety than the mathematical interludes of the introductory chapters. A reader who prefers that the material increase monotonically in difficulty should start with the introductory chapters and digress into the appropriate appendix sections as needed. When using the text to support a course, an instructor should follow a similar path.

As noted earlier, the intended audience is computer science students. Once past colored balls in numbered urns, which constitute the traditional examples in combinatoric problems, the text uses examples that reflect this readership. Client-server performance evaluation, for instance, offers many opportunities for probabilistic analysis. These examples should provide no difficulty for other readers, such as students from the engineering sciences, because the examples make no profound references to advanced concepts, but rather use generally accessible quantities, such as terminal response time, server queue length, error count per 1000 programming statements, or operation count in an algorithm. These examples are no more difficult than those in a more general probability text that ventures beyond the traditional urns, colored balls, and dice.

The requirements of the Computer Science Accreditation Board and the Accreditation Board of Engineering Technology (CSAB/ABET) include a one-semester course in probability and statistics. This text satisfies that requirement. In truth, it is sufficient for a full-year course because it not only develops the traditional introductory probability concepts but also includes considerable material on mathematical reasoning. For a one-semester course, the following selection is appropriate. Note that the topics lie along an acceptable dependency chain in the earlier diagram.

Appendix. Sections as referenced in the items below

Chapter 1. Combinatorics and Probability

Chapter 2. Discrete Distributions

Chapter 3-4. Simulation, Sections 3.1 to 3.3, or Discrete Decision Theory, Sections 4.1 and 4.2

Chapter 6. Continuous distributions, Sections 6.1, 6.3, and 6.4

Chapter 7. Parameter Estimation, Sections 7.1, 7.2, and 7.4

The one-semester abbreviation is possible because Chapters 3 and 4 present major applications of discrete probability and, in the interest of time, need only be sampled. Chapter 5 is advanced material that elaborates the difficulties in extending discrete probability to uncountable sample spaces. It is present for logical completeness and to answer the nagging question that occurs to many students: Is the introduction of a sigma-algebra really necessary in the general definition of a probability space? Consequently, the proposed one-semester course omits Chapter 5 with minimal impact on subsequent material. Finally, Chapter 7 undertakes major applications of continuous probability and also admits partial coverage.

At the time of this writing, many computer science curriculums include only a first probability course. However, there is a recognized need for further study, at least in the form of an elective second course, if not in a required sequel to the introductory course. Anticipating that this increased attention will also expose the need for a more complete mathematical treatment of the material, I have provided unusually detailed excursions into supporting topics, such as estimation arguments with limits, properties of power series, and Markov processes.

Buttressed by these mathematical excursions, the text provides a thorough introduction to probability and statistics—concepts, techniques, and applications. Consequently, it offers a continuing discussion of the real-world meaning of probabilities, particularly when the frequency-of-occurrence interpretation becomes somewhat strained. Any science that uses probability must face the interpretation challenge. How can you apply a result that holds only in a probabilistic sense to a particular data set? The text also discusses competing interpretations, such as the credibility-of-belief interpretation, which might appear more appropriate to history or psychology. The goal is, of course, to remain continually in touch with the real-world meaning of the concepts.

Probability as frequency of occurrence over many trials provides the most compelling interpretation of the phenomenon. It is intuitively plausible, for example, that a symmetric coin should have equal chances of landing heads or tails. The text attempts to carry this interpretation as far as possible. Indeed, the first chapter treats the combinatorics arising from symmetric situations, and this treatment serves as a prelude to the formal definitions of discrete probability. As the theory accumulates layer upon layer of reasoning, however, this viewpoint becomes difficult to sustain in certain cases. When testing a hypothesis, for example, we attempt to infer the prevailing state of nature from sampled data. What does it mean to assign *a priori* probabilities to the possible states? This practice allows statisticians to incorporate expert judgment into the decision rules, but the assigned probabilities do not admit a frequency-of-occurrence interpretation. Rather, they reflect relative strength-of-belief statements about the possible states. As necessary, the text interrupts the technical development to comment on the precise real-world interpretation of the model. Although beautiful as abstract theory, probability and statistics are also rightly praised for their ability to deliver meaningful statements about the real world. Interpreting the precise intent of these statements should be a primary goal of any text.

A trend in modern textbooks, particularly those not addressed specifically to a mathematics curriculum, is to avoid the theorem-proof presentation style. This style can be sterile and detached, in the sense that it provides sparse context for the motivation or application of the theorems. Without the theorem-proof style, on the other hand, arguments lose some precision, and there is a blurring of the line between the general result and its specific applications. I have adopted what I consider a middle ground. I maintain a running prose commentary on the material, but I punctuate the dialog with frequent theorems. Often, the theorem's proof is a simple statement: "See discussion above." This serves to set off the general results, and it also provides reference points for later developments. The ambient prose remains connected with applications and with the questions that motivate the search for new general results.

Plentiful examples, displayed in a contrasting typographical style, play a major role in compensating for the perceived coldness of the theorems. Incidentally, I should say that I do not find the theorems cold, even in isolation. But I am responding to the spirit of the age, which suggests that a theorem wrapped in an example is more digestible than a naked theorem.

The theorems also further a second ambition, noted above, which is to involve the reader more extensively in precise mathematical argument. An aspect of proofs that attracts major criticism is the tendency to display, out of thin air, an expression that magically satisfies all the required constraints and invites the algebraic manipulations necessary to complete the proof. I have tried to avoid this practice by including some explanation of the mysterious expression's origin. I must admit, however, that I am not always successful in this ploy. Sometimes an explanation adds nothing to a careful contemplation of the expression. In such cases, I am tempted to suggest that the reader reflect on the beauty of the expression, note how one part attaches to the known information while another extends toward the desired result, and view the expression as a unifying link, growing naturally from a study of the context of the problem in question. Instead, however, I fall back on the age-old practice: "Consider the following expression ...." The reader should take these words as an invitation to pause and ponder the situation.

In summary, the text develops a main theme of probability and statistics, together with the mathematical techniques needed to support it. Since it is not practical to start with the Peano axioms, there are, however, some prerequisites. Specifically, the text's mathematical level assumes that the reader has mastered differential and integral calculus and has some exposure to matrix algebra. Nevertheless, acknowledging the mechanical fashion in which these subjects are taught these days, the text provides considerable detail in all arguments. It does assume, nevertheless, that readers have some familiarity with limiting operations, even if they do not have significant experience with the concept. For example, readers should be comfortable with l'Hôpital's rule for evaluating limits that initially appear to produce indeterminate results. By contrast, the text does develop the theory of absolutely convergent series to the point of justifying the interchange of summation order in double summations.

As another example, the readers should be generally conversant with power series, although the text develops this topic sufficiently to justify the term-by-term differentiation that is needed to recover parameters from a random variable's moment generating function. The background appendix and the mathematical interludes should bridge the gap between prerequisite knowledge and that needed to establish all probability and statistical concepts encountered. They also serve to present a self-contained book, which is my preference when learning new material. A reader can always skip sections that are peripheral to the main point, but cannot as easily fill in omissions.

Some expositions consist of step-by-step procedures for solving a probabilistic or statistical problem. That is, they involve algorithms. For example, algorithms appear in sections concerned with computer simulations of probabilistic situations. In any case, algorithms in this text appear as mutilated C code, in the sense that I vary the standard syntax as necessary to describe the algorithm most clearly to a human reader. For instance, I use only two iterators, the while-loop and the for-loop, and in each case, the indentation serves to delineate the body of statements under repetition. The left fragment below, intentionally meaningless to focus attention on the code structure, must be reformulated as shown on the right to actually compile properly.

```
while (X > Y)                     while (X > Y) {
    for (j = 1; j <=; j++)            for (j = 1; j <=; j++) {
        Y = Y - t[j];                    Y = Y - t[j];
        t[j] = t[j] + 1.0;               t[j] = t[j] + 1.0;
    if (Y < 0.0)                     }
        Y = 0.0;                     if (Y < 0.0)
                                         Y = 0.0;
                                  }
```

This practice amounts to omitting the opening and closing braces, and it is surprisingly efficient in reducing the code length. The price, of course, is that the code will not execute as written. I also omit variable declarations and simply use any variables that I need in an untyped manner. So a variable, or a function return, can be a scalar, a vector, a matrix, or a list. A particular usage is always clear from context, and these omissions, like those of scope enclosures, produce more compact code. In summary, the algorithms in this text are intended for human consumption. The emphasis is on the concept, not the detailed syntax of a programming language. Nevertheless, I have constructed all the algorithms in functional code and executed them to ensure that they perform as promised.

I emphasize that this text requires no C programming background. A C-based style finds frequent use in books and journals as a concise and precise algorithm presentation method for audiences that are not C literate. This widespread usage confirms the opinion that loose C structures are easily accessible to readers who do not have a background knowledge of C. Consequently, a reader with no previous C experience should not feel anxious about the algorithm descriptions in this text. Although readers must want to understand the algorithm and must expend the time to attend carefully the iterative processes in the C expression, they will find the algorithm comprehensible. Moreover, explanatory prose accompanies all C-structure descriptions.

Another feature that will be appreciated by today's students is the text's early emphasis on discrete distributions. Most introductory probability books quickly proceed to the normal distribution and the central limit theorem because these tools enable certain computational approximations. I certainly do not wish to imply that the normal distribution or the central limit theorem is not important. These topics are developed in detail in the latter half of the book. However, much reasoning with discrete distributions can take place without replacing the discrete distribution with a more tractable normal distribution. This is especially true today when computers complete the mechanical, but sometimes tedious, computations. Working directly with discrete distributions develops a feeling for the subtleties of probability theory that can be overlooked in a rush to obtain approximate results. The many examples involving discrete distributions will sharpen a reader's capabilities for combinatorial reasoning and will exercise his or her skills in discrete mathematical argument. These skills are especially important in today's scientific world, which grows ever more digital, and therefore ever more discrete.

In keeping with this approach, the text develops simulation techniques and certain aspects of statistical inference directly after introducing a repetoire of discrete distributions but before developing continuous distributions. This means that certain proofs, which admit more elegant expositions with continuous tools, contain detailed arguments with limits. These arguments are conceptually similar to those associated with calculus (e.g., passing to an integral limit from a summation), but the text nevertheless presents more intermediate steps than one normally finds in a textbook

All definitions, theorems, and examples share a common numbering sequence within each chapter. To find Theorem 3.14, for example, you can use any definition, theorem, or example to direct your search forward or backward. I find it a considerable annoyance when these items are numbered separately. Of course, this means that there may be a Theorem 3.14 even though there is no Theorem 3.13. I try to make this dissonance more acceptable by placing the number before the theorem, definition, or example. Having located Theorem 3.14, for example, you find that it starts "3.14 Theorem. ...." In this reading, Theorem 3.14 is an item, which happens to be a theorem. There is indeed an item 3.13, which may or may not be a theorem. In the traditional manner, tables and figures exhibit separate numbering sequences within each chapter.

Chapter 1

Combinatorics and Probability

The computer and engineering sciences investigate many nondeterministic phenomena. In computer science, examples include an algorithm's running time, the queue length encountered by a request arriving at a database server, and the delay in transferring a data block from disk to main memory. Currents in semiconductor devices, an important concern in electrical engineering, depend on the energy distributions of charge carriers. Biological evolution appears to involve random mutations in cells' offspring. In any applied science, measurement error adds an uncontrollable variation to the simplest data-gathering activity. Therefore, it is common laboratory practice to average a sequence of measurements to reduce random effects.

We also use probabilistic terms to describe certain ordinary events of everyday life. The chance of rain today is 40%. The probability of heads on a coin toss is 1/2. The odds that team X will win a sports event are 3 : 1. Once in every thirty-six throws, approximately, a pair of dice will exhibit the minimum total of two spots on the upturned faces. In all these cases, we understand that some repetitive process generates outcomes that are not deterministic but nevertheless exhibit a rational pattern. For example, each rotation of the planet brings a new day, on which rain may or may not occur. We interpret a 40% chance of rain as meaning that over an extended run of days, 40% of them will be rainy.

In this initial informal discussion, we use the term *probability* in the ordinary sense implied by the examples above. That is, the probability of an event is its relative frequency of occurrence over many observations. Probability is simply a synonym for chance. The probability of rain is 40%. The chance of rain is 40%. Both mean that similar meteorological conditions in the past have produced rain in 40% of the observations. More exact mathematical definitions appear later in the chapter, and they are consistent with the intuitive meanings discussed here.

When faced with a repeating phenomenon that generates varying outcomes, we can adopt one of two viewpoints. On the one hand, we can argue that the phenomenon does not actually repeat. It only appears to do so because our knowledge is not sufficiently detailed to distinguish between iterations. On the other hand, we can accept the variations and plan accordingly. With the latter approach, we are indifferent as to whether the variations arise because we are not sufficiently informed or because the phenomenon is fundamentally nondeterministic. Consider again the weather forecast that predicts a 40% chance of rain under certain meteorological conditions. The first viewpoint suggests that the meteorological conditions are merely approximations of the true state of nature. It implies that we should direct our efforts toward attaining complete knowledge of nature, which would allow a prediction with 100% certainty. The second attitude suggests that we should live with the uncertainty. We should lay plans that make provisions for alternative outcomes. Indeed, an adherent of the first viewpoint must frequently follow the second course because complete knowledge of nature is inaccessible. Most persons find it practical to take a decision about carrying an umbrella without waiting for meteorologists to acquire complete knowledge of the weather.

Probability theory provides a rigorous mathematical framework for investigating phenomena that exhibit repeatable patterns, even though individual outcomes may appear random. As Hamlet put it, "Though this be madness, yet there is method in't." Probability theory provides a lens that enables us to see the patterns underlying an apparently chaotic surface, a lens that tames the madness to reveal the method.

Knowledge of repeatable patterns can be very useful, even if there are occasional disappointments. If I know that team X will play an extended sequence of games and that the chances of winning remain 3 to 1 throughout the series, I might find a betting office that will accept the following proposition. I will pay the office $10 for each game. When team X wins a game, the office will pay me $15; when team X loses, I get nothing back. Suppose that the team plays 12 games and in keeping with the odds, wins 3 for each loss. That is, the teams wins 9 games and loses 3. I pay $120, and I receive $135, for a net gain of $15. Of course, the betting office can run this calculation as well as I can, and therefore it would not accept my proposition. In truth, the betting office can run the same calculation only if it also knows that the 3 : 1 odds are an accurate description of the situation. So informed bets are made only between parties who have different perceptions of the underlying odds.

Long study and experience can produce credible estimates of the odds for a sports event, or for the chance of rain on a given day, but the results cannot be completely accurate. Other probabilistic descriptions, however, can be precise, at least in an idealized situation, because they describe certain favorable combinations as a fraction of all possible outcomes. Only two outcomes, for example, are possible in a fair coin toss, heads and tails. Therefore, the probability of heads in a fair toss is 1/2.

Similarly, rolling a pair of dice will produce one of the thirty-six patterns in the grouping that follows, where the ordered pair (x, y) means that the first die shows x spots on its upturned face and the second die shows y. We can infer the probabilities of certain simple events from direct observation of the display.

For example, noticing that the outcomes summing to seven are the diagonal entries, running from the lower left to the upper right, we conclude that a seven total has probability 6/36. The calculations for other results are equally

$$
\begin{array}{cccccc}
(1,1) & (1,2) & (1,3) & (1,4) & (1,5) & (1,6) \\
(2,1) & (2,2) & (2,3) & (2,4) & (2,5) & (2,6) \\
(3,1) & (3,2) & (3,3) & (3,4) & (3,5) & (3,6) \\
(4,1) & (4,2) & (4,3) & (4,4) & (4,5) & (4,6) \\
(5,1) & (5,2) & (5,3) & (5,4) & (5,5) & (5,6) \\
(6,1) & (6,2) & (6,3) & (6,4) & (6,5) & (6,6)
\end{array}
$$

straightforward. The probability of a three total is 2/36. The probability that the absolute difference between the two dice is greater than two is 12/36 because the favorable outcomes are the upper-right and lower-left triangles, each containing six entries.

Probabilistic quantities appear frequently in the computer and engineering sciences. To the examples noted earlier we can add the time required for a computer to respond to a command from an interactive terminal, the number of errors in a program module, the number of polls to find a receptive device, the time to access a transmission bus, the number of memory bits flipped by cosmic radiation, or the number of compare operations executed by a sort algorithm. In many cases, it is difficult to calculate meaningful probability weights. For example, suppose that you measure the system response time T to a terminal command by directly observing a number of such operations. You then have an empirical estimate of the fraction of cases for which $T = 2$ seconds. But is it possible to compute this fraction in advance from your knowledge of the system parameters? Such a computation is not likely; it depends on circumstances that are neither repeatable nor easily controlled, such as parallel activity from other system users. By contrast, the number of compare operations in a sort algorithm is subject to analysis. The possible inputs of size N are the permutations of the numbers $1, 2, \ldots, N$, and it is reasonable to assume that all such inputs are equally likely. For a specific input, the algorithm determines exactly the number of compare operations, and we can therefore proceed by direct count to enumerate the fraction of cases for which the count is a particular n. This chapter opens our study of probability by considering just such situations, where we can compute the relevant probabilities by exhaustive counting.

Each observation of a nondeterministic phenomenon is called a *trial*. In symmetric situations, such as coin tosses or dice rolls, we can reasonably assume that each elementary outcome appears with equal frequency in a long sequence of trials. However, as we associate trials in various ways, the equal frequencies quickly disappear. For example, a tossed coin displays either a head or a tail, and we expect a sequence of tosses to exhibit both results with approximately equal frequencies. But, if we change the definition of a trial to cover all tosses necessary to obtain a head, each observation is a number of tails followed by a head. It is not likely that all such observations are equally likely. Do you feel that 6 tails followed by a head will occur as often as 2 tails followed by a head? This chapter's early sections develop the tools to answer such questions. In the process, we obtain an intuitive appreciation for the probability of an event, such as 6 tails followed by a tail, as its relative frequency of occurrence in an extended sequence of observations.

In a particular application, the usual difficulty is the systematic counting of the total possible outcomes and of the outcomes favorable to a certain result. Unlike the simple examples cited above, most applications provide a very large field of possible outcomes, and it is not feasible to list them explicitly. Before launching into the details, we take the time to elaborate three counting principles. All are immediately obvious, so you may feel that there is no need to state them explicitly. In complicated situations, however, when you cannot seem to find a starting point for a solution, you might well return to these simple principles. The first principle suggests breaking the mass to be counted into disjoint pieces and then summing the subcounts from these pieces. That is, if you want to count rooms within a building, you first perform subcounts on each floor and then add the subcounts. In applying this principle, you need to ascertain that there is no overlap among the pieces, which could lead to overcounting some elements. The second principle, known as the multiplicative principle, is actually a special case of the first principle. It again breaks the mass into disjoint pieces, but now such that each piece contains the same number of elements. The total count is then the number of pieces times the count of any one piece. If, for example, you want the number of houses on a rectangular site, you count the houses in one row and multiply by the number of rows. The third principle, known as the pigeonhole principle, states that the allocation of more than n pigeons to n pigeonholes results in multiple occupancy for at least one pigeonhole. If we randomly allocate four balls among three containers, what is the probability that *all* containers receive at most one ball? By the pigeonhole principle, it is zero. Some container must receive at least two balls. Consider a related question: What is the probability that the leftmost container receives at most one ball? Now, that is a more complicated story, which we now begin.

1.1   Combinatorics

Many situations call for counting the number of ways to select, under specified constraints, objects from a given set. When we say that a set contains n objects, we understand, in keeping with the usual definition of a set, that the objects are distinct. We may wish, nevertheless, to disregard distinctions among certain elements. For example, in a set of 7 balls, we can have 4 reds and 3 blacks. In this case, we can adopt a notation that both captures the elements' distinct identities and emphasizes the two internal classes of red and black. The designation $\{r_1, r_2, r_3, r_4, b_1, b_2, b_3\}$ serves this purpose, but other schemes are equally valid. The best notation varies with the context of the problem at hand.

In some manner or other, a problem context always specifies three gen-

Comb.	Sequences
abc	$abc, acb, bac, bca, cab, cba$
abd	$abd, adb, bad, bda, dab, dba$
acd	$acd, adc, cad, cda, dac, dca$
bcd	$bcd, bdc, cbd, cdb, dbc, dcb$

(a) Sampling without replacement

Comb.	Sequences	Comb.	Sequences
abc	$abc, acb, bac, bca, cab, cba$	bbc	bbc, bcb, cbb
abd	$abd, adb, bad, bda, dab, dba$	bcc	bcc, cbc, ccb
acd	$acd, adc, cad, cda, dac, dca$	bbd	bbd, bdb, dbb
bcd	$bcd, bdc, cbd, cdb, dbc, dcb$	bdd	bdd, dbd, ddb
aab	aab, aba, baa	ccd	ccd, cdc, dcc
abb	abb, bab, bba	cdd	cdd, dcd, ddc
aac	aac, aca, caa	aaa	aaa
acc	acc, cac, cca	bbb	bbb
aad	aad, ada, daa	ccc	ccc
add	add, dad, dda	ddd	ddd

(b) Sampling with replacement

TABLE 1.1. 3-combinations and their related 3-sequences from a field of 4 symbols

eral constraints and perhaps further particular constraints. The general constraints answer the questions: How many objects are selected? Does the selection order make a difference? After drawing an object and recording the result, do we return the object to the set before drawing the next item? The responses to some questions affect the possible answers to others. For example, if you return objects to the common pool after each selection, then the total number of objects drawn can exceed the pool size.

1.1 Definition: An ordered selection of k objects is a k-*sequence*. An unordered selection of k objects is a k-*combination*. In the process of *sampling with replacement*, we note the identity of each object as it is drawn, but we return it to the common pool before the next draw. In the alternative process, *sampling without replacement*, we do not return the drawn object to the pool. ∎

From the set $\{a, b, c, d\}$, we generate four 3-combinations and twenty-four 3-sequences when sampling without replacement. The upper tabulation of Table 1.1 suggests a 6-to-1 relationship between the sequences and the combinations. Sampling with replacement, on the other hand, yields twenty 3-combinations and sixty-four 3-sequences, as illustrated in the lower tabulation. If there is a relationship between a combination and the sequences involving the same choices, it is less apparent. Duplications appear in both sequences and combinations when choosing with replacement, and this complicates the count.

As noted earlier, we use the term *probability* for relative frequency of occurrence. Suppose, for example, that we are generating sequences without replacement in the context of Table 1.1. Note that 4 of the 24 possible sequences contain the ordered pair "ab." Consequently, we say that the probability of "ab" occurring as a subsequence is $4/24 = 1/6$. This policy agrees with our intuition when all sequences are equally likely. When, in due course, we come to a formal definition of probability, we will find that it conforms with this frequency-of-occurrence notion.

1.1.1   Sampling without replacement

Under sampling without replacement, neither a sequence nor a combination exhibits a duplicate entry. Sequences, however, have more structure because the insertion order is important. Our goal is to find systematic methods for counting sequences and combinations under sampling without replacement, but also under a variety of further constraints.

We start with sequences. Suppose that we have seven tiles, labeled with the letters: a b c d e f g. Suppose further that we consider a word to be any 3-sequence. How many such words can we compose with the seven tiles? Possible words are abc, cag, def, fed, and the like. We imagine a repetitive process. On each trial, the process selects three tiles from the initial seven and assembles them in the order selected to form a word. There is no replacement as the tiles are selected. However, all seven tiles are reinstated at the beginning of each trial. With this clarification, the desired count is the number of 3-sequences from a set of seven symbols, where the selection process is without replacement. We apply the multiplicative counting principle that breaks the possibilities into groups of the same size. In particular, a systematic count arises from partitioning the words into seven groups—those beginning with a, those beginning with b, and so forth. Within the first group, where the first letter is a, we distinguish six subgroups: those with second letter b, those with second letter c, and so forth through those with second letter g. In the second group, where the first letter is b, we again distinguish six subgroups: those with second letter a, those with second letter c, and so forth through those with second letter g. An important observation is that each group divides into the same number of subgroups. Table 1.2 organizes these subdivisions.

Within each subgroup, where the first two letters are now specified, there remains a choice of one of the five remaining letters for the last tile. Hence each of the 42 subgroups divides naturally into five further pieces, each distinguished by one of the remaining five letters. So, if we choose three tiles randomly from the group of seven and lay them out in the order chosen, we will form one of exactly $7 \cdot 6 \cdot 5 = 210$ three-letter words.

In terms of Definition 1.1, this demonstration shows that there are exactly 210 possible ways of choosing, without replacement, a sequence of length 3 from a group of size 7. Now suppose that we want the probability that a three-letter sequence chosen without replacement from a field of seven letters has its first two letters in alphabetical order. In other words, we want the fraction of the 210 sequences that have the first two letters in order. Referring

First letter	Second letter	Possible third letters
a	b	c d e f g
	c	b d e f g
	d	b c e f g
	e	b c d f g
	f	b c d e g
	g	b c d e f
b	a	c d e f g
	c	a d e f g
	d	a c e f g
	e	a c d f g
	f	a c d e g
	g	a c d e f
c	a	b d e f g
	b	a d e f g
	d	a b e f g
	e	a b d f g
	f	a b d e g
	g	a b d e f
⋮	⋮	⋮

TABLE 1.2. Organizing the three-letter words from seven distinct tiles

to the organization above, we see that the first group, containing $6 \cdot 5 = 30$ sequences, meets the criterion. Since the first letter is a, the first two letters must be in order, regardless of the remaining letters. Sequences in the second group, however, begin with b, so the first subgroup, where the second letter is a, must be excluded. The second group then contributes $5 \cdot 5 = 25$ sequences that meet the criterion. The third group starts with c, so we must exclude its first two subgroups, where the second letter is a or b. Hence, we obtain only $4 \cdot 5 = 20$ sequences here. The pattern is clear: Each group must exclude one more subgroup than its predecessor. The total number of sequences with the first two letters in order is then $6 \cdot 5 + 5 \cdot 5 + 4 \cdot 5 + 3 \cdot 5 + 2 \cdot 5 + 1 \cdot 5 + 0 \cdot 5 = 105$, which gives a probability of $105/210 = 1/2$.

Is this result surprising? The number of sequences with the first two letters in order is exactly one-half of the total number of three-letter sequences. A so-called "sanity check" is always advisable. This means that we should try to confirm by some other method that the result is correct or at least reasonable. An obvious check in this case is that the result is a proper fraction, between zero and 1, because it represents the ratio of some number of selected possibilities to the total number of possibilities. So the result is reasonable in this sense. However, we can argue further that it should be $1/2$ exactly. Suppose that we reorganize the three-letter sequences into two different groups: G_1 contains those sequences with the first two letters in order and G_2 contains those with the first two letters out of order. We can define a function $f : G_1 \to G_2$ by $f(xyz) = yxz$, where xyz is a three-letter sequence from G_1. This means that $x < y$ alphabetically, so yxz is indeed a sequence in G_2. This function establishes a one-to-one correspondence between the two sets, so they must have the same number of elements. Using the notation $|X|$ to indicate the number of elements in a set, we have that the probability of a three-letter sequence having its first two letters in order is $|G_1|/|G_1 \cup G_2| = |G_1|/(2|G_1|) = 1/2$.

With some additional notation, we can extend the example to a general result. Recall that $n!$ means the product of the integers $n \cdot (n-1) \cdot (n-2) \cdots 1$ and that we define $0! = 1$. The notation $P_{n,k}$ will mean the first k factors in the expansion for $n!$.

1.2 Definition: For integers $n > 0, k \geq 0$, $P_{n,k}$ denotes the kth *falling factorial* of n. It is the product $n \cdot (n-1) \cdot (n-2) \cdots (n-k+1)$. For the moment, we think of the "P" as an abbreviation for "product." As a special boundary case, we define $P_{n,0} = 1$. ∎

For $k > n$ the expansion contains a zero factor, and therefore $P_{n,k} = 0$. For $0 \leq k \leq n$, we have

$$
\begin{aligned}
P_{n,k} &= n \cdot (n-1) \cdot (n-2) \cdots (n-k+1) \\
&= \frac{n \cdot (n-1) \cdots (n-k+1) \cdot (n-k) \cdots 2 \cdot 1}{(n-k) \cdot (n-k-1) \cdots 2 \cdot 1} = \frac{n!}{(n-k)!} \\
P_{n,n} &= n \cdot (n-1) \cdot (n-2) \cdots 2 \cdot 1 = n!.
\end{aligned}
$$

1.3 Theorem: Let $n > 0$ and $k \geq 0$. Drawing without replacement from a set S containing n symbols, we can construct exactly $P_{n,k}$ different k-sequences. PROOF: If $k = 0$, then $P_{n,k} = 1$, and there is indeed just one 0-sequence—the empty sequence. If $k > n$, then $P_{n,k} = 0$, but there are no possible k-sequences that we can select without replacement from the pool of n symbols. The theorem is therefore correct for $k = 0$ and for $k > n$. For k in the intermediate range, we proceed by induction.

For $k = 1$, we count exactly n 1-sequences, each being one of the symbols from S. Since $P_{n,1} = n$, the theorem is correct in this base case. If $k > 1$, divide the possible k-sequences into n groups, each distinguished by its starting symbol. The sequences within a group start with a specified first symbol and continue with some sequence of length $k - 1$ from the remaining $n - 1$ symbols. Hence the number of sequences in the group is the number of continuation sequences, which is $P_{n-1,k-1}$ from the induction hypothesis. The total number of k-sequences is then

$$
n \cdot P_{n-1,k-1} = \frac{n(n-1)!}{[n-1-(k-1)]!} = \frac{n!}{(n-k)!} = P_{n,k}. \blacksquare
$$

1.4 Definition: A rearrangement of the ascending sequence $1, 2, 3, \ldots, n$ is called a *permutation* of these numbers. ∎

An immediate application of Theorem 1.3 shows that there are exactly $n!$ permutations of the numbers $1, 2, 3, \ldots, n$. The next example provides a more extensive application and also suggests a more useful formula—one that deals with combinations rather than sequences.

Card			In-order hands	Sub-total
1	2	3		
2	3	4–A	11	
	4	5–A	10	
	5	6–A	9	
	6	7–A	8	
	7	8–A	7	
	8	9–A	6	
	9	10–A	5	
	10	J–A	4	
	J	Q–A	3	
	Q	K–A	2	
	K	A	1	
	A	–	0	66
3	4	5–A	10	
	5	6–A	9	
	6	7–A	8	
	7	8–A	7	
	8	9–A	6	
	9	10–A	5	
	10	J–A	4	
	J	Q–A	3	
	Q	K–A	2	
	K	A	1	
	A	–	0	45
⋮	⋮	⋮	⋮	

Card			In-order hands	Sub-total
1	2	3		
⋮	⋮	⋮	⋮	
8	9	10–A	5	
	10	J–A	4	
	J	Q–A	3	
	Q	K–A	2	
	K	A	1	
	A	–	0	15
9	10	J–A	4	
	J	Q–A	3	
	Q	K–A	2	
	K	A	1	
	A	–	0	10
10	J	Q–A	3	
	Q	K–A	2	
	K	A	1	
	A	–	0	6
J	Q	K–A	2	
	K	A	1	
	A	–	0	3
Q	K	A	1	
	A	–	0	1	
K	A	–	0	0
A	–	–	0	0
			Grand total:	286

TABLE 1.3. In-order three-card hands from thirteen spades (see Example 1.5)

1.5 Example: A dealer removes all cards from a standard deck, except the spades, shuffles the remaining cards, and deals you three cards. What is the probability that you receive your cards in a sequence of increasing value?

As you receive your cards, you keep them. Therefore, you receive a 3-sequence, chosen without replacement from a set of 13. Theorem 1.3 then says that you receive one of the $P_{13,3} = 1716$ possible sequences. To count the in-order sequences, we apply the multiplicative counting principle to elaborate several disjoint categories. If the deal starts $(2,3)$, then any of the remaining 11 cards completes an in-order hand. If the deal starts $(2,4)$, then only 10 of the remaining cards complete an in-order hand because we must exclude the 3. Similarly, only 9 cards successfully complete a hand that starts with $(2,5)$. Thus, the number of in-order hands starting with 2 is $11 + 10 + 9 + \ldots + 1 + 0 = 66$. Continuing with the hands that start with 3, we organize the in-order hands as shown in Table 1.3, which produces a grand total of 286 in-order hands. The required probability is then $286/1716 = 1/6$. So you should receive your three cards in increasing order about once in every six deals. Can you think of a sanity check for why this should be so?

Suppose that we ignore card order for the moment. We divide all possible 3-sequences into groups that contain the same cards. Suppose that there are N such groups. Each group contains 3-sequences that are rearrangements of the same three cards, so according to Theorem 1.3, each group contains $P_{3,3} = 3! = 6$ sequences, and only one of them can be in increasing order. So there must be N sequences in increasing order and a total of $6N$ sequences all together. The probability of receiving an in-order hand is then $N/(6N) = 1/6$. □

We have learned that a set of n distinct symbols admits $P_{n,k}$ k-sequences when they are chosen without replacement. These sequences are, of course, ordered displays of k symbols. A combination, however, is an *unordered* display, which we can visualize as a *subset* of the original set. Suppose that we want to count instead the number of k-combinations. That is, how many k-combinations can we construct if we choose without replacement from n distinct symbols? We follow the suggestion at the end of Example 1.5 to prove the following theorem. First, however, we introduce a notational convenience.

1.6 Definition: For integers $n > 0, k \geq 0$, $C_{n,k}$, pronounced "n choose k," is defined as $P_{n,k}/k!$. For the moment, think of the "C" as denoting "choose." ∎

If $k > n$, then $P_{n,k} = 0$ and consequently $C_{n,k} = 0$. If $0 \leq k \leq n$, then $C_{n,k} = n!/[k!(n-k)!]$. From this formula, it follows that $C_{n,k} = C_{n,n-k}$ for $0 \leq k \leq n$.

1.7 Theorem: Let $n > 0$ and $0 \leq k \leq n$. Choosing without replacement, we can construct exactly $C_{n,k}$ different k-combinations from a set of n symbols. PROOF: By Theorem 1.3 the set admits $P_{n,k}$ k-sequences. If we collect together all sequences that are merely different orderings of the same symbols, we partition the $P_{n,k}$ sequences into N collections. Each collection represents a different k-combination, and, moreover, each k-combination must appear as one of the collections. Therefore, N counts the number of k-combinations. How large are these collections? Suppose that C is one of the collections. Every sequence in C contains the same k symbols. The sequences differ only in the ordering of these symbols. We apply Theorem 1.3 again to conclude that C contains exactly $P_{k,k} = k!$ sequences. This same reasoning applies to the other collections, so each of them is of size $k!$. Then $P_{n,k} = N \cdot k!$, or $N = P_{n,k}/k! = n!/[(n-k)!k!] = C_{n,k}$. ∎

The theorem applies when $k = 0$ because then $C_{n,0} = n!/[0!(n-0)!] = 1$ and there is indeed just one subset of size zero, namely the empty set. It is not possible to select a subset with more than n objects, so $C_{n,k} = 0$ is correct for $k > n$.

A useful interpretation envisions $C_{n,k}$ as the number of patterns obtained by darkening k slots from a field of n. For example, $C_{4,2} = 6$ produces the patterns to the right. In choosing 2 slots from a field of 4, you may, for example, choose slot 1 and then slot 4, or you may choose slot 4 followed by slot 1. In either case, you get the fourth pattern. The pattern is sensitive only to the particular 2-combination, $\{1,4\}$, not to the order in which the slots were chosen. In this sense, exactly 6 patterns are possible. The next example shows how to use this interpretation to count the possibilities from a large field of values.

1.8 Example: You receive 13 cards from a well-shuffled deck. What is the probability that your hand contains 2 or more aces? Because you keep your cards as you receive them, the process unfolds without replacement. Moreover, you are concerned only with the particular 13-combination that you receive. The order in which you receive the cards is not important. Therefore, the total possible outcomes number $C_{52,13}$, the number of 13-combinations from the deck of 52. This is the number of patterns over 52 slots in which 13 are darkened. To count the number of 13-combinations that contain exactly two aces, we imagine the 52 cards arranged linearly over 52 slots, but with some further structure. Specifically, we put the four aces over the first four slots (spades, hearts, diamonds, and then clubs). The remaining 48 cards occupy slots 5 through 52, in an arbitrary order. The following diagram illustrates some sample choices that yield a 13-card hand with exactly 2 aces.

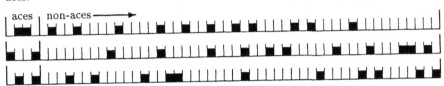

In the top pattern, aces fill slots 2 and 3, while the remaining 11 cards come from slots 6, 9, 14, 19, 22, 25, 28, 30, 35, 37, and 42. The next pattern takes aces from slots 1 and 4 with the other 11 cards from slots 13, 19, 25, 29, 32, 34, 40, 44, 48, 49, and 51. The last pattern shows aces from slots 2 and 4 and another selection for the remaining cards. In general, a valid pattern arises from any of $C_{4,2}$ subpatterns in the first four slots associated with any of $C_{48,11}$ subpatterns from the remaining slots. Hence, the number of 13-card hands that contain exactly 2 aces is $C_{4,2}C_{48,11}$. Similarly, the number of hands containing exactly 3 aces is $C_{4,3}C_{48,10}$. The number for 4 aces is $C_{4,4}C_{48,9}$. The probability of obtaining a hand with 2 or more aces is then $[C_{4,2}C_{48,11} + C_{4,3}C_{48,10} + C_{4,4}C_{48,9}]/C_{52,13} = 0.2573$.

    In terms of the counting principles, we first decided to count separately the hands with 2, 3, and 4 aces, which breaks the field into three disjoint components of varying sizes. We then invoke the multiplicative principle to count the component with k aces as a regular arrangement of $C_{4,k}$ possible ace combinations by $C_{48,13-k}$ possible combinations for the other 11 cards. $\square$

1.9 Example: A poker hand contains five cards. The hand "two-pair" contains cards of three distinct values. Two cards share one value, another two cards share a second value, and the final card assumes the third value. For example, the 4 of spades and diamonds, the king of clubs and diamonds, and the 7 of clubs constitute a two-pair hand. What is the probability of receiving a two-pair hand in a five-card deal from a well-shuffled deck?

    The total number of five-card hands is $C_{52,5} = 2,598,960$. This is the number of 5-combinations from a set of 52. We now need the number of combinations that result in two-pairs. As in the preceding example, we envision the 52 cards aligned linearly over slots, this time in descending value. The four aces occupy the first four slots, then the four kings, then the four queens, and so forth. The final four slots contain the four 2s. We view this arrangement as thirteen consecutive groups of four slots each. Each such group contains cards of a single value. A two-pair hand must involve two cards from each of two distinct groups and a final card from the 44 possibilities outside the two groups. There are $C_{13,2}$ patterns corresponding to the choice of the two groups. The upper bars in the diagram below show one such pattern where the pairs come from the Q-group and the 4-group. The odd card comes from the 9-group.

| A | K | Q | J | 10 | 9 | 8 | 7 | 6 | 5 | 4 | 3 | 2 |

Within each of the two groups chosen, a pair arises from any of $C_{4,2}$ possible combinations. Assuming that the suits are arrayed within groups as spades, hearts, diamonds, clubs, the diagram shows the Q-group generating the queen of spades and the queen of diamonds, one of the six possible pairs. The 4-group generates the four of hearts and the four of clubs, again one of six possibilities. Finally, the 44 remaining cards generate the 9 of hearts, one of $C_{44,1}$ possibilities. The diagram exemplifies only one possible pattern. The total number of two-pair patterns is $C_{13,2}C_{4,2}{}^2C_{44,1} = 123,552$. The desired probability is then $C_{13,2}[C_{4,2}]^2C_{44,1}/C_{52,5}$, which evaluates to 0.0475. □

Following the strategy of the last two examples, you should now be able to verify the probabilities of any poker hand. For a flush, which is five cards of the same suit, it is $C_{4,1}C_{13,5}/C_{52,5} = 0.00198$. This interpretation of a flush includes a royal flush, which is the ace through ten of the same suit. For a straight, which is five cards of consecutive value, it is $C_{9,1}[C_{4,1}]^5/C_{52,5} = 0.00355$. This interpretation includes pure straights as well as straight flushes (consecutive values from the same suit) and royal flushes. The strategy in all cases is to construct the favored hand through a series of choices, where the total number of possibilities divides equally across the choices. The desired count is then the product of the number of subdivisions at each step. This process amounts to the repeated application of the multiplicative counting principle. To choose a straight, for example, you first choose the lowest value, which must be one of the nine values 2 through 10. Any higher value for the lowest would not allow four further cards in increasing value. For each such choice, you then have the option of 4 cards (spades, hearts, diamonds, or clubs) for the lowest value. Then you must choose one of 4 cards for the second lowest value. Indeed, for each of the five sequential values, you must choose one of the 4 available suits. Multiplying the sizes of the subdivisions at each step gives $C_{9,1}[C_{4,1}]^5$, which leads to the probability calculated above.

Although the examples to this point have featured card games, the general ideas of Theorem 1.7 are relevant to a wide range of problems, some of which appear in the following examples. The examples introduce the terms *random sample* and *random selection*. For the moment, we adopt intuitive meanings for these terms. A random sample, or selection, of size k means a subset of k objects, selected without replacement by a process that affords each potential k-combination an equal chance of being chosen. For small objects, we achieve a random sample by mixing all the objects in a jar and drawing k of them without looking at the jar. For larger objects, or for objects that have distinguishable shapes or textures, we must imagine a process that removes these clues.

1.10 Example: Suppose that a bin of 100 components contains 4 defective ones. In a random sample of size 6 from this bin, what is the probability that none will be defective? What is the probability that exactly 1 will be defective?

Let $p_k, k = 0, 1, 2, 3$, or 4, be the probability of k defects in the sample. We imagine an alignment with the 4 defective components in the first 4 slots and the properly functioning components in the subsequent 96 slots. A pattern of 6 slots that exhibits no entry from the first four must necessarily choose all 6 from the remaining 96. Consequently, we calculate p_0 as shown below. In a similar manner, a pattern with one defective entry must have 1 slot from the first four and 5 slots from the remaining 96, which explains the p_1 computation. At this point, a sanity check should be stirring in the back of your mind. Between no defects and one defect, we have used up $0.778 + 0.205 = 0.983$ of the total probability. There are three remaining possibilities, and they must share the final 0.017 probability. The remaining calculations follow as shown, and we note that the values add up properly, within round-off error.

$$p_0 = \frac{C_{4,0}C_{96,6}}{C_{100,6}} = 0.778 \qquad p_2 = \frac{C_{4,2}C_{96,4}}{C_{100,6}} = 0.0167$$

$$p_1 = \frac{C_{4,1}C_{96,5}}{C_{100,6}} = 0.205 \qquad p_3 = \frac{C_{4,3}C_{96,1}}{C_{100,6}} = 3.22 \times 10^{-7}$$

$$p_4 = \frac{C_{4,4}C_{96,0}}{C_{100,6}} = 8.39 \times 10^{-10} \ \square$$

1.11 Example: Suppose that a 10-person committee contains 5 Republicans and 5 Democrats. Random selection produces a 3-person subcommittee. What is the probability that the Democrats will dominate the subcommittee?

Democrats will dominate if they number 2 or 3 on the subcommittee. $C_{10,3} = 120$ subcommittees are possible. $C_{5,2}C_{5,1} = 50$ of them will contain exactly 2 Democrats, and $C_{5,3}C_{5,0} = 10$ of them will contain exactly 3 Democrats. Therefore, Democrats will dominate 60 of the 120 possible subcommittees. The probability of a Democratically dominated committee is $60/120 = 1/2$.

Is this answer reasonable? A subcommittee with an odd number of members must be dominated by one group or the other. Since Democrats and Republicans are equally represented in the candidates, you should suspect from the symmetry of the situation that each group should have the same chance to dominate the subcommittee. That is, each group's chance should be $1/2$. $\square$

The common theme across the last several examples expresses the desired probability as a ratio, in which the numerator is the number of combinations under some constraint and the denominator is the total number of combinations. In the preceding example, the denominator was the number of 3-person combinations from the set of ten, while the numerator was the number of 3-person combinations that contained 2 or more Democrats. In the earlier example, the denominator was the number of 6-component combinations from the field of 100, and the numerator was the number of 6-component combinations with a specified number of defects. With these kinds of problems, the most difficult part is identifying the relevant subsets. Consider the following examples. The first arises in computer science and communications engineering. The second introduces a mathematical tool with widespread application.

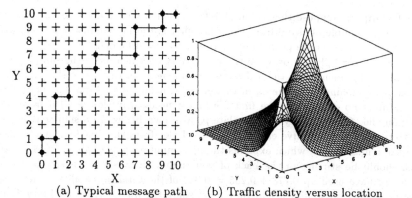

(a) Typical message path   (b) Traffic density versus location

Figure 1.1. Node loading in a switching computer network (see Example 1.12)

1.12 Example: A network has switching computers arrayed in a square pattern as shown in Figure 1.1(a). The computers occupy the grid points (x, y) for integer values of $0 \leq x \leq 10$ and $0 \leq y \leq 10$. Computers communicate only with their immediate neighbors. All message traffic arrives at node $(0,0)$ and departs from node $(10, 10)$ after passing through some path across the intermediate computers. Moreover, each computer forwards messages either up or to the right, so that each link makes progress toward the final destination at $(10, 10)$. Figure 1.1(a) shows a sample valid path. Beyond this restriction, all paths are equally likely. What is the probability that the computer at location $(2, 6)$ participates in a message transfer from $(0, 0)$ to $(10, 10)$?

This question is a particular instance of a more general consideration, which asks how various nodes participate in the overall traffic flow. Figure 1.1(b) gives a graphical depiction of the traffic density, and we now proceed to verify that pattern.

A path must proceed from an intermediate node either upward or to the right. It follows that every path must make 10 steps up and 10 steps to the right as it winds from $(0,0)$ to $(10, 10)$. Every path, therefore, consists of exactly 20 links: 10 horizontal links and 10 vertical links. Paths differ only in the choice of which 10 links are horizontal. So there are $C_{20,10}$ possible paths from $(0,0)$ to $(10, 10)$. Now, consider a path that passes through node $(2, 6)$, such as the one in Figure 1.1(a). The initial segment, from $(0, 0)$ to $(2, 6)$ must contain 8 links: 2 horizontal and 6 vertical. Different paths will choose different locations for the horizontal links. The path illustrated in the figure proceeds horizontally on links 2 and 6 and vertically on the remaining links. Hence there must be $C_{8,2}$ possible segments from $(0, 0)$ to $(2, 6)$. Similarly, there must be $C_{12,8}$ possible continuations from $(2, 6)$ to $(10, 10)$. Since any initial segment can pair with any of these continuations to form a path through $(2, 6)$, the number of such complete paths is $C_{8,2}C_{12,8}$. The required probability is then $p_{(2,6)} = C_{8,2}C_{12,8}/C_{20,10} = 0.075$.

Node (i, j) participates with probability $p_{(i,j)} = C_{i+j,i}C_{20-(i+j),10-i}/C_{20,10}$ in the general case. For a peripheral point, such as $(1, 9)$, the probability is 0.00054, while a central point, such as $(5, 5)$, participates with probability 0.344. If paths are chosen randomly, the central computer at node $(5, 5)$ must bear 34.4% of the traffic, while the peripheral computer at $(1, 9)$ sees very little traffic. These results are important because they reveal where to concentrate the fastest computers and the largest message buffers. Table 1.4 shows the probabilities associated with selected

	$x = 0$	1	2	3	4	5
$y = 0$	**1.0000**	**0.5000**	0.2368	0.1053	0.0433	0.0163
1	**0.5000**	**0.5263**	**0.3947**	0.2477	0.1354	0.0650
2	0.2368	**0.3947**	**0.4180**	**0.3483**	0.2438	0.1463
3	0.1053	0.2477	**0.3483**	**0.3715**	**0.3251**	0.2401
4	0.0433	0.1354	0.2438	**0.3251**	**0.3501**	**0.3151**
5	0.0163	0.0650	0.1463	0.2401	**0.3151**	**0.3437**
6	0.0054	0.0271	0.0750	0.1500	0.2387	**0.3151**
7	0.0015	0.0095	0.0322	0.0779	0.1500	0.2401
8	0.0004	0.0027	0.0110	0.0322	0.0750	0.1463
9	0.0001	0.0005	0.0027	0.0095	0.0271	0.0650
10	0.0000	0.0001	0.0004	0.0015	0.0054	0.0163

TABLE 1.4. Traffic density in a switching network (see Example 1.12)

nodes. The source node is at the upper left corner; the destination node (not shown) is at the lower right. The boldface entries highlight the probabilities greater than 0.3 and show that the traffic load is highest near the source and destination and that a bottleneck persists through the center of the network. Horizontal positions 6–10, which do not appear in the table, are reflections of columns 4–0. That is, column 6 is a copy of column 4, installed upside down, and so forth. The diagonal bottleneck therefore continues to the destination node. □

1.13 Example: Verify the binomial theorem: $(x + y)^n = \sum_{k=0}^{n} C_{n,k} x^k y^{n-k}$.

To multiply two polynomials, we first multiply every term of the first by every term of the second. For example, $(a+2b-2c)(b+2c) = ab+2ac+2b^2+4bc-2bc-4c^2$. Then we combine terms to obtain a more compact expression: $ab + 2ac + 2b^2 + 2bc - 4c^2$. We can view the initial product, before the final simplification, as the sum of all possible two-factor terms in which one factor comes from the first polynomial and the other from the second polynomial. If the two polynomials have n_1 and n_2 terms, respectively, then the product, before simplification, has $n_1 n_2$ terms. If we now multiply the product by a third polynomial with n_3 terms, we obtain a sum of $n_1 n_2 n_3$ three-factor terms. In each such term, each of the three polynomials contributes one factor. Applying this observation to $(x + y)^n$, we get 2^n terms. Each term, moreover, contains n factors, one from each copy of $(x + y)$ in the multiplication. A given term might contain n x-factors and no y-factors and so appear as x^n. It might contain $n - 1$ x-factors and one y and take the form $x^{n-1}y$. The sum of the x and y exponents must always be n, so the general form of $(x+y)^n$, after combining terms, must be $\sum_{k=0}^{n} C_k x^k y^{n-k}$, for some set of constants C_k.

To determine the C_k, we imagine the n factors of $(x+y)^n$ aligned over n slots. To produce a term $x^k y^{n-k}$, we must choose an x-factor from k of the n slots and a y-factor from each of the remaining slots. The diagram to the right illustrates the point for the case $n = 5, k = 2$. The number of patterns is, of course, $C_{5,2} = 10$. That is, $(x + y)^5$ contains the term $10x^2 y^3$. We can calculate the coefficient associated with other values of k in a similar manner. In general, it is $C_{n,k}$, and therefore $(x + y)^n = \sum_{k=0}^{n} C_{n,k} x^k y^{n-k}$. □

The examples to this point illustrate the usefulness of counting the number of combinations available from a set of n objects. There is, however, another viewpoint that extends this concept to a different kind of problem.

Consider a set of n objects, some of which are indistinguishable at a higher level of abstraction. The traditional example consists of n balls, of which r are red and $b = n - r$ are black. For the moment, let us assume that $n = 5$ and $r = 3$, which means, of course, that $b = 2$. Because a set can have no duplicates, the balls must be distinguishable through some further characteristic, so we label the red balls as r_1, r_2, r_3 and the blacks as b_1, b_2. Now if we ask how many 5-sequences we can form, without replacement, from the five balls, Theorem 1.3 immediately answers $5! = 120$. However, if we ignore the subscripts and consider only the colors, there will be fewer sequences because some of them will contain the same colors in the same places and differ only in the subscripts. So, how many sequences are possible when only color is considered?

Suppose that we group the 120 permutations into N collections, each containing the same color pattern. That is, each sequence in a given collection contains red balls in exactly the same positions, differing only in the subscripts on the red balls. By default, all sequences in a given collection must then contain black balls in exactly the same positions, but again differing in the subscripts.

For example, the sequences in Figure 1.2(a) all lie in one collection. The reds occupy positions $1, 2$, and 4, and the three red balls can appear in these slots in any arrangement of $\{r_1, r_2, r_3\}$. Similarly, the blacks appear in slots 3 and 5 in any arrangement of $\{b_1, b_2\}$. So any of the $3! = 6$ 3-sequences of the reds (within the red slots, of course) can associate with any of the $2! = 2$ 2-sequences of the blacks to form a sequence in the collection. The collection then contains $3! \cdot 2! = 12$ sequences, as exemplified by the display above. The same analysis shows that each of the N collections contains 12 sequences. Therefore, $N = 120/12 = 10$, and N counts the number of sequences that are distinguishable only by color.

In the general case, we have n balls, of which r are red and $b = n - r$ are black. If we partition the $n!$ sequences into collections such that all members of a collection exhibit the same color pattern, we find $r!b! = r!(n - r)!$ sequences in each collection. This follows because the r reds can be redistributed in the same slots in any of the $r!$ permutations, and the blacks can be redistributed in their slots in any of $b! = (n - r)!$ permutations. The number of distinguishable color patterns is the number of collections: $n!/[r!(n - r)!] = C_{n,r}$.

In the earlier examples, the result $C_{n,r}$ occurred when we arrayed the objects linearly above a set of slots and looked at the slot patterns formed by choosing r objects from the field of n. In the current context, however, we are choosing *all* the objects, and our interest shifts from which objects are chosen to where the r "special" objects appear. We can still use the slot-pattern technique, with one small variation. We now align the balls' potential *positions* over the slots. It should be clear that a color sequence is determined as soon as the positions of the three red balls are known; the remaining positions must, by default, contain black balls.

For the case $n = 5, r = 3$, we envision the arrangements of Figure 1.2(b).

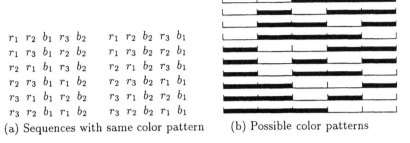

(a) Sequences with same color pattern   (b) Possible color patterns

r_1 r_2 b_1 r_3 b_2 r_1 r_2 b_2 r_3 b_1
r_1 r_3 b_1 r_2 b_2 r_1 r_3 b_2 r_2 b_1
r_2 r_1 b_1 r_3 b_2 r_2 r_1 b_2 r_3 b_1
r_2 r_3 b_1 r_1 b_2 r_2 r_3 b_2 r_1 b_1
r_3 r_1 b_1 r_2 b_2 r_3 r_1 b_2 r_2 b_1
r_3 r_2 b_1 r_1 b_2 r_3 r_2 b_2 r_1 b_1

Figure 1.2. Color patterns exhibited by 3 red balls in a field of 5

The slots are aligned with the potential *positions* of the red balls, not with the balls themselves. The 10 patterns correspond to the possible positions of the red balls, which in turn correspond to the possible color patterns. Since there are $C_{5,3} = 10$ patterns, there are 10 distinguishable color sequences. This analysis verifies the more direct computation above and provides a further interpretation of the $C_{n,k}$ formula.

We can capture the new interpretation in the following theorem. The theorem refers to "classes," which may be colors, as in the example at hand, or any other attribute in the context of a particular problem. In the next example, we interpret class as gender.

1.14 Theorem: Suppose that a set of n objects contains two classes with k objects in the first class and the remaining $n - k$ objects in the second class. Consider n-sequences, chosen without replacement, in which objects of the same class are indistinguishable. The number of such sequences is $C_{n,k}$.

PROOF: See discussion above. ∎

The interpretation accorded $C_{n,k}$ in this theorem is really the same as that which visualizes $C_{n,k}$ as the number of slot patterns in a field of n under the constraint that k slots must be shaded. Objects of the first class occupy the shaded slots, and by default, objects of the second class occupy the remaining slots. Given that objects within a class are indistinguishable, the number of distinguishable n-sequences is precisely the number of slot patterns.

1.15 Example: Consider a dinner party of 10 people—5 men and 5 women. If they randomly choose places at a linear table, what is the probability that there will be a strict alternation of men and women?

The positions of the 5 men determines a gender pattern. Of the $C_{10,5} = 252$ possible patterns, only two exhibit a strict alternation of men and women: one starting with a man and the other starting with a woman. The probability is then $2/252 = 0.00862$. □

At this point, we have two interpretations of $C_{n,k}$, which we will call the *selection* and *allocation* solutions. It is convenient to add a third interpretation to form the following list.

1. $C_{n,k}$ is the number of subsets of size k drawn from a pool of n objects. Stated differently, it is the number of ways to select k *distinguishable*

objects from a pool of n.

2. $C_{n,k}$ is the number of slot patterns in which k of the n slots are darkened. Equivalently, it is the number of ways to allocate k *indistinguishable* objects to n containers in such a manner that no container receives more than one object. The darkened slots indicate the chosen containers.

3. $C_{n,k}$ is the number of binary n-vectors whose components sum to k.

A binary vector is a vector whose components are zeros and 1s. A binary n-vector whose components sum to k must contain exactly k 1s and $n - k$ zeros. These vectors correspond to slot patterns in which the darkened slots represent 1s. Hence the number of binary n-vectors whose components sum to k is the number of slot patterns in a field of n with the requirement that k are dark. This count is $C_{n,k}$.

We conclude this section with a generalization of Theorem 1.14. We start by extending the example of the colored balls. Suppose now that the balls come in three colors: red, black, and white. Specifically, suppose that we have a set of n balls, of which k_1 are red, k_2 are black, and k_3 are white. Of course, $k_1 + k_2 + k_3 = n$. How many n-ball color patterns are possible?

Reasoning as before, we have $n!$ n-sequences when the balls are individually recognizable. We collect sequences that contain the same color balls in the same positions. Suppose that N collections result. Then N is the required number of color patterns. To determine the number of sequences in each collection, consider an arbitrary member of a collection C. This sequence contains k_1 red balls in specific positions, and the $k_1!$ permutations of the red balls among the fixed set of positions result in new sequences with the same color pattern. So, for each fixed arrangement of the black and white balls, there are $k_1!$ sequences that exhibit red balls in identical locations. Of course, each of the $k_1!$ sequences exhibits the same color pattern. Continuing this line of reasoning, we see that each of these $k_1!$ sequences gives rise to $k_2!$ further sequences by permuting the black balls. Consequently, we have $k_1!k_2!$ sequences that display the same color pattern. Moreover, each of the $k_1! \cdot k_2!$ obtained to this point gives rise to $k_3!$ additional sequences by permuting the white balls. Because the chosen collection was arbitrary, we conclude that each collection contains $k_1! \cdot k_2! \cdot k_3!$ of the $n!$ sequences. So $N = n!/[k_1! \cdot k_2! \cdot k_3!]$.

The general case concerns p colors, or classes, which appear $k_1, k_2, \ldots, k_p$ times in the original set of n objects. Of course, $k_1 + k_2 + \ldots + k_p = n$. Theorem 1.14 becomes a special case of the following theorem.

1.16 Theorem: Suppose that a set of n objects contains p classes with $k_i, 1 \leq i \leq p$, objects in class i. $n = \sum_{i=1}^{p} k_i$. Consider n-sequences, chosen without replacement, in which objects of the same class are indistinguishable. The number of such sequences is $n!/[k_1! \cdot k_2! \cdots k_p!]$.

PROOF: The discussion above applies equally well to p classes. ∎

1.17 Example: Verify the multinomial expansion:

$$(x_1 + x_2 + \ldots + x_p)^n = \sum_{k_1+k_2+\ldots+k_p=n} \left(\frac{n!}{k_1! \cdot k_2! \cdots k_p!} x_1^{k_1} \right) x_2^{k_2} \cdots x_p^{k_p}.$$

The notation under the summation symbol implies that the sum extends over all p-sequences of nonnegative integers, chosen with replacement, that sum to n. We will learn later, in Theorem 1.22, that there are precisely $C_{p+n-1,n}$ summands in this expression, but that detail is not important at this point.

Anticipating an induction on p, we denote the left side by L_p^n. The case L_1^n is a straightforward verification, while the case L_2^n reduces to the binomial theorem. Both computations are illustrated below.

$$L_1^n = x_1^n = \frac{n!}{n!} \cdot x_1^n = \sum_{k_1=n} \frac{n!}{k_1!} \cdot x_1^{k_1}$$

$$L_2^n = (x_1 + x_2)^n = \sum_{k_1=1}^{n} \frac{n!}{k_1!(n-k_1)!} \cdot x_1^{k_1} x_2^{n-k_1} = \sum_{k_1+k_2=n} \frac{n!}{k_1!k_2!} \cdot x_1^{k_1} x_2^{k_2}$$

The final reduction for L_2^n occurs because the pairs (k_1, k_2) which sum to n must have the form $(k_1, n - k_1)$. Moreover, as k_1 runs from 0 to n, the expression $(k_1, n - k_1)$ generates all such pairs.

If $p > 2$, we proceed by induction, assuming that

$$L_q^n = \sum_{k_1+k_2+\ldots+k_q=n} \frac{n!}{k_1! \cdot k_2! \cdots k_q!} \cdot x_1^{k_1} x_2^{k_2} \cdots x_q^{k_q},$$

for $q = 1, 2, \ldots, p-1$. We then consider the multinomial of p terms to be a binomial by grouping the first $p-1$ terms as a unit. At that point we can apply the binomial theorem.

$$L_p^n = [(x_1 + x_2 + \ldots + x_{p-1}) + x_p]^n = \sum_{k_p=0}^{n} C_{n,k_p} L_{p-1}^{n-k_p} x_p^{k_p}$$

$$= \sum_{k_p=0}^{n} C_{n,k_p} \left[\sum_{k_1+k_2+\ldots+k_{p-1}=n-k_p} \frac{(n-k_p)!}{k_1! \cdot k_2! \cdots k_{p-1}!} \cdot x_1^{k_1} x_2^{k_2} \cdots x_{p-1}^{k_{p-1}} \right] x_p^{k_p}$$

The double sum effectively enumerates all the p-sequences that sum to n, so we can continue the computation as follows.

$$L_p^n = \sum_{k_1+k_2+\ldots+k_{p-1}+k_p=n} \frac{n!}{k_p!(n-k_p)!} \cdot \frac{(n-k_p)!}{k_1! \cdot k_2! \cdots k_{p-1}!} \cdot x_1^{k_1} x_2^{k_2} \cdots x_{p-1}^{k_{p-1}} x_p^{k_p}$$

$$= \sum_{k_1+k_2+\ldots+k_{p-1}+k_p=n} \frac{n!}{k_1! \cdot k_2! \cdots k_{p-1}! \cdot k_p!} \cdot x_1^{k_1} x_2^{k_2} \cdots x_{p-1}^{k_{p-1}} x_p^{k_p}$$

This calculation verifies the multinomial expansion, but an alternative approach is worth consideration, one that draws on the interpretation of $n!/[k_1! \cdot k_2! \cdots k_p!]$ as the number of color patterns, each of length n, that we can assemble from a group of $n = k_1 + k_2 + \ldots + k_p$ balls of p different colors. From the discussion of Example 1.13, we know that $(x_1 + x_2 + \ldots + x_p)^n$ will generate p^n terms, each containing n factors.

Each copy of $(x_1 + x_2 + \ldots + x_p)$ contributes one of the n factors in each term. Since each copy must draw its contribution from the possibilities $x_1, x_2, \ldots, x_p$, we see that each term in the expansion must have the form $x_1^{k_1} x_2^{k_2} \cdots x_p^{k_p}$ for some set of nonnegative integers $k_1, k_2, \ldots, k_p$ that sum to n. The question is: How many terms have the same set of exponents and are thus eligible for combination? Stated differently, we know that the expansion contains terms of the form $C x_1^{k_1} x_2^{k_2} \cdots x_p^{k_p}$, but we do not yet know the value of C associated with a particular exponent set $k_1, k_2, \ldots, k_p$.

Let $k_1, k_2, \ldots, k_p$ be a fixed set of nonnegative integers that sum to n. Now identify each of $x_1, x_2, \ldots, x_p$ with a distinct color. Each of the n copies of $(x_1 + x_2 + \ldots + x_p)$ now hovers over one of n slots. Each drops one of its colors into the slot below it, under the overall constraint that k_1 copies must drop color x_1, k_2 copies must drop color x_2, and so forth. The number of possible color patterns is $n!/[k_1! \cdot k_2! \cdots k_p!]$. But this is also the number of terms in the expansion that involve $x_1^{k_1} x_2^{k_2} \cdots x_p^{k_p}$. That is, every choice that produces an n-factor term with exactly k_1 x_1-factors, k_2 x_2-factors, and so forth through k_p x_p-factors must correspond to k_1 slots contributing the value x_1, k_2 slots contributing the value x_2, and so forth. Hence $C = n!/[k_1! \cdot k_2! \cdots k_p!]$, for a particular set $k_1, k_2, \ldots, k_p$, and the expansion is verified. $\square$

Example 1.13 shows that the terms $C_{n,k}$ appear as coefficients in the expansion of $(x + y)^n$. For this reason, the $C_{n,k} = n!/[k!(n - k)!], 1 \leq k \leq n$, are called *binomial coefficients*. Similarly, Example 1.17 shows that the terms $n!/[k_1! \cdot k_2! \cdots k_p!]$ appear as coefficients in the expansion of $(x_1 + x_2 + \ldots + x_p)^n$. The expressions $n!/[k_1! \cdot k_2! \cdots k_p!]$, for $k_i \geq 0$ and $k_1 + k_2 + \ldots + k_p = n$, are therefore known as *multinomial coefficients*.

1.18 Example: Suppose that we have a collection of objects, $S = \{x_1, x_2, \ldots, x_n\}$, each with two attributes, $f(x_i)$ and $g(x_i)$. For example, $f(x_i)$ might be the size of object x_i, while $g(x_i)$ might be its weight. In any case, we assume that both $f(x_i)$ and $g(x_i)$ are nonnegative. The goal is to choose a subset, $\{x_{i_1}, x_{i_2}, \ldots, x_{i_p}\}$, such that the total g-values are as large as possible, subject to the constraint that the total f-values do not exceed a prescribed limit, F. Suppose further that our chosen subset must contain at least 2 objects. The code of Figure 1.3 conducts a search for the best subset.

Since this is the first algorithm description in the book, we comment on the C structure's readability, even for a reader with no previous experience in C. Note carefully the convention for marking the scope of while-loop, for-loop, and conditional execution (if) constructions. The body is set off as an indented mass under the control statement. The while-loop controls all statements at the indentation level 1 greater than the while-statement itself [i.e., through the statement X = subset(S, X)]. The while-loop is simply a device for repeatedly executing a step sequence until a specified condition prevails. This example also illustrates the for-loop, which executes a step sequence a prescribed number of times. Executions use consecutive values for an index, which is the variable i in this case. Finally, the code contains if-statements, each with an indented body. The body contains a step sequence that is executed if and only if the attendant condition is true.

This algorithm assumes the availability of a subroutine, subset(S, Y), which cyclically returns the subsets of its first argument. The routine constructs, in a manner that is not of interest here, the next subset beyond that given by the second

```
function binpack (S, F)  // inputs are the set S and the f-limit F
    X = phi;             // X cycles through subsets of S, starting with the empty set
    Y = phi; y = 0.0;    // Y stores the best subset to date; y stores its g-value
    while (X ≠ S)
        if (size(X) ≥ 2)
            a = 0.0; b = 0.0;
            for (i = 1 to size(X))
                a = a + f(X[i]); b = b + g(X[i]);
            if ((a ≤ F) and (b > y))
                y = b; Y = X;
        X = subset(S, X);
    return Y;
```

Figure 1.3. Algorithm for optimal bin packing (see Example 1.18)

---

argument. Successive calls eventually recover all subsets, and the last subset to be returned is S itself. For starting the process, we assume that the empty set is available under the name phi. We also assume a routine that returns the number of elements in a set: size(X). X[i] extracts an element from the subset X.

How many times does the code call the function f if the input set S is of size 10? How many times if the input set is of size n? When X is a subset of size k, the for-loop makes k calls to f, and this loop executes for all subsets of size 2 or greater. Let $t(n)$ denote the number of calls when the input set is of size n. Since $C_{10,k}$ is the number of subsets of size k, we have the following summation.

$$t(10) = \sum_{k=2}^{10} C_{10,k} \cdot k = 45(2) + 120(3) + \ldots + 1(10) = 5110$$

For general n, we obtain the following summation.

$$t(n) = \sum_{k=2}^{n} C_{n,k} \cdot k = \left[\sum_{k=0}^{n} C_{n,k} \cdot k \right] - C_{n,0} \cdot 0 - C_{n,1} \cdot 1 = n \cdot 2^{n-1} - n$$

The last simplification involves summing $kC_{n,k}$, which you may or may not know how to do at this time. Section 1.2 will review these matters, but for the moment, even if you are not able to perform the reduction here, you can easily verify that the general formula generates the correct result for $n = 10$. $\square$

Exercises

1.1 The binary coded decimal (BCD) code assigns a four-bit binary pattern to each digit as follows.

0	1	2	3	4	5	6	7	8	9
0000	0001	0010	0011	0100	0101	0110	0111	1000	1001

This is just one of N possible codes that express the ten digits as distinct four-bit binary patterns. What is N? The BCD code leaves six patterns unused: $1010, 1011, 1100, 1101, 1110$, and 1111. Other codes involve different sets of unused patterns, although each such set will always contain six members. Assume that the first step in constructing a code specifies the set of unused patterns. How many choices are available at this step?

1.2 What is the probability of receiving a full-house poker hand? A full-house hand contains 2 cards of one value and 3 cards of a different value.

1.3 What is the probability of straight-flush poker hand? A straight-flush hand contains 5 cards of consecutive value, all from the same suit.

1.4 In the game of bridge, point cards are aces, kings, queens, and jacks. What is the probability of receiving a 13-card bridge hand containing 3 or more point cards?

1.5 A 5-person committee is formed from a group of 6 men and 7 women. What is the probability that the committee has 3 or more men?

1.6 Seventeen 10-kilohm resistors are inadvertently mixed with a batch of eighty 100-kilohm resistors. Five resistors are drawn from the combined group and wired in series. What is the probability that the series resistance will be less than 500-kilohms?

1.7 If you randomly choose 3 days from the calendar, what is the probability that they all fall in June?

1.8 In a group of 1000 persons, 513 plan to vote Republican and 487 plan to vote Democrat. What is the probability that a random sample of 25 persons will erroneously indicate a Democratic victory?

*1.9 Suppose that 8 balls are tossed toward a linear arrangement of 10 open boxes. What is the probability that the leftmost box receives no balls?

*1.10 You arrive at the gym to find a single free locker, number 4 in a row of 10. You take that locker. When you finish your workout, you find 6 lockers, including yours, still occupied. What is the probability that the neighboring lockers on either side of yours are now free?

1.11 Calculate $\sum_{k_1+k_2+k_3+k_4=7} 7!/[k_1! \cdot k_2! \cdot k_3! \cdot k_4!]$. The notation means the sum over all sequences of four nonnegative integers, chosen with replacement, that add to 7.

1.12 A football fan is sitting high in the stadium and can barely make out the cheerleaders far below on the field. These cheerleaders are arranged in a line and are waving colored fabrics. If 3 cheerleaders have red fabrics, 2 have green fabrics, and 2 have blue fabrics, how many color patterns are discernible by the faraway fan?

1.13 Bob and Alice are dinner guests at a party of eight, 4 male and 4 female. The hostess arranges the guests linearly along a table with the men on one side and the women on the other. What is the probability that Bob and Alice will be facing each other or be within one position of facing each other?

1.14 Six devices $d_1, d_2, \ldots, d_6$ connect with a central processor, which polls them in order to identify one that is ready to communicate. Suppose that three are ready to communicate when the polling cycle starts. What is the probability that the central processor issues exactly two polling requests before communicating.

1.15 If you rearrange the letters of "Mississippi," how many distinguishable patterns are possible?

*1.16 In the context of Example 1.12, what fraction of paths from $(0,0)$ to $(10,10)$ never rise above the diagonal? That is, what fraction of paths pass only through nodes (x,y) with $x \geq y$?

1.1.2 Sampling with replacement

Let us now consider the situation where repetitions are permitted in choosing from a set of n objects. That is, each chosen object, after being recorded in the result, rejoins the set before another object is drawn. This allows an object to appear multiple times in the k-sequence or k-combination under construction. It is an easy task to determine the number of k-sequences from a set of size n, but it is somewhat less obvious how to calculate the number of k-combinations. We start with the easy theorem.

1.19 Theorem: Let $n > 0$ and $k \geq 0$. Under sampling with replacement, a set of n objects admits n^k k-sequences.

PROOF: Let S be the set in question. $|S| = n$. Let $s(n,k)$ denote the number of k-sequences that arise from S. There is just one empty sequence, and $1 = n^0$. Also, there are $n = n^1$ 1-sequences, each consisting of a single object from S. Therefore, $s(n,0) = 1$ and $s(n,1) = n$. We now proceed by induction on k. Assume that $s(n,j) = n^j$, for $j = 1, 2, \ldots, k-1$. Now consider all k-sequences arising from S. Divide them into groups according to the first symbol. There are n such groups. Within each group, the sequence following the initial symbol is a $(k-1)$-sequence from S because S reacquires the initial symbol before any subsequent choices. By the induction hypothesis, the number of such continuation sequences is $s(n, k-1) = n^{k-1}$. Hence there are n groups of n^{k-1} sequences, giving $n \cdot n^{k-1} = n^k$ sequences in total. ∎

1.20 Example: A sequence of heads and tails results from tossing a fair coin 10 times. What is the probability that heads and tails alternate in the sequence?

The sequence is one of the $2^{10} = 1024$ 10-sequences arising from the set $\{H, T\}$ when sampling with replacement. Of these 1024 possibilities, only two provide a strict alternation of heads and tails—the one starting with a head and alternating

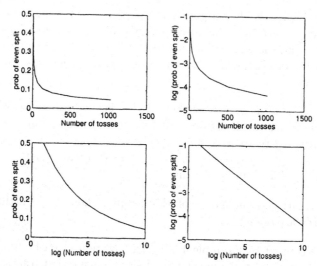

Figure 1.4. Probability of an equal head-tail split in a sequence of coin tosses. (see Example 1.21). Logarithms are base 2.

thereafter, and the one starting with a tail and alternating thereafter. So the probability is $2/1024 = 0.001953$. The coin may be fair, but strict alternation of the two possibilities is not likely. □

1.21 Example: A sequence of heads and tails results from tossing a fair coin 10 times. What is the probability that the sequence has exactly five heads in it?

The sequence is again one of $2^{10} = 1024$ possible 10-sequences that arise from the set $\{H, T\}$ when sampling with replacement. To determine the number of sequences with exactly 5 heads, we note that such a sequence is fixed once we determine the positions of the 5 heads. Hence, we must choose 5 positions, *without* replacement, from the set $\{1, 2, 3, 4, 5, 6, 7, 8, 9, 10\}$. Theorem 1.7 states that such choices number $C_{10,5} = 252$. The desired probability is then $252/1024 = 0.246$. Again, the coin may be fair, but the chance that such fairness will be manifest as an equal split over 10 tosses is rather smaller than 0.5. Indeed, if we were looking for an even split in 100 tosses, the probability would be $C_{100,50}/2^{100} = 0.0796$. This suggests, correctly it turns out, that the chances of an even split become less as the number of tosses increases. Figure 1.4 provides some credibility for this suggestion. Since the probability of an even head-tail split is zero when the number of coin tosses is odd, all the graphs in Figure 1.4 provide probability points only for even numbers of coin tosses. The upper-left graph shows how the probability of an even split drops rapidly as the number of tosses grows larger, but it appears to level out. Switching to a logarithmic scale on the probability axis, we can discern a bit more detail about the transition, as shown in the graph in the upper right. However, the flattening of the decline rate persists in this representation. Using logarithmic scales on both axes provides the clearest indication of the overall trend, as shown in the lower-right graph. □

1.22 Theorem: Let $n > 0$ and $k \geq 0$. Under sampling with replacement, $C_{n+k-1,k}$ k-combinations arise from a set with n objects.

PROOF: Let the elements of the set be $x_1, \ldots, x_n$. In a given k-combination,

Pattern	Repetitions	3-comb
||*|	$(1,1,1,0)$	abc
|||*	$(1,1,0,1)$	abd
|||*	$(1,0,1,1)$	acd
|*|*|*	$(0,1,1,1)$	bcd
**|*||	$(2,1,0,0)$	aab
*|**||	$(1,2,0,0)$	abb
**||*|	$(2,0,1,0)$	aac
*||**|	$(1,0,2,0)$	acc
**|||*	$(2,0,0,1)$	aad
*|||**	$(1,0,0,2)$	add

Pattern	Repetitions	3-comb
|**|*|	$(0,2,1,0)$	bbc
|*|**|	$(0,1,2,0)$	bcc
|**||*	$(0,2,0,1)$	bbd
|*||**	$(0,1,0,2)$	bdd
||**|*	$(0,0,2,1)$	ccd
||*|**	$(0,0,1,2)$	cdd
***||||	$(3,0,0,0)$	aaa
|***||	$(0,3,0,0)$	bbb
||***|	$(0,0,3,0)$	ccc
|||***	$(0,0,0,3)$	ddd

TABLE 1.5. Star-bar patterns and 3-combinations from a field of 4 symbols (see Theorem 1.22)

each x_i is present some integer number of times, say $k_i \geq 0$. Furthermore, these repetition numbers, provided that their sum is k, completely specify the particular k-combination. So the set $\{(k_1, k_2, \ldots, k_n)|k_i \geq 0, \sum k_i = k\}$ is in one-to-one correspondence with the k-combinations. Therefore, an equivalent question is: How many such ordered sets of n nonnegative integers add to k? These are n-sequences chosen from $\{0, 1, 2, 3, \ldots, k\}$ with replacement. Although we know from Theorem 1.19 that the total number of such n-sequences is $(k+1)^n$, not all of these n-sequences sum to k.

Instead, consider the following indirect approach to counting just those n-sequences that add to k. Imagine k asterisks and $n-1$ vertical bars, arranged linearly. Each such display contains $n + k - 1$ symbols. The possibilities for $n = 4, k = 3$ appear as Table 1.5. (The table depicts 3-combinations from the set $\{abcd\}$ rather than from $\{x_1, x_2, x_3, x_4\}$ in order to facilitate comparison with Table 1.1. That earlier table breaks out all the 3-combinations from a set of size 4 but does not offer a method for counting them systematically.)

The number of asterisks to the left of the first bar specifies the repetition count for the first symbol from the set. The number of asterisks between the first and second bar specifies the repetition count for the second symbol, and so forth. Because exactly k asterisks appear in each pattern, the sum of the repetition counts as separated by the bars is always k. Each distinct sequence of repetition counts corresponds to a k-combination, and vice versa. The total number of k-combinations is then the number of ways of choosing $n - 1$ positions for the bars out of the total of $n + k - 1$ possible positions. Theorem 1.7 gives that value as $C_{n+k-1,n-1} = C_{n+k-1,k}$, which proves the theorem. ∎

1.23 Theorem: The number of n-sequences $(x_1, x_2, x_3, \ldots, x_n)$ with nonnegative integer components such that $\sum_{i=0}^{n} x_i = k$ is $C_{n+k-1,k}$.

PROOF: The proof of Theorem 1.22 establishes a one-to-one correspondence between these n-sequences and the k-combinations from a set of size n. The theorem also asserts that the latter count is $C_{n+k-1,k}$. ∎

The preceding section noted three interpretations of $C_{n,k}$, and there are three parallel interpretations for $C_{n+k-1,k}$. They again involve a selection process, an allocation process, and a count of constrained vectors.

1. $C_{n+k-1,k}$ counts the number of ways to select k *distinguishable* objects from a pool of n, given that each object may be chosen zero, one, or more times.

2. $C_{n+k-1,k}$ counts the number of ways to allocate k *indistinguishable* objects to n containers, given that a container may acquire zero, one, or more objects.

3. $C_{n+k-1,k}$ counts the number of n-vectors whose components are nonnegative integers summing to k.

Theorem 1.23 notes the equivalence of the selection solution and the number of n-vectors with nonnegative integer components summing to k. The equivalence of the allocation solution follows because each such n-vector specifies an allocation pattern. The vector $(k_1, k_2, \ldots, k_n)$ specifies k_1 objects in the first container, k_2 in the second, and so forth.

In discussing the pigeonhole principle, the chapter's introduction noted the impossibility of randomly allocating four balls among three containers such that all containers receive at most 1 ball. It then asked the probability of an allocation in which the leftmost container receives at most 1 ball. We can now answer that question. Total allocations number $C_{3+4-1,4} = 15$, while those favoring the leftmost container give it zero or 1 ball. If the leftmost container receives zero balls, the remaining two containers must accommodate four balls, which they can do in $C_{2+4-1,4} = 5$ ways. If the leftmost container receives 1 ball, the remaining two must receive 3 balls, which can happen in $C_{2+3-1,3} = 4$ ways. The probability is then $(5+4)/15 = 0.6$ that the leftmost container receives at most 1 ball.

In this small example, we can tabulate the possible allocations. In the scheme to the right, (k_1, k_2, k_3) means that the left container receives k_1 balls, the center container receives k_2, and the right container k_3. Of course, the k_i must sum to four. The asterisks mark the allocations in which the left container receives at most 1 ball.

∗ $(0,0,4)$	∗ $(0,3,1)$	∗ $(0,2,2)$
∗ $(0,4,0)$	∗ $(1,3,0)$	$(2,2,0)$
$(4,0,0)$	$(3,0,1)$	∗ $(1,1,2)$
∗ $(0,1,3)$	$(3,1,0)$	∗ $(1,2,1)$
∗ $(1,0,3)$	$(2,0,2)$	$(2,1,1)$

When choosing k-sequences or k-combinations with replacement, it is possible for k to be much larger than n, the size of the pool. The following example illustrates this possibility and also investigates the difference between k-sequences and k-combinations in probability calculations.

1.24 Example: Ten tosses of a fair coin produces an ordered pair (h, t), where h is the number of heads over the 10 tosses and t is the number of tails. Of course, $h + t = 10$. How many ordered pairs can result from the experiment? What is the probability of getting $(3, 7)$?

The elements h and t can assume any of the values from $\{0, 1, \ldots, 10\}$, provided that their sum is 10. According to Theorem 1.23, the number of such 2-sequences that sum to 10 is $C_{2+10-1,10} = C_{11,10} = 11$. Such a high-powered formula is not necessary in this simple case. We see that all possible (h, t) pairs must be of the form $(h, 10 - h)$, with $0 \leq h \leq 10$. Hence, there are 11 such pairs. Also, a given (h, t) pair corresponds to a 10-combination from $\{H, T\}$, so Theorem 1.22 also provides the correct answer: $C_{10+2-1,10} = 11$.

Now, since there are 11 possible 10-combinations and just 1 with 3 heads, it might appear that the probability of getting $(3, 7)$ is $1/11$. This is not correct. When calculating a probability as a fraction of some number of equally likely possibilities, we must check carefully that the denominator equitably represents the field of possibilities. Of the 11 possible 10-combinations, are all equally likely in a physical experiment with ten coins? Does $(1, 9)$, in which just 1 head occurs, appear as often as $(5, 5)$, in which 5 heads occur? No, these outcomes are not equally likely.

What actually occurs are 10-sequences, such as $HTTHHHHTTHT$. By Theorem 1.19, there are $2^{10} = 1024$ such sequences., and they are equally likely. Also, since a sequence with 3 heads is determined once the positions of the heads are known, we must have as many such sequences with 3 heads as there are 3-combinations, without replacement, from the positions $\{1, 2, \ldots, 10\}$. Using Theorem 1.7, we then compute

$$\Pr(3, 7) = \frac{1}{1024} C_{10,3} = \frac{120}{1024} = 0.117$$

$$\Pr(5, 5) = \frac{1}{1024} C_{10,5} = \frac{252}{1024} = 0.246$$

$$\Pr(1, 9) = \frac{1}{1024} C_{10,1} = \frac{10}{1024} = 0.0098. \; \square$$

When calculating a probability as a fraction of cases favorable to some constraint out of all possible cases, we must be more careful when the process is sampling with replacement. In the simpler situation, sampling *without* replacement, we can normally use either sequences or combinations, as long as we are consistent in the numerator and denominator. As suggested by Table 1.1, each k-combination, without replacement, gives rise to $k!$ k-sequences. Therefore, a solution using k-sequences in numerator and denominator is the same solution using k-combinations, after multiplying numerator and denominator by $k!$. The value of the fraction does not change. If you want, for example, the probability of receiving 4 hearts when dealt 4 cards from a well-shuffled deck, you can form the appropriate 4-sequence ratio, $P_{13,4}/P_{52,4} = 0.00264$, or the corresponding 4-combination ratio, $C_{13,4}/C_{52,4} = 0.00264$. The numerator and denominator of the first fraction are just $4! = 24$ times those of the second. Because combinations are simpler and extend more easily to complicated situations, they are more frequently used in situations where the underlying process is sampling without replacement.

In sampling with replacement, each k-combination does not give rise to $k!$ k-sequences; sometimes the number of sequences is much less, as illustrated in Table 1.1. Therefore, the ratio of k-combinations is not the same as the ratio of the corresponding k-sequences. In sampling with replacement, it is normally the ratio of the appropriate k-sequences that corresponds to the reality of the experiment. The following example provides another situation where a simple sanity check shows that a proposed solution involving k-combinations cannot be correct.

1.25 Example: Suppose that 10 people are gathered in a room. What is the probability that two or more persons have the same birthday? By having the same birthday, we mean born on the same day of the year, not necessarily being the same age.

We imagine each person announcing a birthday from the 365 possible choices. The birthdays constitute a 10-combination from the set of days: $\{1, 2, 3, \ldots, 365\}$. The 10-combination arises from sampling with replacement because an announcement can duplicate a birthday already used. The number of such 10-combinations is $D = C_{365+10-1,10}$. The combinations corresponding to no two persons having the same birthday are just those that happen to have no duplicate elements. That is, they correspond to the 10-combinations that you could draw *without* replacement. Hence they number $C_{365,10}$. The rest of the combinations, $N = C_{374,10} - C_{365,10}$, then count the sequences for which at least two persons share a birthday. It might appear that the probability of two or more persons having the same birthday is N/D. This is not correct.

Consider the following sanity check. Suppose that the room contains only two persons. You then expect the probability of a shared birthday to be $1/365$. (You ask one person for his birthday, and you have 1 chance in 365 that it will match that of the other person.) The reasoning above, however, gives

$$\frac{N}{D} = \frac{C_{365+2-1,2} - C_{365,2}}{C_{365+2-1,2}} = \frac{1}{183}.$$

In the physical experiment, the persons are distinguishable entities. If you line them up and demand birthdays, from left to right, you will extract a 10-sequence from the set $\{1, 2, 3, \ldots, 365\}$, not a 10-combination. Order does count. A different 10-sequence, even if it represents the same 10-combination, corresponds to a different assignment of birthdays. So it is the 10-sequences that are equally likely. Reordering the elements in a 10-combination does not always produce the same number of 10-sequences. For example, if the 10-combination has all distinct elements, then the reordering will produce 10! 10-sequences. If, on the other hand, it has a single element, repeated 10 times, then reordering produces a single 10-sequence. A 10-combination of distinct elements occurs 10! times as often as the 10-combination containing a single repeated entry.

Returning to the field of equally likely possibilities, there are 365^{10} possible 10-sequences, of which $P_{365,10}$ have no duplicate elements. The following calculation then gives the correct probability.

$$\frac{365^{10} - (365)(364)\ldots(356)}{365^{10}} = 1 - \frac{365}{365} \cdot \frac{364}{365} \cdot \frac{363}{365} \cdots \frac{356}{365} = 0.117$$

If the room contains only two persons, the appropriately modified calculation gives the expected result:

$$\frac{365^2 - (365)(364)}{365^2} = 1 - \frac{365}{365} \cdot \frac{364}{365} = \frac{1}{365}. \quad \square$$

The use of k-sequences or k-combinations arises in a class of problems called *occupancy problems*. The common abstraction is that k objects must be distributed across n containers. The distribution is with replacement, in the sense that a given container can acquire multiple objects. The containers have distinguishable labels; the objects, however, may or may not be distinguishable.

If the objects are distinguishable, then our interest lies with the k-sequences, chosen with replacement from the container labels. A given k-sequence then tells us which containers contain which objects. In particular, the first item in the sequence is the (label of) the container where the first object resides. The second item denotes the location of the second object, and so forth.

If the physical reality of the situation is such that all such k-sequences are equally likely, the consequent probabilities are said to follow *Maxwell-Boltzmann statistics*. In the examples above, we feel that the distribution of 10 coin tosses into 2 bins (heads or tails) follows Maxwell-Boltzmann statistics. Consequently, a sequence, such as $HHTTTHTTHT$, is just as likely as any other sequence. However, two counts, such as $(H = 4, T = 6)$ and $(H = 2, T = 8)$ are not equally likely. The same remarks apply to the distribution of 10 birthdays into 365 containers (the calendar days). We must work with the 10-sequences from the calendar to derive probabilities that comport with experience.

There are cases when it is appropriate to use k-combinations as a basis for probability calculations. One such instance is when the sampling protocol is without replacement, as discussed earlier. When sampling with replacement, however, we must analyze the situation to ensure that the k-combinations properly count the sets under consideration. If we are distributing indistinguishable objects across n bins, a k-combination tells us the *number* of objects in the first bin, the second bin, and so forth. It provides this information in the following manner. The k-combination contains zero or more repetitions of each of the bin labels. So, as noted earlier, each such k-combination corresponds to exactly one set of repetition counts, $k_1, k_2, \ldots, k_n$ that total k. The number of objects in the first bin is k_1. The second bin contains k_2 objects, and so forth.

If physical reality is such that these repetition-count sets are equally likely, then probabilities should be ratios of favored counts to all possible counts. If probabilities follow this law, they are said to follow *Bose-Einstein statistics*. Physicists study certain distributions of particles to labeled energy containers, and they have found, perhaps contrary to intuition, that some particle distributions do obey Bose-Einstein statistics. These examples are beyond the scope of this text. We do, however, use k-combinations to analyze sampling with replacement when the repetition counts themselves participate directly in the probability computations. The following examples illustrates the point.

1.26 Example: Suppose that you have available, in unlimited amounts, food packets of 6 types. The types are labeled $1, 2, 3, 4, 5$, and 6. You must make up a shipment of 20 packets, and there is no limit on the number of repetitions of any

```
function optimumShipment ()
    k = (0, 0, 0, 0, 0, 20);
       // initial assignment: 20 of type 6, 0 of types 1-5
    choice = k; value = 0.0; searching = true;
    while (searching)
        x = 0.0;
        for (j = 1 to 6)
            x = x + k[j] * weight (j);
        y = expert (k) / x;
        if (y > value)
            value = y; choice = k;
        if (k = (20, 0, 0, 0, 0, 0))
            searching = false;
        else
            k = next(k);
    return choice;
```

Figure 1.5. Algorithm to determine optimal food pack (see Example 1.26)

---

one type. The weight of a packet depends only on its type. The nutritional value, however, varies when mixed with other packets. That is, some packets are more nutritional when served in combination with other packets. An expert is available to determine the nutritional value of any 20 packets chosen from the 6 types. The expert charges 1 dollar for each consultation when the shipment includes no packets of type 1. Because packets of type 1 require more expensive testing, the expert charges 2 dollars to evaluate a shipment that includes packets of type 1. Your task is to ship 20 packets to the scene of a catastrophe, and you want to maximize the nutrition delivered per unit weight. What is the maximum that you should have to pay the expert to attain your goal?

Consider the algorithm of Figure 1.5, where the subroutine expert (k) returns the nutritional value of a shipment containing k[1] packets of type 1, k[2] packets of type 2, and so forth, through k[6] packets of type 6. We also assume a routine weight (j) that specifies the weight of a packet of type j. The routine next(k) acquires a new assignment of 20 items in some indeterminate manner. Of interest here is simply the fact that all assignments are eventually processed and that the final assignment is the one that terminates the while-loop: (20, 0, 0, 0, 0, 0).

We need to determine how many times the algorithm invokes the routine expert and what fraction involve packets of type 1. Since the code systematically steps through all possible 6-sequences of nonnegative integers that sum to 20, we can appeal to Theorem 1.23 for the number of such sequences. It is $C_{6+20-1,20} = 53130$. If the first element of the sequence is zero, then the remaining elements are precisely those 5-sequences that sum to 20, and again by Theorem 1.23, they number $C_{5+20-1,20} = 10626$. So our expert costs 1 dollar for each of 10626 calls plus 2 dollars for each of $53130 - 10626 = 42504$ calls, for a total of 95634 dollars. This is the most we will have to pay the expert because this approach investigates every possible assignment. □

1.27 Example: A hash table is a storage structure that calculates the disk storage location of a record as a function of the record key value. The disk storage areas are usually called buckets, and each bucket can hold some maximum number of records, say b. Suppose that there are N buckets. The calculation that operates on the record key to produce the bucket number is called the hash function. A good hash function spreads the records uniformly across the buckets. This is important because when a bucket fills, any further records assigned to that bucket overflow onto new buckets, beyond the original N. Each bucket in the original collection is potentially the beginning of an overflow chain of additional buckets needed to accommodate such inappropriate assignments.

The hash function, for example, could be $f(k) = k \mod 100$, where k is the key value. The key might be a 7-digit part number, for instance. This hash function then returns the last two digits of the part number, which represent a number in the range 0 to 99. Records are stored in the corresponding buckets. If the bucket capacity is $b = 5$, this hash function may or may not generate overflow chains, depending on the specific part numbers that participate. Even though the system has room for 500 records in the 100 buckets, it could still generate overflows on the first six storage operations if the part numbers of these six records all have the same last two digits.

The appeal of a hash table lies in the subsequent retrieval of the stored records. When presented with a key, the hash function can calculate the bucket that contains the record, which allows us to retrieve just that bucket, instead of sequentially searching through all buckets in the structure. Of course, we are occasionally disappointed to find an overflow chain attached to the home bucket, and that unhappy circumstance necessitates further retrievals. Nevertheless, if the overflows are infrequent, most retrievals will be efficient.

For this example, assume that the hash function does spread the records uniformly across the N buckets in the sense that each bucket is equally likely every time a storage assignment is made. A random hash function would be of small worth, because we must be able to locate the storage bucket deterministically when a later retrieval request presents the key. So we assume that just the storage assignment is made in a uniform random fashion. Thereafter, the function remembers the buckets that have been assigned to particular keys, so it can perform retrievals. This is a bit much to expect of a real hash function, but an analysis of this case allows us to infer the approximate behavior of a real hash function that does manage to spread its storage assignments uniformly across the available buckets.

Under this assumption, we want the probability $D(p)$ that a bucket receives exactly p records, including those that might produce overflows, for $p = 0, 1, 2, \ldots, n$. The relevant parameters are: N, the number of buckets; n, the number of records assigned; b, the number of records a bucket can hold without overflowing; and $\alpha = n/Nb$, the load factor, which is the fraction of the system capacity (without overflows) occupied by records when all have been assigned. For the case $N = 10000, b = 10$, the results appear as Table 1.6, and we now proceed to justify these values.

After receiving an assignment, a bucket remains in the pool of choices for subsequent assignments. So the process draws n samples, with replacement, from the buckets $\{1, 2, \ldots, N\}$. Theorem 1.19 asserts that there are N^n possible n-sequences, and our assumption about the uniform nature of the hash function means that each of these sequences is equally likely to occur. If x is an arbitrarily selected bucket, how many of these sequences assign exactly p records to bucket x? That is, how many sequences contain exactly p elements labeled x? Applying the multiplicative

	Number of records					
			Load factor			
	10000	30000	50000	70000	90000	99000
p	0.1	0.3	0.5	0.7	0.9	0.99
0	0.3679	0.0498	0.0067	0.0009	0.0001	0.0001
1	0.3679	0.1494	0.0337	0.0064	0.0011	0.0005
2	0.1839	0.2240	0.0842	0.0223	0.0050	0.0025
3	0.0613	0.2240	0.1404	0.0521	0.0150	0.0081
4	0.0153	0.1680	0.1755	0.0912	0.0337	0.0201
5	0.0031	0.1008	0.1755	0.1277	0.0607	0.0398
6	0.0005	0.0504	0.1462	0.1490	0.0911	0.0656
7	0.0001	0.0216	0.1044	0.1490	0.1171	0.0928
8	0.0000	0.0081	0.0653	0.1304	0.1318	0.1148
9	0.0000	0.0027	0.0363	0.1014	0.1318	0.1263
10	0.0000	0.0008	0.0181	0.0710	0.1186	0.1250
>10	**0.0000**	**0.0003**	**0.0137**	**0.0985**	**0.2940**	**0.4045**

TABLE 1.6. $D(p)$, the probability of a hash table bucket receiving p records (see Example 1.27)

counting principle, we reason that for each of the $C_{n,p}$ ways of choosing locations for the elements labeled x, there are $(N-1)^{n-p}$ ways of completing the sequence by assigning the remaining $n-p$ locations to any one of the remaining $(N-1)$ buckets that are *not* labeled x. In other words, the remaining $n-p$ positions form a $(n-p)$-sequence chosen with replacement from the buckets other than x, and there are $(N-1)$ such buckets. The counting principle then asserts that the number of sequences that contain exactly p elements labeled x is $C_{n,p}(N-1)^{n-p}$. The required probability is then $D(p) = C_{n,p}(N-1)^{n-p}/N^n$. This result is independent of x and is therefore the same for all buckets. Each bucket receives exactly p records with probability $D(p)$. We can therefore interpret $D(p)$ as the fraction of the N buckets that receive exactly p records. Note that these fractions sum properly to 1.

$$\sum_{p=0}^{n} \frac{C_{n,p}(N-1)^{n-p}}{N^n} = \frac{\sum_{p=0}^{n} C_{n,p} 1^p (N-1)^{n-p}}{N^n} = \frac{[1+(N-1)]^n}{N^n} = 1$$

For $p > b$, this expression gives the fraction of buckets that overflow. Consider again Table 1.6, which illustrates the situation with $10,000$ buckets, each of capacity 10. Each column corresponds to a different number of records stored in the hash structure. These values run from $n = 10000$ through $n = 99000$. The corresponding load factors appear below the number of records. The column then gives the probability that a bucket will receive $p = 0, 1, 2, \ldots, 10$ record assignments. The last value, shown in boldface type gives the probability that a bucket will receive more than 10 assignments and will therefore overflow. You can see that that probability of overflow is rather small until the load factor exceeds 0.7. The chance of overflow does rise rapidly after this threshold; there is a 40% chance of overflow when the load factor reaches 0.99. Figure 1.6 depicts the same information in a graphical format, using interpolation to produce smooth curves for each of the load factors $\alpha = 0.1$ through $\alpha = 0.99$. Each curve, therefore, corresponds to one column of data from Table 1.6. Overflow occurs when significant probability exists for $p > 10$. As you

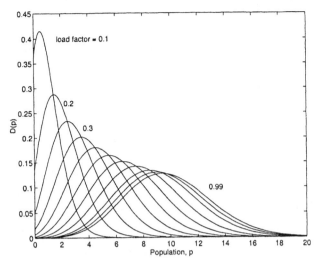

Figure 1.6. Probability that a hash table bucket receives exactly p records (see Example 1.27)

can see, the curves for low load factors, up through about $\alpha = 0.7$, have reasonably small $D(p)$ values, for $p > 10$.

Hash table manipulations occur often in database applications, and bucket fetching is usually the most time-intensive operation because it involves disk accesses. If a hash table contains no overflow chains, each retrieval requires just 1 bucket fetch. Performance deteriorates when overflow chains appear. We can use the $D(p)$ values just obtained to calculate the average number of bucket fetches per retrieval, assuming that the hash function spreads the records in a uniform manner. □

Exercises

1.17 You receive 5 cards from a standard card deck, but you return each card immediately after noting its value and suit. What is the probability that you receive two or more aces? Should this probability be larger or smaller than the probability of receiving two or more aces when you keep the cards?

1.18 Three dice are thrown simultaneously. What is the probability that the sum of the spots on the upturned faces is less than six?

*1.19 You shuffle a deck of cards and then deal all 52 cards in a row. What is the probability that the display contains no adjacent spades?

1.20 In 10 tosses of a fair coin, what is the probability that the number of heads will be greater than twice the number of tails?

1.21 You receive three cards from a well-shuffled standard deck. You note the values and suits and then return the cards. You then receive two further cards. What is the probability that you see two or more aces during the course of the experiment?

1.22 From an unlimited supply of pennies, nickels, dimes, and quarters, you are to select 20 coins. How many different combinations are possible? What fraction consists of monetary amounts that are divisible by 5?

1.23 Consider the equation $\sum_{i=1}^{7} x_i = 30$, where all the x_i are nonnegative integers. What fraction of the solutions have both $x_1 = 0$ and $x_7 = 0$?

1.24 Consider the equation $\sum_{i=1}^{7} x_i = 60$, where each x_i satisfies $x_i \geq i$. What fraction of the solutions also satisfy $x_i \geq 2i$?

*1.25 Suppose that $k > n$ and consider the equation $\sum_{i=1}^{n} x_i = k$, where the x_i are *positive* integers. How many solutions exist?

*1.26 Suppose that $\sum_{i=1}^{n} x_i \leq k$, where the x_i are nonnegative integers. How many solutions exist?

1.27 How many people must assemble to have a probability greater than one-half that two or more persons share the same birthday? How many people are necessary to raise the probability to 0.8?

1.28 A restaurant occasionally announces that the owner will pick up the tab for any patron whose birthday is tonight. If the owner wants a 50% chance of escaping without giving away any free dinners, what is the largest crowd to which he can make the announcement?

*1.29 $\left(\sum_{k=0}^{\infty} \beta^k\right)^n = \sum_{k=0}^{\infty} a_k \beta^k$ for $|\beta| < 1$. Find the coefficients $\{a_k\}$.

*1.30 Example 1.27 analyzes a hash table with N buckets of capacity b to which n records are allocated in a uniform manner. It computes $D(p)$, the fraction of buckets that receive exactly p records as $D(p) = C_{n,p}(N-1)^{n-p}/N^n$. For large N and n and small p, show that $D(p)$ is approximately $(\alpha b)^p e^{-\alpha b}/p!$, where $\alpha = n/(Nb)$ is the table load factor.

1.2   Summations

Consider the code segment to the right, in which we want to count the number of calls to procedure inner. When $i = 1$, j runs from 2 to $n + 1$, which gives n procedure calls. When

```
for (i = 1 to n)
    for (j = i + 1 to n + 1)
        inner(i, j);
```

$i = 2$, j runs from 3 to $n + 1$, producing $n - 1$ calls. The number of calls decreases by 1 for each successive value of i, until the final loop, when $i = n$, produces just 1 call. We can write the total by summing up from 1 to n or by summing down from n to 1. The following manipulation produces the simple expression $S = n(n + 1)/2$ for the sum.

$$
\begin{array}{ccccccccccc}
S &=& 1 &+& 2 &+& 3 &+ \cdots +& n-1 &+& n \\
S &=& n &+& n-1 &+& n-2 &+ \cdots +& 2 &+& 1 \\
\hline
2S &=& (n+1) &+& (n+1) &+& (n+1) &+ \cdots +& (n+1) &+& (n+1)
\end{array}
$$

In summation notation, we have shown that $\sum_{i=1}^{n} i = n(n + 1)/2$. In this expression, the right side is called a *closed form* because it involves no summation symbols. An expression with a summation symbol cannot be expanded without using an ellipsis. The left side, for example, expands as $1 + 2 + 3 + \cdots + n$. The ellipsis (the three dots) is difficult to manipulate in further algebraic simplification. The closed-form on the right, by contrast, easily participates in further algebra.

Although this device quickly provides a closed-form expression for $\sum i$, it is not very helpful for $\sum i^2$ or $\sum i^3$, which also occur in counting operations. What we need is a general expression for $\sum i^k$, for arbitrary k. These sums become increasingly complicated as k grows larger, but there is a systematic approach to the problem. Before launching into a calculation of $\sum_{i=1}^{n} i^2$, we illustrate a useful transformation called index shifting. This technique is helpful in manipulating indices that are slightly offset from those we would find convenient. For example, consider the following two sums, which we can show to be equal.

$$\sum_{i=1}^{n}(i + 1)i = \sum_{i=2}^{n+1} i(i - 1)$$

The left side generates $(2)(1), (3)(2), (4)(3), \ldots, (n + 1)(n)$, as i runs from 1 to n. But these are exactly the terms generated by the right side as i runs from 2 to $n + 1$. The same number of terms result when we shift the sum's lower and upper limits up by 1, and we can ensure that the value of each term remains unchanged by adjusting each occurrence of i to $i - 1$. Observing an index shift of this sort, we quickly verify the conditions needed to maintain equality: (1) the lower and upper indices are shifted by the same amount in the same direction, and (2) the first term generated in the transformed expression is the same as the first term in the old expression. The latter condition holds when the summation variable is adjusted by the same amount as the lower and upper limits, but in the opposite direction. In the example above, the lower and upper limits advance by 1, and each occurrence of the summation index decreases by 1. We need not check terms other than the first. If the initial terms of the two expressions are the same, all succeeding terms will follow along in parallel as the summation index increases uniformly on both sides. The following formula, in which the shift s may be positive or negative, generalizes this rule.

$$\sum_{i=j}^{k} f(i) = \sum_{i=j+s}^{k+s} f(i - s) \tag{1.1}$$

Keeping this transformation in mind, consider the following manipulation, which manages to recover the starting expression plus some other terms. The other terms must, of course, be equal to zero. Note that not all transformations are index shifts; some are obtained by breaking off or adding on a term at the sum's extreme.

$$\sum_{i=1}^{n} i(i-1) = \sum_{i=2}^{n+1}(i-1)(i-2) = \sum_{i=2}^{n+1} i(i-1) - 2\sum_{i=2}^{n+1}(i-1)$$

We adjust the first summation by breaking out the first and last terms; we adjust the second summation by shifting the index.

$$\sum_{i=1}^{n} i(i-1) = -(1)(0) + \sum_{i=1}^{n} i(i-1) + (n+1)(n) - 2\sum_{i=1}^{n} i$$

$$= \sum_{i=1}^{n} i(i-1) + (n+1)(n) - 2\sum_{i=1}^{n} i$$

Canceling the left side with its counterpart on the right and transposing the remaining sum gives $\sum_{i=1}^{n} i = (n+1)n/2$. Of course, we already know this result from the previous calculation. But it is comforting to get the same answer, and besides, this technique generalizes to calculate $\sum_{i=1}^{n} i^k$.

1.28 Theorem: $\sum_{i=1}^{n} P_{i,k} = P_{n+1,k+1}/(k+1)$.

PROOF: To see the general pattern, we repeat the previous discussion with the first three factors of $i!$.

$$\sum_{i=1}^{n} P_{i,3} = \sum_{i=1}^{n} i(i-1)(i-2) = \sum_{i=2}^{n+1}(i-1)(i-2)(i-3)$$

$$= \sum_{i=2}^{n+1} i(i-1)(i-2) - 3\sum_{i=2}^{n+1}(i-1)(i-2)$$

$$= -(1)(0)(-1) + \sum_{i=1}^{n} i(i-1)(i-2) + (n+1)n(n-1)$$

$$-3\sum_{i=1}^{n} i(i-1)$$

The left side again cancels with a right term, giving

$$\sum_{i=1}^{n} P_{i,2} = \sum_{i=1}^{n} i(i-1) = \frac{(n+1)n(n-1)}{3} = \frac{P_{n+1,3}}{3}.$$

So we have established the theorem by direct calculation for $k=1$ and $k=2$, and we now need to handle the general case.

$$\sum_{i=1}^{n} P_{i,k+1} = \sum_{i=1}^{n} i(i-1)(i-2)\cdots(i-k+1)(i-k)$$

$$\sum_{i=1}^{n} P_{i,k+1} = \sum_{i=2}^{n+1} (i-1)(i-2)\cdots(i-k)(i-k-1)$$

$$= \sum_{i=2}^{n+1} P_{i-1,k}[i-(k+1)] = \sum_{i=2}^{n+1} iP_{i-1,k} - (k+1)\sum_{i=2}^{n+1} P_{i-1,k}$$

We adjust the first sum to run from 1 to n by subtracting the summand for $i = 1$ and breaking off the last term:

$$\sum_{i=2}^{n+1} iP_{i-1,k} = -(1)(0)(-1)\cdots(1-k) + \sum_{i=1}^{n} iP_{i-1,k} + (n+1)P_{n,k}$$

$$= \sum_{i=1}^{n} P_{i,k+1} + P_{n+1,k+1}.$$

Consequently,

$$\sum_{i=1}^{n} P_{i,k+1} = \sum_{i=1}^{n} P_{i,k+1} + P_{n+1,k+1} - (k+1)\sum_{i=1}^{n} P_{i,k}.$$

Canceling the left side with its right counterpart and dividing by $(k+1)$ gives the desired result. $\blacksquare$

Using the initial cases already established, we can now compute $\sum_{i=1}^{n} i^k$, for $k = 1, 2, 3, \ldots$. For example, the computation for $\sum_{i=1}^{n} i^2$ is as follows, and the next theorem collects the first several such results.

$$\frac{P_{n+1,3}}{3} = \sum_{i=1}^{n} P_{i,2}$$

$$\frac{(n+1)n(n-1)}{3} = \sum_{i=1}^{n} i(i-1) = \sum_{i=1}^{n}(i^2 - i) = \sum_{i=1}^{n} i^2 - \sum_{i=1}^{n} i$$

$$\sum_{i=1}^{n} i^2 = \frac{(n+1)n(n-1)}{3} + \frac{n(n+1)}{2} = \frac{n(n+1)(2n+1)}{6}$$

1.29 Theorem: $\displaystyle\sum_{i=1}^{n} i^p = \begin{cases} n(n+1)/2, & p = 1 \\ n(n+1)(2n+1)/6, & p = 2 \\ n^2(n+1)^2/4, & p = 3. \end{cases}$

PROOF: The proofs for $\sum_{i=1}^{n} i$ and $\sum_{i=1}^{n} i^2$ appear in the discussion above. For $\sum_{i=1}^{n} i^3$ we again proceed from Theorem 1.28.

$$\frac{P_{n+1,4}}{4} = \sum_{i=1}^{n} P_{i,3}$$

$$\frac{(n+1)n(n-1)(n-2)}{4} = \sum_{i=1}^{n} i(i-1)(i-2) = \sum_{i=1}^{n}(i^3 - 3i^2 + 2i)$$

$$= \sum_{i=1}^{n} i^3 - \frac{3n(n+1)(2n+1)}{6} + \frac{2n(n+1)}{2}$$

Solving for $\sum i^3$ now gives the desired expression: $\sum_{i=1}^{n} i^3 = n^2(n+1)^2/4.$ ∎

1.30 Example: The procedure to the right below accepts a list (i.e., a vector) $A[1..n]$ and sorts it into ascending order. Find $f(n)$, the number of executions of the comparison in line 4.

The algorithm loops give the following nested sums, which after some manipulation, provide an opportunity to involve the first summation formula of Theorem 1.29. Toward the end, the calculation uses the fact that $\sum_{i=1}^{n-1} i = \sum_{i=1}^{n-1}(n-i)$. This is true because the second sum generates the same terms as the first, but in reverse order.

```
function slowsort(A, n)
    for (i = 1 to n - 1)
        for (j = i + 1 to n)
            if (A[j] < A[i])
                x = A[i];
                A[i] = A[j];
                A[j] = x;
```

$$f(n) = \sum_{i=1}^{n-1} \sum_{j=i+1}^{n} (1) = \sum_{i=1}^{n-1} [n - (i+1) + 1] = \sum_{i=1}^{n-1} (n-i) = \sum_{i=1}^{n-1} i = \frac{(n-1)n}{2} \quad \square$$

Besides sums of integers and powers of integers, we also encounter geometric sums and their variations. A plain geometric sum is one in which each term is a constant multiple of its predecessor. We shall see that we can easily convert such sums into closed-form expressions. We shall also find closed-form expressions for certain systematic variations of the plain geometric sum, where the multiple that generates each term from its predecessor is not constant.

First, consider the plain geometric sum $\sum_{i=0}^{n} ar^i$. The first term is a, and each remaining term is r times its predecessor. We can expand the sum and employ a trick similar that used to calculate $\sum_{i=1}^{n} i$. This is one of the few instances where the ellipsis notation allows further manipulation. Letting S represent the sum, the following array immediately implies the subsequent closed form for the geometric sum.

$$
\begin{array}{rl}
S = & a + ar + ar^2 + \ldots + ar^{n-1} + ar^n \\
rS = & \quad\quad ar + ar^2 + \ldots + ar^{n-1} + ar^n + ar^{n+1} \\
\hline
(1-r)S = & a + 0 + 0 + \ldots + 0 + 0 - ar^{n+1}
\end{array}
$$

$$S = \sum_{i=0}^{n} ar^i = \frac{a(1 - r^{n+1})}{1 - r} \qquad (1.2)$$

Equation 1.2 is valid only when $r \neq 1$, but the sum is particularly easy to evaluate directly when $r = 1$. Now consider the following arrangement, where the sum of the first n integers is complicated with a geometric factor. We immediately add zero at the beginning of the sum in the form $0r^0$. Of course, this leaves the sum unchanged.

$$\sum_{i=1}^{n} ir^i = r + 2r^2 + 3r^3 + \cdots + nr^n = 0r^0 + r + 2r^2 + 3r^3 + \cdots + nr^n$$

$$= \sum_{i=0}^{n} ir^i = r \sum_{i=0}^{n} ir^{i-1} = r\frac{d}{dr} \sum_{i=0}^{n} r^i = r\frac{d}{dr}\left(\frac{1 - r^{n+1}}{1 - r}\right)$$

$$= r\left(\frac{nr^{n+1} - (n+1)r^n + 1}{(1-r)^2}\right) = \frac{r[nr^{n+1} - (n+1)r^n + 1]}{(1-r)^2} \qquad (1.3)$$

The expression for an infinite geometric sum finds frequent use, and we can obtain it by considering the limiting form of Equation 1.2.

$$\sum_{i=0}^{\infty} r^i = \lim_{n \to \infty} \sum_{i=0}^{n} r^i = \lim_{n \to \infty} \frac{1 - r^{n+1}}{1 - r} = \frac{1}{1 - r}, \tag{1.4}$$

provided that $|r| < 1$. The first two sections of Appendix A, which discusses sets, functions, and limits, should be reviewed at this time. The following examples provide further illustrations of limiting summations but assume the background familiarity with limit processes provided by the appendix.

1.31 Example: Evaluate $\sum_{i=0}^{\infty} i/2^i$.

For any finite n, Equation 1.3 shows that

$$\sum_{i=1}^{n} ir^i = \frac{r[nr^{n+1} - (n+1)r^n + 1]}{(1-r)^2}.$$

Substituting $r = 1/2$ and adding the inconsequential term for $i = 0$ gives

$$\sum_{i=0}^{n} \frac{i}{2^i} = 2 - \frac{n+2}{2^n}$$

$$\sum_{i=0}^{\infty} \frac{i}{2^i} = \lim_{n \to \infty} \sum_{i=0}^{n} \frac{i}{2^i} = \lim_{n \to \infty} \left(2 - \frac{n+2}{2^n}\right) = 2.$$

This last reduction used l'Hôpital's rule to evaluate the indeterminate form ∞/∞. The rule stipulates that a ratio $f(n)/g(n)$, in which both $f(n)$ and $g(n) \to \infty$, approaches $\lim_{n \to \infty} f'(n)/g'(n)$, provided that the latter is well defined. In the case at hand,

$$\frac{n+2}{2^n} \to \frac{1}{2^n \ln 2} \to 0. \,\square$$

1.32 Example: Show that the geometric variant $\sum_{i=1}^{\infty} ir^i$ converges for $0 \le r < 1$.

If $r = 0$, all partial sums are also zero, and the convergence is trivial. Consequently, assume that $0 < r < 1$. From Equation 1.3 we have

$$\sum_{i=1}^{n} ir^i = \frac{r[nr^{n+1} - (n+1)r^n + 1]}{(1-r)^2} = \frac{r}{(1-r)^2} \cdot [nr^n(r-1) - r^n + 1].$$

Since $r < 1$, $\lim_{n \to \infty} r^n = 0$, so assuming for the moment that nr^n also goes to zero, we have that

$$\sum_{i=1}^{\infty} ir^i = \lim_{n \to \infty} \sum_{i=1}^{n} ir^i = \frac{r}{(1-r)^2}.$$

To show that nr^n actually does go to zero, note that $nr^n = n/(1/r)^n$, which approaches the indeterminate form ∞/∞. Consequently,

$$\lim_{n \to \infty} nr^n = \lim_{n \to \infty} \frac{n}{(1/r)^n} = \lim_{n \to \infty} \frac{1}{(1/r)^n \ln(1/r)} = \lim_{n \to \infty} \frac{r^n}{\ln(1/r)} = 0. \,\square$$

1.33 Example: To the right below is a procedure that searches a list $A[1..n]$ for a target X. The unsophisticated algorithm simply compares each list element with the target, starting at the first element, until it finds a match or falls off the end of the list. If it finds a match, it returns the index of the matching element; otherwise, it returns a 0.

Suppose that this procedure runs in an application environment where the target appears in the first position half the time, in the second position one-fourth of the time, and so forth. Over a large number of executions, what is the average number of comparisons of the target against a list value?

```
int slowsearch(A, n, X)
    for (i = 1 to n)
        if (A[i] == X)
            return i;
    return 0;
```

If we run the algorithm a large number of times, say N, we will find the target in the first position $N/2$ times, in the second position $N/4$ times, and so forth, through finding the target in nth position $N/2^n$ times. The remaining times, which will be $N/2^n$, the target is not in the list. Note that these fractions properly add to N.

$$\left(\frac{N}{2} + \frac{N}{4} + \frac{N}{8} + \cdots + \frac{N}{2^n}\right) + \frac{N}{2^n} = N\left(\frac{1}{2^n} + \sum_{i=1}^{n}\frac{1}{2^i}\right) = N$$

The last equality follows because Equation 1.2 gives

$$\frac{1}{2^n} + \sum_{i=1}^{n}\frac{1}{2^i} = \frac{1}{2^n} + \sum_{i=0}^{n-1}\frac{1}{2^{i+1}} = \frac{1}{2^n} + \frac{1}{2}\sum_{i=0}^{n-1}\left(\frac{1}{2}\right)^i = \frac{1}{2^n} + \frac{1}{2}\left(\frac{1-(1/2)^n}{1-(1/2)}\right) = 1.$$

We are concerned with the number of comparisons in line 3, where the for-loop compares list element $A[i]$ with target X. In particular, we want to know the average number of comparisons over a large number of procedure runs N. So we add the comparisons generated in each run and divide the total by N. (We will develop a more exact definition of average, as a probability concept, later in the text. For now, let us be content with this intuitive meaning of average.) Because $N/2$ runs find the target in the first position, they each do 1 comparison. $N/4$ runs find the target in the second position, and they therefore do two comparisons. The $N/2^n$ runs that find the target in the last position must do n comparisons. The $N/2^n$ runs that do not find the target in the list also do n comparisons each. So we can calculate our required average $f(n)$ as follows, where the second line invokes Equation 1.3 with $r = 1/2$.

$$f(n) = \frac{1}{N}\left(\frac{N}{2}(1) + \frac{N}{4}(2) + \cdots + \frac{N}{2^n}(n) + \frac{N}{2^n}(n)\right)$$

$$= \frac{n}{2^n} + \sum_{i=1}^{n}i\left(\frac{1}{2}\right)^i = \frac{n}{2^n} + \frac{(1/2)[n(1/2)^{n+1} - (n+1)(1/2)^n + 1]}{[1-(1/2)]^2}$$

$$= \frac{n}{2^n} + 2\left[\frac{1}{2}n\left(\frac{1}{2}\right)^n - n\left(\frac{1}{2}\right)^n - \left(\frac{1}{2}\right)^n + 1\right] = 2\left(1 - \frac{1}{2^n}\right)$$

As n becomes large, $f(n)$ approaches 2, which reflects the bias in the application. Recall that most of the time the target is among the early list elements, and it takes, on the average, only about 2 probes to find it. □

The technique used to derive $\sum_{i=0}^{n} ir^i$ is worth remembering because it generalizes to $\sum_{i=0}^{n} i^k r^i$ for other values of k. The method involves differentiating a closed-form expression of $\sum_{i=0}^{n} r^i$. The exercises pursue this line of thought. Here, however, we employ the same general approach for expressions involving binomial coefficients. Summing the coefficients themselves is easy.

$$\sum_{k=0}^{n} C_{n,k} = \sum_{k=0}^{n} C_{n,k} 1^k 1^{n-k} = (1+1)^n = 2^n \tag{1.5}$$

How might we obtain a closed-form expression for $\sum_{k=0}^{n} k \cdot C_{n,k}$? Consider the following differentiable function of x.

$$f(x) = (x+1)^n = \sum_{k=0}^{n} C_{n,k} x^k 1^{n-k} = \sum_{k=0}^{n} C_{n,k} x^k$$

$$f'(x) = n(x+1)^{n-1} = \sum_{k=0}^{n} k C_{n,k} x^{k-1} = \frac{1}{x} \sum_{k=0}^{n} k C_{n,k} x^k$$

We let $x = 1$ to obtain

$$\sum_{k=0}^{n} k \cdot C_{n,k} = n \cdot 2^{n-1}. \tag{1.6}$$

Example 1.18 used this closed-form expression. The technique again generalizes to provide closed-form expressions for $\sum_{k=0}^{n} k^p C_{n,k}$ for higher values of p.

The binomial coefficients exhibit a large number of interrelationships. The following two prove useful in upcoming derivations, and the exercises provide an opportunity to investigate further relationships.

1.34 Theorem: (Pascal's identity) For $0 < k < n$, $C_{n,k} = C_{n-1,k} + C_{n-1,k-1}$. PROOF: Let $S = \{s_1, s_2, \ldots, s_n\}$. Then $C_{n,k}$ counts the number of k-combinations available without replacement from S. We divide these combinations into two groups: those that contain s_n and those that do not contain s_n. Each k-combination in the first group contains s_n plus $k-1$ further choices from $s_1, s_2, \ldots, s_{n-1}$. Since there are $C_{n-1,k-1}$ such choices, the first group must contain $C_{n-1,k-1}$ elements. The second group contains k-combinations that do not include s_n and so must therefore include all k choices from the remaining $s_1, s_2, \ldots, s_{n-1}$. There are $C_{n-1,k}$ such choices, so there are $C_{n-1,k}$ elements in the second group. Adding the two group sizes completes the theorem.

By expressing the right side in factorial terms, we can construct an algebraic proof.

$$C_{n-1,k} + C_{n-1,k-1} = \frac{(n-1)!}{k!(n-k-1)!} + \frac{(n-1)!}{(k-1)!(n-k)!}$$

$$= \frac{(n-k)}{(n-k)} \cdot \frac{(n-1)!}{k!(n-k-1)!} + \frac{k}{k} \cdot \frac{(n-1)!}{(k-1)!(n-k)!}$$

$k \longrightarrow$

n	0	1	2	3	4	5	6	7	8	9	10
0	1										
1	1	1									
2	1	2	1								
3	1	3	3	1							
4	1	4	6	4	1						
5	1	5	10	10	5	1					
6	1	6	15	20	15	6	1				
7	1	7	21	35	35	21	7	1			
8	1	8	28	56	70	56	28	8	1		
9	1	9	36	84	126	126	84	36	9	1	
10	1	10	45	120	210	256	210	120	45	10	1

TABLE 1.7. Pascal's triangle (see Theorem 1.34)

Rearranging to force the factor $(n-1)!/[k!(n-k)!]$ into both terms, we have

$$C_{n-1,k} + C_{n-1,k-1} = (n-k) \cdot \frac{(n-1)!}{k!(n-k)!} + k \cdot \frac{(n-1)!}{k!(n-k)!}$$

$$= (n-k+k) \cdot \frac{(n-1)!}{k!(n-k)!} = \frac{n!}{k!(n-k)!} = C_{n,k}.$$

The first proof is more satisfying for the insight into how to decompose the original $C_{n,k}$ k-combinations into two groups. ∎

Pascal's triangle arranges the binomial coefficients to emphasize the relationship of Theorem 1.34. As shown in Table 1.7, each row corresponds to a particular n and contains the entries $C_{n,k}$, for $k = 0, 1, \ldots, n$. Each row starts and ends with a 1, and Pascal's identity computes each remaining entry as the sum of its north and northwest neighbors.

1.35 Theorem: Suppose that $m \leq n$. Then $\sum_{j=0}^{k} C_{m,j} C_{n-m,k-j} = C_{n,k}$.

PROOF: Recall the convention that $C_{a,b} = 0$ if $b > a$. Hence the sum produces nonzero terms only while $j \leq m$ and $k - j \leq n - m$ [i.e., $j \geq k - (n-m)$]. So although j formally runs from 0 to k, some of the early terms can be zero if $k > (n-m)$, and some of the final terms can be zero if $k > m$.

Consider a set containing m elements of one type and $n - m$ elements of a second type. That is, consider $S = \{x_1, x_2, \ldots, x_m, y_1, y_2, \ldots, y_{n-m}\}$. Now imagine the set elements aligned over n slots. The situation for $n = 9, m = 4$ appears as follows.

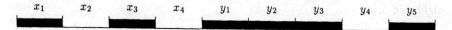

The diagram shows one of $C_{9,6}$ patterns. In this one, 2 elements come from the prefix (the first four slots), and 4 come from the suffix (the last five slots). There are $C_{4,2} C_{5,4}$ patterns of this subtype. Other subtypes contain 1 element from the prefix and 5 from the suffix, which contributes $C_{4,1}C_{5,5}$ patterns, 3 from the prefix and 3 from the suffix, which contributes $C_{4,3}C_{5,3}$ patterns, and finally, 4 from the prefix and 2 from the suffix, which contributes $C_{4,4}C_{5,2}$ patterns. These correspond to $1 \leq j \leq 4$ in the summation formula of the theorem. Notice that the terms for $j = 0, 5$, and 6 produce zero patterns. When $j = 0$, the suffix does not have enough slots to provide $6 - 0 = 6$ choices, and when $j > 4$, the prefix does not have enough slots to provide 5 or 6 choices.

In the general case, let $X = \{x_1, \ldots, x_m\}$ and $Y = \{y_1, \ldots, y_{n-m}\}$. X then refers to the leftmost m slots, while Y refers to the remaining $n - m$ slots. We decompose the $C_{n,k}$ patterns into $k + 1$ groups: those with no chosen slots from X and k chosen slots from Y, those with 1 slot from X and $k - 1$ slots from Y, and so forth, through the final group, which consists of patterns with k slots from X and none from Y. These groups are disjoint, and by the second counting principle, their sizes are as follows, assuming that $k - (n - m) \leq j \leq m$. If j is outside this range, the group size is zero.

Group (j)	Slots from X	Slots from Y	Size of group
0	0	k	$C_{m,0}C_{n-m,k}$
1	1	$k-1$	$C_{m,1}C_{n-m,k-1}$
$\vdots$	$\vdots$	$\vdots$	$\vdots$
k	k	0	$C_{m,k}C_{n-m,0}$

The sum of the last column must then be $C_{n,k}$, which completes the theorem's proof. An algebraic proof of this theorem is not as simple as was the case with Pascal's identity. ∎

Exercises

1.31 Obtain a closed-form expression for $\sum_{i=1}^{n} i^4$.

1.32 Obtain a closed-form expression for $\sum_{i=1}^{n} i^2/2^i$.

1.33 Obtain a closed-form expression for $\sum_{i=6}^{n} i^3/3^i$, for $n > 6$.

*1.34 Obtain a closed-form expression for $\sum_{i=1}^{\infty} \dfrac{1}{i \cdot 2^i}$.

1.35 Obtain a closed-form expression for $\sum_{k=0}^{n}(-1)^k C_{n,k}$.

1.36 Obtain a closed-form expression for $\sum_{k=0}^{n} k^2 C_{n,k}$.

1.37 Prove Newton's identity: $C_{n,k}C_{k,l} = C_{n,l}C_{n-l,k-l}$, for $0 \leq l \leq k \leq n$.

*1.38 Show that $\sum_{k=0}^{m} C_{n+k,k} = C_{n+m+1,m}$.

*1.39 Show that $\sum_{k=0}^{n} C_{n,k}^2 = C_{2n,n}$.

1.40 Obtain a closed-form expression for $\sum_{k_1+k_2+\ldots+k_p=n} n!/[k_1!k_2!\cdots k_p!]$.

1.41 Obtain closed-form expressions for the following:

- $\sum_{k_1+k_2+\ldots+k_p=n} k_1 \cdot \dfrac{n!}{k_1!k_2!\cdots k_p!}$

- $\sum_{k_1+k_2+\ldots+k_p=n} k_1 \cdot k_2 \cdot \dfrac{n!}{k_1!k_2!\cdots k_p!}$.

1.3   Probability spaces and random variables

Although nondeterministic, the outcomes of a random process nevertheless obey certain restrictions. In dice experiments, the outcomes are numbers corresponding to the upturned faces. For two dice, the outcomes are the integers 2 through 12. Other integer outcomes are not possible, nor are outcomes such as heads, tails, blue, or green. The term *sample space* denotes the set of possible outcomes. For experiments with two dice, the sample space is $\{2, 3, \ldots, 12\}$. For coin-tossing situations, it is {heads, tails}. For card deck processes, it is the 52 individual cards. Clearly, the sample space need not contain numbers. It is convenient, nevertheless, to associate numbers with the outcomes in order to involve them in further computations. A functional association of a number with each possible outcome is a random variable, which we will formally define shortly. At this point, however, we note that the domains of such functions are sample spaces. Of course, the sample space outcomes have varying probabilities of occurrence, and when we wish to emphasize this additional feature, we use the term *probability space*. The next definition formalizes the concept for the discrete case.

1.36 Definition: A *discrete probability space* is a countable set Ω (the sample space), together with a function $\Pr : \Omega \to [0, 1]$, such that $\sum_{\omega \in \Omega} \Pr(\omega) = 1$. The elements of Ω are called *outcomes*. For $\omega \in \Omega$, the number $\Pr(\omega)$ is the *probability* of ω. The term *probability distribution* refers to the overall assignment of varying probabilities to the outcomes. Subsets of Ω are called *events*. The probability of an event E is the sum of its constituent probabilities: $\Pr(E) = \sum_{\omega \in E} \Pr(\omega)$. Note that $\Pr(\cdot)$ can refer to an outcome or to an event. The second form represents an extension of the original function to the collection of all subsets of Ω. In any given application, the distinction will always be clear from context. ∎

As the definition implies, probability measures the "size" of sets, where we consider a point to be a singleton set. Consequently, we will find frequent occasion to manipulate sets through unions, intersections, and complements. A quick review of the set theory presented in Appendix A would therefore be appropriate at this point. In any case, we emphasize the notation $A \cup B = A + B$ as alternative expressions for set union. Similarly, $A \cap B = AB$ both represent set intersection. An overbar indicates complement, as in $\overline{A}$.

Later, when we consider uncountable sample spaces, we will update the probability space definition to include a privileged collections of events, called a sigma-algebra, whose members admit probability definitions. For the countable case under discussion here, we can always take this privileged collection to be the set of all subsets of the sample space. That is, all possible events admit probability definitions through the simple expedient of summing the probabilities of their constituent outcomes.

We write $(\Omega, \Pr(\cdot))$ to denote a discrete probability space. If several spaces are under analysis at the same time, we use subscripts on the probability function to clarify its domain. For example, $(X, \Pr_X(\cdot))$ and $(Y, \Pr_Y(\cdot))$ might occur in the same discussion.

Although Definition 1.36 describes an abstract mathematical object, we need to connect it with real-world situations. For example, suppose that $C_1, C_2, \ldots$ are client requests arriving at a database server. In this context, Ω can be the number of requests in the waiting queue when client C_n arrives. Individual outcomes are then natural numbers: $\omega \in \{0, 1, 2, \ldots\}$, each with an associated probability. Of course, if $n = 1$, the first client always finds an empty waiting queue, which is hardly a random outcome. For large n, however, the waiting queue can contain from zero to $n - 1$ previous clients, depending on the time needed to service their requests. Some requests require long service times, while others need only brief attention. Moreover, the arrival rate of clients depends on a number of external circumstances and appears random to an observer. From these variations, we expect the waiting queue size, as seen by client n, to exhibit random lengths.

To be specific, we fix $n = 1000$ to give time for the influence of the initially empty queue to dissipate. Under a frequency-of-occurrence interpretation, what then is the intuitive meaning of $\Pr(2)$? We imagine that the server opens for business with an empty queue at the beginning of each day. At some point in each day, client 1000 arrives and finds some number of requests in the queue. On some days, this number is zero. Sometimes it is 1; sometimes it is 2, and so forth. $\Pr(2)$ is the fraction of days that client n finds two requests in the waiting queue.

In general, the frequency-of-occurrence interpretation requires a repeating environment that, on each cycle, provides opportunities for the outcomes to occur. $\Pr(\omega)$ is then the fraction of cycles in which outcome ω occurs. In the client-server context, the subset $\{\omega | \omega < 5\}$, or more simply $(\omega < 5)$, is an event. Its probability is $\sum_{i=0}^{4} \Pr(i)$, which measures the fraction of cycles in which the client 1000 encounters fewer than five waiting requests. It is reasonable to assume that these probabilities are nearly identical for clients $1001, 1002, \ldots$.

In keeping with the interpretation that outcomes occur in a repeating environment, we extend the frequency-of-occurrence interpretation to events. If event A contains outcome ω, we say that A occurs whenever ω occurs. For the client-server situation, the events $(\omega < 5), (\omega > 1)$, and $(\omega = 0 \bmod 2)$ all occur simultaneously with the outcome $\omega = 4$. Also, $A \subset B$ means that if A occurs, then B occurs. If $AB = \phi$, the empty set, then A and B cannot occur simultaneously. In this case, A and B are said to be *mutually exclusive*. If $A + B$ occurs, then at least one of A and B must occur, because the actual outcome ω must be in one or the other. Since $A\overline{A} = \phi$, if A occurs, then $\overline{A}$ cannot.

Definition 1.36 generalizes the intuitive probability of the first section. There we investigated certain counting mechanisms that enabled us to assign probabilities to events. Specifically, the assigned probability for event E was the number of outcomes favorable to E divided by the total number of possible outcomes. In these cases, the probability assigned to each individual outcome was $1/N$, where N was the total number of possible outcomes. The new definition merely allows arbitrary assignments to the individual outcomes, provided that each assignment is nonnegative and all assignments sum to 1. The earlier probabilities then become special cases of the current definition.

For example, the set of all 5-card hands becomes a probability space with $\Pr(\omega) = 1/C_{52,5}$, for each hand ω. It is important to note that all outcomes have nonnegative probabilities. Moreover, each event E, being a collection of outcomes, has $0 \leq \Pr(E) \leq 1$. Because an event's probability is the sum of the probabilities associated with its constituent outcomes, we have $\Pr(E_1 + E_2 + \ldots) = \sum_i \Pr(E_i)$ when the E_i are pairwise disjoint. This last observation leads to the following simple properties for probability assignments.

1.37 Theorem: Let $(\Omega, \Pr(\cdot))$ be a probability space, in which A, B are events and ϕ is the empty event. Then

$$\Pr(\Omega) = 1$$
$$\Pr(\phi) = 0$$
$$\Pr(\overline{A}) = 1 - \Pr(A)$$
$$\Pr(A + B) = \Pr(A) + \Pr(B) - \Pr(AB)$$
$$\Pr(A + B) \leq \Pr(A) + \Pr(B).$$

PROOF: We have $\Pr(\Omega) = \sum_{\omega \in \Omega} \Pr(\omega) = 1$ directly from the definition. The disjoint union $\Omega = \Omega + \phi$ immediately gives $\Pr(\Omega) = \Pr(\Omega) + \Pr(\phi)$, from which $\Pr(\phi) = 0$ follows. Similarly, $\Omega = A + \overline{A}$ is a disjoint union. Therefore, $1 = \Pr(A) + \Pr(\overline{A})$, which implies that $\Pr(\overline{A}) = 1 - \Pr(A)$. For the remaining assertions, we note that $A + B = AB + \overline{A}B + A\overline{B}$ is the traditional disjoint union of A and B. Moreover, A and B admit the following disjoint unions.

$$A = AB + A\overline{B}$$
$$B = AB + \overline{A}B$$

Therefore,

$$\Pr(A + B) = \Pr(AB) + \Pr(\overline{A}B) + \Pr(A\overline{B})$$
$$\Pr(A) = \Pr(AB) + \Pr(A\overline{B})$$
$$\Pr(B) = \Pr(AB) + \Pr(\overline{A}B).$$

Adding the last two equations and identifying the expression for $\Pr(A + B)$ on the right side, we have

$$\Pr(A) + \Pr(B) = \Pr(AB) + \Pr(AB) + \Pr(\overline{A}B) + \Pr(A\overline{B})$$
$$= \Pr(AB) + \Pr(A + B).$$

It then follows that $\Pr(A + B) = \Pr(A) + \Pr(B) - \Pr(AB)$. Moreover, since $0 \leq \Pr(AB) \leq 1$, we also have $\Pr(A + B) \leq \Pr(A) + \Pr(B)$. ∎

A discrete probability space is, of course, a model of some repetitive process which produces outcomes ω with relative frequencies $\Pr(\omega)$. The process could be, for example, ten tosses of a fair coin. The space Ω then contains all possible 10-sequences from the set $\{H, T\}$. There are 2^{10} such outcomes, and they all have the same probability: $\Pr(\omega) = 1/2^{10} = 1/1024$, for all $\omega \in \Omega$. The collection of outcomes with 4 heads constitutes an event, say the event H_4. The probability of this event is the sum of the probabilities of its constituent outcomes: $\Pr(H_4) = C_{10,4}/1024 = 0.2051$.

1.38 Definition: Given a discrete probability space, $(\Omega, \Pr(\cdot))$, a *discrete random variable* is a function that maps Ω into $\mathcal{R}$, the real numbers. ∎

Because we will deal only with discrete random variables for the next several chapters, the unqualified term *random variable* will mean discrete random variable unless otherwise indicated. Tradition demands capital letters for random variables, but we emphasize that a random variable is, in fact, a *function*. To further overload the notation, we also use the random variable's name to denote its range. Suppose that X is a random variable. Then X denotes a function in the expressions $X(\omega)$, or $X^{-1}(S)$ where $S \subset \mathcal{R}$, or $X : \Omega \to \mathcal{R}$. However, when used as a set, X refers to the range of X. For example, suppose that we have another function that maps range(X) into $\mathcal{R}$. Using X to denote range(X), we write $g : X \to \mathcal{R}$, and we also speak of the image of X under g as $g(X)$. The intended meaning should be clear from context. The next example suggests why such a twofold interpretation of X is appropriate. It shows how probability transfers from events in the probability space to numbers in the random variable's range. Indeed, once this transfer is well understood, it is convenient in many cases to deemphasize the original probability space and think of the random variable's range as the relevant outcomes.

1.39 Example: Let Ω be the set of 10-sequences from $\{H, T\}$, and let $\Pr_\Omega(\omega) = 1/2^{10}$ for each $\omega \in \Omega$. This probability space models ten tosses of a fair coin. Let the random variable $H : \Omega \to \mathcal{R}$ assign $H(\omega)$ the number of heads in the sequence ω. The range of H is $\{0, 1, 2, 3, 4, 5, 6, 7, 8, 9, 10\}$. For i in this range, $H^{-1}(i) = \{\omega \in \Omega \mid H(\omega) = i\}$ is an event in Ω. We can think of the probability of this event as transferring to the number i in the range. Because there are $C_{10,i}$ 10-sequences that exhibit i heads, we see that this probability is $C_{10,i}/2^{10}$. If $i = 4$, for instance, the probability of $H^{-1}(4) = C_{10,4}/2^{10} = 0.2051$. We can often proceed from this point as if the probability space were $H = \{0, 1, 2, \dots, 10\}$ with probabilities as follows.

h	$\Pr_H(h) = C_{10,h}/2^{10}$	h	$\Pr_H(h) = C_{10,h}/2^{10}$
0	0.000977	6	0.2051
1	0.00977	7	0.1172
2	0.04395	8	0.04395
3	0.1172	9	0.00977
4	0.2051	10	0.000977
5	0.2461		

The table shows that the elements of H are not equally likely, even though the ele-

ments of the underlying Ω all have the same probability. Note that the probabilities
of the range elements sum to 1.

$$\sum_{i=0}^{10} C_{10,i}/2^{10} = \frac{1}{2^{10}} \sum_{i=0}^{10} C_{10,i} 1^i 1^{10-i} = \frac{1}{2^{10}}(1+1)^{10} = 1$$

In the new probability space, we speak of $\mathrm{Pr}_H(i)$ or $\mathrm{Pr}(H = i)$ to describe the
probability of the event in Ω that maps to i. Similarly, we refer to $\mathrm{Pr}(H < i)$
or $\mathrm{Pr}(i < H < j)$ when we mean the probability of the event in Ω that maps to
numbers less than i or to numbers between i and j. It frequently happens that the
only events of interest are those with descriptions of this sort. If this is the case,
subsequent analysis need not refer to the original probability space. Rather, the
analysis works with the new space $(H, \mathrm{Pr}_H(\cdot))$. $\square$

In Example 1.39, H becomes a new discrete probability space because
the sum of the elements' probabilities is 1. The next theorem assures us that
this will always happen.

1.40 Theorem: Let X be a random variable over the discrete probabil-
ity space $(\Omega, \mathrm{Pr}_\Omega(\cdot))$. Then $(X, \mathrm{Pr}_X(\cdot))$ is a discrete probability space, with
$\mathrm{Pr}_X(x) = \mathrm{Pr}_\Omega(\{\omega \in \Omega \mid X(\omega) = x\})$.

PROOF: Let $\{\omega_1, \omega_2, \dots\}$ be an enumeration of Ω. Then $\{X(\omega_1), X(\omega_2), \dots\}$
sequentially elaborates all of X, although it may contain some repetitions. (X
may map several Ω-elements to the same value.) If we proceed through the
sequence, eliminating those elements that duplicate elements already passed,
then we obtain an enumeration of X. This shows that X is a countable
set. Let $X = \{x_1, x_2, \dots\}$ be that reduced enumeration. The x_i are now all
distinct.

Now we claim that each $\omega \in \Omega$ belongs to precisely one of the sets
$X^{-1}(x_i)$. Indeed, for $i \neq j$, Theorem A.6 asserts $X^{-1}(x_i) \cap X^{-1}(x_j) =
X^{-1}(\{x_i\} \cap \{x_j\}) = X^{-1}(\phi) = \phi$. So no ω can belong to more than one
$X^{-1}(x_i)$. Also, if $\omega \in \Omega$, then $X(\omega) = x_i$, for some i, which places $\omega \in
X^{-1}(x_i)$. It then follows that

$$\sum_i \mathrm{Pr}_X(x_i) = \sum_i \mathrm{Pr}_\Omega(X^{-1}(x_i)) = \sum_{\omega \in \Omega} \mathrm{Pr}_\Omega(\omega) = 1$$

because $X^{-1}(x_1), X^{-1}(x_2), \dots$ are disjoint subsets whose union is Ω. $\blacksquare$

In the case where the underlying Ω consists of numerical outcomes, we
can define a random variable with the function $X(\omega) = \omega$. The random
variable then retains the same probability distribution as the underlying space.
Another consequence of the theorem is that a function of a random variable
is another random variable. $Y = g(X)$, for example, maps the range of X
to another countable set of real numbers, and the probability transfers from
the elements of X to the new range. That is, $\mathrm{Pr}_Y(Y = y) = \mathrm{Pr}_X(g^{-1}(\{y\}))$.
If $g(t) = t^2 - 3$, for instance, we speak interchangeably of the transformed
random variable as $g(X)$ or as $X^2 - 3$. A functional expression in a random
variable is simply a new random variable.

In view of Theorem 1.40, most analyses start with a statement that a
random variable has a particular probability distribution and continue with-

out further reference to the underlying process that generates the actual outcomes. Further consequences depend only on the probabilities attached to the elements in random variable's range. In Example 1.39, for instance, we say that the random variable H has a *binomial distribution* as given by the table in that example. The next chapter treats the binomial and other discrete distributions in detail. In circumstances where the underlying outcome space is unknown, we simply work with the transferred probabilities: $\Pr_X : X \to [0,1]$. In this case, it is sometimes convenient to extend the probability function to all of $\mathcal{R}$. We do so by defining $\Pr_X(y) = 0$ for all y outside range(X). We then have $\Pr_X : \mathcal{R} \to [0,1]$, but the probability function is zero except on a countable subset of $\mathcal{R}$.

1.41 Example: Let Ω be the collection of possible 5-card hands from a standard card deck. Assume that all hands are equally likely. That is, assume that $\Pr_\Omega(\omega) = 1/C_{52,5}$, for each $\omega \in \Omega$. Let X be the random variable defined by $X(\omega)$ equals 2 times the number of aces plus the number of kings in the hand ω. If, for instance, ω contains 1 ace and 1 king, then $X(\omega) = 3$. What is the probability distribution of X?

The largest value that X can assume is 9, which occurs when ω contains 4 aces and a king. The smallest value is 0, which occurs when ω contains no aces or kings. We must first tabulate the events in Ω that correspond to the inverse images of the numbers 0 through 9. The techniques of Section 1.1.1 are available to calculate the probabilities of all these inverse images. The results appear in Table 1.8.

As a sample calculation, consider the case where $X = 2$. This can arise from two disjoint kinds of hands: those with no aces and two kings and those with 1 ace and no kings. We calculate these values separately. For the case with no aces and two kings, we use the multiplicative counting principle to break the total hands into three successive categories: those that differ in the choice of aces, those that differ in the choice of kings, and those that differ in the final choice from the remaining 44 cards. So the number of hands with no aces and two kings is $C_{4,0}C_{4,2}C_{44,3}$. This gives 79464 hands with no aces and two kings. Similarly, we calculate that 543004 hands have 1 ace and no kings. So $X = 2$ occurs in $79464 + 543004 = 622468$ hands out of $C_{52,5} = 2598960$ possible hands for a probability of $622468/2598960 = 0.2395066$. The final column of Table 1.8 thus gives the probability distribution of the random variable X. $\square$

1.42 Definition: Let $f : \mathcal{R} \to \mathcal{R}$. Let X be a random variable with range $x_1, x_2, \ldots$. The *expected value* of $f(\cdot)$, denoted $E[f(\cdot)]$ or simply $E[f]$, is the sum $E[f(\cdot)] = \sum_i f(x_i)\Pr_X(x_i)$, provided that the sum converges absolutely. ∎

A series $\sum_i a_i$ converges absolutely if $\sum_i |a_i|$ converges, and we insist that the defining sum in the expected value computation possess this property. Sections A.2 and A.3 in Appendix A review convergence and absolute convergence of series. In view of Theorem A.45, this condition allows the summation to proceed over the range of X in any order. Also note that the definition depends on the random variable in question. If there is any ambiguity, we write $E_X[f(\cdot)]$ to clarify that it is the range of X and the corresponding probabilities that appear in the sum. The following theorem gives two useful properties of the expected value operator.

x	$X^{-1}(x)$ components aces	kings	Size of components	$\Pr_\Omega(X^{-1}(x))$ = $\Pr(X = x)$
0	0	0	$C_{4,0}C_{4,0}C_{44,5} = 1086008$	0.4178625
1	0	1	$C_{4,0}C_{4,1}C_{44,4} = 543004$	0.2089313
2	0	2	$C_{4,0}C_{4,2}C_{44,3} = 79464$	
	1	0	$C_{4,1}C_{4,0}C_{44,4} = 543004$	0.2395066
3	0	3	$C_{4,0}C_{4,3}C_{44,2} = 3784$	
	1	1	$C_{4,1}C_{4,1}C_{44,3} = 211904$	0.0829901
4	0	4	$C_{4,0}C_{4,4}C_{44,1} = 44$	
	1	2	$C_{4,1}C_{4,2}C_{44,2} = 22704$	
	2	0	$C_{4,2}C_{4,0}C_{44,3} = 79464$	0.0393280
5	1	3	$C_{4,1}C_{4,3}C_{44,1} = 704$	
	2	1	$C_{4,2}C_{4,1}C_{44,2} = 22704$	0.0090067
6	1	4	$C_{4,1}C_{4,4}C_{44,0} = 4$	
	2	2	$C_{4,2}C_{4,2}C_{44,1} = 1584$	
	3	0	$C_{4,3}C_{4,0}C_{44,2} = 3784$	0.0020670
7	2	3	$C_{4,2}C_{4,3}C_{44,0} = 24$	
	3	1	$C_{4,3}C_{4,1}C_{44,1} = 704$	0.0002801
8	3	2	$C_{4,3}C_{4,2}C_{44,0} = 24$	
	4	0	$C_{4,4}C_{4,0}C_{44,1} = 44$	0.0000262
9	4	1	$C_{4,4}C_{4,1}C_{44,0} = 4$	0.0000015

TABLE 1.8. Random variable X is 2 times aces plus kings in a 5-card hand (see Example 1.41)

1.43 Theorem: $E[\cdot]$ is a linear operator. That is, if $E[f]$ and $E[g]$ exist, we have, for constants a, b,

$$E[af + bg] = aE[f] + bE[g].$$

Also, if $h(t) = K$, a constant, then $E[h] = K$.

PROOF: Let X be the random variable in question, and let $x_1, x_2, \ldots$ enumerate its range. Since $E[f]$ and $E[g]$ exist, their defining sums converge absolutely. Let

$$A = \sum_{i=1}^{\infty} |f(x_i)|\Pr(x_i) \text{ and } B = \sum_{i=1}^{\infty} |g(x_i)|\Pr(x_i).$$

By the triangle inequality,

$$|af(x_i) + bg(x_i)| \leq |af(x_i)| + |bg(x_i)| = |a| \cdot |f(x_i)| + |b| \cdot |g(x_i)|.$$

Consequently,

$$\sum_{i=1}^{n} |af(x_i) + bg(x_i)|\Pr(x_i) \leq |a| \sum_{i=1}^{n} |f(x_i)|\Pr(x_i) + |b| \sum_{i=1}^{n} |g(x_i)|\Pr(x_i)$$

$$\to |a|A + |b|B,$$

which shows that the defining sum for $E[af + bg]$ converges absolutely. Consequently, $E[af + bg]$ exists. Therefore,

$$E[af + bg] = \sum_i (af(x_i) + bg(x_i)) \cdot \Pr(x_i)$$

$$= a \sum_i f(x_i)\Pr(x_i) + b \sum_i g(x_i)\Pr(x_i) = aE[f] + bE[g].$$

Also, if $h(t) = K$, then $E[h(X)] = \sum_i K \cdot \Pr(x_i) = K \sum_i \Pr(x_i) = K.$ ∎

For simple polynomial functions, such as $f(t) = t, g(t) = t^2$, we frequently write the corresponding expression of the random variable as the argument of the expectation operator. That is, we write $E[X]$ for $E[f]$ and $E[X^2]$ for $E[g]$. Slightly more complicated expressions also appear in this abbreviated notation, such as $E[(X - a)^2]$ for $E[h]$ where $h(t) = (t - a)^2$.

When you know the probability distribution of a random variable, you theoretically know everything about it. Practically, however, the distribution is not as easy to manipulate as certain summary values. Two of these values are the mean and variance, which measure, respectively, the central value, around which the probabilities balance, and the dispersion about that balance point.

1.44 Definition: Let X be a random variable with probability distribution $\Pr()$. The *mean* of X, denoted μ or μ_X, and the *variance* of X, denoted σ^2 or σ_X^2, are defined as follows, provided that the defining expected values exist.

$$\mu = E[X] = \sum_i x_i \cdot \Pr(x_i)$$

$$\sigma^2 = E[(X - \mu)^2] = \sum_i (x_i - \mu)^2 \cdot \Pr(x_i),$$

where $x_1, x_2, \ldots$ is the range of X. The square root of the variance, denoted σ or σ_X, is the *standard deviation* of the random variable. ∎

Using Theorem 1.43, we can derive a more useful formula for the variance.

$$\sigma^2 = E[(X - \mu)^2] = E[X^2] - 2\mu E[X] + \mu^2 = E[X^2] - 2\mu^2 + \mu^2$$

$$= E[X^2] - \mu^2 = \left(\sum_i x_i^2 \cdot \Pr(x_i) \right) - \mu^2 \tag{1.7}$$

1.45 Example: What are the mean and variance of the random variable X in Example 1.41?

Table 1.9 excerpts the probability data from the last column of Table 1.8 and tabulates the terms needed for the mean and variance calculations. From the summations in the table's last line, we calculate $\mu = 1.154$ and $\sigma^2 = 2.858 - (1.154)^2 = 1.526$. The standard deviation is $\sqrt{1.526} = 1.235$. Figure 1.7 shows how the mean represents a balance point for the probabilities, in the sense that the x-axis balances at the mean when loaded at the points $x_1, x_2, \ldots$ with weights proportional to the corresponding probabilities. The figure also shows two boundary lines, at the

x_i	$\Pr(X = x_i)$	$x_i \cdot \Pr(X = x_i)$	$x_i^2 \cdot \Pr(X = x_i)$
0	0.4178625	0.0	0.0
1	0.2089313	0.2089313	0.2089313
2	0.2395066	0.4790132	0.9580264
3	0.0829901	0.2489703	0.7469109
4	0.0393280	0.1573120	0.6292480
5	0.0090067	0.0450335	0.2251675
6	0.0020670	0.0124020	0.0744120
7	0.0002801	0.0019607	0.0137249
8	0.0000262	0.0002096	0.0016768
9	0.0000015	0.0000135	0.0001215
$\sum$	1.0000000	1.1538461	2.8582193

TABLE 1.9. Sample calculation of the mean and variance (see Example 1.45)

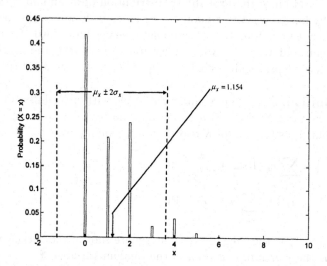

Figure 1.7. The mean as the balance point of probability weights (see Example 1.45)

mean plus or minus two times the standard deviation. The significance of these lines will become clear after the next two theorems. □

1.46 Example: Over a wide range of conditions, the operator of an airline reservation terminal has determined that the response time T, rounded to the nearest 1/2 second, follows the distribution given in the first two columns to the right. The operator obtained these data by appealing to the frequency-of-occurrence meaning of probability. The data reflect the fraction of responses that entail a wait of $0.0, 0.5, \ldots, 4.0$ seconds. What are the mean and variance of T?

t	$\Pr(t)$	$t \cdot \Pr(t)$	$t^2 \cdot \Pr(t)$
0.0	0.15	0.0000	0.0000
0.5	0.05	0.0250	0.0125
1.0	0.20	0.2000	0.2000
1.5	0.30	0.4500	0.6750
2.0	0.05	0.1000	0.2000
2.5	0.05	0.1250	0.3125
3.0	0.10	0.3000	0.9000
3.5	0.05	0.1750	0.6125
4.0	0.05	0.2000	0.8000
$\sum$ 18.0	1.00	1.5750	3.7125

The summary line of the third column gives the mean, $\mu = \sum t_i \Pr(T = t_i) = 1.5750$. The corresponding entry at the bottom of the last column gives $E[T^2]$. The variance is $\sigma^2 = E[T^2] - \mu^2 = 3.7125 - (1.5750)^2 = 1.232.$ $\square$

If $Y = f(X)$, we can compute the mean and variance of Y in the probability space range(Y) or in range(X). Let $y_1, y_2, \ldots$ be an enumeration of range(Y), and let $x_1, x_2, \ldots$ enumerate range(X). Range(X) splits into disjoint sets: those elements that map to y_1, those that map to y_2, and so forth. These sets are, of course, $f^{-1}(y_1), f^{-1}(y_2), \ldots$.

$$\mu_Y = E_Y[Y] = \sum_i y_i \Pr_Y(y_i) = \sum_i y_i \Pr_X(f^{-1}(y_i)) = \sum_i y_i \sum_{j | f(x_j) = y_i} \Pr_X(x_j)$$

$$= \sum_i \sum_{j | f(x_j) = y_i} f(x_j) \Pr_X(x_j) = \sum_j f(x_j) \Pr_X(x_j) = E_X[f]$$

The last line follows because the separate sums over the partition components $f^{-1}(y_1), f^{-1}(y_2), \ldots$ total precisely the sum over range(X), A similar computation gives

$$\sigma_Y^2 = E_Y[(Y - \mu_Y)^2] = E_Y[Y^2] - \mu_Y^2 = E_X[f^2] - (E[f])^2$$
$$= E_X[(f(\cdot) - \mu_Y)^2].$$

1.47 Example: Suppose that X has the probability distribution below, and let $Y = f(X) = 2X^2 + 6$. Find the distribution of Y and its mean and variance.

x	-2	-1	0	1	2
$\Pr(X = x)$	0.25	0.20	0.15	0.35	0.05

Under the specified mapping, Y assumes the values $14, 8,$ and 6. Below we tabulate the inverse images and associated probabilities, which give rise to the mean and variance that follow.

$$Y^{-1}(14) = \{-2, 2\} \rightarrow \Pr(Y = 14) = \Pr(X = -2) + \Pr(X = 2) = 0.30$$
$$Y^{-1}(8) = \{-1, 1\} \rightarrow \Pr(Y = 8) = \Pr(X = -1) + \Pr(X = 1) = 0.55$$
$$Y^{-1}(6) = \{0\} \quad \rightarrow \Pr(Y = 6) = \Pr(X = 0) = 0.15$$

$$\mu_Y = 14(0.30) + 8(0.55) + 6(0.15) = 9.50$$
$$\sigma_Y^2 = 14^2(0.30) + 8^2(0.55) + 6^2(0.15) - 9.50^2 = 9.15$$

Equivalently, we can compute the mean and variance of $Y = f(X)$ using the X distribution.

$$\mu_Y = f(-2)(0.25) + f(-1)(0.20) + f(0)(0.15) + f(1)(0.35) + f(2)(0.05)$$
$$= 14(0.25) + 8(0.20) + 6(0.15) + 8(0.35) + 14(0.05) = 9.50$$
$$\sigma_Y^2 = [f(-2) - 9.5]^2(0.25) + [f(-1) - 9.5]^2(0.20) + [f(0) - 9.5]^2(0.15)$$
$$+ [f(1) - 9.5]^2(0.35) + [f(2) - 9.5]^2(0.05)$$
$$= (14 - 9.5)^2(0.25) + (8 - 9.5)^2(0.20) + (6 - 9.5)^2(0.15) + (8 - 9.5)^2(0.35)$$
$$+ (14 - 9.5)^2(0.05) = 9.15 \ \square$$

Suppose that a random variable X assumes values $x_1, x_2, \ldots$ with corresponding probabilities $p_1 = \Pr(x_1), p_2 = \Pr(x_2), \ldots$. If we repeat the underlying random process a large number of times, say N, and let $N_1, N_2, \ldots$ be the number of times that X generates, respectively, the values $x_1, x_2, \ldots$, then we expect $N_1/N \approx p_1, N_2/N \approx p_2$, and so forth. So the average X value over the course of the N trials is

$$\frac{1}{N} \sum_k x_k \cdot N_k = \sum_k x_k \cdot \frac{N_k}{N} \approx \sum_k x_k p_k = E[X] = \mu.$$

In this sense, the mean is frequently called the *average* value of the random variable. The frequency-of-occurrence interpretation provides a strong connection between the mean, which is a mathematical concept, and the average over a long sequence of observations, which is a real-world construction. This connection enables many applications of probability to real-world problems, and we consequently want no ambiguity about the mean. In particular, we want the same mean regardless of how the defining sum is executed, and this invariance comes with the insistence on absolute convergence. The next example illustrates the difficulties that can arise when absolute convergence does not hold. It uses certain properties of convergent series, which may call for a review of Section A.3 in Appendix A.

1.48 Example: Note that

$$\sum_{i=1}^{n} \frac{1}{i^2} = 1 + \sum_{i=2}^{n} \frac{1}{i^2} \leq 1 + \sum_{i=2}^{n} \frac{1}{i(i-1)} = 1 + \sum_{i=2}^{n} \left(\frac{1}{i-1} - \frac{1}{i} \right)$$

$$= 1 + \sum_{i=1}^{n-1} \frac{1}{i} - \sum_{i=2}^{n} \frac{1}{i} = 1 + 1 - \frac{1}{n} < 2.$$

These partial sums form a nondecreasing sequence with upper bound 2. Hence the sum converges. Let $A = \sum_{i=1}^{\infty} 1/i^2$. A is actually $\pi^2/6$, but that detail is not necessary for this example. Now consider the random variable X with distribution $\Pr(X = i) = 1/(2Ai^2)$, for integer $i \neq 0$. Understanding that $i = 0$ is to be omitted, we sum these weights as follows.

$$\sum_{-\infty}^{\infty} \frac{1}{2Ai^2} = \lim_{M,N \to \infty} \sum_{i=-M}^{N} \frac{1}{2Ai^2} = \frac{1}{2A} \lim_{M,N \to \infty} \left(\sum_{i=1}^{M} \frac{1}{i^2} + \sum_{i=1}^{N} \frac{1}{i^2} \right)$$

$$= \frac{1}{2A}(A + A) = 1$$

Thus we have a proper discrete probability distribution. Consider now the expression for the mean, again with the understanding that $i = 0$ is omitted from the sum.

$$\mu_X = \sum_{-\infty}^{\infty} \frac{i}{2Ai^2} = \lim_{M,N \to \infty} \sum_{i=-M}^{N} \frac{1}{2Ai}$$

$$= \frac{1}{2A} \left(\cdots - \frac{1}{3} - \frac{1}{2} - 1 + 1 + \frac{1}{2} + \frac{1}{3} + \cdots \right)$$

This sum has no definitive limit. (Appendix A investigates this particular sum in the argument accompanying Figure A.1.) By carefully choosing interleaving spans of positive terms with compensating spans of negative terms, we can drive the partial sum back and forth across any prespecified interval. That is, we can force the partial sums to oscillate continually with infinitely many high points greater than 1 and infinitely many low points less than -1. The limit of such an oscillating sequence of partial sums does not exist. Since the terms $1/i$ are approaching zero in magnitude, each positive or negative excursion uses an increasingly large span of terms, but there are always infinitely many remaining to continue the process. We can even force the partial sums to converge to an arbitrary number. For example, we sum a span of positive terms to drive the partial sum to a value in $(42, 43)$. We follow with a span of negative terms to force it back into the interval $(41, 42)$. Alternating positive spans with negative ones, we successively force the partial sum into the ranges $(42, 42+1), (42-1, 42), (42, 42+1/2), (42-1/2, 42), (42, 42+1/4), (42-1/4, 42) \cdots$. These partial sums converge to 42.

It is hardly helpful to have a mean when the value depends on the summation order. We intend to interpret the mean in real-world applications as the average value over many observations. It would be rather artificial to suggest that we sum a run of positive observations, balance it with a run of negative observations, and continue in this manner to achieve some prespecified fraction. Rather, we must take the observations as they occur. We are trying to describe and explain the nondeterministic process; we are not trying to control it. The ambiguous behavior of the expected value operation in this example arises because the sum is not absolutely convergent.

$$\sum_{-\infty}^{\infty} \frac{1}{2A|i|} = \frac{1}{2A} \lim_{M,N\to\infty} \sum_{i=-M}^{N} \frac{1}{|i|} = \frac{1}{2A} \lim_{M,N\to\infty} \sum_{i=1}^{M} \frac{1}{i} + \sum_{i=1}^{N} \frac{1}{i} = \infty$$

In cases like this, we say that the mean does not exist. It is clear that the variance does not exist either, because the defining sum for $E[X^2]$ is not absolutely convergent. $\square$

The variance measures dispersion about the mean, with a larger variance corresponding to a larger dispersion. For example, we can construct the probability distribution of a random variable X by placing probability $1/2$ at $x = 1$ and $1/2$ at $x = -1$. The mean and variance are, respectively, $(1)(1/2) + (-1)(1/2) = 0$ and $(1 - 0)^2(1/2) + (-1 - 0)^2(1/2) = 1$. Now construct a competing random variable Y by placing probability $1/2$ at $y = 10$ and $1/2$ at $y = -10$. The mean is again $(10)(1/2) + (-10)(1/2) = 0$, but the variance is now much larger: $(10 - 0)^2(1/2) + (-10 - 0)^2(1/2) = 100$. To continue the mechanical analogy, if the mean corresponds to the center of gravity, the balance point, then the variance corresponds to the moment of inertia. The following theorems show how the variance places limits on how far from the mean we can find significant probability masses.

1.49 Theorem: (Markov's inequality) Let X be a nonnegative random variable with mean μ. Then, for $x > 0$, $\Pr(X \geq x) \leq \mu/x$.

PROOF: Let $x_1, x_2, \ldots$ be an enumeration of range(X). By assumption, $x_i \geq 0$ for all i.

$$\mu = E[X] = \sum_i x_i \Pr(x_i) = \sum_{\{i|x_i < x\}} x_i \Pr(X_i) + \sum_{\{i|x_i \geq x\}} x_i \Pr(x_i)$$

A smaller value results if we omit the first summation.

$$\mu \geq \sum_{\{i \mid x_i \geq x\}} x_i \Pr(x_i) \geq x \left(\sum_{\{i \mid x_i \geq x\}} \Pr(x_i) \right) = x \cdot \Pr(X \geq x)$$

Dividing by x establishes the theorem. ∎

1.50 Theorem: (Chebyshev's inequality) Let X be a random variable with mean μ and standard deviation $\sigma > 0$. Then, for any $r > 0$, $\Pr(\mu - r\sigma < X < \mu + r\sigma) \geq 1 - 1/r^2$.

PROOF: The result is, of course, trivial for $r \leq 1$. In this case $1 - 1/r^2 \leq 0$, and the probability of any event must be nonnegative. Hence, we expect interesting applications of this theorem to involve $r > 1$. In any case, let $Y = (X - \mu_X)^2 / \sigma_X^2$.

$$\mu_Y = E_Y[Y] = E_X[(X - \mu_X)^2 / \sigma_X^2] = \frac{1}{\sigma_X^2} E_X[(X - \mu_X)^2] = 1$$

Because $Y \geq 0$, we can apply Markov's inequality to obtain $\Pr_Y(Y \geq r^2) \leq 1/r^2$. Then

$$\frac{1}{r^2} \geq \Pr_Y(Y \geq r^2) = 1 - \Pr_Y(Y < r^2) = 1 - \Pr_X\left(\frac{(X - \mu_X)^2}{\sigma_X^2} < r^2 \right)$$

$$= 1 - \Pr_X[(X - \mu_X)^2 < (r\sigma_X)^2]$$

$$= 1 - \Pr_X(\mu_X - r\sigma_X < X < \mu_X + r\sigma_X).$$

Transposing terms gives the expression asserted by the theorem. ∎

1.51 Example: Continuing Examples 1.41 and 1.45, we verify Markov's and Chebyshev's inequalities. Recall that we compute the random variable X as twice the number of aces plus the number of kings in a five-card hand. We have computed $\mu = 1.154$ and $\sigma = 1.235$. Reading the required probabilities from Table 1.8, we construct the following probability values and their Markov bounds.

x	$\Pr(X \geq x)$	bound (μ/x)	x	$\Pr(X \geq x)$	bound (μ_X/x)
1	0.5821375	1.154	6	0.0023748	0.192
2	0.3732062	0.577	7	0.0003078	0.165
3	0.1336996	0.385	8	0.0000277	0.144
4	0.0507095	0.289	9	0.0000015	0.128
5	0.0113815	0.231	10	0	0.115

As you can see, the actual $\Pr(X \geq x)$ is always less than the corresponding Markov bound, μ/x. Indeed, it is usually very much less, which suggests that the Markov bound is not very precise. The Chebyshev inequality states that $\Pr(\mu - 2\sigma < X < \mu + 2\sigma) = \Pr(-1.316 < X < 3.624) \geq 1 - (1/2)^2 = 0.75$. A glance at Figure 1.7 shows that this span includes $X = 0, 1, 2, 3$ for a total probability of 0.949, which again beats the bound by a large margin. Despite this example, the Chebyshev bound cannot be improved in the general case. One of the exercises pursues this idea. □

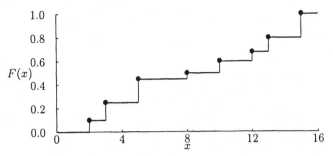

Figure 1.8. A cumulative distribution function

We introduce one final aspect of a random variable in this section, the cumulative distribution function. This new description of the random variable turns out to be equivalent to its probability distribution. However, it is more useful in certain circumstances, such as the computer simulations of Chapter 3.

1.52 Definition: Let X be a random variable with range $x_1 < x_2 < x_3 < \cdots$. The *cumulative distribution function* of X, denoted $F(x)$ or $F_X(x)$, is defined for all real x by

$$F(x) = \Pr(X \le x) = \sum_{\{i \mid x_i \le x\}} \Pr(x_i). \ \blacksquare$$

If, for instance, X has the probability distribution in the third row below, we can calculate $F(x_i)$ by summing the values at or to the left of x_i, as shown in the fourth row.

i	1	2	3	4	5	6	7	8
x_i	2	3	5	8	10	12	13	15
$\Pr(X = x_i)$	0.1	0.15	0.2	0.05	0.1	0.08	0.12	0.2
$F(x_i)$	0.1	0.25	0.45	0.50	0.60	0.68	0.80	1.00

For other values of x, $F(x)$ remains constant at the value of the largest x_i that does not exceed x. That is, $F(x) = F(x_i)$, for $x_i \le x < x_{i+1}$. Also, $F(x) = 0$ for $x < x_1$, and if there is a largest value in the range of X, say x_n, then $F(x) = 1.0$ for $x \ge x_n$. Figure 1.8 shows this staircase behavior of $F(x)$ for the current example. Note that $F(x)$ is continuous from the right, even at the jump points. That is, $\lim_{y \to x^+} F(y) = F(x)$. Recall that the notation $y \to x^+$ means that y approaches x through values greater than x. Except at points in range(X), $F(x)$ is also continuous from the left.

1.53 Example: Suppose that a disk server provides file access services for a collection of clients. The server takes a fixed 10 milliseconds to read or to write a disk sector, which is a fixed-length character string. The server maintains a buffer, which can hold at most two requests. Therefore, if client requests arrive during the 10-millisecond interval while the server is busy with a prior request, the new requests, up to a maximum of 2, will not be lost. Empirical studies for this particular client-server arrangement indicate that the probability of k client requests arriving during any 10-millisecond interval is given by $(0.9)^k e^{-0.9}/k!$. What are the mean and variance of this distribution? What is the probability that the server will lose a client request if it starts its 10-millisecond service interval with an empty buffer? What size buffer would reduce this probability below 0.05?

Let N be the random variable that counts the number of arrivals during a 10-millisecond interval. We then have $\Pr(N = k) = (0.9)^k e^{-0.9}/k!$. Since $\sum_{k=0}^{\infty} x^k/k!$ is the expansion for e^x, we see that

$$\sum_{k=0}^{\infty} \Pr(N = k) = e^{-0.9} \sum_{k=0}^{\infty} \frac{(0.9)^k}{k!} = e^{-0.9} e^{0.9} = 1.$$

Therefore, N is indeed a discrete random variable with range $0, 1, 2, \ldots$. Computation of the mean and variance uses the summation techniques of the preceding section.

$$\mu = E[N] = \sum_{k=0}^{\infty} k \cdot \Pr(N = k) = e^{-0.9} \sum_{k=0}^{\infty} \frac{k \cdot (0.9)^k}{k!} = e^{-0.9} \sum_{k=1}^{\infty} \frac{(0.9)^k}{(k-1)!}$$

$$= e^{-0.9} \sum_{k=0}^{\infty} \frac{(0.9)^{k+1}}{k!} = 0.9 e^{-0.9} \sum_{k=0}^{\infty} \frac{(0.9)^k}{k!} = 0.9 e^{-0.9} e^{0.9} = 0.9$$

$$\sigma^2 = E[N^2] - \mu^2 = \left(\sum_{k=0}^{\infty} k^2 \Pr(N = k) \right) - (0.9)^2$$

$$= -(0.9)^2 + e^{-0.9} \sum_{k=1}^{\infty} \frac{k(0.9)^k}{(k-1)!} = -(0.9)^2 + e^{-0.9} \sum_{k=0}^{\infty} \frac{(k+1)(0.9)^{k+1}}{k!}$$

$$= -(0.9)^2 + (0.9) e^{-0.9} \sum_{k=0}^{\infty} \frac{k(0.9)^k}{k!} + (0.9) e^{-0.9} \sum_{k=0}^{\infty} \frac{(0.9)^k}{k!}$$

$$= -(0.9)^2 + (0.9) E[N] + (0.9) e^{-0.9} e^{0.9} = -(0.9)^2 + (0.9)^2 + (0.9) = 0.9$$

From the formula $\Pr(N = k) = (0.9)^k e^{-0.9}/k!$, we can calculate the cumulative distribution function, $F(x)$, at the first several jump points. To four decimal places, the probability is zero for $k > 6$.

k	0	1	2	3	4	5	6
$\Pr(N = k)$	0.4066	0.3659	0.1647	0.0494	0.0111	0.0020	0.0003
$F(k)$	0.4066	0.7725	0.9372	0.9866	0.9977	0.9997	1.0000

Since the server starts its service interval with an empty buffer, it will not lose a client request as long as no more than 2 such requests arrive in the 10-millisecond interval. The probability of 2 or fewer arrivals is $F(2) = 0.9372$. Hence, the probability is 0.9372 that the server will not lose a request. Equivalently, the probability is $1.0 - 0.9372 = 0.0628$ that it will lose a request. In the sense that the mean of a random variable reflects its average value, we expect an average of 0.9 requests to arrive while the server is busy. Because fewer than 1 request arrives, on the average, during the time necessary to service a prior request, we see that the server has a chance of staying ahead of its clients. Moreover, there are enough buffers to cover the average number of incoming requests during the service interval. Nevertheless, there is still a 6.3% chance of losing a request through buffer overflow. To decrease this value below 0.05, we must be able to buffer sufficient requests, say n, such that

$F(n) > 0.95$. Consulting our tabulation above, we see that we must have $n \geq 3$. That is, we need 1 additional buffer. Of course, if the server starts its 10-millisecond cycle with a partially filled buffer, it will incur a higher probability of an overflow. Analysis of that situation will have to wait until we have developed more tools.

The probability distribution of this example is called a *Poisson* distribution, and it occurs frequently in client-server performance calculations. The next chapter takes up the Poisson distribution in detail. It also develops less tedious methods for determining the mean and variance. $\square$

Exercises

1.42 Let $(\Omega, \Pr(\cdot))$ be a discrete probability space, in which A, B, C are events. Prove the following expression of the probability of the union, and deduce the proper expression for $\Pr(\cup_{i=1}^{n} A_i)$.

$$\Pr(A + B + C) = \Pr(A) + \Pr(B) + \Pr(C)$$
$$-\Pr(AB) - \Pr(AC) - \Pr(BC) + \Pr(ABC).$$

*1.43 Suppose that $A_1 \subset A_2 \subset A_3 \subset \cdots$ is an event sequence. Prove that $\Pr(\cup_{n=1}^{\infty} A_n) = \lim_{n \to \infty} \Pr(A_n)$.

*1.44 Suppose that $B_1 \supset B_2 \supset B_3 \supset \cdots$ is an event sequence. Prove that $\Pr(\cap_{n=1}^{\infty} B_n) = \lim_{n \to \infty} \Pr(B_n)$.

1.45 Two dice are rolled. Let X be the sum of the spots on the upturned faces. What is the range of X? What are the corresponding probabilities? What are μ and σ^2?

1.46 You receive 5 cards from a well-shuffled deck. Let X be twice the number of queens plus the number of jacks in the hand. What is probability distribution of X? What are μ and σ^2? What is the relationship between this problem and that discussed in Examples 1.41 and 1.45?

1.47 As in the preceding problem, you receive 5 cards from a well-shuffled deck. The hand contains n_q queens and n_j jacks. Let $Y = (n_q - n_j)^2$. What is the probability distribution of Y? What are μ_Y and σ_Y^2?

1.48 In 5 tosses of a fair coin, let X be the number of heads minus the number of tails. What is the probability distribution of X? What is the cumulative probability distribution? What are μ and σ^2?

1.49 In a game you receive three cards from a well-shuffled deck. You then receive \$10 if the hand contains an ace and a face card. The face cards are the kings, queens, and jacks. All cards are then returned to the deck before the next game begins. How much would you be willing to pay, per game, to play a large number of hands?

1.50 A hat contains 4 red and 6 black balls. Without looking in the hat, you draw 5 balls. Let X be the number of red balls in the draw. What is the probability distribution of X? What are μ and σ^2?

1.51 Prove the following generalization of Markov's inequality. Suppose that $g(\cdot)$ is a nonnegative, nondecreasing, real-valued function, and let X be a random variable. Then, for $g(x) > 0$, $\Pr(X \geq x) \leq E[g(X)]/g(x)$.

1.52 Let X be a random variable that assumes the values -1 and 1, each with probability $1/2$. Show that the Chebyshev inequality gives strict equality when $r = 1$. This shows that the inequality cannot be improved unless something further is known about the distribution of the random variable.

*1.53 Suppose that a random variable X has the probability distribution $\Pr(X = k) = C_{6,k}/2^6$, for $0 \leq k \leq 6$. What are μ and σ^2? What are the mean and variance if the distribution is $\Pr(X = k) = C_{n,k}p^k(1-p)^{n-k}$, for $0 \leq k \leq n$, where p is a fixed parameter such that $0 < p < 1$?

1.54 Let H be the number of heads in 10 tosses of a fair coin. Compute the mean and variance of H, and verify the Markov and Chebyshev inequalities in this case.

1.55 Generalize Example 1.53 in the following sense. If the probability of k client requests during the 10-millisecond service interval is $e^{-m}m^k/k!$, determine the probability of a lost request if the server starts the service interval with an empty buffer. Express the solution in terms of m. As in the example, the server's buffer is large enough to hold two requests.

1.56 Consider the code to the right, which searches a list L of length N for a target X. The target is always in the list, but it can appear with equal probability in any of the n positions. Let Y be the number of times the algorithm compares the target against a member of the list. What are μ and σ^2?

```
function linearsearch (L, X, N)
    i = 1;
    while (i ≤ N)
        if (L[i] == X)
            return i;
        else
            i = i + 1;
    return 0;
```

1.57 In the preceding problem, suppose that the target is in the list with probability p, for some $0 < p < 1$. If the target is in the list, it is equally likely to be in any of the n positions. Define Y, as before, to be the number of comparisons of the target against a list member. In terms of p, what are μ and σ^2?

1.58 Prove the parallel axis theorem: The variance is the smallest value over the collection $E[(X-a)^2]$ for various a-values. Specifically, $\sigma^2 \leq E[(X-a)^2]$ for any a.

*1.59 Let Ω be the permutations of $\{1, 2, \ldots, n\}$. Suppose that each $\tau \in \Omega$ is equally likely. That is, $\Pr(\tau) = 1/n!$ for each $\tau \in \Omega$. Let $X(\tau)$ be the random variable equal to the number of inversions in τ. Compute $E[X]$.

1.4 Conditional probability

Suppose that we are interested in 5-card hands that contain exactly 2 aces, not as a fraction of all possible 5-card hands, but rather as a fraction of those 5-card hands that contain face cards. If A is the event "contains exactly 2 aces," and F is the event "contains 1 or more face cards," we compute the number of outcomes in these events as follows: $|A| = C_{4,2}C_{48,3} = 103776$ and $|F| = \sum_{k=1}^{5} C_{12,k}C_{40,5-k} = 1940952$. Since there are $C_{52,5}$ possible 5-card hands, we can easily determine the fraction occupied by A or by F. However, the fraction we now want is $|AF|/|F|$. Since an outcome in AF must contain exactly 2 aces and $1, 2,$ or 3 face cards, we compute

$$|AF| = C_{4,2} \sum_{k=1}^{3} C_{12,k}C_{36,3-k} = 60936.$$

Hence the fraction of hands with 1 or more face cards that contains exactly 2 aces is $60936/1940952 = 0.03139$. This is called the conditional probability of A, given F, and we will get to a formal definition shortly. Note that it is different from the unconditional probability of A, which is $|A|/C_{52,5} = 0.03993$. This is not always the case. When the conditional and unconditional probabilities are equal, we have an important special situation, independent events. A precise definition of this concept also appears in due course. Note that

$$\frac{|AF|}{|F|} = \frac{|AF|/C_{52,5}}{|F|/C_{52,5}} = \frac{\Pr(AF)}{\Pr(F)}.$$

We can, therefore, define conditional probability directly in terms of the unconditional probabilities of the components.

1.54 Definition: Let A, B be events in a probability space. For $\Pr(B) > 0$, the *conditional probability* of A given B is $\Pr(A|B) = \Pr(AB)/\Pr(B)$. ∎

For a frequency-of-occurrence interpretation, we again envision a repeating experiment, but we discard all outcomes that do not yield the conditioning event B. The conditional probability $\Pr(A|B)$ is then the fraction of A-occurrences among the remaining outcomes. In the example above, the conditioning event F is "one or more face cards," while A is "exactly two aces." In an experiment that repeatedly generates five-card hands, each from a newly shuffled deck, we disregard all hands that do not have 1 or more face cards. From the remaining hands, we calculate the fraction that contain precisely two aces. This fraction is $\Pr(A|F)$.

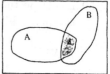

If we envision the outcomes of the probability space as spread out in two dimensions, then we can identify event probabilities with the areas of the corresponding outcome subsets. In the sketch to the right, the rectangle represents all possible outcomes, and the overlapping areas A and B represent two events that share some outcomes. The conditional probability, $\Pr(A|B)$, is the probability of the shared outcomes, divided by the probability of B. It is the fraction of B probability occupied by the AB probability. In effect, we emphasize B as a new probability space, and we enlarge the probability of the outcomes in B such that their sum becomes 1. We compensate by assigning zero probability to outcomes outside of B.

1.55 Example: Example 1.46 examined the distribution of terminal response time T for an airline reservation system. Rounded to the nearest half-second, the response time varies from 0 to 4 seconds, with probabilities as follows.

t_i	0.0	0.5	1.0	1.5	2.0	2.5	3.0	3.5	4.0
$\Pr(T = t_i)$	0.15	0.05	0.20	0.30	0.05	0.05	0.10	0.05	0.05

What are the probabilities $\Pr(T = t | T > 1.0)$?

$$\Pr(T > 1.0) = \sum_{t_i > 1.0} \Pr(t_i) = 0.30 + 0.05 + 0.05 + 0.10 + 0.05 + 0.05 = 0.60.$$

Moreover,

$$(T = t_i) \cap (T > 1.0) = \begin{cases} \phi, & t_i \in \{0.0, 0.5, 1.0\} \\ (T = t_i), & t_i \in \{1.5, 2.0, 2.5, 3.0, 3.5, 4.0\}. \end{cases}$$

We then calculate the probabilities $\Pr(T = t_i | T > 1.0)$ as

$$\Pr(T = t_i | T > 1.0) = \frac{\Pr((T > 1.0) \cap (T = t_i))}{\Pr(T > 1.0)} = \begin{cases} 0, & t_i \leq 1.0 \\ \dfrac{\Pr(T = t_i)}{0.6}, & t_i > 1.0. \end{cases}$$

The new tabulation appears below with row total in the rightmost column.

t_i	0.0	0.5	1.0	1.5	2.0
$\Pr((T = t_i) \cap (T > 1.0))$	0.00	0.00	0.00	0.30	0.05
$\Pr(T = t_i | T > 1.0)$	0.0000	0.0000	0.0000	0.5000	0.0833

	2.5	3.0	3.5	4.0	18.0
	0.05	0.10	0.05	0.05	0.60
	0.0833	0.1667	0.0833	0.0833	1.0000

Because they sum to 1, the probabilities $\Pr(T = t_i | T > 1.0)$ comprise a probability distribution on the range $\{1.5, \ldots, 4.0\}$. In a frequency-of-occurrence interpretation, the conditional probabilities correspond to ratios from a reduced experiment that discards all outcomes with $T \leq 1.0$. Of the remaining outcomes, $\Pr(T = t_i | T > 1.0)$ gives the fraction that equal t_i. $\square$

1.56 Definition: Let A, B be events in a probability space. We say that A and B are *independent* when $\Pr(AB) = \Pr(A) \cdot \Pr(B)$. ∎

Except when the conditioning event has probability zero, independence is equivalent to equality between conditional and unconditional probabilities. This observation is an immediate consequence of the definition, but we will, nevertheless, elevate it to theorem status.

1.57 Theorem: Suppose that $\Pr(B) > 0$. Then A and B are independent events if and only if $\Pr(A|B) = \Pr(A)$.

PROOF: Suppose that A and B are independent. Then

$$\Pr(A|B) = \frac{\Pr(AB)}{\Pr(B)} = \frac{\Pr(A) \cdot \Pr(B)}{\Pr(B)} = \Pr(A).$$

Conversely, suppose that $\Pr(A|B) = \Pr(A)$. Then, multiplying the definition of $\Pr(A|B)$ by $\Pr(B)$, we have $\Pr(AB) = \Pr(A|B) \cdot \Pr(B) = \Pr(A) \cdot \Pr(B)$. ∎

Returning to the card example, we see that the event "exactly two aces" is not independent of the event "one or more face cards." In general, the conditional probability can be greater than, equal to, or less than the unconditional probability. When they are equal, the events are independent. This situation occurs frequently when each outcome is actually an ordered collection with components obtained from unrelated activities. For example, suppose that the experiment involves flipping three coins and rolling a die. A typical outcome is $(HHT, 3)$. In this case, it is a ordered pair that records the result of the coin flip separately from that of the die roll.

Let A be the event "two heads from the coins," and let B be "less than four from the die." By the multiplicative counting principle, the total number of outcomes is $8 \times 6 = 48$. Each of the eight coin possibilities can occur with each of the 6 die possibilities. Figure 1.9 displays this regular arrangement in a rectangle that represents all possible combined outcomes. The 8 coin sequences appear across the bottom of the rectangle and the 6 die possibilities appear along the left side.

The defining condition for event A involves only the coins. Consequently, it appears as two vertical strips that run across all outcomes associated with the die. Similarly, event B involves only the die, and it appears as a horizontal strip that cuts across all coin possibilities. Events that involve a restriction on one component of a composite outcome are called *cylinder events*. (In geometry, the space above a cross-section in the XY-plane is called a cylinder. For example, the solid constructed directly above a circle is a right circular cylinder.) In the case at hand, the geometry of Figure 1.9 immediately reveals that AB occupies the same fraction of B as A occupies of the entire rectangle. Also, BA occupies the same fraction of A as B occupies of the entire rectangle. Consequently, $\Pr(A|B) = \Pr(A)$ and $\Pr(B|A) = \Pr(B)$. The events are independent.

In terms of the individual outcomes, note that the base of A contains those 3-coin sequences that contain 2 heads and the base of B contains the die outcomes that are less than 4. A then contains $4 \cdot 6$ points, which is $1/2$ of the total $8 \cdot 6$ points. AB contains $4 \cdot 3$ points, which is $1/2$ of the $8 \cdot 3$ points in B. Provided that the individual cells are equally likely, the perpendicular arrangement forces AB to occupy the same fraction of B as A occupies of the whole grid. Thus $\Pr(A|B) = \Pr(A)$, and A is independent of B.

This mathematical independence is compatible with our intuitive sense that the coin toss and the die roll are independent. That is, the outcome of the coin toss is not influenced in any way by the result of the die roll, and

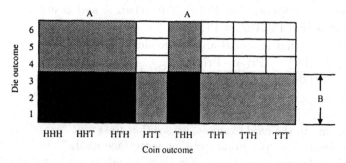

Figure 1.9. Conditional probability in the case of independent events

vice versa. We frequently decompose a compound process into two or more subprocesses that operate without influence on one another. We then *define* the probabilities of the compound events as the products of the corresponding probabilities from the subprocesses. It is then no surprise that events related to distinct subprocesses should prove independent. For example, if you flip a coin, toss a die, and choose a card, you reasonably expect the probability of the outcome (head, four, queen of spades) to be $(1/2) \cdot (1/6) \cdot (1/52)$. You also expect event A, "head on the coin and less than three on the die," to be independent of event B, "spade suit on the card." Indeed, since there are $2 \cdot 6 \cdot 52$ compound outcomes,

$$\Pr(A) = \frac{1 \cdot 2 \cdot 52}{2 \cdot 6 \cdot 52} = \frac{1 \cdot 2}{2 \cdot 6}$$

$$\Pr(B) = \frac{2 \cdot 6 \cdot 13}{2 \cdot 6 \cdot 52} = \frac{13}{52}$$

$$\Pr(AB) = \frac{1 \cdot 2 \cdot 13}{2 \cdot 6 \cdot 52} = \left(\frac{1 \cdot 2}{2 \cdot 6}\right) \cdot \left(\frac{13}{52}\right) = \Pr(A) \cdot \Pr(B).$$

Suppose that $A_0, A_1, \ldots, A_n$ is a partition of the probability space Ω into disjoint subsets, each of nonzero probability. For any event F, it follows [see Figure 1.10(a)] that

$$\Pr(F) = \sum_{\omega \in F} \Pr(\omega) = \sum_{i=0}^{n} \sum_{\omega \in F A_i} \Pr(\omega) = \sum_{i=0}^{n} \Pr(F A_i)$$

$$= \sum_{i=0}^{n} \Pr(F|A_i) \cdot \Pr(A_i). \tag{1.8}$$

This is the *law of total probability*, and it is just a reformulation of the counting principle that justifies breaking an ensemble into convenient pieces to facilitate the count. In terms of probability, the event F occurs in n disjoint ways: It occurs simultaneously with exactly one of the A_i.

Returning to our example of aces and faces in a 5-card hand, we can think of the set A_0 through A_4 in Figure 1.10(a) as the hands containing zero

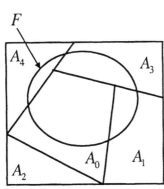

Set	Calculation	Size
A_0	$C_{4,0}C_{48,5}$	1712304
A_1	$C_{4,1}C_{48,4}$	778320
A_2	$C_{4,2}C_{48,3}$	103776
A_3	$C_{4,3}C_{48,4}$	4512
A_4	$C_{4,4}C_{48,1}$	48
F	$C_{12,2}C_{40,3}$	652080
FA_0	$C_{12,2}C_{4,0}C_{36,3}$	471240
FA_1	$C_{12,2}C_{4,1}C_{36,2}$	166320
FA_2	$C_{12,2}C_{4,2}C_{36,1}$	14256
FA_3	$C_{12,2}C_{4,3}C_{36,0}$	264
FA_4		0

(a) An event F intersecting a partition   (b) Intersection sizes

Figure 1.10. Probability of an event as the sum of its disjoint components

through four aces, respectively. This is a disjoint partition of the space of all possible hands. Let the event F be those hands containing exactly two faces cards. For this example, we can work out all the probabilities in the formula above. The probability of two face cards and 1 ace, for instance, is

$$\Pr(FA_1) = \frac{C_{4,1}C_{12,2}C_{36,2}}{C_{52,5}}.$$

Figure 1.10(b) displays the sizes of the various outcome sets involved and allows you to verify that $\Pr(F)$ is indeed equal to the sum of its conditional components, each multiplied by the probability of the conditioning event. That is,

$$\Pr(F) = \sum_{i=0}^{4} \Pr(F|A_i) \cdot \Pr(A_i) = \sum_{i=0}^{4} \left(\frac{\Pr(FA_i)}{\Pr(A_i)} \right) \cdot \Pr(A_i)$$

$$= \sum_{i=0}^{4} \Pr(FA_i) = \sum_{i=0}^{4} \frac{|FA_i|}{T},$$

where $T = C_{52,5}$, the total number of 5-card hands. Substituting the values from Figure 1.10(b), we have

$$\sum_{i=0}^{4} \frac{|FA_i|}{T} = \frac{1}{T}(471240 + 166320 + 14256 + 264 + 0) = \frac{652080}{T} = \Pr(F).$$

The conditional components tell us, for example, that in a hand with two aces there is a

$$\frac{|FA_2|}{|A_2|} = \frac{14256}{103776} = 0.137$$

chance of two face cards also appearing. What this means is that if someone tells you that the hand contains two aces, but does not reveal further details, you have a 13.7% chance of being correct if you guess that the hand also contains exactly two face cards.

The frequency-of-occurrence interpretation is also appropriate. If you receive a large number of hands, each from a freshly shuffled deck, and discard all hands that do not contain precisely two aces, you discover that 13.7% of the remaining hands have two face cards. Note that this probability is different from the unconditional probability of two face cards, which is $652080/C_{52,5} = 0.251$. The latter corresponds to your chance of being correct if you guess that the hand contains exactly two face cards *without* knowing the number of aces. This result accords with intuition. If two of the five positions are reserved for aces, there should be a smaller chance of getting two face cards.

Figure 1.10(b), therefore, contains much useful information for a person engaged in such guessing games. Suppose that the game now turns to a different sort of betting proposition. That is, suppose that you are told that the hand contains two face cards, and you are asked to guess whether it contains exactly two aces. If you reply that it does contain exactly two aces, what are your chances of being correct? Is this information available from the data of Figure 1.10(b)?

Of course, we could calculate another table with the event "two aces" cutting across a partition corresponding to the number of face cards. It turns out, however, that the required information is available in Figure 1.10(b), and the following theorem shows how to extract it.

1.58 Theorem: (Bayes' theorem) Suppose that disjoint events $B_i, 1 \leq i \leq n$, all have positive probability and partition the probability space Ω. Let $A \subset \Omega$ be such that $\Pr(A) > 0$. Then

$$\Pr(B_i|A) = \frac{\Pr(A|B_i) \cdot \Pr(B_i)}{\sum_{j=1}^n \Pr(A|B_j) \cdot \Pr(B_j)}.$$

PROOF: From the definition of conditional probability, we have

$$\Pr(B_i|A) = \frac{\Pr(B_i A)}{\Pr(A)} = \frac{\Pr(A|B_i) \cdot \Pr(B_i)}{\Pr(A)}.$$

The law of total probability asserts that $\Pr(A) = \sum_{j=1}^n \Pr(A|B_j) \cdot \Pr(B_j)$. Substituting this expression in the denominator completes the proof. ∎

In applying Bayes' theorem, we speak of the probabilities $\Pr(B_j)$ as the *prior* probabilities of the partition components. The quantities $\Pr(B_j|A)$ are the *posterior* probabilities. In practice, a prior probability is the weight you place on an event's occurrence when you have no further information. In the example discussed before the theorem, the chance of a five-card hand containing two aces is a prior probability. By contrast, the posterior probability is the revised weight you place on an event's happening after you receive some clarifying information. If you know that the five-card hand in question contains two face cards, you revise (downward) the probability that it contains two aces.

Observing the data of Figure 1.10(b), we find Bayes' theorem rather transparent. We again let $T = C_{52,5}$, the total number of 5-hands. Directly from the definition of conditional probability, we have

$$\Pr(A_2|F) = \frac{\Pr(A_2F)}{\Pr(F)} = \frac{14256/T}{652080/T} = 0.0219.$$

To use Bayes' theorem, we compute the denominator $\sum_{j=0}^{4} \Pr(F|A_j) \cdot \Pr(A_j)$ as

$$\frac{471240/T}{1712304/T} \cdot \frac{1712304}{T} + \frac{166320/T}{778320/T} \cdot \frac{778320}{T} + \ldots + \frac{0/T}{48/T} \cdot \frac{48}{T},$$

which reduces to $(471240 + 166320 + 14256 + 264 + 0)/T$. Consequently,

$$\frac{\Pr(F|A_2) \cdot \Pr(A_2)}{\sum_{j=0}^{4} \Pr(F|A_j) \cdot \Pr(A_j)} = \frac{\frac{14256/T}{103776/T} \left(\frac{103776}{T} \right)}{\frac{471240 + 166320 + 14256 + 264 + 0}{T}}$$

$$= \frac{14256}{652080} = 0.0219.$$

It frequently occurs that we are given conditional probabilities in one direction, but the problem at hand requires conditioning in the reverse direction. Bayes' theorem provides the necessary tool, as illustrated in the next example.

1.59 Example: A manufacturing process produces computer chips, of which 1% are defective. The plant engineers develop a testing procedure T that behaves as follows.

When applied to functional chips, T indicates "good" 99.9% of the time. The remaining 0.1% are false alarms, each of which condemns a functioning chip. When applied to defective chips, T indicates "bad" 99.99% of the time. The remaining 0.01% are misses, which allow defective chips to escape.

$\Pr(T = t|C = c)$		
	c	
t	functional	defective
good	0.999	0.0001
bad	0.001	0.9999

Suppose that we now apply T to a chip of unknown quality. If T indicates "bad," what is the probability that the chip is actually functional?

c	$\Pr(C = c)$
functional	0.99
defective	0.01

Let (C, T) denote the chip quality and the test result. The possible outcomes for (C, T) are (functional, good), (functional, bad), (defective, good), and (defective, bad). We tabulate the known information as shown to the right above. Using Bayes' Theorem, and abbreviating good, bad, functional, and defective with g, b, f, d, we compute the following probabilities.

$$\Pr(C = f|T = b) = \frac{\Pr(T = b|C = f) \cdot \Pr(C = f)}{\Pr(T = b|C = f) \cdot \Pr(C = f) + \Pr(T = b|C = d) \cdot \Pr(C = d)}$$

$$= \frac{(0.001)(0.99)}{(0.001)(0.99) + (0.9999)(0.01)} = 0.09$$

Thus 9% of the condemned chips are actually functional. If you base your judgment on how T performs on chips of known quality, it appears that the test is very effective. However, T is less attractive after you realize that a large fraction of its rejects are functional. This result may run counter to intuition, and the following explanation may help. Consider a long run of chip tests, from which all the good tests are excluded. The remaining tests involve both functional and defective chips, but they all test bad. If the tests total N, they involve about $0.01N$ defective chips and $0.99N$ functional ones. Consequently, the tests that are retained, since they all test bad, must number about $0.001(0.99N) + 0.9999(0.01N) \approx 0.01N$. Among these retained tests are the $0.001(0.99N) \approx 0.001N$ that involved functional chips. These form an approximate fraction $0.001/0.01 = 0.1$ of the retained tests. In all these cases, the test condemns a functional chip. $\square$

Exercises

1.60 An urn contains 8 black, 4 red, and 2 green balls. Four balls are drawn at random. What is the probability that the draw contains exactly 2 red balls given that it contains 1 or more black balls?

1.61 In the preceding problem, let R be the event "the draw contains exactly 2 red balls." Let B_i, for $0 \leq i \leq 4$, be the events "the draw contains exactly i black balls." Compute the probabilities necessary to verify that $\Pr(R) = \sum_{i=0}^{4} \Pr(R|B_i) \cdot \Pr(B_i)$.

1.62 Continuing the analysis of the preceding two problems, calculate directly the probability that the draw contains exactly 2 black balls given that it contains exactly 2 red balls. Verify Bayes' theorem in this setting.

1.63 Construct a probability distribution and define events A and B such that $\Pr(A|B) > \Pr(A)$. Construct another distribution such that $\Pr(A|B) < \Pr(A)$.

1.64 Within a certain population, 1 person in a thousand has virus X, which constitutes a serious health hazard. A test is available. When applied to persons with virus X, the test correctly identifies the sickness 95% of the time. The virus goes undetected in the remaining 5% of the cases. When applied to persons without virus X, the test falsely indicates sickness 0.5% of the time. Suppose that you take the test, and it says that you have the virus. What is the probability that you actually have it?

1.65 Suppose that you toss a fair coin and throw a fair die. Let D be the number of spots on the die's upturned face, and let C be the upturned side of the coin. Compute the conditional probabilities for each of the six possible values of D, given that C is heads. Repeat the calculation given that C is tails. Compute $\Pr(C = \text{heads}|D = 3)$ in two ways, once from the sizes of the outcome sets and once using Bayes' theorem.

1.66 After receiving the first two cards of a 5-card hand, you notice that you have a seven and an eight. What is the probability that the remaining three cards will complete a straight, that is, five cards of consecutive value? Should this probability be higher or lower than the unconditional probability of drawing a straight? Calculate the unconditional probability of a straight to check your intuition.

1.67 Your poker hand contains three kings, a four, and a seven. You have the option of discarding 1 or 2 cards and receiving the same number of new cards from the dealer. The discarded cards do not rejoin the dealer's deck before he deals your new cards. Which gives you a higher chance of filling a full house (3 of one value and 2 of a different value): discarding the four, discarding the seven, or discarding the four and seven?

1.68 A disk server receives requests from many client machines and requires 10 milliseconds to respond to each request. The probability of k additional requests in the 10-millisecond service interval is $e^{-0.9}(0.9)^k/k!$, for $k = 0, 1, 2, \ldots$. If two new calls arrive while the service interval is only partially complete, what is the probability that a third new call will arrive before the server is ready to respond?

*1.69 A voting district contains 1000 voters. In voting for candidate A or candidate B, each voter flips a fair coin. If the result is heads, he votes for A; otherwise, he votes for B. What is the probability that A will win by 100 votes given that he is behind by 100 votes after 500 voters have cast their ballots?

1.70 Suppose that disjoint sets $A_1, \ldots, A_n$ form a partition of the sample space Ω and $\Pr(A_i) > 0$ for $0 \le i \le n$. For B such that $\Pr(B) > 0$ and for an arbitrary event C, show that $\Pr(C|B) = \sum_{i=1}^{n} \Pr(C|BA_i) \cdot \Pr(A_i|B)$.

1.5   Joint distributions

As mentioned earlier, a probabilistic analysis often starts with random variables, rather than with the underlying physical process and its elementary outcomes. Hence we need to extend the conditional probability concept to random variables. That is, we want to know the distribution of a random variable, given that some other random variable lies in a prescribed set. Let us begin by extending the initial example of the preceding section. Let X be the number of aces in a 5-card hand, and let Y be the number of face cards. The event A, "exactly two aces," corresponds to $X = 2$. More precisely, $A = X^{-1}(2)$. The event F, "two or more face cards," corresponds to $Y > 1$. That is, $F = Y^{-1}(\{2, 3, 4, 5\})$. For $0 \le x \le 4$ and $0 \le y \le 5$, we can calculate the probability that transfers to the pair (x, y). For instance,

$$\Pr_{XY}(2, 2) = \Pr((X = 2) \cap (Y = 2)) = \frac{C_{4,2}C_{12,2}C_{36,1}}{C_{52,5}}.$$

After calculating all such combinations, we have Table 1.10. In the upper portion, a cell in row x and column y gives the number of 5-card hands for which $X = x$ and $Y = y$. The lower part then derives $\Pr_{XY}(x, y)$ by dividing these entries by $C_{52,5}$. You can more effectively verify that the probabilities

Number of 5-card hands							
	$Y = 0$	$Y = 1$	$Y = 2$	$Y = 3$	$Y = 4$	$Y = 5$	Totals
$X = 0$	376992	706860	471240	138600	17820	792	1712304
$X = 1$	235620	342720	166320	31680	1980	0	778320
$X = 2$	42840	45360	14256	1320	0	0	103776
$X = 3$	2520	1728	264	0	0	0	4512
$X = 4$	36	12	0	0	0	0	48
Totals	658008	1096680	652080	171600	19800	792	2598960

$\Pr_{XY}(x, y)$							
	$Y = 0$	$Y = 1$	$Y = 2$	$Y = 3$	$Y = 4$	$Y = 5$	Totals
$X = 0$	0.1451	0.2720	0.1813	0.0533	0.0069	0.0003	0.6588
$X = 1$	0.0907	0.1319	0.0640	0.0122	0.0008	0	0.2995
$X = 2$	0.0165	0.0175	0.0055	0.0005	0	0	0.0399
$X = 3$	0.0010	0.0007	0.0001	0	0	0	0.0017
$X = 4$	0.0000	0.0000	0	0	0	0	0.0000
Totals	0.2532	0.4220	0.2509	0.0660	0.0076	0.0003	1.0000

TABLE 1.10. The joint distribution of the number of aces, X, and the number of face cards, Y

sum to 1 by working with the upper table. Adding the row sums in the last column or the column sums in the last row produces the grand total in the lower right, which is indeed $C_{52,5}$. Note that some entries are particularly easy to compute. For example, it is not possible to have 2 aces and 4 face cards in a 5-card hand. So, $\Pr_{XY}(2, 4) = 0$.

1.60 Definition: Let X and Y be random variables over some underlying probability space. The grid of probabilities, $\Pr_{XY}(x, y) = \Pr((X = x) \cap (Y = y))$, for $(x, y) \in \text{range}(X) \times \text{range}(Y)$ is called the *joint probability distribution* of X and Y. ∎

Enumerating the ranges of X and Y as $\text{range}(X) = x_1, x_2, \ldots$ and $\text{range}(Y) = y_1, y_2, \ldots$, we find that the events $((X = x_i) \cap (Y = y_j))$ partition the outcome space. Therefore, $\sum_i \sum_j \Pr_{XY}(x_i, y_j) = 1$. Sometimes the joint probability function arises from an analysis of an underlying probability space, as was the case in the example with aces and faces tabulated in Table 1.10. Sometimes, however, the joint probability function is just given, without any reference to an underlying physical process.

In the former case, you can verify that the sum of the probabilities across all the grid points is 1. These grid points collect the underlying probability of the inverse images, $\Pr_{\Omega}(X^{-1}(x_i) \cap Y^{-1}(y_j))$, which, in total, transfer all the probability from Ω to $\text{range}(X) \times \text{range}(Y)$.

In the latter case, when the probability function makes no reference to an underlying space, the summation requirement must be part of the given information. That is, the given probability grid must exhibit total probability equal to 1. In either case, however, we can treat the grid points themselves as the probability space. Regardless of how $\Pr_{XY}(x, y)$ arises, it immediately spawns related probability distributions, as given in the following definitions.

1.61 Definition: Let $\Pr_{XY}(x_i, y_j)$ be the joint probability distribution of X and Y for $(x_i, y_j) \in \{x_1, x_2, \dots\} \times \{y_1, y_2, \dots\}$. For each x_i, define $\Pr(X = x_i)$ as the joint probability that $(X = x_i)$ and Y assumes any value whatsoever. That is,

$$\Pr_X(x_i) = \Pr(X = x_i) = \Pr((X = x_i) \cap (-\infty < Y < \infty))$$
$$= \sum_j \Pr_{XY}(x_i, y_j).$$

$\Pr_X(\cdot)$ is the *marginal probability distribution* of X. Similarly, the marginal probability distribution of Y is $\Pr_Y(y_j) = \sum_i \Pr_{XY}(x_i, y_j)$. ∎

It is easy to verify that the marginal distributions are true probability distributions because the values associated with the x_i or y_j sum to 1. In Figure 1.10, the marginal distribution of X appears as the row sums in the rightmost column of the lower table. The marginal distribution of Y appears as the column sums in the last row.

When the grid of $\Pr_{XY}$ values is tabulated in the manner of Figure 1.10, the row sums provide the calculations specified by Definition 1.61 and therefore always produce the marginal distribution of the variable on the table's vertical dimension. A similar comment applies to the column sums, which produce the marginal distribution of the variable displayed on the table's horizontal dimension. The location of these sums, on the margins of the tabulation, is the source of the name *marginal* distributions.

1.62 Definition: Random variables X and Y are *independent* if their joint distribution function is the product of the marginals. That is, $\Pr_{XY}(x_i, y_j) = \Pr_X(x_i) \cdot \Pr_Y(y_j)$, for all $(x_i, y_j) \in \text{range}(X) \times \text{range}(Y)$. ∎

If random variables X and Y are independent, then for any sets $A \in \text{range}(X)$ and $B \in \text{range}(Y)$, the events $(X \in A)$ and $(Y \in B)$ are also independent, in the sense of the preceding section. This follow from the independence of X and Y because, letting $Z = (X \in A) \cap (Y \in B)$,

$$\Pr(Z) = \sum_{(x_i, y_j) \in A \times B} \Pr_{XY}(x_i, y_j) = \sum_{x_i \in A} \sum_{y_j \in B} \Pr_X(x_i) \cdot \Pr_Y(y_j)$$
$$= \sum_{x_i \in A} \Pr(X = x_i) \cdot \sum_{y_j \in B} \Pr(Y = y_j) = \Pr(X \in A) \cdot \Pr(Y \in B).$$

1.63 Example: A software shop has extensive records covering many product releases. For each release, the records show the number of kiloblocks (1000 lines) of code, the number of errors reported over the first year of service, and the maintenance cost incurred to repair the errors. Let Ω be the set of software releases. For $\omega \in \Omega$, define $X(\omega)$ as the number of first-year errors per kiloblock of code, rounded to the nearest ten. Define $Y(\omega)$ as the maintenance cost per kiloblock of code, rounded to the nearest \$10,000. Each cell in the upper tabulation of Table 1.11 corresponds to a particular (X, Y) value. The cell contains the number of software releases that have this error count and maintenance cost. Taking a frequency-of-occurrence interpretation of probability, we divide each cell by the grand total, 1586,

Releases by error count (X) and maintenance cost (Y)							
			Y				
X	0	10000	20000	30000	40000	50000	Totals
0	24	0	0	0	0	0	24
10	0	72	141	108	224	4	549
20	0	0	94	280	150	38	562
30	14	21	83	50	62	110	340
40	0	0	21	62	18	10	111
Totals	38	93	339	500	454	162	1586

$\Pr_{XY}(x,y)$							
			y				
x	0	10000	20000	30000	40000	50000	Totals
0	0.0151	0.0000	0.0000	0.0000	0.0000	0.0000	0.0151
10	0.0000	0.0454	0.0889	0.0681	0.1412	0.0025	0.3462
20	0.0000	0.0000	0.0593	0.1765	0.0946	0.0240	0.3544
30	0.0088	0.0132	0.0523	0.0315	0.0391	0.0694	0.2144
40	0.0000	0.0000	0.0132	0.0391	0.0113	0.0063	0.0700
Totals	0.0240	0.0586	0.2137	0.3153	0.2863	0.1021	1.0000

TABLE 1.11. Joint distribution of software errors (X) versus maintenance costs (Y) (see Example 1.63)

to obtain the joint probability distribution in the lower table. The marginal distribution of X appears as the column of row totals on the far right. These entries
give $\Pr(X = x)$, for $x = 0, 10, 20, 30, 40$. The marginal distribution of Y appears
as the row of column totals on the bottom. These entries give $\Pr(Y = y)$, for
$y = 0, 10000, 20000, 30000, 40000, 50000$. To be independent, the distribution table
must have the property that the probability in each interior cell is the product of
its row and column totals. Clearly, X and Y are not independent. Indeed, the table
contains cells with zero entries, but the marginals are nowhere zero. □

1.64 Definition: $F_{XY}(x,y) = \Pr(X \leq x, Y \leq y)$ is the *joint cumulative
distribution* function of X and Y. ∎

Consulting a tabulation of the joint distribution function, such as Table 1.10, you can compute the joint cumulative distribution function at any
grid point by summing all entries above and to the left, that is, in the rectangle whose southeast corner lies at the point in question. Consider again the
random variables that count the aces and faces in a 5-card hand. If X reports
the number of aces and Y the number of face cards, as tabulated in Table 1.10,
the corresponding cumulative distribution function appears in Figure 1.11(a).
The graphical depiction of Figure 1.11(b) shows the same stairstep behavior as the cumulative distribution function of a single random variable. The
only difference is that the stairsteps are now positioned in two dimensions. The
function includes the edges at the tops of the cliffs, but not those at the
bottoms. F_{XY} is therefore continuous as (x, y) descends to (x_0, y_0) through
values such that $x > x_0$ and $y > y_0$.

$F_{XY}(x, y)$						
			y			
x	0	1	2	3	4	5
0	0.1451	0.4171	0.5984	0.6517	0.6586	0.6588
1	0.2358	0.6397	0.8850	0.9505	0.9582	0.9583
2	0.2523	0.6737	0.9245	0.9905	0.9982	0.9982
3	0.2533	0.6754	0.9262	0.9922	0.9999	1.0000
4	0.2533	0.6754	0.9262	0.9922	0.9999	1.0000

(a) Joint cumulative probability

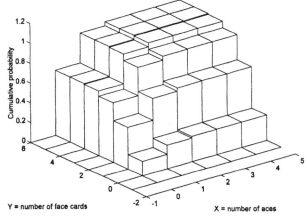

(b) Staircase cumulative distribution function

Figure 1.11. Probability of aces and faces in a 5-card hand

The expected value operator, and therefore the concepts of mean and variance, generalize to joint probability distributions.

1.65 Definition: Let $f : \mathcal{R} \times \mathcal{R} \to \mathcal{R}$. Let X and Y be random variables with ranges $x_1, x_2, \ldots$, and $y_1, y_2, \ldots$, respectively, and with joint probability distribution $\Pr_{XY}(\cdot, \cdot)$. The *expected value* of f, denoted $E[f]$, or $E_{XY}[f]$ if necessary, is the following sum, provided that the sum converges absolutely.

$$E_{XY}[f] = \sum_i \sum_j f(x_i, y_j) \Pr_{XY}(x_i, y_j) \quad \blacksquare$$

As in the case of a single random variable, this definition requires absolute convergence. The reason is the same as before. The absolute convergence allows us to compute the double infinite sum in either order. Furthermore, we may sum across the range of X (or Y) in any convenient order. Section A.3, specifically Theorem A.45, proves these properties of absolute convergence.

In terms of the expected value, we define the mean and variance of X and Y through their marginal distributions. We also introduce a new feature, the covariance of X and Y. The definitions are as follows.

1.66 Definition: Let random variables X and Y have joint probability distribution $\Pr_{XY}(\cdot,\cdot)$. Let the ranges of X and Y be $x_1, x_2, \ldots$ and $y_1, y_2, \ldots$, respectively. The mean and variance of X, the mean and variance of Y, and the covariance of X and Y are calculated as follows.

$$\mu_X = E_X[X] = \sum_i x_i \cdot \Pr_X(x_i)$$

$$\sigma_X^2 = E_X[(X - \mu_X)^2] = \sum_i (x_i - \mu_X)^2 \cdot \Pr_X(x_i)$$

$$\mu_Y = E_Y[Y] = \sum_j y_j \cdot \Pr_Y(y_j)$$

$$\sigma_Y^2 = E_Y[(Y - \mu_Y)^2] = \sum_j (y_j - \mu_Y)^2 \cdot \Pr_Y(y_j)$$

$$\sigma_{XY} = E_{XY}[(X - \mu_X)(Y - \mu_Y)]$$
$$= \sum_i \sum_j (x_i - \mu_X)(y_j - \mu_Y) \cdot \Pr_{XY}(x_i, y_j) \blacksquare$$

As with a single random variable, the expected value operator remains linear in its extension to two random variables. That is, for constants a and b and functions f and g, each a function of two variables, we have

$$E_{XY}[af + bg] = aE_{XY}[f] + bE_{XY}[g].$$

The single-variable formulas for the variance as the expected value of the square minus the square of the expected value also extend to the two-variable case.

$$\sigma_X^2 = E_X[X^2] - \mu_X^2$$
$$\sigma_Y^2 = E_Y[Y^2] - \mu_Y^2$$
$$\sigma_{XY} = E_{XY}[XY] - \mu_X \mu_Y$$

These expressions are straightforward extensions of similar formulas proved for the one-variable case. The exercises ask that you establish some of them.
1.67 Example: Let X be the number of aces in a 5-card hand, and let Y be the number of face cards. Determine the means, variances, and the covariance.

The joint probability distribution of X and Y appears in Table 1.10, together with the marginal distributions. Using these values, we calculate the following expected values for X, Y, X^2, Y^2, and XY.

$E[X]$	$E[Y]$	$E[X^2]$	$E[Y^2]$	$E[XY]$
0.3844	1.1537	0.4744	2.1487	0.3624

The required values are then

$$\mu_X = 0.3844$$
$$\mu_Y = 1.1537$$
$$\sigma_X^2 = 0.4744 - (0.3844)^2 = 0.3266$$
$$\sigma_Y^2 = 2.1487 - (1.1537)^2 = 0.8177$$
$$\sigma_{XY} = 0.3624 - (0.3844)(1.1537) = -0.0811.$$

The interpretation of the negative covariance is that the larger values of X tend to occur with the smaller values of Y, and vice versa. We have already seen that the probability of two face cards is lower when two aces are present than when no information is available about the aces. The negative covariance extends this specific result to general numbers of aces and faces. The negative covariance is intuitively plausible because a larger number of aces reduces the slots available for face cards. $\square$

We have seen how the marginal distributions appear as the row and column sums in the tabulation of a joint probability distribution. These marginal distributions must necessarily sum correctly to 1 because the sum over the entire table is 1. However, we can also produce a new probability distribution from *any* row or column by dividing the entries by the row or column total. This operation generates conditional probability distributions, defined below. The process of dividing a row or column by its sum is called *normalizing* the row or column. Normalization of a nonzero row or column obviously produces a valid probability distribution because the new values must sum to 1.

1.68 Definition: Let X, Y have the joint probability distribution $\Pr_{XY}(\cdot, \cdot)$. For any fixed y_j such that $\Pr_Y(y_j) > 0$, we define the *conditional probability* of $(X = x_i)$ given $(Y = y_j)$ by

$$\Pr_{X|Y=y_j}(x_i) = \frac{\Pr_{XY}(x_i, y_j)}{\Pr_Y(y_j)}.$$

Similarly, for fixed x_i such that $\Pr_X(x_i) > 0$, we define the conditional probability of $(Y = y_i)$ given $(X = x_i)$ by

$$\Pr_{Y|X=x_i}(y_j) = \frac{\Pr_{XY}(x_i, y_j)}{\Pr_X(x_i)}. \blacksquare$$

If we rewrite the definitions in terms of events, we find that this definition of conditional probability is consistent with that developed earlier. That is,

$$\Pr_{X|Y=y_j}(x_i) = \frac{\Pr_{XY}(x_i, y_j)}{\Pr_Y(y_j)} = \frac{\Pr((X = x_i) \cap (Y = y_j))}{\Pr(Y = y_j)}$$
$$= \Pr(X = x_i | Y = y_j).$$

Each conditional distribution has its own expected value operator and produces its own mean and variance. For instance,

$$E_{X|Y=y_j}[f] = \sum_i f(x_i) \cdot \Pr_{X|Y=y_j}(x_i)$$

$$\mu_{X|Y=y_j} = E_{X|Y=y_j}[X] = \sum_i x_i \cdot \Pr_{X|Y=y_j}(x_i)$$

$$\sigma^2_{X|Y=y_j} = E_{X|Y=y_j}[(X - \mu_{X|Y=y_j})^2] = E_{X|Y=y_j}[X^2] - \mu^2_{X|Y=y_j}$$
$$= \sum_i x_i^2 \Pr_{X|Y=y_j}(x_i) - \mu^2_{X|Y=y_j}.$$

The notation becomes cumbersome, and the subscripts are frequently dropped when there is no ambiguity about the probability distribution function in question. Note that the conditional expected value, and consequently the

$\Pr_{X|Y=y}(x)$						
			y			
x	0	1	2	3	4	5
0	0.5729	0.6445	0.7227	0.8077	0.9000	1.0000
1	0.3581	0.3125	0.2551	0.1846	0.1000	0
2	0.0651	0.0414	0.0219	0.0077	0	0
3	0.0038	0.0016	0.0004	0	0	0
4	0.0001	0.0000	0	0	0	0
$\mu_{X|Y=y}$	0.5000	0.4000	0.3000	0.2000	0.1000	0.0000
$\sigma^2_{X|Y=y}$	0.4038	0.3323	0.2562	0.1754	0.0900	0.0000
$\sigma_{X|Y=y}$	0.6355	0.5765	0.5061	0.4188	0.3000	0

TABLE 1.12. Conditional probabilities of ace count X, given face count Y (see Example 1.69)

conditional mean and variance, is no longer a constant; it depends on the conditioning value. That is, $\mu_{X|Y=y_j}$ depends on the parameter y_j. Similarly, $\sigma^2_{X|Y=y_j}$ depends on its conditioning parameter. When we refer to the conditional mean as a function, we write $\mu_{X|Y}(\cdot)$. We understand that this function takes on the values $\mu_{X|Y}(y_1) = \mu_{X|Y=y_1}, \mu_{X|Y}(y_2) = \mu_{X|Y=y_2}, \dots$ at the points $y_1, y_2, \dots$, respectively, and that it is zero at all other points. Similarly, we write $\sigma^2_{X|Y}(\cdot)$ or $\sigma^2_{X|Y}(y_j)$ when we want to emphasize the functional aspect of the conditional variance.

1.69 Example: Example 1.67 computed the means and variances from the distribution of Table 1.10, in which X was the number of aces and Y was the number of face cards in a 5-card hand. Compute the conditional probability distributions of X given the various Y values. Compute the corresponding means and variances.

The desired conditional distributions are the normalized columns from the joint distribution of Table 1.10. Because this table gives the probabilities to four decimal places, certain entries appear to be zero when they are actually just too small to register in the fourth decimal place. The entry $\Pr_{XY}(x_i, y_j)$ is the number of 5-card hands that yield $X = x_i$ and $Y = y_j$ divided by $C_{52,5}$. Therefore, the probabilities are just a constant, $1/C_{52,5}$, times the outcome counts given in the upper part of Table 1.10. It follows that we can obtain the conditional distributions by normalizing the columns of the upper table, where we will not be troubled by zeros in the row or column sums. Following this plan, we arrive at Table 1.12. Each column is a separate conditional distribution, conditioned on the Y value at the top of the column. Below each conditional distribution appears its mean, variance, and standard deviation.

A graphical depiction appears in Figure 1.12. The bullets mark the means $\mu_{X|Y}(\cdot)$ associated with the various Y values. The vertical boxes, centered on the bullets, indicate one-half $\sigma_{X|Y}(\cdot)$ on each side of the mean. As you can see, the means decline linearly with increasing Y value, and the variances shrink steadily. The variance finally disappears when $Y = 5$. Indeed, five face cards in a 5-card hand means no aces—with certainty.

The conditional means vary both above and below the unconditional mean, which was determined to be $\mu_X = 0.3844$ in Example 1.67. This might lead you

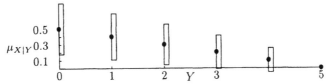

Figure 1.12. Migration of the conditional mean $\mu_{X|Y}$ with increasing Y (see Example 1.69)

to suspect that the expected value of the conditional means, which constitute a function of the conditioning value, is the unconditional mean. This is indeed true, as shown by the next theorem. $\square$

1.70 Theorem: $E_Y[\mu_{X|Y}(\cdot)] = \mu_X.$

PROOF: Let $x_1, x_2, \ldots$ be the range of X. Let $y_1, y_2, \ldots$ be the range of Y.

$$\mu_{X|Y}(y_j) = \mu_{X|Y=y_j} = \sum_i x_i \mathrm{Pr}_{X|Y=y_j}(x_i) = \sum_i x_i \cdot \frac{\mathrm{Pr}_{XY}(x_i, y_j)}{\mathrm{Pr}_Y(y_j)}$$

$$E_Y[\mu_{X|Y}(\cdot)] = \sum_j \mu_{X|Y=y_j} \cdot \mathrm{Pr}_Y(y_j) = \sum_j \sum_i x_i \cdot \frac{\mathrm{Pr}_{XY}(x_i, y_j)}{\mathrm{Pr}_Y(y_j)} \cdot \mathrm{Pr}_Y(y_j)$$

$$= \sum_i x_i \sum_j \mathrm{Pr}_{XY}(x_i, y_j) = \sum_i x_i \mathrm{Pr}_X(x_i) = \mu_X \ \blacksquare$$

The relationship between the conditional and unconditional variances is a bit more complicated. The unconditional variance is the expected value of the conditional variances *plus* the variance of the conditional mean.

1.71 Theorem: $\sigma_X^2 = E_Y[\sigma_{X|Y}^2(\cdot)] + E_Y[(\mu_{X|Y}(\cdot) - \mu_X)^2].$

PROOF: We compute the first term on the right as follows.

$$E_Y[\sigma_{X|Y}^2(\cdot)] = \sum_j \mathrm{Pr}_Y(y_j) \cdot \sigma_{X|Y=y_j}^2$$

$$= \sum_j \mathrm{Pr}_Y(y_j) \sum_i (x_i - \mu_{X|Y=y_j})^2 \cdot \mathrm{Pr}_{X|Y=y_j}(x_i)$$

$$= \sum_j \sum_i x_i^2 \mathrm{Pr}_Y(y_j) \mathrm{Pr}_{X|Y=y_j}(x_i)$$

$$- 2 \sum_j \mathrm{Pr}_Y(y_j) \mu_{X|Y=y_j} \sum_i x_i \mathrm{Pr}_{X|Y=y_j}(x_i)$$

$$+ \sum_j \sum_i \mu_{X|Y=y_j}^2 \mathrm{Pr}_Y(y_j) \mathrm{Pr}_{X|Y=y_j}(x_i)$$

We note that $\mathrm{Pr}_Y(y_j) \mathrm{Pr}_{X|Y=y_j}(x_i)$ is simply the joint probability of $(X = x_i)$ and $(Y = y_j)$. Also, the inner sum of the negative term introduces a second copy of $\mu_{X|Y=y_j}$. Making these substitutions, we have

$$
\begin{aligned}
E_Y[\sigma^2_{X|Y}(\cdot)] &= \sum_i x_i^2 \sum_j \mathrm{Pr}_{XY}(x_i, y_j) - 2\sum_j \mathrm{Pr}_Y(y_j)\mu^2_{X|Y=y_j} \\
&\quad + \sum_j \mathrm{Pr}_Y(y_j)\mu^2_{X|Y=y_j} \sum_i \mathrm{Pr}_{X|Y=y_j}(x_i) \\
&= \sum_i x_i^2 \mathrm{Pr}_X(x_i) - 2E_Y[\mu^2_{X|Y}(\cdot)] + E_Y[\mu^2_{X|Y}(\cdot)] \\
&= E_X[X^2] - E_Y[\mu^2_{X|Y}(\cdot)].
\end{aligned}
$$

Making use of Theorem 1.70, we expand the second term as follows.

$$
\begin{aligned}
E_Y[(\mu_{X|Y}(\cdot) - \mu_X)^2] &= E_Y[\mu^2_{X|Y}(\cdot)] - 2\mu_X E_Y[\mu_{X|Y}(\cdot)] + \mu^2_X \\
&= E_Y[\mu^2_{X|Y}(\cdot)] - \mu^2_X
\end{aligned}
$$

Adding the two expressions gives $E_X[X^2] - \mu^2_X = \sigma^2_X$, which completes the proof. ∎

1.72 Example: Verify the relationships between conditional and unconditional means and variances in Examples 1.67 and 1.69.

Excerpting the marginal distribution for Y from Table 1.10 and the conditional means and variances from Table 1.12, we obtain the following functions of Y.

Y	0	1	2	3	4	5
$\mathrm{Pr}_Y(\cdot)$	0.2532	0.4220	0.2509	0.0660	0.0076	0.0003
$\mu_{X|Y}(\cdot)$	0.5000	0.4000	0.3000	0.2000	0.1000	0.0000
$(\mu_{X|Y}(\cdot) - \mu_X)^2$	0.0134	0.0002	0.0071	0.0340	0.0809	0.1478
$\sigma^2_{X|Y}(\cdot)$	0.4038	0.3323	0.2562	0.1754	0.0900	0.0000

$\sum_j \mu_{X|Y}(y_j)\mathrm{Pr}_Y(y_j)$ yields 0.3846. Due to round-off errors associated with maintaining the probability tables to only four decimal places, this calculation differs from the actual value of $\mu_X = 0.3844$, as computed in Example 1.67. From the tabulation above, we also obtain the two components needed by Theorem 1.71 to compute the variance σ^2_X.

$$
E_Y[\sigma^2_{X|Y}(\cdot)] = \sum_j \sigma^2_{X|Y}(y_j)\mathrm{Pr}_Y(y_j) = 0.3190
$$

$$
E_Y[(\mu_{X|Y}(\cdot) - \mu_X)^2] = \sum_j (\mu_{X|Y}(y_j) - \mu_X)^2 \mathrm{Pr}_Y(y_j) = 0.0082
$$

The sum, 0.3271, differs slightly from $\sigma^2_X = 0.3266$ computed in Example 1.67 due to round-off errors. □

If X and Y are independent random variables, $\mathrm{Pr}_{XY}(x_i, y_j) = \mathrm{Pr}_X(x_i) \cdot \mathrm{Pr}_Y(y_j)$. That is, the jth column of the joint distribution is the column sum, $\mathrm{Pr}(y_j)$, times the marginal distribution on the tabulation's right edge. Normalizing the column then simply recovers the marginal distribution. So $\mathrm{Pr}_{X|Y=y_j}(x_i) = \mathrm{Pr}_X(x_i)$, for any y_j. Similarly, for any x_i, $\mathrm{Pr}_{Y|X=x_i}(y_j) = \mathrm{Pr}_Y(y_j)$. For independent random variables, all conditional distributions reduce to the appropriate marginal distribution.

1.73 Definition: The quantity $\rho_{XY} = \sigma_{XY}/[\sqrt{\sigma_X^2}\sqrt{\sigma_Y^2}]$ is the *correlation coefficient* between X and Y. Random variables X and Y are called *uncorrelated* if $\rho_{XY} = 0$. Equivalently, they are uncorrelated if $\sigma_{XY} = 0$. ∎

Besides having all conditional distributions reduce to one of the marginal distributions, independent random variables are also uncorrelated, as asserted by the next theorem.

1.74 Theorem: Independent random variables are uncorrelated.

PROOF: Suppose that X, Y are independent with respective ranges $x_1, x_2, \cdots$ and $y_1, y_2, \cdots$. Then

$$
\begin{aligned}
\sigma_{XY} &= E_{XY}[(X - \mu_X)(Y - \mu_Y)] = \sum_i \sum_j (x_i - \mu_X)(y_j - \mu_Y)\Pr_{XY}(x_i, y_j) \\
&= \sum_i (x_i - \mu_X)\Pr_X(x_i) \sum_j (y_j - \mu_Y)\Pr_Y(y_j) \\
&= \sum_i (x_i - \mu_X)\Pr_X(x_i)\left(\sum_j y_j \Pr_Y(y_j) - \mu_Y \sum_j \Pr_Y(y_j)\right) \\
&= \sum_i (x_i - \mu_X)\Pr_X(x_i)(\mu_Y - \mu_Y) = 0. \ \blacksquare
\end{aligned}
$$

The converse of Theorem 1.74 is not true. Random variables are uncorrelated if their distribution is symmetric about the means. If the random variables are independent, then the required symmetry is present. However, the symmetry can easily be present without independence, as shown in the next example.

1.75 Example: Construct a joint distribution of two uncorrelated but dependent random variables.

Suppose that $\Pr_{XY}(\cdot, \cdot)$ is symmetrically distributed on integer grid points as shown in Figure 1.13(b). Each of the thirteen marked points receives probability 1/13. The tabulation of the joint distribution, which appears in Figure 1.13(a) exhibits the same pattern. You can see immediately that X and Y are not independent. The marginal distributions are never zero on the sets $\{-2, -1, 0, 1, 2\}$, although zeros do appear at certain grid points. Hence the probability of a zero grid point cannot be the product of its two marginal entries. However, X and Y are uncorrelated.

$$
\begin{aligned}
\mu_X &= (-2)(1/13) + (-1)(3/13) + 0(5/13) + (1)(3/13) + (2)(1/13) = 0 \\
\mu_Y &= (-2)(1/13) * (-1)(3/13) + 0(5/13) + (1)(3/13) + (2)(1/13) = 0 \\
E_{XY}[XY] &= (1/13)[0(2) + (-1)(1) + (0)(1) + (1)(1) + (-2)(0) + (-1)(0) + 0(0) + \\
&\quad (1)(0) + (2)(0) + (-1)(-1) + 0(-1) + (1)(-1) + (-2)(0)] = 0 \\
\sigma_{XY} &= E_{XY}[XY] - \mu_X\mu_Y = 0 \ \square
\end{aligned}
$$

Our discussion has involved just two random variables, but the concepts generalize to a larger collection $X_1, X_2, \ldots$. Of particular interest is the

		Y				
X	-2	-1	0	1	2	
-2	0	0	1/13	0	0	1/13
-1	0	1/13	1/13	1/13	0	3/13
0	1/13	1/13	1/13	1/13	1/13	5/13
1	0	1/13	1/13	1/13	0	3/13
2	0	0	1/13	0	0	1/13
	1/13	3/13	5/13	3/13	1/13	13/13

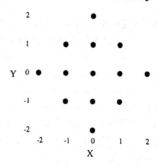

(a) Probability tabulation   (b) Probability configuration

Figure 1.13. Dependent but uncorrelated joint distribution (see Example 1.75)

case when all members of the collection are independent and have the same distribution. That is, all share the distribution of a common X and

$$\Pr(X_1 = x_1, X_2 = x_2, \ldots, X_n = x_n) = \Pr(X = x_1) \cdots \Pr(X = x_n).$$

This situation occurs frequently in the upcoming chapters, and we will often need to compute the distribution of some function of the random variables, such as $X_1 + X_2$, or $X_1 + X_2 + \ldots + X_n$, or $\max(X_1, X_2, \ldots, X_n)$. For the most part, we will develop the tools for such calculations as the need arises. However, as a general illustration, consider the function $Y = X_1 + X_2$, where X_1, X_2 are independent, identically distributed random variables. Suppose that the common distribution is $\Pr(X = i) = p_i$, for $i = 0, 1, 2, \ldots$. What is the distribution of Y? For any outcome, Y is the sum of two nonnegative integers and is therefore itself a nonnegative integer. Moreover, we obtain $Y = i$ precisely when $X_1 = j$ and $X_2 = i - j$. Therefore,

$$\Pr(Y = i) = \sum_{j=0}^{i} \Pr((X_1 = j) \cap (X_2 = i - j))$$

$$= \sum_{j=0}^{i} \Pr(X = j) \cdot \Pr(X = i - j) = \sum_{j=0}^{i} p_j p_{i-j}.$$

This last operation is called a *convolution* of the sequence $p_0, p_1, p_2, \ldots$ with itself. The first several entries are

$$\Pr(Y = 0) = p_0 p_0 = p_0^2$$
$$\Pr(Y = 1) = p_0 p_1 + p_1 p_0 = 2p_0 p_1$$
$$\Pr(Y = 2) = p_0 p_2 + p_1 p_1 + p_2 p_0 = p_1^2 + 2p_0 p_2$$
$$\Pr(Y = 3) = p_0 p_3 + p_1 p_2 + p_2 p_1 + p_3 p_0 = 2(p_0 p_3 + p_1 p_2)$$

1.76 Example: Example 1.46 presented an empirically determined distribution for terminal response times in an airline reservation system. Rounded to the nearest 1/2 second, the response time T exhibited the following distribution.

t	0.0	0.5	1.0	1.5	2.0	2.5	3.0	3.5	4.0
$\Pr(T = t)$	0.15	0.05	0.20	0.30	0.05	0.05	0.10	0.05	0.05

Find the probability that two consecutive response times total 2 seconds or less.

We can reasonably assume that the two response times, T_1 and T_2, are independent and that each has the distribution displayed above. If we count time in half-second units, the possible responses are $0, 1, 2, \ldots, 8$ with probabilities $p_0 = \Pr(T = 0) = 0.15$, $p_1 = \Pr(T = 1) = 0.05$, and so forth, through $p_8 = \Pr(T = 8) = 0.05$. Let $X = T_1 + T_2$. Then still measuring in half-second units, we perform a convolution of the initial terms of the p_i sequence to obtain

$$\Pr(X = 0) = p_0 p_0 = (0.15)^2 = 0.0225$$
$$\Pr(X = 1) = p_0 p_1 + p_1 p_0 = 2(0.15)(0.05) = 0.015$$
$$\Pr(X = 2) = p_0 p_2 + p_1 p_1 + p_2 p_0 = 2(0.15)(0.20) + (0.05)^2 = 0.0625$$
$$\Pr(X = 3) = p_0 p_3 + p_1 p_2 + p_2 p_1 + p_3 p_0 = 2[(0.15)(0.30) + (0.05)(0.20)] = 0.11$$
$$\Pr(X = 4) = p_0 p_4 + p_1 p_3 + p_2 p_2 + p_3 p_1 + p_4 p_0$$
$$= 2[(0.15)(0.05) + (0.05)(0.30)] + (0.20)^2 = 0.085.$$

Summing these values gives the probability that $X \leq 4$ half-seconds: 0.295. $\square$

Exercises

1.71 Prove that the expected value operator remains linear in its extension to joint random variables. That is, $E_{XY}[af + bg] = aE_{XY}[f] + bE_{XY}[g]$.

1.72 Prove that $\sigma_{XY} = E_{XY}[XY] - \mu_X \cdot \mu_Y$.

*1.73 Prove that $|\rho_{XY}| = \left| \sigma_{XY} / \left(\sqrt{\sigma_X^2} \sqrt{\sigma_Y^2} \right) \right| \leq 1$.

1.74 Two dice are rolled. Let X be the sum of the spots on the upturned faces. Let Y be 4 times the sum of the spots on the upturned faces minus 6. Compute σ_X^2, σ_Y^2, and σ_{XY}. Verify the inequality of the preceding problem in this case.

1.75 Two dice are rolled. Let X be the sum of the spots on the upturned faces. Let Y be the square of that sum. Compute σ_X^2, σ_Y^2, and σ_{XY}. Verify in this case that $|\rho_{XY}| = \left| \sigma_{XY} / \left(\sqrt{\sigma_X^2} \sqrt{\sigma_Y^2} \right) \right| < 1$.

1.76 Define the small cards to be those with values two through five, regardless of suit. Let X be the number of face cards in a 5-card hand. Let Y be the number of small cards. Compute the joint probability distribution of X and Y. Also compute the marginal distributions. Are X and Y independent?

1.77 With X and Y as in the preceding problem, compute the conditional probability distributions of X given the various values of Y. Compute the conditional means and variances and verify the relationships among them as asserted by Theorems 1.70 and 1.71.

Number of possible selections of k objects from a pool of n		
Replacement	k-sequences	k-combinations
Without	$P_{n,k} = n!/(n-k)!$	$C_{n,k} = n!/[k!(n-k)!]$
With	n^k	$C_{n+k-1,k} = (n+k-1)!/[k!(n-1)!]$

TABLE 1.13. Counting k-sequences and k-combinations under selection with and without replacement

---

1.78 With X and Y again as in the preceding problem, compute the conditional probability distributions of Y given the various values of X. Compute the conditional means and variances and verify the relationships among them as asserted by Theorems 1.70 and 1.71.

1.79 Again define the small cards as those with values two through five, regardless of suit. Suppose that an experiment consists of dealing a 5-card hand from a well-shuffled deck and then dealing another 5-card hand from a separate well-shuffled deck. Let X be the number of face cards in the first hand. Let Y be the number of small cards in the second hand. Compute the joint probability distribution of X and Y. Also compute the marginal distributions. Are X and Y independent?

1.80 With X and Y as in the preceding problem, compute the conditional probability distributions of X given the various values of Y. Compute the conditional means and variances and verify the relationships among them as asserted by Theorems 1.70 and 1.71.

1.81 From an urn containing 5 black and 4 red balls, you randomly select 3 balls. Let R be the number of red balls in the selection; let B be the number of blacks. Compute the joint probability distribution of R and B. Compute the marginal distributions. Are R and B independent?

1.6   Summary

In selecting k objects from a population of n, we distinguish two replacement protocols and two attitudes toward order. The first distinction concerns whether we do or do not return a selected object to the common pool before making a subsequent selection. The second clarifies whether the selected objects constitute a sequence, in which order of appearance is important, or a combination, in which order of appearance is irrelevant. The number of possibilities varies accordingly, as shown in Table 1.13.

    We can also view $C_{n,k}$ as the number of ways to choose k slots from a field of n, or equivalently, as the number of ways to allocate k indistinguishable objects to n distinguishable containers. This viewpoint is useful in problems where we want to calculate the fraction of possibilities that meet a certain constraint. In such cases, we can construct slot patterns so that the constrained possibilities appear as a product of subpatterns. The chapter repeatedly used

Expression	Interpretation (order of selection or allocation is irrelevant)
$C_{n,k}$	• The number of selections of size k drawn without replacement from a pool of n distinguishable objects • The number of allocations of k indistinguishable objects to n distinguishable containers, such that no container receives more than 1 object • The number of binary n-vectors with components summing to k
$C_{n+k-1,k}$	• The number of selections of size k drawn with replacement from a pool of n distinguishable objects • The number of allocations of k indistinguishable objects to n distinguishable containers • The number of n-vectors with nonnegative integer components summing to k

TABLE 1.14. Interpretations of $C_{n,k}$ and $C_{n+k-1,k}$

5-card hands from a well-shuffled deck to illustrate this point. For example, the probability of a 5-card hand containing 3 cards of one value, 1 card of a different value, and 1 card of yet a third value is $C_{13,3}C_{3,1}C_{4,3}C_{4,1}C_{4,1}/C_{52,5}$.

The first selection in the numerator chooses the 3 different values from the 13 possible values. The second chooses which of the values is to occur three times. The third chooses 3 of the 4 suits for that thrice-repeated value. The fourth and fifth selections choose one suit each for the remaining singleton cards. Using yet another interpretation, $C_{n,k}$ is the number of binary n-vectors in which the components sum to k. The quantity $C_{n+k-1,k}$ admits parallel interpretations, as shown in Table 1.14.

Although we approach probability through such intuitive examples, we define a discrete probability space as an abstract, countable collection of outcomes, each with an assigned probability. The assigned probabilities may or may not come from a combinatorial analysis of the fraction of outcomes associated with certain constraints. In any case, the allocated probabilities must be nonnegative and sum to 1. An event is a subset of outcomes. The probability of an event is the sum of the probabilities associated with its constituent elements. Consequently, the probability of a disjoint union is the sum of the probabilities of its components.

A random variable is a function from a probability space to the real numbers. Probability transfers from the outcomes to the corresponding numbers in the range of the random variable. If X is the random variable defined on probability space $(\Omega, \text{Pr}_\Omega)$, then $\text{Pr}(x) = \text{Pr}(X = x) = \text{Pr}_\Omega(X^{-1}(x))$. The set of values, $\text{Pr}(x)$, for the countable $x \in \text{range}(X)$, is called the probability distribution of X. Once we have the distribution of a random variable, we frequently dispense with further consideration of the original probability space. Instead, we consider the range of the random variable to be the probability space.

Each random variable has an expected value operator, which accepts single-argument functions and returns a number. It is $E[f] = \sum_i f(x_i)\text{Pr}(x_i),$

where $x_1, x_2, \ldots$ is the range of X. While its probability distribution gives complete information about a random variable, certain summary values are often useful. Two of these features are the mean, $\mu = E[X]$, and the variance, $\sigma^2 = E[(X - \mu)^2]$. The mean locates the center of the distribution, in the sense of a balance point, while the variance indicates how widely the probability values are dispersed about the mean. For a fixed variance, the Markov and Chebyshev inequalities place limits on how far a significant probability mass can venture from the mean.

In the intuitive setting where probability arises as the fraction of out-comes favorable to a specified event, conditional probability refers to the fraction within a constrained subset of outcomes rather than the fraction of all possible outcomes. For example, the number of 5-card hands containing exactly 2 queens is $C_{4,2}C_{48,3}$, and therefore the unconditional probability of such a hand is $p = C_{4,2}C_{48,3}/C_{52,5}$. However, the hand's probability, given that it contains 2 aces, is the fraction of hands containing 2 aces and 2 queens out of the hands containing 2 aces. That is, the conditional probability is the fraction of occurrences in a scaled-down experiment, in which outcomes outside the 2-ace hands are not considered. For events A and B, with $\Pr(B) > 0$, this scaling amounts to defining the conditional probability of A, given B, as $\Pr(A|B) = \Pr(AB)/\Pr(B)$. An important result in this context is Bayes' theorem, which allows us to reverse the roles of the two sets. Specifically, for disjoint sets $B_1, B_2, \ldots$ that partition the outcome space, we have

$$\Pr(B_i|A) = \frac{\Pr(A|B_i)\Pr(B_i)}{\sum_j \Pr(A|B_j)\Pr(B_j)},$$

provided that the computation does not involve division by zero. For two random variables, we have the parallel concept of the joint probability distribution, which is a countable tabulation of $(x_i, y_j) \in \text{range}(X) \times \text{range}(Y)$, each with an allocated probability. The allocated probabilities must again be nonnegative and sum to 1. These joint probabilities may come from a combinatorial analysis. For example, X can be the number of aces in a 5-card hand, while Y is the number of kings. For each (x_i, y_j) pair, we work out the fraction of hands that yield $X = x_i$ and $Y = y_j$. That fraction is $\Pr_{XY}(x_i, y_j)$. It is also possible that the probabilities are simply allocated without reference to an underlying experiment. In the latter case, it is still required that the allocations sum to 1.

The joint distribution produces two marginal distributions, which appear as the row and column sums in the tabulation. That is, the marginal probability of X is $\Pr_X(x_i) = \sum_j \Pr_{XY}(x_i, y_j)$. When each cell in the joint tabulation contains the product of its row and column sum, the two random variables are independent. For each function of two real variables, the joint distribution returns an expected value: $E_{XY}[f] = \sum_i \sum_j f(x_i, y_j)\Pr_{XY}(x_i, y_j)$. The mean and variance of X, or of Y, are calculated with the corresponding marginal distribution. The covariance of X and Y is $\sigma_{XY} = E_{XY}[(X - \mu_X)(Y - \mu_Y)] = E_{XY}[XY] - \mu_X\mu_Y$. It is calculated with the joint distribution. The conditional distribution of X, given a particular value of Y, is denoted $\Pr(X = x_i | Y = y_j)$ or $\Pr_{X|Y=y_j}(x_i)$ and is calculated as $\Pr_{XY}(x_i, y_j)/\Pr_Y(y_j)$, provided that $\Pr_Y(y_j) > 0$. This distribution corresponds to a row, or column, in the joint distribution divided by the row, or column, sum. In the case of independent random variables, this normalization process merely recreates the marginal distribution. That is, $\Pr_{X|Y=y_j}(x_i) = \Pr_X(x_i)$ for independent X and Y. For nonindependent random variables the process produces a potentially different distribution for each value of the conditioning variable. Consequently, the expected value operator associated with a conditional distribution, as well as the conditional mean and variance, depend on the conditioning value. Various notations are used for these quantities, including

$$E_{X|Y=y_j}[f] = \sum_i f(x_i)\Pr_{X|Y}(x_i|y_j)$$

$$\mu_{X|Y=y_j} = \mu_{X|Y}(y_j) = E_{X|Y=y_j}[X]$$

$$\sigma^2_{X|Y=y_j} = \sigma^2_{X|Y}(y_j) = E_{X|Y=y_j}[(X - \mu_{X|Y=y_j})^2].$$

Since $\mu_{X|Y}(\cdot)$ is now a function, we can apply the expectation operator E_Y to it. This produces the unconditional mean. That is, $E_Y[\mu_{X|Y}(\cdot)] = \mu_X$. We can also apply E_Y to the conditional variance, but this does not recover the unconditional variance. Instead, the unconditional variance is the expected value of the conditional variance plus the variance of the conditional mean.

Historical Notes

The concept of probability originated with gamblers, possibly in prehistoric times. The astragalus, a small mammalian foot bone, has the features of a primitive die. When tossed, it can come to rest in one of four positions. Archeologists have found these bones around prehistoric sites in numbers that suggest they were used as instruments in some sort of activity. Gaming is a possibility, since it is known that later civilizations (e.g., Egypt around 3500 B.C.) used astragali for board games. These ancient dice and their successors found frequent use in recreational, religious, and even military situations for thousands of years. The choice of a mate or the decision to commit an army to battle could hinge on the dice. Playing cards, which first appeared in the mid-fourteenth century, gradually replaced dice as the preferred instrument for play and fortune-telling. Historians have advanced competing explanations of the long gap between probabilistic activity and any systematic study of that activity: an overwhelming sense of determinism that precluded an objective look at chance phenomena, a religious taboo that forbade any investigation of the connection between chance events and the will of the gods, the lack of truly symmetrical devices that provide the introductory notions of equally likely events, and finally the lack of economic incentives. None of these explanations bears close scrutiny, but the matter of economic incentive has the advantage of good timing.

It is plausible that gamblers from ancient times developed accurate estimates of the probabilities associated with gaming outcomes. However, there is no evidence of a systematic study of probability until commercial needs appeared in the fifteenth and sixteenth centuries. There arose in Florence, for example, a flourishing practice in the insurance of merchant vessels, a practice for which the systematic assessment of risk is indispensable. The seventeenth century saw the emergence of several themes that brought definitive economic pressure to bear on the development of probability. Perhaps motivated by the devastation wrought by the Great Plague in London, John Graunt published the first mortality statistics. John de Witt and John Hudde provided a much-needed analysis of the expected return from annuity contracts, which were used in Holland to finance public works. Leibniz, normally associated with the calculus, worked at this time to apply probability concepts to the law. In short, when the economic incentive appeared, so did the much-delayed analysis.

Although the seventeenth century marks the time when progress in probability theory accelerated quickly, the preceding century saw some notable work, of which one instance is particularly relevant to this chapter. We introduced probabilities here as combinatoric ratios derived from equally likely outcomes. This approach is compelling, and its obvious advantages in the analysis of games makes one wonder why the idea was so late in arriving.

Gerolamo Cardano, in about 1525, proposed that probabilities be assigned in this manner. Some historians conclude that Cardano was an unsavory character, given to lying and dishonesty, who participated fully in the plagiaristic practices of his day. Scientific journals first appeared in the seventeenth century, and most scholars relied on private communications prior to that time. This practice invited frequent disputes as to the originality of new ideas, and perhaps Cardano exploited opportunities of this sort. Other biographers, Ore[58], for example, see Cardano as a capable mathematician, worthy of consideration as the father of probability, an honor traditionally accorded, depending on the source, to Pascal, Fermat, and/or Laplace.

In any case, it appears that Cardano, perhaps with the aid of his student Ludovico Ferrari, first envisioned probabilities as ratios of favored cases to all possible outcomes. Despite this accomplishment, Cardano's major occupation was not mathematics, but rather, medicine. He also dabbled in, and wrote about, physics, music, ethics, morals, horoscopes, and the divination of a person's character from the arrangement of warts on the face.

Some historians trace the origins of probability to a famous correspondence between Pascal and Fermat in the seventeenth century. The subject is the proper disposition of bets when a wager must conclude prematurely. This analysis depends on equally likely elementary outcomes that combine to form events with varying probabilities of occurrence. Laplace contributed many mathematical techniques to the developing art of probability, and he also defined probability as the ratio of favored outcomes to all possible outcomes. These contributions over the centuries attest to the durability of the intuitive approach followed in this chapter.

The lives of Fermat and Pascal make interesting reading. Fermat was a government bureaucrat, for whom mathematics constituted a recreational diversion. Nevertheless, he anticipated Newton's calculus, extended Descartes analytic geometry to higher dimensions, and initiated the field that has become number theory. He posed conjectures in number theory that have occupied mathematicians for the past 350 years. Fermat's life was quiet and peaceful compared to that of his contempory Pascal. Besides being cursed with poor health, Pascal suffered deep anguish over the religious questions of his day. Some of his probabilistic investigations led him to infer the existence of God by showing that the expected value of a life lived under that assumption was infinitely greater than the alternative. Yet, before a second religious conversion removed him from the study of mathematics, he invented projective geometry and settled many issues concerning the cycloid, a curve generated by a fixed point on a rolling circle. Pascal also extended the counting techniques that we now call combinatorics.

As the counting examples of this chapter amply suggest, an application of the symmetry principle can involve much hard work. The various formulas for combinations and permutations, with and without replacement, evolved slowly. Pascal's triangle, which he called the arithmetic triangle, predates Pascal and is of uncertain origin. Boyer[8] notes that the triangle appears in the *Precious Mirror*, a Chinese book dating from the thirteenth century. There is also evidence that it was known to Persian philosophers in the thirteenth century. Pascal, however, used it in the mid-seventeenth century to obtain to obtain combinatorial solutions.

A discussion of combinations appears in an algebra published by the English mathematician John Wallis in the late seventeenth century. Christian Huygens, who authored the first printed textbook on probability in 1657, introduced the concept of expectation.

With the work of James Bernoulli, the pace quickened. He investigated combinations and permutations, their applications to games, and their connection with the binomial theorem. His publications include all the formulas developed in this chapter for the selection of k objects from a field of n, with and without replacement, ordered and unordered. Bernoulli's influence spread through the mathematical community and seeded the efforts of Montmort and de Moivre. Consequently, Bernoulli's major work, *Ars Conjectandi* (The Art of Conjecture), appeared posthumously in 1713 after publications by Pierre Remond de Montmort and Abraham de Moivre had covered much of the same material.

Driven from his native France by religious persecution, de Moivre worked in England as a contemporary of Newton. Although de Moivre was always burdened with financial difficulties, he nevertheless found the time to write several texts that brought organization and clarity to the evolving theory of probability. In an accessible pedagogic style, he presented the concepts of expectation, independence, joint and conditional probability, and a practical method for expanding factorials that eventually became Stirling's formula. He is also credited with one of the first proofs of the central limit theorem, which we will take up in Chapter 6.

When the Reverend Thomas Bayes died in 1761, he had published no mathematics. Nevertheless, his posthumously published "Essay Towards Solving a Problem in the Doctrine of Chances" has generated three centuries of controversy. This work contained a version of Bayes' theorem, which, according to Stigler[84], provides one of the earliest attacks on the inverse probability problem.

This chapter's combinatoric methods permit the calculation of event probabilities, given the outcome distribution. The inverse problem attempts to infer the outcome distribution from observed events. If, for example, we know the probability p of heads associated with a biased coin, we can readily compute the probability of any given sequence. However, suppose that we have observed a particular sequence and want to know what we can infer about an unknown p. This is the inverse problem, and it presents a number of difficulties, the most controversial of which concerns the interpretation of a result. Suppose that some approach declares that the given sequence supports the conclusion $p = 0.3$ with 90% certainty. It is difficult to assign a frequency-of-occurrence meaning to the statement because the actual p value is a constant. Hence, p is either equal to 0.3 or it is not. It is not meaningful to say that $p = 0.3$ in 9 out of 10 trials. Nevertheless, some interpretation is possible in the sense of credibility of belief. That is, the observed sequence may move a person who initially believes that the coin is fair to give greater credibility to the bias $p = 0.3$. Earman[18] gives the matter a thorough examination and demonstrates the Bayes approach to modern statistical problems.

Bayes' contemporary, Richard Price, arranged the posthumous publication of Bayes' mathematical work. During his lifetime, Bayes published only religiously oriented works, such as "Divine Benevolence, or an Attempt to Prove That the Principal End of Divine Providence and Government Is the Happiness of His Creatures." Some historians infer that his attacks on the inverse probability problem were attempts to assign a probability to a first cause for the universe.

Further Reading

The application of probability concepts to everyday situations provides many interesting insights into the intuitive feeling that most people have about the subject. Paulos[63] notes, for example, that a minority ethnic group may feel oppressed even if the fraction of racists is the same in both the majority and the minority. Specifically, suppose that the minority is 13% of the population, while the majority comprises 87%. Moreover, 10% of each group are racists. If the groups are professionally and residentially integrated, a minority person will have a 89.7% chance of finding at least 1 racist in 25 encounters. Under the same circumstances, a majority person will have a 27.9% chance of meeting at least 1 racist in 25 encounters.

The point is that personal assessments of probabilities are frequently different from those derived from a dispassionate analysis of the mathematical data. This is particularly so when the subject matter is an emotional one. Paulos provides many further accounts in which a narrative description, heavily interwoven with the context of the moment, either supports or contradicts mathematical probabilities. The intuitive viewpoint advanced in this chapter emphasizes the ratio of favored outcomes to all possible outcomes. It encounters counterintuitive consequences only when the sizes of the corresponding sets are surprising.

The derivation of probability assignments as ratios of favored outcomes to all possible outcomes is sometimes called the symmetry principle. It is also known as the principle of insufficient reason or the principle of indifference. If we find no reason to attribute a bias to one or more outcomes in a symmetric situation, we assign equal probability to all. Applebaum[3] discusses these matters and introduces the maximum entropy principle, a generalization of the symmetry principle, that systematically accounts for known deviations from perfect symmetry. The symmetry principle is persuasive where it is applicable, but it encounters difficulties when the symmetry is absent or not apparent and when the underlying system does not provide for repeated trials.

There are several alternative approaches. Kolmogorov's[48] axiomatic theory is the least controversial. He simply defines a distribution as a consistent assignment of probability values to subsets of outcomes, such that all subsets receive a probability between zero and 1 inclusive, the collection of all outcomes receives probability 1, and a subset composed of a countable number of disjoint components receives probability equal to the sum of that assigned to the components. The theory then proceeds by developing the consequences of these axioms in an abstract setting or in the context of a particular probability distribution. By simply ignoring any real-world interpretations, this theory avoids controversy. Its usefulness, however, depends on its close parallel with approaches that do attempt to model real-world situations. It is in perfect agreement, for example, with those situations (e.g., cards, dice, urns with red and white balls) where the symmetry principle suggests the probability assignments.

When we apply the symmetry principle to assign probabilities, we expect an event to occur in repeated trials in proportion to those probabilities. This argument appears reasonable, but the connection between probabilities and relative frequencies is difficult to establish conceptually. The frequency viewpoint postulates that the limiting relative frequencies exist, even in situations where no symmetry is available for theoretical computations. A loaded die is an example. Thus, even if there is no apparent symmetry, we can use relative frequencies for probability assignment, provided that we can obtain them. There are practical obstacles here because the frequency will vary across any finite number of trials, and we are hard pressed to complete an infinite number. Nevertheless, with a careful interpretation of probability-zero events, the axiomatic theory is compatible with the relative frequency theory in the limit of infinitely many trials.

A third viewpoint holds that probability assignments are always conditional; they measure the degree of support afforded one statement by another. These measurements may be difficult or impossible to obtain. Indeed, they may even be subjective, depending on the mental state of the individual assessing the situation. Despite the obvious problems that this approach invites, it has the advantage of permitting a probabilistic assessment of a situation that is not embedded in a repetitive process. For example, one can speak of the probability that global war will ensue when country X develops a nuclear weapon. See Gillies[26] or Weatherford[90] for an extended comparison of these different viewpoints.

Combinatorics has a rich literature. Appropriate texts for further study are Cohen[11] and Vilenkin[88], each of which contains many examples of an elementary nature, and Constantine[12], which presents more advanced results.

The two selection protocols, with and without replacement, constitute just two options in a wide continuum of constraints. Sampling without replacement constrains each object to appear a minimum of zero times and a maximum of once in the selection. Sampling with replacement constrains each object to appear a minimum of zero times but does not limit the maximum number of occurrences. A more general approach constrains the ith object to appear a minimum of n_i and a maximum of N_i times. More baroque constraints may also appear, such as constraining an object to an even number of appearances or to a prime number of appearances. The possibilities are endless, and many clever techniques are available to count the number of selections under such constraints. Mott et al.[54] pursue some of these methods at a level about the same as this text. Nelson[56] provides a more advanced treatment, primarily as background material for a detailed analysis of client-server relationships. Riordan[69] is frequently cited as the preferred source for an extensive treatment of combinatorics.

Several examples in this chapter introduced probabilistic analysis of algorithms. Example 1.26 described how to count the executions of a particular statement in a program; Example 1.27 showed how to analyze the performance of a hash table. Such examples occur frequently in computer science, and a large selection of techniques appears in Corman et al.[13]. Brassard and Brately[9] investigate both deterministic and probabilistic algorithm analysis methods. The hash table example in this chapter was adapted from Johnson[38].

Probability measures, random variables, and summarizing features, such as the mean and variance, are primary topics in all texts on probability or statistics. Carlson[10], Fraser[21], Lindgren[51], Papoulis[59], and Ross[72] provide appropriate introductions and expand into concepts beyond those discussed in this text. Allen[1] is of particular interest; he develops the material in a computer science context and provides many constructive examples.

Chapter 2

Discrete Distributions

The preceding chapter described discrete random variables in general terms, in the sense that a distribution is a loosely constrained allocation of probabilities to a countable subset of real numbers. The only constraint in the general case is that the assigned probabilities be nonnegative and sum to 1. More detailed analyses are possible when the distribution assumes a regular form, and this chapter develops the most commonly occurring formats. You may have heard some of the names: Bernoulli, binomial, geometric, Poisson. These random variables are characterized by specific ranges and distributions, and they occur repeatedly in computer science and engineering studies.

In this chapter, all discrete distributions have ranges that are subsets of the nonnegative integers. Moreover, they are all interrelated, in the sense that if one aspect of a process exhibits a certain distribution in the family, another aspect is likely to assume the distribution of a related family member. For example, if the interarrival times of server requests follow a geometric distribution, then the number of arrivals in a fixed time span is binomially distributed. But we a somewhat ahead of ourselves in citing such an example. We must first define the distributions, derive their properties, and catalog their relationships.

We start with the uniform distribution, which was used implicitly in combinatorial analyses of the preceding chapter. For example, when we calculated the probability of receiving two aces in a five-card hand, we assumed that all five-card hands were equally probable. This represents a uniform distribution across the space of all five-card hands, or equivalently, a uniform distribution across the range $1, 2, 3, \ldots, C_{52,5}$. The formal definition is as follows.

2.1 Definition: The *uniform* discrete random variable U_n assumes the integer values $1, 2, \ldots, n$. $\Pr(U_n = k) = 1/n$ for k in the range $1 \leq k \leq n$. We use the abbreviated notation U when the range is understood from context. ∎

In the combinatorial analyses of the preceding chapter, we investigated an event of interest by dividing all possible outcomes into two categories. One category, say C_1, contained all outcomes that satisfied the event-defining constraint. A second category, say C_2, contained those outcomes that did not satisfy the constraint. The probability of the event was then the ratio of category sizes: $|C_1|/(|C_1| + |C_2|)$. We can write this ratio as $|C_1| \cdot [1/(|C_1| + |C_2|)]$, where the second factor is just the reciprocal of the total number of outcomes, say N. If the event is a single outcome, then $|C_1| = 1$, and the event probability is $1/N$. Because this reasoning applies to any outcome, we conclude that we have implicitly established a discrete uniform probability distribution across the outcome space in the sense of the definition above.

Working directly from the definition, we can easily calculate the mean and variance of the uniform distribution. We use Theorem 1.29 to sum powers of the integers.

$$\mu = E[U] = \sum_{k=1}^{n} k \cdot \frac{1}{n} = \frac{1}{n} \cdot \frac{n(n+1)}{2} = \frac{n+1}{2}$$

$$E[U^2] = \sum_{k=1}^{n} k^2 \cdot \frac{1}{n} = \frac{1}{n} \cdot \frac{n(n+1)(2n+1)}{6} = \frac{2n^2 + 3n + 1}{6}$$

$$\sigma^2 = E[U^2] - \mu^2 = \frac{2n^2 + 3n + 1}{6} - \frac{n^2 + 2n + 1}{4} = \frac{n^2 - 1}{12} \tag{2.1}$$

Returning to the example of five-card hands, there are $C_{52,5} = 2598960$ possible outcomes, and we can reasonably assume that they are all equally likely. If we number the hands, in a fixed but arbitrary fashion, from 1 to 2598960, this assumption places probability $1/2598960$ on each integer in this span. This is just the uniform discrete distribution with parameter 2598960. When we calculate the probability of two aces, we are identifying the subset of $1, 2, \ldots, 2598960$ that corresponds to two-ace hands. The answer is the sum of the probabilities across this subset, which is equal to the size of the subset times the common probability value $1/2598960$. This calculation agrees with the approach in the preceding chapter. There we argued that the probability should be the ratio of the number of two-ace hands to the number of all possible hands. In general, the intuitive definition of probability as the ratio of favored cases to all possible cases is always equivalent to a uniform distribution across a set of integers. The integers correspond to an enumeration of the cases.

In an extreme case, a random variable may lose its nondeterministic property. If D is a random variable with a single-point range, then this single point occurs with probability 1. Equivalently, the random variable is actually a constant. We call this extreme case a *degenerate* random variable. Our primary interest is, of course, in random variables that do exhibit nondeterministic behavior, but we nevertheless occasionally encounter degenerate random variables when a parameter takes on an extreme value. We will point out this feature as it occurs.

Many situations demand probability distributions that are neither uniform nor degenerate, and this chapter's goal is to investigate certain well-known cases. The plan is as follows. We start with the Bernoulli distribution, which is a generalization of the fair coin from the preceding chapter. We then show that the sum of a fixed number of independent Bernoulli random variables exhibits a binomial distribution. A special case of the binomial distribution also appeared in the preceding chapter—as the number of heads in n tosses of a fair coin. Some examples from the client-server domain then lead to the Poisson distribution, which is a limiting form of the binomial distribution. The Poisson distribution also made its initial appearance in the examples of the preceding chapter, where it was simply accepted as a given distribution of k requests in a certain time interval. This chapter shows why this distribution arises in such circumstances. When events occur in a nondeterministic but repetitive fashion, we are frequently interested in the interval between events. For example, the number of tosses before a coin lands heads up is a random variable, as is the number of tosses between the ith and $(i + 1)$th heads. This question leads to the geometric distribution and its limiting form, the exponential distribution. Because we want to know the means and variances of these distributions, we will need computational methods that relieve tedious summations. Consequently, after achieving some momentum with the Bernoulli and binomial distributions, we again pause for a mathematical interlude, which develops some interesting techniques for computing many, actually infinitely many, summations simultaneously.

2.1 The Bernoulli and binomial distributions

The Bernoulli random variable generalizes the fair coin of Chapter 1 to the case where the probability of heads can be any value between zero and one, inclusive.

2.2 Definition: For a fixed p, with $0 \leq p \leq 1$, let X be a random variable that takes on the values 1 and 0, with probabilities p and $1 - p$ respectively. X is called a Bernoulli random variable with parameter p. ∎

When $p = 0$ or $p = 1$, the corresponding Bernoulli random variable is degenerate. If $p = 0$, it assumes the constant value 0; if $p = 1$, it assumes the constant value 1. For a nondegenerate Bernoulli random variable X with parameter p, we can conveniently summarize the probabilities of the two possibilities in one formula: $\Pr(x) = p^x(1 - p)^{1-x}$, for $x = 0$ or 1.

2.3 Theorem: Let X be a Bernoulli random variable with parameter p. Then $\mu = p$ and $\sigma^2 = p(1 - p)$.

PROOF:

$$\mu = 1 \cdot p + 0 \cdot (1 - p) = p$$
$$E[X^2] = 1^2 \cdot p + 0^2 \cdot (1 - p) = p$$
$$\sigma^2 = E[X^2] - \mu^2 = p - p^2 = p(1 - p) \ \blacksquare$$

For a fixed $n \geq 1$, suppose that $X_1, \ldots, X_n$ are independent Bernoulli random variables, all with the same parameter p. Consider the sum $S =$

$\sum_{i=1}^{n} X_i$. S can assume all values between 0 and n inclusive. Because the X_i are independent, we have, for any vector $(x_1, x_2, \ldots, x_n)$ of ones and zeros,

$$
\begin{aligned}
\Pr_{X_1,\ldots,X_n}(x_1,\ldots,x_n) &= \Pr_{X_1}(x_1) \cdot \Pr_{X_2}(x_2) \cdots \Pr_{X_n}(x_n) \\
&= p^{x_1}(1-p)^{1-x_1} p^{x_2}(1-p)^{1-x_2} \cdots p^{x_n}(1-p)^{1-x_n} \\
&= p^{x_1+\cdots+x_n}(1-p)^{n-x_1-x_2-\cdots-x_n} = p^y(1-p)^{n-y},
\end{aligned}
$$

where y is the number of ones among the $x_1, x_2, \ldots, x_n$. For a particular y between 0 and n, inclusive, there are $C_{n,y}$ vectors of the form $(x_1, x_2, \ldots, x_n)$ that contain exactly y ones, and each such vector has the same probability: $p^y(1-p)^{n-y}$. The random variable S, moreover, assumes the value y on exactly these outcomes for $X_1, X_2, \ldots, X_n$. Hence $\Pr_S(y) = C_{n,y} p^y (1-p)^{n-y}$. The binomial theorem allows us to verify quickly that this expression is indeed a discrete probability distribution, and the subsequent definition formalizes this distribution as the binomial distribution with parameters n and p.

$$
\sum_{i=0}^{n} \Pr_S(i) = \sum_{i=0}^{n} C_{n,i} p^i (1-p)^{n-i} = [p + (1-p)]^n = 1^n = 1
$$

2.4 Definition: For fixed $n \geq 1$ and $0 \leq p \leq 1$, let X be a random variable with probabilities $\Pr(X = k) = C_{n,k} p^k (1-p)^{n-k}$ for $0 \leq k \leq n$. Then X is called a *binomial random variable* with parameters (n, p). ∎

2.5 Theorem: The sum of $n \geq 1$ independent Bernoulli random variables, all with parameter p, is a binomial random variable with parameters (n, p). PROOF: See the discussion above. ∎

Note that the binomial distribution is degenerate if the underlying Bernoulli distribution is degenerate. That is, the sum of n degenerate Bernoulli random variables with parameter $p = 0$ produces a degenerate binomial random variable that assumes the constant value zero. If the underlying degenerate Bernoulli exhibits $p = 1$, the resulting degenerate binomial assumes the constant value n.

Although the exercise is tedious, we can employ the summation techniques of Section 1.2 to compute the mean and variance of a binomial distribution. We do so here primarily to provide motivation for studying the new computational methods of the next section.

2.6 Theorem: Let X be a binomial random variable with parameters (n, p). Then $\mu = np$ and $\sigma^2 = np(1-p)$. PROOF: For an arbitrary but fixed q, consider the expression $f(p) = (p+q)^n$, which we manipulate in the spirit of Section 1.2. Expanding $f(p)$, we have

$$
f(p) = (p+q)^n = \sum_{k=0}^{n} C_{n,k} p^k q^{n-k}.
$$

Then,

$$f'(p) = n(p+q)^{n-1} = \sum_{k=0}^{n} k C_{n,k} p^{k-1} q^{n-k}$$

$$pf'(p) = np(p+q)^{n-1} = \sum_{k=0}^{n} k C_{n,k} p^{k} q^{n-k}.$$

Since this derivation is valid for any q, the result holds when $q = 1 - p$, in which case the last equality gives us the necessary sum for computing the mean:

$$\mu = \sum_{k=0}^{n} k C_{n,k} p^{k} (1-p)^{n-k} = np[p + (1-p)]^{n-1} = np.$$

Continuing with the second derivative of $f(p)$, we have

$$f'' = n(n-1)(p+q)^{n-2} = \sum_{k=0}^{n} k(k-1) C_{n,k} p^{k-2} q^{n-k}$$

$$p^{2} f'' = n(n-1)p^{2}(p+q)^{n-2} = \sum_{k=0}^{n} k^{2} C_{n,k} p^{k} q^{n-k} - \sum_{k=0}^{n} k C_{n,k} p^{k} q^{n-k}.$$

Letting $q = 1 - p$ and using the summation obtained from the first derivative, we get

$$E[X^2] = \sum_{k=0}^{n} k^{2} C_{n,k} p^{k} (1-p)^{n-k} = n(n-1)p^{2} + np$$

$$\sigma^{2} = E[X^2] - \mu^{2} = n(n-1)p^{2} + np - n^{2}p^{2} = np - np^{2} = np(1-p). \ \blacksquare$$

2.7 Example: A database file has $1,000,000$ records, which occupy disk storage at a density of 10/block. A weekly update modifies 3% of the file, and we assume that the changes are distributed uniformly across the records. The system must rewrite any block with a change to any of its records. What is the probability that the system rewrites block i during the update? What is the average number of blocks rewritten?

For $1 \le j \le 10$, define $Y_{ij} = 1$ if record j in block i is modified, and zero otherwise. Let $X_i = \sum_{j=1}^{10} Y_{ij}$. Under the assumption that each record has a 3% chance of update, Y_{ij} is a Bernoulli random variable with parameter $p = 0.03$. Consequently, X_i is a binomial random variable with parameters $(n = 10, p = 0.03)$, which measures the number of updated records in block i. Because the system must rewrite block i if $X_i > 0$, we compute the probability of such a rewrite as

$$\Pr(X_i > 0) = 1 - \Pr(X_i = 0) = 1 - C_{10,0}(0.03)^{0}(1 - 0.03)^{10} = 0.2626.$$

This result is independent of i, the block number, which means that each block is rewritten with probability 0.2626. The density of 10 records per block implies that there are 100,000 blocks. Let $B_i = 1$, if block i is rewritten, and zero otherwise. Define $Z = \sum_{i=1}^{100,000} B_i$. Each B_i is a Bernoulli random variable with parameter $p = 0.2626$, and we can reasonably assume that they are independent. Rewriting a particular block does not change the probability that another block must be rewritten: $\Pr(B_i = x|B_j = y) = \Pr(B_i = x)$ for all binary values x, y, which implies independence. Therefore, Z has a binomial distribution with parameters $(n = 100,000, p = 0.2626)$. The average number of blocks rewritten is the mean of this distribution, which according to Theorem 2.6, is $np = (100,000)(0.2626) = 26,260$ blocks.

Although the system must rewrite $26,260$ blocks, on the average, for each weekly update, we realize that the actual number will vary. We may be interested then in the probability that the actual number will lie within 1% of this average. That is, what is the probability that Z lies between $26260 - 262.6 \approx 25997$ and $26260 + 262.6 \approx 26523$? The variance of Z is $np(1-p) = (100,000)(0.2626)(1 - 0.2626) = 19364$. Chebyshev's inequality (Theorem 1.50) permits the following estimate, where I is the interval $(0.99\mu_Z \leq Z \leq 1.01\mu_Z)$.

$$\Pr(I) = \Pr(|Z - \mu_Z| \leq 0.01\mu_Z) = \Pr\left(|Z - \mu_Z| \leq \frac{0.01\mu_Z \sigma_Z}{\sigma_Z}\right)$$

$$\geq 1 - \frac{1}{(0.01\mu_Z/\sigma_Z)^2} = 1 - \frac{1}{(262.6/\sqrt{19364})^2} = 0.719$$

In 71.9% or more of the weekly updates, the number of rewritten blocks will be within 1% of the mean value 26260. This bound is actually low. Because Z has a binomial distribution with known parameters, we can express an exact solution as

$$\Pr(Z = k) = C_{100,000,k}(0.2626)^k(1 - 0.2626)^{100,000-k}$$

$$\Pr(25997 \leq Z \leq 26523) = \sum_{k=25997}^{26523} C_{100,000,k}(0.2626)^k(1 - 0.2626)^{100,000-k}.$$

Unfortunately this computation involves large intermediate values, arising from the factorials in $C_{100,000,k}$, which overflow the registers of most computer systems. Later chapters develop more powerful tools for handling this summation. $\square$

We now show that we can express the mean and variance of a sum of random variables in terms of the means and variances of the summands. We can then derive the mean and variance of the binomial random variable by considering it as the sum of independent Bernoulli random variables.

2.8 Theorem: Let $\Pr_{XY}(\cdot, \cdot)$ be the probability distribution of random variables X and Y. Let $S = aX + bY$, for constants a and b. Then $\mu_S = a\mu_X + b\mu_Y$ and $\sigma_S^2 = a^2\sigma_X^2 + 2ab\sigma_{XY} + b^2\sigma_Y^2$.

PROOF: Let $x_1, x_2, \ldots$ and $y_1, y_2, \ldots$ be the ranges of X and Y, respectively. The existence of means and variances implies that the defining sums converge absolutely, and Theorems A.36 and A.45 then allow us to rearrange the summands in any convenient order. We then calculate as follows.

$$\mu_S = \sum_i \sum_j (ax_i + by_j)\Pr_{XY}(x_i, y_j)$$

$$= a\sum_i x_i \sum_j \Pr_{XY}(x_i, y_j) + b\sum_j y_j \sum_i \Pr_{XY}(x_i, y_j)$$

The inner sums are, of course, the marginal probabilities, so we continue with

$$\mu_S = a \sum_i x_i \Pr_X(x_i) + b \sum_j y_j \Pr_Y(y_j) = a\mu_X + b\mu_Y.$$

Similarly,

$$\sigma_S^2 = \sum_i \sum_j [(ax_i + by_j) - (a\mu_X + b\mu_Y)]^2 \Pr_{XY}(x_i, y_j)$$

$$= \sum_i \sum_j [a(x_i - \mu_X) + b(y_j - \mu_Y)]^2 \Pr_{XY}(x_i, y_j)$$

$$= a^2 \sum_i (x_i - \mu_X)^2 \sum_j \Pr_{XY}(x_i, y_j) + b^2 \sum_j (y_j - \mu_Y)^2 \sum_i \Pr_{XY}(x_i, y_j)$$

$$\qquad + 2ab \sum_i \sum_j (x_i - \mu_X)(y_j - \mu_Y) \Pr_{XY}(x_i, y_j)$$

$$= a^2 \sum_i (x_i - \mu_X)^2 \Pr_X(x_i) + b^2 \sum_j (y_j - \mu_Y)^2 \Pr_Y(y_j) + 2ab\sigma_{XY}$$

$$= a^2 \sigma_X^2 + 2ab\sigma_{XY} + b^2 \sigma_Y^2.$$

The result extends to a sum of n random variables. That is, if $S = \sum_{i=1}^n a_i X_i$, where the X_i have means μ_i, variances σ_i^2, and covariances σ_{ij}, then

$$\mu_S = \sum_{i=1}^n a_i \mu_i \qquad\qquad \sigma_S^2 = \sum_{i=1}^n a_i^2 \sigma_i^2 + 2 \sum_{i<j} a_i a_j \sigma_{i,j}.$$

The computation breaks the sum into two parts: $(a_1 X_1 + a_2 X_2 + \ldots + a_{n-1} X_{n-1})$ and $a_n X_n$. The answer follows by applying the two-part result already proved and invoking an induction hypothesis to expand the first part. An exercise asks you to fill in the details. ∎

We can now verify Theorem 2.6 with the new technique. Let the binomial random variable X be the sum of n independent Bernoulli random variables, all with parameter p. That is, $X = \sum_{i=1}^n B_i$. Then X has the binomial distribution with parameters (n, p). Invoking Theorem 2.3 to obtain the mean and variance of the constituent Bernoulli summands, we compute as follows.

$$\mu_X = \sum_{i=1}^n \mu_{B_i} = \sum_{i=1}^n p = np$$

$$\sigma_X^2 = \sum_{i=1}^n \sigma_{B_i}^2 + 2 \sum_{i=1}^n \sum_{j=i+1}^n \sigma_{B_i, B_j} = \sum_{i=1}^n p(1-p) = np(1-p)$$

Notice that all the covariance terms are zero because the Bernoulli random variables are independent, and therefore, according to Theorem 1.74, they are uncorrelated.

2.9 Example: A loaded coin lands heads up with probability p. Let the random variable N be the number of heads in n tosses. What is the probability distribution of N?

Let X_i be the random variable that produces a 1 if the ith toss lands heads up and 0 if the ith toss is tails. X_i is a Bernoulli random variable with parameter p. Furthermore, a given toss is not affected by the results of the other tosses, so it is reasonable to model the joint probability of $X_1, X_2, \ldots, X_n$ as the product of the marginal probabilities of the X_i. In other words, the X_i are independent random variables. Since $N = \sum_{i=0}^{n} X_i$, we see that N is binomial with parameters (n, p). That is, the probability that exactly k heads appear among the n tosses is $\Pr_N(k) = C_{n,k} p^k (1 - p)^{n-k}$. If the coin is fair, then $p = 1 - p = 1/2$, and $\Pr_N(k) = C_{n,k}/2^n$, which is the probability of k heads in n tosses as it appeared in the preceding chapter. $\square$

2.10 Example: A digital communication channel has an error rate of 10^{-7}, which means that the chance of a bit (a one or a zero) arriving reversed at the destination is 10^{-7}. What is the probability that a transmission of a million bits arrives without error?

Let X_i be the Bernoulli random variable that assumes the value 1 when bit i is corrupted. Assuming that an error on a given bit does not affect errors on other bits, the X_i are independent Bernoulli random variables, all with parameter $p = 10^{-7}$. The total number of errors in a transmission of n bits is $S = \sum_{i=0}^{n} X_i$. Hence S has a binomial distribution with parameters $(n = 10^6, p = 10^{-7})$, and an error-free transmission occurs with probability $\Pr(S = 0)$, which we calculate as

$$\Pr(S = 0) = C_{10^6, 0} (10^{-7})^0 (1 - 10^{-7})^{10^6} = 0.905.$$

In other words, the probability of one or more transmission errors is $1 - 0.905 = 0.095$.

Now suppose that we try to reduce this error probability with a parity scheme. Specifically, we break the transmission into 10-bit sequences and follow each of them with a parity bit. The parity bit is a one or zero, whichever is necessary to ensure that the 11-bit sequence of data plus parity contains an odd number of 1s. Into the million-bit data stream we insert 100,000 parity bits, so the total transmission now requires 1.1 million bits, each subject to corruption with probability 10^{-7}.

The receiver performs a parity check on the arriving data stream by verifying that each 11-bit block exhibits an odd number of ones. If a parity violation occurs, the receiver asks for a retransmission. Hence we assume that the transmission is error-free if it contains no errors or if it contains only detectable errors. For $1 \leq j \leq 100,000$, define

$$T_j = \sum_{i=11(j-1)+1}^{11j} X_i.$$

T_j is the number of errors in the jth 11-bit block, and T_j has a binomial distribution with parameters $(11, p)$. An undetectable error occurs in the block if there is an even number, greater than zero, of corrupted bits, since each such scenario leaves an odd number of 1s. Let

$$Y_j = \begin{cases} 1, & \text{block } j \text{ has an undetectable error} \\ 0, & \text{otherwise.} \end{cases}$$

Then

$$\Pr(Y_j = 1) = \sum_{i \text{ even}} C_{11,i} p^i (1-p)^{11-i} = \sum_{i \text{ even}} C_{11,i} (10^{-7})^i (1 - 10^{-7})^{11-i}$$

$$= 5.5 \times 10^{-13} + 3.3 \times 10^{-26} + 4.6 \times 10^{-40} + 1.6 \times 10^{-54} + 1.1 \times 10^{-69}$$

$$\approx 5.5 \times 10^{-13}.$$

We treat the Y_j as independent Bernoulli random variables with parameter $q = 5.5 \times 10^{-13}$. Because there are 10^5 11-bit blocks we define $S = \sum_{j=1}^{10^5} Y_j$. S is then the number of undetectable errors in the transmission, and it has a binomial distribution with parameters $(10^5, q)$. The probability of no undetectable errors is now

$$\Pr(S = 0) = C_{10^5,0} q^0 (1-q)^{10^5} = (1 - 5.5 \times 10^{-13})^{10^5} = 0.999999945,$$

which is, of course, a dramatic improvement over the situation without parity protection. Indeed, you might think that 10^5 parity bits is too much overhead, because you might be willing to settle for, say, 0.9999 as the probability of no undetectable errors. One of the exercises asks you to pursue the option of fewer parity bits. □

Suppose that X_1 has a binomial distribution with parameters (n_1, p); X_2 is binomial with parameters (n_2, p). Assume further that X_1, X_2 are independent, and let $Y = X_1 + X_2$. What is the distribution of Y? Evidently, Y can assume values 0 through $n_1 + n_2$, and for k in this range, $\Pr_Y(k)$ is the sum over the disjoint events that comprise $Y = k$.

$$\Pr_Y(k) = \sum_{i=0}^{n_1} \Pr_{X_1}(i) \cdot \Pr_{X_2}(k-i)$$

$$= \sum_{i=0}^{n_1} C_{n_1,i} p^i (1-p)^{n_1-i} C_{n_2,k-i} p^{k-i} (1-p)^{n_2-k+i}$$

$$= p^k (1-p)^{n_1+n_2-k} \sum_{i=0}^{n_1} C_{n_1,i} C_{(n_1+n_2)-n_1, k-i}$$

$$= C_{n_1+n_2,k} p^k (1-p)^{n_1+n_2-k}$$

The last reduction makes use of Theorem 1.35. Thus Y has a binomial distribution with parameters $(n_1 + n_2, p)$, and we record this fact as the next theorem.

2.11 Theorem: If X_1, X_2 are independent binomial random variables with parameters (n_1, p) and (n_2, p), respectively, then $X_1 + X_2$ has a binomial distribution with parameters $(n_1 + n_2, p)$.

PROOF: See discussion above. ■

The result is not surprising. If X_i has a binomial distribution with parameters (n_i, p), then it is the sum of n_i independent Bernoulli random variables with parameter p. If X_1, X_2 are independent, then $X_1 + X_2$ is the sum of $n_1 + n_2$ independent Bernoulli random variables with parameter p and thus has the derived binomial distribution. Note that this simplification occurs

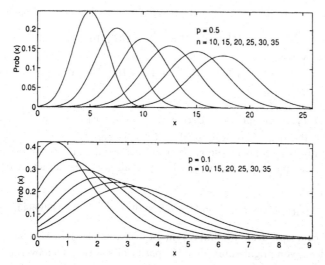

Figure 2.1. Smoothed curves for various binomial distributions.

only when the p parameter is common to the two binomial distributions. If X_i has the binomial distribution with parameters n_i, p_i, then the best we can deduce from the derivation above is the much less tractable expression

$$\mathrm{Pr}_{X_1 + X_2}(k) = \sum_{i=0}^{n_1} C_{n_1,i} C_{n_2,k-i} p_1^i p_2^{k-i} (1 - p_1)^{n_1 - i} (1 - p_2)^{n_2 - k + i}. \qquad (2.2)$$

The binomial distribution with parameters (n, p) is a discrete distribution with nonzero probabilities only at the points $0, 1, 2, \ldots, n$. Figure 2.1 connects these points with smooth curves to depict the general tendency of the distribution to spread out and shift to the right as n increases. The upper curves shows this behavior for $p = 0.5$; the lower curves are for $p = 0.1$. The balance point of each curve, the mean, is at np, as expected. The spreading with larger values of n is also expected since the variance is $np(1 - p)$. This variance is $(0.5)^2 n = 0.25n$ for the upper curves, while it is $(0.1)(0.9)n = 0.09n$ for the lower curves. Thus the curves for $p = 0.5$ are spreading more rapidly with increasing n, although this effect is masked somewhat by the different horizontal scales.

Besides the mean migration to the right and the spreading with increasing n, the curves of Figure 2.1 suggest that the *shape* of the distribution may be tending to a bell-shaped form. This is indeed true, although we will not be in a position to prove it until later in the book. If X_n is the binomial random variable with parameters (n, p), then the probabilities for $X_n = 0, 1, \ldots, n$ tend, with increasing n, to lie on the curve

$$f(x) = \frac{1}{\sigma_n \sqrt{2\pi}} \cdot e^{-(x - \mu_n)^2 / 2\sigma_n^2} = \frac{1}{\sqrt{2\pi np(1 - p)}} \cdot e^{-(x - np)^2 / 2np(1 - p)}, \qquad (2.3)$$

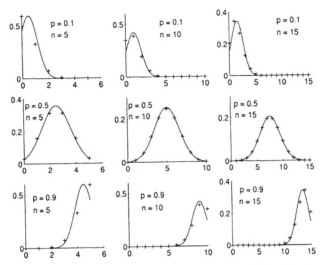

Figure 2.2. Approximating curves for binomial distributions

where $\mu_n = np$ and $\sigma_n^2 = np(1-p)$ are the mean and variance of X_n. Figure 2.2 illustrates the point for several values of n and p. The crosses are the actual probabilities for the binomial random variables, while the smooth curve is the alleged limit. Note that for p near 0.5, the approximation is good even for small n (e.g., $n = 5$). For p close to 0 or 1, the approximation becomes good eventually, but larger n values are required. We will return to this important matter when we take up continuous distributions. For the moment, however, we accept the approximation on faith and use it to complete the calculation that was abandoned in Example 2.7 because of intractably large factorials.

2.12 Example: Example 2.7 determined that the number of block rewrites Z during a database update was binomially distributed with parameters ($n = 100,000, p = 0.2626$). This distribution has a mean of 26260, and a question arose as to the probability that the actual number of rewritten blocks is within 1% of this mean value. The exact solution is

$$\Pr(25997 \leq Z \leq 26523) = \sum_{k=25997}^{26523} C_{100,000,k} (0.2626)^k (1 - 0.2626)^{100,000-k},$$

but it is not computationally tractable because of the large factorials. The approximation of Equation 2.3 is helpful here. Figure 2.3 illustrates the crucial insight for a binomial random variable with a smaller range. The figure shows the limiting curve for the random variable B, which is binomially distributed with parameters ($n = 50, p = 1/2$). The rectangles have areas equal to the individual probabilities $\Pr(B = k)$ for selected k values. Each rectangle has unit width and height $\Pr(B = k) = C_{n,k} p^k (1 - p)^{n-k}$. The combined areas of the rectangles is the probability that $B \leq j$, where j is the horizontal location of the rightmost rectangle. The figure suggests that this area is nearly equal to the area under the limiting curve to

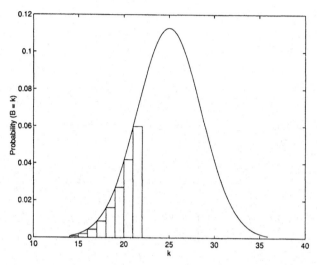

Figure 2.3. Limiting distribution of binomial ($n = 50, p = 0.5$) probabilities (see Example 2.12)

the left of the vertical line at $x = j$. That is,

$$\Pr(B \le j) \approx \int_{-\infty}^{j} \frac{1}{\sigma_B \sqrt{2\pi}} \cdot e^{-(x-\mu_B)^2/2\sigma_B^2} \, dx.$$

The approximation becomes more accurate as n grows larger. Moreover, a better
estimate results if we shift the limiting curve to the right by one-half a unit on the
horizontal axis. The curve then bisects the top edges of the rectangles rather than
skirting the upper left corners. We will take up such subtleties later in the text. For
the moment, we accept the approximation and apply it to the current problem.

The variance of Z is $\sigma_Z^2 = (100,000)(0.2626)(1 - 0.2626) = 19364$. We calculate the probability that the number of rewritten blocks is within 1% of the mean
as follows. We first simplify the integral through the substitution $y = (x - \mu_Z)/\sigma_Z$.
Denoting the event $(0.99\mu_Z \le Z \le 1.01\mu_Z)$ by E, we have

$$\Pr(E) \approx \int_{-\infty}^{1.01\mu_Z} \frac{1}{\sigma_Z \sqrt{2\pi}} e^{-(x-\mu_Z)^2/2\sigma_Z^2} \, dx - \int_{-\infty}^{0.99\mu_Z} \frac{1}{\sigma_Z \sqrt{2\pi}} e^{-(x-\mu_Z)^2/2\sigma_Z^2} \, dx$$

$$= \int_{0.99\mu_Z}^{1.01\mu_Z} \frac{1}{\sigma_Z \sqrt{2\pi}} e^{-(x-\mu_Z)^2/2\sigma_Z^2} \, dx = \frac{1}{\sqrt{2\pi}} \int_{-0.01\mu_Z/\sigma_Z}^{0.01\mu_Z/\sigma_Z} e^{-y^2/2} \, dy$$

$$= \frac{1}{\sqrt{2\pi}} \int_{-1.887}^{1.887} e^{-y^2/2} \, dy.$$

This last integral is still difficult to evaluate, but fortunately the simplification
provides a form that does not depend on the details of this particular problem.
Because it occurs frequently, the function $\Phi(x) = (1/\sqrt{2\pi}) \int_{-\infty}^{x} e^{-y^2/2} \, dy$ is available
from tabulations. These tables were originally constructed from a series expansion of
the integrand. Today, most computer systems provide the function, or some constant

x	0	1	2	3	4	5	6	7	8	9
0.00	.5000	.5040	.5080	.5120	.5160	.5199	.5239	.5279	.5319	.5359
0.10	.5398	.5438	.5478	.5517	.5557	.5596	.5636	.5675	.5714	.5753
0.20	.5793	.5832	.5871	.5910	.5948	.5987	.6026	.6064	.6103	.6141
0.30	.6179	.6217	.6255	.6293	.6331	.6368	.6406	.6443	.6480	.6517
0.40	.6554	.6591	.6628	.6664	.6700	.6736	.6772	.6808	.6844	.6879
0.50	.6915	.6950	.6985	.7019	.7054	.7088	.7123	.7157	.7190	.7224
0.60	.7257	.7291	.7324	.7357	.7389	.7422	.7454	.7486	.7517	.7549
0.70	.7580	.7611	.7642	.7673	.7704	.7734	.7764	.7794	.7823	.7852
0.80	.7881	.7910	.7939	.7967	.7995	.8023	.8051	.8078	.8106	.8133
0.90	.8159	.8186	.8212	.8238	.8264	.8289	.8315	.8340	.8365	.8389
1.00	.8413	.8438	.8461	.8485	.8508	.8531	.8554	.8577	.8599	.8621
1.10	.8643	.8665	.8686	.8708	.8729	.8749	.8770	.8790	.8810	.8830
1.20	.8849	.8869	.8888	.8907	.8925	.8944	.8962	.8980	.8997	.9015
1.30	.9032	.9049	.9066	.9082	.9099	.9115	.9131	.9147	.9162	.9177
1.40	.9192	.9207	.9222	.9236	.9251	.9265	.9279	.9292	.9306	.9319
1.50	.9332	.9345	.9357	.9370	.9382	.9394	.9406	.9418	.9429	.9441
1.60	.9452	.9463	.9474	.9484	.9495	.9505	.9515	.9525	.9535	.9545
1.70	.9554	.9564	.9573	.9582	.9591	.9599	.9608	.9616	.9625	.9633
1.80	.9641	.9649	.9656	.9664	.9671	.9678	.9686	.9693	.9699	.9706
1.90	.9713	.9719	.9726	.9732	.9738	.9744	.9750	.9756	.9761	.9767
2.00	.9772	.9778	.9783	.9788	.9793	.9798	.9803	.9808	.9812	.9817
2.10	.9821	.9826	.9830	.9834	.9838	.9842	.9846	.9850	.9854	.9857
2.20	.9861	.9864	.9868	.9871	.9875	.9878	.9881	.9884	.9887	.9890
2.30	.9893	.9896	.9898	.9901	.9904	.9906	.9909	.9911	.9913	.9916
2.40	.9918	.9920	.9922	.9925	.9927	.9929	.9931	.9932	.9934	.9936
2.50	.9938	.9940	.9941	.9943	.9945	.9946	.9948	.9949	.9951	.9952
2.60	.9953	.9955	.9956	.9957	.9959	.9960	.9961	.9962	.9963	.9964
2.70	.9965	.9966	.9967	.9968	.9969	.9970	.9971	.9972	.9973	.9974
2.80	.9974	.9975	.9976	.9977	.9977	.9978	.9979	.9979	.9980	.9981
2.90	.9981	.9982	.9982	.9983	.9984	.9984	.9985	.9985	.9986	.9986

TABLE 2.1. Partial tabulation of $\Phi(x) = (1/\sqrt{2\pi}) \int_{-\infty}^{x} e^{-y^2/2}\, dy$ (see Example 2.12)

multiple of it, as part of the mathematical library. In any case, Table 2.1 exhibits a partial listing of $\Phi(x)$. To use the table, you find the argument to one decimal place in the leftmost column. You then proceed across horizontally to the column headed by the argument's second decimal place. If, for example, you want $\Phi(2.41)$, you locate 2.4 in the leftmost column and proceed horizontally to the column headed 1. The cell so identified contains $\Phi(2.41) = 0.9920$. Because the integrand in the defining equation is symmetric about $x = 0$, we have

$$\frac{1}{\sqrt{2\pi}} \int_{-\infty}^{0} e^{-y^2/2}\, dy = \frac{1}{\sqrt{2\pi}} \int_{0}^{\infty} e^{-y^2/2}\, dy.$$

From the table, this value is 1/2, which means that $(1/\sqrt{2\pi}) \int_{-\infty}^{\infty} e^{-y^2/2}\, dy = 1$. For $x < 0$, we use the substitution $v = -y$ to obtain

$$\Phi(x) = \frac{1}{\sqrt{2\pi}} \int_{-\infty}^{x} e^{-y^2/2}\, dy = 1 - \frac{1}{\sqrt{2\pi}} \int_{x}^{\infty} e^{-y^2/2}\, dy$$

$$= 1 + \frac{1}{\sqrt{2\pi}} \int_{-x}^{-\infty} e^{-v^2/2}\, dv = 1 - \frac{1}{\sqrt{2\pi}} \int_{-\infty}^{-x} e^{-v^2/2}\, dv = 1 - \Phi(-x).$$

This observation extends the domain of Table 2.1 to the span $-2.99 \le x \le 2.99$.

In the problem at hand, we have

$$\Pr(E) \approx \frac{1}{\sqrt{2\pi}} \int_{-1.887}^{1.887} e^{-y^2/2}\, dy$$

$$= \frac{1}{\sqrt{2\pi}} \int_{-\infty}^{1.887} e^{-y^2/2}\, dy - \frac{1}{\sqrt{2\pi}} \int_{-\infty}^{-1.887} e^{-y^2/2}\, dy$$

$$= \Phi(1.887) - \Phi(-1.887) = \Phi(1.887) - (1 - \Phi(1.887))$$

$$= 2\Phi(1.887) - 1 = 2(0.9704) - 1 = 0.9408.$$

This results assures us that the number of rewritten blocks in 94% of the weekly up-dates will fall within 1% of the mean value 26260. This fraction is significantly higher than the bound that we obtained in Example 2.7 with the Chebyshev inequality. That bound merely noted that the fraction would be larger that 71.9%. □

The rest of this chapter investigates other discrete distributions in a manner similar to that just employed for the binomial distribution. To fa-cilitate the summations required by the mean and variance calculations, we first consider a technique for performing many such summations simultane-ously. Recalling the series expansion for the exponential function, we calculate $E[e^{tX}]$ for some real parameter t for which the defining sum converges abso-lutely. X is a random variable with discrete distribution $\Pr(x_i)$ on a range $x_1, x_2, \ldots$.

$$E[e^{tX}] = \sum_i e^{tx_i} \Pr(x_i) = \sum_i \sum_{j=0}^{\infty} \frac{(tx_i)^j \Pr(x_i)}{j!}$$

Now, because the double sum is assumed absolutely convergent for the t value in question, we can invoke Theorem A.45 to reverse the summation order. This gives

$$E[e^{tX}] = \sum_{j=0}^{\infty} \frac{t^j}{j!} \sum_i x_i^j \Pr(x_i) = \sum_{j=0}^{\infty} \left(\frac{E[X^j]}{j!} \right) t^j.$$

This result suggests that the quantities $E[X^j]$ are available from the series ex-pansion of the single expected value $E[e^{tX}]$, provided that the aforementioned absolute convergence holds. Two definitions are appropriate at this point.

2.13 Definition: Let X be a discrete random variable. The values $E[X^i]$, for $i = 0, 1, 2, \ldots$, are called the *moments* of X. Specifically, $E[X^i]$ is the ith moment of X. The values $E[(X - \mu)^i]$ are the *central moments* of X. ∎

2.14 Definition: Let X be a discrete random variable. The function $\Psi(t)$, or $\Psi_X(t)$ if necessary, is the *moment generating* function for X. Its calculation is $\Psi(t) = E[e^{tX}]$, and it is defined only for those values of t for which the expected value summation is convergent. ∎

Provided that the double sum noted above is absolutely convergent, we now have, in terms of the new definition,

$$\Psi(t) = \sum_{j=0}^{\infty} \left(\frac{E[X^j]}{j!} \right) t^j,$$

which implies that the ith moment of X is $j!$ times the coefficient of t^j in the expansion of $\Psi(t)$. Also notice that successive derivatives of $\Psi(t)$ can produce these values. We justify the term-by-term differentiation in the next section.

$$\Psi(t) = \sum_{j=0}^{\infty} \frac{E[X^j]}{j!} \cdot t^j = E[X^0] + \sum_{j=1}^{\infty} \frac{E[X^j]}{j!} \cdot t^j$$

$$E[X^0] = \Psi(0) = 1$$

$$\Psi'(t) = \sum_{j=1}^{\infty} \frac{E[X^j]}{j!} \cdot jt^{j-1} = E[X^1] + \sum_{j=2}^{\infty} \frac{E[X^j]}{j!} \cdot jt^{j-1}$$

$$E[X^1] = \Psi'(0)$$

$$\Psi''(t) = \sum_{j=2}^{\infty} \frac{E[X^j]}{j!} \cdot j(j-1)t^{j-2} = E[X^2] + \sum_{j=3}^{\infty} \frac{E[X^j]}{j!} \cdot j(j-1)t^{j-2}$$

$$E[X^2] = \Psi''(0)$$

Evidently, this process continues to yield the following theorem, where $f^{(i)}$ denotes the ith derivative of $f(\cdot)$.

2.15 Theorem: Suppose that $\Psi(t)$, the moment generating function for the random variable X, is defined for $t \in (-a, a)$, a nonempty interval about zero. Then the ith moment of X is $i!$ times the coefficient of t^i in the expansion of $\Psi(t)$. It is also equal to $\Psi^{(i)}(0)$.

PROOF: The upcoming mathematical interlude shows that all derivatives of $\Psi(t)$ exist on nonempty $(-a, a)$ when $\Psi(t)$ itself exists there. Moreover, the double sum

$$E[e^{tX}] = \sum_{i=1}^{\infty} e^{tx_i} \Pr(x_i) = \sum_{i=1}^{\infty} \sum_{n=0}^{\infty} \left(\frac{x_i^n \Pr(x_i)}{n!} \right) t^n$$

must be absolutely convergent for $t \in (-a, a)$. (If $0 \le t < a$, the summand is positive and absolute convergence is the same as ordinary convergence, which is given in that range. If $-a < t < 0$, we obtain the absolute value of the summand by substituting $|t|$ for t. Since $0 < |t| < a$, the transformed series converges.) We can therefore reverse the summation order, when $t \in (-a, a)$. The theorem's assertions then follow from the discussion above, provided that the term-by-term differentiation is valid. The next section's mathematical interlude proves the validity of term-by-term differentiation in this circumstance. ∎

The theorem is useful because we can often obtain $\Psi(t)$ without individually considering the terms of the exponential expansion. We can then expand $\Psi(t)$ and read off the moments from the resulting coefficients. We can also evaluate successive derivatives of $\Psi(t)$ at zero to obtain the moments. We can check out, for example, the simple case when X is a Bernoulli random variable with parameter p. In this case, we know that $E[X] = \mu = p$ and that $E[X^2] = \sigma^2 + \mu^2 = p(1-p) + p^2 = p$. We calculate $\Psi(t) = E[e^{tX}]$ directly from the definition of the expected value operator.

$$\Psi(t) = p \cdot e^{1t} + (1-p) \cdot e^{0t} = (1-p) + pe^t = (1-p) + p \sum_{k=0}^{\infty} \frac{t^k}{k!}$$

$$= (1-p) + p + pt + \frac{pt^2}{2!} + \frac{pt^3}{3!} + \ldots = 1 + pt + \frac{pt^2}{2!} + \frac{pt^3}{3!} + \ldots$$

As an application of Theorem 2.22 in the upcoming mathematical interlude, we note that the expansion converges for all t, and we read off $E[X]$ as 1! times the coefficient of t, which gives p, as anticipated. Also, 2! times the coefficient of t^2 gives p, the correct value for $E[X^2]$. Alternatively, we can apply the derivative method to the closed-form expression for $\Psi_X(t)$.

$$\Psi'(t) = pe^t \qquad E[X] = \Psi'(0) = p$$
$$\Psi''(t) = pe^t \qquad E[X^2] = \Psi''(0) = p$$

Indeed, we can see that all the derivatives of $\Psi(t)$ are pe^t, which implies that $E[X^k] = p$ for all $k \geq 1$.

In a similar fashion, we can verify the moments for the binomial distribution with parameters (n, p). Suppose that Y is such a random variable. We know that

$$E[Y] = \mu = np$$
$$E[Y^2] = \sigma^2 + \mu^2 = np(1-p) + (np)^2 = n(n-1)p^2 + np.$$

As in the Bernoulli case, we compute $\Psi(t)$ directly from the definition of expected value and then evaluate its derivatives.

$$\Psi(t) = \sum_{k=0}^{n} e^{kt} C_{n,k} p^k (1-p)^{n-k} = \sum_{k=0}^{n} C_{n,k} (pe^t)^k (1-p)^{n-k}$$

$$= [pe^t + (1-p)]^n$$
$$\Psi'(t) = n[pe^t + (1-p)]^{n-1} pe^t$$
$$\Psi''(t) = n(n-1)[pe^t + (1-p)]^{n-2}(pe^t)^2 + n[pe^t + (1-p)]^{n-1} pe^t$$

$\Psi'(0) = np$, which is the correct value for $E[Y]$, and $\Psi''(0) = n(n-1)p^2 + np$, which is the correct value for $E[Y^2]$.

We close with an example where $E[e^{tX}]$ is not defined in any open interval about zero. In this case, we cannot use the moment generating function to generate moments.

2.16 Example: Consider the distribution on the positive integers. $\Pr(X = k) = 1/[Ak^2]$, where $A = \sum_{k=1}^{\infty}(1/k^2)$. Example 1.48 proved the existence of $A < 2$. For any $t > 0$, l'Hôpital's rule gives

$$\lim_{k\to\infty} \frac{e^{tk}}{k^2} = \lim_{k\to\infty} \frac{(e^t)^k}{k^2} = \lim_{k\to\infty} \frac{(e^t)^k \ln e^t}{2k}$$

$$= \lim_{k\to\infty} \frac{t(e^t)^k}{2k} = \lim_{k\to\infty} \frac{t(e^t)^k \ln e^t}{2} = \lim_{k\to\infty} \frac{t^2(e^t)^k}{2} = \infty.$$

Consequently, there exists N such that $k \geq N$ implies that $e^{tk}/k^2 > 1$. This means that

$$E[e^{tX}] = \sum_{k=1}^{\infty} \frac{e^{tk}}{Ak^2} = \frac{1}{A} \sum_{k=1}^{\infty} \frac{e^{tk}}{k^2}$$

does not exist for positive t. The sum does converge, however, for nonpositive t, because $0 < e^{tk} \leq 1$ for $k = 1, 2, \ldots$. In this case,

$$E[e^{tX}] = \sum_{k=1}^{\infty} \frac{e^{tk}}{Ak^2} \leq \sum_{k=1}^{\infty} \frac{1}{Ak^2} = 1,$$

and Theorem A.35 (dominated convergence) asserts the convergence of $E[e^{tX}]$. Nevertheless, even though $E[e^{tX}]$ exists for nonpositive t, the double sum

$$E[e^{tX}] = \sum_{k=1}^{\infty} \frac{e^{tk}}{Ak^2} = \sum_{k=1}^{\infty} \sum_{n=0}^{\infty} \frac{(tk)^n}{n!Ak^2}$$

is not absolutely convergent. (If it were so, we would have $E[e^{tX}]$ existing for positive t, contradicting our earlier conclusion.) Therefore, in this case, we cannot reverse the summation order to express $E[e^{tX}]$ as

$$E[e^{tX}] = \sum_{n=0}^{\infty} \left(\frac{k^n}{n!Ak^2} \right) t^n = \sum_{n=0}^{\infty} \left(\frac{E[X^n]}{n!} \right) t^n.$$

Consequently, we cannot obtain the moments of X from the expansion of $E[e^{tX}]$. $\Box$

Exercises

2.1 Let X_i, for $1 \leq i \leq n$, be independent Bernoulli random variables with parameter p. Let $Y = X_1 \cdot X_2 \cdots X_n$. What is the distribution of Y?

2.2 Example 2.10 shows how a parity scheme can increase the probability of an error-free transmission across an imperfect channel. The channel corrupts a bit with probability 10^{-7}. Consider a modified scheme that inserts a parity bit after each 100th data bit. What is the probability of an undetectable error in a transmission of a million data bits?

2.3 The random variable X assumes the values $-1, 0, +1$ with probabilities $p/2$, $1 - p$, and $p/2$, respectively. What are μ and σ^2? What is $\Psi(t)$? Verify that the derivatives of $\Psi(t)$, evaluated at zero, deliver the correct values for $E[X]$ and $E[X^2]$.

2.4 Let $X_1, X_2, \ldots, X_n$ be independent random variables, each distributed as in the preceding problem. Let $Y = \sum_{k=0}^{n} X_k^2$. What is the distribution of Y? What are μ and σ^2?

2.5 Let $X_1, X_2, \ldots, X_n$ be independent Bernoulli random variables. Let $S = \sum_{i=1}^{n} X_i$. Let t be a specific vector of k ones and $n - k$ zeros. What is $\Pr((X_1, X_2, \ldots, X_n) = t | S = k)$?

2.6 Let $X_1, X_2, \ldots, X_n$ be independent random variables, each assuming the value $+1$ with probability p and -1 with probability $1 - p$. Let $Y = \sum_{k=0}^{n} X_k$. What is the distribution of Y? What are μ and σ^2? What is $\Psi(t)$? Verify that the derivatives of $\Psi(t)$, evaluated at zero, deliver the correct values for $E[Y]$ and $E[Y^2]$.

2.7 Let X have the distribution to the right. Compute $\Psi(t)$, the first and second moments of X, the mean of X, and the variance of X.

k	0	1	5	6
$\Pr(k)$	0.3	0.2	0.2	0.3

2.8 Let Y be the sum of two independent random variables, each distributed as in the preceding problem. Compute $\Psi_Y(t)$. What is the relationship between $\Psi_Y(t)$ and $\Psi_X(t)$ in the preceding problem?

2.9 Let X_1, X_2 be identically distributed independent random variables, each assuming the values $0, 1, 2$ with probabilities $p, q, 1 - p - q$, respectively. Let $Y = X_1 - X_2$. Calculate $\Psi_Y(t)$ and thereby deduce the mean and variance of Y.

*2.10 Suppose that the random variable X has the distribution $\Pr(k) = 1/2^k$ for $k = 1, 2, \ldots$. Calculate $\Psi(t)$ and subsequently, $E[X]$ and $E[X^2]$. Derive μ and σ^2 from these moments.

*2.11 Let the random variable X have nonzero probabilities on a subset of the nonnegative integers. Find the probability distribution of X given the moment generating function $\Psi(t) = e^t/2 + e^{3t}/(3 - e^{3t})$.

*2.12 Show that $\Psi_X(t) \geq e^{tE[X]}$.

2.13 Complete the extension of Theorem 2.8 to a linear combination of n random variables. That is, let X_i, for $1 \leq i \leq n$, be random variables with means μ_i, variances σ_i^2, and covariances σ_{ij}. Define $S = \sum_{i=1}^{n} a_i X_i$, and show that

$$\mu_S = \sum_{i=1}^{n} a_i \mu_i \qquad\qquad \sigma_S^2 = \sum_{i=1}^{n} a_i^2 \sigma_i^2 + 2 \sum_{i<j} a_i a_j \sigma_{i,j}.$$

2.2   Power series

Certain infinite sequences and series accompany a probability distribution. These closely related constructions provide useful tools for reasoning about the associated distribution. At the conclusion of the preceding section, for example, we saw how the series expansion for $E[e^{tX}]$ can provide values for the moments $E[X^k]$. Consequently, it is important to understand how to manipulate sequence and series expansions. This section's goal is to develop the elementary properties of power series. In particular, we wish to prove that the series expansion of the moment generating function allows term-by-term differentiation and integration. This will justify the argument that established Theorem 2.15, which asserts that the moments of a random variable are available from the derivatives of its moment generating function. The term-by-term operations in the theorem's proof are valid because the expansion of the moment generating function is an especially docile series, called a power series, which admits term-by-term evaluation of both integrals and derivatives. These properties of power series are proved below, following the definition.

2.17 Definition: A series of the form $f(t) = \sum_{n=0}^{\infty} a_n t^n$ is called a *power series.* ∎

For each t, the series $\sum_{n=0}^{\infty} a_n t^n$ is an ordinary series of real numbers, with properties as discussed in Appendix A. We assume familiarity with such properties, so it might now be appropriate to review Section A.3. The questions of convergence and absolute convergence arise for each t. For a set S of real numbers, there also arises the question of uniform convergence over S. A power series is, in many ways, very similar to a simple geometric series, whose properties we studied in Section 1.2. From this perspective, it is easy to settle convergence questions. The next few theorems illustrate the point.

It is clear that a power series $f(t) = \sum_{n=0}^{\infty} a_n t^n$ always converges absolutely at $t = 0$. How far can we venture away from $t = 0$ and maintain the convergence property? If the sequence $\{|a_n|^{1/n}\}$ is well-behaved, we can answer this question in terms of its limit. Suppose that $|a_n|^{1/n} \to 0$. For a fixed $t \neq 0$, there exists N such that $n \geq N$ forces $|a_n|^{1/n} < 1/(2|t|)$. Then, for $N \leq n < m$, we have

$$\left| \sum_{i=n+1}^{m} a_i t^i \right| \leq \sum_{i=n+1}^{m} |a_i t^i| < \sum_{i=n+1}^{m} \frac{|t|^i}{2^i |t|^i} = \frac{1}{2^{n+1}} \sum_{i=0}^{m-n-1} \frac{1}{2^i}$$

$$= \left(\frac{1}{2^{n+1}} \right) \frac{1 - (1/2)^{m-n}}{1 - (1/2)} < \frac{1}{2^n} \to 0,$$

as $n, m \to \infty$. For this fixed t, the partial sums for both $\sum a_n t^n$ and $\sum |a_n t^n|$ are Cauchy sequences and therefore converge. Because t was arbitrary, we have shown that the sequence converges absolutely for all t.

Now suppose that $|a|^{1/n} \to a > 0$. In this case, fix t such that $0 < |t| < 1/a$. Then $a|t| < 1$ and $b = (a|t| + 1)/2 < 1$. There exists N such that $n \geq N$ forces

$$|a_n|^{1/n} - a \leq \left| |a_n|^{1/n} - a \right| < \frac{1}{2} \cdot \left(\frac{1}{|t|} - a \right)$$

$$|a_n|^{1/n} < a + \frac{1}{2} \cdot \left(\frac{1}{|t|} - a \right) = \frac{1}{2} \cdot \left(\frac{1}{|t|} + a \right)$$

$$|a_n t^n| = \left(|a_n|^{1/n} |t| \right)^n < \left[\frac{1}{2} \cdot \left(\frac{1}{|t|} + a \right) \cdot |t| \right]^n = \left(\frac{a|t| + 1}{2} \right)^n = b^n.$$

Because $0 < b < 1$, we continue for $N \le n < m$ to obtain

$$\left| \sum_{i=n+1}^{m} a_i t^i \right| \le \sum_{i=n+1}^{m} |a_i t^i| < \sum_{i=n+1}^{m} b^i = b^{n+1} \sum_{i=0}^{m-n-1} b^i$$

$$= b^{n+1} \frac{1 - b^{m-n}}{1-b} < \frac{b^{n+1}}{1-b} \to 0,$$

as $m, n \to \infty$. Again, the partial sums for both $\sum a_n t^n$ and $\sum |a_n t^n|$ are Cauchy, and the sequence converges absolutely for $t \in (-1/a, 1/a)$.

Finally, suppose that $|a_n|^{1/n} \to \infty$. For any nonzero t, there exists N such that $n \ge N$ forces $|a_n|^{1/n} > 1/|t|$. Then, for $n \ge N$,

$$|a_n t^n| = \left(|a_n|^{1/n} |t| \right)^n > [(1/|t|) \cdot |t|]^n = 1.$$

Theorem A.34 requires $|a_n t^n| \to 0$ for a convergent series, so we conclude that $\sum_{n=0}^{\infty} a_n t^n$ does not converge except at $t = 0$.

2.18 Theorem: Let $f(t) = \sum_{n=0}^{\infty} a_n t^n$ be a power series for which $|a_n|^{1/n} \to a$. If $a = 0$, the series converges absolutely for all t. If a is a real number and $a > 0$, the series converges absolutely for $-1/a < t < 1/a$. Finally, if $a = \infty$, the series converges only for $t = 0$.

PROOF: See discussion above. ∎

2.19 Example: Consider the expansion for the exponential $e^x = \sum_{n=0}^{\infty} x^n / n!$. By the following computation, the sequence $\{(1/n!)^{1/n}\} \to 0$.

$$n! > n(n-1)(n-2) \cdots \lceil n/2 \rceil > \left[\frac{n}{2} \right]^{\lceil n/2 \rceil} \ge \left(\frac{n}{2} \right)^{n/2}$$

$$\left(\frac{1}{n!} \right)^{1/n} < \left(\frac{1}{(n/2)^{n/2}} \right)^{1/n} = \sqrt{2/n} \to 0$$

The power series expansion for e^x then converges absolutely for all x.

Consider the expansion $f(x) = \sum_{n=0}^{\infty} x^n$, in which the coefficients are all 1. Clearly, $|1|^{1/n} \to 1$, so the series converges absolutely for $-1 < x < 1$. Because it is a geometric series, we know that the limit is $1/(1-x)$ for x in this range. Now consider the new series obtained from term-by-term differentiation: $g(x) = \sum_{n=1}^{\infty} n x^{n-1} = \sum_{n=0}^{\infty} (n+1) x^n$. Using l'Hôpital's Rule, we compute

$$b_n = \ln \left[(n+1)^{1/n} \right] = \frac{\ln(n+1)}{n} \to \frac{1/(n+1)}{1} \to 0$$

$$(n+1)^{1/n} \to e^0 = 1.$$

Hence $g(x)$ also converges absolutely for $-1 < x < 1$.

Finally, consider

$$h(x) = \left(\frac{3n+5}{n} \right)^n x^n$$

$$\left[\left(\frac{3n+5}{n} \right)^n \right]^{1/n} = \frac{3n+5}{n} \to 3.$$

Therefore, $h(x)$ converges absolutely for $-1/3 < x < 1/3$. $\square$

If $f(t) = \sum_{n=0}^{\infty} a_n t^n$ and $|a_n|^{1/n}$ neither converges to a real number nor diverges to infinity, then $|a_n|^{1/n}$ must oscillate in a manner that precludes settling down to any definitive limit. In this case, we can use a generalized form of the limit that captures the largest permanent trend in an oscillating sequence.

2.20 Definition: Suppose that the set $\{a_n\}$ has an upper bound. In this case, we define the *upper limit* or the *limit superior* of the sequence $\{a_n\}$ by $\limsup a_n = \lim_{n \to \infty} \sup\{a_i | i \geq n\}$. If the set has no upper bound, we set $\limsup a_n = \infty$. Similarly, if the set $\{a_n\}$ has a lower bound, we define the *lower limit* or the *limit inferior* as $\liminf a_n = \lim_{n \to \infty} \inf\{a_i | i \geq n\}$. If it has no lower bound, we set $\liminf a_n = -\infty$. ∎

Suppose that the set $\{a_n\}$ has upper and lower bounds a and b. That is, $b \leq a_n \leq a$ for all n. Theorem A.40 then asserts that each subset has a least upper bound. (Recall that $\sup S$ and $\text{lub } S$ are equivalent notations for the least upper bound and greatest lower bound of S.) Let $A_n = \text{lub } \{a_i | i \geq n\}$. Then $a_n, a_{n+1}, a_{n+2}, \cdots \leq A_n$, which implies that A_n is an upper bound for $\{a_i | i \geq n + 1\}$. Therefore, A_{n+1}, which is the least upper bound of $\{a_i | i \geq n + 1\}$, cannot exceed A_n. That is, $A_{n+1} \leq A_n$. The sequence $\{A_n\}$ is then nonincreasing. Moreover $A_n \geq a_n \geq b$, so $\{A_n\}$ is bounded from below. According to Theorem A.41, it must converge to a real number. That is, $\limsup a_n = \lim_{n \to \infty} A_n = \overline{A}$, a well-defined real number. Similarly, $\liminf a_n = \underline{A}$, another real number. A bounded sequence always has an upper and a lower limit, even when it does not converge in the ordinary sense.

If the sequence has only an upper bound, then $\limsup a_n = \overline{A}$, a real number, or $\limsup a_n = -\infty$. This follows because the sequence A_n is nonincreasing and therefore cannot oscillate. Similarly, if the sequence has only a lower bound, then $\liminf a_n = \underline{A}$, a real number, or $\liminf a_n = \infty$. We conclude that the upper and lower limits are always either real numbers or one of the two infinities. Retaining $A_n = \text{lub } \{a_i | i \geq n\}$ and defining $B_n = \text{glb } \{a_i | i \geq n\}$, we have $B_n \leq a_n \leq A_n$ for all n. We distinguish several cases.

- $\{A_n\}$ and $\{B_n\}$ both converge. In this case, $A_n \leq B_n$ forces $\liminf a_n \leq \limsup a_n$. Moreover, if equality holds, then $B_n \leq a_n \leq A_n$ forces $\{a_n\}$ to converge to the common limit.

- $B_n \to \infty$. Because $B_n \leq A_n$, we must have $A_n \to \infty$ also. So, given any R, there exists N such that $n \geq N$ forces $R < B_n \leq A_n$. Since $B_n \leq a_n \leq A_n$, we have $R < a_n$ when $n \geq N$. That is, $a_n \to \infty$.

- $A_n \to -\infty$. Reasoning as in the case above, we must have $B_n \to -\infty$ and $a_n \to -\infty$.

Table 2.2 catalogs these and other cases and provides examples of each. We record the properties just discussed and some others in the following theorem.

Case	Example			
	Sequence	lim inf	lim sup	lim
$-\infty = \liminf a_n = \limsup a_n$	$a_n = -n$	$-\infty$	$-\infty$	$-\infty$
$-\infty = \liminf a_n < \limsup a_n = a$	$a_n = \begin{cases} -n, & n \text{ odd} \\ 1/n, & n \text{ even} \end{cases}$	$-\infty$	0	—
$-\infty = \liminf a_n < \limsup a_n = \infty$	$a_n = \begin{cases} -n, & n \text{ odd} \\ n, & n \text{ even} \end{cases}$	$-\infty$	∞	—
$a = \liminf a_n < \limsup a_n = b$	$a_n = (-1)^n$	-1	$+1$	—
$\liminf a_n = a = \limsup a_n$	$a_n = (-1)^n/n$	0	0	0
$a = \liminf a_n < \limsup a_n = \infty$	$a_n = \begin{cases} n, & n \text{ odd} \\ 1/n, & n \text{ even} \end{cases}$	0	∞	—
$\liminf a_n = \limsup a_n = \infty$	$a_n = n$	∞	∞	∞

TABLE 2.2. Possible relationships among upper, lower, and ordinary limits

These properties will prove very useful, both in the current discussion of power series and in the next chapter's treatment of nondeterministic simulations.
2.21 Theorem: The following properties hold for any sequence $\{a_n\}$ of real numbers.

- If the set $\{a_n\}$ has an upper bound, then $\limsup a_n = a$, a real number, or $\limsup a_n = -\infty$. If the set $\{a_n\}$ has a lower bound, then $\liminf a_n = b$, a real number, or $\liminf a_n = +\infty$. If $\{a_n\}$ has both upper and lower bounds, then its upper and lower limits are both real numbers.

- $\liminf a_n \leq \limsup a_n$ and equality implies that a_n converges to the common limit or diverges to the common infinity.

- Suppose that $\limsup a_n = a$, a real number. For any $\epsilon > 0$, we have $a_i > a - \epsilon$ for infinitely many i, and we have $a_i > a + \epsilon$ for only finitely many i. An analogous observation holds for the lower limit. That is, suppose that $\liminf a_i = b$, a real number. For any $\epsilon > 0$, we have $a_i < b + \epsilon$ for infinitely many i, and we have $a_i < b - \epsilon$ for only finitely many i.

- Suppose that $\limsup a_n = \infty$. Then, for any R, we have $a_i > R$ for infinitely many i. Similarly, if $\liminf a_n = -\infty$, then for any R, we have $a_i < R$ for infinitely many i.

- There exist subsequences $\{a_{i_k}\}$ and $\{a_{j_k}\}$ of the original a_n such that $\lim_{k \to \infty} a_{i_k} = \limsup a_n$ and $\lim_{k \to \infty} a_{j_k} = \liminf a_n$.

- Suppose that $b_n \to b > 0$. Then
$$\limsup a_n b_n = \begin{cases} ab, & \text{if } \limsup a_n = a, \text{ a real number} \\ -\infty, & \text{if } \limsup a_n = -\infty \\ \infty, & \text{if } \limsup a_n = \infty. \end{cases}$$

PROOF: The discussion prior to the theorem establishes the first two properties. The remaining proofs all use similar limit arguments that parallel the developments in Appendix A. Consequently, we will prove only a few of them. Continuing with the notation $A_n = \sup\{a_i | i \geq n\}$ and $B_n = \inf\{a_i | i \geq n\}$, we attack the third item as follows.

Suppose that $a_i > a - \epsilon$ for only finitely many i. Then there exists N such that $i \geq N$ forces $a_i \leq a - \epsilon$. This implies that $a - \epsilon$ is an upper bound for any set $\{a_i | i \geq n\}$, provided that $n \geq N$. Consequently, $A_n = \sup\{a_i | i \geq n\} \leq a - \epsilon$ for $n \geq N$, which forces $a \leq a - \epsilon$. Reaching this contradiction, we conclude that $a_i > a - \epsilon$ for infinitely many i. Now suppose that $a_i > a + \epsilon$ for infinitely many i, then every set of the form $\{a_i | i \geq n\}$ must contains some elements $a_i > a + \epsilon$. Consequently, $A_n \geq a + \epsilon$ for all n, which forces another contradiction: $a = \lim_{n \to \infty} A_n \geq a + \epsilon$. We conclude that $a_i > a + \epsilon$ for only finitely many a_i.

In the fourth item, we assume that $\limsup a_n = \infty$. If there are only finitely many $a_i > R$, then there exists N such that $a_i \leq R$ for $i \geq N$. Consequently, $A_n \leq R$ for $n \geq N$, which implies that $\limsup a_n \leq R$. This contradiction establishes that there must be infinitely many $a_i > R$.

To construct the subsequences asserted by the fifth item, we first handle the case where $\liminf a_n = \limsup a_n$. We have shown that this constraint forces the sequence itself to converge to the common limit or diverge to the common infinity. The desired subsequence is then the full sequence itself.

Next consider the case where $\limsup a_n = a$, a real number. The theorem's third part, proved above, provides an infinite number of a_i terms in the span $(a - 1, a + 1)$. Choose from this group the first term of the subsequence a_{i_1}. Among the infinitely many a_i with $a - 1/2 < a_i < a + 1/2$, there must be infinitely many with subscripts greater than i_1. Choosing a_{i_2} from this group, we have with $i_2 > i_1$. From the infinitely many a_i with $a - 1/3 < a_i < a + 1/3$, choose the third term a_{i_3} such that $i_3 > i_2 > i_1$. Continuing in this fashion, we obtain a subsequence $\{a_{i_k}\}$ such that $|a_{i_k} - a| < 1/k$. Clearly, the subsequence converges to a.

The only remaining possibility is that $\liminf a_n < \limsup a_n = \infty$. For each $R \in \{1, 2, 3, \ldots\}$, we invoke the earlier result that there are infinitely many $a_i > R$. We choose $a_{i_1} > 1$ from the first collection. From the infinitely many choices in the second collection, we choose $a_{i_2} > 2$ with $i_2 > i_1$. From the infinitely many choices in the third collection, we choose $a_{i_3} > 3$, taking care to ensure $i_3 > i_2 > i_1$. At the nth stage, we have infinitely many $a_i > n$, and we must avoid only the finite set of indices $i < i_{n-1}$. Consequently, there are infinitely many choices with both $a_i > n$ and $i > i_{n-1} > i_{n-2} > \ldots > i_1$. We choose a_{i_n} from this group. The process continues in this manner, and the resulting subsequence diverges to infinity. ∎

By using the upper limit, we can obtain the convergence interval for a power series, even when the coefficients oscillate. The proof closely follows that of Theorem 2.18. Where that theorem uses some aspect of the ordinary limit, the current theorem uses the corresponding feature of the upper limit.

2.22 Theorem: The power series $f(t) = \sum_{n=0}^{\infty} a_n t^n$ converges absolutely in the interval $-1/a < t < 1/a$, where $a = \limsup |a_n|^{1/n}$. If $a = 0$, the series converges absolutely for all t. If $a = \infty$, the series converges only for $t = 0$.

PROOF: Suppose that $a = 0$. The terms $|a_n|^{1/n}$ are all nonnegative, so $0 \leq \liminf a_n \leq \limsup a_n = a = 0$. Consequently, $a_n \to 0$, and we can invoke Theorem 2.18 to conclude that the power series converges absolutely for all t.

Next, suppose that $\limsup |a_n|^{1/n} = \infty$. For any nonzero t, there exist infinitely many a_n such that $|a_n|^{1/n} > 1/|t|$. For each such a_n,

$$|a_n t^n| = \left(|a_n|^{1/n}|t|\right)^n > \left(\frac{1}{|t|} \cdot |t|\right)^n = 1.$$

With infinitely many terms larger than 1, we cannot have $\lim_{n\to\infty} a_n t^n = 0$, which is necessary for a convergent series. So we conclude that $\sum_{n=0}^{\infty} a_n t^n$ does not converge except at $t = 0$.

Since the $|a_n|^{1/n}$ are all nonnegative, we cannot have $\limsup a_n < 0$. Therefore, the only case remaining is $\limsup |a|^{1/n} = a$, a positive real number. For t such that $0 < |t| < 1/a$, we have $a|t| < 1$ and $b = (a|t| + 1)/2 < 1$. Moreover $a < 1/|t|$, so there are only finitely many $|a_n|^{1/n} > a + [1/|t| - a]/3$. Equivalently, there exists N such that $n \geq N$ forces

$$|a_n|^{1/n} \leq a + \frac{1}{3}\left(\frac{1}{|t|} - a\right) < a + \frac{1}{2}\left(\frac{1}{|t|} - a\right) = \frac{1}{2}\left(\frac{a|t| + 1}{|t|}\right)$$

$$|a_n t^n| = \left(|a_n|^{1/n}|t|\right)^n < \left[\frac{1}{2} \cdot \left(\frac{a|t| + 1}{|t|}\right) \cdot |t|\right]^n = \left(\frac{a|t| + 1}{2}\right)^n = b^n.$$

The rest follows as in the preceding proof. Because $0 < b < 1$, $N \leq n < m$ implies that

$$\left|\sum_{i=n+1}^{m} a_i t^i\right| \leq \sum_{i=n+1}^{m} |a_i t^i| < \sum_{i=n+1}^{m} b^i = b^{n+1} \sum_{i=0}^{m-n-1} b^i$$

$$= b^{n+1}\frac{1 - b^{m-n}}{1 - b} < \frac{b^{n+1}}{1 - b} \to 0,$$

as $m, n \to \infty$. The power series converges absolutely. ∎

2.23 Example: If the sequence of coefficients fails to converge because the power series omits certain terms, we can still determine the convergence interval by using the upper limit of the coefficient sequence. Consider the power series $f(t) = 1 + t^2 + t^4 + \dots$. That is, $f(t) = \sum_{n=0}^{\infty} a_n t^n$, where

$$a_n = \begin{cases} 1, & n \text{ even} \\ 0, & n \text{ odd.} \end{cases}$$

$|a_n|^{1/n}$ is also zero for odd n and 1 for even, so it does not converge. However, $\text{lub}\{|a_i|^{1/i} | i \geq n\} = 1$ for each n. Consequently, $\limsup |a_n|^{1/n} = 1$, and the power series converges for $-1 < t < 1$. □

If $0 < s = \limsup |a_n|^{1/n} < \infty$, we refer to the open interval $(-1/s, 1/s)$ as the *convergence interval* of the power series $f(t) = \sum_{n=0}^{\infty} a_n t^n$. In the special case where $\limsup |a_n|^{1/n} = 0$, the convergence interval is $(-\infty, \infty)$. Theorem 2.22 asserts absolute convergence in this interval. However, it leaves as an exercise the proof that the series diverges for $t \notin [-s, s]$.

We are now in a position to derive the primary characteristics of a power series. The series presents no surprises in its convergence interval. It converges absolutely and uniformly on any closed subinterval. Its limit is a continuous differentiable function, which also admits a power series expansion convergent on the same interval. Term-by-term differentiation of the power series yields the power series for the derivative of the limit function. Term-by-term integration is also valid, and it yields the integral of the limit function. One could hardly ask more of these functions. We establish these properties with a collection of theorems.

2.24 Theorem: For s a positive real number or $s = \infty$, let $(-s, s)$ be the convergence interval of the power series $f(t) = \sum_{n=0}^{\infty} a_n t^n$. Then $\sum_n a_n t^n$ converges absolutely for every $t \in (-s, s)$. Moreover, both $\sum_n a_n t^n$ and $\sum_n |a_n t^n|$ converge uniformly on any closed subinterval $[a, b]$ contained in $(-s, s)$.

PROOF: By Theorem 2.22, we have absolute convergence in $(-s, s)$. For the uniform convergence, note that $[a, b] \subset (-s, s)$ allows us to choose s_1 such that $-s < -s_1 \le a \le b \le s_1 < s$. $\sum_n a_n s_1^n$ is then absolutely convergent. Given $\epsilon > 0$, there exists N such that $\sum_{i=n+1}^{m} |a_i s_1^i| < \epsilon/2$ for $N \le n < m$. For any $t \in [-s_1, s_1]$, we then have

$$\left| \sum_{i=n+1}^{m} a_i t^i \right| \le \sum_{i=n+1}^{m} |a_i t^i| \le \sum_{i=n+1}^{m} |a_i s_1^i| < \frac{\epsilon}{2},$$

provided that $N \le n < m$. It follows that

$$\left| f(t) - \sum_{i=0}^{n} a_i t^i \right| = \left| \lim_{m \to \infty} \sum_{i=0}^{m} a_i t^i - \sum_{i=0}^{n} a_i t^i \right| = \lim_{m \to \infty} \left| \sum_{i=n+1}^{m} a_i t^i \right| \le \frac{\epsilon}{2} < \epsilon,$$

provided that $n \ge N$. Because N is independent of $t \in [-s_1, s_1]$, we conclude that $\sum_n a_n t^n$ converges uniformly on $[-s_1, s_1]$. Because $[a, b] \subset [-s_1, s_1]$, we also have uniform convergence on $[a, b]$.

Finally, to show that the series of absolute values also converges uniformly, let $F(t) = \sum_{n=0}^{\infty} |a_n t^n|$. Then

$$\left| F(t) - \sum_{i=0}^{n} |a_i t^i| \right| = \left| \lim_{m \to \infty} \sum_{i=0}^{m} |a_i t^i| - \sum_{i=0}^{n} |a_i t^i| \right| = \lim_{m \to \infty} \sum_{i=n+1}^{m} |a_i t^i| \le \frac{\epsilon}{2} < \epsilon,$$

again provided that $n \ge N$. Because N is independent of the particular $t \in [-s_1, s_1]$, we conclude that $\sum_n |a_n t^n|$ converges uniformly on $[-s_1, s_1]$ and consequently on $[a, b]$. ∎

2.25 Theorem: For s a positive real number or $s = \infty$, let $(-s, s)$, be the convergence interval for the power series $f(t) = \sum_{n=0}^{\infty} a_n t^n$. Then the power series $\sum_{n=0}^{\infty} (n + 1)a_{n+1} t^n$ and $\sum_{n=1}^{\infty} a_{n-1} t^n / n$, obtained via term-by-term differentiation and integration, share the same convergence interval.

PROOF: The series obtained via term-by-term differentiation is

$$g_1(t) = \sum_{n=0}^{\infty} (n+1)a_{n+1}t^n = \begin{cases} 0, & t = 0 \\ \dfrac{1}{t} \sum_{n=1}^{\infty} n a_n t^n, & t \neq 0, \end{cases}$$

while that obtained via term-by-term integration is

$$g_2(t) = \sum_{n=1}^{\infty} \frac{a_{n-1}t^n}{n} = t \sum_{n=0}^{\infty} \frac{a_n t^n}{n+1}.$$

We investigate their coefficients. By l'Hôpital's rule, we have

$$\ln\left(n^{1/n}\right) = \frac{\ln n}{n} \rightarrow \frac{1/n}{1} \rightarrow 0$$

$$\ln\left(\frac{1}{n+1}\right)^{1/n} = \frac{\ln[1/(n+1)]}{n} \rightarrow \frac{(n+1)[-1/(n+1)^2]}{1} \rightarrow 0,$$

which imply that $n^{1/n} \rightarrow 1$ and $[1/(n+1)]^{1/n} \rightarrow 1$. By Theorem 2.21,

$$\limsup |n a_n|^{1/n} = \limsup \left(n^{1/n}|a_n|^{1/n}\right) = (1)\limsup |a_n|^{1/n}$$

$$\limsup \left|\frac{a_n}{n+1}\right|^{1/n} = \limsup \left[\left(\frac{1}{n+1}\right)^{1/n}|a_n|^{1/n}\right] = (1)\limsup |a_n|^{1/n},$$

which implies that the same convergence intervals for the derived series. ∎

2.26 Theorem: For s a positive real number or $s = \infty$, let $(-s, s)$ be the convergence interval for the power series $f(t) = \sum_{n=0}^{\infty} a_n t^n$. Then $f(t)$ is continuous and differentiable on $(-s, s)$, and for $x \in (-s, s)$,

$$f'(t) = \sum_{n=0}^{\infty} (n+1)a_{n+1}t^n$$

$$\int_0^x f(t)\,dt = \sum_{n=1}^{\infty} \left(\frac{a_{n-1}}{n}\right) x^n.$$

PROOF: We will first show that f is continuous at an arbitrary point $x \in (-s, s)$. Choose x_0 such that $|x| < x_0 < s$. By Theorem 2.24, $\sum_i a_i t^i$ converges uniformly to $f(t)$ for $t \in [-x_0, x_0]$. That is, the sequence of continuous polynomial functions $f_n(t) = \sum_{i=0}^{n} a_i t^i$ converges uniformly to $f(t)$ on $[-x_0, x_0]$. Theorem A.20 then asserts the continuity of $f(t)$ in $(-x_0, x_0)$ and in particular at the point x. This argument remains valid for any arbitrary power series, which means that the derived series obtained by term-by-term differentiation and integration also have continuous limits over the same convergence interval.

We next validate term-by-term integration for $\int_0^x f(t)\,dt$. We need the integral properties developed in Section A.4, so it is appropriate to review that material at this time. In particular, we use Theorem A.52, which expresses the integral of a continuous function as the limit of approximating sums as the partition mesh approaches zero. If $x = 0$, the integral and its alleged series are both zero and therefore equal. So, assume that $x > 0$. (The case for $x < 0$ admits a parallel development, and one of the exercises asks you to fill in the details.) We·again exploit the uniform convergence on the interval $[-a, a]$, where a is chosen such that $-s < -a < 0 < x < a < s$. That is, given $\epsilon > 0$, there exists N such that $n \geq N$ implies that $|f(t) - f_n(t)| < \epsilon/a$ simultaneously for all $t \in [-a, a]$. For such an $n \geq N$, suppose that $0 = x_0 < x_1 < x_2 < \ldots < x_m = x$ is a partition of $[0, x]$. Then $f(x_i) - \epsilon/a \leq f_n(x_i) \leq f(x_i) + \epsilon/a$ for $i = 1, 2, \ldots, m$. Working with the right side of this inequality, we compute

$$\sum_{i=1}^{m} f_n(x_i)(x_i - x_{i-1}) \leq \left(\sum_{i=1}^{m} f(x_i)(x_i - x_{i-1})\right) + \frac{\epsilon}{a}\sum_{i=1}^{m}(x_i - x_{i-1}).$$

The partition widths, $(x_i - x_{i-1})$, sum to $x - 0 = x$, and the fraction $x/a < 1$. Therefore,

$$\sum_{i=1}^{m} f_n(x_i)(x_i - x_{i-1}) \leq \left(\sum_{i=1}^{m} f(x_i)(x_i - x_{i-1})\right) + \frac{\epsilon x}{a}$$

$$\leq \left(\sum_{i=1}^{m} f(x_i)(x_i - x_{i-1})\right) + \epsilon.$$

A similar calculation with the left side of the inequality yields a corresponding bound on the left:

$$\left(\sum_{i=1}^{m} f(x_i)(x_i - x_{i-1})\right) - \epsilon \leq \sum_{i=1}^{m} f_n(x_i)(x_i - x_{i-1}).$$

As the partition mesh, the width of the largest component, approaches zero, the sums approach their respective integrals, which gives

$$\int_0^x f(t)\, dt - \epsilon \leq \int_0^x f_n(t)\, dt \leq \int_0^x f(t)\, dt + \epsilon.$$

Equivalently, $\left|\int_0^x f(t)\, dt - \int_0^x f_n(t)\, dt\right| \leq \epsilon$, which implies that

$$\int_0^x f(t)\, dt = \lim_{n \to \infty} \int_0^x f_n(t)\, dt = \lim_{n \to \infty} \sum_{i=0}^{n} \frac{a_i t^{i+1}}{i+1}\bigg|_{t=0}^{x} = \sum_{i=0}^{\infty} \frac{a_i x^{i+1}}{i+1}$$

$$= \sum_{i=1}^{\infty} \frac{a_{i-1} x^i}{i}.$$

This validates term-by-term integration, and we approach term-by-term differentiation via the fundamental relationship between differentiation and integration. Let $g(t) = \sum_{n=0}^{\infty}(n+1)a_{n+1}t^n$. We know that this series converges on $(-s, s)$ to a continuous limit. The validity of term-by-term integration now gives, for any $x \in (-s, s)$,

$$\int_0^x g(t)\, dt = \sum_{i=0}^{\infty} a_{i+1} t^{i+1}\bigg|_{t=0}^{x} = \sum_{i=0}^{\infty} a_{i+1} x^{i+1} = \sum_{i=1}^{\infty} a_i x^i = f(x) - a_0.$$

This shows that f is a constant plus the integral of a continuous function. Theorem A.51 then asserts that f is differentiable and $f'(x) = g(x)$, which has the form asserted by the theorem. This establishes term-by-term differentiation. ■

This theorem justifies the term-by-term differentiation that recovers the moments of a random variable X from the derivatives of its moment generating function. The technique is especially useful because, as the next theorem demonstrates, we can sometimes obtain the moment generating function without knowing the underlying probability distribution.

2.27 Theorem: Let $X_1, \ldots, X_n$ be independent random variables with corresponding $\Psi_1(t), \ldots, \Psi_n(t)$ moment generating functions. Let $Y = \sum_{i=1}^n X_i$. Then $\Psi_Y(t) = \Psi_1(t) \cdot \Psi_2(t) \cdots \Psi_n(t)$.

PROOF:

$$
\begin{aligned}
\Psi_Y(t) &= E_Y[e^{tY}] = E_{X_1, X_2, \ldots, X_n}[e^{tX_1 + tX_2 + \ldots + tX_n}] \\
&= \sum_{(x_1, x_2, \ldots, x_n)} e^{tx_1} e^{tx_2} \cdots e^{tx_n} \Pr{}_{X_1}(x_1) \Pr{}_{X_2}(x_2) \cdots \Pr{}_{X_n}(x_n) \\
&= \left(\sum_{x \in X_1} e^{xt} \Pr{}_{X_1}(x) \right) \cdot \left(\sum_{x \in X_2} e^{xt} \Pr{}_{X_2}(x) \right) \cdots \left(\sum_{x \in X_n} e^{xt} \Pr{}_{X_n}(x) \right) \\
&= \Psi_1(t) \cdot \Psi_2(t) \cdots \Psi_n(t) \quad ■
\end{aligned}
$$

2.28 Example: Let Y be the sum of two independent binomial random variables with parameters $(n_1, p_1), (n_2, p_2)$. What are the mean and variance of Y?

A glance back at Equation 2.2 reveals the uninviting appearance of the distribution function for Y. So we proceed via the moment generating functions:

$$
\begin{aligned}
\Psi_{X_1}(t) &= [(1 - p_1) + p_1 e^t]^{n_1} \\
\Psi_{X_2}(t) &= [(1 - p_2) + p_2 e^t]^{n_2} \\
\Psi_Y(t) &= [(1 - p_1) + p_1 e^t]^{n_1} \cdot [(1 - p_2) + p_2 e^t]^{n_2} \\
\Psi'_Y(t) &= n_1[(1 - p_1) + p_1 e^t]^{n_1 - 1} p_1 e^t [(1 - p_2) + p_2 e^t]^{n_2} \\
&\quad + [(1 - p_1) + p_1 e^t]^{n_1} n_2 [(1 - p_2) + p_2 e^t]^{n_2 - 1} p_2 e^t \\
\mu_Y &= E_Y[Y] = \Psi'_Y(0) = n_1 p_1 + n_2 p_2 \\
\Psi''(t) &= n_1 p_1 e^t [(1 - p_1) + p_1 e^t]^{n_1 - 1}[(1 - p_2) + p_2 e^t]^{n_2} \\
&\quad + n_1(n_1 - 1)(p_1 e^t)^2[(1 - p_1) + p_1 e^t]^{n_1 - 2}[(1 - p_2) + p_2 e^t]^{n_2} \\
&\quad + n_1 n_2 p_1 p_2 (e^t)^2 [(1 - p_1) + p_1 e^t]^{n_1 - 1}[(1 - p_2) + p_2 e^t]^{n_2 - 1} \\
&\quad + n_2 p_2 e^t [(1 - p_1) + p_1 e^t]^{n_1}[(1 - p_2) + p_2 e^t]^{n_2 - 1} \\
&\quad + n_1 n_2 p_1 p_2 (e^t)^2 [(1 - p_1) + p_1 e^t]^{n_1 - 1}[(1 - p_2) + p_2 e^t]^{n_2 - 1} \\
&\quad + n_2(n_2 - 1)(p_2 e^t)^2[(1 - p_1) + p_1 e^t]^{n_1}[(1 - p_2) + p_2 e^t]^{n_2 - 2} \\
E_Y[Y^2] &= \Psi''_Y(0) = n_1 p_1 + n_2 p_2 + n_1(n_1 - 1)p_1^2 + n_2(n_2 - 1)p_2^2 + 2n_1 n_2 p_1 p_2 \\
&= n_1 p_1 (1 - p_1) + n_2 p_2 (1 - p_2) + (n_1 p_1 + n_2 p_2)^2 \\
\sigma_Y^2 &= E_Y[Y^2] - \mu_Y^2 = n_1 p_1 (1 - p_1) + n_2 p_2 (1 - p_2)
\end{aligned}
$$

Two comments spring to mind. First, evaluating the derivatives of Ψ_Y is tedious, so you may wonder if this approach is easier than dealing with the complicated distribution function. Well, there's no free lunch. Second, we already have a method, Theorem 2.8, for evaluating the mean and variance of a linear combination of random variables, and it is much easier to use. This is true, but the moment generating function can give us higher moments, while Theorem 2.8 provides only the mean and variance. $\square$

Theorem 2.27 finds frequent use when the X_i are independent, identically distributed random variables. In this case, the moment generating function of the sum is the nth power of the common moment generating function of the summands. The following example illustrates the point.

2.29 Example: Suppose that X_i assumes values $1, 2, 3$ with respective probabilities $1/2, 1/3, 1/6$. Let $Y = \sum_{i=1}^{10} X_i$. What are the mean and variance of Y?

$$\Psi_{X_i}(t) = (1/2)e^t + (1/3)e^{2t} + (1/6)e^{3t}$$
$$\Psi_Y(t) = [(1/2)e^t + (1/3)e^{2t} + (1/6)e^{3t}]^{10}$$
$$\Psi_Y'(t) = 10 \cdot [(1/2)e^t + (1/3)e^{2t} + (1/6)e^{3t}]^9 [(1/2)e^t + (2/3)e^{2t} + (1/2)e^{3t}]$$
$$\mu_Y = E_Y[Y] = \Psi_Y'(0) = 10[(1/2) + (2/3) + (1/2)] = 50/3$$
$$\Psi_Y'' = 10 \cdot 9[(1/2)e^t + (1/3)e^{2t} + (1/6)e^{3t}]^8 [(1/2)e^t + (2/3)e^{2t} + (1/2)e^{3t}]^2$$
$$\quad + 10 \cdot [(1/2)e^t + (1/3)e^{2t} + (1/6)e^{3t}]^9 [(1/2)e^t + (4/3)e^{2t} + (3/2)e^{3t}]$$
$$E_Y[Y^2] = \Psi_Y''(0) = 10 \cdot 9[(1/2) + (2/3) + (1/2)]^2 + 10[(1/2) + (4/3) + (3/2)]$$
$$\quad = 250 + (100/3)$$
$$\sigma_Y^2 = E_Y[Y^2] - \mu_Y^2 = 250 + (100/3) - (2500/9) = 50/9$$

We can directly calculate $\mu_{X_i} = 1(1/2) + 2(1/3) + 3(1/6) = 5/3$ and use Theorem 2.8 to find $\mu_Y = 10(5/3) = 50/3$. Similarly, $\sigma_{X_i}^2 = (1 - 5/3)^2(1/2) + (2 - 5/3)^2(1/3) + (3 - 5/3)^2(1/6) = 5/9$. Then $\sigma_Y^2 = 10(5/9) = 50/9$. The moment generating function, however, does give us access to higher moments.

Note that the sum of 10 identically distributed random variables is not the same as 10 times one of them. If we let $Z = 10X_1$, then $\mu_Z = 10\mu_{X_1} = 50/3$, the same as μ_Y. However, $\sigma_Z^2 = 100\sigma_{X_1}^2 = 500/9$, which is much larger than σ_Y^2. Moreover, Z can assume only the values $10, 20, 30$, whereas Y assumes values $10, 11, 12, \ldots, 30$. $\square$

It is a natural question to ask if you can recover the probability distribution from the moment generating function. The answer is yes, but such inversion formulas are difficult to derive. Here is one that works for the special case of a random variable that assumes values on the nonnegative integers. It is useful in this text, and in the computer and engineering sciences in general, because such random variables frequently occur.

2.30 Theorem: Suppose that random variable X has positive probabilities on the nonnegative integers $0 \leq k_0 < k_1 < k_2 < \ldots$. Then $\Psi(t)$ converges for $t \leq 0$, $\lim_{t \to -\infty} e^{-yt}\Psi(t) = 0$ for $y < k_0$, and $\lim_{t \to -\infty} e^{-yt}\Psi(t) = \infty$ for $y > k_0$. Moreover, $\lim_{t \to -\infty} e^{-k_0 t}\Psi(t) = \Pr(k_0)$.

PROOF: Because the k_i are nonnegative, $e^{k_i t} \leq 1$ when $t \leq 0$. Therefore, for $t \leq 0$, we have $|e^{k_i t}\Pr(k_i)| \leq \Pr(k_i)$, and $\sum_i \Pr(k_i)$ converges. By Theorem A.35, $\Psi(t) = \sum_i e^{k_i t}\Pr(k_i)$ converges.

If $y < k_0$, then $k_i - y > 0$ for all i. Actually, $0 < k_0 - y < k_i - y$ for $i = 1, 2, \ldots$. For $t < 0$ it then follows that

$$0 \le e^{-yt}\Psi(t) = \sum_i e^{(k_i - y)t}\Pr(k_i) < e^{(k_0 - y)t}\sum_i \Pr(k_i) = e^{(k_0 - y)t}.$$

Since $e^{(k_0 - y)t} \to 0$ as $t \to -\infty$, $e^{-yt}\Psi_X(t)$ must also have limiting value 0.

If $y > k_0$, we write the expansion for $e^{-ty}\Psi_X(t)$ in three parts:

$$e^{-ty}\Psi(t) = \sum_{k_i < y} e^{(k_i - y)t}\Pr(k_i) + \sum_{k_i = y} e^{(k_i - y)t}\Pr(k_i) + \sum_{k_i > y} e^{(k_i - y)t}\Pr(k_i)$$

$$\ge \sum_{k_i < y} e^{(k_i - y)t}\Pr(k_i) \ge e^{(k_0 - y)t}\Pr(k_0).$$

Because $k_0 - y$ is negative, the last expression becomes arbitrarily large as $t \to -\infty$, so $\lim_{t \to -\infty} e^{-yt}\Psi(t) = \infty$.

Finally, if $y = k_0$,

$$e^{-k_0 t}\Psi(t) = e^{-k_0 t + k_0 t}\Pr(k_0) + \sum_{i>0} e^{(k_i - k_0)t}\Pr(k_i)$$

$$= \Pr(k_0) + \sum_{i>0} e^{(k_i - k_0)t}\Pr(k_i). \tag{2.4}$$

For $t < 0$, the remaining summation involves only negative exponents and, as in the case for $y < x_0$, fades to zero as $t \to -\infty$. Specifically, for $t < 0$, the terms $e^{(k_i - k_0)t}$ for $i = 2, 3, \ldots$ are all smaller than $e^{(k_1 - k_0)t}$. Therefore,

$$0 \le \sum_{i>0} e^{(k_i - k_0)t}\Pr(k_i) < e^{(k_1 - k_0)t}\sum_{i>0} \Pr(k_i) < e^t.$$

The last inequality follows because $(k_1 - k_0)$ is a nonzero integer and therefore at least 1. Hence the sum in Equation 2.4 has limiting value zero, which establishes the theorem. ∎

We can use this theorem to recover the probability distribution from the moment generating function, provided that the random variable has nonzero probabilities on a subset of the nonnegative integers.. We move y in from $-\infty$, watching $\lim_{t \to -\infty} e^{-yt}\Psi(t)$ as we go. When the limit jumps from zero to infinity, we have just crossed k_0, the leftmost point of nonzero probability. We evaluate $\Pr(k_0)$ as the limit for $y = k_0$. Let us call this process "scanning across k_0." Knowing both k_0 and $\Pr(k_0)$, we can then replace $\Psi(t)$ with $\Psi(t) - e^{x_0 t}\Pr(x_0)$ and scan across k_1 to recover the next probability jump. Continuing this operation, we discover the range of X and the associated probabilities, one point at a time.

2.31 Example: Let the random variable X have nonzero probabilities on a subset of the nonnegative integers. Suppose that $\Psi(t) = e^t/2 + e^{3t}/[3 - e^{3t}]$. What is the probability distribution of X?

To start the process, let $g_0(t) = \Psi(t)$. We investigate the limiting behavior, as $t \to -\infty$, of $e^{-yt}g_0(t)$.

$$\lim_{t \to -\infty} e^{-yt}g_0(t) = \lim_{t \to -\infty} \left[\frac{e^{(1-y)t}}{2} + \frac{e^{(3-y)t}}{3 - e^{3t}} \right] = \begin{cases} 0, & y < 1 \\ 1/2, & y = 1 \\ \infty, & y > 1 \end{cases}$$

The first probability jump, of size $1/2$, occurs at $X = 1$. We then let $g_1(t) = g_0(t) - (1/2)e^t$ and continue the process.

$$g_1(t) = \frac{e^t}{2} + \frac{e^{3t}}{3 - e^{3t}} - \frac{e^t}{2} = \frac{e^{3t}}{3 - e^{3t}}$$

$$\lim_{t \to -\infty} e^{-yt}g_1(t) = \lim_{t \to -\infty} \frac{e^{(3-y)t}}{3 - e^{3t}} = \begin{cases} 0, & y < 3 \\ 1/3, & y = 3 \\ \infty, & y > 3 \end{cases}$$

The second nonzero probability then occurs at $X = 3$. It is $\Pr(3) = 1/3$.

$$g_2(t) = g_1(t) - \frac{e^{3t}}{3} = \frac{e^{6t}}{3(3 - e^{3t})}$$

$$\lim_{t \to -\infty} e^{-yt}g_2(t) = \lim_{t \to -\infty} \frac{e^{(6-y)t}}{3(3 - e^{3t})} = \begin{cases} 0, & y < 6 \\ 1/9, & y = 6 \\ \infty, & y > 6 \end{cases}$$

We conclude that $\Pr(6) = 1/9$. Evidently, the process continues to produce additional distribution points. For example, the next jump occurs at $X = 9$:

$$g_3(t) = g_2(t) - \frac{e^{6t}}{9} = \frac{e^{9t}}{9(3 - e^{3t})}$$

$$\lim_{t \to -\infty} e^{-yt}g_3(t) = \lim_{t \to -\infty} \frac{e^{(9-y)t}}{9(3 - e^{3t})} = \begin{cases} 0, & y < 9 \\ 1/27, & y = 9 \\ \infty, & y > 9. \end{cases}$$

The general pattern is

$$g_i(t) = \frac{e^{3it}}{3^{i-1}(3 - e^{3t})}$$

$$\lim_{t \to -\infty} e^{-yt}g_i(t) = \lim_{t \to -\infty} \frac{e^{(3i-y)t}}{3^{i-1}(3 - e^{3t})} = \begin{cases} 0, & y < 3i \\ 1/3^i, & y = 3i \\ \infty, & y > 3i. \end{cases}$$

The probability distribution for X is therefore as follows.

k	1	3	6	9	12	15	$\cdots$
$\Pr(k)$	1/2	1/3	$1/3^2$	$1/3^3$	$1/3^4$	$1/3^5$	$\cdots$

Note that

$$\sum_k \Pr(k) = 1/2 + (1/3)\sum_{k=0}^{\infty}(1/3)^k = 1/2 + (1/3)(1/(1-1/3)) = 1/2 + (1/3)(3/2) = 1.$$

Because this process recovers the probability distribution, under certain conditions, from the moment generating function, we are confident that this moment generating function can represent just one distribution. Therefore, if we can guess a distribution that yields the moment generating function, then that distribution must be the correct inversion. In the example at hand, we can take advantage of the convergent geometric series $\sum_{i=0}^{\infty} x^i = 1/(1-x)$, for $0 \le x < 1$. If $t < (\ln 3)/3$, then $e^{3t}/3 < 1$, and we can write

$$
\begin{aligned}
\Psi(t) &= \frac{e^t}{2} + \frac{e^{3t}}{3 - e^{3t}} = \frac{e^t}{2} + \frac{e^{3t}}{3} \cdot \frac{1}{1 - e^{3t}/3} \\
&= \frac{e^t}{2} + \frac{e^{3t}}{3}[1 + (e^{3t}/3) + (e^{3t}/3)^2 + \ldots] \\
&= \frac{e^t}{2} + \frac{e^{3t}}{3} + \frac{e^{6t}}{3^2} + \frac{e^{9t}}{3^3} + \ldots = \frac{e^t}{2} + \sum_{i=1}^{\infty} \frac{e^{3it}}{3^i}.
\end{aligned}
$$

Examining the weights accorded $e^t, e^{3t}.e^{6t}, \ldots$, we obtain the same distribution tabulated above. $\square$

If a function $f(\cdot)$ possesses derivatives of all orders, then it frequently has a power series $\sum_{k=0}^{\infty} f^{(k)}(0)x^k/k!$ over some convergence interval. Indeed, this condition is sufficient to invoke Taylor's formula for the following exact representation, where y lies between zero and x.

$$
f(x) = \left(\sum_{k=0}^{n} \frac{f^{(k)}(0)}{k!} x^k \right) + \frac{f^{(n+1)}(y)}{(n+1)!} x^{n+1}
$$

To show that $f(x) = \sum_{k=0}^{\infty} f^{(k)}(0)x^k/k!$, we need only show that the remainder term approaches zero. The following example illustrates the point with $f(x) = \ln(1-x)$, yielding a power series that will be useful in the upcoming discussion of geometric random variables.

2.32 Example: Express $f(x) = \ln(1-x)$ as a power series and determine the convergence interval.

The derivatives at zero are

$$
\begin{aligned}
f^{(0)}(0) &= \ln(1-x)|_{x=0} = 0 \\
f^{(1)}(0) &= -(1-x)^{-1}\big|_{x=0} = -1 \\
f^{(2)}(0) &= -(1-x)^{-2}\big|_{x=0} = -1 \\
f^{(3)}(0) &= -2(1-x)^{-3}\big|_{x=0} = -2 \\
f^{(4)}(0) &= -(3)(2)(1-x)^{-4}\big|_{x=0} = -3! \\
&\vdots \qquad \vdots \\
f^{(k)}(0) &= -(k-1)!.
\end{aligned}
$$

Using Taylor's formula with remainder, we have

$$
\ln(1-x) = -\left(\sum_{k=1}^{n} \frac{(k-1)!}{k!} x^k \right) + \frac{f^{(n+1)}(y)}{(n+1)!} x^{n+1},
$$

with y lying between zero and x. For the remainder term,

$$
\left| \frac{f^{(n+1)}(y)}{(n+1)!} x^{n+1} \right| = \left| \frac{-n!(1-y)^{-(n+1)}}{(n+1)!} x^{n+1} \right| = \frac{1}{n+1} \left| \frac{x}{1-y} \right|^{n+1}.
$$

If we restrict $x \in (-1, 1/2)$, then $|x/(1-y)| < 1$ and the remainder approaches zero with increasing n. Consequently,

$$\ln(1-x) = -\sum_{k=0}^{\infty} \frac{x^k}{k},$$

for $-1 < x < 1/2$. The convergence interval of the series is actually $-1 < x < 1$, as we can easily verify $\limsup(1/n)^{1/n} = 1$. So the series converges on $-1 < x < 1$ and is equal to $\ln(1-x)$ for $-1 < x < 1/2$. Is the series equal to $\ln(1-x)$ on the balance of the convergence interval $1/2 \le x < 1$? It is. Because we can differentiate and integrate term by term, we can reason as follows on the entire interval $-1 < x < 1$.

$$\hat{f}(x) = -\sum_{k=1}^{\infty} \frac{x^k}{k}$$

$$\hat{f}'(x) = -\sum_{k=1}^{\infty} x^{k-1} = -(1 + x + x^2 + \dots) = \frac{-1}{1-x}$$

$$\hat{f}(x) = \int_0^x \frac{-dt}{1-t} = \ln(1-x) \quad \square$$

Formal arithmetic with power series produces new power series. By formal arithmetic, we mean arithmetic operations without regard for convergence consequences. Suppose that $f(t) = \sum_{n=0}^{\infty} a_n t^n, g(t) = \sum_{n=0}^{\infty} b_n t^n$ have convergence intervals $(-a, a)$ and $(-b, b)$, respectively. Assume that both a and b are nonzero and one or both may be $+\infty$. If we let $c = \min(a, b)$, then the formal sum does indeed converge to $f(t) + g(t)$. That is,

$$\sum_{n=0}^{\infty} (a_n + b_n)t^n = f(t) + g(t),$$

for $t \in (-c, c)$. This follows immediately from the corresponding operation on partial sums:

$$\left| \sum_{n=0}^{N} (a_n + b_n)t^n - [f(t) + g(t)] \right| = \left| \left(\sum_{n=0}^{N} a_n t^n - f(t) \right) + \left(\sum_{n=0}^{N} b_n t^n - g(t) \right) \right|$$

$$\le \left| \sum_{n=0}^{N} a_n t^n - f(t) \right| + \left| \sum_{n=0}^{N} b_n t^n - g(t) \right| \to 0$$

as $N \to \infty$. A similar argument applies to the formal difference $\sum_{n=0}^{\infty} (a_n - b_n)t^n = f(t) - g(t)$, for $t \in (-c, c)$. The formal product is more complicated.

$$\left(\sum_{n=0}^{\infty} a_n t^n \right) \left(\sum_{n=0}^{\infty} b_n t^n \right) = (a_0 + a_1 t + a_2 t^2 + \dots)(b_0 + b_1 t + b_2 t^2 + \dots)$$

$$= a_0 b_0 + (a_0 b_1 + a_1 b_0)t + (a_0 b_2 + a_1 b_1 + a_2 b_0)t^2$$
$$+ (a_0 b_3 + a_1 b_2 + a_2 b_1 + a_3 b_0)t^3 + \dots$$

$$= \sum_{n=0}^{\infty} \left(\sum_{k=0}^{n} a_k b_{n-k} \right) t^n$$

This formal product also converges, for $t \in (-c, c)$, to the expected $f(t)g(t)$. We reason as follows. First, note that the product of partial sums exhibits the same pattern for coefficients.

$$
\left(\sum_{n=0}^{N} a_n t^n \right) \left(\sum_{n=0}^{N} b_n t^n \right) = (a_0 b_0) + (a_0 b_1 + a_1 b_0)t + (a_0 b_2 + a_1 b_1 + a_2 b_0)t^2
$$
$$
+ \ldots + (a_0 b_N + \ldots + a_{N-1} b_1 + a_N b_0)t^N
$$
$$
+ (a_1 b_N + a_2 b_{N-1} + \ldots + a_{N-1} b_2 + a_N b_1)t^{N+1}
$$
$$
+ \ldots + (a_N b_N)t^{2N}
$$

In this expression, the coefficient of t^n is $\sum_{k=0}^{n} a_k b_{n-k}$, provided that $0 \leq n \leq N$. For subsequent powers, $\sum_{k=0}^{n} a_k b_{n-k}$ generates too many terms. To obtain the correct coefficient, we must discard the formula-generated terms for which k or $n - k$ exceeds N. That is,

$$
\left(\sum_{n=0}^{N} a_n t^n \right) \left(\sum_{n=0}^{N} b_n t^n \right) = \sum_{n=0}^{2N} \left(\sum_{k=0}^{n} a_k b_{n-k} \right) t^n - \sum_{n=N+1}^{2N} \left(\sum_{k \in I_n} a_k b_{n-k} \right) t^n,
$$

where $I_n = \{k \mid 0 \leq k \leq n \text{ and } (k > N \text{ or } n - k > N)\}$. In a similar fashion, we obtain

$$
\left(\sum_{n=0}^{2N} a_n t^n \right) \left(\sum_{n=0}^{2N} b_n t^n \right) = \sum_{n=0}^{2N} \left(\sum_{k=0}^{n} a_k b_{n-k} \right) t^n + \sum_{n=2N+1}^{4N} \left(\sum_{k \in J_n} a_k b_{n-k} \right) t^n,
$$

where $J_n = \{k \mid 0 \leq k \leq n \text{ and } k \leq 2N \text{ and } n - k \leq 2N\}$. For the absolute series, we observe the same adjustments, except the coefficient of $|t|^n$ is $\sum_{k=0}^{n} |a_k b_{n-k}|$, provided that $0 \leq n \leq N$. Moreover, the summands are all nonnegative for the absolute series. Therefore,

$$
\left(\sum_{n=0}^{N} |a_n||t|^n \right) \left(\sum_{n=0}^{N} |b_n||t|^n \right) \leq \sum_{n=0}^{2N} \left(\sum_{k=0}^{n} |a_k b_{n-k}| \right) |t|^n
$$
$$
\leq \left(\sum_{n=0}^{2N} |a_n||t|^n \right) \left(\sum_{n=0}^{2N} |b_n||t|^n \right).
$$

Being power series, $\sum a_n t^n$ and $\sum b_n t^n$ converge absolutely, say to $F(t)$ and $G(t)$, in their convergence intervals. For $t \in (-c, c)$, both extremes of the inequality above converge to $F(t)G(t)$, which forces convergence of the center expression to the same value. This demonstrates the absolute convergence of $\sum_{n=0}^{\infty} \left(\sum_{k=0}^{n} a_k b_{n-k} \right) t^n$, which means that the series also converges in the ordinary sense. It remains to show that it converges to $f(t)g(t)$. Note that the difference

$$D_1 = \left| \sum_{n=0}^{2N} \left(\sum_{k=0}^{n} a_k b_{n-k} \right) t^n - \left(\sum_{n=0}^{N} a_n t^n \right) \left(\sum_{n=0}^{N} b_n t^n \right) \right|$$

$$= \left| \sum_{n=N+1}^{2N} \left(\sum_{k \in I_n} a_k b_{n-k} \right) t^n \right| \leq \sum_{n=N+1}^{2N} \left(\sum_{k \in I_n} |a_k b_{n-k}| \right) |t|^n$$

$$= \sum_{n=0}^{2N} \left(\sum_{k=0}^{n} |a_k b_{n-k}| \right) |t|^n - \left(\sum_{n=0}^{N} |a_n| |t|^n \right) \left(\sum_{n=0}^{N} |b_n| |t|^n \right)$$

approaches zero with increasing N because both the terms in the final reduction approach $F(t)G(t)$. Moreover,

$$\left(\sum_{n=0}^{N} a_n t^n \right) \left(\sum_{n=0}^{N} b_n t^n \right) \rightarrow f(t)g(t),$$

and therefore the difference

$$D_2 = \left| \sum_{n=0}^{2N} \left(\sum_{k=0}^{n} a_k b_{n-k} \right) t^n - f(t)g(t) \right|$$

behaves as follows with increasing N.

$$D_2 \leq \left| \sum_{n=0}^{2N} \left(\sum_{k=0}^{n} a_k b_{n-k} \right) t^n - \left(\sum_{n=0}^{N} a_n t^n \right) \left(\sum_{n=0}^{N} b_n t^n \right) \right|$$

$$+ \left| \left(\sum_{n=0}^{N} a_n t^n \right) \left(\sum_{n=0}^{N} b_n t^n \right) - f(t)g(t) \right| \rightarrow 0$$

We have shown that the formal sum, difference, and product of two power series are all new power series that converge on the common convergence interval of their components. The next theorem summarizes these results and contributes two additional arithmetic properties.

2.33 Theorem: Suppose that $f(t) = \sum a_n t^n$ and $g(t) = \sum b_n t^n$ are power series with convergence intervals $(-a, a)$ and $(-b, b)$, respectively. Suppose further that a and b are both nonzero; one or both may be $+\infty$. The following related power series then converge as indicated.

- Let $c = \min(a, b)$. For $t \in (-c, c)$, we have

$$\sum_{n=0}^{\infty} (a_n + b_n) t^n = f(t) + g(t)$$

$$\sum_{n=0}^{\infty} (a_n - b_n) t^n = f(t) - g(t)$$

$$\sum_{n=0}^{\infty} \left(\sum_{k=0}^{n} a_k b_{n-k} \right) t^n = f(t)g(t).$$

- For $K > 0$ and $-a/K < t < a/K$, we have $\sum_{n=0}^{\infty} a_n(Kt)^n = f(Kt)$.

- Let

$$\sum_{k=0}^{\infty} c_{nk} t^k = \left(\sum_{k=0}^{\infty} a_k t^k \right)^n = (f(t))^n$$

be the power series obtained by repeated formal multiplication. From the result above, this series converges absolutely on $(-a, a)$. Then, provided $F(t) = \sum_{k=0}^{\infty} |a_k||t|^k < b$,

$$\sum_{k=0}^{\infty} \left(\sum_{n=0}^{\infty} b_n c_{nk} \right) t^k = g(f(t)).$$

PROOF: The discussion above establishes the proper convergence for the formal sum, difference, and product of two power series. For the second item, we simply note that $-a/K < t < a/K$ implies that $-a < Kt < a$, and therefore the new series converges as stated. For the final item, first note that the series for $f(t)$ and $g(t)$ are both absolutely convergent, say to $F(t)$ and $G(t)$ in their convergence intervals. So, if $t \in (-a, a)$ and $F(t) < b$,

$$G(F(t)) = \sum_{n=0}^{\infty} |b_n|[F(t)]^n = \sum_{n=0}^{\infty} \sum_{k=0}^{\infty} |b_n||c_{nk}||t|^k.$$

Because the summands are all nonnegative, the convergence is absolute, which implies that the associated sequence obtained by removing the absolute values also converges. Moreover, it converges under the rearrangement

$$\sum_{k=0}^{\infty} \sum_{n=0}^{\infty} b_n c_{nk} t^k = \sum_{n=0}^{\infty} b_n \left(\sum_{k=0}^{\infty} c_{nk} t^k \right)$$

$$= \sum_{n=0}^{\infty} b_n[f(t)]^n = g(f(t)). \blacksquare$$

2.34 Example: Verify that $(e^t)^2 = e^{2t}$ as the product of two power series.

From the expansion $e^x = \sum_{n=0}^{\infty} x^n/n!$, which is valid for all x, we know in advance that

$$e^{2t} = \sum_{n=0}^{\infty} \frac{(2t)^n}{n!} = \sum_{n=0}^{\infty} \left(\frac{2^n}{n!} \right) t^n.$$

We need only verify that the formal product generates the same series.

$$(e^t)^2 = \left(\sum_{n=0}^{\infty} \frac{t^n}{n!} \right) \left(\sum_{n=0}^{\infty} \frac{t^n}{n!} \right) = \sum_{n=0}^{\infty} c_n t^n,$$

where

$$c_n = \sum_{k=0}^{n} \left(\frac{1}{k!}\right)\left(\frac{1}{(n-k)!}\right) = \frac{1}{n!}\sum_{k=0}^{n}\frac{n!}{k!(n-k)!} = \frac{1}{n!}\sum_{k=0}^{n}C_{n,k} = \frac{2^n}{n!}. \ \Box$$

2.35 Example: Suppose that $0 < p < 1$. Show that $h(t) = p/[1 - (1-p)e^t]$ has a power series expansion convergent in a nonempty interval around zero.

We construct $h(t)$ from components with known power series expansions. The factor $(1-p)e^t$ has the familiar expansion $\sum(1-p)t^n/n!$, absolutely convergent for all t. Also, $g(x) = p/(1-x) = p\sum_{n=0}^{\infty}x^n$, absolutely convergent for $|x| < 1$. The desired composition is then $h(t) = g((1-p)e^t)$. Consulting Theorem 2.33, we see that $h(t)$ has a convergent power series for

$$F(t) = \sum_{n=0}^{\infty}\left|\frac{(1-p)t^n}{n!}\right| = (1-p)\sum_{n=0}^{\infty}\frac{|t|^n}{n!} = (1-p)e^{|t|} < 1,$$

or equivalently, for $|t| < \ln[1/(1-p)]$. This range constitutes a nonempty interval around zero. Knowing where convergence is valid, we obtain the coefficients by formally composing the component power series.

$$h(t) = p\sum_{n=0}^{\infty}[(1-p)e^t]^n = p + p\sum_{n=1}^{\infty}(1-p)^n e^{nt} = p + \sum_{n=1}^{\infty}p(1-p)^n\sum_{k=0}^{\infty}\left(\frac{n^k}{k!}\right)t^k$$

$$= p + \sum_{n=1}^{\infty}p(1-p)^n\left[1 + \sum_{k=1}^{\infty}\left(\frac{n^k}{k!}\right)t^k\right]$$

$$= p + p\sum_{n=1}^{\infty}(1-p)^n + \sum_{n=1}^{\infty}\sum_{k=1}^{\infty}\left(\frac{p(1-p)^n n^k}{k!}\right)t^k$$

$$= 1 + \sum_{k=1}^{\infty}\left(\sum_{n=1}^{\infty}\frac{p(1-p)^n n^k}{k!}\right)t^k$$

Consequently, $c_0 = 1$ and $c_k = (p/k!)\sum_{n=1}^{\infty}n^k(1-p)^n$ for $k > 0$. $\Box$

Exercises

*2.14 Let $f(t) = \sum_{n=0}^{\infty}a_n t^n$ be a power series for which $|a_n|^{1/n} \to a$, a real number. Show that the sum diverges for $t > a$.

2.15 Suppose that $\liminf a_i = b$, a real number, and let $\epsilon > 0$ be chosen arbitrarily. Show that $a_i < b + \epsilon$ for infinitely many i and $a_i < b - \epsilon$ for only finitely many i.

2.16 Suppose that $\liminf a_n = +\infty$. Show, for an arbitrary number R, that there are only finitely many $a_i \leq R$.

2.17 Determine the convergence interval of $f(t) = \sum_{n=0}^{\infty}a_n t^n$, where

$$a_n = \begin{cases} 2^n n^{5/2}/(n^2 + 3n), & n \text{ odd} \\ 0, & n \text{ even}. \end{cases}$$

*2.18 Determine the convergence interval of $f(t) = \sum_{n=0}^{\infty} a_n t^n$, where

$$a_n = \begin{cases} 3^n, & n \text{ odd} \\ (-1)^{n/2} 2^{2n}, & n \text{ even.} \end{cases}$$

2.19 Suppose that $\limsup a_n = \infty$ and $\lim_{n \to \infty} b_n = b$, a negative real number. Show that $\liminf a_n b_n = -\infty$.

2.20 Theorem 2.26 justifies term-by-term integration of a power series. That is, if $f(t) = \sum_{i=0}^{\infty} a_i t^i$ for $t \in (-s, s)$, then $\int_0^x f(t)\,dt = \sum_{i=1}^{\infty} a_{i-1} t^i / i$ for $x \in (-s, s)$. The text establishes this fact for $x \geq 0$. Prove it for $x < 0$.

*2.21 Random variable X has nonzero probabilities on a subset of the nonnegative integers. $\Psi(t) = (2e^{-t} - 1)^{-1}$. What is the probability distribution of X?

*2.22 Random variable Y has nonzero probabilities on a subset of the nonnegative integers. $\Psi(t) = (2e^{-t} - 1)^{-2}$. What is the probability distribution of Y?

*2.23 Let X have the uniform discrete distribution with parameter n. Show that $\Psi(t) = (e^{nt} - 1)/[n(1 - e^{-t})]$, for $t \leq 0$. Verify that $\Psi'(0)$ and $\Psi''(0)$ correctly generate μ and σ^2 as given in Equation 2.1.

2.24 Suppose that the random variable X has moment generating function $\Psi_X(t)$. Let $Y = aX + b$, for constants $a \neq 0$ and b. Express the moment generating function for Y in terms of $\Psi_X(t)$.

2.25 Derive the power series expansion for $f(x) = \cos x + \sin x$. What is the convergence interval?

2.26 Let $h(x) = \sum_{n=1}^{\infty} x^n / n$. Find the convergence interval. Find a closed-form expression for $h(x)$.

2.27 By formal multiplication of the component power series, find the power series for the following function. What is its convergence interval?

$$h(x) = \frac{1}{x^2 - 3x + 1} = \left(\frac{1}{1-x}\right)\left(\frac{1}{1-2x}\right)$$

2.3 Geometric and negative binomial forms

Before giving a formal definition, we illustrate a geometric distribution by constructing a probability space and assigning probabilities to its outcomes. We envision a nondeterministic process that generates the outcomes $0, 1, 2, \ldots$ as follows. Each observation involves executing a subprocess that repeatedly flips a coin until heads occurs. The trial's outcome is the number of tails prior to the first head. Outcomes $0, 1, 2, 3, \ldots$ correspond to subprocess results $H, TH, TTH, \ldots$. Suppose that the probability of heads on a given toss is p. Over a large number of observations, we note the fraction for which the subprocess produces $H, TH, TTH, \ldots$. We expect a fraction p for H, $(1-p)p$ for TH, $(1-p)^2 p$ for TTH, and so forth. That is, for $k = 0, 1, 2, \ldots$, we have $\Pr(k) = (1-p)^k p$. Provided that $p > 0$, we have $\sum_{k=0}^{\infty} (1-p)^k p = 1$. There remains one outcome that we have not considered. The subprocess can, theoretically, return an infinite sequence of tails. In this case, we add a new outcome, which we designate with the symbol ∞. Of course, we have used up all the available probability with the other outcomes, so we must assign $\Pr(\infty) = 0$.

The example generalizes to any random process where an outcome is the number of "failures" prior to the first "success" in repeated observations of a Bernoulli random variable. This provides a connection to applications beyond simple coin tossing. For example, if a storage structure can accommodate N records and currently holds n, a fraction $p = n/N$ of its storage slots are occupied. Equivalently, each slot is occupied with probability p. If we systematically probe the structure for an empty slot, what is the probability that we will encounter k occupied slots before finding an empty one? This is precisely the probability of k failures before the first success in repeated observations of a Bernoulli random variable with parameter p. The desired probability is then $(1-p)^k p$.

2.36 Definition: A random variable X has a *geometric* distribution with parameter p if $0 < p < 1$ and $\Pr(X = k) = (1-p)^k p$, for $k = 0, 1, 2, \ldots$. ∎

Before deriving the mean, variance, and moment generating function for the geometric distribution, we investigate another viewpoint on repeated observations of a Bernoulli random variable. In the opening example, we used a subprocess to sample the Bernoulli random variable. Now, we imagine that the main process simply returns an infinite binary sequence by simultaneously sampling infinitely many independent Bernoulli random variables with common parameter p. Let $X_1, X_2, \ldots$ be independent Bernoulli random variables, all with common parameter p. For each X_k, we have $\Pr(X_k = x) = p^x (1-p)^{1-x}$, where $x = 0$ or 1. By independent, we mean that any finite subset $X_{i_1}, X_{i_2}, \ldots, X_{i_n}$ exhibits

$$\Pr(X_{i_1} = x_1, \ldots, X_{i_n} = x_n) = \prod_{k=1}^{n} \Pr(X_{i_k} = x_k) = \prod_{k=1}^{n} p^{x_k} (1-p)^{1-x_k}.$$

Define $S = \min\{i \mid X_i = 1\} - 1$. Thus $S + 1$ is the position of the first one; S is the number of zeros preceding the first one. We can imagine that the underlying process consists of simultaneously flipping an infinite sequence of loaded coins, each of which lands heads up with probability p. After all coins come to rest—remember this is just a mental experiment—we examine the results, starting with the first coin in the sequence. If the first head occurs on coin k, then $S = k - 1$ for this trial. The process produces just one possible outcome in a large probability space. The space actually contains all possible infinite sequences with heads and tails as elements. We need a meaningful assignment of probabilities to these outcomes, which we can then transfer to S. Replacing each head with a one and each tail with a zero, we have an equivalent outcome space, which now contains all infinite binary sequences. For a particular outcome ω, $S(\omega)$ is the number of zeros prior to the first one.

We have two immediate difficulties. First, S is not defined for one outcome, the sequence with all zeros. Second, the space of all sequences is not countable. This follows from the standard diagonalization argument. Suppose that the space is countable. We first arrange the sequences according to the supposed enumeration, following the pattern below.

$$\omega_1 = x_{11}x_{12}x_{13}\cdots$$

$$\omega_2 = x_{21}x_{22}x_{23}\cdots$$

$$\omega_3 = x_{31}x_{32}x_{33}\cdots$$

We then construct the sequence $\omega' = y_1 y_2 y_3 \cdots$ such that y_k is the complement of x_{kk}. That is, $y_k = 1 - x_{kk}$. Clearly, ω' differs from ω_k in the kth bit, and therefore ω' is not in the enumeration. This contradiction establishes the uncountability of the collection of all binary sequences.

We bypass both problems through the simple expedient of assigning probabilities directly to *subsets* of sequences. If we can accomplish this task such that the subsets include the range of S, we will have a probability distribution for S, and we can then ignore the underlying set of sequences. Since S is the number of zeros prior to the first 1, the range of S is $0, 1, 2, \ldots$, which is obviously countable. What is the appropriate assignment for $S = 0$? The outcomes yielding $S = 0$ are all binary sequences that start with a 1. The initial 1 comes from the event $(X_1 = 1)$, which has probability p, so we assign $\Pr(S = 0) = p$. This allocation is intuitively compelling. In the coin-tossing experiment, if we ignore all aspects of the outcome except the status of the first coin, we expect heads with probability p. That is, the events $E = \{x_1, x_2, x_3, \ldots | x_1 = 1\}$ and $(X_1 = 1)$ are coincident. Following the same reasoning, we assign $\Pr(S = 1) = (1 - p)p$ because $S = 1$ coincides with the event $(X_1 = 0) \cap (X_2 = 1)$. In general, the probability that $S = k$ is the probability that X_1 through X_k are all zero and $X_{k+1} = 1$, which is $(1 - p)^k p$. We obtain the same geometric distribution that was generated through repeated observation of the same Bernoulli random variable. Now, however, we observe the pattern associated with a single observation of infinitely many independent Bernoulli random variables.

Despite the added complexity, we often encounter probability spaces in which the outcomes are infinite sequences. Consequently, we probe a bit further into the mechanism for consistently assigning probability directly to subsets of outcomes. We assemble the infinite sequences in blocks, such that the members of a given block have the same finite prefix. The finite prefix serves as an identifier of the block. For example, the prefix 01101 is the block $\{01101\ldots\}$. This notation blurs the differences between sequences in the same block, but we can tolerate this impreciseness if the only outcomes of interest involve finite prefixes. In defining the random variable S, we need only distinguish the blocks $1, 01, 001, \ldots$. Note that we still have infinitely many outcomes, but each outcome is now a block of infinite binary sequences, which is identified with a finite prefix. Most important, the new space of prefixes is countable.

2.37 Definition: A sequence $X_1, X_2, \ldots$ of random variables is a *stochastic process*. Outcomes are infinite sequences $x_1, x_2, \ldots$, such that $x_i \in \text{range}(X_i)$. The process description must include a consistent probability assignment to all subsets associated with finite prefixes. Consistent means that all assignments are nonnegative, the total assigned probability is 1, and the assignment to the union of a countable collection of disjoint components is the sum of the assignments to the components. A *Bernoulli process* is a stochastic process in which the X_i are independent Bernoulli random variables with a common parameter. ∎

To describe a Bernoulli process with parameter p, we must complete a consistent probability assignment to all subsets associated with finite prefixes. We established assignments to prefixes $1, 01, 001, 0001, \ldots$ while discussing the geometric distribution. These prefixes represent disjoint sequence collections, and their probabilities sum to 1. Their union is the entire sequence space, except the sequence $00000\ldots$. If we assign probability zero to this exceptional sequence, then we have a consistent assignment to this point. However, there remain many prefixes which still lack a probability assignment. We approach the task systematically by assigning probabilities in order of prefix length: $\Pr(1) = p, \Pr(0) = 1 - p, \Pr(00) = (1 - p)^2$, and so forth. In general,

$$\Pr(x_1 x_2 \ldots x_n) = \prod_{i=1}^{n} p^{x_i} (1 - p)^{1 - x_i}.$$

This assignment is compatible with that already established for prefixes that have the forms $1, 01, 001, 0001$, and so forth. Prefixes of length $n+1$ subdivide parent prefixes of length n. In the assignment list above, we see, for example, that $1 = 10 \cup 11$ and

$$\Pr(10) + \Pr(11) = p(1 - p) + p^2 = p(1 - p + p) = p = \Pr(1).$$

Each such subdivision allocates probability to two components such that the total is the probability of the parent prefix. If we complete the assignments through prefixes of length N, we have a probability space in which the elementary outcomes are N-prefixes. Shorter prefixes designate unions of selected N-prefixes, such as, for example,

$$101 = 10100 \cup 10101 \cup 10110 \cup 10111.$$

If we let an asterisk stand for an arbitrary bit (0 or 1), N-patterns of ones, zeros, and asterisks also constitute unions of N-prefixes. For example,

$$*10** = 010** \cup 110**$$
$$= 01000 \cup 01001 \cup 01010 \cup 01011 \cup 11000 \cup 11001 \cup 11010 \cup 11011.$$

Because we can consistently allocate probability to prefixes through any desired length, we avoid the necessity of assigning probability directly to the outcomes. Consequently, we avoid the complications associated with uncountably many outcomes. We will revisit this interpretation in connection with Markov processes, which appear in the next chapter. For the moment, we accept Definition 2.36 as the criterion for a geometric distribution, regardless of the circumstances in which it arises, and we continue with computations of its mean and variance.

2.38 Theorem: Let X have a geometric distribution with parameter $p, 0 < p < 1$. Then

$$\mu = \frac{1-p}{p}$$

$$\sigma^2 = \frac{1-p}{p^2}$$

$$\Psi(t) = \frac{p}{1-(1-p)e^t}.$$

PROOF:

$$\Psi(t) = E[e^{tX}] = \sum_{k=0}^{\infty} e^{kt} p(1-p)^k = p \sum_{k=0}^{\infty} [(1-p)e^t]^k$$

This is a geometric series, which converges for $(1-p)e^t < 1$. Consequently,

$$\Psi(t) = p \cdot \frac{1}{1-(1-p)e^t} = \frac{p}{1-(1-p)e^t}.$$

We recognize this expression as a composition of $g(x) = 1/(1-x)$ and $f(t) = (1-p)e^t$. Both components have familiar power series, and the composition series converges for $|t| < -\ln(1-p)$. See Example 2.35 for the details. Theorem 2.15 then gives the desired moments.

$$\Psi'(t) = -p[1-(1-p)e^t]^{-2}[-(1-p)e^t] = \frac{p(1-p)e^t}{[1-(1-p)e^t]^2}$$

$$\Psi''(t) = p(1-p)e^t \left(\frac{2(1-p)e^t}{[1-(1-p)e^t]^3} + \frac{1}{[1-(1-p)e^t]^2} \right)$$

The mean and variance then follow.

$$\mu = E[X] = \Psi'(0) = \frac{p(1-p)}{p^2} = \frac{1-p}{p}$$

$$E[X^2] = \Psi''(0) = p(1-p) \left[\frac{2(1-p)}{p^3} + \frac{1}{p^2} \right] = \frac{(1-p)(2-p)}{p^2}$$

$$\sigma^2 = E[X^2] - \mu^2 = \frac{(1-p)(2-p) - (1-p)^2}{p^2} = \frac{1-p}{p^2} \quad \blacksquare$$

If X is geometric with parameter p, then $Y = X + 1$ gives the number of Bernoulli trials up to and including the first success. $\mu_Y = \mu_X + 1 = (1 -$

$p)/p + 1 = 1/p$. This is a plausible result. Suppose that you are watching for the first time a die lands with two spots on the upturned face. Each roll defines a Bernoulli random variable with $p = 1/6$. You then expect, on the average, 6 rolls to encounter the two. There can be, however, significant variation in the values observed because the standard deviation is $\sqrt{(5/6)/(1/6)^2} = 5.48$.

The following two examples illustrate the use of geometric distributions in computer science problems.

2.39 Example: Example 1.27 introduced the study of hash tables. A hash table, you will recall, is a storage structure, together with an algorithm that calculates the storage address from a record key. For a record with key k, a hash function $f_1(k)$ returns the storage address. The hash function attempts to spread the storage uniformly across the table, but there are probabilistic variations from this ideal performance. Sometimes, therefore, the hash function returns the same address for several keys. We say that the keys *collide* at the computed address. We can design the hash table so that each address accommodates several records, but that remedy only delays the problem. There can still occur sufficient collisions at a particular address to overrun its capacity. Another approach, which can be used with or without an enlarged address capacity, is called secondary hashing. This scheme uses a sequence of independent hash functions, $f_1, f_2, \ldots$, each of which attempts to spread the keys uniformly across the table. If $f_1(k)$ produces a collision, then the algorithm tries $f_2(k)$, and so forth. If a record insertion requires the services of hash functions $f_1, f_2, \ldots, f_j$ to successfully find space for a new record, we say that the insertion requires j *probes*.

Suppose that we have a hash table with address capacity 1, served by an algorithm that uses secondary hashing. Suppose that the fraction α of the table is filled when we make an attempt to store a new record. α is the table *load factor*, and we assume that $\alpha < 1$. Let X be the random variable corresponding to the number of probes necessary to insert the new record. What is the distribution of X? What are its mean and variance? If $\alpha = 0.8$, what is the average number of probes to complete an insertion? What is the probability that the insertion will require more than twice the average?

If the table contains N addresses, then αN are filled. The probability that f_1 produces an empty address is then $(N - \alpha N)/N = 1 - \alpha$. Indeed, if further probes are necessary, each f_i will find an empty address with probability $1 - \alpha$ because each successive probe operates on the same table and all the hash functions give an equal chance to each address. Let Y_j be a Bernoulli random variable associated with the jth hash function; Y_j is 1 if the jth hash function finds an empty slot. So, $\Pr(Y_j = 1) = 1 - \alpha$ for all j. Because the hash functions operate independently, the probability of the jth probe finding an empty address is not influenced by the previous probes. In other words, the joint probability of $Y_1, Y_2, \ldots, Y_j$ is the product of the component probabilities. Then $X = \min\{i | Y_i = 1\}$ is the number of probes required for an insertion.

If we let $Z = X - 1$, then Z is a geometric random variable with parameter $1 - \alpha$. Using Theorem 2.38, we calculate as follows.

$$\Pr_Z(j) = [1 - (1 - \alpha)]^j (1 - \alpha) = \alpha^j (1 - \alpha)$$

$$\mu_X = \mu_Z + 1 = \frac{1 - (1 - \alpha)}{1 - \alpha} + 1 = \frac{1}{1 - \alpha}$$

$$\sigma_X^2 = \sigma_Z^2 = \frac{1 - (1 - \alpha)}{(1 - \alpha)^2} = \frac{\alpha}{(1 - \alpha)^2}$$

If $\alpha = 0.8$, the average number of probes is $\mu_X = 1/(1 - 0.8) = 5$. The probability that the insertion will require more than 10 probes is

$$\Pr(X > 10) = \Pr(Z > 9) = \sum_{i=10}^{\infty} \Pr_Z(i) = \sum_{i=10}^{\infty} \alpha^i(1 - \alpha) = (1 - \alpha)\alpha^{10} \sum_{i=0}^{\infty} \alpha^i$$

$$= (1 - \alpha)\alpha^{10} \cdot \frac{1}{1 - \alpha} = \alpha^{10} = (0.8)^{10} = 0.107. \quad \Box$$

2.40 Example: At the beginning of each millisecond, a file server simultaneously observes four request queues. If the server is idle and if it finds one or more nonempty queues, it randomly selects a request and begins working on it. In any given millisecond, the event "new arrival in queue i" is independent of similar events in the other queues and has probability p_i. For the four queues, the p_i are $0.02, 0.04, 0.08$, and 0.10. Suppose that the server is idle at time $t = 0$, just prior to checking the queue status, and that the queues were all empty on the previous check. What is the average time that the server will remain idle?

Let $A_{ij} = 1$ if a request arrives in queue i in millisecond j. Otherwise $A_{ij} = 0$. If a request appears in queue i on the first check after $t = 0$, we have $A_{i1} = 1$. If another request appears on the second check, we have $A_{i2} = 1$. These events are all independent, and they occur whether or not the server is busy. The A_{ij} are independent Bernoulli random variables with parameters p_i. For each i, let G_i be the number of $A_{ij} = 0$ prior to the first $A_{ij} = 1$. G_i is then a geometric random variable with parameter p_i, and since queue i is empty at $t = 0$, it measures the number of milliseconds before the first arrival. The distribution of G_i is $\Pr(G_i = j) = (1 - p_i)^j p_i$, for $j = 0, 1, 2, \ldots$. Let T be the random variable that measures the milliseconds before the server finds a request. Clearly, $T = \min\{G_1, G_2, G_3, G_4\}$. We want to know μ_T.

For any $j \geq 0$, the minimum G_i exceeds j precisely when all the G_i exceed j. That is,

$$\Pr(T \geq j) = \Pr((G_1 \geq j) \cap (G_2 \geq j) \cap (G_3 \geq j) \cap (G_4 \geq j)).$$

Because the events $(G_i \geq j)$ are independent, we have

$$\Pr(T \geq j) = \prod_{i=1}^{4} \Pr(G_i \geq j) = \prod_{i=1}^{4} \sum_{k=j}^{\infty} (1 - p_i)^k p_i = \prod_{i=1}^{4} p_i(1 - p_i)^j \sum_{k=0}^{\infty} (1 - p_i)^k$$

$$= \prod_{i=1}^{4} p_i(1 - p_i)^j \frac{1}{1 - (1 - p_i)} = \prod_{i=1}^{4} (1 - p_i)^j = \left[\prod_{i=1}^{4} (1 - p_i) \right]^j.$$

Let $1 - q = \prod(1 - p_i) = (0.98)(0.96)(0.92)(0.90) = 0.7790$, or $q = 0.2210$.

$$\Pr(T \geq j) = (1 - q)^j$$

$$\Pr(T = j) = \Pr(T \geq j) - \Pr(T \geq j + 1) = (1 - q)^j - (1 - q)^{j+1}$$

$$= (1 - q)^j[1 - (1 - q)] = (1 - q)^j q.$$

We conclude that T is also a geometric random variable with parameter $q = 0.2210$. Hence $\mu_T = (1 - q)/q = 3.525$ milliseconds. Having observed all empty queues, the server can expect to wait an average 3.525 milliseconds before a request arrives. Note that this argument remains valid even if the server has been idle for an arbitrary amount of time. For example, if the server has been idle for 10 milliseconds, it can still expect, on the average, to remain idle for an additional 3.525 milliseconds. This is an interesting property, unique to geometric distributions, that we will now study in more detail. $\square$

As noted in the example, the geometric distribution possesses an interesting feature, called the *memoryless* property. Intuitively, this property asserts that the number of trials before the next success in a sequence of Bernoulli trials is independent of the number of failures already observed. Suppose that you are watching the tosses of a fair coin. The number of tosses to obtain the first heads is a geometric random variable with parameter 0.5. If you have observed 10 tails in a row, you might think that the number of additional tosses necessary to obtain a head would be less than the number required if the long sequence of tails had not been observed. This intuition is not correct. That is, when the tosses start, the probabilities that the first head will arrive on the first, second, third toss are $(1 - 1/2)^0(1/2) = 1/2$, $(1 - 1/2)^1(1/2) = 1/4$, and $(1 - 1/2)^2(1/2) = 1/8$, respectively, with similar statements for the first head arriving on tosses beyond three. It turns out that after observing 10 tails, the probabilities that the first head will arrive on the subsequent first, second, third toss are still $1/2, 1/4, 1/8$. The process is oblivious to a long string of failures because it starts anew with each toss. The next theorem formalizes this result.

2.41 Theorem: Suppose that X is a geometric random variable with parameter p. Then for any $i = 0, 1, 2, \ldots$, the conditional probability that $X = i+j$, given that $X \geq i$, is the same as the unconditional probability that $X = j$. That is, $\Pr(X = i + j | X \geq i) = \Pr(X = j)$.

PROOF: Invoking Definition 1.54 for conditional probability, we compute as follows. For a fixed i, and $j \geq 0$,

$$\Pr(X = i + j | X \geq i) = \frac{\Pr((X = i + j) \cap (X \geq i))}{\Pr(X \geq i)} = \frac{\Pr(X = i + j)}{\sum_{k=i}^{\infty} \Pr(X = k)}$$

$$= \frac{p(1 - p)^{i+j}}{\sum_{k=i}^{\infty} p(1 - p)^k} = \frac{p(1 - p)^{i+j}}{p(1 - p)^i \sum_{k=0}^{\infty}(1 - p)^k}$$

$$= \frac{(1 - p)^j}{1/[1 - (1 - p)]} = p(1 - p)^j = \Pr(X = j).$$

The geometric distribution is the only distribution on the nonnegative integers that possesses the memoryless property. One of the exercises asks you to prove this converse theorem. ∎

Because the Bernoulli process involves independent Bernoulli random variables at each step, this result should not seem surprising. From any point in the sequence of trials, the probability of $0, 1, 2, \ldots$ further failures before the next success is just as if the sequence had started at that point. In particular, when a success occurs, the distribution of the number of failures before the next success is the same as the distribution of the number of failures before the first success. Therefore, in a Bernoulli process with parameter $p > 0$, the number of failures *between* successes is geometric with parameter p.

Before verifying this conjecture, we pause to acquire an important formula. The nth power of a convergent geometric series is

$$\left(\sum_{k=0}^{\infty} \beta^k\right)^n = \sum_{k=0}^{\infty} C_{n+k-1,k}\beta^k. \qquad (2.5)$$

Consider the partial sum through $k = N$ of the original series. Each term is a power of β, and the exponents range from 0 to N. Consequently, the nth power of the partial sum will also contain powers of β, with exponents in the range 0 through nN. However, different powers of β acquire different coefficients in the multiplication process. The products that contribute to β^k in the nth power have the form $\beta^{k_1}\beta^{k_2}\cdots\beta^{k_n}$, where $k_1 + k_2 + \ldots k_n = k$. Considering the nth power as the product of n copies of the series, the β^{k_1} comes from the first copy, the β^{k_2} from the second copy, and so forth. Theorem 1.23 counts the number of solutions to $k_1 + k_2 + \ldots k_n = k$. It is $N(k,n) = C_{n+k-1,k}$, and it follows that $C_{n+k-1,k}$ is the coefficient of β^k in the nth power. This reasoning assumes that each copy can indeed deliver a β^{k_i} factor as required. Of course, the copy can comply only if $k_i \le N$ because each copy is a partial sum extending only through β^N. So, as k nears its upper limit nN, the formula overestimates the size of the coefficient. For any fixed k, the formula is eventually correct as N tends to infinity. So, let us denote the actual coefficient as $C^*_{n+k-1,k}$, with the understanding that $C^*_{n+k-1,k} \le C_{n+k-1,k}$ and the inequality reverts to an equality for sufficiently large N. That is,

$$\left(\sum_{k=0}^{\infty} \beta^k\right)^n = \left(\lim_{N\to\infty} \sum_{k=0}^{N} \beta^k\right)^n = \lim_{N\to\infty}\left(\sum_{k=0}^{N} \beta^k\right)^n = \lim_{N\to\infty}\sum_{k=0}^{nN} C^*_{n+k-1,k}\beta^k$$

$$= \sum_{k=0}^{\infty} C_{n+k-1,k}\beta^k.$$

This establishes Equation 2.5.

We return to the main discussion, where we have a Bernoulli process with parameter $p > 0$. We wish to show that the number of failures between any two successes has a geometric distribution with parameter p. For this purpose, we fix an integer $j > 0$ and let X be the number of zeros between the jth and the $(j + 1)$th ones. We interpret the Bernoulli process as a probability space of infinite binary sequences, as discussed at the beginning of this section. Recall that we have a mechanism for consistently assigning probability to *subsets* of sequences, each associated with a finite prefix. For prefix $(x_1 x_2 \cdots x_n)$, this assignment is

$$\Pr(x_1 x_2 \ldots x_n) = p^{\sum x_i}(1 - p)^{n - \sum x_i}.$$

We consider constrained prefixes that terminate with the $(j + 1)$th one. Of course, the prefix must have length at least $j + 1$, but there is no upper bound on its length. A constrained prefix of length $n \geq j+1$ is completely specified when the locations of the j nonterminal ones are known. Consequently, if N_n is the number of such prefixes of length n, we have $N_n = C_{n-1,j}$. Because each prefix contains exactly $j+1$ ones, all prefixes of length n have the same probability assignment: $p^{j+1}(1-p)^{n-j-1}$. Because the prefixes represent disjoint collections of sequences, the total probability assignment is then

$$T = \sum_{n=j+1}^{\infty} N_n p^{j+1}(1-p)^{n-j-1},$$

which we can simplify because we know N_n.

$$T = p^{j+1} \sum_{n=j+1}^{\infty} C_{n-1,j}(1-p)^{n-j-1} = p^{j+1} \sum_{n=0}^{\infty} C_{n+j,j}(1-p)^n$$

$$= p^{j+1} \sum_{n=0}^{\infty} C_{j+n+1-1,n}(1-p)^n = p^{j+1} \left(\sum_{n=0}^{\infty} (1-p)^n \right)^{j+1}$$

$$= p^{j+1} \left(\frac{1}{1-(1-p)} \right)^{j+1} = 1$$

The subset of sequences outside the union of the selected prefixes receives probability assignment zero. This subset contains sequences that have j or fewer ones. The selected prefixes, however, represent all ways in which $X = s$, for $s = 0, 1, 2, \ldots$. The event $(X = s)$ is comprised of those prefixes that terminate with the pattern $\ldots 1000 \ldots 0001$, where there are exactly s zeros in the terminating pattern. Clearly, a prefix must have length at least $s+j+1$ to accommodate this terminating pattern plus the remaining $j-1$ ones. A prefix of sufficient length n is completely specified when the locations of the remaining $j-1$ ones are located among the $n-(s+2)$ positions prior to the terminating pattern. Consequently,

$$\Pr(X = s) = \sum_{n=s+j+1}^{\infty} C_{n-s-2,j-1} p^{j+1}(1-p)^{n-j-1}$$

$$= p^{j+1} \sum_{n=0}^{\infty} C_{j+n-1,j-1}(1-p)^{n+s}$$

$$= p^{j+1}(1-p)^s \sum_{n=0}^{\infty} C_{j+n-1,n}(1-p)^n$$

$$= p^{j+1}(1-p)^s \left(\sum_{n=0}^{\infty} (1-p)^n \right)^j$$

$$= p^{j+1}(1-p)^s \left(\frac{1}{1-(1-p)} \right)^j = p(1-p)^s.$$

This is precisely the geometric distribution with parameter p. Thus we have proved part of the following theorem.

2.42 Theorem: Let $B_1, B_2, \ldots$ be a sequence of independent Bernoulli random variables, each with parameter p. In a binary sequence from this process, let X_1 be the number of zeros prior to the first 1 and X_i be the number of zeros between the $(i-1)$th and ith 1s, for $i = 2, 3, \ldots$. Then the X_i are independent geometric random variables with parameter p.

PROOF: The discussion above proved that the X_i have identical geometric distributions with parameter p. Because the B_i are independent, it seems reasonable that the X_i should be independent. Nevertheless, it is comforting to establish this fact mathematically, so we will show that the X_i in any finite collection are independent. Let $i_1 < i_2 < \cdots < i_n$ index a finite collection of the X_i. Set $N = i_n$ and consider the prefixes that terminate with the Nth one. Just as in the argument above, we can show that these selected prefixes represent disjoint subsets of sequences and consume all the probability assignment. We want to identify the prefixes that constitute the event $E = (X_{i_1} = s_1, X_{i_2} = s_2, \ldots, X_{i_n} = s_n)$, for a given set of nonnegative integers $s_1, s_2, \ldots, s_n$. Let $s = \sum_{k=1}^{n} s_{i_k}$. A prefix satisfying this condition must have length $k \geq s + N$ to accommodate the s zeros and N ones. All qualifying prefixes of length k have probability $p^N (1-p)^{k-N}$. Moreover, the prefix has n blocks of zeros, of sizes $s_1, s_2, \ldots, s_n$, that fit into fixed positions before their corresponding 1s. We group the remaining bits into $N - n$ blocks, each containing a one and its preceding zeros. These must share the remaining $k - (n + s)$ slots. Running from left to right, these blocks can be of sizes $t_1 + 1, t_2 + 1, \ldots, t_{N-n} + 1$. The t_j represent the unspecified number of zeros preceding the 1 in block j. Consequently, the t_j are nonnegative and

$$\sum_{j=1}^{N-n} (t_j + 1) = k - n - s$$

$$\sum_{j=1}^{N-n} t_j = k - s - N.$$

Invoking Theorem 1.23, we count such nonnegative $\{t_1, t_2, \ldots, t_{N-n}\}$ vectors as

$$C_{N-n+k-s-N-1, k-s-N} = C_{k-n-s-1, k-s-N} = C_{k-n-s-1, (k-n-s-1)-(k-s-N)}$$
$$= C_{k-n-s-1, N-n-1}.$$

Each such vector specifies a different prefix in the event E. Consider the following example, in which we justify the counts in the following array.

Block	1	2	3	4	5	6	7	8	9
Leading zeros	0	4	1	2	0	0	2	0	3
Why?	$t_1 = 0$	$X_2 = 4$	$t_2 = 1$	$t_3 = 2$	$X_5 = 0$	$t_4 = 0$	$t_5 = 2$	$t_6 = 0$	$X_9 = 3$

To contribute to event E, a prefix must exhibit a length greater than or equal to $N + s = 9 + 7 = 16$. Suppose that we count the prefixes of length $k = 21$

that contribute to E. Since $N - n = 6$, we seek vectors $(t_1, t_2, \ldots, t_6)$ such that

$$\sum_{j=1}^{6} t_j = k - s - N = 21 - 7 - 9 = 5.$$

One such vector is $(0, 1, 2, 0, 2, 0)$, and it determines the prefix completely. Specifically, the prefix contains nine blocks, each containing a 1 with some number of preceding zeros. The blocks appear in the left-to-right order tabulated above. Note that the leading zeros total 12, which, together with the 9 ones, gives the correct length $k = 21$. So, we can assess the probability of event E by summing the probabilities of the participating prefixes.

$$\Pr(E) = \Pr_{X_{i_1}, \ldots, X_{i_n}}(s_1, \ldots, s_n) = \sum_{k=s+N}^{\infty} C_{k-n-s-1, N-n-1} p^N (1-p)^{k-N}$$

$$= p^N \sum_{k=0}^{\infty} C_{k+N-n-1, N-n-1} (1-p)^{k+s}$$

$$= p^N (1-p)^s \sum_{k=0}^{\infty} C_{(N-n)+k-1, k} (1-p)^k$$

$$= p^N (1-p)^s \left(\sum_{k=0}^{\infty} (1-p)^k \right)^{N-n} = p^N (1-p)^s \left(\frac{1}{1-(1-p)} \right)^{N-n}$$

$$= p^n (1-p)^s = p^n (1-p)^{s_1+s_2+\ldots+s_n} = \Pi_{i=1}^{n} [p(1-p)^{s_i}]$$

$$= \Pi_{i=1}^{n} \Pr_{X_i}(s_i)$$

Hence, $X_{i_1}, X_{i_2}, \ldots, X_{i_n}$ are independent. ∎

2.43 Example: A job consists of two tasks, and the processing time, in milliseconds, for each task is a geometric random variable with parameter $p = 0.3$. Since the range of the geometric random variable is $0, 1, 2, \ldots$, there is the possibility that a task requires no process time. Suppose that both tasks are started simultaneously on separate processors. What is the average time to complete the job?

Let G_1 and G_2 be the respective process times for the tasks, and let T be the process time for the full job. The G_i are independent and have identical distributions $\Pr(G_i = j) = (1-p)^j p$. The job finishes when all tasks are complete. Therefore, $T = \max\{G_1, G_2\}$, and We seek μ_T. We can approach this computation directly by first deriving the distribution of T. Note the following rearrangement of the series expansion for the mean.

$$
\begin{array}{llll}
\sum_{j=0}^{\infty} \Pr(T > j) = \Pr(T = 1) & + \Pr(T = 2) & + \Pr(T = 3) & + \ldots \\
& + \Pr(T = 2) & + \Pr(T = 3) & + \ldots \\
& & + \Pr(T = 3) & + \ldots \\
\hline
= (1)\Pr(T = 1) + (2)\Pr(T = 2) + (3)\Pr(T = 3) + \ldots = \mu_T
\end{array}
$$

This formula is known as the *survivor expression for the mean*. Now, $\Pr(T > j) = 1 - \Pr(T \leq j)$, and we can calculate the latter from the observation that $(T \leq j)$ if and only if $(G_1 \leq j)$ and $(G_2 \leq j)$. Consequently,

$$\Pr(T > j) = 1 - \Pr(T \leq j) = 1 - \prod_{i=1}^{2}\sum_{k=0}^{j}(1-p)^k p = 1 - p^2\left(\sum_{k=0}^{j}(1-p)^k\right)^2$$

$$= 1 - p^2\left[\frac{1-(1-p)^{j+1}}{1-(1-p)}\right]^2 = 1 - 1 + 2(1-p)^{j+1} - (1-p)^{2(j+1)}$$

$$= 2(1-p)^{j+1} - [(1-p)^2]^{j+1}$$

$$\mu_T = 2\sum_{j=0}^{\infty}(1-p)^{j+1} - \sum_{j=0}^{\infty}[(1-p)^2]^{j+1}$$

$$= 2(1-p)\left[\frac{1}{1-(1-p)}\right] - (1-p)^2\left[\frac{1}{1-(1-p)^2}\right] = \frac{(1-p)(3-p)}{p(2-p)}.$$

Substituting $p = 0.3$, we have $\mu_T = 3.706$.

We might attempt the following clever interpretation of the geometric distribution's memoryless property. The first completion occurs in time $T_1 = \min\{G_1, G_2\}$. In Example 2.40 we found that the minimum of n geometric random variables with parameters $p_1, p_2, \ldots, p_n$ is also a geometric random variable with parameter $q = 1 - \prod_{i=1}^{n}(1 - p_i)$. So, $\mu_{T_1} = (1-p)^2/[1-(1-p)^2] = 0.961$. What is the distribution of the remaining process time for the unfinished task? Suppose that $T_1 = t$ and that T_2 is the remaining process time for the other process, say G. We have that $\Pr(T_2 = j) = \Pr(G = t + j | G \geq t) = \Pr(G = j)$, which is geometrically distributed with the same parameter. The mean time interval between the first and second completions is then $\mu_{T_2} = (1-p)/p = 2.333$. Consequently, $\mu_T = \mu_{T_1} + \mu_{T_2} = 2.294$, which is incorrect. One of the exercises asks you to investigate this discrepancy. $\square$

Suppose now that $X_1, X_2, \ldots, X_n$ are independent geometric random variables with common parameter p. Let $Y = \sum_{i=1}^{n} X_i$. In terms of a Bernoulli process, X_i is the number of zeros after the $(i-1)$st and before the ith 1. Y then measures the total number of zeros before the nth 1. We seek the distribution of Y.

We first observe that Y assumes the values $0, 1, 2, \ldots$. Then, knowing the common moment generating function of the X_i, $\Psi_X = p[1-(1-p)e^t]^{-1}$, we can invoke Theorem 2.27 to obtain $\Psi_Y(t) = p^n[1-(1-p)e^t]^{-n}$. The limiting process of Section 2.2 can then sequentially recover each of the probabilities: $\Pr_Y(0), \Pr_Y(1), \ldots$. For example,

$$\Pr(Y = 0) = \lim_{t\to-\infty} e^{-0t}\Psi_Y(t) = \lim_{t\to-\infty}\frac{p^n}{[1-(1-p)e^t]^n} = p^n$$

$$\Pr(Y = 1) = \lim_{t\to-\infty} e^{-1t}[\Psi_Y(t) - p^n e^{0t}] = \lim_{t\to-\infty} e^{-t}p^n\left[\frac{1-(1-(1-p)e^t)^n}{[1-(1-p)e^t]^n}\right]$$

$$= \lim_{t\to-\infty}\frac{p^n}{[1-(1-p)e^t]^n} \cdot \lim_{t\to-\infty}\frac{1-(1-(1-p)e^t)^n}{e^t}$$

$$= p^n\lim_{t\to-\infty}\frac{-n(1-(1-p)e^t)^{n-1}[-(1-p)e^t]}{e^t} = n(1-p)p^n.$$

A simple observation shortens this tedious process. Now that we know there is only one possible distribution corresponding to this $\Psi_Y(t)$, we can manipulate the expression to reveal the coefficients of e^{kt}. These coefficients must be $\Pr(Y = k)$. For $(1-p)e^t < 1$, we calculate as follows.

$$\Psi_Y(t) = \frac{p^n}{[1-(1-p)e^t]^n} = p^n \left(\frac{1}{1-(1-p)e^t}\right)^n$$

$$= p^n(1 + (1-p)e^t + (1-p)^2 e^{2t} + (1-p)^3 e^{3t} + \ldots)^n$$

$$= p^n \sum_{k=0}^{\infty} C_{n+k-1,k}(1-p)^k e^{kt} = \sum_{k=0}^{\infty} C_{n+k-1,k} p^n (1-p)^k e^{kt} \qquad (2.6)$$

Consequently, $\Pr(Y = k) = C_{n+k-1,k} p^n (1-p)^k$. Note that these values describe a valid discrete probability distribution:

$$\sum_{k=0}^{\infty} C_{n+k-1,k} p^n (1-p)^k = p^n \left(\sum_{k=0}^{\infty}(1-p)^k\right)^n = p^n \left(\frac{1}{1-(1-p)}\right)^n = 1.$$

This distribution is the negative binomial distribution, and we will state a formal definition shortly. First, however, we entertain a short digression to explain the distribution's name. Definition 1.2 established the falling factorial $P_{n,k}$ for positive n. It defines $P_{n,k}$ as the product of k factors. The first factor is n, and each succeeding factor is one less. This process extends naturally to negative n: $P_{-n,k} = (-n)(-n-1)(-n-2)\cdots(-n-(k-1))$. For positive n the falling factorial is zero for $k > n$ because the expansion contains a zero factor. For negative n, the factors are nonzero for all k. Note that

$$P_{-n,k} = (-1)^k P_{n+k-1,k} = \frac{(-1)^k (n+k-1)!}{(n-1)!}.$$

This extension allows us to use binomial coefficients with negative numbers in the upper position. Recalling that the binomial coefficient for positive upper elements is $C_{n,k} = P_{n,k}/k! = n!/[k!(n-k)!]$, we adopt a similar definition for the case when n is negative.

2.44 Definition: The *negative binomial coefficient*, $C_{-n,k}$, is defined by

$$C_{-n,k} = \frac{P_{-n,k}}{k!} = (-1)^k \cdot \frac{(n+k-1)!}{k!(n-1)!} = (-1)^k C_{n+k-1,k}. \blacksquare$$

Now consider the following expression, where $|a| > |b|$.

$$(a+b)^{-n} = \frac{1}{(a+b)^n} = \frac{1}{a^n[1+(b/a)]^n} = a^{-n}\left(\frac{1}{1-(-b/a)}\right)^n$$

$$= a^{-n}\left(\sum_{k=0}^{\infty}(-b/a)^k\right)^n = a^{-n}\sum_{k=0}^{\infty} C_{n+k-1,k}(-b/a)^k$$

$$= \sum_{k=0}^{\infty}(-1)^k C_{n+k-1,k}(-b)^k a^{-n-k} = \sum_{k=0}^{\infty} C_{-n,k} b^k a^{-n-k}.$$

So the binomial expansion is valid for negative powers, provided that $|a| > |b|$. This latter condition is easily satisfied by exchanging the roles of a and b if necessary. [If $a = b$, then $(a + b)^{-n} = 2^{-n}a^{-n}$, a particularly simple series expansion.] The only difference in the negative binomial expansion is that an infinite sum replaces the finite sum that appears ordinary binomial expansion. In terms of negative binomial coefficients, we can rewrite the Y probabilities derived above as

$$\Pr{}_Y(k) = C_{n+k-1,k}p^n(1-p)^k = (-1)^k C_{-n,k}p^n(1-p)^k.$$

Because the probabilities involve the negative binomial coefficients, this distribution is known as the negative binomial distribution.

2.45 Definition: Let integer $n \geq 1$ and $0 < p < 1$. A *negative binomial* random variable Y with parameters (n, p) has the distribution

$$\Pr{}_Y(k) = C_{-n,k}(-1)^k p^n(1-p)^k$$

for $k = 0, 1, 2, \ldots$. This is the distribution of the sum of n independent geometric random variables with common parameter p. ∎

2.46 Theorem: Let X have a negative binomial distribution with parameters (n, p), both greater than zero. Then

$$\Psi(t) = \frac{p^n}{[1 - (1-p)e^t]^n}$$

$$\mu = \frac{n(1-p)}{p}$$

$$\sigma^2 = \frac{n(1-p)}{p^2}.$$

PROOF: The discussion above establishes the moment generating function, from which we can calculate the mean and variance.

$$\Psi'(t) = p^n(-n)[1 - (1-p)e^t]^{-n-1}[-(1-p)e^t]$$
$$= np^n(1-p)e^t[1 - (1-p)e^t]^{-n-1}$$

$$\mu = E[X] = \Psi'(0) = np^n(1-p)p^{-n-1} = \frac{n(1-p)}{p}$$

$$\Psi'' = np^n(1-p)\{e^t[1 - (1-p)e^t]^{-n-1}$$
$$+ e^t(-n-1)[1 - (1-p)e^t]^{-n-2}[-(1-p)e^t]\}$$

$$E[X^2] = \Psi''(0) = np^n(1-p)\left(\frac{1}{p^{n+1}} + \frac{(n+1)(1-p)}{p^{n+2}}\right)$$

$$= \frac{n(1-p)}{p} + \frac{n(n+1)(1-p)^2}{p^2}$$

$$\sigma^2 = E[X^2] - \mu^2 = \frac{n(1-p)}{p} + \frac{n(n+1)(1-p)^2}{p^2} - \frac{n^2(1-p)^2}{p^2}$$

$$= \frac{n(1-p)}{p^2}$$

These are the anticipated values because as a sum of n independent geometric random variables with common parameter p, X must have, according to Theorem 2.8, a mean and variance that are each n times the corresponding values for the geometric distribution. ∎

2.47 Example: You repeatedly receive cards from a well-shuffled deck, and you replace each card before receiving the next. What is the chance that you will see your fourth ace on or before the tenth card?

Let X_i be the random variable that reports 1 if you receive an ace on the ith card and zero otherwise. Then X_i is a Bernoulli random variable with parameter $p = 1/13$. The number of deals before the first ace is a geometric random variable with parameter p. The number of additional deals before the second ace is another independent geometric random variable with parameter p. Hence the number of deals before the second ace is a negative binomial distribution with parameters $(2, p)$. Continuing in this fashion, we see that the number of deals, excluding the aces themselves, up to the fourth ace is a negative binomial distribution with parameters $(4, p)$. Let X be the number of "failures" before the fourth ace. Then $T = X + 4$ is the number of deals up to and including the fourth ace. We seek that probability that $T \leq 10$, or equivalently, the probability that $X \leq 6$.

$$\Pr(T \leq 10) = \Pr(X \leq 6) = \sum_{k=0}^{6} C_{4+k-1,k}(1/13)^4(12/13)^k = 0.005 \qquad (2.7)$$

We can verify this result by assuming that we see a total of 10 cards, even if the fourth ace occurs before the last card. We are then sampling, with replacement, from a set of 52. Four aces occur at or before the tenth card if four or more aces appear in the sample. Out of $(52)^{10}$ possible sequences, four aces appear in $C_{10,4} \cdot 4^4 \cdot 48^6$ such sequences. This is because each of the $C_{10,4}$ positions where the aces can occur admits 4 possibilities (the 4 suits) for each of the four ace slots. It also admits 48 possibilities for each of the non-ace slots. Five aces appear in $C_{10,5} \cdot 4^5 \cdot 48^5$ sequences, and so forth. The probability of 4 or more aces in 10 cards is then

$$\frac{1}{52^{10}} \sum_{k=4}^{10} C_{10,k} 4^k 48^{10-k} = \frac{1}{4^{10} 13^{10}} \sum_{k=4}^{10} C_{10,k} \frac{4^k (4 \cdot 12)^{10}}{(4 \cdot 12)^k}$$

$$= \frac{1}{13^{10}} \sum_{k=4}^{10} C_{10,k} \cdot 12^{10-k} = 0.005.$$

Since

$$E[T] = E[X] + 4 = \frac{4(1 - 1/13)}{1/13} + 4 = \frac{4(12/13) + 4(1/13)}{1/13} = 52,$$

we expect a significant probability of seeing the fourth ace at or before card 52. Indeed, if we extend the sum in Equation 2.7 to 52, we obtain a probability of 0.5744. ☐

2.48 Example: Example 2.39 showed that the number of probes needed to insert a record into a hash table is 1 plus a geometric random variable with parameter $1 - \alpha$, where α is the table load factor, the fraction of its capacity that is filled. Suppose that each probe requires a 10-millisecond disk access. Assume that $\alpha = 0.8$ and the table is so large that adding a small number of records does not change α significantly. What is the average time needed for a transaction that inserts 10 new records? Find an upper bound on the probability that this time will exceed 1 second. What can you do to decrease this probability?

Let T be the time needed to insert 10 records. In milliseconds, T is $10N$, where N is the number of probes. N, in turn, is 10 plus X, where X is the sum of 10 independent geometric random variables, each with parameter $1 - \alpha = 0.2$. So X has a negative binomial distribution with parameters $(10, 0.2)$. Accordingly, $\mu_X = 10(1 - 0.2)/0.2 = 40$. Then $\mu_N = 40 + 10 = 50$, and $\mu_T = 10(50) = 500$ milliseconds. Also, $\sigma_N^2 = \sigma_X^2 = 10(0.8)/(0.2)^2 = 200$, so $\sigma_T^2 = (10)^2(200) = 20000$. Therefore, $\sigma_T = 141.4$ milliseconds. An insertion delay of 1 second represents a deviation of $1000 - 500 = 3.536\sigma_T$ from the mean. Chebyshev's inequality asserts that the probability of so large a deviation is less than $1/(3.536)^2 = 0.08$. An exact value is available from the negative binomial distribution, since the probability that $T > 1.0$ is the same as the probability that $N > 100$ or $X > 90$. Although difficult to evaluate without the approximation tools from later in the book, the latter value is

$$\Pr_X(X > 90) = \sum_{k=91}^{\infty} C_{10+k-1,k}(0.2)^{10}(1 - 0.2)^k.$$

If 0.08 is too large a bound for the probability that the insertion time will exceed 1 second, we can use a larger hash table, which will lower the load factor. If α is reduced to 0.5, for instance, then μ_T becomes $(10)[10 + 10(1 - 0.5)/0.5] = 200$ milliseconds, and σ_T becomes $\sqrt{(10)^2(10)(1 - 0.5)/(0.5)^2} = 44.27$ milliseconds. An insertion delay of 1 second now represents a deviation of $1000 - 200 = 17.9\sigma_T$ from the mean. The probability of this deviation is less than $1/(17.9)^2 = 0.003$. $\square$

Figure 2.4 illustrates several negative binomial distributions, all with $p = 0.2$ and varying n. The distributions are discrete, of course, but the smoothed curves show how, just as in the case for the binomial distribution, they tend to a bell-shaped curve centered on the mean. We discuss this limiting tendency in a later chapter.

Figure 2.5 shows the relationships among the discrete distributions discussed to this point. Two arrows, labeled "conditional," connect the binomial and negative binomial distributions to the discrete uniform. To understand these relationships, consider first a binomial random variable X with parameters (n, p). X is the sum of n independent Bernoulli random variables, $(B_1, B_2, \ldots, B_n)$, and $(X = k)$ is the event in which exactly k of the B_i are 1. As such, X presents a summary of the activity among the B_i. We can keep the complete observation with a random vector $B = (B_1, B_2, \ldots, B_n)$, which assumes 2^n values, from $(000\ldots0)$ through $(111\ldots1)$. What is the conditional distribution of B, given $X = k$? For a binary n-vector $(b_1, b_2, \ldots, b_n)$, we have

$$\Pr(B = (b_1, \ldots, b_n)|X = k) = \frac{\Pr((B = (b_1, \ldots, b_n) \cap (\sum b_i = k))}{\Pr(\sum b_i = k)}$$

$$= p^k(1 - p)^{n-k}/C_{n,k}p^k(1 - p)^{n-k}$$

$$= 1/C_{n,k}, \tag{2.8}$$

if $\sum b_i = k$. Of course, the probability is zero if $\sum b_i \neq k$. The conditional

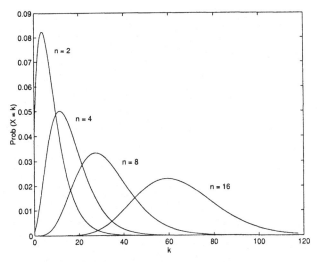

Figure 2.4. Smoothed curves for negative binomial distributions, all with $p = 0.2$

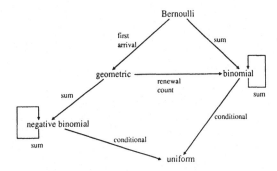

Figure 2.5. Relationships among discrete distributions

distribution is therefore a discrete uniform, which places probability $1/C_{n,k}$ on each of the B vectors having components that sum to k.

A similar interpretation explains the conditional arrow connecting the negative binomial distribution to the discrete uniform. If X has the negative binomial distribution with parameters (n, p), we envision X as the sum of n independent geometric random variables, each with parameter p. Each such geometric random variable represents the number of zeros preceding the first 1 in a sequence of independent Bernoulli random variables. X is therefore the number of zeros preceding the nth 1. When $X = k$, the corresponding Bernoulli sequence must start with an $(n + k)$-prefix consisting of n ones and k zeros, and the last bit of the prefix must be a 1. Let B_{n+k} be the prefix that actually occurs when $X = k$. There are precisely $C_{n+k-1,k}$ possible values for B_{n+k}, corresponding to the number of ways we can choose locations for the k zeros from the $n + k - 1$ nonterminal positions. The probability of each

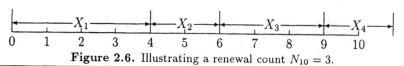

Figure 2.6. Illustrating a renewal count $N_{10} = 3$.

such prefix is $p^n(1-p)^k$. If $(b_1, b_2, \ldots, b_{n+k})$ is a binary vector with exactly n ones, we have

$$\Pr(B_{n+k} = (b_1, b_2, \ldots, b_{n+k})|X = k) = \frac{p^n(1-p)^k}{C_{n+k-1,k}p^n(1-p)^k} = \frac{1}{C_{n+k-1,k}}.$$

This conditional distribution is discrete uniform. On each possible prefix, it places probability $1/C_{n+k-1,k}$.

The remaining unexplained connection in Figure 2.5 is the renewal count connection between the geometric and binomial distributions. For the discrete case, a renewal count is the number of events that occur along a fixed linear span, given that the intervals between events are a specified sequence of random variables. The following definition describes a renewal count more precisely.

2.49 Definition: Let $X_1, X_2, \ldots$ be a sequence of random variables. For a fixed x, the random variable $N_x = \max\{n| \sum_{i=1}^{n} X_i \leq x\}$, is a *renewal count* for the sequence. ∎

The probability space is the set of all infinite sequences $(x_1, x_2, \ldots)$, in which $x_i \in \text{range}(X_i)$. When an outcome occurs, it simultaneously specifies the values of all the x_i components and therefore determines N_x. If we think of the X_i as the interarrival distances or times between events, then N_x measures the number of events in the span $[0, x]$. Figure 2.6 illustrates an outcome that exhibits $N_{10} = 3$. The outcomes that produce $(N_x \leq n)$ are precisely those that produce $\sum_{i=1}^{n+1} X_i > x$, and this observation leads to the distribution of N_x. The theorem below derives the distribution of N_n, the renewal counts for a sequence of independent geometric random variables with the same parameter.

2.50 Theorem: Let $G_1, G_2, \ldots$ be a sequence of independent geometric random variables, each with parameter p. Then N_n, the renewal counts for the sequence $(G_1 + 1, G_2 + 1, \ldots)$, are binomial random variables with parameters (n, p).

PROOF: The distribution of each G_i is $\Pr(G_i = j) = (1-p)^j p$. By definition, we have $N_n = \max\{k| \sum_{i=1}^{k}(G_i + 1) \leq n\}$. Since each $(G_i + 1)$ assumes the values $1, 2, \ldots$, we must have $N_n \leq n$. Moreover, $N_n = n$ if and only if the first n of the G_i are zero. Hence $\Pr(N_n = n) = p^n$. For the remaining possibilities, let $0 \leq k < n$ and compute as follows. We use the fact that the sum of independent geometric random variables has a negative binomial distribution.

$$\Pr(N_n \leq k) = \Pr\left(\sum_{i=1}^{k+1}(G_i + 1) > n\right) = \Pr\left(\sum_{i=1}^{k+1} G_i > n - k - 1\right)$$

Substituting the negative binomial probabilities, we obtain

$$\Pr(N_n \le k) = \sum_{j=n-k}^{\infty} C_{(k+1)+j-1,j} p^{k+1} (1-p)^j$$

$$= p^{k+1} \sum_{j=0}^{\infty} C_{n+j,n-k+j} (1-p)^{n-k+j}$$

$$= p^{k+1} (1-p)^{n-k} \sum_{j=0}^{\infty} C_{n+j,k} (1-p)^j.$$

Theorem 1.35 asserts $C_{n+j,k} = \sum_{i=0}^{k} C_{n,i} C_{j,k-i}$, and because all the summands are nonnegative, we can exploit the absolute convergence to rearrange the summands to suit our convenience.

$$\Pr(N_n \le k) = p^{k+1} (1-p)^{n-k} \sum_{i=0}^{k} C_{n,i} \sum_{j=0}^{\infty} C_{j,k-i} (1-p)^j$$

$$= p^{k+1} (1-p)^{n-k} \sum_{i=0}^{k} C_{n,i} \sum_{j=k-i}^{\infty} C_{j,k-i} (1-p)^j$$

The last reduction follows because $C_{j,k-i} = 0$ for $j < k-i$. We now invoke a useful trick for evaluating sums of this sort, where the upper member of the combinatorial symbol runs with the summation while the lower member remains constant. Consider the power series $f(x) = \sum_{j=0}^{\infty} x^j = 1/(1-x)$, which is convergent for $|x| < 1$. Using term-by-term differentiation, we compute the mth derivative.

$$f^{(m)}(x) = \sum_{j=m}^{\infty} j(j-1) \cdots (j-m+1) x^{j-m} = \sum_{j=m}^{\infty} P_{j,m} x^{j-m} = \frac{m!}{(1-x)^{m+1}}$$

This observation is pertinent to the computation in progress if we let $m = k-i$ and $x = (1-p)$. That is,

$$\sum_{j=k-i}^{\infty} C_{j,k-i} (1-p)^j = \frac{(1-p)^{k-i}}{(k-i)!} \sum_{j=k-i}^{\infty} P_{j,k-i} (1-p)^{j-(k-i)}$$

$$= \frac{(1-p)^{k-i}}{(k-i)!} \cdot \frac{(k-i)!}{[1-(1-p)]^{k-i+1}} = \frac{(1-p)^{k-i}}{p^{k-i+1}}.$$

Substituting in the earlier expression for $\Pr(N_n \le k)$, we have

$$\Pr(N_n \le k) = p^{k+1} (1-p)^{n-k} \sum_{i=0}^{k} C_{n,i} \frac{(1-p)^{k-i}}{p^{k-i+1}} = \sum_{i=0}^{k} C_{n,i} p^i (1-p)^{n-i}.$$

Since N_n is nonnegative, $\Pr(N_n = 0) = \Pr(N_n \le 0) = C_{n,0} p^0 (1-p)^n =$

$(1-p)^n$. For $0 < k < n$, we have

$$
\begin{aligned}
\Pr(N_n = k) &= \Pr(N \leq k) - \Pr(N \leq k - 1) \\
&= \sum_{i=0}^{k} C_{n,i} p^i (1-p)^{n-i} - \sum_{i=0}^{k-1} C_{n,i} p^i (1-p)^{n-i} \\
&= C_{n,k} p^k (1-p)^{n-k}.
\end{aligned}
$$

Therefore, N_n is binomially distributed with parameters (n, p). ∎

2.51 Example: A system contains a processor, a memory module, and a clock. The memory and processor can communicate on each clock pulse, and the pulses occur every 2 nanoseconds. The processor issues store/retrieve commands to the memory such that the interval, measured in clock pulses, between commands is a geometric random variable with parameter $p = 0.05$. That is, the probability that k clock pulses appear between commands is $(1-p)^k p$. Furthermore, the interval between any two given commands is independent of the interval between any other two commands. What is the probability that the memory receives no commands in a 40-nanosecond interval? What is the probability that it receives less than three commands?

Let G_1 be the number of clock pulses prior to the first command, and for $i > 1$, let G_i count the pulses between the commands $i - 1$ and i. Command 1 occurs on pulse $G_1 + 1$. Command 2 arrives on pulse $(G_1 + 1) + (G_2 + 1)$, and so forth. The 40 nanosecond span is 20 pulses, and letting N_{20} be the number of commands in this time span, we have $N_{20} = \max\{n | \sum_{i=1}^{n}(G_i + 1) \leq 20\}$. By the theorem above, N_{20} has a binomial distribution with parameters $(20, 0.05)$. Therefore,

$$
\Pr(N_n = 0) = C_{20,0}(0.05)^0 (1 - 0.05)^{20} = 0.3585
$$

$$
\Pr(N_n < 3) = \sum_{k=0}^{2} C_{20,k}(0.05)^k (1 - 0.05)^{20-k} = 0.9245. \ \square
$$

Exercises

*2.28 Consider a Bernoulli process as a probability space of infinite binary sequences. A finite prefix identifies the set of sequences beginning with the prefix. The probability assignment is

$$
\Pr(x_1 x_2 \ldots x_n) = p^{\sum x_i}(1-p)^{n - \sum x_i},
$$

for prefix $(x_1 x_2 \ldots x_n)$. Let S_k be the set of prefixes containing exactly k 1s, including a terminating 1. If $j \geq k$ is the length of $\omega \in S_k$, then $\Pr(\omega) = p^k(1-p)^{j-k}$. Consider the prefix sets S_k and S_n for $k < n$. For $\omega \in S_k$, define $Q(\omega)$ as all the prefixes in S_n that are continuations of ω. $0101 \in S_2$, for example, and the extended prefixes $0101101, 01010000011, 010111 \in S_4$ are continuations. Show that $\Pr(\omega) = \Pr(Q(\omega))$.

2.29 Compute the mean and variance of a geometric random variable with parameter p by directly summing the series $\sum_n n(1-p)^n p$ and $\sum_n n^2(1-p)^n p$. Compare your results with Theorem 2.38.

2.30 Let $X_1, X_2, \ldots$ be a sequence of independent Bernoulli random variables, all with parameter $p > 0$. The geometric random variable is the number of zeros prior to the first 1. Let Y be the location of the first 1. Find the moment generating function, the mean, and the variance of Y. This related distribution is also called a geometric distribution.

2.31 Suppose that a system contains three devices connected in series such that the system fails if any one of the devices fails. Starting from an arbitrary $t = 0$, let E_{ij} be the event "device i fails in hour j." Here $1 \leq i \leq 3$ and $j = 1, 2, 3, \ldots$. Assuming the E_{ij} to be independent Bernoulli random variables with the following parameters, what is the mean time to system failure?

$$\Pr(E_{ij} = 1) = \begin{cases} 0.0001, & i = 1, j = 1, 2, 3 \ldots \\ 0.0003, & i = 2, j = 1, 2, 3, \ldots \\ 0.004, & i = 3, j = 1, 2, 3, \ldots . \end{cases}$$

*2.32 Suppose that X is a discrete random variable with range contained in the nonnegative integers. Suppose further that X has the memoryless property: $\Pr_X(X = i + j)/\Pr_X(X \geq i) = \Pr_X(j)$ for all nonnegative integers i and j. Show that X must have a geometric distribution.

*2.33 Example 2.43 considered the random variable $T = \max\{G_1, G_2\}$, where each G_i is geometric with parameter $p = 0.3$. The example computed $\mu_T = 3.706$ correctly, but speculated that one might take a shortcut by exploiting the memoryless property of the geometric distribution. In particular, the first completion occurs in time $T_1 = \min\{G_1, G_2\}$, which is a geometric random variable with parameter $q = 1 - (1 - p)^2$. So, $\mu_{T_1} = (1 - p)^2/[1 - (1 - p)^2] = 0.961$. The remaining process time is the excess from the other process, which is geometric with parameter p by the memoryless property. Consequently, $\mu_{T_2} = (1 - 0.3)/0.3 = 2.333$, and $\mu_T = \mu_{T_1} + \mu_{T_2} = 3.294$, which is incorrect. What is wrong with this reasoning?

*2.34 From its distribution function, compute directly the mean and variance of a negative binomial distribution with parameters (n, p). Compare the results with those asserted by Theorem 2.46.

2.35 In a sequence of Bernoulli trials, the negative binomial distribution with parameters (n, p) describes the number of failures prior to the nth success. If X is this negative binomial random variable, let $Y = X + n$, which is the number of trials up to and including the nth success. Find the moment generating function, the mean, and the variance of Y. This related distribution is also called a negative binomial distribution.

2.36 What is the average number of times that you must roll a die to observe 10 even numbers? What is the probability that twice as many rolls are necessary?

2.37 An urn contains 8 black balls and two red balls. You repeatedly draw one ball from the urn, replacing it before the next draw. What is the probability that you will encounter a red ball before the third draw? What is the probability that you will encounter two red balls before the tenth draw?

2.38 An urn contains 8 black balls and two red balls. You repeatedly draw two balls from the urn, replacing them before the next draw. What is the probability that you will encounter a red-black pair before the third draw? What is the probability that you will encounter two red-red pairs before the tenth draw?

*2.39 Show that $C_{-n_1-n_2,k} = \sum_{i=0}^{k} C_{-n_1,i} C_{-n_2,k-i}$.

*2.40 Since a negative binomial random with parameters (n,p) is the sum of n independent geometric random variables with parameter p, the sum of two independent negative binomial random variables, with parameters (n_1,p) and (n_2,p), should be another negative binomial random variable with parameters (n_1+n_2,p). Prove this directly from the distributions. That is, if X_1, X_2 are independent and

$$\Pr(X_i = k) = C_{-n_i,k} p^{n_i} (p-1)^k, \text{ for } k = 0,1,2,\dots \text{ and } i = 1,2,$$

then

$$\Pr(X_1 + X_2 = k) = C_{-n_1-n_2,k} p^{n_1+n_2} (p-1)^k, \text{ for } k = 0,1,2,\dots.$$

2.41 A freeway entrance ramps admits cars and measures the number of car lengths between admissions. Suppose that this measure exhibits a geometric distribution with $p = 0.04$. Suppose further that the cars continue on the freeway at the same constant speed with which they pass the entrance gate. What is the average number of cars on a freeway segment that can hold 100 cars bumper to bumper in a single lane?

2.4   The Poisson distribution

In this section we derive the limiting form of the binomial distribution with parameters (n,p) as $n \to \infty$ and $p \to 0$ in such a manner that the product np converges to a positive constant λ. First note that l'Hôpital's rule implies the following limit.

$$y = (1-p)^{1/p}$$

$$\lim_{p\to 0} \ln y = \lim_{p\to 0} \frac{\ln(1-p)}{p} = \lim_{p\to 0} \frac{-1/(1-p)}{1} = -1$$

$$\lim_{p\to 0} y = \lim_{p\to 0} (1-p)^{1/p} = e^{-1} \qquad (2.9)$$

Now let X_i be a sequence of binomial distributions with parameters (n_i, p_i) such that $\lim_{i \to \infty} n_i = \infty$, $\lim_{i \to \infty} p_i = 0$, and $\lim_{i \to \infty} n_i p_i = \lambda > 0$. For a fixed k and $n_i > k$, we have

$$\Pr_{X_i}(k) = C_{n_i,k} p_i^k (1-p_i)^{n_i-k} = \frac{n_i(n_i-1)\cdots(n_i-k+1)}{k!} \cdot \frac{p_i^k(1-p_i)^{n_i}}{(1-p_i)^k}$$

$$= \frac{(n_i p_i)^k}{k!} \cdot \left(\frac{n_i-1}{n_i}\right) \cdots \left(\frac{n_i-k+1}{n_i}\right) \cdot \frac{[(1-p_i)^{1/p_i}]^{n_i p_i}}{(1-p_i)^k}.$$

Consequently,

$$\lim_{i \to \infty} \Pr_{X_i}(k) = \frac{\lambda^k}{k!} \cdot (e^{-1})^\lambda = \frac{\lambda^k e^{-\lambda}}{k!}. \tag{2.10}$$

For $k = 0, 1, 2, \ldots$, the expressions in Equation 2.10 form a valid discrete probability distribution because

$$\sum_{k=0}^{\infty} \frac{\lambda^k e^{-\lambda}}{k!} = e^{-\lambda} \sum_{k=0}^{\infty} \frac{\lambda^k}{k!} = e^{-\lambda} e^{\lambda} = 1.$$

This limiting distribution is the Poisson distribution. We provide a formal definition, derive the summarizing features, and then discuss the distribution's utility.

2.52 Definition: The Poisson random variable P_λ has distribution $\Pr(P_\lambda = k) = \lambda^k e^{-\lambda}/k!$, for $k = 0, 1, 2, \ldots$. ∎

2.53 Theorem: Let X have a Poisson distribution with parameter λ. Then

$$\Psi(t) = e^{-\lambda(1-e^t)}$$
$$\mu = \lambda$$
$$\sigma^2 = \lambda.$$

PROOF:

$$\Psi(t) = \sum_{k=0}^{\infty} \frac{e^{kt} e^{-\lambda} \lambda^k}{k!} = e^{-\lambda} \sum_{k=0}^{\infty} \frac{(\lambda e^t)^k}{k!} = e^{-\lambda} e^{\lambda e^t} = e^{-\lambda(1-e^t)}$$

$$\Psi'(t) = e^{-\lambda(1-e^t)} \cdot \lambda e^t = \lambda e^t \Psi(t)$$

$$\Psi''(t) = \lambda e^t \Psi(t) + \lambda e^t \Psi'(t) = \lambda e^t (\Psi(t) + \lambda e^t \Psi(t)) = \Psi(t)[\lambda e^t + (\lambda e^t)^2]$$

$$\mu = E[X] = \Psi'(0) = \lambda$$

$$E[X^2] = \Psi''(0) = \lambda + \lambda^2$$

$$\sigma^2 = E[X^2] - \mu^2 = \lambda \quad \blacksquare$$

The Poisson distribution realistically describes the pattern of requests over time in many client-server situations. Examples are incoming customers at a bank teller, calls into a company's telephone exchange, requests for storage/retrieval services from a database server, and interrupts to a central processor. It also has higher-dimensional applications, such as the spatial distribution of defects on integrated circuit wafers and the volume distribution of contaminants in well water. In such cases, the "events," which are request arrivals or defect occurrences, are independent. Customers do not conspire to achieve some special pattern in their access to a bank teller; rather they operate as independent agents. The manufacture of hard disks or integrated circuits introduces unavoidable defects because the process pushes the limits of geometric tolerances. Therefore, a perfectly functional process will still occasionally produce a defect, such as a small area on the disk surface where the magnetic material is not spread uniformly or a shorted transistor on an integrated circuit chip. These errors are independent in the sense that a defect at one point does not influence, for better or worse, the chance of a defect at another point. Moreover, if the time interval or spatial area is small, the probability of an event is correspondingly small. This is a characterizing feature of a Poisson distribution: event probability decreases with the window of opportunity and is linear in the limit. A second characterizing feature, negligible probability of two or more events in a small interval, is also present in the examples mentioned here.

This reasoning suggests that we can approximate the total number of events over some macro-interval as the sum of many independent Bernoulli random variables with small p parameters. In the limit, this distribution becomes the Poisson distribution. To buttress the argument's plausibility, consider another viewpoint. It involves a time-based situation, but the analysis remains valid for spatial distributions, such as the number of defects in a square centimeter of an integrated circuit. We start with a base time interval, $[0, 1]$, during which events occur according to some underlying random process. This interval can represent one minute, one hour, one day, or some larger unit, as long as the nature of the process remains unchanged over that time span.. Let $0 = t_0 < t_1 < t_2 < \ldots < t_n = 1$ be a partition of the interval, and let $\Delta t_i = t_i - t_{i-1}$, for $i = 1, 2, \ldots, n$. As $n \to \infty$, we insist that all the Δt_i approach zero. That is, $\lim_{n \to \infty} \max_{i=1}^{n} \Delta t_i = 0$. As before, we assume that event counts in disjoint subintervals are independent random variables $X_1, X_2, \ldots, X_n$. Furthermore, we now assume that there exists a proportionality constant λ, which for small Δt_i describes the probabilities of events in the subintervals as follows.

$$\Pr(X_i = 1) = g_1(\Delta t_i) = \Pr(1 \text{ event in } \Delta t_i) = \lambda \Delta t_i + \eta_1(\Delta t_i)$$
$$\Pr(X_i > 1) = g_2(\Delta t_i) = \Pr(2 \text{ or more events in } \Delta t_i) = \eta_2(\Delta t_i)$$
$$\Pr(X_i = 0) = g_0(\Delta t_i) = \Pr(0 \text{ events in } \Delta t_i) = 1 - g_1(\Delta t_i) - g_2(\Delta t_i), \quad (2.11)$$

where $\eta_j(x)$ has the property: $\lim_{x \to 0} \eta_j(x)/x = 0$, for $j = 1, 2$. That is, the probability of an event is proportional to the "exposure" Δt_i plus other terms that behave like Δt_i^2 or higher powers as the exposure window shrinks to zero. These probability assignments are consistent with the earlier derivation. That is, if

$$\Delta t = \Delta t_i = 1/n, \text{ for } 1 \le i \le n,$$

then as $n \to \infty$,

$$p = g_1(\Delta t) = \frac{\lambda}{n} + \eta_1(1/n) = \frac{1}{n}\left[\lambda + \frac{\eta_1(1/n)}{1/n}\right] \to 0.$$

Moreover,

$$np = \lambda + \frac{\eta_1(1/n)}{1/n} \to \lambda.$$

In this case, we can use g_1 to associate a Bernoulli random variable with each subinterval. We then have n such random variables, and the earlier discussion shows that the limiting probability of k events in $[0, 1]$ is $\lambda^k e^{-\lambda}/k!$. There is, however, the complication that a subinterval can experience more than one event, which renders the Bernoulli sequence an inexact model. We will show that this complication vanishes in the limit, and the distribution is indeed a Poisson. Before launching into the proof, we note certain similarities with a Bernoulli *process*.

A Bernoulli process is a sequence of independent Bernoulli random variables with the same parameter. The intervals between successive ones is geometrically distributed, and the renewal counts are binomial. Our goal is to establish similar results in the limiting case, but with less restrictive constraints on the random variable sequence. In the discussion above, the sequence $X_1, X_2, \ldots$ comprises independent random variables, but they need not be Bernoulli and they need not have the same parameters. Rather, they must act like Bernoulli random variables only in the limit, as specified in Equations 2.11, with parameter p proportional to the subinterval width Δt. Thus, while compatible with the earlier treatment, in which the subintervals associate with Bernoulli random variables under a common parameter, these new probability assignments are more flexible. In particular, the new probabilities associated with the subintervals are not necessarily equal; they must merely be proportional, in the limit, to the size of the subinterval. This feature has intuitive appeal because it seems reasonable that the probability of receiving a client request, or of encountering a defect, should be proportional to the exposure time, at least for very small exposure windows. These assumptions may then leave us somewhat more comfortable in applying the results to a particular situation, but as we shall presently see, they lead to the same Poisson distribution as was obtained with the apparently more rigid assumptions. Equations 2.11 also specify precisely the sense in which the probability of two or more events in a given subinterval is negligible: Such probabilities behave like the square, or higher power, of the subinterval size. In the limit, we will have the Poisson process, whose precise definition we defer until a bit later in the argument. For the moment, we use the term *pre-Poisson process* to denote the random variable sequence $X_1, X_2, \ldots, X_n$, in which X_i counts the number of events in $[t_{i-1}, t_i)$ and satisfies Equations 2.11. We will now develop the ramifications of these assumptions.

Let $f(t; k)$ be the probability of k events in the interval $[0, t]$, for $0 \le t < 1$. For a fixed $t \in [0, 1)$ and a small subinterval $[t, t + \Delta t)$, the subinterval is a member of some partition $0 = t_1 < t_2 < \ldots < t_n$, when n is sufficiently large. Because the event counts in disjoint subintervals are independent, the probability of zero events in the span $[0, t + \Delta t)$ is the product of the probabilities of zero events in the two pieces.

$$f(t + \Delta t; 0) = f(t; 0)[1 - \lambda \Delta t - \eta_1(\Delta t) - \eta_2(\Delta t)]$$

$$\frac{f(t + \Delta t; 0) - f(t; 0)}{\Delta t} = -\lambda f(t; 0) - f(t; 0)\left[\frac{\eta_1(\Delta t)}{\Delta t} + \frac{\eta_2(\Delta_t)}{\Delta t}\right]$$

$$f'(t; 0) = \lim_{\Delta t \to 0} \frac{f(t + \Delta t; 0) - f(t; 0)}{\Delta t} = -\lambda f(t; 0)$$

Consequently, $f'(t; 0)/f(t; 0) = [\ln f(t; 0)]' = -\lambda$, and the solution is then $f(t; 0) = Ke^{-\lambda t}$. Since it is certain that zero events will occur in an interval of size zero, we have $f(0; 0) = 1$. This boundary condition forces the constant $K = 1$. So $f(t; 0) = e^{-\lambda t}$. As t approaches 1, this value becomes $e^{-\lambda}$. Definition 2.52 asserts the same value for the Poisson distribution at $k = 0$. That is, $\Pr(P_\lambda = 0) = e^{-\lambda}$.

We now proceed by induction. Assume that $f(t; j) = e^{-\lambda t}(\lambda t)^j / j!$, for $j = 0, 1, \ldots, k - 1$. Consider $f(t + \Delta t; k)$, for $k > 0$.

$$f(t + \Delta t; k) = \sum_{j=0}^{k} \Pr(j \text{ events in } [0, t)) \cdot \Pr(k - j \text{ events in } [t, t + \Delta t))$$

$$= \sum_{j=0}^{k} f(t; j) \cdot \Pr(k - j \text{ events in } [t, t + \Delta t))$$

$$= f(t; k)[1 - \lambda \Delta t - \eta_1(\Delta t) - \eta_2(\Delta t)]$$

$$+ f(t; k - 1)[\lambda \Delta t + \eta_1(\Delta t)] + \sum_{j=0}^{k-2} f(t; j)\eta_2(\Delta t)$$

We can now rearrange this expression to obtain a ratio that leads to the derivative of f.

$$\frac{f(t + \Delta t; k) - f(t; k)}{\Delta t} = -\lambda f(t; k) - f(t; k)\left[\frac{\eta_1(\Delta t)}{\Delta t} + \frac{\eta_2(\Delta t)}{\Delta t}\right]$$

$$+ \lambda \cdot \frac{(\lambda t)^{k-1} e^{-\lambda t}}{(k - 1)!} + f(t; k - 1) \cdot \frac{\eta_1(\Delta t)}{\Delta t}$$

$$+ \frac{\eta_2(\Delta t)}{\Delta t} \sum_{j=0}^{k-2} f(t; j)$$

Letting Δt approach zero, we obtain the derivative expression

$$f'(t; k) = -\lambda f(t; k) + \frac{e^{-\lambda t} \lambda^k t^{k-1}}{(k - 1)!},$$

which we can manipulate into a simple differential equation.

$$f'(t;k) + \lambda f(t;k) = \frac{e^{-\lambda t} \lambda^k t^{k-1}}{(k-1)!}$$

$$e^{\lambda t}[f'(t;k) + \lambda f(t;k)] = \frac{\lambda^k t^{k-1}}{(k-1)!}$$

The left side is the derivative of $e^{\lambda t} f(t;k)$, so

$$\left[e^{\lambda t} f(t;k)\right]' = \frac{\lambda^k t^{k-1}}{(k-1)!}$$

$$e^{\lambda t} f(t;k) = \frac{(\lambda t)^k}{k!} + K.$$

Since the probability of $k > 0$ events in the interval $[0,0)$ is 0, we have $f(0;k) = 0$, and it follows that the integration constant $K = 0$. For all $k \geq 0$ and $t \in [0,1)$, we can now assert that $f(t;k) = e^{-\lambda t}(\lambda t)^k/k!$. Letting $t \to 1$, the probability of k events in the interval $[0,1)$ is $f(1,k) = e^{-\lambda}\lambda^k/k!$, which is precisely $\Pr(P_\lambda = k)$, the Poisson distribution. The probability of an event in an interval of width zero is zero, so we can assert the same distribution for the closed interval. That is, the probability of k events in the interval $[0,1]$ is $e^{-\lambda}\lambda^k/k!$. Furthermore, we have established the same format for any fraction of the interval. That is, the probability that k events occur in the interval $[0,t]$ is $e^{-\lambda t}(\lambda t)^k/k!$. We capture our observations in the following theorem.

2.54 Theorem: For each $n = 1, 2, \ldots$, let $0 = t_0^{(n)} < t_1^{(n)} < \cdots < t_n^{(n)} = 1$ be a partition of $[0,1]$, and define $\Delta t_i^{(n)} = t_i^{(n)} - t_{i-1}^{(n)}$. Suppose further that $\lim_{n\to\infty} \left(\max_{1 \leq i \leq n} \Delta t_i^{(n)} \right) = 0$. Let $X_1^{(n)}, X_2^{(n)}, \ldots, X_n^{(n)}$ be independent random variables that are "almost Bernoulli" in the following sense:

$$\Pr(X_i^{(n)} = 1) = g_1(\Delta t_i^{(n)}) = \lambda \Delta t_i^{(n)} + \eta_1(\Delta t_i^{(n)})$$

$$\Pr(X_i^{(n)} > 1) = g_2(\Delta t_i^{(n)}) = \eta_2(\Delta t_i^{(n)})$$

$$\Pr(X_i^{(n)} = 0) = 1 - g_1(\Delta t_i^{(n)}) - g_2(\Delta t_i^{(n)}),$$

where $\lim_{x\to 0} \eta_i(x)/x = 0$ for $i = 1, 2$. Let $S_n = \sum_{j=1}^n X_j^{(n)}$. Then

$$\lim_{n\to\infty} \Pr(S_n = k) = \Pr(P_\lambda = k) = \frac{e^{-\lambda}\lambda^k}{k!}.$$

PROOF: See discussion above. ∎

2.55 Example: Suppose that commercial airplane accidents occur, on the average, once in every 10^{12} passenger-miles. What is the probability that there will be an accident in the first 10^9 passenger miles? What is the probability of an accident in the first 10^{13} passenger-miles?

This situation matches the Poisson distribution assumptions: a large number of potential events, independent of each other and each with a very small probability of occurrence. The long term average is $\lambda = 1$. The base interval is $[0, 10^{12}]$ passenger-miles. The first question concerns the fraction $t = 10^9/10^{12} = 0.001$

of the base interval. The probability of an accident in this fractional interval is 1 minus the probability of no accidents. That is, if $X = k$ is the event "k accidents in 10^9 passenger-miles," then $\Pr(X = k) = e^{-\lambda t}(\lambda t)^k/k!$. In particular, the probability of one or more accidents in this window is

$$\Pr(X) = 1 - e^{-\lambda t}(\lambda t)^0/0! = 1 - e^{-1(0.001)} = 0.00099.$$

If the fraction t changes to $10^{13}/10^{12} = 10.0$, then the probability of an accident becomes

$$1 - e^{-\lambda t} = 1 - e^{-1(10)} = 0.999955.$$

An accident rate of one per 10^{12} passenger-miles means that there is essentially no chance of an accident in a span of 10^9 passenger-miles, but an accident is almost certain in 10^{13} passenger-miles. $\square$

As the example illustrates, the fraction t of the base interval can be larger than 1. However, we must know, or be able to assume with some credibility, that the underlying random process that produces the events is the same over the longer interval. To see that this extension is valid, notice that we can always choose a larger base interval. Suppose, for example, that the original base interval $[0, 1]$ represents one hour and that $t = 2$, a fraction larger than 1. If we expand the base interval so that $[0, 1]$ now represents two hours, then λ must double because it is the average number of events in the base interval. Also, t now changes to 1.0 because the fraction 1.0 of the new interval is the same time as the fraction 2.0 of the original interval. The parameter λt then remains unchanged; it is 2 times the original λ or 1 times the new λ. The distribution, of course, depends only on the product λt. The same argument holds for an arbitrary fraction t, greater than or less than 1, of the original base interval.

2.56 Example: A hard disk manufacturing process deposits a thin coating of magnetic material on a plastic substrate. Empirical measurement has shown that the process introduces 20 defects per 15000 cm². A disk pack contains 800 cm² of this material. A disk pack with one defect is marketable because the disk controller can compensate for one error by using a spare location. What is the process yield, which is the percentage of marketable disk packs?

The situation constitutes a Poisson process. The base interval is 15000 cm², and the fraction in question is $t = 800/15000 = 0.053$. The overall average defect rate is $\lambda = 20$. The probability of zero defects or one defect is

$$\frac{e^{-\lambda t}(\lambda t)^0}{0!} + \frac{e^{-\lambda t}(\lambda t)^1}{1!} = e^{-20(0.053)}(1 + 20(0.053)) = 0.711.$$

The probability of a given disk pack being marketable is 0.711, so the yield, the percentage of marketable disk packs, is 71.1%. $\square$

2.57 Example: A central database server receives, on the average, 25 requests per second from its clients. What is the probability that the server will receive no requests in a 10-millisecond interval? What is the probability that it will receive more than 2 requests in a 10-millisecond interval?

It is reasonable to assume that the client requests are independent and that the probability of a request occurring in a small subinterval is proportional to the width of that subinterval. In other words, the Poisson distribution applies to this situation. The base interval is 1 second, and the fraction in question is $t = 0.01$ second. The average event rate is $\lambda = 25$ per second. Hence the probability of no requests in 0.01 seconds is $e^{-\lambda t} = e^{-0.01(25)} = 0.779$. The probability of more than 2 requests is 1 minus the probabilities of 0, 1, or 2 requests:

$$1 - \sum_{k=0}^{2} \frac{e^{-\lambda t}(\lambda t)^k}{k!} = 1 - e^{-0.25}\left[1 + 0.25 + \frac{0.25^2}{2}\right] = 0.00216. \ \square$$

Because the Poisson assumptions (Equations 2.11) assign probabilities that depend only on the subinterval size, not its position, the resulting Poisson distribution applies to any fraction of the base interval, starting at any point. If we need to know, for example, the distribution of events in the span $[0.25, 0.75]$ of the base interval, the probability of k events is $(0.5\lambda)^k e^{-0.5\lambda}/k!$.

Although we derived the Poisson distribution from Equations 2.11, it is comforting to close the loop by demonstrating that the distribution does indeed satisfy the Poisson assumptions. The probability of one event in a small interval Δt is $e^{-\lambda\Delta t}(\lambda\Delta t)^1/1!$. So

$$g_1(\Delta t) = e^{-\lambda\Delta t}(\lambda\Delta t) = \lambda\Delta t + \lambda\Delta t(e^{-\lambda\Delta t} - 1) = \lambda\Delta t + \eta_1(\Delta t)$$

$$\frac{\eta_1(\Delta t)}{\Delta t} = \lambda(e^{-\lambda\Delta t} - 1),$$

and this last expression approaches zero as $\Delta t \to 0$. The verification that $g_2(\Delta t)$ has the proper form for small Δt is one of the exercises.

In parallel with the Bernoulli process, we now define the Poisson process by exploiting our knowledge of the renewal count. That is, we know that the renewal counts of a Bernoulli process, in the case of independent interarrival times, are binomial. Because a renewal count gives the number of events in a fixed span, we can define the Poisson process as a sequence of random variables that deliver Poisson renewal counts.

2.58 Definition: A Poisson *process* with parameter λ is a sequence $T_1, T_2, \ldots$ of independent identically distributed nonnegative random variables such that the renewal counts $N_t = \max\{n| \sum_{i=1}^{n} T_i \leq t\}$ have Poisson distributions. That is, $\Pr(N_t = k) = e^{-\lambda t}(\lambda t)^k/k!$. Note that t is a nonnegative real number, which implies an uncountable number of renewal counts. The T_i are called the *interarrival times* for the process. ∎

In a particular application, events of interest occur at the conclusion of each T_i. In this interpretation, T_i measures the spacing between events i and $i - 1$. Frequently, this spacing is a time measure, which explains the terminology interarrival time. For this discussion, we assume that the T_i represent time intervals, although the concept is more general. Consider T_1, the time before the first event. For some fixed t, what is the probability that $T_1 \leq t$? From the definition of the renewal count N_t, we see that $T_1 > t$ if and only if $N_t = 0$. The known distribution of N_t then forces

$$\Pr(T_1 \leq t) = 1 - \Pr(T_1 > t) = 1 - \Pr(N_t = 0) = 1 - \frac{e^{-\lambda t}(\lambda t)^0}{0!} = 1 - e^{-\lambda t}.$$

This is not the cumulative distribution function of a discrete random variable. It does not exhibit staircase jumps at discrete points. We conclude that T_1 does not take on a discrete range of values, but rather that it can assume any nonnegative real value. T_1 is an example of a continuous random variable, for which we defer detailed study to a later chapter. For the moment, it is sufficient to note that a continuous random variable's distribution is not an assignment of probability masses to a countable range, but rather a consistent assignment to *intervals*.

Specifically, we describe a continuous random variable X with a *continuous* cumulative distribution function $F(t) = \Pr(X \leq t)$. This specification has two immediate consequences. First, it determines $\Pr(X \in (a, b])$ for all intervals $(a, b]$. Since the event $(X \leq b)$ is the disjoint union of the two events $(X \leq a) \cup (a < X \leq b)$, we have

$$\begin{aligned}
F(b) &= \Pr(X \leq b) = \Pr(X \leq a) + \Pr(a < X \leq b) \\
&= F_X(a) + \Pr(X \in (a, b]) \\
\Pr(X \in (a, b]) &= F(b) - F(a).
\end{aligned}$$

$$(2.12)$$

Second, $\Pr(X = t) = 0$ for any single point t. This follows because, for all $\Delta t > 0$, the event $(X \in (t - \Delta t, t])$ contains the event $(X = t)$. Hence, for all $\Delta t > 0$, the continuity of F entails

$$\begin{aligned}
\Pr(X = t) &\leq \Pr(X \in (t - \Delta t, t]) = F(t) - F(t - \Delta t) \\
\Pr(X = t) &\leq \lim_{\Delta t \to 0} [F(t) - F(t - \Delta t)] = 0.
\end{aligned}$$

A continuous random variable X with the cumulative distribution function $F(t) = 1 - e^{-\lambda t}$ is called an *exponential* random variable with parameter λ. Chapter 6 will take up its advanced properties, but we can conclude here that T_1 is an exponential random variable with parameter λ. Because the T_i have identical distributions, they must all be exponential with parameter λ. That is, they are all nondiscrete random variables.

This discovery invites a reconsideration of the Poisson process definition, which involves the sequence $T_1, T_2, \ldots$ of independent random variables. Because we have defined independence only for discrete random variables, the definition remains vague until we have extended the concept to continuous random variables. Although we must wait for Chapter 6 to develop the details, the correct extension is as follows. We have discrete X and Y independent if and only if $\Pr((X = x_i) \cap (Y = y_j)) = \Pr(X = x_i) \cdot \Pr(Y = y_j)$ for all (x_i, y_j) in the joint range. Some reflection reveals that this condition is equivalent to $\Pr((X \leq x) \cap (Y \leq y)) = \Pr(X \leq x) \cdot \Pr(Y \leq y)$ for all (x, y). The original independence criterion encounters difficulties with continuous random variables, because it is always trivially true. The probabilities of points are zero. However, the second equivalent condition circumvents these problems, and therefore we adopt it as the more general definition of independent random variables. That is, X and Y are independent if $\Pr((X \leq x) \cap (Y \leq y)) = F_X(x)F_Y(y)$. The sequence $T_1, T_2, \ldots$ in the Poisson process definition is then independent in the sense that

$$\Pr((T_{i_1} \leq x_1) \cap (T_{i_2} \leq x_2) \cap \ldots \cap (T_{i_k} \leq x_k)) = \prod_{j=1}^{k} F(x_j),$$

for any finite collection $\{T_{i_1}, T_{i_2}, \ldots, T_{i_k}\}$. Here $F(\cdot)$ is the common cumulative distribution function of the T_i.

One important property of the exponential distribution is immediately available. The conditional probability that the first event occurs after time $t + s$, given that it occurs after time t is

$$\Pr(T_1 > t + s | T_1 > t) = \frac{e^{-\lambda(t+s)}}{e^{-\lambda t}} = e^{-\lambda s} = \Pr(T_1 > s).$$

When we establish the zero point for measuring time, there is a certain probability that s time units will elapse before the first event. After observing t time units in which no event occurs, the probability of an additional s time units with no event remains the same as that for s uneventful time units at the beginning of the experiment. In a Poisson process, the first arrival time is therefore memoryless in the sense that $\Pr(T_1 > t + s | T_1 > t) = \Pr(T_1 > s)$. Recall that Theorem 2.41 asserts a similar property for the geometric distribution, and one of that section's exercises claims that the geometric distribution is the only discrete distribution with the memoryless property. We do not have a contradiction here because T_1 is not a *discrete* random variable. However, this fact should arouse some suspicion that the first arrival time in a Poisson process might be a limiting case of the geometric distribution. This is true in the following sense. If Y is geometric with parameter p, then Y gives the number of failures prior to the first success in a Bernoulli process with parameter p. The probability of at least n failures before the first success is

$$\Pr(Y \geq n) = \sum_{k=n}^{\infty}(1-p)^k p = p\sum_{k=0}^{\infty}(1-p)^{k+n} = (1-p)^n p\sum_{k=0}^{\infty}(1-p)^k$$

$$= (1-p)^n p \cdot \frac{1}{1 - (1-p)} = (1-p)^n. \tag{2.13}$$

Now, if $p \to 0$ and $n \to \infty$ in such a manner that $np \to \lambda t$, Equation 2.9 asserts that $(1-p)^n = [(1-p)^{1/p}]^{np}$ approaches $e^{-\lambda t}$. We formalize this observation in the following theorem.

2.59 Theorem: Let $G_1, G_2, \ldots$ be a sequence of independent geometric random variables with parameters $p_1, p_2, \ldots$. Suppose that $p_n \to 0$ and $np_n \to \lambda > 0$. Let $E_n = G_n/n$. Then $\lim_{n \to \infty} \Pr(E_n > t) = e^{-\lambda t}$.

PROOF:

$$\Pr(E_n > t) = \Pr(G_n > nt) = \sum_{k=\lfloor nt+1 \rfloor}^{\infty} p_n(1-p_n)^k$$

$$\Pr(E_n > t) = p_n(1 - p_n)^{\lfloor nt+1 \rfloor} \sum_{k=0}^{\infty} (1 - p_n)^k.$$

Evaluating the sum, we have

$$\Pr(E_n > t) = \frac{p_n(1 - p_n)^{\lfloor nt+1 \rfloor}}{1 - (1 - p_n)} = (1 - p_n)^{\lfloor nt+1 \rfloor}.$$

Consequently, $(1 - p_n)^{nt+1} \le \Pr(E_n > t) \le (1 - p_n)^{nt}$. We will argue that both extremes in this expression converge to $e^{-\lambda t}$, which will force $\Pr(E_n > t)$ to converge to the same value. Note that

$$(1 - p_n)^{nt+1} = (1 - p_n)(1 - p_n)^{nt}.$$

Because $p_n \to 0$, the first factor approaches 1. It therefore suffices to show $(1-p_n)^{nt} \to e^{-\lambda t}$ to prove that both extremes bracketing $\Pr(E_n > t)$ approach $e^{-\lambda t}$. Recall that Example 2.32 showed $\ln(1 - x) = -\sum_{k=1}^{\infty} \frac{x^k}{k}$, for $-1 < x < 1$. Exploiting this representation, we have

$$\lim_{n\to\infty} \ln(1 - p_n)^n = \lim_{n\to\infty} n \ln(1 - p_n) = -\lim_{n\to\infty} n \left(p_n + \frac{p_n^2}{2} + \frac{p_n^3}{3} + \cdots \right)$$

$$= -\lim_{n\to\infty} \left(np_n + \frac{(np_n)p_n}{2} + \frac{(np_n)p_n^2}{3} + \cdots \right) = -\lambda$$

$$\lim_{n\to\infty} (1 - p_n)^n = e^{-\lambda}$$

$$\lim_{n\to\infty} (1 - p_n)^{nt} = \lim_{n\to\infty} [(1 - p_n)^n]^t = e^{-\lambda t}. \ \blacksquare$$

2.60 Example: The database server of Example 2.57 receives, on the average, 25 requests per second from its clients. What is the average time between incoming requests?

As in Example 2.57, we assume a Poisson distribution for the number of incoming requests over an interval of width t seconds: $\Pr(k) = (25t)^k e^{-25t}/k!$. The interarrival times, T, follow $\Pr(T > s) = e^{-25s}$. Since T is not a discrete random variable, we have not yet studied how to determine its features, such as its average or variance. We can, however, make a limiting argument, which will be further developed in a later chapter. In keeping with our earlier notation, let μ_T be the average interarrival time.

Let L be a temporary upper limit on the interarrival time. It is temporary because we will let it approach infinity near the conclusion of the argument. Divide the span $[0, L]$ into n small subintervals, each of width $L/n = \Delta s$. The partition is then $0 = s_0 < s_1 < s_2 < \cdots < s_n = L$. For a particular s_i, the outcome sets $(T \le s_i)$ and $(s_i < T \le s_i + \Delta s)$ are disjoint, and their union is the outcome set $(T \le s_i + \Delta s)$. Consequently,

$$\Pr(s_i < T \le s_i + \Delta s) = \Pr(T \le s_i + \Delta s) - \Pr(T \le s_i)$$

$$= [1 - e^{-25(s_i+\Delta s)}] - [1 - e^{-25s_i}] = e^{-25s_i}(1 - e^{-25\Delta s}).$$

In the interval $s_i < T \le s_i + \Delta s$, $T_i \approx s$, and the approximation becomes exact as $\Delta s \to 0$. Let μ_n be the following approximating expression for the average interarrival time. It treats T as a constant over the small subintervals, and it cuts off the possible s values at L.

$$\mu_n = \sum_{i=0}^{n-1} s_i \Pr(s_i < T \le s_i + \Delta s) = \sum_{i=0}^{n-1} s_i e^{-25 s_i} (1 - e^{-25 \Delta s})$$

$$= (1 - e^{-25 \Delta s}) \cdot \sum_{i=0}^{n-1} s_i e^{-25 s_i}$$

For small Δs,

$$1 - e^{-25 \Delta s} = 1 - [1 + (-25 \Delta s) + (-25 \Delta s)^2 / 2! + (-25 \Delta s)^3 / 3! + \dots)]$$
$$= 25 \Delta s [1 + (-25 \Delta s) / 2! + (-25 \Delta s)^2 / 3! + \dots] = 25 \Delta s \cdot R(\Delta s),$$

where $R(\Delta s) \to 1$ as $\Delta s \to 0$. Because $\Delta s = L/n \to 0$ as $n \to \infty$, we have

$$\mu_n = 25 R(\Delta s) \sum_{i=0}^{n-1} s_i e^{-25 s_i} \Delta s$$

$$\lim_{n \to \infty} \mu_n = 25 \int_0^L s e^{-25 s} \, ds.$$

Finally, we let $L \to \infty$ to remove the upper constraint on the interarrival time.

$$\mu_T = 25 \int_0^\infty s e^{-25 s} \, ds = \frac{1}{25}$$

So the average interarrival time is $1/25$ second. This result is intuitively plausible. If the server receives 25 requests per second, on the average, it seems reasonable that the average time between requests should be $1/25$ second. $\square$

Definition 2.52 describes the number of events in a base interval in terms of the average number, λ. As we have seen, the expression scales to any multiple t of the base interval $[0, 1]$. The probability of k events in an interval $[0, t]$ is then $(\lambda t)^k e^{-\lambda t} / k!$. These probabilities depend on the combined parameter $m = \lambda t$. The Poisson distributions for various values of $m = \lambda t$ appear in Figure 2.7. Although the distributions are discrete, the figure depicts smoothed curves to emphasize the trend toward a bell-shaped limit as the parameter increases. Note that each curve is symmetric about $k = m$, which is indeed the mean of the distribution $(\lambda t)^k e^{-\lambda t} / k!$.

We conclude this section with a theorem that describes the particularly simple distribution that results from adding independent Poisson random variables.

2.61 Theorem: Let X and Y be independent Poisson random variables with parameters λ_X and λ_Y, respectively. Let $Z = X + Y$. Then Z has a Poisson distribution with parameter $\lambda_X + \lambda_Y$.

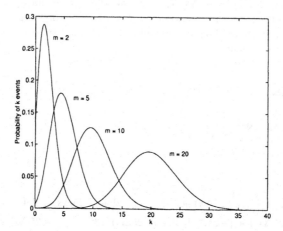

Figure 2.7. Smoothed curves for Poisson distributions where $m = \lambda t$

PROOF:

$$\Pr_Z(k) = \sum_{i=0}^{k} \Pr_X(i)\Pr_Y(k-i) = \sum_{i=0}^{k} \frac{\lambda_X^i e^{-\lambda_X}}{i!} \cdot \frac{\lambda_Y^{k-i} e^{-\lambda_Y}}{(k-i)!}$$

$$= \frac{e^{-(\lambda_X+\lambda_Y)}}{k!} \sum_{i=0}^{k} \frac{k!}{i!(k-i)!} \cdot \lambda_X^i \lambda_Y^{k-i} = \frac{e^{-(\lambda_X+\lambda_Y)}}{k!}(\lambda_X + \lambda_Y)^k$$

This last expression is the Poisson distribution with parameter $\lambda_X + \lambda_Y$. ∎

Figure 2.8 updates the earlier relationship map to include the limiting distributions discussed in this section. The renewal count connection between the exponential and Poisson distributions reflects our discovery that Poisson renewal counts force exponential interarrival times. To complete the relationship, we need to show the converse. That is, the renewal counts of a sequence of independent exponential random variables have Poisson distributions. We defer this proof until Chapter 6.

We opened this section with a proof that $\Pr_{X_i}(k) \to \lambda^k e^{-\lambda}/k!$, where the X_i are binomial random variables with parameters (n_i, p_i) such that $n_i \to \infty$ and $n_i p_i \to \lambda > 0$. This result appears in Figure 2.8 as the arrow from binomial to Poisson. The limit is useful for evaluating a binomial (n, p) probability for large n and small p. In particular, suppose that random variable X is binomially distributed with parameters (n, p). If n is large and p is small, we define $\lambda = np$ and envision X as one of the advanced X_i in the limiting process. The following expression is then the *Poisson approximation* for a binomial distribution.

$$\Pr(X = k) = C_{n,k} p^k (1-p)^{n-k} \approx \frac{\lambda^k e^{-\lambda}}{k!} = \frac{(np)^k e^{-np}}{k!}$$

2.62 Example: In a large population, it is known that 1 in a 100 carries a particular

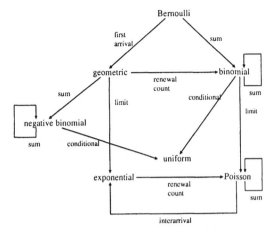

Figure 2.8. Limiting relationships among discrete distributions

virus. In a random sample of 200 persons, what is the chance of encountering at most one infected person?

From the information given, we infer that a randomly chosen person has the infection with probability 0.01. Letting X be the number of infections in the 200 person sample, we conclude that X is the sum of 200 independent Bernoulli random variables, each with parameter $p = 0.01$. Consequently, X is binomially distributed with parameters $(n = 200, p = 0.01)$. The exact solution is

$$\Pr(X \leq 1) = \sum_{k=0}^{1} C_{200,k}(0.01)^k(0.99)^{200-k} = (0.99)^{200} + 200(0.01)(0.99)^{199}$$

$$= 0.4046.$$

Defining $\lambda = np = 2$, the Poisson approximation is

$$\Pr(X \leq 1) \approx \sum_{k=0}^{1} \frac{2^k e^{-2}}{k!} = e^{-2} + 2e^{-2} = 3e^{-2} = 0.4060. \ \square$$

Exercises

2.42 A secretary receives, on the average, 50 phone calls per hour. What is the probability that she will enjoy a 5-minute break without any calls? What is the probability that she will receive more than 10 calls in 5 minutes? What is the probability of quiet span of 2 minutes or more between calls?

2.43 A manufacturing process for magnetic tape introduces flaws on the average of 3 defects per 1000 feet of tape. What is the probability that a 2400-foot reel will contain fewer than 5 defects? What is the probability that the first defect will occur in the first 100 feet of the tape?

2.44 A substance undergoes radioactive decay by emitting particles at random times. Observation shows that the probability of an emission is proportional to the observation time, in the limit of small observation times. Suppose that empirical data show that the substance emits 200 particles per minute, on the average. What is the probability that a 1-minute count will be exactly 200?

2.45 Bacteria develop on a laboratory plate at an average rate of 2 colonies on 10 square centimeters. What is the probability of 7 or fewer colonies on 30 square centimeters?

2.46 Suppose that events come from two sources, one averaging λ_1 events per unit time and the other averaging λ_2 events per unit time. What is the distribution of the number of events in the interval $[0, 1.5]$ from both sources simultaneously?

2.47 Suppose that a central processor receives interrupts under circumstances that justify a Poisson distribution. On the average, the processor receives 150 interrupts in a 10-millisecond period. What is the average time between interrupts?

2.48 A Web server overloads if it receives more than 25 page requests in a 1-second interval. If traffic averages 1000 requests per minute, what is the probability of an overload?

2.49 Consider a Poisson distribution over a small fraction, Δt, of the base interval. Show that, in accordance with the Poisson probability assumptions, the probability of two or more events in the subinterval Δt has the form $\eta_2(\Delta t)$, where $\lim_{\Delta t \to 0} \eta_2(\Delta t)/\Delta t = 0$.

*2.50 Evaluate $\sum_{k=0}^{\infty} k^2/k!$.

2.5   The hypergeometric distribution

A Bernoulli process occurs when sampling, with replacement, a population with two features. Suppose, for example, that an urn contains 8 white balls and 5 black balls. A trial consists of drawing a ball from the urn, noting its color, and returning it to the urn before the next draw. Let $B_i = 1$ when a black ball appears on the ith draw and zero otherwise. Then B_i is a Bernoulli random variable with parameter $p = 5/13$, the B_i are independent, and $B_1, B_2, \ldots$ is a Bernoulli process. From our earlier work, we know many interesting features of this process. For example, we know that its interarrival times are independent geometric random variables, and we know that its renewal counts are binomially distributed. The template of colored balls in an urn obviously generalizes to a variety of situations. The set could contain people, who are either male or female. It could contain components, which are either defective or not. It could contain numbers, which are either even or odd. The important point is that each observation returns to the pool before the next sample is taken. This maintains the independence of successive trials.

In this section we want to study the distribution that occurs when the samples are not replaced. That is, suppose that we have a set of $n_1 + n_2$ objects, with n_1 possessing a defining characteristic (e.g., a special color) and the remaining n_2 not possessing that feature. If an object possesses the defining characteristic, we will call it "special." We repeatedly draw objects from the pool, and we let $X_i = 1$ if the ith object drawn is special. We let $X_i = 0$ otherwise. We do not replace the objects, so the pool shrinks in size with each successive draw. Consequently, this sequence can contain at most $n_1 + n_2$ entries. Each X_i is a Bernoulli random variable, for which we seek the parameter p_i. In contrast with the situation using replacement, the X_i are not independent, and it is not immediately apparent that they share a common parameter. So we first establish that they do indeed share a common parameter.

The random vector $(X_1, \ldots, X_{n_1+n_2})$ assumes binary vector values in which exactly n_1 components are 1. There are $C_{n_1+n_2,n_1}$ such vectors, corresponding to the number of location choices for the n_1 ones from the field of $n_1 + n_2$. Symmetry considerations suggest that all such vectors have the same probability. That is, there is no reason why the selection protocol should encounter the special objects in any particular order. The probability that $X_j = 1$ then derives from the number of $(n_1 + n_2)$-vectors in which the jth entry is a 1. This count is the number of ways of selecting locations for the remaining $n_1 - 1$ ones from the available $n_1 + n_2 - 1$ positions. Consequently,

$$\Pr(X_j = 1) = \frac{C_{n_1+n_2-1,n_1-1}}{C_{n_1+n_2,n_1}} = \frac{(n_1+n_2-1)!}{(n_1-1)!n_2!} \cdot \frac{n_1!n_2!}{(n_1+n_2)!} = \frac{n_1}{n_1+n_2}.$$

In the final analysis, then, X_j is a Bernoulli random variable with parameter $p = n_1/(n_1 + n_2)$, which is the fraction of special objects in the collection. You might find this result surprising. Intuition suggests that the probability of $X_1 = 1$ is $n_1/(n_1 + n_2)$, but you might think that the probability of $X_2 = 1$ would be different because the collection contains one less object when the second selection is taken. It is true that the *conditional* probabilities, $\Pr(X_2 = 1 | X_1 = 0)$ and $\Pr(X_2 = 1 | X_1 = 1)$, do differ from $n_1/(n_1 + n_2)$. But the *unconditional* probability that $X_2 = 1$ remains exactly the same as the probability that $X_1 = 1$. In other words, if you do not know the result of the first draw, the probability of getting a special object on the second draw is the same as on the first. A short example should clarify the situation.

2.63 Example: Suppose that an urn contains 5 black and 3 white balls. The black balls are special. We draw two balls, without replacement, and let $X_i = 1$ if the ith draw is black. Table 2.3 tabulates the four possibilities. Simple calculations establish the table's entries. In particular, $\Pr(X_1 = 0, X_2 = 0)$ is $C_{5,0}C_{3,2}/C_{8,2} = 6/56$, while $\Pr(X_1 = X_2 = 1)$ is $C_{5,2}C_{3,0}/C_{8,2} = 20/56$. The 30/56 probability for a mixed outcome splits evenly between the two possibilities. That is,

$$\Pr(X_1 = 0, X_2 = 1 \text{ or } X_1 = 1, X_2 = 0) = \frac{C_{5,1}C_{3,1}}{C_{8,2}} = 30/56.$$

X_1	X_2 0	1	
0	6/56	15/56	3/8
1	15/56	20/56	5/8
	3/8	5/8	8/8

X_2	0	1
$\Pr(X_2\|X_1 = 0)$	2/7	5/7
$\Pr(X_2\|X_1 = 1)$	3/7	4/7

TABLE 2.3. Joint and marginal probabilities for a two-ball selection from a field of black and white balls

In the leftmost tabulation, the sums on the right margin and bottom row give the marginal (unconditional) probabilities of X_1 and X_2, respectively. As you can see, they are exactly the same. Moreover, $\Pr(X_i = 1)$ is 5/8 in both cases, which is the fraction of black balls in the collection. Note also that the joint probability in a given cell is *not* equal to the product of the unconditional entries from the right and bottom margins. Therefore, X_1 and X_2 are not independent. We can nevertheless use Theorem 2.8 to determine the mean and variance of the sum of such Bernoulli random variables. However, we cannot neglect the covariance terms. We will pursue this calculation after a formal definition of the hypergeometric random variable. The conditional probabilities of $X_2 = 1$ are different, however, being $5/7 > 5/8$ if $X_1 = 0$ and $4/7 < 5/8$ if $X_1 = 1$. That is, the chances of a black ball on the second draw increase if the first ball is white and decrease if it is black. $\square$

Returning to the general situation, where X_i is 1 when the ith draw encounters one of the n_1 special objects in the field of $n_1 + n_2$, we have already noted that the number of ones in the observation vector $(x_1, x_2, \ldots, x_{n_1+n_2})$ must be n_1. This is because selecting all the objects necessarily selects all the special objects. However, if we shorten the sequence, we introduce nondeterministic outcomes. Accordingly, let $n \leq n_1 + n_2$, and consider $Y = \sum_{i=1}^n X_i$. The probability of exactly k special objects among the first n choices is $C_{n_1,k}C_{n_2,n-k}/C_{n_1+n_2,n}$. Of course, we can obtain k special objects only when $0 \leq k \leq n_1$ and $n - k \leq n_2$, but the probability expression remains valid for larger k because $C_{n_1,k} = 0$ for $k > n_1$ and $C_{n_2,n-k} = 0$ for $k < n - n_2$. This situation produces the hypergeometric random variable.

2.64 Definition: For a fixed parameter set (n_1, n_2), and $n \leq n_1+n_2$, let Y assume the values $0, 1, \ldots, n$ with probabilities $\Pr(k) = C_{n_1,k}C_{n_2,n-k}/C_{n_1+n_2,n}$. Then Y is a *hypergeometric* random variable with parameters (n_1, n_2, n). ∎

If Y is hypergeometric with parameters (n_1, n_2, n), we have

$$\sum_{k=0}^n \Pr_Y(k) = \sum_{k=0}^n \frac{C_{n_1,k}C_{n_2,n-k}}{C_{n_1+n_2,n}} = \sum_{k=0}^n \frac{C_{n_1,k}C_{(n_1+n_2)-n_1,n-k}}{C_{n_1+n_2,n}},$$

which Theorem 1.35 reduces to

$$\sum_{k=0}^n \Pr_Y(k) = \frac{C_{n_1+n_2,n}}{C_{n_1+n_2,n}} = 1.$$

This establishes the validity of the probability assignment. As noted above, $\Pr_Y(k)$ is nonzero only for $n - n_2 \leq k \leq \min(n, n_1)$. Nevertheless, the formula of Definition 2.64 is valid for any nonnegative k, even though the range of nonzero probabilities is always a finite set.

With parameters (n_1, n_2, n), the hypergeometric random variable Y corresponds to the sum of the n Bernoulli random variables that arise in choosing n objects, without replacement, from a collection of $n_1 + n_2$ objects, of which n_1 are special. In particular, if X_i, for $1 \leq i \leq n$, is 1 when the ith choice is special and zero when it is not, then $Y = \sum_{i=1}^{n} X_i$ is hypergeometric.

2.65 Theorem: Let Y have the hypergeometric distribution with parameters n_1, n_2, and $n \leq n_1 + n_2$. Let $p = n_1/(n_1 + n_2)$. Then

$$\mu_Y = np$$

$$\sigma_Y^2 = np(1-p)\left(\frac{n_1 + n_2 - n}{n_1 + n_2 - 1}\right).$$

PROOF: We will use the interpretation of the hypergeometric random variable as the sum of Bernoulli random variables associated with selections from a two-category population. Accordingly, let S be a collection of n_1 special objects and n_2 nonspecial objects. Draw n objects at random from S. Define $X_i = 1$ if the ith draw is special; let $X_i = 0$ otherwise. Then $Y = \sum_{i=1}^{n} X_i$. Each X_i is a Bernoulli random variable with parameter p. Therefore, if μ_i, σ_i^2 denote the mean and variance of X_i, we know that $\mu_i = p$ and $\sigma_i^2 = p(1-p)$.

From Theorem 2.8, we conclude immediately that $\mu_Y = \sum_{i=1}^{n} \mu_i = np$. To compute the variance, however, we must first calculate the covariance, σ_{ij}, of X_i and X_j. For a binary n-vector $t = (t_1, t_2, \ldots, t_n)$, let q_t be the probability that $X_k = t_k$ for $1 \leq k \leq n$. Then $E[X_i X_j] = \sum_t t_i t_j q_t$, where the sum extends over all n-vectors t. As noted earlier, $q_t = 0$ when t contains more than n_1 or fewer than $n - n_2$ ones. The group of vectors containing exactly k ones constitute the event $(Y = k)$ and therefore have total probability

$$\Pr(Y = k) = \frac{C_{n_1,k} C_{n_2, n-k}}{C_{n_1+n_2, n}}.$$

There are $C_{n,k}$ vectors in the group, and symmetry considerations again suggest that they are equally likely. That is, there is no reason why one particular choice of k locations for the ones should have a higher probability than the other choices. Therefore,

$$q_t = \frac{C_{n_1,k} C_{n_2, n-k}}{C_{n_1+n_2, n} C_{n,k}},$$

for each binary n-vector t with exactly k ones.

When either $t_i = 0$ or $t_j = 0$, the summand in $E[X_i X_j] = \sum_t t_i t_j q_t$ is zero. So we need consider only those n-vectors for which $t_i = t_j = 1$. These vectors always have at least these two 1s, and they may have as many as n ones. (Again, if $n > n_1$, vectors with more than n_1 ones will have zero probability.) We sum by group, starting with the vectors having 2 ones, continuing with those having 3 ones, and so forth. A group with k ones contains $C_{n-2,k-2}$ vectors, corresponding to the choices for positioning the $k-2$ ones among the $n-2$ slots not equal to i or j. Hence,

$$E[X_iX_j] = \sum_{k=2}^{n} C_{n-2,k-2} \cdot \frac{C_{n_1,k}C_{n_2,n-k}}{C_{n_1+n_2,n}C_{n,k}}.$$

A bit of algebra provides a reasonably compact result, from which we obtain σ_{ij} and eventually σ_Y^2. Note that a summation over all $i < j$ includes a total of $n(n-1)/2$ terms, and $1 - p = 1 - n_1/(n_1 + n_2) = n_2/(n_1 + n_2)$.

$$E[X_iX_j] = \sum_{k=2}^{n} \frac{(n-2)!}{(k-2)!(n-k)!} \cdot \frac{n_1!}{k!(n_1-k)!} \cdot \frac{n_2!}{(n-k)!(n_2-n+k)!}$$

$$\cdot \frac{n!(n_1+n_2-n)!}{(n_1+n_2)!} \cdot \frac{k!(n-k)!}{n!}$$

$$= \frac{n_1(n_1-1)(n-2)!(n_1+n_2-n)!}{(n_1+n_2)!}$$

$$\cdot \sum_{k=0}^{n-2} \frac{(n_1-2)!}{k!((n_1-2)-k)!} \cdot \frac{n_2!}{((n-2)-k)!(n_2-(n-2)+k)!}$$

$$= \frac{n_1(n_1-1)(n-2)!(n_1+n_2-n)!}{(n_1+n_2)!} \sum_{k=0}^{n-2} C_{n_1-2,k}C_{n_2,(n-2)-k}$$

$$= \frac{n_1(n_1-1)(n-2)!(n_1+n_2-n)!}{(n_1+n_2)!} C_{n_1+n_2-2,n-2}$$

$$= \frac{n_1(n_1-1)(n-2)!(n_1+n_2-n)!}{(n_1+n_2)!} \cdot \frac{(n_1+n_2-2)!}{(n-2)!(n_1+n_2-n)!}$$

$$= \frac{n_1(n_1-1)}{(n_1+n_2)(n_1+n_2-1)}$$

$$\sigma_{ij} = E[X_iX_j] - \mu_i\mu_j = \frac{n_1(n_1-1)}{(n_1+n_2)(n_1+n_2-1)} - \left(\frac{n_1}{n_1+n_2}\right)^2$$

$$= \frac{-n_1n_2}{(n_1+n_2)^2(n_1+n_2-1)} = \frac{-p(1-p)}{n_1+n_2-1}$$

$$\sigma_Y^2 = \sum_{i=1}^{n}\sigma_i^2 + 2\sum_{i<j}\sigma_{ij} = np(1-p) + 2 \cdot \frac{n(n-1)}{2} \cdot \frac{-p(1-p)}{n_1+n_2-1}$$

$$= np(1-p)\left[1 - \frac{n-1}{n_1+n_2-1}\right] = np(1-p) \cdot \frac{n_1+n_2-n}{n_1+n_2-1} \quad\blacksquare$$

2.66 Example: A carnival worker, the game master, invites you to draw three balls, without replacement, from a hat that contains 4 black and 10 white balls. He will pay you \$10 for each black ball. What is a fair charge to play this game? If the game master wants to earn \$400 in a 5-hour shift, while playing 40 games per hour, what should he charge per game?

The number of black balls, N, is a random variable. A fair charge for the game is $10 \cdot E[N]$ because, on the average, that is the amount the carnival worker can expect to pay out per game. The fair charge allows him to break even. N has a hypergeometric distribution with parameters $n_1 = 4, n_2 = 10, n = 3$. Let $p = n_1/(n_1 + n_2) = 4/14$. We calculate $E[N] = np = 3 \cdot 4/14 = 6/7$, and the fair charge is then $\$60/7 = \$8.57$. Playing $40 \cdot 5 = 200$ games in his five-hour shift, the game master can expect to pay out $200(8.57) = 1714.29$. To make $\$400$, he must charge a total of $\$2114.29$ for the 200 games, which amounts to $\$10.57$ per game.

Assuming the game master does charge $\$10.57$ per game, what is the probability that he will lose money over the course of 200 games?

Let R_i be the game master's gain on the ith game. The probabilities of this random variable appear to the right. $T = \sum_{i=1}^{200} R_i$ is the total gain over 200 games. We want the probability that $T < 0$.

x	$\Pr(R_i = x)$	blacks
10.57	$C_{4,0}C_{10,3}/C_{14,3} = 0.3297$	0
0.57	$C_{4,1}C_{10,2}/C_{14,3} = 0.4945$	1
−9.43	$C_{4,2}C_{10,1}/C_{14,3} = 0.1648$	2
−19.43	$C_{4,3}C_{10,0}/C_{14,3} = 0.0110$	3

It is reasonable to assume that the R_i are independent because one game should have no effect on the others. (All balls are returned to the hat to begin a new game.) Each $R_i = 10.57 - 10N$, so $\mu_{R_i} = 10.57 - 10np = 2.00$ and

$$\sigma_{R_i}^2 = 100np(1 - p)(n_1 + n_2 - n)/(n_1 + n_2 - 1) = 51.81.$$

Applying Theorem 2.8, we calculate $\mu_T = 200(2.00) = 400$, as expected, and $\sigma_T^2 = 200(51.81) = 10362$. An outcome $T < 0$ then represents a divergence of $400/\sqrt{10362} = 3.93$ standard deviations from the mean. Chebyshev's inequality tells us that such an event has probability less than $1/(3.93)^2 = 0.0648$. Actually, Chebyshev's inequality says that $\Pr_T(|T - \mu_T|/\sigma_T > k) < 1/k^2$, so the figure 0.0648 includes both the probability that $T < 0$ and the probability that $T > 800$. A more reasonable bound on the probability that $T < 0$ is then $0.0648/2 = 0.0324$. The Chebyshev inequality is normally very conservative. That is, it gives an upper bound that may be generously high. Better estimation techniques, which appear in a later chapter, give $\Pr_T(T < 0) \approx 1.89 \cdot 10^{-15}$. The game master is very unlikely to have a bad day. $\Box$

If we interpret the hypergeometric random variable as the sum of n dependent Bernoulli random variables, we see that its mean is the same as a sum of n independent Bernoulli random variables with the same parameter. Its variance, in general, is somewhat less. If $n = 1$, the variance is $p(1 - p)$, which is the variance for one trial. This is a reasonable result because the replacement policy is not relevant in this case. As a further check, if $n = n_1 + n_2$, then the entire population constitutes the sample, and the number of special objects must always be n_1. That is, if $n = n_1 + n_2$, there is no random variation, and the variance should be zero. The variance formula of Theorem 2.65 does indeed return zero in this case.

Suppose that we keep n constant, but let $n_1, n_2 \to \infty$ in such a manner that $n_1/(n_1 + n_2)$ remains constant at p. This corresponds to selecting, without replacement, n objects from an ever-increasing population, in which the fraction of special objects remains constant. It seems reasonable, in the limit, that the replacement policy should not matter. If we replace an object in a large population, it is unlikely to be chosen again. Indeed, the variance, $np(1 - p)(n_1 + n_2 - n)/(n_1 + n_2 - 1)$, approaches $np(1 - p)$, the value associated with a policy of replacing each object prior to the next draw. Moreover, for any fixed k, the probability of observing k special objects approaches the binomial value $C_{n,k} p^k (1 - p)^{n-k}$. We can verify this fact as follows.

$$\frac{C_{n_1,k} C_{n_2,n-k}}{C_{n_1+n_2,n}} = \frac{n_1!}{k!(n_1 - k)!} \cdot \frac{n_2!}{(n - k)!(n_2 - n + k)!} \cdot \frac{n!(n_1 + n_2 - n)!}{(n_1 + n_2)!}$$

We write this product as $A_1 A_2 C_{n,k}$, where

$$A_1 = \frac{n_1(n_1 - 1) \cdots (n_1 - k + 1)}{(n_1 + n_2)(n_1 + n_2 - 1) \cdots (n_1 + n_2 - k + 1)}$$

$$A_2 = \frac{n_2(n_2 - 1) \cdots (n_2 - n + k + 1)}{(n_1 + n_2 - k)(n_1 + n_2 - k - 1) \cdots (n_1 + n_2 - n + 1)}.$$

A_1 is a product of k factors, each of the form $(n_1 - i)/(n_1 + n_2 - i)$ for $i = 0, 1, \ldots, k - 1$. Consequently, each factor approaches p as $n_1, n_2 \to \infty$ such that $n_1/(n_1 + n_2)$ remains p. Similarly, A_2 contains $n - k$ factors, each approaching $n_2/(n_1 + n_2) = 1 - p$. Therefore, $C_{n_1,k} C_{n_2,n-k}/C_{n_1+n_2,n} \to C_{n,k} p^k (1 - p)^{n-k}$. So, as the population grows larger, provided that the fraction of special objects remains constant, the hypergeometric distribution approaches the binomial distribution. In other words, the fact that the observations are not replaced in the population becomes increasingly irrelevant. We record this fact as our last theorem.

2.67 Theorem: Let Y_{n_1,n_2} be a sequence of hypergeometric random variables with parameters (n_1, n_2, n). Define $p = n_1/(n_1 + n_2)$ and let $n_1, n_2 \to \infty$ in such a manner that the ratio $n_1/(n_1 + n_2)$ remains p. Then, for $0 \le k \le n$,

$$\Pr(Y_{n_1,n_2} = k) \to C_{n,k} p^k (1 - p)^{n-k}.$$

That is, the limiting distribution is binomial with parameters (n, p).

PROOF: See discussion above. ∎

Exercises

2.51 The rules of a card game allow the player to receive five cards, without replacement, from a standard deck. The player receives one dollar for each ace or face card. To break even on the average, how much should the dealer charge per game? What charge would allow the dealer to turn an average profit of $100 over the course of 200 games? Assuming that the dealer charges this latter amount, find an upper bound on the probability that the dealer will lose money over the course of 200 games.

2.52 An urn contains 4 black, 6 red, and 8 white balls. You draw balls without replacement. The random variable X_i is 2 if the ith ball is black; it is zero otherwise. Let $Y = \sum_{i=1}^{5} X_i$. What is the probability distribution of Y? What are its mean and variance?

2.53 An unordered list of 1000 persons contains 20% teenagers. The teenagers are equally likely to occupy any given subset of 200 list positions. The algorithm to the right accepts the list and returns the locations of the first two teenagers that it finds. The function teen tests a list item; it returns true if the item is a teenager. The vector

```
function findTeen(L, N)
  found = 0; i = 0;
  location = (0, 0);
  while ((found < 2) and (i < N))
    if (teen (L[i]))
      location[found] = i;
      found = found + 1;
    i = i + 1;
  return location;
```

L contains entries in locations L[0] through L[N - 1]. The returns are in location [0] and location [1]. What is the average number of iterations of the while-loop?

*2.54 When reading a hard disk file, we access the file one sector at a time. Each access is called a probe and incurs a time penalty of 10 milliseconds. A hash file resides on the disk, and the corresponding hash function converts a record key into a sector address. The hash file contains 1000 sectors, which provide 1000 home addresses. Each sector can hold 10 records, and an attempt to store more records creates an overflow chain that spills onto other sectors that are outside the hash file's home sectors. Performance deteriorates as overflow chains accumulate because of the time penalty associated with reading sectors beyond a record's home address. Assume that overflow chains show no preference for attaching to particular home sectors. If there are 100 overflowing home addresses, for example, they are equally likely to occur at any subset of 100 addresses.

To test the severity of the overflow problem, we have written a routine that chooses a home address at random and probes 100 sectors, starting at the chosen point. If the process encounters the last sector, it continues from the beginning sector. Suppose that there are actually 150 overflowing addresses. What is the probability that the test routine will report 10 addresses with overflow chains? What is the probability that five executions will report the counts $10, 12, 9, 8, 11$?

2.55 An urn contains 10 objects, of which 4 are special. We draw objects without replacement and observe the random variable X, which is the number of objects before the first special object. What is the probability distribution of X? What are its mean and variance?

*2.56 Generalize the preceding problem to a collection of $n_1 + n_2$ objects, of which n_1 are special. X is the number of objects drawn, without replacement, before the first special object. What is the probability distribution of X? What are its mean and variance? Acceptable expressions for the mean and variance can involve summations.

*2.57 Consider again the urn with 10 objects, of which 4 are special. Drawing objects without replacement, let X be the number of objects between

the first and second special objects. What is the probability distribution of X? What are its mean and variance?

2.58 Investigate the case $n_1 = n_2 = n = 1$ to show that the sum of two independent hypergeometric random variables, with the same parameters, is not a hypergeometric random variable with doubled parameters.

2.59 Let random variable A be the number of aces in a 13-card bridge hand. What are the expected value and variance of A?

2.60 Let random variable P be 4 times the number of aces plus 3 times the number of kings plus 2 times the number of queens plus the number of jacks in a 13-card bridge hand. What are the expected value and variance of P?

2.61 Suppose, in a population of 100 voters, that 46% favor a given proposition. Among 10 voters, chosen at random without replacement, what is the probability that 6 or more favor the proposition?

2.62 An urn contains 4 black balls and 6 white balls. Let random variable B be the number of black balls drawn, without replacement, before a white ball appears. What are the expected value and variance of B?

2.63 In the context of the preceding exercise, let W be the number of white balls drawn, without replacement, before a black ball appears. What are the expected value and variance of W? Intuitively, should the expected value of W be larger or smaller than the expected value of B?

2.6   Summary

This chapter's discrete distributions all stem from a Bernoulli process, in which we repeatedly draw an object at random from a two-category population. We designate the two categories as special and nonspecial. The Bernoulli random variable assumes just two values, one or zero, and corresponds to drawing a special or nonspecial object, respectively, from the population. The fraction p of special objects in the population is a parameter of the Bernoulli random variable. If we draw a total of n objects from the population, the number of special objects drawn constitutes another random variable, which is the sum of the Bernoulli random variables associated with the individual draws. This sum follows a binomial distribution if we replace each object before drawing the next. It follows a hypergeometric distribution if we do not replace the drawn objects. In the former case, the Bernoulli summands are independent; in the latter case they are not. The binomial distribution then requires two parameters, (n, p), while the hypergeometric requires three (n_1, n_2, n), where n_1 and n_2 are the number of special and nonspecial objects, respectively, in the population. In the limit of a large population, the replacement policy does not matter, so the hypergeometric distribution approaches

Distribution (parameters)	Probabilities	Mean and variance	Moment-generating function
Bernoulli (p)	$p^x(1-p)^{1-x}$ $x = 0, 1$	$\mu = p$ $\sigma^2 = p(1-p)$	$\Psi(t) = (1-p) + pe^t$
binomial (n, p)	$C_{n,k}\, p^k(1-p)^{n-k}$ $k = 0, 1, \ldots, n$	$\mu = np$ $\sigma^2 = np(1-p)$	$\Psi(t) = [(1-p) + pe^t]^n$
hypergeometric (n_1, n_2, n) $[p = n_1/(n_1 + n_2)]$	$\dfrac{C_{n_1,k} C_{n_2, n-k}}{C_{n_1+n_2, n}}$ $k = 0, 1, \ldots, n$	$\mu = np$ $\sigma^2 = np(1-p)\cdot$ $\left(\dfrac{n_1 + n_2 - n}{n_1 + n_2 - 1}\right)$	
geometric (p)	$(1-p)^k p$ $k = 0, 1, 2, \ldots$	$\mu = (1-p)/p$ $\sigma^2 = (1-p)/p^2$	$\Psi(t) = p[1 - (1-p)e^t]^{-1}$
negative binomial (n, p)	$C_{n+k-1, k}\, p^n (1-p)^k$ $k = 0, 1, 2, \ldots$	$\mu = n(1-p)/p$ $\sigma^2 = n(1-p)/p^2$	$\Psi(t) = p^n[1 - (1-p)e^t]^{-n}$
Poisson (λ)	$\dfrac{\lambda^k}{k!} \cdot e^{-\lambda}$ $k = 0, 1, 2, \ldots$	$\mu = \lambda$ $\sigma^2 = \lambda$	$\Psi(t) = e^{-\lambda(1-e^t)}$

TABLE 2.4. Features of discrete distributions

the corresponding binomial distribution. The characteristics for Bernoulli, binomial, and hypergeometric distributions appear in Table 2.4.

Two further random variables arise from a Bernoulli process with parameter p. The geometric random variable, with the same parameter p, is the number of trials, with replacement, before the first special object appears. This is frequently termed the number of failures before the first success. In terms of the independent Bernoulli random variables associated with successive trials, it is the number of zeros prior to the first 1. The geometric distribution has the distinction of being memoryless. That is, if X has a geometric distribution, then $\Pr(X = i + j | X > i) = \Pr(X = j)$ for any $i, j \geq 0$. After we observe i failures, we find the probability of j additional failures before the first success is the same as the probability of j failures at the onset of the experiment. The geometric distribution is the only discrete distribution with this property. It follows that the number of failures *between* successes in a Bernoulli process also have geometric distributions. That is, if $G_1, G_2, \ldots$ count the number of failures between successes $i - 1$ and i, then the G_i are independent geometric random variables with the common parameter p.

For a sequence $X_1, X_2, \ldots$ of random variables, we define the renewal counts N_t as $N_t = \max\{n | \sum_{i=1}^n X_i \leq t\}$. Utilizing the fact that the events $(N_t \leq n)$ and $(\sum_{i=1}^{n+1} X_i > t)$ comprise the same outcomes, we showed that the renewal count for a sequence of geometric random variables has a binomial distribution.

Although the sum of independent binomial random variables with parameters $(n_1, p), (n_2, p), \ldots, (n_k, p)$ is simply another binomial random variable with parameters $(n_1 + n_2 + \ldots + n_k, p)$, this property does not hold for geometric random variables. The sum of n geometric random variables with common parameter p produces a random variable with a new distribution: the negative binomial distribution with parameters (n, p). In terms of the Bernoulli process, the negative binomial distribution describes the number of failures prior to the nth success. If X_i, for $1 \leq i \leq k$, have negative binomial distributions with parameters $(n_1, p), (n_2, p), \ldots, (n_k, p)$, then their sum is another negative binomial random variable with parameters $(n_1 + n_2 + \ldots + n_k, p)$. This is intuitively plausible because the sum represents the number of failures before success number $n_1 + n_2 + \ldots + n_k$ in the same Bernoulli process. The characteristics of the geometric and negative binomial distributions also appear in Table 2.4.

A sequence of binomial distributions B_i with parameters (n_i, p_i) has a limiting form as $n_i \to \infty, p_i \to 0$ in such a manner that $n_i p_i \to \lambda$, a positive constant. In particular, $\Pr(B_i = k)$ approaches $\lambda^k e^{-\lambda}/k!$, which is the Poisson distribution with parameter λ.

We can interpret this limit in a useful way as follows. If the average number of events in some base interval (of time or distance or area) is λ, and if these events originate independently with small probabilities in narrow slices of the interval, we can approximate the number of events in the interval as the sum of a large number of independent Bernoulli random variables. Each such random variable corresponds to one narrow slice and produces 1 if an event occurs in the slice and zero otherwise. If we have n_i slices with probability p_i of an event in each slice, then the number of events has a binomial distribution with parameters (n_i, p_i). The average of this distribution is $n_i p_i$, which must approximate the known value λ.

As the number of slices increases toward infinity, the slice widths decrease toward zero. If we assign the probabilities p_i values that are proportional to the slice widths, they also will decrease toward zero, even though the product $n_i p_i$ remains near λ. This is precisely the limiting situation that produces the Poisson distribution.

The same distribution arises from the Poisson assumptions: (1) event counts in disjoint intervals are independent random variables, (2) the probability of one event in a small interval is λ times the interval width plus other terms that involve higher powers of the interval width, and (3) the probability of more than one event in a small interval is negligible in comparison with the probability of one event.

The Poisson distribution scales to any multiple, greater than or less than 1, of the base interval for which λ is the average. So long as the underlying random process is unchanged, the number of events in a multiple t of the base interval has the distribution $\Pr(k \text{ events in } [0, t]) = (\lambda t)^k e^{-\lambda t}/k!$. This distribution depends only on the product λt.

The Poisson distribution applies to many situations in which events arrive independently from a multitude of potential sources, each with a small probability of contributing in any given time slice. The number of customers arriving at a teller's window, for example, certainly increases with the time that the window is open. In the limit, for small time intervals, it is reasonable to assume that the probability of an arrival in the interval is proportional to the interval width. We can then approximate the number of arrivals over a larger time span by a binomial distribution with a large n and a small p.

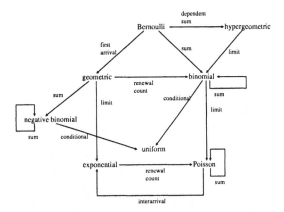

Figure 2.9. Complete map of relationships among discrete distributions

In the limit, provided that the np product approaches the overall average λ, this distribution becomes the Poisson distribution. The characteristics of the Poisson distribution also appear in Table 2.4.

Just as the number of Bernoulli trials between events have independent geometric distribution, so the time between events in a Poisson process have independent descriptions. This description is not a discrete distribution. Rather, if T is the interarrival time between events (or before the first event), we have $\Pr(T > t) = e^{-\lambda t}$. Figure 2.9 shows the relationships among all the discrete distributions discussed in this chapter. The exponential entry is not a discrete distribution, but the diagram includes it to demonstrate the analogy between the number of trials before the first success (Bernoulli process) and the time before the first event (Poisson process).

For a random variable X, the values $E[X^i]$, for $i = 1, 2, \ldots$, are the moments of X. Moments are useful in mean and variance calculations: $\mu_X = E[X], \sigma_X^2 = E[X^2] - \mu_X^2$. Higher moments provide further information about the shape of the probability distribution. The moment generating function $\Psi_X(t) = E_X[e^{tX}]$ involves all the moments of X. Indeed, the linearity of the expectation operator and the series expansion of the exponential give $\Psi(t) = 1 + E[X]t + E[X^2]t^2/2! + E[X^3]t^3/3! + \ldots$. Since $\Psi(t)$ is a power series, we can disentangle the components corresponding to the different moments.

To support calculations with the moment generating function, this chapter's mathematical interlude develops the elementary properties of power series. These properties include absolute convergence, uniform convergence inside a closed subinterval, and term-by-term differentiation and integration. These features ensure that convergence surprises seldom interfere with arguments based on power series.

In particular, we apply power series arguments to the moment generating function. We conclude that $E[X^i] = \Psi^{(i)}(0)$, where $\Psi^{(i)}$ denotes the ith derivative of Ψ. This observation is especially useful for obtaining the moments of new distributions that arise as the sum of independent components. In this case, the moment generating function of the new distribution is the product of the component moment generating functions. The binomial random variable, for example, is the sum of n independent Bernoulli random variables with a common parameter p. Hence, the binomial moment generating function is the nth power of the Bernoulli moment generating function. These relationships among the moment generating functions are apparent in the last column of Table 2.4.

Finally, just as we obtain the moment generating function from a calculation on the probability distribution, we can reverse the process and obtain the probability distribution from a calculation on the moment generating function. We proved this for the special case where the nonzero probabilities occur on some subset of nonnegative integers. Chapter 6 will develop a more general inversion formula. The importance of the inversion is not so much the inversion process itself, but the fact that the moment generating function determines the location and size of the probability jumps. Consequently, when confronted with a moment generating function of an unknown distribution, we can then guess a distribution that produces that function. The guess must then be the correct probability distribution. We employed this technique to obtain the negative binomial distribution.

Historical Notes

James Bernoulli developed the properties of the distribution that bears his name, as well as the binomial distribution. James was the eldest of three generations of Bernoullis who contributed significantly to the mathematical and scientific discoveries of their age. James appeared in 1654 in a family with a long history as successful merchants. James' father envisioned the prosperous merchant life for his sons, but each drifted into mathematics as though it were a genetic calling. At the conclusion of the preceding chapter, we noted James Bernoulli's role in the development of combinatorial formulas, and in a subsequent chapter we will investigate certain laws of large numbers, for which he furnished the first rigorous proofs. In addition to the Bernoulli distribution, there are the Bernoulli differential equation and the lemniscate of Bernoulli. The first is a particular form of nonlinear equation that Bernoulli solved through a felicitous substitution. It is an extension of Newton's calculus. The lemniscate is the polar curve $r^2 = a\cos 2\theta$. Another curve, the logarithmic spiral, so captivated his attention that it appears in his epitaph.

His brothers, John and Nicolaus, were also highly competent mathematicians. John had a curious financial arrangement with the Marquis de l'Hôpital whereby the marquis obtained all rights to John Bernoulli's mathematical work. The famous l'Hôpital's rule was actually discovered by John Bernoulli and transmitted to l'Hôpital under this arrangement. John's son Daniel continued the family tradition, particularly in applied mathematics, where he made fundamental contributions to fluid mechanics. The Bernoulli effect, in which a low pressure results from constricting a gaseous flow, bears Daniel's name.

Simeon Denis Poisson was a prolific researcher who apparently felt life's purpose was to create and teach mathematics. The Poisson distribution appears in his *Recherches sur la probabilité des jugements en matière criminelle et en matière civile,* published in 1837. As the name implies, this work applies probabilistic argument to court decisions. The application of probability in such social context was very controversial at the time, and his text garnered Poisson a great deal of criticism. History, however, has vindicated his efforts, and probabilistic concepts now form a crucial part of the economic, political, and social sciences. Poisson's name also attaches to an equation in electrostatics that relates electric potential with spatial charge distribution and to certain physical constants in the theories of electricity and elasticity. In one of his less glorious moments, Poisson rejected as incomprehensible a paper by the young Évariste Galois. That paper contained much of what was later known as Galois theory, which settled longstanding questions about solving polynomial equations. Bell[5] and Boyer[8] provide more information on both Bernoulli and Poisson.

Further Reading

The properties of power series are among the first concepts encountered in the analysis of real functions. Indeed, the development in this chapter is a continuation of the real line structure discussed in Appendix A. Consequently, the sources cited there, Rudin[75] and [76], remain the recommended avenues for further study. The moment generating function is just one of a variety of generating functions that facilitate computations with probability distributions. The characteristic function, for example, is a seemingly small variation, $\phi_X(t) = E_X[e^{itX}] = \sum_k \Pr_X(k)e^{itk}$, where $i = \sqrt{-1}$. Whereas the moment generating function, $\Psi_X(t) = E_X[e^{tX}] = \sum_k \Pr_X(k)e^{tk}$, may diverge for certain distributions, the characteristic function does not. This stability follows because $|e^{itk}| \leq 1$. Another example is the factorial moment generating function, $\eta_X(t) = E_X[t^X] = \sum_k \Pr_X(k) \cdot t^k$. An analysis similar to that undertaken for $\Psi_X(t)$ reveals that $E[X(X-1)\cdots(X-k+1)] = \eta_X^{(k)}(1)$. These latter expressions are called the factorial moments of X, and they are also useful in deriving summary characteristics, such as the mean and variance. Finally, when range(X) is a subset of the nonnegative integers, the expression $\zeta(z) = \sum_{k=0}^{\infty} \Pr_X(k)z^k$ is called the z-transform of X. It is useful for certain combinatorial problems. Zehna[94] and Lindgren[51] provide introductory treatments of these functions; Riordan[69] and Nelson[56] are appropriate texts for further study. Graham et al.[27] bring a novel perspective to the subject of generating functions in their excellent text, *Concrete Mathematics,* so named because it treats a blend of *con*tinuous and dis*crete* mathematics.

This chapter just opens the door on a deep subject, stochastic processes, which are sequences of random variables exhibiting various influences on one another. Our use of this concept was limited to Bernoulli processes, in which the random variables are independent, and Poisson processes, for which the renewal counts are Poisson random variables. Karlin and Taylor[42] provide a more systematic introduction to the topic.

Bell's classic anthology[5] provided some material for the historical reflections above and in subsequent chapters. For many of the mathematicians mentioned in this text, Bell portrays interesting lives, replete with anecdotes in which seemingly accidental circumstances lead to important discoveries. Some more recent mathematical historians (e.g., Yandell[93]) suggest that Bell may give undue emphasis to these circumstances. Nevertheless, Bell's book has proved an inspiration for several mathematical generations, and it remains an entertaining and informative historical source—both for the details about individual mathematicians and for a feeling about the times when they worked.

Chapter 3

Simulation

The preceding two chapters introduced the basic concepts of probability theory and developed several discrete distributions. This chapter and the next take up two important applications, simulations and decision theory, that demonstrate the practical utility of the theory. We start with simulations, which are widely used throughout science and engineering. In these fields, problems arise that are very difficult, or impossible, to solve analytically. The operation count of a complex algorithm or the queue lengths associated with producer-consumer processes are examples of these intractable problems. In many cases, however, we can quickly obtain rather accurate estimates of certain unknown parameters by simulating the problem. That is, we build a mathematical model of the problem and simulate its operation in the context of specified probability distributions on the inputs.

For example, consider three bank tellers: a slow one, who takes an average of 10 minutes to serve a customer; a fast one, who finishes with a customer in 4 minutes, on the average; and one of intermediate speed, who requires 8 minutes. Suppose that the variation is a uniform plus or minus 2 minutes in all three cases. For purposes of this simple example, we quantize time in 1-minute increments. That is, all services require an integral number of minutes. The fast teller, for instance, complete a service in 2, 3, 4, 5, or 6 minutes, and all these times are equally probable. Customers arrive at an average rate of 25 per hour. What is the average waiting time that a customer can expect? Waiting time varies due to the probabilistic nature of the arrival and disposition processes. The waiting time, T, for an individual customer is a random variable, but with an unknown distribution. Indeed, the distribution may not even be constant. The first customer simply engages the fastest teller and depart in 2 to 6 minutes—4 minutes on the average. A subsequent arrival faces a more complex environment. He may find all tellers busy, in which case his waiting time extends to include some time in a service queue. Furthermore, he may find himself at a faster or slower teller, depending on which service window opens when he arrives at the front of the queue.

For ease of analysis, we frequently assume that the distribution does

179

stabilize to a constant probability profile some short time after the system commences operation. In this chapter's mathematical interlude, we prove that such stability ensues under rather general conditions. Even with this simplification, however, we still do not know the steady-state distribution in the general case, so we cannot analytically compute μ_T, the expected waiting time. We can, however, configure a computer program that tracks the progress of customers through the system. By averaging the waiting times of the virtual customers in the simulation, we obtain an estimate of μ_T. The program simply generates arrivals from a Poisson process with parameter $\lambda = 25$ and disposes of them via one of the three tellers. The accuracy of such simulations is a topic for a later chapter. Here we are interested in the simulation details.

Accordingly, this chapter's two main goals are to develop algorithms that generate random numbers from prescribed distributions and to construct a general simulation template that adapts to a wide range of client-server situations similar to the example above. We start with a discussion of the uniform random number generator supplied with most computer systems. We then show how to use this generator to obtain random samples from any given discrete distribution. Because random number generation is a basic, and frequently called, feature of all simulations, we investigate some optimization techniques to improve the running time of the corresponding algorithms. We then introduce client-server examples that use discrete probability distributions to describe client request arrivals and their corresponding service times. We tailor our first simulation algorithms to address particular problems, but we soon realize the need for a more general approach that can respond to varying configurations of input streams, queue policies, and multiple servers. After constructing the generic algorithm and exercising it with several examples, we consider how we might verify the simulation results. The last section is a mathematical interlude that introduces the Markov chain, an analytical tool that enables such verification in many cases of practical importance.

3.1   Random number generation

A random variable associates a number with each outcome of some underlying probability space. As discussed in Chapter 1, we can consider the random variable's range to be the probability space of interest, provided that we know the probabilities associated with each value. In either case, we envision some nondeterministic process that we can cycle as often as we want. Each repetition either produces an outcome that determines the values of one or more random variables or directly fixes the random variables' values. We do not distinguish between these two approaches because both conclude with an *observation* of the random variables. We can repeat the process to obtain further observations, and unless otherwise noted, we assume that an observation does not influence in any way the chances of obtaining particular values on subsequent observations. It is convenient to extend the notion of observation to a finite sequence of simple observations. We refer to this process as sampling, and the formal definition is as follows.

3.1 Definition: A *simple sample* from a random variable X is a value in the range of X that we obtain by observing an outcome in the underlying probability space. If the range of X is the probability space in question, then we envision an anonymous process that delivers values from the range of X according to the prescribed probabilities. Any two simple samples are always independent in the sense that observing one sample does not change the probabilities associated with the other sample. A *sample of size N* is a vector of N such independent simple samples. ∎

If, for example, X has the distribution to the right, then we expect samples from X to contain numbers from the set $\{2, 4, 7\}$. For a vector of

x	2	4	7
$\Pr(X = x)$	0.4	0.1	0.5

samples, we expect the components to vary unpredictably over the set, but for lengthy vectors, we expect about 40% of the components to be 2, 10% to be 4, and 50% to be 7. That is, we expect the relative frequencies of the sample values to approximate the corresponding probabilities rather closely.

3.2 Definition: A random variable U with range $[a, b)$, for $a < b$, has a *continuous uniform distribution* if its cumulative distribution function is

$$F_U(u) = \Pr(U \leq u) = \begin{cases} 0, & u < a \\ (u - a)/(b - a), & a \leq u < b \\ 1, & u \geq b. \end{cases} \; \blacksquare$$

This definition requires some explanation because if U has a continuous uniform distribution, then U is not a discrete random variable. At this point, we have defined random variables only when they map the outcomes of a discrete probability space to the real numbers. Hence, all discrete random variables have a countable range, whereas U has an uncountable range: the real numbers in $[a, b)$. In Chapter 5, we will generalize the random variable concept to include those with uncountable ranges. For the moment, however, we continue to work with the discrete definition. We encountered a similar situation in Chapter 2 with the exponential random variable, which arose in connection with the interarrival times of a Poisson process. At that point, we discovered that we can work with such random variables through their continuous cumulative distribution functions. Referring to Equation 2.12, which relates the probability of intervals to the cumulative distribution function, we conclude for $a \leq x \leq y < b$ that

$$\Pr(U \in (x, y]) = F_U(y) - F_U(x) = \frac{y - a - (x - a)}{b - a} = \left(\frac{1}{b - a} \right)(y - x).$$

The continuous uniform distribution then assigns probability to subintervals within $[a, b)$ in proportion to the subinterval width. Consequently, a single point, being a subinterval of zero width, receives probability zero. That is, for each x, we have $\Pr(U = x) = 0$.

We assume that our computer system provides a utility that generates samples from a random variable U_0 with a uniform distribution on $[0, 1)$. On

UNIX systems, this function is called rand(). For any $u \in [0, 1)$, the fraction of rand() returns that are less than or equal to u is equal to u. That is, we expect 20% of an extended sequence of such returns to be less than or equal to 0.2. We expect 70% of the returns to be less than or equal to 0.7, and so forth. This is in accordance with Definition 3.2 with $a = 0$ and $b = 1$. In this case, the definition says that $\Pr(U_0 \le u) = (u - 0)/(1 - 0) = u$. The utility function may have a different name on your system, but that is not important. What is important is that this function, no matter how cleverly implemented, can be only an approximation of a true continuous uniform random variable.

An implementation will suffer from two deficiencies. The first arises from the way a computer represents real numbers; the second appears because a deterministic algorithm generates the samples. A computer represents a real number in the same way that it represents any structure—as a finite string of zeros and ones. The ones and zeros are called bits, which is an abbreviation of *binary digit* and is itself a bit of an oxymoron. The size of the string is normally 32 or 64 bits, corresponding to the size of the hardware storage units that hold and manipulate the strings. We assume 32 bits for this discussion. Since there are only 2^{32} patterns of 32 bits, the computer can represent only finitely many numbers. Each execution of rand() returns one of these patterns, one that happens to represent a number in the interval $[0, 1)$. The interval $[0, 1)$, however, contains uncountably many real numbers. Therefore, many, actually most, of the real numbers in $[0, 1)$ never appear in the computer's calculations, nor do they appear among the values returned by rand(). According to Definition 3.2, a continuous uniform random variable's range must be a real interval $[a, b)$. It follows that rand() cannot faithfully return samples from a continuous uniform random variable; it omits most of the required range.

If rand() does not exactly correspond to a continuous uniform random variable, how serious is the discrepancy? The answer depends on how the machine represents real numbers as bit strings. There are a variety of schemes, but they all suffer from the same shortcoming: bit strings of finite length can represent only finitely many numbers. Suppose that the machine represents real numbers in $[0, 1)$ as 32-bit strings in which the assumed binary point is at the extreme left. The leftmost bit then contributes zero or 1 times 2^{-1}. Its neighbor to the right contributes zero or 1 times 2^{-2}, and so forth. The rightmost bit contributes zero or 1 times 2^{-32}. The sum of these contributions is the numerical value. For example, the bit pattern

$$B = 10110001 \quad 00000000 \quad 00000000 \quad 00000000$$
$$= 1 \times 2^{-1} + 0 \times 2^{-2} + 1 \times 2^{-3} + 1 \times 2^{-4} + \ldots + 0 \times 2^{-32}$$
$$= \frac{1}{2} + \frac{1}{8} + \frac{1}{16} + \frac{1}{256} = 0.69140625.$$

There are $2^{32} = 4294967296$ such patterns, ranging in numerical value from 0 to $1 - 2^{-32}$. They are equally spaced in $[0, 1)$; the difference between adjacent patterns is 2^{-32}. The best that rand() can hope to accomplish is to return these 2^{32} bit patterns with equal probabilities. That is, rand() attempts to simulate a *discrete* uniform distribution on the integers $0, 1, 2, \ldots, 4294967295$, which we interpret as numbers in $[0, 1)$ by shifting the binary point to the extreme left (i.e., by multiplying by 2^{-32}). Assume for the moment that rand() succeeds in producing this discrete uniform distribution, which we will call U_0'. Then

$$\text{range}(U_0') = \left\{ 0, \frac{1}{2^{32}}, \frac{2}{2^{32}}, \frac{3}{2^{32}}, \ldots, \frac{4294967295}{2^{32}} \right\}.$$

For the true continuous uniform random variable U_0, we have $\Pr(U_0 \le u) = \Pr(U_0 < u) = u$ because the point u itself has probability zero. How much does $\Pr(U_0' < u)$ differ from $\Pr(U_0 < u)$? If $0 \le u < 1$, then $\Pr(U_0' < u)$ is the number of range points to the left of u divided by 2^{32}. It follows that $\Pr(U_0' < u) = u$ only when u is one of the range points. When $u = 3/2^{32}$, for example, there are three range points less than u: $0, 1/2^{32}$, and $2/2^{32}$, which yields $\Pr(U_0' < 3/2^{32}) = 3/2^{32}$. For u between range points, say $i/2^{32} < u < (i+1)/2^{32}$, the points $0/2^{32}, 1/2^{32}, \ldots, i/2^{32}$ are all less than u. Because there are $i + 1$ such points, we have $\Pr(U_0' < u) = (i+1)/2^{32} > u$. Hence $\Pr(U_0' < u)$ is too high by a small amount. The error is at most 2^{-32}, which occurs when u is just greater than a given range point. The error decreases linearly to zero as u approaches the next range point.

To use computer simulations, we must tolerate this small error. So long as we do not expect accurate probabilities for extremely small intervals, the error remains negligible. If, for example, we divide $[0, 1)$ into 100 or 1000 intervals of equal size, we can expect the returns from rand() to accumulate about equally in all the subintervals. By contrast, if we divide $[0, 1)$ into 2^{32} subintervals, we should not expect the subintervals to receive equitable returns from rand(). As long as we keep this ultimate limitation in mind, we can ignore the small error in $\Pr(U_0' < u)$. Consequently, for the rest of this book, we assume that rand() simulates a random variable U_0 with the property that $\Pr(U_0 < u) = u$, for $0 \le u < 1$.

There is, nevertheless, the second deficiency: A deterministic algorithm will not produce unpredictable samples. In the final analysis, it is possible to predict the next sample from a sequence of observed samples. Since this is hardly random behavior, we must be content with the appearance of randomness over a large number of returns. Algorithm rand() asks the user to choose a starting sample, X_0, which is called the *seed*. It then generates subsequent samples from the following formula: $X_{n+1} = (a_1 X_n + a_2) \mod a_3$. The X_n are integers in the range 0 to 4294967295; they are later scaled to the interval $[0, 1)$ by shifting the binary point. The a_i are carefully chosen constants. This technique is called the linear congruential method, and its performance depends critically on the choice of constants a_i. Because there are only 2^{32} possible X values, the sequence must eventually repeat itself. Suppose that $j > i$ and $X_j = X_i$. Because the formula is deterministic, X_{j+1} will then equal X_{i+1}, which will force $X_{j+2} = X_{i+2}$, and so forth. The subsequence

```
                                   function Bernoulli (p)
                                      if (rand() < p)
                                         return 1;
      function uniform(a, b)          else
         return a + (b - a) * rand();         return 0;
         (a) Uniform over [a, b)         (b) Bernoulli with parameter p
```

Figure 3.1. Algorithms for sampling uniform and Bernoulli distributions

between X_i and X_{j-1} will repeat over and over in an endless cycle. The goal then is to choose the a_i to produce a very long cycle, one approaching the maximum length of 2^{32}. This problem has generated reams of research. Every choice for the a_i seems to reveal some flaw if someone is willing to examine the situation closely enough. We will take the position that rand() is sufficient for our purposes. A frequent practice is to choose the initial seed, X_0, from the system real-time clock. Each simulation run then starts with a different initial value, one that is difficult or impossible to ascertain in advance.

Our first improvement scales rand() to provide a uniform distribution on an arbitrary interval $[a, b)$. The simple code appears in Figure 3.1(a). The call U = uniform(a, b) simulates an observation of the random variable U, for which the range is $[a, b)$. We refer to the algorithm in Figure 3.1(a) simply as Algorithm 3.1(a), and similarly for algorithms appearing in subsequent figures.

3.3 Theorem: Algorithm 3.1(a) correctly simulates a continuous uniform distribution on $[a, b)$.

PROOF: Let U be the random variable computed by the algorithm. Examining the computation, we see that rand() returns a number $r \in [0, 1)$, so the computation in uniform produces a result in $[a, b)$. Moreover, $a \leq U < u < b$ is equivalent to $0 \leq r < (u - a)/(b - a)$. Therefore,

$$\Pr(a \leq U < u) = \Pr\left(0 \leq r < \frac{u - a}{b - a}\right) = \frac{u - a}{b - a},$$

for $a \leq u < b$. Hence, by Definition 3.2, U has a uniform distribution on $[a, b)$. ∎

The computation in Algorithm 3.1(a) is actually an evaluation of the inverse of the cumulative distribution function $F_U(u) = u$. This technique generalizes nicely to the more complicated situations in the next section, so a closer look is in order. Examining $F_U(u)$ in Definition 3.2, we note that F_U is monotone nondecreasing. That is, $u_1 < u_2$ implies that $F_U(u_1) \leq F_U(u_2)$. On the crucial section, $a < u < b$, F_U is monotone increasing and therefore has a well-defined inverse: $F_U^{-1}(y) = a + (b - a)y$, for $0 < y < 1$. This is precisely the computation of Algorithm 3.1(a), which suggests that a general method for obtaining samples from a given distribution is to evaluate the inverse of its cumulative distribution function at a point chosen from a uniform distribution on $[0, 1)$. Except for adjustments to accommodate flat sections in the cumulative distribution function, where the inverse is not well-defined, this

```
function integrateExp()
    N = 5000;
    s = 0.0;
    for (i = 0 to N - 1)
        x = uniform (0, 2.0);
        y = uniform (0, exp (2.0));
        if (y < exp (x))
            s = s + 1.0;
    z = 2.0 * exp (2.0) * (s / N);
    return z;
```

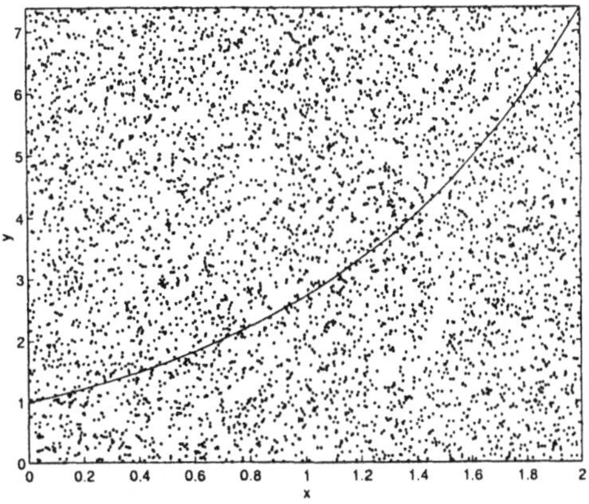

(a) Integration algorithm   (b) Distribution of random points

Figure 3.2. Integration of e^t by sampling a uniform distribution (see Example 3.4)

---

method does indeed generalize. The next section will show how this approach can provide an algorithm to sample from any discrete distribution. Before taking up that discussion, however, we illustrate an immediate application of the continuous uniform distribution.

3.4 Example: Approximate $\int_0^2 e^t \, dt$.

This integral is easy enough to evaluate analytically: $\int_0^2 e^t \, dt = e^t \big|_0^2 = e^2 - 1 = 6.389$. However, this is not really the point of the example. The integrand could pose a more difficult challenge. We have chosen a simple integrand, with a known solution, to emphasize the correctness of a more novel method. From Figure 3.2(b), we see that the desired integral is the area within a certain rectangle and below the curve $f(t) = e^t$. If we uniformly cover the rectangle with random points, the fraction that lie below the curve, multiplied by the area of the rectangle, should approximate the desired area. The cloud of points in the figure represents 5000 (x, y) points, each with x chosen from uniform $(0, 2)$ and y from uniform $(0, 7.389)$. That is, the x-coordinates come from a uniform distribution across the base of the rectangle, while the y-coordinates come from a uniform distribution across its height. For each random point (x, y), we can check its position by comparing y with e^x. If y is less, the point is below the line $f(t) = e^t$; otherwise, it is on or above the line. The algorithm of Figure 3.2(a) supplies the details. Five executions provided solutions $6.4048, 6.2955, 6.4994, 6.3014$, and 6.3768. These values average to 6.376, which is reasonably close to the 6.389 computed analytically above. This integration method is called the *Monte Carlo technique*. □

Using the uniform random number generator rand(), we can easily construct an algorithm that returns samples from a Bernoulli distribution with parameter p. The simple code appear in Figure 3.1(b). Let B be the random variable obtained via repeated calls to Algorithm 3.1(b), and recall that rand() simulates U_0, the continuous uniform random variable on $[0, 1)$. Then

$\Pr(B = 1) = \Pr(U_0 < p) = p$. So the algorithm performs as expected.

3.5 Example: A dealer engages you in a card game with the following rules. For each game, you pay the dealer $1 and receive two cards from a well-shuffled deck. If the cards are an ace and a face card, you receive $25. Otherwise, you receive nothing. You return your cards to the dealer's deck before beginning a new game. Construct a simulation that tracks your winnings over a long sequence of games. Approximate your average winnings per game. Comment on any long-term trend.

Let X_i be a Bernoulli random variable that assumes the value 1 when you win the ith game. The parameter p is the probability of receiving an ace and face card, which is $p = C_{4,1}C_{12,1}C_{36,0}/C_{52,2} = 0.036$. We use Algorithm 3.1(b) to simulate each game. Each time the algorithm returns a 1, we increment the winnings by $24. This represents the $25 payoff less the $1 game fee. For each zero, we decrement by $1.

The code of Figure 3.3(a), which accepts the number of games as input parameter n, implements this scheme. If y is the zero or 1 returned by the Bernoulli call, then the expression $24y - (1 - y) = 25y - 1$ correctly calculates the winnings in either case. The algorithm returns a 3-vector containing the high point in the accumulated winnings, the low point, and the average winnings per game over the full series. Although the algorithm does not report the history of the accumulated winnings over the course of many games, you can accomplish this modification by keeping the winnings information as a vector, in which the ith component stores the accumulated winnings through game i. Figure 3.3(b) presents such a history over the course of 10000 games. The trend clearly shows that this game is biased in the dealer's favor. The graph does exhibit occasional upward segments in which your winnings accumulate. These temporary setbacks for the dealer may encourage you to continue playing the game. In the long term, however, it appears that you should not do so. How representative is the particular series represented in the graph? Twenty simulations, each 10,000 games, returned the results of Table 3.1.

These multiple runs confirm the negative trend. To gain further insight into the results, we can perform some analysis. Redefine X_i as the the winnings on the ith game. Then X_i assumes the value 24 with probability 0.036 and -1 with probability 0.964. Accordingly, all X_i share the mean $0.036(24) + 0.964(-1) = -0.1$ and variance $0.036(24+0.1)^2 + 0.964(-1+0.1)^2 = 21.69$. Let H be the high point in the accumulated winnings over n games, L be the low point, and W be the average winnings per game. Thus, $H = \max_{1 \le j \le n}\{\sum_{i=1}^{j} X_i\}$, $L = \min_{1 \le j \le n}\{\sum_{i=1}^{j} X_i\}$, and $W = (1/n)\sum_{i=1}^{n} X_i$. For compatibility with the tabulated observations above, let $n = 10000$. Under the reasonable assumption that the X_i are independent, we invoke Theorem 2.8 to compute

$$\mu_W = \frac{1}{10000} \sum_{i=1}^{10000} (-0.1) = -0.1$$

$$\sigma_W^2 = \frac{1}{(10000)^2} \sum_{i=1}^{10000} 21.69 = 0.002169$$

$$\sigma_W = \sqrt{0.002169} = 0.047.$$

Chebyshev's inequality then bounds the chance of observing a W more than $3\sigma_W$ from the mean, that is, outside the interval $(-0.241, 0.041)$, as less than 1/9.

Indeed, none of the average winnings from the 20 sample series is outside this range. Actually, most runs cluster rather closely to the expected value of -0.1.

```
function aceface (N)        // input is the number of games, N
    p = 0.036; w = 0.0; high = 0.0; low = 0.0;
    for (i = 0 to N - 1)
        w = w + 25.0 * Bernoulli (p) - 1.0;
        if (w > high)
            high = w;
        if (w < low)
            low = w;
    average = w / N;
    return (high, low, average);
```
                    (a) Simulation algorithm

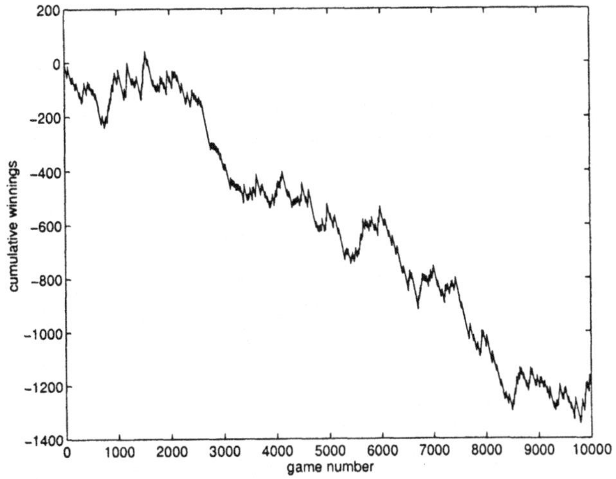

(b) History of accumulated winnings

Figure 3.3. Simulation of the ace-face game (see Example 3.5)

Therefore, any one of the runs is a rather accurate assessment of the average winnings per game, because W has such a small variance.

Because we can so easily obtain this analytic average, the simulation is hardly necessary to assess this parameter. The simulation's real contributions are the estimates for H and L. These observations vary more widely over the 20 runs, and unlike W, we do not have a simple formula to compute their mean and variance.

Consider L, the low point of the accumulated winnings over n games. It has a probability distribution because there are only 2^n possible sequences of wins and losses, each of which produces a particular low point. So there are a μ_L and σ_L^2, even though we can obtain their values only with some difficulty. The calculation is certainly beyond the methods of this chapter.

Consider, however, $\overline{L}$, the average of L over m series. From Theorem 2.8, $\mu_{\overline{L}} = \mu_L$ and $\sigma_{\overline{L}}^2 = \sigma_L/m$. Thus the variance of $\overline{L}$ becomes very small when the average encompasses a large number of runs. For the 20 data points of Table 3.1, $\overline{L} = -644.5$. We still do not know if 20 runs is sufficient to drive the variance to an acceptably small value, but we can always make further runs and watch the change in $\overline{L}$.

Further analysis of the simulation results must await the developments of a

Simulation	Accumulated winnings			Simulation	Accumulated winnings		
	Low	High	Average		Low	High	Average
1	-777	104	0.000	11	-1383	17	-0.100
2	-1045	94	-0.100	12	-893	21	-0.100
3	-544	167	0.000	13	-1421	107	-0.100
4	-572	65	0.000	14	-1556	-1	-0.200
5	-1599	55	-0.200	15	-1175	12	-0.100
6	-759	43	-0.100	16	-1950	-1	-0.200
7	-546	213	0.000	17	-1068	-1	-0.100
8	-1265	59	-0.100	18	-811	125	-0.100
9	-325	231	0.000	19	-950	134	-0.100
10	-767	46	-0.100	20	-788	155	-0.100

TABLE 3.1. The ace-face game over many simulations, each 10,000
games (see Example 3.5)

---

later chapter. □

Exercises

3.1 Test rand(), or its equivalent, on your computer system. Generate
10,000 samples and count the fraction that lie in the subintervals $[(i - 1)/100, i/100)$, for $i = 1, 2, \ldots, 100$. Repeat the experiment several
times and average the counts for each subinterval.

3.2 For each $s = 0, 1, 2, \ldots, 31$, compute the sequence $x_0, x_1, x_2, \ldots$ until
a repetition occurs. The recurrence formula is $x_0 = s$ and $x_{i+1} = (13x_i + 5) \mod 32$. Let t be the value of s that produces the longest
sequence prior to a repetition. Let $Y = \{y_0, y_1, y_2, \ldots\}$ be the infinite
sequence associated with the seed t. What are the relative frequencies
of $0, 1, \ldots, 31$ in the sequence Y?

3.3 Use the Monte Carlo integration technique of Example 3.4 to approximate $\int_0^1 te^{2t}\, dt$. Compare the result with an analytical solution.

3.4 Use the Monte Carlo integration technique of Example 3.4 to approximate $\int_0^1 [(e^t - 1)/t]\, dt$.

*3.5 Consider the modified cumulative distribution function below, which
characterizes a continuous random variable V. Modify the algorithm of
Figure 3.1(a) to generate returns with relative frequencies matching the
new cumulative distribution. That is, the fraction of returns less than
or equal to u should be $F_V(u)$. Prove that your algorithm delivers this

result.

$$F_V(u) = \begin{cases} 0, & u < 0 \\ \dfrac{u}{2}, & 0 \leq u < \dfrac{1}{2} \\ \dfrac{3u-1}{2}, & \dfrac{1}{2} \leq u < 1 \\ 1, & u \geq 1 \end{cases}$$

*3.6 Repeat the preceding exercise for the cumulative distribution $F_W(u) = u^2$ for $0 \leq u < 1$.

*3.7 Stretch the continuous uniform distribution to cover the range $[0, 1) \cup [2, 3)$. That is, repeat the preceding exercise for the following cumulative distribution.

$$F_Z(u) = \begin{cases} 0, & u < 0 \\ \dfrac{u}{2}, & 0 \leq u < 1 \\ \dfrac{1}{2}, & 1 \leq u < 2 \\ \dfrac{u-1}{2}, & 2 \leq u < 3 \\ 1, & u \geq 3 \end{cases}$$

3.8 If X, Y are independent random variables, each uniformly distributed on $[-1, 1)$, then the pair (X, Y) plots in the square $\{(x, y) | -1 \leq x < 1, -1 \leq y < 1\}$. A large set of samples (X_i, Y_i) should scatter uniformly across the square. Consequently, the fraction of samples for which $X_i^2 + Y_i^2 < 1$ should approximate the area ratio of the inscribed circle to the square. This ratio is $\pi(1)^2/2^2 = \pi/4$. Use this observation in a simulation to approximate π.

3.2 Inverse transforms and rejection filters

The discussion following Theorem 3.3 suggests the following generalization. To obtain returns that simulate a desired distribution, we evaluate the inverse of the cumulative distribution function at a point chosen from a continuous uniform distribution on $[0, 1)$. That is, the algorithm should return $x = F^{-1}(y)$, where y is chosen from a continuous uniform distribution on $[0, 1)$. For a discrete random variable, the cumulative distribution function exhibits a staircase graph, and consequently, the inverse is not well-defined. In particular, $F^{-1}(y)$ is a set of values when y corresponds to one of the horizontal segments of the graph. Also, $F^{-1}(y)$ is the empty set when y does

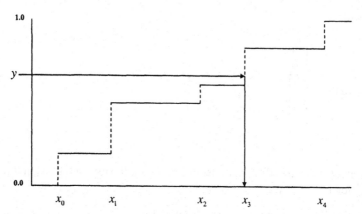

Figure 3.4. Generating a random number from its cumulative distribution function

not correspond to such a horizontal segment. The following mechanism circumvents this difficulty. Because it is still an evaluation of the inverse of the cumulative distribution function, it retains the title of inverse transform method.

Suppose that the discrete random variable X has the range $x_0 < x_1 < x_2 < \cdots$. These are the points where X has nonzero probability, so the cumulative distribution function F jumps by an amount $\Pr(X = x_i) = \Pr(x_i)$ at each x_i. Figure 3.4 shows the general appearance of F, and the following piecewise description captures its staircase behavior.

$$F(x) = \begin{cases} 0, & x < x_0 \\ \Pr(x_0), & x_0 \leq x < x_1 \\ \Pr(x_0) + \Pr(x_1), & x_1 \leq x < x_2 \\ \Pr(x_0) + \Pr(x_1) + \Pr(x_2), & x_2 \leq x < x_3 \\ \vdots \end{cases}$$

Consider an algorithm that maps y, obtained from a uniform distribution on $[0, 1)$, to x_i by choosing i as the smallest index such that $F(x_i) > y$. What is the probability that the algorithm returns x_0? This occurs whenever $y < F(x_0) = \Pr(x_0)$. Equivalently, this occurs when $y \in [0, \Pr(x_0))$. Since y comes from a uniform distribution on $[0, 1)$, this probability is just $\Pr(x_0)$. The algorithm therefore returns x_0 with the correct probability. Consider another x_i. The algorithm returns x_i, for $i > 0$, when $F(x_{i-1}) \leq y < F(x_i)$. We rewrite this event using the description of F above to obtain

$$y \in [\Pr(x_0) + \ldots + \Pr(x_{i-1}), \Pr(x_0) + \ldots + \Pr(x_{i-1}) + \Pr(x_i)).$$

Because y comes from a uniform distribution on $[0, 1)$, the probability is the interval width, which is $\Pr(x_i)$.

The algorithm requires a uniform generator on $[0, 1)$, and it needs access to the successive values in the random variable's range, together with their

```
function genericInverseTransform()        function binomial1(n, p)
    y = rand();                               y = rand();
    x = firstx();                             k = 0;
    cum = prob(x);                            cum = C_{n,0} * (1 - p)^n;
    while (cum <= y)                          while (cum <= y) {
        x = nextx();                              k = k + 1;
        cum = cum + prob(x);                      cum = cum + C_{n,k} * p^k * (1 - p)^{n-k};
    return x;                                 return k;
```

(a) Generic algorithm   (b) Binomial application

Figure 3.5. Inverse transform algorithm for random observations from
a discrete distribution

probabilities. In the algorithm of Figure 3.5(a), these sources are rand(),
firstx(), nextx(), and prob(). This generic algorithm becomes more specific
when we are generating a particular distribution. In the general case, however,
we have proved the following theorem.

3.6 Theorem: Repeated calls to Algorithm 3.5(a) return values x with probabilities (i.e., relative frequencies) prob(x).

PROOF: The algorithm scans sequentially through increasing values of the
random variable's range. By accumulating the probabilities across these entries, it builds the cumulative distribution function as it proceeds. The while-
loop terminates when the cumulative distribution function exceeds y, which
was chosen from a uniform distribution on $[0, 1)$. The x value at that point is
the smallest range entry where the cumulative distribution function exceeds
y. Referring to the discussion above, we conclude that the algorithm returns
x with probability prob(x). ■

We can adapt Algorithm 3.5(a) to a particular distribution by substituting specific expressions for the functions that obtain the range values and
their probabilities. The algorithm of Figure 3.5(b), for example, returns values
from a binomial distribution with parameters (n, p).

In an analysis of algorithm performance, we normally ignore setup operations and to concentrate on that part of the algorithm that changes from
one execution to the next. Typically, the analysis emphasizes operation counts
that vary with the program input. In the case of probabilistic algorithms, in
which random numbers play a significant role, the focus must also include
operation counts that vary with the particular random numbers chosen in an
execution. This approach reflects our concern with the running time when the
inputs are large. For large inputs, the setup operations become negligible in
comparison with the loops whose running time increases in proportion to the
input. We are not very anxious about the complementary situation, because
short inputs normally produce short running times.

In Algorithm 3.5(b), the execution time, excepting setup operations, is
proportional to the number of while-loop iterations. If we let T be the number
of while-loop iterations, then T becomes a random variable. Examining the

code, we see that when the algorithm returns 0, the loop is not executed at all. Therefore, $T = 0$ in this case. Since the loop increases k by one for each iteration, we conclude that the returned value measures the number of loop iterations. Therefore, the probability that $T = k$ is just the probability that the algorithm returns k. But this latter value is $C_{n,k}p^k(1-p)^{n-k}$ because the routine is simulating the binomial random variable with parameters (n, p). This means that the distribution of T is also binomial with parameters (n, p). On the average, we can expect np loop iterations per call.

If np is large, we have an inefficient generator. Recall that most of the probability is concentrated near the mean. Chebyshev's inequality asserts this fact. Most of the time, therefore, the while-loop of Algorithm 3.5(b) must count from zero to the vicinity of np before it terminates. This suggests an improvement to the algorithm. We should start the search at np, actually at an integer close to np, and then search forward or backward as appropriate. Suppose that we truncate np to obtain the starting point $s = \lfloor np \rfloor$. We then need the cumulative distribution function at this point: $F(s) = \sum_{k=0}^{s} C_{n,k}p^k(1-p)^{n-k}$. This calculation involves a sum over the same range that we are trying to avoid. However, $F(s)$ need not be recalculated on each call. As long as the parameters (n, p) are unchanged, we can store $F(s)$ for future use.

Since we are optimizing the algorithm, we should also investigate the subsidiary routines that deliver binomial coefficients and powers of p. For binomial coefficients, we can shorten the calculation by computing adjacent binomial coefficients in terms of current ones. That is, if we have $C_{n,k}$, we can move forward or backward with the formulas

$$C_{n,k+1} = \frac{n!}{(k+1)!(n-k-1)!} = \frac{n!}{k!(n-k)!} \cdot \frac{n-k}{k+1} = C_{n,k} \cdot \frac{n-k}{k+1}$$

$$C_{n,k-1} = \frac{n!}{(k-1)!(n-k+1)!} = \frac{n!}{k!(n-k)!} \cdot \frac{k}{n-k+1}$$

$$= C_{n,k} \cdot \frac{k}{n-k+1}. \tag{3.1}$$

We can also move forward and backward in the powers $p^k(1-p)^{n-k}$ as follows.

$$p^{k+1}(1-p)^{n-k-1} = p^k(1-p)^{n-k} \cdot \frac{p}{1-p}$$

$$p^{k-1}(1-p)^{n-k+1} = p^k(1-p)^{n-k} \cdot \frac{1-p}{p} \tag{3.2}$$

These formulas can substitute for the calculation that delivers powers of p, provided that we keep track of the current powers of p and $1 - p$. Figure 3.6 incorporates these changes and optimizations to produce a more efficient algorithm for generating random observations from a binomial distribution. The algorithm starts with a preamble that executes only if the parameters (n, p) have changed from the previous call. The static declarations establish variables that survive from one execution to the next. The preamble sets up

```
function binomial (n, p)
    static np, lastn = 0;
    static startprob, startcum, lastp = 0.0;

    if ((n ≠ lastn) or (p ≠ lastp))      // initialize if new parameters
        lastn = n; lastp = p; np = floor (n * p);
        startcum = (1 - p)^n; startprob = startcum;
        for (k = 0 to np - 1)
            startprob = startprob * (p / (1 - p)) * ((n - k) / (k + 1));
            startcum = startcum + startprob;
    // main segment assumes initialized startcum and startprob
    k = np; prob = startprob; cum = startcum;
    y = rand();
    if (cum ≤ y)      // search among higher k
        while (cum <= y)
            prob = prob * (p / (1 - p)) * ((n - k) / (k + 1));
            cum = cum + prob;
            k = k + 1;
        return k;
    else      // return starting value or search among lower k
        cum = cum - prob;
        while (cum > y)
            prob = prob * ((1 - p) / p) * (k / (n - k + 1));
            cum = cum - prob;
            k = k - 1;
        return k;
```

Figure 3.6. Improved algorithm for random observations from a binomial distribution

a starting point at $\lfloor np \rfloor$, called simply np in the algorithm, and calculates the cumulative distribution, called startcum, at that point. In preparation for using Equations 3.1 and 3.2, it also establishes $\Pr(\lfloor np \rfloor)$ in the variable startprob.

In the absence of parameter changes, the routine skips immediately to the main segment, where it initializes k to the starting value $\lfloor np \rfloor$. After obtaining a sample y from a uniform distribution on $[0, 1)$, the algorithm tests y against the starting value of the cumulative distribution function. If y is larger or equal, the correct intercept with the cumulative distribution function lies with $k > \lfloor np \rfloor$. Otherwise, the intercept occurs at $\lfloor np \rfloor$ or lower. Accordingly, one of the two while-loops searches as in the preceding algorithm by adjusting the cumulative probability function up or down with increasing or decreasing k. Note that neither while-loop executes if the intercept occurs precisely at $\lfloor np \rfloor$.

Figure 3.7 illustrates a test of the algorithm for $n = 15, p = 0.5$. The test uses Algorithm 3.6 to obtain 1000 samples from the binomial distribution and then calculates the fraction that are equal to i, for $0 \leq i \leq 15$. These relative frequencies appear as plus signs in the figure. The solid curve is the

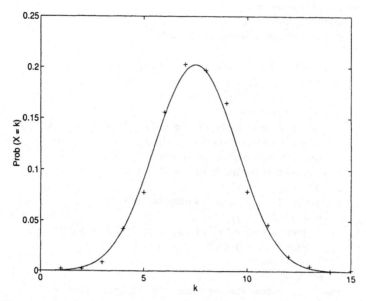

Figure 3.7. Performance of the inverse transform in generating samples of a binomial random variable X

true binomial distribution $y(k) = C_{15,k}(0.5)^k(1 - 0.5)^{15-k}$, which has been smoothed to facilitate comparison with the relative frequencies. The graph verifies that the algorithm delivers the expected results.

To analyze the algorithm's efficiency, we measure the number of while-loop iterations. The other operations constitute a fixed overhead that will not contribute materially to the running time when the number of while-loop iterations is large. Let $k_0 = \lfloor np \rfloor$. We see that returns $k_0 + 1, k_0 + 2, \ldots, n$ correspond to $1, 2, \ldots, n - k_0$ iterations of the upper while-loop. Similarly, returns $k_0 - 1, k_0 - 2, \ldots, 0$ correspond to $1, 2, \ldots, k_0$ iterations of the lower while-loop. The returned value k_0 means that neither loop was exercised. Because the algorithm generates each return k with probability $C_{n,k}p^k(1 - p)^{n-k}$, we can easily express the mean of the random variable N that measures the number of while-loop iterations. It is

$$\mu_N = \sum_{0 \le k \le \lfloor np \rfloor} C_{n,k}p^k(1 - p)^{n-k}(\lfloor np \rfloor - k)$$

$$+ \sum_{\lfloor np \rfloor < k \le n} C_{n,k}p^k(1 - p)^{n-k}(k - \lfloor np \rfloor)$$

$$= \sum_{k=0}^{n} |k - \lfloor np \rfloor| C_{n,k}p^k(1 - p)^{n-k} = E[|B - \lfloor np \rfloor|],$$

where B is a binomial random variable with parameters (n, p). The final expression is nevertheless difficult to evaluate in the general case. More advanced techniques, to be studied in a later chapter, allow us to establish

$$\mu_N = \sqrt{\frac{2np(1-p)}{\pi}} \leq \sqrt{\frac{n}{2\pi}} = 0.399\sqrt{n} \tag{3.3}$$

for large n. The bound follows because $p(1-p) \leq 1/4$ for $0 \leq p \leq 1$. Algorithm 3.6 is then significantly more efficient than our initial effort, Algorithm 3.5(b). The first algorithm, you will recall, has running time proportional to np, which is linear in n. Although we have not yet studied the techniques needed to establish Equation 3.3, we can verify it to some extent by simulating the average value of $X = |B - \lfloor np \rfloor|$. That is, for a given (n, p), let $X_i = |B_i - \lfloor np \rfloor|$, where B_i is the ith random number generated with Algorithm 3.6. For a large number, M, of such samples, we define the sample average by $\overline{X} = (1/M) \sum_{i=1}^{M} X_i$. We are a little ahead of ourselves in this exercise; the next chapter's goal is to analyze the inferences that we can draw from repeatedly sampling a population. For the moment, however, let us rely on the intuitive association between probability and relative frequency, which suggests that $\mu_N = E[|B - \lfloor np \rfloor|] \approx \overline{X}$.

The data to the right come from an experiment in which Algorithm 3.6 provided 500 samples for each of several n values. The parameter p was fixed at 0.5, which

n	8	16	32	64	128
$\ln n$	2.0794	2.7726	3.4657	4.1589	4.8520
$\ln \overline{X}$	0.1467	0.4331	0.7683	1.1594	1.4951

is the point where $p(1-p)$ attains its maximum value of $1/4$. Figure 3.8 plots $\ln n$ versus $\ln \overline{X}$; each plus symbol corresponds to one data point from the table above. Except for small deviations, the points lie along a straight line. We can approximate the slope m from the extreme data points, and we can compute the intercept b where $\ln n = 0$.

$$m \approx \frac{1.4951 - 0.1467}{4.8520 - 2.0794} = 0.49$$

$$b \approx 1.4951 - (0.49)(4.8520) = -0.88$$

So $\ln \overline{X} \approx -0.88 + 0.49 \ln n$ or equivalently, $\mu_N \approx \overline{X} \approx 0.41 n^{0.49}$. This provides an empirical verification of Equation 3.3. The solid line in Figure 3.8 graphs the natural logarithm of Equation 3.3.

This situation shows how a simulation can quickly provide a rather accurate approximation for a parameter that is difficult to compute analytically. This is, of course, the main reason for simulations, and further examples appear later in the chapter.

At this point, however, it should be clear that the inverse transform method can produce samples from any discrete distribution. We simply adapt the generic code of Algorithm 3.5(a) to the details of the given distribution. The following example illustrates the method in the case of a Poisson distribution.

3.7 Example: Design and test an algorithm that provides random samples of a Poisson random variable P with parameter m. That is, $\Pr(P = k) = m^k e^{-m}/k!$.

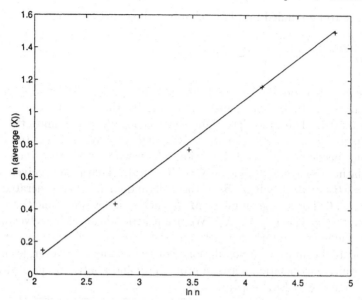

Figure 3.8. Operation counts from the binomial distribution algorithm

The algorithm must return an integer k in the range $0, 1, 2, \ldots$ such that $\Pr(k) = m^k e^{-m}/k!$. Since the expected value of this distribution is m, the search starts at $k = \lfloor m \rfloor$ and proceeds into larger or smaller k, as appropriate, to find the smallest k such that $F(k) > y$. Here $F(k)$ is the cumulative distribution function $\sum_{j=0}^{k} m^j e^{-m}/j!$, and y is chosen from a uniform distribution on $[0,1)$. Imitating the optimization used for the binomial distribution (Algorithm 3.6), we note that we can move forward or backward in the probability distribution with a multiplicative factor.

$$\Pr(P = k+1) = \frac{m^{k+1}e^{-m}}{(k+1)!} = \frac{m^k e^{-m}}{k!} \cdot \frac{m}{k+1} = \Pr(P = k) \cdot \frac{m}{k+1}$$

$$\Pr(P = k-1) = \frac{m^{k-1}e^{-m}}{(k-1)!} = \frac{m^k e^{-m}}{k!} \cdot \frac{k}{m} = \Pr(P = k) \cdot \frac{k}{m}$$

The algorithm appears in Figure 3.9. A preamble executes only when a new parameter m is detected. It computes the starting values. The variables mean, startprob, and cumprob take the values $\lfloor m \rfloor$, $\Pr(\lfloor m \rfloor)$, and $F(\lfloor m \rfloor)$ respectively. The search proceeds from k equal to mean into higher or lower values as determined by the random variable y, which comes from a uniform distribution on $[0,1)$. The results of 3000 executions with $m = 10$ appear in Figure 3.10. The plus symbols mark the relative frequency of $k = 0, 1, 2, \ldots$ among the returns. The solid line, smoothed to facilitate comparison with the plus signs, graphs the true Poisson distribution with parameter 10. □

The algorithms developed above for the binomial and Poisson distributions involve an iterative search along the cumulative distribution staircase. These algorithms illustrate the general case, which applies even if there is no pattern across the probability distribution. However, if there is a pattern,

```
function poisson (m) {
    static mean, lastm = 0;
    static startprob, startcum;
    if (m ≠ lastm)      // initialize if new m parameter
        lastm = m; mean = floor(m);
        startcum = exp(-m); startprob = startcum;
        for (k = 0 to mean - 1)
            startprob = startprob * m / (k + 1);
            startcum = startcum + startprob;
    // main segment assumes initialized startcum and startprob
    k = mean; prob = startprob; cum = startcum;
    y = rand();
    if (cum ≤ y)       // search among higher k
        while (cum <= y)
            prob = prob * m / (k + 1);
            cum = cum + prob;
            k = k + 1;
        return k;
    else       // return starting value or search among lower k
        cum = cum - prob;
        while (cum > y)
            prob = prob * k / m;
            cum = cum - prob;
            k = k - 1;
        return k;
```

Figure 3.9. Algorithm for random observations from a Poisson distribution (see Example 3.7)

---

sometimes it is possible to calculate the return value directly without a search.
The following theorem illustrates the point for two discrete distributions: a
discrete uniform distribution on $1, 2, \ldots, n$ and a geometric distribution with
parameter $0 < p < 1$. In these cases, the simplification is possible because the
sums that characterize the cumulative distribution have an easily accessible
closed form. For the binomial or Poisson distribution, the corresponding sum
does not admit a closed form, so the search method is necessary.

3.8 Theorem: Let $0 < p < 1$, and let $n > 1$ be an integer. If U_0 is the
continuous uniform random variable on $[0, 1)$ and

$$X = \left\lfloor \frac{\ln(1 - U_0)}{\ln(1 - p)} \right\rfloor$$
$$Y = \lfloor nU_0 \rfloor + 1,$$

then X is a geometric random variable with parameter p and Y is a discrete
uniform random variable on $1, 2, \ldots, n$.

PROOF: Consider X first. If $U_0 = 0$, then $X = 0$. Otherwise, X is the
integer part of the ratio of two natural logarithms, both negative. Hence X

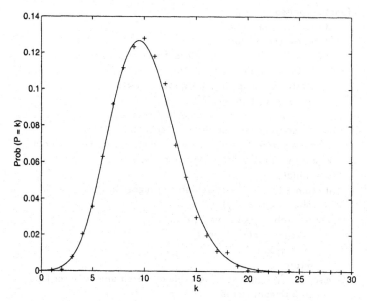

Figure 3.10. Inverse transform performance generating Poisson samples
(see Example 3.7)

is a nonnegative integer. The following events are then equivalent.

$$X = j$$

$$j \leq \frac{\ln(1 - U_0)}{\ln(1 - p)} < j + 1$$

$$1 - (1 - p)^j \leq U_0 < 1 - (1 - p)^{j+1}$$

Therefore, $\Pr_X(j) = \Pr(U_0 \in [1 - (1 - p)^j, 1 - (1 - p)^{j+1})) = p(1 - p)^j$.
Thus X has a geometric distribution with parameter p. Now consider Y.
Using a similar analysis, we reason that $0 \leq nU_0 < n$, which implies that
$Y \in \{1, 2, 3, \ldots, n\}$. Furthermore,

$$\Pr_Y(j) = \Pr(j - 1 \leq nU_0 < j) = \Pr\left(\frac{j-1}{n} \leq U_0 < \frac{j}{n}\right)$$

$$= \frac{j - (j - 1)}{n} = \frac{1}{n}.$$

Thus Y has a uniform discrete distribution on $1, 2, \ldots, n$. ■

Note that U_0 and $T = 1 - U_0$ have the same distribution. Specifically,
for $u \in (0, 1]$ we have

$$\Pr(T \leq u) = \Pr(1 - U_0 \leq u) = \Pr(U_0 \geq 1 - u) = 1 - \Pr(U_0 < 1 - u)$$

$$= 1 - (1 - u) = u.$$

We can then mimic the proof of Theorem 3.8 to show that $\lfloor \ln U_0 / \ln(1 - p) \rfloor$
has a geometric distribution with parameter p. Unfortunately, this small

simplification opens the possibility that U_0 can be zero and consequently, the computation $\ln U_0$ will produce a run-time error. As described in Section 3.1, the implementation of U_0 via rand() cannot assume the value 1. Therefore, $1 - U_0$ cannot be zero, and $\ln(1 - U_0)$ is always well defined. For this technical reason, we will forego the simplification to $\lfloor \ln U_0 / \ln(1 - p) \rfloor$ and generate our geometric random variables with $\lfloor \ln(1 - U_0) / \ln(1 - p) \rfloor$.

In light of Theorem 3.8, the algorithms for the geometric and uniform discrete distributions each contain one line, a computation on the uniform value returned by rand(). From these examples we see that a simpler algorithm results when we are able to calculate the return value, as compared with the case when we must search along the cumulative distribution function. The computations in Theorem 3.8 suggest that the simpler case occurs only when the cumulative distribution function has a manageable closed form. Another method, however, allows the computational approach for certain irregular distributions, provided that the distribution function is not too different from that of another, more easily simulated random variable. This is the rejection filter method, which we will now develop in detail.

Suppose that we have a reasonable simulation for a random variable X. The values $\Pr_X(j)$, for $j = 0, 1, 2, \ldots$, comprise the probability distribution of X. Our need, however, is for a simulation of another random variable Y with the same range as X but with a different probability distribution $\Pr_Y(j)$. If $\Pr_X(j) > \Pr_Y(j)$ for some j, then we expect a higher fraction of j-returns from the X simulation than from a Y simulation. This suggests the possibility of running the X simulation and rejecting some of the j-returns. In particular, we want to reject just enough j-returns to bring the frequency down to $\Pr_Y(j)$. The matter is further complicated because we need to accomplish this balancing act simultaneously for all j.

Consider the situation in Figure 3.11(a). Here both X and Y share the range $\{0, 1, 2, 3\}$. The light bars, on the left at each point, show the probabilities for X; the dark bars show the corresponding probabilities for Y. If we generate N returns from the X simulation, we will find that $0.4N$ are zeros, $0.3N$ are ones, and so forth. We can modify these proportions as follows. When we receive the return 0 from the X simulation, we can accept or reject it in a random fashion by choosing a value u from the continuous uniform distribution on $[0, 1)$. If $u < 1/2$, we accept; otherwise, we reject. In this manner one-half of the 0 returns are ignored, so the accepted returns constitute $0.2N$ of the total. This is, of course, just the right fraction for a Y simulation. Moving to the next range value, we see that we need to reject $2/3$ of the 1 returns to reduce the $0.3N$ fraction to $0.1N$. The next range value, unfortunately, presents a problem. Here we need to generate more 2 returns than are provided by the X simulation. Indeed, we need 1.5 times as many. The situation is even worse with the last range point, where we need 4.0 times as many 3 returns.

We cannot expect the X probabilities to be larger than the Y probabilities across the entire range. Because both probability lists must sum to

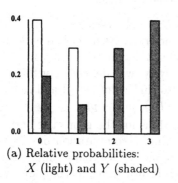

(a) Relative probabilities:
X (light) and Y (shaded)

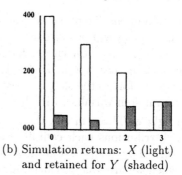

(b) Simulation returns: X (light)
and retained for Y (shaded)

Figure 3.11. Rejection filter concept

1, an X deficit at some range points must appear to compensate for an X excess at other points. We can circumvent this difficulty by keeping all the X returns at the point where the $\Pr_Y(j)/\Pr_X(j)$ is the largest and trimming at the other points to keep the desired final proportions. Figure 3.11(b) implements this idea. The vertical axis now measures the number of returns, rather than the fraction of returns at each range point. The total number of returns is normalized to 1000 to illustrate the calculations.

The highest $\Pr_Y(j)/\Pr_X(j)$ occurs at $j = 3$, where the ratio is 4. Let $K = 4$, a value that will appear in our subsequent calculations. We keep all the X returns of value 3. This policy appears in the figure as a dark bar of the same height as the light bar for range value 3. Consequently, the Y simulation returns 100 threes. Next consider the twos. The X simulation provides 200 of them, but the Y simulation needs only 75 to keep the proper proportion with the 100 threes. The Y probabilities demand that the twos be 3/4 as numerous as the threes, so we accept only 75/200 of these X returns. We accomplish this filtering in a random fashion by choosing a value u from the continuous uniform distribution on $[0, 1)$ and accepting when $u < 75/200$. Notice that $75/200 = \Pr_Y(2)/(K \cdot \Pr_X(2))$. The X simulation returns 300 ones, of which we retain only 25. That is, we keep $\Pr_Y(1)/(K \cdot \Pr_X(1)) = 0.1/1.2 = 1/12$ of them. Continuing this pattern for the remaining range point, 0, we obtain the following acceptance pattern.

Range point, j	0	1	2	3
Number returned by X simulation	400	300	200	100
Number retained for Y simulation	50	25	75	100
$\Pr_Y(j)/(K\Pr_X(j))$	1/8	1/12	3/8	1
Fraction of all Y retentions	0.2	0.1	0.3	0.4

Of the 1000 X returns, 750 are rejected. The remaining 250 distribute across the range points in accordance with the Y probabilities. Although we reject most of the X returns, we do manage to modify the distribution of the accepted points to conform to a new distribution. We will now capture this idea in a more general algorithm.

```
function simulateY ( )
    while (true)
        j = simulateX();
        u = rand();
        if (u < Yprob(j) / (K * Xprob(j)))
            return j;
```

```
function fourpoint ( )
    probratio = (0.5, 0.25, 0.75, 1.0);
    while (true)
        j = floor (4 * rand());
        u = rand();
        if (u < probratio [j])
            return j;
```

(a) Generic algorithm   (b) Adapted to Example 3.10

Figure 3.12. Rejection filter algorithm

The required condition is that there exist a constant $K > 0$ such that $\Pr_Y(j) \leq K \cdot \Pr_X(j)$ for all j. Summing this inequality over all j, we see that $K \geq 1$. We also assume that $\Pr_X(j)$ is positive for all range points, which ensures $0 \leq \Pr_Y(j)/[K\Pr_X(j)] \leq 1$ for all j. Under these conditions, consider the algorithm of Figure 3.12(a). The subroutines Xprob() and Yprob() access the X and Y probability values. The routine simulateX() obtains a sample j from the X distribution. The algorithm performs a simple calculation involving the two probability distributions at the point j, the bounding constant K, and a random number u chosen from a continuous uniform distribution on $[0, 1)$. Sometimes the algorithm returns the sample j, and sometimes it recycles in the while-loop for a new sample. Note that the while-loop condition is always true, so the loop cycles indefinitely until a j sample is deemed worthy of return. (The return statement breaks out of the loop and terminates the algorithm.) It is apparent that the returns from the X simulation are filtered to reduce the frequency of some returns as compared with others. The next theorem shows that the reduction is precisely the amount needed to match the Y probabilities.

3.9 Theorem: Suppose that random variables X and Y share the range $j = 0, 1, 2, \ldots$. Suppose further that $\Pr_X(j) > 0$ for all j and that there exists $K \geq 1$ such that $\Pr_Y(j)/\Pr_X(j) \leq K$ for all j. Then Algorithm 3.12(a) correctly simulates the random variable Y.

PROOF: Let Z be the random variable simulated by the algorithm. We need to show that $\Pr_Z(j) = \Pr_Y(j)$ for $j = 0, 1, 2, \ldots$. Recall that U_0 is the continuous uniform random variable on $[0, 1)$ simulated by rand(). The choice of u is not influenced by the choice of j. Moreover, each while-loop iteration operates independently of any previous iterations; each iteration chooses new random numbers. Consequently, once an iteration starts, its probability of terminating with return value j is the same as that of any other iteration that executes. Let p_j denote the probability that an executing iteration terminates and delivers the return value j.

$$p_j = \Pr_X(j) \cdot \Pr\left(U_0 < \frac{\Pr_Y(j)}{K\Pr_X(j)}\right) = \Pr_X(j) \cdot \frac{\Pr_Y(j)}{K\Pr_X(j)} = \frac{\Pr_Y(j)}{K}$$

For any executing iteration, moreover, the event "iteration terminates algorithm" is the disjoint union of the events "iteration terminates algorithm with return value j," for $j = 0, 1, 2, \ldots$. Let q_T be the probability that an executing iteration terminates the algorithm. Then $q_T = \sum_j \Pr_Y(j)/K = 1/K$.

While p_j and q_T apply to any executing iteration, the probability that a given iteration executes at all depend on its iteration number. The first iteration certainly executes. The second executes only if the first fails to terminate the algorithm. In general, iteration $i+1$ executes only if the previous i iterations all failed to satisfy the termination condition. Therefore, the algorithm terminates after iteration i with probability $(1 - q_T)^{i-1} q_T$. Also, it terminates after iteration i and delivers return value j with probability $(1 - q_T)^{i-1} p_j$. Now, the event $(Z = j)$ is the disjoint union of the events "algorithm terminates with return value j after iteration i," for $i = 1, 2, 3, \ldots$. The following calculation then shows that the algorithm's return values follow the probability distribution for Y.

$$\Pr_Z(j) = \sum_{i=1}^{\infty} (1 - q_T)^{i-1} p_j = p_j \sum_{i=0}^{\infty} \left(1 - \frac{1}{K}\right)^i = K p_j = \Pr_Y(j) \ \blacksquare$$

3.10 Example: Implement a rejection filter to simulate the random variable Y with distribution to the right below.

We assume access to the floor(x) routine (used in earlier examples) to perform the round-down operation $\lfloor x \rfloor$. From Theorem 3.8 we know that floor (4 * rand())

j	0	1	2	3
$\Pr_Y(j)$	0.2	0.1	0.3	0.4

returns a discrete uniform random variable with range $\{0, 1, 2, 3\}$. We will use this as our reference distribution X in the generic algorithm. Since $\Pr_X(j) = 0.25$ for all range values, the largest $\Pr_Y(j)/\Pr_X(j)$ ratio occurs at $j = 3$. Therefore, $K = \Pr_Y(3)/\Pr_X(3) = 0.4/0.25 = 1.6$. $\Pr_Y(j)/(K \Pr_X(j)$ at the four range points constitutes the vector $\{0.5, 0.25, 0.75, 1.0\}$. We incorporate these values directly into the code as the vector probratio in Figure 3.12(b).

A test, involving 10000 calls to Algorithm 3.12(b), produced the frequencies tabulated to the

Range point	0	1	2	3
Return fraction	0.1973	0.1004	0.3055	0.3968

right. Note that they closely match the given Y probabilities. $\Box$

Recall that the inverse transform method requires a search along the cumulative distribution function for the Poisson case, although the geometric case admits a simple calculation. The next example shows that we can use a rejection filter to transform a geometric simulation into a Poisson simulation.

3.11 Example: Using a rejection filter on a geometric simulation, construct an algorithm to simulate X, a Poisson random variable with parameter λ.

Let G be a geometric random variable with parameter p. Then X and G share the range $\{0, 1, 2, \ldots\}$. We will use G as the reference distribution because it is easy to simulate. We will decide on a p parameter later. At the range points, the probability ratios are

$$r_j = \frac{\Pr_X(j)}{\Pr_G(j)} = \frac{e^{-\lambda} \lambda^j}{j! p (1-p)^j}.$$

We need a constant K that exceeds this ratio for all j. Note that

$$\frac{r_{j+1}}{r_j} = \frac{e^{-\lambda} \lambda^{j+1} j! p (1-p)^j}{(j+1)! p (1-p)^{j+1} e^{-\lambda} \lambda^j} = \frac{\lambda}{(j+1)(1-p)}.$$

```
function rejectFilterPoisson (lambda)
    k = ceiling (2.0 * lambda) - 1.0;
    while (true)
        j = floor (ln (1.0 - rand()) / ln (0.5));      // geometric j
        if (rand() < exp ((j - k) * ln (2.0 * lambda)) * factorial (k) / factorial (j))
            return j;
```

Figure 3.13. Poisson distribution via a rejection filter over a geometric distribution (see Example 3.11)

This means that the sequence $r_0, r_1, r_2, \ldots$ is nondecreasing while $\lambda/[(j+1)(1-p)] \geq 1$. This latter condition is equivalent to $j \leq [\lambda/(1-p)] - 1$. The largest ratio then occurs at $k = \lceil \lambda/(1-p) \rceil - 1$. Let $p = 0.5$, which gives $k = \lceil 2\lambda \rceil - 1$. The bound K that we need to adapt the generic algorithm is then

$$K = r_k = \frac{e^{-\lambda}\lambda^k}{k!(1/2)^{k+1}} = \frac{2^{k+1}e^{-\lambda}\lambda^k}{k!}.$$

The rejection bound when the G simulation returns j is then

$$\frac{\Pr_X(j)}{K \cdot \Pr_G(j)} = \frac{e^{-\lambda}\lambda^j}{j!} \cdot \frac{1}{(1/2)(1/2)^j} \cdot \frac{k!}{e^{-\lambda}\lambda^k 2^{k+1}}$$

$$= \frac{(2\lambda)^{j-k}k!}{j!} = \frac{e^{(j-k)\ln(2\lambda)}k!}{j!}.$$

j	Simulation frequency	True Poisson probability
0	0.0808	0.0821
1	0.2147	0.2052
2	0.2545	0.2565
3	0.2109	0.2138
4	0.1323	0.1336
5	0.0636	0.0668
6	0.0288	0.0278
7	0.0106	0.0099
8	0.0028	0.0031
9	0.0009	0.0009
10	0.0001	0.0002

Using these values, we adapt the generic algorithm as shown in Figure 3.13. We assume a ceiling function to round up a real value to the next largest integer. Furthermore, with $p = 1/2$, Theorem 3.8 states that $\lfloor [\ln(1 - U_0)]/[\ln(1/2)] \rfloor$ is geometrically distributed when U_0 is the continuous uniform distribution on $[0, 1)$. The program computes $(2\lambda)^{j-k}$ as $e^{(j-k)\ln(2\lambda)}$. A test run exercised Algorithm 3.13 10000 times to simulate a Poisson random variable with parameter $\lambda = 2.5$. The relative frequencies of the returned values, shown to the right, compare favorably with the expected probabilities. The more complicated computations, repeated exponentials and logarithms, result in slower performance than the inverse transform algorithm (see Example 3.7). $\square$

Exercises

3.9 Develop an algorithm that searches along the cumulative distribution function to generate samples from a geometric distribution with parameter p. For $p = 0.25$, perform 1000 executions and compare the relative frequencies of the returned values with the corresponding probabilities $(1 - p)^k p$.

3.10 Develop an algorithm that returns samples from the negative binomial distribution with parameters (n, p). Perform 1000 executions and compare the relative frequencies of the returned values with the corresponding probabilities $C_{n+k-1,k}p^n(1-p)^k$.

3.11 Example 3.7 develops an algorithm that returns samples from a Poisson distribution with parameter m. Let the random variable N be the number of while-loop iterations in a given execution. If P is the Poisson random variable, show that $\mu_N = E_P[|P-m|]$. Develop a simulation that approximates $E_P[|P-m|]$. Run the simulation and show that $\mu_N \approx \sqrt{2m/\pi}$.

3.12 Let random variable X have the distribution to the right. Construct an algorithm similar to Figure 3.5(a) that samples from this distribution. Let N be the number of while-loop iterations in a given run. Show that N has the same range as X, and furthermore, $\Pr_N(k) = \Pr_X(k)$ for all k in the common range.

k	0	1	2	3	4
$\Pr_X(k)$	0.3	0.1	0.4	0.1	0.1

3.13 Use Theorem 3.8 to construct an algorithm that generates samples from a geometric distribution. The parameter p should be an input variable. Test the algorithm for $p = 0.25$ by generating 10000 samples and computing the relative frequencies of returns $0, 1, 2, \ldots$. Compare these relative frequencies with the the probabilities expected for a geometric distribution with parameter $p = 0.25$.

3.14 Use Theorem 3.8 and some additional scaling to construct an algorithm that generates samples from a discrete uniform distribution on the even integers $6, 8, 10, 12, \ldots, 20$. Test the algorithm by generating 1000 samples and computing the relative frequencies of the returns.

3.15 Let X be a geometric random variable with parameter p. The range of X is the nonnegative integers. Theorem 3.8 asserts that $Z = \lfloor \ln(1 - U_0)/(1-p) \rfloor$ has the same distribution. What is the actual range of Z when U_0 is simulated by rand() on a 32-bit machine?

3.16 Suppose that X has the following distribution.

x	0	1	2	3	4	5	6
$\Pr_X(x)$	0.01	0.02	0.04	0.08	0.16	0.32	0.37

Devise an algorithm that generates samples from this distribution without searching along the cumulative distribution function. That is, the algorithm should compute the return in the spirit of Theorem 3.8. Prove that your algorithm delivers the expected result.

3.17 Implement a rejection filter over a discrete uniform distribution to simulate the random variable X with the following distribution. Test the algorithm by generating 10000 samples and computing the relative frequencies of the range points.

j	0	1	2	3	4	5	6
$\Pr_X(j)$	0.3	0.15	0.05	0.2	0.1	0.05	0.15

3.18 Implement a rejection filter over a discrete uniform distribution to simulate a binomial random variable with parameters $(n = 15, p = 0.4)$. Test the algorithm by generating 10000 samples and computing the relative frequencies of the range points.

3.19 Referring to Algorithm 3.12(a), the generic rejection filter algorithm, let the random variable N be the number of while-loop iterations. Find the distribution of N. Find the mean and variance of N.

3.20 Random variable X has the distribution tabulated to the right. Implement a rejection filter over a discrete uniform distribution to simulate X. Test the algorithm by generating 10000 samples and computing the relative frequencies of the range points.

j	0	3	8	12	15
$\Pr_X(j)$	0.3	0.15	0.4	0.07	0.08

*3.21 Let U_0 be the continuous uniform random variable on $[0, 1)$. Find a function f such that $Y = f(U_0)$ has the property $\Pr(Y < y) = 1 - e^{-y}$ for $y \geq 0$.

3.3 Client-server systems

At this point we have algorithms for simulating binomial and Poisson distributions, either of which can model the number of events in some simulation interval. However, the simulation may need more detail. In particular, it may need to know the event locations within the simulation interval in order to trigger internal system changes. For example, a customer arrival in a bank lobby requires the simulation to update the queue length for at least one teller or, in the event of an idle teller, to engage that service. Suppose that the simulation interval contains n slots, each the locus of a potential event. Assuming sufficiently narrow slots, at most one event occurs in each slot, say with probability p, and events in distinct slots are independent. This input assumption corresponds to a Bernoulli process with parameter p. The total number of events is then a random variable N having a binomial distribution with parameters (n, p). The event *locations* comprise a random binary n-vector V, in which the 1s mark the event locations. There are a number of approaches to simulating this vector.

The simulation can generate n Bernoulli random variables $X_i, 1 \leq i \leq n$. It then computes $V = (X_1, \ldots, X_n)$ and $N = \sum_{i=1}^{n} X_i$. This is clearly the most direct approach because it reflects the action, or lack thereof, at each slot in the physical situation. Using a different approach, the simulation can first generate the number of events N and then generate the locations V. Equation 2.8 notes the uniform conditional distribution across the $C_{n,k}$ vectors V, given $N = k$. That is, all valid location vectors are equally likely, given a

```
function BernoulliVector (n, p )
    for (i = 1 to n)
        v[i] = 0;
    i = 1 + floor (ln(1.0 - rand()) / ln(1.0 - p));
    while (i ≤ n)
        v[i] = 1;
        i = i + 1 + floor (ln(1.0 - rand()) / ln(1.0 - p));
    return v;
```

Figure 3.14. Simulation of event locations in a Bernoulli process

---

specific total number of events. So, when the first phase generates $N = k$ as the total number of events, the second phase must generate the location vector from a discrete uniform distribution over the $C_{n,k}$ candidates. To implement the second phase, we can list all n-vectors with k ones in lexicographical order and generate a uniform index $i = \lfloor C_{n,k} U_0 \rfloor + 1$ into the list. Recall that U_0 is the continuous uniform random variable on $[0, 1)$, for which we use the system routine rand().

A third approach simulates the number of uneventful slots before each event. This technique generalizes to the Poisson case, so we develop it in greater detail. The situation at hand is a Bernoulli process, and we know that the number of failures before the first event has a geometric distribution with parameter p. The number of failures between subsequent events follows the same distribution. Let $G_1, G_2, \ldots$ be a sequence of independent geometric random variables, all with parameter p. We can sample from any one of these distributions with the computation $\lfloor \ln(1 - U_0) / \ln(1 - p) \rfloor$. The simulation fills the vector V sequentially with G_1 zeros followed by a 1, then G_2 zeros followed by a 1, and so forth until the n slots are specified. The algorithm of Figure 3.14 provides the details. Because N, the number of 1s in V is the renewal count for the sequence $G_1 + 1, G_2 + 1, \ldots$, Theorem 2.50 assures us that N has the binomial distribution with parameters (n, p).

We are now ready to simulate a small system. Although we will present a more generic client-server simulation shell at a later point, this example is not complicated, and we can easily assemble a simulation by tracking the arrivals and departures at the server.

3.12 Example: Suppose that a server admits requests from a single anonymous client in a quantized fashion. In particular, the server opens a turnstile at the beginning of each millisecond to admit a request with probability $p = 0.1$. If the server is busy at the time, the request enters a waiting queue. The service policy is first-in, first out (FIFO) from the queue. The server takes a random amount of time to respond to a request. The service time is a discrete uniform random variable on $\{7, 8, 9, 10, 11\}$ milliseconds. We are interested in the following averages over a 5-second period: the average time a request waits in the queue, the average time a request resides in the system, and the total time that the server is idle.

An 5-second interval contains 5000 milliseconds, so the number of requests is binomially distributed with parameters $(n = 5000, p = 0.1)$. The server can expect $np = 500$ requests on the average. The average response time is $(1/5)(7 + 8 + 9 + 10 + 11) = 9.0$ milliseconds, which means that the server can handle $5000/9 =$

555.56 requests, on the average. So we do not expect the receiving queue to grow indefinitely. The queue will simply buffer probabilistic variations from the average. To estimate the desired averages, we resort to the simulation of Figure 3.15.

The simulation maintains a list of pending events, each a vector of the form (type, arrive, exitQ, exitSys), in the variable eventList. The type is zero or 1, designating an arrival event or a departure event, respectively. The remaining vector components contain the millisecond slot number in which the request arrives, exits the wait queue, or leaves the system. The eventList maintains pending events in chronological order; at the front of the list is the event with the earliest timestamp. The timestamp is the arrive time for an arrival event; it is the exitSys time for a departure event.

Two routines interface with the eventList; the insert operation places a new event in the list, and first removes the earliest event. After some initialization, the algorithm seeds the eventList with a first arrival and then commences a loop that repeatedly removes the first event for processing. The simulation halts when the eventList becomes empty. When the algorithm processes an arrival, it generates the next arrival for the eventList, provided that its arrival slot is not beyond N, an input parameter that controls the simulation length. For this example, we run the simulation with N = 5000.

If the server is busy (serverBusy $= 1$) when the algorithm processes an arrival, it places the request in a waitQueue. The two routines enqueue and dequeue insert and remove requests from this queue. If, after processing a departure, the algorithm finds a waiting entry, it does not free the server but rather removes the head of the waitQueue, changes it to a departure event, and inserts it into the eventList. Instead of computing the entire N-vector of arrival requests, the algorithm acquires them as needed by generating the number of fallow slots between arrivals. This number is, of course, a geometric random variable with parameter $p = 0.1$, which we simulate with the calculation $\lfloor \ln(1 - U_0) / \ln(0.9) \rfloor$. The random variable U_0, continuous and uniform on $[0, 1)$, is available with rand(), and we use it to generate departure times via $\lfloor 5U_0 \rfloor + 7$. Further details appear in Figure 3.15. This is a systematic approach to an event-driven simulation, but it is not the most generic template. For the moment, however, it suffices to introduce the subject.

Let us follow the eventList and waitQueue through several cycles. Table 3.2 provides this information for a short simulation of 100 milliseconds. The values reflect the status at the top of the while-loop. On the first cycle, the eventList contains the first arrival (at time 1) and an empty waitQueue. Upon removing the arrival event, the algorithm generates the following arrival (at time 11) and places it in the eventList. Since the server is available, the exitQ time becomes the arrive time, the difference of zero then signifying that the request spent no time in the waitQueue. The algorithm adds the service time, calculated from a uniform distribution on $\{7, 8, 9, 10, 11\}$, to the exitQ time to produce the exitSys time. After changing the type to 1, the algorithm places this departure event in the eventList. Because the departure time, 12, exceeds the arrival time of the next request, 11, the departure record appears second in the eventList, as we see at the beginning of the second cycle.

The second cycle processes the second arrival, (0, 11, 0, 0), and generates another, (0, 15, 0, 0). Because the server is now busy, (0, 11, 0, 0) enters the wait-Queue, as shown at the beginning of cycle 3. Cycle 3 processes the departure event. It then removes the waiting request from the waitQueue, calculates its departure

```
function clientServer1 (N)        // input N is total milliseconds to simulate
    totQTime = 0.0;        // initialize to accumulate total time in required categories
    totSysTime = 0.0; totIdleTime = 0.0;
    n = 0;        // number of requests processed
    serverBusy = 0;        // zero denotes server is free
    lastService = 0;        // start of an idle span of slots
    clear(eventList); clear(waitQueue);
    j = floor (ln(1.0 - rand()) / ln(0.9)) + 1;        // generate first arrival time
    type = 0; arrive = j; exitQ = 0, exitSys = 0;        // first event, arrival = type 0
    if (j ≤ N)
        insert(eventList, (type, arrive, exitQ, exitSys));        // schedule first arrival
    while (notEmpty(eventList))        // loop until no further events
        (type, arrive, exitQ, exitSys) = first (eventList);        // earliest pending event
        if (type = 0)        // process arrival
            n = n + 1;
            j = arrive + floor(ln(1.0 - rand()) / ln(0.9)) + 1;        // next arrival time
            if (j ≤ N)        // conditionally schedule next arrival
                insert(eventList, (0, j, 0, 0));
            if (serverBusy)        // new arrival to wait queue
                enqueue(waitQueue, (type, arrive, exitQ, exitSys));
            else
                totIdleTime = totIdleTime + arrive - lastService - 1;
                serverBusy = 1;
                type = 1;        // change to departure event
                exitQ = arrive;        // no wait queue time
                exitSys = exitQ + floor (5.0 * rand()) + 7;        // departure time
                insert(eventList, (type, arrival, exitQ, exitSys));        // schedule departure
        else        // process departure
            totQTime = totQTime + exitQ - arrive;        // accumulate wait queue time
            totSysTime = totSysTime + exitSys - arrive;        // accumulate system time
            now = exitSys;        // save departure time
            if (notEmpty(waitQueue))
                (type, arrive, exitQ, exitSys) = dequeue(waitQueue);
                type = 1;        // change to departure event
                exitQ = now;
                exitSys = now + floor(5.0 * rand()) + 7;        // generate departure time
                insert(eventList, (type, arrive, exitQ, exitSys));        // schedule departure
            else
                serverBusy = 0;
                lastService = now - 1;
    avgQTime = totQTime / n; avgSysTime = totSysTime / n;
    return (avgQTime, avgSysTime, totIdleTime);
```

Figure 3.15. Client-server simulation: Bernoulli arrivals and uniform
                service times (see Example 3.12)

Cycle	eventList				waitQueue			
	type	arrive	exitQ	exitSys	type	arrive	exitQ	exitSys
1	0	1	0	0	—empty—			
2	0	11	0	0	—empty—			
	1	1	1	12				
3	1	1	1	12	0	11	0	0
	0	15	0	0				
4	0	15	0	0	—empty—			
	1	11	12	21				
5	1	11	12	21	0	15	0	0
	0	29	0	0				
6	0	29	0	0	—empty—			
	1	15	21	30				
7	1	15	21	30	0	29	0	0
	0	32	0	0				
8	0	32	0	0	—empty—			
	1	29	30	38				
9	0	37	0	0	0	32	0	0
	1	29	30	38				
10	1	29	30	38	0	32	0	0
	0	39	0	0	0	37	0	0
11	0	39	0	0	0	37	0	0
	1	32	38	47				
12	1	32	38	47	0	37	0	0
	0	61	0	0	0	39	0	0
13	1	37	47	58	0	39	0	0
	0	61	0	0				
14	0	61	0	0	—empty—			
	1	39	58	65				
15	1	39	58	65	0	61	0	0
	0	84	0	0				
16	1	61	65	74	—empty—			
	0	84	0	0				
17	0	84	0	0	—empty—			
18	0	91	0	0	—empty—			
	1	84	84	92				
19	1	84	84	92	0	91	0	0
20	1	91	92	101	—empty—			

TABLE 3.2. The eventList and waitQueue for the client-server simulation (see Example 3.12)

time, and places it in the eventList. This gives the data at the beginning of cycle 4. This pattern continues until the processing of an arrival event generates a new arrival beyond the simulation time limit. This new arrival is discarded, which allows the eventList to empty.

While processing a departure, the algorithm calculates the waitQueue time (exitQ - arrive) and the system time (exitSys - arrive) for the request. It updates accumulators for both these variables. The algorithm detects server idle time while processing an arrival request when the server is not busy. It maintains another accumulator for this variable. Finally, the algorithm returns a 3-vector that specifies the average queue time and the average system time per request and the total server idle time—all in milliseconds.

Run	Average queue time (milliseconds)	Average system time (milliseconds)	Total server idle time (milliseconds)
1	37.5	46.5	236.0
2	23.0	32.0	688.0
3	29.4	38.4	496.0
4	48.4	57.5	602.0
5	43.4	52.4	487.0
6	42.5	51.6	623.0
7	29.1	38.1	613.0
8	29.9	38.9	581.0
9	25.9	35.0	1016.0
10	68.4	77.4	155.0

TABLE 3.3. Client-server simulation results (see Example 3.12)

How accurate are these returns? Table 3.3 shows the results of 10 simulation runs, each of 5000-millisecond duration. Let T be the average queue time per request. The simulations provides samples of T that vary from 29 to 68 milliseconds over the 10 runs, with an overall mean of 37.8 milliseconds. In general, let T_i be the average wait queue time of the ith simulation and define $\overline{T_n} = (1/n)\sum_{i=1}^{n} T_i$. Following the argument of Example 3.5, if we assume that n simulation runs deliver independent samples from a fixed but unknown distribution for T, then the mean of $\overline{T_n}$ is μ_T and the variance is σ_T^2/n. By making yet more simulation runs, if necessary, we can eventually drive the variance of $\overline{T_n}$ to a very small value, which means that there is only a small probability that our observed $\overline{T_n}$ will differ from μ_T by a significant amount.

Is $n = 10$ sufficient? We are not yet in a position to answer that question. However, we can draw some confidence from the fact that $n = 20$ produces $\overline{T_{20}} = 39.8$ milliseconds, which is not very different from $\overline{T_{10}} = 37.8$. We can express our lack of confidence in a precise value by stating that the average wait queue time is about 38 milliseconds per request. $\square$

We turn now to the situation where the arrivals follow a Poisson distribution with parameter λ. We know how to simulate the number of arrivals in the base interval (Algorithm 3.9), but this process does not give us the actual arrival times. For the binomial case, we exploited the fact that the number of uneventful slots between events is a geometric random variable. We have a similar situation in the Poisson case because the interarrival intervals are independent random variables E_i, each characterized by $\Pr(E_i > s) = e^{-\lambda s}$. Letting, as usual, U_0 denote the continuous uniform random variable on $[0, 1)$, consider the expression $E = (-1/\lambda)\ln(1 - U_0)$. We have

$$\Pr(E > s) = \Pr(\ln(1 - U_0) < -\lambda s) = \Pr((1 - U_0) < e^{-\lambda s})$$
$$= 1 - \Pr(U_0 \leq 1 - e^{-\lambda s}) = 1 - (1 - e^{-\lambda s}) = e^{-\lambda s}. \qquad (3.4)$$

That is, $-\ln(1 - U_0)/\lambda$ follows the proper probability law for E_i. Consequently, we can simulate an interarrival span by simulating $-\ln(1 - U_0)/\lambda$. Of course, rand() is available to simulate U_0. These considerations lead to the algorithm of Figure 3.16, which delivers a vector of event times within

```
function poissonVector (lambda)
    v = ();      // initialize to empty vector – no events
    T = -ln (1.0 - rand()) / lambda;      // time of first event
    i = 0;
    while (T ≤ 1)
        i = i + 1;
        v[i] = T;
        T = T - ln (1.0 - rand()) / lambda;      // time of next event
    return v;
```

Figure 3.16. Simulation of event times in a Poisson process

---

the base interval $[0, 1]$. The length of the vector returned is itself a random variable, which depends on when the accumulating interarrival times exceed 1. We will show that this vector length is a Poisson random variable with parameter λ. The proof needs the following technical result concerning the sum of independent exponential random variables.

3.13 Theorem: Let $X_1, X_2, \ldots$ be independent continuous random variables with common cumulative distribution function $F(t) = 1 - e^{-\lambda t}$. Let $S_n = \sum_{i=1}^{n} X_i$. Then S_n is a continuous random variable with cumulative distribution function $F_n(t) = 1 - e^{-\lambda t} \sum_{j=0}^{n-1} (\lambda t)^j / j!$.

PROOF: Let $T_n(t) = 1 - e^{-\lambda t} \sum_{j=0}^{n-1} (\lambda t)^j / j!$. We want to show $F_n(t) = T_n(t)$ for $n = 1, 2, \ldots$. For $n > 1$, we have

$$
T_n'(t) = \lambda e^{-\lambda t} \sum_{j=0}^{n-1} \frac{(\lambda t)^j}{j!} - e^{-\lambda t} \sum_{j=1}^{n-1} \frac{j(\lambda t)^{j-1}\lambda}{j!}
$$

$$
= \lambda e^{-\lambda t} \left[\sum_{j=0}^{n-1} \frac{(\lambda t)^j}{j!} - \sum_{j=1}^{n-1} \frac{(\lambda t)^{j-1}}{(j-1)!} \right]
$$

$$
= \lambda e^{-\lambda t} \left[\sum_{j=0}^{n-1} \frac{(\lambda t)^j}{j!} - \sum_{j=0}^{n-2} \frac{(\lambda t)^j}{j!} \right] = \frac{\lambda e^{-\lambda t}(\lambda t)^{n-1}}{(n-1)!}.
$$

Noting that $T_1'(t) = \lambda e^{-\lambda t}$, we conclude that this expression holds for $n = 1$ as well. We now proceed by induction toward the main result. For the base case, note that $S_1 = X_1$. Therefore, $F_1(t) = F(t) = 1 - e^{-\lambda t} = T_1(t)$. Now suppose that $n > 1$ and $F_i(t) = T_i(t)$ for $1 \le i < n$. We will show that $F_n(t) = T_n(t)$, which will complete the proof. Fix $t > 0$ and consider the partition $0 = y_0 < y_1 < y_2 < \cdots < y_m = t$ of the interval $[0, t]$, where $y_k - y_{k-1} = \Delta y = t/m$ for $k = 1, 2, \ldots, m$.

The disjoint events of Table 3.4 are subsets of the event $(S_n > t)$. The probability of the first-line event is just $1 - F_1(t) = e^{-\lambda t}$. We calculate the probabilities for the m events on the second line as follows. We use the induction hypothesis to reduce the F_1 difference to a T_1 difference and then the mean value theorem of calculus to obtain the final expression.

Events $(0 \le k \le m-1)$	Probability
$(X_1 > t)$	$e^{-\lambda t}$
$(S_1 \in (y_k, y_k + \Delta y]) \cap (X_2 > t - y_k)$	$T_1'(\tau_{1k})\Delta y e^{-\lambda(t-y_k)}$
$(S_2 \in (y_k, y_k + \Delta y]) \cap (X_3 > t - y_k)$	$T_2'(\tau_{2k})\Delta y e^{-\lambda(t-y_k)}$
$\vdots$	$\vdots$
$(S_{n-1} \in (y_k, y_k + \Delta y]) \cap (X_n > t - y_k),$	$T_{n-1}'(\tau_{n-1,k})\Delta y e^{-\lambda(t-y_k)}$

TABLE 3.4. Disjoint events comprising $(S_n > t)$ (see Theorem 3.13)

$$\begin{aligned} \Pr(S_1 \in (y_k, y_k + \Delta y]) &= F_1(y_k + \Delta y) - F_1(y_k) \\ &= T_1(y_k + \Delta y) - T_1(y_k) = T_1'(\tau_{1k})\Delta y \end{aligned}$$

Finally, by independence of S_1 and X_2, we have

$$\Pr(S_1 \in (y_k, y_k + \Delta y) \cap (X_2 > t - y_k) = T_1'(\tau_{1k})\Delta y \cdot e^{-\lambda(t-y_k)}.$$

The points τ_{1k} lie in the intervals $(y_k, y_k + \Delta y)$. The remaining lines follow from similar calculations. Since all these disjoint events lie within the event $(S_n > t)$, we have

$$\begin{aligned} 1 - F_n(t) = \Pr(S_n > t) &\ge e^{-\lambda t} + \sum_{j=1}^{n-1}\sum_{k=0}^{m-1} T_j'(\tau_{jk})\Delta y \cdot e^{-\lambda(t-y_k)} \\ &= e^{-\lambda t} + \sum_{j=1}^{n-1}\sum_{k=0}^{m-1} \lambda e^{-\lambda \tau_{jk}} \cdot \frac{(\lambda \tau_{jk})^{j-1}}{(j-1)!} \cdot \Delta y e^{-\lambda(t-y_k)} \end{aligned}$$

$$1 - F_n(t) = e^{-\lambda t} + e^{-\lambda t}\sum_{j=1}^{n-1} \frac{\lambda^j}{(j-1)!} \cdot \sum_{k=0}^{m-1} \tau_{jk}^{j-1} e^{-\lambda(\tau_{jk}-y_k)}\Delta y$$

$$\ge e^{-\lambda t}\left[1 + \sum_{j=1}^{n-1} \frac{\lambda^j}{(j-1)!} \cdot e^{-\lambda \Delta y} \cdot \sum_{k=0}^{m-1} \tau_{jk}^{j-1}\Delta y\right].$$

The last line follows because $\tau_{jk} \in (y_k, y_k + \Delta y)$ and therefore cannot differ by more than Δy from the endpoint y_k. So $e^{-\lambda(\tau_{jk}-y_k)} \ge e^{-\lambda \Delta y}$. We now let $m \to \infty$, which forces $\Delta y \to 0$ and converts the innermost summation to an integral.

$$\begin{aligned} 1 - F_n(t) &\ge e^{-\lambda t}\left[1 + \sum_{j=1}^{n-1} \frac{\lambda^j}{(j-1)!} \cdot \int_0^t y^{j-1}\,dy\right] \\ &= e^{-\lambda t}\left[1 + \sum_{j=1}^{n-1} \frac{\lambda^j}{(j-1)!} \cdot \frac{t^j}{j}\right] = e^{-\lambda t}\sum_{j=0}^{n-1} \frac{(\lambda t)^j}{j!} \end{aligned}$$

$$F_n(t) \le 1 - e^{-\lambda t}\sum_{j=0}^{n-1} \frac{(\lambda t)^j}{j!} = T_n(t)$$

The reverse inequality involves a similar computation. Consider the same partition of $[0, t]$ and note that, for $k = 0, 1, \ldots, m - 1$, the following disjoint events are all contained in the event $(S_n \le t)$. As before, $\tau_k \in (y_k, y_k + \Delta y]$.

Event(s)	$(S_{n-1} \in (y_k, y_k + \Delta y]) \cap (X_n \le t - y_k - \Delta y), k = 0, 1, 2, \ldots, m-1$
Probability	$T'_{n-1}(\tau_k)\Delta y[1 - e^{-\lambda(t-y_k-\Delta y)}]$

Consequently,

$$F_n(t) = \Pr(S_n \le t) \ge \sum_{k=0}^{m-1} \lambda e^{-\lambda \tau_k} \frac{(\lambda \tau_k)^{n-2}}{(n-2)!} \Delta y \cdot (1 - e^{-\lambda(t-y_k-\Delta y)})$$

$$= \frac{\lambda^{n-1}}{(n-2)!} \sum_{k=0}^{m-1} \tau_k^{n-2} e^{-\lambda \tau_k} \Delta y$$

$$- \frac{\lambda^{n-1} e^{-\lambda(t-\Delta y)}}{(n-2)!} \sum_{k=0}^{m-1} \tau_k^{n-2} e^{-\lambda(\tau_k - y_k)} \Delta y$$

Because $\tau_k - y_k \ge 0$, we have $e^{-\lambda(\tau_k - y_k)} \le 1$. Consequently,

$$F_n(t) \ge \frac{\lambda^{n-1}}{(n-2)!} \sum_{k=0}^{m-1} \tau_k^{n-2} e^{-\lambda \tau_k} \Delta y - \frac{\lambda^{n-1} e^{-\lambda(t-\Delta y)}}{(n-2)!} \sum_{k=0}^{m-1} \tau_k^{n-2} \Delta y.$$

Letting $m \to \infty$, we have $\Delta y \to 0$, and

$$F_n(t) \ge \frac{\lambda^{n-1}}{(n-2)!} \int_0^t y^{n-2} e^{-\lambda y} \, dy - \frac{\lambda^{n-1} e^{-\lambda t}}{(n-2)!} \int_0^t y^{n-2} \, dy.$$

The integrals involve only mechanical calculus, so we will leave it to the exercises to show that, for $n \ge 2$,

$$\frac{\lambda^{n-1}}{(n-2)!} \int_0^t y^{n-2} e^{-\lambda y} \, dy = 1 - e^{-\lambda t} \sum_{j=0}^{n-2} \frac{(\lambda t)^j}{j!}.$$

Then

$$F_n(t) \ge \left(1 - e^{-\lambda t} \sum_{j=0}^{n-2} \frac{(\lambda t)^j}{j!} \right) - \frac{\lambda^{n-1} e^{-\lambda t}}{(n-2)!} \cdot \frac{t^{n-1}}{(n-1)}$$

$$= 1 - e^{-\lambda t} \sum_{j=0}^{n-1} \frac{(\lambda t)^j}{j!} = T_n(t).$$

Combining this result with the earlier inequality, we have $F_n(t) = T_n(t)$. $\blacksquare$

Theorem 3.13 has a more elegant proof, but it involves techniques with continuous random variables that we have not yet developed. We will revisit the matter in a later chapter when we have more powerful tools at our disposal. In any case, it is only an intermediate result. The next theorem uses it to show that the vector returned by Algorithm 3.16 has a Poisson distribution.

3.14 Theorem: Let N be the length of the vector returned by Algorithm 3.16 when called with argument λ. Then N has a Poisson distribution with parameter λ.

PROOF: We need to show that $\Pr(N = k) = e^{-\lambda}\lambda^k/k!$, for $k = 0, 1, 2, \ldots$. We proceed by induction on k. The algorithm returns a zero-length vector when the initial assignment to variable T exceeds one. Invoking Equation 3.4, we compute

$$\Pr(N = 0) = \Pr\left(\frac{-\ln(1 - U_0)}{\lambda} > 1\right) = e^{-\lambda} = \frac{e^{-\lambda}\lambda^0}{0!}.$$

For $k \geq 1$, we accept the induction hypothesis $\Pr(N = j) = \lambda^j e^{-\lambda}/j!$ for $j = 0, 1, 2, \ldots, k - 1$. Then

$$\Pr(N < k) = \sum_{j=0}^{k-1} \frac{e^{-\lambda}\lambda^j}{j!}$$

$$\Pr(N \geq k) = 1 - \Pr(N < k) = 1 - \sum_{j=0}^{k-1} \frac{e^{-\lambda}\lambda^j}{j!}.$$

Let successive executions of the computation $-[\ln(1 - U_0)]/\lambda$ yield the random variables $X_1, X_2, \ldots$, and let $S_n = \sum_{i=1}^{n} X_i$. Theorem 3.13 gives the cumulative distribution function of S_n:

$$F_n(t) = 1 - e^{-\lambda t}\sum_{j=0}^{n-1} \frac{(\lambda t)^j}{j!}.$$

Note that $N > k$ if and only if $S_{k+1} \leq 1$, so the events $(S_{k+1} \leq 1)$ and $(N > k)$ contain the same outcomes. Therefore,

$$\Pr(N > k) = \Pr(S_{k+1} \leq 1) = F_{k+1}(1) = 1 - e^{-\lambda}\sum_{j=0}^{k} \frac{\lambda^j}{j!}$$

$$\Pr(N = k) = \Pr(N \geq k) - \Pr(N > k)$$

$$= 1 - \sum_{j=0}^{k-1} \frac{e^{-\lambda}\lambda^j}{j!} - \left(1 - e^{-\lambda}\sum_{j=0}^{k} \frac{\lambda^j}{j!}\right) = \frac{e^{-\lambda}\lambda^k}{k!}.$$

We conclude that N has the Poisson distribution with parameter λ. ∎

Algorithm 3.16 generates a vector of arrival times that follow a Poisson distribution with parameter λ. It computes the next arrival time by adding an interarrival increment, $-\ln(1 - U_0)/\lambda$, to the previous arrival time. In a simulation, however, we can generate the arrivals as needed, rather than generating the entire vector at the outset. This is the same policy that we followed with the Bernoulli process. We generated the next event by adding a randomly generated number of uneventful slots to the position of the last event. Another useful observation derives from our earlier discussion of Poisson distributions, in which we determined that we can scale the base interval to suit our convenience (see Section 2.4). If we double the base interval, for instance, then we simply halve λ. The next example illustrates the point.

3.15 Example: Revisit the client-server system of Example 3.12 with the new assumptions. Client requests now arrive according to a Poisson distribution at an average rate of 100 per second, and the service time is a continuous uniform random variable on the interval [7, 11) milliseconds.

We refer to Example 3.12 as the quantized example. We keep the total simulation time at 5 seconds as in the quantized example. We therefore expect an average of 500 requests, each of which requires, on the average, about 9 milliseconds to discharge. Nine milliseconds is the midpoint of the service interval, which spans 7 to 11 milliseconds. The server can then expect to be busy for 4500 milliseconds of the 5000-millisecond total. There will be, of course, probabilistic variations that expand or contract the waiting queue, but our general expectation is that the server can keep up with the requests over the long term. As in the quantized example, we do not expect the wait queue to grow indefinitely. Indeed, the averages are the same as in the quantized example, so we expect similar results for the average time a request waits in the queue, the average time it spends in the system, and the total idle time for the server.

If we adopt 1 millisecond as the base interval, the λ for the Poisson arrival process is 0.1. The computation $-\ln(1 - U_0)/\lambda$ then produces interarrival times in milliseconds. We use Algorithm 3.1(a) to simulate the server time for a given request. We invoke this algorithm with the call uniform (7, 11). The logistics for handling the event list and the wait queue are the same as for the quantized example, so the code in Figure 3.17 requires little additional explanation. The various time markers and accumulators (e.g., totQtime, totSystime, totIdleTime, lastService) now hold continuous time values, in milliseconds, rather than slot counts.

The results of 20 simulation runs appear in Figure 3.18. Over the 20 runs, the average queue time per request has a mean of 41.2 milliseconds, which compares closely with the 39.8 milliseconds obtained from the quantized example. □

Examples 3.12 and 3.15 illustrate quantized and continuous client-server applications with one input process, one wait queue, and one server. The situation generalizes to multiple input streams, multiple servers, and arbitrary queue protocols. We can have, for example, several sources of requests, each operating under its own probabilistic rule (e.g., Poisson or Bernoulli process) with its own particular parameters. The buffering component can contain a separate queue for each input stream, a single queue for all streams, or some combination of these extremes. A server can deal with a particular input stream, or with a specified group of streams, or with all comers. The servers can differ in the probabilistic distribution of service time. The parameters for the input streams, the queues, and the servers can change during the course of the simulation. Customer arrivals at a bank teller, for example, can follow a Poisson distribution with $\lambda = 10$ per hour in the morning and change to $\lambda = 20$ per hour in the afternoon.

Figure 3.19 illustrates a generic organizational plan that can accommodate many of these variations. A set of input processes, each governed by

```
function clientServer2 (T, lambda)       // input T is total milliseconds to simulate;
    totQTime = 0.0;       // initialize to accumulate total time in required categories
    totSysTime = 0.0; totIdleTime = 0.0;
    n = 0;       // number of requests processed
    serverBusy = 0;       // zero denotes server is free
    lastService = 0;       // start of an idle period
    clear(eventList); clear(waitQueue);
    t = -ln(1.0 - rand() / lambda;       // generate first arrival time
    type = 0; arrive = t; exitQ = 0, exitSys = 0;       // first event, arrival = type 0
    if (t ≤ T)
        insert(eventList, (type, arrive, exitQ, exitSys));       // schedule first arrival
    while (notEmpty(eventList))       // loop until no further events
        (type, arrive, exitQ, exitSys) = first (eventList);       // earliest pending event
        if (type = 0)       // process arrival
            n = n + 1;
            t = arrive - ln(1.0 - rand() / lambda;       // next arrival time
            if (t ≤ T)       // conditionally schedule next arrival
                insert(eventList, (0, t, 0, 0));
            if (serverBusy)       // new arrival to wait queue
                enqueue(waitQueue, (type, arrive, exitQ, exitSys));
            else
                totIdleTime = totIdleTime + arrive - lastService;
                serverBusy = 1;
                type = 1;       // change to departure event
                exitQ = arrive;       // no wait queue time
                exitSys = exitQ + uniform (7, 11);       // generate departure time
                insert(eventList, (type, arrival, exitQ, exitSys));       // schedule departure
        else       // process departure
            totQTime = totQTime + exitQ - arrive;       // accumulate wait queue time
            totSysTime = totSysTime + exitSys - arrive;       // accumulate system time
            now = exitSys;       // save departure time
            if (notEmpty(waitQueue))
                (type, arrive, exitQ, exitSys) = dequeue(waitQueue);
                type = 1;       // change to departure event
                exitQ = now;
                exitSys = now + uniform (7, 11);       // generate departure time
                insert(eventList, (type, arrive, exitQ, exitSys));       // schedule departure
            else
                serverBusy = 0;
                lastService = now;
    avgQTime = totQTime / n; avgSysTime = totSysTime / n;
    return (avgQTime, avgSysTime, totIdleTime);
```

Figure 3.17. Client-server simulation algorithm with Poisson arrivals
and uniform service (see Example 3.15)

Run	Average queue time (msec)	Average system time (msec)	Total server idle time (msec)	Run	Average queue time (msec)	Average system time (msec)	Total server idle time (msec)
1	38.45	47.46	417.25	11	97.12	106.13	360.29
2	28.87	37.90	621.72	12	38.50	47.55	502.18
3	38.84	47.80	583.33	13	30.27	39.30	513.20
4	24.08	33.11	577.01	14	26.39	35.42	526.71
5	30.68	39.64	615.61	15	14.24	23.23	756.19
6	31.62	40.60	260.98	16	34.29	43.27	461.04
7	95.94	104.91	172.28	17	43.82	52.84	349.79
8	20.94	29.95	732.17	18	74.48	83.52	228.74
9	45.50	54.51	176.30	19	25.92	34.93	852.58
10	55.65	64.68	371.07	20	28.41	37.47	549.14

Figure 3.18. Client-server simulation results (see Example 3.15)

different probabilistic rules, feeds a queue farm. The queue farm implements some particular policy for distributing new arrivals into waiting configurations. Finally, the queue farm feeds the servers, which request new inputs from particular queues according to a prescribed procedure. The database component collects the history of each request as it passes through the system. This organization, a collection of interacting components, suggests a control algorithm to coordinate their activities. That algorithm, which appears in Figure 3.20, uses the now-familiar event list to schedule arrival and departure events. The event list now stores only the request identifier (requestID) and the arrival or departure time. Any further information about a request is available from the database. At the beginning of the simulation, the algorithm primes the event list with an arrival from each of the input processes. Some input processes may begin operation long after the beginning of the simulation interval. This is not problematic; the priming event from such a process merely lies dormant in the event list until simulation time has advanced to expose the event.

The algorithm relies on other modules for specific tasks associated with the input processes, the buffer queues, the servers, and the database. Specifically, the algorithm communicates with the module inputProcess with the calls of Table 3.5. The control algorithm uses these procedures to prime an event list with one request from each input process. It then enters a while-loop that repeatedly processes the earliest event in the list. After processing an arrival, it loads the event list with another arrival from the corresponding input process, provided that the new request arrives before closing time. Thus the event list will eventually become depleted as the simulation time moves beyond the closing time of all the input processes.

To process an arrival, the algorithm simply hands it to the bufferQueue module. That module applies its policies, as specified in the initialization parameter qParams, to route the request to the appropriate queue. The control algorithm communicates with the bufferQueue module through the interfaces

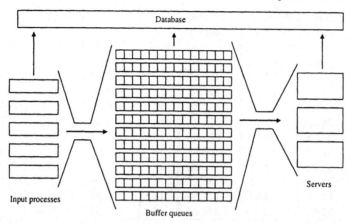

Figure 3.19. Generic organization for client-server systems

noted in Table 3.5. There is even less activity associated with a departure; the algorithm simply frees the associated server, using an interface from the table. After checking for arrivals and departures, the while-loop offers the bufferQueue an opportunity to provide work for available servers. The support routines write all database entries. InputProcess starts a history record for each requestID that it passes to the control algorithm. When the bufferQueue removes a request, it updates the history record to reflect the time spent in the queue. When the serverFarm accepts a request, it informs the database of the request's status change because the request is now a departure item. It also notes the service time in the history.

The supporting routines can be complex, depending on the policies under simulation, but they are basically logistical functions that do not add to the concept under discussion. Therefore, we will not develop these supporting algorithms in detail. Rather, we consider an example that illustrates their formats in a particular instance. Similar comments apply to the database interface and its subsequent interrogation. Because the database maintains a complete history of each request, we can systematically accumulate any performance measure of interest. We can, for instance, determine the average queue times associated with requests that exit from different servers or the fraction of work performed by a particular server. The next example illustrates some possibilities.

3.16 Example: In the following situation, we wish to simulate 5 seconds of operation. We take 1 millisecond as the unit time increment. A server group handles request from three clients. The first client's requests appear as a Poisson process with parameter $\lambda_1 = 0.02$ per unit time through the first 1000 time units of operation. This client then slows to a rate of $\lambda_2 = 0.01$ per unit time for the balance of the operational period. The second client is a synchronous source that provides a request opportunity every 50 time units. The probability is $p = 0.1$ that any given opportunity will actually produce a request. The third client is a priority source from which requests appear according to a Poisson process with $\lambda_3 = 0.001$ per unit

```
function genericClientServer (inParams, qParams, serverParams)
      // parameters to configure auxiliary modules
   inputProcess.initialize (inParams);
   bufferQueue.initialize (qParams);
   serverFarm.initialize (serverParams);
   database.clear (); clearEvents (eventList);
   i = 1;
   while (inputProcess.exists (i))       // prime event list from input processes
      (valid, requestID, arriveTime) = inputProcess.next (i, 0.0);
         // next arrival after time 0.0
      if (valid)
         insertEvent (eventList, requestID, arriveTime);
      i = i + 1;
   time = 0.0;       // simulation time to enable event list withdrawals
   while (eventList ≠ φ)
      requestID = nextEvent (eventList);       // acquire next event
      status = database.status (requestID);
      if (status = arrival)       // process arrival
      time = database.arriveTime (requestID);       // update simulation time
         bufferQueue.insert(requestID);       // release request to queues
         i = database.inputID (requestID);    // identify input; acquire new arrival
         (valid, requestID, arriveTime) = inputProcess.next (i, time);
         if (valid)
            insertEvent (eventList, requestID, arriveTime);
      else       // process departure
         time = database.departTime (requestID);       // update simulation time
         j = database.serverID (requestID);       // identify and free server
         serverFarm.free (j);
      freeList = serverFarm.freeList ();       // list of available servers
      if (bufferQueue.exists (freeList))       // requests available for a free server?
         (j, requestID) = bufferQueue.remove (freeList, time);       // request server j
         departTime = serverFarm.accept (j, requestID, time);
         insertEvent (eventList, requestID, departTime);       // schedule departure
```

Figure 3.20. Control program for a generic client-server simulation

---

time. A request from the third client bypasses any requests from the other sources that are waiting in the queues, but it does not interrupt a server that is already working on a lower-priority request.

Two servers are available. The slower one responds to a request with a service time that is uniformly distributed on the interval $[130, 170)$ time units. The service time for the faster server is a discrete uniform random variable on $\{90, 91, 92, \ldots, 110\}$ time units.

Except for the priority associated with the third client, all requests proceed through a first-come, first-serve queue. The simulation is to estimate the average queue time for the high-priority and ordinary requests and the percent idle time for the two servers.

We can handle the rate variation from the first client by modeling it as two sources. One opens at time 0 and closes at time 1000. The second opens at time

Call	Purpose
inputProcess.initialize (inParams)	Establish number of input streams, their opening and closing times, and their probabilistic properties.
inputProcess.exists (i)	Return true if input i is defined.
inputProcess.next (i, lastTime)	Return (valid, requestID, arriveTime), corresponding to the next request arrival from input process i. The lastTime parameter is necessary because the input process operates by generating interarrival times, which add to the time of the last arrival. If the next arrival falls beyond the closing time of the process, valid is false. Otherwise, requestID identifies a valid request registered in the database.
bufferQueue.initialize (qParams)	Establish the number of queues, their input processes, destination servers, and priority-handling schemes.
bufferQueue.insert (requestID)	Place requestID in the correct queue, referring to the database as necessary.
bufferQueue.exists (freeList)	Return true if bufferQueue can provide a request for a server in freeList.
bufferQueue.remove (freeList, time)	Remove a request from a queue feeding a server in freeList. Return the requestID and the identity of the chosen server. Using the current simulation time, update the database history by noting its queue exit time.
serverFarm.initialize (serverParams)	Establish the number of servers and their probabilistic parameters.
serverFarm.free (j)	Free server j.
serverFarm.freeList ()	Return a vector of free servers.
serverFarm.accept (j, requestID, time)	Load server j with a request at the specified simulation time. The server generates a departure time and updates the request history in the database.

TABLE 3.5. Support routines needed by the generic event-driven simulation shell

```
function inputProcess.initialize (inParams)
    param = inParams;    // local storage of param for other inputProcess routines

function inputProcess.exists (i)
    if (i ≤ nrows (param))    // compare i with number of rows in param
        return true;
    else
        return false;

function inputProcess.next (i, lastTime)
    (openTime, closeTime, type, genParam) = param (i, -);    // get param row i
    if (type = Poisson)
        delta = - ln (1.0 - rand ()) / genParam;    // exponential interarrival time
    else
        (n, p) = genParam;
        delta = n * floor (ln (1.0 - rand ()) / ln (1 - p));
            // generate number of uneventful slots, each n time units
    arriveTime = max (lastTime, openTime) + delta;
    if (arriveTime > closeTime)
        return (false, 0, 0.0);
    else
        requestID = database.create ();    // start history record in database
        database.setInputID (requestID, i);
        database.setStatus (requestID, arrival);
        database.setArriveTime (requestID, arriveTime);
        return (true, requestID, arriveTime);
```

Figure 3.21. Input process module to support the generic client-server
simulation (see Example 3.16)

1000 and closes at time 5000, the simulation end time. We therefore consider four
input processes, parameterized as follows.

inputID	open time	close time	generator	generator parameters
1	0.0	1000.0	Poisson	0.02
2	1000.0	5000.0	Poisson	0.01
3	0.0	5000.0	Bernoulli	(50.0, 0.1)
4	0.0	5000.0	Poisson	0.001

This 4×4 matrix, everything except the first column, comprises the inParams input for the generic algorithm of Figure 3.20. The inputProcess module, shown in
Figure 3.21, must respond to each of the interface calls listed in Table 3.5. The
functions are straightforward manipulations of the input parameters.

We accommodate the high-priority client by providing a separate queue for
calls from this source. When the control algorithm asks for a waiting client request,
the bufferQueue module tries to satisfy the demand from the high-priority queue. We
establish this policy simply by scanning the queues in order of decreasing priority.
So we list the high-priority queue first. For this example, the qParams parameter
is then the rightmost two columns of the following matrix. Note that both queues
feed both servers, but they receive from different input processes. Recall that the
high-priority client arrives through input process 4.

```
function bufferQueue.initialize (qParams)
    param = qParams;      // local storage of param for other bufferQueue routines
    for i = 1 to nrows (param)      // clear all queues
        clearQueue (queue (i));

function bufferQueue.exists (freeList)
    for i = 1 to nrows (param)
        (source, destination) = param (i, –);      // extract row i from param
        if ((destination ∩ freeList ≠ φ) and (queue (i) ≠ φ))
            return true;      // queue i can fill request
    return false;      // no queue can respond

function bufferQueue.remove (freeList, simTime)
    for i = 1 to nrows (param)
        (source, destination) = param (i, –);      // extract row i from param
        if ((destination ∩ FreeList ≠ φ) and (queue (i) ≠ φ))
            requestID = dequeue (queue (i));
            j = choose (destination ∩ freeList);      // choose an available server
            database.setServerID (requestID, j);      update database history
            database.setExitQTime (requestID, simTime);
            return (j, requestID);

function bufferQueue.insert (requestID)
    j = database.inputID (requestID);      // identify source process
    for i = 1 to nrows (param)
        (source, destination) = param (i, –);      // extract row i from param
        if (j ∈ source)
            enqueue (queue (i), requestID);
            database.setQueueID (requestID, i);
            return;
```

Figure 3.22. Buffer queue module to support the generic client-server
simulation (see Example 3.16)

queueID	source processes	destination servers
1	(4)	(1, 2)
2	(1, 2, 3)	(1, 2)

The algorithms of Figure 3.22 implement the bufferQueue interface routines as required by the control program. Note that the queue insertion routine places a
request in the first queue that accepts from the request's source process. This example involves only two queues, and they have disjoint source requirements. The
code handles the more general situation, where the source requirements of several
queues may overlap, but not in a particularly sophisticated manner. You might
want, for example, to choose the shortest queue when there are several possibilities.
This policy would require only minor changes to the algorithm.

The algorithms for the final supporting module, serverFarm, appear in Figure 3.23. The following matrix provides the server descriptions. This matrix, except
the first column, comprises the input parameter serverParams.

```
function serverFarm.initialize (serverParams)
    param = serverParams;      // local storage of param for other serverFarm routines
    for i = 1 to nrows (param)      // all servers start in not busy status
        serverBusy (i) = 0;

function serverFarm.free (j)
    serverBusy (j) = 0;

function serverFarm.freeList ()
    j = 0;
    for i = 1 to nrows (param)
        if (serverBusy (i) = 0)
            j = j + 1;
            v (j) = i;
    return v;

function serverFarm.accept (j, requestID, simTime)
    (dist, (low, high)) = param (j, -);      // extract row j from param
    if (dist = continuous uniform)
        delta = (high - low) * rand() + low;      // scaled continuous uniform
    else
        delta = low + floor ((high - low + 1) * rand ());      // scaled discrete uniform
    departTime = simTime + delta;
    database.setExitSysTime (requestID, departTime);
    database.setStatus (requestID, departure);
    serverBusy (j) = 1;
    return departTime;
```

Figure 3.23. Server module to support the generic client-server simulation (see Example 3.16)

serverID	service distribution	distribution parameters
1	continuous uniform	(130, 170)
2	discrete uniform	(90, 110)

A typical run produced 73 requests from the three clients during the 5000-millisecond simulation interval. They arrived through the 4 input processes, progressed through the 2 queues, and emerged from the 2 servers. The database maintained history records of all these events, as illustrated in Figure 3.6 for the first 20 client requests and the first 10 service intervals for the two servers. The database histories allow us to check that the high-priority third client, who arrives on input process 4, has exclusive use of queue 1. We can also spot check the service times by subtracting the queue exit time from the server exit time. In this case, we verify that server 1 responds in 130–170 milliseconds and server 2 requires 90–110 milliseconds, as expected from the probability distributions. We also note that arrivals from input process 2 occur later than 1000 milliseconds because this input process simulates the first client after a rate change. Similarly, input process 1 presents arrivals only in the span 0–1000 milliseconds.

We can manipulate the database to sample any measurement of interest. We can compute the average queue time by averaging the difference, queue exit time

Client request	Input process	Arrival time	Queue	Queue exit time	Server	Server exit time
1	1	69.00	2	69.00	1	231.62
2	2	1289.40	2	1305.51	2	1395.51
3	3	550.00	2	809.51	2	908.51
4	4	2427.84	1	2433.88	2	2528.88
5	1	78.39	2	78.39	2	178.39
6	1	95.02	2	178.39	2	282.39
7	1	215.89	2	231.62	1	368.35
8	1	241.37	2	282.39	2	376.39
9	1	357.76	2	368.35	1	513.26
10	1	397.51	2	397.51	2	503.51
11	1	412.78	2	503.51	2	602.51
12	1	426.23	2	513.26	1	667.54
13	1	480.01	2	602.51	2	703.51
14	1	500.45	2	667.54	1	827.39
15	1	537.37	2	703.51	2	809.51
16	1	592.37	2	908.51	2	1002.51
17	3	550.00	2	827.39	1	973.66
18	3	1350.00	2	1395.51	2	1486.51
19	1	616.05	2	973.66	1	1119.94
20	1	790.88	2	1002.51	2	1099.51

TABLE 3.6. Timing (milliseconds) for the multiclient, multiserver simulation (see Example 3.16)

minus arrival time, across all requests. We can compute separate average queue times for the high-priority and ordinary requests by separating the histories for input process 4 from the other input processes. We can compute the work time for a given server by isolating the records for that server and summing the time differences between server exit and queue exit. The idle time is the maximum server exit time minus the work time. Figure 3.7 presents the results of such calculations for 20 simulation runs. Each simulation run samples four measurements, each from an unknown distribution. We know that the average over the 20 runs will have a variance that is 1/20 of the variance of the corresponding distribution. This means that, as the number of simulation runs increases, the measurement averages will be less and less likely to deviate from the true means of their underlying distributions. We can with some confidence, therefore, report that a priority request will encounter an average queue wait of 26.8 milliseconds, while an ordinary request will typically suffer a 297.3 millisecond wait. The slower server, server 1, is idle an average of 6.6% of the time, while the faster server is typically idle 11.9%. □

Example 3.16, together with the control algorithm of Figure 3.20, provides a rather general client-server simulation. It can handle all systems in which the input processes, queue policies, and server characteristics can be captured in simple matrix parameters. There are, of course, systems that lie beyond such parameterization, such as those in which requests move among several queues between arrival and disposition. Another shortcoming of this simulation template is that it mixes programming logistics (e.g., event list

Simulation	Queue time (priority requests)	Queue time (ordinary requests)	Fraction Idle (Server 1)	Fraction Idle (Server 2)
1	36.60	278.90	0.0214	0.0686
2	36.40	203.69	0.0852	0.1328
3	26.70	119.43	0.0933	0.2313
4	24.25	304.69	0.1092	0.1520
5	33.54	361.02	0.0073	0.0099
6	14.94	180.80	0.0894	0.0948
7	45.45	558.37	0.0197	0.0239
8	51.81	261.97	0.0098	0.0195
9	0.00	130.68	0.0794	0.1924
10	9.12	198.51	0.1635	0.3114
11	31.17	173.66	0.1286	0.1340
12	12.02	303.77	0.0103	0.0748
13	41.22	762.04	0.0000	0.0000
14	10.61	164.29	0.0385	0.1401
15	24.27	846.46	0.0072	0.0135
16	35.82	366.28	0.0956	0.1457
17	19.39	110.64	0.1454	0.3186
18	31.84	251.69	0.0000	0.0097
19	39.63	188.57	0.0919	0.1275
20	11.70	181.14	0.1242	0.1881
Averages	26.82	297.33	0.0660	0.1194

TABLE 3.7. Selected statistics for the multi-client, multi-server simulation (see Example 3.16)

---

maintenance) with application concepts (e.g., probability distributions on the input processes). Commercial simulation languages are available that effectively suppress the programming detail and allow the user to deal exclusively with the application intricacies. The suggestions for further reading, at the end of the chapter, provide references for such languages.

Exercises

*3.22 Let V be an n-vector, whose ith component is B_i, a Bernoulli random variable with parameter p. The B_i, for $1 \leq i \leq n$, are independent. Let $N = \sum_{i=1}^{n} B_i$ be the number of ones in V. Let L be the location of the first 1 in V. That is, $L = \min\{i|B_i = 1\}$.

  – Suppose that $n = 5, k = 2$. What is the distribution of L given that $N = k$? That is, what is $\Pr(L = j|N = k)$, for $j = 1, 2, 3, 4, 5$?

  – Find the distribution of L, given $N = k$, in the general case.

3.23 Modify the simulation of Example 3.12 by changing the service time to a discrete uniform distribution on $\{8, 9, 10, 11, 12\}$ milliseconds. Is there enough time, on the average, for the server to accommodate the

500 expected arrivals in the 5-second simulation span? Make several simulation runs to confirm your expectations as to the average time that a request spends in the wait queue.

3.24 Modify the simulation of Example 3.12 by changing the probability to 0.15 that a request arrives in each 1-millisecond interval. Leave the service time as a discrete uniform distribution on $\{7, 8, 9, 10, 11\}$. Can the server keep up, on the average, with the new arrival stream? Make several simulations, of increasing lengths, to verify your conjecture. Let W_i, for $i \geq 10$, be the average wait time experienced by requests $i - 9, i - 8, \ldots, i$. Instrument the algorithm to measure W_i and thereby expose the general trend in the wait queue length.

3.25 Use induction to show that

$$\frac{\lambda^{n-1}}{(n-2)!} \int_0^t y^{n-2} e^{-\lambda y} \, dy = 1 - e^{-\lambda t} \sum_{j=0}^{n-2} \frac{(\lambda t)^j}{j!},$$

for $n \geq 2$, as required in the proof of Theorem 3.13.

3.26 Modify the simulation of Example 3.15 by changing the service time to a continuous uniform distribution on the interval $(8, 12]$ milliseconds. What change do you expect in the average time that a request spends in the wait queue? Make several simulation runs to confirm your conjecture.

3.27 Modify the simulation of Example 3.15 by changing the parameter on the input process to $\lambda = 150$ requests per second. Let W_i, for $i \geq 10$, be the average wait time experienced by requests $i - 9, i - 8, \ldots, i$. Instrument the algorithm to measure W_i and thereby expose the general trend in the wait queue length.

*3.28 Implement the simulation of Example 3.16. Use it to estimate the average queue time for priority and ordinary requests and the fraction of idle time associated with the two servers.

*3.29 Working with an implementation of Example 3.16, explore the effect on server idle time of shifting a server's probability distribution to smaller values. Suppose, for instance, that the time for server 1 to complete a request is a continuous uniform random variable on the interval $[40, 80)$ milliseconds, instead of $[130, 170)$ milliseconds.

*3.30 Design and implement a simulation for the following system. Use 1 millisecond as the unit time increment, and simulate 10 seconds of operation. A server group handles requests from three clients: A, B, and C. Client A has the highest priority, followed by client B and then client C. When an eligible server becomes free, it accepts a client A request, if one is available. Otherwise, it looks to the other clients in priority order. A server always finishes a request before accepting a new one. It does not interrupt service just because a higher-priority request arrives while it is busy. Multiple requests from the same source queue up for service in a first-come, first-serve manner. Clients A, B, and C each generate requests according to a Poisson process. The rates are $\lambda_A = 5$ requests per second for client A, $\lambda_B = 10$ for client B, and $\lambda_C = 20$ for client C. Three servers are available: $\alpha, \beta,$ and γ. The fastest, α, accepts requests only from clients A and B; the midrange unit, β, serves only clients B and C; and the slowest, γ, serves all comers. The service times for the three servers are all continuous uniform random variables, on the interval $[45, 55)$ milliseconds for α, $[90, 110)$ milliseconds for β, and $[180, 220)$ milliseconds for γ. Let $T_\alpha, T_\beta, T_\gamma$ be the total work time logged by the three servers. Let $T = T_\alpha + T_\beta + T_\gamma$ be the total working time for all servers combined. Use the simulation to estimate $W_\alpha = T_\alpha/T, W_\beta = T_\beta/T,$ and $W_\gamma = T_\gamma/T$, the fractional loads carried by the servers. Make 20 simulation runs and report the average of the desired measurements.

*3.31 Referring to the preceding exercise, calculate the effect on the fractional loads if client A ceases operation 5 seconds into the simulation.

3.4 Markov chains

The preceding section's simulations are useful for obtaining estimates of certain parameters, such as the mean time in a wait queue, under circumstances where the underlying probability distributions are difficult or impossible to specify. This section's purpose is to demonstrate an analytical tool that permits direct calculation in simple cases. Unfortunately, the method becomes increasingly complex in large systems, so the simulation technique remains valuable. The new method is the finite Markov chain, which applies to systems with a finite number of states.

A Markov chain is a stochastic process, but unlike those considered to this point, it allows certain dependencies among the random variables. A Bernoulli process, you will recall, is an infinite sequence $B_1, B_2, \ldots$ of independent Bernoulli random variables with the same parameter. A Poisson process is an infinite sequence $T_1, T_2, \ldots$ of independent exponential interarrival times, all with the same parameter, which generate Poisson renewal counts. A Markov chain $X_1, X_2, \ldots$ is more akin to a Bernoulli process, except the range of each X_i is $\{1, 2, \ldots, n\}$ rather than $\{0, 1\}$ and the probability assignment is slightly more complicated. In the Bernoulli process, a prefix is a collection of infinite sequences that agree in a certain finite number of initial values. For example, the prefix 011001 represents all sequences $011001\ldots$, and because the B_i are independent, we assign probability

$$\Pr(011001) = (1-p)pp(1-p)(1-p)p = p^3(1-p)^3.$$

In general, a prefix $b = b_1 \cdots b_k$ has probability $p^{S(b)}(1 - p)^{k-S(b)}$, where $S(b) = \sum_{j=1}^{k} b_i$. In discussing the Bernoulli process, we showed that this is a consistent probability assignment to the countable space of finite prefixes. We establish a Markov chain with similar probability assignments to prefixes.

3.17 Definition: Let α be a row vector of n nonnegative components that sum to 1; let $P = [p_{ij}]$ be an $n \times n$ matrix of nonnegative entries, in which each row sums to 1. Let $X_0, X_1, \ldots$ be an infinite sequence of random variables, each with range $\{1, 2, \ldots, n\}$. Outcomes are infinite sequences $X_0 = x_0, X_1 = x_1, X_2 = x_2, \ldots$. For each finite prefix $x_0, x_1, \ldots, x_k$, representing a collection of outcomes, we assign probability

$$\Pr(x_0 x_1 \ldots x_k) = \alpha_{x_0} p_{x_0 x_1} p_{x_1 x_2} \cdots p_{x_{k-1} x_k}.$$

A union of distinct k-prefixes necessarily represents disjoint outcome collections, and they receive probability equal to the sum of the constituent assignments. An n-state *Markov chain* comprises the random variables $X_0, X_1, \ldots$, the initial distribution α, the transition matrix P, and the above specified probability assignments to prefix set unions. ∎

Note that the definition does not directly specify the probability distributions of the constituent X_i. When we defined a Poisson process, we did not directly fix the distributions of the T_i. Rather, we provided an overall constraint that the renewal counts must all have a common Poisson distribution. From that constraint, we discovered that the T_i are exponentially distributed. The situation here is similar. We have a global constraint on the probability of k-prefix unions, and the marginal distributions of the individual X_i remains to be discovered. In contrast with the Poisson process, note that the X_i are *not* asserted to be independent or identically distributed. We proceed to investigate the consequences of the definition.

We first need to show that the Markov probability assignment is consistent. That is, all assignments are nonnegative, the total assignment is 1, and the sum of disjoint component assignments is the assignment to their union. The definition emphasizes the nonnegativity of the factors that contribute to the probability of a prefix. The entire space Ω is the disjoint union of the prefixes $1, 2, \ldots, n$. Consequently,

$$\Pr(\Omega) = \sum_{j=1}^{n} \Pr(j) = \sum_{j=1}^{n} \alpha_j = 1.$$

Indeed, for any k, the entire space is the disjoint union of prefixes of length k.

$$\Pr(\Omega) = \sum_{j_1=1}^{n} \sum_{j_2=1}^{n} \cdots \sum_{j_k=1}^{n} \Pr(j_1 j_2 \ldots j_k) = \sum_{j_1=1}^{n} \sum_{j_2=1}^{n} \cdots \sum_{j_k=1}^{n} \alpha_{j_1} p_{j_1 j_2} \cdots p_{j_{k-1} j_k}$$

$$= \sum_{j_1=1}^{n} \alpha_{j_1} \sum_{j_2=1}^{n} p_{j_1 j_2} \sum_{j_3=1}^{n} p_{j_2 j_3} \cdots \sum_{j_{k-1}=1}^{n} p_{j_{k-2} j_{k-1}} \sum_{j_k=1}^{n} p_{j_{k-1} j_k}$$

The last summation is the sum of row j_{k-1} of P and is therefore 1. The number of summations thus decreases by 1 to reveal another final summation that reduces to 1. This reduction continues until the entire expression becomes $\sum_{j=1}^{n} \alpha_j = 1$. In a given application, we endeavor to describe all events of interest in terms of prefixes through some fixed maximum length. Consequently, we can treat these disjoint prefixes as the atomic outcomes, and their probabilities properly sum to 1.

3.18 Example: Define a 3-state Markov chain with the following initial distribution and transition matrix. Using an asterisk to denote unspecified entries, what is the probability of the event $* * 2 * 1$?

$$\alpha = \begin{bmatrix} 0.2 & 0.3 & 0.5 \end{bmatrix} \qquad P = \begin{bmatrix} 0.1 & 0.2 & 0.7 \\ 0.2 & 0.5 & 0.3 \\ 0.4 & 0.2 & 0.4 \end{bmatrix}$$

The event in question contains all infinite sequences that begin $ab2c1$, where a, b, c can be any member of $\{1, 2, 3\}$. Consequently, we can work in the space of prefixes of maximum length 5.

$$\Pr(* * 2 * 1) = \sum_{a=1}^{3}\sum_{b=1}^{3}\sum_{c=1}^{3} \Pr(ab2c1) = \sum_{a=1}^{3}\sum_{b=1}^{3}\sum_{c=1}^{3} \alpha_a p_{ab} p_{b2} p_{2c} p_{c1}$$

$$= \sum_{a=1}^{3} \alpha_a \sum_{b=1}^{3} p_{ab} p_{b2} \sum_{c=1}^{3} p_{2c} p_{c1}$$

We see that it is helpful to have the components of $Q = P^2$.

$$Q = [q_{ij}] = P^2 = \begin{bmatrix} 0.33 & 0.26 & 0.41 \\ 0.24 & 0.35 & 0.41 \\ 0.24 & 0.26 & 0.50 \end{bmatrix}$$

Continuing,

$$\Pr(* * 2 * 1) = \sum_{a=1}^{3} \alpha_a \sum_{b=1}^{3} p_{ab} p_{b2} q_{21} = 0.24 \sum_{a=1}^{3} \alpha_a \sum_{b=1}^{3} p_{ab} p_{b2} = 0.24 \sum_{a=1}^{3} \alpha_a q_{a2}$$

$$= 0.24[(0.2)(0.26) + (0.3)(0.35) + (0.5)(0.26)] = 0.06888. \ \square$$

When a prefix of length k splits into n prefixes of length $k + 1$, the probability of the parent k-prefix distributes across its n disjoint children according to a row of the transition matrix P. That is, child $(x_0 x_1 \ldots x_k x_{k+1})$ of prefix $(x_0 x_1 \ldots x_k)$ has probability

$$\Pr(x_0 x_1 \ldots x_k x_{k+1}) = \alpha_{x_0} p_{x_0 x_1} p_{x_1 x_2} \cdots p_{x_{k-1} x_k} p_{x_k x_{k+1}}$$

$$= \Pr(x_0 x_1 \ldots x_k) \cdot p_{x_k x_{k+1}}.$$

The child that extends the prefix with the entry x_{k+1} receives the fraction $p_{x_k x_{k+1}}$ of the parent probability. These fractions constitute row x_k of the transition matrix and therefore sum to 1. It follows that the probabilities of the children sum to that of the parent k-sequence. Probability is exactly conserved when the prefix depth increases by 1. Consequently, when a prefix is the union of longer prefixes, the assigned probability as a short prefix is exactly the sum of the assigned probabilities to the longer prefixes.

To understand the development to come, it is important to appreciate the dual nature of a Markov chain's sample space. Directly from the definition, the sample space is the set of all infinite sequences $(x_0, x_1, x_2, \dots)$, where each $x_i \in S = \{1, 2, \dots, n\}$. There is an assumed, anonymous, non-deterministic process that produces these outcomes. With each cycle, a new vector $(x_0, x_1, x_2, \dots)$ appears. Each X_i now has an observed value: $X_0 = x_0, X_1 = x_1, \dots$. Under the frequency-of-occurrence interpretation of probability, the fraction α_i of the vectors will exhibit $x_0 = i$. If we isolate the vectors in which $x_t = i$, then the fraction p_{ij} of this collection will exhibit $x_{t+1} = j$. This is a static perspective because we envision an outcome as an entire infinite sequence that appears all at once.

An equivalent dynamic viewpoint is frequently more appealing in certain application contexts. This alternative considers a clocked system that moves among the n states $\{1, 2, \dots, n\}$, generating an outcome in the process. The system's initial state, as well as its transitions, are nondeterministic. It originates in state i with probability α_i. On the kth clock pulse, it moves from its current state, say u, to its next state, say v, with probability p_{uv}. After t clock pulses, the values $(X_0, X_1, \dots, X_t) = (x_0, x_1, \dots, x_t)$ are determined, but sequence values beyond t are as yet unspecified. The time-oriented process examines an outcome in the process of happening. However, our probability assignment mechanism can handle this ambiguity because it needs only finite prefixes to determine probability. The probability of the system describing the trajectory $x_0 x_1 \dots x_t$ through the first t clock pulses is precisely the probability of the prefix $x_0 x_1 \dots x_t$ in the static sense. All time-oriented probabilities involving a finite history are computations on finite prefixes and therefore have well-defined probabilities. Suppose that we are interested in the probability p that the system will enter state j on clock pulse $t + m$, given that it is in state i after clock pulse t. We work with prefixes of length $t + m$ and compute

$$p = \frac{\Pr(* * * \cdots * i * * * \cdots * j)}{\Pr(* * * \cdots * i * * * \cdots * *)},$$

where the numerator describes prefixes with fixed components i and j in locations t and $t + m$, respectively, and arbitrary entries elsewhere. The denominator includes prefixes with a fixed component i in location t only. Example 3.18 shows how to calculate these constituent probabilities, although we will discover more expeditious methods shortly. The point at the moment is that time-oriented interpretations of a Markov chain are possible, and events depending on a finite history correspond to combinations of finite prefixes in the static interpretation. Under the dynamic interpretation, we imagine a new random experiment with each state transition. From this perspective, each outcome is a state, in the range $\{1, 2, \dots, n\}$, which specifies the next system configuration. The first random experiment determines X_0, the initial state. The next experiment establishes X_1, and so forth. If the system is currently in state i, we can imagine that the next state comes from rolling an n-sided die, for which the probabilities of sides 1 through n correspond to row i of the transition matrix. This mental model contrasts with the static interpretation, in which we imagine a single random experiment that establishes all states simultaneously: $X_0, X_1, X_2, \ldots$. The outcomes are then infinite sequences of states $\{i_0, i_1, i_2, \ldots\}$. An outcome may again be generated with an n-sided die that changes its probability structure from one roll to the next. Now, however, we require infinitely many rolls to specify a single outcome.

Knowing how a parent t-prefix distributes probability to its $(t+1)$-prefix children, we have $\Pr(X_{t+1} = j | X_t = i) = p_{ij}$. The conditional probability of entering state j from current state i is p_{ij}, regardless t. This property is called *homogeneity*. Because there are only finitely many possible states and because the conditional probabilities $\Pr(X_{t+1} = j | X_t = i)$ are independent of t, the Markov chain under description is more accurately called a *finite, homogeneous Markov chain*. The computation above also shows that the conditional probability of entering state j from current state i is independent of the states visited prior to entering the current state. The next theorem generalizes this property.

3.19 Theorem: Let $P = [p_{ij}]$ be the transition matrix of a Markov chain with states $S = \{1, 2, \ldots, n\}$, and let $S_0, S_1, \ldots, S_{t-1}$, for $t \geq 1$, be arbitrary nonempty subsets of S. Then for all $1 \leq i, j \leq n$,
$$\Pr(X_{t+1} = j | (X_t = i) \cap (X_{t-1} \in S_{t-1}) \cap (X_{t-2} \in S_{t-2}) \cap \ldots \cap (X_0 \in S_0))$$
$$= \Pr(X_{t+1} = j | X_t = i) = p_{ij}.$$

PROOF: Define event C and probability q by

$$C = (X_{t-1} \in S_{t-1}) \cap (X_{t-2} \in S_{t-2}) \cap \ldots \cap (X_0 \in S_0)$$
$$q = \Pr(X_{t+1} = j | (X_t = i) \cap C).$$

We are to show that $q = p_{ij}$. Let $Y = (X_1, X_2, \ldots, X_{t-1})$, and let v denote a vector $(v_0, v_1, \ldots, v_{t-1}) \in S' = S_0 \times S_1 \times \cdots \times S_{t-1}$. We can then write event C as a disjoint union:

$$C = \bigcup_{v \in S'} (Y = v).$$

For a particular $v = (v_0, v_1, \ldots, v_{t-1})$, the event $(Y = v)$ constitutes a specific trajectory through the prior states $(v_0, v_1, \ldots, v_{t-1})$. Working with prefixes through position $t + 1$, we have

$$\Pr(X_{t+1} = j | (X_t = i) \cap (Y = v)) = \frac{\alpha_{v_0} p_{v_0 v_1} \cdots p_{v_{t-1} i} p_{ij}}{\sum_{k=1}^{n} \alpha_{v_0} p_{v_0 v_1} \cdots p_{v_{t-1} i} p_{ik}} = p_{ij}.$$

Therefore,

$$q = \frac{\Pr((X_{t+1} = j) \cap (X_t = i) \cap C)}{\Pr((X_t = i) \cap C)}$$
$$= \frac{\sum_{v \in S'} \Pr((X_{t+1} = j) \cap (X_t = i) \cap (Y = v))}{\Pr((X_t = i) \cap C)},$$

where we have used the law of total probability to expand the expression across the elements of C. Continuing, we have

$$q = \frac{\sum_{v \in S'} \Pr((X_{t+1} = j)|(X_t = i) \cap (Y = v)) \cdot \Pr((X_t = i) \cap (Y = v))}{\Pr((X_t = i) \cap C)}$$

$$= p_{ij} \cdot \frac{\sum_{v \in S'} \Pr((X_t = i) \cap (Y = v))}{\Pr((X_t = i) \cap C)} = p_{ij} \cdot \frac{\Pr((X_t = i) \cap C)}{\Pr((X_t = i) \cap C)} = p_{ij}.$$

Of course, if we set all the $S_i = S$, we obtain $\Pr(X_{t+1} = j|X_t = i) = p_{ij}$. ∎

Given this theorem, we can now write $p_{ij} = \Pr(X_{t+1} = j|X_t = i)$, or if it suits our purposes, we can install arbitrary conditions on the the positions prior to t. That is, $p_{ij} = \Pr(X_{t+1} = j|(X_t = i) \cap C)$, where C involves arbitrary constraints on components $X_0, X_1, \ldots, X_{t-1}$. We emphasize again that these probabilities apply independently of t. For example, the probability that X_{12} is j, given that X_{11} is i, is the same as the probability that X_{102} is j given that X_{101} is i. The value is p_{ij} in either case.

From the conditional probabilities, we derive the distributions of the X_t.

$$\Pr(X_0 = j) = \alpha_j$$

$$\Pr(X_1 = j) = \sum_{i=1}^{n} \Pr(X_1 = j|X_0 = i) \cdot \Pr(X_0 = i) = \sum_{i=1}^{n} \alpha_i p_{ij}$$

$$\Pr(X_2 = j) = \sum_{k=1}^{n} \Pr(X_2 = j|X_1 = k) \cdot \Pr(X_1 = k) = \sum_{k=1}^{n} \alpha_i \sum_{i=1}^{n} p_{ik} p_{kj}$$

The last inner sum constitutes component (i, j) of P^2. Hence, row vector α elaborates the probability distribution for X_0. Similarly, row vectors αP and αP^2 elaborate the distributions for X_1 and X_2. Continuing in this manner, we obtain αP^t as the vector of probabilities for X_t, for $t > 0$. If we define P^0 to be the $n \times n$ identity matrix, then the formula applies for all $t \geq 0$. We formalize these results in the following theorem.

3.20 Theorem: Let $X_0, X_1, \ldots$ be the random variables of an n-state Markov chain with initial distribution α and transition matrix P. Then $\Pr(X_t = i)$ is the ith component of αP^t for $1 \leq i \leq n$ and $t \geq 0$.

PROOF: See discussion above. ∎

The following example shows the usefulness of this framework for modeling a simple client-server system. The situation is essentially that of Example 3.12 with the parameters changed to produce a Markov chain model with fewer states.

3.21 Example: A server admits requests from a single anonymous client in a quantized fashion. In particular, the server opens a turnstile at the beginning of each millisecond to admit a request with probability $p = 0.3$. If the server is busy at the time, the request enters a waiting queue, provided that the queue contains less than three waiting entries. If the queue contains three waiting entries, the new request is lost. The queue policy is, as usual, first-in, first-out (FIFO). The server takes a random amount of time to respond to a request; the service time is

State		1 (0,0)	2 (1,1)	3 (1,2)	4 (1,3)	5 (2,1)	6 (2,2)	7 (2,3)	8 (3,1)	9 (3,2)	10 (3,3)	11 (4,1)	12 (4,2)	13 (4,3)
1	(0,0)	0.70	0.10	0.10	0.10									
2	(1,1)	0.70	0.10	0.10	0.10									
3	(1,2)		0.70			0.30								
4	(1,3)			0.70			0.30							
5	(2,1)		0.23	0.23	0.23	0.10	0.10	0.10						
6	(2,2)					0.70			0.30					
7	(2,3)						0.70			0.30				
8	(3,1)					0.23	0.23	0.23	0.10	0.10	0.10			
9	(3,2)								0.70			0.30		
10	(3,3)									0.70			0.30	
11	(4,1)								0.23	0.23	0.23	0.10	0.10	0.10
12	(4,2)											1.00		
13	(4,3)												1.00	

TABLE 3.8. Transition matrix for the client-server system (see Example 3.21)

a discrete uniform random variable on $\{1, 2, 3\}$ milliseconds. Use a Markov chain model to determine the average time that a request spends in the waiting queue.

It is a simple matter to reprogram the simulation of Example 3.12 to reject new requests when the wait queue contains three entries. We also change the parameters on the input and service distribution to match the current situation. The resulting simulation, averaged over 20 runs, yields 0.9379 milliseconds for the average time that a request spends in the waiting queue. We expect the Markov model to produce a similar value.

To this end, we characterize the system state with a pair (n, s), where n is the number of requests in the system, including the one currently under service, and s is the time needed by the server to complete the current request. State $(2, 3)$, for example, indicates that 2 requests are in the system. One is in the queue and the other is at the server, which requires 3 more milliseconds to finish with it. Table 3.8 lists the possible states and the corresponding transition probabilities. State $(0, 0)$ indicates an empty system; states (n, s) for $1 \leq n \leq 4, 1 \leq s \leq 3$ indicate system occupancies with the server needing s milliseconds to complete its current charge. The total number of states is then 13.

Consider the first row of the transition matrix. These are the probabilities of moving from the empty state $(0, 0)$. If no arrival occurs with the next clock tick, the system stays empty. So state $(0, 0)$ follows itself with probability 0.7. The remaining 0.3 probability divides equally among states $(1, 1), (1, 2)$, and $(1, 3)$ because the service time associated with the new arrival comes from the uniform discrete distribution on $\{1, 2, 3\}$. Now consider the second row, which elaborates the probabilities from state $(1, 1)$. The one request at the server will depart on the next clock tick, so if no new arrival appears, the system will be empty. This happens with probability 0.7. If a new arrival appears, it proceeds directly to the server, which will require 1, 2, or 3 milliseconds to respond. The remaining transition probabilities follow from similar considerations, with some local variations. States $(4, 2)$ and $(4, 3)$ cannot move to states with a higher system occupancy, so regardless of whether a new arrival appears or not, they must decay to $(4, 1)$ and $(4, 2)$, respectively. For

State		Expected queue time
1	(0,0)	0.0
2	(1,1)	0.0
3	(1,2)	1.0
4	(1,3)	2.0
5	(2,1)	2.0
6	(2,2)	3.0
7	(2,3)	4.0

State		Expected queue time
8	(3,1)	4.0
9	(3,2)	5.0
10	(3,3)	6.0
11	(4,1)	6.0
12	(4,2)	0.0
13	(4,3)	0.0

TABLE 3.9. Conditional mean (milliseconds) queue time by exiting state (see Example 3.21)

state $(4,1)$, if a new arrival appears, it will happen just as the queue shifts a new request to the server. The new arrival can then take up a position in the newly shortened queue.

Table 3.8 gives transition probabilities to two decimal places. The 0.23 entries are actually $7/30$, and further calculations will use the more accurate representation. In the first such calculation, we determine the expected queue time for a request that arrives on the clock tick when the system moves to a specific state. Arriving requests played a part in the scenarios used to construct the transition matrix, so we need to revisit those scenarios and note the expected queue time for such a request. If a request arrives just as the system leaves the empty state, the new state will be $(1,s)$, where s is the service time assigned the new request. The request itself suffers no queue time. If the request arrives as the system leaves state $(1,1)$, the new state will again be $(1,s)$, and the request suffers no queue time. These observations accounts for the first two entries in Table 3.9, which tabulates the expected queue time for a request given that the request coincides with a transition from a particular state. The queue times calculated to this point are deterministic, but most of the rest are mean values associated with a random variable. For these calculations, let U be the discrete random variable on $\{1,2,3\}$. Its mean is $\mu_U = 2$.

If an incoming request occurs as the system leaves state $(1,s)$ with $s > 1$, the new state will be $(2, s-1)$, and the new request must wait in the queue for $s-1$ milliseconds. This explains rows 3 and 4 of the table. If the request arrives as the system leaves state $(2,1)$, the new state will $(2,s)$ as the single queue entry engages the server and the new request becomes the sole queue member. The new request's queue time is now a random variable, depending on the service time assigned to the previous queue entry. The expected queue time is $\mu_U = 2$. If the new request appears as the system leaves state $(2,s)$ with $s > 1$, the new state will be $(3, s-1)$. In this case the new request suffers a queue wait of $s-1$ milliseconds while the server finishes its current task plus, on the average, $\mu_U = 2$ milliseconds while the server deals with the other queue element. These cases, therefore, incur waits of $s+1$ milliseconds.

If a new request occurs as the system leaves state $(3,1)$, the new state will be $(3,s)$, with the new request waiting $2\mu_U = 4$ milliseconds on the average. If the new request arrives as the system leaves state $(3,s)$ with $s > 1$, the new state will be $(4, s-1)$, and the new request will suffer a mean queue time of $s-1+2\mu_U = s+3$ milliseconds. Finally, if the new request occurs as the system exits state $(4,1)$, the new request must wait $3\mu_U = 6$ milliseconds in the queue. If it arrives as the system leaves state $(4, s)$ with $s > 1$, it is lost and does not contribute to the wait time scenarios. We have now accounted for all the entries of Table 3.9.

Now let X be the random variable that indicates the state the system is leaving when a request arrives. X assumes the values 1 through 13, corresponding to states $(0, 0)$ through $(4, 3)$. Let Q be the random variable that measures the queue time of a request. Then, invoking Theorem 1.70, we have

$$\mu_Q = \sum_{i=1}^{13} \mu_{Q|X=i} \cdot \Pr(X = i).$$

The conditional means, $\mu_{Q|X=i}$, are the values in Table 3.9, but the state probabilities, $\Pr(X = i)$, are still unknown. To determine these values, we define a Markov chain for the system. Let X_0 denote the initial state, and let $X_1, X_2, \ldots$ indicate the system state just after to the ith clock tick. The transition matrix, P, is the matrix of Table 3.8, and the initial distribution, α, is $(1, 0, 0, \ldots, 0)$ because the system starts in state $(0, 0)$. We know that the distributions of the X_k are the vectors αP^k, so it is reasonable to investigate powers of P. We are perhaps surprised to find that the powers P^k start to stabilize around $k = 35$. Beyond that point, the entries change by less than 0.0001 from one power to the next. Moreover, we find that the stable value consists of 13 nearly identical rows, the average of which we will call β. Therefore, P^k, for $k > 35$, is very nearly the following row repeated 13 times.

$\beta = (0.4020, 0.1723, 0.1338, 0.0787, 0.0911, 0.0503, 0.0157, 0.0283, 0.0124,$

$\qquad\qquad\qquad\qquad\qquad 0.0045, 0.0073, 0.0028, 0.0007)$

This means that αP^k, the distribution of X_k, is β for k sufficiently large. Since all X_k have the same distribution for large k, an arriving request will encounter state i with probability β_i. In other words, β provides the missing $\Pr(X = i)$ values needed to complete the computation of μ_Q. Substituting the numbers, we have

$$\mu_Q = \sum_{i=1}^{13} \mu_{Q|X=i} \cdot \Pr(X = i) = \sum_{i=1}^{13} \mu_{Q|X=i} \cdot \beta_i = 0.9337.$$

This value confirms the 0.9379 milliseconds obtained from the simulation. $\square$

The example illustrates the usefulness of Markov chains in modeling client-server systems, but it also reveals how large the state space becomes for problems that are still quite elementary. If we return to the original quantized client-server system of Example 3.12, the input Bernoulli process with $p = 0.1$ and the uniform service time on the integers $\{7, 8, 9, 10, 11\}$ give rise to states $(0, 0)$ and (n, s) for $1 \le s \le 11$ and n ranging to a system occupancy that safely exceeds the queue lengths that can occur. We can instrument the simulation of Figure 3.15 to report the maximum queue length that develops over the simulation duration. Unfortunately, we find queue lengths in the high 20s, which produces large transition matrices in the corresponding Markov model. We can, nevertheless, limit the system occupancy and verify that the modified simulation and the Markov prediction are in agreement. One of the exercises pursues this program.

Despite the shortcomings associated with bounded system occupancy and large transition matrices, the Markov chain model does provide an analytical approach to the average values of certain client-server parameters. The example, however, raises two questions. When can we be sure that the powers of the transition matrix will converge to repeated rows, each representing the long-term probability of the system occupying the various states? Second, can we compute these long-term probabilities without computing large powers of the transition matrix? We proceed by further specializing the Markov chains to forms that admit satisfactory answers to these questions.

Exercises

3.32 Write a computer program to verify that the transition matrix of Table 3.8, when raised to increasingly higher powers, does stabilize to the following repeated row as asserted in Example 3.21.

$$\beta = (0.4020, 0.1723, 0.1338, 0.0787, 0.0911, 0.0503, 0.0157, 0.0283,$$
$$0.0124, 0.0045, 0.0073, 0.0028, 0.0007)$$

3.33 Extend the Markov model of Example 3.21 to calculate the average system time per request and the total server idle time. Compare the computed values with simulated results from the algorithm of Figure 3.12, suitably modified to reject requests that arrive when the queue contains three waiting entries.

3.34 Modify the simulation of Figure 3.15 to limit the system occupancy to 5 requests, 4 in the wait queue and 1 at the server. Using $p = 0.1$ for the input Bernoulli process and a uniform discrete service time on 7–11 milliseconds, make 20 simulation runs, each of 10000-millisecond duration. Each run reports the average queue time per request. Use the average of the 20 reports as the simulated value of the average queue time. Compose the corresponding Markov transition matrix and verify that its powers converge to repeated rows β, where the left-to-right components of β are as follows.

> 0.1519, 0.0168, 0.0187, 0.0208, 0.0231, 0.0256, 0.0285, 0.0316,
> 0.0266, 0.0209, 0.0147, 0.0077, 0.0242, 0.0248, 0.0252, 0.0254,
> 0.0254, 0.0251, 0.0243, 0.0192, 0.0141, 0.0091, 0.0044, 0.0218,
> 0.0215, 0.0211, 0.0206, 0.0200, 0.0195, 0.0189, 0.0149, 0.0110,
> 0.0073, 0.0036, 0.0175, 0.0171, 0.0167, 0.0162, 0.0158, 0.0154,
> 0.0150, 0.0119, 0.0088, 0.0058, 0.0029, 0.0139, 0.0122, 0.0105,
> 0.0089, 0.0074, 0.0058, 0.0043, 0.0029, 0.0017, 0.0008, 0.0003

The states are, of course, $(0,0)$ for the empty state and (n, s) for occupied states, where $1 \leq n \leq 5, 1 \leq s \leq 11$. We interpret β_i as the limiting probability that the system is in state i. In particular, let $(0,0)$ be state 1, and let (n, s) be state $11(n-1) + s + 1$. Calculate q_i, the expected queue time for a request, given that it arrives just as the system is leaving state i. Verify that $\sum_{i=1}^{56} q_i \beta_i$ agrees with the simulated value of the average queue time.

3.35 Suppose that a quantized client-server system contains a single server that deals with a single input stream. The stream delivers requests from a Bernoulli process with $p = 0.2$, and the server responds with service times that are uniformly distributed on the set $\{3, 4, 5\}$. Describe the system as a Markov chain with a specific transition matrix. Using the matrix and the methods of Example 3.21, calculate the average queue time per request.

3.36 Suppose that a quantized client-server system contains a single server that deals with two input streams: stream 1, a Bernoulli process with $p = 0.1$, and stream 2, a Bernoulli process with $p = 0.2$. Requests from the two streams share a single wait queue, where they enter on a first-come, first-in basis when the server is busy. If requests appear simultaneously on both input streams, stream 1 takes priority on entering the queue or engaging the server. The service time is a random variable on the set $\{1, 2, 3\}$ with $\Pr_S(1) = 0.25, \Pr_S(2) = 0.5, \Pr_S(3) = 0.25$. Describe the system as a Markov chain with a specific transition matrix. Using the matrix and the methods of Example 3.21, calculate the average queue time per request.

3.37 The random variable T, measured in seconds, has the distribution to the right, which has been developed empirically by testing adept players

T	2.0	2.5	3.0	3.5
$\Pr(T)$	0.2	0.3	0.3	0.2

on the following video game. T represents the time required to dispatch a ball through the game's exit gate. The game itself proceeds as follows.

Numbered spheres enter at random from the upper edge of the display screen. A random mechanism allows an entrance opportunity once every 0.5 seconds. At each such opportunity, a sphere actually materializes with probability 0.2. The spheres fall into a queue, from which they emerge in the same order as they entered. On leaving the queue, a sphere encounters a series of gates and barriers that impede its general progress toward the exit. The player controls the positions of these obstacles in an attempt to accelerate the sphere's departure. When the sphere exits, the player's score increases by the numbers of points on the sphere. To add urgency to the matter, the sphere's numerical value decreases by ten for each second that it remains on the screen, either negotiating the barriers or waiting in the queue. A new sphere departs the queue only when the previous sphere has reached the exit. If the sphere in play reaches zero value, the player must nevertheless dispatch it through the exit before another sphere can leave the queue. Expedited play is thus encouraged. Moreover, not only do spheres in the queue lose value as they wait, but a newly entering sphere immediately decays to zero value and vanishes if it encounters a full queue of three waiting spheres.

When no player is at the controls, the machine idles by self-adjusting its barriers to achieve passage times T in accordance with the distribution tabulated above. The numerical value of all entering spheres, however, is zero, and therefore the scoreboard remains at zero. To start a game, a player inserts some integral number of dollars, which buys an equal number of spheres. At this point, newly entering spheres are worth 100 points each until the money is exhausted. The player must dispatch any zero-point spheres that remain in the queue or in the barriers before he deals with his spheres. The player now controls the gates.

Simulate this game as a Markov chain to determine the average score of an adept player. Use this information to determine how the machine could pay off in a manner that leaves a 20% profit for the house.

3.4.1   Irreducible aperiodic Markov chains

We investigate here some restrictions that render Markov chains particularly appropriate for client-server models, such as those considered earlier in simulation examples. We start with a useful interpretation of the powers of the transition matrix. Suppose that $P = [p_{ij}]$ is the transition matrix of an n-state Markov chain. We know that p_{ij} is the probability that the system will, after one transition, be in state j given that it is currently in state i. The next theorem shows how to extend these probabilities to cover multiple transitions. To simplify the notation, we define $p_{ij}^{(t)}$ as the (i,j)th element of P^t for $t \geq 0$. Recalling that P^0 is the $n \times n$ identity matrix, we have $p_{ii}^{(0)} = 1$ and $p_{ij}^{(0)} = 0$ for $i \neq j$.

3.22 Theorem: Let $P = [p_{ij}]$ be the transition matrix of an n-state Markov chain, $X_0, X_1, \ldots$. Then $p_{ij}^{(\tau)}$ is the probability that the system will be in state j after τ transitions given that it is currently in state i. That is, $\Pr(X_{t+\tau} = j | X_t = i) = p_{ij}^{(\tau)}$, for all $t \geq 0$.

PROOF: Let $S = \{1, 2, \ldots, n\}$ be the states, and let $q = \Pr(X_{t+\tau} = j | X_t = i)$. Theorem 3.19 establishes the theorem's truth for $\tau = 1$, and we proceed by induction for $\tau > 1$. Specifically, suppose that $\tau > 1$ and $\Pr(X_{t+(\tau-1)} = j | X_t = i) = p_{ij}^{(\tau-1)}$. Because the events $(X_{t+(\tau-1)} = m)$, for $1 \leq m \leq n$, form a disjoint partition of all outcomes, it follows that

$$q = \Pr(X_{t+\tau} = j | X_t = i) = \frac{\Pr((X_{t+\tau} = j) \cap (X_t = i))}{\Pr(X_t = i)}$$

$$= \frac{\sum_{m=1}^{n} \Pr((X_{t+\tau} = j) \cap (X_{t+(\tau-1)} = m) \cap (X_t = i))}{\Pr(X_t = i)} = \sum_{m=1}^{n} q_m,$$

where the q_m can be further reduced as follows.

$$q_m = \frac{\Pr(X_{t+\tau} = j | (X_{t+(\tau-1)} = m) \cap (X_t = i)) \cdot \Pr((X_{t+(\tau-1)} = m) \cap (X_t = i))}{\Pr(X_t = i)}$$

$$= \Pr(X_{t+(\tau-1)} = m | X_t = i) \cdot p_{mj} = p_{im}^{(\tau-1)} p_{mj}$$

On the last line, we equated $\Pr(X_{t+\tau} = j | (X_{t+(\tau-1)} = m) \cap (X_t = i))$ with p_{mj} as allowed by Theorem 3.19. Summing the q_m, we have

$$q = \Pr(X_{t+\tau} = j | X_t = i) = \sum_{m=1}^{n} q_m = \sum_{m=1}^{n} p_{im}^{(\tau-1)} p_{mj} = p_{ij}^{(\tau)}. \quad \blacksquare$$

3.23 Definition: We refer to $p_{ij}^{(k)}$ as the *k-step transition probabilities*. $\blacksquare$

A Markov chain is termed irreducible if there is nonzero probability of cycling between any two given states. Since $p_{ij}^{(t)}$ is the probability of entering state j after t transitions from state i, the formal definition is as follows.

3.24 Definition: Let $P = [p_{ij}]$ be the transition matrix of an n-state Markov chain. The chain is *irreducible* if for every i, j there exist $k_1, k_2 \geq 0$ such that $p_{ij}^{(k_1)} > 0$ and $p_{ji}^{(k_2)} > 0$. $\blacksquare$

Client-server systems as typified in Example 3.21 always give rise to irreducible Markov chains. The states correspond to combinations of queue length and remaining service time for the current task. The transition matrix allows the queue length to grow and shrink within its allowed limit, so there is nonzero probability of moving from states associated with one queue length to those associated with the next larger or next smaller queue length. Within states corresponding to a given queue length, the system moves successively down along states associated with the remaining service time for the task at the server. Hence it is always possible to envision a sequence of events that will move the system from a given state to any specified destination state.

The intuitive idea of irreducibility is that there is no proper subset of inescapable states. A wandering trajectory must eventually visit every state in the system. Client-server systems do have this property, provided that we do not include dead states in the model. A realistic model must provide for a return to the empty state, and all useful states are reachable from there.

A consequence of irreducibility is that the convergence properties of sequences $\{p_{ii}^{(k)}\}$ and $\{p_{jj}^{(k)}\}$, for $k = 1, 2, \ldots$, are interrelated. This property is useful in the coming development, so we record it as a theorem.

3.25 Theorem: Let i, j be two states of an irreducible Markov chain with transition matrix $P = [p_{ij}]$. Then there exist a constant $A > 0$ and a positive integer M such that $m > M$ implies that $p_{ii}^{(m)} \geq A p_{jj}^{(m-M)}$ and $p_{jj}^{(m)} \geq A p_{ii}^{(m-M)}$. In particular, $M = \min\{k | p_{ij}^{(k)} > 0\} + \min\{k | p_{ji}^{(k)} > 0\}$ suffices.

PROOF: The irreducibility means there exist positive integers K_1 and K_2 such that $p_{ij}^{(K_1)} > 0$ and $p_{ji}^{(K_2)} > 0$. Let K_1 and K_2 be the smallest such values, and choose $M = K_1 + K_2$. Now, for any $m > K_1 + K_2$,

$$P^m = P^{K_1} P^{m - K_1 - K_2} P^{K_2} = P^{K_2} P^{m - K_1 - K_2} P^{K_1}.$$

So, with $A = p_{ij}^{(K_1)} p_{ji}^{(K_2)} > 0$, we have, for $m > M$,

$$p_{ii}^{(m)} = \sum_{k=1}^{n} p_{ik}^{(K_1)} \left[\sum_{l=1}^{n} p_{kl}^{(m - K_1 - K_2)} p_{li}^{(K_2)} \right] \geq p_{ij}^{(K_1)} p_{jj}^{(m-M)} p_{ji}^{(K_2)} = A p_{jj}^{(m-M)}$$

$$P = P^5$$

$$\begin{bmatrix} 0 & 1/2 & 1/2 & 0 & 0 \\ 0 & 0 & 0 & 1 & 0 \\ 0 & 0 & 0 & 1 & 0 \\ 0 & 0 & 0 & 0 & 1 \\ 1 & 0 & 0 & 0 & 0 \end{bmatrix}$$

$$P^2$$

$$\begin{bmatrix} 0 & 0 & 0 & 1 & 0 \\ 0 & 0 & 0 & 0 & 1 \\ 0 & 0 & 0 & 0 & 1 \\ 1 & 0 & 0 & 0 & 0 \\ 0 & 1/2 & 1/2 & 0 & 0 \end{bmatrix}$$

$$P^3$$

$$\begin{bmatrix} 0 & 0 & 0 & 0 & 1 \\ 1 & 0 & 0 & 0 & 0 \\ 1 & 0 & 0 & 0 & 0 \\ 0 & 1/2 & 1/2 & 0 & 0 \\ 0 & 0 & 0 & 1 & 0 \end{bmatrix}$$

$$P^4$$

$$\begin{bmatrix} 1 & 0 & 0 & 0 & 0 \\ 0 & 1/2 & 1/2 & 0 & 0 \\ 0 & 1/2 & 1/2 & 0 & 0 \\ 0 & 0 & 0 & 1 & 0 \\ 0 & 0 & 0 & 0 & 1 \end{bmatrix}$$

Figure 3.24. Transition matrix powers for Example 3.24

$$p_{jj}^{(m)} = \sum_{k=1}^{n} p_{jk}^{(K_2)} \left[\sum_{l=1}^{n} p_{kl}^{(m-K_1-K_2)} p_{lj}^{(K_1)} \right] \geq p_{ji}^{(K_2)} p_{ii}^{(m-M)} p_{ij}^{(K_1)} = A p_{ii}^{(m-M)}. \blacksquare$$

This result relates to convergence. Suppose that P is the transition matrix of an irreducible Markov chain, and suppose that $\lim_{m \to \infty} p_{ii}^{(m)} = v_i$. The theorem asserts a positive constant A and an integer M such that, for large m,

$$A p_{ii}^{(m-M)} \leq p_{jj}^{(m)} \leq \frac{p_{ii}^{(m+M)}}{A}.$$

In the limit, the variation in the $p_{jj}^{(m)}$ is essentially confined to $A v_i \leq p_{jj}^{(m)} \leq v_i/A$. In particular, if $\lim_{m \to \infty} p_{ii}^{(m)} = 0$, then $p_{jj}^{(m)}$ must also converge to zero.

A second constraint on Markov chains arising from simple client-server models is aperiodicity, defined as follows. The notation $a|b$, for integers a and b, means that a divides b evenly.

3.26 Definition: Let $P = [p_{ij}]$ be the transition matrix of an n-state Markov chain. An integer $r \geq 1$ is a *candidate period* for state i if the nonzero entries in the sequence $p_{ii}^{(1)}, p_{ii}^{(2)}, \ldots$ occur only at multiples of r. That is, $p_{ii}^{(t)} > 0$ implies that $r|t$. The integer $\tau_i = \max\{r| \ r$ is a candidate period for state $i \}$ is the *period* of state i. If $\tau_i = 1$, state i is called *aperiodic*. A Markov chain is *aperiodic* if all its states are aperiodic. The chain is *periodic* with period $\tau > 1$ if all its states have the same period τ. $\blacksquare$

3.27 Example: Determine the periodicity of the states in the Markov chain with transition matrix P shown in Figure 3.24.

Figure 3.24 also elaborates the transition matrix powers, and we note that $P^5 = P$. So higher powers just repeat the first four. The sequences $p_{ii}^{(t)}$ then exhibit the following pattern, repeating the first four entries. For any state i, if t is not a

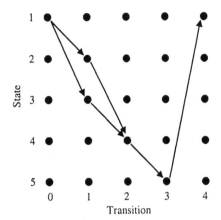

Figure 3.25. Evolution of state 1 in a periodic Markov chain (see Example 3.27)

multiple of $r = 4$, then $p_{ii}^{(t)} = 0$. This is also true for $r = 1$ and $r = 2$. Therefore, 1, 2, and 4 are the candidate periods for state i. The period is 4 because 4 is the largest candidate. Since all states have period 4, the Markov chain has period 4.

	$t = 1$	2	3	4	5	6	7	8	9	
$p_{11}^{(k)}$	0	0	0	1	0	0	0	1	0	...
$p_{22}^{(k)}$	0	0	0	1/2	0	0	0	1/2	0	...
$p_{33}^{(k)}$	0	0	0	1/2	0	0	0	1/2	0	...
$p_{44}^{(k)}$	0	0	0	1	0	0	0	1	0	...
$p_{55}^{(k)}$	0	0	0	1	0	0	0	1	0	...

The next theorem will show that all states in an irreducible Markov chain have the same period. Figure 3.25 illustrates the possible evolution of the system starting from state 1. The arrows indicate nonzero probabilities. You can see that all paths (there are only two) emanating from state 1 return to state 1 after 4 transitions. Further evolution of the system merely repeats these paths. Note that this Markov chain is irreducible because the path from each state visits all other states via transitions of nonzero probability. □

As suggested by the preceding example, all states in an irreducible Markov chain have the same period. Indeed, suppose that $P = [p_{ij}]$ is the transition matrix of an irreducible n-state Markov chain, in which τ_i and τ_j are the periods of states i and j, respectively. The irreducibility means there exist positive integers K_1 and K_2 such that $p_{ij}^{(K_1)} > 0$ and $p_{ji}^{(K_2)} > 0$. Letting K_1 and K_2 be the smallest such positive integers, and invoking Theorem 3.25, we obtain $A > 0$ such that

$$p_{ii}^{(m)} \geq A p_{jj}^{(m-M)}$$
$$p_{jj}^{(m)} \geq A p_{ii}^{(m-M)},$$

for $m > M = K_1 + K_2$. Also

$$p_{ii}^{(M)} = \sum_{k=1}^{n} p_{ik}^{(K_1)} p_{ki}^{(K_2)} \geq p_{ij}^{(K_1)} p_{ji}^{(K_2)} > 0$$

$$p_{jj}^{(M)} = \sum_{k=1}^{n} p_{jk}^{(K_2)} p_{kj}^{(K_1)} \geq p_{ji}^{(K_2)} p_{ij}^{(K_1)} > 0,$$

which means that $\tau_i | M$ and $\tau_j | M$. We can use these facts to relate the periods of states i and j as follows. Suppose that $m \geq 1$ and $\tau_i \nmid m$. Then, since $\tau_i | M$, we must have that $\tau_i \nmid (m + M)$. Consequently, $p_{ii}^{(m+M)} = 0$. So,

$$0 = p_{ii}^{(m+M)} \geq A p_{jj}^{(m)} \geq 0,$$

which implies that $p_{jj}^{(m)} = 0$. So τ_i is a candidate period for state j, and therefore $\tau_i \leq \tau_j$. We reason similarly if $\tau_j \nmid m$ to derive $\tau_j \leq \tau_i$. Therefore $\tau_i = \tau_j$; the two states have the same period. We record this result as the following theorem, which also asserts a simple condition for determining when an irreducible Markov chain is aperiodic.

3.28 Theorem: Let P be the transition matrix of an irreducible Markov chain. All states in the chain have the same period. Moreover, if $p_{ii} > 0$ for one or more i, then the period is 1, and the chain is aperiodic.

PROOF: The discussion above shows that all states have the same period, so we need only find a single state with period 1. That state i is, of course, one of those with $p_{ii} > 0$. Let τ_i be the period of such a state. Then $p_{ii}^{(1)} = p_{ii} > 0$ means that $\tau_i | 1$. Therefore, $\tau_i = 1$. ∎

The simple client-server systems illustrated in this chapter all give rise to irreducible Markov chains satisfying the condition $p_{ii} > 0$ for the empty state i. With a probabilistic interarrival time, there is a nonzero probability that an empty system will remain empty in the next time slot. So our earlier questions about the convergence of P^t to a single repeated row become questions about the transition matrix of an irreducible aperiodic Markov chain. To show this convergence, we introduce a new matrix of probabilities—the first-passage probabilities.

3.29 Definition: For an n-state Markov chain, let

$$q_{ij}^{(k)} = \Pr((X_{t+1} \neq j) \cap \ldots \cap (X_{t+k-1} \neq j) \cap (X_{t+k} = j) | X_t = i),$$

for $1 \leq i, j \leq n$ and $k \geq 1$. These are the *first-passage* probabilities. $q_{ij}^{(k)}$ is the probability that the kth transition from state i enters state j for the first time. That is, state j does not appear as an intermediate state on the trajectory from i to j. This probability is independent of the position t. ∎

This definition is meaningful only if $q_{ij}^{(k)}$ is indeed independent of the starting location t. Before proving this fact, we consider an example that shows how to compute a first-passage probability in a simple case.

3.30 Example: Consider a game that repeatedly flips a fair coin. We receive one dollar for each head, and we must pay one dollar for each tail. Suppose that we start with zero dollars, and we stop when our fortune rises to two dollars or our debt reaches two dollars. We construct a Markov chain for this situation by identifying state 1 with a 2-dollar debt, state 2 with a 1-dollar debt, state 3 with a fortune of zero dollars, state 4 with a 1-dollar fortune, and state 5 with a 2-dollar fortune. The probability of a transition from state 1 or 5 is zero, because we cease playing when we reach either state. From state 3, where we start, we proceed to state 2 or 3, each with probability 0.5. In general, from any state $i \in \{2, 3, 4\}$ we progress to either state $i - 1$ or $i + 1$, each with probability 0.5. The transition matrix is then as follows.

$$P = \begin{bmatrix} 1.0 & 0.0 & 0.0 & 0.0 & 0.0 \\ 0.5 & 0.0 & 0.5 & 0.0 & 0.0 \\ 0.0 & 0.5 & 0.0 & 0.5 & 0.0 \\ 0.0 & 0.0 & 0.5 & 0.0 & 0.5 \\ 0.0 & 0.0 & 0.0 & 0.0 & 1.0 \end{bmatrix}$$

The k-step transition probabilities are easy to obtain, because we need only compute powers of the transition matrix. Suppose, for example, that we are one dollar ahead at some point, and we want to know the probability of breaking even after exactly seven more coin tosses. We compute

$$P^7 = \begin{bmatrix} 1.0000 & 0.0000 & 0.0000 & 0.0000 & 0.0000 \\ 0.7188 & 0.0000 & 0.0625 & 0.0000 & 0.2188 \\ 0.4375 & 0.0625 & 0.0000 & 0.0625 & 0.4375 \\ 0.2188 & 0.0000 & 0.0625 & 0.0000 & 0.7188 \\ 0.0000 & 0.0000 & 0.0000 & 0.0000 & 1.0000 \end{bmatrix}$$

and observe the entry in row 4, column 3. This probability is 0.0625. The first-passage probabilities, however, are more difficult to compute. Suppose instead that we want the probability of breaking even *for the first time* after exactly seven coin tosses, given that we currently have a fortune of one dollar. The probability 0.0625 is too high, because it includes all trajectories from one dollar to zero dollars in seven transition, including those that might pass through zero dollars at some intermediate point. In this small example, we can apply ad hoc reasoning to conclude that the actual first-passage probability $q_{43}^{(7)} = 0$. Indeed, starting at state 4, the next transition must lead either to state 5 or to state 3. If it leads to state 5, then all future transitions cycle deterministically back to state 5. If it leads to state 3, then the path is disqualified for contribution to the first-passage probability because it enters state 3 too soon. Consequently, there is no scenario that can progress from state 4 to state 3 in exactly seven transitions without encountering state 3 at some intermediate point.

Suppose that we want $q_{35}^{(6)}$, the probability of leaving the game with 2 dollars after exactly six coin flips from a neutral position. The only valid trajectories are (3232345), (3234345), (3432345), and (3434345). All others hit state 5 too soon or do not hit it at all. In all four paths, the transition probabilities are 0.5 at each step. Consequently, each path has probability $(1/2)^6$. So, $q_{35}^{(6)} = 4/2^6 = 0.0625$. By computing P^6, we note that $p_{35}^{(6)} = 0.4375$, a much larger value that includes paths that hit state 5 earlier than the sixth transition and therefore remain trapped there. These additional paths are (3234555), (3434555), (3455555). Along these paths, the transition probabilities are 0.5 until the path enters state 5, where the

subevent E	$\Pr(E \cap (X_t = i) \cap (X_{t+\tau} = j))$
first state j after X_t occurs at X_{t+1}	$\Pr(X_t = i) \cdot q_{ij}^{(1)} p_{jj}^{(\tau-1)}$
first state j after X_t occurs at X_{t+2}	$\Pr(X_t = i) \cdot q_{ij}^{(2)} p_{jj}^{(\tau-2)}$
first state j after X_t occurs at X_{t+3}	$\Pr(X_t = i) \cdot q_{ij}^{(3)} p_{jj}^{(\tau-3)}$
$\vdots$	$\vdots$
first state j after X_t occurs at $X_{t+\tau-1}$	$\Pr(X_t = i) \cdot q_{ij}^{(\tau-1)} p_{jj}^{(1)}$
first state j after X_t occurs at $X_{t+\tau}$	$\Pr(X_t = i) \cdot q_{ij}^{(\tau)}$

TABLE 3.10. Disjoint components of the event $(X_t = i) \cap (X_{t+\tau} = j)$

transition probability changes to 1.0. The probabilities of these three paths are, respectively, $(0.5)^4(1.0)^2 = 0.0625$, $(0.5)^4(1.0)^2 = 0.0625$, and $(0.5)^2(1.0)^4 = 0.2500$. These contributions, added to the 0.0625 associated with the paths for $q_{35}^{(6)}$, raise $p_{35}^{(6)}$ to 0.4375. Although this exhaustive elaboration of paths is feasible for a small example, we need a more systematic way of dealing with first-passage probabilities. Accordingly, we continue with the theory. $\square$

We now verify, for the general case, that the first-passage probability $q_{ij}^{(\tau)}$ is indeed independent of the position t where state i occurs. The event $(X_t = i) \cap (X_{t+\tau} = j)$ decomposes into the disjoint components of Table 3.10. Hence,

$$p_{ij}^{(\tau)} = \frac{\Pr(X_t = i)\left[q_{ij}^{(\tau)} + \sum_{k=1}^{\tau-1} q_{ij}^{(k)} p_{jj}^{(\tau-k)}\right]}{\Pr(X_t = i)} = q_{ij}^{(\tau)} + \sum_{k=1}^{\tau-1} q_{ij}^{(k)} p_{jj}^{(\tau-k)}.$$

A rearrangement gives a recursive equation for the first-passage probabilities.

$$q_{ij}^{(\tau)} = p_{ij}^{(\tau)} - \sum_{k=1}^{\tau-1} q_{ij}^{(k)} p_{jj}^{(\tau-k)} \tag{3.5}$$

We see that $q_{ij}^{(1)} = p_{ij}^{(1)} = p_{ij}$, which is independent of t. For purposes of an induction argument, assume that $q_{ij}^{(m)}$ is independent of t for $m = 1, 2, \ldots, k-1$. Now note that $q_{ij}^{(k)}$ is an expression in $q_{ij}^{(m)}$ and $p_{jj}^{(m)}$, for $m < k$. The $p_{jj}^{(m)}$ are independent of t, and by the induction hypothesis, so are the $q_{ij}^{(m)}$. We conclude that the first-passage probabilities, like the transition probabilities, depend only on the initial and final states, not where they occur in the sequence.

3.31 Example: Using Equation 3.5, verify the first-passage probabilities of the preceding example. Recall that we are flipping a fair coin and that our winnings increase by one dollar for each head and decrease by one dollar for each tail. We cease playing when we have won or lost two dollars. We reasoned in the preceding example that $q_{43}^{(7)} = 0$ and $q_{35}^{(6)} = 0.0625$. To apply Equation 3.5 in calculating $q_{43}^{(7)}$, we need $p_{43}^{(k)}$ and $p_{33}^{(k)}$, for $1 \le k \le 6$. These are available from the powers of P, the transition matrix, and we install them in the following tabulation to organize the computation. The right portion of the table organizes the similar calculation for $q_{35}^{(6)}$.

k	$p_{43}^{(k)}$	$p_{33}^{(k)}$	$\sum_{j=1}^{k-1} q_{43}^{(j)} p_{33}^{(k-j)}$	$q_{43}^{(k)}$	$p_{35}^{(k)}$	$p_{55}^{(k)}$	$\sum_{j=1}^{k-1} q_{35}^{(j)} p_{55}^{(k-j)}$	$q_{35}^{(k)}$
1	0.5000	0.0000	0.0000	0.5000	0.0000	1.0000	0.0000	0.0000
2	0.0000	0.5000	0.0000	0.0000	0.2500	1.0000	0.0000	0.2500
3	0.2500	0.0000	0.2500	0.0000	0.2500	1.0000	0.2500	0.0000
4	0.0000	0.2500	0.0000	0.0000	0.3750	1.0000	0.2500	0.1250
5	0.1250	0.0000	0.1250	0.0000	0.3750	1.0000	0.3750	0.0000
6	0.0000	0.1250	0.0000	0.0000	0.4375	1.0000	0.3750	**0.0625**
7	0.0625	0.0000	0.0625	**0.0000**				

The boldfaced entries confirm the values obtained earlier. □

3.32 Example: Examples 3.12 and 3.21 treat a client-server configuration, in which a single client generates a request with probability p at the beginning of each millisecond. The single server handles the requests in a first-come, first-serve fashion while any additional arrivals accumulate in a bounded queue. The server response time is a discrete uniform distribution on $\{a, a+1, a+2, \dots, b\}$ milliseconds. In Example 3.21, we used $p = 0.3, a = 1$, and $b = 3$, and we showed that a Markov chain analysis revealed the same value as a simulation for the expected time a request spends in the waiting queue. Continuing with these parameters, investigate the random variable T, which is the time, in milliseconds, between idle system states.

Extracting the context of Example 3.21, we find that the system has 13 states, each corresponding to a pair (n, s), where n denotes the number of requests in the system and s is the time remaining for the request currently under service. The transition matrix appears in Table 3.8, where we note that state 1, corresponding to $(n = 0, s = 0)$, is the idle system state. T is then the first-passage time from state 1 to itself and has probabilities $q_{11}^{(k)}$, for $k = 1, 2, \dots$. We organize the calculations as in the preceding example but enlist some computer assistance to complete the entries. The results appear below.

k	$p_{11}^{(k)}$	$\sum_{j=1}^{k-1} q_{11}^{(j)} p_{11}^{(k-j)}$	$q_{11}^{(k)}$
1	0.7000	0.0000	0.7000
2	0.5600	0.4900	0.0700
3	0.4760	0.4410	0.0350
4	0.4304	0.3969	0.0335
5	0.4089	0.3777	0.0312
6	0.4026	0.3737	0.0289
7	0.4021	0.3792	0.0229
8	0.4021	0.3855	0.0166
9	0.4021	0.3891	0.0130
10	0.4021	0.3917	0.0104

The tabulated $q_{11}^{(k)}$, for $k = 1, 2, \dots, 10$, account for 96% of the probability associated with T and provide the approximation $\mu_T = 1.92$. This means that the system reenters the idle state, on the average about once every 2 milliseconds. Note also that the average service time is 2 milliseconds. That is, the service time is uniformly distributed between 1 and 3 milliseconds, which places the average at the center of that span. The input process with $p = 0.3$ delivers about 30 requests in 100 milliseconds and the system can handle, on the average, 50 requests in that time. So we do expect frequent returns to the idle state. □

Exercises

3.38 In a board game, you move a token to the left or right depending on a card drawn from a well-shuffled standard deck. The chosen card returns to the deck before each new draw. The token starts at square 0 and moves to the left, into squares with negative numbers, when certain cards appear. It moves to the right, into squares with positive numbers when other cards appear. The size of the moves is as follows, where L, R denote left and right, and x designates the value of a card in the range $[2, 10]$.

Card	Ace	King	Queen	Jack	Other
Move	R4	R3	R2	R1	Lx

You win the game when your token lands in square 4 or in square -4; you lose if your token travels to the right of square 4 without stopping on square 4 or if you travel to the left of square -4 without stopping there. Model this game as a Markov chain with states $0, \pm1, \pm2, \pm3$, win, and lose. Find the probability distribution of the states after three draws.

3.39 Show that the Markov chain associated with the preceding exercise is not irreducible. Change the game rules to produce an irreducible Markov chain.

3.40 A 4-state Markov chain has the transition matrix to the right. Theorem 3.25 says that there exist a constant $A > 0$ and an integer M such that

$$p_{11}^{(m)} \geq A p_{22}^{(m-M)}$$
$$p_{22}^{(m)} \geq A p_{11}^{(m-M)},$$

$$\begin{bmatrix} 1/2 & 0 & 1/2 & 0 \\ 0 & 1/2 & 1/2 & 0 \\ 0 & 0 & 1/2 & 1/2 \\ 1/2 & 1/2 & 0 & 0 \end{bmatrix}$$

for $m > M$. Find A and M in this case.

*3.41 Let $P = [p_{ij}]$ be the transition matrix of an irreducible Markov chain. Imitate the proof of Theorem 3.25 to show that if $\lim_{k \to \infty} p_{ij}^{(k)} = 0$, then $\lim_{k \to \infty} p_{ji}^{(k)} = 0$ also.

3.42 At any given time, the token in a board game occupies one of six squares. Each move occurs with the toss of a single die. If the die shows i spots on the upturned face, the token moves from its current square to square i. Note that it is possible for the token to remain stationary if the die returns the number of its current square. When the game starts, the token is in square 1. If the token lands in square 5 for the first time on the fifth die toss, you win \$10. With any other scenario, you lose. To break even over the long run, how much should the game operator charge to play each game? How much should he charge to average one dollar per game in profit?

3.43 Consider a coin-flipping game in which you win one dollar for each head and lose one dollar for each tail. The game ends when you have won five dollars or lost four dollars. Assuming that the coin is fair, what is the probability that you will win the five dollars? What is the probability that you will lose four dollars?

3.44 Repeat the preceding exercise with a biased coin for which the probability of heads is 0.4.

3.45 In the context of the preceding two exercises, suppose that the coin is fair, but that you win one dollar on a head and lose two dollars on a tail. The game concludes when the winnings total five dollars or when the losses equal or exceed four dollars. What is the probability of a win when the game ends? What is the probability of a loss?

3.46 Example 3.32 investigates the time span between idle states for a system containing one server and a single request queue. Time flows in a quantized manner. At the beginning of each millisecond, a request arrives with probability 0.3. The server responds in a first-come, first-serve fashion, and the response time is a discrete uniform random variable on the set $\{1, 2, 3\}$ milliseconds. Change the input rate such that a request arrives with probability 0.5 at the beginning of each millisecond. Let state 1 be the system idle state and compute the first-passage probabilities $q_{11}^{(k)}$ for $k = 1, 2, \ldots, 20$.

3.4.2   Convergence properties

This subsection's discussion uses only the analysis tools from Appendix A and those developed in the earlier mathematical interludes. The arguments, however, are demanding. The proofs here require more than simple awareness of the tools. To master these proofs, the reader must attend carefully to the details, and this task becomes more digestible with experience. Consequently, depending on his preparation, the reader may want to omit this section; the material is not necessary for the remainder of the book. For a reader with sufficient preparation, however, the insights available from a close study of the convergence properties of Markov chains are well worth the effort.

We continue with the notation of the preceding subsection. In particular, the $q_{ij}^{(k)}$, for $k = 1, 2, \ldots$, are the first-passage probabilities from state i to state j. The $p_{ij}^{(k)}$, also for $k = 1, 2, \ldots$, are the k-step transition probabilities. If $X_t = i$, the probability that $X_{t+\tau} = i$ for some $\tau \geq 1$ is $\sum_{\tau=1}^{\infty} q_{ii}^{(\tau)}$. This is the probability that the system ever returns to state i given that it is currently in state i. Accordingly, this sum may be 1 or less that 1, and that provides a criterion for distinguishing among states.

Terms	Bound
$q_{ij}^{(1)} p_{jj}^{(1)}$	$= p_{ij}^{(2)} - q_{ij}^{(2)}$
$q_{ij}^{(1)} p_{jj}^{(2)} + q_{ij}^{(2)} p_{jj}^{(1)}$	$= p_{ij}^{(3)} - q_{ij}^{(3)}$
$q_{ij}^{(1)} p_{jj}^{(3)} + q_{ij}^{(2)} p_{jj}^{(2)} + q_{ij}^{(3)} p_{jj}^{(1)}$	$= p_{ij}^{(4)} - q_{ij}^{(4)}$
$\vdots$	$\vdots$
$q_{ij}^{(1)} p_{jj}^{(N)} + q_{ij}^{(2)} p_{jj}^{(N-1)} + \ldots + q_{ij}^{(N)} p_{jj}^{(1)}$	$= p_{ij}^{(N)} - q_{ij}^{(N)}$
$q_{ij}^{(2)} p_{jj}^{(N)} + q_{ij}^{(3)} p_{jj}^{(N-1)} + \ldots + q_{ij}^{(N)} p_{jj}^{(2)}$	$\leq p_{ij}^{(N+1)} - q_{ij}^{(N+1)}$
$q_{ij}^{(3)} p_{jj}^{(N)} + q_{ij}^{(4)} p_{jj}^{(N-1)} + \ldots + q_{ij}^{(N)} p_{jj}^{(3)}$	$\leq p_{ij}^{(N+2)} - q_{ij}^{(N+2)}$
$\vdots$	$\vdots$
$q_{ij}^{(N-1)} p_{jj}^{(N)} + q_{ij}^{(N)} p_{jj}^{(N-1)}$	$\leq p_{ij}^{(2N-1)} - q_{ij}^{(2N-1)}$
$q_{ij}^{(N)} p_{jj}^{(N)}$	$\leq p_{ij}^{(2N)} - q_{ij}^{(2N)}$

Figure 3.26. Bounds involving first passage and multistep transition probabilities (see Theorem 3.34)

3.33 Definition: Let i be a state in an n-state Markov chain, and let $\{q_{ij}^{(\tau)}\}$ be the first-passage probabilities. State i is *recurrent* if $\sum_{\tau=1}^{\infty} q_{ii}^{(\tau)} = 1$; otherwise, it is *transient*. If i is a recurrent state, let T_i be the random variable that gives the number of transitions for the system to reenter state i for the first time, given that it starts in state i. T_i is called the *first return time* for state i. If $E(T_i) = \sum_{k=1}^{\infty} k q_{ii}^{(k)} < \infty$, then state i is said to be *positive recurrent*; otherwise it is a *null recurrent* state. An *ergodic* state is an aperiodic, positive recurrent state. ∎

We have seen that the Markov chains arising from simple client-server systems are irreducible and aperiodic. We will now show that they are also ergodic. In light of the definition above, we need only show that all states are positive recurrent. Equation 3.5 gives the relationship between first-passage and transition probabilities. It is complicated because $q_{ij}^{(\tau)}$ depends on all the previous elements in its sequence: $q_{ij}^{(1)}, q_{ij}^{(2)}, \ldots, q_{ij}^{(\tau-1)}$. It also depends on all the elements in the sequence $p_{jj}^{(1)}, p_{jj}^{(2)}, \ldots, p_{jj}^{(\tau-1)}$. We will develop a simpler condition for a recurrent state, a condition that depends only on the k-step probabilities $p_{jj}^{(k)}$. Consider the following interesting product of first-passage and k-step probabilities.

$$\left(\sum_{k=1}^{N} q_{ij}^{(k)} \right) \left(\sum_{k=1}^{N} p_{jj}^{(k)} \right) = \left(q_{ij}^{(1)} + q_{ij}^{(2)} + \ldots + q_{ij}^{(N)} \right) \left(p_{jj}^{(1)} + p_{jj}^{(2)} + \ldots + p_{jj}^{(N)} \right)$$

Figure 3.26 organizes the terms of this product and uses Equation 3.5 to obtain a bound on each component group. The first line contains those terms in which the superscripts on the q_{ij} and p_{jj} factors sum to 2. The second line contains the terms in which the superscripts sum to 3, and so forth.

Therefore, because $q_{ij}^{(1)} = p_{ij}^{(1)}$,

$$\sum_{k=1}^{N} p_{ij}^{(k)} - \sum_{k=1}^{N} q_{ij}^{(k)} \leq \left(\sum_{k=1}^{N} q_{ij}^{(k)}\right)\left(\sum_{k=1}^{N} p_{jj}^{(k)}\right) \leq \sum_{k=1}^{2N} p_{ij}^{(k)} - \sum_{k=1}^{2N} q_{ij}^{(k)} \qquad (3.6)$$

Now suppose that i is a recurrent state. Letting $j = i$ in the rightmost inequality of Equation 3.6, we have

$$\left(\sum_{k=1}^{N} q_{ii}^{(k)}\right)\left(\sum_{k=1}^{N} p_{ii}^{(k)}\right) + \sum_{k=1}^{2N} q_{ii}^{(k)} \leq \sum_{k=1}^{2N} p_{ii}^{(k)}.$$

Because i is recurrent, we know $\sum_{k=1}^{\infty} q_{ii}^{(k)} = 1$. Suppose that $\sum_{k=1}^{\infty} p_{ii}^{(k)} = c < \infty$. Then, letting $N \to \infty$ in the equation above, we have $c + 1 \leq c$. This is not possible for a finite number c. We conclude that i recurrent implies that $\sum_{k=1}^{\infty} p_{ii}^{(k)}$ diverges. Conversely, suppose that $\sum_{k=1}^{\infty} p_{ii}^{(k)}$ diverges. The left part of Equation 3.6, again with $j = i$, gives

$$\sum_{k=1}^{N} p_{ii}^{(k)} - \sum_{k=1}^{N} q_{ii}^{(k)} \leq \left(\sum_{k=1}^{N} q_{ii}^{(k)}\right)\left(\sum_{k=1}^{N} p_{ii}^{(k)}\right)$$

$$1 - \frac{\sum_{k=1}^{N} q_{ii}^{(k)}}{\sum_{k=1}^{N} p_{ii}^{(k)}} \leq \sum_{k=1}^{N} q_{ii}^{(k)}.$$

Letting $N \to \infty$, we obtain $\sum_{k=1}^{\infty} q_{ii}^{(k)} \geq 1$. Because it represents the probability of an eventual return to state i, we know that $\sum_{k=1}^{\infty} q_{ii}^{(k)} \leq 1$. We conclude that the sum is exactly 1 and the state is recurrent. The following theorem summarizes these observations.

3.34 Theorem: A state i in a Markov chain is recurrent if and only if $\sum_{k=1}^{\infty} p_{ii}^{(k)}$ diverges.

PROOF: See discussion above. ∎

3.35 Theorem: All states in a finite irreducible Markov chain are recurrent.

PROOF: Suppose that all states are transient. That is, $q_i = \sum_{k=1}^{\infty} q_{ii}^{(k)} < 1$ for all states i. Each time the system enters state i, there is a nonzero probability $1 - q_i$ that it will never return. Thus, starting from state i, the probability of zero reentries is $1 - q_i > 0$. The probability of exactly one reentry is $q_i(1 - q_i)$. The probability of exactly two reentries is $q_i^2(1 - q_i)$. This is the familiar geometric distribution with parameter $1 - q_i$. Suppose that we start the system from an arbitrary state. Then $X_k = i$ for $m + 1$ positions k means that there have been m reentries. The probability of m or more reentries is

$$\sum_{k=m}^{\infty} q_i^k (1 - q_i) = q_i^m (1 - q_i) \sum_{k=0}^{\infty} q_i^k = q_i^m (1 - q_i) \cdot \frac{1}{1 - q_i} = q_i^m.$$

Because $q_i < 1$, this probability approaches zero with increasing m. As there are only n states, we can choose m such that $q_i^m < 1/(2n)$ simultaneously for

$i = 1, 2, \ldots, n$. Then, if we let E_i be the event that state i occurs more than m times among the X_k, we have $\Pr(E_i) < 1/(2n)$. Let E be the event that one or more states occurs more than m times among the X_k. Since there are infinitely many X_k, it is clear that E is a certain event. That is, $\Pr(E) = 1$. However, $E \subset \cup_{i=1}^{n} E_i$, which implies that $\Pr(E) \leq \sum_{i=1}^{n} \Pr(E_i) < n[1/(2n)] = 1/2$. This contradiction proves that there must be at least one recurrent state.

Let j be the recurrent state, and let i be another state. By Theorem 3.34, $\sum_{k=1}^{\infty} p_{jj}^{(k)}$ diverges. Because the chain is irreducible, Theorem 3.25 applies. Hence there exist $A > 0$ and positive integer M such that $p_{ii}^{(m)} \geq A p_{jj}^{(m-M)}$, for $m > M$. Consequently, $\sum_{k=1}^{\infty} p_{ii}^{(k)}$ diverges, which means that state i is also recurrent. We conclude that all states are recurrent. ∎

For a recurrent state i, we have, by definition, $\sum_{k=1}^{\infty} q_{ii}^{(k)} = 1$. Starting from state i, a return to state i is certain. It turns out that the probability of $2, 3, 4, \ldots$ returns is also certain.

3.36 Theorem: Let i be a recurrent state in Markov chain. Fix a position τ, and for any integer $s \geq 1$, let E_s be the event that s or more occurrences of state i occur after position τ. Then $\Pr(E_s | X_\tau = i) = 1$.

PROOF: By the definition of a recurrent state, $\Pr(E_1 | X_\tau = i) = \sum_{k=1}^{\infty} q_{ii}^{(k)} = 1$, and we can proceed by induction. Assume that $\Pr(E_s | X_\tau = i) = \sum_{k=0}^{\infty} b_k = 1$, where b_k is the probability that the sth return occurs in position $\tau + k$. Clearly, $b_k = 0$ for $k = 0, 1, 2, \ldots, s - 1$, and we let $q_{ii}^{(0)} = 0$ for convenience. Denote by c_k the probability that return $s + 1$ occurs at position $\tau + k$. Then $c_k = 0$ for $k = 0, 1, 2, \ldots, s$ and

$$
\begin{aligned}
c_{s+1} &= b_s q_{ii}^{(1)} \\
c_{s+2} &= b_s q_{ii}^{(2)} + b_{s+1} q_{ii}^{(1)} \\
c_{s+3} &= b_s q_{ii}^{(3)} + b_{s+1} q_{ii}^{(2)} + b_{s+2} q_{ii}^{(1)} \\
&\vdots \qquad\qquad \vdots \\
c_t &= b_s q_{ii}^{(t-s)} + b_{s+1} q_{ii}^{(t-s-1)} + \ldots + b_{t-1} q_{ii}^{(1)}
\end{aligned}
$$

$$\sum_{k=s+1}^{t} c_k = \sum_{k=s+1}^{t} \sum_{m=1}^{k-s} q_{ii}^{(m)} b_{k-m}.$$

We can add the zero terms to obtain summations that start at zero. The full expansion then exhibits the pattern to the left in the display below.

$$\sum_{k=0}^{t} c_k = \sum_{k=0}^{t} \sum_{m=0}^{k} q_{ii}^{(m)} b_{k-m} \qquad\qquad \left(\sum_{k=0}^{t} q_{ii}^{(k)} \right) \cdot \left(\sum_{k=0}^{t} b_k \right)$$

$q_{ii}^{(0)} b_0 +$
$q_{ii}^{(0)} b_1 + q_{ii}^{(1)} b_0 +$
$q_{ii}^{(0)} b_2 + q_{ii}^{(1)} b_1 + q_{ii}^{(2)} b_0 +$
$\vdots$
$q_{ii}^{(0)} b_t + q_{ii}^{(1)} b_{t-1} + \ldots + q_{ii}^{(t)} b_0.$

$q_{ii}^{(0)} b_0 +$

$q_{ii}^{(0)} b_1 + q_{ii}^{(1)} b_0 +$

$q_{ii}^{(0)} b_2 + q_{ii}^{(1)} b_1 + q_{ii}^{(2)} b_0 +$

$\vdots$

$q_{ii}^{(0)} b_t + q_{ii}^{(1)} b_{t-1} + \ldots + q_{ii}^{(t)} b_0 +$

$q_{ii}^{(1)} b_t + q_{ii}^{(2)} b_{t-1} + \ldots + q_{ii}^{(t)} b_1 +$

$\vdots$

$q_{ii}^{(t)} b_t$

The right side above displays terms from the product of two series. Comparing the two columns, we find that the product of series includes all the terms for $\sum_{k=0}^{t} c_k$ and some of the terms for $\sum_{k=0}^{2t} c_k$. Because all these terms are nonnegative, we conclude that

$$\sum_{k=0}^{t} c_k \leq \left(\sum_{k=0}^{t} q_{ii}^{(k)}\right) \cdot \left(\sum_{k=0}^{t} b_k\right) \leq \sum_{k=0}^{2t} c_k.$$

Letting $t \to \infty$, we obtain $\sum_{k=0}^{\infty} c_k \leq 1 \leq \sum_{k=0}^{\infty} c_k$, which forces $\sum_{k=0}^{\infty} c_k = \Pr(E_{s+1}|X_\tau = i) = 1$. Completing the induction, we have $\Pr(E_s|X_\tau = i) = 1$ for all $s \geq 1$. ∎

Given an irreducible, aperiodic Markov chain, we now know that all states are recurrent, and moreover, each state recurs infinitely often. We have yet to show that the states are *positive* recurrent. To this end, let i be one of the n states. To prove positive recurrence, we need to show that $\sum k q_{ii}^{(k)}$ converges.

As a first step in that direction, we will connect recurrence with convergence of the sequence $\{p_{ii}^{(k)}\}$. Because i will be fixed throughout the argument, we use the simpler notation $p_k = p_{ii}^{(k)}$ and $q_k = q_{ii}^{(k)}$. Because the state is recurrent, $\sum_{k=1}^{\infty} q_k = 1$. If we compute $\sum_{k=1}^{t} p_k$ from Equation 3.5 and collect terms in the various p_k, we obtain the following, in which our interest rests with the sum of the leading terms on the right-hand sides. Recall that $p_0 = 1$ because P^0 is the identity matrix.

$$p_1 = q_1 p_0$$
$$p_2 = q_2 p_0 + q_1 p_1$$
$$p_3 = q_3 p_0 + q_2 p_1 + q_1 p_2$$
$$\vdots \qquad \vdots$$
$$p_t = q_t p_0 + q_{t-1} p_1 + \ldots + q_1 p_{t-1}$$

Summing and solving for the total of the first column after the equal sign, we have

$$p_0 \sum_{k=1}^{t} q_k = p_1 \left(1 - \sum_{k=1}^{t-1} q_k\right) + p_2 \left(1 - \sum_{k=1}^{t-2} q_k\right) + \ldots + p_{t-1}(1 - q_1) + p_t(1).$$

To obtain a uniform pattern, we transpose the left side and add 1 to both sides. This produces

$$1 = 1 - p_0 \sum_{k=1}^{t} q_k + p_1 \left(1 - \sum_{k=1}^{t-1} q_k\right) + p_2 \left(1 - \sum_{k=1}^{t-2} q_k\right)$$
$$+ \ldots + p_{t-1}(1 - q_1) + p_t(1)$$

$$1 = p_0 \left(1 - \sum_{k=1}^{t} q_k\right) + p_1 \left(1 - \sum_{k=1}^{t-1} q_k\right) + p_2 \left(1 - \sum_{k=1}^{t-2} q_k\right)$$

$$+ \ldots + p_{t-1}(1 - q_1) + p_t(1)$$

$$1 = \sum_{j=0}^{t} p_j \left(1 - \sum_{k=1}^{t-j} q_k\right).$$

Now, define $w_s = 1 - \sum_{k=1}^{s} q_k = \sum_{k=1}^{\infty} q_k - \sum_{k=1}^{s} q_k = \sum_{k=s+1}^{\infty} q_k$, for $s \geq 0$. Note that $w_0 = 1$. In these new terms, The derivation above is

$$1 = w_0 p_t + w_1 p_{t-1} + w_2 p_{t-2} + \ldots + w_t p_0. \tag{3.7}$$

Recall that the first return time, T_i, is the random variable that counts the transitions, starting from state i, to return the system to state i for the first time. In our simplified notation, which suppresses mention of the state i, we let $T = T_i$. Accordingly, $\mu_T = \sum_{k=1}^{\infty} k q_k$, provided that the sum converges. This expression appears when we sum the newly created sequence $\{w_k\}$:

$$
\begin{array}{rcccccc}
w_0 &=& q_1 &+& q_2 &+& q_3 &+& \cdots = 1 \\
w_1 &=& && q_2 &+& q_3 &+& \cdots \\
w_2 &=& &&&& q_3 &+& \cdots \\
&& \vdots && \vdots && \vdots && \vdots && \vdots \\
\hline
\sum_{k=0}^{\infty} w_k &=& q_1 &+& 2q_2 &+& 3q_3 &+& \cdots = \sum_{k=1}^{\infty} k q_k.
\end{array}
$$

We know that $\sum p_k$ diverges because we are working with a recurrent state. However, we now want to draw a further distinction between null-recurrent and positive-recurrent states. In particular, we want to show that null recurrence, which is characterized by $\sum_{k=1}^{\infty} k q_k = \sum_{k=0}^{\infty} w_k = \infty$, occurs precisely when p_k converges to zero.

Suppose, to the contrary, that p_k converges to zero and $\sum_{k=0}^{\infty} w_k = \mu_T < \infty$. Let $0 < \epsilon < 1$. Then there exists N such that $\sum_{k=N+1}^{\infty} w_k < \epsilon/2$, which implies that, for any $t > N$, $\sum_{k=N+1}^{t} w_k p_{t-k} < \epsilon/2$. Let $W = \max\{w_0, w_1, w_2, \ldots, w_N\}$. Because $w_0 = 1$, we have $W \geq 1$. Turning now to the p_k, there exists M such that $p_k < \epsilon/[2W(N+1)]$ when $k \geq M$. Applying these constraints to Equation 3.7, for $t > N + M$, we have

$$1 = \sum_{k=0}^{t} w_k p_{t-k} = \sum_{k=0}^{N} w_k p_{t-k} + \sum_{k=N+1}^{t} w_k p_{t-k} < W \sum_{k=t-N}^{t} p_k + \frac{\epsilon}{2}$$

$$< W(N+1) \left(\frac{\epsilon}{2W(N+1)}\right) + \frac{\epsilon}{2} = \epsilon < 1.$$

This contradiction establishes that p_k convergent to zero implies that $\sum_{k=0}^{\infty} w_k = \sum_{k=1}^{\infty} k q_k$ divergent. That is, if state k is recurrent, which forces $\sum p_k \to \infty$, but nevertheless $p_k \to 0$, then the state is null-recurrent. To see that the converse is also true, we adapt an argument from Feller[19], which

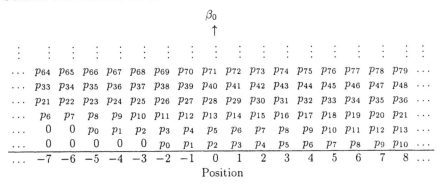

Figure 3.27. Arrangement of multistep transition probabilities p_k to obtain convergence

involves a detailed look at multiple copies of the sequence p_k and, in particular, how Equation 3.5 relates the sequence to the first-passage probabilities q_k. Let $\beta_0 = \limsup p_k$. Because the p_k are all in the bounded range $[0, 1]$, this limit is a well-defined number, and $0 \leq \beta_0 \leq 1$. There is then a subsequence p_{i_k} such that $\lim_{k\to\infty} p_{i_k} = \beta_0$. We now align multiple copies of sequences $\{\dots, 0, 0, p_0, p_1, p_2, p_3, \dots\}$ such that the first copy has the term p_{i_1} in the center. The next copy has p_{i_2} in the center. We continue in this manner to produce a display similar to Figure 3.27. The figure illustrates the case where the subsequence converging to β_0 is $p_2, p_5, p_{13}, p_{28}, p_{40}, \dots$. These terms comprise the central column of the display—over position number 0. Each line contains the complete sequence, with most of it to the right. An infinite number of zeros precedes the finite number of sequence terms to the left of the center column, and therefore each line is a sequence that stretches infinitely in both directions.

The column over position zero converges to β_0 because the construction places the terms of the convergent subsequence over this position. Immediately to the right, over position 1, is a column of terms from the original sequence. Call this column, starting from the bottom $\{a_1, a_2, a_3, \dots\}$. It may not converge, but $\beta_1 = \limsup a_k$ is a well-defined number. Since the a_k are a subset of the original p_k, we must have $\beta_1 \leq \beta_0 \leq 1$. (Otherwise, we would have infinitely many a_k, and hence infinitely many p_k, greater than β_0, which would contradict $\beta_0 = \limsup p_k$.) A subsequence $\{a_{i_k}\}$ then converges to β_1, and we can indicate this convergence in Figure 3.27 by crossing out lines so as to leave the a_{i_k} vertically over position 1. In order to leave some part of the display stable, let us agree to leave the first line of the original display, regardless of whether or not the term over position 1 is a_{i_1} or not. That is, we leave the bottom of column 1 as p_{i_1+1}. We then cross out higher lines until levels $2, 3, \dots$ coincide with $a_{i_2}, a_{i_3}, \dots$. The net result is that the display shifts downward when intermediate lines are deleted. The new display looks just like the old, except two columns now converge. Column 0 converges to β_0, and column 1 converges to β_1.

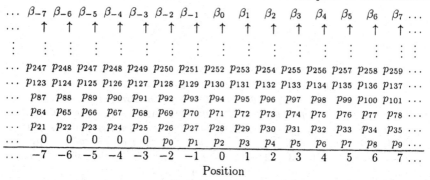

Figure 3.28. Final arrangement of multistep transition probabilities p_k to obtain convergence

We now repeat this process with the focus on position -1 to obtain a display of the same overall appearance but with columns $-1, 0, 1$ converging to $\beta_{-1}, \beta_0, \beta_1$. While forcing convergence on column -1, we maintain the first two lines of the previous display, and we also have $\beta_{-1} \le \beta_0 \le 1$. We continue with columns $2, -2, 3, -3, 4, -4, \ldots$. The original subsequence over column 0 becomes thinner and thinner but still converges to β_0. After the first step, the bottom of column 0 does not change. After the second step, the first two terms at the bottom of column 0 do not change. After three steps, the first three terms of the column are stable. The process is therefore producing a new subsequence of p_k, which maintains its convergence to β_0 while affording convergence to its neighboring columns. For any fixed column, say s, the process eventually reaches column s and thins it to a subsequence approaching β_s. Columns already converging remain convergent; they just lose some of their intermediate terms. The applicable display is now Figure 3.28. Each line contains the entire p_k sequence with a left shift as required to locate the center element in accordance with the process just described.

We will now argue that all the β_j are the same, so that we have infinitely many subsequences of p_k converging to the same β_0. Eventually we have p_k itself converging to β_0. The first step is to rewrite Equation 3.5 in our simpler notation. Recalling that $p_{ii}^{(0)} = p_0 = 1$, we have

$$p_t = \sum_{s=1}^{t} q_s p_{t-s}.$$

This equation defines each p_t in terms of the preceding $p_0, p_1, \ldots, p_{t-1}$. We can apply this formula to the display of Figure 3.28. Let the jth line in the display be $\ldots, r_{j,-3}, r_{j,-2}, r_{j,-1}, r_{j,0}, r_{j,1}, r_{j,2}, r_{j,3}, \ldots$. We have, for each line j,

$$r_{j,t} = \sum_{s=1}^{\infty} q_s r_{j,t-s},$$

for $t \in \{\ldots, -3, -2, -1, 0, 1, 2, 3, \ldots\}$. (Recall that the $r_{j,t-s}$ are zero for large s because line j is just the original sequence p_k shifted to put a particular value over position 0. A finite number of terms from the original sequence extends to the left of position 0, and all terms are zero yet farther to the left.) We want to extend this relation to the limits $\ldots, \beta_{-2}, \beta_{-1}, \beta_0, \beta_1, \beta_2, \ldots$. Accordingly, consider a fixed position t, and let $\epsilon > 0$ be an arbitrary small number. Because $\sum_{s=1}^{\infty} q_s = 1$, there exists N such that $\sum_{s=N+1}^{\infty} q_s < \epsilon/4$. For column m in the range $t-N$ through t, we have terms $r_{j,m} \to \beta_m$ as $j \to \infty$. Columns $t-N$ through t constitute $N+1$ convergent sequences, so there exists an integer M such that $j \geq M$ implies that $|r_{j,m} - \beta_m| < \epsilon/[4(N+1)]$ simultaneously for $t - N \leq m \leq t$. Now, remembering that the r, q, and β values are all in the range $[0, 1]$, we have, for $j > M$,

$$\left| \beta_t - \sum_{s=1}^{\infty} q_s \beta_{t-s} \right| = \left| \beta_t - r_{j,t} + \sum_{s=1}^{\infty} q_s r_{j,t-s} - \sum_{s=1}^{\infty} q_s \beta_{t-s} \right|$$

$$\leq \left| \beta_t - r_{j,t} \right| + \left| \sum_{s=1}^{N} q_s (r_{j,t-s} - \beta_{t-s}) \right| + \left| \sum_{s=N+1}^{\infty} q_s r_{j,t-s} \right|$$

$$+ \left| \sum_{s=N+1}^{\infty} q_s \beta_{t-s} \right|$$

$$\leq \frac{\epsilon}{4(N+1)} + \sum_{s=1}^{N} q_s |r_{j,t-s} - \beta_{t-s}| + 2 \sum_{s=N+1}^{\infty} q_s.$$

Invoking the available bounds, we conclude that

$$\left| \beta_t - \sum_{s=1}^{\infty} q_s \beta_{t-s} \right| \leq \frac{\epsilon}{4(N+1)} + N \cdot \frac{\epsilon}{4(N+1)} + 2 \cdot \frac{\epsilon}{4} = \epsilon.$$

Therefore $\beta_t = \sum_{s=1}^{\infty} q_s \beta_{t-s}$ for $t \in \{\ldots, -3, -2, -1, 0, 1, 2, 3, \ldots\}$. Since $\beta_s \leq \beta_0$ for all s, we can argue that

$$\beta_0 = \sum_{s=1}^{\infty} q_s \beta_{-s} \leq \beta_0 \cdot \sum_{s=1}^{\infty} q_s = \beta_0$$

$$\sum_{s=1}^{\infty} q_s \beta_{-s} = \sum_{s=1}^{\infty} q_s \beta_0$$

$$\sum_{\{s|q_s>0\}} q_s (\beta_0 - \beta_{-s}) = 0.$$

The last expression is a sum of nonnegative terms, and therefore $\beta_{-s} = \beta_0$ for all s such that $q_s > 0$. At this point, we have a partial success. If we let $Z' = \{s | \beta_s = \beta_0\}$, we now know that $-s \in Z'$ whenever $q_s > 0$. Our goal is to show that Z' actually contain all the integers. If $a \in Z'$,

$$\beta_0 = \beta_a = \sum_{s=1}^{\infty} q_s \beta_{a-s} \leq \beta_0 \cdot \sum_{s=1}^{\infty} q_s = \beta_0$$

$$\sum_{s=1}^{\infty} q_s \beta_{a-s} = \sum_{s=1}^{\infty} q_s \beta_0$$

$$\sum_{\{s|q_s>0\}} q_s (\beta_0 - \beta_{a-s}) = 0.$$

So $\beta_{a-s} = \beta_0$ for all s such that $q_s > 0$. This means that Z' must include $-2s = -s - s, -3s = -2s - s$, and so forth for all s such that $q_s > 0$. In general, if $c_1, c_2, \ldots, c_m$ are positive integers and $s_1, s_2, \ldots, s_m$ are such that $q_{s_j} > 0$, we can argue that $-(c_1 s_1 + c_2 s_2 + \ldots + c_m s_m) \in Z'$. We just build up the expression term by term. Letting $a = -s_1$ in the equations above, we obtain $-s_1 - s_1 = -2s_1 \in Z'$. Then letting $a = -2s_1$, we obtain $-2s_1 - s_1 = -3s_1 \in Z'$. Continuing in this manner, we obtain $-c_1 s_1 \in Z'$. Then we let $a = -c_1 s_1$ to derive $-c_1 s_1 - s_2 \in Z'$. Another iteration yields $-c_1 s_1 - 2s_2 \in Z'$, and several more passes give $-c_1 s_1 - c_2 s_2 \in Z'$. In this fashion we eventually discover $-c_1 s_1 - c_2 s_2 - \ldots - c_m s_m \in Z'$. So we have $\beta_{-t} = \beta_0$ whenever t is a linear combination, via position integers, of the set $\{s|q_s > 0\}$. We now show that the aperiodicity of the Markov chain (remember the Markov chain?) forces these linear combinations to encompass a great many integers.

The aperiodicity means that the largest integer that evenly divides the positions k of all the nonzero p_k is 1. This property carries over to the q_k. Suppose, to the contrary, that $m > 1$ evenly divides the positions k of all the nonzero q_k. Then $q_1 = q_2 = \ldots = q_{m-1} = 0$, which forces

$$p_1 = q_1 = 0$$
$$p_2 = q_2 + q_1 p_1 = 0$$
$$\vdots \quad \vdots$$
$$p_{m-1} = q_{m-1} + q_{m-2} p_1 + q_{m-3} p_2 + \ldots + q_1 p_{m-2} = 0.$$

Therefore, p_m is the first of the p_k sequence that can be nonzero. Continuing by induction, suppose that $p_j = 0$ except when $m|j$ for $j = 1, 2, \ldots, (k-1)m$. For j in the range $(k-1)m + 1$ through $km - 1$, we have $q_j = 0$ since $m \nmid j$. Therefore,

$$p_j = q_j + \sum_{v=1}^{j-1} q_{j-v} p_v = \sum_{v=1}^{(k-1)m} q_{j-v} p_v + \sum_{v=(k-1)m+1}^{j-1} q_{j-v} p_v$$

$$= \sum_{v=1}^{k-1} q_{j-vm} p_{vm} + \sum_{v=0}^{j-(k-1)m-2} p_{v+(k-1)m+1} q_{j-v-(k-1)m-1} = 0.$$

The first sum above is zero because $m \nmid j$ implies that $m \nmid (j - vm)$. The second sum is vacuous if $j = (k - 1)m + 1$, and if $(k - 1)m + 2 \leq j \leq km - 1$, the q-term subscripts are in the range 1 to $m - 2$ and are consequently zero.

We have shown that if $m > 1$ evenly divides the positions k of all the nonzero q_k, then m also evenly divides the positions of all the nonzero p_k. The chain's aperiodicity means that no such $m > 1$ can exist. Therefore, the largest integer that evenly divides the positions k of all the nonzero q_k is 1. In other words, there are integers $s_1, s_2, \ldots, s_r$ such that $q_{s_j} > 0$ for $j = 1, 2, \ldots, r$ and the greatest common divisor of $\{s_1, s_2, \ldots, s_r\}$ is 1. Euclid's algorithm then asserts the existence of integers $d_1, d_2, \ldots, d_r$, not necessarily positive, with $\sum_{j=1}^{r} d_j s_j = 1$. Let $D = \max\{|d_j|\} + 1$, and let $M = D(s_1 + s_2 + \ldots + s_r)^2$. For $m \geq M$, we divide m by $(s_1 + s_2 + \ldots + s_r)$ and distinguish the quotient $Q \geq D(s_1 + s_2 + \ldots + s_r)$ and remainder $0 \leq R < (s_1 + s_2 + \ldots + s_r)$:

$$m = Q(s_1 + s_2 + \ldots + s_r) + R = Q(s_1 + s_2 + \ldots + s_r) + R(1)$$

$$= Q(s_1 + s_2 + \ldots + s_r) + R(d_1 s_1 + d_2 s_2 + \ldots + d_r s_r) = \sum_{j=1}^{r}(Q + Rd_j)s_j.$$

We now have m as a linear combination of $s_1, s_2, \ldots, s_r$. Moreover, each coefficient is a positive integer:

$$Q + Rd_j \geq D(s_1 + s_2 + \ldots + s_r) - R|d_j|$$

$$\geq D(s_1 + s_2 + \ldots + s_r) - |d_j|(s_1 + s_2 + \ldots + s_r)$$

$$= (s_1 + s_2 + \ldots + s_r)(D - |d_j|) > 0.$$

Since all linear combinations t, via positive integers, of locations $s_1, s_2, \ldots, s_r$ where $q_{s_j} > 0$ must have $\beta_{-t} = \beta_0$, we conclude that $\beta_t = \beta_0$ for all $t \leq -M$. This provides the opening wedge of an induction argument that forces all $\beta_t = \beta_0$. We know that each β_t is related to those that precede it. The demonstration proceeds as follows.

$$\beta_t = \sum_{s=1}^{\infty} q_s \beta_{t-s}$$

$$\beta_{-M+1} = \sum_{s=1}^{\infty} q_s \beta_{-M+1-s} = \sum_{s=1}^{\infty} q_s \beta_0 = \beta_0$$

$$\beta_{-M+2} = \sum_{s=1}^{\infty} q_s \beta_{-M+2-s} = \sum_{s=1}^{\infty} q_s \beta_0 = \beta_0$$

$$\vdots \qquad \vdots$$

We conclude that $\beta_t = \beta_0 = \limsup p_k$, for all $t \in \{\ldots, -2, -1, 0, 1, 2, \ldots\}$. Referring to Figure 3.28, we see that infinitely many subsequences of the p_k converge to β_0.

The columns of our convergent display, as exemplified by Figure 3.28, are subsequences of the p_k. If we let t assume the ascending subscripts of the p_k that appear in the center column, Equation 3.7 asserts a relationship between the center term and the t terms to its left. Since the terms yet farther to the left are zero, we can extend the sum to infinity. Specifically,

$$1 = w_0 r_{j,0} + w_1 r_{j,-1} + w_2 r_{j,-2} + \ldots = \sum_{k=0}^{\infty} w_k r_{j,-k},$$

for each line j of the convergent display. Recall that we have shown if p_k converges to zero, then $\sum_{k=0}^{\infty} w_k = \sum_{k=1}^{\infty} k q_k$ diverges. This long digression into convergent columns represents our attempt to prove the converse. To this end, suppose that $\sum_{k=0}^{\infty} w_k$ diverges and $\beta_0 > 0$. Under these conditions, there exists an M such that $\sum_{k=0}^{M} w_k > 2/\beta_0$. Consider the $M+1$ columns $0, -1, -2, \ldots, -M$ of the convergent display. Since these columns converge to β_0, there exists N such that $j \geq N$ implies that $|r_{j,-s} - \beta_0| < \beta_0/2$, which in turn implies that $r_{j,-s} > \beta_0 - \beta_0/2 = \beta_0/2$, simultaneously for all $0 \leq s \leq M$. Then, for any line $j \geq N$ of the convergent display,

$$1 = \sum_{k=0}^{\infty} w_k r_{j,-k} \geq \sum_{k=0}^{M} w_k r_{j,-k} > \sum_{k=0}^{M} w_k \cdot \frac{\beta_0}{2} > \left(\frac{\beta_0}{2}\right) \cdot \left(\frac{2}{\beta_0}\right) = 1.$$

This contradiction means that if $\sum_{k=0}^{\infty} w_k$ diverges, then $\beta_0 = \limsup p_k = 0$. Because all p_k are in the range $[0,1]$, we have $0 \leq \liminf p_k \leq \limsup p_k = 0$. Hence $\lim_{k \to \infty} p_k = 0$. We record our results to this point in the following theorem and then use it to show, as promised, that an irreducible Markov chain must be ergodic.

3.37 Theorem: Let $P = [p_{ij}]$ be the transition matrix of an irreducible aperiodic Markov chain. For any state i, $\sum_{k=1}^{\infty} k q_{ii}^{(k)}$ diverges if and only if $\lim_{k \to \infty} p_{ii}^{(k)} = 0$.

PROOF: See discussion above. ∎

3.38 Theorem: All states of an irreducible aperiodic finite Markov chain are ergodic.

PROOF: Let the chain have n states. From Theorem 3.35, we know that all states are recurrent. Therefore, $\sum_{k=1}^{\infty} q_{ii}^{(k)} = 1$ for each state i. Moreover, $\sum_{k=1}^{\infty} q_{ji}^{(k)} \leq 1$ for any state i, j because it represents that probability that state i will follow state j at some point in the future.

Suppose that all states are null-recurrent. That is, $\sum_{k=1}^{\infty} k q_{ii}^{(k)}$ diverges for all states i. Theorem 3.37 then asserts that $\lim_{k \to \infty} p_{ii}^{(k)} = 0$ for each state i. Fix states i and j, and let $\epsilon > 0$ be arbitrary. Choose N such that $\sum_{k=N+1}^{\infty} q_{ji}^{(k)} < \epsilon/2$. Choose M such that $k \geq M$ implies that $0 \leq p_{ii}^{(k)} < \epsilon/(2N)$. From Equation 3.5 we have, for $t > M + N$,

$$p_{ji}^{(t)} = q_{ji}^{(t)} p_{ii}^{(0)} + \ldots + q_{ji}^{(1)} p_{ii}^{(t-1)} = \sum_{k=1}^{N} q_{ji}^{(k)} p_{ii}^{(t-k)} + \sum_{k=N+1}^{t} q_{ji}^{(k)} p_{ii}^{(t-k)}$$

$$p_{ji}^{(t)} \le \sum_{k=1}^{N} p_{ii}^{(t-k)} + \sum_{k=N+1}^{t} q_{ji}^{(k)} < N \cdot \frac{\epsilon}{2N} + \frac{\epsilon}{2} = \epsilon.$$

Hence $\lim_{k \to \infty} p_{ji}^{(k)} = 0$. Since there are only n^2 entries, we can choose N' such that $k \ge N'$ implies that $0 \le p_{ij}^{(k)} < 1/(2n)$ for all $1 \le i, j \le n$. For $k \ge N'$, the sum of row i in P^k is then $\sum_{j=1}^{n} p_{ij}^{(k)} \le n(1/(2n)) = 1/2$. But this is not possible; each row in P^k must sum to 1. There must exist, therefore, at least one positive-recurrent state.

Let i be the positive-recurrent state, and let j be another state. Suppose that j is null-recurrent. Then $p_{jj}^{(k)} \to 0$. From Theorem 3.25, there exists $A > 0$ and M such that $p_{jj}^{(m)} \ge A p_{ii}^{(m-M)}$ for $m > M$. That is, $p_{ii}^{(m-M)} \le p_{jj}^{(m)}/A$ for $m > M$. Hence, $p_{ii}^{(k)} \to 0$, and state i is null-recurrent—a contradiction. We conclude that all states are positive recurrent. Given that the chain is aperiodic, this means all states are ergodic. ∎

At this point, we have $p_{ii}^{(k)}$ convergent (to zero) in the null-recurrent case. However, we also now know that irreducible aperiodic Markov chains have only positive recurrent states, and it is these types of chains that appear in client-server simulations. So the convergence question is still open. To settle the issue, we return to our simplified notation: $p_k = p_{ii}^{(k)}, q_k = q_{ii}^{(k)}$, for a fixed state i. We continue the argument that we interrupted for the last two theorems. In particular, we use the convergent display, as exemplified by Figure 3.28, and the relationship between the w_k and the p_k as given by Equation 3.7. In the positive-recurrent case, we have $\sum_{k=0}^{\infty} w_k = \mu_T < \infty$. Moreover, since p_k cannot converge to zero, we have $\beta_0 = \limsup p_k > 0$. We know that

$$1 = w_0 p_t + w_1 p_{t-1} + w_2 p_{t-2} + \ldots + w_t p_0,$$

for any $t \ge 0$. In terms of the row j $(\ldots, r_{j,-2}, r_{j,-1}, r_{j,0}, r_{j,1}, r_{j,2}, \ldots)$ of the convergent display, this translates to

$$1 = w_0 r_{j,0} + w_1 r_{j,-1} + w_2 r_{j,-2} + \ldots = \sum_{k=0}^{\infty} w_k r_{j,-k}.$$

Given $\epsilon > 0$, there exists M such that $\sum_{k=M+1}^{\infty} w_k < \epsilon/2$, which means, of course, that $\sum_{k=0}^{M} w_k > \mu_T - \epsilon/2$. Since the $r_{j,k}$ are all in the range $[0, 1]$, we have $\sum_{k=M+1}^{\infty} w_k r_{j,-k} < \epsilon/2$. Arguing as before, we can determine N such that the $M + 1$ columns $m \in \{0, -1, -2, -3, \ldots, -M\}$ all exhibit $|r_{j,m} - \beta_0| < \epsilon/(2\mu_T)$ whenever $j > N$. Equivalently, $j > N$ implies that $\beta_0 - \epsilon/(2\mu_T) < r_{j,m} < \beta_0 + \epsilon/(2\mu_T)$ simultaneously for all $m \in \{0, -1, -2, \ldots, -M\}$. Consequently, for $j > N$,

$$1 = \sum_{k=0}^{\infty} w_k r_{j,-k} = \sum_{k=0}^{M} w_k r_{j,-k} + \sum_{k=M+1}^{\infty} w_k r_{j,-k} \le \sum_{k=0}^{M} w_k r_{j,-k} + \frac{\epsilon}{2}.$$

Continuing, we have

$$1 < \sum_{k=0}^{M} w_k \left(\beta_0 + \frac{\epsilon}{2\mu_T} \right) + \frac{\epsilon}{2}$$

$$\leq \left(\beta_0 + \frac{\epsilon}{2\mu_T} \right) \sum_{k=0}^{\infty} w_k + \frac{\epsilon}{2} = \beta_0 \mu_T + \frac{\epsilon}{2\mu_T} \cdot \mu_T + \frac{\epsilon}{2} = \beta_0 \mu_T + \epsilon.$$

Also,

$$1 = \sum_{k=0}^{\infty} w_k r_{j,-k} \geq \sum_{k=0}^{M} w_k r_{j,-k} > \sum_{k=0}^{M} w_k \left(\beta_0 - \frac{\epsilon}{2\mu_T} \right)$$

$$> \left(\beta_0 - \frac{\epsilon}{2\mu_T} \right) \left(\mu_T - \frac{\epsilon}{2} \right) = \beta_0 \mu_T - \frac{\beta_0 \epsilon}{2} - \frac{\epsilon}{2} + \frac{\epsilon^2}{4\mu_T} > \beta_0 \mu_T - \epsilon.$$

Thus, for arbitrary $\epsilon > 0$, we have $1 - \epsilon < \beta_0 \mu_T < 1 + \epsilon$, which implies that $\beta_0 = 1/\mu_T$. Thus all the columns of the convergent display tend to $1/\mu_T$. That is, infinitely many subsequences of $\{p_k\}$ converge to $1/\mu_T$, and we are now in a position to show that p_k itself converges to $1/\mu_T$.

Let $\gamma = \liminf p_k$. Then $\gamma \leq \beta_0$, and there exists a subsequence p_{i_k} converging to γ. Let $\epsilon > 0$ be arbitrary. Choose N such that $\sum_{k=N+1}^{\infty} w_k < \epsilon \beta_0 / 2$. Because $\beta_0 = \limsup p_k$, there can be only finitely many $p_k \geq \beta_0 (1 + \epsilon/2)$. Accordingly, choose M such that $p_k < \beta_0 (1 + \epsilon/2)$ for $k > M$. For $t > M + N$ we then have

$$1 = w_0 p_t + w_1 p_{t-1} + \ldots + w_t p_0 = w_0 p_t + \sum_{k=1}^{N} w_k p_{t-k} + \sum_{k=N+1}^{t} w_k p_{t-k}$$

$$< p_t + \beta_0 \left(1 + \frac{\epsilon}{2} \right) \sum_{k=1}^{N} w_k + \sum_{k=N+1}^{t} w_k$$

$$= p_t + \beta_0 \left(1 + \frac{\epsilon}{2} \right) \left(-w_0 + \sum_{k=0}^{N} w_k \right) + \sum_{k=N+1}^{\infty} w_k$$

$$< p_t + \beta_0 \left(1 + \frac{\epsilon}{2} \right) \left(-1 + \sum_{k=0}^{\infty} w_k \right) + \frac{\beta_0 \epsilon}{2}$$

$$= p_t + \beta_0 \left(1 + \frac{\epsilon}{2} \right) \left(-1 + \frac{1}{\beta_0} \right) + \frac{\beta_0 \epsilon}{2}$$

$$= p_t - \beta_0 \left(1 + \frac{\epsilon}{2} \right) + \left(1 + \frac{\epsilon}{2} \right) + \frac{\epsilon \beta_0}{2} = p_t - \beta_0 + 1 + \frac{\epsilon}{2},$$

from which we conclude that $\beta_0 - p_t < \epsilon/2$. Now choose $i_k > M + N$ such that $|p_{i_k} - \gamma| < \epsilon/2$. For $t = i_k$, we continue the argument above.

$$|\beta_0 - \gamma| - |\gamma - p_{i_k}| \leq |\beta_0 - \gamma + \gamma - p_{i_k}| = \beta_0 - \gamma + \gamma - p_{i_k} < \frac{\epsilon}{2}$$

$$|\beta_0 - \gamma| < \frac{\epsilon}{2} + |p_{i_k} - \gamma| < \frac{\epsilon}{2} + \frac{\epsilon}{2} = \epsilon$$

Hence $\beta_0 = \gamma$. Equivalently, $\limsup p_k = \liminf p_k = \lim p_k = \beta_0$. Finally, we have the main sequence convergent, and we can now prove that the powers of the transition matrix stabilize to a well-defined limit.

3.39 Theorem: Let P be the transition matrix of an irreducible, aperiodic, n-state Markov chain. Then the rows of P^k all converge to $(1/\mu_{T_1}, \ldots, 1/\mu_{T_n})$.

PROOF: From Theorem 3.38 we know that all states are positive recurrent. Hence $\sum_{t=1}^{\infty} q_{ii}^{(t)} = 1$ and $\sum_{t=1}^{\infty} t q_{ii}^{(t)} = \mu_{T_i} < \infty$ for all states i. The lengthy argument above shows that $\lim_{t\to\infty} p_{ii}^{(t)} = 1/\mu_{T_i}$. We will also need the fact that $\sum_{k=1}^{\infty} q_{ji}^{(k)} = 1$. That is, we need that the eventual passage from state j to state i is certain. Suppose that it is not. That is, suppose that $q = \sum_{k=1}^{\infty} q_{ji}^{(k)} < 1$. Because the chain is irreducible, there exists K_1 such that $p_{ij}^{(K_1)} > 0$. Consequently, for an arbitrary position t,

$$\Pr((X_{t+K_1} = j) \cap [\cap_{k>t+K_1}(X_k \neq i)]|X_t = i) = p_{ij}^{(K_1)}(1-q) > 0.$$

That is, the probability of at most $K_1 - 1$ returns to state i is positive, which implies that the probability of $K_1, K_1 + 1, \ldots$ returns to state i is less than 1. This contradicts Theorem 3.36, which asserts that the probability of s returns is 1, for any $s \geq 1$. Therefore, $\sum_{k=1}^{\infty} q_{ji}^{(k)} = 1$. From Equation 3.5, we have

$$p_{ji}^{(t)} = q_{ji}^{(t)} p_{ii}^{(0)} + q_{ji}^{(t-1)} p_{ii}^{(1)} + \ldots + q_{ji}^{(1)} p_{ii}^{(t-1)} = \sum_{k=1}^{t} q_{ji}^{(k)} p_{ii}^{(t-k)}.$$

Let $\epsilon > 0$ be arbitrary. Choose N such that $\sum_{k=N+1}^{\infty} q_{ji}^{(k)} < \epsilon/[3(1 + 1/\mu_{T_i})]$ and also $\sum_{k=N+1}^{\infty} q_{ji}^{(k)} < \epsilon \mu_{T_i}/3$. Then choose M such that $t > M$ implies that $|p_{ii}^{(t)} - 1/\mu_{T_i}| < \epsilon/3$. For $t > M + N$, we argue that

$$\left| p_{ji}^{(t)} - \frac{1}{\mu_{T_i}} \right| = \left| \sum_{k=1}^{t} q_{ji}^{(k)} p_{ii}^{(t-k)} - \left(\frac{1}{\mu_{T_i}} \right) \left(\sum_{k=1}^{\infty} q_{ji}^{(k)} \right) \right|$$

$$= \left| \sum_{k=1}^{t} q_{ji}^{(k)} \left(p_{ii}^{(t-k)} - \frac{1}{\mu_{T_i}} \right) - \frac{1}{\mu_{T_i}} \sum_{k=t+1}^{\infty} q_{ji}^{(k)} \right|$$

$$\leq \left| \sum_{k=1}^{t} q_{ji}^{(k)} \left(p_{ii}^{(t-k)} - \frac{1}{\mu_{T_i}} \right) \right| + \frac{1}{\mu_{T_i}} \sum_{k=N+1}^{\infty} q_{ji}^{(k)}$$

$$\left| p_{ji}^{(t)} - \frac{1}{\mu_{T_i}} \right| < \left| \sum_{k=1}^{N} q_{ji}^{(k)} \left(p_{ii}^{(t-k)} - \frac{1}{\mu_{T_i}} \right) \right| + \left| \sum_{k=N+1}^{t} q_{ji}^{(k)} \left(p_{ii}^{(t-k)} - \frac{1}{\mu_{T_i}} \right) \right|$$

$$+ \frac{1}{\mu_{T_i}} \cdot \frac{\epsilon \mu_{T_i}}{3}$$

$$< \frac{\epsilon}{3} \sum_{k=1}^{N} q_{ji}^{(k)} + \left(1 + \frac{1}{\mu_{T_i}} \right) \sum_{k=N+1}^{t} q_{ji}^{(k)} + \frac{\epsilon}{3}$$

$$< \frac{\epsilon}{3} + \left(1 + \frac{1}{\mu_{T_i}} \right) \cdot \frac{\epsilon}{3(1 + 1/\mu_{T_i})} + \frac{\epsilon}{3} = \epsilon.$$

Hence $p_{ji}^{(k)} \to 1/\mu_{T_i}$, for $j = 1, 2, \ldots, n$. That is, the ith column of P^k converges to $1/\mu_{T_i}$, repeated n times. Equivalently, the rows of P^k converge to $(1/\mu_{T_1}, \ldots, 1/\mu_{T_n})$. ∎

For a final theorem, we show how to obtain the limiting value of P^k without computing large powers of the transition matrix.

3.40 Theorem: Let P be the transition matrix of an irreducible, aperiodic, n-state Markov chain. Then the rows of P^k all converge to the row vector u, which is the unique nonzero solution of the matrix equation $uP = u$ satisfying the constraints $u_i \geq 0$ and $\sum_{i=1}^{n} u_i = 1$.

PROOF: Suppose that $u = (u_1, u_2, \ldots, u_n)$ satisfies $uP = u$, all $u_i \geq 0$, and $\sum_{i=1}^{n} u_i = 1$. Postmultiplying by P, we have $uP^2 = uP = u, uP^3 = u$, and so forth. That is, $uP^k = u$ for all $k \geq 0$. From the preceding theorem, we know that all rows of P^k converge to a common value, which we will call $\beta = (\beta_1, \beta_2, \ldots, \beta_n)$. Given $\epsilon > 0$, choose N such that $k \geq N$ implies that $|p_{ij}^{(k)} - \beta_j| < \epsilon$ for all $1 \leq i, j \leq n$. For $k > N$ we then have

$$|u_j - \beta_j| = \left| \sum_{i=1}^{n} u_i p_{ij}^{(k)} - \beta_j \right| = \left| \sum_{i=1}^{n} u_i p_{ij}^{(k)} - \beta_j \sum_{i=1}^{n} u_i \right| = \left| \sum_{i=1}^{n} u_i (p_{ij}^{(k)} - \beta_j) \right|$$

$$< \epsilon \sum_{i=1}^{n} |u_i| = \epsilon \sum_{i=1}^{n} u_i = \epsilon.$$

So $u_j = \beta_j$ for $1 \leq j \leq n$. We have shown that any solution satisfying the restrictions $u_i \geq 0$ and $\sum_{i=1}^{n} u_i = 1$ must be β. Of course, β satisfies the constraints, and we now show that it is a solution. Let $(\beta P)_j$ denote the jth component of βP. Then, for any k,

$$|(\beta P)_j - \beta_j| \leq |(\beta P)_j - p_{1j}^{(k+1)}| + |p_{1j}^{(k+1)} - \beta_j|$$

$$= \left| \sum_{i=1}^{n} \beta_i p_{ij} - \sum_{i=1}^{n} p_{1i}^{(k)} p_{ij} \right| + \left| p_{1j}^{(k+1)} - \beta_j \right|$$

$$\leq \sum_{i=1}^{n} p_{ij} |\beta_i - p_{1i}^{(k)}| + |p_{1j}^{(k+1)} - \beta_j|$$

$$|(\beta P)_j - \beta_j| < \sum_{i=1}^{n} |\beta_i - p_{1i}^{(k)}| + |p_{1j}^{(k+1)} - \beta_j|.$$

This terminal expression approaches zero as $k \to \infty$. Therefore, $(\beta P)_j = \beta_j$ for each j, or equivalently, $\beta P = \beta$. We conclude that β is the unique solution to $uP = u$ satisfying the constraints $\sum_{i=1}^{n} u_i = 1$ and $0 \le u_i \le 1$ for each i. ∎

We close this lengthy interlude with a final example that utilizes the last theorem to determine the long-term occupancy distribution of the Markov states.

3.41 Example: A controller handles read-write requests from three processes directed toward a common memory. Traffic conditions are such that an unlimited number of requests are always pending for each process. Upon completing a request from process i, the controller takes the next request from process j with probability p_{ij}, where $P = [p_{ij}]$ is the following matrix.

$$P = \begin{bmatrix} 0.100 & 0.400 & 0.500 \\ 0.300 & 0.200 & 0.500 \\ 0.700 & 0.200 & 0.100 \end{bmatrix}$$

Assuming that read-write requests require a constant fixed time to service, what fraction of time does the controller spend with each process?

We can model this situation as a 3-state Markov chain. The current state, at any given time, is the number of the process whose request is with the controller. The transition matrix is P. Because there is nonzero probability of a transition between any two states, the chain is irreducible. It is also aperiodic, since it has at least one state, actually all three, with $p_{ii}^{(1)} > 0$. The powers of P will therefore converge to identical rows, each representing the steady-state occupancy probabilities for the three states. Let us call the common row $\beta = (\beta_1, \beta_2, \beta_3)$, where $\beta_1 + \beta_2 + \beta_3 = 1$. From the preceding theorem, we know that β satisfies the matrix equation $\beta = \beta P$. Hence

$$\beta_1 = 0.1\beta_1 + 0.3\beta_2 + 0.7\beta_3 = 0.1\beta_1 + 0.3\beta_2 + 0.7(1 - \beta_1 - \beta_2)$$
$$\beta_1 = -0.25\beta_2 + 0.4375$$
$$\beta_2 = 0.4\beta_1 + 0.2\beta_2 + 0.2\beta_3 = 0.4\beta_1 + 0.2\beta_2 + 0.2(1 - \beta_1 - \beta_2)$$
$$\beta_2 = 0.2\beta_1 + 0.2 = 0.2(-0.25\beta_2 + 0.4375)$$
$$\beta_2 = 0.2738$$
$$\beta_1 = -0.25(0.2738) + 0.4375 = 0.3691$$
$$\beta_3 = 1 - 0.2738 - 0.3691 = 0.3571.$$

With the matrices P^k converging as follows,

$$P^k \to \begin{bmatrix} \beta_1 & \beta_2 & \beta_3 \\ \beta_1 & \beta_2 & \beta_3 \\ \beta_1 & \beta_2 & \beta_3 \end{bmatrix} = \begin{bmatrix} 0.3691 & 0.2738 & 0.3571 \\ 0.3691 & 0.2738 & 0.3571 \\ 0.3691 & 0.2738 & 0.3571 \end{bmatrix},$$

we have, for any initial state α, that $\alpha P^k \approx \beta$ when k is large. That is, $\Pr(X_k = i) = \beta_i$, independent of the initial conditions. We can, therefore, interpret β_i as the fraction of time the controller devotes to process i. □

Exercises

---

*3.47 Find two different 3×3 transition matrices whose powers converge to the same values.

3.48 Exhibit the transition matrix of a finite Markov chain that is not irreducible.

3.49 Exhibit the transition matrix of an irreducible finite Markov chain with period 3.

3.50 Let $P = [p_{ij}]$ be the transition matrix of an n-state irreducible Markov chain. Suppose that state i is transient. Show that $\lim_{k \to 0} p_{ii}^{(k)} = 0$.

3.51 The transition matrix $P = [p_{ij}]$ of a 2-state Markov chain appears to the right, where $0 < p < 1$. Show by direct calculation that $\sum_{k=1}^{\infty} p_{11}^{(k)}$ diverges and consequently state 1 is recurrent.
$$\begin{bmatrix} p & (1-p) \\ p & (1-p) \end{bmatrix}$$

3.52 Let $P = [p_{ij}]$ be the transition matrix of an n-state Markov chain, in which state i is transient. As usual, the $q_{ij}^{(k)}$ are the first-passage probabilities. Prove that $\lim_{k \to \infty} p_{ii}^{(k)} = \left(\sum_{k=1}^{\infty} q_{ii}^{(k)} \right) / \left[1 - \sum_{k=1}^{\infty} q_{ii}^{(k)} \right]$.

3.53 Let $P = [p_{ij}]$ be the transition matrix of an n-state irreducible Markov chain. Let the random variable X_i denote the state at position i, and let $E_s(t, i)$ be the event $(X_k = i$ for s positions $k > t)$. Show that $\Pr(E_s(t, i) | X_t = i) = 1$.

3.54 For an n-state irreducible Markov chain. Prove that $\sum_{i=1}^{n} 1/\mu_{T_i} = 1$. T_i is the first return time for state i.

3.55 The transition matrix of 4-state Markov chain appears to the right. If T_i is the first return time for state i, find μ_{T_i} for $1 \leq i \leq 4$.
$$\begin{bmatrix} 1/2 & 0 & 1/2 & 0 \\ 0 & 1/2 & 1/2 & 0 \\ 0 & 0 & 1/2 & 1/2 \\ 1/2 & 1/2 & 0 & 0 \end{bmatrix}$$

3.5   Summary

To simulate systems containing nondeterministic components, we must be able to sample the probability distributions of these components. That is, we must be able to cycle the underlying process that generates outcomes in accordance with a given distribution. Because many such processes are physical, such as customers arriving at a service queue, we resort to a mathematical computation to generate equivalent information. Instead of observing customers, we compute a sequence of successive arrival times that exhibits the same random pattern. The mathematical algorithms are known as random number generators. We assume that our computer operating system provides a utility function rand() that delivers random numbers in the range $[0, 1)$ and matches the cumulative distribution $\Pr(U \le u) = u$, for $0 \le u \le 1$. The linear congruential formula, $X_{n+1} = (a_1 X_n + a_2) \mod a_3$, provides the usual mechanism to bootstrap from one random number to the next. The quality of the approximation depends on the choice of the constants a_i.

The rand() utility provides samples from a uniform distribution on $[0, 1)$, which is easily extended to a uniform distribution on an arbitrary interval $[a, b)$. We simply generate a + (b - a) * rand(). We can also extend rand() to other discrete distribution through inverse transforms and rejection filters.

If $F_Y()$ is an invertible cumulative distribution function for the random variable Y, then the numbers generated by $F_Y^{-1}(\text{rand}())$ follow the distribution for Y. This is the inverse transform calculation. The actual algorithm is complicated by the fact that F_Y jumps in a staircase fashion, which necessitates an iterative approach in computing $F_Y^{-1}(\text{rand}())$. Nevertheless, the inverse transform provides a generic method for sampling any discrete distribution. We adapted it to the binomial, Poisson, and geometric distributions, and we showed how to improve the convergence of the iterative algorithm by the judicious choice of a starting point.

The rejection filter is based on the realization that we can thin returns from a related distribution to obtain a different, desired distribution. Specifically, if we have, or can easily realize, a return sequence that matches a random variable X, we can systematically discard some of the elements and produce a subsequence that matches the related random variable Y. We used this technique to filter a geometric generator so as to produce a Poisson generator.

Discrete client-server systems are those in which events occur at discrete time intervals. In the simplest case, the client request inputs appear as a Bernoulli process. Each time slot admits an input with some fixed probability. The servers handle the requests in time spans chosen from some discrete distribution. Intermediate queues buffer requests that arrive while the servers are busy. The model exhibits little conceptual difference when a Poisson process generates the inputs.

We first considered a single-queue, single-server system to illustrate the use of the random number generators. We expanded this example to a generic shell that can accommodate multiple input streams, queues, and servers. Simulators patterned after these examples can estimate performance parameters, such as the average time that a request waits in a queue, the servers' idle times, and the system throughput in requests serviced per unit time.

The last section introduced the Markov chain. Using this analytical tool, we can derive the expected values of performance parameters of client-server systems, although even small systems can produce serious computational difficulties. A Markov chain is a sequence of random variables, $X_0, X_1, X_2, \ldots$, each of which assumes values in a finite state space $S = \{1, 2, \ldots, n\}$. The probability, p_{ij}, that X_{t+1} assumes state j depends only on the state i assumed by X_t.

Two viewpoints are compatible with this probability rule. We can envision a single system that moves in time across successive states. At time 0, it is in state X_0. At time 1, it is in state X_1, and so forth. Each transition is nondeterministic, but depends only on the current state. From this viewpoint, typical questions ask about the probability of achieving state j in k transitions from state i.

The second viewpoint envisions an ensemble of trajectories as the outcome space. A particular trajectory $(i_0, i_1, i_2, \ldots)$ means that $X_0 = i_0, X_1 = i_1, X_2 = i_2, \ldots$. We assign probability to this outcome space such that $\Pr(X_{t+1} = j | X_t = i) = p_{ij}$. The interpretation now, however, is not that state i evolves to state j with probability p_{ij}, but rather that the collection of trajectories with i in position t and j in position $t+1$ has the fraction p_{ij} of the probability allocated to the collection with i in position t.

The interpretations are equivalent, and we exploit this fact in the applications. However, it is the second interpretation that we use to develop the properties of Markov chains.

We introduce the matrix sequences $p_{ij}^{(k)}$ and $q_{ij}^{(k)}$ for the k-step transition probabilities and first-passage probabilities, respectively. The first gives the probability that the system will be in state j after k transitions from state i. This is the dynamic interpretation. As necessary, we also interpret $p_{ij}^{(k)}$ as the fraction of the probability allocated to trajectories with i in position t that is carried by the subcollection with j in position $t+k$. The $q_{ij}^{(k)}$ gives the probability that the system enters state j for the first time k transitions after leaving state i. It is, of course, smaller than the corresponding $p_{ij}^{(k)}$ because the latter includes the possibility of multiple intermediate visits to state j. We derive relationships between these two matrix sequences and exploit them to discover useful properties of Markov chains.

The Markov chains that arise in the analysis of client-server systems are frequently irreducible and aperiodic. Irreducibility means that there is a nonzero probability of a multistep transition from any state to any other. Aperiodicity means that the greatest common divisor of the nonzero $p_{ij}^{(k)}$ locations is 1. In other words, there is no pattern that says states j follow state i only at multiples of 2 transitions or 3 transitions or any higher multiple. With the initial definition, aperiodicity applies to a particular state. That is, a state i is aperiodic if the greatest common divisor of $\{k | p_{ii}^{(k)} > 0\}$ is 1. We showed, however, that all states in an irreducible Markov chain have the same status and that the property extends to the $p_{ij}^{(k)}$ and also to the $q_{ij}^{(k)}$. Depending on the convergence properties of $p_{ii}^{(k)}$ and $\sum p_{ii}^{(k)}$, we further categorize a state i as transient, null-recurrent, or positive-recurrent. Table 3.11 organizes these categories and also presents the behavior of certain functions of the first-passage probabilities in the aperiodic case.

All states in an irreducible Markov chain are positive-recurrent. Using these properties, we showed, for an irreducible, aperiodic Markov chain, that

Category	$\sum_{k=1}^{\infty} p_{ii}^{(k)}$	$p_{ii}^{(k)}$	$\sum_{k=1}^{\infty} q_{ii}^{(k)}$	$\mu_{T_i} = \sum_{k=1}^{\infty} k q_{ii}^{(k)}$
Transient	Converges	Approaches 0	Less than 1	
Null-recurrent	Diverges	Approaches 0	Equal 1	Diverges
Positive-recurrent	Diverges	Approaches $1/\mu_{T_i}$	Equal 1	Converges

TABLE 3.11. Markov chain state categories

the powers of its transition matrix, P^k, converge to a matrix with repeated rows. The common row vector is the unique solution to the matrix equation $uP = u$ with $u_i \geq 0$ for all i and $\sum u_i = 1$. The values u_i correspond to the probability that $X_k = i$ for large k, regardless of the initial state. This insight allows us to calculate the expected value of a system parameter, queue length, for example, from the conditional expected values, given that requests enter when the system is in a specific state. We used this technique to verify the earlier simulation results.

Historical Notes

In 1948, D. H. Lehmer[50] proposed the linear congruential method for generating samples from a uniform distribution. Although this method remains the most widely used today, it is still a controversial item. Various researchers have found patterns, such as consecutive pairs or triples lying on a small number of hyperplanes, that belie the alleged random distribution. See Park and Miller[60] and Gentle[25] for further details.

The Monte Carlo integration method illustrated in Example 3.4 is an application of a more general technique that uses random numbers in deterministic computations. An early use of the method occurred in the late eighteenth century, when G. L. L. Buffon applied it in the calculation of π, although the random aspect entered through physical experiment: tossing a needle on a grid of parallel lines. Subsequent applications graduated to random number tables, which were normally generated from document lists. Although this certainly represented an improvement over physical machinery, the technique remained somewhat impractical until the advent of computers, which made their initial appearances in the mid-1940s. The name Monte Carlo comes from the code name of a secret mathematical project run by John von Neumann and Stanislaus Ulam in connection with the atomic bomb development in the 1940s. The term reflects a perceived similarity between the method and certain gambling techniques.

Simulation has a long history. In some sense, the herdsman who maintained an inventory of his sheep with a collection of pebbles was engaged in a simulation. In modern times, but before the widespread availability of suitable digital computers, simulations exploited similarities between the differential equations describing a system of interest (e.g., an aircraft in flight) and those of electrical circuits. Using analog simulators, an engineer could observe the evolution of a carefully constructed circuit and interpret the results in terms of parameters in the real system. The rise and fall of a particular voltage, for example, could be scaled to represent the altitude of an aircraft.

These simulators found extensive use in the aerospace industry well into the 1960s. Although ingenious plug-boards relieved some of the burden of special-purpose wiring, these devices were extremely cumbersome in comparison with digital simulations, where software control permits an almost instantaneous switch from one problem context to another.

The first simulation languages for digital computers started to appear in the 1950s. In his review of the history of such languages, Nance[55] presents a genealogical tree relating some 30 languages over a period of 30 years. He credits K. D. Tocher with the first simulation language, General Simulation Program (GSP), and notes that IBM's General Purpose Simulation System (GPSS) appeared shortly thereafter. These early products addressed the non-programmer's need to construct simulations by introducing features that automate the data flow logistics and thereby free the user to concentrate on the application at hand. This division of labor remains apparent in all simulations languages in use today.

Early simulation languages also foreshadowed the current emphasis on object-oriented software because early products used interacting entities and processes. SIMULA was, as its name suggests, a simulation language, although its general-purpose constructs allowed the widest possible interpretation of "simulation." SIMULA, which introduced abstract data types, class, inheritance, co-routines, and virtual parallelism, is regarded as the first object-oriented language.

Some other early languages include the General Activity Simulation Program (GASP), developed at U.S. Steel, the General Electric Manufacturing Simulator (GEMS), developed at General Electric, SIMSCRIPT, developed at the RAND Corporation, and the Control and Simulation Language (CSL), developed by Esso Petroleum. These origins betray the intense interest in simulation aroused by problems in large industrial plants. In the scientific community, FORTRAN was the most popular general-purpose computer language in this era. Therefore, these simulation languages all took the form of a convenience wrapper around a FORTRAN core. Many were simply collections of FORTRAN routines for handling the logistics of event lists, process cycles, random number generation, statistical reduction, and report writing. Some of these lines became extinct; others merged and evolved into products that are available today. The suggestions for further reading below provide some pointers to information about current simulation languages.

A. A. Markov, a student of Chebyshev, introduced the Markov chain, which initiated the theory of stochastic processes in the early twentieth century. In the 1920s, Norbert Wiener provided a rigorous extension to continuous processes, and Andrei Kolmogorov developed the foundations of a general theory in the 1930s.

Further Reading

The linear congruential method is one of many techniques proposed for random number generation. Some alternatives include normalized digit sequences from the expansion of π; the midsquare method, which computes the next random number by extracting the middle digits from the square of the previous one; and digitized noise from an electronic component. Except for the last, these methods are more appropriately termed pseudo-random number generators. The output, being the result of a deterministic calculation, is naturally deterministic. However, an examination of a large collection of numbers generated with these algorithms gives the appearance of a uniform probability distribution. Many simulation texts (e.g., Karian and Dudewicz[41]) include a discussion of random number generation, including tests that you can use to verify the effectiveness of your system's rand() routine. See Knuth[47] for the mathematical details of the linear congruential method.

Gentle[25] reviews the current state-of-the-art in computer random number generation. His treatment includes the concerns that arise in choosing constants for the linear congruential formula, the practice of randomly shuffling the output streams from several generators to improve performance, and programming subtleties needed to avoid unintentional overflows. He also describes generators that employ a feedback shift register, a method that generates the next random number in sequence by shifting the current number 1 bit to the left and replacing the rightmost bit with an exclusive-or between the removed bit and one or more of the retained bits.

Gentle also investigates algorithms for generating random samples from a specified distribution. His list includes the inverse transform and acceptance/rejection methods of this chapter, as well as some alternative methods. This text also discusses the quality of commercially available random number generation software.

Simulation programs, as presented in this chapter, address two aspects of the underlying problem. The first concerns the logistics of creating arrivals according to input probability distributions, tracking their progress through queues, and allocating service times at various stations, again according to specified probability distributions. The event list and the program structure that supports it deal with these logistics. The second aspect is more particular to given simulation. It involves setting parameters that properly describe the probability distributions and the flow of arrivals past the service stations. A variety of specialized programming languages are available to handle the logistical aspects. The user is then free to concentrate on the particulars of the specific application. The user manipulates the parameters of the probability distributions and alternative traffic flow patterns and thereby gains the insight necessary to implement a real-world system. Simulation languages typically provide block primitives, such as arrival creators, queues, and delays, that the user can arrange to reflect the flow patterns of the problem at hand.

The General Purpose Simulation System (GPSS), released by IBM in 1961, was the first widely used simulation language. Its descendants remain available today. In GPSS/H, for instance, the single-server situation of Example 3.12 appears approximately as follows.

```
Simulate
    generate 1     // create a candidate arrival every millisecond
    transfer 0.1, real, virtual     // immediately transfer 90% out of the system
real:
    queue waitQueue
    seize server
    depart waitQueue     // waitQueue provides information about time waiting for server
    advance 9, 2     // server holds for 9 ± 2 milliseconds
    release server
    terminate 1     // request leaves system decreasing simulation count by one
virtual:
terminate 1     // terminate virtual request, decreasing simulation count by one
    start 5000     // start above process with simulation count of 5000
```

Some simulation languages and their associated references are GPSS (Schriber[78], Karian and Dudewicz[41]), SIMAN (Pegden et al.[65]), and Arena (Kelton et al.[44]). SIMAN code is very similar to the GPSS sample illustrated above. Pegden develops general simulation techniques and describes the SIMAN language as a specific vehicle. Arena is an extension of SIMAN, in the sense that low-level SIMAN constructs are available, but it allows you to build a simulation by interconnecting panels that represent higher-level system units. This visual approach replaces the linear code. From the visual specification, Arena compiles the linear SIMAN code, which you may view and edit if you wish. You specify parameters for each panel by editing icons, menus, and selection boxes in a manner that has become standard for most personal computer software. Arena also provides animation facilities, through which you can, for example, watch queue lengths grow and shrink as the simulation progresses. Kelton's text provides an extensive introduction to Arena. Banks and Carson[4] provides a more comprehensive view of discrete-event simulation that includes queuing system examples similar to those illustrated in this chapter.

The Markov chains considered here are finite, homogeneous, irreducible, aperiodic, and hence ergodic. A more general study takes up Markov chains that are not so restricted. Some references, in order of accessibility, are Ross[72], Cox and Miller[14], and Feller[19]. Ross assembles the needed convergence results without proof but illustrates the concepts with numerous examples. Cox and Miller prove most of the results derived here but defer to · Feller for the supporting theorems needed to show convergence of the rows of the transition matrix. Feller, as usual, exhaustively completes every aspect of Markov chains. Taylor and Karlin[86] is an accessible introduction to stochastic modeling, which treats Markov chains in some detail.

Chapter 4

Discrete Decision Theory

The preceding chapter illustrated the considerable power of the discrete probability model in connection with client-server simulations. This chapter continues with a second application area, decision theory. It is a common fact of life that we must frequently make decisions before we have complete information on all the contextual parameters. The choice of carrying an umbrella in uncertain weather conditions is a simple example. In many such cases, however, we can analyze the situation probabilistically and derive the *expected* consequences associated with our choices. Acting to maximize the expected profit, or to minimize the expected loss, is then appropriate if we have the opportunity to repeat the experience many times. The frequency-of-occurrence interpretation of probability then reassures us that our calculated expected profit will approximate our average real profit over the many repetitions. The following example should clarify the point.

Suppose that a computer needs to communicate with a remote machine across a network. A local cable connects the workstation with a gateway device, which in turn connects to a common carrier. The carrier uses some combination of land lines and satellite links to reach a destination gateway, which provides the final leg to the remote machine. At any given time, various malfunctions along the network result in an overall service characterization of good, fair, or poor. Network services, such as parity checks, redundant packets, and retransmission requests, are available to ensure message delivery despite component failures and traffic conditions. These advanced services, however, are expensive. For a low price, the network offers a "best effort" delivery, which is satisfactory when network conditions are good. Also available are two higher-priced options, which provide satisfactory performance under fair and poor network conditions. We waste resources if we choose a high-priced delivery option when a lower-priced one can handle the task. We also suffer some loss if we opt for the low-priced service and find that the message does not arrive at the destination. In short, we face a decision problem.

This chapter investigates the decision problem under increasingly complex circumstances. In the general case, a problem requires a decision, but the optimal decision depends on an unknown state of nature. In the example here, the state of nature is the network condition: fair, good, or poor. Our initial approach makes no attempt to ascertain the true state of nature. It simply quantifies the problem and develops a rational, conservative decision procedure based on the expected losses associated with each choice. We then consider the case where the state of nature, although unknown in a particular case, nevertheless exhibits a known probability distribution. In the example, this corresponds to knowing that network conditions are good, fair, and poor with probabilities, say, $0.6, 0.3$, and 0.1. With this knowledge, we can refine our decision procedure to lower the expected loss, provided that nature obligingly presents its states in a random fashion with the specified probabilities. Finally, we develop an approach that samples the state of nature and integrates that new knowledge into the decision procedure. In the current example, a sample might be a probe message to determine current conditions. Unfortunately, the probe will not provide certain knowledge of the current network conditions. Rather, it will follow a probability distribution that varies with the actual state of the network.

This final stage, characterized by samples from an unknown distribution, leads to a definition of statistical inference and a general mechanism for formulating an optimal decision rule. In this context, this chapter's mathematical interlude investigates sufficient statistics, a concept that shows how to reduce the sample without discarding relevant information. The chapter closes with an application to hypothesis testing, a special case of a decision procedure where the available actions are acceptance or rejection of a conjecture about the unknown state of nature.

4.1   Decision methods without samples

The following example quantifies the situation discussed in the introduction, although it reduces the number of natural states to enable illustrative graphs. It also provides a conservative decision principle. Note that no observations are undertaken in an attempt to determine the true state of nature.

4.1 Example: Continuing the network example, suppose that we need to transmit a message between two computers. Faced with the prospect of paying higher rates for ensured delivery across an unreliable network, we discern four possible actions. For various network conditions, Table 4.1 elaborates the available actions and their associated costs. In most cases, a correct guess of the network conditions still incurs some cost. The cost incurred for an incorrect guess varies, depending on the cost of repairing the effect of a partially inaccurate delivery. The costs associated with message abandonment represent the resources expended in constructing a message that is not sent. In any case, the exact source of these data are not of interest here. We are concerned with a decision procedure in the face of these known costs.

The last column tabulates the maximum cost incurred for each action, with the maximum taken over the possible states of nature. A conservative approach assumes that the worst will happen and chooses the action that minimizes that maximum cost. The action so chosen is called the minimax action. Here, this

| Action | Network condition | | Maximum cost |
	θ_1: Good	θ_2: Poor	
a_1: Request deluxe service	9	0	9
a_2: Request intermediate service	6	3	6
a_3: Request economy service	1	4	4
a_4: Abandon message	6	6	6

TABLE 4.1. Loss versus network state for various actions (see Example 4.1)

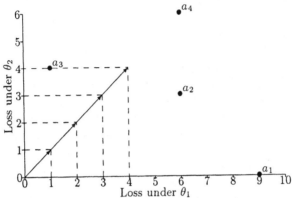

Figure 4.1. The minimax rule for a loss configuration of actions and states of nature (see Example 4.1)

approach chooses action a_3, which advocates economy service. The maximum cost for this action, 4, is the smallest of the maxima over the range of actions. Figure 4.1 locates each action in a two-dimensional space. The horizontal coordinate gives the loss when the state of nature is θ_1; the vertical coordinate is the loss when the state of nature is θ_2. An expanding wedge, originating at the origin, travels along the diagonal. The expansion stops when the horizontal or vertical edge first touches an action point. The point so encountered is the minimax action. If the wedge's vertical edge strikes the point, then the action's loss under θ_1 is larger than that under θ_2, but the maximum losses for other actions are larger still. Hence the point meets the minimax criterion. A similar remark applies if the wedge's horizontal edge first strikes an action point. If the wedge strikes two or more points simultaneously, there is a tie for the action that minimizes the maximum loss. Since the maximum loss is the same for all tied points, we can choose the minimax action arbitrarily from the tied points. □

In the general situation, a loss function, $L(a, \theta)$, describes the loss incurred by action a when the state of nature is θ. For a specific loss function, we define the no-data pure minimax decision rule as follows. This is a no-data decision rule because it involves no sample that might indicate the prevalent state of nature.

4.2 Definition: Suppose that one of the n possibilities $\theta_1, \theta_2, \ldots, \theta_n$ is the current state of nature. Suppose further that m actions are available: $a_1, a_2, \ldots, a_m$. Let $L(a_i, \theta_j)$ describe the loss associated with action a_i when θ_j prevails as the state of nature. In this context, the *no-data pure minimax* *decision rule* advocates action a_k where

$$\max_{1 \le i \le n} L(a_k, \theta_i) = \min_{1 \le j \le m} \left[\max_{1 \le i \le n} L(a_j, \theta_i) \right]. \ \blacksquare$$

A pure decision rule is one that determines a specific action from among the available choices. A mixed decision rule, by contrast, advocates a probabilistic mixture of several actions. That is, if the available actions are $a_1, \ldots, a_m$, then a mixed decision rule is a vector of probabilities, $p_1, \ldots, p_m$, that recommends action a_1 with probability p_1, action a_2 with probability p_2, and so forth. Of course, $\sum_{j=1}^m p_j = 1$. The user needs an independent random device, an m-sided coin, for example, with probability p_i for face i, to complete the decision process. At each decision point, he consults the random device and takes the action so determined. With this definition, we see that a pure decision rule is simply a mixed rule for which the probability vector contains a single 1. Consequently, the following discussion includes pure decision rules as a special case.

For the rest of the chapter, we use the terms *weighted sum, weighted average,* and *convex combination* to refer to expressions of the form $\sum_{i=1}^n p_i x_i$, where $\sum_{i=1}^n p_i = 1$. We use the term *probability vector* for the coefficients $(p_1, p_2, \ldots, p_n)$. Under this terminology, we understand that the coefficients are nonnegative and that they sum to 1. Therefore, we will frequently omit mention of this constraint.

Under a mixed decision rule, the actual loss is a random variable. Let $p = (p_1, p_2, \ldots, p_m)$ denote a mixed decision rule. For each state of nature θ_i, the expected loss is the weighted sum of the losses associated with the various actions:

$$E_{a;p}[L(\cdot, \theta_i)] = \sum_{j=1}^m p_j L(a_j, \theta_i).$$

The subscript on the expectation operator indicates that the probability distribution $p = (p_1, p_2, \ldots, p_m)$ lies over the available actions $a = (a_1, a_2, \ldots, a_m)$. The notation also uses a dot to replace the first argument in $L(a_j, \theta_i)$. This means that L is considered as a function of its first argument in the normal expansion of an expected value (see Definition 1.42). We now test a candidate probability vector in much the same manner as we tested a candidate action in arriving at the no-data pure minimax rule. Over the possible states of nature θ_i, candidate p exhibits varying expected losses $E_{a;p}[L(\cdot, \theta_i)]$. Continuing with the loss avoidance approach, we associate candidate p with the largest expected loss across this range of natural states. Then we choose among the candidates to select one with the smallest such association.

4.3 Definition: Suppose that one of the n possibilities $\theta_1, \theta_2, \ldots, \theta_n$ is the current state of nature. Suppose further that m actions are available: $a_1, a_2, \ldots, a_m$. Let $L(a_i, \theta_j)$ describe the loss associated with action a_i when

Cost as a function of action and prevailing network state				
		Network condition		Maximum
Strategy	Action	θ_1: Good	θ_2: Poor	cost
Pure	a_1: Deluxe service	9	0	9
	a_2: Intermediate service	6	3	6
	a_3: Economy service	1	4	4
	a_4: Abandon message	6	6	6
Mixed	(p_1, p_2, p_3, p_4)	$9p_1 + 6p_2$ $+1p_3 + 6p_4$	$0p_1 + 3p_2$ $+4p_3 + 6p_4$	— —

TABLE 4.2. Loss table when mixed strategies are available (see Example 4.4)

θ_j prevails. The *no-data mixed minimax decision rule* advocates the probability vector $p = (p_1, p_2, \ldots, p_m)$ where

$$\max_{1 \le i \le n} E_{a;p}[L(\cdot, \theta_i)] = \min_q \max_{1 \le i \le n} E_{a;q}[L(\cdot, \theta_i)].$$

The minimum is taken over all probability vectors $q = (q_1, q_2, \ldots, q_m)$. ∎

In the preceding example, there were only four pure decision rules, each advocating one of the actions a_1, a_2, a_3, a_4. There are, however, infinitely many mixed decision rules corresponding to all (p_1, p_2, p_3, p_4) probability vectors. Over the possible states of nature, it can happen that the maximum expected losses for various mixed decision rules admit a lower minimum than the maximum losses for the pure decision rules. We continue the example to see if a mixed minimax rule can outperform the pure minimax strategy.

4.4 Example: Example 4.1 derived the no-data pure minimax decision rule for choosing a message transmission option when faced with an unreliable network. Table 4.2 extends the earlier table of options to include mixed decision rules.

Under θ_i, the expected loss for a mixed strategy is just a weighted sum of the losses under θ_i for the pure strategies. Referring to Figure 4.1, which locates the pure strategies on a two-dimensional plot according to the losses under θ_1 (x-axis) and θ_2 (y-axis), we can interpret the expected loss for a mixed strategy as a weighted sum of the pure action points. That is,

$$\begin{pmatrix} E_{a;p}[L(\cdot, \theta_1)] \\ E_{a;p}[L(\cdot, \theta_2)] \end{pmatrix} = p_1 \begin{pmatrix} 9 \\ 0 \end{pmatrix} + p_2 \begin{pmatrix} 6 \\ 3 \end{pmatrix} + p_3 \begin{pmatrix} 1 \\ 4 \end{pmatrix} + p_4 \begin{pmatrix} 6 \\ 6 \end{pmatrix},$$

where the upper and lower elements are the respective x and y coordinates. Consequently, each mixed strategy corresponds to a point in the space enclosed by the outermost pure action points. In this case, that space is a triangle, as illustrated in Figure 4.2.

We can justify this conclusion as follows. Let P be the point computed in the equation above for the mixed strategy (p_1, p_2, p_3, p_4). The x-coordinate of P is then the weighted average of the x-coordinates of the four fixed points. Hence it must lie between the minimum and maximum x-coordinates of those points. Imagine two vertical lines, one starting near $x = -\infty$ and the other near $x = \infty$, that move inward toward each other. They stop when they encounter points in the set $\{a_1, a_2, a_3, a_4\}$. One will stop at the minimum x-coordinate; the other will stop at the maximum. So the x-coordinate of P must lie between these two lines. For

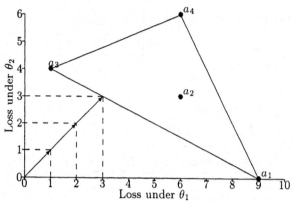

Figure 4.2. Applying the mixed minimax rule to a convex loss configuration (see Example 4.4)

a similar reason, two horizontal lines, converging from $-\infty$ and ∞, will stop at the minimum and maximum y-coordinates of the set $\{a_1, a_2, a_3, a_4\}$. This exercise confines P to a rectangle containing the four points. However, we can rotate the coordinate system and repeat the exercise to show that P is actually confined by two parallel lines of *any* orientation that converge toward the set $\{a_1, a_2, a_3, a_4\}$. By choosing lines parallel to the sides of the triangle of Figure 4.2, we confine P to that triangle. The set of all weighted sums of the fixed points is called the *convex hull* of those fixed points. You can envision the convex hull as the persistent space remaining after the fixed points are pinched from all directions by parallel lines. In a higher-dimensional situation, parallel hyperplanes pinch the set.

Given that any mixed strategy is a point in the triangle, the strategy's worst performance corresponds to its larger coordinate. If the x-coordinate is larger, it returns a larger expected loss under θ_1 than under θ_2. If the y-coordinate is larger, it fares worse under θ_2. We can therefore use an expanding wedge, as we did in searching for the pure minimax rule, to find the mixed minimax rule. The wedge stops when its horizontal or vertical edge encounters a mixed strategy point. Figure 4.2 illustrates the point. We see that the first point encountered lies on the line between a_1 and a_3. Hence $p_2 = p_4 = 0$ in the mixed minimax strategy. Moreover, the strategy point lies at the intersection of the line $x = y$ with the line between a_1 and a_3. The latter has equation $x + 2y = 9$, and the intersection point is $(3, 3)$. Consequently, $p_1(9) + (1 - p_1)(1) = 3$, which yields $p_1 = 1/4$. The required no-data mixed minimax rule is then $(p_1, p_2, p_3, p_4) = (1/4, 0, 3/4, 0)$. It advocates deluxe service one time out of four, and economy service three times out of four. The corresponding expected loss is 3 under either θ_1 or θ_2. Hence 3 is the minimum over the maximum coordinate of all mixed strategies. This is less than the 4 obtained with the pure minimax rule. $\square$

The pure minimax rule in Example 4.1 chooses economy service a_3. This is the same as the mixed strategy with probability vector $(0, 0, 1, 0)$. Example 4.4 showed that this is not the optimal mixed strategy, but it could be under different circumstances. If the convex hull of the available actions is oriented so that the expanding wedge from the origin first strikes it at a vertex, then the mixed minimax rule will be a pure strategy.

In the example above, the only serious contenders for the minimax strategy are combinations of actions a_1 and a_3. The figures show how one of these contenders intercepts the expanding wedge before any of the remaining points can enter the competition. This is a frequently occurring situation, where we say that certain strategies dominate the field.

4.5 Definition: Suppose that one of the n possibilities $\theta_1, \theta_2, \ldots, \theta_n$ is the current state of nature. Suppose further that m actions are available: $a_1, a_2, \ldots, a_m$. Let $L(a_j, \theta_i)$ describe the loss associated with action a_j when θ_i prevails. We say that action a_j *dominates* action a_k if $L(a_j, \theta_i) \leq L(a_k, \theta_i)$, for $1 \leq i \leq n$, and $L(a_j, \theta_{i_0}) < L(a_k, \theta_{i_0})$ for some i_0. We say that the mixed strategy $(p_1, \ldots, p_m)$ *dominates* the mixed strategy $(q_1, \ldots, q_m)$ if $E_{a;p}[L(\cdot, \theta_i)] \leq E_{a;q}[L(\cdot, \theta_i)]$, for $1 \leq i \leq n$, and $E_{a;p}[L(\cdot, \theta_{i_0})] < E_{a;q}[L(\cdot, \theta_{i_0})]$ for some i_0. ∎

Suppose that action a_j dominates action a_k. In a graphical rendition, where actions are placed at coordinates corresponding to the losses under the various states of nature, the a_k-coordinates are consistently larger than or equal to the corresponding a_j-coordinates across all states of nature. In the two-dimensional case, a_k lies to the northeast of a_j. In any case, $\max_{1 \leq i \leq n} L(a_j, \theta_i) \leq \max_{1 \leq i \leq n} L(a_k, \theta_i)$, and the minimax selection mechanism need never choose a_k. If $\max_{1 \leq i \leq n} L(a_j, \theta_i) < \max_{1 \leq i \leq n} L(a_k, \theta_i)$, it cannot choose a_k. In the boundary case where equality holds, and a better choice is not available, the selection can still avoid a_k by breaking the tie in favor of a_j. We can, therefore, exclude dominated points from consideration in the pure minimax procedure.

Similar reasoning shows that dominated actions need not contend in the mixed minimax procedure. Indeed, suppose that a_j and a_k are two distinct actions with a_j dominant. Consider a mixed strategy involving a_k: $p = (p_1, p_2, \ldots, p_m)$ with $p_k > 0$. Construct a new mixed strategy, $q = (q_1, q_2, \ldots, q_m)$ as follows. Set $q_i = p_i$ when $i \neq j$ and $i \neq k$, and set $q_j = p_j + p_k, q_k = 0$. That is, we obtain the new strategy by reassigning the probability associated with a_k to a_j. Given that the p_i sum to one, the q_i also sum to 1. Moreover, for any state of nature θ_i,

$$q_j L(a_j, \theta_i) + q_k L(a_k, \theta_i) = (p_j + p_k) L(a_j, \theta_i) = p_j L(a_j, \theta_i) + p_k L(a_j, \theta_i)$$
$$\leq p_j L(a_j, \theta_i) + p_k L(a_k, \theta_i).$$

To the left side of this inequality, we add the terms $q_t L(a_t, \theta_i)$ for $t \neq j$ or k. To the right side, we add the terms $p_t L(a_t, \theta_i)$, which are equal to those added on the left. Therefore,

$$E_{a;q}[L(\cdot, \theta_i)] = \sum_{t=1}^{m} q_t L(a_t, \theta_i) \leq \sum_{t=1}^{m} p_t L(a_t, \theta_i) = E_{a;p}[L(\cdot, \theta_i)].$$

Because a_j dominates a_k, the strict inequality $L(a_j, \theta_{i_0}) < L(a_k, \theta_{i_0})$ holds for some state i_0. For i_0, therefore, the inequalities above are strict. Consequently, the mixed strategy $q = (q_1, q_2, \ldots, q_m)$, for which $q_k = 0$, dominates

$p = (p_1, p_2, \ldots, p_m)$. We then have

$$\max_{1 \leq i \leq n} E_{a;q}[L(\cdot, \theta_i)] \leq \max_{1 \leq i \leq n} E_{a;p}[L(\cdot, \theta_i)].$$

In the contention for the mixed minimax strategy, any probability vector $p = (p_1, p_2, \ldots, p_m)$ with $p_k > 0$ has a competitor $q = (q_1, q_2, \ldots, q_m)$ with $q_k = 0$, such that the maximum expected loss under q is no larger than that under p. If there is a unique p that minimizes the numbers $\max_{1 \leq i \leq n} E_{a;p}[L(\cdot, \theta_i)]$, then p_k must be 0. If there are several vectors that tie for the minimum, then one of them must have $p_k = 0$, and we can break the tie by choosing that vector. We conclude that the dominated action a_k will receive zero weight in the mixed minimax rule. This observation provides a practical method of calculating the mixed minimax decision rule in situations where there are many possibilities for the state of nature. In this case, it is difficult to visualize the actions' convex hull in the higher-dimensional space. We start the calculation by placing zeros in the probability vector for all dominated actions. If a single action remains, then it receives weight 1, and the mixed minimax strategy is a pure strategy. If several actions remain, then the minimax point is the intersection of the convex hull of these points with the expanding wedge from the origin. The next example illustrates the technique.

4.6 Example: The losses, $L(a_j, \theta_i)$, for five states of nature and six actions appear to the right below. We seek the pure and mixed minimax decision strategy in the absence of any data that might indicate the prevailing state of nature.

The no-data pure minimax decision rule chooses action a_2, which exhibits the smallest value in the rightmost column. No consideration of dominating strategies can make this calculation any simpler. For the mixed minimax computation, however, we can benefit by reducing the contenders. Note that a_2 dominates a_1 because a_2 shows a smaller (or equal) loss consistently across all states of nature. Similarly, a_4 dominates a_3,

Action	States of nature					Max
	θ_1	θ_2	θ_3	θ_4	θ_5	Loss
a_1	3	6	5	7	5	7
a_2	2	4	5	5	4	5
a_3	6	7	5	1	2	7
a_4	6	6	5	1	2	6
a_5	7	8	4	4	4	8
a_6	3	8	1	4	4	8

and a_6 dominates a_5. Accordingly, the no-data mixed minimax procedure takes the form $p = (0, p_2, 0, p_4, 0, p_6)$. Of the remaining three actions, none dominates another. In comparing losses for any pair of them, we find states of nature in which one shows the smaller loss and states of nature in which the other is superior. Consequently, we must solve for the combination $p_2 a_2 + p_4 a_4 + p_6 a_6$ that first encounters the expanding wedge.

Here we have identified the remaining action points with their locations in 5-space. That is, $a_2 = (2, 4, 5, 5, 4)$, and similarly for a_4 and a_6. We now show how to analyze the situation in the two-dimensional space of Figure 4.3. We seek values for p_2, p_4, p_6 that minimize the leftmost expression below. Each line gives the expected loss under strategy p for a particular state of nature.

$$\max \left\{ \begin{array}{l} 2p_2 + 6p_4 + 3p_6 \\ 4p_2 + 6p_4 + 8p_6 \\ 5p_2 + 5p_4 + 1p_6 \\ 5p_2 + 1p_4 + 4p_6 \\ 4p_2 + 2p_4 + 4p_6 \end{array} \right\} \qquad \max \left\{ \begin{array}{l} L_1 \\ L_2 \\ L_3 \\ L_4 \\ L_5 \end{array} \right\} = \max \left\{ \begin{array}{l} -p_2 + 3p_4 + 3 \\ -4p_2 - 2p_4 + 8 \\ 4p_2 + 4p_4 + 1 \\ p_2 - 3p_4 + 4 \\ -2p_4 + 4 \end{array} \right\}$$

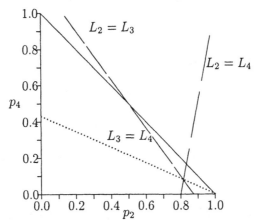

Figure 4.3. Permissible p_2, p_4 values for the mixed minimax search (see Example 4.6)

In general, this is a difficult problem. We can reduce it a bit further by substituting $1 - p_2 - p_4$ for p_6. We then seek to minimize $\max\{L_1, L_2, L_3, L_4, L_5\}$, where $L_i = E_{a;p}[L(\cdot, \theta_i)]$. That is, we seek to minimize the rightmost expression above. Because $p_2 + p_4 + p_6 = 1$, the permissible values for p_2 and p_4 form a triangle in the first quadrant to the southwest of the line $p_2 + p_4 = 1$, as shown in Figure 4.3. The third nonzero element of the probability vector, p_6, makes up the difference so that the sum remains 1. Since $p_2 + p_4 \leq 1$ in the triangle, we can argue that loss L_2 will always exceed losses L_1 and L_5 in this area:

$$L_2 - L_1 = -4p_2 - 2p_4 + 8 - (-p_2 + 3p_4 + 3) = -3p_2 - 5p_4 + 5$$
$$\geq -3p_2 - 5(1 - p_2) + 5 = 2p_2 \geq 0$$
$$L_2 - L_5 = -4p_2 - 2p_4 + 8 - (-2p_4 + 4) = -2p_2 - 2p_4 + 4 = -2(p_2 + p_4) + 4$$
$$\geq -2 + 4 = 2 > 0.$$

The competition, therefore, is among L_2, L_3, and L_4. Each of these components assumes the maximum over some portion of the triangle. The broken and dotted lines in Figure 4.3 show the intersections where $L_2 = L_3, L_2 = L_4$, and $L_3 = L_4$. These lines subdivide the triangle into areas, in each of which one of the three components is largest. To minimize the largest component, therefore, we are sometimes working with L_2, sometimes with L_3, and sometimes with L_4. The minimum cannot occur in the interior of a subarea because movement toward a border can decrease the maximum component. In the interior of an area where $L_2 = -4p_2 - 2p_4 + 8$ is the largest component, for example, a smaller value ensues from moving to larger p_2 or larger p_4 values. Consequently, the minimum must occur on a boundary. On a boundary line, where two components are equal, we can use either as the maximum. Because this maximum is a linear function in p_2 and p_4, it will increase, decrease, or remain constant along the entire boundary line. It cannot pass through a local minimum as we move along the boundary. The boundary where $L_2 = L_3$, for example, is the line $p_2 = (-3/4)p_4 + (7/8)$. The maximum loss along this line is the common value of L_2 and L_3 there, which is $p_4 + (9/2)$. We can decrease this value by moving along the line toward the extremity associated with smaller p_4 values.

We conclude that the minimum must occur at one of the intersections, of which there are eight. The intersections, together with the component values at the intersections, appear to the right. The minimum in the last column occurs at $(p_2, p_4) =$ $(0.82, 0.08)$. More precisely, it occurs at $(p_2, p_4) = (31/38, 3/38)$. The no-data minimax mixed strategy is then $(0, 31/38, 0, 3/38, 0, 4/38)$. Under this strategy, the maximum expected loss

	Loss L_i under θ_i					Max
Intersection	L_1	L_2	L_3	L_4	L_5	Loss
$(0.00, 1.00)$	6.00	6.00	5.00	1.00	2.00	6.00
$(0.00, 0.43)$	4.29	7.14	2.71	2.71	4.86	7.14
$(0.00, 0.00)$	3.00	8.00	1.00	4.00	4.00	8.00
$(0.80, 0.00)$	2.20	4.80	4.20	4.80	4.00	4.80
$(0.88, 0.00)$	2.13	4.50	4.50	4.88	4.00	4.88
$(1.00, 0.00)$	2.00	4.00	5.00	5.00	4.00	5.00
$(0.83, 0.17)$	2.67	4.33	5.00	4.33	3.67	5.00
$(0.82, 0.08)$	2.42	4.58	4.58	4.58	3.84	4.58
$(0.50, 0.50)$	4.00	5.00	5.00	3.00	3.00	5.00

is 4.58, which occurs when the state of nature is θ_2, θ_3, or θ_4. This is less than 5.00, which is the maximum loss incurred with the no-data pure minimax strategy. $\square$

The no-data minimax rule, either the pure or the mixed version, operates without any knowledge of the actual state of nature. It simply determines, for each candidate action or random mixture of actions, the worst loss across the states of nature. The minimax rule specifies the action, or random mixture of actions, that minimizes these pessimistic values. If, in Examples 4.1 and 4.4 above, we know that poor network conditions (θ_2) prevail 80% of the time, we might consistently choose the deluxe service. We then incur a loss of 9 with probability 0.2 and a loss of 0 with probability 0.8. This is a pure strategy with an expected loss of $0.2(9) + 0.8(0) = 1.8$, which is significantly better than the minimax rule, pure or mixed version. Note that the expected value is now computed with respect to a distribution across the states of nature. This leads to the definition of our next decision rule. It presupposes a knowledge of the distribution of the states of nature. Because this distribution is not based on a sample, it is called a *prior distribution*. The prior distribution may reflect our past experience with the states of nature, but we conduct no sample with the express purpose of "spying" on nature before taking our decision.

If nature assumes states $\theta_1, \ldots, \theta_n$ with probabilities $q = (q_1, \ldots, q_n)$, then the expected loss for action a_j is

$$E_{\theta;q}[L(a_j, \cdot)] = \sum_{i=1}^{n} q_i L(a_j, \theta_i).$$

The subscript indicates that the expectation calculation uses distribution q over states of nature θ.

4.7 Definition: Suppose that one of the n possibilities $\theta_1, \theta_2, \ldots, \theta_n$ is the current state of nature. Suppose further that m actions are available: $a_1, a_2, \ldots, a_m$. Also, it is known that θ_i prevails with $\Pr(\theta_i) = q_i$. Let $L(a_j, \theta_i)$ describe the loss associated with action a_j when θ_i prevails. In this context, the *no-data pure Bayes decision rule* advocates action a_k, where

$$E_{\theta;q}[L(a_k, \cdot)] = \min_{1 \le j \le m} E_{\theta;q}[L(a_j, \cdot)]. \blacksquare$$

Action	Network condition		$E_{\theta;q}[L(a_j,\cdot)]$
	θ_1: Good $q_1 = \Pr(\theta_1) = 0.2$	θ_2: Poor $q_2 = \Pr(\theta_2) = 0.8$	
a_1: Deluxe service	9	0	1.8
a_2: Intermediate service	6	3	3.6
a_3: Economy service	1	4	3.4
a_4: Abandon message	6	6	6.0

TABLE 4.3. Network losses and an associated prior distribution of network states (see Example 4.8)

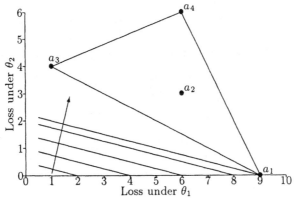

Figure 4.4. The pure Bayes rule: the first action encountered by expanding parallel lines (see Example 4.8)

Let us return to Example 4.1 with the additional information provided by a known distribution across the states of nature.

4.8 Example: Table 4.3 recapitulates the costs associated with various actions under two possible states of nature. The data are as in Example 4.1, except the last column now gives the expected cost for each action. The expectation is with respect to the distribution across the states of nature: $q_1 = \Pr(\theta_1) = 0.2, q_2 = \Pr(\theta_2) = 0.8$.

The no-data pure Bayes decision rule chooses action a_1, which with an expected loss of 1.8 incurs less cost than the other options under the given distribution across θ_1 and θ_2. The graphical interpretation of Figure 4.4 again locates the action points on the L_1, L_2 plane, where L_i is the loss associated with state θ_i. Any strategy in the plane, (x, y), has expected loss $0.2x + 0.8y$. The points of constant loss K form lines with equations $0.2x + 0.8y = K$. Lines in this family exhibit slope $-1/4$ and y-intercept $K/0.8$. They are therefore parallel and displaced increasingly to the northeast with increasing K. The arrow in the figure shows the direction of increasing K. As we successively plot lines with increasing K, we eventually encounter one of the action points. The first such encounter determines the action for the no-data pure Bayes decision procedure because it identifies the smallest K actually achieved by an available action. The figure shows how this construction identifies the action a_1 under the current circumstances.

It is clear from the figure that an expanding family of parallel lines, of any given slope, will first encounter the convex hull of the action points at a vertex. If the parallel lines are exactly parallel to the line between a_1 and a_3, they will first touch the convex hull simultaneously at all points on that line. In this boundary case, all points on the line a_1, a_3 exhibit the same expected loss, and we can choose a vertex to break the tie. This means that a mixed strategy, which corresponds to some point in the convex hull of $\{a_1, a_2, a_3, a_4\}$, cannot deliver a lower expected loss than the chosen pure strategy. The following discussion investigates this point more thoroughly. $\square$

4.9 Definition: Suppose that one of the n possibilities $\theta_1, \theta_2, \ldots, \theta_n$ is the current state of nature. Suppose further that m actions are available: $a_1, a_2, \ldots, a_m$. Let $q = (q_1, q_2, \ldots, q_n)$, where $q_i = \Pr(\theta_i)$, be the prior probabilities of the states. Let $L(a_j, \theta_i)$ describe the loss associated with action a_j when θ_i prevails. In this context, the *no-data mixed Bayes decision rule* is a probability vector $p = (p_1, p_2, \ldots, p_m)$ that advocates action a_k with probability p_j, where p is chosen such that

$$\sum_{i=1}^{n} q_i \sum_{j=1}^{m} p_j L(a_j, \theta_i) = \min_r \sum_{i=1}^{n} q_i \sum_{j=1}^{m} r_j L(a_j, \theta_i).$$

The minimum is taken over all probability vectors $r = (r_1, r_2, \ldots, r_m)$. ∎

In this definition, the inner sum $\sum_{j=1}^{m} r_j L(a_j, \theta_i)$ is the expected loss from the mixed strategy r when θ_i prevails. The outer sum then calculates the expected value of this function with respect to the distribution q across the possible states of nature. In the final step, the mixed rule calculates the same expected value as the pure rule, but it does so for many more candidates. The candidates are all weighted sums of the available actions, with losses defined in proportion to the weights. As suggested by the preceding example, however, the mixed Bayes strategy provides no advantage of the pure Bayes strategy. In the example, this follows because the expanding parallel lines, which represent increasing expected losses, must first touch the convex hull of the action points at a vertex. We can also make this argument analytically, which relieves us of a difficult graphical interpretation in higher dimensions.

For purposes of deriving a contradiction, suppose that a mixed Bayes strategy, $p = (p_1, p_2, \ldots, p_m)$, provides a lower expected loss with respect to the distribution $q = (q_1, q_2, \ldots, q_n)$ across the states of nature than any pure Bayes strategy. That is,

$$\sum_{i=1}^{n} q_i \sum_{j=1}^{m} p_j L(a_j, \theta_i) < \sum_{i=1}^{n} q_i L(a_k, \theta_i),$$

for all $1 \le k \le m$. We multiply both sides of these inequalities by p_k and sum to obtain

$$\sum_{k=1}^{m} p_k \left(\sum_{i=1}^{n} q_i \sum_{j=1}^{m} p_j L(a_j, \theta_i) \right) < \sum_{k=1}^{m} p_k \sum_{i=1}^{n} q_i L(a_k, \theta_i).$$

Of course, the parenthesized sum on the left is independent of k, which allows further reduction:

$$\left(\sum_{i=1}^{n} q_i \sum_{j=1}^{m} p_j L(a_j, \theta_i) \right) \sum_{k=1}^{m} p_k < \sum_{i=1}^{n} \sum_{k=1}^{m} q_i p_k L(a_k, \theta_i)$$

$$\sum_{i=1}^{n} \sum_{j=1}^{m} q_i p_j L(a_j, \theta_i) < \sum_{i=1}^{n} \sum_{j=1}^{m} q_i p_j L(a_j, \theta_i).$$

This last equation is, of course, impossible. We therefore conclude the following theorem, which refers to "a" Bayes strategy, rather than "the" Bayes strategy because the selection procedure does not specify a particular choice in the case of ties.

4.10 Theorem: A no-data mixed Bayes strategy cannot deliver a lower expected cost than a no-data pure Bayes strategy.

PROOF: See discussion above. ∎

The minimax and Bayes decision rules of this section are no-data solutions, in the sense that they operate in the absence of any data that might reveal the actual state of nature. Returning to the examples that deal with a computer network under various operating conditions, we might send a short test packet to ascertain the network state before choosing the transmission option. The normal circumstance is that the information so obtained does not specify the exact state of nature, although it may indicate that some states are more probable than others. That is, the test data may lead to a revision of the prior distribution across the states of nature. The test packet, even in poor network conditions, may find a window in the traffic and quickly scurry to its destination and back. That is, the return time for the packet is a random variable, whose distribution depends on the unknown state of nature. Under good network conditions, a fast return time has higher probability than under poor conditions. However, it is unfortunately the case that the distributions usually overlap. It is then possible to experience a fast return time even when network conditions are poor. The test packet constitutes a sample of a random variable with an unknown distribution. That is, the distribution when θ_1 prevails is different than when θ_2 prevails. These samples, or functions of the samples, are called statistics. This chapter's goal is to show how to use statistics to make meaningful decisions in the face of uncertainty. The next section takes the first step, which is a systematic study of the properties of statistics. It also develops sample-driven variations of the minimax and Bayes decision rules, which, in general, deliver lower expected costs than their no-data counterparts.

Exercises

4.1 Modify the data in Examples 4.1 and 4.4 such that action a_3 (economy service) becomes the pure strategy advocated by the mixed minimax decision rule.

Action	Nature states	
	θ_1	θ_2
a_1	4	8
a_2	5	2
a_3	10	1

(a)

Action	Nature states		
	θ_1	θ_2	θ_3
a_1	4	0	8
a_2	3	2	7
a_3	6	5	2
a_4	4	4	1

(b)

Action	Nature states			
	θ_1	θ_2	θ_3	θ_4
a_1	1	2	3	4
a_2	4	3	2	1
a_3	3	2	1	4
a_4	2	1	4	3
a_5	1	4	3	2

(c)

Figure 4.5. Loss functions for the exercises

4.2 Figure 4.5(a) gives the losses associated with three actions and two states of nature. Draw a graph that illustrates the mixed minimax strategy $p = (p_1, p_2, p_3)$. Set up the appropriate equations and solve analytically for p.

4.3 For the loss function of Figure 4.5(a), suppose the prior distribution for the states of nature is $q_1 = \Pr(\theta_1) = 0.6, q_2 = \Pr(\theta_2) = 0.4$. What is the pure Bayes strategy?

4.4 Find the pure and mixed minimax strategies for the loss configuration of Figure 4.5(b).

4.5 For the loss function of Figure 4.5(b), determine the pure Bayes strategy when the prior probabilities are $q_1 = \Pr(\theta_1) = 0.1, q_2 = \Pr(\theta_2) = 0.2$, and $q_3 = \Pr(\theta_3) = 0.7$.

4.6 Find the pure and mixed minimax strategies for the loss configuration of Figure 4.5(c).

4.7 For the loss function of Figure 4.5(c), determine the pure Bayes strategy for the preceding problem when prior probabilities are $q_1 = \Pr(\theta_1) = 0.1, q_2 = \Pr(\theta_2) = 0.2, q_3 = \Pr(\theta_3) = 0.3$, and $q_4 = \Pr(\theta_4) = 0.4$.

4.8 Suppose that $L(a_j, \theta_i)$ characterizes the loss associated with action a_j $(1 \le j \le m)$ when the state of nature is $\theta_i (1 \le i \le n)$. Consider a new loss function, obtained by subtracting the minimum value for each θ_i. That is, $L'(a_j, \theta_i) = L(a_j, \theta_i) - \min_{1 \le t \le m} L(a_t, \theta_i)$, for $1 \le i \le n$. If we measure the loss with L', every state of nature has a zero-loss action. A loss function with this property is called a *regret function*. Show that the pure Bayes strategy remains unchanged when the calculation uses L' in place of L. Provide an example to show that the mixed minimax strategy under L' can differ from that under L.

*4.9 Let S be the convex hull of the points $\{a_1, a_2, \ldots, a_n\}$ in three-space. Suppose that $x \in S$, and let T be the convex hull of $\{x, a_1, a_2, \ldots, a_n\}$. Show that $S = T$.

*4.10 Let S be the convex hull of the points $\{a_1, a_2, \ldots, a_n\}$ in three-space. Suppose that x_i, for $i = 1, 2, \ldots$, is a sequence of points in S such that $\lim_{i \to \infty} x_i = x$. Show that $x \in S$.

4.2 Statistics and their properties

As established in Definition 3.1, a sample of a random variable X is one or more values obtained by observing independent outcomes in the underlying probability space. We denote the individual observations as $X_1, X_2, \ldots, X_N$, where N is the size of the sample. When we have no need to emphasize N, we use the boldface $\mathbf{X}$ for the vector $(X_1, X_2, \ldots, X_N)$. Similarly, we use boldface $\mathbf{x} = (x_1, x_2, \ldots, x_N)$ for a value assumed by $\mathbf{X}$. The distribution of each X_i is the distribution of X, and by definition, X_i is independent of X_j for $i \neq j$. Therefore, the joint probability of a particular observed value $\mathbf{x}$ is the product of the probabilities of the components. That is,

$$\Pr(\mathbf{X} = \mathbf{x}) = \Pr(X_1 = x_1)\Pr(X_2 = x_2)\cdots\Pr(X_N = x_N).$$

4.11 Definition: A *statistic* is a function of a sample. The underlying random variable is called the *population*. ∎
   If $\mathbf{X} = (X_1, X_2, \ldots, X_N)$ is a sample from population X, the following are statistics. In each case, we can view the statistic as a reduction of the sample. The reduction may produce a single number or a vector of numbers.

- the sample mean $\overline{X} = \left(\sum_{i=1}^{N} X_i\right)/N$

- the sample sum $S = \sum_{i=1}^{N} X_i = N\overline{X}$

- the sample variance $s_X^2 = \left(\sum_{i=1}^{N}(X_i - \overline{X})^2\right)/N$

- the sample standard deviation $\sqrt{s_X^2}$

- the order statistic $X_{(1)}, X_{(2)}, \ldots, X_{(N)}$, where $X_{(1)} = \min_{1 \leq i \leq N} X_i$ is the smallest of the sample values, $X_{(2)}$ is the next largest, and so forth, through $X_{(N)} = \max_{1 \leq i \leq N} X_i$

- the sample maximum $X_{(N)}$

- the sample minimum $X_{(1)}$

- the range $R = X_{(N)} - X_{(1)}$

- the median $m = \begin{cases} X_{((N+1)/2)}, & \text{if } N \text{ is odd} \\ [X_{(N/2)} + X_{((N/2)+1)}]/2, & \text{if } N \text{ is even} \end{cases}$

- the sample itself $(X_1, X_2, \ldots, X_N)$

Action	Network condition	
	θ_1: Good	θ_2: Poor
a_1: Deluxe service	9	0
a_2: Intermediate service	6	3
a_3: Economy service	1	4
a_4: Abandon message	6	6

	Network condition	
	θ_1: Good	θ_2: Poor
$\Pr_X(1)$	0.8	0.3
$\Pr_X(0)$	0.2	0.7
μ_X	0.8	0.3
σ_X^2	0.16	0.21

(a) Loss for state θ_i (b) Bernoulli traffic probe X

Figure 4.6. Network loss functions and traffic probe distributions (see Example 4.12)

A statistic from population X is a random variable and therefore possesses a distribution, which depends on the population distribution. Consequently, each statistic has a mean and a variance. If the statistic's name involves the term mean or variance, a discussion of the computed mean or variance may appear confusing. For example, we speak of the mean of the sample mean or the variance of the sample variance. Remember that "sample mean" is just a compound title for a random variable, and the reference is to the mean of this random variable, computed in the normal fashion as an expected value. Similarly, "sample variance" is simply a random variable, perhaps with an ill-chosen name, for which the variance is an acceptable computation. To compute the mean or variance, we need the random variable's distribution.

In a decision problem, the distribution of the population X is generally not known. Instead, X may have many candidate distributions. In this case, a statistic from X also has several candidate distributions, each of which enables the computation of different means and variances. The following example illustrates the point in the case of the unreliable computer network of the preceding section.

4.12 Example: We return to the context of Example 4.1, where we must choose among transmission services with varying costs, depending on network conditions. Figure 4.6(a) repeats the information on losses as a function of network state. Suppose that we sample the state of nature with a test packet. Let $X = 1$ if the packet indicates good network conditions, and let $X = 0$ if it indicates poor conditions. X is a random variable because it cannot determine the network condition with certainty. In poor conditions, there is still some possibility that a packet may find a hole in the traffic and give a good indication. Also, in generally good conditions, there is some possibility that the packet encounters congestion and gives a poor indication. To be useful, X must have a known distribution under each state of nature. Suppose that X has one the distributions of Figure 4.6(b), according to the prevailing state of nature. These are rational distributions. Under good network conditions, $X = 1$ with probability 0.8; under poor conditions, $X = 0$ with probability 0.7. Now suppose that we take a sample (X_1, X_2) of size two and consider some statistics of this sample.

- The sample itself (X_1, X_2) has distributions as shown below. Under θ_1, for example, the probability of $(X_1, X_2) = (1, 1)$ is $\Pr(X_1 = 1) \cdot \Pr(X_2 = 1) = (0.8)^2$. The remaining entries admit similar calculations.

Distributions of (X_1, X_2)		
(X_1, X_2)	θ_1 : Good	θ_2 : Poor
$(1,1)$	$(0.8)^2 = 0.64$	$(0.3)^2 = 0.09$
$(1,0)$	$(0.8)(0.2) = 0.16$	$(0.3)(0.7) = 0.21$
$(0,1)$	$(0.2)(0.8) = 0.16$	$(0.7)(0.3) = 0.21$
$(0,0)$	$(0.2)^2 = 0.04$	$(0.7)^2 = 0.49$

- The sample sum, $S = X_1 + X_2$, being the sum of two independent Bernoulli random variables, is binomially distributed. Under θ_1, the contributing Bernoulli components have common parameter 0.8, while under θ_2 the parameter is 0.3. S assumes the values $0, 1, 2$, while the sample mean, $\overline{X} = S/2$, assumes the values $0, 1/2, 1$ with corresponding probabilities. The distributions appear below.

Distribution of $S = X_1 + X_2$ and $\overline{X} = (X_1 + X_2)/2$			
S	$\overline{X}$	θ_1 : Good	θ_2 : Poor
0	0	$C_{2,0}(0.8)^0(0.2)^2 = 0.04$	$C_{2,0}(0.3)^0(0.7)^2 = 0.49$
1	1/2	$C_{2,1}(0.8)^1(0.2)^1 = 0.32$	$C_{2,0}(0.3)^1(0.7)^1 = 0.42$
2	1	$C_{2,2}(0.8)^2(0.2)^0 = 0.64$	$C_{2,2}(0.3)^2(0.7)^0 = 0.09$

Under θ_1 we have $\mu_{\overline{X}} = 0(0.04) + (1/2)(0.32) + 1(0.64) = 0.8 = \mu_X$, and a similar computation holds when the state of nature is θ_2. That is, the mean of the statistic $\overline{X}$ is the mean of X under either state of nature.

- $s_X^2 = \left(\sum_{i=1}^2 (X_i - \overline{X})^2 \right) /2$ is the sample variance, which requires the calculation to the right for each possible (X_1, X_2) vector. We see that the sample variance assumes just two values, and consequently the sample standard deviation assumes just two values. We obtain the distributions by summing the probabilities associated with a particular value. For example, the probability under θ_1 that $s_X^2 = 0.00$ is $0.64 + 0.04 = 0.68$. The corresponding distributions appear in the lower table. Under θ_1, mean$(s_X^2) = (0.00)(0.68) + (0.25)(0.32) = 0.08$, while

				Probability	
(X_1, X_2)	$\overline{X}$	s_X^2	θ_1	θ_2
$(1,1)$	1.0	0.00	0.64	0.09
$(1,0)$	0.5	0.25	0.16	0.21
$(0,1)$	0.5	0.25	0.16	0.21
$(0,0)$	0.0	0.00	0.04	0.49

		Probability	
s_X^2	$\sqrt{s_X^2}$	θ_1	θ_2
0.00	0.00	0.68	0.58
0.25	0.50	0.32	0.42

under θ_2, it is mean$(s_X^2) = (0.00)(0.58) + (0.25)(0.42) = 0.105$. Thus, the means of the sample variances, 0.08 under θ_1 and 0.105 under θ_2, are not equal to the X-variances, which are 0.16 and 0.21. Curiously, they are just half as much.

- The order statistic, $X_{(1)}, X_{(2)}$, permits a further reduction to obtain the range R, the median m, and the sample maximum and minimum. For all these statistics, the calculations are tabulated below. Note that order statistic assumes three values, as does the median, but the range, sample maximum, and sample minimum assume only two.

(X_1,X_2)	$(X_{(1)},X_{(2)})$	R	m	$X_{(2)}$	$X_{(1)}$	Probability θ_1	Probability θ_2
$(1,1)$	$(1,1)$	0	1.0	1	1	0.64	0.09
$(1,0)$	$(0,1)$	1	0.5	1	0	0.16	0.21
$(0,1)$	$(0,1)$	1	0.5	1	0	0.16	0.21
$(0,0)$	$(0,0)$	0	0.0	0	0	0.04	0.49

$(X_{(1)},X_{(2)})$	m	Probability θ_1	Probability θ_2
$(1,1)$	1.0	0.64	0.09
$(0,1)$	0.5	0.32	0.42
$(0,0)$	0.0	0.04	0.49

R	$\Pr_R$ θ_1	$\Pr_R$ θ_2	$X_{(2)}$	$\Pr_{X_{(2)}}$ θ_1	$\Pr_{X_{(2)}}$ θ_2	$X_{(1)}$	$\Pr_{X_{(1)}}$ θ_1	$\Pr_{X_{(1)}}$ θ_2
0	0.68	0.58	0	0.04	0.49	0	0.36	0.91
1	0.32	0.42	1	0.96	0.51	1	0.64	0.09

For all the statistics illustrated in this example, we obtain the distribution in the same manner. For each possible statistical value y, we examine the outcomes in the underlying population that produce y. The sum of the associated probabilities is $\Pr(y)$. $\square$

We see that it is a simple, if computationally tedious, task to calculate the distribution of a statistic. We also see that certain statistics, such as the sample mean, are single numbers. We will eventually use statistics in our decision rules, and we are therefore interested in those situations where a single-valued statistic can enable a good decision. A function over the sample that does not discard information necessary for an optimal decision, even when it reduces the sample to a single number, is called a sufficient statistic. A later section will investigate the characteristics of sufficient statistics.

The example also shows that, having computed the distribution, we can calculate other descriptive features of a statistic, such as its mean and variance. For simple statistics, we can derive general formulas for these features in terms of the corresponding features of the underlying population. The calculations involve expected values, and significant simplification follows from the following simple observation. For arbitrary functions f and g, if $i \neq j$,

$$E[f(X_i)g(X_j)] = E[f(X_i)] \cdot E[g(X_j)]. \tag{4.1}$$

This follows from the independence of the samples:

$$E[f(X_i)g(X_j)] = \sum_k \sum_l f(x_k)g(x_l)\Pr_{X_i,X_j}(x_k,x_l)$$

$$= \sum_k \sum_l f(x_k)g(x_l)\Pr_{X_i}(x_k)\Pr_{X_j}(x_l)$$

$$= \left(\sum_k f(x_k)\Pr_{X_i}(x_k)\right)\left(\sum_l g(x_l)\Pr_{X_j}(x_l)\right)$$

$$= E[f(X_i)] \cdot E[g(X_j)], \tag{4.2}$$

where $x_1, x_2, \ldots$ is the range of X.

Suppose that $(X_1, X_2, \ldots, X_N)$ is a sample from population X, and consider the sample sum $S = \sum_{i=1}^{N} X_i$. Because the X_i are independent, and hence uncorrelated, the expectation operator's linearity (Theorem 1.43) provides a quick calculation of the mean and variance. Since each X_i has the same distribution as X, we have

$$\text{mean}(S) \;=\; \mu_S = E\left[\sum_{i=1}^{N} X_i\right] = \sum_{i=1}^{N} E[X_i] = \sum_{i=1}^{N} E[X] = N\mu_X.$$

The variance computation invokes Equation 4.1 to eliminate most of the terms from the square of a sum.

$$\text{var}(S) \;=\; E[(S - N\mu_X)^2] = E\left[\left(\sum_{i=1}^{N}(X_i - \mu_X)\right)^2\right]$$

$$= E\left[\sum_{i=1}^{N}(X_i - \mu_X)^2 + \sum_{i\neq j}(X_i - \mu_X)(X_j - \mu_X)\right]$$

$$= \sum_{i=1}^{N} E[(X_i - \mu_X)^2] + \sum_{i\neq j} E[X_i - \mu_X]E[X_j - \mu_X]$$

$$= \sum_{i=1}^{N} E[(X - \mu_X)^2] = N\sigma_X^2$$

Since $\overline{X} = S/N$, we have

$$\text{mean}(\overline{X}) \;=\; \mu_{\overline{X}} = \frac{\mu_S}{N} = \mu_X$$

$$\text{var}(\overline{X}) \;=\; E\left[\left(\frac{S}{N} - \mu_X\right)^2\right] = E\left[\frac{(S - N\mu_X)^2}{N^2}\right] = \frac{\text{var}(S)}{N^2} = \frac{\sigma_X^2}{N}.$$

To compute the mean of the sample variance, we need the covariance of X_i and $\overline{X}$. Since X_i and $\overline{X}$ both have mean μ_X, we have

$$\text{cov}(X_i, X_j) \;=\; E[(X_i - \mu_X)(\overline{X} - \mu_X)] = E\left[\frac{(X_i - \mu_X)(S - N\mu_X)}{N}\right]$$

$$= \frac{1}{N}E\left[(X_i - \mu_X)\sum_{j=1}^{N}(X_j - \mu_X)\right]$$

$$= \frac{1}{N}\left(E[(X_i - \mu_X)^2] + \sum_{j\neq i} E[(X_i - \mu_X)(X_j - \mu_X)]\right)$$

$$= \frac{1}{N}E[(X - \mu_X)^2] = \frac{\sigma_X^2}{N}.$$

Then,

$$\text{mean}(s_X^2) = E\left[\frac{1}{N}\sum_{i=1}^{N}(X_i - \overline{X})^2\right] = \frac{1}{N}E\left[\sum_{i=1}^{N}[(X_i - \mu_X) - (\overline{X} - \mu_X)]^2\right]$$

$$= \frac{1}{N}E\left[\sum_{i=1}^{N}[(X_i - \mu_X)^2 - 2(X_i - \mu_X)(\overline{X} - \mu_X) + (\overline{X} - \mu_X)^2]\right]$$

$$= \frac{1}{N}\left(\sum_{i=1}^{N}E[(X_i - \mu_X)^2] - 2\sum_{i=1}^{N}E[(X_i - \mu_X)(\overline{X} - \mu_X)]\right.$$
$$\left. + \sum_{i=1}^{N}E[(\overline{X} - \mu_X)^2]\right)$$

$$= \frac{1}{N}\left(N\sigma_X^2 - 2N\cdot\frac{\sigma_X^2}{N} + N\cdot\frac{\sigma_X^2}{N}\right) = \left(\frac{N-1}{N}\right)\sigma_X^2.$$

This formula explains why, in the example above, the mean of the sample variance was just one-half of the population variance; the fraction $(N - 1)/N$ is $1/2$ when $N = 2$.

The variance of the sample variance is somewhat more difficult to compute. We first establish the intermediate results of Table 4.4, which appear in the expansion of $\text{var}(s_X^2)$. Because higher-order moments of X appear, we switch to the notation

$$\delta_2 = E[(X - \mu_X)^2] = \sigma_X^2$$
$$\delta_4 = E[(X - \mu_X)^4].$$

The table's first entry is relatively easy to verify. If $i = j$,

$$E[(X_i - \mu_X)^2(X_j - \mu_X)^2] = E[(X_i - \mu_X)^4] = E[(X - \mu_X)^4] = \delta_4.$$

If $i \neq j$, Equation 4.1 implies that

$$E[(X_i - \mu_X)^2(X_j - \mu_X)^2] = E[(X_i - \mu_X)^2]E[(X_j - \mu_X)^2] = \delta_2^2.$$

Consider the third entry.

$$E[(X_i - \mu_X)^2(\overline{X} - \mu_X)^2] = E\left[(X_i - \mu_X)^2\left(\frac{1}{N}\sum_{j=1}^{N}(X_j - \mu_X)\right)^2\right]$$

$$= \frac{1}{N^2}\sum_{j=1}^{N}\sum_{k=1}^{N}E[(X_i - \mu_X)^2(X_j - \mu_X)(X_k - \mu_X)]$$

$$= \frac{1}{N^2}\sum_{j=1}^{N}E[(X_i - \mu_X)^2(X_j - \mu_X)^2]$$

The last reduction follows because if $j \neq k$ in the double summation, then one of them is different from i. Suppose that it is j. Then $E[(X_j - \mu_X)] = 0$

F_{ijk}	$E[F_{ijk}]$	$\sum_{i=1}^{N} \sum_{j=1}^{N} E[F_{ijk}]$
$F_{ij1} = (X_i - \mu_X)^2(X_j - \mu_X)^2$	$\begin{cases} \delta_4, \; i = j \\ \delta_2^2, \; i \neq j \end{cases}$	$N\delta_4 + N(N-1)\delta_2^2$
$F_{ij2} = (X_i - \mu_X)^2(X_j - \mu_X)(\overline{X} - \mu_X)$	$\begin{cases} \dfrac{\delta_4}{N}, \; i = j \\ \dfrac{\delta_2^2}{N}, \; i \neq j \end{cases}$	$\delta_4 + (N-1)\delta_2^2$
$F_{ij3} = (X_i - \mu_X)^2(\overline{X} - \mu_X)^2$	$\dfrac{\delta_4 + (N-1)\delta_2^2}{N^2}$	$\delta_4 + (N-1)\delta_2^2$
$F_{ij4} = (X_i - \mu_X)(X_j - \mu_X)(\overline{X} - \mu_X)^2$	$\begin{cases} \dfrac{\delta_4 + (N-1)\delta_2^2}{N^2}, \; i = j \\ \dfrac{2\delta_2^2}{N^2}, \qquad\quad\; i \neq j \end{cases}$	$\dfrac{\delta_4 + 3(N-1)\delta_2^2}{N}$
$F_{ij5} = (X_i - \mu_X)(\overline{X} - \mu_X)^3$	$\dfrac{3(N-1)\delta_2^2 + \delta_4}{N^3}$	$\dfrac{\delta_4 + 3(N-1)\delta_2^2}{N}$
$F_{ij6} = (\overline{X} - \mu_X)^4$	$\dfrac{3(N-1)\delta_2^2 + \delta_4}{N^3}$	$\dfrac{\delta_4 + 3(N-1)\delta_2^2}{N}$

TABLE 4.4. Intermediate results needed to compute $\mathrm{var}(s_X^2)$

---

stands alone as a factor when the expectation operation is factored according to Equation 4.1. Therefore, a summand in the double summation is zero when $j \neq k$. The same reasoning holds if it is k that differs from i. The remaining terms collapse to the single summation, which yields a single δ_4, when $j = i$, and a total of $(N-1)\delta_2^2$, across the remaining $(N-1)$ summands where $j \neq i$. The exercises ask that you verify the remaining table entries.

Now, we start the computation of the variance of the sample variance in the usual manner.

$$\mathrm{var}(s_X^2) = E\left[\left(s_X^2 - \frac{(N-1)\sigma_X^2}{N}\right)^2\right]$$

$$= E[(s_X^2)^2] - 2\left(\frac{N-1}{N}\right)\delta_2 E[s_X^2] + \left(\frac{N-1}{N}\right)^2 \delta_2^2$$

At this point, all the terms are known except for $E[(s_X^2)^2]$. Continuing the simplification, we have

$$\mathrm{var}(s_X^2) = E[(s_X^2)^2] - 2\left(\frac{N-1}{N}\right)^2 \delta_2^2 + \left(\frac{N-1}{N}\right)^2 \delta_2^2$$

$$= E[(s_X^2)^2] - \left(\frac{N-1}{N}\right)^2 \delta_2^2. \qquad (4.3)$$

There remains the expansion of $E[(s_X^2)^2]$:

$$E[(s_X^2)^2] = E\left[\left(\frac{1}{N}\sum_{i=1}^{N}(X_i - \overline{X})^2\right)^2\right].$$

Persevering with the expansion, we find the expressions of Table 4.4. We also abbreviate $\mu_X = \mu$.

$$E[(s_X^2)^2] = \frac{1}{N^2}E\left[\left(\sum_{i=1}^{N}[(X_i - \mu) - (\overline{X} - \mu)]^2\right)^2\right]$$

$$= \frac{1}{N^2}\sum_{i=1}^{N}\sum_{j=1}^{N}E\left[\left((X_i - \mu) - (\overline{X} - \mu)\right)^2\left((X_j - \mu) - (\overline{X} - \mu)\right)^2\right]$$

$$E[(s_X^2)^2] = \frac{1}{N^2}\sum_{i=1}^{N}\sum_{j=1}^{N}E\left[[(X_i - \mu)^2 - 2(X_i - \mu)(\overline{X} - \mu) + (\overline{X} - \mu)^2]\right.$$

$$\left.\cdot\,[(X_j - \mu)^2 - 2(X_j - \mu)(\overline{X} - \mu) + (\overline{X} - \mu)^2]\right]$$

$$= \frac{1}{N^2}\sum_{i=1}^{N}\sum_{j=1}^{N}(F_{ij1} - 2F_{ij2} + F_{ij3} - 2F_{ij2} + 4F_{ij4} - 2F_{ij5}$$

$$+ F_{ij3} - 2F_{ij5} + F_{ij6})$$

Substituting the table values yields

$$E[(s_X^2)^2] = \delta_2^2 + \frac{\delta_4 - 3\delta_2^2}{N} + \frac{-2\delta_4 + 5\delta_2^2}{N^2} + \frac{\delta_4 - 3\delta_2^2}{N^3},$$

and further substitution in Equation 4.3 gives

$$\mathrm{var}(s_X^2) = E[(s_X^2)^2] - \left(\frac{N-1}{N}\right)^2\delta_2^2 = \frac{\delta_4 - \delta_2^2}{N} - \frac{2(\delta_4 - 2\delta_2^2)}{N^2} + \frac{\delta_4 - 3\delta_2^2}{N^3}.$$

We summarize the results to this point as the following theorem.

4.13 Theorem: Let $X_1, X_2, \ldots, X_N$ be a sample from the population X, which has moments $\mu_X, \delta_2 = \sigma_X^2$, and $\delta_4 = E_X[(X - \mu_X)^4]$. The sample sum S, sample mean $\overline{X}$, and sample variance s_X^2 have the following characteristics.

Statistic	Mean	Variance		
S	$N\mu_X$	$N\sigma_X^2$		
$\overline{X}$	μ_X	$\dfrac{\sigma_X^2}{N}$		
s_X^2	$\dfrac{(N-1)\sigma_X^2}{N}$	$\dfrac{\delta_4 - \delta_2^2}{N}$	$-\dfrac{2(\delta_4 - 2\delta_2^2)}{N^2}$	$+\dfrac{\delta_4 - 3\delta_2^2}{N^3}$

PROOF: See discussion above. ∎

4.14 Example: Verify Theorem 4.13 for the data of Example 4.12.

The earlier example features a test probe X that provides a probabilistic reading of the network state. Figure 4.7(a) gathers the distributions of X and of various statistics computed from a sample of size $N = 2$. These data are excerpted from the earlier example. Figure 4.7(b) lists the means and variances, computed as expected values directly from the distributions. Under θ_1, Theorem 4.13 provides

Distributions			
		Under	
Quantity	Range	θ_1	θ_2
X	0	0.2	0.7
	1	0.8	0.3
S	0	0.04	0.49
	1	0.32	0.42
	2	0.64	0.09
$\overline{X}$	0.0	0.04	0.49
	0.5	0.32	0.42
	1.0	0.64	0.09
s_X^2	0.00	0.68	0.58
	0.25	0.32	0.42

Moments						
	Under θ_1			Under θ_2		
	mean	variance	δ_4	mean	variance	δ_4
X	0.8	0.16	0.0832	0.3	0.21	0.0777
S	1.6	0.32		0.6	0.42	
$\overline{X}$	0.8	0.08		0.3	0.105	
s_X^2	0.08	0.0136		0.105	0.015225	

Figure 4.7. Verification of the mean and variance of common statistics
(see Example 4.14)

---

the following values for the mean and variance of the three statistics.

$$\text{mean}(S) = N\mu_X = 2(0.8) = 1.6$$
$$\text{var}(S) = N\sigma_X^2 = 2(0.16) = 0.32$$
$$\text{mean}(\overline{X}) = \mu_X = 0.8$$
$$\text{var}(\overline{X}) = \frac{\sigma_X^2}{N} = \frac{0.16}{2} = 0.08$$
$$\text{mean}(s_X^2) = \frac{(N-1)\sigma_X^2}{N} = \frac{0.16}{2} = 0.08$$
$$\text{var}(s_X^2) = \frac{\delta_4 - \delta_2^2}{N} - \frac{2(\delta_4 - 2\delta_2^2)}{N^2} + \frac{\delta_4 - 3\delta_2^2}{N^3}$$
$$= \frac{0.0832 - (0.16)^2}{2} - \frac{2(0.0832 - 2(0.16)^2)}{4} + \frac{0.0832 - 3(0.16)^2}{8}$$
$$= 0.0136$$

These computations verify the entries under θ_1 in Figure 4.7(b), and similar calculations show that the theorem's formulas also correctly calculate the entries under θ_2. □

A statistic's mean and variance depend on its distribution, which in turn depends on the distribution of the underlying population X. We have seen that the mean and variance of the sample mean are simple functions of the mean and variance of X. The same is true for the sample sum. The variance of the sample variance, however, introduces the higher-order moment, δ_4, of the population. In general, the mean and variance of more complicated statistics will be functions of the X distribution, rather than functions of summarizing characteristics, such as μ_X or σ_X^2. Various reductions of the order statistic, such as the sample maximum $X_{(N)}$, the sample minimum $X_{(1)}$, the sample range R, and the sample median m, fall into this category.

We will, therefore, show how to obtain the *cumulative* distributions for the order statistic components and leave further reductions to specific cases.

The sample maximum can assume any value from range(X), but it is less than some fixed value if and only if all sample components are less than that value. That is, the event $(X_{(N)} \leq t)$ is the intersection of the events $(X_i \leq t)$ for $i = 1, 2, \ldots, N$. The independence of the X_i then implies that

$$F_{X_{(N)}}(t) = \Pr(X_{(N)} \leq t) = \Pr(X_1 \leq t) \cdot \Pr(X_2 \leq t) \cdots \Pr(X_N \leq t)$$
$$= (\Pr(X \leq t))^N = [F_X(t)]^N .$$

We can generalize the argument for an arbitrary component $X_{(k)}$ of the order statistic. We note that $X_{(k)} \leq t$ if and only if k or more of the components are less than or equal to t. For a fixed t, if we let

$$Y_i = \begin{cases} 1, & X_i \leq t \\ 0, & X > t, \end{cases}$$

then Y_i is a Bernoulli random variable with parameter $F_X(t)$. Consequently, $Y = Y_1 + Y_2 + \ldots + Y_N$ is binomially distributed with parameters $(N, F_X(t))$. Therefore, the probability that $Y = j$, which is the probability that exactly j components are less than or equal to t, is $C_{N,j}[F_X(t)]^j[1 - F_X(t)]^{N-j}$. Hence

$$F_{X_{(k)}}(t) = \sum_{j=k}^{N} C_{N,j} [F_X(t)]^j [1 - F_X(t)]^{N-j} .$$

Letting $k = 1$, we obtain the distribution for the sample minimum $X_{(1)}$:

$$F_{X_{(1)}}(t) = \sum_{j=1}^{N} C_{N,j} [F_X(t)]^j [1 - F_X(t)]^{N-j}$$
$$= \left(\sum_{j=0}^{N} C_{N,j} [F_X(t)]^j [1 - F_X(t)]^{N-j} \right) - [1 - F_X(t)]^N$$
$$= (F_X(t) + 1 - F_X(t))^N - [1 - F_X(t)]^N = 1 - [1 - F_X(t)]^N .$$

To deal with the range $R = X_{(N)} - X_{(1)}$, we need the joint distribution of $X_{(1)}$ and $X_{(N)}$. Note that we can write, for any s, the event $(X_{(N)} \leq t)$ as the following disjoint union:

$$(X_{(N)} \leq t) = [(X_{(N)} \leq t) \cap (X_{(1)} \leq s)] \cup [(X_{(N)} \leq t) \cap (X_{(1)} > s)] .$$

It follows that

$$F_{X_{(1)}, X_{(N)}}(s, t) = \Pr((X_{(1)} \leq s) \cap (X_{(N)} \leq t))$$
$$= \Pr(X_{(N)} \leq t) - \Pr((X_{(N)} \leq t) \cap (X_{(1)} > s)).$$

The event $((X_{(N)} \leq t) \cap (X_{(1)} > s))$ occurs precisely when all sample components lie in the interval $(s, t]$. If $s > t$, the interval is empty, and the corresponding event has probability zero. If $s \le t$, the probability that X_i lies in the interval is $F_X(t) - F_X(s)$. Because the samples are independent, the probability that all sample components lie in the interval is $[F_X(t) - F_X(s)]^N$. Substituting these results in the equation above, we obtain

$$F_{X_{(1)}, X_{(N)}}(s, t) = \begin{cases} [F_X(t)]^N - [F_X(t) - F_X(s)]^N, & s \le t \\ [F_X(t)]^N, & s > t. \end{cases}$$

For future reference, we record these observations as the next theorem. We then take up some examples that illustrate how to pursue further analysis with a specific F_X.

4.15 Theorem: Let $X_1, X_2, \ldots, X_N$ be a sample from the population X, and let $X_{(1)}, X_{(2)}, \ldots, X_{(N)}$ be the corresponding order statistic. The cumulative distribution of component $X_{(k)}$ and the joint distribution of the smallest and largest are, respectively,

$$F_{X_{(k)}}(t) = \sum_{j=k}^{N} C_{N,j} [F_X(t)]^j [1 - F_X(t)]^{N-j}$$

$$F_{X_{(1)}, X_{(N)}}(s, t) = \begin{cases} [F_X(t)]^N - [F_X(t) - F_X(s)]^N, & s \le t \\ [F_X(t)]^N, & s > t. \end{cases}$$

PROOF: See discussion above. ∎

4.16 Example: Verify that Theorem 4.15 correctly computes the cumulative distributions for the order statistic components in Example 4.12.

The example used a sample of size $N = 2$: X_1, X_2. We verify the order statistic distribution under θ_1; the computation under θ_2 is similar. Recall that the X distribution under θ_1 is $\Pr_X(0) = 0.2, \Pr_X(1) = 0.8$. The cumulative distribution is therefore

$$F_X(t) = \begin{cases} 0, & t < 0 \\ 0.2, & 0 \le t < 1 \\ 1.0, & t \ge 1. \end{cases}$$

According to Theorem 4.15,

$$F_{X_{(1)}}(0) = \sum_{j=1}^{2} C_{2,j} [F_X(0)]^j [1 - F_X(0)]^{2-j}$$
$$= C_{2,1}(0.2)^1 (1 - 0.2)^{2-1} + C_{2,2}(0.2)^2 (1 - 0.2)^{2-2}$$
$$= 2(0.2)(0.8) + (1)(0.04) = 0.36$$

$$F_{X_{(1)}}(1) = \sum_{j=1}^{2} C_{2,j} [F_X(1)]^j [1 - F_X(1)]^{2-j}$$
$$= C_{2,1}(1.0)^1 (1 - 1.0)^{2-1} + C_{2,2}(1.0)^2 (1 - 1.0)^{2-2}$$
$$= 2(1.0)(0.0) + (1)(1.0)(1.0) = 1.00$$

$$F_{X_{(2)}}(0) = \sum_{j=2}^{2} C_{2,j} [F_X(0)]^j [1 - F_X(0)]^{2-j}$$
$$= C_{2,2}(0.2)^2 (1 - 0.2)^{2-2} = (1)(0.04)(1.0) = 0.04$$

$$F_{X_{(2)}}(1) = \sum_{j=2}^{2} C_{2,j} [F_X(1)]^j [1 - F_X(1)]^{2-j}$$

$$= C_{2,2}(1.0)^2(1 - 1.0)^{2-2} = (1)(1.0)(1.0) = 1.00.$$

Consequently,

$$\Pr_{X_{(1)}}(0) = F_{X_{(1)}}(0) - 0 = 0.36$$
$$\Pr_{X_{(1)}}(1) = F_{X_{(1)}}(1) - F_{X_{(1)}}(0) = 1.00 - 0.36 = 0.64$$
$$\Pr_{X_{(2)}}(0) = F_{X_{(2)}}(0) - 0 = 0.04$$
$$\Pr_{X_{(2)}}(1) = F_{X_{(2)}}(1) - F_{X_{(2)}}(0) = 1.00 - 0.04 = 0.96.$$

These values are identical to those obtained in Example 4.12. $\square$

4.17 Example: Let X have the distribution of Figure 4.8(a). Consider a sample X_1, X_2, X_3, and determine the mean and variance of the sample minimum, the sample maximum, the sample median, and the sample range.

Theorem 4.15, applied to the cumulative distribution $F_X(\cdot)$, yields the results of Figure 4.8(b). From these cumulative distributions, we obtain the corresponding probability distributions by subtraction. Denoting the range of X as $x_1 < x_2 < x_3 < x_4$, we have

$$\Pr_{X(k)}(x_t) = F_{X_{(k)}}(x_t) - F_{X_{(k)}}(x_{t-1})$$
$$\Pr_{X_{(1)}, X_{(3)}}(x_s, x_t) = F_{X_{(1)}, X_{(3)}}(x_s, x_t) - F_{X_{(1)}, X_{(3)}}(x_{s-1}, x_t) - F_{X_{(1)}, X_{(3)}}(x_s, x_{t-1})$$
$$+ F_{X_{(1)}, X_{(3)}}(x_{s-1}, x_{t-1}).$$

With these formulas, we can transform the cumulative distribution tables into probability tables. For the left table of Figure 4.8(b), the new probability entry is the old cumulative distribution entry minus its old neighbor to the north. The neighbors of the top entries are zero. For the right table, the new probability entry is the sum of the old cumulative distribution entry with its old northwest neighbor, minus its old north and west neighbors. As before, zeros substitute for missing neighbors. Accordingly, the tables transform to those of Figure 4.8(c). The parenthesized entries in the right table give the range associated with the cell. The sample minimum is $X_{(1)}$, the sample maximum is $X_{(3)}$, and the sample median is $X_{(2)}$. The means and variances of these statistics are immediately available from the probability entries of Figure 4.8(c).

$$\text{mean}(X_{(1)}) = 1(0.784) + 4(0.091) + 6(0.117) + 9(0.008) = 1.922$$
$$\text{var}(X_{(1)}) = 1^2(0.784) + 4^2(0.091) + 6^2(0.117) + 9^2(0.008) - (1.922)^2 = 3.406$$
$$\text{mean}(X_{(2)}) = 1(0.352) + 4(0.148) + 6(0.396) + 9(0.104) = 4.256$$
$$\text{var}(X_{(2)}) = 1^2(0.352) + 4^2(0.148) + 6^2(0.396) + 9^2(0.104) - (4.256)^2 = 7.286$$
$$\text{mean}(X_{(3)}) = 1(0.064) + 4(0.061) + 6(0.387) + 9(0.488) = 7.022$$
$$\text{var}(X_{(3)}) = 1^2(0.064) + 4^2(0.061) + 6^2(0.387) + 9^2(0.488) - (7.022)^2 = 5.192$$

From the right table of Figure 4.8(c), we see that the range R assumes values 0, 2, 3, 5, and 8. Summing probabilities from the relevant cells, we have $\Pr(R = 0) = 0.064 + 0.001 + 0.027 + 0.008 = 0.100$. Similar calculations yield (0.100, 0.036, 0.150,

x	1	4	6	9
$\Pr_X(x)$	0.4	0.1	0.3	0.2
$F_X(x)$	0.4	0.5	0.8	1.0

(a) Population distribution

x	$F_{X_{(1)}}(x)$	$F_{X_{(2)}}(x)$	$F_{X_{(3)}}(x)$
1	0.784	0.352	0.064
4	0.875	0.500	0.125
6	0.992	0.896	0.512
9	1.000	1.000	1.000

$F_{X_{(1)},X_{(3)}}(x,y)$	$y=1$	4	6	9
$x=1$	0.064	0.124	0.448	0.784
4	0.064	0.125	0.485	0.875
6	0.064	0.125	0.512	0.992
9	0.064	0.125	0.512	1.000

(b) Derived cumulative distributions for order statistic components

x	$\Pr_{X_{(1)}}(x)$	$\Pr_{X_{(2)}}(x)$	$\Pr_{X_{(3)}}$
1	0.784	0.352	0.064
4	0.091	0.148	0.061
6	0.117	0.396	0.387
9	0.008	0.104	0.488

$\Pr(X_{(1)}=x, X_{(2)}=y)$	$y=1$	4	6	9
$x=1$	0.064	0.060	0.324	0.336
	(0)	(3)	(5)	(8)
4	0.000	0.001	0.036	0.054
		(0)	(2)	(5)
6	0.000	0.000	0.027	0.090
			(0)	(3)
9	0.000	0.000	0.000	0.008
				(0)

(c) Probabilities for order statistic components

Figure 4.8. Distributions for Example 4.17

0.378, 0.336) for the full set of range value probabilities. The mean and variance are then

$$\text{mean}(R) = 0(0.100) + 2(0.036) + 3(0.150) + 5(0.378) + 8(0.336) = 5.1$$
$$\text{var}(R) = 0^2(0.100) + 2^2(0.036) + 3^2(0.150) + 5^2(0.378) + 8^2(0.336) - (5.1)^2$$
$$= 6.438. \ \square$$

Now we know what statistics are and how to compute their distributions. Our intentions, however, are more ambitious. We want to use statistics to improve our decision procedures. To illustrate the possibilities, we further develop the situation started in Example 4.1, which deals with service options available for an unreliable network. This approach introduces yet another expected loss calculation, so we first pause to reflect on the multiple meanings of the term "expected loss" encountered to this point.

1. For the no-data minimax decision rule, we consider mixed strategies of the form $p = (p_1, p_2, \ldots, p_m)$, for which the expected loss is

$$\overline{L}(\theta_i) = E_{a;p}[L(\cdot, \theta_i)] = \sum_{j=1}^{m} p_j L(a_j, \theta_i).$$

We calculate this expected value over actions, and we use a particular state of nature as a parameter. A frequency interpretation of this expected value is as follows. If θ_i prevails as the constant state of nature while we make many repeated decisions, each utilizing a random selection that chooses action a_j with probability p_j, then our average cost will be $\overline{L}(\theta_i)$ per decision. For each mixed strategy, the no-data minimax procedure first maximizes $\overline{L}(\theta_i)$ over θ_i, which gives the most pessimistic assessment for each strategy. The procedure then minimizes these results over the possible strategies $(p_1, p_2, \ldots, p_m)$. This policy acts as though nature knows when we use the probability vector p and perversely presents the least favorable θ_i for that vector. It is important to understand that nature does not know our particular choice of action for each decision; nature knows only that we are randomly choosing among the actions using probabilities $p = (p_1, p_2, \ldots, p_m)$. Accordingly, nature chooses the θ_i that make this procedure most expensive for us. We respond with the mixed minimax decision rule, which minimizes our expected losses when we confront such a knowledgeable adversary.

2. When contemplating a no-data pure Bayes decision rule, we consider pure strategies of the form a_j, for which the expected loss is

$$\overline{L}(a_j) = E_{\theta;q}[L(a_j, \cdot)] = \sum_{i=1}^{n} q_i L(a_j, \theta_i),$$

where q_i is the prior probability of θ_i. In contrast with the preceding entry, this expected value is calculated over states of nature and is parameterized by a particular action. The frequency interpretation is that repeated choices of action a_j incur an average cost of $\overline{L}(a_j)$ per decision, provided that nature assumes random states with the single constraint that state θ_i appear with probability q_i. We can view nature as an adversary, who chooses a state for each decision. Nature uses a randomizing device such that state θ_i appears with frequency q_i. Unlike the previous interpretation, the pure Bayes decision rule does not assume that nature knows our strategy and chooses a fixed state to exact the highest price. Nature simply presents random states in keeping with the prior distribution. In response, the no-data Bayes decision rule minimizes $\overline{L}(a_j)$ over the available actions.

3. When working with a no-data mixed Bayes decision rule, we again consider mixed strategies of the form $(p_1, p_2, \ldots, p_m)$, for which the expected loss is

$$\overline{L} = E_{a;p}[E_{\theta;q}[L(\cdot, \cdot)]] = \sum_{j=1}^{m} \sum_{i=1}^{n} p_j q_i L(a_j, \theta_i).$$

The q_i are again the prior probabilities of the states of nature θ_i. The expected loss here is an expected value of an expected value, calculated over both actions and states of nature. The frequency interpretation involves randomization both by us (the decision makers) and by nature. If we make repeated decisions, each utilizing a selection mechanism that chooses a_j with probability p_j, and if nature independently chooses a state for each decision such that state θ_i appears with frequency q_i, then our average cost per decision will be $\overline{L}$. The no-data Bayes rule minimizes this quantity over the possible strategies $(p_1, p_2, \ldots, p_m)$. We have seen that this complication yields no advantage over the no-data pure Bayes rule.

When we introduce sample data into the process, further variations of expected loss appear, conditioned on the sample observations. Before considering this problem in its general form, we present an illustrative example.

4.18 Example: Examples 4.1, 4.4, 4.8, and 4.12 analyze the costs and benefits associated with various service options for message transmission across an unreliable network. Example 4.12 computes the distributions of certain statistics associated with test packets. For each of the two possible states of nature, the tables below summarize the costs across the available services, the distributions of a test packet X, and the derived distribution of the sample sum $S = X_1 + X_2$, the sum of two such test packet results.

Cost $L(a_j, \theta_i)$		
	Network condition	
Action	θ_1: Good	θ_2: Poor
a_1: Deluxe service	9	0
a_2: Intermediate service	6	3
a_3: Economy service	1	4
a_4: Abandon message	6	6

X distribution		
	θ_1	θ_2
$\mathrm{Pr}_X(1)$	0.8	0.3
$\mathrm{Pr}_X(0)$	0.2	0.7

S distribution		
S	θ_1	θ_2
0	0.04	0.49
1	0.32	0.42
2	0.64	0.09

To specify a pure decision rule based on S, we must designate an action when $S = 0$, an action when $S = 1$, and an action when $S = 2$. Because each entails four choices, there are $4^3 = 64$ possible decision procedures. We represent each rule as a triple (i, j, k), which commands action a_i if $S = 0$, a_j if $S = 1$, and a_k if $S = 2$. Table 4.5 lists the possible rules.

Some rules simply ignore the data. For example, the rule $(2, 2, 2)$ commands action a_2 (intermediate service) regardless of S. This strategy degenerates to the no-data pure strategy that advocates a_2. Other rules use the data irrationally. The rule $(3, 2, 1)$, for example, advocates a_3 (economy service) when $S = 0$. But $S = 0$ means that both test packets have reported poor network conditions, for which deluxe service is more appropriate. It also advocates a_1 (deluxe service) when $S = 2$, which is again contrary to intuition. How do we choose among the rules?

For each state of nature, we know the distribution of S. We can therefore compute an expected loss, given a particular state of nature, for each decision rule. Consider $d_9 = (1, 3, 2)$, for example, which advocates actions a_1 when $S = 0$, a_3 when $S = 1$, and a_2 when $S = 2$. Under θ_1, the sample sum assumes values $0, 1, 2$ with probabilities $0.04, 0.32, 0.64$, respectively, and the corresponding actions a_1, a_3, a_2 incur losses of $9, 1, 6$, respectively. A loss of 9, for example, occurs when a_1 is chosen, which happens when $S = 0$, which occurs with probability 0.04. The

Rule		Expected Loss			Rule		Expected Loss			Rule		Expected Loss		
		θ_1	θ_2	max			θ_1	θ_2	max			θ_1	θ_2	max
d_0	$(1,1,1)$	9.00	0.00	9.00	d_{21}	$(2,2,2)$	6.00	3.00	6.00	d_{42}	$(3,3,3)$	1.00	4.00	4.00
d_1	$(1,1,2)$	7.08	0.27	7.08	d_{22}	$(2,2,3)$	2.80	3.09	3.09	d_{43}	$(3,3,4)$	4.20	4.18	4.20
d_2	$(1,1,3)$	3.88	0.36	3.88	d_{23}	$(2,2,4)$	6.00	3.27	6.00	d_{44}	$(3,4,1)$	7.72	4.48	7.72
d_3	$(1,1,4)$	7.08	0.54	7.08	d_{24}	$(2,3,1)$	6.32	3.15	6.32	d_{45}	$(3,4,2)$	5.80	4.75	5.80
d_4	$(1,2,1)$	8.04	1.26	8.04	d_{25}	$(2,3,2)$	4.40	3.42	4.40	d_{46}	$(3,4,3)$	2.60	4.84	4.84
d_5	$(1,2,2)$	6.12	1.53	6.12	d_{26}	$(2,3,3)$	1.20	3.51	3.51	d_{47}	$(3,4,4)$	5.80	5.02	5.80
d_6	$(1,2,3)$	2.92	1.62	2.92	d_{27}	$(2,3,4)$	4.40	3.69	4.40	d_{48}	$(4,1,1)$	8.88	2.94	8.88
d_7	$(1,2,4)$	6.12	1.80	6.12	d_{28}	$(2,4,1)$	7.92	3.99	7.92	d_{49}	$(4,1,2)$	6.96	3.21	6.96
d_8	$(1,3,1)$	6.44	1.68	6.44	d_{29}	$(2,4,2)$	6.00	4.26	6.00	d_{50}	$(4,1,3)$	3.76	3.30	3.76
d_9	$(1,3,2)$	4.52	1.95	4.52	d_{30}	$(2,4,3)$	2.80	4.35	4.35	d_{51}	$(4,1,4)$	6.96	3.48	6.96
d_{10}	$(1,3,3)$	1.32	2.04	2.04	d_{31}	$(2,4,4)$	6.00	4.53	6.00	d_{52}	$(4,2,1)$	7.92	4.20	7.92
d_{11}	$(1,3,4)$	4.52	2.22	4.52	d_{32}	$(3,1,1)$	8.68	1.96	8.68	d_{53}	$(4,2,2)$	6.00	4.47	6.00
d_{12}	$(1,4,1)$	8.04	2.52	8.04	d_{33}	$(3,1,2)$	6.76	2.23	6.76	d_{54}	$(4,2,3)$	2.80	4.56	4.56
d_{13}	$(1,4,2)$	6.12	2.79	6.12	d_{34}	$(3,1,3)$	3.56	2.32	3.56	d_{55}	$(4,2,4)$	6.00	4.74	6.00
d_{14}	$(1,4,3)$	2.92	2.88	2.92	d_{35}	$(3,1,4)$	6.76	2.50	6.76	d_{56}	$(4,3,1)$	6.32	4.62	6.32
d_{15}	$(1,4,4)$	6.12	3.06	6.12	d_{36}	$(3,2,1)$	7.72	3.22	7.72	d_{57}	$(4,3,2)$	4.40	4.89	4.89
d_{16}	$(2,1,1)$	8.88	1.47	8.88	d_{37}	$(3,2,2)$	5.80	3.49	5.80	d_{58}	$(4,3,3)$	1.20	4.98	4.98
d_{17}	$(2,1,2)$	6.96	1.74	6.96	d_{38}	$(3,2,3)$	2.60	3.58	3.58	d_{59}	$(4,3,4)$	4.40	5.16	5.16
d_{18}	$(2,1,3)$	3.76	1.83	3.76	d_{39}	$(3,2,4)$	5.80	3.76	5.80	d_{60}	$(4,4,1)$	7.92	5.46	7.92
d_{19}	$(2,1,4)$	6.96	2.01	6.96	d_{40}	$(3,3,1)$	6.12	3.64	6.12	d_{61}	$(4,4,2)$	6.00	5.73	6.00
d_{20}	$(2,2,1)$	7.92	2.73	7.92	d_{41}	$(3,3,2)$	4.20	3.91	4.20	d_{62}	$(4,4,3)$	2.80	5.82	5.82
										d_{63}	$(4,4,4)$	6.00	6.00	6.00

TABLE 4.5. Decision rules based on the sample sum for the unreliable network (see Example 4.18)

expected loss is then $0.04(9) + 0.32(1) + 0.64(6) = 4.52$. Under θ_2, the probabilities are $0.49, 0.42, 0.09$, and the losses are $0, 4, 3$ for an expected loss of $0.49(0) + 0.42(4) + 0.09(3) = 1.95$. In this manner, we can calculate the expected loss associated with each rule, and these data also appear in Table 4.5.

As with the no-data analysis, we resort again to a graphical layout to clarify the choices. Figure 4.9 shows the 64 possible decision rules as points in a two-dimensional space. The horizontal coordinate measures the expected loss under θ_1; the vertical coordinate measures the expected loss under θ_2. Rules d_0, d_{21}, d_{42}, and d_{63} choose actions a_1, a_2, a_3, and a_4, respectively, regardless of the sample sum. We can therefore identify these rules with their no-data counterparts. The expected loss in these cases is just the loss associated with the chosen action, so the triangular pattern of the previous examples describes the convex hull of these four points. As shown clearly in the figure, rules that do use the data can produce points outside this triangle. Those to the southwest of the triangle provide lower expected losses than any pure or mixed strategy in the triangle. Those to the northeast of the triangle provide higher losses and correspond to using the data in an irrational manner. To the northeast, for example, we find rule 60, which advocates a_4, a_4, a_1 when $S = 0, 1, 2$, respectively. This corresponds to message abandonment when $S \leq 1$ and deluxe service when $S = 2$. But $S = 2$ means that both test packets have indicated good network conditions—a situation when expensive deluxe service is probably not necessary. By contrast, rule 10 resides to the southwest of the triangle

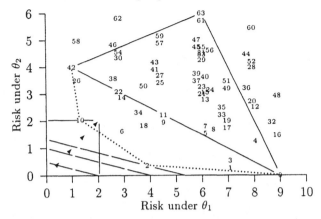

Figure 4.9. Expected losses from various decision rules over the sample sum statistic (see Example 4.18)

and advocates a_1, a_3, a_3 when $S = 0, 1, 2$, respectively. This is a more intelligent use of the data. It advocates deluxe service when $S = 0$. That is, it chooses deluxe service when both test packets have found poor network conditions. Moreover, it advocates economy service when both packets have reported good conditions.

The situation is similar to that associated with the previous examples, except that decision rules, each a function of the chosen statistic S, now replace the actions. Accordingly, we can employ the earlier techniques. As illustrated in Figure 4.9, an expanding wedge from the origin first encounters rule d_{10}. This is the pure minimax solution, which chooses the rule that minimizes the maximum loss. Indeed, a glance at Table 4.5 shows that d_{10} has the smallest value in the rightmost column. That column lists the maximum, across the possible states of nature, of the expected loss. If θ_1 prevails and we repeatedly make decisions using rule d_{10}, then we will incur an average cost of 1.32 per decision. If θ_2 prevails and we use d_{10} to make repeated decisions, then we will incur an average cost of 2.04. Hence, if nature is most unkind, we lose 2.04 per decision. Other rules also perform well under some states of nature and poorly under others. For some rules, nature is unkind in choosing θ_1; for others, the worse performance comes under θ_2. If we pessimistically assume that nature contrives to be unkind, regardless of our chosen rule, then we should choose d_{10}. Under d_{10}, nature's obstinacy costs us, on the average, 2.04 per decision. Under any other rule, we would pay more. In the face of nature's most unfavorable state, θ_2, for rule d_{10}, we will still encounter varying costs because the statistic S is a random variable. Rule d_{10} sometimes chooses a_1 for a loss of 0, and it sometimes chooses a_3 for a loss of 4. The distribution of S under θ_2 is such that the average is 2.04.

Figure 4.9 also shows that the expanding wedge encounters the convex hull of the decision rules, the dotted line, before it reaches any particular rule. Specifically, it encounters the edge between rules d_{10} and d_2. This suggests a mixed strategy, in which we choose between rules d_{10} and d_2 with carefully chosen probabilities. Denote by $(\hat{L}_1, \hat{L}_2)$ the expected losses under θ_1 and θ_2, respectively. Then d_{10} is at $(\hat{L}_1, \hat{L}_2) = (1.32, 2.04)$, while d_2 is at $(3.88, 0.36)$. The line between them is $\hat{L}_2 = -0.65625\hat{L}_1 + 2.90625$. The point of the expanding wedge is at $\hat{L}_1 = \hat{L}_2$. Solving these two equations for the intersection gives $\hat{L}_1 = \hat{L}_2 = 1.75$, which is indeed less than the 2.04 from the pure minimax solution. The fraction $(1.75 - 1.32)/(3.88 - 1.32) = 0.17$ indicates that the intersection is displaced 17% from d_{10}. This means that we should choose d_{10} with probability $1.0 - 0.17 = 0.83$. With this policy, our loss under θ_1 is $0.83(1.32) + 0.17(3.88) = 1.75$. Under θ_2 it is $0.83(2.04) + 0.17(0.36) = 1.75$. This is the mixed minimax solution: Use rule d_{10} with probability 0.83 and rule d_2 with probability 0.17. With this procedure, we are ambivalent about the state of nature because we incur an average cost of 1.75 per decision, regardless of the state of nature. In other words, even if nature knows our mixed strategy and schemes to maximize out expense, nature finds no advantage in one state over the other.

Finally, if we have prior probabilities for the states of nature, we can approach the convex hull of the decision rules with parallel lines representing constant expected loss. This is yet another usage of the term expected loss. Now it means the expected value, with respect to the prior distribution across the states of nature, of the expected losses calculated previously with the statistic's distributions. The earlier calculations, you will recall, produced values in which θ_i appears as a parameter. We will discuss this matter in more detail outside the example. For the moment, suppose we know that θ_1 occurs with probability 0.2, while θ_2 occurs with probability 0.8. For any point in the $(\hat{L}_1, \hat{L}_2)$ plane of Figure 4.9, we expect an average cost of $\hat{L}_1$ per decision when θ_1 prevails and $\hat{L}_2$ per decision when θ_2 prevails. Over many decisions, we expect 20% of them to take place under θ_1 and 80% under θ_2. The overall average cost per decision is then $0.2\hat{L}_1 + 0.8\hat{L}_2$. Lines of constant $0.2\hat{L}_1 + 0.8\hat{L}_2$ have slope -0.25. This family of parallel lines first touches the convex hull of the decision rules at d_2, as shown in the figure. This means that d_2 will incur the lowest average cost per decision, provided that nature obligingly generates θ_1 and θ_2 with probabilities 0.2 and 0.8, respectively. The average cost in that case is $0.2(3.88) + 0.8(0.36) = 1.064$.

Before leaving the example, it is worthwhile to note that the four rules on the southwest boundary, d_{42}, d_{10}, d_2, d_0, dominate the field. That is, for any other rule, one of these four has lower expected losses for *both* states of nature. $\square$

We will now formally define the concepts illustrated in the example. The definitions closely parallel those of the preceding section. The difference is that the earlier discussion used the raw actions where the current treatment uses decision rules based on the sample statistic. In all the definitions, we assume the following context. There are n states of nature $(\theta_1, \theta_2, \ldots, \theta_n)$ and m actions $(a_1, a_2, \ldots, a_m)$. A random variable X provides indirect evidence of the prevailing state of nature, in the sense that X has a known distribution under each state of nature. That is, if $\{x_1, x_2, \ldots\}$ is the range of X, then the $\Pr_{X|\theta_i}(x_i)$ are known. $\mathbf{X} = (X_1, X_2, \ldots, X_N)$ is a sample from the population X, which means that the X_i are independent random variables, each with the same distribution as X. Y is a statistic over the sample, and the distribution of Y is known for each state of nature. That is, $\Pr_{Y|\theta_i}(y_i)$ is available. Here $\{y_1, y_2, \ldots\}$ is the range of Y. The earlier discussions in this section have shown how to calculate these distributions from the distributions of X. Finally, $L(a_j, \theta_i)$ is the loss incurred with action a_j when state θ_i prevails.

As the example above suggests, our analysis will deal with losses associated with decision rules, rather than with specific actions. With decision rules, the loss is an expected value, which varies with the statistical value obtained from the sample. Even though we have worked with expected losses in the no-data problem, where we tolerated the possible confusion associated with different probability distributions, we attempt to avoid yet greater confusion by changing the terminology somewhat for the data-based problem. Specifically, we will refer to *risk*, rather than loss, when dealing with decision rules. The first definition should clarify this point.

4.19 Definition: A *decision rule* based on statistic Y is a function from range(Y) to the available actions. If $d(\cdot)$ is a decision rule and $y_k \in$ range(Y), then $d(y_k)$ is the action advocated by the rule when the sample produces $Y = y_k$. For each combination of a decision rule $d(\cdot)$ and a state of nature θ_i, we define the *risk* as

$$\hat{L}(d,\theta_i) = E_{Y|\theta_i}[L(d(\cdot),\theta_i)] = \sum_k \Pr_{Y|\theta_i}(y_k)L(d(y_k),\theta_i). \blacksquare$$

Risk is evidently the expected loss discussed in the preceding example. The new terminology, however, frees us to introduce further expected values based on distributions across the risk arguments. That is, if we have a distribution across the state of nature, in which $q_i = \Pr(\theta_i)$, we can construct an expected risk $E_{\theta;q}[\hat{L}(d,\cdot)] = \sum_i q_i \hat{L}(d,\theta_i)$. If we have a distribution across the decision rules, in which $p_d = \Pr(d)$, we can construct an expected risk $E_{d;p}[\hat{L}(\cdot,\theta_i)] = \sum_d p_d \hat{L}(d,\theta_i)$. The first expected value contains the decision rule d as a parameter; the second contains the state of nature θ_i.

4.20 Definition: The *statistical pure minimax decision rule*, based on statistic Y, advocates rule $d_*(\cdot)$, where

$$\max_{1\leq i\leq n} \hat{L}(d_*,\theta_i) = \min_d \max_{1\leq i\leq n} \hat{L}(d,\theta_i).$$

The minimum is taken over all decision rules that map range(Y) to the available actions. $\blacksquare$

4.21 Definition: Let $d_1, d_2, \ldots$ be an enumeration of the decision rules, each of which maps the range of Y to the available actions. A *mixed decision rule* is a probability vector $q = (q_1, q_2, \ldots)$ that advocates rule d_k with probability q_k. The *statistical mixed minimax decision rule*, based on statistic Y, is the mixed rule $p = (p_1, p_2, \ldots)$ such that

$$\max_{1\leq i\leq n} E_{d;p}[\hat{L}(\cdot,\theta_i)] = \min_q \max_{1\leq i\leq n} E_{d;q}[\hat{L}(\cdot,\theta_i)].$$

The minimum is taken over all mixed rules $q = (q_1, q_2, \ldots)$. $\blacksquare$

4.22 Definition: Let $q = (q_1, q_2, \ldots, q_n)$ be a prior distribution over the states of nature. That is, $\Pr(\theta_i) = q_i$. In this context, the *statistical pure Bayes decision rule* is rule $d'(\cdot)$, where

$$E_{\theta;q}[\hat{L}(d',\cdot)] = \min_d E_{\theta;q}[\hat{L}(d,\cdot)].$$

The minimum is taken over all decision rules that map range(Y) to the available actions. ∎

4.23 Definition: We say that decision rule d_j *dominates* rule d_k if $\hat{L}(d_j, \theta_i) \leq \hat{L}(d_k, \theta_i)$, for $1 \leq i \leq n$ and $\hat{L}(d_j, \theta_{i_0}) < \hat{L}(d_k, \theta_{i_0})$ for some i_0. In the same sense, we say that the mixed strategy $(p_1, p_2, \dots)$ *dominates* the mixed strategy $(q_1, q_2, \dots)$ if $E_{d;p}[\hat{L}(\cdot, \theta_i)] \leq E_{d;q}[\hat{L}(\cdot, \theta_i)]$ for $1 \leq i \leq n$ and for some i_0, $E_{d;p}[\hat{L}(\cdot, \theta_{i_0})] < E_{d;q}[\hat{L}(\cdot, \theta_{i_0})]$. ∎

The new decision rules are analogous to the corresponding no-data selections of Definitions 4.2, 4.3, and 4.7. The difference is that risk, which involves an expected value over a statistical distribution, substitutes for loss. The statistical pure minimax rule is the decision rule that minimizes the risk when nature somehow manages to present the least favorable state. The statistical mixed minimax rule is a mixed strategy over all decision rules, but it also seeks to minimize the risk when nature presents the least favorable state. This mixed strategy is a probability vector $p = (p_1, p_2, \dots)$. To use it, we generate a random integer k from the distribution p over $\{1, 2, \dots\}$ and then apply rule $d_k(\cdot)$ to the sample statistic. Finally, the statistical pure Bayes rule is the rule that minimizes the expected risk when nature presents random states consistent with a given prior distribution.

When using a statistic over a sample, we can, as in the no-data case, eliminate dominated decision rules from contention when searching for the pure minimax, mixed minimax, or Bayes rule. The argument is analogous to that given for the no-data case, with decision rules and their risks replacing actions and their losses. (See page 278.) This provides a useful simplification because the number of decision rules can quickly escalate. The next example illustrates the point.

4.24 Example: Continuing Example 4.18, identify the statistical minimax rules, both pure and mixed, for the sample sum statistic. Also identify the statistical pure Bayes rule, given the prior distribution $\Pr(\theta_1) = 0.2, \Pr(\theta_2) = 0.8$ across the two states of nature. Compare the costs with those of the corresponding no-data rules. Repeat the exercise with the sample mean, the sample range, and the sample itself.

For the sample sum, the first row in Table 4.6 collects the calculations of the preceding example. The corresponding information for the no-data case comes from Examples 4.1, 4.4, and 4.8. This example develops the remaining entries.

Recall that risks are expected losses incurred because the chosen statistic does not give an infallible reading on the prevailing state of nature. Nevertheless, over many decisions the risk gives the average cost per decision, provided that nature presents a constant state. Furthermore, the expected losses and expected risks that appear for the minimax and Bayes rules also approximate the average cost per decision, provided that nature maintains a constant maximally injurious state (minimax) or provides random states according to the given prior distribution. Therefore, all the elements in Table 4.6 are comparable in this sense. They all represent the average cost per decision. The assumptions under which they deliver this average across repeated trials are slightly different.

Using the sample sum as evidence of the prevailing state of nature, we can adopt the pure minimax rule, $(1, 3, 3)$, and thereby cut our average cost to a maximum of 2.04 per decision. We may pay less, if nature is cooperative, but if nature

	Statistical					
	Pure minimax		Mixed minimax		Pure Bayes	
Statistic	Rule	Maximum Risk	Rule	Maximum Expected Risk	Rule	Expected Risk
S	$(1,3,3)$	2.04	$0.17(1,1,3) + 0.83(1,3,3)$	1.750	$(1,1,3)$	1.064
$\overline{X}$	$(1,3,3)$	2.04	$0.17(1,1,3) + 0.83(1,3,3)$	1.750	$(1,1,3)$	1.064
R	$(3,1)$	3.56	$0.293(3,3) + 0.707(3,1)$	2.811	$(1,1)$	1.800
(X_1, X_2)	$(1,3,3,3)$	2.04	$0.17(1,1,1,3) + 0.83(1,3,3,3)$	1.750	$(1,1,1,3)$	1.064

	No-data					
	Pure minimax		Mixed minimax		Pure Bayes	
—	Action	Maximum Loss	Action	Maximum Expected Loss	Action	Expected Loss
	a_3	4.00	$0.25a_1 + 0.75a_3$	3.000	a_1	1.800

TABLE 4.6. For the unreliable network, comparison of rules based on various statistics (see Example 4.24)

perversely assumes the least favorable stance for this rule, we still hold the average cost to 2.04. Variation occurs, nevertheless, due to the fluctuation in the reported value of the sample sum, even though nature maintains the fixed state that is most expensive for our chosen rule.

Continuing with the sample sum as evidence, we can adopt the mixed minimax rule, $(1,1,3)$ with probability 0.17 and $(1,3,3)$ with probability 0.83, to reduce the maximum cost to 1.75 per decision. This means that a long run of decisions, taken during a period of least favorable θ_i, will incur an average cost of 1.75 per decision In both these cases, the average cost might be less, if nature does not present the least favorable θ_i.

If we know, however, that θ_i assumes different values, randomly and independently of our decisions, with probabilities $\Pr(\theta_1) = 0.2, \Pr(\theta_2) = 0.8$, then we can use the sample sum in a pure Bayes rule to reduce the average cost to 1.064 per decision. These figures improve on the corresponding no-data losses as tabulated in the lower portion of the table. In the absence of data, the pure minimax rule holds the loss to a maximum of 4.00 per decision, and the mixed minimax rule reduces this value to 3.00. The no-data Bayes rule incurs an average cost of 1.8, which is still greater than its statistical counterpart.

Now consider the sample mean $\overline{X} = S/2$. Let us refer to the risks calculated above for the sample sum as $\hat{L}_S$, in order to distinguish them from those to be computed now for $\overline{X}$. The latter will be denoted by $\hat{L}_{\overline{X}}$. The sample mean assumes three values, so there are again $4^3 = 64$ possible decision rules. We can use the list of Table 4.5, but we now associate the triple (j,k,l) with the rule that commands action a_j when $\overline{X} = 0$, a_k when $\overline{X} = 0.5$, and a_l when $\overline{X} = 1$. Suppose that (j,k,l) represents rule d. Under θ_i, we have

$$\hat{L}_{\overline{X}}(d, \theta_i) = \Pr_{\overline{X}}(0) \cdot L(a_j, \theta_i) + \Pr_{\overline{X}}(0.5) \cdot L(a_k, \theta_i) + \Pr_{\overline{X}}(1) \cdot L(a_k, \theta_i)$$
$$= \Pr_S(0) \cdot L(a_j, \theta_i) + \Pr_S(1) \cdot L(a_k, \theta_i) + \Pr_S(2) \cdot L(a_k, \theta_i) = \hat{L}_S(d, \theta_i).$$

Thus the risks, under both states of nature, are the same for the sample mean as for the sample sum. The decision rules then appear in the same locations in the $\hat{L}_1$–$\hat{L}_2$ plane, and Figure 4.9 is equally applicable to the sample mean. The expanding wedge operation then finds the same pure and mixed minimax rules, and the family

Rule		θ_1	θ_2	max
		Expected Loss		
d_0	$(1,1)$	9.00	0.00	9.00
d_1	$(1,2)$	8.04	1.26	8.04
d_2	$(1,3)$	6.44	1.68	6.44
d_3	$(1,4)$	8.04	2.52	8.04
d_4	$(2,1)$	6.96	1.74	6.96
d_5	$(2,2)$	6.00	3.00	6.00
d_6	$(2,3)$	4.40	3.42	4.40
d_7	$(2,4)$	6.00	4.26	6.00

Rule		θ_1	θ_2	max
		Expected Loss		
d_8	$(3,1)$	3.56	2.32	3.56
d_9	$(3,2)$	2.60	3.58	3.58
d_{10}	$(3,3)$	1.00	4.00	4.00
d_{11}	$(3,4)$	2.60	4.84	4.84
d_{12}	$(4,1)$	6.96	3.48	6.96
d_{13}	$(4,2)$	6.00	4.74	6.00
d_{14}	$(4,3)$	4.40	5.16	5.16
d_{15}	$(4,4)$	6.00	6.00	6.00

TABLE 4.7. Sample range decision rules for the unreliable network (see Example 4.24)

of parallel lines for the prior distribution $(\Pr(\theta_1) = 0.2, \Pr(\theta_2) = 0.8)$ acquires the same Bayes rule. The second line of Table 4.6 reflects these observations. The sample mean gives rules and average costs per decision that are identical with those obtained from the sample sum.

The sample range R, by contrast, is not so tractable. We assemble the information needed to compute the risks from Examples 4.1, 4.4, 4.8, and 4.12.

Costs as a function of action and network condition		
	Network condition	
Action	θ_1: Good	θ_2: Poor
a_1: Deluxe service	9	0
a_2: Intermediate service	6	3
a_3: Economy service	1	4
a_4: Abandon message	6	6

$\Pr(X = x)$		
x	θ_1	θ_2
1	0.8	0.3
0	0.2	0.7

$\Pr(R = r)$		
r	θ_1	θ_2
0	0.68	0.58
1	0.32	0.42

Because R assumes just two values, there are only $4^2 = 16$ possible decision rules. We represent each with a pair (j, k), which commands a_j when $R = 0$ and a_k when $R = 1$. If d is the rule (j, k), then under θ_i we compute

$$\hat{L}(d, \theta_i) = \Pr(R = 0|\theta_i) \cdot L(a_j, \theta_i) + \Pr(R = 1|\theta_i) \cdot L(a_k, \theta_i).$$

The data for these computations are all available in the tables above. Table 4.7 lists the results, and Figure 4.10 plots the risks for the 16 rules. The graph is disappointing, compared with that for the sample sum. Only one rule, d_8, appears to the southwest of the no-data triangle, and it is only slightly displaced from the triangle. The expanding wedge finds the pure minimax rule to be d_8, with a maximum risk of 3.56. For the mixed minimax rule, the expanding wedge encounters the convex hull on the edge between d_{10} and d_8. Solving for the intersection gives the mixed rule $0.293 d_{10} + 0.707 d_8$ with a maximum expected risk of 2.811. Finally, the pure Bayes rule, assuming $\Pr(\theta_1) = 0.2, \Pr(\theta_2) = 0.8$, is d_0 with an expected risk of $0.2(9.00) + 0.8(0.00) = 1.8$. The Bayes rule is no better than its no-data counterpart, and the pure and minimax rules are only marginally better than their no-data partners. The reason for this poor performance is that the distribution of the sample range under θ_1 is nearly the same as under θ_2. In this case, the sample range has little discriminating power.

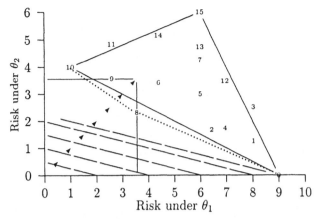

Figure 4.10. Expected losses from various decision rules over the sample range (see Example 4.24)

Keeping the full sample, we have four possible outcomes: $(X_1, X_2) = (0,0), (0,1), (1,0), (1,1)$. A decision rule must prescribe an action for each outcome, which gives $4^4 = 256$ possible decision rules. We describe each such rule as a quadruple (i, j, k, l). The rule advocates action a_i if $(X_1, X_2) = (0,0)$, a_j if $(0,1)$, and so forth. The distribution of the full sample is available from Example 4.12 and is repeated to the right above.

$\Pr((X_1, X_2) = (x_1, x_2))$		
(x_1, x_2)	θ_1 : Good	θ_2 : Poor
$(0,0)$	0.04	0.49
$(0,1)$	0.16	0.21
$(1,0)$	0.16	0.21
$(1,1)$	0.64	0.09

From our analysis of other statistics, we see the contention for the minimax and Bayes rules is among candidates on the southwest frontier of the convex hull. With some computer assistance, we calculate the risks for the 256 possible rules and thereby ascertain the minimax solution to be $(1,3,3,3)$ with a maximum risk of 2.04. We continue the program to delete all rules that are dominated. The southwest frontier of the convex hull arises from the undominated rules. Table 4.8 lists the undominated rules, and the pure minimax rule appears among them. Figure 4.11 graphs the 256 rules, identifying by number only those in Table 4.8. The other rules appear as anonymous points. We can see that the relevant portion of the convex hull extends through rules d_8, d_6, d_2, and d_0. Point d_4 is very close to the critical southwest frontier, but it is actually just inside the dotted line. We determine the mixed minimax rule by noting where the expanding wedge intersects the line between d_6 and d_2. This is at the point $(1.75, 1.75)$, which is 17% removed from d_6. The mixed minimax rule is then $0.83d_6 + 0.17d_2 = 0.83(1,3,3,3) + 0.17(1,1,1,3)$, with a maximum expected risk of 1.75. Finally, the Bayes rule is d_2, which marks where the family of parallel lines first encounters the convex hull. The expected risk is $0.2(3.88) + 0.8(0.36) = 1.064$.

Table 4.6 collects the results for the four statistics (sample sum, mean, range, and whole sample) and also for the no-data decision rules. We see that the average cost per decision is lowered by including the sample information in the decision process. Some statistics, however, fare better than others. The sample sum and sample mean give results that are the same as those that use the entire sample. Evidently, reducing the sample to a single number via the sum or mean computation

Rule		Expected Loss		
		θ_1	θ_2	max
d_0	$(1,1,1,1)$	9.00	0.00	9.00
d_1	$(1,1,1,2)$	7.08	0.27	7.08
d_2	$(1,1,1,3)$	3.88	0.36	3.88
d_3	$(1,2,1,3)$	3.40	0.99	3.40
d_4	$(1,1,3,3)$	2.60	1.20	2.60

Rule		Expected Loss		
		θ_1	θ_2	max
d_5	$(1,2,3,3)$	2.12	1.83	2.12
d_6	$(1,3,3,3)$	1.32	2.04	2.04
d_7	$(2,3,3,3)$	1.20	3.51	3.51
d_8	$(3,3,3,3)$	1.00	4.00	4.00

TABLE 4.8. Undominated decision rules over the whole sample for the unreliable network (see Example 4.24)

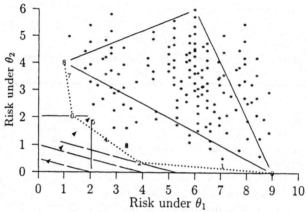

Figure 4.11. Expected losses from various decision rules over the whole sample (see Example 4.24)

does not degrade the quality of the decision rules. Indeed, if you examine closely the rules for the entire sample, you find that they advocate the same actions as the corresponding rules for the sample sum. For example, under the sample sum, the pure minimax rule is $(1,3,3)$, while under the entire sample it is $(1,3,3,3)$. The two middle values specify the action when $(X_1, X_2) = (0,1)$ or $(1,0)$, which both correspond to $S = 1$. The commanded action, a_3 in either case, is the same as that advocated under the sample sum. Similar observations hold for the mixed minimax and Bayes rules. By contrast, reducing the sample to the sample range does degrade the performance of the rules. This poor performance is due to the fact that the distribution for the sample range under θ_1 differs very little from its distribution under θ_2. One of the exercises asks you to show that a statistic with identical distributions under θ_1 and θ_2 produces decision rules whose risks all lie inside the no-data convex hull. Such a statistic cannot improve the average cost per decision over that obtained with the no-data rules. □

As illustrated in the example, minimax and Bayes decision procedures have an advantage when they use information from a sample. In general, the statistical versions reduce the expected cost per decision from that associated with the no-data versions. The extent of the advantage depends on the ability of the statistic to separate the states of nature. That is, if the statistic has markedly different distributions for the different states of nature, then the statistical rules can exploit this difference. The upcoming mathematical interlude develops the notion of sufficient statistics, which categorizes those reductions of the sample that retain the ability to distinguish among the states of nature. Before engaging that topic, however, we consider some further analysis of the statistical Bayes procedure.

As noted earlier, the Bayes procedure operates under different assumptions than the minimax procedures. Under the minimax rule, we focus our attention on the worst state nature can present. When we calculate the expected risk, we assume that nature presents this worst state consistently. That is, we endow nature with a malicious personality that knows our minimax strategy. Recall that a pure minimax strategy is just a special case within the mixed strategies. Nature knows how we plan to choose randomly among the available decision rules and then contrives to present the state that will make this strategy most expensive. Nature maintains this fixed state while we undertake repeated decisions. Under these circumstances, our average loss per decision will be the expected risk of the minimax rule.

By contrast, the Bayes procedure operates under the assumption that the state of nature is a random variable, which can change with each decision, as long as it conforms to the known distribution. With this approach, we do not endow nature with a malicious personality. Nature does not know our decision rule and does not attempt to increase our expense. The expected loss per decision associated with the Bayes rule is the average we can expect over repeated decisions, provided that nature presents random states according to the known distribution.

Although we may frequently feel that nature intends to exact the largest cost for our rewards, the more likely circumstance is that nature is indifferent to our concerns. Therefore, the Bayes approach is more rational given that we have a credible distribution for the states of nature. Accordingly, we emphasize the Bayes procedure for the remainder of the chapter. The Bayes selection procedure involves two steps.

First, we calculate the risks for each decision rule over the chosen statistic. Actually, for each decision rule we calculate a *set* of risks, one for each possible state of nature. This risk set locates the decision rule in an n-dimensional space with coordinate axes measuring the risks for the various states of nature. In the example above, there were only two possible states of nature, and the risk calculations deployed the decision rules in the plane.

Second, we establish a family of parallel hyperplanes, corresponding to the known prior distribution across the states of nature, and determine where these first strike the convex hull of the risks. Example 4.24 illustrates how quickly the number of decision rules can mount. If the range of the statistic has r values and there are m actions, then there are m^r decision rules. If there are n possible states of nature, we must then calculate nm^r risk coordinates. Is there a way we can avoid calculating the risks for all these decision rules?

Consider this alternative approach. Suppose that T is the statistic chosen over the sample $\mathbf{X} = (X_1, X_2, \dots, X_N)$. We use the sample data, $\mathbf{X} = \mathbf{x}$, to update the prior distribution across the states of nature to a *posterior distribution*. That is, we move from $\Pr_\theta(\cdot)$ to $\Pr_{\theta|T=t}(\cdot)$, where $t = T(\mathbf{x})$ is the T-value actually observed. We now revert to the no-data Bayes rule, but we use the posterior distribution across the states of nature. That is, choose the action that minimizes the expected loss with respect to the posterior distribution.

This approach still gives a decision rule, because it provides an unambiguous path from the statistical value to the action. That is, we observe a T-value, and it fixes the posterior distribution, which we can call the modified prior distribution across the states of nature. We then apply the no-data Bayes rule, using this modified prior distribution, to obtain an action. The process is somewhat indirect, but it is nevertheless a decision rule because it is a unambiguous mapping from range(T) to the space of actions.

It turns out that this decision rule is identical with the statistical Bayes rule as defined previously. Moreover, it has the advantage that we need not calculate the risks for the multitude of potential decision rules. To verify this claim, we start with definitions of the new concepts. As before, the states of nature are $\theta_1, \theta_2, \dots, \theta_n$ and the chosen statistic is T. The available actions are $a_1, a_2, \dots, a_m$, and the loss function is $L(a_j, \theta_i)$. The prior distribution across $\theta_1, \theta_2, \dots, \theta_n$ is $q_1, q_2, \dots, q_n$.

4.25 Definition: A *posterior distribution* $\Pr_{\theta|T=t}(\cdot)$ across the states of nature is a new distribution calculated from a given prior distribution $\Pr_\theta(\cdot)$ and the conditional distributions of the statistic T, $\Pr_{T|\theta=\theta_i}(\cdot)$. The computation is essentially an application of Bayes' theorem (Theorem 1.58):

$$\Pr_{\theta|T=t}(\theta_i) = \frac{\Pr((\theta = \theta_i) \cap (T = t))}{\Pr(T = t)} = \frac{\Pr_{T|\theta=\theta_i}(t) \cdot \Pr_\theta(\theta_i)}{\Pr_T(t)}$$

$$= \frac{\Pr_{T|\theta=\theta_i}(t) \cdot \Pr_\theta(\theta_i)}{\sum_{k=1}^n \Pr_{T|\theta=\theta_k}(t) \cdot \Pr_\theta(\theta_k)}. \quad\blacksquare$$

We use the notation $q_1^{(t)}, \dots, q_n^{(t)}$ for the posterior distribution across $\theta_1, \theta_2, \dots, \theta_n$ conditioned on $T = t$. The definition above then entails

$$q_i^{(t)} \cdot \Pr_T(t) = q_i \cdot \Pr_{T|\theta=\theta_i}(t). \tag{4.4}$$

4.26 Definition: The *posterior Bayes decision rule* $d_*(\cdot)$ is given by $d_*(t) = a_j$, where

$$E_{\theta;q^{(t)}}[L(a_j, \cdot)] = \min_{1 \le k \le m} E_{\theta;q^{(t)}}[L(a_k, \cdot)]. \quad\blacksquare$$

We reason as follows to show that the posterior Bayes decision rule is identical with that obtained from the normal Bayes process using the prior distribution across the states of nature. For an arbitrary decision rule d, we can use Equation 4.4 to rearrange the expected risk r under the prior distribution across θ. Let $t_1, t_2, \ldots$ be the range of the statistic T.

$$R = E_{\theta;q}[\hat{L}(d, \cdot)] = \sum_{i=1}^{n} q_i \hat{L}(d, \theta_i) = \sum_{i=1}^{n} \sum_{k} q_i \Pr_{T|\theta=\theta_i}(t_k) \cdot L(d(t_k), \theta_i)$$

$$= \sum_{i=1}^{n} \sum_{k} q_i^{(t_k)} \Pr_T(t_k) L(d(t_k), \theta_i) = \sum_{k} \Pr_T(t_k) E_{\theta;q^{(t_k)}}[L(d(t_k), \cdot)]$$

We obtain the prior Bayes rule by choosing the $d(\cdot)$ that minimizes the left side. Equivalently, we can choose the $d(\cdot)$ that minimizes the final expression on the right. That expression is the sum of nonnegative entries, so if we can adjust each term separately, the minimum will occur for the $d(\cdot)$ that brings all summands to their smallest values simultaneously. Each summand involves one point $t_k \in \text{range}(T)$. We can make it smallest by choosing $d(t_k)$ to be the action that minimizes $E_{\theta;q^{(t_k)}}[L(d(t_k), \cdot)]$. Since we define $d(\cdot)$ by assigning a value at each t_k independently, these operations do not conflict; we can choose the minimizing action at each t_k without regard to the other range points. However, Definition 4.26 prescribes just this operation to specify the posterior Bayes rule. Hence the $d(\cdot)$ chosen by the prior Bayes process is the same rule as that chosen by the posterior Bayes process.

4.27 Theorem: The prior and posterior Bayes procedures select the same decision rule.

PROOF: See discussion above. ∎

To illustrate the computational advantage of the posterior Bayes procedure, we revisit Example 4.24 for the sample sum case. You will recall that we computed the risks under two states of nature for 64 decision rules.

4.28 Example: Continuing the unreliable network example, the following tables restate the relevant information for a two-point sample over the population X, from which we have extracted the sample sum statistic. As in the previous examples, we assume a prior distribution $\Pr_\theta(\theta_1) = 0.2$, $\Pr_\theta(\theta_2) = 0.8$ across the states of nature.

Cost as a function of action and network condition		
	Network condition	
Action	θ_1: Good	θ_2: Poor
a_1: Deluxe service	9	0
a_2: Intermediate service	6	3
a_3: Economy service	1	4
a_4: Abandon message	6	6

$\Pr(X = x)$		
x	θ_1	θ_2
1	0.8	0.3
0	0.2	0.7

$\Pr(S = s)$		
s	θ_1	θ_2
0	0.04	0.49
1	0.32	0.42
2	0.64	0.09

We wish to calculate the posterior Bayes decision rule and verify that it accords with the prior Bayes rule. Example 4.24 calculated the latter to be rule $(1, 1, 3)$. We need to calculate the posterior probabilities $\Pr_{\theta|S=t}(\theta_i)$, for $t = 0, 1, 2$ and $i = 1, 2$. Using Definition 4.25, we illustrate the calculation for $a = \Pr_{\theta|S=0}(\theta_1)$.

$$a = \frac{\Pr_{S|\theta=\theta_1}(0) \cdot \Pr_\theta(\theta_1)}{\Pr_{S|\theta=\theta_1}(0) \cdot \Pr_\theta(\theta_1) + \Pr_{S|\theta=\theta_2}(0) \cdot \Pr_\theta(\theta_2)} = \frac{(0.04)(0.2)}{(0.04)(0.2) + (0.49)(0.8)} = 0.02$$

This establishes the top entry in the second column of the following table, and similar calculations determine the rest of first three columns.

			Posterior expected loss under specified action				
S	$\Pr_{\theta\mid S}(\theta_1)$	$\Pr_{\theta\mid S}(\theta_2)$	a_1	a_2	a_3	a_4	Minimizing action
0	0.02	0.98	0.18	3.06	3.94	6.00	a_1
1	0.16	0.84	1.44	3.48	3.52	6.00	a_1
2	0.64	0.36	5.76	4.92	2.08	6.00	a_3

We can now compute the posterior expected losses for the actions. If $S = 0$, for example, the calculation for a_1 is

$$E_{\theta;q^{(t)}}[L(a_1, \cdot)] \; = \; 9(0.02) + 0(0.98) = 0.18.$$

Similar calculations establish the remaining entries in the table. The last column chooses the action that produces the smallest expected loss given that the sample sum is a particular observed value. The chosen actions constitute rule $(1, 1, 3)$, which matches the prior Bayes rule. $\square$

Exercises

*4.11 Let $\mathbf{X} = (X_1, X_2, \ldots, X_N)$ be a sample from the population X. Verify the following entries from Table 4.4.

$$E[(X_i - \mu_X)^2(X_j - \mu_X)(\overline{X} - \mu_X)] = \begin{cases} \dfrac{\delta_4}{N}, & i = j \\[2mm] \dfrac{\delta_2^2}{N} & i \neq j \end{cases}$$

$$E[(X_i - \mu_X)(X_j - \mu_X)(\overline{X} - \mu_X)^2] = \begin{cases} \dfrac{\delta_4 + (N-1)\delta_2^2}{N^2}, & i = j \\[2mm] \dfrac{2\delta_2^2}{N^2}, & i \neq j \end{cases}$$

$$E[(X_i - \mu_X)(\overline{X} - \mu_X)^3] = \dfrac{3(N-1)\delta_2^2 + \delta_4}{N^3}$$

$$E[(\overline{X} - \mu_X)^4] = \dfrac{3(N-1)\delta_2^2 + \delta_4}{N^3}$$

4.12 Show that the sample minimum and the sample maximum are not independent random variables.

4.13 Example 4.17 calculates the means and variances for the sample minimum, maximum, median, and range for a sample of size three over a population X with distribution as shown to

x	1	4	6	9
$\Pr_X(x)$	0.4	0.1	0.3	0.2
$F_X(x)$	0.4	0.5	0.8	1.0

the right. The example uses Theorem 4.15 to obtain the results. Verify the answers by calculating the probabilities for each possible sample

Loss Table		
	State of nature	
Action	θ_1	θ_2
a_1	8	2
a_2	7	10
a_3	3	6

Loss Table		
	State of nature	
Action	θ_1	θ_2
a_1	3	1
a_2	6	4
a_3	8	3

Loss Table			
		State of nature	
Action	θ_1	θ_2	θ_3
a_1	9	3	1
a_2	7	6	4
a_3	3	5	8
a_4	1	4	9

Distribution Table		
	State of nature	
X	θ_1	θ_2
0	0.8	0.1
1	0.1	0.2
2	0.1	0.7

Distribution Table		
	State of nature	
X	θ_1	θ_2
0	0.8	0.1
1	0.1	0.2
2	0.1	0.7

Distribution Table			
		State of nature	
X	θ_1	θ_2	θ_3
0	0.8	0.1	0.1
1	0.1	0.8	0.1
2	0.1	0.1	0.8

| (a) | (b) | (c) |

Figure 4.12. Loss functions and population distributions for the exercises

directly. There are $4^3 = 64$ such samples. Combine these probabilities with the corresponding statistical values for the minimum, maximum, median, and range to compute their means and variances.

4.14 Suppose that decision rule $d_j(\cdot)$ dominates rule $d_k(\cdot)$. Prove that the search for the statistical pure minimax decision rule need not consider $d_k(\cdot)$ in the contention. Show that this exclusion also applies to the search for the statistical mixed minimax decision rule.

4.15 The loss functions and population distributions of Figure 4.12(a) involve three actions under two possible states of nature. A sample of size three (X_1, X_2, X_3) is taken over the population X. Using the sample sum as a statistic, find the pure minimax, mixed minimax, and Bayes decision rules, together with their expected cost per decision. For the Bayes rule, use $\Pr(\theta_1) = 0.3, \Pr(\theta_2) = 0.7$ as the prior distribution across the states of nature.

4.16 Repeat the preceding exercise using the sample range.

4.17 For the situation in the preceding two exercises, verify that the posterior Bayes rule is the same as the prior Bayes rule.

4.18 Suppose that the losses associated with three actions under two possible states of nature are as shown in Figure 4.12(b). Obviously, action a_1 dominates the other two, so the optimal decision rule should advocate a_1, regardless of the sample data. Call this rule d_*. Consider a sample of size three from the random variable X, whose population distribution appears in the figure, and show that d_* dominates all other decision rules over the sample sum statistic. This shows that the mechanism for choosing the minimax or Bayes rule does not contradict common sense. If the least expensive option is obvious, the mechanism will choose it.

4.19 Suppose that the losses associated with four actions under three possible states of nature are as shown in Figure 4.12(c). A sample of size three (X_1, X_2, X_3) is taken over the population X. Using the sample sum as a statistic, find the pure minimax, mixed minimax, and Bayes decision rules, together with their expected cost per decision. For the Bayes rule, use $\Pr(\theta_1) = 0.3, \Pr(\theta_2) = 0.4, \Pr(\theta_3) = 0.3$ as the prior distribution across the states of nature.

4.20 Repeat the preceding exercise with the sample range.

4.21 For the situation in the preceding two exercises, verify that the posterior Bayes rule is the same as the prior Bayes rule.

4.22 Suppose that actions $a_1, \ldots, a_m$ are available, with losses given by $L(a_j, \theta_i)$ under states of nature θ_1 and θ_2. A sample $\mathbf{X} = (X_1, \ldots, X_N)$ gives an indirect reading of the state of nature, in the sense that the distribution of the sample varies from θ_1 to θ_2. A statistic $S = f(\mathbf{X})$, however, has a distribution under θ_1 that is identical to that under θ_2. For any decision rule over this statistic, show that the risk lies within the convex hull of the points $(L(a_j, \theta_1), L(a_j, \theta_2)), j = 1, 2, \ldots, m$.

4.3 Sufficient statistics

This mathematical interlude develops the notion of a sufficient statistic, which intuitively is a statistic that is as good as the entire sample for purposes of taking decisions.

As in the preceding section, we denote by $\mathbf{X} = (X_1, X_2, \ldots, X_N)$ a sample from a population X. Let C be the collection of all potential sample points: $C = \{(x_1, x_2, \ldots, x_N) | x_i \in \text{range}(X), 1 \le i \le N\}$. A statistic T is a function of the sample. Suppose that $T = f(\mathbf{X})$. The values assumed by T need not be simple scalars. The order statistic, for example, is the vector obtained by sorting the sample values in increasing order. In any case, let $t_1, t_2, \ldots$ denote the possible values for T. The sets $C_k = f^{-1}(t_k)$ constitute a partition of the space of all sample points. That is, $C_j \cap C_k = \phi$ when $j \ne k$, and $\cup_k C_k = C$.

When the event $(\mathbf{X} = \mathbf{x})$ occurs, so does the event $(T = f(\mathbf{x}))$. This implies $(\mathbf{X} = \mathbf{x}) \subset (T = f(\mathbf{x}))$, and therefore for any state of nature θ_i, we have

$$(\mathbf{X} = \mathbf{x}) = (\mathbf{X} = \mathbf{x}) \cap (T = f(\mathbf{x}))$$
$$(\mathbf{X} = \mathbf{x}) \cap (\theta = \theta_i) = (\mathbf{X} = \mathbf{x}) \cap (T = f(\mathbf{x})) \cap (\theta = \theta_i).$$

We then obtain the following decomposition.

$$\Pr_{\mathbf{X}|\theta_i}(\mathbf{x}) = \Pr(\mathbf{X} = \mathbf{x}|\theta = \theta_i) = \frac{\Pr((\mathbf{X} = \mathbf{x}) \cap (\theta = \theta_i))}{\Pr(\theta = \theta_i)}$$

$$\Pr_{\mathbf{X}|\theta_i}(\mathbf{x}) = \frac{\Pr((\mathbf{X} = \mathbf{x}) \cap (T = f(\mathbf{x})) \cap (\theta = \theta_i))}{\Pr(\theta = \theta_i)}$$

$$= \frac{\Pr((\mathbf{X} = \mathbf{x}) \cap (T = f(\mathbf{x})) \cap (\theta = \theta_i))}{\Pr(\theta = \theta_i)} \cdot \frac{\Pr((T = f(\mathbf{x})) \cap (\theta = \theta_i))}{\Pr((T = f(\mathbf{x})) \cap (\theta = \theta_i))}$$

$$= \Pr(\mathbf{X} = \mathbf{x}|(T = f(\mathbf{x})) \cap (\theta = \theta_i)) \cdot \Pr(T = f(\mathbf{x})|\theta = \theta_i) \qquad (4.5)$$

This calculation shows that the probability of a sample point $\mathbf{x}$, given a specific state of nature θ_i, factors into two components. The first is the probability of $\mathbf{x}$, given θ_i and the corresponding statistical value $f(\mathbf{x})$. The second is the probability of the sample's statistical value $f(\mathbf{x})$ given θ_i. If the first component is independent of θ_i, some interesting consequences entail when formulating decision rules based on the statistic T.

Recall that we can obtain the statistical Bayes decision rule by selecting, for each value of the chosen statistic, the action that minimizes the posterior loss. (See Theorem 4.27 and its preceding discussion.) Consider the entire sample as one statistic, and let T be a second statistic. The expressions to minimize, by choosing the appropriate action a_k, are

$$\sum_{i=1}^{n} \Pr(\theta = \theta_i|\mathbf{X} = \mathbf{x}) \cdot L(a_k, \theta_i)$$

$$\sum_{i=1}^{n} \Pr(\theta = \theta_i|T = f(\mathbf{x})) \cdot L(a_k, \theta_i).$$

Now if it happens that $\Pr(\theta = \theta_i|\mathbf{X} = \mathbf{x}) = \Pr(\theta = \theta_i|T = f(\mathbf{x}))$ for all $1 \le i \le n$, then the minimizing action for the first expression is the same as for the second. In other words, the Bayes decision rule based on the entire sample advocates the same action as the Bayes decision rule based on statistic T. All sample points $\mathbf{x}$ with the same $f(\mathbf{x})$ value command the same minimizing action as does that $f(\mathbf{x})$. We encountered this situation in Example 4.24, in which the actions commanded by the sample sum decision rule were the same as those commanded by the decision rule over the entire sample. All sample points corresponding to a particular sample sum commanded the same action as the given sample sum. We will show that this is no accident—if the first component of the factorization of the sample distribution above is indeed independent of θ_i. This criterion will then form the basis for the definition of a sufficient statistic, which carries the intuitive meaning that it is as good as the full sample for use in decision rules.

Assume that $\Pr(\mathbf{X} = \mathbf{x}|(T = f(\mathbf{x})) \cap (\theta = \theta_i))$ is independent of θ_i. Then,

$$\Pr(\mathbf{X} = \mathbf{x}|T = f(\mathbf{x})) = \frac{\Pr((\mathbf{X} = \mathbf{x}) \cap (T = f(\mathbf{x})))}{\Pr(T = f(\mathbf{x}))}$$

$$= \frac{\sum_{i=1}^{n} \Pr((\mathbf{X} = \mathbf{x}) \cap (T = f(\mathbf{x})) \cap (\theta = \theta_i))}{\Pr(T = f(\mathbf{x}))} = \sum_{i=1}^{n} q_i.$$

Each q_i admits the further reduction. Specifically, for an arbitrarily chosen θ_j, we can replace θ_i with θ_j in the expression $\Pr(\mathbf{X} = \mathbf{x} | (T = f(\mathbf{x})) \cap (\theta = \theta_i))$. This gives

$$
\begin{aligned}
q_i &= \frac{\Pr(\mathbf{X} = \mathbf{x} | (T = f(\mathbf{x})) \cap (\theta = \theta_i)) \cdot \Pr((T = f(\mathbf{x})) \cap (\theta = \theta_i))}{\Pr(T = f(\mathbf{x}))} \\
&= \frac{\Pr(\mathbf{X} = \mathbf{x} | (T = f(\mathbf{x})) \cap (\theta = \theta_j))}{\Pr(T = f(\mathbf{x}))} \cdot \Pr((T = f(\mathbf{x})) \cap (\theta = \theta_i)).
\end{aligned}
$$

Consequently,

$$
\begin{aligned}
\sum_{i=1}^{n} q_i &= \Pr(\mathbf{X} = \mathbf{x} | T = f(\mathbf{x})) \\
&= \frac{\Pr(\mathbf{X} = \mathbf{x} | (T = f(\mathbf{x})) \cap (\theta = \theta_j))}{\Pr(T = f(\mathbf{x}))} \sum_{i=1}^{n} \Pr((T = f(\mathbf{x})) \cap (\theta = \theta_i)) \\
&= \Pr(\mathbf{X} = \mathbf{x} | (T = f(\mathbf{x})) \cap (\theta = \theta_j)) \cdot \frac{\Pr(T = f(\mathbf{x}))}{\Pr(T = f(\mathbf{x}))} \\
&= \Pr(\mathbf{X} = \mathbf{x} | (T = f(\mathbf{x})) \cap (\theta = \theta_j)).
\end{aligned}
$$

We can therefore drop the θ_i restriction from the first factor of Equation 4.5 and write the probability of a sample point $\mathbf{x}$ given state of nature θ_i as follows.

$$
\Pr(\mathbf{X} = \mathbf{x} | \theta = \theta_i) = \Pr(\mathbf{X} = \mathbf{x} | T = f(\mathbf{x})) \cdot \Pr(T = f(\mathbf{x}) | \theta = \theta_i)
$$

The equivalence of the posterior probabilities for the θ_i, given either the sample point $\mathbf{x}$ or the statistical value $f(\mathbf{x})$, is a direct consequence of this factorization. Using Bayes' theorem, at the beginning and again at the end of the following derivation, proves the point.

$$
\begin{aligned}
\Pr(\theta = \theta_i | T = f(\mathbf{x})) &= \frac{\Pr(T = f(\mathbf{x}) | \theta = \theta_i) \cdot \Pr(\theta = \theta_i)}{\Pr(T = f(\mathbf{x}))} \\
&= \frac{\Pr(\mathbf{X} = \mathbf{x} | \theta = \theta_i) \cdot \Pr(\theta = \theta_i)}{\Pr(T = f(\mathbf{x})) \cdot \Pr(\mathbf{X} = \mathbf{x} | T = f(\mathbf{x}))} \\
&= \frac{\Pr(\mathbf{X} = \mathbf{x} | \theta = \theta_i) \cdot \Pr(\theta = \theta_i)}{\Pr((\mathbf{X} = \mathbf{x}) \cap (T = f(\mathbf{x})))} \\
&= \frac{\Pr(\mathbf{X} = \mathbf{x} | \theta = \theta_i) \cdot \Pr(\theta = \theta_i)}{\Pr(\mathbf{X} = \mathbf{x})} = \Pr(\theta = \theta_i | \mathbf{X} = \mathbf{x})
\end{aligned}
$$

We conclude that the Bayes rule over the statistic T is that same as that over the entire sample, *if* the component $\Pr(\mathbf{X} = \mathbf{x} | (T = f(\mathbf{x})) \cap (\theta = \theta_i))$ is independent of the state of nature θ_i.

4.29 Definition: Let $\Pr(\mathbf{X} = \mathbf{x} | \theta = \theta_i)$ be a family of distributions for the sample $\mathbf{X}$. If $\Pr(\mathbf{X} = \mathbf{x} | (T = f(\mathbf{x})) \cap (\theta = \theta_i))$ is independent of θ_i, then we say that statistic $T = f(\mathbf{X})$ is *sufficient* for the family. That is, sufficiency requires, for all i, j,

$$\Pr(\mathbf{X} = \mathbf{x} | (T = f(\mathbf{x})) \cap (\theta = \theta_i)) = \Pr(\mathbf{X} = \mathbf{x} | (T = f(\mathbf{x})) \cap (\theta = \theta_j))$$
$$= \Pr(\mathbf{X} = \mathbf{x} | T = f(\mathbf{x})). \ \blacksquare$$

From our previous discussion, we conclude that $T = f(\mathbf{X})$ sufficient for the family $\Pr(\mathbf{X} = \mathbf{x} | \theta = \theta_i)$ implies that

$$\Pr(\mathbf{X} = \mathbf{x} | \theta = \theta_i) = \Pr(\mathbf{X} = \mathbf{x} | T = f(\mathbf{x})) \cdot \Pr(T = f(\mathbf{x}) | \theta = \theta_i).$$

4.30 Theorem: Let $T = f(\mathbf{X})$ be a sufficient statistic for the family of sample distributions $\Pr(\mathbf{X} = \mathbf{x} | \theta = \theta_i)$. Denote the statistical Bayes decision rule using the entire sample by $d(\mathbf{x})$; denote by $d_T(f(\mathbf{x}))$ the statistical Bayes decision rule using statistic T. Then the rules command identical actions. That is, $d(\mathbf{y}) = d_T(f(\mathbf{x}))$, for all $\mathbf{y} \in f^{-1}(f(\mathbf{x}))$.

PROOF: See discussion above. $\blacksquare$

We now revisit Example 4.24 to show how the sample sum is a sufficient statistic and thereby explain why its Bayes decision rule is the same as that for the entire sample.

4.31 Example: In the context of the two-point sample from the unreliable network, verify that the sample sum is a sufficient statistic and that the sample range is not.

In Example 4.24, we calculated the distribution of the sample, given each of the two possible states of nature. In the vector notation of this section, $\mathbf{X} = (X_1, X_2)$, and these distributions appear to the upper right. We also know the priori probabilities for the θ_i: $\Pr(\theta_1) = 0.2, \Pr(\theta_2) = 0.8$. The sample sum $S = X_1 + X_2$ induces a three-component partition on the space of sample values. $S = 0$ corresponds to the single point $(0, 0)$, and $S = 2$ corresponds to the single point $(1, 1)$. $S = 1$ encompasses the two points $(0, 1)$ and $(1, 0)$. We can compute the conditional probabilities for $S = s$, given a particular θ_i, by aggregating the probabilities masses of the corresponding partition components. For example, $\Pr(S = 1 | \theta = \theta_1) = 0.16 + 0.16 = 0.32$. The table to the lower right summarizes these computations.

$\Pr(\mathbf{X} = \mathbf{x} | \theta = \theta_i)$		
$\mathbf{x}$	θ_1	θ_2
$(0,0)$	0.04	0.49
$(0,1)$	0.16	0.21
$(1,0)$	0.16	0.21
$(1,1)$	0.64	0.09

$\Pr(S = s | \theta = \theta_i$		
s	θ_1	θ_2
0	0.04	0.49
1	0.32	0.42
2	0.64	0.09

The event $(S = x_1 + x_2)$ contains the event $(\mathbf{X} = (x_1, x_2))$. Therefore, we have

$$q_i = \Pr(\mathbf{X} = (x_1, x_2) | (S = x_1 + x_2) \cap (\theta = \theta_i))$$
$$= \frac{\Pr((\mathbf{X} = (x_1, x_2)) \cap (\theta = \theta_i))}{\Pr((S = x_1 + x_2) \cap (\theta = \theta_i))}.$$

Dividing numerator and denominator by $\Pr(\theta = \theta_i)$, we obtain a solution in terms of conditional probabilities.

$$q_i = \frac{\Pr((\mathbf{X} = (x_1, x_2)) \cap (\theta = \theta_i))/\Pr(\theta = \theta_i)}{\Pr((S = x_1 + x_2) \cap (\theta = \theta_i))/\Pr(\theta = \theta_i)} = \frac{\Pr(\mathbf{X} = (x_1, x_2) | \theta = \theta_i)}{\Pr(S = x_1 + x_2 | \theta = \theta_i)}$$

The two tables above, therefore, enable the computations of the q_i, which in turn determine the sufficiency of S. The results are as follows.

$\mathbf{x} = (x_1, x_2)$	$(0,0)$	$(0,1)$	$(1,0)$	$(1,1)$
$q_1 = \Pr(\mathbf{X} = (x_1, x_2)\|(S = x_1 + x_2) \cap (\theta = \theta_1))$	1.00	0.50	0.50	1.00
$q_2 = \Pr(\mathbf{X} = (x_1, x_2)\|(S = x_1 + x_2) \cap (\theta = \theta_2))$	1.00	0.50	0.50	1.00

Because the two rows are identical, we conclude that $\Pr(\mathbf{X} = (x_1, x_2)|S = x_1 + x_2)$ is independent of θ_i. The sample sum is a sufficient statistic. Under the prior Bayes decision rule based on the entire sample, those points having the same sample sum all command the same action. Moreover, this action is the same as that commanded by the prior Bayes rule based on the sample sum. We lose nothing pertinent to the decision problem by reducing the sample to the sample sum.

Example 4.24 showed that the sample range does not perform as well as the entire sample, in the sense that the Bayes rule over the sample range has a higher average cost per decision. We expect, therefore, that the sample range is not a sufficient statistic. Following the pattern established with the sample sum, we compute the conditional distributions of the sample range as shown to the right. We then compute

$\Pr(R = r\|\theta = \theta_i)$		
r	θ_1	θ_2
0	0.68	0.58
1	0.32	0.42

$$\Pr(\mathbf{X} = (x_1, x_2)|(R = |x_1 - x_2|) \cap (\theta = \theta_i)) = \frac{\Pr(\mathbf{X} = (x_1, x_2)|\theta = \theta_i)}{\Pr(R = |x_1 - x_2| \,|\theta = \theta_i)}$$

to obtain the following unequal probability displays.

$\mathbf{x} = (x_1, x_2)$	$(0,0)$	$(0,1)$	$(1,0)$	$(1,1)$
$\Pr(\mathbf{X} = (x_1, x_2)\|(R = |x_1 - x_2|) \cap (\theta = \theta_1))$	0.0588	0.5000	0.5000	0.9412
$\Pr(\mathbf{X} = (x_1, x_2)\|(R = |x_1 - x_2|) \cap (\theta = \theta_2))$	0.8448	0.5000	0.5000	0.1552

Consequently, the sample range is not a sufficient statistic. $\square$

Recall that the statistic $T = f(\mathbf{X})$ induces a partition in the space of sample points. For a given sample point $\mathbf{x}$, the set $f^{-1}(f(\mathbf{x}))$ is the partition component containing $\mathbf{x}$. Any decision rule based on the statistic T expands naturally to a decision rule based on the entire sample. We simply let the expanded rule command the same action for all sample points in a given component. If $d(\cdot)$ is the rule based on T, we denote by $\bar{d}(\cdot)$ the expanded rule that operates on the individual sample points. The formal relationship is

$$\bar{d}(\mathbf{x}) = d(f(\mathbf{x})). \tag{4.6}$$

Any full-sample decision rule that is constant on partition components contracts to a rule that uses T. To specify an action for a statistical value $T = y$, the contracted rule simply assigns the common action of the component $f^{-1}(y)$. Clearly, a contraction of a previously expanded rule recovers the original rule. In these terms, Theorem 4.30 states that the Bayes decision rule over a sufficient statistic expands to the Bayes decision rule over the entire sample. Of course, this forces the latter rule to be constant over the partition components induced by the sufficient statistic.

Consider the convex hull of the full-sample decision rules. In selecting the mixed minimax rule, the expanding wedge first encounters a point in the southwest face of the convex hull. That point is a convex combination of vertices, each of which is the Bayes decision rule for some prior distribution $q = (q_1, q_2, \ldots, q_n)$. That is, the distribution components determine the orientation of the hyperplane family $q_1 \hat{L}_1 + q_2 \hat{L}_2 + \ldots + q_n \hat{L}_n = K$, which advances toward the convex hull with increasing K. By varying the distribution coefficients, we can arrange contact with the convex hull at any chosen vertex on the southwest side. Each of these vertices, being a Bayes decision rule for the full sample, is therefore an expansion of a Bayes decision rule for the sufficient statistic T. The mixed minimax rule is then a convex combination of rules that are constant over partitions induced by T. In other words, the mixed minimax rule that uses the sufficient statistic T produces the same maximum expected cost over the states of nature as the minimax rule that uses the full sample. Moreover, it commands the same actions, since it is constant on sample points corresponding to the same statistical value.

4.32 Theorem: Let $T = f(\mathbf{X})$ be a sufficient statistic for the family of sample distributions $\Pr(\mathbf{X} = \mathbf{x} | \theta = \theta_i)$. Denote the statistical mixed minimax decision rule using the entire sample by $d(\mathbf{x})$; denote by $d_T(f(\mathbf{x}))$ the corresponding rule that uses statistic T. Then the rules command identical actions. That is, $d(\mathbf{y}) = d_T(f(\mathbf{x}))$, for all $\mathbf{y} \in f^{-1}(f(\mathbf{x}))$.

PROOF: See discussion above. ∎

The last two theorems announce the importance of a sufficient statistic. It allows a reduction of the sample and a corresponding radical decrease in the number of potential decision rules. It does so in such a manner that neither the mixed minimax nor the Bayes procedure suffers any degradation. The optimal rule remains in the reduced set, and the selection mechanisms find it. In these circumstances, it is natural to ask how far these reductions can proceed. For example, the full sample is obviously a sufficient statistic because

$$\Pr(\mathbf{X} = \mathbf{x} | (\mathbf{X} = \mathbf{x}) \cap (\theta = \theta_i)) = \frac{\Pr((\mathbf{X} = \mathbf{x}) \cap (\theta = \theta_i))}{\Pr((\mathbf{X} = \mathbf{x}) \cap (\theta = \theta_i))} = 1.0,$$

which is clearly independent of the state of nature. At the other extreme, a statistic that maps the sample to a constant is most unlikely to be sufficient. If $T(\mathbf{x}) = K$ for all $\mathbf{x}$, then $(T = K)$ is the certain event. It contains all outcomes. Consequently,

$$\Pr(\mathbf{X} = \mathbf{x} | (T = K) \cap (\theta = \theta_i)) = \Pr(\mathbf{X} = \mathbf{x} | \theta = \theta_i),$$

which is independent of θ only if the underlying population X has the same distribution under all states of nature. We can trace sufficiency through functional relationships, however, in the following sense.

Each statistic corresponds to a partition of the sample points, in which a component corresponds to those points that map to the same statistical value. Viewing a sample through a statistic removes the distinction among individual points and retains only the partition component associated with the observed statistical value. Hence less information is available from the statistical value than from the sample itself. This loss of information is smallest when the statistic is the sample itself. Indeed, the loss is zero in this case because viewing the statistic is the same as viewing the sample. Any function that assigns a different value to each potential sample point also suffers no loss of information. Viewing such a statistical value remains the same as viewing the sample because, knowing the mapping, we can recover the observed sample from the computed statistical value. However, a function that collapses several potential sample points into a single value loses information. Knowing only the reported statistical value, we cannot reconstruct the sample. If this loss is not harmful to our decision-making process, then we characterize the function as a sufficient statistic.

The sufficiency property actually attaches to the *partition* and extends to any statistic that induces that partition. Starting with the finest partition, in which each sample point occupies its own separate component, we apply a function that aggregates points with the same functional value. We can apply a second function to the values returned by the first. This process aggregates lumps from the first partition into larger components, thereby forming a coarser partition. We can continue with further functions until all potential sample points reside in a single component, which means that the final function is constant across the values delivered by its predecessor. As noted above, the initial fine partition is sufficient, while the final single-lump partition is not, except in the trivial case where the sample distributions for the various states of nature are all the same. Somewhere in the sequence the sufficiency property is lost. After it is lost, it can never again be recovered. Equivalently, if a new sufficient statistic is a function of an old statistic, then the old statistic must also be sufficient.

Before embarking on a proof of this fact, consider the following visualization of a sufficient statistic T. Suppose that T induces a four-component partition on the sample points, as suggested by Figure 4.13. A pair of vertical bars represents each potential sample point. The height of the left bar is the probability of the sample point under state of nature θ_i; the height of the right bar is the probability under state of nature θ_j. Because T is a sufficient statistic, the bars within a given component all adjust by the same factor when moving from θ_i to θ_j. In component 1, each probability decreases by about 50%. In component 2, each increases by about 10%. This constant adjustment within each component is the defining characteristic of a sufficient statistic. We can derive this result analytically as follows.

Let $\mathbf{X} = (X_1, X_2, \ldots, X_N)$ be a sample over the population X. Let T be a statistic, and consider a point $\mathbf{x}$ for which $T(\mathbf{x}) = t$. The point resides in component $T^{-1}(t)$. Since event $(\mathbf{X} = \mathbf{x}) \subset (T = t)$, we have

$$\mathrm{Pr}_{\mathbf{X}|\theta_i}(\mathbf{x}) = \mathrm{Pr}(\mathbf{X} = \mathbf{x}|\theta = \theta_i) = \frac{\mathrm{Pr}((\mathbf{X} = \mathbf{x}) \cap (\theta = \theta_i))}{\mathrm{Pr}(\theta = \theta_i)}$$

$$= \frac{\mathrm{Pr}((\mathbf{X} = \mathbf{x}) \cap (T = t) \cap (\theta = \theta_i))}{\mathrm{Pr}(\theta = \theta_i)}.$$

Because $\mathrm{Pr}((T = t) \cap (\theta = \theta_i))$ is the sum of contributions from the individual

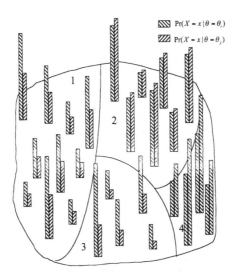

$$\boxed{\begin{array}{ll} \text{\SSS} & \Pr(X=x|\theta=\theta_i) \\ \text{\ZZZ} & \Pr(X=x|\theta=\theta_j) \end{array}}$$

Figure 4.13. Illustrating proportional probability shifts within components of a sufficient statistic

sample points z that deliver $T(z) = t$, we can simplify this expression as follows.

$$\Pr_{\mathbf{X}|\theta_i}(\mathbf{x}) = \frac{\Pr((\mathbf{X}=\mathbf{x})\cap(T=t)\cap(\theta=\theta_i))}{\Pr((T=t)\cap(\theta=\theta_i))} \cdot \frac{\Pr((T=t)\cap(\theta=\theta_i))}{\Pr(\theta=\theta_i)} \quad (4.7)$$

$$= \Pr(\mathbf{X}=\mathbf{x}|(T=t)\cap(\theta=\theta_i)) \cdot \sum_{z\in T^{-1}(t)} \frac{\Pr((\mathbf{X}=z)\cap(\theta=\theta_i))}{\Pr(\theta=\theta_i)}$$

$$= \Pr(\mathbf{X}=\mathbf{x}|(T=t)\cap(\theta=\theta_i)) \cdot \sum_{z\in T^{-1}(t)} \Pr(\mathbf{X}=z|\theta=\theta_i) \quad (4.8)$$

On the right side, the second factor is independent of the individual sample point $\mathbf{x}$; it depends only on the component $T^{-1}(t)$ and the state of nature θ_i. Consequently, denote this factor by $K(t, \theta_i)$. If T is a sufficient statistic, the first factor remains unchanged when the state of nature switches to θ_j. Therefore, T sufficient implies that

$$\frac{\Pr(\mathbf{X}=\mathbf{x}|\theta=\theta_i)}{\Pr(\mathbf{X}=\mathbf{x}|\theta=\theta_j)} = \frac{K(t, \theta_i)}{K(t, \theta_j)}.$$

Because the right side is independent of the sample point $\mathbf{x}$, any two points in the same component exhibit the same proportional probability adjustment when nature switches from θ_i to θ_j. That is, T sufficient implies that

$$\frac{\Pr(\mathbf{X}=\mathbf{x}|\theta=\theta_i)}{\Pr(\mathbf{X}=\mathbf{x}|\theta=\theta_j)} = \frac{K(t, \theta_i)}{K(t, \theta_j)} = \frac{\Pr(\mathbf{X}=\mathbf{y}|\theta=\theta_i)}{\Pr(\mathbf{X}=\mathbf{y}|\theta=\theta_j)},$$

when $\mathbf{x}, \mathbf{y} \in T^{-1}(t)$.

4.33 Theorem: Let $\mathbf{X}$ be a sample from a population X, which has distributions $\Pr(\mathbf{X} = \mathbf{x}|\theta = \theta_i)$ for various states of nature θ_i. T is a sufficient statistic if and only if, for any two states of nature θ_i and θ_j, the quantity

$$C(\mathbf{x}, \theta_i, \theta_j) = \frac{\Pr(\mathbf{X} = \mathbf{x}|\theta = \theta_i)}{\Pr(\mathbf{X} = \mathbf{x}|\theta = \theta_j)}$$

is constant across $\mathbf{x}$ in a particular component of the T-partition.

PROOF: The discussion above shows that T sufficient implies that $C(\mathbf{z}, \theta_i, \theta_j)$ is constant for $\mathbf{z} \in T^{-1}(T(\mathbf{x}))$. For the converse, suppose that $C(\mathbf{z}, \theta_i, \theta_j)$ is constant as $\mathbf{z}$ varies across the component $T^{-1}(T(\mathbf{x}))$. We then have

$$\sum_{\mathbf{z} \in T^{-1}(T(\mathbf{x}))} \Pr(\mathbf{X} = \mathbf{z}|\theta = \theta_i) = C(\mathbf{x}, \theta_i, \theta_j) \cdot \sum_{\mathbf{z} \in T^{-1}(T(\mathbf{x}))} \Pr(\mathbf{X} = \mathbf{z}|\theta = \theta_j).$$

Letting $t = T(\mathbf{x})$ and applying Equation 4.8 then gives the ratio r_{ij}:

$$r_{ij} = \frac{\Pr(\mathbf{X} = \mathbf{x}|(T = t) \cap (\theta = \theta_i))}{\Pr(\mathbf{X} = \mathbf{x}|(T = t) \cap (\theta = \theta_j))}$$

$$= \frac{\Pr(\mathbf{X} = \mathbf{x}|\theta = \theta_i)}{\sum_{\mathbf{z} \in T^{-1}(t)} \Pr(\mathbf{X} = \mathbf{z}|\theta = \theta_i)} \cdot \frac{\sum_{\mathbf{z} \in T^{-1}(t)} \Pr(\mathbf{X} = \mathbf{z}|\theta = \theta_j)}{\Pr(\mathbf{X} = \mathbf{x}|\theta = \theta_j)}$$

$$= C(\mathbf{x}, \theta_i, \theta_j) \cdot \frac{\sum_{\mathbf{z} \in T^{-1}(t)} \Pr(\mathbf{X} = \mathbf{z}|\theta = \theta_j)}{C(\mathbf{x}, \theta_i, \theta_j) \sum_{\mathbf{z} \in T^{-1}(t)} \Pr(\mathbf{X} = \mathbf{z}|\theta = \theta_j)} = 1.$$

Therefore, for any states θ_i and θ_j, we have the ratio $r_{ij} = 1$, which proves that T is sufficient. ■

The following example illustrates the probability adjustments within components that occur when the state of nature changes. It again uses the unreliable network situation, for which the sample sum is a known sufficient statistic.

4.34 Example: As you will no doubt recall by this time, the unreliable network offers four service options with varying costs depending on the prevailing state of nature. A test packet X returns a probabilistic indication of the state of nature. The cost data and the distribution of X are as follows.

Cost $L(a_j, \theta_i)$		
	Network condition	
Action	θ_1: Good	θ_2: Poor
a_1: Deluxe service	9	0
a_2: Intermediate service	6	3
a_3: Economy service	1	4
a_4: Abandon message	6	6

X distribution		
	θ_1	θ_2
$\Pr_X(1)$	0.8	0.3
$\Pr_X(0)$	0.2	0.7

From the population X, we take a sample of size three, $\mathbf{X} = (X_1, X_2, X_3)$, and form the sample sum, $S = X_1 + X_2 + X_3$. A potential sample point $\mathbf{x} = (x_1, x_2, x_3)$ is a 3-vector of zeros and ones with probability

$$q_{\mathbf{x}} = \Pr(\mathbf{X} = \mathbf{x}|\theta = \theta_i)$$

$$= \begin{cases} (0.8)^{x_1}(0.2)^{1-x_1}(0.8)^{x_2}(0.2)^{1-x_2}(0.8)^{x_3}(0.2)^{1-x_3} = (0.8)^s(0.2)^{3-s}, & \text{if } \theta = \theta_1 \\ (0.3)^{x_1}(0.7)^{1-x_1}(0.3)^{x_2}(0.7)^{1-x_2}(0.3)^{x_3}(0.7)^{1-x_3} = (0.3)^s(0.7)^{3-s}, & \text{if } \theta = \theta_2, \end{cases}$$

where $s = x_1 + x_2 + x_3$ is the sample sum associated with a particular observation. Using these formulas, we calculate the probabilities for the potential sample points and group them according to their sample sums.

Probabilities within components of the sample sum statistic											
Points within	State of nature		Points within	State of nature		Points within	State of nature		Points within	State of nature	
$S = 0$	θ_1	θ_2	$S = 1$	θ_1	θ_2	$S = 2$	θ_1	θ_2	$S = 3$	θ_1	θ_2
$(0,0,0)$	0.008	0.343	$(0,0,1)$	0.032	0.147	$(0,1,1)$	0.128	0.063	$(1,1,1)$	0.512	0.027
			$(0,1,0)$	0.032	0.147	$(1,0,1)$	0.128	0.063			
			$(1,0,0)$	0.032	0.147	$(1,1,0)$	0.128	0.063			
adjustment: 42.875			adjustment: 4.594			adjustment: 0.492			adjustment: 0.0527		

Within a given partition component, the adjustment noted on the bottom line is the factor by which $\Pr(\mathbf{X} = \mathbf{x}|\theta = \theta_1)$ expands or contracts to $\Pr(\mathbf{X} = \mathbf{x}|\theta = \theta_2)$ when shifting states of nature. Clearly, it is constant over points within a partition component, as predicted by the preceding theorem. Indeed, it demonstrates, in this example at least, that the points within a component are equiprobable under both states of nature. This equiprobability is not, however, a necessary characteristic of a sufficient statistic, as shown in the next example. □

4.35 Example: Construct a situation in which a sufficient statistic partition contains points with different probabilities in the same component.

If the sample distribution does not change with the state of nature, then all statistics are sufficient. We need simply choose a statistic that has varying probabilities over the points within a component. Consequently, consider a sample of size two, $\mathbf{X} = (X_1, X_2)$ from population X with the distribution to the right.

$\Pr(X = x \mid \theta = \theta_i)$		
x	θ_1	θ_2
0	0.2	0.2
1	0.8	0.8

Because the distribution does not vary with the state of nature, all statistics are sufficient (Theorem 4.33). Let statistic $T(\mathbf{x}) = x_1$. The component $T = 0$ contains points $(0,0)$ and $(0,1)$, while component $T = 1$ contains $(1,0)$ and $(1,1)$.

Probabilities within components of statistic $T(\mathbf{x}) = x_1$					
Points within	State of nature		Points within	State of nature	
$T = 0$	θ_1	θ_2	$T = 1$	θ_1	θ_2
$(0,0)$	0.04	0.04	$(1,0)$	0.16	0.16
$(0,1)$	0.16	0.16	$(1,1)$	0.64	0.64
adjustment: 1.00			adjustment: 1.00		

The probability within components is not equally distributed. □

We are now in a position to justify the earlier remark that successive coarsenings of the initial fine partition never recover sufficiency once it is lost.

4.36 Theorem: Let $\mathbf{X}$ be a sample from population X, which has a family of distributions indexed by θ_i: $\Pr(\mathbf{X} = \mathbf{x}|\theta = \theta_i)$, for $1 \leq i \leq n$. Let $T_1 = f(\mathbf{X})$ and $T_2 = g(T_1)$. That is, $T_2(\mathbf{x}) = g(f(\mathbf{x}))$. Then, if T_2 is a sufficient statistic for the family of distributions, so is T_1.

PROOF: If $\mathbf{x}, \mathbf{y}$ are two points in a component of the T_1 partition, then $f(\mathbf{x}) = f(\mathbf{y})$. Therefore, $T_2(\mathbf{x}) = g(f(\mathbf{x})) = g(f(\mathbf{y})) = T_2(\mathbf{y})$, which implies that $\mathbf{x}, \mathbf{y}$ reside in the same component of the T_2 partition. That is, each component of

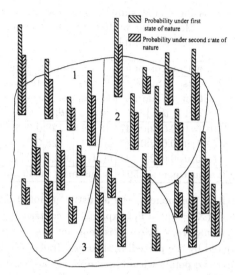

Figure 4.14. Probability shifts within components of two functionally
related statistics (see Theorem 4.36)

the T_2 partition is comprised of one or more components of the T_1 partition.
Equivalently, the T_2 partition coarsens the T_1 partition, or T_1 refines T_2.
Refer to Figure 4.14, and suppose that the larger enclosure is a component
of the T_2 partition. The subareas are the T_1 components that aggregate to
form the T_2 component. Because T_2 is sufficient, when the state of nature
changes from θ_i to θ_j, all points in the large enclosure shift probability by a
common factor. Since any given T_1 component resides entirely within a T_2
component, all points in the T_1 component must shift by the same common
factor associated with its enclosing T_2 component. With this observation, the
preceding theorem then asserts the sufficiency of T_1. ■

According to this theorem, when a function of a statistic is sufficient, so
is the original statistic. Moreover, the theorem's proof shows that sufficiency
is actually a property of the partition. Each component of the partition
corresponds to a different value in the statistic's range. From this standpoint,
we can consider the statistical values as tags for the partition components. If
we create a new statistic by assigning new (distinct) tags to the components,
the sufficiency status of the old statistic migrates to the new. If the conditional
distribution of the sample points within a particular component is independent
of the state of nature, it will remain so when we relabel the components. In
terms of the preceding theorem, the new tags are a function of the old tags,
and vice versa. Hence either both statistics are sufficient, or both are not.
The question is then: How far can we enlarge the partition components while
maintaining the conditional distributions independent of the state of nature?

If we start with the finest possible partition, each sample point constitutes a distinct component. This corresponds to the full-sample statistic,

which is sufficient. We then aggregate points to enlarge the components. In the earlier examples dealing with the unreliable network, we found that we could aggregate points that produce the same sample mean value. In some cases, we may be able to coalesce the lumps of a previous aggregation and still maintain the independence property. In general, however, the process must terminate. The ultimate aggregation would be that which encompasses all sample points in a single component. This corresponds to a statistic that is identically constant for all sample points. We argued above that it is unlikely that we can achieve this extreme. Typically, the process terminates at an earlier stage. In any case, the final partition then corresponds to a minimal sufficient statistic. The formal definition follows.

4.37 Definition: Let $\mathbf{X}$ be a sample from population X, which has distributions $\Pr(\mathbf{X} = \mathbf{x}|\theta = \theta_i)$ for $1 \leq i \leq n$. A sufficient statistic T is a *minimal sufficient statistic* for the distribution family if, for any sufficient statistic S, there exists a function $f(\cdot)$ with $T = f(S)$. ∎

We saw in the proof of the preceding theorem that a functional reduction of a statistic coarsens its partition by coalescing components. Because sufficiency status attaches to the partition, rather than the corresponding statistic, we can speak of minimal sufficient partitions. A minimal sufficient partition is then one that can be obtained by combining components from any other sufficient partition. If we have two minimal sufficient partitions, we must map one into the other without actually coalescing any components. (If we were to combine components, the finer partition would not be minimal sufficient. It could not be obtained by aggregating components of the coarser partition.) This means that the two minimal sufficient partitions must have the same components. Their statistical tags can be different, but there must be a one-to-one correspondence between the tag sets. So, while there may be many minimal sufficient statistics, they all induce the same partition. For example, if the sample mean $\overline{X}$ is minimal sufficient, so are $S = N\overline{X}$ and $\overline{X} + 42$. Both simply attach new tags to the $\overline{X}$ partition components.

4.38 Theorem: All minimal sufficient statistics induce the same partition in the space of sample points.

PROOF: See discussion above. ∎

A minimal sufficient statistic T enjoys two advantages. The Bayes and mixed minimax decision rules over T coincide with those over the full sample, and no other sufficient statistic has a smaller range. The latter property greatly reduces the number of rules under consideration in the Bayes and minimax selection processes. In theory, we can construct the minimal sufficient partition with an equivalence relation over the sample points. We declare points $\mathbf{x}, \mathbf{y}$ equivalent and write $\mathbf{x} \equiv \mathbf{y}$ when there exists a factor $h(\mathbf{x}, \mathbf{y}) > 0$, independent of θ, such that for all states of nature θ_i

$$\Pr(\mathbf{X} = \mathbf{x}|\theta = \theta_i) = h(\mathbf{x}, \mathbf{y}) \cdot \Pr(\mathbf{X} = \mathbf{y}|\theta = \theta_i).$$

Obviously $\mathbf{x} \equiv \mathbf{x}$; we simply take $h(\mathbf{x}, \mathbf{x}) = 1$. If $\mathbf{x} \equiv \mathbf{y}$, then independent of

θ_i, we have

$$\Pr(\mathbf{X} = \mathbf{x}|\theta = \theta_i) = h(\mathbf{x}, \mathbf{y})\Pr(\mathbf{X} = \mathbf{y}|\theta = \theta_i)$$

$$\Pr(\mathbf{X} = \mathbf{y}|\theta = \theta_i) = \left(\frac{1}{h(\mathbf{x}, \mathbf{y})}\right)\Pr(\mathbf{X} = \mathbf{x}|\theta = \theta_i).$$

So, with $h(\mathbf{y}, \mathbf{x}) = 1/h(\mathbf{x}, \mathbf{y})$, we have $\mathbf{y} \equiv \mathbf{x}$. Finally, if $\mathbf{x} \equiv \mathbf{y}$ and $\mathbf{y} \equiv \mathbf{z}$, we set $h(\mathbf{x}, \mathbf{z}) = h(\mathbf{x}, \mathbf{y}) \cdot h(\mathbf{y}, \mathbf{z})$ to verify that $\mathbf{x} \equiv \mathbf{z}$. Hence we have a true equivalence relation. Let $C_1, C_2, \ldots$ be the equivalence classes of this relation, and define the statistic $T(\mathbf{x}) = k$, where $\mathbf{x} \in C_k$. Then, since $(\mathbf{X} = \mathbf{x}) \subset (T = T(\mathbf{x}))$, we have the following for any state of nature θ_i.

$$q_i = \Pr(\mathbf{X} = \mathbf{x}|(T = T(\mathbf{x})) \cap (\theta = \theta_i)) = \frac{\Pr((\mathbf{X} = \mathbf{x}) \cap (T = T(\mathbf{x})) \cap (\theta = \theta_i))}{\Pr((T = T(\mathbf{x})) \cap (\theta = \theta_i))}$$

Now, the event $(T = T(\mathbf{x}))$ is the disjoint union $\cup(\mathbf{X} = \mathbf{y})$ across all $\mathbf{y}$ equivalent to $\mathbf{x}$. Consequently,

$$q_i = \frac{\Pr((\mathbf{X} = \mathbf{x}) \cap (\theta = \theta_i))}{\sum_{\mathbf{y} \equiv \mathbf{x}} \Pr((\mathbf{X} = \mathbf{y}) \cap (\theta = \theta_i))} = \frac{\Pr(\mathbf{X} = \mathbf{x}|\theta = \theta_i) \cdot \Pr(\theta = \theta_i)}{\sum_{\mathbf{y} \equiv \mathbf{x}} \Pr(\mathbf{X} = \mathbf{y})|\theta = \theta_i) \cdot \Pr(\theta = \theta_i)}$$

$$= \frac{\Pr(\mathbf{X} = \mathbf{x}|\theta = \theta_i)}{\sum_{\mathbf{y} \equiv \mathbf{x}} h(\mathbf{y}, \mathbf{x})\Pr(\mathbf{X} = \mathbf{x})|\theta = \theta_i)} = \frac{1}{\sum_{\mathbf{y} \equiv \mathbf{x}} h(\mathbf{y}, \mathbf{x})}.$$

Hence $\Pr(\mathbf{X} = \mathbf{x}|(T = T(\mathbf{x})) \cap (\theta = \theta_i))$ is independent of θ_i, and T is a sufficient statistic. To show that T is minimal sufficient, we must show that T is a function of any other sufficient statistic. To this end, let T' be a sufficient statistic. Let $D(\mathbf{x})$ be the T' component containing $\mathbf{x}$. For $\mathbf{x}, \mathbf{y} \in D(\mathbf{x})$, we have $T'(\mathbf{x}) = T'(\mathbf{y}) = t'$, and both $\Pr(\mathbf{X} = \mathbf{x}|(T' = t') \cap (\theta = \theta_i))$ and $\Pr(\mathbf{X} = \mathbf{y}|(T' = t') \cap (\theta = \theta_i))$ are independent of θ. We can argue as follows to show that $\mathbf{x} \equiv \mathbf{y}$.

$$\Pr_{\mathbf{X}|\theta_i}(\mathbf{x}) = \Pr(\mathbf{X} = \mathbf{x}|\theta = \theta_i) = \frac{\Pr((\mathbf{X} = \mathbf{x}) \cap (\theta = \theta_i))}{\Pr(\theta = \theta_i)}$$

$$= \frac{\Pr((\mathbf{X} = \mathbf{x}) \cap (T' = t') \cap (\theta = \theta_i))}{\Pr(\theta = \theta_i)}$$

$$= \frac{\Pr((\mathbf{X} = \mathbf{x}) \cap (T' = t') \cap (\theta = \theta_i))}{\Pr(\theta = \theta_i)} \cdot \frac{\Pr((T' = t') \cap (\theta = \theta_i))}{\Pr((T' = t') \cap (\theta = \theta_i))}$$

$$= \Pr(\mathbf{X} = \mathbf{x}|(T' = t') \cap (\theta = \theta_i)) \cdot \Pr(T' = t'|\theta = \theta_i)$$

Analogous calculations show a similar result for $\mathbf{y}$. That is,

$$\Pr(\mathbf{X} = \mathbf{y}|\theta = \theta_i) = \Pr(\mathbf{X} = \mathbf{y}|(T' = t') \cap (\theta = \theta_i)) \cdot \Pr(T' = t'|\theta = \theta_i).$$

Solving for $\Pr(T' = t'|\theta = \theta_i)$ in the latter expression and then substituting in the former gives

$$\Pr(\mathbf{X} = \mathbf{x}|\theta = \theta_i) = \frac{\Pr(\mathbf{X} = \mathbf{x}|(T' = t') \cap (\theta = \theta_i))}{\Pr(\mathbf{X} = \mathbf{y}|(T' = t') \cap (\theta = \theta_i))} \cdot \Pr(\mathbf{X} = \mathbf{y}|\theta = \theta_i)$$

$$= h(\mathbf{x}, \mathbf{y}) \cdot \Pr(\mathbf{X} = \mathbf{y}|\theta = \theta_i).$$

Because T' is sufficient, the factor $h(\mathbf{x}, \mathbf{y})$ is independent of θ_i. Therefore, $\mathbf{x} \equiv \mathbf{y}$, which implies that $\mathbf{x}, \mathbf{y}$ both reside in the same component of the T partition. Since $\mathbf{y}$ was an arbitrary member of $D(\mathbf{x})$, we have $D(\mathbf{x})$ contained entirely within a T component.

Now, for $t' \in \text{range}(T')$, define $f(t') = k$, where $D(\mathbf{x})$ is the T' component with $T'(\mathbf{x}) = t'$ and $D(\mathbf{x}) \subset C_k$. Then $T'(\mathbf{x}) = t'$ if and only if $T(\mathbf{x}) = k$. Therefore, $T(\mathbf{x}) = k = f(t') = f(T'(\mathbf{x}))$. This proves that $T = f(T')$, and therefore T is minimal sufficient.

4.39 Theorem: The partition of a minimal sufficient statistic is the set of equivalence classes induced by the relation $\mathbf{x} \equiv \mathbf{y}$.

PROOF: See discussion above. ∎

The discussion above provides a method for finding a minimal sufficient statistic. We simply start with the finest partition, in which each sample point is a distinct component, and coalesce points for which the probabilities retain the same proportion across all states of nature. We can then assign arbitrary distinct tags to each partition component to obtain a minimal sufficient statistic. In truth, this procedure is impractical because of the large number of point pairs that must be compared. The next result provides a more direct approach.

4.40 Theorem: Let $\mathbf{X}$ be a sample from population X, which has distributions $\Pr(\mathbf{X} = \mathbf{x} | \theta = \theta_i)$ for $1 \leq i \leq n$. T is a sufficient statistic for the distribution family if and only if the distributions factor: $\Pr(\mathbf{X} = \mathbf{x} | \theta = \theta_i) = f(T(\mathbf{x}), \theta_i) \cdot g(\mathbf{x})$.

PROOF: Suppose that T is a sufficient statistic. As in the preceding proof, we have

$$
\begin{aligned}
\Pr_{\mathbf{X}|\theta_i}(\mathbf{x}) = \Pr(\mathbf{X} = \mathbf{x} | \theta = \theta_i) &= \frac{\Pr((\mathbf{X} = \mathbf{x}) \cap (\theta = \theta_i))}{\Pr(\theta = \theta_i)} \\
&= \frac{\Pr((\mathbf{X} = \mathbf{x}) \cap (T = T(\mathbf{x})) \cap (\theta = \theta_i))}{\Pr(\theta = \theta_i)} \\
&= \frac{\Pr((\mathbf{X} = \mathbf{x}) \cap (T = T(\mathbf{x})) \cap (\theta = \theta_i))}{\Pr(\theta = \theta_i)} \cdot \frac{\Pr((T = T(\mathbf{x})) \cap (\theta = \theta_i))}{\Pr((T = T(\mathbf{x})) \cap (\theta = \theta_i))} \\
&= \Pr(\mathbf{X} = \mathbf{x} | (T = T(\mathbf{x})) \cap (\theta = \theta_i)) \cdot \Pr(T = T(\mathbf{x}) | \theta = \theta_i) \\
&= \Pr(\mathbf{X} = \mathbf{x} | T = T(\mathbf{x})) \cdot \Pr(T = T(\mathbf{x}) | \theta = \theta_i) = g(\mathbf{x}) \cdot f(T(\mathbf{x}), \theta_i).
\end{aligned}
$$

Conversely, suppose that $\Pr(\mathbf{X} = \mathbf{x} | \theta = \theta_i) = f(T(\mathbf{x}), \theta_i) \cdot g(\mathbf{x})$. We can sum probability over all $\mathbf{y}$ for which $T(\mathbf{y}) = T(\mathbf{x})$ to obtain $\Pr(T = T(\mathbf{x}) | \theta = \theta_i)$:

$$
\begin{aligned}
\Pr(T = T(\mathbf{x}) | \theta = \theta_i) &= \sum_{\mathbf{y} \in T^{-1}(T(\mathbf{x}))} \Pr(\mathbf{X} = \mathbf{y} | \theta = \theta_i) \\
&= f(T(\mathbf{x}), \theta_i) \sum_{\mathbf{y} \in T^{-1}(T(\mathbf{x}))} g(\mathbf{y}).
\end{aligned}
$$

Then, because $(\mathbf{X} = \mathbf{x}) \subset (T = T(\mathbf{x}))$, we have

$$
q_i = \Pr(\mathbf{X} = \mathbf{x} | (T = T(\mathbf{x}) \cap (\theta = \theta_i)) = \frac{\Pr((\mathbf{X} = \mathbf{x}) \cap (T = T(\mathbf{x})) \cap (\theta = \theta_i))}{\Pr((T = T(\mathbf{x})) \cap (\theta = \theta_i))},
$$

which we manipulate to exploit the given factorization.

$$q_i = \frac{\Pr((\mathbf{X} = \mathbf{x}) \cap (\theta = \theta_i))}{\Pr((T = T(\mathbf{x})) \cap (\theta = \theta_i))} \cdot \frac{\Pr(\theta = \theta_i)}{\Pr(\theta = \theta_i)} = \frac{\Pr(\mathbf{X} = \mathbf{x} | \theta = \theta_i)}{\Pr(T = T(\mathbf{x}) | \theta = \theta_i)}$$

$$= \left(\frac{f(T(\mathbf{x}), \theta_i)}{\Pr(T = T(\mathbf{x}) | \theta = \theta_i)} \right) \cdot g(\mathbf{x}) = \frac{f(T(\mathbf{x}), \theta_i) g(\mathbf{x})}{f(T(\mathbf{x}), \theta_i) \cdot \sum_{\mathbf{y} \in T^{-1}(T(\mathbf{x}))} g(\mathbf{y})}$$

$$= \frac{g(\mathbf{x})}{\sum_{\mathbf{y} \in T^{-1}(T(\mathbf{x}))} g(\mathbf{y})}$$

The last expression is evidently independent of θ_i. Therefore, T is sufficient. ∎

    This theorem says that a statistic is sufficient if and only if the θ variation is confined to the partition components, as opposed to the individual sample points. If we can factor $\Pr(\mathbf{X} = \mathbf{x} | \theta = \theta_i)$ as required by the theorem, we can sometimes use this form to discover a minimal sufficient statistic. Example 4.31 showed that the sample mean was sufficient in the unreliable network situation. We can now show that it is minimal sufficient.

4.41 Example: To this point, examples with the unreliable network have used samples of size two, and they have presented the distributions in a tabular form. Now we consider a sample of size N, and we will work with analytic expressions for the distributions. Recall that the underlying population X has the distributions to the right below.

    To factor these distributions as required by Theorem 4.40, we need algebraic expressions. Under θ_1, X is a Bernoulli random variable with parameter $p_1 = 0.8$; under θ_2, it is a Bernoulli random variable with parameter $p_2 = 0.3$. It is convenient to denote the states of nature as p_1, p_2 instead

$\Pr(X = x \vert \theta = \theta_i)$		
x	θ_1	θ_2
0	0.2	0.7
1	0.8	0.3

of θ_1, θ_2. Indeed, p_1, p_2 provide a more immediate indexing of the probability distributions. Let the sample be $\mathbf{X} = (X_1, X_2, \dots, X_N)$, and let $\mathbf{x} = (x_1, x_2, \dots, x_N)$ be a generic sample point. The vector $\mathbf{x}$ consists of zeros and ones. The sample distribution then assumes the following form. As usual, $S = \sum_{i=1}^{N} X_i$ is the sample sum statistic.

$$\Pr(\mathbf{X} = \mathbf{x} | p_j) = p_j^{x_1}(1 - p_j)^{1-x_1} p_j^{x_2}(1 - p_j)^{1-x_2} \cdots p_j^{x_N}(1 - p_j)^{1-x_N}$$

$$= p_j^{S(\mathbf{x})}(1 - p_j)^{N - S(\mathbf{x})},$$

for $j = 1, 2$. Taking $f(S(\mathbf{x}), p_j) = p_j^{S(\mathbf{x})}(1 - p_j)^{N-S(\mathbf{x})}$ and $g(\mathbf{x}) = 1$, we have the factorization needed to show that the sample sum is a sufficient statistic. Since $S(\mathbf{x}) = N\overline{X}(\mathbf{x})$, we can write

$$\Pr(\mathbf{X} = \mathbf{x} | p_j) = p_j^{N\overline{X}(\mathbf{x})}(1 - p_j)^{N(1 - \overline{X}(\mathbf{x}))},$$

which shows that the sample mean is also sufficient. Now suppose that T is a sufficient statistic. Then the factorization criterion asserts that

$$\Pr(\mathbf{X} = \mathbf{x} | p_j) = f(T(\mathbf{x}), p_j) \cdot g(\mathbf{x}),$$

for some functions $f(\cdot, \cdot)$ and $g(\cdot)$. Suppose that $\mathbf{x}, \mathbf{y}$ are two points in the same component of the T partition. We have $T(\mathbf{x}) = T(\mathbf{y})$, which yields

$$p_j^{S(\mathbf{x})}(1 - p_j)^{N - S(\mathbf{x})} = f(T(\mathbf{x}), p_j)g(\mathbf{x})$$
$$p_j^{S(\mathbf{y})}(1 - p_j)^{N - S(\mathbf{y})} = f(T(\mathbf{y}), p_j)g(\mathbf{y}) = f(T(\mathbf{x}), p_j)g(\mathbf{y})$$

$$\frac{g(\mathbf{x})}{g(\mathbf{y})} = p_j^{S(\mathbf{x}) - S(\mathbf{y})}(1 - p_j)^{-(S(\mathbf{x}) - S(\mathbf{y}))} = \left(\frac{p_j}{1 - p_j}\right)^{S(\mathbf{x}) - S(\mathbf{y})}.$$

The left side is independent of p_j, and therefore the right side must also be independent of p_j. Hence, either $p_1/(1 - p_1) = p_2/(1 - p_2)$ or $S(\mathbf{x}) = S(\mathbf{y})$. The former alternative implies that $p_1 = p_2$, which means that there is really only one state of nature. (There may actually be two states of nature in the physical situation, corresponding to good and poor network conditions, but if $p_1 = p_2$, the distributions of the test packet X do not distinguish between them. As far as the family of distributions is concerned, there is but one state of nature in this case.) We conclude that $S(\mathbf{x}) = S(\mathbf{y})$ and that the S component containing $S(\mathbf{x})$ includes the entire T component associated with $\mathbf{x}$. As before, this implies that $S = h(T)$, for some function $h(\cdot)$, which proves that S is a minimal sufficient statistic. Also, $\overline{X} = S/N = h(T)/N = h'(T)$, which implies that $\overline{X}$ is minimal sufficient. $\Box$

Suppose that we have in hand both the distributions of X under all possible states of nature and a sample observation $\mathbf{x} = (x_1, x_2, \ldots, x_N)$. We are asked to guess the prevailing state of nature. We look at the values $\Pr(\mathbf{X} = \mathbf{x} | \theta = \theta_i)$ as a function of θ_i. Let $\hat{\theta}$ be the state of nature that maximizes these values. In other words, the observation $\mathbf{x}$ is a more probable event under $\hat{\theta}$ than under any other state of nature. $\hat{\theta}$ is called the *maximum likelihood estimate* for θ.

4.42 Example: Continuing the unreliable network example, suppose that X has the following distributions under states of nature $p_1 = 0.8, p_2 = 0.3$.

$$\Pr(X = x | p_j) = p_j^x(1 - p_j)^{1 - x},$$

for $x = 0, 1$ and $j = 1, 2$. Suppose that $\mathbf{x} = (1, 1, 0, 1, 1, 0, 0, 1, 1, 1)$ is an observed sample of size 10. What is the probability of this observation under the two states of nature?

If $p_1 = 0.8$ prevails, the probability is $0.8^7 \cdot 0.2^3 = 0.001678$; if $p_2 = 0.3$ prevails, the probability is $0.3^7 \cdot 0.7^3 = 0.000075$. The observation is $0.001678/0.000075 = 22.4$ times as likely under p_1 as it is under p_2. The maximum likelihood estimate for the state of nature is then p_1. $\Box$

4.43 Definition: Suppose that $\mathbf{X} = (X_1, X_2, \ldots, X_N)$ is a sample over a population X having distributions $\Pr(\mathbf{X} = \mathbf{x} | \theta = \theta_i)$ for various states of nature θ_i. For a given sample value $\mathbf{x} = (x_1, x_2, \ldots, x_N)$, the *likelihood function* for $\mathbf{x}$ is $\mathcal{L}_{\mathbf{x}}(\theta_i) = \Pr(\mathbf{X} = \mathbf{x} | \theta = \theta_i)$. $\blacksquare$

4.44 Definition: The *maximum likelihood equivalence relation* on the space of sample points is that which considers $\mathbf{x}$ equivalent to $\mathbf{y}$ (written $\mathbf{x} \sim \mathbf{y}$) when there exists $K(\mathbf{x}, \mathbf{y}) > 0$ such that $\mathcal{L}_{\mathbf{x}}(\cdot) = K(\mathbf{x}, \mathbf{y})\mathcal{L}_{\mathbf{y}}(\cdot)$. That is, two points are equivalent when their likelihood functions are scalar multiples of one another. The scalar multiple can vary with the points. A *maximum* *likelihood statistic* is a statistic T such that $T(\mathbf{x}) = T(\mathbf{y})$ if and only if $\mathbf{x} \sim \mathbf{y}$. The partition induced by T is called the *maximum likelihood partition*, and it is, of course, the set of equivalence classes determined by the maximum likelihood equivalence relation. ∎

If $\mathbf{x} \sim \mathbf{y}$, we have, for all θ_i,

$$\mathcal{L}_{\mathbf{x}}(\theta_i) = K(\mathbf{x}, \mathbf{y}) \cdot \mathcal{L}_{\mathbf{y}}(\theta_i)$$
$$\Pr(\mathbf{X} = \mathbf{x} | \theta = \theta_i) = K(\mathbf{x}, \mathbf{y}) \cdot \Pr(\mathbf{X} = \mathbf{y} | \theta = \theta_i)$$
$$\mathbf{x} \equiv \mathbf{y}.$$

The steps are reversible, so $\mathbf{x} \sim \mathbf{y}$ if and only if $\mathbf{x} \equiv \mathbf{y}$. The equivalence classes of the two relations then describe the same partition, which means that a maximum likelihood statistic is minimal sufficient.

4.45 Theorem: A maximum likelihood statistic is minimal sufficient.

PROOF: See argument above. ∎

If we categorize the possible likelihood functions into groups, where the functions within a particular group differ by a multiplicative constant, then the values assumed by a maximum likelihood statistic index these groups. As emphasized in the discussion above, two likelihood functions, $\mathcal{L}_{\mathbf{x}}(\cdot)$ and $\mathcal{L}_{\mathbf{y}}(\cdot)$ are proportional if and only if the corresponding probabilities $\Pr(\mathbf{X} = \mathbf{x} | \theta = \theta_i)$ and $\Pr(\mathbf{X} = \mathbf{y} | \theta = \theta_i)$ are also proportional with the same proportionality constant for all θ. The maximum likelihood viewpoint is sometimes more effective in identifying a minimal sufficient statistic. The next example illustrates the point.

4.46 Example: Continuing with the unreliable network of Example 4.41, use likelihood functions to ascertain a minimal sufficient statistic.

For the two states of nature, $p_1 = 0.8, p_2 = 0.3$, we have the following distributions, where $S(\mathbf{x})$ is the sample sum.

$$\Pr(\mathbf{X} = \mathbf{x} | p_j) = p_j^{S(\mathbf{x})}(1 - p_j)^{N - S(\mathbf{x})}$$

Therefore,

$$\mathcal{L}_{\mathbf{x}}(\theta_i) = p_j^{S(\mathbf{x})}(1 - p_j)^{N - S(\mathbf{x})}.$$

$\mathbf{x} \sim \mathbf{y}$ means $\mathcal{L}_{\mathbf{x}}(\cdot) = K(\mathbf{x}, \mathbf{y})\mathcal{L}_{\mathbf{y}}(\cdot)$, which implies that

$$K(\mathbf{x}, \mathbf{y}) = \frac{p_j^{S(\mathbf{x})}(1 - p_j)^{N - S(\mathbf{x})}}{p_j^{S(\mathbf{y})}(1 - p_j)^{N - S(\mathbf{y})}} = \left(\frac{p_j}{1 - p_j}\right)^{S(\mathbf{x}) - S(\mathbf{y})}$$
$$S(\mathbf{x}) = S(\mathbf{y})$$
$$K(\mathbf{x}, \mathbf{y}) = 1.$$

Because this argument is reversible, we conclude that $\mathbf{x} \sim \mathbf{y}$ if and only if $S(\mathbf{x}) = S(\mathbf{y})$. The sample sum is therefore a maximum likelihood statistic. Consequently, it is minimal sufficient.

While the perspective of this example is slightly different in its emphasis on likelihood functions, the calculations concerning the minimal sufficient partition are very similar to those employed in Example 4.41, which emphasized the proportionality of the conditional probabilities $\Pr(\mathbf{X} = \mathbf{x} | \theta = \theta_i)$ and $\Pr(\mathbf{X} = \mathbf{y} | \theta = \theta_i)$. □

Exercises

4.23 Suppose, as in Example 4.28, that X has the distributions to the right. For a two-point sample from population X, show that the order statistic is a sufficient statistic.

$\Pr(X = x \mid \theta = \theta_i)$		
x	θ_1	θ_2
0	0.2	0.7
1	0.8	0.3

4.24 Suppose that $\mathbf{X} = (X_1, X_2, \ldots, X_N)$ is a sample from a population X with distributions as shown to the right. Show that the order statistic is sufficient but not minimal sufficient.

$\Pr(X = x \mid \theta = \theta_i)$		
x	θ_1	θ_2
0	0.1	0.7
1	0.9	0.3

4.25 For $j \in \{1, 2, 3\}$, let $0 < p_j < 1$ be three distinct probability values. Let n be a fixed positive integer. Suppose that the population X has the binomial distribution with parameter (n, p_j) under state of nature p_j. That is, suppose that $\Pr(X = k \mid p_j) = C_{n,k} p_j^k (1 - p_j)^{n-k}$, for $k = 0, \ldots, n$. Let $\mathbf{X} = (X_1, \ldots, X_N)$ be a sample. Show that the sample sum $S(\mathbf{X}) = \sum_{i=1}^{N} X_i$ is a sufficient statistic.

4.26 Repeat the preceding exercise when X has a negative binomial distribution: $\Pr(X = k \mid p_j) = (-1)^k C_{-n,k} p_j^n (1 - p_j)^k$.

4.27 When a server comes online, it receives requests from a Poisson process with an average rate of $\lambda_1 = 0.5, \lambda_2 = 5.0$, or $\lambda_3 = 50.0$ requests per second. The three rates correspond to three possible states of nature. The server has 10 seconds to bind a processor for handling its load. Three classes of processor are available: α, β, and γ with average process rates of $1.0, 10.0$, and 100.0 requests per second, respectively. An α processor charges $10.00 per hour, while the β and γ processors charge $100.00 and $1000.00, respectively. Once the server has chosen a processor, it must retain that processor for the duration of a duty cycle, which is 4 hours. If the server cannot keep up with the request stream, it can enlist help from a farm of high-speed emergency servers, which charge $0.01 per request.

If the server chooses an α processor and $\lambda_1 = 0.5$ prevails as the state of nature, the expected cost for the duty cycle is $4(10.00) = \$40.00$ because the α can handle the 0.5 requests per second with capacity to spare.

Cost for given processor P and state of nature λ_i			
P	$\lambda_1 = 0.5$	$\lambda_2 = 5.0$	$\lambda_3 = 50.0$
α	40	616	7096
β	400	400	6160
γ	4000	4000	4000

If λ_2 prevails, however, the server will call for help on $4/5$ of the requests, which will arrive at 5 per second. The cost will then be $40.00 + (4/5)(5)(60)(60)(4)(0.01) = \$616.00$. If λ_3 prevails, the cost will be $40.00 + (49/50)(50)(60)(60)(4)(0.01) = \$7096.00$. Similar calculations for the remaining cases give the matrix to the right above.

In the 10 seconds before it must irrevocably bind a processor, the server monitors the input line to sample the request rate. Specifically, it counts the number of requests in 5 consecutive 2-second intervals. This sample is $\mathbf{X} = (X_1, \ldots, X_5)$ from a population X with distributions $\Pr(X = k | \lambda_j) = (2\lambda_j)^k e^{-2\lambda_j} / k!$, for $k = 0, 1, 2, \ldots$ and $j = 1, 2, 3$. A particular sample value is $\mathbf{x} = (k_1, k_2, k_3, k_4, k_5)$. Use the factorization criterion to show that $S(\mathbf{x}) = \sum_{i=1}^{5} k_i$ is a sufficient statistic.

4.28 In the context of the preceding exercise, use the likelihood function approach to show that $S(\mathbf{x})$ is minimal sufficient.

4.29 Continuing Exercise 4.27, assume a prior distribution across the states of nature with $\Pr(\lambda_i) = 1/3$ for $i = 1, 2, 3$. Compute the Bayes decision rule using the minimal statistic $S(\mathbf{x})$.

4.30 Remaining in the context of Exercise 4.27, suppose that the sample is $\mathbf{x} = (4, 10, 6, 3, 5)$. What is the maximum likelihood estimate for the prevailing state of nature?

*4.31 Given a uniform prior distribution across the states of nature θ, show that the maximum likelihood estimate of θ is that which maximizes the posterior probability $\Pr(\theta = \theta_i | \mathbf{X} = \mathbf{x})$.

*4.32 Suppose that the population X has distinct distributions $\Pr(X = x | \theta = \theta_i)$ for various states of nature θ_i. Let $\mathbf{X} = (X_1, X_2, \ldots, X_N)$ be a sample, and consider the vector-valued statistic $T(\mathbf{x}) = (x_1, x_2, \ldots, x_{N-1})$. That is, the statistic ignores the last sample value. Show that T cannot be a sufficient statistic.

*4.33 Let $\mathbf{X} = (X_1, X_2, \ldots, X_N)$ be a sample from the population X, which has the following distributions under the three possible states of nature N_1, N_2, N_3. The N_j are distinct positive integers. Accordingly, the jth distribution is the discrete uniform distribution on the integers $[1..N_j]$. Use the factorization criterion to obtain a sufficient statistic for this family of distributions.

$$\Pr(X = k | N_j) = \begin{cases} \dfrac{1}{N_j}, & 1 \le k \le N_j \\ 0, & k > N_j \end{cases}$$

*4.34 Consider a sample $\mathbf{X} = (X_1, X_2, \ldots, X_N)$ from a population X, which has varying distributions under two possible states of nature, θ_1 and θ_2. Show that T is a sufficient statistic, where

$$T(\mathbf{x}) = \frac{\Pr(\mathbf{X} = \mathbf{x} | \theta = \theta_1)}{\Pr(\mathbf{X} = \mathbf{x} | \theta = \theta_2)}.$$

4.4   Hypothesis testing

An important subclass of decision problems involves just two possible actions: accept a hypothesis or reject it. The hypothesis itself is a conjecture about the prevailing state of nature, and it can normally be worded as a statement that nature presents a specific state. For example, suppose that we have a biased coin, for which the probability of heads is either $p_1 = 1/4$ or $p_2 = 3/4$. The states of nature are then p_1 and p_2. The statement "The state of nature is p_1" is a hypothesis, as is the statement "In a long sequence of coin flips, the number of tails will dominate by approximately $3 : 1$." The second statement is equivalent to the first, in the sense that tails will dominate if and only if p_1 prevails as the state of nature.

4.47 Definition: A *simple hypothesis* is a statement that specifies a particular state of nature. A *composite hypothesis* is a statement that the state of nature lies in a prescribed set. In either case, the *alternative hypothesis* is the statement that the state of nature lies in the complement of the set specified in the hypothesis. H_0 denotes the hypothesis, either simple or composite, and H_1 denotes the alternative. ∎

In the biased coin example, the simple hypothesis H_0 asserts that the state of nature is p_1. The alternative is also simple; it asserts that the state of nature is p_2. When we collect sample data and reach a conclusion in this context, we say that we are testing a simple hypothesis against a simple alternative. Consider a more complicated situation, in which p, the probability of heads, can be any value in the range $[0, 1]$. If we let H_0 be the assertion: "The state of nature is $p = 0.4$," then the alternative, H_1, is "$p \in [0, 1]$, but $p \neq 0.4$." This implies a test of a simple hypothesis against a composite alternative.

It is traditional to call H_0 the null hypothesis. The terminology derives from a frequent application in which H_0 asserts that a particular treatment has no effect. Suppose, for example, that someone proposes a treatment for our biased coin, which we know to have $p = 1/4$. The treatment purports to remove the bias. The null hypothesis, H_0, then asserts that the treatment in reality has no effect. That is, the probability remains $p = 1/4$. A test, involving sample flips of the treated coin, either accepts or rejects the null hypothesis. Depending on other claims made by the treatment provider, the alternative may be simple ($p = 1/2$) or composite ($p \neq 1/4$). That is, if the treatment does have some effect, does it perform as advertised to deliver an unbiased coin, or does it simply change the existing bias?

We think of H_0 as a subset of the states. It is a singleton subset if the hypothesis is simple. H_1 is the complement, in the sense that $H_0 \cup H_1$ encompass all possible states of nature as determined by the application. It is also traditional to assume zero loss when we arrive at the correct decision. That is, we suffer no loss in accepting H_0 if $\theta \in H_0$, and we have no loss in rejecting H_0 if $\theta \in H_1$. To each state of nature θ, we can then attach a cost, $c(\theta)$, which is the cost of taking the wrong decision when θ prevails as the actual state. The loss function $L(\text{accept/reject}, \theta)$ is then

$$L(\text{accept}, \theta) = \begin{cases} 0, & \theta \in H_0 \\ c(\theta), & \theta \in H_1 \end{cases} \qquad L(\text{reject}, \theta) = \begin{cases} c(\theta), & \theta \in H_0 \\ 0, & \theta \in H_1 \end{cases}$$

4.48 Definition: If a test rejects H_0, when in reality $\theta \in H_0$, the test makes a *type I error*. If a test accepts H_0 when $\theta \in H_1$, the test makes a *type II error*. ∎

The error type terminology is well established and reflects the frequently occurring situation in which it is more serious to reject the null hypothesis when, in truth, there is no effect. In this context, a type I error is more serious, and our first concern in designing a decision procedure is to maintain a low probability of a type I error, even at the expense of an elevated probability of a type II error, if necessary. Because reversing the roles of H_0 and H_1 swaps the meanings of the two error types, we can select H_0 and H_1 to place the type I error where we choose.

As in previous sections, we assume that we have a random variable X, which exhibits different distributions depending on the state of nature. A sample is again a random vector $\mathbf{X} = (X_1, X_2, \ldots, X_N)$ of independent observations, and a particular sample value is $\mathbf{x} = (x_1, x_2, \ldots, x_N)$. In the general exposition we work with a statistic $T(\mathbf{x})$, which includes, in the extreme case, the sample itself. That is, $T(\mathbf{x}) = \mathbf{x}$ is an acceptable statistic. A decision rule maps $\text{range}(T)$ to the actions $\{\text{accept, reject}\}$. Each such rule $d(\cdot)$ has a natural expansion $\overline{d}(\cdot)$ that maps $\text{range}(\mathbf{X})$ to $\{\text{accept, reject}\}$. As noted earlier (Equation 4.6), $\overline{d}(\mathbf{x}) = d(T(\mathbf{x}))$, and $\overline{d}(\cdot)$ is constant on components of the T partition. We identify each decision rule $d(\cdot)$ with the subset that $\overline{d}(\cdot)$ maps to "reject." This subset is called the rule's *critical region*. Because $\overline{d}(\cdot)$ is constant on any given component of the T partition, that component is either entirely inside or entirely outside the critical region. Consequently, the critical region is the union of selected components of the T partition. Specifically, if we let $R = \{t \in \text{range}(T) | d(t) = \text{reject}\}$, then the critical region C is the union of the components corresponding to $t \in R$. That is,

$$C = \bigcup_{t \in R} T^{-1}(t) = \bigcup_{t \in R} \{\mathbf{x} \in \text{range}(\mathbf{X}) | T(\mathbf{x}) = t\}.$$

Exploiting this correspondence, we have $d(\mathbf{x}) = \text{reject}$ if and only if $\mathbf{x} \in C$. Of course, this implies that $d(\mathbf{x}) = \text{accept}$ if and only if $\mathbf{x} \notin C$. So, instead of denoting a rule by $d(\cdot)$ as in previous sections, we will use its critical region C. In this terminology, a rule C advocates rejecting H_0 if $\mathbf{x} \in C$ and accepting H_0 otherwise.

4.49 Definition: A *critical region* C in the space of potential sample points corresponds to a decision rule in hypothesis testing. The rule advocates rejecting H_0 if the sample point $\mathbf{x} \in C$. If the decision rule uses statistic T, then the critical region is the union of selected components of the T partition. ∎

For a given decision rule C, we compute the risk $\hat{L}(C, \theta)$ in the normal manner. Here, C' is the complement of C in the space of potential sample points.

$$\hat{L}(C,\theta) = E_{\mathbf{X}|\theta}[L(\cdot,\theta)]$$
$$= \begin{cases} c(\theta)\Pr(C|\theta) + 0 \cdot \Pr(C'|\theta) = c(\theta)\Pr(C|\theta), & \theta \in H_0 \\ 0 \cdot \Pr(C|\theta) + c(\theta)\Pr(C'|\theta) = c(\theta)(1 - \Pr(C|\theta)), & \theta \in H_1 \end{cases} \quad (4.9)$$

Except for identifying a test with its critical region, the general context is no different from that considered in previous sections. We may therefore proceed with the minimax and Bayes algorithms for selecting a decision rule. The next example illustrates the point.

4.50 Example: For a biased coin, suppose we know that the probability of heads is either $p_0 = 0.4$ or $p_1 = 0.6$. Letting the Bernoulli random variable X be 1 if a coin flip shows heads, we probe the situation by flipping the coin $N = 6$ times to produce the sample $\mathbf{X} = \mathbf{x}$. Let $H_0 = \{p_0\}$, and investigate the risks associated with various decision rules over the sample sum.

In this case, both H_0 and its alternative, $H_1 = \{p_1\}$, are simple hypotheses. Let $\mathbf{x} = (x_1, x_2, \ldots, x_6)$, a vector of ones and zeros, and let $S(\mathbf{x}) = \sum_{i=1}^{6} x_i$ be the observed sample sum. The distributions are

$$\Pr(\mathbf{X} = \mathbf{x}|p_j) = p_j^{S(\mathbf{x})}(1 - p_j)^{6-S(\mathbf{x})}$$
$$\Pr(S = s|p_j) = C_{6,s} p_j^s (1 - p_j)^{6-s},$$

for $j = 0, 1$. There are 7 components in the S partition, corresponding to the values $S = 0, 1, 2, \ldots, 6$. Consequently, there are $2^7 = 128$ distinct critical regions, each corresponding to a different subset of these components. We identify each decision rule with a subset of $\{0, 1, 2, \ldots, 6\}$. The rule $\{1, 4, 6\}$, for example, advocates rejecting H_0 when the sample sum equals 1, 4, or 6. To calculate the risks associated with a rule $C = (i_1, i_2, \ldots, i_t)$, we need

$$\Pr(S \in C|p_j) = \sum_{k=1}^{t} \Pr(S = i_k|p_j) = \sum_{k=1}^{t} C_{6,i_k} p_j^{i_k} (1 - p_j)^{6-i_k}.$$

We can use cost penalties to specify the relative severities of type I and type II errors. Suppose that we specify $c_0 = c(p_0) = 10$ as the cost of erroneously rejecting H_0, while we set $c_1 = c(p_1) = 1$ as the cost of erroneously accepting. For a given rule C, the risk under p_0 is 10 times the probability of a type I error, or $10 \cdot \Pr(C|p_0)$. Under p_1, it is 1 times the probability of a type II error, or $1 - \Pr(C|p_1)$. The rule $\{5, 6\}$, for example, has risks

$$\hat{L}_0 = \hat{L}(\{5, 6\}, p_0) = (10)\Pr(\{5, 6\}|p_0) = 10\left(C_{6,5}(0.4)^5(0.6)^1 + C_{6,6}(0.4)^6(0.6)^0\right)$$
$$= 0.4096$$
$$\hat{L}_1 = \hat{L}(\{5, 6\}, p_1) = (1)(1 - \Pr(\{5, 6\}|p_1))$$
$$= 1 - \left(C_{6,5}(0.6)^5(0.4)^1 + C_{6,6}(0.6)^6(0.4)^0\right) = 0.7667.$$

After calculating the risks for all such rules, we proceed in the established tradition to plot them in the $\hat{L}_0$-$\hat{L}_1$ plane and to use an expanding wedge to locate the mixed minimax strategy. The calculations are tedious, because of the large number of rules, but we can offload much of the burden to a computer program. Table 4.9 lists the undominated rules and indicates the minimax rule $\{5, 6\}$.

Rule	$\hat{L}(\cdot, p_0)$	$\hat{L}(\cdot, p_1)$	Maximum risk
1 ϕ	0.0000	1.0000	1.0000
2 $\{6\}$	0.0410	0.9533	0.9533
3 $\{5\}$	0.3686	0.8134	0.8134
4 $\{5,6\}$	0.4096	0.7667	0.7667 ← minimax rule
5 $\{4\}$	1.3824	0.6890	1.3824
6 $\{4,6\}$	1.4234	0.6423	1.4234
7 $\{4,5\}$	1.7510	0.5023	1.7510
8 $\{4,5,6\}$	1.7920	0.4557	1.7920
9 $\{3,4\}$	4.1472	0.4125	4.1472
10 $\{3,4,6\}$	4.1882	0.3658	4.1882
11 $\{3,4,5\}$	4.5158	0.2259	4.5158
12 $\{3,4,5,6\}$	4.5568	0.1792	4.5568
13 $\{2,3,4,5\}$	7.6262	0.0876	7.6262
14 $\{2,3,4,5,6\}$	7.6672	0.0410	7.6672
15 $\{1,4,5,6\}$	3.6582	0.4188	3.6582
16 $\{1,3,4,5,6\}$	6.4230	0.1423	6.4230
17 $\{1,2,3,4,5,6\}$	9.5334	0.0041	9.5334
18 $\{0,5,6\}$	0.8762	0.7626	0.8762
19 $\{0,4,5,6\}$	2.2586	0.4516	2.2586
20 $\{0,3,4,5,6\}$	5.0234	0.1751	5.0234
21 $\{0,2,3,4,5,6\}$	8.1338	0.0369	8.1338
23 $\{0,1,3,4,5,6\}$	6.8896	0.1382	6.8896
24 $\{0,1,2,3,4,5,6\}$	10.0000	0.0000	10.0000

TABLE 4.9. Risks for undominated decision rules in the coin-flipping hypothesis test (see Example 4.50)

A risk plot of the 24 undominated rules appears in Figure 4.15, where it is apparent that the expanding wedge first encounters the convex hull at a point between rules 4 and 8 (i.e., between $\{5,6\}$ and $\{4,5,6\}$). Note the unequal scales on the horizontal and vertical axes, which alter the normally symmetric appearance of the expanding wedge. Here, it rises more quickly in the vertical direction. Also, rules 22 and 9 plot in approximately the same location, which is identified only by rule 22.

Returning to the expanding wedge and its encounter with the line between rules 4 and 8, we compute that line's equation to be $\hat{L}_1 = -0.2250\hat{L}_0 + 0.8588$, which intersects $\hat{L}_1 = \hat{L}_0$ at $(\hat{L}_0, \hat{L}_1) = (0.7011, 0.7011)$. This point divides the distance from rule 4 to rule 8 such that the fraction 0.211 lies between the point and rule 4. The mixed minimax strategy then advocates rule $\{5,6\}$ with probability 0.789 and rule $\{4,5,6\}$ with probability 0.211. This rule delivers an average cost per decision of 0.7011, which is slightly better than the 0.7667 of the pure minimax rule.

Both these costs assume that nature maintains the fixed state that is most injurious to our chosen strategy, while we make repeated decisions. These conditions hold, for example, if a coin-minting factory contains either a "0.4" machine or a "0.6" machine, chosen by a malicious operator who is aware of our coin-testing strategy. We pass judgment on each output coin, using our mixed minimax rule.

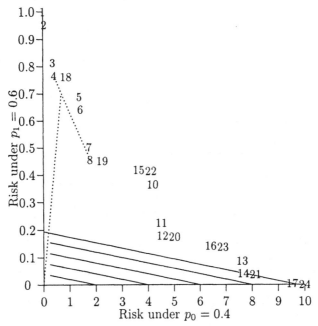

Figure 4.15. Risk plot of undominated rules in the coin-flipping hypothesis test (see Example 4.50)

We occasionally reach the wrong conclusion, but our average risk per decision is 0.7011 or less. Notice that we are not testing whether the factory has a "0.4" machine. Instead, we are repeatedly testing $H_0 = \{p_0\}$ on separate coins. This allows us to talk rationally about the average risk per decision.

If we have credible knowledge of the prior probabilities of p_0 and p_1, we can generate a Bayes decision rule. In hypothesis testing, it is not normally the case that nature switches among states. Rather, nature maintains a fixed state, but we do not know what it is. We use our knowledge of the sample and its distributions to reach a credible conclusion about the truth of the null hypothesis. Under these circumstances, the prior probabilities cannot represent frequency of occurrence. Rather, $\Pr(p_0) = 1/49$, for example, must represent a subjective belief that state p_0 is much less likely that state p_1. Indeed, it places a quantitative measure on this judgment. Such probabilities are more difficult to quantify than those that reflect frequency counts. Nevertheless, some theorists argue that they exist, despite the problems encountered in measuring them, and that decision rules that incorporate them are more consistent with the biases of the person holding the beliefs.

Consider again the situation in which a coin-minting factory contains either a "0.4" machine or a "0.6" machine. We may have strong reasons to suspect that an "0.4" machine is unlikely. We may, for example, know from our experience with coin manufacturing that "0.4" machines are much more expensive than "0.6" machines. Moreover, we know of only a few factories in the world that use "0.4" machines, and those are located in foreign countries and are operated at a net loss. Certain statisticians, known as Bayesians, argue that we should incorporate this knowledge into the decision procedure.

If we conclude that the assignment $\Pr(p_0) = 1/49, \Pr(p_1) = 48/49$ properly reflects this knowledge, then we can construct a Bayes rule that lays lines of constant risk on the plot of Figure 4.15. These lines, with increasing expected risk, first touch the convex hull at rule 14: $\{2, 3, 4, 5, 6\}$. The figure's resolution may not be sufficient to distinguish between rules $14, 17,$ and 21. In that case, we must calculate the quantity $K = \hat{L}_0/49 + 48\hat{L}_1/49$ for each of the three rules and select the rule with the smallest K-value. That rule is rule 14, which advocates rejecting H_0, except when $S = 0$ or $S = 1$. The strong bias against p_0 forces a rule that overcomes the relatively severe penalty associated with an erroneous rejection.

While we can say that this approach incorporates the "expert" testimony that p_0 is unlikely, what meaning do we attach to the expected risk? That risk is $(1/49)(7.6672) + (48/49)(0.0410) = 0.1966$. A fanciful interpretation postulates a large number of parallel universes, in which we are making simultaneous decisions. In one universe out of 49, p_0 holds, and the coin-minting machine is a "0.4" machine, despite the strong expert evidence against the fact. In the remaining universes, p_1 holds, and the coin-minting machine is a "0.6" machine, in keeping with the expert evidence.

Repeated testing of coins in the p_0 universe costs, on the average, 7.6672 per decision. This is still an average figure because the sample is not a foolproof guide. In the more numerous p_1 universes, the average cost is 0.0410. The expected Bayes risk is the weighted average of these two possibilities.

The construction ensures that this weighted average is less than that associated with any other rule. Therefore, a god-like entity, operating simultaneously in all the universes, can minimize expense by using rule 14, which almost always rejects p_0. We, however, stranded in our single universe, can only argue that it is far more likely that we are in one of the more numerous p_1 universes, where it behooves us to slant our decision procedure toward rejecting $H_0 = \{p_0\}$. $\square$

The example introduces no new concepts; it merely casts the now-familiar decision problem in the language of hypothesis testing. We still have a loss function that measures the cost of each action, and this cost still varies with the unknown state of nature. The new constraint is that the action space is binary. We are to accept or reject a specified hypothesis. Nevertheless, to this point we have simply employed our established statistical methods that attempt to control the risks.

We now consider an alternative approach that emphasizes controlling the type I and type II errors, rather than the risks. For a test C on sample $\mathbf{X}$, the probability of a type I error is $\Pr(\mathbf{X} \in C | \theta \in H_0)$; for a type II error, the probability is $1 - \Pr(\mathbf{X} \in C | \theta \in H_1)$. Both expressions involve the chance of observing a sample in the critical region, as a function of the state of nature. This quantity also appears in the risk calculations (Equation 4.9) and so deserves a special name.

4.51 Definition: For a decision rule C over a sample $\mathbf{X}$, the quantity $\mathcal{P}_C(\theta) = \Pr(\mathbf{X} \in C | \theta)$ is the *power function* of the rule. For the mixed rule C, which is a combination of the critical regions $C_1, C_2, \ldots, C_k$ via the probability vector $p_1, p_2, \ldots, p_k$, the power function is $\mathcal{P}_C(\theta) = \sum_{i=1}^{k} p_i \mathcal{P}_{C_i}(\theta)$. ∎

The power function provides a figure of merit to distinguish among tests. A good test C exhibits a power function $\mathcal{P}_C(\theta)$ that is small, ideally zero, when

$\theta \in H_0$ and large, ideally 1, when $\theta \in H_1$. The following subsections develop this idea.

4.4.1   Simple hypothesis versus simple alternative

In this subsection, we treat the restricted case of a simple hypothesis versus a simple alternative. The definitions and theorems will note the constraint, but the running commentary will often omit it. Each reference to a decision rule C includes the case where C is a mixed rule. The constituent critical areas $C_1, C_2, \ldots, C_k$ and the combining probability vector $p_1, p_2, \ldots, p_k$ will not be mentioned unless necessary for the argument at hand.

4.52 Definition: When using decision rule C over sample X to test simple hypothesis $H_0 = \{\theta_0\}$ against simple alternative $H_1 = \{\theta_1\}$, we call $\alpha_C = \mathcal{P}_C(\theta_0)$ the *type I error size*. Similarly, we call $\beta_C = 1 - \mathcal{P}_C(\theta_1)$ the *type II error size*. ∎

4.53 Example: What are the error sizes of the minimax and Bayes rules developed in Example 4.50?

We recall that the question involves a biased coin, which has probability of heads $p_0 = 0.4$ or $p_1 = 0.6$. The null hypothesis is $H_0 = \{p_0\}$, which we test by flipping the coin 6 times. The sample gives a binary vector $\mathbf{x} = (x_1, \ldots, x_6)$, in which $x_i = 1$ if the ith toss is heads. We consider decision rules over the sample sum statistic $S(\mathbf{x}) = \sum_{i=1}^{6} x_i$. A rule's critical section is set of nonnegative integers $i_1, i_2, \ldots, i_k$, which advocates rejecting H_0 if $S(\mathbf{x}) \in \{i_1, i_2, \ldots, i_k\}$. Excerpting from the example, the power function is

$$\mathcal{P}_{\{i_1, i_2, \ldots, i_k\}}(p_j) = \Pr(S(\mathbf{x} \in \{i_1, i_2, \ldots, i_k\} | H_0) = \sum_{t=1}^{k} \Pr(S = i_t | p_j)$$

$$= \sum_{t=1}^{k} C_{6, i_t} p_j^{i_t} (1 - p_j)^{6 - i_t}.$$

The example cited obtained $\{5, 6\}$ for the minimax rule. Its error sizes are

$$\alpha_{\{5,6\}} = \mathcal{P}_{\{5,6\}}(p_0) = C_{6,5}(0.4)^5(0.6)^1 + C_{6,6}(0.4)^6(0.6)^0 = 0.04096$$

$$\beta_{\{5,6\}} = 1 - \mathcal{P}_{\{5,6\}}(p_1) = 1 - \left[C_{6,5}(0.6)^5(0.4)^1 + C_{6,6}(0.6)^6(0.4)^0 \right] = 0.76672.$$

While the type I error size is reassuringly small, the type II error size is not. Recall, however, the circumstance under which this rule was selected. A type I error incurred a cost of 10, compared with a cost of 1 for a type II error. It is reasonable then to expect a much lower type I error probability. A similar comment applies to the mixed minimax rule $D = (0.789)\{5, 6\} + (0.211)\{4, 5, 6\}$, for which the expected error sizes are

$$\alpha_D = (0.789)(0.04096)$$
$$+ (0.211) \left[C_{6,4}(0.4)^4(0.6)^2 + C_{6,5}(0.4)^5(0.6)^1 + C_{6,6}(0.4)^6(0.6)^0 \right]$$
$$= 0.03736$$

$$\beta_D = (0.789)(0.76672)$$
$$+ (0.211) \left[1 - C_{6,4}(0.6)^4(0.4)^2 - C_{6,5}(0.6)^5(0.4)^1 - C_{6,6}(0.6)^6(0.4)^0 \right]$$
$$= 0.70109.$$

The Bayes rule $\{2,3,4,5,6\}$ resulted from the further assumption that p_0 was a very unlikely state of nature. The prior distribution was $\Pr(p_0) = 1/49, \Pr(p_1) = 48/49$. The Bayes rule's error sizes are

$$\alpha_{\{2,3,4,5,6\}} = \sum_{k=2}^{6} C_{6,k}(0.4)^k (0.6)^{6-k} = 0.76672$$

$$\beta_{\{2,3,4,5,6\}} = 1 - \sum_{k=2}^{6} C_{6,k}(0.6)^k (0.4)^{6-k} = 0.04096,$$

which is just the reverse of those for the minimax rule. The prior assumption about the state of nature has forced a rule that allows a high probability of a type I error. After all, if p_0 occurs very infrequently, we will seldom have occasion to reject it erroneously. Therefore, we will seldom encounter the situation where the high α works to our disadvantage. $\square$

4.54 Definition: When using rule C over sample $\mathbf{X}$ to test a simple hypothesis, $H_0 = \{\theta_0\}$, against a simple alternative, $H_1 = \{\theta_1\}$, we refer to the type I error size, α_C, as the *significance level* of the test. We call the complement of the type II error size, $1 - \beta_C$, the *power* of the test. Note the distinction between the *power function*, which is a function of θ, and the *power* of the text, which is the value of the power function at θ_1. ∎

In testing a simple hypothesis, $H_0 = \{\theta_0\}$, against a simple alternative, $H_1 = \{\theta_1\}$, a decision rule C has $\alpha_C = \mathcal{P}_C(\theta_0)$ and $1 - \beta_C = \mathcal{P}_C(\theta_1)$. Assuming that a type I error is the more serious, we first guard against it by insisting on a low significance level, typically $0.05, 0.01$, or 0.005. If a decision rule uses a significance level of 0.05, for example, and rejects the null hypothesis, we say that it rejects at the 5% level. This means that there is a 5% chance that the rejection is erroneous. Equivalently, we can assert that either our rejection is correct or we have observed a sample point that has only one chance in 20 of occurring. We normally set the significance level such that we reject the null hypothesis when the only other possibility is the occurrence of a very improbable event. The quantity $1 - \beta_C$ measures the power of the test to correctly reject the null hypothesis when its alternative actually prevails. Having set α_C to obtain the desired assurance against a type I error, we then select for the highest power (lowest type II error) across the remaining qualified decision rules.

4.55 Definition: Suppose that we are testing a simple hypothesis against a simple alternative. A test C is called a *most significant test* if the significance level of C, α_C, is less than or equal to the significance level of D, α_D, for all tests D with $\beta_D \le \beta_C$. That is, among all tests that do not exceed the type II error size of C, none has a smaller type I error size. ∎

4.56 Definition: Suppose that we are testing a simple hypothesis against a simple alternative. A test C is called a *most powerful test* if the power of C, $1 - \beta_C$, is greater than or equal to the power of D, $1 - \beta_D$, for all tests D with $\alpha_D \le \alpha_C$. That is, among all tests that do not exceed the type I error size of C, none has a smaller type II error size. ∎

A most powerful, most significant test occupies a privileged position on the risk plot. This follows because the risk coordinates are direct functions of α_C and β_C. In particular, assuming that C is a mixed rule arising from pure rules $C_1, C_2, \ldots, C_k$ via probability vector $p_1, p_2, \ldots, p_k$, we have

$$\hat{L}(C, \theta_0) = \sum_{i=1}^{k} p_i \hat{L}(C_i, \theta_0) = \sum_{i=1}^{k} p_i A \alpha_{C_i} = A \alpha_C$$

$$\hat{L}(C, \theta_1) = \sum_{i=1}^{k} p_i \hat{L}(C_i, \theta_1) = \sum_{i=1}^{k} p_i B \beta_{C_i} = B \beta_C,$$

where $A, B > 0$ are the costs for incorrect decisions under $H_0 = \theta_0$ and $H_1 = \theta_1$, respectively. The significance levels and powers functions of competing tests then specify the risk plot, which in turn permits construction of the minimax and Bayes decision rules. Because undominated decision rules are the only true contenders in these constructions, it is helpful to know that a most significant and most powerful test is undominated. Indeed, if D is such a test, appearing at $(A\alpha_D, B\beta_D)$ on the risk plot, a dominating test C must exhibit no larger risks under either θ_0 or θ_1, and it must show a smaller risk for at least one of the two states of nature. Suppose that $A\alpha_C < A\alpha_D$ and $B\beta_C \leq B\beta_D$. Then $\alpha_C < \alpha_D$, and $\beta_C \leq \beta_D$, which contradicts the fact that D is a most significant test. A similar conclusion follows if the smaller risk occurs under θ_1. Consequently, no test dominates D.

4.57 Theorem: In the case of a simple hypothesis versus a simple alternative, no test or convex combination of tests dominates a most significant, most powerful test.

PROOF: See discussion above. ∎

Our goal now is to identify an appropriate sufficient statistic, to construct most significant and most powerful tests over it, and to relate these tests to the minimax and Bayes decision rules.

A statistic aggregates the sample points into lumps, and each candidate critical area contains selected lumps. The number of candidate critical areas (i.e., the number of candidate decision rules) is then much less than the number of subsets of sample points. A sufficient statistic has the further advantage that it retains all information relevant to an optimal decision. The following statistic is appropriate for testing a simple hypothesis against a simple alternative.

4.58 Definition: Let X have two distributions $\text{Pr}_{X|\theta_i}(\cdot)$, corresponding to two states of nature θ_0 and θ_1, and let $\mathbf{X} = (X_1, X_2, \ldots, X_N)$ be a sample. The quantity

$$R(\mathbf{x}) = \frac{\text{Pr}(\mathbf{X} = \mathbf{x} | \theta = \theta_0)}{\text{Pr}(\mathbf{X} = \mathbf{x} | \theta = \theta_1)}$$

is called the *likelihood ratio statistic*. ∎

It is convenient to have the likelihood ratio a well-defined positive number for each sample point. Sample points are vectors $(x_1, x_2, \ldots, x_N)$, whose components come from the range of X, the underlying random variable. Because X has discrete distributions under both states of nature, the range of X will not include points that exhibit zero probability for both states. Moreover, if a point has zero probability for just one state, then an observation of that point identifies the state of nature with certainty, and we can take an optimal decision without embarking on a statistical study. For the purposes of developing a statistical rule, we can therefore remove these points from the range. (We must, of course, adjust the distributions to the conditional distributions given that these problematic points have not occurred.) The remaining points give rise to samples $\mathbf{x} = (x_1, x_2, \ldots, x_N)$ for which

$$\Pr(\mathbf{X} = \mathbf{x}|\theta_i) = \Pr(X = x_1|\theta_i) \cdot \Pr(X = x_2|\theta_i) \cdots \Pr(X = x_N|\theta_i) > 0.$$

The likelihood ratio is a well-defined positive number on these points, and we will assume this context whenever the statistic occurs.

4.59 Theorem: The likelihood ratio statistic is minimal sufficient.

PROOF: Let R be the likelihood ratio statistic over sample $\mathbf{X} = (X_1, \ldots, X_N)$ over a population X having distributions for two states of nature, θ_0 and θ_1. For notational convenience, we abbreviate $\Pr(\mathbf{X} = \mathbf{x}|\theta = \theta_i)$ as $\Pr(\mathbf{x}|\theta_i)$. Then $R(\mathbf{x}) = \Pr(\mathbf{x}|\theta_0)/\Pr(\mathbf{x}|\theta_1)$. Let

$$f_1(R(\mathbf{x}), \theta) = \begin{cases} [R(\mathbf{x})]^{1/2}, & \text{if } \theta = \theta_0 \\ [R(\mathbf{x})]^{-1/2}, & \text{if } \theta = \theta_1 \end{cases}$$

$$f_2(\mathbf{x}) = [\Pr(\mathbf{x}|\theta_0) \cdot \Pr(\mathbf{x}|\theta_1)]^{1/2}.$$

Then

$$f_1(R(\mathbf{x}), \theta) \cdot f_2(\mathbf{x}) = \begin{cases} \left[\dfrac{\Pr(\mathbf{x}|\theta_0)}{\Pr(\mathbf{x}|\theta_1)}\right]^{1/2} \cdot [\Pr(\mathbf{x}|\theta_0) \cdot \Pr(\mathbf{x}|\theta_1)]^{1/2}, & \text{if } \theta = \theta_0 \\[3ex] \left[\dfrac{\Pr(\mathbf{x}|\theta_1)}{\Pr(\mathbf{x}|\theta_0)}\right]^{1/2} \cdot [\Pr(\mathbf{x}|\theta_0) \cdot \Pr(\mathbf{x}|\theta_1)]^{1/2}, & \text{if } \theta = \theta_1 \end{cases}$$

$$= \begin{cases} \Pr(\mathbf{x}|\theta_0), & \text{if } \theta = \theta_0 \\ \Pr(\mathbf{x}|\theta_1), & \text{if } \theta = \theta_1. \end{cases}$$

By the factorization criterion, Theorem 4.40, R is sufficient. Now suppose that T is another sufficient statistic. Since $(\mathbf{X} = \mathbf{z}) \subset (T = T(\mathbf{z}))$, we have, for any sample point $\mathbf{z}$,

$$\begin{aligned} R(\mathbf{z}) &= \frac{\Pr(\mathbf{z}|\theta_0)}{\Pr(\mathbf{z}|\theta_1)} = \frac{\Pr((\mathbf{X} = \mathbf{z}) \cap (T = T(\mathbf{z})) \cap (\theta = \theta_0))/\Pr(\theta = \theta_0)}{\Pr((\mathbf{X} = \mathbf{z}) \cap (T = T(\mathbf{z})) \cap (\theta = \theta_1))/\Pr(\theta = \theta_1)} \\[2ex] &= \frac{\Pr(\mathbf{X} = \mathbf{z}|(T = T(\mathbf{z})) \cap (\theta = \theta_0)) \cdot \Pr((T = T(\mathbf{z})) \cap (\theta = \theta_0))/\Pr(\theta = \theta_0)}{\Pr(\mathbf{X} = \mathbf{z}|(T = T(\mathbf{z})) \cap (\theta = \theta_1)) \cdot \Pr((T = T(\mathbf{z})) \cap (\theta = \theta_1))/\Pr(\theta = \theta_1)} \\[2ex] &= \frac{\Pr(T = T(\mathbf{z})|\theta = \theta_0)}{\Pr(T = T(\mathbf{z})|\theta = \theta_1)}. \end{aligned}$$

The last reduction follows because T is sufficient, which forces the expression $\Pr(\mathbf{X} = \mathbf{z}|(T = T(\mathbf{z})) \cap (\theta = \theta_i))$ to be independent of θ_i.

Now, if $T(\mathbf{x}) = T(\mathbf{y})$, we have

$$R(\mathbf{x}) = \frac{\Pr(T = T(\mathbf{x})|\theta = \theta_0)}{\Pr(T = T(\mathbf{x})|\theta = \theta_1)} = \frac{\Pr(T = T(\mathbf{y})|\theta = \theta_0)}{\Pr(T = T(\mathbf{y})|\theta = \theta_1)} = R(\mathbf{y}).$$

This means that each component of the T partition lies entirely within some component of the R partition, which implies that $R = h(T)$ for some function h. Therefore, R is minimal sufficient. ∎

4.60 Definition: A *threshold test*, in the context of a simple hypothesis versus a simple alternative, is a test whose critical area is $\{\mathbf{x}|T(\mathbf{x}) < t\}$, for some real-valued statistic T and some threshold value t. ∎

Example 4.50 used the sample sum, which assumed integral values between 0 and 6. There were 128 potential decision rules, each corresponding to points with sample sums in a particular subset of these 7 values. Among these rules, however, are just 8 threshold tests: $S < t$ for $t = 0, 1, 2, \ldots, 7$. If we can justify restricting our search to threshold rules, we find a much smaller field of contenders. This implies, of course, that our procedure for selecting the optimal rule is much less difficult. Our first result in this direction shows that a threshold test on the likelihood ratio statistic is a most significant and most powerful test. Eventually, we show that only threshold rules need be considered in a search for the optimal rule.

4.61 Theorem: (Neyman-Pearson) In the context of a simple hypothesis, $H_0 = \{\theta_0\}$, versus a simple alternative, $H_1 = \{\theta_1\}$, the threshold test with critical area $R(\mathbf{x}) = \Pr(\mathbf{X} = \mathbf{x}|\theta_0)/\Pr(\mathbf{X} = \mathbf{x}|\theta_1) < t$ is most significant and most powerful.

PROOF: Because R assumes only positive values, we can, without loss of generality, take $t > 0$. Let D be the threshold test $R < t$. We first show that it is most powerful. To this end, consider an arbitrary test C, which is a convex combination of $C_1, C_2, \ldots, C_k$ via probability vector $p_1, p_2, \ldots, p_k$. Note that C could be a pure test, if a single p_i carries all the probability, so the argument below compares the threshold test D with both pure and mixed competitors. We must verify that $\alpha_C \leq \alpha_D$ implies that $\beta_D \leq \beta_C$. Note that $(D - C_i)$ and $(D \cap C_i)$ are disjoint events with union D. Similarly, $(C_i - D)$ and $(C_i \cap D)$ are disjoint events with union C_i. Consequently,

$$\begin{aligned}
\beta_C - \beta_D &= \sum_{i=1}^{k} p_i(1 - \beta_D) - \sum_{i=1}^{k} p_i(1 - \beta_{C_i}) \\
&= \sum_{i=1}^{k} p_i[\Pr(\mathbf{X} \in D|\theta_1) - \Pr(\mathbf{X} \in C_i|\theta_1)] \\
&= \sum_{i=1}^{k} p_i\,[\Pr(\mathbf{X} \in D - C_i|\theta_1) + \Pr(\mathbf{X} \in C_i \cap D|\theta_1) \\
&\qquad - \Pr(\mathbf{X} \in C_i - D|\theta_1) - \Pr(\mathbf{X} \in C_i \cap D|\theta_1)]
\end{aligned}$$

$$\beta_C - \beta_D = \sum_{i=1}^{k} p_i \left[\sum_{\mathbf{x}|\mathbf{x}\in D-C_i} \Pr(\mathbf{X}=\mathbf{x}|\theta_1) - \sum_{\mathbf{x}|\mathbf{x}\in C_i-D} \Pr(\mathbf{X}=\mathbf{x}|\theta_1) \right].$$

In the first inner sum, $\mathbf{x} \in D$, so $R(\mathbf{x}) = \Pr(\mathbf{X}=\mathbf{x}|\theta_0)/\Pr(\mathbf{X}=\mathbf{x}|\theta_1) < t$. Equivalently, the summand is greater than $(1/t)\Pr(\mathbf{X}=\mathbf{x}|\theta_0)$. In the second inner sum, $\mathbf{x} \notin D$, so $\Pr(\mathbf{X}=\mathbf{x}|\theta_0) \geq t \cdot \Pr(\mathbf{X}=\mathbf{x}|\theta_1)$, which renders the summand less than or equal to $(1/t)\Pr(\mathbf{X}=\mathbf{x}|\theta_0)$. Consequently,

$$\beta_C - \beta_D \geq \frac{1}{t} \sum_{i=1}^{k} p_i \left[\sum_{\mathbf{x}|\mathbf{x}\in D-C_i} \Pr(\mathbf{X}=\mathbf{x}|\theta_0) - \sum_{\mathbf{x}|\mathbf{x}\in C_i-D} \Pr(\mathbf{X}=\mathbf{x}|\theta_0) \right]$$

$$= \frac{1}{t} \sum_{i=1}^{k} p_i [\Pr(\mathbf{X} \in D - C_i|\theta_0) - \Pr(\mathbf{X} \in C_i - D|\theta_0)]$$

$$= \frac{1}{t} \sum_{i=1}^{k} p_i \Big\{ [\Pr(\mathbf{X} \in D - C_i|\theta_0) + \Pr(\mathbf{X} \in C_i \cap D|\theta_0)]$$

$$- [\Pr(\mathbf{X} \in C_i - D|\theta_0) + \Pr(\mathbf{X} \in C_i \cap D|\theta_0)] \Big\}$$

$$= \frac{1}{t} \sum_{i=1}^{k} p_i [\Pr(\mathbf{X} \in D|\theta_0) - \Pr(\mathbf{X} \in C_i|\theta_0)]$$

$$= \frac{1}{t} \left[\Pr(\mathbf{X} \in D|\theta_0) \left(\sum_{i=1}^{k} p_i \right) - \sum_{i=1}^{k} p_i \Pr(\mathbf{X} \in C_i|\theta_0) \right]$$

$$= \frac{1}{t}(\alpha_D - \alpha_C) \geq 0.$$

Hence $\beta_C \geq \beta_D$, which proves the likelihood ratio threshold test most powerful.

Reversing the roles of α and β in the argument above provides a proof that it is also most significant. In abbreviated form, the argument runs as follows.

$$\alpha_C - \alpha_D = \sum_{i=1}^{k} p_i [\Pr(\mathbf{X} \in C_i|\theta_0) - \Pr(\mathbf{X} \in D|\theta_0)]$$

$$= \sum_{i=1}^{k} p_i \left[\sum_{\mathbf{x}|\mathbf{x}\in C_i-D} \Pr(\mathbf{X}=\mathbf{x}|\theta_0) - \sum_{\mathbf{x}|\mathbf{x}\in D-C_i} \Pr(\mathbf{X}=\mathbf{x}|\theta_0) \right]$$

$$\geq t \cdot \sum_{i=1}^{k} p_i \left[\sum_{\mathbf{x}|\mathbf{x}\in C_i-D} \Pr(\mathbf{X}=\mathbf{x}|\theta_1) - \sum_{\mathbf{x}|\mathbf{x}\in D-C_i} \Pr(\mathbf{X}=\mathbf{x}|\theta_1) \right]$$

$$\geq t \cdot \sum_{i=1}^{k} p_i [\Pr(\mathbf{X} \in C_i - D|\theta_1) - \Pr(\mathbf{X} \in D - C_i|\theta_1)]$$

$$\alpha_C - \alpha_D \geq t \cdot \sum_{i=1}^{k} p_i[\Pr(\mathbf{X} \in C_i|\theta_1) - \Pr(\mathbf{X} \in D|\theta_1)]$$

Adding and subtracting 1 allows us to express these results in terms of the known β difference.

$$\alpha_C - \alpha_D \geq t \cdot \left[(1 - \Pr(\mathbf{X} \in D|\theta_1)) - \sum_{i=1}^{k} p_i(1 - \Pr(\mathbf{X} \in C_i|\theta_1))\right]$$

$$= t(\beta_D - \beta_C) \geq 0.$$

Consequently, if β_C does not exceed β_D, we must have $\alpha_C - \alpha_D \geq 0$, or equivalently, D exhibits a lower type I error probability than C. Hence D, the likelihood ratio threshold test, is most significant. ∎

Invoking Theorem 4.57, we conclude that the southwest rectangle below and to the right of a likelihood ratio threshold test, including the boundary lines, contains no convex combination of decision rules. These threshold tests must, therefore, lie on the boundary of the convex hull of the decision rules. Because we are working with discrete distributions, $R(\mathbf{x})$ assumes discrete values: $t_1 < t_2 < t_3 < \cdots$. Let $D_i = \{\mathbf{x}|R(\mathbf{x}) < t_i\}$. Then $\phi = D_1 \subset D_2 \subset D_3 \subset \ldots$. Since $\alpha_{D_i} = \Pr(\mathbf{x} \in D_i|\theta_0)$, it follows that $0 = \alpha_{D_1} < \alpha_{D_2} < \alpha_{D_3} < \cdots$. Similarly, $\beta_{D_i} = 1 - \Pr(\mathbf{x} \in D_i|\theta_1)$ implies that $1 = \beta_{D_1} > \beta_{D_2} > \beta_{D_3} > \cdots$. The tests $D_1, D_2, \ldots$ therefore form a piecewise linear frontier that stretches from the northwest to the southeast on the risk plot. We shall see that this frontier is the boundary of the convex hull of all decision rules, which means that the mixed minimax and Bayes rules will always be threshold tests.

First, however, we present an example to illustrate the central role of threshold rules. The example also demonstrates that a statistic that rises or falls monotonically with the likelihood ratio serves as a proxy for that ratio. Such a proxy statistic may be more easily accessible than the likelihood ratio.

4.62 Example: Recall the coin-flipping test of Example 4.50, in which a biased coin has the probability of heads either $p_0 = 0.4$ or $p_1 = 0.6$. We again probe the situation by flipping the coin $N = 6$ times, obtaining a sample $\mathbf{X} = \mathbf{x}$, but we now use the likelihood ratio statistic to pursue the analysis. As before, let $H_0 = \{p_0\}, H_1 = \{p_1\}$. Writing the sample as $\mathbf{x} = (x_1, x_2, \ldots, x_6)$, a vector of ones and zeros, we have

$$R(\mathbf{x}) = \frac{\Pr(\mathbf{X} = \mathbf{x}|\theta_0)}{\Pr(\mathbf{X} = \mathbf{x}|\theta_1)} = \frac{(0.4)^{x_1+x_2+\cdots+x_6}(0.6)^{6-x_1-x_2-\cdots-x_6}}{(0.6)^{x_1+x_2+\cdots+x_6}(0.4)^{6-x_1-x_2-\cdots-x_6}}$$

$$= \left(\frac{0.4}{0.6}\right)^{S(\mathbf{x})}\left(\frac{0.6}{0.4}\right)^{6-S(\mathbf{x})} = \left(\frac{2}{3}\right)^{2S(\mathbf{x})-6},$$

where S is the sample sum. We can therefore compute the R values from the S values and vice versa, noting that the smallest R arises from the largest S. Let $r_0, \ldots, r_6$ be the likelihood ratios corresponding to $S = 0, \ldots, 6$. Then $\Pr(R <$

$r_i | \theta_j) = \Pr(S > i | \theta_j)$, and

$$\alpha_{R < r_i} = \Pr(R < r_i | \theta_0) = \Pr(S > i | \theta_0) = \sum_{j=i+1}^{6} C_{6,j} (0.4)^j (0.6)^{6-j}$$

$$\beta_{R < r_i} = 1 - \Pr(R < r_i | \theta_1) = 1 - \Pr(S > i | \theta_1) = \sum_{j=0}^{i} C_{6,j} (0.6)^j (0.4)^{6-j}.$$

These values appear in the tabulation to the right, where we include the case $i = -1$ to capture the threshold rule that includes all sample values in its critical area. $S > -1$ includes all sample values, as does $R < 12.0000$. The threshold rules $D_i = \{R < r_i\}$ are identical with the threshold rules $\{S > i\}$ from the earlier example. In order of increasing α, they appear in Table 4.9 as entries $0, 2, 4, 8, 12, 14, 17$, and 24. Figure 4.15 shows how these rules form the southwest boundary of the convex hull of the 128

i	r_i	$\alpha_{R<r_i}$	$\beta_{R<r_i}$
6	0.0878	0.0000	1.0000
5	0.1975	0.0041	0.9533
4	0.4444	0.0410	0.7667
3	1.0000	0.1792	0.4557
2	2.2500	0.4557	0.1792
1	5.0625	0.7667	0.0410
0	11.3906	0.9533	0.0041
-1	12.0000	1.0000	0.0000

possible decision rules. This explains why the mixed minimax rule, a linear combination of $S > 4$ and $S > 3$, and the Bayes rule, $S > 1$, both assume the threshold form. □

Our goal is to show that we need consider only likelihood ratio threshold rules in constructing the mixed minimax or Bayes test in the case of a simple hypothesis against a simple alternative. At this point, we know that such rules lie on the southwest boundary of the convex hull of all rules, but we have not eliminated the possibility that some nonthreshold rule also appears as a vertex on that boundary. The next theorem wraps up this last detail.

4.63 Theorem: In the case of a simple hypothesis, $H_0 = \{p_0\}$, versus a simple alternative, $H_1 = \{p_1\}$, the mixed minimax and Bayes decision rules constructed with the likelihood ratio statistic are threshold tests.

PROOF: Let $0 < t_1 < t_2 < \ldots$ be the range of the likelihood ratio, and let D_i, with critical area $(R(\mathbf{x}) < t_i)$ be the corresponding likelihood ratio threshold test. Then $\phi = D_1 \subset D_2 \subset \ldots$, which implies that the corresponding α values are increasing and the β values are decreasing.

If $A, B > 0$ are the costs of type I and type II errors, respectively, the D_i appear on the risk plot at points $(A\alpha_{D_i}, B\beta_{D_i})$, which, being undominated, are vertex points of the convex hull of all decision rules over the likelihood ratio statistic. To prove the theorem, it suffices to show that the piecewise linear trace through these vertices is the southwest boundary of the convex hull.

Figure 4.16 illustrates the situation. We must show that all nonthreshold tests lie above the linear segments connecting the threshold tests. The critical area D_i contains points $\mathbf{x}$ with $R(\mathbf{x}) < t_i$. That is, it contains points with $R(\mathbf{x}) \in \{t_1, t_2, \ldots, t_{i-1}\}$.

Consider D_{i+1}, the next threshold test along the boundary. We obtain critical area D_{i+1} by adding to D_i those points with $R(\mathbf{x}) = t_i$. For these new points, $\Pr(\mathbf{x}|\theta_0) = t_i \Pr(\mathbf{x}|\theta_1)$. In adding these points, we increase α by the

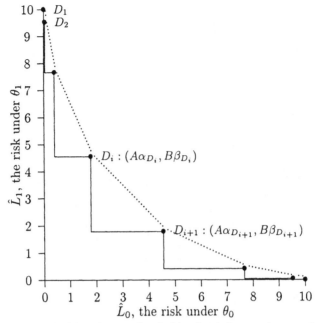

Figure 4.16. Likelihood ratio threshold rules define vertices on the convex hull of all rules (see Theorem 4.63)

θ_0 probabilities of the new points, and we decrease β by the θ_1 probabilities. That is,

$$\alpha_{D_{i+1}} = \alpha_{D_i} + \sum_{\mathbf{x}|R(\mathbf{x})=t_i} \Pr(\mathbf{X} = \mathbf{x}|\theta_0)$$

$$\beta_{D_{i+1}} = \beta_{D_i} - \sum_{\mathbf{x}|R(\mathbf{x})=t_i} \Pr(\mathbf{X} = \mathbf{x}|\theta_1).$$

The slope of the line connecting D_i and D_{i+1} is then

$$\frac{B(\beta_{D_{i+1}} - \beta_{D_i})}{A(\alpha_{D_{i+1}} - \alpha_{D_i})} = \frac{-B \sum_{\mathbf{x}|R(\mathbf{x})=t_i} \Pr(\mathbf{X} = \mathbf{x}|\theta_1)}{A \sum_{\mathbf{x}|R(\mathbf{x})=t_i} \Pr(\mathbf{X} = \mathbf{x}|\theta_0)}$$

$$= \frac{-B \sum_{\mathbf{x}|R(\mathbf{x})=t_i} \Pr(\mathbf{X} = \mathbf{x}|\theta_1)}{At_i \sum_{\mathbf{x}|R(\mathbf{x})=t_i} \Pr(\mathbf{X} = \mathbf{x}|\theta_1)} = \frac{-B}{At_i}.$$

Since $0 < t_1 < t_2 < \ldots$, the slopes of the lines connecting $D_1, D_2, \ldots$ become increasingly shallow, as suggested by Figure 4.16. The argument generalizes to critical areas for tests that are not necessarily threshold tests. Suppose that a test E contains points $\mathbf{x}$ such that $R(\mathbf{x}) \in \{t_{i_1}, t_{i_2}, \ldots\}$. The t_{i_j} are a subset of range(R) but not necessarily a contiguous one starting with t_1. If we generate a new test E' by adding points with $R(\mathbf{x}) = t$, for some new value $t \notin \{t_{i_1}, t_{i_2}, \ldots\}$, we obtain a test E' on the risk plot that is $B \cdot \Pr(R = t|\theta_1)$ below and $At \cdot \Pr(R = t|\theta_1)$ to the right of E. Graphically, we suspend a triangle from the old test such that the sides are parallel with the axes. The end of the hypotenuse locates the new test. The available triangles appear in Figure 4.16, descending from the point $(0, B)$ in order of increasing values from range(R). Suppose that we construct E by adding points to an initially empty critical area. We first add the points associated with the smallest range(R) value in E. Then we add the points with the next larger range value, and so forth. This process deploys the triangles as shown in the figure, as long as we do not skip any range values. If E is not a threshold test, however, we must eventually skip a value, which omits one of the triangles. Subsequent triangles must have shallower hypotenuse slopes than the omitted one, so they displace E to the right of the boundary curve. We conclude that nonthreshold tests appear above the southwest boundary defined by the threshold tests.

Analytically, we proceed as follows. Let p_n and q_n be the amounts by which the type I error increases and the type II error decreases when the points $\{\mathbf{x} | R(\mathbf{x}) = t_n\}$ join the critical area. That is,

$$p_n = \Pr(R = t_n | \theta_0)$$
$$q_n = \Pr(R = t_n | \theta_1).$$

From the definition of the likelihood ratio,

$$p_n = \sum_{\mathbf{x} | R(\mathbf{x}) = t_n} \Pr(\mathbf{X} = \mathbf{x} | \theta_0) = \sum_{\mathbf{x} | R(\mathbf{x}) = t_n} t_n \Pr(\mathbf{X} = \mathbf{x} | \theta_1) = t_n q_n.$$

Threshold test D_i then appears at $(A \sum_{j=1}^{i-1} p_j, B(1 - \sum_{j=1}^{i-1} q_j))$ on the risk plot. Consequently, the line between D_i and D_{i+1} satisfies the equation

$$\hat{L}_1 = B \left(1 - \sum_{j=1}^{i-1} q_j \right) - \frac{Bq_i}{Ap_i} \left(\hat{L}_0 - A \sum_{j=1}^{i-1} p_j \right).$$

An arbitrary test C contains points $\mathbf{x}$ with $R(\mathbf{x}) \in \{t_{i_1}, t_{i_2}, \ldots\}$, where $t_{i_1}, t_{i_2}, \ldots$ comprise an arbitrary subset from range(R). This test appears on the risk plot at $(A \sum_k p_{i_k}, B(1 - \sum_k q_{i_k}))$. Hence, C is above the line between D_i and D_{i+1} if

$$B \left(1 - \sum_k q_{i_k} \right) \geq B \left(1 - \sum_{j=1}^{i-1} q_i \right) - \frac{Bq_i}{Ap_i} \left(A \sum_k p_{i_k} - A \sum_{j=1}^{i-1} p_j \right).$$

Equivalently,

$$\sum_{j=1}^{i-1} q_j - \sum_k q_{i_k} \geq \frac{1}{t_i} \left(\sum_{j=1}^{i-1} t_j q_j - \sum_k t_{i_k} q_{i_k} \right)$$

$$\sum_{j=1}^{i-1} (t_i - t_j) q_j \geq \sum_k (t_i - t_{i_k}) q_{i_k}.$$

This last inequality is certainly true. If we let $I = \{i_1, i_2, \ldots\} \cap \{1, 2, \ldots, i-1\}$, we can rewrite it as

$$\sum_{j \in \{1,2,\ldots,i-1\}-I} (t_i - t_j) q_j \geq \sum_{i_k \geq i} (t_i - t_{i_k}) q_{i_k}.$$

The left side is nonnegative because $j < i$ implies that $t_i > t_j$; the right side is nonpositive because $i_k \geq i$ implies that $t_{i_k} \geq t_i$. We conclude that the test C lies above, or on, the line defined by D_i and D_{i+1}. Because i was arbitrary, C lies above or on all such boundary line segments. Finally, if C is the convex combination of pure tests $C_1, C_2, \ldots, C_k$, it appears on the risk plot inside the polygon whose vertices are the C_j risks. Therefore, C also lies on or above all lines determined by D_i and D_{i+1}. This verifies that the piecewise linear trajectory connecting the threshold tests in order of increasing type I error is the southwest boundary of the convex hull. ∎

The analysis of a simple hypothesis versus a simple alternative reduces to a threshold test of the likelihood ratio statistic. We can substitute another statistic, T, for the likelihood ratio if it is monotone in the sense that $T(\mathbf{x}) < T(\mathbf{y})$ if and only if $R(\mathbf{x}) < R(\mathbf{y})$. A threshold test using R then corresponds to a threshold test using T.

4.64 Example: Recall the situation of Examples 4.50 and 4.62. We have a biased coin, for which the probability of heads is either $p_0 = 0.4$ or $p_1 = 0.6$. The Bernoulli random variable X is 1 if a coin flip shows heads. We take a sample, $\mathbf{x} = (x_1, x_2, \ldots, x_6)$, which we use to test $H_0 = \{p_0\}$ versus the simple alternative $H_1 = \{p_1\}$. Given that the costs of type I and type II errors are 10 and 1, respectively, we seek the mixed minimax rule and the Bayes rules under a prior distribution with $\Pr(p_0) = 1/49$. Example 4.62 calculated the relationship between the likelihood ratio, $R(\mathbf{x})$, and its monotone counterpart $S(\mathbf{x})$, the sample sum: $R(\mathbf{x}) = (2/3)^{2S(\mathbf{x})-6}$.

The negation of the sample sum, $-S(\mathbf{x})$, defines the same partition as the likelihood ratio, and it is monotone in the sense that $S(\mathbf{x}) > S(\mathbf{y})$ if and only if $-S(\mathbf{x}) < -S(\mathbf{y})$ if and only if $R(\mathbf{x}) < R(\mathbf{y})$. The Bayes rule will therefore be a test of the form $-S < k$, or equivalently, $S > k'$. The mixed minimax rule will be a convex combination of such tests. The important point is that there are only eight candidate tests, corresponding to $S > k'$ for $k' \in \{-1, 0, 1, 2, 3, 4, 5, 6\}$, whereas there were 128 candidates in our original formulation. From this point the search proceeds as in the preceding example, but the smaller field of candidates allows manual computations, in contrast with the preceding example, which used a computer program to identify the undominated rules. Hence, we find the same mixed minimax and Bayes rules as before. Specifically, the minimax rule rejects with probability 0.789 when $S > 4$ and with probability 0.211 when $S > 3$. The Bayes rule rejects when $S > 1$. □

Sometimes the error costs and the prior probabilities are not part of the problem. In a frequently occurring situation, the statistician must assess the *statistical significance* of the evidence against a null hypothesis. In this context, called significance testing, the error costs do not enter the analysis directly. Instead, we design a sampling test with critical area corresponding to a prespecified, acceptably low α, the probability of a type I error. Among tests with this α or lower, we naturally select one that has the lowest probability of a type II error. In other words, we choose a most powerful test, which will be a threshold test on the likelihood ratio or on some equivalent statistic. We then observe the sample, and if it falls in the critical area, we characterize it as statistically significant at the α level. Error costs can enter indirectly. If a type I error is more costly than a type II error, we can lower the initial α, which renders the more costly error less likely. In designing the test, we can adjust the size of the sample and the threshold value. Because the distributions are discrete, we have available only a discrete collection of threshold values. The resulting candidates each exhibit a particular α, and we might not find the desired value among them. In that case, we use a convex combination of threshold tests to obtain the required α. The next example illustrates the point. In a significance test, the alternative, H_1, is usually composite, but this example manages to remain in the realm of simple hypothesis versus simple alternative.

4.65 Example: Our coin mint produces biased coins, which, when flipped, land heads up with probability $p_0 = 0.4$. We are replacing it with a new one that will produce unbiased coins ($p_1 = 0.5$). The company that performs the installation insists on working in absolute secrecy to protect its proprietary technology. When the installation is complete, there is no apparent physical change. The new mint looks exactly like the old mint. By repeatedly flipping a coin from the allegedly new mint, we want to test the null hypothesis that no installation actually took place. In other words, we want to test $H_0 = \{p_0\}$ versus $H_1 = \{p_1\}$.

We select a significance level, $\alpha = 0.01$. Suppose that we flip the coin N times to produce the sample $\mathbf{x} = (x_1, x_2, \dots, x_N)$. The likelihood ratio is

$$R(\mathbf{x}) = \frac{\Pr(\mathbf{X} = \mathbf{x}|p_0)}{\Pr(\mathbf{X} = \mathbf{x}|p_1)} = \frac{p_0^{S(\mathbf{x})}(1 - p_0)^{N - S(\mathbf{x})}}{p_1^{S(\mathbf{x})}(1 - p_1)^{N - S(\mathbf{x})}} = (1.2)^N \cdot \left(\frac{2}{3}\right)^{S(\mathbf{x})},$$

where $S(\mathbf{x})$ is the sample sum. The negation of the sample sum then defines the same partition as the likelihood ratio, and it is monotone with it. Threshold tests of the form $S > t$, for $t \in \{-1, 0, 1, \dots, N\}$ are the candidates. Each is a most powerful test with

$$\alpha_t = \Pr(S > t|p_0) = \sum_{i=t+1}^{N} C_{N,i} p_0^i (1 - p_0)^{N-i}$$

$$\beta_t = 1 - \Pr(S > t|p_1) = 1 - \sum_{i=t+1}^{N} C_{N,i} p_1^i (1 - p_1)^{N-i} = \sum_{i=0}^{t} C_{N,i} p_1^i (1 - p_1)^{N-i}.$$

It follows that $1 = \alpha_{-1} > \alpha_0 > \alpha_1 > \cdots > \alpha_N = 0$ and $0 = \beta_{-1} < \beta_0 < \beta_1 < \cdots < \beta_N = 1$. For various N, we tabulate the thresholds that provide α values bracketing the desired 0.01. We also note the β values. The results appear in Table 4.10.

For $N = 50$, for example, the test that rejects when $S > 27$ has $\alpha = 0.0160$, while the test that rejects when $S > 28$ has $\alpha = 0.0076$. We can construct a mixed

N	t	α_t	β_t	α_{t+1}	β_{t+1}
10	7	0.0123	0.9453	0.0017	0.9893
20	12	0.0210	0.8684	0.0065	0.9423
30	17	0.0212	0.8192	0.0083	0.8998
40	22	0.0189	0.7852	0.0083	0.8659
50	27	0.0160	0.7601	0.0076	0.8389
60	32	0.0133	0.7405	0.0066	0.8169
70	37	0.0109	0.7248	0.0056	0.7985
80	41	0.0158	0.6312	0.0088	0.7118
90	46	0.0126	0.6240	0.0071	0.7008
100	51	0.0100	0.6178	0.0058	0.6914

TABLE 4.10. For various sample sizes, thresholds that bracket the desired significance level $\alpha = 0.01$ (see Example 4.65)

test to obtain an exact $\alpha = 0.01$:

$$q(0.0160) + (1 - q)(0.0076) = 0.0100$$
$$q = 0.286.$$

To obtain $\alpha = 0.01$, we apply critical area $S > 27$ with probability 0.286 and $S > 28$ with probability 0.714. In practice, we observe the sample, reject H_0 if $S > 28$, accept H_0 if $S < 28$, and consult a uniform random number generator if $S = 28$. Assuming that the generator dispenses values uniformly in the range $[0, 1)$, we reject on $S = 28$ if the generator returns a value less than 0.286. Otherwise, we accept on $S = 28$. Observations $S > 28$, and some proportion of the observations $S = 28$, are deemed statistically significant at the 0.01 level. Equivalently, they are called significant at the 1% level. Upon observing a significant sample, we reject the null hypothesis. In the case at hand, we conclude that the new mint is now in operation, and we pay the installation fee. If the new mint has not been installed, there is only one chance in one hundred that this conclusion is erroneous. However, if the new mint has indeed been installed, the error probability is between 0.76 and 0.84 that we will incorrectly conclude the contrary and withhold the rightfully due payment.

For $N = 50$, Figure 4.17 shows the approximate distributions of the sample sum under p_0 and p_1. The distributions are actually discrete, assuming values only on the integers, but the smoothed curve provides a more easily interpretable picture. The vertical line at $S = 28$ delineates the critical area for the threshold test discussed above. Note that only a small part of the p_0 distribution lies to the right of the line. This corresponds to the small type I error size, calculated above to be in the 0.01 range. Unfortunately, a large part of the p_1 distribution lies to the left of the line. This corresponds to the large type II error size, calculated above to be in the 0.76 to 0.84 range.

The β tabulation above shows a monotone decrease with increasing sample size. Distribution graphs for these varying sample sizes all have the same general appearance as Figure 4.17 but with the distinguishing difference that the curves for p_0 and p_1 are increasingly separated. A large enough sample size reduces $\beta < 0.01$ while retaining α in the same range. To compute the required sample size, we take advantage of the approximation (Equation 2.3):

$$\Pr(S = k) \approx [2\pi Np(1 - p)]^{-1/2} e^{-(k - Np)^2/2Np(1-p)}.$$

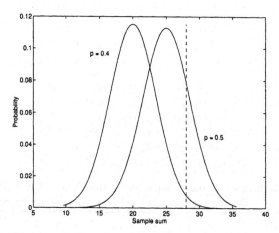

Figure 4.17. Approximate sample sum distributions for sample size $N = 50$ (see Example 4.65)

For large N, the threshold test $S > t$ has

$$\alpha = \sum_{k=t+1}^{N} \Pr(S = k|p_0) \approx \int_{t}^{\infty} \frac{1}{\sqrt{2\pi N p_0 (1 - p_0)}} \cdot e^{-(x - N p_0)^2 / 2 N p_0 (1 - p_0)} \, dx$$

$$\beta = 1 - \sum_{k=t+1}^{N} \Pr(S = k|p_1) \approx \int_{-\infty}^{t} \frac{1}{\sqrt{2\pi N p_1 (1 - p_1)}} \cdot e^{-(x - N p_1)^2 / 2 N p_1 (1 - p_1)} \, dx.$$

We set $\alpha = \beta = 0.01$ to obtain two equations in t and N. Chapter 6 develops the methods to solve such equations. For the moment, we merely note the answer: $N = 531.98, t = 239.11$. Since N must be an integer and increasing N lowers β for a fixed α, we choose $N = 532$. In this example, we flip a coin from the new process 532 times, and we note a statistically significant observation if $S > 239$. We pay the installation fee if $S > 239$, and we withhold it otherwise. Regardless of the installation status, we have only a 0.01 error probability. Using the larger sample, $N = 532$, Figure 4.18 illustrates the threshold position in the two sample sum distributions. In contrast with the smaller sample (Figure 4.17), the two distributions show little overlap. Moreover, the p_0 area to the right of the threshold, which is the type I error size, appears comparable to the p_1 area left of the threshold, which is the type II error size. □

Exercises

4.35 Write a computer program to identify the undominated decision rules in Example 4.50. Also report their risks, assuming that a type I error costs 10 and a type II error costs 1. This exercise verifies the results of Table 4.50.

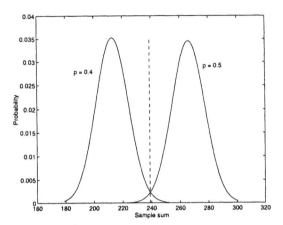

Figure 4.18. Separation of distributions achieved with sample size $N = 532$ (see Example 4.65)

4.36 Traffic conditions in a computer network are either good or bad. A probe produces the random variable X with the distributions to the right. Let H_0 be the hypothesis that traffic conditions are good. Assume that a type I error is 4 times

	$\Pr(X = x \mid \text{good/bad})$	
x	good	bad
0	0.2	0.9
1	0.8	0.1

as expensive as a type II error. For a sample of size 3, find the undominated decision rules over the sample mean statistic, and verify that the mixed minimax rule is a combination of threshold rules. Given the prior probability $\Pr(\text{good}) = 0.7$, verify that the Bayes rule is also a threshold rule.

4.37 Suppose that X has the following distributions for $i = 1, 2$.

$$\Pr(X = k) = \frac{\lambda_i^k e^{-\lambda}}{k!}, \qquad \text{for } k = 0, 1, 2, \dots$$

For a sample $\mathbf{X} = (X_1, X_2, \dots, X_N)$ from this population, construct the likelihood ratio statistic R, and find another statistic T that is monotone with it in the sense that $T(\mathbf{x}) < T(\mathbf{y})$ if and only if $R(\mathbf{x}) < R(\mathbf{y})$.

4.38 Let $R(\mathbf{x})$ be the likelihood ratio statistic. Show that $R'(\mathbf{x}) = 1/R(\mathbf{x})$ is minimal sufficient.

4.39 Suppose that X is binomially distributed with parameters $(10, p_0)$ or $(10, p_1)$, where $p_0 = 0.1$ and $p_1 = 0.9$. Devise a most powerful, most significant test of $H_0 = \{p_0\}$ versus $H_1 = \{p_1\}$ with $\alpha = 0.05$ and $\beta \leq 0.1$.

4.40 Let Y be geometrically distributed with parameter p_0 or p_1, where $p_0 = 0.2$ and $p_1 = 0.3$. Devise a most powerful, most significant test of $H_0 = \{p_0\}$ versus $H_1 = \{p_1\}$ with $\alpha = 0.05$ and $\beta \leq 0.1$.

4.41 Suppose that X has a negative binomial distribution with parameters $(10, p_0)$ or $(10, p_1)$, where $p_0 = 0.3$ and $p_1 = 0.4$. Devise a most powerful, most significant test of $H_0 = \{p_0\}$ versus $H_1 = \{p_1\}$ with $\alpha = 0.05$ and $\beta \leq 0.1$.

4.42 A laboratory device produces random pulses such that $N(t)$, the number of pulses in time interval t, has a Poisson distribution with parameter $\lambda = 5/\text{second}$. That is, $\Pr(N(t) = k) = (5t)^k e^{-5t}/k!$. The device operates through the radioactive decay of an internal element. You ask a technician to replace the element to boost λ to $10/\text{second}$. After a few days, you test the device. Let H_0 be the assertion that the technician ignored your request, and let H_1 be the alternative that the device now delivers pulses with $\lambda = 10$. Let the random variable X be the number of pulses observed in a 1-second interval. A sample from this population is then a vector $\mathbf{X} = (X_1, X_2, \ldots, X_N)$ in which each component is an observed count in a distinct 1-second interval. Devise a most significant, most powerful decision rule that maintains the probability of type I and type II errors below 0.1 and uses the smallest sample size.

*4.43 Let H_0 be the hypothesis that X has a discrete uniform distribution on the integers $1, 2, \ldots, 10$. Under the alternative, H_1, the distribution is a discrete uniform one on the integers $1, 2, \ldots, 8$. Devise a most powerful, most significant test with $\alpha = 0.05$ and $\beta \leq 0.1$.

4.44 You have a coin, which is either normal or magic. With a normal coin, the probability of a flip landing heads is 0.5. With a magic coin, the probability starts at 0.4 but changes with each toss. If p is the probability of a head on toss i and toss i actually delivers heads, then the probability of a head on toss $i + 1$ is $1 - p$. If toss i delivers tails, the probability remains unchanged for toss $i + 1$. Let H_0 be the hypothesis that the coin is normal. Devise a most powerful, most significant test with $\alpha = 0.5$ and $\beta \leq 0.1$.

4.4.2   Composite hypotheses

We have seen that the test of a simple hypothesis against a simple alternative always reduces to a threshold test on the likelihood ratio statistic, or on some equivalent statistic. If H_0 or H_1 is composite, we find our earlier definitions of error size inapplicable, so the first step is to broaden these definitions. The probability of a type I error, which erroneously rejects the null hypothesis, varies with the state of nature that prevails when the hypothesis is true. Similarly, the probability of a type II error, which erroneously accepts the null hypothesis, varies with the prevailing state of nature in H_1. In each case, we now define the error size to be the largest probability over the available states. As in the preceding subsection, we assume that each reference to a decision rule C includes the case where C is a mixed rule. The constituent critical areas $C_1, C_2, \ldots, C_k$ and the combining probability vector $p_1, p_2, \ldots, p_k$ will not be mentioned unless necessary for the argument at hand.

4.66 Definition: When using decision rule C over sample X to test hypothesis H_0 against alternative H_1, we call $\alpha_C = \sup_{\theta \in H_0} \mathcal{P}_C(\theta)$ the *type I error size*. Similarly, we call $\beta_C = \sup_{\theta \in H_1}(1 - \mathcal{P}_C(\theta))$ the *type II error size*. ∎

These definitions are compatible with those given earlier when both the null hypothesis and its alternative are simple. As before, we first guard against a type I error by establishing an upper bound for α. Since α is now a supremum over the error probabilities across all states in H_0, this step ensures that the probability of an erroneous rejection will not exceed α. Indeed, it may be much lower, depending on the particular state within H_0 that prevails. Among the tests that meet the established α, we naturally want to choose one that minimizes the probability of a type II error. Since this error varies across the states in H_1, we choose, if possible, a test that simultaneously minimizes the type II error probability across all states in H_1.

4.67 Definition: In testing H_0 versus H_1, C is a *uniformly most powerful test* if for all tests D with $\alpha_D \leq \alpha_C$, we have $\mathcal{P}_C(\theta) \geq \mathcal{P}_D(\theta)$ for all $\theta \in H_1$. ∎

Suppose that C is a uniformly most powerful test and D is a test with $\alpha_D \leq \alpha_C$. Then, for any $\theta \in H_1$, we have

$$\mathcal{P}_C(\theta) \geq \mathcal{P}_D(\theta)$$
$$1 - \mathcal{P}_C(\theta) \leq 1 - \mathcal{P}_D(\theta)$$

Pr(accept under $C|\theta$) $\leq$ Pr(accept under $D|\theta$).

Hence the probability of a type II error is smaller under C than under D, regardless of the prevailing state within H_1. A uniformly most powerful test then satisfies our requirements. No test with the same or lower type I error probability can deliver a smaller type II error probability, regardless of the state of nature. Unfortunately, as shown in the next example, a uniformly most powerful test may not exist—except perhaps at unrealistic α levels. The example needs the following property, which asserts that a uniformly most powerful test exhibits a lower power at $\theta \in H_0$ than at any point $\theta' \in H_1$.

4.68 Theorem: For a uniformly most powerful test C, $\mathcal{P}_C(\theta) \leq \mathcal{P}_C(\theta')$, whenever $\theta \in H_0$ and $\theta' \in H_1$.

PROOF: Let D_1, D_2 be the tests that never reject and that always reject, respectively, the null hypothesis. Let D be the mixed test that combines D_1, D_2 with probability vector $(1 - \alpha_C, \alpha_C)$. Then, for any $\theta \in H_0 \cup H_1$, we have

$$\mathcal{P}_{D_1}(\theta) = 0$$
$$\mathcal{P}_{D_2}(\theta) = 1$$
$$\mathcal{P}_D(\theta) = (1 - \alpha_C) \cdot 0 + \alpha_C \cdot 1 = \alpha_C.$$

For $\theta' \in H_1$, the uniformly most powerful test C must exhibit $\mathcal{P}_C(\theta') \geq \mathcal{P}_D(\theta')$. Therefore, for $\theta \in H_0, \theta' \in H_1$, we have

$$\mathcal{P}_C(\theta) \leq \sup_{\xi \in H_0} \mathcal{P}_C(\xi) = \alpha_C = \mathcal{P}_D(\theta') \leq \mathcal{P}_C(\theta'). \blacksquare$$

4.69 Example: We have a coin for which the probability of heads is an unknown parameter p. To test the null hypothesis $H_0 = \{0.5\}$ against the composite alternative $H_1 = \{p|p \neq 0.5\}$, we flip the coin N times and observe the number of heads. This corresponds to a sample $\mathbf{X} = (X_1, X_2, \dots, X_N)$ from a Bernoulli population with parameter p. The number of heads is the sample sum S. Show that no uniformly most powerful test exists with an α in the range $(0, 1)$.

The factorization criterion, Theorem 4.40, asserts the sufficiency of the sample sum statistic:

$$\Pr(\mathbf{X} = \mathbf{x}|p) = p^{x_1}(1-p)^{1-x_1} \cdot p^{x_2}(1-p)^{1-x_2} \cdots p^{x_N}(1-p)^{1-x_N}$$
$$= p^{S(\mathbf{x})}(1-p)^{N-S(\mathbf{x})} = f(S(\mathbf{x}), p) \cdot 1 = f(S(\mathbf{x}), p) \cdot g(\mathbf{x}).$$

The sample sum assumes integer values from 0 through N with probabilities

$$\Pr(S = k|p) = C_{N,k} p^k (1-p)^{N-k}. \tag{4.10}$$

The pure tests have critical areas $\{\mathbf{x}|S(\mathbf{x}) \in \{i_1, i_2, \dots, i_t\}\}$, where $\{i_1, i_2, \dots, i_t\}$ is a subset of $0, 1, 2, \dots, N$. The power functions of these pure tests are sums in which each term takes the form of Equation 4.10. The test with critical area $S \in \{2, 4\}$, for example, has the power function

$$\mathcal{P}_{\{2,4\}}(p) = C_{N,2} p^2 (1-p)^{N-2} + C_{N,4} p^4 (1-p)^{N-4}.$$

These power functions are simply polynomials in p and are therefore continuously differentiable functions of p. Mixed tests are convex combinations of pure tests, and consequently, their power functions are also continuously differentiable functions of p.

Now suppose that C is a uniformly most powerful test with $0 < \alpha_C < 1$. By Theorem 4.68, $\mathcal{P}_C(0.5) \leq \mathcal{P}_C(p)$ for all $p \in [0, 1]$. Since the derivative $\mathcal{P}'_C(0.5)$ exists, we can compute it as a limit either with p ascending from below 0.5 or with p descending from above 0.5. These two approaches give

$$\mathcal{P}'_C(0.5) = \lim_{p \to 0.5+} \frac{\mathcal{P}_C(p) - \mathcal{P}_C(0.5)}{p - 0.5} \geq 0$$

$$\mathcal{P}'_C(0.5) = \lim_{p \to 0.5-} \frac{\mathcal{P}_C(p) - \mathcal{P}_C(0.5)}{p - 0.5} \leq 0.$$

We conclude that $\mathcal{P}'_C(0.5) = 0$. This means that $\mathcal{P}_C(p)$ is flat or opening upward in the near vicinity of $p = 0.5$. To force a contradiction, we will construct a test D with $\alpha_D = \alpha_C$ but with $\mathcal{P}'_D(0.5) > 0$. This means that the D power function will intersect the C power function at $p = 0.5$ on a left-to-right upward trajectory. This will force $\mathcal{P}_D(p) > \mathcal{P}_C(p)$ for some p slightly greater than 0.5, which contradicts the alleged uniformly most powerful status of C. Figure 4.19 illustrates the situation. The bowl-shaped curve is the power function of C, while the rising curve is the power function of D. Of course, this contradicts the fact that C is a uniformly most powerful test.

It remains to construct D. We distinguish two cases: $\alpha_C > 0.5^N$ and $\alpha_C \leq 0.5^N$. For the first case, $\alpha_C > 0.5^N$, let D_1 be the test that always rejects H_0, regardless of the sample value, and let D_2 be the test that rejects H_0 if $S = N$. Let D be the mixed test that combines D_1 and D_2 with the probability vector $(q, 1 - q)$,

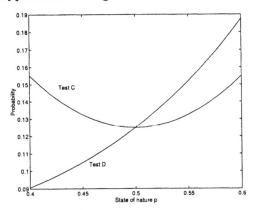

Figure 4.19. Power functions of an alleged uniformly most powerful test
and a competitor (see Example 4.69)

where $q = (\alpha_C - 0.5^N)/(1 - 0.5^N)$. Note that $0 < \alpha_C - 0.5^N < 1 - 0.5^N$, so $0 < q < 1$.
We have

$$\mathcal{P}_{D_1}(p) = 1.0$$

$$\mathcal{P}_{D_2}(p) = C_{N,N} p^N (1-p)^{N-N} = p^N$$

$$\mathcal{P}_D(p) = \frac{\alpha_C - 0.5^N}{1 - 0.5^N} \cdot 1 + \left(1 - \frac{\alpha_C - 0.5^N}{1 - 0.5^N}\right) \cdot p^N = \frac{\alpha_C(1 - p^N) + p^N - 0.5^N}{1 - 0.5^N}$$

$$\alpha_D = \mathcal{P}_D(0.5) = \alpha_C = \mathcal{P}_C(0.5).$$

The $\mathcal{P}_D$ curve intersects the $\mathcal{P}_C$ curve at $p = 0.5$. Moreover, for all $p > 0$,

$$\mathcal{P}'_D(p) = \frac{N p^{N-1}(1 - \alpha_C)}{1 - 0.5^N} > 0.$$

In particular, the derivative is positive at $p = 0.5$, which means that the $\mathcal{P}_D$ curve
has a tangent sloping to the northeast at $p = 0.5$. Because the $\mathcal{P}_C$ curve has
a horizontal tangent at that point, the intersection must appear as depicted in
Figure 4.19. Consequently, for a point slightly greater than $p = 0.5$, the $\mathcal{P}_D$ curve
must lie above the $\mathcal{P}_C$ curve. As noted above, this configuration contradicts the
uniformly most powerful status of C.

For the second case, $\alpha_C \leq 0.5^N$, let D_1 be the test that always accepts H_0,
regardless of the sample value, and let D_2 remain as in the first case. Let D be the
mixed test that combines D_1 and D_2 with the probability vector $(q, 1 - q)$, where
$q = (0.5^N - \alpha_C)/0.5^N$. Again $0 \leq q < 1$, and we have

$$\mathcal{P}_{D_1}(p) = 0$$

$$\mathcal{P}_{D_2}(p) = C_{N,N} p^N (1-p)^{N-N} = p^N$$

$$\mathcal{P}_D(p) = \frac{0.5^N - \alpha_C}{0.5^N} \cdot 0 + \left(1 - \frac{0.5^N - \alpha_C}{0.5^N}\right) \cdot p^N = \frac{\alpha_C p^N}{0.5^N}$$

$$\alpha_D = \mathcal{P}_D(0.5) = \alpha_C = \mathcal{P}_C(0.5).$$

As in the first case, the $\mathcal{P}_D$ curve intersects the $\mathcal{P}_C$ curve at $p = 0.5$, and

$$\mathcal{P}'_D(p) = \frac{N p^{N-1} \alpha_C}{0.5^N} > 0.$$

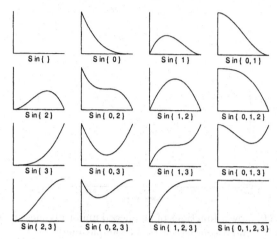

Figure 4.20. Power functions for pure tests over the sample sum (see Example 4.69)

The intersection is again as depicted in Figure 4.19. The existence of test D undermines the alleged uniformly most powerful claim for C.

Having established that no uniformly most powerful test exists, except in the unrealistic cases where $\alpha = 0$ or 1, we still have not identified the most appropriate test. The power functions of the pure tests over the sample sum appear in Figure 4.20 for the case $N = 3$. In each graph, the horizontal axis elaborates the states of nature running from $p = 0$ to $p = 1$. The vertical axis measures the probability of the critical area. Of course, the ideal power function is zero at $p = 0.5$ and 1 at all other points. Unfortunately, no combination of the entries in the figure can produce this ideal function. Notice that most of the graphs dip to the horizontal axis at one or more extreme p values. These excursions correspond to a high probability of a type II error, if nature presents the injurious p value. Assuming that we must guard against large type II errors both when $p < 0.5$ and when $p > 0.5$, we find our choices reduced to the three critical regions: $\{0, 3\}$, $\{0, 1, 3\}$, and $\{0, 2, 3\}$. The latter two are asymmetric in the probability of type II errors. Therefore, if the losses associated with an erroneous decision when $p < 0.5$ are comparable with those when $p > 0.5$, we prefer the test with critical region $\{0, 3\}$

We can mix the chosen test $\{0, 3\}$ with one of the extreme tests $\{\}$ or $\{0, 1, 2, 3\}$ to acquire the desired α. Without such an adjustment, we have

$$\alpha_{\{0,3\}} = \mathcal{P}_{\{0,3\}}(0.5) = \Pr(S \in \{0, 3\}|p = 0.5)$$
$$= C_{3,0}(0.5)^0(1 - 0.5)^{3-0} + C_{3,3}(0.5)^3(1 - 0.5)^{3-3} = 0.25.$$

Suppose that we want a test E with power function shaped like that of test $\{0, 3\}$ but with $\alpha_E = 0.05$. We note that the extreme test, $S \in \phi$, has a power function that is identically zero. We mix in the proper proportion of this test to obtain the desired α_E.

$$0.05 = \alpha_E = q \cdot 0 + (1 - q) \cdot (0.25)$$
$$q = 0.8$$

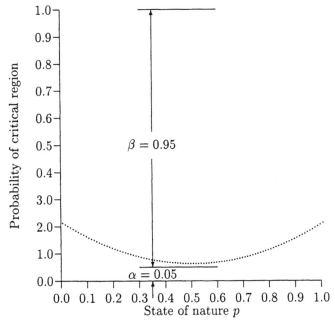

Figure 4.21. Power function of mixed test that achieves $\alpha = 0.05$ (see Example 4.69)

The mixed test then operates as follows. Twenty percent of the time it rejects if $S \in \{0, 3\}$. The remaining 80% of the time it accepts regardless of the sample. With this test, we have a 5% chance of making a type I error. Our probability of making a type II error varies with the prevailing $p \in H_1$. For $p \neq 0.5$, this error probability is

$$1 - \mathcal{P}_E(p) = 1 - \left[(0.8) \cdot 0 + 0.2 \left(C_{3,0} p^0 (1-p)^{3-0} + C_{3,3} p^3 (1-p)^{3-3} \right) \right]$$
$$= 1 - 0.2[p^3 + (1-p)^3].$$

Over H_1, the supremum of this expression occurs as p approaches 0.5, where it attains the value $\beta = 0.95$. Figure 4.21 shows the power function of the mixed test and the close relationship between the chosen α and the resulting β. Because $\alpha + \beta = 1$, it is not possible to achieve a low α except at the expense of a large β. This is intuitively plausible because H_1 includes such states as $p = 0.501$, which is very close to H_0. For states in such close proximity to H_0, the test is understandably prone to error. $\square$

The example shows that the general case may not admit a uniformly most powerful test. Special circumstances, however, may produce a different result. Suppose that a single real parameter indexes the states of nature and that the states in H_0 lie entirely to one side of the states for H_1. This condition does not prevail in the preceding example because $H_1 = \{p|p \neq 0.5\}$ surrounds $H_0 = \{0.5\}$. If we test $H_0 = \{p|p \leq 0.5\}$ versus $H_1 = \{p|p > 0.5\}$, however, we find that a uniformly most powerful test does exist. Without loss of generality, we assume that H_0 lies to the left of H_1. That is, $\theta_0 \in H_0$

and $\theta_1 \in H_1$ imply that $\theta_0 < \theta_1$. If we construct a test based on statistic T, which delivers the desired α and has a monotone increasing power function, we can often show that the test is actually uniformly most powerful. The next example illustrates the technique.

4.70 Example: The context here is the same as in the preceding example, except that we are now testing the null hypothesis $H_0 = \{p|p \leq 0.5\}$ against the composite alternative $H_1 = \{p|p > 0.5\}$. We again flip the coin N times and observe the number of heads. Choose arbitrary $p_0 \in H_0, p_1 \in H_1$ and show that a sample sum threshold test for p_0 versus p_1 is actually a uniformly most powerful test for H_0 versus H_1.

As in the preceding example, the sample sum is a sufficient statistic. The probabilities at hand are

$$\Pr(\mathbf{X} = \mathbf{x}|p) = p^{S(\mathbf{x})}(1 - p)^{N - S(\mathbf{x})}$$
$$\Pr(S = k|p) = C_{N,k}p^k(1 - p)^{N-k}.$$

Let $p_0 \leq 0.5 < p_1$ be two arbitrary states. The likelihood ratio threshold test $\Pr(\mathbf{X} = \mathbf{x}|p_0)/\Pr(\mathbf{X} = \mathbf{x}|p_1) < K$ is most powerful for the simple hypothesis p_0 versus the simple alternative p_1. Note that $p_0 < p_1$ implies that $p_0/p_1 < 1$ and $(1 - p_1)/(1 - p_0) < 1$. Therefore, the equivalence

$$\frac{\Pr(\mathbf{X} = \mathbf{x}|p_0)}{\Pr(\mathbf{X} = \mathbf{x}|p_1)} = \frac{p_0^S(1 - p_0)^{N-S}}{p_1^S(1 - p_1)^{N-S}} = \left(\frac{1 - p_0}{1 - p_1}\right)^N \left(\frac{p_0}{p_1} \cdot \frac{1 - p_1}{1 - p_0}\right)^S$$

means that larger S values are monotonically associated with smaller likelihood ratios. Since S assumes only integer values, it is convenient to express this equivalence in the reverse direction. For integer c, the test $S > c$ is equivalent to the likelihood ratio test

$$\frac{\Pr(\mathbf{X} = \mathbf{x}|p_0)}{\Pr(\mathbf{X} = \mathbf{x}|p_1)} < \left(\frac{1 - p_0}{1 - p_1}\right)^N \left(\frac{p_0}{p_1} \cdot \frac{1 - p_1}{1 - p_0}\right)^c = c'(p_0, p_1, c).$$

The tests $S > c$, for $c = -1, 0, 1, 2, \ldots, N$, are then most powerful tests of p_0 versus p_1, for any $p_0 < p_1$. Their power functions are

$$\mathcal{P}_c(p) = \Pr(S > c|p) = \sum_{k=c+1}^{N} C_{N,k}p^k(1 - p)^{N-k}.$$

Figure 4.22 illustrates the monotone behavior of this function for $N = 6, c = 3$. This is a general property, which we can establish by investigating the derivative.

$$\mathcal{P}_c'(p) = \sum_{k=c+1}^{N} C_{N,k}[kp^{k-1}(1 - p)^{N-k} + p^k(N - k)(1 - p)^{N-k-1}(-1)]$$

$$= \sum_{k=c+1}^{N} C_{N,k}p^{k-1}(1 - p)^{N-k-1}[k(1 - p) - (N - k)p]$$

$$= \sum_{k=c+1}^{N} C_{N,k}(k - Np)p^{k-1}(1 - p)^{N-k-1}$$

$$= \frac{1}{p(1 - p)} \sum_{k=c+1}^{N} (k - Np)C_{N,k}p^k(1 - p)^{N-k}$$

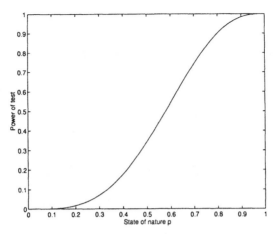

Figure 4.22. Power function of test $(S > 3)$ for $N = 6$ (see Example 4.70)

If $c \geq Np$, the summands are all positive, and hence so is $\mathcal{P}'_c(p)$. If $c < Np$, then extending the lower summation limit to zero adds terms that are either zero or negative. In this case,

$$\mathcal{P}'_c(p) \geq \frac{1}{p(1-p)} \sum_{k=0}^{N} (k - Np) C_{N,k} p^k (1-p)^{N-k}$$

$$= \frac{1}{p(1-p)} \left[\sum_{k=0}^{N} k C_{N,k} p^k (1-p)^{N-k} - Np \sum_{k=0}^{N} C_{N,k} p^k (1-p)^{N-k} \right]$$

$$= \frac{1}{p(1-p)} (Np - Np) = 0.$$

This follows because the first sum, according to Theorem 2.6, is the expected value of a binominal random variable with parameters (N, p) and the second sum is the binomial expansion of $[p + (1-p)]^N = 1$. In either case, the derivative is nonnegative, which means that the power function is monotone nondecreasing.

Because $S > c$ is most powerful for any p_0 versus any p_1, as long as $p_0 < p_1$, no test can exhibit a power function that crosses that of $S > c$ in an ascending sense. That is, for all tests D, if $\mathcal{P}_D(p_0) < \mathcal{P}_c(p_0)$, then $\mathcal{P}_D(p)$ remains below $\mathcal{P}_c(p)$ for all $p > p_0$. Indeed, if the contrary were to occur, $\mathcal{P}_D(p_0) \leq \mathcal{P}_c(p_0)$ and $\mathcal{P}_D(p_1) > \mathcal{P}_c(p_1)$ for some $p_1 > p_0$, this would contradict that fact that $S > c$ is a most powerful test for p_0 versus p_1.

We are now ready to show that $S > c$, which arose from consideration of simple hypothesis p_0 versus simple p_1 for an arbitrary choice $p_0 < p_1$, is actually a uniformly most powerful test of H_0 versus H_1. Suppose that D is a test with $\alpha_D \leq \alpha_c$. Because the power function of $S > c$ is nondecreasing,

$$\alpha_c = \sup_{p \in H_0} \Pr(S > c | p) = \sup_{p \leq 0.5} \mathcal{P}_c(p) = \mathcal{P}_c(0.5).$$

Hence,

$$\mathcal{P}_D(0.5) \leq \sup_{p \leq 0.5} \mathcal{P}_D(p) = \alpha_D \leq \alpha_c = \mathcal{P}_c(0.5).$$

This says that the power function of D lies below that of $S > c$ at $p = 0.5$. Therefore, it must lie below it for all larger p. That is, $\mathcal{P}_D(p) \le \mathcal{P}_c(p)$ for all $p \in H_1$. We conclude that the test $S > c$ is a uniformly most powerful test of H_0 versus H_1. $\square$

4.71 Example: Continuing the preceding example, construct a uniformly most powerful test of $H_0 = \{p|p \le 0.5\}$ versus $H_1 = \{p|p > 0.5\}$ with $\alpha = 0.05$.

We now know that the test will have the form $S > c$ for some appropriate threshold c and sample size N. Since the power function will be monotone nondecreasing, the supremum over the probabilities of a type I error will occur at $p = 0.5$ That is, $\alpha_c = \mathcal{P}_c(0.5)$. Similarly, the supremum over the probabilities of a type II error will occur as $p \to 0.5^+$. That is, $\beta_c = 1 - \mathcal{P}_c(0.5)$. Hence the type I and type II error probabilities sum to 1, which means that achieving $\alpha = 0.05$ will entail $\beta = 0.95$. As noted earlier, this is an expected consequence of the fact that H_0 and H_1 come arbitrarily close to one another. A test with a tight guard against erroneously rejecting a true $p = 0.5 \in H_0$ necessarily performs poorly in accepting H_0 when $p = 0.501 \in H_1$.

Because $\alpha_c = \mathcal{P}_c(0.5) = \sum_{k=c+1}^{N} C_{N,k}(0.5)^k (0.5)^{N-k} = (1/2^N) \sum_{k=c+1}^{N} C_{N,k}$, we can calculate the maximum probability of a type I error for various N and c.

For selected N, the table to the right lists the threshold values that bracket the desired $\alpha = 0.05$ and the proportions for a mixed test that achieves $\alpha = 0.05$ exactly. The line for $N = 10$, for example, specifies a threshold $c = 7$. The next two columns give the α values for the tests $S > 8$ and $S > 7$. These are 0.0107 and 0.0547, which properly bracket the desired $\alpha = 0.05$. The next column gives the proportion, q, with which to mix the lower α test, $S > 8$, and the higher

N	c	α_{c+1}	α_c	q	$\mathcal{P}_{mix}(0.6)$
3	2	0.0000	0.1250	0.6000	0.0864
4	3	0.0000	0.0625	0.2000	0.1037
5	3	0.0312	0.1875	0.8800	0.1089
6	4	0.0156	0.1094	0.6333	0.1151
7	5	0.0078	0.0625	0.2286	0.1288
8	5	0.0352	0.1445	0.8643	0.1347
9	6	0.0195	0.0898	0.5667	0.1404
10	7	0.0107	0.0547	0.1067	0.1544

α test, $S > 7$, to achieve exactly $\alpha = 0.05$. Specifically, we should use the test $S > 8$ with probability 0.1067 and test $S > 7$ with probability $1 - 0.1067 = 0.8933$. The final column gives the power function of the mixed test at $p = 0.6$. Although all the mixed tests have $\alpha = 0.05$, by construction, they differ in how quickly the power function climbs as p increases into H_1. As expected, the larger the sample size, the steeper the climb. A large sample size can therefore reduce the range of $p \in H_1$ with a high probability of a type II error. Figure 4.23 shows the power functions for the sample sizes in the table.

If we choose $N = 10$, for example, the probability of a type II error is less than $1 - \mathcal{P}(0.6) = 0.8456$, provided that the prevailing $p > 0.6$. Even larger sample sizes, assuming that we can afford the sampling expense, can reduce the range of $p \in H_1$ with high type II error probabilities to a small sliver. $\square$

Exercises

4.45 Let X be a Poisson random variable with unknown parameter λ, and let $S = X_1 + X_2 + \ldots + X_N$ be the sample sum. What is the power function of the test $S > 3$?

4.46 Suppose that X is a Bernoulli random variable with parameter p. Based on a sample $\mathbf{X} = (X_1, X_2, \ldots, X_N)$ over the population X, devise a

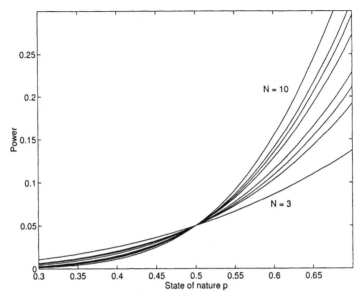

Figure 4.23. For various sample sizes, power functions of the mixed test
with $\alpha = 0.05$ (see Example 4.71)

uniformly most powerful test of $H_0 = \{p|p \geq 0.6\}$ versus $H_1 = \{p|p < 0.6\}$.with $\alpha = 0.1$.

4.47 Suppose that X is binomially distributed with parameters $(10, p)$. Devise a uniformly most powerful test of $H_0 = \{p|p \leq 0.4\}$ versus $H_1 = \{p|p > 0.4\}$ with $\alpha = 0.05$.

4.48 Let Y be geometrically distributed with parameter p. Devise a uniformly most powerful test of $H_0 = \{p|p \geq 0.6\}$ versus $H_1 = \{p|p < 0.6\}$ with $\alpha = 0.05$.

4.49 Suppose that X has a negative binomial distribution with parameters $(10, p)$. Devise a uniformly most powerful test of $H_0 = \{p|p \geq 0.8\}$ versus the alternative $H_1 = \{p|p < 0.8\}$ with $\alpha = 0.05$.

4.5   Summary

A decision problem demands a choice among certain actions, each of which incurs a cost that varies with an unknown state of nature. Sample data, if available, may indirectly suggest the prevailing state, in the sense that different states correspond to different probability distributions for the sample. In the absence of such data, however, it is still possible to develop a rational decision procedure. Two such strategies are the no-data minimax and the no-data Bayes decision rules. The minimax approach seeks to minimize the cost under the pessimistic assumption that the most injurious state of nature prevails. This viewpoint endows nature with an adversarial personality. That is, nature knows the decision procedure and presents the most expensive state for that decision. Faced with such an omniscient adversary, the minimax rule notes that each candidate action has a maximum loss across the states of nature and advocates the action with the smallest maximum loss. This policy places an upper bound on the loss per decision. Over repeated trials, we can actually expect somewhat lower losses because it is unlikely that nature consistently presents the most expensive state. Indeed, if nature were consistent, the problem would become deterministic.

A mixed version of this rule advocates a random selection among the m actions according to a probability vector $p = (p_1, p_2, \ldots, p_m)$. That is, the decision maker employs an independent random number generator to assist in each decision. The generator produces an integer $i \in [1, m]$ with probability p_i, and the decision maker then takes action a_i. Under the pessimistic assumption that nature knows the probability vector and adjusts its state to exact the largest average cost, the no-data mixed minimax rule minimizes that expense. Averaged over many decisions, the mixed minimax rule can deliver a lower cost than the pure minimax rule, even if nature assumes the least favorable state for this strategy. Both the pure and mixed minimax policies assume that nature repeatedly presents the same state as occasions arise for repeated decisions. This state may be the worst possible for the strategy chosen, but it does not vary from one decision to the next.

By contrast, the Bayes approach seeks to exploit a known distribution across the states of nature. This does not mean that sample data are used to infer the state of nature. Rather, it means that the decision maker has some reason to feel that the states occur randomly according to a particular distribution. For each action, the loss then constitutes a random variable, which takes potentially different values for different states of nature. The Baysian decision maker then chooses the action that minimizes the expected loss. In a repeated sequence of decisions, he takes the same action each time, although he expects the state of nature to vary randomly from one decision opportunity to the next. His rationale is that his costs, averaged over many decisions, will then be lower than they would be with any other action, provided, of course, that nature obligingly presents random states according to the known prior distribution.

A mixed Bayes rule specifies a probability vector $p = (p_1, p_2, \ldots, p_m)$, which advocates action a_i with probability p_i. This again requires an independent random number generator to assist in each decision. The decision maker treats the expected loss for each action as a second random variable, with a discrete distribution across the actions. He then chooses the distribution p that minimizes the expected value of this second random variable. The mixed Bayes rule is, however, an academic exercise because it cannot deliver a lower expected loss than the pure version.

Both the minimax and Bayes rules admit graphical constructions, in

Statistic	Mean	Variance		
S	$N\mu_X$	$N\sigma_X^2$		
$\overline{X}$	μ_X	$\dfrac{\sigma_X^2}{N}$		
s_X^2	$\dfrac{(N-1)\sigma_X^2}{N}$	$\dfrac{\delta_4 - \delta_2^2}{N}$	$-\dfrac{2(\delta_4 - 2\delta_2^2)}{N^2}$	$+\dfrac{\delta_4 - 3\delta_2^2}{N^3}$

TABLE 4.11. Mean and variance of selected statistics

which the losses under the different states of nature constitute the coordinates of an action. In the two-dimensional case, the pure minimax action is the one first encountered by an expanding square from the origin. The expanding square's first encounter with the action set's convex hull defines the mixed minimax rule. The Bayes rule specifies the action first encountered by a family of parallel lines, with slopes determined by the prior probabilities of the states of nature. The first encounter with the convex hull defines the mixed Bayes rule, but this encounter will always occur at a vertex of the convex hull, which is one of the actions. With n states of nature, an expanding n-cube replaces the expanding square in locating the minimax rules, and parallel hyperplanes replace the parallel lines in locating the Bayes rule.

The decision problem may involve a random variable, called the population, which has differing distributions depending on the state of nature. In this case, a sample provides indirect evidence of the prevailing state. A sample is a set of observations drawn from the population. A statistic is a function of the sample. It may reduce the sample to a single number, as is the case with the sample sum, for example, or at the other extreme, it may be the vector of observed values without any reduction. Scalar statistics of interest are the sample sum, mean, variance, maximum, minimum, range, and median. Two useful vector statistics are the sample itself and the order statistic. A statistic is a random variable because it assumes different values when presented with different sample observations. Consequently, a statistic has a distribution and summarizing characteristics, such as a mean, a variance, and higher-order moments. Sometimes we can calculate these quantities from the corresponding population parameters. Table 4.11 lists some of the results, which assume a sample of size N: $\mathbf{X} = (X_1, X_2, \ldots, X_N)$.

For more complicated statistics, we must first calculate the distribution and then compute the mean and variance directly. The kth component of the order statistic, for example, has cumulative distribution function

$$F_{X_{(k)}}(t) = \sum_{j=k}^{N} C_{N,j}[F_X(t)]^j[1 - F_X(t)]^{N-j}.$$

To incorporate statistic T into the decision problem, we replace actions with decision rules and losses with risks. A decision rule is a function from range(T) to the available actions. For each decision rule $d(\cdot)$ and state of nature θ_i, the risk is the expected loss $\hat{L}(d, \theta_i) = E[L(d(\cdot), \theta_i)]$, computed with respect to the statistic's distribution. If nature maintains a constant state θ_i and the decision maker repeatedly uses rule d on the varying T values obtained from successive samples, then the risk is the average loss per decision. From this point, we proceed as in the no-data case to obtain minimax and Bayes rules. The advantage of the sample-driven approach is most apparent when there are just two states of nature. The graphs then reveal decision rules well outside the convex hull of the raw actions. Those to the southwest of the convex hull provide opportunities for reduced average cost, and the expanding square or advancing parallel lines do indeed first intercept these more advantageous decision rules.

Using Bayes' theorem, we can compute a posterior distribution across the states of nature, conditioned on the appearance of a particular value $T = t$. If we then apply the no-data Bayes approach to locate the optimal action for t, we have in effect determined a decision rule, which turns out to be identical with the Bayes rule obtained by examining the risk spectrum of the collection of all possible rules. The posterior method has the advantage that it eliminates the necessity of calculating risks for all possible rules. One merely waits until the statistical observation is available and then determines the optimal action for that statistical value.

There remains the question of which statistic to use in a particular decision problem. A sufficient statistic is one that retains all relevant information pertinent to the decision at hand. That is, a decision rule based on a sufficient statistic incurs losses no greater than if the entire sample were used. The formal definition requires that $\Pr(\mathbf{X} = \mathbf{x} | (T = f(\mathbf{x}) \cap (\theta = \theta_i))$ be independent of θ_i. Because a decision rule d over statistic T expands naturally to a decision rule d' over the entire sample via $d'(\mathbf{x}) = d(T(\mathbf{x}))$, we can compare the mixed minimax and Bayes rules from these two contexts. When T is a sufficient statistic, we find them identical, and therefore the rules over T deliver the same expected costs as those over the entire sample.

Every statistic, whether sufficient or not, induces a partition on the space of possible sample points. The components are those subsets in which every point yields the same statistical value. If the statistic is sufficient, the conditional distribution within a component is independent of the state of nature. Stated differently, the probability weights of sample points in the same component shift by the same factor when the state of nature changes. Indeed, this property is a defining characteristic of a sufficient statistic.

Creating a new statistic as a function of an existing one coarsens the existing partition by aggregating selected components. It follows that $T_2 = f(T_1)$ sufficient implies T_1 sufficient. T is minimal sufficient if T is a functional reduction of every other sufficient statistic. While there are many minimal sufficient statistics, they all induce the same partition in the space of sample points. This partition is the coarsest for which the conditional probabilities within components are independent of the state of nature. Theoretically, we can obtain this coarsest partition as the equivalence classes of the relation $\mathbf{x} \equiv \mathbf{y}$, which identifies two points when the ratio of their probability weights is independent of the state of nature. Any assignment of numeric tags to the resulting components then defines a minimal sufficient statistic.

The likelihood function for a particular sample point $\mathbf{x}$ is $\mathcal{L}_{\mathbf{x}}(\theta_i) = \Pr(\mathbf{X} = \mathbf{x}|\theta = \theta_i)$. The equivalence relation $\mathbf{x} \sim \mathbf{y}$ identifies two points when their likelihood functions are scalar multiples of one another, and it also yields the minimal sufficient partition. To find a sufficient statistic, it is helpful to know that T is sufficient if and only if the probability distributions $\Pr(\mathbf{X} = \mathbf{x}|\theta = \theta_i)$ factor as $f(T(\mathbf{x}), \theta_i) \cdot g(\mathbf{x})$. This observation frequently allows us to determine a sufficient statistic by inspection.

Hypothesis testing is a special class of decision problem in which the states of nature segregate into two classes, H_0 and H_1. The null hypothesis is the statement that the state of nature resides in H_0; its alternative is that the state of nature is in H_1. The decision problem is then to accept or reject the null hypothesis. If H_0 contains a single state, it is termed a simple hypothesis; otherwise it is composite. Similarly, H_1 may be a simple or composite alternative. As in the general case, we have a random variable with known distributions for the various states.

A sample from this population produces a point in the space of all possible sample values, and we can therefore characterize each pure decision rule by the subset of sample points that command rejection of the null hypothesis. This subset is the rule's critical region. If we reduce the sample with a statistic T, then the critical region is the union of selected components of the T partition. The probability of the critical region C, as a function of the state of nature, is the power function $\mathcal{P}_C(\theta)$ of the decision rule. In terms of the power function, we can elaborate the risks associated with each state, and we can then identify the minimax and Bayes decision rules in the traditional manner.

We can also pursue an analysis that emphasizes the probabilities of erroneously rejecting or accepting the null hypothesis. A type I error is an erroneous rejection, whose probability is traditionally denoted by α. A type II error is an erroneous acceptance with probability denoted by β. The simplest case involves a simple hypothesis $H_0 = \{p\}$ versus a simple alternative $H_1 = \{q\}$, where test C has $\alpha = \mathcal{P}_C(p)$ and $\beta = 1 - \mathcal{P}_C(q)$. α is also called the significance level of the test. In this context, we define a most powerful test as one whose β is less than or equal to that of any test with a lower or equal α. That is, C is a most powerful test if among all tests with smaller or equal type I error probabilities, none has a smaller type II error probability. Similarly, a most significant test is one whose α is less than or equal to that of any test with a lower or equal β. A most powerful, most significant test occupies an undominated position on the risk graph. The likelihood ratio statistic $R(\mathbf{x}) = \Pr(\mathbf{x}|p)/\Pr(\mathbf{x}|q)$ is minimal sufficient, and the threshold test $R < t$ is most powerful and most significant. Tests of this form then constitute the southwest boundary of the convex hull of all decision rules, which means that the mixed minimax and Bayes solutions are of this form. Thus, regardless of the optimality criterion (minimax, Bayes, or α-β error sizes), the optimal test is a threshold test on the likelihood ratio statistic, or on some statistic T that is monotone with it in the sense that $T(\mathbf{x}) < T(\mathbf{y})$ if and only if $R(\mathbf{x}) < R(\mathbf{y})$.

If one or both hypotheses are composite, the type I and type II error probabilities vary with the prevailing state in H_0 or H_1. Consequently, we redefine α and β to be the suprema of $\mathcal{P}_C(p)$ and $1 - \mathcal{P}_C(q)$ over $p \in H_0$ and $q \in H_1$ respectively. If the sample has a distribution under state p that is nearly identical to that under state q, we say that states p and q are close together. With composite hypotheses, H_0 often contains states that are very close to states in H_1. This proximity makes it impossible for a test to achieve simultaneously a low α and a low β. In an attempt to obtain the optimal β while maintaining a specified α threshold, we define a uniformly most powerful test. A uniformly most powerful test C has $\mathcal{P}_C(q) \geq \mathcal{P}_D(q)$ for all $q \in H_1$, when compared with any test D that has the same or smaller α. That is, C is uniformly most powerful if it never exhibits a larger type II error probability, regardless of the prevailing $q \in H_1$, than any test with the same or lower type I error probability. Unfortunately, a uniformly most powerful test may not exist in a particular situation. However, in the case where a real parameter indexes the states of nature and H_0 lies completely to one side of H_1, a test with a monotone increasing power function sometimes satisfies the criteria for uniformly most powerful.

Historical Notes

Implicit in the inference methods of this chapter is the assumption that probability assignments reflect reality in the sense of relative frequency of occurrence across many repetitions of a random experiment. For example, when we ask if a manufacturing process has deteriorated from a certain tolerable rate of defective products, we are assuming that defects appear in the process output stream in a random fashion, either at the tolerable rate or at an augmented rate. The probability distribution that governs the sporadic appearance of defects is related to the relative frequency of defects in the process output. Inference theory is on fairly solid ground in these applications.

There are, however, situations in which no repetitive process exists to connect the probability distribution with observable relative frequencies. The historian, for example, attempts to reconstruct the past by selecting, balancing, and interpreting the available evidence. Because the evidence is incomplete, and sometimes contradictory, the reconstruction is an exercise in probabilistic inference. The nature of the probabilities is different from cases in which an underlying repetitive process supplies the random variation. The past is a unique, nonrepeatable story. The events of this story either did or did not happen. The probabilities are then subjective judgments that attempt to quantify the level of certainty ascribed to the events by witnesses, participants, and eventually historians. For example, did Napoleon pass through village X on his return from Elba or not? What does it mean when a historian says that there is an 80% probability that he did pass through village X?

When a jury convicts an individual of a crime, it is making a probabilistic judgment, and the chain of inferences supporting that judgment can literally be a matter of life and death. Among the eminent personalities of the nineteenth century who grappled with the so-called jury problem were two noted mathematicians, Laplace and Poisson. What is the chance of error if a jury convicts with a vote of 10 : 2? Is a conviction on a vote of 7 : 5 less compelling? Laplace concluded that 7 : 5 convictions have an error rate in excess of 1/3, "a terrifying figure." His analysis included assumptions that a defendant's prior probability of guilt is 1/2 and that the reliability of an individual juror is uniformly distributed between 1/2 and 1. In this matter, Laplace's notions of probability were of the level-of-belief, as opposed to the relative frequency, variety. His analysis used no data on conviction rates. Poisson, however, took the other approach and appealed to his law of large numbers and actual conviction rates to infer juror reliability. He concluded that the error rate in 7 : 5 convictions ranges between 0.03 and 0.17, depending on whether the issue is a crime against property or against a person. The issue has not been settled. In 1831, about the time of Poisson's work, the French law was changed so that an 8 : 4 vote, as opposed to the earlier 7 : 5, became necessary for conviction. The law was reversed in 1835. See Hacking[31] for a detailed account.

More recently, Kadane and Schum[40] apply a probabilistic framework, initially proposed by the judicial scholar Wigmore[92], to systematize the chain of evidence in a celebrated criminal case of the 1920s. They discuss the significance of probability measures that do not have a relative frequency interpretation, and from this standpoint, they conclude that the verdict invites a large chance of error.

The minimax and Bayes rules are special cases of a more general problem that has generated a large literature. It concerns optimizing an objective function over a convex set. When the objective function is linear, as it is in the Bayes procedure, the process is termed linear programming. It prescribes a method for systematically moving among the vertices of the constraint space in search of the optimal location. The primary difficulty is the exponential growth of vertices with problem size.

When the objective function is not linear, the process is termed nonlinear programming, and a systematic attack is much more difficult. The French mathematician Fourier worked with the linear form in the early nineteenth century, and the Russian mathematician Kantorovich considered some industrial applications in the early twentieth century. The large logistical problems associated with World War II gave birth to an area of optimization techniques called operations research, which includes linear and nonlinear programming. The most well-known contributor was George Danzig, who formulated, in 1947, the simplex algorithm for linear programming.

As shown in this chapter, statistics are helpful tools in optimizing a decision-making process in the face of uncertainty, and of all potential users, political institutions are perhaps the most active participants in such processes. Indeed, governments, for better or for worse, have long held an abiding interest in statistics. Some governments use the information to enhance the well-being of their subjects; others use the data to tax and exploit them. Most governments appear interested in both possibilities. Reflecting their service to the state, early statistical methods were known as political arithmetic (see Roberts[70]) and stretch back nearly 1000 years. Shortly after his victory at Hastings (1066), William the Conqueror commissioned the Domesday Book, so called because one could not contest its contents. The ambitious aim of this book was to count the English population and, more important, to list each person's assets. Government, even when engaged in noble practice, is essentially parasitic in nature, and it needs to know the strength of the host on which it feeds.

English churches began the systematic recording of births, marriages, and deaths in the sixteenth century, and John Graunt compiled the first mortality tables in the mid-seventeenth century. He also developed inference methods (e.g., estimating the number of men suitable for military service), together with cross-checks to confirm their accuracy. People were becoming aware that large data collections can reveal patterns and trends that are hidden by random fluctuations in individual observations. David[16] and Hacking[30] give detailed accounts.

Further Reading

Linear and nonlinear programming, of which the minimax and Bayes procedures are special applications, have been studied extensively. Best and Ritter[6] and Schrijver[79] are introductory texts on the subject. The latter includes a discussion of Karmakar's polynomial-time algorithm for linear programming. A more advanced treatment appears in Sposito[82], and Rockafellar[71] is the definitive reference for convex analysis.

It has perhaps occurred to the reader that the prior distribution across the states of nature, which is necessary for the Bayes procedure, may be somewhat difficult to obtain. We have avoided this controversy here by simply assuming that the distribution is known, but the literature on statistical decision theory is concerned extensively with this problem. Pratt et al.[67] provide a complete introduction to the subject, including a discussion of the subjective assessments needed to estimate prior probabilities. Weiss[91] gives a more mathematically concise treatment of the topic.

Chapter 5

Real Line-Probability

Although discrete probability distributions are useful in many applications, as demonstrated in the preceding two chapters, there remain problems that demand a different type of distribution. The characterizing feature of a discrete distribution is that the outcome space is a countable collection of points. Frequently, the range of a random variable substitutes for the underlying outcome space, and this range is then a countable subset of the real numbers. We can extend the outcomes to include all real numbers by specifying probability zero for those points that are not in the countable subset. Of course, these extended points do not enter into expected value computations, which remain sums over countable sets. This method, unfortunately, is not applicable in all cases. We can have a distribution that assigns probability zero to all individual points in such a manner that nonzero probability accumulates on uncountable collections. To deal with such distributions, we must consistently assign probability directly to subsets of outcomes. Because we use random variables to map outcome subsets to real number subsets, a consistent probability assignment is equivalent to allocating probability to subsets of real numbers.

We have used this approach successfully in earlier chapters. The outcome spaces for Poisson processes and Markov chains, for example, were infinite sequences, and we managed a consistent probability assignment to groups of sequences. In both cases, the groups selected were those identified by finite prefixes, and the consequent assignment was sufficient to derive many useful results. This chapter outlines how to extend this technique to uncountable outcome spaces of real numbers. We let $\mathcal{R}$ denote the real numbers, and we use $\mathcal{R}^n$ for the collection of n-vectors $(x_1, x_2, \ldots, x_n)$, in which each $x_i \in \mathcal{R}$.

To illustrate the difficulties that arise with probability assignments to individual outcomes, consider a nondeterministic situation, in which all real numbers are possible outcomes. For example, the outcome may be the speed of a particular vehicle as it passes a checkpoint on the highway. The highway is bidirectional, and we assign positive speeds to vehicles traveling in a particular direction and negative speeds to those traveling the other way. Of course,

371

there are upper and lower limits, such as ± 120 miles per hour, beyond which we expect to find few observations. Nevertheless, it is convenient to consider all possible real numbers as outcomes in this experiment. In situations such as this, we want to assign probabilities to the outcomes as a prelude to calculating useful features, such as the mean and variance. There are certain constraints on the assignment. It must extend meaningfully to events, which in this context are subsets of real numbers, and it must observe the three basic axioms of a probability distribution: nonnegativity, normalization, and countable additivity. That is, all assigned probabilities must be nonnegative, the total must sum to 1, and the probability of a countable union of disjoint events must be the sum of the component probabilities. Suppose that $\Pr : \mathcal{R} \to [0, 1]$ is a tentative assignment. Let $A_n = \{x \mid \Pr(x) > 1/n\}$. Then the set of points with nonzero probability is $A = \cup_{n=1}^{\infty} A_n$. For purposes of establishing a contradiction, assume that some A_n contains more than n points. Consider a set $A_n' = \{x_1, x_2, \ldots, x_{n+1}\} \subset A_n$. The disjoint union $A_n' = \{x_1\} \cup \{x_2\} \cup \ldots \cup \{x_{n+1}\}$ implies that $\Pr(A_n') = \sum_{i=1}^{n+1} \Pr(x_i) > (n+1)(1/n) > 1$, which violates the normalization property. Therefore, A_n contains at most n points, and A, being a countable collection of countable sets, is also countable.

This conclusion means that a probability assignment can specify nonzero probability only for a countable collection of points. The collection may be finite or countably infinite, but in either case, the remaining outcomes, which comprise a large uncountable set, must all receive probability zero.

This is not a promising start toward a more general probability distribution, but it is not an insurmountable obstacle. We can assign part of the probability weight to a countable collection of individual points, and we can spread the remaining probability across certain events in such a manner that the constituent outcomes individually receive zero weight. How is this possible? Suppose, for example, that we first distribute 0.4 probability over a countable collection of individual points, $\mathcal{C} = \{x_1, x_2, \ldots\}$. This constitutes a component $\Pr_1(\cdot)$. We then spread the remaining 0.6 across the interval $[0, 1]$ such that

$$\Pr_2([a, b]) = \Pr_2((a, b]) = \Pr_2([a, b)) = \Pr_2((a, b)) = 1.8 \int_a^b x^2 \, dx$$

for any event of the form $[a, b], (a, b], [a, b)$, or $(a, b) \subset [0, 1]$. For any such event, the total probability is that contributed by the integral plus that contributed by the discrete points. For the closed interval, for example, we have

$$\Pr([a, b]) = \Pr_2([a, b]) + \sum_{x_i \in [a,b] \cap \mathcal{C}} \Pr_1(x_i) = 1.8 \int_a^b x^2 \, dx + \sum_{x_i \in [a,b] \cap \mathcal{C}} \Pr_1(x_i).$$

Note that all such events receive nonnegative probability assignments, and

$$\Pr(\mathcal{R}) = 1.8 \int_0^1 x^2 \, dx + \sum_{i=1}^{\infty} \Pr_1(x_i) = \left. \frac{1.8 x^3}{3} \right|_0^1 + 0.4 = 0.6 + 0.4 = 1,$$

which ensures that the normalization property holds. We can extend the assignment to unions of intervals, and the nature of the integral maintains the countable additivity property for disjoint unions. Moreover, $\text{Pr}_2(\cdot)$ assigns zero weight to an individual point, even though an event containing the point may accumulate nonzero probability. This solution is not without its problems. For example, a major concern is that probability is not specified for events other than unions of intervals. Nevertheless, the approach shows that progress is possible toward a more general probability assignment, even when the outcomes with nonzero probability must constitute a countable set.

Distributing probability directly to events, with an integral as in the example above or in some other manner, offers an avenue of promise that has some connection with certain issues in the preceding chapters. There we developed discrete distributions and their applications, but we did encounter two occasions that introduced nondiscrete random variables. The first was the time between events in a Poisson process; the second was the continuous uniform distribution that serves as a model for computerized random number generation. In each case, we used a *cumulative* distribution function to capture enough of the random variable's behavior to complete the discussion.

For the interarrival time T in a Poisson process with parameter λ and the continuous uniform random variable U on $[0,1)$, the cumulative distribution functions were

$$F_T(t) \;=\; \text{Pr}(T \le t) = \text{Pr}(T \in (-\infty, t]) = \begin{cases} 0, & t < 0 \\ 1 - e^{-\lambda t}, & t \ge 0 \end{cases}$$

$$F_U(t) \;=\; \text{Pr}(U \le t) = \text{Pr}(U \in (-\infty, t]) = \begin{cases} 0, & t < 0 \\ t, & 0 \le t < 1 \\ 1, & t \ge 1. \end{cases}$$

These functions are continuous, in contrast with the staircase cumulative distribution functions of discrete random variables. In common with the integral approach above, they also assign probability directly to certain intervals—those of the form $(-\infty, t]$. These observations suggest that we can describe a probability assignment over $\mathcal{R}$ by directly specifying the cumulative distribution function.

The first section below brings this approach to a successful conclusion and develops the promised extension from a discrete probability space to a general probability space. It continues with an extended definition of a random variable, which can assume values throughout $\mathcal{R}$, and develops analogs of the features already familiar from our earlier study: mean, variance, and moment generating function. Also discussed are joint, marginal, and conditional distributions of several random variables of the more general type. This first section is rather lengthy because it addresses a general extension of discrete probability.

The second section shows how this general extension applies to joint random variables, although it omits the detailed constructions employed in the first section to establish the one-dimensional case. The final section then considers a subclass in which the cumulative distribution functions are differentiable. With this property available, we can exploit ordinary calculus to obtain results that are derivable in a more general context only with the Lebesgue integral, a topic that is beyond the scope of this text. This chapter does use, however, the Riemann-Stieltjes integral, which is a straightforward extension of the ordinary Riemann integral. Appendix A develops the properties of the Riemann-Stieltjes integral, including its evaluation via an ordinary integral in particular cases of interest.

This chapter appears here because a rigorous extension of discrete probability concepts to uncountable outcome spaces requires no small measure of explanation. However, it is not strictly necessary to understand the particular distributions and applications in the upcoming chapters. Indeed, the next chapter's nondiscrete distribution are all of a common form. Specifically, a differentiable cumulative distribution function F completely describes the distribution. If $F' = f$, we refer to f as the distribution density and calculate all expected values with an ordinary integral: $E[g] = \int_{-\infty}^{\infty} g(x) f(x)\, dx$. Consequently, skipping this chapter may be appropriate for a first reading. The concepts, nevertheless, are very interesting.

5.1   One-dimensional real distributions

We continue the thread started in the introduction. If we attempt to assign probability directly to events, which are subsets of $\mathcal{R}$, via a cumulative distribution function F, what properties must the function possess? Clearly, F directly assigns probability only to events of the form $(-\infty, x]$. By carefully restricting F, can we extend the assignment such that it also specifies the probability of other events? Can we find an extension that consistently assigns probability to all subsets of R? We use the notation 2^W for the collection of all subsets of W. This may seem an obscure notation, but you can show by induction that a set W of size $|W| = n$ has 2^n subsets, including the empty set ϕ and the set W itself. Think of 2^W as an extension of this observation, because we frequently use it when W is infinite or even uncountable. In these terms, our conjecture is: Does F determine a probability function $\Pr : 2^R \to [0, 1]$? It turns out that nonnegativity, normalization, and countable additivity are sufficiently restrictive that an extension to $2^{\mathcal{R}}$ is not possible. However, with proper constraints on the function, it can specify probability for many subsets of interest.

If we have interest in an event A associated with some application, it is natural that we also deal with the event $\overline{A}$. Similarly, if the sequence of events $A_1, A_2, \ldots$ arise in the application, so do the events corresponding to the occurrence of any or all of the A_n. These derived events are $\cup_{n=1}^{\infty} A_n$ and $\cap_{n=1}^{\infty} A_n$ respectively. Consequently, we hope to use the cumulative distribution function F to specify probability for as many subsets as possible, including at the very least those subsets that have these natural relationships to the intervals $(-\infty, x]$. Although our primary interest is $\mathcal{R}$ as an outcome space, we place our initial definitions in a more general setting.

5.1 Definition: Let X be a set of points, and let $\mathcal{A}$ be a collection of subsets of X. $\mathcal{A}$ is a σ-algebra over X if it satisfies the following three properties.

(1) $X \in \mathcal{A}$

(2) If $A \in \mathcal{A}$, then $\overline{A} = X - A \in \mathcal{A}$

(3) If $A_i \in \mathcal{A}$, for $i = 1, 2, \ldots$,, then $\cup_{i=1}^{\infty} A_i \in \mathcal{A}$. ∎

A σ-algebra is a collection of subsets that contains the base set and is closed under complements and countable unions. Since $X \in \mathcal{A}$, $\phi = \overline{X} \in \mathcal{A}$. Consequently, for any set X, the collection $\{\phi, X\}$ constitutes a σ-algebra. We are normally interested in more profound σ-algebras. Now that we know the empty set must be a member of any σ-algebra, we can infer that the collection is closed under finite unions. When considering $\cup_{i=1}^{n} A_i$, we simply set $A_i = \phi$ for $i > n$. Moreover, since

$$\bigcap_{i=1}^{\infty} A_i = \overline{\bigcup_{i=1}^{\infty} \overline{A_i}},$$

we see that the σ-algebra remains closed under countable intersections. As for finite intersections $\cap_{i=1}^{n} A_i$, we set $A_i = X$ for $i > n$ to show that $\cap_{i=1}^{n} A_i = \cap_{i=1}^{\infty} A_i$ is in the σ-algebra.

5.2 Theorem: A σ-algebra is closed under countable unions and intersections, both finite and infinite.

PROOF: See discussion above. ∎

5.3 Definition: Let X be a set of points, and let $\mathcal{C}$ be a collection of subsets of X. The *smallest σ-algebra* containing $\mathcal{C}$ is a σ-algebra $\mathcal{A}$ over X with the following properties.

(a) $\mathcal{A}$ contains $\mathcal{C}$.

(b) If $\mathcal{A}'$ is a σ-algebra over X containing $\mathcal{C}$, then $\mathcal{A} \subset \mathcal{A}'$. ∎

If it exists, the smallest σ-algebra containing $\mathcal{C}$ is unique. Indeed, if $\mathcal{A}$ and $\mathcal{A}'$ are both smallest σ-algebras containing $\mathcal{C}$, then the definition forces $\mathcal{A} \subset \mathcal{A}'$ and $\mathcal{A}' \subset \mathcal{A}$, which means that they are equal. Moreover, the smallest σ-algebra containing $\mathcal{C}$ does exist; it is simply the intersection of all σ-algebras containing $\mathcal{C}$. There is at least one such σ-algebra, the set of all subsets of X, and it is a straightforward exercise to show that the intersection of all σ-algebras containing $\mathcal{C}$ satisfies the defining properties of a σ-algebra.

When we seek to characterize the subsets of $\mathcal{R}$ that are likely candidates for events in probability calculations, we immediately encounter a σ-algebra. Clearly, this collection should contain the intervals because many situations call for the probability that a random variable lies between two bounds. Moreover, it should remain closed under complements and countable unions because these operations define events that are naturally related to their component events.

5.4 Definition: The *Borel sets* comprise the smallest σ-algebra over $\mathcal{R}$ containing all intervals of the form $(-\infty, x]$, where x is an arbitrary real number. We use $\mathcal{B}$ to designate the collection of Borel sets. ∎

Beyond subsets of the form $(-\infty, x]$, what further subsets appear in $\mathcal{B}$? Of course, ϕ and $\mathcal{R}$ are in $\mathcal{B}$ by the definition of a σ-algebra. Because it must be closed under complements, sets of the form (x, ∞) appear in $\mathcal{B}$. Then, because $[x, \infty) = \cap_{n=1}^{\infty} (x - 1/n, \infty)$, we have all intervals of the form $[x, \infty) \in \mathcal{B}$. Complement operations then produce all $(-\infty, x)$ forms. Via $(a, b) = (-\infty, b) \cap (a, \infty)$, we obtain all open intervals in $\mathcal{B}$. This exercise continues to obtain all open, closed, and half-open intervals in $\mathcal{B}$, as well as all singleton points $\{x\}$.

Our goal is now somewhat more clear. If we assert F as a cumulative distribution function, we want to extend the initial probability assignments of the form $\Pr((-\infty, x]) = F(x)$ to all the Borel sets, because this larger collection contains events that are naturally related to the intervals. Let us first assume that this process has been completed successfully and investigate what conditions F must *necessarily* satisfy to maintain non-negativity, normalization, and countable additivity. We first record a consequence of these properties for increasing and decreasing sequences of sets.

5.5 Theorem: Suppose that X is a set of points, $\mathcal{A}$ is a σ-algebra over X, and $\Pr : \mathcal{A} \to [0, 1]$ such that non-negativity, normalization, and countable additivity hold for sets in $\mathcal{A}$. Let $\{A_n\}$ and $\{B_n\}$ be sequences of sets, all members of $\mathcal{A}$. Furthermore, $A_1 \subset A_2 \subset A_3 \subset \ldots$ is an increasing sequence, and $B_1 \supset B_2 \supset B_3 \supset \ldots$ is a decreasing sequence. Then

$$\Pr(\cup_{n=1}^{\infty} A_n) = \lim_{n \to \infty} \Pr(A_n)$$
$$\Pr(\cap_{n=1}^{\infty} B_n) = \lim_{n \to \infty} \Pr(B_n).$$

PROOF: Let $C_n = A_n - \cup_{i=1}^{n-1} A_i$. The C_n are disjoint, and $\cup_{n=1}^{\infty} C_n = \cup_{n=1}^{\infty} A_n$. Also, for any finite m, $\cup_{n=1}^{m} C_n = A_m$. So,

$$\sum_{n=1}^{m} \Pr(C_n) = \Pr(\cup_{n=1}^{m} C_n) = \Pr(A_m)$$

$$\Pr(\cup_{n=1}^{\infty} A_n) = \Pr(\cup_{n=1}^{\infty} C_n) = \sum_{n=1}^{\infty} \Pr(C_n) = \lim_{m \to \infty} \sum_{n=1}^{m} \Pr(C_n) = \lim_{m \to \infty} \Pr(A_m).$$

We treat the decreasing sequence by complements. That is, $\overline{B_1} \subset \overline{B_2} \subset \ldots$ is an increasing sequence. Applying the result just proved to this sequence yields

$$\Pr(\cup_{n=1}^{\infty} \overline{B_n}) = \lim_{n \to \infty} \Pr(\overline{B_n}) = \lim_{n \to \infty} (1 - \Pr(B_n)) = 1 - \lim_{n \to \infty} \Pr(B_n)$$

$$\Pr(\cap_{n=1}^{\infty} B_n) = \Pr(\overline{\cup_{n=1}^{\infty} \overline{B_n}}) = \lim_{n \to \infty} \Pr(B_n). ∎$$

Now, assume that we start with a function F and successfully define a probability assignment $\Pr(\cdot)$ such that $F(x) = \Pr((-\infty, x])$ for all x and $\Pr(\cdot)$

enjoys the nonnegativity, normalization, and countable additivity properties on the Borel sets. We can reason as follows to obtain certain properties that F must possess. The disjoint union $\mathcal{R} = (-\infty, x] \cup (x, \infty)$ means that $\Pr((-\infty, x]) + \Pr((x, \infty)) = 1$. Consequently,

$$0 \leq \Pr((-\infty, x]) = F(x) = 1 - \Pr((x, \infty)) \leq 1.$$

So, $F(\cdot)$ is bounded between zero and one. Also, if $x_1 < x_2$, then $(-\infty, x_2] = (-\infty, x_1] \cup (x_1, x_2]$ as a disjoint union, which implies that

$$\Pr((-\infty, x_2]) = \Pr((-\infty, x_1]) + \Pr((x_1, x_2])$$
$$F(x_2) = F(x_1) + \Pr((x_1, x_2]) \geq F(x_1).$$

This shows that $F(\cdot)$ must be monotone nondecreasing. We now want to investigate the one-sided limits of F, such as $\lim_{y \to x+} F(y)$. Exploiting Theorem A.28, we can probe this behavior with monotone approach sequences. Specifically, suppose that $x_1 \leq x_2 \leq \ldots < x$ is an arbitrary increasing sequence approaching x from below. Let $A_n = (-\infty, x_n]$. Then $A_1 \subset A_2 \subset \ldots$ is an expanding sequence of sets, and $\cup_{n=1}^{\infty} A_n = (-\infty, x)$. Invoking Theorem 5.5, we have

$$\Pr((-\infty, x)) = \lim_{n \to \infty} \Pr(A_n) = \lim_{n \to \infty} \Pr((-\infty, x_n]) = \lim_{n \to \infty} F(x_n).$$

We conclude that

$$\lim_{y \to x^-} F(y) = \Pr((-\infty, x)) \leq \Pr((-\infty, x]) = F(x).$$

Using a similar argument based on a decreasing set sequence, we obtain

$$\lim_{y \to x^+} F(y) = \Pr((-\infty, x]) = F(x).$$

Therefore, $F(y)$ has a limit as y approaches x from below or from above. Moreover, the limit from above is $F(x)$. We conclude that $F(\cdot)$ is continuous from above at each point x.

Now let the increasing sequence $x_1 \leq x_2 \leq \ldots \to \infty$. The corresponding sets $A_1 = (-\infty, x_1], A_2 = (-\infty, x_2], \ldots$ form an expanding sequence with $\cup_{n=1}^{\infty} A_n = \mathcal{R}$. This implies that

$$1 = \Pr(\mathcal{R}) = \lim_{n \to \infty} \Pr(A_n) = \lim_{n \to \infty} F(x_n).$$

Consequently, $\lim_{y \to \infty} F(y) = 1$. Similarly, a decreasing sequence, $x_1 \geq x_2 \geq \ldots \to -\infty$, defines a contracting sequence of sets, $A_1 = (-\infty, x_1], A_2 = (-\infty, x_2], \ldots$, with $\cap_{n=1}^{\infty} A_n = \phi$. Evoking the same argument once again, we have

$$0 = \Pr(\phi) = \lim_{n \to \infty} \Pr(A_n) = \lim_{n \to \infty} F(x_n),$$

which implies that $\lim_{y \to -\infty} F(y) = 0$. The following theorem summarizes these necessary properties of a true cumulative distribution function.

5.6 Theorem: If $F(x) = \Pr((-\infty, x])$ extends to $\Pr(\cdot)$, which determines probability weights for the Borel sets and preserves the nonnegativity, normalization, and countable additivity properties, then F must obey the following constraints.

- $0 \leq F(x) \leq 1$, for all $x \in \mathcal{R}$.

- $x_1 < x_2$ implies that $F(x_1) \leq F(x_2)$. That is, $F(\cdot)$ is monotone nondecreasing.

- $\lim_{x \to -\infty} F(x) = 0$; $\lim_{x \to \infty} F(x) = 1$.

- At any point x, $\lim_{y \to x^-} F(y)$ and $\lim_{y \to x^+} F(y)$ both exist. Moreover,

$$\lim_{y \to x^-} F(y) \leq \lim_{y \to x^+} F(y) = F(x).$$

- $F_X(\cdot)$ is continuous except at a countable number of points.

PROOF: The discussion above establishes all these properties except the last. At any point x, the limits from either side exist, so discontinuity can occur only when these two limits are unequal. At any such point x, we have

$$\Pr(x) = \Pr(-\infty, x]) - \Pr(-\infty, x) = F(x) - \lim_{y \to x^-} F(y)$$

$$= \lim_{y \to x^+} F(y) - \lim_{y \to x^-} F(y) > 0.$$

The points of discontinuity correspond to individual points with nonzero probability. As was demonstrated in the introduction, these points constitute a countable set. ∎

5.7 Definition: A *cumulative distribution function* is a function $F : \mathcal{R} \to [0, 1]$ that satisfies the properties of Theorem 5.6. ∎

Definition 5.7 allows a concise reference to a function that satisfies the theorem's properties, although we have not yet demonstrated that such a function corresponds to the cumulative distribution function of a real random variable. We continue toward that goal. At this point, we know that *if* F extends to a nonnegative, normalized, countably additive probability assignment on the Borel sets, *then* it must have the properties elaborated in Theorem 5.6. We do not yet know that we can construct such an extension. In truth, however, the extension is possible, and we now undertake the appropriate constructions. We adopt a more concise notation for the one-sided limits of F: $F(x^-) = \lim_{y \to x^-} F(y)$, and similarly for $F(x^+)$. Consider the following process, which assigns probability to every subset of $\mathcal{R}$. Of course, we need to verify that the process is unambiguous and that is satisfies nonnegativity, normalization, and countable additivity on the Borel sets.

5.8 Definition: Given a cumulative distribution function F, define $\Pr : 2^{\mathcal{R}} \to [0, 1]$ as follows. Let $\mathcal{O}$ be the open intervals. That is, $\mathcal{O}$ contains all intervals of the form (a, b), where $a = -\infty$ and $b = \infty$ are possible. $\mathcal{O}$

also contains the empty set. Define $\mathrm{Pr}_F(\phi) = 0$ and, for $I = (x, y) \in \mathcal{O}$,
$\mathrm{Pr}_F(I) = F(y^-) - F(x^+) = F(y^-) - F(x)$.

For any countable $\mathcal{G} = \{I_1, I_2, \ldots\} \subset \mathcal{O}$, define $s(\mathcal{G}) = \sum_{n=1}^{\infty} \mathrm{Pr}_F(I_n)$.
Finally, for any $A \subset \mathcal{R}$, *extend* F to a pseudo-probability assignment with

$$\mathrm{Pr}(A) = \inf \left\{ s(\mathcal{G}) \mid \mathcal{G} = \{I_1, I_2, \ldots\} \subset \mathcal{O} \text{ and } A \subset \bigcup_{n=1}^{\infty} I_n \right\}. \blacksquare$$

The properties of F guarantee the nonnegativity and normalization of $\mathrm{Pr}_F(\cdot)$ over $\mathcal{O}$. Also, there is only one way to construct an interval $I \in \mathcal{O}$ as a countable union of disjoint intervals $I_n \in \mathcal{O}$. One of the I_n must equal I and the rest of the I_n must be empty. Consequently, $\mathrm{Pr}_F(\cdot)$ is countably additive over the restricted subcollection $\mathcal{O}$. We use the term *pseudo-probability assignment* for the derived measure $\mathrm{Pr}(\cdot)$, for which we have not yet established nonnegativity, normalization, and countable additivity over $2^{\mathcal{R}}$.

For a general subset $A \in 2^{\mathcal{R}}$, Definition 5.8 establishes $\mathrm{Pr}(A)$ through a competition among all possible countable sequences of open intervals, each of which covers A with its union. Within a given competing sequence, the open intervals are not necessarily disjoint. For each competitor $\mathcal{G}$, we think of $s(\mathcal{G})$ as a score, computed by summing the $\mathrm{Pr}_F(\cdot)$ values of its constituent open intervals. The sequence with the smallest score wins, and the winning score becomes $\mathrm{Pr}(A)$. It is conceivable that no sequence actually exhibits a winning score. The infimum is the greatest lower bound of the set of all scores, and it is possible that all scores are larger.

Note that $s(\mathcal{G}) = \infty$ for many competitors $\mathcal{G}$. A simple example is a sequence that repeats infinitely often an interval (a, b) for which $\mathrm{Pr}_F((a, b)) = F(b^-) - F(a) > 0$. However, these competitors are all soundly beaten by the sequence $\mathcal{G}_0 = \{\mathcal{R}, \phi, \phi, \ldots\}$, which certainly covers the target A with its union and which has score $s(\mathcal{G}_0) = \mathrm{Pr}_F(\mathcal{R}) = 1$. The competition for the infimum is then essentially limited to those sequences with scores less than or equal to 1.

We first relieve our anxiety concerning the consistency of the process. If we apply the competitive process to an interval in $\mathcal{O}$, for which we already have a probability value directly from $\mathrm{Pr}_F(\cdot)$, we want the process to deliver this same answer. That is, we want $\mathrm{Pr}(K) = \mathrm{Pr}_F(K)$ for $K \in \mathcal{O}$. Accordingly, suppose that $K \in \mathcal{O}$ and $\mathrm{Pr}_F(K) = p$. Of course, $0 \le p \le 1$. We nevertheless invoke the competition, which delivers the value q. We want to show that $q = p$. If $K = \phi$, then $p = 0$, but also $q = 0$ because the competitor $\phi, \phi, \phi, \ldots$ delivers a score of zero, which must equal the infimum. Assume then that K is a nonempty open interval. One competitor is the sequence $K, \phi, \phi, \ldots$, for which the score is p. Therefore, the infimum over all competitors must be less than or equal to p. That is, $q \le p$. To obtain the reverse inequality, consider more closely the composition of a typical competitor. It is a sequence $J_1, J_2, \ldots$, and each J_n is an open interval. If two members admit an inclusion, say $J_i \subset J_j$, then we can remove J_i while still maintaining coverage of K. This removal creates a new competitor with a lower score. If two members exhibit an overlap, say $J_i \cap J_j \neq \phi$, we can replace both J_i and J_j with $J_i \cup J_j$, which is another open interval. A typical case is $J_i = (a, b), J_j = (c, d)$ with $a \leq c < b \leq d$. Because F is monotone nondecreasing, $F(b^-) - F(c^+) \geq 0$, and consequently,

$$
\begin{aligned}
\Pr_F(J_i \cup J_j) &= \Pr((a, d)) = F(d^-) - F(a+) \\
&\leq F(d^-) - F(a^+) + F(b^-) - F(c^+) \\
&= (F(b^-) - F(a^+)) + (F(d^-) - F(c^+)) \\
&= \Pr_F(J_i) + \Pr_F(J_j).
\end{aligned}
$$

The replacement again lowers the score. So any competitor has a score that is larger than that of a competitor with no inclusions or overlaps. But a competitor with no inclusions or overlaps must consist of disjoint intervals. To cover K with disjoint open intervals, one of the intervals must include K. Consequently, the score of any such competitor cannot be less than p. It follows that all competitors must have score greater than or equal to p, which forces $q \geq p$. We conclude that $q = p$.

We have shown that $\Pr(I) = \Pr_F(I)$, for all open intervals $I \in \mathcal{O}$. The same technique suffices to determine the probability assignments for related closed, half-open, and half-closed intervals. For example, $\Pr([x, \infty)) = 1 - F(x^-)$. We leave these extensions as exercises and state the following theorem, which summarizes the properties of $\Pr(\cdot)$.

5.9 Theorem: Let $\Pr(\cdot)$ extend cumulative distribution function F. Then for all $x \leq y$ and for all sets A, B, A_n, we have

$$
\begin{array}{ll}
0 \leq \Pr(A) \leq 1 & \Pr(\phi) = 0 \\
\Pr((x, y)) = F(y^-) - F(x^+) & \Pr(\mathcal{R}) = 1 \\
\Pr((x, y]) = F(y) - F(x^+) & \Pr(\cup_{n=1}^{\infty} A_n) \leq \sum_{n=1}^{\infty} \Pr(A_n) \\
\Pr([x, y)) = F(y^-) - F(x^-) & \Pr(B) \leq \Pr(A), \text{ when } B \subset A. \\
\Pr([x, y]) = F(y) - F(x^-)
\end{array}
$$

PROOF: The discussion above established all points except the last two. To deal with a countable union, suppose that $\epsilon > 0$ is an arbitrary small number. Choose $\mathcal{G}_1, \mathcal{G}_2, \ldots$, each a collection of open intervals in $\mathcal{O}$, such that

$$
A_n \subset \bigcup_{I \in \mathcal{G}_n} I
$$

$$
s(\mathcal{G}_n) < \Pr(A_n) + \epsilon/2^n
$$

$$
\sum_{n=1}^{\infty} s(\mathcal{G}_n) < \sum_{n=1}^{\infty} \Pr(A_n) + \epsilon(1/2^1 + 1/2^2 + \ldots) = \sum_{n=1}^{\infty} \Pr(A_n) + \epsilon.
$$

Now, $\mathcal{G} = \cup_{n=1}^{\infty} \mathcal{G}_n$ is also a countable collection of open intervals, and we have $\cup_{n=1}^{\infty} A_n \subset \cup_{I \in \mathcal{G}} I$. Moreover, each $I \in \mathcal{G}$ appears in one or more of the $\mathcal{G}_n$. Consequently,

$$
s(\mathcal{G}) = \sum_{I \in \mathcal{G}} \Pr_F(I) \leq \sum_{n=1}^{\infty} \sum_{I \in \mathcal{G}_n} \Pr_F(I) = \sum_{n=1}^{\infty} s(\mathcal{G}_n),
$$

and therefore

$$\Pr\left(\bigcup_{n=1}^{\infty} A_n\right) \le s(\mathcal{G}) \le \sum_{n=1}^{\infty} s(\mathcal{G}_n) < \sum_{n=1}^{\infty} \Pr(A_n) + \epsilon.$$

We conclude that $\Pr(\cup_{n=1}^{\infty} A_n) \le \sum_{n=1}^{\infty} \Pr(A_n)$.

Finally, suppose that $B \subset A$. If $\mathcal{G}$ is a countable collection of open intervals with $A \subset \cup_{I \in \mathcal{G}} I$, then $B \subset A \subset \cup_{I \in \mathcal{G}} I$. Consequently, every competitor for $\Pr(A)$ is also a competitor for $\Pr(B)$. The B competition may, however, have additional competitors, which forces $\Pr(B) \le \Pr(A)$. ∎

The theorem asserts that $\Pr(\cdot)$ is nonnegative, normalized, and *subadditive*, where subadditive means that the probability of a countable union is no greater than the sum of its constituent probabilities. All that remains is to show that it is countably additive on the Borel sets. You might feel inclined to go for total victory and try to show that $\Pr(\cdot)$ is countably additive on $2^{\mathcal{R}}$. Unfortunately, the next example shows that countable additivity over arbitrary disjoint subsets is not possible.

5.10 Example: Let $Q = \{q_1, q_2, \dots\}$ enumerate the rational numbers in the interval $[0, 1)$. Define an equivalence relation on $[0, 1)$ by $x \approx y$ if and only if $|x - y| \in Q$. Clearly, this relation is reflexive, symmetric, and transitive, so it partitions $[0, 1)$ into disjoint equivalence classes. Construct set A by selecting one point from each equivalence class, and let

$$A_n = \{x + q_n \mid x \in A, x + q_n < 1\} \cup \{x + q_n - 1 \mid x \in A, x + q_n \ge 1\}.$$

In constructing A_n, each point in A is displaced q_n to the right. If the displacement lies outside $[0, 1)$, it wraps around such that the surplus appears to the right of 0. Now consider the following simple cumulative distribution function and its extension to a probability assignment $\Pr(\cdot)$ via Definition 5.8.

$$F(x) = \begin{cases} 0, & x < 0 \\ x, & 0 \le x < 1 \\ 1, & x \ge 1 \end{cases}$$

We have $\Pr([0, 1)) = F(1^-) - F(0^+) = 1 - 0 = 1$. But, what are $\Pr(A)$ and $\Pr(A_n)$? First, note that F is continuous, so it assigns zero probability to individual points. Consequently, the probability of any subinterval of $[0, 1)$ is simply the difference of its endpoints, regardless of whether or not the endpoints are included in the subinterval. So, for any set B, if $\mathcal{G}$ is a competitor in the contest to determine $\Pr(B)$, we can alter the open intervals in $\mathcal{G}$ by including endpoints as suits our needs. These alterations do not change the score of $\mathcal{G}$. Second, note that F assigns zero probability to intervals outside $[0, 1)$. So, if $B \subset [0, 1)$, we can modify the intervals in a competitor $\mathcal{G}$ by intersecting them with $[0, 1)$. The score of $\mathcal{G}$ again remains unchanged. Finally, we argue that a modified competitor $\mathcal{G} = \{I_1, I_2, \dots\}$ includes A in its union if and only if its translated associate $\mathcal{G}'$ includes A_n. Here, $\mathcal{G}' = \{I_1', I_2', \dots\}$, where

$$I_j' = \{x + q_n \mid x \in I_j, x + q_n < 1\} \cup \{x + q_n - 1 \mid x \in I_j, x + q_n \ge 1\}.$$

If, for example, $I_j = (a, b) \subset [0, 1)$, then $I_j' = (a + q_n, b + q_n)$ if $b + q_n < 1$. Otherwise, $I_j' = (a + q_n, 1) \cup [0, b + q_n - 1)$. Since the score of a competitor is the sum of its subinterval widths, $\mathcal{G}$ and $\mathcal{G}'$ have identical scores. At this point, we have shown that for every competitor in the competition for $\Pr(A)$, there is a competitor in the competition for $\Pr(A_n)$ with the same score, and vice versa. We conclude that $\Pr(A_n) = \Pr(A)$ for all n.

Now, we claim that the $\{A_n\}$ are disjoint. Indeed, if $x \in A_i \cap A_j$, then x is $a + q_i$ or $a + q_i - 1$ for some $a \in A$. Similarly, $x = b + q_j$ or $x = b + q_j - 1$ for some $b \in A$. Whichever case obtains, we must have $|a - b| \in Q$, which places a and b in the same equivalence class. Because A contains one element from each class, we conclude that $a = b$, which forces $q_i = q_j$. Equivalently, $i = j$. That is, the $\{A_n\}$ are disjoint.

Finally, we assert that $[0, 1) = \cup_{n=1}^{\infty} A_n$. Since $A_n \subset [0, 1)$ for all n, we immediately have $\cup_{n=1}^{\infty} A_n \subset [0, 1)$. For the reverse inclusion, suppose that $x \in [0, 1)$. Then x is in one of the equivalence classes, which means that there exists $y \in A$ such that $|x - y| \in Q$. If $x - y \geq 0$, we have $x - y = q_n$, for some n, which places $x \in A_n$. If $x - y < 0$, then $1 + x - y = q_n$, for some n, which still places $x = y + q_n - 1 \in A_n$. In either case, $x \in \cup_{n=1}^{\infty} A_n$. This proves the reverse inclusion and $[0, 1) = \cup_{n=1}^{\infty} A_n$. If countable additivity over disjoint subsets holds for arbitrary subsets of $\mathcal{R}$, we must have

$$1 = \Pr([0, 1)) = \sum_{n=1}^{\infty} \Pr(A_n) = \lim_{N \to \infty} \sum_{n=1}^{N} \Pr(A) = \lim_{N \to \infty} N\Pr(A).$$

If $\Pr(A) = 0$, this equation cannot be true. If $\Pr(A) > 0$, it still cannot be true. Because we know that the probability assignment of Definition 5.8 is non-negative, there are no further alternatives. We conclude that countable additivity over disjoint sets does not hold in the general case. $\square$

Despite the preceding example, we can salvage some utility from Definition 5.8's probability assignment if we can show that it is countably additive over a large and useful collection of subsets. We will now show that it exhibits the proper behavior when confined to the Borel sets. The proof involves some intricate and demanding set constructions, which we attempt to render more intuitively persuasive with the following definition.

5.11 Definition: For a given $\Pr(\cdot)$, which extends the cumulative distribution function F, define the collection of *perfect slicers* as

$$\mathcal{B}' = \{A \subset \mathcal{R} \mid \Pr(B) = \Pr(B \cap A) + \Pr(B \cap \overline{A}) \text{ for all } B \subset \mathcal{R} \}. \blacksquare$$

For each fixed A, we have $B = (B \cap A) \cup (B \cap \overline{A})$ for all B. The subadditivity of $\Pr(\cdot)$ then implies that $\Pr(B) \leq \Pr(B \cap A) + \Pr(B \cap \overline{A})$ for all B. We place in $\mathcal{B}'$ only those sets A for which this inequality is actually an equality for all sets B. That is, $A \in \mathcal{B}'$ if it slices every set into two disjoint components while conserving probability; the probability of the whole is exactly the sum of the probabilities of the two components. We will first show that the perfect slicers constitute a σ-algebra containing all intervals of the form $(-\infty, x]$. This implies that all Borel sets are perfect slicers. Then we will show that the condition on $\mathcal{B}'$ membership implies countable additivity for unions of disjoint subsets.

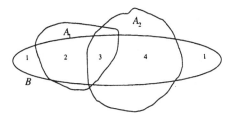

Figure 5.1. Slicing a set B with sets known to maintain probability measure

If $A \in \mathcal{B}'$, we have

$$\Pr(B) = \Pr(B \cap A) + \Pr(B \cap \overline{A}) = \Pr(B \cap \overline{A}) + \Pr(B \cap \overline{\overline{A}}),$$

for any set B. Consequently, $\overline{A} \in \mathcal{B}'$, which implies that $\mathcal{B}'$ is closed under complements. Also, for any B,

$$\Pr(B) = \Pr(B \cap \mathcal{R}) + \Pr(\phi) = \Pr(B \cap \mathcal{R}) + \Pr(B \cap \overline{\mathcal{R}}),$$

which confirms that $\mathcal{R} \in \mathcal{B}'$. Now, consider two sets $A_1, A_2 \in \mathcal{B}'$. Let B be an arbitrary set, and envision the decomposition suggested by Figure 5.1. We decompose B into four parts, the segments outside both A_1 and A_2, inside A_1 but outside A_2, inside both A_1 and A_2, and inside A_2 but outside A_1. These components bear the labels $1, 2, 3$, and 4 in the figure. Set A_1 is a perfect slicer. That is, as an element of $\mathcal{B}'$, it conserves probability when it divides another set into a component in A_1 and a component outside A_1. Consequently, $\Pr(B)$ is the sum of the probabilities of segments 2–3 and 1–4. Then, because A_2 is a perfect slicer, segment 1–4 breaks into segments 1 and 4 without any loss, which gives $\Pr(B)$ as the sum of the probabilities of segments 2–3, 4, and 1. Finally, we decompose segment 2–3–4 with set A_1 into segments 2–3 and 4, which shows $\Pr(B)$ to be the lossless sum of the probabilities of segments 2–3–4 and 1. Segment 2–3–4 is the intersection of B with $A_1 \cup A_2$; segment 1 is the intersection with $\overline{A_1 \cup A_2}$. That is, $A_1 \cup A_2$ functions as a perfect slicer.

This shows that A_1, A_2 in $\mathcal{B}'$ implies that $A_1 \cup A_2 \in \mathcal{B}'$. An induction argument now shows that $A_1, A_2, \dots, A_n \in \mathcal{B}'$ implies that $\cup_{j=1}^{n} A_j \in \mathcal{B}'$ for any n. That is, $\mathcal{B}'$ is closed under finite unions. Because $\mathcal{B}'$ is closed under complement operations, the rearrangement $\cap_{j=1}^{n} A_j = \overline{\cup_{j=1}^{n} \overline{A_j}}$ implies that it is also closed under finite intersections.

5.12 Theorem: $\mathcal{B}'$ contains $\mathcal{R}$, and it is closed under complements, finite unions, and finite intersections. Moreover, if $A_1, A_2, \dots, A_n$ are *disjoint* elements of $\mathcal{B}'$, then for any set B, we have

$$\Pr\left(B \cap \left(\cup_{i=1}^{n} A_i\right)\right) = \sum_{i=1}^{n} \Pr(B \cap A_i).$$

PROOF: The discussion above established these claims except the last expression, which we prove by induction on n. For $n = 1$, it reads $\Pr(B \cap A_1) =$

$\Pr(B \cap A_1)$, which is clearly true. Assume that it holds through $n-1$. Because the $\{A_i\}$ are disjoint, $i < n$ implies that $A_i \cap A_n = \phi$ and $A_i \cap \overline{A_n} = A_i$. Then, because A_n is a perfect slicer, we have

$$\Pr\left(B \cap \left(\cup_{i=1}^n A_i\right)\right) = \Pr\left(B \cap \left(\cup_{i=1}^n A_i\right) \cap A_n\right) + \Pr\left(B \cap \left(\cup_{i=1}^n A_i\right) \cap \overline{A_n}\right)$$

$$= \Pr(B \cap A_n) + \Pr\left(B \cap \left(\cup_{i=1}^{n-1} A_i\right)\right) = \sum_{i=1}^n \Pr(B \cap A_i). \blacksquare$$

We now want to show that $\mathcal{B}'$ is closed under countably infinite unions. Suppose that $A = \cup_{n=1}^\infty A_n$, with each $A_n \in \mathcal{B}'$. Let $B_n = A_n - \cup_{i=1}^{n-1} A_i$. Since $X - Y = X \cap \overline{Y}$, each $B_n \in \mathcal{B}'$, and the B_n are disjoint. Moreover, $\cup_{n=1}^\infty B_n = \cup_{n=1}^\infty A_n = A$. We need to show that $\cup_{n=1}^\infty B_n$ is a perfect slicer. For any n, $\cup_{i=1}^\infty B_i \subset \cup_{i=1}^n B_i$. From Theorem 5.12, finite unions of the B_i are perfect slicers. So, for any set C, the preceding theorem allows the reduction

$$\Pr(C) = \Pr\left(C \cap \left(\cup_{i=1}^n B_i\right)\right) + \Pr\left(C \cap \left(\overline{\cup_{i=1}^n B_i}\right)\right)$$

$$= \sum_{i=1}^n \Pr(C \cap B_i) + \Pr\left(C \cap \left(\overline{\cup_{i=1}^n B_i}\right)\right)$$

$$\geq \sum_{i=1}^n \Pr(C \cap B_i) + \Pr\left(C \cap \left(\overline{\cup_{i=1}^\infty B_i}\right)\right).$$

Letting $n \to \infty$ and invoking the known subadditivity on the first term, we have

$$\Pr(C) \geq \sum_{i=1}^\infty \Pr(C \cap B_i) + \Pr\left(C \cap \left(\overline{\cup_{i=1}^\infty B_i}\right)\right)$$

$$\geq \Pr\left(\cup_{i=1}^\infty (C \cap B_i)\right) + \Pr\left(C \cap \left(\overline{\cup_{i=1}^\infty B_i}\right)\right)$$

$$= \Pr\left(C \cap \left(\cup_{i=1}^\infty B_i\right)\right) + \Pr\left(C \cap \left(\overline{\cup_{i=1}^\infty B_i}\right)\right).$$

The subadditivity of the $\Pr(\cdot)$ function provides the reverse inequality, so we conclude that $\cup_{i=1}^\infty B_i = \cup_{i=1}^\infty A_i \in \mathcal{B}'$. At this point, we finally know that $\mathcal{B}'$ is a σ-algebra, and we proceed to show that $\Pr(\cdot)$ is countably additive on $\mathcal{B}'$.

For any $A \in \mathcal{B}'$, we have $1 = \Pr(\mathcal{R}) = \Pr(\mathcal{R} \cap A) + \Pr(\mathcal{R} \cap \overline{A}) = \Pr(A) + \Pr(\overline{A})$, which means that $\Pr(\overline{A}) = 1 - \Pr(A)$, as expected. Now, if $A = \cup_{i=1}^\infty A_i$, and the A_i are disjoint members of $\mathcal{B}'$, we have, for any n,

$$1 = \Pr(\mathcal{R}) = \Pr\left(\mathcal{R} \cap \left(\cup_{i=1}^n A_i\right)\right) + \Pr\left(\mathcal{R} \cap \left(\overline{\cup_{i=1}^n A_i}\right)\right)$$

$$= \sum_{i=1}^n \Pr(A_i) + \Pr\left(\overline{\cup_{i=1}^n A_i}\right) \geq \sum_{i=1}^n \Pr(A_i) + \Pr\left(\overline{\cup_{i=1}^\infty A_i}\right)$$

$$= \sum_{i=1}^n \Pr(A_i) + \Pr(\overline{A}) = \sum_{i=1}^n \Pr(A_i) + 1 - \Pr(A)$$

$$\Pr(A) \geq \sum_{i=1}^n \Pr(A_i).$$

Letting $n \to \infty$, we have $\Pr(A) \geq \sum_{i=1}^{\infty} \Pr(A_i)$. The subadditivity of $\Pr(\cdot)$ again provides the reverse inequality, so we conclude that $\Pr(\cdot)$ is countably additive on $\mathcal{B}'$.

Now we will show that intervals of the form $(-\infty, x]$ are in $\mathcal{B}'$. To this end, we show that $(-\infty, x]$ perfectly slices an arbitrary set B. Given any $\epsilon > 0$, choose a collection $\mathcal{G} = \{I_1, I_2, \ldots\}$ of open intervals in $\mathcal{O}$ such that $B \subset \cup_{n=1}^{\infty} I_n$ and $s(\mathcal{G}) < \Pr(B) + \epsilon$. For each $I_n \in \mathcal{G}$, define

$$I_n^+ = I_n \cap (x, \infty)$$
$$I_n^- = I_n \cap (-\infty, x].$$

Let $I_n = (a_n, b_n)$, where a_n may be $-\infty$ and b_n may be ∞. Then $\Pr(I_n) = F(b_n^-) - F(a_n^+)$, and we distinguish the following possibilities. Recall that F is continuous from the right: $F(t^+) = F(t)$.

Case	I_n^-	I_n^+	$\Pr(I_n^-)$	$\Pr(I_n^+)$	$\Pr(I_n^-) + \Pr(I_n^+)$
$a_n < x < b_n$	$(a_n, x]$	(x, b_n)	$F(x) - F(a_n^+)$	$F(b_n^-) - F(x^+)$	$F(b_n^-) - F(a_n^+)$
$a_n < b_n \leq x$	(a_n, b_n)	ϕ	$F(b_n^-) - F(a_n^+)$	0	$F(b_n^-) - F(a_n^+)$
$x \leq a_n < b_n$	ϕ	(a_n, b_n)	0	$F(b_n^-) - F(a_n^+)$	$F(b_n^-) - F(a_n^+)$

In all cases, we have $\Pr(I_n^-) + \Pr(I_n^+) = \Pr(I_n)$, which enables the following argument.

$$B \cap (-\infty, x] \subset \cup_{n=1}^{\infty} I_n^-$$

$$\Pr(B \cap (-\infty, x]) \leq \Pr(\cup_{n=1}^{\infty} I_n^-) \leq \sum_{n=1}^{\infty} \Pr(I_n^-)$$

$$B \cap ((x, \infty)) \subset \cup_{n=1}^{\infty} I_n^+$$

$$\Pr(B \cap ((x, \infty))) \leq \Pr(\cup_{n=1}^{\infty} I_n^+) \leq \sum_{n=1}^{\infty} \Pr(I_n^+)$$

$$\Pr(B \cap ((-\infty, x])) + \Pr(B \cap (x, \infty)) \leq \sum_{n=1}^{\infty} (\Pr(I_n^-) + \Pr(I_n^+))$$

$$= \sum_{n=1}^{\infty} \Pr(I_n) < \Pr(B) + \epsilon$$

Because this last inequality holds for an arbitrary $\epsilon > 0$, we conclude that

$$\Pr(B \cap ((-\infty, x])) + \Pr(B \cap (x, \infty)) \leq \Pr(B).$$

The subadditivity of $\Pr(\cdot)$ provides the reverse inequality, so we have shown that $(-\infty, x]$ is a perfect slicer. Because $\mathcal{B}'$ is a σ-algebra containing all intervals of the form $(-\infty, x]$, it must contain all the Borel sets. We summarize the discussion with the following theorem.

5.13 Theorem: $\mathcal{B}'$, the collection of perfect slicers, is a σ-algebra containing the Borel sets. Moreover, $\Pr(\cdot)$ is countably additive on $\mathcal{B}'$.

PROOF: See discussion above. ∎

We now have a practical method for spreading probability measure over events, provided that the events are Borel sets of real numbers. Following the tradition associated with discrete distributions, we define a probability space in a more abstract setting and use a random variables to map the outcomes, events, and probabilities into the real numbers. However, we must now take care that mapped events are Borel sets.

5.14 Definition: A general *probability space* contains three components $(\Omega, \mathcal{A}, \Pr(\cdot))$. Ω is a point set, possibly uncountable, $\mathcal{A}$ is a σ-algebra over Ω, and $\Pr : \mathcal{A} \to [0, 1]$ is a probability assignment function that satisfies

$$\Pr(\Omega) = 1$$

$$\Pr(\cup_{n=1}^{\infty} A_n) = \sum_{n=1}^{\infty} \Pr(A_n), \quad \text{when the } A_n \text{ are disjoint members of } \mathcal{A}. \blacksquare$$

If Ω is countable, we can take $\mathcal{A} = 2^{\Omega}$, and the new definition then accords with Definition 1.36 for a discrete probability space. In the normal manner, we can now define a random variable X over a general probability space. X maps outcomes into the real numbers and transfers probability to selected subsets of $\mathcal{R}$. We insist that all intervals, which are the most frequently used events in range(X), receive a probability assignment. Specifically, the definition is as follows.

5.15 Definition: Let $(\Omega, \mathcal{A}, \Pr(\cdot))$ be a general probability space. A general *random variable* is a function $X : \Omega \to \mathcal{R}$ such that $X^{-1}((-\infty, x]) \in \mathcal{A}$ for every $x \in \mathcal{R}$. $\blacksquare$

When Ω is countable and $\mathcal{A} = 2^{\Omega}$, the restriction is superfluous because $X^{-1}(C) \in 2^{\Omega}$ for any set C. Therefore, the definition accords with Definition 1.38, which describes a discrete random variable. In the general case, $X^{-1}((-\infty, x])$ always has a well-defined probability because it always lies in the σ-algebra $\mathcal{A}$. The restriction then ensures that all events of the form $(X \leq x)$ have a well-defined probability:

$$\Pr(X \leq x) = \Pr(X \in (-\infty, x]) = \Pr(X^{-1}((-\infty, x])).$$

However, the question arises: What other events in $\mathcal{R}$ receive probability assignments? In general, a set $C \subset \mathcal{R}$ receives a probability assignment if and only if $X^{-1}(C) \in \mathcal{A}$. Consequently, we examine the collection

$$\mathcal{C} = \{C \in \mathcal{R} | X^{-1}(C) \in \mathcal{A}\}.$$

We will show that $\mathcal{C}$ is a σ-algebra containing all intervals of the form $(-\infty, x]$. Indeed, the definition of a random variable forces $\mathcal{C}$ to contain all such intervals. Clearly, $\mathcal{R} \in \mathcal{C}$ because $X^{-1}(\mathcal{R}) = \Omega \in \mathcal{A}$. Because inverse images preserve complements (Theorem A.6), we have

$$X^{-1}(\overline{C}) = \overline{X^{-1}(C)} \in \mathcal{A},$$

when $C \in \mathcal{C}$, which implies that $\overline{C} \in \mathcal{C}$. Similarly, if $C = \cup_{n=1}^{\infty} C_i$, with each $C_i \in \mathcal{C}$, we have

$$X^{-1}(C) = X^{-1}(\cup_{n=1}^{\infty} C_i) = \cup_{n=1}^{\infty} X^{-1}(C_i) \in \mathcal{A},$$

which implies that $C \in \mathcal{C}$. This proves that $\mathcal{C}$ is a σ-algebra containing the intervals $(-\infty, x]$, which implies that $\mathcal{B} \subset \mathcal{C}$. Thus every Borel set receives a probability assignment when the random variable X transfers the probability weights to $\mathcal{R}$.

5.16 Theorem: A general random variable specifies probability assignments for all Borel sets.

PROOF: See discussion above. ∎

Once a random variable transfers outcomes and probabilities to the real numbers, all analysis takes place in the new setting. That is, we work with the induced probability space consisting of real outcomes, Borel events, and probabilities of these Borel events. The distribution of the random variable then refers to the induced probability assignments to certain subsets of real numbers. We describe such a distribution by giving its cumulative distribution function F, with the understanding that a compatible $\Pr(\cdot)$ function consistently assigns probability to all Borel sets. The next example illustrates the point.

5.17 Example: Suppose that the lifetime of a disk drive is a random variable T with the cumulative distribution function

$$F_T(t) = \begin{cases} 0, & t < 0 \\ 0.1 + 0.9(1 - e^{-0.03t}), & t \geq 0, \end{cases}$$

where t is measured in months. Note that $F_T(\cdot)$ satisfies the properties of a cumulative distribution function. Let $\Pr(\cdot)$ be the extension to the Borel sets. We consider some events of interest in this situation.

$\Pr(T = 0) = \Pr([0, 0]) = F(0) - F(0^-) = 0.1 - 0 = 0.1$. That is, there is a 10% chance that the drive will fail immediately upon being placed in service. If the drive starts successfully, however, then

$$\begin{aligned} \Pr(0 < a < T < b) &= F(b^-) - F(a) = 0.9(e^{-0.03a} - e^{-0.03b}) \\ &= 0.9e^{-0.03a}(1 - e^{-0.03(b-a)}). \end{aligned}$$

The probability that the drive will last longer than a year is $\Pr(T \in (12, \infty))$, which is

$$\Pr(T \in (12, \infty)) = 0.9e^{-0.36} = 0.628.$$

The probability that it will fail in the first six months is

$$\Pr(T \in [0, 6]) = F(6) - F(0^-) = F(6) = 0.1 + 0.9(1 - e^{0.18}) = 0.248. \ \square$$

For general random variables, we naturally want descriptive features, such as the mean, variance, and higher moments. For the discrete case, these features appear as expected values of certain elementary functions, and we expect similar definitions in the general case. Accordingly, we extend the expected value operation as a Riemann-Stieltjes integral. This integral is a simple extension of the ordinary integral encountered in elementary calculus. In particular, suppose that F is a nondecreasing function on $[a, b]$, such as a cumulative distribution function. We define the Riemann-Stieltjes integral of a bounded f with respect to F through a limiting process similar to that used for the ordinary integral. For each partition P of $[a, b]$, $a = x_1 \leq x_2 \leq \ldots \leq x_n = b$, we define upper and lower sums:

$$S_{f,F}(P) = \sum_{i=1}^{n} [F(x_i) - F(x_{i-1})] \cdot \sup\{f(x) \mid x_{i-1} \leq x \leq x_i\}$$

$$s_{f,F}(P) = \sum_{i=1}^{n} [F(x_i) - F(x_{i-1})] \cdot \inf\{f(x) \mid x_{i-1} \leq x \leq x_i\}.$$

Letting P range over $\mathcal{P}$, the collection of all partitions, we extract the infimum of the upper sums and the supremum of the lower sums:

$$U_f = \inf\{S_f(P) \mid P \in \mathcal{P}\}$$
$$L_f = \sup\{s_f(P) \mid P \in \mathcal{P}\}.$$

If $U_f = L_f$, we say that f is integrable with respect to F on (a, b) and write $\int_a^b f \, dF = U_f = L_f$. This process differs from that associated with the ordinary integral only in that the F gain over $[x_{i-1}, x_i]$, which is $F(x_i) - F(x_{i-1})$, replaces the cell width $(x_i - x_{i-1})$ in the sums. Indeed, if we let $F(x) = x$, we recover the ordinary integral as a special case of the Riemann-Stieltjes integral. Appendix A develops the properties of the Riemann-Stieltjes integral in detail, including the extensions when (a, b) is an infinite interval and when f is unbounded. These extensions are limits of approximating integrals as defined above. At this point, we assume familiarity with the properties elaborated in Appendix A. In particular, for the next definition, we need the concept of absolute convergence for the integral. We say that $\int_{-\infty}^{\infty} f \, dF$ converges absolutely if

$$\int_{-\infty}^{\infty} |f| \, dF = \lim_{M,N \to \infty} \int_{-M}^{N} |f| \, dF = A < \infty.$$

5.18 Definition: Let X be a general random variable with cumulative distribution function F. The *expected value* of $g(\cdot)$ is $E_X[g(\cdot)] = E[g] = \int_{-\infty}^{\infty} g \, dF$, provided that the integral converges absolutely. ∎

The absolute convergence condition is necessary to avoid pathologies similar to those associated with lack of absolute convergence in the discrete case. Consequently, an expected value is not defined for certain troublesome $g(\cdot)$. As before, we drop the X subscript on the expectation operator when the underlying random variable is clear from context. Also, we frequently specify the functional form as the argument for the operator. That is, for example, we write $E[X^2]$ instead of $E[g(\cdot)]$, $g(t) = t^2$.

Suppose that X is a discrete random variable with nonzero probability only on a countable set. That is, $\Pr(X = x_i) = p_i$, for $i = 1, 2, \ldots$. F_X is then a staircase function with a jump $p_i > 0$ at each x_i. In this case, assuming that g is integrable with respect to F, the integral reduces to a summation:

$\int_a^b g \, dF = \sum_{i=1}^\infty p_i g(x_i)$. Consequently, the new definition coincides with the old.

To facilitate expected value computations, we assume from this point forward that the cumulative distribution function F is differentiable everywhere except on a countable collection of points, called discontinuities, where $F(x) - F(x^-) > 0$. For any interval (a, b) free of discontinuities, we then have (Theorem A.53)

$$\int_a^b g \, dF = \int_a^b g(x) F'(x) \, dx.$$

We now have an ordinary integral, whose value is unchanged if we modify the integrand on an isolated set of points. Consequently, we can define $F'(x) = 0$, or any other convenient value, for those points where F' does not exist. With this modification, we need only add the contributions from the discontinuities to obtain a working formula for arbitrary intervals. That is, if the discontinuities are at points $\{x_1, x_2, \dots\}$, we have

$$\int_a^b g \, dF = \int_a^b g(x) F'(x) \, dx + \sum_{x_i \in (a, b]} g(x_i) \Pr(x_i).$$

The evaluation mechanism includes a discontinuity contribution from the right endpoint of an integration span, but it excludes any such contribution from the left endpoint. This policy allows the assembly of an integral over consecutive spans, such as

$$\int_a^c g \, dF = \int_a^b g \, dF + \int_b^c g \, dF,$$

without counting an endpoint contribution twice. If, for some reason, we specifically want to include a left endpoint or exclude a right endpoint, we adopt special notation. For example,

$$\int_{a^-}^b g \, dF = \lim_{c \to a^-} \int_c^b g \, dF = \int_a^b g(x) F'(x) \, dx + \sum_{x_i \in [a, b]} g(x_i) \Pr(x_i).$$

With such a piecewise differentiable F, we can define the distribution with two components: a countable collection of points where $F(x) - F(x^-) > 0$ and a density function, whose integral advances the cumulative distribution function between the discontinuous points. Suppose, for example, that we are describing the cumulative distribution function of a random variable X. We first specify a countable collection $D_X = \{x_1, x_2, \dots\}$ and a mapping $f_d(x_i) \geq 0$ that gives the jump in the cumulative distribution function at each $x_i \in D_X$. That is, $F(x_i) - F(x_i^-) = f_d(x_i)$. We complete the description with a density function $f_c(x) \geq 0$ such that

$$F_X(x) = \int_{-\infty}^x dF = \sum_{x_i \leq x} f_d(x_i) + \int_{-\infty}^x f_c(y) \, dy.$$

For the example above, which dealt with the service life of a disk drive, we can show that the components

$$f_d(0) = 0.1$$

$$f_c(x) = \begin{cases} 0, & x < 0 \\ 0.027e^{-0.03x}, & x \geq 0 \end{cases}$$

deliver the given cumulative distribution function. Indeed, for $t < 0$, we compute $F_T(t) = 0$, and for $t \geq 0$,

$$F_T(t) = \sum_{x_i \leq t} f_d(x_i) + \int_{-\infty}^{t} f_c(y)\,dy = 0.1 + \int_0^t 0.027e^{-0.03y}\,dy$$

$$= 0.1 + 0.9(1 - e^{-0.03t}).$$

When we describe the cumulative distribution function with these two constituents, we refer to $f_d(\cdot)$ and $f_c(\cdot)$ as the discrete and continuous density components. The precise definition is as follows.

5.19 Definition: Let X be a random variable with cumulative distribution function

$$F_X(x) = \int_{-\infty}^{x} dF = \sum_{x_i \leq x} f_d(x_i) + \int_{-\infty}^{x} f_c(y)\,dy,$$

where $D_X = \{x_1, x_2, \dots\}$, $f_d : D_X \to [0,1]$, and $f_c : \mathcal{R} \to [0, \infty)$. $D_X, f_d(\cdot)$, and $f_c(\cdot)$ constitute the *density* of X. If $f_c(x) = 0$ for all x, we say that X is a *discrete random variable*; if $D_X = \phi$ or $f_d(x_i) = 0$ for all $x_i \in D_X$, we say that X is a *continuous random variable*. Otherwise, we call X a *mixed random variable* and refer to its $f_d(\cdot)$ and $f_c(\cdot)$ as the discrete and continuous density components. ∎

Note that this definition of a discrete random variable agrees with that used in preceding chapters. In that case, the integral portion of the $F_X(\cdot)$ computation vanishes, leaving $F_X(x) = \sum_{x_i \leq x} f_d(x_i)$, which is just the sum of the probability weights for all range points less than or equal to x. For the Poisson random variable with parameter λ, for example, $D_X = \{0, 1, 2, \dots\}$, and $f_d(i) = \lambda^i e^{-\lambda}/i!$. Of course, $f_c(\cdot)$ is identically zero. If X is a Bernoulli random variable with parameter p, then $D_X = \{0, 1\}$, and $f_d(x) = p^x(1 - p)^{1-x}$ for $x = 0, 1$. Again, $f_c(\cdot)$ is identically zero.

For each $x_i \in D_X$, we have

$$\Pr(X = x_i) = \Pr([x_i, x_i]) = F(x_i) - F(x_i^-) = \int_{-\infty}^{x_i} dF - \int_{-\infty}^{x_i^-} dF$$

$$= \sum_{x_j \leq x_i} f_d(x_j) + \int_{-\infty}^{x_i} f_c(y)\,dy - \sum_{x_j < x_i} f_d(x_j) - \int_{-\infty}^{x_i} f_c(y)\,dy$$

$$= f_d(x_i),$$

as anticipated. If $x \notin D_X$, the set $(x_i \leq x)$ coincides with the set $(x_i < x)$, and the computation above gives $\Pr(X = x) = 0$. This result is also anticipated because F differentiable on $\overline{D_X}$ implies that it is continuous there. Because the only points that have nonzero probability lie in D_X, we frequently say that these points possess *probability lumps*, as opposed to the points in $\overline{D_X}$, which acquire nonzero probability only in aggregate sets.

We now proceed to establish the mean, variance, and moments of a general random variable with a piecewise differentiable cumulative distribution function. These properties are straightforward extensions of the corresponding concepts for discrete random variables. We first note the linearity of the more general expected value operator.

5.20 Theorem: The extended expectation operator remains linear. That is, we have $E[ag_1(\cdot) + bg_2(\cdot)] = aE[g_1(\cdot)] + bE[g_2(\cdot)]$ and $E[a] = a$.

PROOF: Theorem A.51 establishes the linearity of the Riemann-Stieltjes integral. ∎

The mean, variance, and the moment generating function have the anticipated definitions, provided that the required expected values exist.

5.21 Definition:

$$
\begin{aligned}
k\text{th moment}(X) &= E[X^k] \\
\text{mean}(X) &= \mu_X = E[X] \\
\text{var}(X) &= \sigma_X^2 = E[(X - \mu_X)^2] \\
\text{standard deviation}(X) &= \sigma_X = \sqrt{\text{var}(X)} \\
\Psi_X(t) &= E[e^{tX}] \ \blacksquare
\end{aligned}
$$

The shortcut for computing the variance as the expected value of the square minus the square of the expected value remain available. It depends only on the linearity of the expected value operator.

$$
\sigma^2 = E[(X - \mu)^2] = E[X^2] - 2\mu E[X] + \mu^2 = E[X^2] - \mu^2
$$

Furthermore, the differentiation process that recovers the higher order moments from the moment generating function remains valid. To verify this property, suppose that $\Psi(t)$ exists for t in an interval around zero.

$$
\Psi(t) = E[e^{tX}] = \int_{-\infty}^{\infty} e^{tx}\, dF = \int_{-\infty}^{\infty} \left(\sum_{n=0}^{\infty} \frac{(tx)^n}{n!} \right) dF
$$

For a fixed t, the summation converges uniformly for x in any bounded interval $[a, b]$. Theorem A.55 then allows the interchange of the summation and integration operations:

$$
\int_a^b \left(\sum_{n=0}^{\infty} \frac{(tx)^n}{n!} \right) dF = \sum_{n=0}^{\infty} \int_a^b \frac{(tx)^n}{n!}\, dF.
$$

The left side approaches $\Psi(t)$ as $[a, b] \rightarrow (-\infty, \infty)$, which means that the right side must also. Consequently,

$$\Psi(t) = \sum_{n=0}^{\infty} \left(\int_{-\infty}^{\infty} x^n \, dF \right) \frac{t^n}{n!} = \sum_{n=0}^{\infty} \frac{E[X^n]t^n}{n!}$$

$$\Psi^{(k)}(t) = \sum_{n=k}^{\infty} \frac{P_{n,k} E[X^n] t^{n-k}}{n!}$$

$$\Psi^{(k)}(0) = E[X^k].$$

5.22 Example: Recall the context of Example 5.17, in which the lifetime T of a disk drive has the following two-part density function.

$$f_d(0) = 0.1$$

$$f_c(t) = \begin{cases} 0, & t < 0 \\ 0.027 e^{-0.03t}, & t \geq 0, \end{cases}$$

where t is measured in months. What are the mean, variance, and moment generating function? Verify that the derivatives of the moment generating function correctly deliver the first and second moments.

For $t \geq 0$, we have

$$F_T(t) = \Pr(T \leq t) = f_d(0) + \int_0^t 0.027 e^{-0.03\tau} \, d\tau = 0.1 + \left[\frac{0.027}{-0.03} e^{-0.03\tau} \right]_0^t$$

$$= 0.1 + 0.9(1 - e^{-0.03t}),$$

which matches that given in the earlier example. Figure 5.2 graphs the density and cumulative distribution. Note that the discrete component of the density appears as a spike of the appropriate height. The required expectation calculations are as follows.

$$\mu_T = (0) f_d(0) + \int_{-\infty}^{\infty} t f_c(t) \, dt = \int_0^{\infty} 0.027 t e^{-0.03t} \, dt$$

$$= \left[t \cdot \frac{0.027}{-0.03} e^{-0.03t} \right]_0^{\infty} + 0.9 \int_0^{\infty} e^{-0.03t} \, dt = \frac{0.9}{-0.03} e^{-0.03t} \Big|_0^{\infty} = 30$$

$$E[T^2] = (0)^2 f_d(0) + \int_{-\infty}^{\infty} t^2 f_c(t) \, dt = \int_0^{\infty} 0.027 t^2 e^{-0.03t} \, dt$$

$$= \left[t^2 \cdot \frac{0.027}{-0.03} e^{-0.03t} \right]_0^{\infty} + 2(0.9) \int_0^{\infty} t e^{-0.03t} \, dt$$

$$= \left[t \cdot \frac{1.8}{-0.03} e^{-0.03t} \right]_0^{\infty} + \frac{1.8}{0.03} \int_0^{\infty} e^{-0.03t} \, dt = \frac{60}{-0.03} e^{-0.03t} \Big|_0^{\infty} = 2000$$

$$\sigma_T^2 = E[T^2] - \mu_T^2 = 2000 - 30^2 = 1100$$

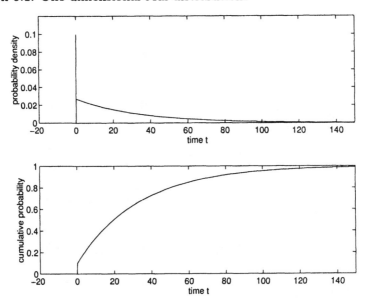

Figure 5.2. The density (upper) and cumulative distribution for disk drive lifetime (see Example 5.22)

Also, for $\tau < 0.03$,

$$\Psi_T(\tau) = E[e^{\tau T}] = e^0 f_d(0) + \int_{-\infty}^{\infty} e^{\tau t} f_c(t)\, dt = 0.1 + \int_0^{\infty} 0.027 e^{\tau t} e^{-0.03t}\, dt$$

$$= 0.1 + 0.027 \int_0^{\infty} e^{-(0.03-\tau)t}\, dt = 0.1 + \left[\frac{0.027}{-(0.03-\tau)} e^{-(0.03-\tau)t} \right]_0^{\infty}$$

$$= 0.1 + \frac{0.027}{0.03 - \tau} = 0.1 - 0.027(\tau - 0.03)^{-1}$$

This expression admits simple derivatives, which deliver the correct results. We can expect a disk drive to last 30 months, on the average, but with a large variance.

$$\Psi_T'(\tau) = 0.027(\tau - 0.03)^{-2}$$

$$\Psi_T'(0) = 0.027(-0.03)^{-2} = \frac{0.027}{0.0009} = 30$$

$$\Psi_T''(\tau) = -0.054(\tau - 0.03)^{-3}$$

$$\Psi_T''(0) = -0.054(-0.03)^{-3} = \frac{-0.054}{-0.000027} = 2000. \ \square$$

The Markov and Chebyshev inequalities remain valid in the general case, as shown in the final theorem below.

5.23 Theorem: If X is a general nonnegative random variable with mean μ, then $\Pr(X \geq x) \leq \mu/x$ for any $x > 0$. If X is a general random variable with mean μ and variance $\sigma^2 > 0$, then $\Pr(\mu - r\sigma < X < \mu + r\sigma) \geq 1 - 1/r^2$ for any $r > 0$.

PROOF: Let F be the cumulative distribution function. First, suppose that X is nonnegative, which means $F(y) = 0$ for $y < 0$. Then, for $x > 0$, we have

$$\mu = \int_{-\infty}^{\infty} y \, dF = \int_{0}^{x} y \, dF + \int_{x}^{\infty} y \, dF \geq \int_{x}^{\infty} y \, dF \geq x \int_{x}^{\infty} dF$$
$$= x \cdot \Pr(X \geq x)$$

$$\Pr(X \geq x) \leq \frac{\mu}{x}.$$

Now remove the nonnegative constraint on X. The related random variable $Y = (X - \mu)^2$ remains nonnegative. The Markov inequality then forces

$$\Pr(Y \geq r^2 \sigma^2) \leq \frac{\mu}{r^2 \sigma^2} = \frac{E[(X - \mu)^2]}{r^2 \sigma^2} = \frac{1}{r^2}$$

$$\Pr(\mu - r\sigma < X < \mu + r\sigma) = \Pr(|X - \mu| < r\sigma) = \Pr(Y < r^2 \sigma^2)$$
$$= 1 - \Pr(Y \geq r^2 \sigma^2) \geq 1 - \frac{1}{r^2}. \blacksquare$$

Exercises

5.1 $|A|$ denotes the number of elements in set A. If $|A| = n$, prove that $|2^A| = 2^n$.

5.2 Let C be a collection of subsets of X. Show that the intersection of all σ-algebras containing C is the smallest σ-algebra containing C.

5.3 For $x \in \mathcal{R}$, the subset $\{x\}$ is a singleton-point set. Show that all singleton-point sets are Borel sets.

*5.4 Let $\Pr(\cdot)$ extend the cumulative distribution function F. Prove the following.

$$\Pr((-\infty, x]) = F(x)$$
$$\Pr([x, \infty)) = 1 - F(x^-)$$
$$\Pr((x, y]) = F(y) - F(x)$$
$$\Pr([x, y)) = F(y^-) - F(x^-)$$
$$\Pr([x, y]) = F(y) - F(x^-)$$

*5.5 Let $\Pr(\cdot)$ extend the cumulative distribution function F and continue to denote the collection of perfect slicers as $\mathcal{B}'$. Show that $\Pr(A) = 0$ implies that $A \in \mathcal{B}'$.

*5.6 Construct the *Cantor ternary set* $C = \cap_{n=1}^{\infty} C_n$ as follows. C_1 results from dropping the open interval comprising the middle third of $[0, 1]$. C_2 follows by dropping the middle third of the two components in C_1. In general, C_{n+1} follows by dropping the middle third from each component in C_n. Show that the Cantor ternary set is closed and uncountable. Extend the cumulative distribution function

$$F(x) = \begin{cases} 0, & x < 0 \\ x, & 0 \le x < 1 \\ 1, & x \ge 1 \end{cases}$$

to a probability assignment $\Pr(\cdot)$ on the Borel sets. Show that the Cantor ternary set is a Borel set that receives probability assignment zero.

5.7 A disk drive has a 10% chance of failing immediately upon startup. If it survives for 36 months, it is destroyed and replaced with a new drive. The probability distribution for the random variable T, the lifetime of the drive, has the form

$$\Pr(T = 0) = 0.1$$
$$\Pr(T = 36) = A$$
$$\Pr(0 < T < 36) = \int_0^t f_T(t) \, dt = \int_0^t Be^{-0.03t} \, dt.$$

Identify the set D_T, and determine the parameters A and B. Derive $F_T(\cdot)$, the cumulative distribution function, and verify that $F_T'(t) = f_T(t)$ for $t \notin D_T$. Compute the probability that a drive survives longer than one year.

5.8 For the random variable T of the preceding exercise, compute μ_T and σ_T^2. Also derive the moment generating function and verify that the derivatives of that function correctly produce $E[T]$ and $E[T^2]$.

5.9 Let X have cumulative distribution function

$$F_X(x) = \begin{cases} 0, & x < 0 \\ 0.1 + 0.7x^2, & 0 \le x < 0.5 \\ 0.475 + 0.7(x^2 - 0.25), & 0.5 \le x < 1.0 \\ 1.0, & x \ge 1.0. \end{cases}$$

Find the corresponding two-part density function $f_d(\cdot)$, $f_c(\cdot)$. Compute μ_X and σ_X^2.

5.10 Verify the Markov and Chebyshev inequalities for the random variable X of the preceding exercise.

5.2   Joint random variables

This section's goal is to describe the distribution of a set of n random variables, and we start modestly with $n = 2$. To describe two real random variables, X and Y, we need a joint distribution over $\mathcal{R}^2 = \{(x,y)|x,y \in \mathcal{R}\}$. We can argue, exactly as in the one-dimensional case, that individual points carrying nonzero probability must form a countable set. However, the remaining probability can spread in a more complicated fashion. Some of it may spread along lines or other curves; some may spread across two-dimensional regions.

The general approach extends the ideas that led to a probability function on the Borel sets. First, semi-infinite rectangles of the form $\{(x,y)| -\infty < x \le a, -\infty < y \le b\}$ substitute for the intervals $(-\infty, a]$, and the smallest σ-algebra containing these rectangles becomes the collection of two-dimensional Borel sets. Second, starting with general conditions on a cumulative distribution function $F(x,y)$, similar to those of Theorem 5.6, we note that this function defines probability allocations to the semi-infinite rectangles. That is, $F(x,y) = \Pr((X \le x) \cap (Y \le y))$. Third, we extend this allocation to a more inclusive probability function on all subsets of the plane, which proves countably additive on the two-dimensional Borel sets. Consequently, we can describe a two-dimensional distribution by giving its cumulative distribution function $F(x,y)$. Finally, we define a two-dimensional integral with respect to F as the lower bound of upper sums, in parallel with the development in Appendix A that establishes the one-dimensional Riemann-Stieltjes integral. This section explores some issues in the process, without attempting the completeness of the one-dimensional case.

If $K = [a,b] \times [c,d]$ is a rectangular region in the plane, we partition it with rectangles obtained from independent partitions of $[a,b]$ and $[c,d]$. That is, if $a = x_0 \le x_1 \le \ldots \le x_m = b$ and $c = y_0 \le y_1 \le \ldots \le y_n = d$ are partitions of $[a,b]$ and $[c,d]$, then

$$P = \{[x_{i-1}, x_i] \times [y_{j-1}, y_j] \mid 1 \le i \le m, 1 \le j \le n\}$$

is a partition of K. For each cell,

$$(X \le x_i) = (X \le x_{i-1}) \cup (x_{i-1} < X \le x_i),$$

and the union is disjoint. Therefore,

$$
\begin{aligned}
(X \le x_i) \cap (Y \le y_j) &= [(X \le x_{i-1}) \cap (Y \le y_j)] \\
&\quad \cup [(x_{i-1} < X \le x_i) \cap (Y \le y_j)] \\
F(x_i, y_j) &= \Pr((X \le x_i) \cap (Y \le y_j)) \\
&= F(x_{i-1}, y_j) + \Pr((x_{i-1} < X \le x_i) \cap (Y \le y_j)).
\end{aligned}
$$

Transposing, we obtain an F-difference:

$$\Pr((x_{i-1} < X \le x_i) \cap (Y \le y_j)) = F(x_i, y_j) - F(x_{i-1}, y_j).$$

We also have the disjoint union

$$(x_{i-1} < X \le x_i) \cap (Y \le y_j) = [(x_{i-1} < X \le x_i) \cap (Y \le y_{j-1})]$$
$$\cup [(x_{i-1} < X \le x_i) \cap (y_{j-1} < Y \le y_j)],$$

which gives

$$0 \le \Pr((x_{i-1} < X \le x_i) \cap (y_{j-1} < Y \le y_j))$$
$$= \Pr((x_{i-1} < X \le x_i) \cap (Y \le y_j)) - \Pr((x_{i-1} < X \le x_i) \cap (Y \le y_{j-1}))$$
$$= F(x_i, y_i) - F(x_{i-1}, y_j) - F(x_i, y_{j-1}) + F(x_{i-1}, y_{j-1}).$$

Accordingly, we write

$$\Delta F_{ij} = F(x_i, y_i) - F(x_{i-1}, y_j) - F(x_i, y_{j-1}) + F(x_{i-1}, y_{j-1}) \ge 0$$

for this expression. Given a function $g(x, y)$, each partition P produces an upper and a lower sum,

$$S_{g,F}(P) = \sum_{i=1}^{m} \sum_{j=1}^{n} \Delta F_{ij} \sup\{g(x, y) \mid x_{i-1} \le x \le x_i, y_{j-1} \le y \le y_j\}$$

$$s_{g,F}(P) = \sum_{i=1}^{m} \sum_{j=1}^{n} \Delta F_{ij} \inf\{g(x, y) \mid x_{i-1} \le x \le x_i, y_{j-1} \le y \le y_j\},$$

just as described in Appendix A for the one-dimensional Riemann-Stieltjes integral. If the lower bound of the upper sums equals the upper bound of the lower sums, we say that g is integrable with respect to F over region K and define $\int_K g \, dF = \inf S_{g,F}$, where the infimum is across all partitions of $K = [a, b] \times [c, d]$. The integral over the entire plane is the limit, if it exists, as $[a, b] \to (-\infty, \infty)$ and $[c, d] \to (-\infty, \infty)$. The expected value is then a calculation involving the newly defined integral. That is, $E[g] = \int_{\mathcal{R}^2} g \, dF$, provided that $\int_{\mathcal{R}^2} |g| \, dF$ exists.

There are a number of calculation techniques for evaluating these integrals, corresponding to the methods available in the one-dimensional case. Appendix A discusses the one-dimensional form of these computations. The random variables in this text, however, all have cumulative distributions for which the integral reduces to familiar forms. Recall the one-dimensional development of the preceding section. There we obtained a general result that expressed expected values with a Riemann-Stieltjes integral. However, we noted that many useful cumulative distribution functions take a special form involving a two-part density. A discrete component specifies probability masses on a countable point set, while an integrable function advances the cumulative distribution function between such points. A similar observation holds in the two-dimensional case, except that the density has three components. A zero-dimensional component specifies probability lumps on a countable point set in the plane. A one-dimensional density accumulates probability through a linear integral along curves in the plane. Finally, a two-dimensional component contributes probability via a two-dimensional integral. Each component

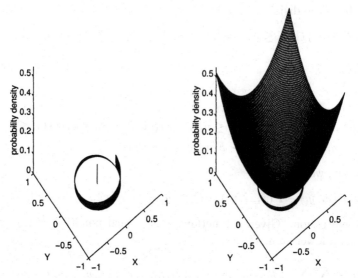

Figure 5.3. Joint density components for two random variables X and Y (see Example 5.24)

requires separate treatment when computing an expected value. The example below illustrates the point.

5.24 Example: Real random variables X and Y have a joint density with three components. A lump of probability 0.1 exists at point $(0,0)$. Probability in the amount 0.2 spreads over the circle of radius 0.5 centered at the origin, with linearly increasing density along the arc originating at $(0.5, 0)$ and continuing counterclockwise. The remaining 0.7 spreads over the square $\{(x,y)| -1 \leq x \leq 1, -1 \leq y \leq 1\}$, such that the density increases in proportion to the distance from the origin. Figure 5.3 illustrates the components. The left panel shows a single spike at the origin of height 0.1 and a fence of steadily increasing height erected on the circle of radius 0.5 in the plane. The right panel combines these lower-dimensional components with a bowl-shaped function that spreads probability over two-dimensional areas. In the right graph, the bowl hides the one-dimensional spike at the origin, but it is still there.

The density $f(\cdot, \cdot)$ then requires three components:

$$f(x, y) = (f_0(x, y), f_1(x, y), f_2(x, y))$$

$$f_0(x, y) = \begin{cases} 0.1, & x = y = 0 \\ 0, & \text{elsewhere} \end{cases}$$

$$f_1(x, y) = \begin{cases} \dfrac{0.4t}{(2\pi)^2}, & x = 0.5\cos t, y = 0.5\sin t, 0 \leq t < 2\pi \\ 0, & \text{otherwise} \end{cases}$$

$$f_2(x, y) = \begin{cases} 0.2625(x^2 + y^2), & -1 \leq x \leq 1, -1 \leq y \leq 1 \\ 0, & \text{otherwise.} \end{cases}$$

The zero-dimensional component contributes a probability lump of 0.1 to any integration that passes over the origin. The one-dimensional component requires a

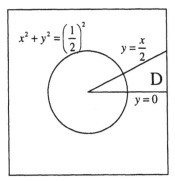

Figure 5.4. Outcomes comprising the event $(0 < 2Y < X)$ (see Example 5.24)

---

parametric expression. It gives zero probability to individual points, but it allows positive probability to accumulate on any nonempty arc of the circle of radius 0.5 centered at the origin. Points on this circle assume the form $(0.5\cos t, 0.5\sin t)$ for $t \in [0, 2\pi)$. Any two-dimensional integration that includes an arc of this circle receives the probability weight of the arc. The computation, however, involves a one-dimensional integral along the arc. Nevertheless, an integration that includes one or more isolated points on the arc receives no contribution from the one-dimensional component. Individual points on the circle have zero probability; probability accumulates only over arc segments. Let F denote the cumulative distribution function. We first verify that the assigned probability totals 1.

$$\int_{\mathcal{R}^2} dF = 0.1 + \int_0^{2\pi} \frac{0.4t}{(2\pi)^2} \, dt + 0.2625 \int_{-1}^1 \int_{-1}^1 (x^2 + y^2) \, dx \, dy$$
$$= 0.1 + 0.2 + 0.7 = 1.0$$

Now suppose that we want $\Pr(0 < 2Y < X)$. As shown in Figure 5.4, outcomes in this event lie in the first quadrant, above the x-axis and below the line $y = x/2$. Both boundary lines are excluded. Hence, the point mass at the origin does not contribute to the probability of the event. However, an arc of the one-dimensional component does fall in the integration area. This arc contains points $(\cos t, \sin t)$ for $t \in (0, \tan^{-1}(0.5))$. Consequently, denoting the integration area as D, we have

$$\Pr(0 < 2Y < X) = \int_D dF$$
$$= \int_0^{\tan^{-1}(0.5)} \frac{0.4t}{(2\pi)^2} \, dt + \int_{x=0}^1 \int_{y=0}^{x/2} 0.2625(x^2 + y^2) \, dy \, dx$$
$$= 0.0366.$$

Expected value computations involve similar integrals, which in the worst case may include three components. As an example of an expected value calculation, let us compute $\int_{\mathcal{R}^2} g(x, y) \, dF$, when $g(x, y) = xy$. Because this function is zero at the origin, the zero-dimensional integral component vanishes. However, the remaining two components both require integrations.

$$E[XY] = \int_{\mathcal{R}^2} xy \, dF = (0)(0)(0.1) + \int_0^{2\pi} (0.5 \cos t)(0.5 \sin t)\frac{0.4t}{(2\pi)^2} \, dt$$

$$+ \int_{-1}^{1} \int_{-1}^{1} 0.2625xy(x^2 + y^2) \, dx \, dy$$

$$= -0.004 \ \square$$

Having some experience now with $n = 2$, we start anew with an arbitrary dimension $n \geq 1$ and develop the basic definitions. With n random variables, the basic outcomes can be n-vectors, which are already members of $\mathcal{R}^n$. It is possible, however, that the random variables create the n-vectors from a more abstract experiment.

Following the usual tradition, we write $\mathbf{x}$ for the point in $\mathcal{R}^n$ with coordinates $(x_1, x_2, \ldots, x_n)$. The upcoming discussion involves convergence properties in $\mathcal{R}^n$, a review of which appears in Appendix A. In particular, $||\mathbf{x}||$ is the norm of $\mathbf{x}$ and denotes $\max_{1 \leq k \leq n} |x_k|$. $\mathbf{x} + \mathbf{y}$ denotes a component-by-component operation: $(x_1 + y_1, x_2 + y_2, \ldots, x_n + y_n)$. Similarly, $\mathbf{x} \leq \mathbf{y}$ denotes a component-by-component relationship: $x_i \leq y_i$, for $1 \leq i \leq n$. Boldface $\mathbf{0}$ is the n-vector with zero in each component; $\boldsymbol{\infty}$ is the n-vector with ∞ in each component. We also extend the notations $(a, b), (a, b], [a, b),$ and $[a, b]$ to boldface equivalents that denote rectangles in $\mathcal{R}^n$. For example, $[\mathbf{a}, \mathbf{b}) = \{\mathbf{x} \mid \mathbf{a} \leq \mathbf{x} < \mathbf{b}\}$. A parenthesized endpoint may be $\pm\boldsymbol{\infty}$.

5.25 Definition: Let $(\Omega, \mathcal{A}, \mathrm{Pr}_\Omega(\cdot))$ be a general probability space in the sense of Definition 5.14. That is, Ω is a point set, $\mathcal{A}$ is a σ-algebra over Ω, and $\mathrm{Pr}_\Omega : \mathcal{A} \to [0, 1]$ is a probability assignment function. The vector-valued function $\mathbf{X} = (X_1, \ldots, X_n)$ is a set of *joint random variables* if each $X_i : \Omega \to \mathcal{R}$ and for every $\mathbf{x} \in \mathcal{R}^n$,

$$\mathbf{X}^{-1}((-\boldsymbol{\infty}, \mathbf{x}]) = X_1^{-1}((-\infty, x_1]) \cap \ldots \cap X_n^{-1}((-\infty, x_n]) \in \mathcal{A} \ \blacksquare$$

In the most frequently occurring case, $\Omega = \mathcal{R}^n$, and the random variable X_i is simply projection on the ith component. That is, $X_i(\mathbf{x}) = x_i$. In any case, $\mathcal{B}^n$, the collection of n-dimensional Borel sets, is the smallest σ-algebra containing all the intervals $(-\boldsymbol{\infty}, \mathbf{x}]$ as $\mathbf{x}$ ranges over all vectors in $\mathcal{R}^n$. Because inverse images preserve complements and unions, we have $\mathbf{X}^{-1}(B) \in \mathcal{A}$ for every Borel set $B \in \mathcal{B}^n$. Consequently, the joint random variables induce a new probability space $(\mathcal{R}^n, \mathcal{B}^n, \mathrm{Pr}_\mathbf{X}(\cdot))$ via $\mathrm{Pr}_\mathbf{X}(B) = \mathrm{Pr}_\Omega(\mathbf{X}^{-1}(B))$, for every $B \in \mathcal{B}^n$. The induced probability function $\mathrm{Pr}_\mathbf{X}(\cdot)$ is the *joint distribution* of the random variables $X_1, X_2, \ldots, X_n$. With the probability distribution in hand, we can obviously compute the joint cumulative distribution, according to the next definition.

5.26 Definition: Let $\mathbf{X} = (X_1, X_2, \ldots, X_n)$ be joint random variables over a probability space $(\Omega, \mathcal{A}, \mathrm{Pr}_\Omega(\cdot))$, and let $\mathrm{Pr}_\mathbf{X}(\cdot)$ be the induced probability on the Borel sets of $\mathcal{R}^n$. The function

$$F_{\mathbf{X}}(\mathbf{x}) = \Pr_{\mathbf{X}}((-\infty, \mathbf{x}]) = \Pr(X_1 \leq x_1, X_2 \leq x_2, \dots, X_n \leq x_n)$$

is the *joint cumulative distribution* of the random variables. ∎

Because $\mathbf{x} \leq \mathbf{y}$ implies that $(-\infty, \mathbf{x}] \subset (-\infty, \mathbf{y}]$, we have F monotone nondecreasing. That is, $\mathbf{x} \leq \mathbf{y}$ implies that $F(\mathbf{x}) \leq F(\mathbf{y})$. Moreover, the countable additivity property forces certain additional conditions analogous to those elaborated in Theorem 5.6 for a single random variable. Indeed, the properties follow from convergence considerations in $\mathcal{R}^n$, just as Theorem 5.6 followed from the corresponding convergence in $\mathcal{R}$.

For example, we can demonstrate $\lim_{\mathbf{x} \to \infty} F(\mathbf{x}) = 1$ with an expanding sequence of sets. Specifically, for integer m, write $\mathbf{m}$ for the n-vector $(m, m, \dots, m)$. Then $\cup_{m=1}^{\infty}(-\infty, \mathbf{m}] = \mathcal{R}^n$, and Theorem 5.5 computes the probability of an expanding set sequence: $\lim_{m \to \infty} \Pr((-\infty, \mathbf{m}]) = \Pr(\mathcal{R}^n) = 1$. Given $\epsilon > 0$, there exists M such that $m \geq M$ implies that $\Pr((-\infty, \mathbf{m}]) \geq 1 - \epsilon$. If $\mathbf{x} \in \mathcal{I}(\infty; M)$, a neighborhood of ∞, then $x_i > M$ for $i = 1, 2, \dots, n$. Consequently, $(-\infty, \mathbf{M}] \subset (-\infty, \mathbf{x}]$, and $F(\mathbf{x}) = \Pr(-\infty, \mathbf{x}]) \geq \Pr((-\infty, \mathbf{M}]) \geq 1 - \epsilon$.

Using appropriate nested sequences of sets, we can use similar arguments to establish

$$\lim_{\mathbf{x} \to -\infty} F(\mathbf{x}) = 0$$

$$F(\mathbf{x}^-) = \lim_{\mathbf{y} \to \mathbf{x}^-} F(\mathbf{y}) \leq F(\mathbf{x})$$

$$F(\mathbf{x}^+) = \lim_{\mathbf{y} \to \mathbf{x}^+} F(\mathbf{y}) = F(\mathbf{x}).$$

The argument is very little changed if we conduct the approach in a selected subset of dimensions, rather than all n. In $\mathcal{R}^5$, for example,

$$F(x_1, x_2^-, x_3, x_4^-, x_5) \leq F(\mathbf{x}) = F(x_1^+, x_2, x_3, x_4^+, x_5).$$

The notation becomes cumbersome, but the idea should be clear. An approach from below, in any specified set of dimensions, yields a limit that does not exceed $F(\mathbf{x})$. An approach from above, in any specified set of dimensions, yields precisely $F(\mathbf{x})$ in the limit.

Now, we develop the probability of the n-dimensional box $\mathbf{a} < \mathbf{x} \leq \mathbf{b}$ in terms of F. We adopt the notation $(a_1 : b_1, t_2, \dots, t_n)$ to mean $a_1 < X_1 \leq b_1$ and $X_i \leq t_i$ for $i = 2, 3, \dots, n$. The double term may also appear in other coordinates, where it means that the corresponding X_i is also constrained to a bounded interval. We then have the following disjoint union and its consequent probability equation.

$$(b_1, t_2, \dots, t_n) = (a_1, t_2, \dots, t_n) \cup (a_1 : b_1, t_2, \dots, t_n)$$

$$\Pr(a_1 : b_1, t_2, \dots, t_n) = F(b_1, t_2, \dots, t_n) - F(a_1, t_2, \dots, t_n) \tag{5.1}$$

Continuing the pattern,

$$(a_1 : b_1, b_2, t_3, \ldots, t_n) = (a_1 : b_1, a_2, t_3, \ldots, t_n)$$
$$\cup (a_1 : b_1, a_2 : b_2, t_3, \ldots, t_n)$$
$$\Pr(a_1 : b_1, a_2 : b_2, t_3, \ldots, t_n) = \Pr(a_1 : b_1, b_2, t_3, \ldots, t_n)$$
$$- \Pr(a_1 : b_1, a_2, t_3, \ldots, t_n).$$

Both terms on the right are available from the first expansion (Equation 5.1).

$$\Pr(a_1 : b_1, a_2 : b_2, t_3, \ldots, t_n)$$
$$= [F(b_1, b_2, t_3, \ldots, t_n) - F(a_1, b_2, t_3, \ldots, t_n)]$$
$$- [F(b_1, a_2, t_3, \ldots, t_n) - F(a_1, a_2, t_3, \ldots, t_n)]$$
$$= F(b_1, b_2, t_3, \ldots, t_n) - F(a_1, b_2, t_3, \ldots, t_n)$$
$$- F(b_1, a_2, t_3, \ldots, t_n) + F(a_1, a_2, t_3, \ldots, t_n) \qquad (5.2)$$

The pattern mixes upper and lower coordinate limits of the box $\mathbf{a} < \mathbf{x} \le \mathbf{b}$ in the first two positions. Specifically, we first see the term that has no lower coordinates; it is positive. Then we see the terms that correspond to one lower coordinate; they are negative. Finally, we have the term that has two lower coordinates, and it is positive. We guess the following form when we bring in the third position.

$$\Pr(a_1 : b_1, a_2 : b_2, a_3 : b_3, t_4, \ldots, t_n) = F(b_1, b_2, b_3, t_4, \ldots, t_n)$$
$$- F(b_1, b_2, a_3, t_4, \ldots, t_n) - F(b_1, a_2, b_3, t_4, \ldots, t_n) - F(a_1, b_2, b_3, t_4, \ldots, t_n)$$
$$+ F(b_1, a_2, a_3, t_4, \ldots, t_n) + F(a_1, b_2, a_3, t_4, \ldots, t_n) + F(a_1, a_2, b_3, t_4, \ldots, t_n)$$
$$- F(a_1, a_2, a_3, t_4, \ldots, t_n)$$

Indeed, this equation follows from the disjoint union

$$(a_1 : b_1, a_2 : b_2, b_3, t_4, \ldots, t_n) = (a_1 : b_1, a_2 : b_2, a_3, t_4, \ldots, t_n)$$
$$\cup (a_1 : b_1, a_2 : b_2, a_3 : b_3, t_4, \ldots, t_n)$$

upon substitution of the expression from Equation 5.2. The pattern continues until we arrive at

$$\Delta F(\mathbf{a}, \mathbf{b}) = \Pr(a_1 : b_1, \ldots, a_n : b_n) = \sum_{k=0}^{n} (-1)^k \sum_{\mathbf{x} \in C_k} F(\mathbf{x}), \qquad (5.3)$$

where $C_k = \{(x_1, \ldots, x_n) | \ x_i = a_i \text{ for } k \text{ positions } i \text{ and } x_i = b_i \text{ for the rest}\}$. Because it is a probability, $\Delta F(\mathbf{a}, \mathbf{b})$ must be nonnegative. The following theorem summarizes these necessary properties of a cumulative distribution function.

5.27 Theorem: Let $F(\mathbf{x})$ be the joint cumulative distribution function of random variables $\mathbf{X} = (X_1, X_2, \ldots, X_n)$. Then F satisfies the following constraints.

- $0 \le F(\mathbf{x}) \le 1$, for all $\mathbf{x} \in \mathcal{R}^n$.

- $\lim_{\mathbf{x} \to -\infty} F(\mathbf{x}) = 0$; $\lim_{\mathbf{x} \to \infty} F(\mathbf{x}) = 1$.

- $F(\mathbf{x})$ is monotone nondecreasing.

- $F(\mathbf{x}^-) \le F(\mathbf{x}) = F(\mathbf{x}^+)$, and F is continuous from above.

- $\Delta F(\mathbf{a}, \mathbf{b}) = \Pr(\mathbf{a} < \mathbf{X} \le \mathbf{b})$. $\Delta F(\mathbf{a}, \mathbf{b})$ is monotone nondecreasing in $\mathbf{b}$ and monotone nonincreasing in $\mathbf{a}$.

- $\Delta F(\mathbf{a}^+, \mathbf{b}) = \Delta F(\mathbf{a}, \mathbf{b}^+) = \Delta F(\mathbf{a}, \mathbf{b})$.

- $\Delta F(\mathbf{a}^-, \mathbf{b}) \le \Delta F(\mathbf{a}, \mathbf{b})$.

- $\Delta F(\mathbf{a}, \mathbf{b}^-) \le \Delta F(\mathbf{a}, \mathbf{b})$.

PROOF: The discussion above establishes all items except the one-sided limits on ΔF, which follow from consideration of expanding and contracting sequences of sets. ∎

Theorem 5.27 says that a cumulative distribution function arising from a probability distribution on $\mathcal{R}^n$ must satisfy certain properties. As in the one-dimensional case, a converse theorem also applies. That is, we can start with a function that satisfies the properties of Theorem 5.27 and construct a probability distribution across the Borel sets in $\mathcal{R}^n$. The construction is lengthy and parallels that of the preceding section. Consequently, we briefly sketch the necessary steps and note those that require some variation on the earlier proofs. The exercises provide hints for completing the details.

Suppose that F satisfies the properties of Theorem 5.27. For open rectangles, we allocate probabilities $\Pr((\mathbf{x}, \mathbf{y})) = \Delta F(\mathbf{x}, \mathbf{y}^-)$. We extend this probability assignment to all subsets of $\mathcal{R}^n$ through a competition, exactly as was done in the preceding section for $\mathcal{R}^1$. That is, we let $\mathcal{O}$ be the collection of all open rectangles, and for any countable subcollection $\mathcal{G} = \{R_1, R_2, \dots\} \subset \mathcal{O}$, we associate a score $s(\mathcal{G}) = \sum_{i=1}^{\infty} \Pr(R_i)$. Finally, for $A \subset \mathcal{R}^n$, define

$$\Pr(A) = \inf\{s(\mathcal{G}) \mid \mathcal{G} = \{R_1, R_2, \dots\} \subset \mathcal{O}, A \subset \cup_{i=1}^{\infty} R_i\}.$$

At this point, we must verify that the competition produces $\Pr((\mathbf{a}, \mathbf{b})) = \Delta F(\mathbf{a}, \mathbf{b}^-)$ for open rectangles $(\mathbf{a}, \mathbf{b})$. The verification generalizes the discussion preceding Theorem 5.9 to higher dimensions. Moreover, the theorem's proof of subadditivity for finite unions applies directly because it is entirely set-theoretic and independent of dimension.

The next step identifies the perfect slicers as those sets B such that $\Pr(A) = \Pr(A \cap B) + \Pr(A \cap \overline{B})$ for all $A \in \mathcal{R}^n$. To show that $\Pr(\cdot)$ is countably additive when restricted to the perfect slicers, we examine the proof of Theorem 5.13, which is also independent of dimension. For the final step, we adapt the argument in Theorem 5.13, which establishes the n-dimensional Borel sets as a subalgebra of the perfect slicers. This requires a demonstration that each $(-\infty, \mathbf{x}]$ is a perfect slicer of open rectangles. We then have the following theorem.

5.28 Theorem: Let $F : \mathcal{R}^n \to \mathcal{R}$ satisfy the properties of Theorem 5.27. Then F defines a probability allocation to the Borel sets for which it is the cumulative distribution function.

PROOF: See discussion above. ∎

In light of this theorem, we identify a probability distribution on $\mathcal{R}^n$ with its cumulative distribution function. We then establish a Riemann-Stieltjes integral to compute expected values. The integral is an exact analog of that developed in the preceding section for $\mathcal{R}^1$, except that the approximating sums involve rectangle partitions, rather than interval partitions, and the mixed difference $\Delta F(\mathbf{a}_i, \mathbf{b}_i)$ replaces the quantity $F(x_i) - F(x_{i-1})$. Specifically, for a function g defined over a rectangle K, we define upper and lower sums for each partition $P = [\mathbf{a}_1, \mathbf{b}_1], \ldots, [\mathbf{a}_m, \mathbf{b}_m]$ of K.

$$S_{g,F}(P) = \sum_{i=1}^{m} \Delta F(\mathbf{a}_i, \mathbf{b}_i) \cdot \sup\{g(\mathbf{x}) \mid \mathbf{a}_i \leq \mathbf{x} \leq \mathbf{b}_i\}$$

$$s_{g,F}(P) = \sum_{i=1}^{m} \Delta F(\mathbf{a}_i, \mathbf{b}_i) \cdot \inf\{g(\mathbf{x}) \mid \mathbf{a}_i \leq \mathbf{x} \leq \mathbf{b}_i\}$$

When $\inf S_{g,F}$, over all partitions, equals $\sup s_{g,F}$, we say that g is integrable with respect to F over K and write $\int_K g\,dF = \inf S_{g,F} = \sup s_{g,F}$.

5.29 Definition: Let F be the cumulative distribution function of a distribution on $\mathcal{R}^n$. For $g : \mathcal{R}^n \to \mathcal{R}$, define the *expected value* of g by

$$E[g(\cdot)] = \int_{\mathcal{R}^n} g\,dF, \quad \text{provided that } \int_{\mathcal{R}^n} |g|\,dF \text{ exists.} \quad \blacksquare$$

Before examining the details of expected value computations, we record another important feature of the cumulative joint distribution function and consider a simple example.

5.30 Theorem: Let F be the cumulative distribution function of a distribution on $\mathcal{R}^n$. The expression

$$F(\infty, \infty, \ldots, x_{i_1}, \ldots, x_{i_2}, \ldots, x_{i_k}, \ldots, \infty),$$

formed by substituting ∞ in the positions not among $i_1, i_2, \ldots, i_k$, is the cumulative distribution function of the joint random variables $X_{i_1}, X_{i_2}, \ldots, X_{i_k}$.

PROOF: The ∞ substitutions mean, of course, that a limit process sends the corresponding variables to ∞. Consider a sequence B_m of expanding rectangles having the form

$$B_m = \{\mathbf{y} \mid y_{i_t} \leq x_{i_t}, 1 \leq t \leq k, \text{ and } y_j \leq m \text{ for } j \notin \{i_1, i_2, \ldots, i_k\}\}.$$

We have

$$(X_{i_1} \leq x_{i_1}, \ldots, X_{i_k} \leq x_{i_k}) = \cup_{m=1}^{\infty} B_m$$

$$F_{X_{i_1}, \ldots, X_{i_k}}(x_{i_1}, \ldots, x_{i_k}) = \Pr(X_{i_1} \leq x_{i_1}, \ldots, X_{i_k} \leq x_{i_k}) = \Pr\left(\cup_{m=1}^{\infty} B_m\right)$$

$$= \lim_{m \to \infty} \Pr(B_m)$$

$$= \lim_{m \to \infty} F(\ldots, x_{i_1}, \ldots, x_{i_2}, \ldots, x_{i_k}, \ldots),$$

where the argument m appears in positions not among the $i_1, i_2, \ldots, i_k$. Because the limit is the same along any such increasing sequence, we have

$$F_{X_{i_1}, \ldots, X_{i_k}}(x_{i_1}, \ldots, x_{i_k}) = \lim_{x_{j_1}, \ldots, x_{j_{n-k}} \to \infty} F(\mathbf{x}),$$

where the j_t are the positions missing from the list $i_1, i_2, \ldots, i_k$. The expression $F(\infty, \ldots, x_{i_1}, \ldots, x_{i_2}, \ldots, x_{i_k}, \infty)$ is an abbreviation of this latter limit. ∎

5.31 Definition: Let F be the cumulative distribution function of the joint random variables $X_1, X_2, \ldots, X_n$, and suppose that $i_1 < i_2 < \cdots < i_k$ is a nonempty subset of $\{1, 2, \ldots, n\}$. Let $j_1 < j_2 < \cdots < j_{n-k}$ be the remaining positions.

$$F_{X_{i_1}, X_{i_2}, \ldots, X_{i_k}}(x_{i_1}, x_{i_2}, \ldots, x_{i_k}) = \lim_{x_{j_1}, x_{j_2}, \ldots, x_{j_{n-k}} \to \infty} F(x_1, x_2, \ldots, x_n)$$

is called the *marginal cumulative distribution* of $X_{i_1}, X_{i_2}, \ldots, X_{i_k}$. ∎

5.32 Example: Let X, Y be joint random variables with cumulative distribution

$$F(x, y) = \begin{cases} (1 - e^{-x})(1 - e^{-y}), & x \geq 0, y \geq 0 \\ 0, & \text{elsewhere.} \end{cases}$$

Verify the difference properties of Theorem 5.27. Compute the marginal cumulative distribution functions for X and Y.

For $(u, v) < (0, 0) \leq (x, y)$, the terms $F(u, y), F(x, v)$, and $F(u, v)$ are all zero. Consequently, $\Delta F[(u, v), (x, y)] = F(x, y)$. Also, for $(0, 0) \leq (u, v) < (x, y)$, we have

$$\begin{aligned} \Delta F[(u, v), (x, y)] &= (1 - e^{-x})(1 - e^{-y}) - (1 - e^{-u})(1 - e^{-y}) - (1 - e^{-x})(1 - e^{-v}) \\ &\quad + (1 - e^{-u})(1 - e^{-v}) \\ &= e^{-u} e^{-v} \left(1 - e^{-(y-v)}\right) \left(1 - e^{-(x-u)}\right). \end{aligned}$$

We note that $\Delta F((u, v), (x, y))$ is continuous and therefore the limiting properties asserted by the theorem are all equalities:

$$\begin{aligned} \Delta F((u, v)^+, (x, y)) &= \Delta F((u, v), (x, y)+) = \Delta F((u, v)^-, (x, y)) \\ &= \Delta F((u, v), (x, y)^-) = \Delta F((u, v), (x, y)). \end{aligned}$$

The marginal cumulative distribution function of X is

$$F_X(x) = F(x, \infty) = \lim_{y \to \infty} (1 - e^{-x})(1 - e^{-y}) = 1 - e^{-x}$$

for $x \geq 0$ and zero elsewhere. Similarly, $F_Y(y) = 1 - e^{-y}$. □

In the example above, the $F_X(x) = 1 - e^{-x}$, for $x \geq 0$, is the marginal distribution of X in the context of the joint distribution of X and Y. If X and Y arise from some physical system, such as, say, the response times and processing costs of database queries, our decision to treat them in a joint analysis reflects our concern with their interaction. Another analysis, dealing only with response time, for example, would use the single random variable X. In the latter case, $F_X(\cdot)$ would not be called the marginal distribution of X, even though it assumes the same expression as that obtained from $F(x, \infty)$.

Each random variable, or subset of random variables, has a distribution, or joint distribution, which is termed marginal only in the context of a larger set of random variables.

5.33 Example: Let $F(x_1, x_2, \ldots, x_n) = x_1 x_2 \cdots x_n$. Show that $\Delta F(\mathbf{x}, \mathbf{y})$ is the n-dimensional volume of the rectangle $[\mathbf{x}, \mathbf{y}]$.

For each k in the range $0, 1, \ldots, n$, the volume expression $\prod_{i=1}^{n}(y_i - x_i)$ contains $C_{n,k}$ terms that have k factors from among the x_i and $n - k$ factors from among the y_i. These terms have the form $t_1 t_2 \cdots t_n$, where k of the $t_i = -x_i$ and the remaining $n - k$ of the $t_i = y_i$. The term's sign is then $(-1)^k$. That is,

$$
\prod_{i=1}^{n}(y_i - x_i) = \sum_{k=0}^{n}(-1)^k \sum_{(t_1,\ldots,t_n) \in C_k} t_1 t_2 \cdots t_n
$$

$$
= \sum_{k=0}^{n}(-1)^k \sum_{(t_1,\ldots,t_n) \in C_k} F(t_1, t_2, \ldots, t_n),
$$

where $C_k = \{(t_1, \ldots, t_n) |\ t_i = x_i$ for k positions i and $t_i = y_i$ for other positions$\}$. Noting Equation 5.3, we see that this is precisely the definition of $\Delta F(\mathbf{x}, \mathbf{y})$.

In this case, the approximating sums for the Riemann-Stieltjes integral reduce to the corresponding sums for the ordinary multidimensional integral. That is,

$$
\int_{\mathbf{x}}^{\mathbf{y}} g \, dF = \int_{x_1}^{y_1} \int_{x_2}^{y_2} \cdots \int_{x_n}^{y_n} g(t_1, \ldots, t_n) \, dt_1 \cdots dt_n. \ \square
$$

For expected value computations, the integral of Definition 5.29 can present some difficulties. However, as illustrated in the example above, certain circumstances reduce the integral to a more familiar form. For our purposes, it suffices to consider only distributions where F, and consequently, $\int g \, dF$, admit such reductions. The example at the beginning of the section illustrated the computations, and a second example appears below. In such cases, the cumulative distribution function arises as the integral of a density, which may involve several components.

5.34 Definition: Let F be the cumulative distribution of joint random variables $\mathbf{X} = (X_1, X_2, \ldots, X_n)$. If there exists a function f such that, for any g integrable with respect to F and any region K,

$$
\int_K g \, dF = \int_K g f,
$$

then f is called the *joint density* of $\mathbf{X}$. ∎

As in Example 5.24, the density may contain lower-dimensional components that are specified parametrically. For those components, the integration shifts to the corresponding lower-dimensional form. In dimension k, for $0 \leq k \leq n$, the corresponding component provides a nonzero contribution only when it is nonzero over a subregion with positive k-dimensional volume. The next example illustrates the point. It also shows that when the distribution arises from a density, the marginal distributions do as well.

5.35 Example: The joint density $f_{X,Y}(\cdot,\cdot)$ has zero-, one-, and two-dimensional components:

$$f_0(x,y) = \begin{cases} 0.1, & x = y = 0 \\ 0, & \text{elsewhere} \end{cases}$$

$$f_1(x,y) = \begin{cases} 0.2, & 0 \leq x \leq 1, y = 0 \\ 0.2, & x = 0, 0 \leq y \leq 1 \\ 0, & \text{elsewhere} \end{cases}$$

$$f_2(x,y) = \begin{cases} 0.5(x+y), & 0 \leq x \leq 1, 0 \leq y \leq 1 \\ 0, & \text{elsewhere.} \end{cases}$$

These expressions describe a probability lump of size 0.1 at $(0,0)$, a one-dimensional fence of probability uniformly distributed from $(0,0)$ to $(1,0)$ on the x-axis, a similar fence on the y-axis, and a two-dimensional contribution that spreads nonzero probability over the unit square in the first quadrant. Find the cumulative distribution and its marginals. Also find the marginal densities of X and Y.

We compute the cumulative distribution in pieces. For (x,y) in the interior of the second, third, or fourth quadrants, $F(x,y) = 0$. Furthermore, $F(0,0) = 0.1$ because the integral's upper limit includes the origin, which carries the point mass 0.1. That is,

$$F(0,0) = \int_{(-\infty,-\infty)}^{(0,0)} dF = f_0(0,0) = 0.1.$$

The one-dimensional components contribute nothing because the integration region contains no interval where their densities are nonzero. Similarly, the two-dimensional component is zero on any two-dimensional volume within the region and contributes nothing to the integral. For $0 < x \leq 1, y = 0$, only the one-dimensional fence along the x-axis adds to the point mass already accumulated. The fence contributes no further weight after x passes the point $(1,0)$. So,

$$F(x,0) = \begin{cases} 0.1 + \int_0^x (0.2)\, dt = 0.1 + 0.2t|_0^x = 0.1 + 0.2x, & 0 < x \leq 1 \\ 0.1 + \int_0^1 (0.2)\, dt = 0.1 + 0.2t|_0^1 = 0.1 + 0.2 = 0.3, & x > 1. \end{cases}$$

Similarly, $F(0,y) = 0.1 + 0.2y$ for $x = 0, 0 < y \leq 1$, and $F(0,y) = 0.3$ for $x = 0, y > 1$. There remains only the interior of the first quadrant, where

$$F(x,y) = 0.1 + \int_0^x (0.2)\, dt + \int_0^y (0.2)\, ds + \int_0^x \int_0^y (0.5)(t+s)\, ds\, dt$$

$$= 0.1 + 0.2x + 0.2y + (0.5) \int_0^x \left[ts + \frac{s^2}{2} \right]_{s=0}^y dt$$

$$= 0.1 + 0.2(x+y) + 0.5 \left[\frac{t^2 y}{2} + \frac{y^2 t}{2} \right]_0^x$$

$$= 0.1 + 0.2(x+y) + 0.5 \left(\frac{x^2 y + xy^2}{2} \right)$$

$$= 0.1 + 0.2(x+y) + 0.25xy(x+y) = 0.1 + (x+y)(0.25xy + 0.2),$$

provided that $0 < x \leq 1, 0 < y \leq 1$. For (x,y) beyond the unit square, the density is zero, and the integral stops accumulating probability. For (x,y) above the unit square, we have

$$F(x,y) = 0.1 + (x+1)[0.25x(1) + 0.2] = 0.3 + y(0.25y + 0.45),$$

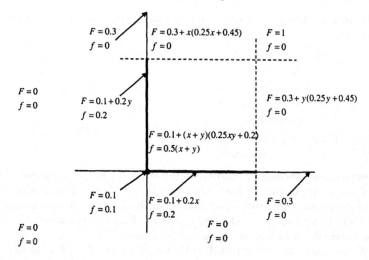

Figure 5.5. Density and cumulative distribution (see Example 5.35)

and a similar expression holds for (x, y) to the right of the unit square. The final result is then

$$F(x,y) = \begin{cases} 0, & x < 0 \text{ or } y < 0 \\ 0.1, & (x, y) = (0,0) \\ 0.1 + 0.2x, & y = 0, 0 < x \le 1 \\ 0.1 + 0.2y, & x = 0, 0 < y \le 1 \\ 0.3, & y = 0, x > 1 \text{ or } x = 0, y > 1 \\ 0.1 + (x + y)(0.25xy + 0.2), & 0 < x \le 1, 0 < y \le 1 \\ 0.3 + y(0.25y + 0.45), & x > 1, 0 < y \le 1 \\ 0.3 + x(0.25x + 0.45), & y > 1, 0 < x \le 1 \\ 1, & x > 1, y > 1. \end{cases}$$

Figure 5.5 shows the various areas with their individual expressions for the density and cumulative distribution. To obtain the marginal distribution for X, we send y to infinity:

$$F_X(x) = \begin{cases} 0, & x < 0 \\ 0.3, & x = 0 \\ 0.3 + x(0.25x + 0.45), & 0 < x \le 1 \\ 1, & x > 1. \end{cases}$$

The marginal density for X is a function $f_X(\cdot)$ with $F_X(x) = \int_{-\infty}^{x} f_X(t)\, dt$. Evidently, $f_X(\cdot)$ must have a point mass of size 0.3 at $x = 0$, and it must have a one-dimensional component that integrates to $x(0.25x + 0.45)$ on $(0, 1]$. The derivative of this expression serves the purpose. Therefore,

$$f_X(x) = (f_0(x), f_1(x))$$

$$f_0(x) = \begin{cases} 0.3, & x = 0 \\ 0, & \text{elsewhere} \end{cases}$$

$$f_1(x) = \begin{cases} 0, & x < 0 \text{ or } x > 1 \\ 0.5x + 0.45, & 0 \le x \le 1. \end{cases}$$

$F_Y(\cdot)$ and $f_Y(\cdot)$ have similar forms. $\square$

5.36 Definition: Let $\mathbf{X} = (X_1, X_2, \ldots, X_n)$ be a set of joint random variables with density $f_{\mathbf{X}}(\cdot)$. Then the means, variances, covariances, and correlations are as follows.

$$
\begin{aligned}
\text{mean}(X_i) &= \mu_{X_i} = E[X_i] \\
\text{var}(X_i) &= \sigma^2_{X_i} = E[(X_i - \mu_{X_i})^2] \\
\text{cov}(X_i, X_j) &= E[(X_i - \mu_{X_i})(X_j - \mu_{X_j})] \\
\text{corr}(X_i, X_j) &= \rho_{X_i, X_j} = \frac{\text{cov}(X_i, X_j)}{\sqrt{\text{var}(X_i) \cdot \text{var}(X_j)}} \quad \blacksquare
\end{aligned}
$$

Even if the integration assumes different forms for density components of different dimensions, each such form is linear. Therefore, the following shortcut formulas remain valid.

$$
\begin{aligned}
\sigma^2_{X_i} &= E[X_i^2] - \mu^2_{X_i} \\
\text{cov}(X_i, X_j) &= E(X_i X_j) - \mu_{X_i}\mu_{X_j}
\end{aligned}
\tag{5.4}
$$

The following example illustrates the point.

5.37 Example: Find the means, variances, covariances, and correlations of the random variables in Example 5.24.

In the example cited, probability appears on the square centered at the origin $\{(x, y)| -1 \le x \le 1, -1 \le y \le 1\}$ according to the multicomponent density $f_{X,Y}(x, y) = (f_0(x, y), f_1(x, y), f_2(x, y))$, where

$$
f_0(x, y) = \begin{cases} 0.1, & x = y = 0 \\ 0, & \text{elsewhere} \end{cases}
$$

$$
f_1(x, y) = \begin{cases} \dfrac{0.4t}{(2\pi)^2}, & x = 0.5\cos t, y = 0.5\sin t, 0 \le t < 2\pi \\ 0, & \text{otherwise} \end{cases}
$$

$$
f_2(x, y) = \begin{cases} 0.2625(x^2 + y^2), & -1 \le x \le 1, -1 \le y \le 1 \\ 0 & \text{otherwise.} \end{cases}
$$

These expressions describe a zero-dimensional point mass at the origin, a one-dimensional fence on a circle, centered at the origin and having radius $1/2$, and a two-dimensional surface over the square. All integrals in the mean and variance calculations include both the point mass and the circular distribution.

$$
\begin{aligned}
\mu_X &= \int_{\mathcal{R}^2} x \, dF = (0)(0.1) + \int_0^{2\pi} (0.5\cos t)\frac{0.4t}{(2\pi)^2}\, dt + \int_{-1}^1 \int_{-1}^1 0.2625x(x^2 + y^2) \, dx \, dy \\
&= 0
\end{aligned}
$$

$$
\begin{aligned}
\mu_Y &= \int_{\mathcal{R}^2} y \, dF = (0)(0.1) + \int_0^{2\pi} (0.5\sin t)\frac{0.4t}{(2\pi)^2}\, dt + \int_{-1}^1 \int_{-1}^1 0.2625y(x^2 + y^2) \, dx \, dy \\
&= \frac{-0.1}{\pi} = -0.0318
\end{aligned}
$$

These results comport with intuition. The point mass and the two-dimensional distribution are symmetric about the origin. The circular distribution, however, exhibits linearly increasing weight as the arc extends counterclockwise from $(1/2, 0)$. For μ_X, the low weights for x-values in the first quadrant supplement by the high weights in the fourth quadrant. The resulting sum just balances the intermediate weights for corresponding negative x-values in the second and third quadrants. Therefore, $\mu_X = 0$. For μ_Y, however, positive y-values in the first and second quadrants all have lower weights than their negative counterparts in the third and fourth quadrants, which means that the center of mass must shift downward. Consequently, $\mu_Y < 0$. The variance computations follow similar patterns.

$$\sigma_X^2 = \int_{\mathcal{R}^2} (x-0)^2 \, dF$$

$$= (0)(0.1) + \int_0^{2\pi} (0.5 \cos t)^2 \frac{0.4t}{(2\pi)^2} \, dt + \int_{-1}^1 \int_{-1}^1 0.2625 x^2 (x^2 + y^2) \, dx \, dy$$

$$= 0.3517$$

$$E[Y^2] = (0)^2(0.1) + \int_0^{2\pi} (0.5 \sin t)^2 \frac{0.4t}{(2\pi)^2} \, dt + \int_{-1}^1 \int_{-1}^1 0.2625 y^2 (x^2 + y^2) \, dx \, dy$$

$$= 0.3517$$

$$\sigma_Y^2 = E[Y^2] - \mu_Y^2 = 0.3517 - (-0.0318)^2 = 0.3517 - 0.0010 = 0.3507$$

Also,

$$\text{cov}(X, Y) = E[XY] - \mu_X \mu_Y$$

$$= (0)(0)(0.1) + \int_0^{2\pi} (0.5 \cos t)(0.5 \sin t) \frac{0.4t}{(2\pi)^2} \, dt$$

$$+ \int_{-1}^1 \int_{-1}^1 0.2625 xy(x^2 + y^2) \, dx \, dy - (0)(-0.0318)$$

$$= -0.00398$$

$$\rho_{X,Y} = \frac{-0.00398}{\sqrt{(0.3517)(0.3507)}} = -0.0113. \; \square$$

If X is one of a group of joint random variables, we expect the same values for μ_X and σ_X^2 when we compute using the marginal density for X as when we calculate with the joint density of the group. The following example demonstrates this equivalence.

5.38 Example: Compute μ_X and σ_X^2 for the density of Example 5.35, both from the joint density and from the marginal density for X.

The example gave the joint density as $f_{XY}(x, y) = (f_0(x, y), f_1(x, y), f_2(x, y))$, where

$$f_0(x, y) = \begin{cases} 0.1, & x = y = 0 \\ 0, & \text{elsewhere} \end{cases}$$

$$f_1(x, y) = \begin{cases} 0.2, & 0 \le x \le 1, y = 0 \\ 0.2, & x = 0, 0 \le y \le 1 \\ 0, & \text{elsewhere} \end{cases}$$

$$f_2(x, y) = \begin{cases} 0.5(x + y), & 0 \le x \le 1, 0 \le y \le 1 \\ 0, & \text{elsewhere,} \end{cases}$$

and calculated the marginal density for X to be

$$f_X(x) = (f_0(x), f_1(x))$$

$$f_0(x) = \begin{cases} 0.3, & x = 0 \\ 0, & \text{elsewhere} \end{cases}$$

$$f_1(x) = \begin{cases} 0, & x < 0 \text{ or } x > 1 \\ 0.5x + 0.45, & 0 \leq x \leq 1. \end{cases}$$

From the former, we have

$$\mu_X = (0)(0.1) + \int_0^1 0.2x \, dx + \int_0^1 (0)0.2y \, dy + \int_0^1 \int_0^1 0.5x(x+y) \, dx \, dy = 0.3917$$

$$\sigma_X^2 = (0)^2(0.1) + \int_0^1 0.2x^2 \, dx + \int_0^1 (0)^2 0.2y \, dy$$

$$+ \int_0^1 \int_0^1 0.5x^2(x+y) \, dx \, dy - (0.3917)^2$$

$$= 0.1216.$$

From the latter, we obtain

$$\mu_X = (0)(0.3) + \int_0^1 x(0.5x + 0.45) \, dx = 0.3917$$

$$\sigma_X^2 = E[X^2] - (0.3917)^2 = (0)^2(0.3) + \int_0^1 x^2(0.5x + 0.45) \, dx - (0.3917)^2 = 0.1216.$$

The values are identical. $\square$

5.39 Definition: Let $\mathbf{X} = (X_1, X_2, \ldots, X_n)$ be a set of joint random variables with cumulative distribution F arising from a density. The X_i are *independent* if

$$F_{\mathbf{X}}(x_1, x_2, \ldots, x_n) = F_{X_1}(x_1) \cdot F_{X_2}(x_2) \cdots F_{X_n}(x_n).$$

They are uncorrelated if $i \neq j$ implies that $\text{cov}(X_i, X_j) = 0$. ∎

5.40 Example: Continuous joint random variables X and Y have density

$$f_{XY}(x, y) = \begin{cases} 1/2, & -1 \leq x \leq 1, -(1 - |x|) \leq y \leq (1 - |x|) \\ 0, & \text{elsewhere.} \end{cases}$$

Show that X, Y are uncorrelated but dependent.

The region of nonzero density is the diamond shown in Figure 5.6. This area is actually a rotated square of side $\sqrt{2}$, over which the density is the constant $1/2$. From the symmetry of the density, it follows that $E[X] = E[Y] = E[XY] = 0$. Consequently, the random variables are uncorrelated.

To see that they are not independent, we compute the joint cumulative distribution and its marginals for points in the region A, the triangular area outside the diamond in the first quadrant, but not yet completely above or to the right of it. Consider a point $(x, y) \in A$, as shown in the figure. The semi-infinite rectangle to the southwest of the point covers the diamond, except for triangular portions in the upper and right corners. Some geometric reasoning reveals that the areas of these triangles are $(1 - y)^2$ and $(1 - x)^2$, respectively. Consequently,

$$F_{XY}(x, y) = \frac{1}{2}[(\sqrt{2})^2 - (1 - x)^2 - (1 - y)^2] = 1 - \frac{(1 - x)^2 + (1 - y)^2}{2},$$

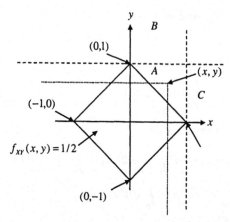

Figure 5.6. Joint density of uncorrelated, dependent random variables
(see Example 5.40)

for $(x,y) \in A$. The marginal distribution is $F_X(x) = F_{XY}(x,\infty)$, and to evaluate this limit, we need the joint distribution in the region B, the corridor above A. A point $(x,y) \in B$ subtends a rectangle that includes the entire diamond, except for the right triangular corner. Hence, for $(x,y) \in B$,

$$F_{XY}(x,y) = \frac{1}{2}[(\sqrt{2})^2 - (1-x)^2] = 1 - \frac{(1-x)^2}{2}.$$

This expression does not depend on y, so $F_X(x) = F_{XY}(x,\infty) = 1 - (1-x)^2/2$. A similar analysis involving region C reveals that $F_Y(y) = F_{XY}(\infty,y) = 1-(1-y)^2/2$. The following calculation, for the point $(1/2,3/4) \in A$, shows that the random variables are not independent.

$$F_{XY}(1/2,3/4) = 1 - \frac{(1/2)^2 + (1/4)^2}{2} = \frac{27}{32} = \frac{216}{256}$$

$$F_X(1/2) \cdot F_Y(3/4) = \left[1 - \frac{(1/2)^2}{2}\right] \cdot \left[1 - \frac{(1/4)^2}{2}\right] = \frac{7}{8} \cdot \frac{31}{32} = \frac{217}{256}. \quad \Box$$

The example shows that uncorrelated random variables need not be independent. However, independent random variables are uncorrelated, just as in the discrete case. We defer the proof to the next section, where we will work with densities that do not have lower-dimensional components. In that context, we can use ordinary calculus results to derive several more interesting properties of densities and cumulative distributions. We conclude this section with a final definition in the general context. For two vectors $\mathbf{x}, \mathbf{y} \in \mathcal{R}^n$, we use the dot product notation, $\mathbf{x} \cdot \mathbf{y}$, for the expression $x_1 y_1 + x_2 y_2 + \ldots + x_n y_n$.

5.41 Definition: Let $\mathbf{X} = (X_1, X_2, \ldots)$ be joint random variables. The *joint moment generating function* is

$$\Psi_{\mathbf{X}}(\mathbf{t}) = E[e^{\mathbf{t} \cdot \mathbf{X}}] = \int_{\mathcal{R}^n} e^{x_1 t_1 + \ldots + x_n t_n} \, dF. \quad \blacksquare$$

As in the discrete case, this function is useful for computing means, variances, and covariances, although we defer the details to the next section, where a simpler density function facilitates the analysis. The actual computation of the moment generating function is straightforward, as illustrated in the example below.

5.42 Example: Compute the moment generating function of random variables X and Y, for which the joint density $f_{XY}(\cdot, \cdot)$ contains the following zero-, one-, and two-dimensional components.

$$f_0(x, y) = \begin{cases} 0.1, & x = y = 0 \\ 0, & \text{elsewhere} \end{cases}$$

$$f_1(x, y) = \begin{cases} 0.2e^{-x}, & x \geq 0, y = 0 \\ 0.2e^{-y}, & x = 0, y \geq 0 \\ 0, & \text{elsewhere} \end{cases}$$

$$f_2(x, y) = \begin{cases} 0.5e^{-(x+y)}, & x > 0, y > 0 \\ 0, & \text{elsewhere} \end{cases}$$

We first verify that the total assigned probability is 1.

$$\int_{\mathcal{R}^2} dF = 0.1 + 0.2 \int_0^\infty e^{-x}\, dx + 0.2 \int_0^\infty e^{-y}\, dy$$
$$+ 0.5 \int_0^\infty \int_0^\infty e^{-(x+y)}\, dx\, dy = 1$$

Now, we compute $\Psi_{XY}(t, s)$ by integrating the separate components. For $t, s < 1$, we have

$$\Psi_{XY}(t, s) = (0.1)e^{(0)t + (0)s} + \int_0^\infty (0.2)e^{-x}e^{xt+(0)s}\, dx + \int_0^\infty (0.2)e^{-y}e^{(0)t+ys}\, dy$$
$$+ \int_0^\infty \int_0^\infty (0.5)e^{-(x+y)}e^{xt+ys}\, dx\, dy$$
$$= 0.1 + \frac{0.2}{1-t} + \frac{0.2}{1-s} + \frac{0.5}{(1-t)(1-s)}.$$

Note that

$$E[X] = (0)(0.1) + 0.2 \int_0^\infty (0)e^{-y}\, dy + 0.2 \int_0^\infty xe^{-x}\, dx + 0.5 \int_0^\infty \int_0^\infty xe^{-(x+y)}\, dx\, dy$$
$$= 0.7$$

and

$$\partial_1 \Psi_{XY}(t, s) = \frac{0.2}{(1-t)^2} + \frac{0.5}{(1-t)^2(1-s)}$$
$$\partial_1 \Psi_{XY}(0, 0) = 0.7.$$

Similar computations verify that

$$E[Y] = \partial_2 \Psi_{XY}(0, 0)$$
$$E[XY] = \partial_{12} \Psi_{XY}(0, 0)$$
$$E[X^2] = \partial_{11} \Psi_{XY}(0, 0),$$

and so forth. $\square$

5.11 Prove the following properties of the cumulative distribution function, which were omitted from the proof of Theorem 5.27.

(a) $\lim_{x \to -\infty} F(x) = 0$

(b) $\lim_{y \to x+} F(y) = F(x)$

5.12 Let $F(x_1, x_2, x_3, x_4)$ be the cumulative distribution function of a probability distribution on $\mathcal{R}^4$. Show that

$$F(x_1^-, x_2^-, x_3, x_4) = \lim_{(y_1, y_2) \to (x_1, x_2)^-} F(y_1, y_2, x_3, x_4) \le F(x_1, x_2, x_3, x_4)$$

by expressing $F(x_1, x_2, x_3, x_4) - F(y_1, y_2, x_3, x_4)$ as a the limiting probability of a nested collection of shrinking rectangles.

*5.13 The discussion preceding Theorem 5.28 sketches the process for allocating probability to arbitrary sets in $\mathcal{R}^n$. In summary, the procedure starts with an appropriate function F, assigns probabilities for open rectangles with $\Pr((\mathbf{x}, \mathbf{y})) = \Delta F(\mathbf{x}, \mathbf{y}^-)$, and then establishes a competitive mechanism for assigning probability to an arbitrary set. If $\mathcal{O}$ is the collection of all open rectangles, each countable subcollection $\mathcal{G} = \{R_1, R_2, \dots\}$ receives the score $s(\mathcal{G}) = \sum_{i=1}^{\infty} \Pr(R_i)$, and

$$\Pr(A) = \inf\{s(\mathcal{G}) \mid \mathcal{G} = \{R_1, R_2, \dots\} \subset \mathcal{O}, A \subset \cup_{i=1}^{\infty} R_i\}.$$

For the case $n = 2$, verify that the competition yields the same probability as the assignment via F. That is, show that

$$\Delta F(\mathbf{a}, \mathbf{b}^-) = \inf\{s(\mathcal{G}) \mid \mathcal{G} = \{R_1, R_2, \dots\} \subset \mathcal{O}, (\mathbf{a}, \mathbf{b}) \subset \cup_{i=1}^{\infty} R_i\}.$$

Let the left and right sides of the equation above be p and q, respectively. One approach first uses the subcollection $(\mathbf{a}, \mathbf{b}), \phi, \phi, \dots$ to argue that $q \le p$ and then establishes the reverse inequality by cases. If $\mathcal{G}$ contains one or more components R_i with $(\mathbf{a}, \mathbf{b}) \subset R_i$, then $s(\mathbf{G}) \ge p$. If no $\mathcal{G}$ component contains $(\mathbf{a}, \mathbf{b})$, then some point in $(\mathbf{a}, \mathbf{b})$ is common to two components R_i and R_j. Because the components are open rectangles, there must be an open rectangle R common to R_i and R_j. This common rectangle contributes twice to the score of $\mathcal{G}$, which drives the score above p. Figure 5.7 illustrates the point for the simple case where two components cover $(\mathbf{a}, \mathbf{b})$. The monotonicity of ΔF forces the probability sum assigned to the two components to equal or exceed that assigned to $(\mathbf{a}, \mathbf{b})$, which then forces $s(\mathbf{G}) \ge p$.

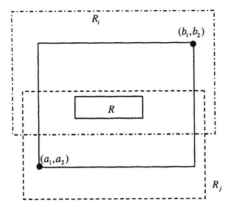

Figure 5.7. Covering an open rectangle with a countable collection of open rectangles

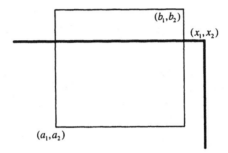

Figure 5.8. Slicing an open rectangle with $(-\infty, \mathbf{x}]$

---

*5.14 A second gap in the proof of Theorem 5.28 concerns the relationship between the Borel sets and the perfect slicers. To show that Borel sets are perfect slicers, it suffices to show that half-open rectangles of the form $(-\infty, \mathbf{x}]$ are perfect slicers of open rectangles. Adapt the argument of Theorem 5.13 to prove this statement for $n = 2$.

First, enumerate the ways the slicer $A = (-\infty, \mathbf{x}]$ can intersect the open rectangle $B = (\mathbf{a}, \mathbf{b})$. Figure 5.8 illustrates one possibility. In this case, $B \cap A$ is the lower part of B, including the upper boundary line. $B \cap \overline{A}$ is the upper part of B, excluding all boundary lines. Consequently,

$$\Pr(B \cap A) = \Delta F((a_1, a_2), (b_1^-, x_2^+))$$
$$\Pr(B \cap \overline{A}) = \Delta F((a_1, x_2), (b_1^-, b_2^-)).$$

We can expand these expressions, and using the fact that ΔF is continuous from above, show that they sum to $\Pr(B)$.

5.15 The joint density of X and Y has a zero-dimensional component that places probability 0.1 at point $(1/2, 1/2)$, a one-dimensional component that spreads probability 0.3 uniformly over the two diagonals of the unit square in the first quadrant, and a two-dimensional component that distributes probability over the first quadrant in proportion to $e^{-(x+y)}$. Write the components of this density and compute the joint cumulative distribution function. Derive the marginal distribution for X and its corresponding density.

5.16   For the joint distribution of the preceding problem, find the means, variances, covariance, and correlation. Verify that the computation of μ_X using the joint density gives the same results as that using the marginal density for X.

5.17   The joint density of X and Y has a one-dimensional component that spreads probability 0.5 uniformly over the parabola $y = x^2$ in the unit square of the first quadrant. It also has a two-dimensional component that spreads the remaining 0.5 across the unit square in proportion to the product xy. Write the components of this density and compute the joint cumulative distribution function. Derive the marginal distribution for Y and its density.

5.18   For the joint distribution of the preceding problem, find the means, variances, covariance, and correlation. Verify that the computation of μ_Y using the joint density gives the same results as that using the marginal density for Y.

5.19   For the density of Example 5.42, verify that

$$\sigma_X^2 = \partial_{11}\Psi_{XY}(0,0) - [\partial_1\Psi_{XY}(0,0)]^2$$
$$\mathrm{cov}(X,Y) = \partial_{12}\Psi_{XY}(0,0) - \partial_1\Psi_{XY}(0,0)\cdot\partial_2\Psi_{XY}(0,0).$$

5.20   Compute the moment generating function for the continuous uniform density

$$f_U(x) = \begin{cases} 1, & 0 \le x \le 1 \\ 0, & \text{elsewhere.} \end{cases}$$

*5.21   For the density $f(x) = (1/\sqrt{2\pi})e^{-x^2/2}$, compute the moment generating function.

5.3   Differentiable distributions

The density function provides primary access to computations involving one or more random variables. We have seen that the general density can contain several components, each requiring separate attention in a calculation. In the case of two random variables, for example, the density is a function over the plane that can involve probability concentrations on points and lines as well as two-dimensional regions. In this case, a typical computation decomposes into three parts, each dealing with one kind of probability contribution. The calculations are considerably easier if the lower-dimensional components are not present. A single continuous random variable X, for example, has a one-component density $f(x)$ with the property $F(x) = \int_{-\infty}^{x} f(t)\, dt$. The integration involves no point masses. Similarly, a set of n continuous joint random variables has a one-component density $f(x_1, x_2, \ldots, x_n)$ such that

$$F(x_1, x_2, \ldots, x_n) = \int_{-\infty}^{x_1} \int_{-\infty}^{x_2} \cdots \int_{-\infty}^{x_n} f(t_1, t_2, \ldots, t_n)\, dt_n\, dt_{n-1} \cdots dt_1.$$

Again, the integration involves no lower-dimensional operations.

5.43 Definition: X is a *continuous random variable* if its cumulative distribution function $F_X(\cdot)$ arises from an integrable, nonnegative density $f_X(\cdot)$. That is, $F(x) = \int_{-\infty}^{x} f_X(t)\, dt$. A set $\mathbf{X} = (X_1, X_2, \ldots, X_n)$ of joint random variables is continuous if the joint cumulative distribution function $F_{\mathbf{X}}(\cdot)$ arises from a single integrable, nonnegative density $f_{\mathbf{X}}(\cdot)$ component based on n-dimensional volume elements. That is,

$$F_{\mathbf{X}}(x_1, x_2, \ldots, x_n) = \int_{-\infty}^{x_1} \int_{-\infty}^{x_2} \cdots \int_{-\infty}^{x_n} f_{\mathbf{X}}(t_1, t_2, \ldots, t_n)\, dt_n \cdots dt_1. \blacksquare$$

We use the term *integrable function* in the sense of Section A.4. In particular, the function has only isolated singular points, in the neighborhood of which the function is unbounded. If the function has n arguments, we assume that it obeys this constraint as a function of any one argument, with the remaining $(n-1)$ arguments held constant. From Theorem A.51, we know that an integrable $f(\cdot)$ delivers a continuous $F(x) = \int_{a}^{x} f(t)\, dt$ and that a continuous integrand delivers a differentiable $F(x)$. If follows immediately that the cumulative distribution function of a continuous random variable is continuous, which, of course, explains the name. The same results hold for the iterated integral of a density function of several variables. That is, the various marginal cumulative distributions are continuous functions. This continuity means that the n-dimensional integral receives no contribution from integrand values on lower-dimensional subsets.

The next chapter develops a catalog of useful continuous distributions, but all enjoy a further property that allows the full use of familiar calculus operations in their analyses. Assuming a set of n joint random variables, the additional property is that the cumulative distribution function possesses continuous partial derivatives through order n at all points, except possibly on a closed set of zero volume. We write $\partial_i F(x_1, x_2, \ldots, x_n)$ for the value of the partial derivative with respect to the ith argument, evaluated at point $\mathbf{x} = (x_1, x_2, \ldots, x_n)$. Conventional notation is $(\partial F/\partial x_i)(x_1, x_2, \ldots, x_n)$, which uses x_i in a double role—once to identify a particular function among the n first partial derivatives and again to specify the ith component of the evaluation point. The jth partial derivative of the ith partial derivative at point $\mathbf{x} = (x_1, x_2, \ldots, x_n)$ is $\partial_{ij} F(x_1, x_2, \ldots, x_n)$, and so forth.

For the random variables $\mathbf{X} = (X_1, X_2, \ldots, X_n)$, suppose that the joint cumulative distribution function F possesses these continuous derivatives. Recalling the notation preceding Definition 5.5.3, we compute ΔF as follows.

$$\Pr(a_1 : b_1, t_2, \ldots, t_n) = F(b_1, t_2, \ldots, t_n) - F(a_1, t_2, \ldots, t_n)$$
$$= \int_{a_1}^{b_1} \partial_1 F(s_1, t_2, \ldots, t_n) \, ds_1$$

$$\Pr(a_1 : b_1, a_2 : b_2, t_3, \ldots, t_n)$$
$$= \Pr(a_1 : b_1, b_2, t_3, \ldots, t_n) - \Pr(a_1 : b_1, a_2, t_3, \ldots, t_n)$$
$$= \int_{a_1}^{b_1} \partial_1 F(s_1, b_2, t_3, \ldots, t_n) \, ds_1 - \int_{a_1}^{b_1} \partial_1 F(s_1, a_2, t_3, \ldots, t_n) \, ds_1$$
$$= \int_{a_2}^{b_2} \int_{a_1}^{b_1} \partial_{12} F(s_1, s_2, t_3, \ldots, t_n) \, ds_1 \, ds_2$$

The pattern continues until we obtain

$$\Delta F(\mathbf{a}, \mathbf{b}) = \Pr(a_1 : b_1, \ldots, a_n : b_n)$$

$$= \int_{a_n}^{b_n} \cdots \int_{a_1}^{b_1} \partial_{1,2,\ldots,n} F(s_1, \ldots, s_n) \, ds_1 \cdots ds_n.$$

We can arbitrarily set $\partial_{1,2,\ldots,n} F(s_1, s_2, \ldots, s_n) = 0$ for the exceptional points $(s_1, s_2, \ldots, s_n)$ where the derivatives are not defined. This modification does not change the probability computations above because integrand values on a set of volume zero are not relevant for the integral. Now, let $\mathbf{a} \to -\infty$ to obtain

$$F(b_1, \ldots, b_n) = \Pr((-\infty, \mathbf{b}]) = \int_{-\infty}^{b_1} \cdots \int_{-\infty}^{b_n} \partial_{1,2,\ldots,n} F(s_1, \ldots, s_n) \, ds_n \ldots ds_1.$$

Therefore, we can take $\partial_{1,2,\ldots,n} F(\cdot)$ as the density of $\mathbf{X}$ at all points where the derivatives exist. The remaining points comprise a set of zero volume, so the value of the density at those points is not relevant for integral calculations. We can set the density at those points to any convenient value.

5.44 Example: Let X be a continuous random variable with cumulative distribution function

$$F(x) = \begin{cases} 0, & x < 0 \\ 1 - e^{-x}, & x \geq 0 \end{cases}$$

$F(\cdot)$ has a continuous derivative at all points except $x = 0$:

$$F'(x) = \begin{cases} 0, & x < 0 \\ e^{-x}, & x > 0. \end{cases}$$

Consequently, the density of X is $F'(\cdot)$, with any convenient value for $f(0)$. For example,

$$f(x) = \begin{cases} 0, & x < 0 \\ e^{-x}, & x \geq 0 \end{cases}$$

puts $f(0) = 1$ and renders $f(\cdot)$ continuous from the right at $x = 0$. If we had some pressing need for continuity from the left, we could have chosen $f(0) = 0$. These choices do not change any calculations that involve the integral of the density.

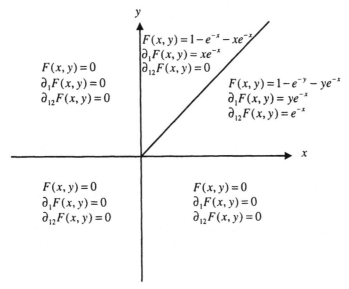

Figure 5.9. Cumulative distribution function (see Example 5.44)

Consider a two-dimensional case with cumulative distribution:

$$F(x,y) = \begin{cases} 0, & x < 0 \text{ or } y < 0 \\ 1 - e^{-x} - xe^{-x}, & 0 \le x < y \\ 1 - e^{-y} - ye^{-x}, & 0 \le y \le x. \end{cases}$$

Figure 5.9 notes the regions where $F(\cdot,\cdot)$ assumes its various forms, together with the derivatives in those regions. The density is then

$$f_{XY}(x,y) = \partial_{12}F(x,y) = \begin{cases} e^{-x}, & 0 < y < x \\ 0, & \text{elsewhere.} \end{cases}$$

The boundary lines separating the regions have two-dimensional volume zero, and the density on these boundaries is not relevant for computations. The expression above sets the density equal to zero for these points. One of the exercises asks that you recover the cumulative distribution function by integrating $\int_{-\infty}^{x}\int_{-\infty}^{y} f_{XY}(t,s)\,ds\,dt$. $\square$

For continuous random variables with a differentiable joint cumulative distribution function, we can obtain the marginal density of a subset by integrating over the remaining components. For two joint random variables X and Y, this means that

$$f_X(x) = \int_{-\infty}^{\infty} f_{XY}(x,y)\,dy$$

$$f_Y(y) = \int_{-\infty}^{\infty} f_{XY}(x,y)\,dx.$$

For three random variables X, Y, Z, it means that

$$f_X(x) = \int_{-\infty}^{\infty} \int_{-\infty}^{\infty} f_{XYZ}(x, y, z) \, dy \, dz$$

$$f_{XY}(x, y) = \int_{-\infty}^{\infty} f_{XYZ}(x, y, z) \, dz,$$

and so forth. For n joint random variables, the notation is somewhat cumbersome, but the idea is the same. This result follows from the observations below. Let $j_1, j_2, \ldots, j_{n-k}$ index the components that are *not* in the marginal set $i_1, i_2, \ldots, i_k$. Then we obtain the marginal cumulative distribution $F_{i_1, \ldots, i_k}(x_{i_1}, \ldots, x_{i_k})$ with n integrations. For each component i_t in the marginal set, the integration limits are $-\infty$ to x_{i_t}; for each component j_t outside the marginal set, the limits are $-\infty$ to ∞. We rearrange the integration order to range first over the components outside the marginal set. We denote $f_{X_1, \ldots, X_n}$ simply as f.

$$K(x_{i_1}, \ldots, x_{i_k}) = \int_{s_{j_1} = -\infty}^{\infty} \cdots \int_{s_{j_{n-k}} = -\infty}^{\infty} f(s_1, \ldots, s_n) \, ds_{j_1} \ldots ds_{j_{n-k}}$$

Then, completing the full integration, we have

$$F_{X_{i_1}, \ldots, X_{i_k}}(x_{i_1}, \ldots, x_{i_k}) = \int_{s_{i_1} = -\infty}^{x_{i_1}} \cdots \int_{s_{i_k} = -\infty}^{x_{i_k}} K(x_{i_1}, \ldots, x_{i_k}) \, ds_{i_1} \ldots ds_{i_k}.$$

It is now apparent that K is the marginal density of $X_{i_1}, X_{i_2}, \ldots, X_{i_k}$. When we computed marginal densities in the preceding section, we first derived the joint cumulative distribution and sent certain arguments to infinity to obtain the marginal cumulative distribution. We then recognized that distribution as the integral of a density, possibly with lower-dimensional components. In the differentiable case, we can proceed directly to the marginal density by integrating out the variables not in the chosen set.

Continuous random variables with a differentiable joint distributions also admit a characterization of independence in terms of densities. Recall that Definition 5.39 establishes independence when the joint cumulative distribution is the product of its marginals. When the differentiability condition prevails, this is equivalent to the joint density equal to the product of the marginal densities. Indeed, suppose that $X_1, X_2, \ldots, X_n$ are independent. Then

$$F_{X_1, X_2, \ldots, X_n}(x_1, x_2, \ldots, x_n) = F_{X_1}(x_1) F_{X_2}(x_2) F_{X_n}(x_n)$$

$$f_{X_1, X_2, \ldots, X_n}(x_1, x_2, \ldots, x_n) = \partial_{1,2,\ldots,n} F_{X_1, X_2, \ldots, X_n}(x_1, x_2, \ldots, x_n)$$

$$= F'_{X_1}(x_1) \cdot F'_{X_2}(x_2) \ldots F'_{X_n}(x_n)$$

$$= f_{X_1}(x_1) f_{X_2}(x_2) \ldots f_{X_n}(x_n).$$

Conversely, if the joint density is the product of the marginals, we have

$$F_{X_1,\ldots,X_n}(x_1,\ldots,x_n) = \int_{-\infty}^{x_1} \cdots \int_{-\infty}^{x_n} f_{X_1}(t_1)\ldots f_{X_n}(t_n)\, dt_1 \ldots dt_n$$

$$= \left(\int_{-\infty}^{x_1} f_{X_1}(t_1)\, dt_1\right) \cdots \left(\int_{-\infty}^{x_n} f_{X_n}(t_n)\, dt_n\right)$$

$$= F_{X_1}(x_1)F_{X_2}(x_2)\cdots F_{X_n}(x_n),$$

and the random variables are independent. We see that the differentiability condition enables simple calculus relationships between densities and distributions and between joint functions and marginals. We summarize these properties in the following theorem.

5.45 Theorem: Let $\mathbf{X} = (X_1, X_2, \ldots, X_n)$ be joint continuous random variables arising from density $f_{\mathbf{X}}(\cdot)$. Suppose that the joint cumulative distribution function $F_{\mathbf{X}}(\cdot)$ has continuous partial derivatives through order n, except on a set of zero n-dimensional volume. Then

- $f_{\mathbf{X}}(\mathbf{x}) = \partial_{1,2,\ldots,n} F_{\mathbf{X}}(\mathbf{x})$, for all $\mathbf{x}$ where the differentiability condition holds. The value of $f_{\mathbf{X}}(\cdot)$ on the exceptional set is not relevant for integral computations and may be assumed zero.

- The marginal density of a subset $X_{i_1}, X_{i_2}, \ldots, X_{i_k}$ is

$$f_{X_{i_1},\ldots,X_{i_k}}(x_{i_1},\ldots,x_{i_n}) =$$
$$\int_{x_{j_1}=-\infty}^{\infty} \cdots \int_{x_{j_{n-k}}=-\infty}^{\infty} f_{X_1,\ldots,X_n}(x_1,\ldots,x_n)\, dx_{j_1}\cdots dx_{j_{n-k}},$$

where $j_1,\ldots,j_{n-k}$ are the components excluded from the set $i_1,\ldots,i_k$.

- The random variables are independent if and only if the joint density is the product of the marginal densities. That is, independence is equivalent to $f_{\mathbf{X}}(\mathbf{x}) = f_{X_1}(x_1)f_{X_2}(x_2)\cdots f_{X_n}(x_n)$.

PROOF: See discussion above. ∎

In the context of differentiable distributions, it is easy to show that independent random variables are uncorrelated, just as in the discrete case.

5.46 Theorem: Let $\mathbf{X} = (X_1, X_2, \ldots, X_n)$ be a set of joint continuous random variables with density $f_{\mathbf{X}}(\cdot)$. Suppose further that the joint cumulative distribution function possesses continuous partial derivatives through order n. If the random variables are independent, then they are uncorrelated.

PROOF: Invoking the preceding theorem, we have a joint density that is the product of its marginals. Therefore,

$$\mathrm{cov}(X_i, X_j) = \int_{-\infty}^{\infty} \cdots \int_{-\infty}^{\infty} (t_i - \mu_{X_i})(t_j - \mu_{X_j})f_{X_1}(t_1)\ldots f_{X_n}(t_n)\, dt_1 \cdots dt_n$$

$$= \left(\int_{-\infty}^{\infty} (t_i - \mu_{X_i})f_{X_i}(t_i)\, dt_i\right)\left(\int_{-\infty}^{\infty} f_{X_j}(t_j)(t_j - \mu_{X_j})\, dt_j\right)$$

$$= 0. \qquad ∎$$

   The expressions for means, variances, covariances, correlations, and moment generating functions remain as given in Definitions 5.36 and 5.41. The only difference is that the integrals have a single n-dimensional component.

5.47 Example: Find the means, variances, covariances, and moment generating functions for the distributions of Example 5.44.

   The cited example first considered a single random variable with density

$$f(x) = \begin{cases} 0, & x < 0 \\ e^{-x}, & x \geq 0. \end{cases}$$

Directly from the definitions, and without the complication of lower-dimensional integrals, we calculate

$$\mu_X = \int_{-\infty}^{\infty} x f(x)\, dx = \int_0^{\infty} x e^{-x}\, dx = -(x+1)e^{-x}\big|_0^{\infty} = 1.0$$

$$E[X^2] = \int_{-\infty}^{\infty} x^2 f(x)\, dx = \int_0^{\infty} x^2 e^{-x}\, dx = -(x^2 + 2x + 2)e^{-x}\big|_0^{\infty} = 2.0.$$

From the latter equation, it follows that $\sigma_X^2 = E[X^2] - \mu_X^2 = 1.0$. Finally,

$$\Psi_X(t) = E[e^{tX}] = \int_{-\infty}^{\infty} e^{tx} f(x)\, dx = \int_0^{\infty} e^{-(1-t)x}\, dx = \frac{1}{1-t}, \text{ for } t < 1.$$

   The example then considered two joint random variables with density

$$f_{XY}(x, y) = \begin{cases} e^{-x}, & 0 < y < x \\ 0, & \text{elsewhere.} \end{cases}$$

We first obtain the marginal densities, which are both zero for negative arguments and assume the following expression for nonnegative ones.

$$f_X(x) = \int_{-\infty}^{\infty} f_{XY}(x, y)\, dy = \int_0^x e^{-x}\, dy = ye^{-x}\big|_{y=0}^x = xe^{-x}$$

$$f_Y(y) = \int_{-\infty}^{\infty} f_{XY}(x, y)\, dx = \int_y^{\infty} e^{-x}\, dx = -e^{-x}\big|_y^{\infty} = e^{-y}$$

We then compute the one-dimensional features as follows.

$$\mu_X = \int_0^{\infty} x f_X(x)\, dx = \int_0^{\infty} x^2 e^{-x}\, dx = 2.0$$

$$\mu_Y = \int_0^{\infty} y f_Y(y)\, dy = \int_0^{\infty} y e^{-y}\, dy = 1.0$$

$$E[X^2] = \int_0^{\infty} x^2 f_X(x)\, dx = \int_0^{\infty} x^3 e^{-x}\, dx = 6.0$$

$$E[Y^2] = \int_0^{\infty} y^2 e^{-y}\, dy = 2.0$$

$$\sigma_X^2 = E[X^2] - \mu_X^2 = 2.0$$

$$\sigma_Y^2 = E[Y^2] - \mu_Y^2 = 1.0$$

$$\Psi_X(t) = E[e^{tX}] = \int_{-\infty}^{\infty} e^{tx} f_X(x)\, dx = \int_0^{\infty} x e^{-(1-t)x}\, dx = \frac{1}{(1-t)^2}$$

$$\Psi_Y(s) = E[e^{sY}] = \int_{-\infty}^{\infty} e^{sy} f_Y(y)\, dy = \int_0^{\infty} e^{-(1-s)y}\, dy = \frac{1}{1-s}$$

For the joint features, we use the joint density.

$$E[XY] = \int_{-\infty}^{\infty} \int_{-\infty}^{\infty} xy f_{XY}(x,y) \, dy \, dx = \int_{0}^{\infty} \int_{0}^{x} xy e^{-x} \, dy \, dx$$

$$= \int_{0}^{\infty} x e^{-x} \left[y^2/2 \right]_{0}^{x} \, dx = \int_{0}^{\infty} (x^3/2) e^{-x} \, dx = 3.0$$

$$\mathrm{cov}(X,Y) = E[XY] - \mu_X \mu_Y = 1.0$$

$$\mathrm{corr}(X,Y) = \frac{\mathrm{cov}(X,Y)}{\sqrt{\sigma_X^2 \sigma_Y^2}} = \frac{1}{\sqrt{2}} = 0.707$$

Because they have nonzero covariance, X and Y are not independent. $\square$

The example computed the marginal means and variances with integrals over the marginal densities. With continuous random variables, it is easy to show that this is permissible in the general case.

$$E[X] = \int_{-\infty}^{\infty} \int_{-\infty}^{\infty} x f_{XY} \, dy \, dx = \int_{-\infty}^{\infty} x \left(\int_{-\infty}^{\infty} f_{XY}(x,y) \, dy \right) dx$$

$$= \int_{-\infty}^{\infty} x f_X(x) \, dx$$

A similar calculation holds for the variance.

As a final topic in this section, we consider conditional distributions. If X, Y are joint random variables with density $f_{XY}(x,y)$, is it meaningful to speak of the random variable X, given that Y has a particular value? Recall the interpretation in the discrete case. Assuming that $\Pr(Y = y) \neq 0$, we defined the conditional distribution $F_{X|Y=y}(\cdot)$ as

$$F_{X|Y=y}(x) = \Pr(X \leq x | Y = y) = \frac{\Pr(X \leq x \text{ and } Y = y)}{\Pr(Y = y)}$$

and interpreted the value as the relative frequency of event $(X \leq x)$ among those trials for which $(Y = y)$. With a continuous distribution, a certain difficulty arises because $\Pr(Y = y) = 0$ for each individual point y. Obtaining an empirical estimate of $\Pr(X \leq x | Y = y)$ then becomes a lengthy task. When we discard all experiments in which $Y \neq y$, the remaining experiments are few or zero in number, which renders the estimate unreliable. We can, however, note the relative frequency of $(X \leq x)$ among those trials for which $(y \leq Y \leq y + \Delta y)$, for smaller and smaller Δy. As $\Delta y \rightarrow 0$, we need an increasingly lengthy process because the qualifying outcomes become fewer and fewer. We hope to ascertain the limiting tendency before this process becomes too long. Actually, we obtain the limit mathematically and interpret it as the limiting relative frequency of $(X \leq x)$ among those trials for which $(y \leq Y \leq y + \Delta y)$ as $\Delta y \rightarrow 0$.

In particular, suppose that the joint random variables X, Y have a continuously differentiable cumulative distribution function. The density is $f_{XY}(\cdot, \cdot)$, which is bounded, nonnegative, and continuous except possibly on a set of zero two-dimensional volume. Assume further that the marginal densities, $f_X(\cdot)$ and $f_Y(\cdot)$, are also continuous, except possibly on a set of zero one-dimensional volume. We define the distribution of X, given $Y = y$, as the limiting distribution of X, given $y \leq Y \leq y + \Delta y$, as $\Delta y \to 0$.

5.48 Definition: Given joint random variables X and Y, the conditional distribution of X, given $Y = y$, is

$$F_{X|Y=y}(x) = \Pr(X \leq x | Y = y) = \lim_{\Delta y \to 0} \Pr(X \leq x | y \leq Y \leq y + \Delta y),$$

provided that the limit exists. For joint random variables $X_1, \ldots, X_n$, let $i_1, \ldots, i_k$ index a nonempty subset, and let $j_1, \ldots, j_{n-k}$ be the remaining components. For $(y_1, y_2, \ldots, y_k) \in \mathcal{R}^k$ and $(z_1, z_2, \ldots, z_{n-k}) \in \mathcal{R}^{n-k}$, the conditional distribution of $X_{j_1}, \ldots, X_{j_{n-k}}$, given $X_{i_1} = y_1, X_{i_2} = y_2, \ldots, X_{i_k} = y_k$, is

$$F_{X_{j_1}, \ldots, X_{j_{n-k}} | X_{i_1} = y_1, \ldots, X_{i_k} = y_k}(z_1, \ldots, z_{n-k})$$
$$= \Pr(X_{j_1} \leq z_1, \ldots, X_{j_{n-k}} \leq z_{n-k} | X_{i_1} = y_1, \ldots, X_{i_k} = y_k)$$
$$= \lim_{\Delta y_1, \ldots, \Delta y_k \to 0} \Pr(X_{j_1} \leq z_1, \ldots, X_{j_{n-k}} \leq z_{n-k} | E(\Delta y_1, \ldots, \Delta y_k)),$$

where

$$E(\Delta y_1, \ldots, \Delta y_k) = (y_1 \leq X_{i_1} \leq y_1 + \Delta y_1, \ldots, y_k \leq X_{i_k} \leq y_k + \Delta y_k). \blacksquare$$

Working in two dimensions, we now show that the limit does exist when $f_Y(y) \neq 0$, and we obtain its value as the integral of the density $f_{XY}(x, y)/f_Y(y)$. Consider a point (x, y) and a small neighborhood containing $[y, y + \Delta y]$, in which the marginal density $f_Y(\cdot)$ is continuous. We assume that $f_Y(y) \neq 0$. Let

$$A(\Delta y) = \left| \Pr(X \leq x | y \leq Y \leq y + \Delta y) - \frac{\int_{-\infty}^{x} f_{XY}(t, y)\, dt}{f_Y(y)} \right|.$$

We want to show that $\lim_{\Delta y \to 0} A(\Delta y) = 0$. Invoking the definition of conditional probability, we first expand it to the following form.

$$A(\Delta y) = \left| \frac{\Pr(X \leq x \text{ and } y \leq Y \leq y + \Delta y)}{\Pr(y \leq Y \leq y + \Delta y)} - \frac{\int_{-\infty}^{x} f_{XY}(t, y)\, dt}{f_Y(y)} \right|$$

Placing this expression over the common denominator $f_Y(y)\Pr(y \leq Y \leq y + \Delta y)$, we reduce the corresponding numerator N.

$$|N| = \left| f_Y(y) \int_{-\infty}^{x} \int_{y}^{y+\Delta y} f_{XY}(t, s)\, ds\, dt \right.$$
$$\left. - \left(\int_{-\infty}^{x} f_{XY}(t, y)\, dt \right) \left(\int_{y}^{y+\Delta y} f_Y(s)\, ds \right) \right|$$
$$= \left| \int_{y}^{y+\Delta y} \int_{-\infty}^{x} [f_Y(y) f_{XY}(t, s) - f_Y(s) f_{XY}(t, y)]\, dt\, ds \right|$$

$$|N| \leq \left| \int_y^{y+\Delta y} \int_{-\infty}^x [f_Y(y)f_{XY}(t,s) - f_Y(y)f_{XY}(t,y)]\, dt\, ds \right|$$

$$+ \left| \int_y^{y+\Delta y} \int_{-\infty}^x [f_Y(y)f_{XY}(t,y) - f_Y(s)f_{XY}(t,y)]\, dt\, ds \right|$$

Consequently,

$$A(\Delta y) \leq \left| \frac{\int_y^{y+\Delta y} \int_{-\infty}^x [f_Y(y)f_{XY}(t,s) - f_Y(y)f_{XY}(t,y)]\, dt\, ds}{f_Y(y)\Pr(y \leq Y \leq y+\Delta y)} \right|$$

$$+ \left| \frac{\int_y^{y+\Delta y} \int_{-\infty}^x [f_Y(y)f_{XY}(t,y) - f_Y(s)f_{XY}(t,y)]\, dt\, ds}{f_Y(y)\Pr(y \leq Y \leq y+\Delta y)} \right|.$$

Naming the two components on the right, we have $A(\Delta y) = A_1(\Delta y) + A_2(\Delta y)$. The simpler component is $A_2(\Delta y)$, so we deal with it first.

$$A_2(\Delta y) = \left| \frac{\left(\int_y^{y+\Delta y}[f_Y(y) - f_Y(s)]\, ds \right) \left(\int_{-\infty}^x f_{XY}(t,y)\, dt \right)}{f_Y(y)\Pr(y \leq Y \leq y+\Delta y)} \right|$$

$$\leq \frac{\left| \int_y^{y+\Delta y}[f_Y(y) - f_Y(s)]\, ds \right| \cdot \int_{-\infty}^\infty f_{XY}(t,y)\, dt}{f_Y(y)\Pr(y \leq Y \leq y+\Delta y)}$$

$$= \frac{\left| \int_y^{y+\Delta y}[f_Y(y) - f_Y(s)]\, ds \right| \cdot f_Y(y)}{f_Y(y)\Pr(y \leq Y \leq y+\Delta y)}$$

$$= \left| \frac{\int_y^{y+\Delta y}[f_Y(y) - f_Y(s)]\, ds}{\Pr(y \leq Y \leq y+\Delta y)} \right| = \left| \frac{f_Y(y)\Delta y - \Pr(y \leq Y \leq y+\Delta y)}{\Pr(y \leq Y \leq y+\Delta y)} \right|$$

Because $F_Y'(y) = f_Y(y)$, we can write

$$\Pr(y \leq Y \leq y+\Delta y) = F_Y(y+\Delta y) - F_Y(y)$$
$$= f_Y(y)\Delta y + [F_Y(y+\Delta y) - F_Y(y)] - f_Y(y)\Delta y$$
$$= f_Y(y)\Delta y + o_1(\Delta y),$$

where $o_1(\Delta y) = [F_Y(y+\Delta y) - F_Y(y)] - f_Y(y)\Delta y$ has the property

$$\lim_{\Delta y \to 0} \frac{o_1(\Delta y)}{\Delta y} = \lim_{\Delta y \to 0} \frac{F_Y(y+\Delta y) - F_Y(y)}{\Delta y} - f_Y(y) = 0.$$

Accordingly,

$$A_2(\Delta y) = \left| \frac{f_Y(y)\Delta y - f_Y(y)\Delta y - o_1(\Delta y)}{f_Y(y)\Delta y + o_1(\Delta y)} \right| = \frac{|o_1(\Delta y)/\Delta y|}{|f_Y(y) + o_1(\Delta y)/\Delta y|} \to 0,$$

as $\Delta y \to 0$. Now consider $A_1(\Delta y)$, after canceling the factor $f_Y(y)$ from numerator and denominator.

$$A_1(\Delta y) = \left| \frac{\int_y^{y+\Delta y} \int_{-\infty}^x [f_{XY}(t,s) - f_{XY}(t,y)]\, dt\, ds}{\Pr(y \le Y \le y + \Delta y)} \right|$$

$$= \left| \frac{\int_y^{y+\Delta y} [\partial_2 F_{XY}(x,s) - \partial_2 F_{XY}(x,y)]\, ds}{\Pr(y \le Y \le y + \Delta y)} \right|$$

$$= \left| \frac{F_{XY}(x, y + \Delta y) - F_{XY}(x,y) - \partial_2 F_{XY}(x,y)\Delta y}{\Pr(y \le Y \le y + \Delta y)} \right|,$$

where we have used the reduction

$$F_{XY}(x,y) = \int_{-\infty}^y \int_{-\infty}^x f_{XY}(t,s)\, dt\, ds$$

$$\partial_2 F_{XY}(x,y) = \int_{-\infty}^x f_{XY}(t,y)\, dt.$$

Because $\partial_2 F_{XY}(x,y) = \lim_{\Delta y \to 0}[F_{XY}(x, y + \Delta y) - F_{XY}(x,y)]/\Delta y$, we can express

$$F_{XY}(x, y + \Delta y) - F_{XY}(x,y) = \partial_2 F_{XY}(x,y)\Delta y + [F_{XY}(x, y + \Delta y)$$
$$- F_{XY}(x,y)] - \partial_2 F_{XY}(x,y)\Delta y$$
$$= \partial_2 F_{XY}(x,y)\Delta y + o_2(\Delta y),$$

where $o_2(\Delta y) = [F_{XY}(x, y+\Delta y) - F_{XY}(x,y)] - \partial_2 F_{XY}(x,y)\Delta y$ has the property

$$\lim_{\Delta y \to 0} \frac{o_2(\Delta y)}{\Delta y} = \lim_{\Delta y \to 0} \frac{F_{XY}(x, y + \Delta y) - F_{XY}(x,y)}{\Delta y} - \partial_2 F_{XY}(x,y) = 0.$$

Using this result and the previously established expression for $\Pr(y \le Y \le y + \Delta y)$, we have

$$A_1(\Delta y) = \left| \frac{\partial_2 F_{XY}(x,y)\Delta y + o_2(\Delta y) - \partial_2 F_{XY}(x,y)\Delta y}{f_Y(y)\Delta y + o_1(\Delta y)} \right|$$

$$= \frac{|o_2(\Delta y)/\Delta y|}{f_Y(y) + o_1(\Delta y)/\Delta y} \to 0,$$

as $\Delta y \to 0$. Hence,

$$F_{X|Y=y}(x) = \frac{\int_{-\infty}^x f_{XY}(t,y)\, dt}{f_Y(y)}.$$

Differentiating, we have $f_{X|Y=y}(x) = f_{XY}(x,y)/f_Y(y)$. The following theorem summarizes these results.

5.49 Theorem: Suppose that $X_1, X_2, \dots, X_n$ have a differentiable joint distribution, arising from a joint density that is continuous except possibly on a set of zero n-dimensional volume. Suppose further that the marginal distributions share this property. The conditional density of a set $X_{j_1}, \dots, X_{j_{n-k}}$, given specific values for the other components, $X_{i_1} = x_{i_1}, \ldots, X_{i_k} = x_{i_k}$, is then

$$f_{X_{j_1}, \ldots, X_{j_{n-k}} | X_{i_1} = x_{i_1}, \ldots, X_{i_k} = x_{i_k}}(x_{j_1}, \ldots, x_{j_{n-k}}) = \frac{f_{X_1, \ldots, X_n}(x_1, \ldots, x_n)}{f_{X_{i_1}, \ldots, X_{i_k}}(x_{i_1}, \ldots, x_{i_k})},$$

provided that the marginal density in the denominator is not zero. For two dimensions, these conditional densities are $f_{X|Y=y}(x) = f_{XY}(x, y)/f_Y(y)$ and $f_{Y|X=x}(y) = f_{XY}(x, y)/f_X(x)$.

PROOF: The discussion above establishes the theorem for $n = 2$. Higher dimensions require a similar argument. ∎

A conditional distribution is a distribution in its own right. The total assigned probability mass is 1, as shown by the following computation for $n = 2$.

$$\int_{-\infty}^{\infty} f_{X|Y=y}(x) \, dx = \frac{\int_{-\infty}^{\infty} f_{XY}(x, y) \, dx}{f_Y(y)} = \frac{f_Y(y)}{f_Y(y)} = 1$$

Consequently, a conditional distribution has a mean and variance, although both depend on the value of the conditioning variable. That is, for example, $\mu_{X|Y=y}$ is a function of y. The definitions use the conditional density.

$$\mu_{X|Y=y} = E_{X|Y=y}[X] = \frac{\int_{-\infty}^{\infty} x f_{XY}(x, y) \, dx}{f_Y(y)}$$

$$\sigma^2_{X|Y=y} = E_{X|Y=y}[(X - \mu_{X|Y=y})^2] = \frac{\int_{-\infty}^{\infty} (x - \mu_{X|Y=y})^2 f_{XY}(x, y) \, dx}{f_Y(y)}$$

Expanding the integral for $\sigma^2_{X|Y=y}$ shows that the traditional shortcut applies:

$$\sigma^2_{X|Y=y} = E_{X|Y=y}[X^2] - \mu^2_{X|Y=y}.$$

5.50 Example: Consider the joint density

$$f_{XY}(x, y) = \begin{cases} e^{-x}, & 0 < y < x \\ 0, & \text{elsewhere}, \end{cases}$$

for which the descriptive features were calculated in Example 5.47. There we determined the marginal densities to be

$$f_X(x) = \begin{cases} xe^{-x}, & x \geq 0 \\ 0, & x < 0 \end{cases}$$

$$f_Y(y) = \begin{cases} e^{-y}, & y \geq 0 \\ 0, & y < 0. \end{cases}$$

The conditional densities are therefore

$$f_{X|Y=y}(x) = \begin{cases} e^{-(x-y)}, & 0 < y < x \\ 0, & 0 \leq x \leq y \\ \text{undefined}, & \text{elsewhere} \end{cases}$$

$$f_{Y|X=x}(y) = \begin{cases} 1/x, & 0 < y < x \\ 0, & 0 < x \leq y \\ \text{undefined}, & \text{elsewhere}. \end{cases}$$

We compute the mean and variance of X, given $Y = y$, for $y \geq 0$.

$$\mu_{X|Y=y} = E_{X|Y=y}[X] = \int_{-\infty}^{\infty} x f_{X|Y=y}(x)\, dx = \int_{y}^{\infty} x e^{-(x-y)}\, dx = y + 1$$

$$E_{X|Y=y}[X^2] = \int_{-\infty}^{\infty} x^2 f_{X|Y=y}(x)\, dx = \int_{y}^{\infty} x^2 e^{-(x-y)}\, dx = y^2 + 2y + 2$$

$$\sigma_{X|Y=y}^2 = y^2 + 2y + 2 - (y+1)^2 = 1.$$

Since $\mu_{X|Y=y}$ is a function of y, we can compute its expected value $E_Y[\mu_{X|Y=y}]$. Using $f_Y(y) = e^{-y}$, for $y \geq 0$, we have

$$E_Y[\mu_{X|Y=y}] = \int_{0}^{\infty} (y+1) e^{-y}\, dy = 2.$$

From Example 5.47, we note that $\mu_X = 2$. This is no coincidence, as shown in the next theorem. $\square$

In the discrete case, we showed that $E_Y[\mu_{X|Y=y_j}] = \mu_X$ (Theorem 1.70), and we established a slightly more complex formula relating the variance and the conditional variance (Theorem 1.71). The following theorem shows that these results hold for continuous distributions.

5.51 Theorem: Let X, Y be joint continuous random variables. They may constitute a subset of a larger collection of joint random variables. Then we have

$$\mu_X = E_Y[\mu_{X|Y=y}]$$
$$\mu_Y = E_X[\mu_{Y|X=x}]$$
$$\sigma_X^2 = E_Y[\sigma_{X|Y=y}^2] + E_Y[(\mu_{X|Y=y} - \mu_X)^2]$$
$$\sigma_Y^2 = E_X[\sigma_{Y|X=x}^2] + E_X[(\mu_{Y|X=x} - \mu_Y)^2].$$

PROOF: We need to evaluate

$$E_Y[\mu_{X|Y=y}] = \int_{-\infty}^{\infty} f_Y(y) \left(\int_{-\infty}^{\infty} x f_{X|Y=y}(x)\, dx \right) dy,$$

but a difficulty arises in that $f_{X|Y=y}(\cdot)$ is defined only when $f_Y(y) > 0$. Note, however, that if $f_Y(y) = 0$, the entire integral involving $f_{X|Y=y}(\cdot)$ has coefficient zero in the larger integral and therefore does not contribute to the larger integral. So, for purposes of this computation, we can set $f_{X|Y=y}(x) \equiv 0$ when $f_Y(y) = 0$.

Also, if $f_Y(y) = 0$, we must have $f_{XY}(x, y) = 0$ for all x. The argument is straightforward. We have $f_Y(y) = \int_{-\infty}^{\infty} f_{XY}(x, y)\, dx$, and the integrand is continuous and nonnegative. If $f_{XY}(x_0, y) > 0$ for some point x_0, then continuity forces

$$f_{XY}(x_0, y) - f_{XY}(x, y) \leq |f_{XY}(x_0, y) - f_{XY}(x, y)| \leq \frac{1}{2} f_{XY}(x_0, y)$$

$$f_{XY}(x, y) \geq \frac{1}{2} f_{XY}(x_0, y),$$

for x in a neighborhood of x_0, say $x \in (x_0 - \delta, x_0 + \delta)$ with $\delta > 0$. But this forces

$$f_Y(y) = \int_{-\infty}^{\infty} f_{XY}(x, y) \, dx \geq \int_{x_0-\delta}^{x_0+\delta} f_{XY}(x, y) \, dx > \frac{1}{2} f_{XY}(x_0, y)(2\delta) > 0,$$

a contradiction. Consequently, taking $f_{X|Y=y}(x) \equiv 0$ when $f(y) = 0$ means that $f_{X|Y=y}(x) \cdot f_Y(y) = f_{XY}(x, y)$ for all y. Accordingly,

$$E_Y[\mu_{X|Y=y}] = \int_{-\infty}^{\infty} \int_{-\infty}^{\infty} x f_{X|Y=y}(x) \cdot f_Y(y) \, dx \, dy$$

$$= \int_{-\infty}^{\infty} x \left(\int_{-\infty}^{\infty} f_{XY}(x, y) \, dy \right) dx = \int_{-\infty}^{\infty} x f_X(x) \, dx = \mu_X.$$

A similar computation establishes $\mu_Y = E_X[\mu_{Y|X=x}]$. For the variance, we use the traditional shortcut:

$$\sigma^2_{X|Y=y} = E_{X|Y=y}[X^2] - \mu^2_{X|Y=y}.$$

The first term in the alleged expression for σ^2_X is then

$$E_Y[\sigma^2_{X|Y=y}] = \int_{-\infty}^{\infty} f_Y(y) \left(E_{X|Y=y}[X^2] - \mu^2_{X|Y=y} \right) dy.$$

We employ the same device as above to accommodate points where $f_{X|Y=y}(\cdot)$ is not defined. In this context, we compute

$$\int_{-\infty}^{\infty} f_Y(y) E_{X|Y=y}[X^2] \, dy = \int_{-\infty}^{\infty} \int_{-\infty}^{\infty} x^2 f_{X|Y=y}(x) f_Y(y) \, dx \, dy$$

$$= \int_{-\infty}^{\infty} \int_{-\infty}^{\infty} x^2 f_{XY}(x, y) \, dx \, dy = \int_{-\infty}^{\infty} x^2 f_X(x) \, dx$$

$$= E[X^2].$$

So,

$$E_Y[\sigma^2_{X|Y=y}] = E[X^2] - E_Y[\mu^2_{X|Y=y}]. \tag{5.5}$$

Using the equivalence already derived for the mean, $\mu_X = E_Y[\mu_{X|Y=y}]$, we have

$$E_Y[(\mu_{X|Y=y} - \mu_X)^2] = E_Y[\mu^2_{X|Y=y}] - 2\mu_X E_Y[\mu_{X|Y=y}] + \mu^2_X$$

$$= E_Y[\mu^2_{X|Y=y}] - \mu^2_X.$$

Combining with Equation 5.5, we have

$$E_Y[\sigma^2_{X|Y=y}] + E_Y[(\mu_{X|Y=y} - \mu_X)^2] = E[X^2] - \mu^2_X = \sigma^2_X.$$

The proof involving σ^2_Y is similar. ∎

5.52 Example: In the context of Example 5.50, verify the relationships asserted by Theorem 5.51.

We have the following joint, marginal, and conditional densities.

$$f_{XY}(x,y) = \begin{cases} e^{-x}, & 0 < y < x \\ 0, & \text{elsewhere} \end{cases}$$

$$f_X(x) = \begin{cases} xe^{-x}, & x \geq 0 \\ 0, & x < 0 \end{cases}$$

$$f_Y(y) = \begin{cases} e^{-y}, & y \geq 0 \\ 0, & y < 0 \end{cases}$$

$$f_{X|Y=y}(x) = \begin{cases} e^{-(x-y)}, & 0 < y < x \\ 0, & 0 \leq x \leq y \\ \text{undefined}, & \text{elsewhere} \end{cases}$$

$$f_{Y|X=x}(y) = \begin{cases} 1/x, & 0 < y < x \\ 0, & 0 < x \leq y \\ \text{undefined}, & \text{elsewhere} \end{cases}$$

The cited example calculated

$$\mu_{X|Y=y} = y + 1$$
$$\sigma^2_{X|Y=y} = 1,$$

and $E_Y[\mu_{X|Y=y}] = 2$. An earlier computation, from Example 5.47, provides $\mu_X = 2$ and $\sigma^2_X = 2$. The verification of the marginal mean of the conditional mean is therefore complete. For the variance, Theorem 5.51 delivers the anticipated result:

$$E_Y[\sigma^2_{X|Y=y}] + E_Y[(\mu_{X|Y=y} - \mu_X)^2] = E_Y[1] + E_Y[(y + 1 - 2)^2]$$
$$= 1 + \int_0^\infty (y-1)^2 e^{-y}\, dy = 2. \ \square$$

Exercises

5.22 Example 5.44 derived the density of the continuous random variables with cumulative distribution

$$F(x,y) = \begin{cases} 0, & x < 0 \text{ or } y < 0 \\ 1 - e^{-x} - xe^{-x}, & 0 \leq x < y \\ 1 - e^{-y} - ye^{-x}, & 0 \leq y \leq x. \end{cases}$$

The density is

$$f_{XY}(x,y) = \partial_{12}F(x,y) = \begin{cases} e^{-x}, & 0 < y < x \\ 0, & \text{elsewhere}. \end{cases}$$

Verify that $F(x,y) = \int_{-\infty}^x \int_{-\infty}^y f_{XY}(t,s)\, ds\, dt$.

5.23 The conditional densities of Example 5.50 are

$$f_{X|Y=y}(x) = \begin{cases} e^{-(x-y)}, & 0 < y < x \\ 0, & 0 \le x \le y \\ \text{undefined}, & \text{elsewhere} \end{cases}$$

$$f_{Y|X=x}(y) = \begin{cases} 1/x, & 0 < y < x \\ 0, & 0 < x \le y \\ \text{undefined}, & \text{elsewhere}. \end{cases}$$

For each of these densities, show that the total assigned probability is 1.

5.24 In the context of Example 5.50, compute $\mu_{Y|X=x}$ and $\sigma^2_{Y|X=x}$ directly from the conditional densities. Verify the relationships between conditional and marginal features, as asserted by Theorem 5.51. That is, verify that

$$\mu_Y = E_X[\mu_{Y|X=x}]$$
$$\sigma^2_Y = E_X[\sigma^2_{Y|X=x}] + E_X[(\mu_{Y|X=x} - \mu_Y)^2].$$

5.25 Show that two continuous random variables are independent if and only if all conditional distributions equal the appropriate marginal distribution. That is, show that independence is equivalent to

$$f_X(x) = f_{X|Y=y}(x),$$

for all y such that $f_Y(y) > 0$.

5.26 X and Y are joint random variables with density

$$f_{XY}(x, y) = \begin{cases} 6e^{-3x+y}, & 0 \le y \le x \\ 0, & \text{elsewhere}. \end{cases}$$

Find the marginal and conditional densities, means, and variances. Verify that

$$\mu_X = E_Y[\mu_{X|Y=y}]$$
$$\sigma^2_X = E_Y[\sigma^2_{X|Y=y}] + E_Y[(\mu_{X|Y=y} - \mu_X)^2].$$

Are X and Y independent?

5.27 X and Y are joint random variables with density

$$f_{XY}(x, y) = \begin{cases} 2, & 0 \le y \le x \le 1 \\ 0, & \text{elsewhere}. \end{cases}$$

Find the marginal and conditional densities, means, and variances. Verify that

$$\mu_X = E_Y[\mu_{X|Y=y}]$$
$$\sigma_X^2 = E_Y[\sigma_{X|Y=y}^2] + E_Y[(\mu_{X|Y=y} - \mu_X)^2].$$

Are X and Y independent?

5.28 X and Y are joint random variables with density

$$f_{XY}(x,y) = \begin{cases} 1/\pi, & x^2 + y^2 \leq 1 \\ 0, & \text{elsewhere.} \end{cases}$$

Show that X and Y are not independent, although $\text{cov}(X,Y) = 0$.

5.29 The joint density of X and Y is

$$f_{XY}(x,y) = \begin{cases} 6e^{-2x+y}, & 0 \leq y \leq x/2 \\ 0, & \text{elsewhere.} \end{cases}$$

Compute the conditional probability of the event $(X + Y < 4)$, given $(X < 3)$.

*5.30 Let X and Y be joint, continuous, independent random variables with densities $f_X(\cdot)$ and $f_Y(\cdot)$. Define $Z = X + Y$. Show that the density of Z is

$$f_Z(z) = \int_{-\infty}^{\infty} f_X(x) f_Y(z - x)\, dx.$$

5.31 The density of random variable X is $f_X(x) = e^{-|x|}/2$ for $-\infty < x < \infty$. Verify that the total assigned probability is 1, and find the mean and variance.

5.32 The density of random variable X is $f_X(x) = 2e^x[\pi(1 + e^{2x})]$ for $-\infty < x < \infty$. Verify that the total assigned probability is 1, and find the mean and variance.

5.4 Summary

This chapter constitutes an introduction to a branch of analysis called measure theory. We start with a discussion of the philosophical problems associated with extending probability concepts to uncountable sets, particularly to the real line. To maintain the countable additivity property, we conclude that we can actually assign probability only to subsets, and we commence a search for a collection of subsets sufficiently large to cover those that occur in typical applications.

If events (subsets) $E_1, E_2, \ldots$ are of interest, it is likely that the events $\cup E_i$ and $\cap E_i$ will also appear in the discussion. This reasoning leads to σ-algebras of subsets, which are collections that are closed under complementation and countable unions. The Borel sets are the smallest σ-algebra containing all intervals of the form $(-\infty, x]$, where x is an arbitrary real number. We can assign probability to the Borel sets in a consistent fashion.

The most straightforward assignment uses a cumulative distribution function, $F(x)$, which we interpret as the probability of the interval $(-\infty, x]$. This function must satisfy a number of constraints. It must be monotone nondecreasing and continuous from above at all points. It must be continuous except on a countable number of points. Of course, $0 \leq F(x) \leq 1$ for all x, $\lim_{x \to -\infty} F(x) = 0$, and $\lim_{x \to \infty} F(x) = 1$.

The cumulative distribution function assigns probability directly to intervals of the form $(-\infty, x]$. Through complements and intersections, we also obtain probabilities for the open intervals. The function then assigns probability to the Borel sets via

$$\Pr(A) = \inf\{s(\mathcal{G}) \mid \mathcal{G} = \{I_1, I_2, \ldots\} \text{ and } A \subset \cup_{n=1}^{\infty} I_n\},$$

where $s(\mathcal{G})$ is the sum of the probabilities assigned to the open intervals comprising $\mathcal{G}$. The measure of the Borel set A is then the lower bound of all competing sums of measures of open intervals, provided that a competitor must contain A within the union of its open intervals. This extension actually extends the measure to all subsets of the real line, but it preserves the countable additivity property only on a proper subcollection called the perfect slicers, a σ-algebra containing the Borel sets. The Borel sets then receive consistent probability assignments.

A general probability space is a triple $(\Omega, \mathcal{A}, \Pr(\cdot))$, where Ω is a point set, $\mathcal{A}$ is a σ-algebra over Ω, and $\Pr : \mathcal{A} \to [0, 1]$ is a probability assignment function. A real random variable is a function $X : \Omega \to \mathcal{R}$ such that $X^{-1}((-\infty, x]) \in \mathcal{A}$ for every real x. As in the discrete case, a real random variable induces a probability assignment on $\mathcal{R}$ via $\Pr(B) = \Pr_\Omega(X^{-1}(B))$ for every Borel set B. Subsequent analysis takes place in this induced space, where many computations involve expected values.

The expected value operator takes the form $E[g] = \int_{-\infty}^{\infty} g \, dF$, where F is the cumulative distribution function of X. This is a Riemann-Stieltjes integral, which is a limit of approximating sums very similar to the ordinary Riemann integral. In this text, all expected value operations reduce to a familiar Riemann integral. For this restricted subclass of distributions, the cumulative distribution function has the form

$$F(x) = \sum_{x_i \leq x} f_d(x_i) + \int_{-\infty}^{x} f_c(y) \, dy,$$

where $D = \{x_1, x_2, \ldots\}$ is a countable collection. The pair (f_d, f_c) are collectively called the density of the distribution. $f_d(\cdot)$ gives a discrete contribution, while $f_c(\cdot)$ gives a contribution that accumulates over intervals while assigning zero to individual points.

For these random variables, expected value computations use only Riemann integrals.

$$E[g] = \sum_{x \in D} g(x) f_d(x) + \int_{-\infty}^{\infty} g(x) f_c(x)\, dx,$$

provided that the sum and the integral are absolutely convergent. This is an extension of expectation operator defined earlier for discrete distributions. It remains linear: $E[c_1 g_1 + c_2 g_2] = c_1 E[g_1] + c_2 E[g_2]$. The mean and variance are then $\mu = E[X]$ and $\sigma^2 = E[(X - \mu)^2]$. The moment generating function is $E[e^{tX}]$.

In higher dimensions, the Borel sets comprise the smallest σ-algebra containing the half-open rectangles $(-\infty, \mathbf{x}]$. To define probability allocations on these sets, we again start with a function F that will eventually become the cumulative distribution function. It and its mixed difference $\Delta F(\mathbf{x}, \mathbf{y})$ must satisfy certain monotonicity and limit constraints. F then directly specifies probability allocations to open rectangles, and a competition among open covers extends the measure to all subsets of $\mathcal{R}^n$. The resulting probability assignment proves countably additive on n-dimensional Borel sets. A higher-dimensional Riemann-Stieltjes integral, in which the approximating sums use the ΔF measure of partition cells, provides an expected value operator.

For actual computations, we restrict our attention to those distributions for which the expected value calculations reduce to ordinary integrals. In practice, this amounts to a multicomponent density, corresponding to probability accumulations on subsets of lower dimension than the full $\mathcal{R}^n$. In $\mathcal{R}^2$, for example, a density has three components: point masses, line masses, and regional masses. The first corresponds to a summation over a countable collection of points, while the last is a two-dimensional Riemann integral. Line masses involve linear integrals along curves in the plane.

Continuous distributions have single component densities that accumulate nonzero probability only over positive volumes in $\mathcal{R}^n$. Lower-dimensional structures, such as boundaries and individual points, receive zero probability under these distributions. If the density is continuous, the corresponding cumulative distribution function is differentiable, and we refer to it as a differentiable distribution.

All distributions considered in the rest of the text have differentiable densities. In these cases, it is easy to derive the marginal distributions for a subset of the variables by integrating out the remaining variables. For the two-dimensional case, this is $f_{X_1}(x) = \int_{-\infty}^{\infty} f_{X_1, X_2}(x, y)\, dy$.

We can establish conditional densities by normalizing the joint density. In two dimensions, this is $f_{X_1 | X_2 = x_2}(x_1) = f_{X_1, X_2}(x_1, x_2) / f_{X_2}(x_2)$, provided that the marginal in the denominator is not zero. As in the discrete case, a conditional distribution defines a probability distribution in its own right, for which the mean and variance are related to their unconditional counterparts.

Historical Notes

Maistrov[52] credits the English mathematician Thomas Simpson with the first investigations of continuous distributions. Simpson is remembered primarily for Simpson's rule, a numerical integration technique. Nevertheless, he published, in 1740 and 1742, two texts on probability theory, in which he explains the role of the arithmetic mean in reducing observational errors. However, Simpson's limiting approach to the mean error was foreshadowed by Bernoulli's limit theorem for the binomial distribution. The concept of a continuous distribution was very much "in the air" in the eighteenth century.

Probability assignments to subsets of real numbers is a special case of a more general problem concerned with extending a measure, such as length, from intervals to more complicated sets. Boyer[8] credits Emile Borel with the idea of restricting the assignment to a σ-algebra that preserves countable additivity. As we have seen, this approach enables the description of a distribution with a cumulative distribution function. We restricted our discussion to those cumulative distribution functions that arise as the ordinary (Riemann) integral of a density. For this case, which covers most distributions of interest, expected value computations are also ordinary integrals. For the more general case, Henri Lebesgue developed, around 1900, a more general integral. Like the Riemann integral, it involves a limiting process but works with increasingly fine subdivisions of the function's range, rather than its domain.

Borel is perhaps most widely known for the Heine-Borel property of the real numbers: if the union of an infinite collection of open intervals contains a closed bounded interval $[a, b]$, then a finite subset suffices to cover $[a, b]$. According to Katz[43], Borel, in 1894, was the first to prove this property for countable collections, although Heine apparently used it in the 1870s. Later, Henri Lebesgue generalized the principle to arbitrary collections. Open sets have a particularly appealing form: they are always unions of intervals. Consequently, it is sometimes possible to derive results for arbitrary sets by covering them with a collection of intervals. This chapter's approach to general probability allocations is a case in point.

Further Reading

When outcomes are elements of $\mathcal{R}$, the set of real numbers, an event is an arbitrary set of such numbers. Ideally, a probability assignment consistently specifies the probability associated with any such event. Nevertheless, we have restricted our development to a special class of events—those constructed from intervals through a finite number of complements or countable unions. This may seem unnecessarily restrictive, but it is not possible to extend the general assignment $F(b^-) - F(a)$, which assigns probability to intervals, to the class of all subsets of $\mathcal{R}$ while preserving countable additivity Recognizing the ultimate impossibility of reaching the class of all subsets as valid events, it is nevertheless possible to develop a probability assignment that is more inclusive than the one considered here. This is the province of measure theory, for which Royden[74] and Halmos[32] are classic references. Indeed, the treatment here is an adaptation of Royden's measure-theory arguments to a probability context. What we have called the perfect slicers are Royden's measurable sets. Royden also provides an example of an unmeasurable set. An adaptation of that example forms the basis of Example 5.10, which demonstrates that the pseudo-probability measure derived from a cumulative distribution is not countably additive for arbitrary subset sequences.

Chapter 6

Continuous Distributions

Certain particularly useful continuous random variables occur repeatedly in computer and engineering sciences applications. This chapter develops a catalog of these distributions. It begins with the normal distribution, which is the limiting form of the sample sum over an arbitrary population. A mathematical interlude proves this remarkable limit and also some associated results of the same nature known as the laws of large numbers. We then consider continuous distributions needed to complete Figure 2.9, which elaborates the relationships among the discrete distributions of Chapter 2. For example, the sum of n independent exponential distributions is an Erlang distribution. We also study distributions that are related to the normal, such as the chi-square, the t, and the F distributions, which prove useful in the parameter estimation techniques of the next chapter. The concluding section discusses computer algorithms that generate samples from continuous distributions.

6.1 The normal distribution

This section has two parts. The first establishes the univariate and bivariate normal distributions. The basic properties of the distribution are apparent in the univariate case, where scalar parameters characterize the features. The bivariate extension illustrates the major differences that appear when matrix parameters replace the scalars. The second part deals with multivariate normal distributions, in which $n \times n$ matrices substitute for the 2×2 bivariate matrices. The extension involves no new concepts but entails greater facility with matrix algebra. The required techniques build on the introductory material of Section A.5. To enhance readability, we frequently write $\exp(t)$ instead of e^t when t involves complicated details.

6.1.1 The univariate and bivariate normal distributions

Our first definition establishes the normal random variable in one dimension. it involves two parameters, which we initially call a and b. While a can be any real number, we must have $b > 0$.

6.1 Definition: A continuous random variable X is called *univariate normal* if its density is

$$f_X(x) \; = \; \frac{1}{b\sqrt{2\pi}} \exp\left(\frac{-(x-a)^2}{2b^2}\right).$$

We often speak of a "normal" distribution without the "univariate" qualifier if the one-dimensional nature of the distribution is clear from context. ∎
    The integral $\int_{-\infty}^{\infty} e^{-x^2}\, dx$ appears in many calculations with a normal density. It is somewhat difficult to compute. However, once this result is known, it facilitates the evaluation of many integrals of a similar nature. The trick to computing $\int_{-\infty}^{\infty} e^{-x^2/2}\, dx$ is a two-dimensional approach, which conveniently allows polar coordinates. Specifically, the Cartesian integral gives

$$\int_{x=-\infty}^{\infty} \int_{y=-\infty}^{\infty} e^{-(x^2+y^2)}\, dy\, dx \; = \; \int_{x=-\infty}^{\infty} e^{-x^2} \left(\int_{y=-\infty}^{\infty} e^{-y^2}\, dy\right) dx$$

$$= \left(\int_{-\infty}^{\infty} e^{-x^2}\, dx\right)^2 ,$$

whereas the polar integral of the same function gives

$$\int_{\theta=0}^{2\pi} \int_{r=0}^{\infty} e^{-r^2} r\, dr\, d\theta \; = \; \int_{0}^{2\pi} \left[\frac{-e^{-r^2}}{2}\right]_{0}^{\infty} d\theta = \frac{1}{2}\int_{0}^{2\pi} d\theta = \pi.$$

Equating the two results, we see that $\int_{-\infty}^{\infty} e^{-x^2}\, dx = \sqrt{\pi}$. We can now verify that the normal density allocates total probability 1, and we can compute the mean and variance. For all these calculations, we transform the integral with the substitution $y = (x-a)/(b\sqrt{2})$, $dy = dx/(b\sqrt{2})$.

$$\int_{-\infty}^{\infty} f_X(x)\, dx \; = \; \frac{1}{b\sqrt{2\pi}} \int_{-\infty}^{\infty} \exp\left(\frac{-(x-a)^2}{2b^2}\right) dx$$

$$= \; \frac{1}{b\sqrt{2\pi}} \int_{-\infty}^{\infty} e^{-y^2} \left(b\sqrt{2}\right) dy = \frac{1}{\sqrt{\pi}} \cdot \sqrt{\pi} = 1$$

$$\mu_X \; = \; \int_{-\infty}^{\infty} x f_X(x)\, dx = \frac{1}{b\sqrt{2\pi}} \int_{-\infty}^{\infty} x \exp\left(\frac{-(x-a)^2}{2b^2}\right) dx$$

$$= \; \frac{1}{b\sqrt{2\pi}} \int_{-\infty}^{\infty} (a + b\sqrt{2}y)e^{-y^2} \left(b\sqrt{2}\right) dy$$

$$= \; \frac{1}{\sqrt{\pi}} \left[a \int_{-\infty}^{\infty} e^{-y^2}\, dy + b\sqrt{2}\int_{-\infty}^{\infty} ye^{-y^2}\, dy\right] = a$$

$$\sigma_X^2 = \int_{-\infty}^{\infty} (x - \mu_X)^2 f_X(x)\, dx$$

$$= \frac{1}{b\sqrt{2\pi}} \int_{-\infty}^{\infty} (x - a)^2 \exp\left(\frac{-(x - a)^2}{2b^2}\right) dx$$

$$= \frac{1}{b\sqrt{2\pi}} \int_{-\infty}^{\infty} (by\sqrt{2})^2 e^{-y^2} \left(b\sqrt{2}\right) dy = \frac{2b^2}{\sqrt{\pi}} \int_{-\infty}^{\infty} y^2 e^{-y^2}\, dy$$

$$= \frac{2b^2}{\sqrt{\pi}} \left[y \cdot \frac{-e^{-y^2}}{2} \Big|_{-\infty}^{\infty} + \int_{-\infty}^{\infty} \frac{e^{-y^2}}{2}\, dy \right] = \frac{2b^2}{\sqrt{\pi}} \cdot \frac{\sqrt{\pi}}{2} = b^2$$

Because the mean and variance are simple functions of a and b, it is traditional to specify them directly as parameters. That is, we say that X is a normal random variable with parameters (μ, σ^2). The normal distribution with $\mu = 0$ and $\sigma^2 = 1$ is particularly useful because problems involving arbitrary means and variances reduce to this standard case.

6.2 Definition: The *standard normal* random variable has a normal distribution with parameters $(\mu = 0, \sigma^2 = 1)$. Its density and cumulative distribution function are as follows.

$$f(x) = \frac{1}{\sqrt{2\pi}} e^{-x^2/2} \qquad \Phi(x) = \frac{1}{\sqrt{2\pi}} \int_{-\infty}^{x} e^{-t^2/2}\, dt \; \blacksquare$$

The symbol $\Phi(\cdot)$ is the common notation for the standard normal's cumulative distribution function. $\Phi(x)$ is the probability of the event $(Z \leq x)$, where Z is standard normal. The computation of $\Phi(x)$ is difficult. It involves a power series expansion of $e^{-x^2/2}$, followed by a term-by-term integration. To avoid repetitive calculations of this sort, we use tables, such as Table 2.1 or the longer version that appears in Appendix B.

The next example shows how to reduce a general problem to one involving the standard normal distribution, for which a solution is available from the table. You may want to review the discussion associated with Table 2.1, in particular the derivation which shows that $\Phi(-x) = 1 - \Phi(x)$. With this observation, we can infer the Φ-value for negative arguments from the positive entries in the table.

6.3 Example: From many long years of experience, the weather bureau knows that the temperature T on the 4th of July in Seattle is a normal random variable with $\mu = 70$ degrees (Fahrenheit) and $\sigma^2 = (30)^2$. What is the probability that the temperature will be below freezing in Seattle on the 4th of July?

We use the substitution $s = (t - 70)/30$, $dt = 30\, ds$ to transform the following integral.

$$\Pr(T < 32) = \Pr(T \leq 32) = \frac{1}{30\sqrt{2\pi}} \int_{-\infty}^{32} \exp\left(\frac{-(t - 70)^2}{2(30)^2}\right) dt$$

$$= \frac{1}{30\sqrt{2\pi}} \int_{-\infty}^{(32-70)/30} e^{-s^2/2} (30)\, ds$$

This expression simplifies to the standard normal form.

$$\Pr(T < 32) = \frac{1}{\sqrt{2\pi}} \int_{-\infty}^{-1.267} e^{-s^2/2} \, ds = \Phi(-1.267) = 1 - \Phi(1.267)$$

Linear interpolation from the table between $\Phi(1.26) = 0.8962$ and $\Phi(1.27) = 0.8980$ gives $\Phi(1.267) = 0.8975$. Consequently, $\Pr(T < 32) = 1 - 0.8975 = 0.1025$. $\square$

The example uses an integral transformation that uncovers the integral expression for $\Phi(x)$, which is then available from the table. Another approach transforms the random variable directly to a standard normal. The next theorem shows that a linear transformation of a normal random variable is also normal, with a mean and variance simply related to the corresponding features of the original distribution.

6.4 Theorem: Let X be a normal random variable with parameters (μ, σ^2). For constants $a \neq 0$ and b, define $Y = aX + b$. Then Y is a normal random variable with parameters $(a\mu + b, a^2\sigma^2)$. In particular, $Z = (X - \mu)/\sigma$ is standard normal.

PROOF: For $a > 0$, with the integral transformation $s = ax + b$, $ds = a \, dx$, we calculate

$$F_Y(t) = \Pr(Y \leq t) = \Pr(aX + b \leq t) = \Pr(X \leq (t - b)/a) = F_X((t - b)/a)$$

$$= \frac{1}{\sigma\sqrt{2\pi}} \int_{-\infty}^{(t-b)/a} \exp\left(\frac{-(x - \mu)^2}{2\sigma^2}\right) dx$$

$$= \frac{1}{\sigma\sqrt{2\pi}} \int_{-\infty}^{t} \exp\left(\frac{-[(s - b)/a - \mu]^2}{2\sigma^2}\right) (1/a) \, ds$$

$$= \frac{1}{(a\sigma)\sqrt{2\pi}} \int_{-\infty}^{t} \exp\left(\frac{-[(s - (a\mu + b)]^2}{2(a\sigma)^2}\right) ds$$

$$f_Y(t) = F_Y'(t) = \frac{1}{(a\sigma)\sqrt{2\pi}} \exp\left(\frac{-[(t - (a\mu + b)]^2}{2(a\sigma)^2}\right).$$

The last expression reveals that Y is normal with parameters $(a\mu + b, a^2\sigma^2)$. The calculation for $a < 0$ uses the same integral transformation and concludes that

$$f_Y(t) = F_Y'(t) = \frac{1}{(|a|\sigma)\sqrt{2\pi}} \exp\left(\frac{-[t - (a\mu + b)]^2}{2(|a|\sigma)^2}\right).$$

In either case, we have Y normal with parameters $(a\mu + b, a^2\sigma^2)$.

$Z = (X - \mu)/\sigma$ corresponds to $a = 1/\sigma$ and $b = -\mu/\sigma$. Consequently, Z is normal with parameters $(0, 1)$. That is, Z is standard normal. $\blacksquare$

6.5 Example: Answer the question in Example 6.3 by directly transforming T to a standard normal random variable.

We have T normal with parameters $(\mu = 70, \sigma^2 = (30)^2)$. Applying Theorem 6.4, we conclude that $Z = (T - 70)/30$ is standard normal. We then calculate

$$\Pr(T < 32) = \Pr(T \leq 32) = \Pr\left(\frac{T - 70}{30} \leq \frac{32 - 70}{30}\right) = \Pr(Z \leq -1.267)$$

$$= \Phi(-1.267) = 1 - \Phi(1.267).$$

The computation then proceeds as in the example cited to obtain $\Pr(T \leq 32) = 0.1025.$ $\square$

As the example suggests, we can relate an outcome interval from a normal distribution to an equivalent interval from the standard normal random variable Z. That is, if X is normal with parameters (μ, σ^2), then

$$\Pr(a < X < b) = \Pr\left(\frac{a - \mu}{\sigma} < \frac{X - \mu}{\sigma} < \frac{b - \mu}{\sigma}\right)$$

$$= \Pr\left(\frac{a - \mu}{\sigma} < Z < \frac{b - \mu}{\sigma}\right).$$

At our convenience, we can replace the "less than" conditions with "less than or equal to" because individual point outcomes have probability zero. In terms of $\Phi(\cdot)$, the probability that X differs from its mean by less than some multiple k of its standard deviation is

$$\Pr(|X - \mu| \leq k\sigma) = \Pr\left(-k \leq \frac{X - \mu}{\sigma} \leq k\right) = \Phi(k) - \Phi(-k) = 2\Phi(k) - 1.$$

Accordingly, the probability that an outcome is within one standard deviation of the mean is $2\Phi(1.0) - 1 = 0.6826$. For two standard deviations, the probability rises to $2\Phi(2.0) - 1 = 0.9544$. Figure 6.1 provides a graph of the standard normal density and shows the spans encompassed by $\pm\sigma$ and $\pm 2\sigma$.

We now derive the moment generating function for a normal random variable X with parameters (μ, σ^2). The defining integral involves an exponential argument, in which we complete the square. We then transform the integral with the substitution $y = [x - (\mu + \sigma^2 t)]/(\sigma\sqrt{2})$, $dy = dx/(\sigma\sqrt{2})$.

$$\Psi_X(t) = E_X[e^{tX}] = \frac{1}{\sigma\sqrt{2\pi}} \int_{-\infty}^{\infty} e^{tx} \exp\left(\frac{-(x - \mu)^2}{2\sigma^2}\right) dx$$

$$= \frac{1}{\sigma\sqrt{2\pi}} \int_{-\infty}^{\infty} \exp\left(\frac{-[x^2 - 2(\mu + \sigma^2 t)x + \mu^2]}{2\sigma^2}\right) dx$$

$$= \frac{1}{\sigma\sqrt{2\pi}} \int_{-\infty}^{\infty} \exp\left(\frac{-[x - (\mu + \sigma^2 t)]^2 + (\mu + \sigma^2 t)^2 - \mu^2}{2\sigma^2}\right) dx$$

$$= \frac{1}{\sigma\sqrt{2\pi}} \exp\left(\frac{2\mu\sigma^2 t + \sigma^4 t^2}{2\sigma^2}\right) \int_{-\infty}^{\infty} \exp\left(\frac{-[x - (\mu + \sigma^2 t)]^2}{2\sigma^2}\right) dx$$

$$= \frac{1}{\sigma\sqrt{2\pi}} \exp\left(\mu t + \frac{\sigma^2 t^2}{2}\right) \int_{-\infty}^{\infty} e^{-y^2} (\sigma\sqrt{2}) \, dy = \exp\left(\mu t + \frac{\sigma^2 t^2}{2}\right)$$

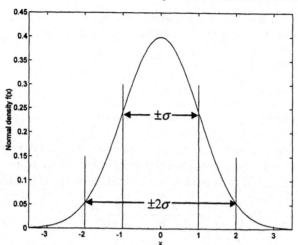

Figure 6.1. The standard normal density

For a standard normal random variable Z, this expression reduces to $\Psi_Z(t) = e^{t^2/2}$, which we can use to derive higher-order moments.

$$\Psi_Z(t) = E[e^{tZ}] = \int_{-\infty}^{\infty} e^{tz} f_Z(z)\, dz = \int_{-\infty}^{\infty} \left(\sum_{n=0}^{\infty} \frac{(tz)^n}{n!} \right) f_Z(z)\, dz$$

Exploiting the interchange of summation and integration, we obtain

$$\Psi_Z(t) = \sum_{n=0}^{\infty} \left(\int_{-\infty}^{\infty} z^n f_Z(z)\, dz \right) \frac{t^n}{n!} = \sum_{n=0}^{\infty} \frac{E[Z^n] t^n}{n!}.$$

Consequently, the nth moment of Z is the coefficient of $t^n/n!$ in the power series expansion of $\Psi_Z(t)$. The latter is

$$\Psi_Z(t) = e^{t^2/2} = 1 + \frac{t^2/2}{1!} + \frac{t^4/4}{2!} + \frac{t^6/8}{3!} + \ldots + \frac{t^{2n}/2^n}{n!} + \ldots$$

$$= 1 + \frac{t^2}{2!} \cdot \frac{2!}{2 \cdot 1!} + \frac{t^4}{4!} \cdot \frac{4!}{2^2 \cdot 2!} + \frac{t^6}{6!} \cdot \frac{6!}{2^3 \cdot 3!}$$

$$+ \ldots + \frac{t^{2n}}{(2n)!} \cdot \frac{(2n)!}{2^n \cdot n!} + \ldots$$

$$= 1 + \frac{t^2}{2!} + \frac{t^4}{4!}(3) + \frac{t^6}{6!}(5)(3) + \frac{t^8}{8!}(7)(5)(3)$$

$$+ \ldots + \frac{t^{2n}}{(2n)!}(2n-1)(2n-3)(2n-5) \cdots (3)(1) + \ldots.$$

From the last reduction, we can read off the moments. All odd moments are zero, and an even moment $E[Z^{2n}]$ is the product of the odd integers from $(2n-1)$ down through (1). The next theorem summarizes these features.

6.6 Theorem: Let X be normal with parameters (μ, σ^2). Let Z be standard normal. Then

$$\Psi_X(t) = \exp\left(\mu t + \frac{\sigma^2 t^2}{2}\right)$$

$$\Psi_Z(t) = e^{t^2/2}$$

$$E[Z^n] = \begin{cases} (n-1)(n-3)\cdots(3)(1), & n = 2,4,6,\ldots \\ 0, & n = 1,3,5,\ldots. \end{cases}$$

PROOF: See discussion above. ∎

The two-dimensional normal distribution is a function of five real parameters, traditionally arranged as a two-component vector and a four-component matrix. The formal definition is as follows.

6.7 Definition: Given the parameters $\mathbf{a} = \begin{bmatrix} a_1 \\ a_2 \end{bmatrix}$ and $\mathbf{A} = \begin{bmatrix} b_1 & c \\ c & b_2 \end{bmatrix}$ with $b_1 > 0$, $b_2 > 0$, and $\det(\mathbf{A}) = b_1 b_2 - c^2 > 0$, we say that continuous random variables $\mathbf{X} = (X_1, X_2)$ are *jointly normal*, or *bivariate normal*, if their joint density is of the form

$$f_{\mathbf{X}}(\mathbf{x}) = \frac{1}{2\pi\sqrt{\det(\mathbf{A})}} \exp\left(\frac{-(\mathbf{x}-\mathbf{a})'\mathbf{A}^{-1}(\mathbf{x}-\mathbf{a})}{2}\right). ∎$$

As in the univariate case, the parameters are closely related to the means, variances, and covariance, and we next derive the appropriate relationships. The term $Q(\mathbf{x}) = (\mathbf{x}-\mathbf{a})'\mathbf{A}^{-1}(\mathbf{x}-\mathbf{a})$ is called a *quadratic form*. Theorem A.68 gives an expression for the inverse of a matrix with nonzero determinant. We invert $\mathbf{A}$ and then expand the quadratic form.

$$\mathbf{A}^{-1} = \begin{bmatrix} b_1 & c \\ c & b_2 \end{bmatrix}^{-1} = \frac{1}{\det(\mathbf{A})} \begin{bmatrix} b_2 & -c \\ -c & b_1 \end{bmatrix}$$

$$Q(\mathbf{x}) = \frac{\begin{bmatrix} x_1 - a_1 & x_2 - a_2 \end{bmatrix} \begin{bmatrix} b_2 & -c \\ -c & b_1 \end{bmatrix} \begin{bmatrix} x_1 - a_1 \\ x_2 - a_2 \end{bmatrix}}{\det(\mathbf{A})}$$

$$= \frac{b_2(x_1 - a_1)^2 - 2c(x_1 - a_1)(x_2 - a_2) + b_1(x_2 - a_2)^2}{\det(\mathbf{A})}$$

To facilitate computing the marginal density $f_{X_1}(\cdot)$, we further transform the quadratic form into a perfect square in $(x_2 - a_2)$ plus residual terms in $(x_1 - a_1)$.

$$Q = (\mathbf{x}-\mathbf{a})'\mathbf{A}^{-1}(\mathbf{x}-\mathbf{a})$$

$$= \frac{b_1}{\det(\mathbf{A})}\left[(x_2 - a_2)^2 - \frac{2c(x_1 - a_1)(x_2 - a_2)}{b_1} + \frac{c^2(x_1 - a_1)^2}{b_1^2}\right.$$

$$\left. + \frac{-c^2(x_1 - a_1)^2}{b_1^2} + \frac{b_2(x_1 - a_1)^2}{b_1}\right]$$

We manipulate this expression into a sum of squares as follows.

$$Q = \frac{b_1}{\det(\mathbf{A})} \left\{ \left[(x_2 - a_2) - \frac{c}{b_1}(x_1 - a_1) \right]^2 + (x_1 - a_1)^2 \left[\frac{b_2}{b_1} - \frac{c^2}{b_1^2} \right] \right\}$$

$$= \frac{b_1 b_2 - c^2}{b_1 \det(\mathbf{A})}(x_1 - a_1)^2 + \frac{b_1}{\det(\mathbf{A})} \left[(x_2 - a_2) - \frac{c}{b_1}(x_1 - a_1) \right]^2$$

$$= \frac{1}{b_1}(x_1 - a_1)^2 + \frac{b_1}{\det(\mathbf{A})} \left[(x_2 - a_2) - \frac{c}{b_1}(x_1 - a_1) \right]^2$$

We integrate the joint density to obtain the marginal $f_{X_1}(\cdot)$:

$$f_{X_1}(x_1) = \int_{-\infty}^{\infty} f_{\mathbf{X}}(x_1, x_2) \, dx_2$$

$$= \frac{1}{2\pi\sqrt{\det(\mathbf{A})}} \int_{-\infty}^{\infty} \exp\left(\frac{-(\mathbf{x} - \mathbf{a})'\mathbf{A}^{-1}(\mathbf{x} - \mathbf{a})}{2} \right) \, dx_2.$$

We now replace the quadratic form in the exponential with the expanded expression above and then employ the now familiar integral transformation to obtain a single square in the exponential argument: $y = [(x_2 - a_2) - (c/b_1)(x_1 - a_1)]\sqrt{b_1/(2\det(\mathbf{A}))}$.

$$f_{X_1}(x_1) = \frac{1}{2\pi\sqrt{\det(\mathbf{A})}} \exp\left(\frac{-(x_1 - a_1)^2}{2b_1} \right)$$

$$\cdot \int_{-\infty}^{\infty} \exp\left(\frac{-b_1[(x_2 - a_2) - (c/b)(x_1 - a_1)]^2}{2\det(\mathbf{A})} \right) \, dx_2$$

$$= \frac{1}{2\pi\sqrt{\det(\mathbf{A})}} \exp\left(\frac{-(x_1 - a_1)^2}{2b_1} \right) \int_{-\infty}^{\infty} e^{-y^2} \sqrt{\frac{2\det(\mathbf{A})}{b_1}} \, dy$$

$$= \frac{1}{\sqrt{2\pi b_1}} \exp\left(\frac{-(x_1 - a_1)^2}{2b_1} \right)$$

This last expression is, of course, the density of a univariate normal random variable with parameters $(\mu = a_1, \sigma^2 = b_1)$. Because we know that the $f_{X_1}(x_1)$ expression integrates to 1, we can quickly verify that the joint density also integrates to 1:

$$\int_{-\infty}^{\infty} \int_{-\infty}^{\infty} f_{\mathbf{X}}(x_1, x_2) \, dx_2 \, dx_1 = \int_{-\infty}^{\infty} f_{X_1}(x_1) \, dx_1 = 1.$$

Similar calculations show that X_2 is normal with parameters $(\mu = a_2, \sigma^2 = b_2)$. Consequently, we can immediately express their means and variances: $\mu_{X_1} = a_1$, $\mu_{X_2} = a_2$, $\sigma_{X_1}^2 = b_1$, and $\sigma_{X_2}^2 = b_2$. There remains the covariance, which computations similar to those illustrated above show to be c. As with the one-dimensional normal distribution, it is traditional to specify the means,

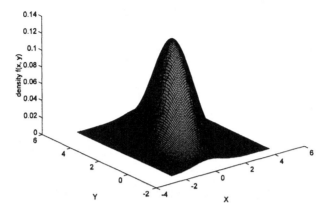

Figure 6.2. Bivariate normal density: $\mu_1 = 1, \mu_2 = 2$; $\sigma_{11} = 1, \sigma_{22} = 2, \sigma_{12} = 0.5$

variances, and covariance directly as parameters. That is, the two-dimensional normal distribution has density

$$f_{\mathbf{X}}(\mathbf{x}) = \frac{1}{2\pi\sqrt{\det(\Sigma)}} \exp\left(\frac{-(\mathbf{x} - \boldsymbol{\mu})'\Sigma^{-1}(\mathbf{x} - \boldsymbol{\mu})}{2}\right),$$

where the *mean vector* $\boldsymbol{\mu}$ and *covariance matrix* Σ are $\boldsymbol{\mu} = \begin{bmatrix} \mu_1 \\ \mu_2 \end{bmatrix}$ and $\Sigma = \begin{bmatrix} \sigma_{11} & \sigma_{12} \\ \sigma_{12} & \sigma_{22} \end{bmatrix}$. The covariance matrix must be invertible. Note that we have adopted a simpler subscript, writing μ_i for μ_{X_i}, σ_{ii} for $\sigma_{X_i}^2$, and σ_{ij} for $\mathrm{cov}(X_1, X_2)$. Figure 6.2 is a density plot of bivariate normal X, Y with parameters $\boldsymbol{\mu} = \begin{bmatrix} 1 \\ 2 \end{bmatrix}$ and $\Sigma = \begin{bmatrix} 1.0 & 0.5 \\ 0.5 & 2.0 \end{bmatrix}$.

Figure 6.3 is a contour plot of the same density. Lines of constant density are concentric ellipses. The variances determine the eccentricity of the ellipses (i.e., their deviations from circles), while the covariance determines the extent of rotation from an uncorrelated position where the ellipse axes are parallel to the coordinate axes. To see how the parameters determine the orientation of the contour plot, we rewrite the density in terms of the correlation coefficient $\rho = \sigma_{12}/\sqrt{\sigma_{11}\sigma_{22}}$.

$$\Sigma = \begin{bmatrix} \sigma_{11} & \rho\sqrt{\sigma_{11}\sigma_{22}} \\ \rho\sqrt{\sigma_{11}\sigma_{22}} & \sigma_{22} \end{bmatrix}$$

$$\Sigma^{-1} = \frac{1}{\sigma_{11}\sigma_{22}(1 - \rho^2)} \begin{bmatrix} \sigma_{22} & -\rho\sqrt{\sigma_{11}\sigma_{22}} \\ -\rho\sqrt{\sigma_{11}\sigma_{22}} & \sigma_{11} \end{bmatrix}$$

Some algebraic reduction then gives the following expression for the joint

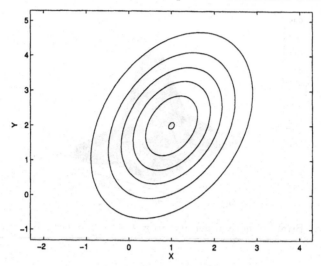

Figure 6.3. Contour plot of bivariate normal density: $\mu_1 = 1, \mu_2 = 2$;
$\sigma_{11} = 1, \sigma_{22} = 2, \sigma_{12} = 0.5$

density:

$$f_{12}(\mathbf{x}) = \frac{1}{2\pi\sqrt{\sigma_{11}\sigma_{22}(1-\rho^2)}} \exp\left(\frac{-Q(\mathbf{x})}{2(1-\rho^2)}\right), \tag{6.1}$$

where

$$Q(\mathbf{x}) = \left[\frac{(x_1-\mu_1)^2}{\sigma_{11}} - \frac{2\rho(x_1-\mu_1)(x_2-\mu_2)}{\sqrt{\sigma_{11}\sigma_{22}}} + \frac{(x_2-\mu_2)^2}{\sigma_{22}}\right].$$

Figure 6.4 shows, from top to bottom and left to right, the contour plots of bivariate normal densities with correlations ranging from $\rho = -0.8$ to $\rho = 0.8$. All have the same means and variances: $\boldsymbol{\mu} = \begin{bmatrix} 0 \\ 0 \end{bmatrix}$ and $\boldsymbol{\Sigma} = \begin{bmatrix} 1.0 & \rho\sqrt{2} \\ \rho\sqrt{2} & 2.0 \end{bmatrix}$.

The center plot, for which $\rho = 0$, clearly exhibits a vertical dispersion that is twice the horizontal value. This reflects $\sigma_{22} = 2\sigma_{11}$. The rotation apparent in the other plots derives from an interplay between the nonzero correlation and the variances. From Equation 6.1, we see that points of constant density lie on ellipses of the form

$$\frac{(x_1-\mu_1)^2}{\sigma_{11}} - \frac{2\rho(x_1-\mu_1)(x_2-\mu_2)}{\sqrt{\sigma_{11}\sigma_{22}}} + \frac{(x_2-\mu_2)^2}{\sigma_{22}} = K.$$

We transform to a new coordinate system, centered at (μ_1, μ_2) and rotated counterclockwise through angle θ. Some geometric reasoning (Figure 6.5) reveals that the new (y_1, y_2) coordinates relate to the old (x_1, x_2) values through

$$\begin{aligned} y_1 &= (x_1-\mu_1)\cos\theta + (x_2-\mu_2)\sin\theta & (x_1-\mu_1) &= y_1\cos\theta - y_2\sin\theta \\ y_2 &= -(x_1-\mu_1)\sin\theta + (x_2-\mu_2)\cos\theta & (x_2-\mu_2) &= y_1\sin\theta + y_2\cos\theta. \end{aligned}$$

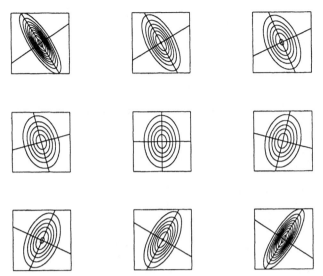

Figure 6.4. Bivariate normal densities with $\rho = -0.8, -0.6, \ldots, 0.6, 0.8$

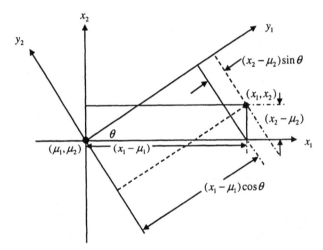

Figure 6.5. Coordinate relationships between rotated systems

Writing the ellipse equation in (y_1, y_2) terms, we find that

$$K = \frac{(y_1 \cos\theta - y_2 \sin\theta)^2}{\sigma_{11}} - \frac{2\rho(y_1 \cos\theta - y_2 \sin\theta)(y_1 \sin\theta + y_2 \cos\theta)}{\sqrt{\sigma_{11}\sigma_{22}}}$$
$$+ \frac{(y_1 \sin\theta + y_2 \cos\theta)^2}{\sigma_{22}}.$$

When we isolate the cross-product term $y_1 y_2$, we find that it has the following coefficient, which we equate to zero to obtain a rotation that eliminates such cross-product terms from the ellipse.

$$\frac{-2\sin\theta\cos\theta}{\sigma_{11}} + \frac{-2\rho(\cos^2\theta - \sin^2\theta)}{\sqrt{\sigma_{11}\sigma_{22}}} + \frac{2\sin\theta\cos\theta}{\sigma_{22}} = 0$$

Solving for θ, we obtain $\theta = (1/2)\tan^{-1}[2\rho\sqrt{\sigma_{11}\sigma_{22}}/(\sigma_{11} - \sigma_{22})]$. For this rotational value, the ellipse equation contains only terms in y_1^2 and y_2^2 and can be rearranged in the form $y_1^2/a^2 + y_2^2/b^2 = 1$. This is the standard form of an ellipse, in the rotated coordinated system, with semimajor and semiminor axes lengths a and b. The contour plots in Figure 6.4 include a line with slope $\tan\theta$ and its perpendicular. As expected, these lines coincide with the major and minor axes of the ellipses. Note that $\theta = 0$ when $\rho = 0$.

We extend the simpler notation for means and variances to density functions. In the development above, we have already used $f_{12}(\cdot,\cdot)$ for $f_{X_1,X_2}(\cdot,\cdot)$, but we now include such notations as $f_1(\cdot) = f_{X_1}(\cdot)$ and $f_2(\cdot) = f_{X_2}(\cdot)$ for the marginal densities of the first and second components. We also use $f_{1|2}(\cdot)$ for the conditional density of X_1, given a particular X_2 value. That is, $f_{1|2}(x_1) = f_{X_1|X_2=x_2}(x_1)$. Moreover, we write $\mu_{1|2}$ for the conditional mean of X_1, given $X_2 = x_2$. That is, $\mu_{1|2} = \mu_{X_1|X_2=x_2}$. With these conditional notations, the argument x_2 does not appear explicitly, but x_2 appears in any expressions derived for these quantities. Continuing this practice, we write $\sigma_{11|2}$ for the variance of X_1, given $X_2 = x_2$.

At this point, we know that the marginal distributions of a bivariate normal pair are themselves normal. It so happens that the conditional distributions are as well. The next theorem summarizes these observations.

6.8 Theorem: Let $\mathbf{X} = (X_1, X_2)$ be bivariate normal with parameters $\mu = \begin{bmatrix} \mu_1 \\ \mu_2 \end{bmatrix}$ and $\Sigma = \begin{bmatrix} \sigma_{11} & \sigma_{12} \\ \sigma_{12} & \sigma_{22} \end{bmatrix}$. Then X_1 is normal with parameters (μ_1, σ_{11}), and X_2 is normal with parameters (μ_2, σ_{22}). Given $X_2 = x_2$, the conditional density of X_1 is normal with parameters

$$\mu_{1|2} = \mu_1 + \frac{\sigma_{12}(x_2 - \mu_2)}{\sigma_{22}} \qquad\qquad \sigma_{11|2} = \sigma_{11} - \frac{\sigma_{12}^2}{\sigma_{22}}.$$

A symmetric statement holds for the conditional density of X_2 given $X_1 = x_1$.
PROOF: The discussion above establishes the marginal densities. Now consider the conditional density of X_1 given $X_2 = x_2$. We have

$$f_{1|2}(x_1) = \frac{f_{12}(x_1, x_2)}{f_2(x_2)} = K \cdot R(x_1, x_2),$$

where $K = \left(2\pi\sqrt{\det(\Sigma)}\right)^{-1} / \left(\sqrt{2\pi\sigma_{22}}\right)^{-1}$ and $R(x_1, x_2)$ is the ratio of two exponentials:

$$R(x_1, x_2) = \frac{\exp\left(\dfrac{-\sigma_{22}(x_1 - \mu_1)^2 + 2\sigma_{12}(x_1 - \mu_1)(x_2 - \mu_2) - \sigma_{11}(x_2 - \mu_2)^2}{2\det(\Sigma)}\right)}{\exp\left(\dfrac{-(x_2 - \mu_2)^2}{2\sigma_{22}}\right)}.$$

This expression reduces to

$$f_{1|2}(x_1) = \frac{e^t}{\sqrt{2\pi[\det(\Sigma)]/\sigma_{22}}},$$

where

$$
\begin{aligned}
t &= \frac{-\sigma_{22}(x_1 - \mu_1)^2}{2\det(\Sigma)} + \frac{\sigma_{12}(x_1 - \mu_1)(x_2 - \mu_2)}{\det(\Sigma)} \\
&\quad + \left[\frac{1}{2\sigma_{22}} - \frac{\sigma_{11}}{2\det(\Sigma)}\right](x_2 - \mu_2)^2 \\
&= \frac{-\sigma_{22}(x_1 - \mu_1)^2}{2\det(\Sigma)} + \frac{\sigma_{12}(x_1 - \mu_1)(x_2 - \mu_2)}{\det(\Sigma)} + \frac{-\sigma_{12}^2(x_2 - \mu_2)^2}{2\sigma_{22}\det(\Sigma)} \\
&= \frac{-\sigma_{22}}{2\det(\Sigma)}\left[(x_1 - \mu_1)^2 - \frac{2\sigma_{12}}{\sigma_{22}}(x_2 - \mu_2)(x_1 - \mu_1) + \frac{\sigma_{12}^2}{\sigma_{22}^2}(x_2 - \mu_2)^2\right] \\
&= \frac{-\sigma_{22}}{2\det(\Sigma)}\left[(x_1 - \mu_1) - \frac{\sigma_{12}}{\sigma_{22}}(x_2 - \mu_2)\right]^2.
\end{aligned}
$$

Consequently,

$$f_{1|2}(x_1) = \frac{1}{\sqrt{2\pi[\det(\Sigma)]/\sigma_{22}}}\exp\left(-\frac{\left[x_1 - \left(\mu_1 + \frac{\sigma_{12}(x_2 - \mu_2)}{\sigma_{22}}\right)\right]^2}{2[\det(\Sigma)]/\sigma_{22}}\right).$$

Since $\det(\Sigma)/\sigma_{22} = (\sigma_{11}\sigma_{22} - \sigma_{12}^2)/\sigma_{22} = \sigma_{11} - \sigma_{12}^2/\sigma_{22}$, we have the expression desired. The density is normal with parameters noted in the theorem. Similar computations produce the desired expressions for $f_{2|1}(\cdot)$ and the conditional quantities $\mu_{2|1}$ and $\sigma_{22|1}$. ∎

This theorem offers several interesting observations. First, the marginal distributions are not influenced by the covariance term σ_{12}. Second, the conditional mean is a linear function of the conditioning variable. That is, $\mu_{1|2}$ is a linear function of x_2. Third, the conditional variance is not a function of the conditioning variable, and it is *less* than the unconditional variance. Finally, if $\sigma_{12} = 0$, we can show that $f_{12} = f_1 \cdot f_2$, which means that uncorrelated normal random variables must be independent. While independent random variables must be uncorrelated, the converse is not true in general, and we have seen several examples of dependent, uncorrelated random variables. For normal random variables, however, independent and uncorrelated are equivalent. The next theorem provides the details.

6.9 Theorem: Let (X_1, X_2) be bivariate normal random variables. Then the following are equivalent.

(a) X_1 and X_2 are independent.

(b) X_1 and X_2 are uncorrelated.

(c) $f_{1|2} = f_1$ and $f_{2|1} = f_2$ whenever the conditional densities are defined.

PROOF: We have that (a) implies (b) from our earlier work; it is true regardless of whether the distributions are normal or not. To show that (b) implies (c), suppose that X_1, X_2 are uncorrelated. Consequently, $\sigma_{12} = 0$, and the last theorem then asserts that $\mu_{1|2} = \mu_1$ and $\sigma_{11|2} = \sigma_{11}$. This means that $f_{1|2}(\cdot)$ is a normal density with parameters (μ_1, σ_{11}), which is precisely the marginal density $f_1(\cdot)$. A similar argument shows that $f_{2|1} = f_2$. Finally, to show that (c) implies (a), we note that $f_{1|2} = f_1$ implies that $f_{12}(x_1, x_2)/f_2(x_2) = f_1(x_1)$, whenever $f_2(x_2) > 0$. For all such points, therefore, we have $f_{12} = f_1 \cdot f_2$. For a point x_2 where $f_2(x_2) = 0$, we have

$$f_2(x_2) = \int_{-\infty}^{\infty} f_{12}(x_1, x_2)\, dx_1 = 0.$$

Since f_{12} is nonnegative, we conclude that $f_{12}(x_1, x_2) = 0$ for all x_1. [Since f_{12} is continuous, if it were greater than zero at a point, it would be greater than zero on a neighborhood of the point, which would force the integral $\int_{-\infty}^{\infty} f_{12}(x_1)\, dx_1 > 0$.] Consequently, $f_{12} = f_1 \cdot f_2$ everywhere, and the random variables are independent. ∎

We have simplified many two-dimensional integral expressions in this section by first transforming and evaluating the inner integral and then operating on the outer one. Recall from multivariate calculus that we can transform both integrals simultaneously. That is, suppose that $\mathbf{f} : D \to D'$, where D is a region in the (x_1, x_2) plane and D' is in the (y_1, y_2) plane. Suppose further that $\mathbf{f}(\cdot)$ is bijective with inverse $\mathbf{g} : D' \to D$. Each of $\mathbf{f}$ and $\mathbf{g}$ must contain two components. For $\mathbf{f}$, one component specifies the y_1 component of $\mathbf{f}(x_1, x_2)$, and the other specifies the y_2 component. So $\mathbf{f} = (f_1, f_2)$ with $y_i = f_i(x_1, x_2)$. Similarly, $\mathbf{g} = (g_1, g_2)$ with $x_i = g_i(y_1, y_2)$. Then

$$\int\int_D h(x_1, x_2)\, dx_1\, dx_2 = \int\int_{D'} h(\mathbf{g}(y_1, y_2)) \cdot \frac{1}{J(y_1, y_2)}\, dy_1\, dy_2,$$

where $J(y_1, y_2)$ is the Jacobian of the transformation given by

$$J(\mathbf{y}) = \left| \det \begin{bmatrix} \partial_1 f_1(\mathbf{g}(\mathbf{y})) & \partial_2 f_1(\mathbf{g}(\mathbf{y})) \\ \partial_1 f_2(\mathbf{g}(\mathbf{y})) & \partial_2 f_2(\mathbf{g}(\mathbf{y})) \end{bmatrix} \right|.$$

6.10 Example: Use a transform to integrate $h(x, y) = xy$ over the annular section D of Figure 6.6.

Consider the transform $\mathbf{f} = (f_1, f_2)$ given by
$$u = f_1(x, y) = (x^2 + y^2)^{1/2}$$
$$v = f_2(x, y) = \tan^{-1}(y/x),$$
which maps D into a region in the (u, v) plane. To determine the shape of the target region, we proceed counterclockwise around D and solve for the corner points. Starting with the corner labeled $(1, \sqrt{3})$, these corners map as tabulated below. In the (u, v) plane, these mappings trace out a rectangular area $D' = \{(u, v) \mid 1 \le u \le 2, \pi/6 \le v \le \pi/3\}$.

(x, y)	$(1, \sqrt{3})$	$(\sqrt{3}, 1)$	$(\sqrt{3}/2, 1/2)$	$(1/2, \sqrt{3}/2)$
(u, v)	$(2, \pi/3)$	$(2, \pi/6)$	$(1, \pi/6)$	$(1, \pi/3)$

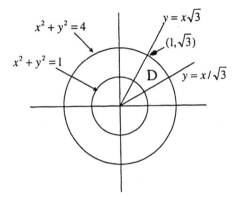

Figure 6.6. Transforming an integration domain (see Example 6.10)

The Jacobian, in terms of (x, y), is

$$J(x, y) = \left| \det \begin{bmatrix} \partial_1 f_1(x, y) & \partial_2 f_1(x, y) \\ \partial_1 f_2(x, y) & \partial_2 f_2(x, y) \end{bmatrix} \right| = \left| \det \begin{bmatrix} \dfrac{x}{\sqrt{x^2 + y^2}} & \dfrac{y}{\sqrt{x^2 + y^2}} \\ \dfrac{-y}{x^2 + y^2} & \dfrac{x}{x^2 + y^2} \end{bmatrix} \right|.$$

The inverse mapping is $\mathbf{g} = (g_1, g_2)$, given by

$$x = g_1(u, v) = u \cos v$$
$$y = g_2(u, v) = u \sin v.$$

In terms of (u, v), we then have

$$J(u, v) = \left| \det \begin{bmatrix} \dfrac{u \cos v}{\sqrt{(u \cos v)^2 + (u \sin v)^2}} & \dfrac{u \sin v}{\sqrt{(u \cos v)^2 + (u \sin v)^2}} \\ \dfrac{-u \sin v}{(u \cos v)^2 + (u \sin v)^2} & \dfrac{u \cos v}{(u \cos v)^2 + (u \sin v)^2} \end{bmatrix} \right|$$

$$= \left| \det \begin{bmatrix} \cos v & \sin v \\ -\dfrac{\sin v}{u} & \dfrac{\cos v}{u} \end{bmatrix} \right| = \frac{\cos^2 v + \sin^2 v}{u} = \frac{1}{u}.$$

Consequently,

$$\iint_D xy \cdot dx\, dy = \iint_{D'} \frac{(u \cos v)(u \sin v)}{1/u} \, du\, dv = \int_{v=\pi/6}^{\pi/3} \int_{u=1}^{2} u^3 \sin v \cos v \, du\, dv$$

$$= \int_{v=\pi/6}^{\pi/3} \left[\frac{u^4}{4} \right]_1^2 \sin v \cos v \, dv = \int_{v=\pi/6}^{\pi/3} \left(\frac{2^4 - 1^4}{4} \right) \sin v \cos v \, dv.$$

Using the identity $\sin 2v = 2 \sin v \cos v$, we have

$$\iint_D xy \cdot dx\, dy = \frac{15}{8} \int_{\pi/6}^{\pi/3} 2 \sin v \cos v \, dv = \frac{15}{8} \int_{\pi/6}^{\pi/3} \sin 2v \, dv$$

$$= \frac{15}{8} \left[\frac{-\cos 2v}{2} \right]_{\pi/6}^{\pi/3} = \frac{15}{16} (\cos \pi/3 - \cos 2\pi/3) = \frac{15}{16}.$$

Of course, (u, v) are thinly disguised polar coordinates, and we could have set up the problem initially in a polar context. However, the example serves to illustrate how the Jacobian transforms the differential area when both variables are mapped simultaneously. $\square$

We use this technique to obtain the joint distribution of a linear combination of normal random variables. Suppose that $\mathbf{X} = (X_1, X_2)$ are bivariate normal with parameters $(\boldsymbol{\mu}, \boldsymbol{\Sigma})$, and

$$\mathbf{Y} = \begin{bmatrix} Y_1 \\ Y_2 \end{bmatrix} = \mathbf{TX} = \begin{bmatrix} t_{11} & t_{12} \\ t_{21} & t_{22} \end{bmatrix} \begin{bmatrix} X_1 \\ X_2 \end{bmatrix},$$

where $\det(\mathbf{T}) \neq 0$. We determine $F_{\mathbf{Y}}(y_1, y_2)$ by integrating the $\mathbf{X}$ distribution over the inverse image of the rectangle $(-\infty, \mathbf{y}]$. To this end, let

$$D' = \{\mathbf{v} \mid \mathbf{v} \leq \mathbf{y}\}$$
$$D = \mathbf{T}^{-1}(D') = \{\mathbf{x} \mid \mathbf{Tx} \in D'\}.$$

Then

$$F_{\mathbf{Y}}(y_1, y_2) = \Pr((Y_1 \leq y_1)(Y_2 \leq y_2)) = \Pr(\mathbf{X} \in D) = \int\!\!\int_D f_{\mathbf{X}}(x_1, x_2)\, dx_2\, dx_1$$

$$= \frac{1}{2\pi\sqrt{\det(\boldsymbol{\Sigma})}} \int\!\!\int_D \exp\left(-\frac{1}{2}(\mathbf{x} - \boldsymbol{\mu})'\boldsymbol{\Sigma}^{-1}(\mathbf{x} - \boldsymbol{\mu})\right)\, dx_1\, dx_2.$$

For the transformation $\mathbf{v} = \mathbf{Tx}$, the Jacobian is simply $|\det(\mathbf{T})|$. Consequently,

$$F_{\mathbf{Y}}(y_1, y_2) = \frac{1}{2\pi\sqrt{\det(\boldsymbol{\Sigma})}} \cdot \frac{1}{|\det(\mathbf{T})|}$$

$$\cdot \int\!\!\int_{D'} \exp\left(-\frac{1}{2}(\mathbf{T}^{-1}\mathbf{v} - \boldsymbol{\mu})'\boldsymbol{\Sigma}^{-1}(\mathbf{T}^{-1}\mathbf{v} - \boldsymbol{\mu})\right)\, dv_1\, dv_2.$$

To simplify this expression, we need some elementary facts about matrices beyond those developed in Appendix A. Specifically, we need $(\mathbf{AB})' = \mathbf{B}'\mathbf{A}'$, $(\mathbf{A}^{-1})' = (\mathbf{A}')^{-1}$, and $(\mathbf{AB})^{-1} = \mathbf{B}^{-1}\mathbf{A}^{-1}$. The entry in row i, column j of $(\mathbf{AB})'$ is

$$[(\mathbf{AB})']_{ij} = [\mathbf{AB}]_{ji} = \sum_k a_{jk}b_{ki} = \sum_k b'_{ik}a'_{kj} = [\mathbf{B}'\mathbf{A}']_{ij},$$

where a, b denote elements of $\mathbf{A}$ and $\mathbf{B}$, respectively, and a', b' are elements of $\mathbf{A}'$ and $\mathbf{B}'$. This establishes the first fact, which we use to obtain the second. Here $\mathbf{I}$ is the identity matrix.

$$\mathbf{A}'(\mathbf{A}^{-1})' = (\mathbf{A}^{-1}\mathbf{A})' = \mathbf{I}' = \mathbf{I}$$
$$(\mathbf{A}^{-1})'\mathbf{A}' = (\mathbf{A}\mathbf{A}^{-1})' = \mathbf{I}' = \mathbf{I}.$$

This shows that $(\mathbf{A}')^{-1} = (\mathbf{A}^{-1})'$. Finally,

$$\mathbf{B}^{-1}\mathbf{A}^{-1}\mathbf{A}\mathbf{B} = \mathbf{B}^{-1}\mathbf{B} = \mathbf{I}$$
$$\mathbf{A}\mathbf{B}\mathbf{B}^{-1}\mathbf{A}^{-1} = \mathbf{A}\mathbf{A}^{-1} = \mathbf{I},$$

which establishes the last fact. With these equivalences available, we continue the computation of $F_{\mathbf{Y}}(y_1, y_2)$. In particular, the exponential argument is $Q/2$, where

$$Q = (\mathbf{T}^{-1}\mathbf{v} - \mu)'\boldsymbol{\Sigma}^{-1}(\mathbf{T}^{-1}\mathbf{v} - \mu) = [\mathbf{T}^{-1}(\mathbf{v} - \mathbf{T}\mu)]'\boldsymbol{\Sigma}^{-1}[\mathbf{T}^{-1}(\mathbf{v} - \mathbf{T}\mu)]$$
$$= (\mathbf{v} - \mathbf{T}\mu)'(\mathbf{T}')^{-1}\boldsymbol{\Sigma}^{-1}\mathbf{T}^{-1}(\mathbf{v} - \mathbf{T}\mu) = (\mathbf{v} - \mathbf{T}\mu)'(\mathbf{T}\boldsymbol{\Sigma}\mathbf{T}')^{-1}(\mathbf{v} - \mathbf{T}\mu),$$

which implies that

$$F_{\mathbf{Y}}(y_1, y_2) = \frac{1}{2\pi|\det(\mathbf{T})|\sqrt{\det(\boldsymbol{\Sigma})}}$$
$$\cdot \int\int_{D'} \exp\left(-\frac{1}{2}(\mathbf{v} - \mathbf{T}\mu)'(\mathbf{T}\boldsymbol{\Sigma}\mathbf{T}')^{-1}(\mathbf{v} - \mathbf{T}\mu)\right) dv_1\, dv_2.$$

We have $|\det(\mathbf{A})| = \sqrt{\det(\mathbf{A})\det(\mathbf{A}')}$ because $\det(\mathbf{A}) = \det(\mathbf{A}')$ (Theorem A.65). Also, Theorem A.68 asserts, among other items, that $\det(\mathbf{A}\mathbf{B}) = \det(\mathbf{A})\det(\mathbf{B})$. Applying these reductions to the case in hand, we have

$$F_{\mathbf{Y}}(y_1, y_2) = \frac{1}{2\pi\sqrt{\det(\mathbf{T}\boldsymbol{\Sigma}\mathbf{T}')}} \cdot$$
$$\int\int_{D'} \exp\left(-\frac{1}{2}(\mathbf{v} - \mathbf{T}\mu)'(\mathbf{T}\boldsymbol{\Sigma}\mathbf{T}')^{-1}(\mathbf{v} - \mathbf{T}\mu)\right) dv_1\, dv_2$$
$$f_{\mathbf{Y}}(y_1, y_2) = \partial_{12}F_{\mathbf{Y}}(y_1, y_2)$$
$$= \frac{1}{2\pi\sqrt{\det(\mathbf{T}\boldsymbol{\Sigma}\mathbf{T}')}}\exp\left(-\frac{1}{2}(\mathbf{y} - \mathbf{T}\mu)'(\mathbf{T}\boldsymbol{\Sigma}\mathbf{T}')^{-1}(\mathbf{y} - \mathbf{T}\mu)\right).$$

This shows that $\mathbf{Y} = (Y_1, Y_2)$ is bivariate normal with mean $\mathbf{T}\mu$ and covariance matrix $\mathbf{T}\boldsymbol{\Sigma}\mathbf{T}'$.

6.11 Theorem: Let $\mathbf{X} = (X_1, X_2)$ be bivariate normal with parameters $\mu, \boldsymbol{\Sigma}$, and let $\mathbf{Y} = \mathbf{T}\mathbf{X}$, where $\mathbf{T}$ is a two-by-two matrix with nonzero determinant. Then $\mathbf{Y}$ is bivariate normal with parameters $(\mathbf{T}\mu, \mathbf{T}\boldsymbol{\Sigma}\mathbf{T}')$.

PROOF: See discussion above. ∎

It follows immediately that a linear combination of bivariate normals is normal. Suppose that we need the combination $aX_1 + bX_2$, taken from the bivariate normal set (X_1, X_2). We simply choose a transformation T, with nonzero determinant, that yields the required combination as one of the transformed pair. For example, choose $T = \begin{bmatrix} a & b \\ -b & a \end{bmatrix}$. This yields the pair $(Y_1 = aX_1 + bX_2, Y_2 = -bX_1 + aX_2)$. The first component is the desired linear combination, and its density is the marginal density of Y_1, which is normal.

6.12 Example: X $= (X_1, X_2)$ is bivariate normal with $\Sigma = \begin{bmatrix} 1 & 1 \\ 1 & 2 \end{bmatrix}$. Find a linear combination $aX_1 + bX_2$ that is independent of X_2.

Consider the linear transformation $\mathbf{Y} = \mathbf{TX}$, where $\mathbf{T} = \begin{bmatrix} a & b \\ 0 & 1 \end{bmatrix}$. Provided that $a \neq 0$, this transformation has nonzero determinant, and it produces $Y_1 = aX_1 + bX_2$ and $Y_2 = X_2$. We have $\mathbf{Y}$ bivariate normal, and its covariance matrix is

$$T\Sigma T' = \begin{bmatrix} a^2 + 2ab + 2b^2 & a + 2b \\ a + 2b & 2 \end{bmatrix}.$$

If we choose $a = -2b$, the $\mathbf{Y}$ components are uncorrelated and therefore independent. Take $b = -1, a = 2$. Then $Y_1 = 2X_1 - X_2$ has variance $2^2 + 2(2)(-1) + 2(-1)^2 = 2$, and it is independent of $Y_2 = X_2$. $\square$

Exercises

6.1 An autopilot attempts to maintain an airplane's altitude at a constant 18000 feet. Because of random variations, the actual altitude A is a normal random variable with parameters $(\mu = 18000, \sigma^2 = 50^2)$. What is the probability that the altitude differs from the assigned 18000 value by more than 100 feet?

6.2 Let U be normal, and let $V = aU + b$, for constants a, b. Assume that $a \neq 0$. Show that $\Psi_V(t) = e^{bt}\Psi_U(at)$. Use this result to derive $\Psi_X(t)$ for a normal random variable X, having parameters (μ, σ^2), from $\Psi_Z(t) = e^{t^2/2}$ for standard normal Z.

6.3 Let X be normal with parameters $(0, \sigma^2)$. Find the moments of X.

*6.4 Let $X_1, X_2, \ldots, X_n$ be a sample from a population X, which is normally distributed with parameters (μ, σ^2). Define the sample mean and variance as in the discrete case:

$$\overline{X} = \frac{1}{n}\sum_{i=1}^{n} X_i \quad \text{and} \quad s_X^2 = \frac{1}{n}\sum_{i=1}^{n}(X_i - \overline{X})^2.$$

Show that $\text{mean}(s_X^2) = (n-1)\sigma^2/n$ and $\text{var}(s_X^2) = 2(n-1)\sigma^4/n^2$.

*6.5 Given the bivariate normal density

$$f_{\mathbf{X}}(\mathbf{x}) = \frac{1}{2\pi\sqrt{|A|}}e^{-(\mathbf{x}-\mathbf{a})'A^{-1}(\mathbf{x}-\mathbf{a})/2},$$

where $\mathbf{a} = \begin{bmatrix} a_1 \\ a_2 \end{bmatrix}$ and $A = \begin{bmatrix} b_1 & c \\ c & b_2 \end{bmatrix}$ satisfy $b_1 > 0, b_2 > 0$, and $\det(A) > 0$, show that

$$f_{X_2}(x_2) = \frac{1}{\sqrt{2\pi b_2}}e^{-(x_2-a_2)^2/(2b_2)}.$$

*6.6 In the context of the preceding exercise, show that $\text{cov}(X_1, X_2) = c$.

6.7 Let X_1, X_2 be bivariate normal with $\mu = \begin{bmatrix} 0 \\ 0 \end{bmatrix}$ and $\Sigma = \begin{bmatrix} 3 & 4 \\ 4 & 6 \end{bmatrix}$. Find the equations of the lines coinciding with the major and minor axes of the ellipses on a contour density plot.

6.8 What is the maximum rotation that can appear on the contour plot of a bivariate normal density? What conditions produce this maximum rotation?

6.9 Let X_1, X_2 be bivariate normal with $\mu = \begin{bmatrix} \mu_1 \\ \mu_2 \end{bmatrix}$ and $\Sigma = \begin{bmatrix} v & \rho v \\ \rho v & v \end{bmatrix}$, for some $v > 0$ and $-1 < \rho < 1$. What rotation aligns a new coordinate system with the major and minor axes of the contour ellipses? What are the lengths of the semimajor and semiminor axes of the ellipse corresponding to $f_{X_1, X_2}(x_1, x_2) = K$? What is the qualitative behavior of this ellipse as $\rho \to \pm 1$?

6.10 Complete the proof of Theorem 6.8 by showing that $f_{2|1}(\cdot)$ is normal with parameters
$$\mu_{2|1} = \mu_2 + \frac{\sigma_{12}(x_1 - \mu_1)}{\sigma_{11}} \text{ and } \sigma_{22|1} = \sigma_{22} - \frac{\sigma_{12}^2}{\sigma_{11}}.$$

6.11 In the context of the bivariate normal distribution of Theorem 6.8, use the marginal density for X_2 to show $E_{X_2}[\mu_{1|2}] = \mu_1$ and $E_{X_2}[(\mu_{1|2} - \mu_1)^2] = \sigma_{12}^2/\sigma_{22}$. Note that these results comply with Theorem 5.51, which provides a general relationship between conditional means and variances and their marginal counterparts.

6.12 Suppose that (X_1, X_2) are independent and bivariate normal. Using the expressions of Theorem 6.8, show that $f_{1|2}(x_1) = f_1(x_1)$, for all x_1 and any conditioning value x_2 such that $f_2(x_2) > 0$.

*6.13 A bolt is to fit in a hole. The bolt measures X inches; the hole measures Y inches. X and Y are independent normal random variables with parameters $(\mu_X = 1.0, \sigma_X^2 = 0.01^2)$ and $(\mu_Y = 1.0, \sigma_Y^2 = 0.02^2)$, respectively. What is the probability that the bolt will not fit in the hole?

*6.14 Let $\mathbf{X} = (X_1, X_2)$ be bivariate normal with $\Sigma = \begin{bmatrix} 1 & 1 \\ 1 & 2 \end{bmatrix}$. Find a matrix $\mathbf{T}$ such that $\mathbf{Y} = \mathbf{TX}$ consists of two independent standard normal random variables.

6.1.2   The multivariate normal distribution

This section extends the bivariate concept to a joint normal distribution of n random variables. Before giving a precise definition, we consider some properties of the covariance matrix that characterizes a bivariate normal distribution.

For $\mathbf{X} = (X_1, X_2)$ with means (μ_1, μ_2), let $\boldsymbol{\Sigma} = \begin{bmatrix} \sigma_{11} & \sigma_{12} \\ \sigma_{21} & \sigma_{22} \end{bmatrix}$ be the covariance matrix. The diagonal elements of $\boldsymbol{\Sigma}$ are the variances of X_1 and X_2 and are therefore positive. Moreover, the joint density expression requires $\boldsymbol{\Sigma}^{-1}$, so we must have $\det \boldsymbol{\Sigma} \neq 0$. Indeed, we can argue that the determinant must be positive. For any λ, we have

$$
\begin{aligned}
0 \leq\ & E\left[((X_1 - \mu_1) - \lambda(X_2 - \mu_2))^2 \right] \\
=\ & \lambda^2 E[(X_2 - \mu_2)^2] - 2\lambda E[(X_1 - \mu_1)(X_2 - \mu_2)] + E[(X_1 - \mu_1)^2] \\
=\ & \sigma_{22}\lambda^2 - 2\sigma_{12}\lambda + \sigma_{11}.
\end{aligned}
$$

For the quadratic to be nonnegative, it must have no real roots, or just one repeated real root. Consequently, $4\sigma_{12}^2 - 4\sigma_{22}\sigma_{11} \leq 0$, which forces $\sigma_{11}\sigma_{22} - \sigma_{12}^2 = \det \boldsymbol{\Sigma} \geq 0$. Since we already know that the determinant cannot equal zero, it must be positive. Finally, consider the quadratic form

$$
\begin{aligned}
Q =\ & \begin{bmatrix} x_1 & x_2 \end{bmatrix} \begin{bmatrix} \sigma_{11} & \sigma_{12} \\ \sigma_{21} & \sigma_{22} \end{bmatrix} \begin{bmatrix} x_1 \\ x_2 \end{bmatrix} = \sigma_{11}x_1^2 + 2\sigma_{12}x_1 x_2 + \sigma_{22}x_2^2 \\
=\ & \sigma_{11}\left(x_1^2 + \frac{2\sigma_{12}x_2}{\sigma_{11}}x_1 + \frac{\sigma_{12}^2 x_2^2}{\sigma_{11}^2} - \frac{\sigma_{12}^2 x_2^2}{\sigma_{11}^2} + \frac{\sigma_{22}x_2^2}{\sigma_{11}} \right) \\
=\ & \sigma_{11}\left[\left(x_1 + \frac{\sigma_{12}x_2}{\sigma_{11}} \right)^2 + \frac{x_2^2}{\sigma_{11}^2}(\sigma_{11}\sigma_{22} - \sigma_{12}^2) \right] \\
=\ & \sigma_{11}\left[\left(x_1 + \frac{\sigma_{12}x_2}{\sigma_{11}} \right)^2 + \frac{x_2^2}{\sigma_{11}^2}\det \boldsymbol{\Sigma} \right] \geq 0.
\end{aligned}
$$

Moreover, $Q = 0$ if and only if $x_1 = x_2 = 0$. In summary, we observe that $\boldsymbol{\Sigma}$ is a symmetric invertible matrix with positive entries on the main diagonal and a positive determinant. Also, the quadratic form $\mathbf{x}'\boldsymbol{\Sigma}\mathbf{x}$ is nonnegative in general and positive if and only if $\mathbf{x} \neq \mathbf{0}$. In higher dimensions, the designation that captures these features is positive definite, as defined below.

6.13 Definition: Let $\mathbf{A}$ be an $n \times n$ matrix. $\mathbf{A}$ is *positive semidefinite* if $\mathbf{x}'\mathbf{A}\mathbf{x} \geq 0$ for all n-vectors $\mathbf{x}$. $\mathbf{A}$ is *positive definite* if it is positive semidefinite and $\mathbf{x}'\mathbf{A}\mathbf{x} = 0$ implies that $\mathbf{x} = \mathbf{0}$. ∎

   To show that positive definiteness captures the desired properties of a covariance matrix, we note some consequences of that feature. Suppose that $\mathbf{D}$ is a symmetric positive definite $n \times n$ matrix. Let d_{ij} denote the elements of $\mathbf{D}$, and let $\mathbf{d}_{i\cdot}$ be the ith row. First, we note that $\mathbf{D}$ has nonzero determinant. If not, the rows of $\mathbf{D}$ are dependent, which implies the existence of constants $c_1, c_2, \ldots, c_n$, not all zero, such that $\sum_{i=1}^{n} c_i \mathbf{d}_{i\cdot} = \mathbf{0}$. Arranging these constants as a column vector $\mathbf{x}$, we have $\mathbf{x} \neq \mathbf{0}$, but $\mathbf{x}'\mathbf{D} = \mathbf{0}$. The latter implies that $\mathbf{x}'\mathbf{D}\mathbf{x} = 0$, which contradicts the positive definite status of $\mathbf{D}$.

   Next, we observe that the main diagonal entries d_{ii} must all be positive. To see this, construct a column vector $\mathbf{x}$ with 1 in component i and zeros in the remaining components. Then $\mathbf{x} \neq \mathbf{0}$, which forces $d_{ii} = \mathbf{x}'\mathbf{D}\mathbf{x} > 0$.

For real t and $i \neq j$, construct $\mathbf{E}_{ij;t}$ by replacing the zero in row j column i of the $n \times n$ identity matrix with t. If we postmultiply $\mathbf{D}\mathbf{E}_{ij;t}$, the result is $\mathbf{D}$ with t times column j added to column i. Since this result has the same determinant as $\mathbf{D}$, we conclude that $\det \mathbf{E}_{ij;t} = 1$. If we premultiply $\mathbf{E}'_{ij;t}\mathbf{D}$, the result is $\mathbf{D}$ with t times row j added to row i. We use these elementary matrices to manipulate $\mathbf{D}$.

Since $d_{11} > 0$, we can multiply column 1 by appropriate constants $t_j = -d_{1j}/d_{11}$ and add the result to columns j to produce zeros in the first row in all positions after column 1. To maintain the symmetry and to clear the off-diagonal entries in column 1, we multiply row 1 by these same constants and add to rows $2, 3, \ldots, n$. In terms of elementary matrices, we form

$$\mathbf{D}_1 = \mathbf{E}'_{n,1;t_n}\mathbf{E}'_{n-1,1;t_{n-1}} \cdots \mathbf{E}'_{2,1;t_2}\mathbf{D}\mathbf{E}_{2,1;t_2}\mathbf{E}_{3,1;t_3} \cdots \mathbf{E}_{n,1;t_n},$$

where

$$t_2 = -d_{12}/d_{11} = -d_{21}/d_{11}$$
$$t_3 = -d_{13}/d_{11} = -d_{31}/d_{11}$$
$$\vdots \qquad \vdots$$
$$t_n = -d_{1n}/d_{11} = -d_{n1}/d_{11}.$$

This clears the nondiagonal entries in row 1 and in column 1, while maintaining the matrix symmetry. For example, suppose that

$$\mathbf{D} = \begin{bmatrix} 1 & 1 & 1 \\ 1 & 3 & 3 \\ 1 & 3 & 6 \end{bmatrix}.$$

To clear the trailing entries in row 1, we multiply column 1 by $t_2 = t_3 = -1$ and add to columns 2 and 3.

$$\mathbf{D}\mathbf{E}_{21;-1}\mathbf{E}_{31;-1} = \begin{bmatrix} 1 & 1 & 1 \\ 1 & 3 & 3 \\ 1 & 3 & 6 \end{bmatrix}\begin{bmatrix} 1 & -1 & 0 \\ 0 & 1 & 0 \\ 0 & 0 & 1 \end{bmatrix}\begin{bmatrix} 1 & 0 & -1 \\ 0 & 1 & 0 \\ 0 & 0 & 1 \end{bmatrix} = \begin{bmatrix} 1 & 0 & 0 \\ 1 & 2 & 2 \\ 1 & 2 & 5 \end{bmatrix}.$$

We follow by multiplying row 1 by the same t_2 and t_3 and adding to rows 2 and 3, respectively.

$$\mathbf{D}_1 = \mathbf{E}'_{31;-1}\mathbf{E}'_{21;-1}\begin{bmatrix} 1 & 0 & 0 \\ 1 & 2 & 2 \\ 1 & 2 & 5 \end{bmatrix}$$

$$= \begin{bmatrix} 1 & 0 & 0 \\ 0 & 1 & 0 \\ -1 & 0 & 1 \end{bmatrix}\begin{bmatrix} 1 & 0 & 0 \\ -1 & 1 & 0 \\ 0 & 0 & 1 \end{bmatrix}\begin{bmatrix} 1 & 0 & 0 \\ 1 & 2 & 2 \\ 1 & 2 & 5 \end{bmatrix} = \begin{bmatrix} 1 & 0 & 0 \\ 0 & 2 & 2 \\ 0 & 2 & 5 \end{bmatrix}.$$

This clears the trailing entries in column 1 and restores the matrix symmetry. Returning to the general case, we set

$$\mathbf{E}_1 = \mathbf{E}_{2,1;t_2}\mathbf{E}_{3,1;t_3} \cdots \mathbf{E}_{n,1;t_r}.$$

Then $D_1 = E_1'DE_1$ has zeros in row 1 and in column 1, except in the leading position. Moreover, D_1 remains positive definite because for any x, we have $x'D_1x = (E_1x)'D(E_1x) \geq 0$. Equality holds if and only if $E_1x = 0$, which since E_1 is invertible, occurs if and only if $x = 0$.

Consequently, the diagonal entries in D_1 are still positive, and we can repeat the process above to clear the remaining nondiagonal elements in row 2 and column 2. In the example started above, we need to multiply column 2 by -1 and add it to column 3. That is,

$$D_2 = E_{32;-1}'D_1E_{32;-1} = \begin{bmatrix} 1 & 0 & 0 \\ 0 & 1 & 0 \\ 0 & -1 & 1 \end{bmatrix} \begin{bmatrix} 1 & 0 & 0 \\ 0 & 2 & 2 \\ 0 & 2 & 5 \end{bmatrix} \begin{bmatrix} 1 & 0 & 0 \\ 0 & 1 & -1 \\ 0 & 0 & 1 \end{bmatrix}$$

$$= \begin{bmatrix} 1 & 0 & 0 \\ 0 & 2 & 0 \\ 0 & 0 & 3 \end{bmatrix}.$$

In the general case, we continue the process to obtain $D_n = E'DE$, where D_n has nonzero entries only on the main diagonal. Moreover, these entries are all positive. The matrix E is invertible because, being the product of the elementary matrices that produce the column transformations, its determinant is 1. Suppose that

$$D_n = \begin{bmatrix} \lambda_1 & 0 & \cdots & 0 \\ 0 & \lambda_2 & \cdots & 0 \\ \vdots & \vdots & & \vdots \\ 0 & 0 & \cdots & \lambda_n \end{bmatrix}.$$

We note that $0 < \det D_n = \det E' \cdot \det D \cdot \det E = \det D$. Defining

$$F = \begin{bmatrix} 1/\sqrt{\lambda_1} & 0 & \cdots & 0 \\ 0 & 1/\sqrt{\lambda_2} & \cdots & 0 \\ \vdots & \vdots & & \vdots \\ 0 & 0 & \cdots & 1/\sqrt{\lambda_n} \end{bmatrix},$$

we have $(EF)'D(EF) = F'D_nF = I$, the $n \times n$ identity matrix. Of course, EF is invertible; its determinant is

$$\det EF = \det E \det F = (1) \left(\prod_{i=1}^{n} 1/\sqrt{\lambda_i} \right) > 0.$$

We have proved the following theorem.

6.14 Theorem: Let D be a symmetric positive definite $n \times n$ matrix. Then $\det D > 0$, and there exists an invertible matrix B such that $B'DB = I$, the $n \times n$ identity matrix.

PROOF: Let B be EF in the discussion above. ∎

6.15 Example: Show that

$$\mathbf{D} = \begin{bmatrix} 32 & -13 & -15/2 & -13/2 \\ -13 & 16 & 5 & 3 \\ -15/2 & 5 & 5/2 & 5/2 \\ -13/2 & 3 & 5/2 & 7/2 \end{bmatrix}$$

is positive definite and identify the matrix $\mathbf{B}$ that produces $\mathbf{B}'\mathbf{DB} = \mathbf{I}$.

We transform $\mathbf{D}$ by adding multiples of rows to other rows and multiples of columns to other columns. This reduces $\mathbf{D}$ to a diagonal matrix with positive diagonal entries, if the original matrix is indeed positive definite. If these diagonal entries are $\lambda_1, \ldots, \lambda_4$, we then premultiply and postmultiply by the diagonal matrix with entries $\lambda_1^{-1/2}, \lambda_2^{-1/2}, \ldots, \lambda_4^{-1/2}$ to obtain the identity. Each such row operation is actually a premultiplication by an elementary matrix. This means that $\mathbf{B}'$ is the product of these elementary matrices $\mathbf{E}_i$ and the diagonal matrix $\mathbf{F}$. That is,

$$\mathbf{B}' = \mathbf{F}\mathbf{E}_k\mathbf{E}_{k-1}\cdots\mathbf{E}_1.$$

This equation is unchanged if we append the identity matrix to the right end, which means that $\mathbf{B}'$ evolves from the identity matrix via the same row operations as were performed on $\mathbf{D}$, followed by the premultiplication of the diagonal matrix $\mathbf{F}$. For convenience, we can align the identity matrix to the left of $\mathbf{D}$ and perform the row operations on the enlarged array.

We proceed to multiply row 1 by $13/32 = 0.4062$ and add to row 2. To maintain the symmetry, we then multiply column 1 by the same constant and add to column 2. That is, we multiply column 1 of $\mathbf{D}$ by the constant and add to column 2 of $\mathbf{D}$. These are columns 5 and 6 in the extended array. These two operations evolve $\mathbf{D}$ and its companion identity matrix from

$$\left[\begin{array}{cccc|cccc} 1 & 0 & 0 & 0 & 32.0 & -13.0 & -7.5 & -6.5 \\ 0 & 1 & 0 & 0 & -13.0 & 16.0 & 5.0 & 3.0 \\ 0 & 0 & 1 & 0 & -7.5 & 5.0 & 2.5 & 2.5 \\ 0 & 0 & 0 & 1 & -6.5 & 3.0 & 2.5 & 3.5 \end{array}\right]$$

to

$$\left[\begin{array}{cccc|cccc} 1.0000 & 0 & 0 & 0 & 32.0000 & 0 & -7.5000 & -6.5000 \\ 0.4062 & 1.0000 & 0 & 0 & 0 & 10.7188 & 1.9531 & 0.3594 \\ 0 & 0 & 1.0000 & 0 & -7.5000 & 1.9531 & 2.5000 & 2.5000 \\ 0 & 0 & 0 & 1.0000 & -6.5000 & 0.3594 & 2.5000 & 3.5000 \end{array}\right].$$

As expected, the array remains symmetric and has zeros in locations $(1, 2)$ and $(2, 1)$ of the $\mathbf{D}$ portion. The next step multiplies row 1 by $15/64$ and adds to row 3. This operation extends across the entire array. The complementary operation multiplies column 1 of the $\mathbf{D}$ part by $15/64$ and adds to column 3. Continuing in this fashion we clear the nondiagonal entries in row 1 and in column 1 of the $\mathbf{D}$ portion. The evolution now appears as follows.

$$\left[\begin{array}{cccc|cccc} 1.0000 & 0 & 0 & 0 & 32.0000 & 0 & 0 & 0 \\ 0.4062 & 1.0000 & 0 & 0 & 0 & 10.7188 & 1.9531 & 0.3594 \\ 0.2344 & 0 & 1.0000 & 0 & 0 & 1.9531 & 0.7422 & 0.9766 \\ 0.2031 & 0 & 0 & 1.0000 & 0 & 0.3594 & 0.9766 & 0.9766 \end{array}\right]$$

We continue to clear the entries below and to the right of $\mathbf{D}$ position $(2,2)$ and finally below and to the right of position $(3,3)$. The final result is

$$
\left[\begin{array}{cccc|cccc}
1.0000 & 0.0000 & 0.0000 & 0.0000 & 32.0000 & 0.0000 & 0.0000 & 0.0000 \\
0.4062 & 1.0000 & 0.0000 & 0.0000 & 0.0000 & 10.7188 & 0.0000 & 0.0000 \\
0.1603 & -0.1822 & 1.0000 & 0.0000 & 0.0000 & 0.0000 & 0.3863 & 0.0000 \\
-0.1887 & 0.3962 & -2.3585 & 1.0000 & 0.0000 & 0.0000 & 0.0000 & 0.0189
\end{array}\right].
$$

We now have $\mathbf{D}$ in diagonal form, and the positive entries show that it is indeed positive definite. The diagonal matrix needed to finalize the transformation to the identity is

$$
\mathbf{F} = \left[\begin{array}{cccc}
1/\sqrt{32} & 0 & 0 & 0 \\
0 & 1/\sqrt{10.7188} & 0 & 0 \\
0 & 0 & 1/\sqrt{0.3863} & 0 \\
0 & 0 & 0 & 1/\sqrt{0.0189}
\end{array}\right].
$$

Consequently,

$$
\mathbf{B}' = \left[\begin{array}{cccc}
1/\sqrt{32} & 0 & 0 & 0 \\
0 & 1/\sqrt{10.7188} & 0 & 0 \\
0 & 0 & 1/\sqrt{0.3863} & 0 \\
0 & 0 & 0 & 1/\sqrt{0.0189}
\end{array}\right]
$$

$$
\cdot \left[\begin{array}{cccc}
1.0000 & 0.0000 & 0.0000 & 0.0000 \\
0.4062 & 1.0000 & 0.0000 & 0.0000 \\
0.1603 & -0.1822 & 1.0000 & 0.0000 \\
-0.1887 & 0.3962 & -2.3585 & 1.0000
\end{array}\right]
$$

$$
= \left[\begin{array}{cccc}
0.1768 & 0.0000 & 0.0000 & 0.0000 \\
0.1241 & 0.3054 & 0.0000 & 0.0000 \\
0.2580 & -0.2932 & 1.6089 & 0.0000 \\
-1.3736 & 2.8846 & -17.1701 & 7.2801
\end{array}\right].
$$

A straightforward multiplication verifies that $\mathbf{B}'\mathbf{DB} = \mathbf{I}$. □

A symmetric positive definite matrix is the required parameter for a multivariate normal distribution, for which a formal definition is now appropriate. Unless otherwise noted, boldface lower case denotes an n-vector, except for a vector of random variables, which continues to use boldface upper case. In general, however, boldface upper case denotes an $n \times n$ matrix. Components of $\mathbf{t}$ are t_i; components of $\mathbf{A}$ are a_{ij}. Components of a transposed vector $\mathbf{t}'$ or matrix $\mathbf{A}'$ are written t_i' or a_{ij}'.

6.16 Definition: Let $\mathbf{X}$ be an n-vector of random variables. Let $\mathbf{a}$ be an n-vector of real numbers, and let $\mathbf{A}$ be a symmetric positive definite $n \times n$ matrix. We say that $\mathbf{X}$ has a *multivariate normal distribution* with parameters $(\mathbf{a}, \mathbf{A})$ if the joint density function is

$$
f_{\mathbf{X}}(\mathbf{t}) = \frac{1}{(2\pi)^{n/2}\sqrt{\det \mathbf{A}}} \exp\left(\frac{-(\mathbf{t}-\mathbf{a})'\mathbf{A}^{-1}(\mathbf{t}-\mathbf{a})}{2}\right). \ \blacksquare
$$

We verify that this density assigns total probability 1 and that the parameters contain the means, variances, and covariances of the $\mathbf{X}$ components. This effort involves n-dimensional integrals, which require multidimensional variable transformations, similar to those used in the two-dimensional case. Specifically, let D be a region in the $(x_1, x_2, \ldots, x_n)$ coordinate system and D' a corresponding region in the $(y_1, y_2, \ldots, y_n)$ system. Let $\mathbf{f} : D \to D'$ be a bijective transformation with inverse $\mathbf{g} : D' \to D$. Both $\mathbf{f}$ and $\mathbf{g}$ require n components, one for each coordinate in the target system. That is, we can write them as $y_i = f_i(x_1, x_2, \ldots, x_n)$ and $x_i = g_i(y_1, y_2, \ldots, y_n)$ for $i = 1, 2, \ldots, n$. The Jacobian $J(\mathbf{y})$ is

$$J(\mathbf{y}) = \left| \det \begin{bmatrix} \partial_1 f_1(\mathbf{g}(\mathbf{y})) & \partial_2 f_1(\mathbf{g}(\mathbf{y})) & \cdots & \partial_n f_1(\mathbf{g}(\mathbf{y})) \\ \partial_1 f_2(\mathbf{g}(\mathbf{y})) & \partial_2 f_2(\mathbf{g}(\mathbf{y})) & \cdots & \partial_n f_2(\mathbf{g}(\mathbf{y})) \\ \vdots & \vdots & & \vdots \\ \partial_1 f_n(\mathbf{g}(\mathbf{y})) & \partial_2 f_n(\mathbf{g}(\mathbf{y})) & \cdots & \partial_n f_n(\mathbf{g}(\mathbf{y})) \end{bmatrix} \right|,$$

and the integral of a function $h(x_1, x_2, \ldots, x_n)$ over region D admits the evaluation

$$\int_D h(x_1, \ldots, x_n) \, dx_1 \cdots dx_n$$
$$= \int_{D'} \frac{h(g_1(y_1, \ldots, y_n), \ldots, g_n(y_1, \ldots, y_n)) \, dy_1 \cdots dy_n}{J(y_1, \ldots, y_n)}.$$

Now suppose that $\mathbf{X}$ has a multivariate normal distribution with parameters $(\mathbf{a}, \mathbf{A})$. Because $\mathbf{A}$ is positive definite, there is an invertible matrix $\mathbf{B}$ such that $\mathbf{B}'\mathbf{AB} = \mathbf{I}$. Under the transformation $\mathbf{y} = \mathbf{f}(\mathbf{x}) = \mathbf{B}'(\mathbf{x} - \mathbf{a})$, we have

$$y_i = f_i(\mathbf{x}) = \sum_{j=1}^{n} b_{ji}(x_j - a_j)$$

$$\partial_j f_i(\mathbf{x}) = b_{ji}$$
$$J(\mathbf{y}) = |\det \mathbf{B}'|.$$

The Jacobian is a constant. Also, $\det(\mathbf{B}'\mathbf{AB}) = \det \mathbf{I} = 1$. Therefore,

$$\det \mathbf{B}' \cdot \det \mathbf{A} \cdot \det \mathbf{B} = 1 \qquad\qquad J(\mathbf{y}) = |\det \mathbf{B}'| = \frac{1}{\sqrt{\det \mathbf{A}}}.$$
$$(\det \mathbf{B}')^2 \cdot \det \mathbf{A} = 1$$

The integral for total probability assignment is then as follows.

$$\int_{\mathcal{R}^n} f_{\mathbf{X}}(\mathbf{x}) \, dx_1 \cdots dx_n$$

$$= \int_{\mathcal{R}^n} \frac{1}{(2\pi)^{n/2} \sqrt{\det \mathbf{A}}} \exp\left(\frac{-(\mathbf{x} - \mathbf{a})' \mathbf{A}^{-1} (\mathbf{x} - \mathbf{a})}{2} \right) dx_1 \cdots dx_n$$

$$= \int_{\mathcal{R}^n} \frac{1}{(2\pi)^{n/2}} \exp\left(\frac{-\mathbf{y}' \mathbf{B}^{-1} \mathbf{A}^{-1} (\mathbf{B}')^{-1} \mathbf{y}}{2} \right) dy_1 \cdots dy_n$$

But, $\mathbf{B}^{-1}\mathbf{A}^{-1}(\mathbf{B}')^{-1} = (\mathbf{B}'\mathbf{A}\mathbf{B})^{-1} = \mathbf{I}^{-1} = \mathbf{I}$. Consequently,

$$
\begin{aligned}
\int_{\mathcal{R}^n} f_{\mathbf{X}}(\mathbf{x})\, dx_1 \cdots dx_n &= \int_{\mathcal{R}^n} \frac{1}{(2\pi)^{n/2}} \exp\left(\frac{-\mathbf{y}'\mathbf{y}}{2}\right) dy_1 \cdots dy_n \\
&= \int_{\mathcal{R}^n} \frac{1}{(2\pi)^{n/2}} \exp\left(\frac{-(y_1^2 + \ldots + y_n^2)}{2}\right) dy_1 \cdots dy_n \\
&= \left(\frac{1}{\sqrt{2\pi}} \int_{-\infty}^{\infty} e^{-y^2/2}\, dy\right)^n = 1.
\end{aligned}
$$

The total probability is thus 1. Also note that the shifted random variables $\mathbf{X} - \mathbf{a}$ have a multivariate normal distribution with parameters $(\mathbf{0}, \mathbf{A})$:

$$
\begin{aligned}
F_{\mathbf{X}-\mathbf{a}}(\mathbf{t}) &= \Pr(X_1 - a_1 \le t_1, \ldots, X_n - a_n \le t_n) \\
&= \Pr(X_1 \le t_1 + a_1, \ldots, X_n \le t_n + a_n) = F_{\mathbf{X}}(\mathbf{t} + \mathbf{a}) \\
f_{\mathbf{X}-\mathbf{a}}(\mathbf{t}) &= f_{\mathbf{X}}(\mathbf{t} + \mathbf{a}) = \frac{1}{(2\pi)^{n/2}\sqrt{\mathbf{A}}} \exp\left(\frac{-\mathbf{t}'\mathbf{A}^{-1}\mathbf{t}}{2}\right).
\end{aligned}
$$

Many properties follow from appropriately chosen linear transformations of a multivariate random variable. To exploit this approach, we need the distribution of $\mathbf{TX}$, a vector of linear combinations of the $\mathbf{X}$ components. The proof of Theorem 6.11 applies here, provided that we give an n-dimensional interpretation to the 2×2 matrices. We restate the result as follows.

6.17 Theorem: Let $\mathbf{X}$ be an n-dimensional vector of random variables with a multivariate normal distribution with parameters $(\mathbf{a}, \mathbf{A})$. Let $\mathbf{T}$ be an $n \times n$ matrix with nonzero determinant. Then $\mathbf{Y} = \mathbf{TX}$ is multivariate normal with parameters $(\mathbf{Ta}, \mathbf{TAT}')$.

PROOF: With an n-dimensional interpretation of the vectors and matrices, the proof of Theorem 6.11 applies verbatim. ∎

From this point, we can continue with linear algebra alone. The desired properties follow from appropriate selections for the transformation $\mathbf{T}$. For $\mathbf{X}$ multivariate normal with parameters $(\mathbf{a}, \mathbf{A})$, we first verify that the parameters contain the means and covariances of the X_i. We again apply $\mathbf{B}$ such that $\mathbf{B}'\mathbf{A}\mathbf{B} = \mathbf{I}$. $\mathbf{Y} = \mathbf{B}'(\mathbf{X} - \mathbf{a})$ is then multivariate normal with parameters $(\mathbf{0}, \mathbf{I})$. The density of $\mathbf{Y}$ is then

$$
f_{\mathbf{Y}}(\mathbf{t}) = (2\pi)^{-n/2} e^{-(y_1^2 + y_2^2 + \ldots + y_n^2)/2}.
$$

Because the integrals separate easily, we can compute the marginal distributions of the Y_i as follows.

$$
f_{Y_i}(y_i) = \frac{e^{-y_i^2/2}}{\sqrt{2\pi}} \left((2\pi)^{-1/2} \int_{-\infty}^{\infty} e^{-y^2/2}\, dy\right)^{n-1} = \frac{e^{-y_i^2/2}}{\sqrt{2\pi}}
$$

$$
f_{Y_i, Y_j}(y_i, y_j) = \frac{e^{-(y_i^2 + y_j^2)/2}}{\sqrt{2\pi}} \left((2\pi)^{-1/2} \int_{-\infty}^{\infty} e^{-y^2/2}\, dy\right)^{n-2} = f_{Y_i}(y_i) f_{Y_j}(y_j)
$$

The Y_i are then independent standard normal random variables. The means and covariances are zero; the variances are 1. We now work backwards to derive the means, variances, and covariances of the X_i. Let $\mathbf{C} = (\mathbf{B}')^{-1}$. Then $\mathbf{B}'\mathbf{A}\mathbf{B} = \mathbf{I}$ implies that $\mathbf{A} = (\mathbf{B}')^{-1}\mathbf{B}^{-1} = \mathbf{C}(\mathbf{B}'')^{-1} = \mathbf{C}((\mathbf{B}')^{-1})' = \mathbf{C}\mathbf{C}'$. Since $\mathbf{X} - \mathbf{a} = (\mathbf{B}')^{-1}\mathbf{Y} = \mathbf{C}\mathbf{Y}$, we have

$$X_i - a_i = \sum_{j=1}^{n} c_{ij}Y_j$$

$$E[X_i - a_i] = \sum_{j=1}^{n} c_{ij}E[Y_j] = 0$$

$$\text{mean}(X_i) = a_i$$

Because the covariances among the Y_j are zero, we also have

$$\text{var}(X_i) = E[(X_i - a_i)^2] = E\left[\left(\sum_{j=1}^{n} c_{ij}Y_j\right)^2\right]$$

$$= \sum_{j=1}^{n} c_{ij}^2 E[Y_j^2] + 2\sum_{j \neq k} c_{ij}c_{ik}E[Y_j Y_k] = \sum_{j=1}^{n} c_{ij}^2 = \sum_{j=1}^{n} c_{ij}c_{ji}' = a_{ii}$$

$$\text{cov}(X_i, X_j) = E[(X_i - a_i)(X_j - a_j)] = E\left[\left(\sum_{k=1}^{n} c_{ik}Y_k\right)\left(\sum_{l=1}^{n} c_{jl}Y_l\right)\right]$$

$$= \sum_{k=1}^{n}\sum_{l=1}^{n} c_{ik}c_{jl}E[Y_k Y_l] = \sum_{k=1}^{n} c_{ik}c_{jk} = \sum_{k=1}^{n} c_{ik}c_{kj}' = a_{ij}.$$

We have shown that the parameters $(\mathbf{a}, \mathbf{A})$ contain the means and variances of the $\mathbf{X}$ components. As in the univariate and bivariate cases, it is traditional to use a notation that emphasizes this fact. Henceforth, we will speak of $\mathbf{X}$ as multivariate normal with parameters $(\boldsymbol{\mu}, \boldsymbol{\Sigma})$, where $\boldsymbol{\mu}$ is the mean vector and $\boldsymbol{\Sigma}$ is the covariance matrix.

A subset of the $\mathbf{X}$ random variables is also multivariate normal with parameters obtained by selecting the corresponding components from $\boldsymbol{\mu}$ and $\boldsymbol{\Sigma}$. Relabeling the X_i if necessary, we can assume that the selected subset is $X_1, X_2, \ldots, X_k$. We then envision the random variable vector, the mean vector, and the covariance matrix partitioned to isolate these entries. We use $\hat{\mathbf{x}}$ to denote the $\mathbf{x}$ vector shortened to the first k components. Similarly, $\check{\mathbf{x}}$ denotes the vector containing the final $n - k$ entries. Expressed as transpose vectors, they have the following structures.

$$\mathbf{X}' = [\, X_1 \ X_2 \ \cdots \ X_k \,|\, X_{k+1} \ \cdots \ X_n \,]$$

$$\boldsymbol{\mu}' = [\, \mu_1 \ \mu_2 \ \cdots \ \mu_k \,|\, \mu_{k+1} \ \cdots \ \mu_n \,]$$

The corresponding covariance matrix partition is

$$
\Sigma = \left[
\begin{array}{ccc|ccc}
\sigma_{11} & \cdots & \sigma_{1k} & \sigma_{1,k+1} & \cdots & \sigma_{1n} \\
\vdots & & \vdots & \vdots & & \vdots \\
\sigma_{k1} & \cdots & \sigma_{kk} & \sigma_{k,k+1} & \cdots & \sigma_{kn} \\
\hline
\sigma_{k+1,1} & \cdots & \sigma_{k+1,k} & \sigma_{k+1,k+1} & \cdots & \sigma_{k+1,n} \\
\vdots & & \vdots & \vdots & & \vdots \\
\sigma_{n1} & \cdots & \sigma_{nk} & \sigma_{n,k+1} & \cdots & \sigma_{nn}
\end{array}
\right] = \left[\begin{array}{c|c} \mathbf{A} & \mathbf{B} \\ \hline \mathbf{B}' & \mathbf{C} \end{array}\right].
$$

We seek an invertible transformation $\mathbf{Y} = \mathbf{TX}$ such that $Y_i = X_i$ for $1 \leq i \leq k$ and the transformed covariance matrix $\mathbf{T\Sigma T'}$ exhibits zero covariances between Y_i and Y_j when $1 \leq i \leq k$ and $k+1 \leq j \leq n$. That is, we want

$$
\mathbf{T} = \left[\begin{array}{c|c} \mathbf{I} & \mathbf{0} \\ \hline \mathbf{E} & \mathbf{F} \end{array}\right] \quad \text{and} \quad \mathbf{T\Sigma T'} = \left[\begin{array}{c|c} \mathbf{G} & \mathbf{0} \\ \hline \mathbf{0} & \mathbf{H} \end{array}\right].
$$

Assume for the moment that we have such a transformation $\mathbf{T}$. Theorem A.80 then enables the following computations.

$$
\mathbf{Tx} = \left[\begin{array}{c} \hat{\mathbf{x}} \\ \hline \mathbf{E}\hat{\mathbf{x}} + \mathbf{F}\check{\mathbf{x}} \end{array}\right]
$$

$$
\det \mathbf{T\Sigma T'} = \det \mathbf{G} \det \mathbf{H}
$$

$$
(\mathbf{T\Sigma T'})^{-1} = \left[\begin{array}{c|c} \mathbf{G}^{-1} & \mathbf{0} \\ \hline \mathbf{0} & \mathbf{H}^{-1} \end{array}\right]
$$

$$
\mathbf{x}'\,(\mathbf{T\Sigma T'})^{-1}\,\mathbf{x} = \hat{\mathbf{x}}'\mathbf{G}^{-1}\hat{\mathbf{x}} + \check{\mathbf{x}}'\mathbf{H}^{-1}\check{\mathbf{x}}
$$

Let $\boldsymbol{\nu} = \mathbf{T}\boldsymbol{\mu}$ and note that $\hat{\boldsymbol{\nu}} = \hat{\boldsymbol{\mu}}$. Now the density of $\mathbf{Y} = \mathbf{TX}$ is

$$
f_{\mathbf{Y}}(\mathbf{y}) = \frac{1}{(2\pi)^{n/2}\sqrt{\det \mathbf{T\Sigma T'}}} \exp\left(\frac{-(\mathbf{x} - \mathbf{T}\boldsymbol{\mu})'(\mathbf{T\Sigma T'})^{-1}(\mathbf{y} - \mathbf{T}\boldsymbol{\mu})}{2}\right),
$$

which we can further manipulate, given the special form of $(\mathbf{T\Sigma T'})^{-1}$. That is,

$$
(\mathbf{x} - \mathbf{T}\boldsymbol{\mu})'(\mathbf{T\Sigma T'})^{-1}(\mathbf{y} - \mathbf{T}\boldsymbol{\mu}) = (\hat{\mathbf{y}} - \hat{\boldsymbol{\mu}})'\mathbf{G}^{-1}(\hat{\mathbf{y}} - \hat{\boldsymbol{\mu}}) + (\check{\mathbf{y}} - \check{\boldsymbol{\nu}})'\mathbf{H}^{-1}(\check{\mathbf{y}} - \check{\boldsymbol{\nu}}).
$$

Consequently,

$$
f_{\mathbf{Y}}(\mathbf{y}) = \left(\frac{1}{(2\pi)^{k/2}\sqrt{\det \mathbf{G}}} \exp\left(\frac{-(\hat{\mathbf{y}} - \hat{\boldsymbol{\mu}})'\mathbf{G}^{-1}(\hat{\mathbf{y}} - \hat{\boldsymbol{\mu}})}{2}\right)\right) \cdot
$$

$$
\left(\frac{1}{(2\pi)^{(n-k)/2}\sqrt{\det \mathbf{H}}} \exp\left(\frac{-(\check{\mathbf{y}} - \check{\boldsymbol{\nu}})'\mathbf{H}^{-1}(\check{\mathbf{y}} - \check{\boldsymbol{\nu}})}{2}\right)\right),
$$

which is simply $f_{\hat{\mathbf{Y}}}(\hat{\mathbf{y}}) \cdot f_{\check{\mathbf{Y}}}(\check{\mathbf{y}})$. We conclude that $\hat{\mathbf{X}} = \hat{\mathbf{Y}}$ is multivariate normal with parameters $(\hat{\boldsymbol{\mu}}, \mathbf{G})$. But, we can perform enough of the multiplication

$\mathbf{T}\boldsymbol{\Sigma}\mathbf{T}'$ to identify $\mathbf{G}$. Specifically,

$$\left[\begin{array}{c|c}\mathbf{G} & \mathbf{0} \\ \hline \mathbf{0} & \mathbf{H}\end{array}\right] = \left[\begin{array}{c|c}\mathbf{I} & \mathbf{0} \\ \hline \mathbf{E} & \mathbf{F}\end{array}\right]\left[\begin{array}{c|c}\mathbf{A} & \mathbf{B} \\ \hline \mathbf{B}' & \mathbf{C}\end{array}\right]\left[\begin{array}{c|c}\mathbf{I}' & \mathbf{E}' \\ \hline \mathbf{0}' & \mathbf{F}'\end{array}\right] = \left[\begin{array}{c|c}\mathbf{A} & \mathbf{B} \\ \hline \mathbf{EA} + \mathbf{FB}' & \mathbf{EB} + \mathbf{FC}\end{array}\right]\left[\begin{array}{c|c}\mathbf{I} & \mathbf{E}' \\ \hline \mathbf{0} & \mathbf{F}'\end{array}\right]$$

$$= \left[\begin{array}{c|c}\mathbf{A} & \mathbf{AE}' + \mathbf{BF}' \\ \hline \mathbf{EA} + \mathbf{FB}' & (\mathbf{EA} + \mathbf{FB}')\mathbf{E}' + (\mathbf{EB} + \mathbf{FC})\mathbf{F}'\end{array}\right].$$

Consequently, $\mathbf{G} = \mathbf{A}$, which is the upper left $k \times k$ corner of $\boldsymbol{\Sigma}$. This shows that $\hat{\mathbf{X}}$ is multivariate normal with means and covariances obtained by selecting rows and columns 1 through k from $(\boldsymbol{\mu}, \boldsymbol{\Sigma})$. The computation also reveals how to construct the required transformation $\mathbf{T}$. We must have all zeros in the off-diagonal segments of $\mathbf{T}\boldsymbol{\Sigma}\mathbf{T}'$. Because $\mathbf{A}' = \mathbf{A}$, these components are transposes of one another. That is,

$$\mathbf{AE}' + \mathbf{BF}' = \mathbf{A}'\mathbf{E}' + \mathbf{BF}' = (\mathbf{EA} + \mathbf{FB}')'.$$

Therefore, if we achieve zero status for one of them, we also zero out the other. The required $\mathbf{T}$ then arises if we can choose $\mathbf{E}$ and $\mathbf{F}$ such that $\mathbf{EA} + \mathbf{FB}' = \mathbf{0}$. If we let $\mathbf{F}$ equal the $(n-k) \times (n-k)$ identity matrix, we ensure that $\det \mathbf{T} = 1$, which assures an invertible transformation. Then we solve for $\mathbf{E}$:

$$\mathbf{EA} + \mathbf{FB}' = \mathbf{0}$$
$$\mathbf{EA} = -\mathbf{IB}' = -\mathbf{B}'$$
$$\mathbf{E} = -\mathbf{B}'\mathbf{A}^{-1}.$$

The required transformation is then $\mathbf{T} = \left[\begin{array}{c|c}\mathbf{I} & \mathbf{0} \\ \hline \mathbf{B}'\mathbf{A}^{-1} & \mathbf{I}\end{array}\right]$.

6.18 Theorem: Let $\mathbf{X}$ be multivariate normal with parameters $(\boldsymbol{\mu}, \boldsymbol{\Sigma})$. If we construct $\mathbf{Y}$ by selecting components from $\mathbf{X}$, then $\mathbf{Y}$ is multivariate normal with parameters obtained by selecting the same components from $(\boldsymbol{\mu}, \boldsymbol{\Sigma})$.

PROOF: See discussion above. ∎

6.19 Example: Suppose that $\mathbf{X}$ is multivariate normal with parameters

$$\boldsymbol{\mu} = \begin{bmatrix} 1 \\ 2 \\ 3 \\ 4 \\ 5 \end{bmatrix} \qquad \boldsymbol{\Sigma} = \begin{bmatrix} 42 & -39 & -21 & 8 & -10 \\ -39 & 54 & 22 & -4 & 15 \\ -21 & 22 & 12 & -4 & 5 \\ 8 & -4 & -4 & 4 & 0 \\ -10 & 15 & 5 & 0 & 5 \end{bmatrix}.$$

Verify that components 2 and 4 are bivariate normal with parameters

$$\mathbf{m} = \begin{bmatrix} 2 \\ 4 \end{bmatrix} \qquad \mathbf{S} = \begin{bmatrix} 54 & -4 \\ -4 & 4 \end{bmatrix}.$$

The parameters $(\mathbf{m}, \mathbf{S})$ are obtained by striking rows and columns $1, 3, 5$ from the original parameters. We first verify that the covariance matrix is positive definite. Using the row and column operations illustrated earlier, we evolve the matrix as follows.

$$\begin{array}{ccccc|ccccc} 1 & 0 & 0 & 0 & 0 & 42 & -39 & -21 & 8 & -10 \\ 0 & 1 & 0 & 0 & 0 & -39 & 54 & 22 & -4 & 15 \\ 0 & 0 & 1 & 0 & 0 & -21 & 22 & 12 & -4 & 5 \\ 0 & 0 & 0 & 1 & 0 & 8 & -4 & -4 & 4 & 0 \\ 0 & 0 & 0 & 0 & 1 & -10 & 15 & 5 & 0 & 5 \end{array} \longrightarrow$$

$$
\begin{array}{ccccc|ccccc}
1.0000 & 0.0000 & 0.0000 & 0.0000 & 0.0000 & 42.0000 & 0.0000 & 0.0000 & 0.0000 & 0.0000 \\
0.9286 & 1.0000 & 0.0000 & 0.0000 & 0.0000 & 0.0000 & 17.7857 & 0.0000 & 0.0000 & 0.0000 \\
0.3695 & -0.1406 & 1.0000 & 0.0000 & 0.0000 & 0.0000 & 0.0000 & 1.1486 & 0.0000 & 0.0000 \\
-0.2145 & -0.2517 & 0.4196 & 1.0000 & 0.0000 & 0.0000 & 0.0000 & 0.0000 & 1.6131 & 0.0000 \\
0.2601 & -0.3468 & 0.5780 & -0.2890 & 1.0000 & 0.0000 & 0.0000 & 0.0000 & 0.0000 & 0.0867
\end{array}
$$

The left half of the result is the $\mathbf{B}'$ that produces $\mathbf{B}'\mathbf{\Sigma}\mathbf{B}$ equal to the right half. The positive diagonal entries in the right half then show that the covariance matrix is positive definite. We could continue the evolution into the identity matrix by premultiplying and postmultiplying by a diagonal matrix containing the square roots of the diagonal entries on the right. This process produces five linear combinations that are decoupled in the sense that their covariances are all zero. However, we want only to decouple components 2 and 4 from the rest. We therefore proceed in the spirit of the discussion prior to Theorem 6.18 and partition the original mean vector and covariance matrix to isolate the desired components. We first rearrange the $\mathbf{X}$ vector so that X_2 and X_4 are the first two components. This gives

$$
\mathbf{X} = \begin{bmatrix} X_2 \\ X_4 \\ X_1 \\ X_3 \\ X_5 \end{bmatrix} \qquad \mu = \begin{bmatrix} 2 \\ 4 \\ 1 \\ 3 \\ 5 \end{bmatrix} \qquad \mathbf{\Sigma} = \left[\begin{array}{cc|ccc} 54 & -4 & -39 & 22 & 15 \\ -4 & 4 & 8 & -4 & 0 \\ \hline -39 & 8 & 42 & -21 & -10 \\ 22 & -4 & -21 & 12 & 5 \\ 15 & 0 & -10 & 5 & 5 \end{array}\right].
$$

If we write the covariance matrix as $\mathbf{\Sigma} = \left[\begin{array}{c|c} \mathbf{A} & \mathbf{B} \\ \hline \mathbf{B}' & \mathbf{C} \end{array}\right]$, then the required transformation

is $\mathbf{T} = \left[\begin{array}{c|c} \mathbf{I} & \mathbf{0} \\ \hline -\mathbf{B}'\mathbf{A}^{-1} & \mathbf{I} \end{array}\right]$. Because

$$
\mathbf{B}'\mathbf{A}^{-1} = \begin{bmatrix} -39 & 8 \\ 22 & -4 \\ 15 & 0 \end{bmatrix} \begin{bmatrix} 54 & -4 \\ -4 & 4 \end{bmatrix}^{-1} = \begin{bmatrix} -0.6200 & 1.3800 \\ 0.3600 & -0.6400 \\ 0.3000 & 0.3000 \end{bmatrix},
$$

we have

$$
\mathbf{T} = \left[\begin{array}{cc|ccc} 1.0000 & 0.0000 & 0.0000 & 0.0000 & 0.0000 \\ 0.0000 & 1.0000 & 0.0000 & 0.0000 & 0.0000 \\ \hline 0.6200 & -1.3800 & 1.0000 & 0.0000 & 0.0000 \\ -0.3600 & 0.6400 & 0.0000 & 1.0000 & 0.0000 \\ -0.3000 & -0.3000 & 0.0000 & 0.0000 & 1.0000 \end{array}\right].
$$

A matrix multiplication then verifies that

$$
\mathbf{T}\mathbf{\Sigma}\mathbf{T}' = \left[\begin{array}{cc|ccc} 54.0000 & -4.0000 & 0.0000 & 0.0000 & 0.0000 \\ -4.0000 & 4.0000 & 0.0000 & 0.0000 & 0.0000 \\ \hline 0.0000 & 0.0000 & 6.7800 & -1.8400 & -0.7000 \\ 0.0000 & 0.0000 & -1.8400 & 1.5200 & -0.4000 \\ 0.0000 & 0.0000 & -0.7000 & -0.4000 & 0.5000 \end{array}\right] \qquad \mathbf{T}\mu = \left[\begin{array}{c} 2.0000 \\ 4.0000 \\ \hline -3.2800 \\ 4.8400 \\ 3.2000 \end{array}\right],
$$

which separates the density of $\mathbf{Y} = \mathbf{T}\mathbf{X}$ into two factors. The first is the density of the first two $\mathbf{Y}$ components, which are X_2 and X_4. These components are now evidently multivariate normal with the anticipated parameters. $\square$

We can now state a more general form of Theorem 6.17. Recall that theorem asserts that $\mathbf{Y} = \mathbf{T}\mathbf{X}$ is multivariate normal if $\mathbf{T}$ is invertible. This requires that $\mathbf{T}$ be square. In truth, $\mathbf{Y} = \mathbf{T}\mathbf{X}$ is multivariate normal for nonsquare $\mathbf{T}$, provided only that the rows of $\mathbf{T}$ are independent.

6.20 Theorem: Let $\mathbf{X}$ be multivariate normal with parameters (μ, Σ). Suppose that $\mathbf{X}$ has n components. Let $\mathbf{T}$ be $m \times n$, where $m \leq n$, and let the rows of $\mathbf{T}$ be independent. Then $\mathbf{Y} = \mathbf{TX}$ is multivariate normal with parameters $(\mathbf{T}\mu, \mathbf{T}\Sigma\mathbf{T}')$.

PROOF: According to Theorem A.77, we can add $n - m$ additional independent rows to $\mathbf{T}$, forming $\left[\dfrac{\mathbf{T}}{\mathbf{W}}\right]$. Then

$$\left[\frac{\mathbf{Y}}{\mathbf{Z}}\right] = \left[\frac{\mathbf{T}}{\mathbf{W}}\right] \mathbf{X}$$

is multivariate normal with parameters

$$\left[\frac{\mathbf{T}}{\mathbf{W}}\right]\mu = \left[\frac{\mathbf{T}\mu}{\mathbf{W}\mu}\right]$$

$$\left[\frac{\mathbf{T}}{\mathbf{W}}\right]\Sigma\,[\,\mathbf{T}'\,|\,\mathbf{W}'\,] = \left[\frac{\mathbf{T}\Sigma}{\mathbf{W}\Sigma}\right][\,\mathbf{T}'\,|\,\mathbf{W}'\,] = \left[\begin{array}{c|c}\mathbf{T}\Sigma\mathbf{T}' & \mathbf{T}\Sigma\mathbf{W}' \\ \hline \mathbf{W}\Sigma\mathbf{T}' & \mathbf{W}\Sigma\mathbf{W}'\end{array}\right].$$

Theorem 6.18 then asserts that any subset of selected components is multivariate normal with parameters selected similarly from the overall mean vector and covariance matrix. We select the first m components, which comprise the $\mathbf{Y}$ vector. The selected parameters are $\mathbf{T}\mu$ and $\mathbf{T}\Sigma\mathbf{T}'$. ∎

Exercises

6.15 Suppose that row and column operations on $\mathbf{D}$ via elementary matrices $\mathbf{E}_{i,j;t}$ yield a matrix with a nonpositive diagonal element. Show that $\mathbf{D}$ cannot be positive definite.

6.16 Example 6.15 illustrates a technique for reducing a positive definite matrix $\mathbf{D}$ to the identity with a transformation of the form $\mathbf{B}'\mathbf{DB}$. The technique simultaneously evolves the identity matrix into $\mathbf{B}'$ by operating on an extended array that starts with the identity matrix to the left of $\mathbf{D}$. Devise a similar technique that evolves $\mathbf{B}$ rather than $\mathbf{B}'$.

6.17 Show that the following matrix is positive definite and identify the matrix $\mathbf{B}$ such that $\mathbf{B}'\mathbf{DB} = \mathbf{I}$.

$$\mathbf{D} = \begin{bmatrix} 16 & 13 & 14 & 2 \\ 13 & 22 & 11 & 7 \\ 14 & 11 & 16 & 0 \\ 2 & 7 & 0 & 4 \end{bmatrix}$$

6.18 Let $\mathbf{D}$ be positive definite. Form $\mathbf{D}_1$ by deleting rows $i_1, i_2, \ldots, i_k$ and columns $i_1, i_2, \ldots, i_k$. Show that $\mathbf{D}_1$ remains positive definite and $\det \mathbf{D}_1 > 0$.

6.19 Suppose that $\mathbf{X}$ is multivariate normal with parameters

$$\mu = \begin{bmatrix} 0 \\ 0 \\ 0 \\ 0 \end{bmatrix} \qquad \Sigma = \begin{bmatrix} 22 & -11 & -3 & 8 \\ -11 & 21 & -3 & -8 \\ -3 & -3 & 3 & 0 \\ 8 & -8 & 0 & 4 \end{bmatrix}.$$

Find an invertible transformation $\mathbf{Y} = \mathbf{TX}$ such that $Y_i = X_i$, for $i = 1, 2$, and $\mathrm{cov}(Y_i, Y_j) = 0$ when $i \in \{1, 2\}$ and $j \in \{3, 4\}$.

6.20 Suppose that $\mathbf{X}$ is multivariate normal with parameters

$$\mu = \begin{bmatrix} 1 \\ -1 \\ 2 \\ 0 \end{bmatrix} \qquad \Sigma = \begin{bmatrix} 11 & 6 & -2 & 17 \\ 6 & 5 & 0 & 12 \\ -2 & 0 & 2 & 0 \\ 17 & 12 & 0 & 33 \end{bmatrix}.$$

Show that $\mathbf{Y} = \mathbf{TX}$ is bivariate normal under the following transformation. Find the mean vector and covariance matrix of $\mathbf{Y}$.

$$\mathbf{T} = \begin{bmatrix} 1 & 4 & 1 & -1 \\ 2 & 0 & 6 & 3 \end{bmatrix}$$

6.21 Suppose that $\mathbf{X}$ is multivariate normal with parameters

$$\mu = \begin{bmatrix} 0 \\ 0 \\ 0 \\ 0 \end{bmatrix} \qquad \Sigma = \begin{bmatrix} 22 & -11 & -3 & 8 \\ -11 & 21 & -3 & -8 \\ -3 & -3 & 3 & 0 \\ 8 & -8 & 0 & 4 \end{bmatrix}.$$

Find an invertible transformation $\mathbf{Y} = \mathbf{TX}$ such that the $Y_1, Y_2, \ldots, Y_4$ are independent standard normal random variables.

6.2   Limit theorems

This section's main objective is to prove certain limiting properties of sequences of distributions. However, we first investigate some questions about integrals that appeared in preceding sections. The analysis extends the integral properties discussed in Section A.4.

Consider the question of absolute convergence that arose in connection with the expected value computation. For a one-component, one-dimensional density $f_X(\cdot)$, we assert $E[g(\cdot)] = \int_{-\infty}^{\infty} g(x) f_X(x)\, dx$, provided that the integral is absolutely convergent. That is, we insist that $\int_{-\infty}^{\infty} |g(x) f_X(x)|\, dx < \infty$. Otherwise, we say that the expected value does not exist. The reason for this constraint is easily demonstrated with an example. Consider the density $f(x) = 1/[\pi(1 + x^2)]$, known as a Cauchy distribution. The total assigned probability is 1, but the calculation for the mean encounters the absolute convergence constraint.

$$\int_{-\infty}^{\infty} f(x)\, dx = \int_{-\infty}^{\infty} \frac{dx}{\pi(1 + x^2)} = \frac{1}{\pi} \tan^{-1} x \Big|_{-\infty}^{\infty} = \frac{1}{\pi} \left[\frac{\pi}{2} + \frac{\pi}{2} \right] = 1$$

$$
\mu = \int_{-\infty}^{\infty} \frac{x \, dx}{\pi(1 + x^2)} = \lim_{M,N \to \infty} \int_{-M}^{N} \frac{x \, dx}{\pi(1 + x^2)}
$$

$$
= \lim_{M,N \to \infty} \left[\int_{-M}^{0} \frac{x \, dx}{\pi(1 + x^2)} + \int_{0}^{N} \frac{x \, dx}{\pi(1 + x^2)} \right]
$$

$$
= \lim_{M,N \to \infty} \left[\frac{1}{2\pi} \ln(1 + x^2) \Big|_{-M}^{0} + \frac{1}{2\pi} \ln(1 + x^2) \Big|_{0}^{N} \right]
$$

$$
= \frac{1}{2\pi} \lim_{M,N \to \infty} \left[\ln(1 + N^2) - \ln(1 + M^2) \right]
$$

Because both $\ln(1 + M^2)$ and $\ln(1 + N^2)$ are unbounded as $M, N \to \infty$, the last expression has no limit. Given any position n_0 in the sequence, we hold $M > n_0$ fixed and let N increase until the expression assumes a large positive value. We then hold N at that point and let M increase until the expression reverses and assumes a large negative value. Therefore, beyond any fixed n_0, there is a pair (m_1, n_1) where the expression is a large negative value and another pair (m_2, n_2) where it is a large positive value. Actually, there are infinitely many such pairs, because we can continue the oscillation process. Consequently, no limit exists. The Cauchy distribution has no mean.

The symmetric limit $\lim_{N \to \infty} \int_{-N}^{N} x f(x) \, dx$ does exist. For each N, the integral is zero, and therefore the limit is zero. However, we do not define the improper integral as a symmetric limit. Recall that the integral is simply the limiting form of a collection of approximating sums. The symmetric limit relies on assembling the contributions in the approximating sum in a particular order so that positive contributions always balance negative ones. This is not a satisfactory approach for determining the mean, which we can envision as the balance point of the density function. That balance point should not depend on a specific order in which the weights to the right and left are deployed.

In the example above, the integral for μ is not absolutely convergent:

$$
\int_{-\infty}^{\infty} \left| \frac{x}{\pi(1 + x^2)} \right| dx = \frac{2}{\pi} \int_{0}^{\infty} \frac{x \, dx}{1 + x^2} = \frac{1}{\pi} \lim_{N \to \infty} \ln(1 + N^2) = \infty.
$$

We now show how absolute convergence protects against the difficulty exhibited in the example. Intuitively, we conclude that absolute convergence keeps the integral from decomposing into positive and negative components that can alternately override each other. To make this notion precise, suppose that $g(\cdot)$ is integrable over all finite spans $[a, b]$ and has a finite absolute integral: $\int_{-\infty}^{\infty} |g(x)| \, dx = S < \infty$. For each $n = 1, 2, \ldots$, there exist a_n and b_n such that

$$
\left| \int_{a}^{b} |g(x)| \, dx - S \right| < \frac{1}{2n},
$$

whenever $a \leq a_n$ and $b \geq b_n$. We can choose the a_n and b_n such that $a_1 > a_2 > \cdots$ and $b_1 < b_2 < \cdots$. Consequently, for $n \geq m$ we have the difference $D_{m,n}$ bounded as follows.

$$D_{m,n} = \left| \int_{a_n}^{b_n} g(x)\,dx - \int_{a_m}^{b_m} g(x)\,dx \right| = \left| \int_{a_n}^{a_m} g(x)\,dx + \int_{b_m}^{b_n} g(x)\,dx \right|$$

$$\leq \left| \int_{a_n}^{a_m} g(x)\,dx \right| + \left| \int_{b_m}^{b_n} g(x)\,dx \right| \leq \int_{a_n}^{a_m} |g(x)|\,dx + \int_{b_m}^{b_n} |g(x)|\,dx$$

$$= \left| \int_{a_n}^{b_n} |g(x)|\,dx - \int_{a_m}^{b_m} |g(x)|\,dx \right|$$

$$\leq \left| \int_{a_n}^{b_n} |g(x)|\,dx - S \right| + \left| S - \int_{a_m}^{b_m} |g(x)|\,dx \right| < \frac{1}{2n} + \frac{1}{2m} \leq \frac{1}{m}.$$

Therefore, $\int_{a_n}^{b_n} g(x)\,dx$ is a Cauchy sequence and possesses a limit L. We claim $\int_{-\infty}^{\infty} g(x)\,dx = L$. Indeed, given $\epsilon > 0$, choose $n > 2/\epsilon$ such that

$$\left| \int_{a_n}^{b_n} g(x)\,dx - L \right| < \frac{\epsilon}{2}.$$

Then, for $a \leq a_n$ and $b \geq b_n$, we have

$$\left| \int_a^b g(x)\,dx - L \right| = \left| \int_a^{a_n} g(x)\,dx + \int_{a_n}^{b_n} g(x)\,dx + \int_{b_n}^b g(x)\,dx - L \right|$$

$$\leq \left| \int_{a_n}^{b_n} g(x)\,dx - L \right| + \int_a^{a_n} |g(x)|\,dx + \int_{b_n}^b |g(x)|\,dx$$

$$\leq \frac{\epsilon}{2} + \left| \int_a^b |g(x)|\,dx - \int_{a_n}^{b_n} |g(x)|\,dx \right|$$

$$\leq \frac{\epsilon}{2} + \left| \int_a^b |g(x)|\,dx - S \right| + \left| S - \int_{a_n}^{b_n} |g(x)|\,dx \right| < \frac{\epsilon}{2} + \frac{2}{2n} < \epsilon.$$

We conclude that $\int_{-\infty}^{\infty} |g(x)|\,dx = S < \infty$ forces $\int_{-\infty}^{\infty} g(x)\,dx$ to exist as a finite number. That is, the limit process for the latter integral does not involve a trade-off between arbitrarily large positive and negative components. Moreover, the analysis is essentially unchanged if the integrand limits approach the boundaries of some finite or half-infinite interval (a, b), $(-\infty, b)$, or (a, ∞). We summarize the more general result as the following theorem.

6.21 Theorem: Suppose that $g(\cdot)$ is integrable on each finite span $[a, b]$ properly contained in (c, d) and has a finite absolute integral on (c, d). Here, c may be $-\infty$, and d may be ∞. Then the integral

$$\int_c^d g(x)\,dx = \lim_{a \to c^+, b \to d^-} \int_a^b g(x)\,dx$$

converges and $|\int_c^d g(x)\,dx| \leq \int_c^d |g(x)|\,dx$.

PROOF: The discussion above shows that $\int_c^d g(x)\,dx$ exists as a real number. Theorem A.51 then asserts that

$$\left| \int_a^b g(x)\,dx \right| \leq \int_a^b |g(x)|\,dx \leq \int_c^d |g(x)|\,dx,$$

for every finite span $[a, b]$ properly contained in (c, d). Consequently, the limit must also obey this bound. ■

6.22 Example: Let $f(x) = x/\sqrt{1 - |x|}$ for $-1 < x < 1$. Evaluate $\int_{-1}^{1} f(x)\, dx$.

From the symmetry of the integrand, we see that the integral, if it exists, must be zero. There is the possibility, however, that the positive and negative contributions are unbounded in such a manner that no limit exists as the integration boundaries approach ± 1. In view of the preceding theorem, it suffices to show that $\int_{-1}^{1} |f(x)|\, dx < \infty$.

$$\int_{-1}^{1} \left| \frac{x}{\sqrt{1 - |x|}} \right| dx = 2 \int_{0}^{1} \frac{x\, dx}{\sqrt{1 - x}}$$

We transform the integral with $y = \sqrt{1 - x}$, $dy = -\, dx/(2\sqrt{1 - x})$ to obtain

$$\int_{-1}^{1} |f(x)|\, dx = 4 \int_{0}^{1} (1 - y^2)\, dy \leq 4 \int_{0}^{1} dy = 4 < \infty. \ \square$$

Exercises

6.22 Show that the second moment of the Cauchy distribution does not exist.

6.23 Suppose that $g(\cdot)$ is integrable on each span $[a, b] \subset (-x_0, x_0)$, but

$$\lim_{x \to -x_0^+} g(x) = -\infty$$
$$\lim_{x \to x_0^-} g(x) = \infty.$$

Suppose further that $\int_{-x_0}^{x_0} |g(x)|\, dx < \infty$. Imitate the proof of Theorem 6.21 to show that $\int_{-x_0}^{x_0} g(x)\, dx$ exists.

6.24 Show that the following $f(x)$ is a valid probability density with a well-defined mean value.

$$f(x) = \begin{cases} \dfrac{1}{3(1 - |x|)^{1/3}}, & -1 < x < 1 \\ 0, & \text{elsewhere} \end{cases}$$

6.2.1 Convergence concepts

We now embark on this section's primary topic, limit relationships among random variables. Suppose that we have a probability space Ω. The infinite sequence of random variables $X_1, X_2, \ldots$ is defined on this space, as is the single random variable Y. With each cycle of the underlying random process, an $\omega \in \Omega$ emerges, which fixes the values of all the random variables. These random variables are simply functions with domain Ω. When ω occurs, $X_i(\omega)$

is a fixed real number for each i. We want to investigate the circumstances that give meaning to the statement $\lim_{n\to\infty} X_n = Y$.

A particularly stringent condition is $\lim_{n\to\infty} X_n(\omega) = Y(\omega)$ for all $\omega \in \Omega$. This is the pointwise convergence of Definition A.18. Under this interpretation, we can infer the value of Y at any given ω by observing the values of $X_n(\omega)$ for large n. The definition of strong convergence is almost this restrictive—but not quite. It insists on convergence only on most of Ω, where we interpret "most" with respect to the probability of the Ω region where pointwise convergence does indeed take place.

6.23 Definition: Let $(\Omega, \mathcal{A}, \Pr(\cdot))$ be a probability space, where $\mathcal{A}$ is the σ-algebra of Ω subsets on which $\Pr(\cdot)$ is defined. Let Y and the sequence $X_1, X_2, \ldots$ be random variables on Ω. With $B = \{\omega \mid \lim_{n\to\infty} X_n(\omega) = Y(\omega)\}$, we say that sequence X_n *converges strongly* to Y if $\Pr(B) = 1$. ■

The $X_i(\cdot)$ are simply functions on Ω. For a given ω, we may have $X_n(\omega) \to Y(\omega)$ or not. We cannot judge strong convergence without consulting the probability allocation across Ω. If the allocation gives zero weight to the nonconverging points, then we can assert strong convergence. The next example considers a single set of functions, but with two different probability allocations.

6.24 Example: Let Ω be the real interval $[0, 1]$. Define

$$Y(\omega) = 1$$

$$X_n(\omega) = \begin{cases} 0, & 0 \leq \omega < 1/2, n \in \{1, 3, 5, 7, \ldots\} \\ 1, & 0 \leq \omega < 1/2, n \in \{2, 4, 6, 8, \ldots\} \\ \dfrac{n-\omega}{n}, & 1/2 \leq \omega \leq 1. \end{cases}$$

On the left half of $[0, 1]$, $X_n(\omega)$ oscillates between 0 and 1 and therefore has no limit. On the right half, the sequence $X_n(\omega) = (n-\omega)/n \to 1$ with increasing n, and that limit happens to be the value of $Y(\omega)$.

Now impose a probability allocation on Ω with the density $f(x) = 1$ for $0 \leq x \leq 1$ and zero elsewhere. This assignment allocates to each subinterval a probability equal to its width. In this case, we cannot assert the strong convergence of X_n. The points where the convergence holds constitutes a set with probability assignment $1/2$. Consider now an alternative probability allocation via the density $g(x) = 2$ for $1/2 \leq x \leq 1$ and zero elsewhere. Under this assignment, each subinterval in $[1/2, 1]$ has probability equal to twice its width. This density still results in a total assignment of 1, but it allocates probability 1 to the set $[1/2, 1]$, where convergence does indeed take place. So, under this second assignment, X_n converges strongly to Y. □

Strong convergence clearly includes the case where $X_n(\omega) \to Y(\omega)$ for all ω. That is, it includes the case where X_n converges to Y everywhere. Following this tradition, if X_n converges strongly to Y, we say that it converges *almost everywhere*, or *almost surely*. The name certainly suggests that there is another form of convergence, which we now define.

6.25 Definition: Let $(\Omega, \mathcal{A}, \Pr(\cdot))$ be a probability space, where $\mathcal{A}$ is the σ-algebra of Ω subsets on which $\Pr(\cdot)$ is defined. Let Y and the sequence

$X_1, X_2, \ldots$ be random variables on Ω. We say that X_n converges weakly to Y if for every $\epsilon > 0$,

$$\lim_{n \to \infty} \Pr(|X_n - Y| \geq \epsilon) = \lim_{n \to \infty} \Pr(\{\omega \mid |X_n(\omega) - Y(\omega)| \geq \epsilon\}) = 0.$$

Weak convergence is also called *convergence in probability*. ▌

With weak convergence, the set where X_n fails to approach Y can move around as n increases, provided that the probability of these mobile sets approaches zero. Consequently, it is possible to construct a scenario in which, for every ω, $X_n(\omega)$ fails to approach $Y(\omega)$ but which still manages to achieve weak convergence. The next example provides the details.

6.26 Example: Let Ω be the real interval $[0,1]$. Allocate probability to intervals with the density $f(x) = 1$ for $0 \leq x \leq 1$. That is, each interval receives probability equal to its width. Define $Y(\omega) = 0$ for all $\omega \in [0,1]$. Define the sequence $X_1, X_2, \ldots$ as follows. Each $X_i(\omega) = 0$, except on subintervals as suggested by the following pattern.

$$
\begin{aligned}
X_1(\omega) &= 1, \ \omega \in [0, 1/2] & X_5(\omega) &= 1, \ \omega \in [1/2, 3/4] \\
X_2(\omega) &= 1, \ \omega \in [1/2, 1] & X_6(\omega) &= 1, \ \omega \in [3/4, 1] \\
X_3(\omega) &= 1, \ \omega \in [0, 1/4] & X_7(\omega) &= 1, \ \omega \in [0, 1/8] \\
X_4(\omega) &= 1, \ \omega \in [1/4, 1/2] & X_8(\omega) &= 1, \ \omega \in [1/8, 1/4]
\end{aligned}
$$

The first two X_i are pulses of height 1 that occupy the left and right halves of $[0,1]$. The next four X_i are pulses of height 1 that occupy successive quarters of the interval. The next eight X_i are pulses of height 1 that occupy successive eighths of the interval. For any fixed ω, the values of $X_n(\omega)$ remain zero for increasing runs of n. However, the mobile pulses revisit ω infinitely often. Each visit boosts the X_n value to 1. That is, beyond any landmark N, there are infinitely many n where $X_n(\omega) = 1$ and infinitely many points where $X_n(\omega) = 0$. Consequently, there is no limit. $X_n(\omega)$ does not converge for any ω. However, we can still say that $X_n \to Y$ weakly because, for any $0 < \epsilon < 1$, the event $|X_n - Y| \geq \epsilon$ corresponds to ω in the subinterval where $X_n(\omega) = 1$. With increasing n, these subinterval widths approach zero. □

The example shows that weak convergence does not necessarily imply strong convergence. The converse, however, is a valid implication, as noted in the following theorem.

6.27 Theorem: If $X_n \to Y$ strongly, then $X_n \to Y$ weakly.

PROOF: Let $\epsilon > 0$ be arbitrary. If we let $D = \{\omega \mid X_n(\omega) \to Y(\omega)\}$, strong convergence ensures that $\Pr(D) = 1$ and consequently, $\Pr(\overline{D}) = 0$. Now let $D_n = \{\omega \mid |X_n(\omega) - Y(\omega)| < \epsilon\}$. If $\omega \in D$, then there exists an N such that $\omega \in D_n$ for all $n \geq N$. Consequently, $\omega \in \cap_{n=N}^{\infty} D_n$. This implies that $D \subset \cup_{N=1}^{\infty} \cap_{n=N}^{\infty} D_n$. Taking complements, we have

$$\bigcap_{N=1}^{\infty} \bigcup_{n=N}^{\infty} \overline{D_n} \subset \overline{D}$$

$$\Pr\left(\bigcap_{N=1}^{\infty} B_N\right) \leq \Pr(\overline{D}) = 0,$$

where $B_N = \cup_{n=N}^{\infty} \overline{D_n}$ is a contracting sequence. That is, $B_1 \supset B_2 \supset B_3 \supset \cdots$. Using Theorem 5.5, we conclude $\lim_{N \to \infty} \Pr(B_N) = 0$. Since $\overline{D_N} \subset \cup_{n=N}^{\infty} \overline{D_n} = B_N$, we also conclude that

$$\lim_{N \to \infty} \Pr(|X_N - Y| \geq \epsilon) = \lim_{N \to \infty} \Pr(\overline{D_N}) = 0,$$

which establishes the weak convergence. ∎

We are now ready to prove the *weak law of large numbers*, also known as Khintchine's theorem.

6.28 Theorem: (Khintchine) Let X_n be a sequence of independent identically distributed random variables with common mean μ and variance σ^2. Define $Y_n = (1/n) \sum_{i=1}^{n} X_i$ and $Y = \mu$. Then Y_n converges weakly to Y.

PROOF: We have $\mu_{Y_n} = (1/n) \sum_{i=1}^{n} \mu_{X_i} = \mu$ and $\sigma_{Y_n}^2 = (1/n^2) \sum_{i=1}^{n} \sigma_{X_i}^2 = \sigma^2/n$. Chebyshev's inequality then asserts that $\Pr(|Y_n - \mu| \geq r\sigma/\sqrt{n}) \leq 1/r^2$ for any $r > 0$.

Given arbitrary $\epsilon > 0$, we need to show $\lim_{n \to \infty} \Pr(|Y_n - \mu| \geq \epsilon) = 0$. Given $\epsilon' > 0$, first fix $r > 0$ such that $1/r^2 < \epsilon'$. Now choose N such that $n \geq N$ implies that $r\sigma/\sqrt{n} < \epsilon$. For $n \geq N$, we then have

$$(|Y_n - \mu| \geq \epsilon) \subset \left(|Y_n - \mu| \geq \frac{r\sigma}{\sqrt{n}} \right)$$

$$\Pr(|Y_n - \mu| \geq \epsilon) \leq \Pr\left(|Y_n - \mu| \geq \frac{r\sigma}{\sqrt{n}} \right) \leq \frac{1}{r^2} < \epsilon',$$

which establishes the desired limit. ∎

6.29 Definition: Let $(\Omega, \mathcal{A}, \Pr(\cdot))$ be a probability space, where $\mathcal{A}$ is the σ-algebra of Ω subsets on which $\Pr(\cdot)$ is defined. Let Y and the sequence $X_1, X_2, \ldots$ be random variables on Ω. We say that X_n *converges in quadratic mean* to Y if $\lim_{n \to \infty} E[(X_n - Y)^2] = 0$. ∎

When Ω is the real numbers and $\Pr(\cdot)$ arises from density f, it is rather straightforward to prove that convergence in quadratic mean implies weak convergence. For arbitrary $\epsilon > 0$, let $B_n = \{\omega \mid |X_n(\omega) - Y(\omega)| \geq \epsilon\}$. Applying Markov's inequality to the nonnegative random variable $(X_n - Y)^2$, we have $E[(X_n - Y)^2] \geq \epsilon^2 \Pr(B_n)$. So, if $E[(X_n - Y)^2] \to 0$, we must also have $\Pr(B_n) \to 0$, which is the defining condition for weak convergence.

However, convergence in quadratic mean is not related to strong convergence. That is, we can have convergence in quadratic mean without strong convergence, or vice versa. Indeed, the sequence X_n of Example 6.26 converges in quadratic mean to the constant zero, but $X_n(\omega)$ does not converge for any ω. Therefore, X_n does not converge strongly. For a counterexample in the reverse direction, consider the following sequence, again on the interval $[0,1]$ with a uniform probability density: $f(x) = 1$ for $0 \leq x \leq 1$, and zero elsewhere.

$$Z_n(\omega) = \begin{cases} \sqrt{n}, & 0 \leq \omega \leq 1/n \\ 0, & \text{elsewhere} \end{cases}$$

Here $Z_n(\omega) \to 0$ for all $\omega \neq 0$. Consequently, Z_n converges strongly to zero. By Theorem 6.27, Z_n converges weakly to zero. Certainly, Z_n cannot converge to zero in quadratic mean because

$$E[(Z_n - 0)^2] = \int_0^1 Z_n^2(\omega)\, d\omega = \int_0^{1/n} \left(\sqrt{n}\right)^2 d\omega = 1.$$

However, we want to show that Z_n does not converge in quadratic mean to any random variable. To establish a contradiction, suppose that Z_n converges in quadratic mean to Y. Then Z_n converges weakly both to Y and to zero. For $k = 1, 2, \ldots$, let $C_k = \{\omega \,||Y(\omega)| > 1/k\}$.

For each k, we will argue that $\Pr(C_k) = 0$. Indeed, given an arbitrary $\epsilon > 0$, the weak convergence allows us to choose n such that

$$\Pr\left(|Z_n - Y| \geq \frac{1}{3k}\right) = \Pr(D_1) < \frac{\epsilon}{2}$$

$$\Pr\left(|Z_n - 0| \geq \frac{1}{3k}\right) = \Pr(D_2) < \frac{\epsilon}{2}.$$

We have $\Pr(D_1 \cup D_2) \leq \Pr(D_1) + \Pr(D_2) < \epsilon$. Also, for $\omega \in \overline{D_1} \cap \overline{D_2}$, we have

$$|Y(\omega)| = |Y(\omega) - Z_n(\omega) + Z_n(\omega) - 0| \leq |Z_n(\omega) - Y(\omega)| + |Z_n(\omega) - 0|$$

$$< \frac{1}{3k} + \frac{1}{3k} < \frac{1}{k}.$$

Consequently, $\omega \in \overline{D_1} \cap \overline{D_2}$ implies that $\omega \in \overline{C_k}$. Equivalently, $C_k \subset \overline{\overline{D_1} \cap \overline{D_2}} = D_1 \cup D_2$, which forces $\Pr(C_k) < \epsilon$. Because $\epsilon > 0$ was arbitrary, we must have $\Pr(C_k) = 0$.

Letting $C = \{\omega \mid |Y(\omega)| > 0\}$, we have $\Pr(C) = \Pr\left(\cup_{k=1}^\infty C_k\right) = 0$. We can exclude a set of width zero from the integral without changing the result.

$$E[(Z_n - Y)^2] = \int_0^1 [Z_n(\omega) - Y(\omega)]^2 f(\omega)\, d\omega = \int_{\overline{C}} [Z_n(\omega) - 0]^2\, d\omega$$

$$= \int_0^1 Z_n^2(\omega)\, d\omega = 1.$$

This contradiction establishes that Z_n does not converge in quadratic mean to any random variable.

Figure 6.7 summarizes the relationships among the three convergence concepts, and Theorem 6.30 restates the valid implications.

6.30 Theorem: Strong convergence implies weak convergence. Convergence in quadratic mean implies weak convergence. However, strong convergence and convergence in quadratic mean are unrelated concepts.

PROOF: See discussion above. ∎

Although the three convergence concepts do exhibit subtle differences, we frequently encounter all three types together, as illustrated in the next example.

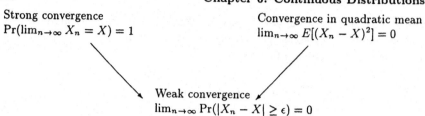

Strong convergence
$\Pr(\lim_{n\to\infty} X_n = X) = 1$

Convergence in quadratic mean
$\lim_{n\to\infty} E[(X_n - X)^2] = 0$

Weak convergence
$\lim_{n\to\infty} \Pr(|X_n - X| \geq \epsilon) = 0$

Figure 6.7. Three convergence concepts for random variables

6.31 Example: Let Ω be the real interval $(0, 1]$ with a uniform probability distribution. Discuss the convergence of the random variable sequence $X_n(\omega) = -(\ln \omega)/n$.

For any $\omega \in (0, 1]$, we have $\lim_{n\to\infty} X_n(\omega) = \lim_{n\to\infty} -(\ln \omega)/n = 0$. Consequently, X_n converges strongly to zero. By Theorem 6.30, X_n also converges weakly to zero. There remains the question of convergence in quadratic mean, which we settle with the computation:

$$E[(X_n - 0)^2] = \int_0^1 \frac{\ln^2 \omega}{n^2}\, d\omega = \left.\frac{\omega(\ln^2 \omega - 2\ln \omega + 2)}{n^2}\right|_0^1 = \frac{2}{n^2} \to 0.$$

Note that X_n has range $[0, \infty)$ and its cumulative distribution is

$$F_{X_n}(x) = \Pr(X_n \leq x) = \Pr\left(\frac{-\ln \omega}{n} \leq x\right) = \Pr(\omega \geq e^{-nx})$$

$$= \Pr(\omega \in [e^{-nx}, 1]) = 1 - e^{-nx}.$$

The density is then $f_{X_n}(x) = ne^{-nx}$. That is, X_n is thus an exponential random variable with parameter n. We conclude that a sequence of such random variables approaches zero under all three modes of convergence. $\square$

Exercises

6.25 Construct an example in which X_n converges weakly to Y, but nevertheless, $\lim_{n\to\infty} \Pr(|X_n - Y| > 0) = 1$. This shows that the ϵ in the definition of weak convergence cannot be replaced with a zero.

6.26 Let Ω be the real interval $(0, 1]$ with probability density $f(\omega) = 1/(2\sqrt{x})$. For the random variable sequence $X_n(\omega) = (-\ln \omega)/n$, discuss strong convergence, weak convergence, and convergence in quadratic mean.

6.27 Let Ω be the real interval $(0, 1]$ with a uniform probability distribution. For the random variable sequence $X_n(\omega) = 1/(n\omega)$, discuss strong convergence, weak convergence, and convergence in quadratic mean.

6.28 Let Ω be the unit square $\{(u, v) \mid 0 < u \leq 1, 0 < v \leq 1\}$ with a uniform probability density. For the random variable sequence $X_n(u, v) = -\ln(uv)/n$, show that X_n approaches zero strongly, weakly, and in quadratic mean.

6.29 Let Ω be the unit square $\{(u, v) \mid 0 < u \le 1, 0 < v \le 1\}$ with a uniform probability density. show that X_n, defined below, approaches zero strongly and weakly but not in quadratic mean.

$$X_n(u, v) = \begin{cases} \sqrt{n}, & 0 < u \le \dfrac{1}{\sqrt{n}}, 0 < v \le \dfrac{1}{\sqrt{n}} \\ \dfrac{-\ln(uv)}{n}, & \dfrac{1}{\sqrt{n}} < u \le 1, \dfrac{1}{\sqrt{n}} < v \le 1 \end{cases}$$

6.2.2 An inversion formula

We define two new generating functions, each closely related to the moment generating function. We call $\Psi_c(t)$ and $\Psi_s(t)$, respectively, the cosine and sine moment generating functions. They are

$$\Psi_c(t) = E[\cos(xt)] = \int \cos(xt)\, dF(x)$$

$$\Psi_s(t) = E[\sin(xt)] = \int \sin(xt)\, dF(x),$$

where F is a cumulative distribution function. The power series expansions of the sine and cosine functions are very similar to that of the exponential.

$$\sin(x) = x - \frac{x^3}{3!} + \frac{x^5}{5!} - \cdots = \sum_{k=0}^{\infty} \frac{(-1)^k x^{2k+1}}{(2k+1)!}$$

$$\cos(x) = 1 - \frac{x^2}{2!} + \frac{x^4}{4!} + \cdots = \sum_{k=0}^{\infty} \frac{(-1)^k x^{2k}}{(2k)!}$$

If the ordinary moment generating function exists as a power series expansion in a nonempty interval around zero, then its sine and cosine relatives do as well. Moreover, the coefficients of the sine and cosine variants are simply selected coefficients from the ordinary moment generating function with some sign changes. This observation follows because the distribution moments determine the coefficients, regardless of the generating function. That is,

$$\Psi_c(t) = \int \left[\sum_{k=0}^{\infty} \frac{(-1)^k (xt)^{2k}}{(2k)!} \right] dF(x) = \sum_{k=0}^{\infty} \frac{(-1)^k t^{2k}}{(2k)!} \int x^{2k}\, dF(x)$$

$$= \sum_{k=0}^{\infty} \frac{(-1)^k t^{2k} E[X^{2k}]}{(2k)!} = \sum_{k=0}^{\infty} \frac{(-1)^k t^{2k} \Psi^{(2k)}(0)}{(2k)!}$$

$$\Psi_s(t) = \sum_{k=0}^{\infty} \frac{(-1)^k t^{2k+1} E[X^{2k+1}]}{(2k+1)!} = \sum_{k=0}^{\infty} \frac{(-1)^k t^{2k+1} \Psi^{(2k+1)}(0)}{(2k+1)!}.$$

Differentiating, we obtain

$$\Psi_c^{(2k)}(0) = (-1)^k \Psi^{(2k)}(0) = (-1)^k E[X^{2k}]$$

$$\Psi_s^{(2k+1)}(0) = (-1)^k \Psi^{(2k+1)}(0) = (-1)^k E[X^{2k+1}],$$

which means that we can also compute the ordinary moment generating function from its sine and cosine relatives.

$$\Psi(t) = \sum_{k=0}^{\infty} \frac{t^k E[X^k]}{k!} = \sum_{k=0}^{\infty} \left[\frac{t^{2k} E[X^{2k}]}{(2k)!} + \frac{t^{2k+1} E[X^{2k+1}]}{(2k+1)!} \right]$$

$$= \sum_{k=0}^{\infty} \left[\frac{(-1)^k t^{2k} \Psi_c^{(2k)}(0)}{(2k)!} + \frac{(-1)^k t^{2k+1} \Psi_s^{(2k+1)}(0)}{(2k+1)!} \right]$$

Because we can compute one from the other, we regard the pair $(\Psi_c(t), \Psi_s(t))$ as equivalent to $\Psi(t)$. This subsection's first goal is to invert $\Psi(t)$. That is, we wish to compute the cumulative distribution function from the moment generating function. In light of the discussion above, it suffices to compute F from the pair $(\Psi_c(t), \Psi_s(t))$.

6.32 Example: For the cumulative distribution function $F(x) = 1 - \lambda e^{-\lambda x}$, $x \geq 0$, recover the moment generating function from the sine and cosine variations.

This distribution has density $f(x) = F'(x) = \lambda e^{-\lambda x}$ for $x \geq 0$. Consequently, we calculate

$$\Psi_c(t) = \int_0^{\infty} \lambda e^{-\lambda x} \cos(tx)\, dx = -e^{-\lambda x} \cos(tx)\Big|_0^{\infty} - t \int_0^{\infty} e^{-\lambda x} \sin(tx)\, dx$$

$$= 1 - \frac{t}{\lambda} \int_0^{\infty} \lambda e^{-\lambda x} \sin(tx)\, dx$$

$$= 1 - \frac{t}{\lambda} \left[-e^{-\lambda x} \sin(tx)\Big|_0^{\infty} + t \int_0^{\infty} e^{-\lambda x} \cos(tx)\, dx \right]$$

$$= 1 - \frac{t^2}{\lambda^2} \int_0^{\infty} \lambda e^{-\lambda x} \cos(tx)\, dx = 1 - \frac{t^2}{\lambda^2} \Psi_c(t)$$

$$\Psi_c(t) = \frac{\lambda^2}{\lambda^2 + t^2} = \frac{1}{1 + (t/\lambda)^2}$$

$$\Psi_s(t) = \int_0^{\infty} \lambda e^{-\lambda x} \sin(tx)\, dx = -e^{-\lambda x} \sin(tx)\Big|_0^{\infty} + t \int_0^{\infty} e^{-\lambda x} \cos(tx)\, dx$$

$$= (t/\lambda) \Psi_c(t) = \frac{(t/\lambda)}{1 + (t/\lambda)^2}.$$

These expressions expand easily to power series, which immediately yield the derivatives at zero.

$$\Psi_c(t) = 1 - \frac{t^2}{\lambda^2} + \frac{t^4}{\lambda^4} - \ldots = \sum_{k=0}^{\infty} \frac{(-1)^k t^{2k}}{\lambda^{2k}}$$

$$\Psi_c^{(2k)}(0) = \frac{(-1)^k (2k)!}{\lambda^{2k}}$$

The odd derivatives are zero: $\Psi_c^{(2k+1)}(0) = 0$. Similarly,

$$\Psi_s(t) = \frac{t}{\lambda} \left[1 - \frac{t^2}{\lambda^2} + \frac{t^4}{\lambda^4} - \ldots \right] = \sum_{k=0}^{\infty} \frac{(-1)^k t^{2k+1}}{\lambda^{2k+1}}$$

$$\Psi_s^{(2k+1)}(0) = \frac{(-1)^k (2k+1)!}{\lambda^{2k+1}}$$

and the even derivatives are zero. We then obtain $\Psi(t)$ from

$$\Psi(t) = \sum_{k=0}^{\infty} \left[\frac{(-1)^k t^{2k} \Psi_c^{(2k)}(0)}{(2k)!} + \frac{(-1)^k t^{2k+1} \Psi_s^{(2k+1)}(0)}{(2k+1)!} \right]$$

$$= \sum_{k=0}^{\infty} \left[\frac{t^{2k}}{\lambda^{2k}} + \frac{t^{2k+1}}{\lambda^{2k+1}} \right] = \sum_{k=0}^{\infty} \frac{t^k}{\lambda^k} = \sum_{k=0}^{\infty} \left(\frac{t}{\lambda} \right)^k = \frac{1}{1 - (t/\lambda)} = \frac{\lambda}{\lambda - t}.$$

We confirm the result with a direct calculation. For $t < \lambda$,

$$\Psi(t) = \int_0^{\infty} e^{tx} \lambda e^{-\lambda x} \, dx = \lambda \int_0^{\infty} e^{-(\lambda - t)x} \, dx = \left. \frac{-\lambda e^{-(\lambda-t)x}}{(\lambda - t)} \right|_0^{\infty} = \frac{\lambda}{\lambda - t}. \; \square$$

The upcoming inversion formula depends on the remarkable ability of the integral $\int_0^{\infty} (1/t) \sin \alpha t \, dt$ to transition sharply when α crosses zero. Because $(1/t) \sin \alpha t$ approaches α as t approaches zero, the integrand does not have a singularity at $t = 0$. Indeed, we define the integrand to be α at $t = 0$ and thereby transform it to a continuous function over the integration span. The integral is clearly zero if $\alpha = 0$. If it exists for nonzero α, then a simple calculation shows that it is independent of the nonzero α. If, $\alpha > 0$, we use the transformation $\tau = \alpha t$, $d\tau = \alpha \, dt$ to obtain

$$\int_0^{\infty} \frac{\sin \alpha t}{t} \, dt = \int_0^{\infty} \frac{\sin \tau}{t/\alpha} \cdot \frac{d\tau}{\alpha} = \int_0^{\infty} \frac{\sin \tau}{\tau} \, d\tau.$$

Similarly, if $\alpha < 0$, the same transformation and the fact that the $(\sin \tau)/\tau$ is an even function recover the symmetric result:

$$\int_0^{\infty} \frac{\sin \alpha t}{t} \, dt = \int_0^{-\infty} \frac{\sin \tau}{\tau} \, d\tau = - \int_{-\infty}^0 \frac{\sin \tau}{\tau} \, d\tau = - \int_0^{\infty} \frac{\sin \tau}{\tau} \, d\tau.$$

We also note that

$$\left| \int_{2n\pi}^{\infty} \frac{\sin t}{t} \, dt \right| = \sum_{k=n}^{\infty} \int_{2k\pi}^{2(k+1)\pi} \frac{\sin t}{t} \, dt$$

$$\leq \sum_{k=n}^{\infty} \left[\frac{1}{2k\pi} \int_{2k\pi}^{(2k+1)\pi} \sin t \, dt + \frac{1}{2(k+1)\pi} \int_{(2k+1)\pi}^{(2k+2)\pi} \sin t \, dt \right]$$

$$= \frac{1}{\pi} \sum_{k=n}^{\infty} \left(\frac{1}{k} - \frac{2}{2k+1} \right) = \frac{1}{\pi} \sum_{k=n}^{\infty} \left(\frac{1}{k(2k+1)} \right)$$

$$< \frac{1}{\pi} \sum_{k=n}^{\infty} \frac{1}{k^2}.$$

Because the tail of the series $\sum 1/k^2$ approaches zero with increasing n, we conclude that the tail of the integral does as well. That is, $\int_0^{\infty} (1/t) \sin t \, dt$ converges. The actual value is $\pi/2$, which can be obtained as a limiting form of the related integral $\int_0^{\infty} (1/t) e^{-\beta t} \sin t \, dt$. The latter integral is somewhat easier to manipulate. In any case, for the inversion formula development to come, it is important only to know that this integral is a definite constant. Nevertheless, we will use the correct constant, $\pi/2$, as we proceed.

6.33 Theorem: $\displaystyle\int_0^\infty \frac{\sin\alpha t}{t}\,dt = \begin{cases} \dfrac{-\pi}{2}, & \alpha < 0 \\[2mm] 0, & \alpha = 0 \\[2mm] \dfrac{\pi}{2}, & \alpha > 0. \end{cases}$

PROOF: See discussion above. ∎

With this technical result in hand, we are ready for the inversion theorem. For a fixed interval $[a,b]$, let $x = (b+a)/2, h = (b-a)/2$. Consider the following integral. Assume throughout the development that the integral involving the cumulative distribution function F spans $-\infty$ to ∞.

$$
\begin{aligned}
I &= \int_0^\infty \frac{2\sin(ht)}{t}\left[\Psi_c(t)\cos(xt) + \Psi_s(t)\sin(xt)\right]\,dt \\[2mm]
&= \int_0^\infty \frac{2\sin(ht)}{t}\left[\cos(xt)\int\cos(yt)\,dF(y) + \sin(xt)\int\sin(yt)\,dF(y)\right]\,dt \\[2mm]
&= \int\left[\int_0^\infty \frac{2\sin(ht)}{t}\left[\cos(xt)\cos(yt) + \sin(xt)\sin(yt)\right]\,dt\right]\,dF(y) \\[2mm]
&= \int\left[\int_0^\infty \frac{2\sin(ht)}{t}\cdot\cos(xt - yt)\,dt\right]\,dF(y) \\[2mm]
&= \int\left[\int_0^\infty \frac{1}{t}[2\sin(ht)\cos(xt - yt)]\,dt\right]\,dF(y).
\end{aligned}
$$

Using the identity $2\sin\theta\cos\phi = \sin(\theta + \phi) + \sin(\theta - \phi)$, we simplify further to obtain

$$
\begin{aligned}
I &= \int\left[\int_0^\infty \frac{1}{t}[\sin(ht + xt - yt) + \sin(ht - xt + yt)]\,dt\right]\,dF(y) \\[2mm]
&= \int\left[\int_0^\infty \frac{1}{t}[\sin(-t(y - x - h)) + \sin(t(y - x + h))]\,dt\right]\,dF(y) \\[2mm]
&= \int\left[\int_0^\infty \frac{\sin[t(y - (x - h))] - \sin[t(y - (x + h))]}{t}\,dt\right]\,dF(y).
\end{aligned}
$$

Evaluating the integrals with Theorem 6.33, we have

$$
\int_0^\infty \frac{\sin[t(y - (x - h))]\,dt}{t} = \begin{cases} -\pi/2, & y < x - h \\ 0, & y = x - h \\ \pi/2, & y > x - h \end{cases}
$$

$$
\int_0^\infty \frac{-\sin[t(y - (x + h))]\,dt}{t} = \begin{cases} \pi/2, & y < x + h \\ 0, & y = x + h \\ -\pi/2, & y > x + h. \end{cases}
$$

Therefore,

$$\int_0^\infty \frac{\sin[t(y-(x-h))] - \sin[t(y-(x+h))]}{t}\, dt = \begin{cases} 0, & y < x-h \\ \pi/2, & y = x-h \\ \pi, & x-h < y < x+h \\ \pi/2, & y = x+h \\ 0, & y > x+h. \end{cases}$$

Now, $x+h = (b+a)/2 + (b-a)/2 = b$ and $x-h = (b+a)/2 - (b-a)/2 = a$, and we conclude that

$$\int_0^\infty \frac{\sin[t(y-(x-h))] - \sin[t(y-(x+h))]}{t}\, dt = \left(\frac{\pi}{2}\right) \left[\chi_{[a,b]}(y) + \chi_{(a,b)}(y)\right],$$

where χ_B is the characteristic function of a set B. That is, $\chi_B(x) = 1$ when $x \in B$ and $\chi_B(x) = 0$ when $x \notin B$. Finally, we have

$$I = \left(\frac{\pi}{2}\right) \int \left[\chi_{[a,b]} + \chi_{(a,b)}\right]\, dF = \left(\frac{\pi}{2}\right) \left[\Pr([a,b]) + \Pr((a,b))\right]$$

$$= \left(\frac{\pi}{2}\right) \left[F(b) - F(a^-) + F(b^-) - F(a)\right].$$

6.34 Theorem: Let F be a cumulative distribution function, and let $\Psi_c(t)$ and $\Psi_s(t)$ be the sine and cosine moment generating functions. Then

$$\frac{F(b) + F(b^-)}{2} = \lim_{a \to -\infty} \frac{1}{\pi} \int_0^\infty \frac{2\sin[(b-a)t/2]}{t}$$
$$\cdot \left[\Psi_c(t)\cos[(b+a)t/2] + \Psi_s(t)\sin[(b+a)t/2]\right]\, dt.$$

If F is continuous at b, the limit is $F(b)$.

PROOF: The first part follows from the integral I developed above. In particular, let $a \to -\infty$ and recall that $h = (b-a)/2$ and $x = (b+a)/x$. Then both $F(a)$ and $F(a^-)$ approach zero, and

$$\frac{F(b) + F(b^-)}{2} = \lim_{a \to -\infty} \frac{I}{\pi}.$$

Noting the definition of I and substituting for h and x in terms of (a,b), the right side assumes the form asserted by the theorem. If F is continuous at b, then $F(b^-) = F(b)$, and consequently, $(F(b) + F(b^-))/2 = F(b)$. ∎

The sine and cosine moment generating function always exist because $\sin(tx)$ and $\cos(tx)$ are bounded functions. So, $\int |\sin(tx)|\, dF \le \int dF = 1$, which implies not only that $\Psi_s(t)$ exists but also that it is bounded. $|\Psi_s(t)| \le 1$. Similar remarks apply to the cosine moment generating function.

6.35 Theorem: The cosine and sine moment generating functions uniquely determine a distribution.

PROOF: $\Psi_c(t)$ and $\Psi_s(t)$ always exist, and the inversion formula determines the corresponding distribution. Indeed, where F is continuous, the formula determines F precisely. Where F jumps, the formula determines the midpoint of the jump, which, of course, determines the jump itself. ∎

For purposes of deriving a distribution from known generating functions, the inversion theorem is not very helpful. The integrals and limit processes are too complicated. To illustrate the difficulty, the next example performs the inversion for a very simple distribution. Nevertheless, the uniqueness theorem is very useful, as will be seen in the upcoming sections. If we can determine the moment generating function of a distribution, it is often the case that we recognize it as belonging to a familiar distribution. The uniqueness theorem then ensures that we infer the correct distribution.

6.36 Example: The Bernoulli distribution with parameter p has a probability mass of size $(1-p)$ at $x = 0$ and a complementary mass of size p at $x = 1$. The cumulative distribution function is then

$$F(x) = \begin{cases} 0, & x < 0 \\ 1 - p, & 0 \le x < 1 \\ 1, & x \ge 1. \end{cases}$$

The moment generating function is simply $\Psi(t) = e^{0 \cdot t}(1 - p) + e^{1 \cdot t}p = 1 - p + pe^t$. Verify that the inversion formula recovers $F(x)$.

From $\Psi(t)$

$$\Psi(t) = (1 - p) + p\left(1 + t + \frac{t^2}{2!} + \frac{t^3}{3!} + \dots\right) = 1 + pt + \frac{pt^2}{2!} + \frac{pt^3}{3!} + \dots$$

$$\Psi^{(k)}(0) = \begin{cases} 1, & k = 0 \\ p, & k > 0, \end{cases}$$

we have

$$\Psi_s(t) = \sum_{k=0}^{\infty} \frac{(-1)^k t^{2k+1} \Psi^{(2k+1)}(0)}{(2k+1)!} = pt - \frac{pt^3}{3!} + \frac{pt^5}{5!} - \dots = p\sin(t)$$

and

$$\Psi_c(t) = \sum_{k=0}^{\infty} \frac{(-1)^k t^{2k} \Psi^{(2k)}(0)}{(2k)!} = 1 - \frac{pt^2}{2!} + \frac{pt^4}{4!} - \dots$$

$$= 1 + p\left(-1 + 1 - \frac{t^2}{2!} + \frac{t^4}{4!} - \dots\right) = (1 - p) + p\cos(t).$$

We simplify the inversion formula before applying it.

$$2\sin[(b - a)t/2]\cos[(b + a)t/2] = \sin[(b - a)t/2 + (b + a)t/2]$$
$$+ \sin[(b - a)t/2 - (b + a)t/2]$$
$$= \sin(bt) - \sin(at)$$
$$2\sin[(b - a)t/2]\sin[(b + a)t/2] = \cos[(b - a)t/2 - (b + a)t/2]$$
$$- \cos[(b - a)t/2 + (b + a)t/2]$$
$$= \cos(at) - \cos(bt)$$

The integrand of the inversion formula is then

$$K(a, b, t) = \frac{\Psi_c(t)[(\sin(bt) - \sin(at)] + \Psi_s(t)[\cos(at) - \cos(bt)]}{t}.$$

Substituting the sine and cosine moment generating functions for the case at hand, we obtain the following:

$$K = \frac{(1-p)[\sin(bt) - \sin(at)] + p\cos(t)[\sin(bt) - \sin(at)] + p\sin(t)[\cos(at) - \cos(bt)]}{t}$$

$$= \frac{1}{t}\Big[(1-p)[\sin(bt) - \sin(at)]$$

$$+ p[\sin(bt)\cos(t) - \cos(bt)\sin(t) - \sin(at)\cos(t) + \cos(at)\sin(t)]\Big]$$

$$= (1-p)\left[\frac{\sin(bt)}{t} - \frac{\sin(at)}{t}\right] + p\left[\frac{\sin(b-1)t}{t} - \frac{\sin(a-1)t}{t}\right].$$

So,

$$\frac{F(b) + F(b^-)}{2} = \frac{1}{\pi}\lim_{a\to-\infty}\int_0^\infty K(a,b,t)\,dt.$$

The integral $\int K(a,b)\,dt$ has four additive components, each of the form

$$\int_0^\infty \frac{\sin(\alpha t)\,dt}{t} = \begin{cases} -\pi/2, & \alpha < 0 \\ 0, & \alpha = 0 \\ \pi/2, & \alpha > 0. \end{cases}$$

Because we want a to approach $-\infty$, we can safely assume that $a < 0$. As b assumes various positions, the components add in different ways.

$$\int_0^\infty K\,dt = \begin{cases} (1-p)(-\pi/2 + \pi/2) + p(-\pi/2 + \pi/2) = 0, & a < b < 0 \\ (1-p)(0 + \pi/2) + p(-\pi/2 + \pi/2) = (1-p)\pi/2, & a < 0 = b \\ (1-p)(\pi/2 + \pi/2) + p(-\pi/2 + \pi/2) = (1-p)\pi, & a < 0 < b < 1 \\ (1-p)(\pi/2 + \pi/2) + p(0 + \pi/2) = (1-p/2)\pi, & a < 0 < 1 = b \\ (1-p)(\pi/2 + \pi/2) + p(\pi/2 + \pi/2) = \pi, & a < 0 < 1 < b \end{cases}$$

Therefore,

$$L(b) = \frac{F(b) + F(b^-)}{2} = \begin{cases} 0, & a < b < 0 \\ (1-p)/2, & a < 0 = b \\ (1-p), & a < 0 < b < 1 \\ (1-p/2), & a < 0 < 1 = b \\ 1, & a < 0 < 1 < b. \end{cases}$$

$L(b)$ is constant on the spans $(-\infty, 0), (0, 1)$, and $(1, \infty)$. Consider one of these spans, say, $0 < b < 1$. Because F is monotone nondecreasing, we have $L(b) = (F(b) + F(b^-))/2 \leq F(b)$. If $0 < b < c < 1$, we have $L(c) \geq F(c^-) \geq F(b)$ for the same reason. Then $F(b) \leq \lim_{c\to b+} L(c) = L(b) \leq F(b)$, which forces $F(b) = L(b)$. That is, F is continuous on the span $(0, 1)$. A similar analysis holds for the spans $(-\infty, 0)$ and $(1, \infty)$. We conclude that

$$F(b) = \begin{cases} 0, & b < 0 \\ (1-p), & 0 \leq b < 1 \\ 1, & b \geq 1, \end{cases}$$

which is precisely the cumulative distribution function of the Bernoulli distribution. $\square$

6.37 Definition: Let X_n be a sequence of random variables with cumulative distribution functions F_n. Let X be a random variable with cumulative distribution function F. Then X_n *converges in distribution* to X if $F_n(x) \to F(x)$ for every x where F is continuous. ∎

Suppose that X_n converges in distribution to X. If g is a bounded continuous function, we will show that $E_{X_n}[g] \to E_X[g]$. We start with the corresponding integrals over finite spans. In particular, suppose that $[a, b]$ is a bounded interval and that F is continuous at the endpoints. We want to establish

$$\int_a^b g \, dF_n \to \int_a^b g \, dF. \tag{6.2}$$

Let M be the bound on g. That is, $|g(x)| \leq M$ for all x. If $F(b) = F(a)$, we have

$$\left| \int_a^b g \, dF_n \right| \leq \int_a^b |g| \, dF_n \leq M \int_a^b dF_n$$

$$= M[F_n(b) - F_n(a)] \to M[F(b) - F(a)] = 0$$

$$\left| \int_a^b g \, dF \right| \leq \int_a^b M[F(b) - F(a)] = 0,$$

from which Equation 6.2 follows immediately. So, we assume that $F(b) > F(a)$. From Theorem A.44, g continuous on $[a, b]$ implies that g is uniformly continuous there. Given $\epsilon > 0$, there exists $\delta > 0$ such that $|x - y| < \delta$ implies that $|g(x) - g(y)| < \epsilon/[4(F(b) - F(a))]$. The points where F is not continuous form a countable set, so we can construct a partition $a = x_0 < x_1 < \ldots < x_k = b$ such that F is continuous at each boundary point and $|x_i - x_{i-1}| < \delta$. Define

$$c_i = \inf\{g(x) | x_{i-1} \leq x \leq x_i\} \leq M$$
$$d_i = \sup\{g(x) | x_{i-1} \leq x \leq x_i\} \leq M.$$

We then have $0 \leq d_i - c_i \leq \epsilon/[4(F(b) - F(a))] < \epsilon/[2(F(b) - F(a))]$ for $i = 1, 2, \ldots, k$ and

$$\sum_{i=1}^{k} c_i \chi_{(x_{i-1}, x_i]}(x) \leq g(x) \leq \sum_{i=1}^{k} d_i \chi_{(x_{i-1}, x_i]}(x),$$

for $x \in (a, b]$. As before, χ_B is the characteristic function of the set B. Consequently, we can bound the difference D between the integral with respect to F_n and that with respect to F.

$$D = \int_a^b g \, dF - \int_a^b g \, dF_n \leq \sum_{i=1}^{k} d_i \int_a^b \chi_{(x_{i-1}, x_i]} \, dF - \sum_{i=1}^{k} c_i \int_a^b \chi_{(x_{i-1}, x_i]} \, dF_n$$

$$D = \sum_{i=1}^{k} (d_i - c_i)[F(x_i) - F(x_{i-1}) + c_i[(F(x_i) - F(x_{i-1})) - (F_n(x_i) - F_n(x_{i-1}))]$$

$$\leq \frac{\epsilon}{2[F(b) - F(a)]}[F(b) - F(a)] + M[F(b) - F_n(b) + F_n(a) - F(a)]$$

Now, choose N such that $n \geq N$ implies both that $|F_n(b) - F(b)| < \epsilon/(4M)$ and that $|F_n(a) - F(a)| < \epsilon/(4M)$. Then $n \geq N$ forces $\int_a^b g\, dF - \int_a^b g\, dF_n < \epsilon$. A similar argument with reversed inequalities forces the difference greater than $-\epsilon$, which establishes Equation 6.2 for finite integration spans. To extend the result to the infinite span $(-\infty, \infty)$, we argue as follows. Given arbitrary $\epsilon > 0$, first choose a and b such that $a < b$, F is continuous at a and at b, $F(a) < \epsilon/(8M)$, and $F(b) > 1 - \epsilon/(8M)$. Now, choose N such that $n \geq N$ implies that $|F_n(a) - F(a)| < \epsilon/(8M)$, $|F_n(b) - F(b)| < \epsilon/(8M)$, and $\left| \int_a^b g\, dF_n - \int_a^b g\, dF \right| < \epsilon/4$. Of course, $n \geq N$ entails $F_n(a) < \epsilon/(4M)$ and $F_n(b) > 1 - \epsilon/(4M)$. Now, for $n \geq N$, we compute

$$D_\infty = \left| \int_{-\infty}^{\infty} g\, dF_n - \int_{-\infty}^{\infty} g\, dF \right| = \left| \int_{-\infty}^{a} g\, dF_n + \int_a^b g\, dF_n + \int_b^{\infty} g\, dF_n \right.$$

$$\left. - \int_{-\infty}^{a} g\, dF - \int_a^b g\, dF - \int_b^{\infty} g\, dF \right|$$

$$\leq MF_n(a) + M[1 - F_n(b)] + \frac{\epsilon}{4} + MF(a) + M[1 - F(b)]$$

$$< \frac{\epsilon}{4} + \frac{\epsilon}{4} + \frac{\epsilon}{4} + \frac{\epsilon}{8} + \frac{\epsilon}{8} = \epsilon.$$

We have proved the following theorem.

6.38 Theorem: Suppose that X_n converges in distribution to X. Then $E_{X_n}[g] \to E_X[g]$, for every bounded continuous function g.

PROOF: Let F_n and F be the cumulative distribution functions of X_n and X, respectively. The discussion above establishes that $E_{X_n}[g] = \int_{-\infty}^{\infty} g\, dF_n \to \int_{-\infty}^{\infty} g\, dF = E_X[g]$. ∎

6.39 Theorem: Suppose that X_n converges to X in distribution. Then the sine and cosine moment generating functions converge. That is, $\Psi_{c(n)}(t) = E_{X_n}[\cos(tx)] \to E_X[\cos(tx)] = \Psi_c(t)$ and similarly for the sine variant. Where the ordinary moment generating function exists as a power series expansion, it also converges in the same sense.

PROOF: Because $\sin(tx)$ and $\cos(tx)$ are bounded continuous functions of x, the preceding theorem asserts the convergence of the sine and cosine moment generating functions. At a t value where the ordinary moment generating function exists as a power series expansion, it is determined by the sine and cosine variants at the same t. Consequently, the convergence carries over to the ordinary moment generating functions. ∎

Theorem 6.39 constitutes half of what is commonly called the continuity theorem. Loosely speaking, it states that convergence in distribution is equivalent to convergence of generating functions.

6.40 Theorem: (Continuity Theorem) Let X_n be a sequence of random variables, for which the cosine and sine moment generating functions converge to the corresponding generating functions for X. That is, $\Psi_{c(n)}(t) \to \Psi_c(t)$ for all t and similarly for the sine variant. Here, the (n) annotation refers to X_n and the unannotated form refers to X. Then, X_n converges in distribution to X.

PROOF: Let F_n and F be the cumulative distribution functions of X_n and X. We first establish a convergent subsequence $F_{n_k} \to G$.

Note that a bounded sequence of real numbers must have at least one limit point. If $|y_n| < M$, then repeatedly divide the interval $[-M, M]$ in half and note that one of the two parts must contain infinitely many y_n. This produces a sequence of nested intervals, and the least upper bound of their lower endpoints must be a limit point of the sequence. See Theorem A.43 for more detail on this sort of construction. We apply this observation to countably many sequences, each associated with a rational number. Specifically, let $x_1, x_2, x_3, \ldots$ be an enumeration of the rational numbers. Choose a limit point z_1 for the set $\{F_n(x_1)\}$ and a subsequence $F_{n_k}(x_1)$ converging to it. Rename this subsequence $F_n^{(1)}$. Now choose a limit point z_2 from the collection $F_n^{(1)}(x_2)$ and a subsequence $F_{n_k}^{(1)}(x_2)$ converging to it. This second subsequence remains a subsequence of the original F_n and it now converges at both x_1 and x_2. Rename it $F_n^{(2)}$. Continuing in this fashion, we establish a two-dimensional grid, which starts with the following pattern.

$$
\begin{array}{llllll}
F_1^{(1)} & F_2^{(1)} & F_3^{(1)} & F_4^{(1)} & \ldots & F^{(1)}(x_1) \to z_1 \\
F_1^{(2)} & F_2^{(2)} & F_3^{(2)} & F_4^{(2)} & \ldots & F^{(2)}(x_i) \to z_i, \text{ for } i = 1, 2 \\
F_1^{(3)} & F_2^{(3)} & F_3^{(3)} & F_4^{(3)} & \ldots & F^{(3)}(x_i) \to z_i, \text{ for } i = 1, 2, 3 \\
F_1^{(4)} & F_2^{(4)} & F_3^{(4)} & F_4^{(4)} & \ldots & F^{(4)}(x_i) \to z_i, \text{ for } i = 1, 2, 3, 4
\end{array}
$$

The first row defines a subsequence of the original F_n. Each row thereafter defines a subsequence of the preceding row. The diagonal selections $F_1^{(1)}, F_2^{(2)}, \ldots$ then constitute a subsequence of the original F_n that converges for all x_i. That is, if the diagonal sequence is F_{n_k}, we have $F_{n_k}(x_i) \to z_i$ for all i. Define

$$
G(x) = \begin{cases} \lim_{k \to \infty} F_{n_k}(x) = z_i, & x = x_i \\ \text{glb}\,\{G(x_i)|x_i > x\}, & x \notin \{x_1, x_2, \ldots\}. \end{cases}
$$

Because the F_n are monotone nondecreasing, if $x_i < x_j$, then $F_{n_k}(x_i) \leq F_{n_k}(x_j)$ for all k. Consequently, $x_i < x_j$ implies that $G(x_i) \leq G(x_j)$. If y is irrational and $x_i < y$, then $x_i < x_j$, for every $x_j > y$, which implies that $G(x_i) \leq G(x_j)$ for all such x_j. Then, of course, $G(x_i) \leq \text{glb}\{G(x_j)|x_j > y\}$, so $G(x_i) \leq G(y)$. A similar argument applies when irrational y is less than rational x_i. Finally, if $x < y$ and both are irrational, there exists a rational x_i between them. Then

$$G(x) = \text{glb } \{G(x_j)|x_j > x\} \leq G(x_i) \leq \text{glb } \{G(x_j)|x_j > y\} = G(y).$$

We conclude that G is monotone nondecreasing.

Because G is the limit of a sequence of cumulative distribution functions, we must have $A = \lim_{x \to \infty} G(x) \leq 1$. Moreover, a monotone function must have right and left limits. For example, $G(x^-) = \text{lub } \{G(y)|y < x\}$. However, it is not necessarily true that $G(x^-) = G(x^+)$. If $E_n = \{x|G(x^+) - G(x^-) > 1/n\}$, then E_n contains finitely many points. Otherwise, the jumps on E_n would push G above 1. The union $E = \cup_{n=1}^{\infty} E_n$ then contains all discontinuities, and it is countable. At $x \in E$, we redefine G, if necessary, to force $G(x) = G(x^+)$. This induces right continuity and does not affect the value of G at any point of continuity. The total probability assigned by G is A, which is potentially less than 1. In all other respects, G is a cumulative distribution function, and subsequence F_{n_k} converges to G in distribution.

Now, let $\Psi_c^*(t)$ and $\Psi_s^*(t)$ denote the cosine and sine moment generating functions of G. From Theorem 6.39, we have $\Psi_{c(n)}(t) \to \Psi_c^*(t)$ and $\Psi_{s(n)}(t) \to \Psi_s^*(t)$, for all t. By hypothesis, we know that $\Psi_{c(n)}(t) \to \Psi_c(t)$, the cosine moment generating function for the distribution F and similarly for the sine variant. This means that $\Psi_c^* = \Psi_c$ and $\Psi_s^* = \Psi_s$. In particular,

$$A = \int_{-\infty}^{\infty} dG = \int_{-\infty}^{\infty} \cos(0 \cdot x) \, dG(x) = \Psi_c^*(0) = \Psi_c(0) = \int_{-\infty}^{\infty} dF = 1.$$

This removes the last doubt about G as a cumulative distribution function. Using the notation $\overline{F}(x) = [F(x) + F(x^-)]/2$, we invoke the inversion theorem to conclude that $\overline{F}(x) = \overline{G}(x)$ for all x. When F is continuous at x, we have $\overline{F}(x) = F(x)$ and consequently $F(x) = \overline{G}(x)$.

Now suppose that F is continuous at b. Given $\epsilon > 0$, there exists $\delta > 0$ such that $|F(x) - F(b)| < \epsilon/2$ for $|x - b| < \delta$. Because the points where F or G is discontinuous constitute a countable set, we can find points a, c such that F and G are continuous at both points and $b - \delta < a < b < c < b + \delta$. Monotonicity implies that $F_{n_k}(a) \leq F_{n_k}(c)$, for all subsequence functions F_{n_k}. Passing to the limit, we have $G(a) \leq G(c)$. But, between $G(a)$ and $G(b)$ lie both $G(b^-)$ and $G(b^+)$. Specifically,

$$F(a) = \overline{F}(a) = \overline{G}(a) = G(a) \leq G(b^-) \leq G(b^+) \leq G(c) = \overline{G}(c) = \overline{F}(c) = F(c).$$

Moreover, $|F(c) - F(a)| \leq |F(c) - F(b)| + |F(b) - F(a)| < 2(\epsilon/2) = \epsilon$. This implies that $|G(b^+) - G(b^-)| < \epsilon$. Because $\epsilon > 0$ was arbitrary, we conclude that $G(b^+) = G(b^-)$ and G is also continuous at b. Therefore, $F(b) = G(b) = \lim_{k \to \infty} F_{n_k}(b)$.

At this point, we have X_{n_k} converges in distribution to X. Now, consider the full sequence F_n. If x_0 is a point of continuity of F such that $F_n(x_0)$ does not converge to $F(x_0)$, then there exists a $B > 0$ such that $|F_n(x_0) - F(x_0)| > B$ for infinitely many n. Therefore, there is a subsequence F_{n_k} such that

$|F_{n_k}(x_0) - F(x_0)| > B > 0$ for all k. Replace the original sequence F_n with this subsequence, which we rename F_n'. It is clear from the construction that no subsequence of F_n' can converge to F at the point x_0. Nevertheless, we reengage the process described above to obtain a subsequence F_{n_k}' that converges to F at all points of continuity of F. This is possible because the new sequence has the same limiting values for the sine and cosine moment generating functions. This contradiction proves that the original sequence must converge to F at x_0. ∎

6.41 Theorem: (Central limit theorem) Let $(X_1, X_2, \ldots, X_n)$ be a sample from a population X having mean μ and variance σ^2. Suppose that moment generating function for X has a convergent power series in a nonempty interval around zero. Then $Y_n = [1/(\sigma\sqrt{n})] \sum_{i=1}^n (X_i - \mu)$ converges in distribution to the standard normal Z.

PROOF: A power series expansion of the ordinary moment generating function for Z yields its cosine and sine variations, which are then everywhere convergent. From Theorem 6.6 we obtain $\Psi_Z(t) = e^{t^2/2}$. In view of the preceding theorem, we need only show that $\Psi_{Y_n}(t)$ converges to $e^{t^2/2}$. The cosine and sine variations then follow in lockstep, which implies that the Y_n converge in distribution to Z.

Note that $Y_n \sigma \sqrt{n} = \sum_{i=1}^n (X_i - \mu)$ and therefore has a moment generating function which is the nth power of the moment generating function for $(X - \mu)$. Because we know that the coefficients of the linear and quadratic terms are $E[X - \mu] = 0$ and $E[(X - \mu)^2]/2! = \sigma^2/2!$, we can write

$$\Psi_{Y_n \sigma \sqrt{n}}(t) = \left(1 + \frac{\sigma^2 t^2}{2!} + c_3 t^3 + c_4 t^4 + \ldots\right)^n = \left(1 + \frac{\sigma^2 t^2}{2} + o(t^2)\right)^n$$

$$\Psi_{Y_n}(t) = E[e^{tY_n}] = E[e^{tY_n \sigma \sqrt{n}/(\sigma\sqrt{n})}] = \Psi_{Y_n \sigma \sqrt{n}}(t/(\sigma\sqrt{n}))$$

$$= \left(1 + \frac{t^2}{2n} + o[t^2/(n\sigma^2)]\right)^n,$$

where $o(y)$ is such that $\lim_{y\to 0} o(y)/y = 0$. Continuing, we have

$$\Psi_{Y_n}(t) = \sum_{k=0}^n C_{n,k} \left[1 + \frac{t^2}{2n}\right]^k \left[o\left(\frac{t^2}{n\sigma^2}\right)\right]^{n-k}$$

$$= \left[1 + \frac{t^2}{2n}\right]^n + \sum_{k=0}^{n-1} C_{n,k} \left[1 + \frac{t^2}{2n}\right]^k \left[o\left(\frac{t^2}{n\sigma^2}\right)\right]^{n-k} = \left[1 + \frac{t^2}{2n}\right]^n + S_n.$$

Disregarding S_n for the moment, we investigate the limit of the primary term as $n \to \infty$. We have

$$\ln\left[1 + \frac{t^2}{2n}\right]^n = n \ln\left[1 + \frac{t^2}{2n}\right] = \frac{\ln[1 + t^2/(2n)]}{1/n}$$

$$\to \frac{[1 + t^2/(2n)]^{-1}(t^2/2)(-1/n^2)}{-1/n^2} \to \frac{t^2}{2},$$

which implies that $[1 + t^2/(2n)]^n \to e^{t^2/2}$. This is a promising development, as we now need only show that $S_n \to 0$. Accordingly, we manipulate S_n as follows.

$$S_n = \left(\frac{o[t^2/(n\sigma^2)]}{t^2/(n\sigma^2)}\right) \sum_{k=0}^{n-1} C_{n,k} \left(1 + \frac{t^2}{2n}\right)^k \left(\frac{t^2}{n\sigma^2}\right)^{n-k} \left(\frac{o[t^2/(n\sigma^2)]}{t^2/(n\sigma^2)}\right)^{n-k-1}$$

$$= \frac{t^2}{\sigma^2} \left(\frac{o[t^2/(n\sigma^2)]}{t^2/(n\sigma^2)}\right) \cdot$$

$$\sum_{k=0}^{n-1} C_{n,k} \left(\frac{1}{n}\right)^{n-k} \left(1 + \frac{t^2}{2n}\right)^k \left(\frac{t^2}{\sigma^2}\right)^{n-k-1} \left(\frac{o[t^2/(n\sigma^2)]}{t^2/(n\sigma^2)}\right)^{n-k-1}$$

There exists N such that $n \geq N$ implies that $|o[t^2/(n\sigma^2)]/[t^2/(n\sigma^2)]| < \sigma^2/t^2$, which in turn forces the product of the last two factors in the summand to be less than 1 in absolute value. Moreover,

$$C_{n,k} \left(\frac{1}{n^{n-k}}\right) = \frac{n!}{k!(n-k)!n^{n-k}} = \left(\frac{n}{(n-k)n^{n-k}}\right) \left(\frac{(n-1)!}{k!(n-k-1)!}\right)$$

$$= \frac{1}{n^{n-k-1}(n-k)} C_{n-1,k} \leq C_{n-1,k},$$

for $0 \leq k \leq n - 1$. Therefore, $n \geq N$ implies that

$$|S_n| \leq \frac{t^2}{\sigma^2} \left(\frac{o(t^2/(n\sigma^2))}{t^2/(n\sigma^2)}\right) \sum_{k=0}^{n-1} C_{n-1,k} \left(1 + \frac{t^2}{2n}\right)^k (1)^{n-k-1}$$

$$= \frac{t^2}{\sigma^2} \left(\frac{o(t^2/(n\sigma^2))}{t^2/(n\sigma^2)}\right) \left(1 + \frac{t^2}{2n}\right)^{n-1}$$

$$= \frac{t^2}{\sigma^2} \left(1 + \frac{t^2}{2n}\right)^{-1} \left(\frac{o(t^2/(n\sigma^2))}{t^2/(n\sigma^2)}\right) \left(1 + \frac{t^2}{2n}\right)^n$$

$$\to \frac{t^2}{\sigma^2}(1)(0)e^{t^2/2} = 0.$$

Therefore, $\lim_{n\to\infty} \Psi_{Y_n}(t) = e^{t^2/2}$. ∎

6.42 Example: The central limit theorem permits approximate calculations for a variety of distributions. Suppose that a biased coin has probability $p = 0.4$ of heads. In 1000 tosses, what is the probability that the number of heads exceeds 410?

The number of heads H in n tosses has a binomial distribution with parameters (n, p). Consequently, the exact solution is

$$\Pr(H = k) = C_{1000,k} p^k (1-p)^{1000-k} = C_{1000,k}(0.4)^k(0.6)^{1000-k}$$

$$\Pr(H > 410) = \sum_{k=411}^{1000} C_{1000,k}(0.4)^k(0.6)^{1000-k}.$$

For an approximation, we note that H is the sum of 1000 independent Bernoulli random variables B_i, each with parameter $p = 0.4$. Each has mean $p = 0.4$ and variance $p(1-p) = 0.24$. That is, $\sigma = \sqrt{0.24} = 0.49$. From the central limit theorem,

$$Z = \frac{\sum_{i=1}^{1000}(B_i - 0.4)}{(0.49)\sqrt{1000}}$$

has approximately a standard normal distribution. Consequently,

$$\Pr(H > 410) = \Pr\left(\sum_{i=1}^{1000} B_i > 410\right) = \Pr\left(\frac{\sum_{i=1}^{1000}(B_i - 0.4)}{(0.49)\sqrt{1000}} > \frac{410 - 1000(0.4)}{(0.49)\sqrt{1000}}\right)$$

$$\approx \Pr(Z > 0.645) = 1 - \Pr(Z \le 0.645) = 1 - \Phi(0.645)$$

$$= 1 - 0.74 = 0.26. \; \square$$

Exercises

*6.30 The Cauchy distribution has density $f(x) = (1/\pi)/(1+x^2)$. Verify that $\Psi_c(t)$ and $\Psi_s(t)$ exist, but that $\Psi(t)$ does not.

*6.31 Show that $\int_0^\infty (1/t)\sin\alpha t\, dt$ is not absolutely convergent.

*6.32 Complete the proof of Theorem 6.33 by showing that $\int_0^\infty (1/t)\sin t\, dt = \pi/2$. The standard approach is to consider limiting configurations of the related integral $\int_0^\infty (1/t)e^{-\beta t}\sin t\, dt$, for $\beta \ge 0$.

6.33 Suppose that $\Psi(t) = p_1 + p_2 e^t + p_3 e^{2t}$ for positive p_i such that $p_1 + p_2 + p_3 = 1$. Use the inversion formula to determine the corresponding cumulative distribution function.

6.34 Left-handed persons comprise approximately 10% of the general population. A crowded athletic stadium contains 5000 people. What is the probability that there are between 750 and 800 left-handers in the stadium?

6.3   Gamma and beta distributions

Figure 2.9 displayed that relationships among the discrete distributions studied in Chapter 2. Figure 6.8 reproduces that summary and adds the gamma distribution in the lower right corner. In this section, we define the gamma distribution and see how it fits in the diagram. We start with the gamma function.

6.43 Definition: The *gamma function* is $\Gamma(t) = \int_0^\infty x^{t-1}e^{-x}\, dx$, defined for $t > 0$. $\blacksquare$

For a fixed $t > 0$, we perform one integration by parts to obtain

$$\Gamma(t) = \frac{x^t}{t}e^{-x}\Big|_{x=0}^{\infty} + \int_0^\infty \frac{x^t}{t}e^{-x}\, dx = 0 + \frac{1}{t}\int_0^\infty x^{(t+1)-1}e^{-x}\, dx$$

$$= \frac{\Gamma(t+1)}{t}.$$

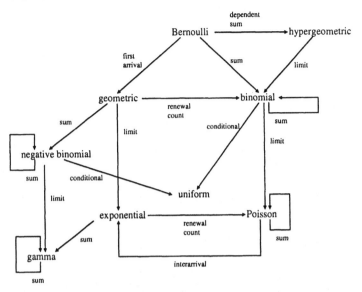

Figure 6.8. Relationships among discrete and continuous distributions

Hence, $\Gamma(t + 1) = t\Gamma(t)$. Because $\Gamma(1) = \int_0^\infty e^{-t}\, dt = 1$, we exploit this recursion to obtain, for integer $n \geq 1$,

$$\Gamma(n) = (n - 1)\Gamma(n - 1) = (n - 1)(n - 2)\Gamma(n - 2) = \cdots$$
$$= (n - 1)(n - 2)(n - 3) \cdots (1)\Gamma(1) = (n - 1)!.$$

Thus $\Gamma(\cdot)$ is a generalization of the factorial function; Figure 6.9 shows its general form. When we make the substitution $y = x^{1/2}$, the computation for $\Gamma(1/2)$ presents a familiar integral.

$$\Gamma(1/2) = \int_0^\infty x^{-1/2} e^{-x}\, dx = \int_0^\infty 2e^{-y^2}\, dy = \frac{1}{2}\int_{-\infty}^\infty 2e^{-y^2}\, dy$$
$$= \int_{-\infty}^\infty e^{-y^2}\, dy = \sqrt{\pi}.$$

Consequently, for any integer $n \geq 0$,

$$\Gamma\left(n + \frac{1}{2}\right) = \left(n - \frac{1}{2}\right)\left(n - \frac{3}{2}\right)\left(n - \frac{5}{2}\right) \cdots \left(\frac{1}{2}\right)\Gamma\left(\frac{1}{2}\right)$$
$$= \frac{(2n - 1)(2n - 3)(2n - 5) \cdots (3)(1)\sqrt{\pi}}{2^n} = \frac{(2n)!\sqrt{\pi}}{2^{2n}n!}.$$

We summarize the properties of the gamma function in the following theorem.
6.44 Theorem: Properties of the gamma function include the following.

- $\Gamma(n + 1) = n!$, for integer $n \geq 0$

- $\Gamma(t + 1) = t\Gamma(t)$, for real $t > 0$

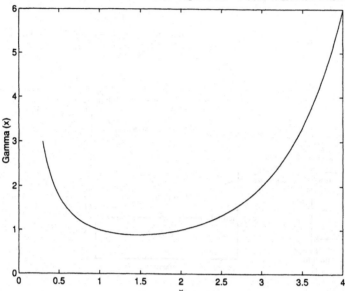

Figure 6.9. The gamma function: $\Gamma(x) = \int_0^\infty t^{x-1} e^{-t}\, dt$

- $\Gamma\left(n + \dfrac{1}{2}\right) = \dfrac{(2n)!\sqrt{\pi}}{2^{2n}n!}$, for integer $n \geq 0$

- $\Gamma(t) = \lambda^t \int_0^\infty x^{t-1} e^{-\lambda x}\, dx$, for $\lambda > 0$

- $\Gamma(t) = \dfrac{1}{2^{t-1}} \int_0^\infty x^{2t-1} e^{-x^2/2}\, dx$

PROOF: The discussion above established the first three properties. The last two items provide equivalent integrals for evaluating $\Gamma(t)$. Each involves a different exponential term in the integrand, which makes them useful in various circumstances. With the substitution $y = \lambda x$, we obtain

$$\int_0^\infty x^{t-1} e^{-\lambda x}\, dx = \int_0^\infty \left(\frac{y}{\lambda}\right)^{t-1} e^{-y} \frac{1}{\lambda}\, dy = \frac{1}{\lambda^t} \int_0^\infty y^{t-1} e^{-y}\, dy = \frac{\Gamma(t)}{\lambda^t}.$$

The form with $e^{-x^2/2}$ in the integrand follows from a similar substitution. ∎

We must resort to numerical integration to evaluate the gamma function for an arbitrary argument. Continued application of the relation $\Gamma(t+1) = t\Gamma(t)$ reduces the argument to the range $0.5 \leq t < 1.5$, for which tables are available. Table 6.1 provides a tabulation.

From the property that involves $e^{-\lambda t}$ in the integrand, we see that $\Gamma(t)$ is precisely the normalizing factor needed to convert $\lambda^t x^{t-1} e^{-\lambda x}$ to a probability density. This produces the gamma distribution.

x	0	1	2	3	4	5	6	7	8	9
0.50	1.7725	1.7384	1.7058	1.6747	1.6448	1.6161	1.5886	1.5623	1.5369	1.5126
0.60	1.4892	1.4667	1.4450	1.4242	1.4041	1.3848	1.3662	1.3482	1.3309	1.3142
0.70	1.2981	1.2825	1.2675	1.2530	1.2390	1.2254	1.2123	1.1997	1.1875	1.1757
0.80	1.1642	1.1532	1.1425	1.1322	1.1222	1.1125	1.1031	1.0941	1.0853	1.0768
0.90	1.0686	1.0607	1.0530	1.0456	1.0384	1.0315	1.0247	1.0182	1.0119	1.0059
1.00	1.0000	0.9943	0.9888	0.9835	0.9784	0.9735	0.9687	0.9642	0.9597	0.9555
1.10	0.9514	0.9474	0.9436	0.9399	0.9364	0.9330	0.9298	0.9267	0.9237	0.9209
1.20	0.9182	0.9156	0.9131	0.9108	0.9085	0.9064	0.9044	0.9025	0.9007	0.8990
1.30	0.8975	0.8960	0.8946	0.8934	0.8922	0.8912	0.8902	0.8893	0.8885	0.8879
1.40	0.8873	0.8868	0.8864	0.8860	0.8858	0.8857	0.8856	0.8856	0.8857	0.8859

TABLE 6.1. The gamma function

6.45 Definition: The *gamma distribution* with parameters (α, λ) has the density $f(x) = \lambda^{\alpha} x^{\alpha-1} e^{-\lambda x} / \Gamma(\alpha)$ for $x \geq 0$. The density is zero for $x < 0$. Both parameters must be positive. ∎

If $\alpha = 1$, we have $f(x) = \lambda e^{-\lambda x}$, for $x \geq 0$, which corresponds to cumulative distribution function $F(x) = \int_0^x \lambda e^{-\lambda t} \, dt = 1 - e^{-\lambda x}$. This is the exponential distribution, which appeared in Chapter 2 as the interarrival time between Poisson events. We now want to show that a gamma distribution with parameters (n, λ) describes the sum of n independent exponentials, each with parameter λ. This development requires the following identity, which we quickly establish by induction. For integer $n \geq 0$ and $\lambda > 0$,

$$\int_0^t x^n e^{-\lambda x} \, dx = \frac{n!}{\lambda^{n+1}} - e^{-\lambda t} \sum_{i=0}^{n} \frac{n!}{i!} \cdot \frac{t^i}{\lambda^{n+1-i}}. \tag{6.3}$$

For $n = 0$, the integral is

$$\int_0^t e^{-\lambda x} \, dx = \left. \frac{-e^{-\lambda x}}{\lambda} \right|_0^t = \frac{1}{\lambda} - e^{-\lambda t} \frac{1}{\lambda},$$

which is also the value of the expression in Equation 6.3. Proceeding by induction, suppose that the equation valid for $n = k$. Then, integrating by parts, we have

$$\int_0^t x^{k+1} e^{-\lambda x} \, dx = \left. x^{k+1} \cdot \frac{-e^{-\lambda x}}{\lambda} \right|_0^t + \frac{k+1}{\lambda} \int_0^t x^k e^{-\lambda x} \, dx$$

$$= \frac{-t^{k+1} e^{-\lambda t}}{\lambda} + \frac{k+1}{\lambda} \left[\frac{k!}{\lambda^{k+1}} - e^{-\lambda t} \sum_{i=0}^{k} \frac{k!}{i!} \cdot \frac{t^i}{\lambda^{k+1-i}} \right]$$

$$\int_0^t x^{k+1} e^{-\lambda x} \, dx = \frac{-t^{k+1} e^{-\lambda t}}{\lambda} + \left[\frac{(k+1)!}{\lambda^{k+2}} - e^{-\lambda t} \sum_{i=0}^{k} \frac{(k+1)!}{i!} \cdot \frac{t^i}{\lambda^{k+2-i}} \right]$$

$$= \frac{(k+1)!}{\lambda^{k+2}} - e^{-\lambda t} \sum_{i=0}^{k+1} \frac{(k+1)!}{i!} \cdot \frac{t^i}{\lambda^{k+2-i}},$$

which is the expected form for $n = k + 1$. We can now establish the following theorem.

6.46 Theorem: Let $X_1, X_2, \ldots, X_n$ be independent exponential random variables, each with parameter λ. Then $Y = \sum_{i=0}^{n} X_i$ is a gamma random variable with parameters (n, λ).

PROOF: As noted above, a gamma density with parameters $(1, \lambda)$ is an exponential distribution. Therefore, the theorem is true for $n = 1$, and we proceed by induction. Now suppose that the sum of k independent λ-exponentials is a (k, λ)-gamma, and consider $Y = X_1 + \ldots + X_{k+1}$, where the X_i are independent λ-exponentials. Then X_{k+1} is independent of $Z = \sum_{i=1}^{k} X_i$, and the induction hypothesis provides the joint density:

$$f_{X_{k+1}, Z}(x, z) = \begin{cases} \lambda e^{-\lambda x} \cdot \dfrac{1}{\Gamma(k)} \lambda^k z^{k-1} e^{-\lambda z}, & x, z \geq 0 \\ 0, & \text{otherwise.} \end{cases}$$

The probability of the event $(Y \leq t)$ is the integral of this joint density over the first-quadrant triangle with vertices $(0,0), (t,0)$, and $(0,t)$.

$$F_Y(t) = \Pr(Y \leq t) = \int_{z=0}^{t} \left(\int_{x=0}^{t-z} \lambda e^{-\lambda x} \, dx \right) \frac{\lambda^k}{\Gamma(k)} z^{k-1} e^{-\lambda z} \, dz$$

$$= \frac{\lambda^k}{(k-1)!} \int_0^t (1 - e^{-\lambda(t-z)}) z^{k-1} e^{-\lambda z} \, dz$$

$$= \frac{\lambda^k}{(k-1)!} \left[\int_0^t z^{k-1} e^{-\lambda z} \, dz - e^{-\lambda t} \int_0^t z^{k-1} \, dz \right]$$

Invoking Equation 6.3 for the left integral, we obtain the required density:

$$F_Y(t) = \frac{\lambda^k}{(k-1)!} \left[\frac{(k-1)!}{\lambda^k} - e^{-\lambda t} \sum_{i=0}^{k-1} \frac{(k-1)!}{i!} \cdot \frac{t^i}{\lambda^{k-i}} \right] - \left[\frac{\lambda^k e^{-\lambda t}}{(k-1)!} \cdot \frac{z^k}{k} \right]_0^t .$$

This expression simplifies as follows, and the derivative then establishes the desired density.

$$F_Y(t) = 1 - e^{-\lambda t} \sum_{i=0}^{k-1} \frac{(\lambda t)^i}{i!} - \left[\frac{(\lambda t)^k e^{-\lambda t}}{k!} \right] = 1 - e^{-\lambda t} \sum_{i=0}^{k} \frac{(\lambda t)^i}{i!}$$

$$f_Y(t) = F_Y'(t) = \lambda e^{-\lambda t} \sum_{i=0}^{k} \frac{(\lambda t)^i}{i!} - e^{-\lambda t} \sum_{i=1}^{k} \frac{i(\lambda t)^{i-1} \lambda}{i!}$$

$$= \lambda e^{-\lambda t} \left[\sum_{i=0}^{k} \frac{(\lambda t)^i}{i!} - \sum_{i=1}^{k} \frac{(\lambda t)^{i-1}}{(i-1)!} \right]$$

$$= \lambda e^{-\lambda t} \left[\sum_{i=0}^{k} \frac{(\lambda t)^i}{i!} - \sum_{i=0}^{k-1} \frac{(\lambda t)^i}{i!} \right] = \lambda e^{-\lambda t} \cdot \frac{(\lambda t)^k}{k!} = \frac{\lambda^{k+1} t^k e^{-\lambda t}}{\Gamma(k+1)} . \blacksquare$$

When α is a positive integer, the corresponding gamma distribution frequently carries an alternative name, as noted in the following definition.

6.47 Definition: The *exponential distribution* with parameter λ is a gamma distribution with parameters $(1, \lambda)$. Its density is $f(x) = \lambda e^{-\lambda x}$ for $x \geq 0$ and zero elsewhere. For integer $n \geq 1$, the *Erlang distribution* with parameters (n, λ) is a gamma distribution with parameters (n, λ). It density is $f(x) = \lambda^n x^{n-1} e^{-\lambda x}/(n-1)!$ for $x \geq 0$ and zero elsewhere. ∎

Theorem 6.46 partially justifies the position of the gamma random variable in Figure 6.8. It is clear that the sum of n independent gamma random variables retains a gamma distribution. If the summands have parameters $(n_1, \lambda), \ldots, (n_k, \lambda)$, then the first is a sum of n_1 exponentials with parameter λ, the second is a sum of n_2 such exponentials, and so forth. The sum of the n gamma random variables is then the sum of $n_1 + n_2 + \ldots + n_k$ exponentials and consequently has a gamma distribution with parameters $(n_1 + \ldots + n_k, \lambda)$. To verify the last feature noted in the Figure 6.8, we must show that the gamma distribution is a limiting form of the negative binomial distribution. Before taking up that discussion, we establish the mean, variance, and moment generating function.

6.48 Theorem: Let X be a gamma random variable with parameters (α, λ). Then

$$\mu_X = \frac{\alpha}{\lambda}$$

$$\sigma_X^2 = \frac{\alpha}{\lambda^2}$$

$$\Psi_X(t) = \left(1 - \frac{t}{\lambda}\right)^{-\alpha} = \left(\frac{\lambda}{\lambda - t}\right)^{\alpha}.$$

PROOF: The computations are straightforward.

$$\mu_X = \frac{1}{\Gamma(\alpha)} \int_0^\infty \lambda^\alpha x^\alpha e^{-\lambda x} \, dx = \frac{1}{\lambda \Gamma(\alpha)} \int_0^\infty \lambda^{\alpha+1} x^\alpha e^{-\lambda x} \, dx$$

$$= \frac{\Gamma(\alpha+1)}{\lambda \Gamma(\alpha)} = \frac{\alpha \Gamma(\alpha)}{\lambda \Gamma(\alpha)} = \frac{\alpha}{\lambda}$$

Most computations of this sort involve manipulating the integral into a form that is recognizable as a gamma function. The remaining calculations provide further illustrations of the pattern.

$$E[X^2] = \frac{1}{\Gamma(\alpha)} \int_0^\infty \lambda^\alpha x^{\alpha+1} e^{-\lambda x} \, dx = \frac{1}{\lambda^2 \Gamma(\alpha)} \int_0^\infty \lambda^{\alpha+2} x^{\alpha+1} e^{-\lambda x} \, dx$$

$$= \frac{\Gamma(\alpha+2)}{\lambda^2 \Gamma(\alpha)} = \frac{(\alpha+1)\alpha \Gamma(\alpha)}{\lambda^2 \Gamma(\alpha)} = \frac{\alpha^2 + \alpha}{\lambda^2}$$

$$\sigma_X^2 = E[X^2] - \mu_X^2 = \frac{\alpha^2 + \alpha}{\lambda^2} - \frac{\alpha^2}{\lambda^2} = \frac{\alpha}{\lambda^2}$$

$$\Psi_X(t) = E[e^{tX}] = \frac{1}{\Gamma(\alpha)} \int_0^\infty e^{tx} \lambda^\alpha x^{\alpha-1} e^{-\lambda x} \, dx$$

$$\Psi_X(t) = \frac{1}{\Gamma(\alpha)} \left(\frac{\lambda}{\lambda - t}\right)^\alpha \int_0^\infty (\lambda - t)^\alpha x^{\alpha - 1} e^{-(\lambda - t)x} \, dx$$

$$= \left(\frac{\lambda}{\lambda - t}\right)^\alpha = \left(1 - \frac{t}{\lambda}\right)^{-\alpha}$$

The computation for $\Psi_X(t)$ holds for $t < \lambda$. ∎

6.49 Example: When a database request arrives at a server, it passes in sequence through five processes. The service time of each process is an exponential random variable with parameter $\lambda = 2.5$ milliseconds^{-1}. If T is the total service time accrued by the request within the server, what are the average and variance of this random variable?

Let T_i be the service time of the ith process. Then $T = \sum_{i=1}^5 T_i$ is an Erlang random variable with parameters $(5, 2.5)$. Consequently, $\mu_T = 5/2.5 = 2.0$ milliseconds. Of course, this is just 5 times the service time associated with one process. The variance is $\sigma_T^2 = 5/(2.5)^2 = 0.8$, which is again 5 times the variance of a single process. □

Using Equation 6.3, we obtain an expression for the cumulative distribution function of an Erlang random variable. For $x \geq 0$, we have

$$F(x) = \int_0^x \frac{\lambda^n t^{n-1} e^{-\lambda t}}{(n-1)!} \, dt = \frac{\lambda^n}{(n-1)!} \int_0^x t^{n-1} e^{-\lambda t} \, dt$$

$$= \frac{\lambda^n}{(n-1)!} \left[\frac{(n-1)!}{\lambda^n} - e^{-\lambda x} \sum_{i=0}^{n-1} \frac{(n-1)!}{i!} \cdot \frac{x^i}{\lambda^{n-i}}\right]$$

$$= 1 - e^{-\lambda x} \sum_{i=0}^{n-1} \frac{(\lambda x)^i}{i!}.$$

This expression proves useful in analyzing a random variable that is the minimum or maximum of several Erlang random variables with the same parameter. The following example illustrates the point.

6.50 Example: A client simultaneously sends the same database request to three servers. The response time of each server is an Erlang random variable with parameters $(5, 2.5)$. The client accepts the response of the fastest server and sends cancellation commands to the other two. What is the average waiting time before the client receives a response? How does this compare with the response time of Exercise 6.49?

Let $\lambda = 2.5$ and $n = 5$. Suppose that T_i is the response time of the ith server, and T is the client's waiting time. Then $T = \min\{T_1, T_2, T_3\}$. Each T_i has the cumulative distribution function $F_{T_i}(x) = 1 - e^{-\lambda x} \sum_{i=0}^{n-1} (\lambda x)^i / i!$ for $x \geq 0$. The event $(T > x)$ occurs precisely when all three $(T_i > x)$ occur. Assuming independent response times for the different servers, we have, for $x \geq 0$,

$$1 - F_T(x) = \Pr(T > x) = \prod_{i=1}^3 (1 - F_{T_i}(x)) = \left[e^{-\lambda x} \sum_{i=0}^{n-1} \frac{(\lambda x)^i}{i!}\right]^3$$

$$f_T(x) = -3 \left[e^{-\lambda x} \sum_{i=0}^{n-1} \frac{(\lambda x)^i}{i!} \right]^2 \cdot \left[e^{-\lambda x} \sum_{i=1}^{n-1} \frac{i(\lambda x)^{i-1} \lambda}{i!} - \lambda e^{-\lambda x} \sum_{i=0}^{n-1} \frac{(\lambda x)^i}{i!} \right]$$

$$= -3 \left[e^{-\lambda x} \sum_{i=0}^{n-1} \frac{(\lambda x)^i}{i!} \right]^2 \cdot \lambda e^{-\lambda x} \left[\sum_{i=0}^{n-2} \frac{(\lambda x)^i}{i!} - \sum_{i=0}^{n-1} \frac{(\lambda x)^i}{i!} \right]$$

$$= -3 \left[e^{-\lambda x} \sum_{i=0}^{n-1} \frac{(\lambda x)^i}{i!} \right]^2 \cdot \left[\frac{-\lambda^n x^{n-1} e^{-\lambda x}}{(n-1)!} \right].$$

That is,

$$f_T(x) = \frac{3\lambda^n}{(n-1)!} e^{-3\lambda x} x^{n-1} \left(\sum_{i=0}^{n-1} \frac{(\lambda x)^i}{i!} \right)^2$$

$$= \frac{3\lambda^n}{(n-1)!} e^{-3\lambda x} \sum_{i=0}^{n-1} \sum_{j=0}^{n-1} \frac{\lambda^{i+j} x^{i+j+n-1}}{i!j!}.$$

The required average is now available as an integral of this density.

$$E[T] = \int_0^\infty x f_T(x) \, dx = \frac{3\lambda^n}{(n-1)!} \sum_{i=0}^{n-1} \sum_{j=0}^{n-1} \frac{\lambda^{i+j}}{i!j!} \int_0^\infty x^{i+j+n} e^{-3\lambda x} \, dx$$

$$= \frac{3\lambda^n}{(n-1)!} \sum_{i=0}^{n-1} \sum_{j=0}^{n-1} \left(\frac{\lambda^{i+j}}{i!j!} \right) \frac{\Gamma(i+j+n+1)}{(3\lambda)^{i+j+n+1}}$$

$$= \frac{1}{3^n \lambda (n-1)!} \sum_{i=0}^{n-1} \sum_{j=0}^{n-1} \frac{(i+j+n)!}{i!j!3^{i+j}}$$

Substituting $\lambda = 2.5$ and $n = 5$, we obtain $E[T] = 1.298$, which is less than the average response time of 2.0 in the preceding exercise. In that problem, the client sent its request to a single server. In the current problem, the client sends the same request simultaneously to three identical servers. Because it accepts the earliest response, we expect the average waiting time to be less.

The client might wait for all three responses. Three identical responses raises the credibility of the answer. In this case, however, we expect the average waiting time to be greater than that associated with a single server. One of the exercises pursues this possibility. $\square$

We now show that the Erlang distribution, which is a special case of a gamma distribution, is the limit of a sequence of negative binomial distributions. This limit occurs under conditions parallel to those in which the exponential distribution appears as the limit of geometric distributions. In particular, Theorem 2.59 justified the following construction. Start with $G_1, G_2, \ldots$, a sequence of geometric random variables with parameters $p_1, p_2, \ldots$. Impose the constraint $p_n \to 0$ in such a manner that $np_n \to \lambda > 0$. Finally, form the sequence $E_n = G_n/n$. The theorem showed that the cumulative distribution of E_n approaches that of an exponential random variable with parameter λ.

Now suppose that $H_1, H_2, \ldots$ is a sequence of negative binomial random variables with parameters (m, p_n), where $p_n \to 0$ in such a manner that

$np_n \to \lambda > 0$. Consider the related sequence $F_n = H_n/n$. We claim that the limiting distribution of the F_n is the Erlang distribution with parameters (m, λ). That is,

$$\lim_{n \to \infty} \Pr(F_n \le t) = \int_0^t \frac{\lambda^m x^{m-1} e^{-\lambda x}}{(m-1)!} \, dx.$$

We prove the claim by showing that the moment generating functions $\Psi_{F_n}(\cdot)$ converge to $\Psi_F(\cdot)$, where the latter is the moment generating function of the Erlang distribution. From Theorem 6.48, we have $\Psi_F(t) = [\lambda/(\lambda - t)]^m$. For each H_n, we obtain the moment generating function from Theorem 2.46: $\Psi_{H_n}(t) = [p_n/(1 - (1 - p_n)e^t)]^m$. Consequently,

$$\Psi_{F_n}(t) = E[e^{tF_n}] = E[e^{tH_n/n}] = \Psi_{H_n}(t/n) = \left(\frac{p_n}{1 - (1 - p_n)e^{t/n}} \right)^m$$

$$= \left(\frac{p_n}{p_n + (1 - p_n)(1 - e^{t/n})} \right)^m$$

$$= \left(\frac{np_n}{np_n + n(1 - e^{t/n}) - np_n(1 - e^{t/n})} \right)^m.$$

The only troublesome term is $n(1 - e^{t/n})$, which formally approaches $\infty \cdot 0$. However, l'Hôpital's helpful procedure gives

$$\lim_{n \to \infty} n(1 - e^{t/n}) = \lim_{n \to \infty} \frac{1 - e^{t/n}}{1/n} = \lim_{n \to \infty} \frac{-e^{t/n}(-t/n^2)}{-1/n^2} = \lim_{n \to \infty} -te^{t/n} = -t.$$

Therefore, $\lim_{n \to \infty} \Psi_{F_n}(t) = [\lambda/(\lambda - t)]^m = \Psi_F(t)$.

6.51 Theorem: Let $H_1, H_2, \ldots$ be a sequence of negative binomial random variables with parameters (m, p_n), where $p_n \to 0$ in such a manner that $np_n \to \lambda > 0$. Then the limiting distribution of H_n/n is an Erlang distribution with parameters (m, λ). That is,

$$\lim_{n \to \infty} \Pr(H_n/n \le t) = \int_0^t \frac{\lambda^m x^{m-1} e^{-\lambda x}}{(m-1)!} \, dx.$$

PROOF: See discussion above. ∎

The theorem holds in the case $m = 1$, which shows that the limiting distribution of scaled geometric distributions, G_n/n, is an exponential distribution with parameter λ. The convergence criteria remain the same. Each G_n is geometric with parameter p_n, and the $p_n \to 0$ in such a manner that $np_n \to \lambda > 0$. Theorem 2.59 established this fact directly, without reference to the moment generating function.

In the context of a Poisson process, we argued that the first arrival time must have an exponential distribution and extended that observation to all interarrival times because the process definition assumes that they are identically distributed. We are now in a position to argue directly that the interarrival times are exponentially distributed.

6.52 Theorem: For a Poisson process with parameter λ, $\Pr(T_j > t) = e^{-\lambda t}$ for all interarrival times T_j.

PROOF: Following Definition 2.58, the Poisson process is a sequence of independent nonnegative random variables, $T_1, T_2, \ldots$, such that the renewal counts, $N_t = \max\{n | \sum_{i=1}^{n} T_i \leq t\}$, for each $t > 0$, have Poisson distributions with parameters λt. That is, $\Pr(N_t = k) = e^{-\lambda t}(\lambda t)^k/k!$. Let $Y_n = \sum_{k=1}^{n} T_k$. The event $(Y_n > x)$ coincides with $\cup_{k=0}^{n-1}(N_x = k)$. Consequently,

$$1 - F_{Y_n}(x) = \Pr(Y_n > x) = \sum_{k=0}^{n-1} \frac{e^{-\lambda x}(\lambda x)^k}{k!}$$

$$F_{Y_n}(x) = 1 - e^{-\lambda x} \sum_{k=0}^{n-1} \frac{(\lambda x)^k}{k!}$$

$$f_{Y_n}(x) = F'_{Y_n}(x) = \lambda e^{-\lambda x} \sum_{k=0}^{n-1} \frac{(\lambda x)^k}{k!} - e^{-\lambda x} \sum_{k=1}^{n-1} \frac{k(\lambda x)^{k-1}\lambda}{k!}$$

$$= \lambda e^{-\lambda x} \left[\sum_{k=0}^{n-1} \frac{(\lambda x)^k}{k!} - \sum_{k=0}^{n-2} \frac{(\lambda x)^k}{k!} \right] = \frac{\lambda^n x^{n-1} e^{-\lambda x}}{(n-1)!}.$$

That is, Y_n is an Erlang random variable with parameters (n, λ). In particular, $T_1 = Y_1$ is exponential with parameter λ. Therefore,

$$\Pr(T_1 > t) = \int_t^\infty \lambda e^{-\lambda x} \, dx = -e^{-\lambda x}\big|_{x=t}^\infty = e^{-\lambda t}.$$

This shows that T_1 has the required distribution. Moreover, for any $n > 1$,

$$\left(\frac{\lambda}{\lambda - t}\right)^n = \Psi_{Y_n}(t) = \Psi_{Y_{n-1}}(t) \cdot \Psi_{T_n}(t) = \left(\frac{\lambda}{\lambda - t}\right)^{n-1} \Psi_{T_n}(t)$$

$$\Psi_{T_n}(t) = \frac{\lambda}{\lambda - t}$$

$$f_{T_n}(x) = \lambda e^{-\lambda x}$$

Therefore, T_n is also exponential with parameter λ, and $\Pr(T_n > t)$ has the desired value. ∎

Figure 6.10 plots a sequence of gamma density functions with parameters $(\alpha, \lambda = 2.0)$ and $1.0 \leq \alpha \leq 6$. Note how the center of gravity shifts to the right, in keeping with $\mu = \alpha/\lambda$. Moreover, the curves exhibit increasing dispersion about the mean, in accordance with $\sigma^2 = \alpha/\lambda^2$.

Let (u, v) be a pair of positive parameters. On the interval $0 < x < 1$, the function $x^{u-1}(1-x)^{v-1}$ is nonnegative and can therefore serve as a probability density if it is properly normalized. The beta function provides that normalization and produces the family of beta random variables.

6.53 Definition: Defined for $u, v > 0$, the *beta function* is

$$\beta(u, v) = \int_0^1 x^{u-1}(1-x)^{v-1} \, dx.$$

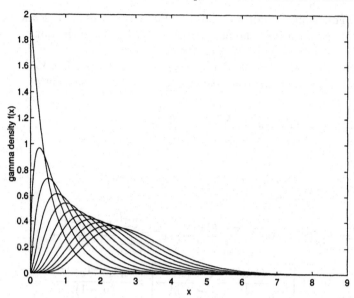

Figure 6.10. Gamma density, $\lambda = 2.0$ and $\alpha = 1.0, 1.5, \ldots, 6.0$, from left to right

---

A *beta random variable* with parameters (u, v), both positive, has density

$$f(x) = \frac{x^{u-1}(1-x)^{v-1}}{\beta(u,v)},$$

for $0 < x < 1$ and zero elsewhere. ∎

6.54 Theorem: The following are features of the beta function.

$$\beta(u,v) = \beta(v,u)$$
$$\beta(u,v) = 2\int_0^{\pi/2} \cos^{2u-1}\theta \sin^{2v-1}\theta \, d\theta$$
$$\beta(u,v) = \frac{\Gamma(u)\Gamma(v)}{\Gamma(u+v)}$$

PROOF: The first property follows easily from the definition and the transformation $y = (1-x)$ in the integral.

$$\beta(u,v) = \int_0^1 x^{u-1}(1-x)^{v-1}\,dx = \int_1^0 (1-y)^{u-1}y^{v-1}(-\,dy)$$
$$= \int_0^1 y^{v-1}(1-y)^{u-1}\,dy = \beta(v,u)$$

For the second property, we transform via $x = \cos^2\theta$, $dx = -2\cos\theta\sin\theta \, d\theta$.

$$\beta(u,v) = \int_0^1 x^{u-1}(1-x)^{v-1}\,dx$$

$$\beta(u,v) = \int_{\pi/2}^{0} (\cos^2\theta)^{u-1}(1-\cos^2\theta)^{v-1}(-2\cos\theta\sin\theta)\,d\theta$$

$$= 2\int_0^{\pi/2} \cos^{2u-1}\theta\sin^{2v-1}\theta\,d\theta$$

For the last property, we use a polar coordinate transformation to evaluate the product of two gamma functions.

$$\Gamma(u)\Gamma(v) = \left(\frac{1}{2^{u-1}}\int_0^\infty x^{2u-1}e^{-x^2/2}\,dx\right)\left(\frac{1}{2^{v-1}}\int_0^\infty y^{2v-1}e^{-y^2/2}\,dy\right)$$

$$= \frac{1}{2^{u-1}2^{v-1}}\int_0^\infty\int_0^\infty x^{2u-1}y^{2v-1}e^{-(x^2+y^2)/2}\,dx\,dy$$

In polar coordinates, $x = r\cos\theta, y = r\sin\theta$ and the differential area $dx\,dy$ becomes $r\,dr\,d\theta$. Consequently,

$$\Gamma(u)\Gamma(v) = \frac{1}{2^{u+v-2}}\int_{\theta=0}^{\pi/2}\left(\int_{r=0}^\infty r^{2u+2v-2}e^{-r^2/2}r\,dr\right)\cos^{2u-1}\theta\sin^{2v-1}\theta\,d\theta$$

$$= 2\int_{\theta=0}^{\pi/2}\left(\frac{1}{2^{u+v-1}}\int_{r=0}^\infty r^{2(u+v)-1}e^{-r^2/2}\,dr\right)\cos^{2u-1}\theta\sin^{2v-1}\theta\,d\theta$$

$$= \Gamma(u+v)\cdot 2\int_0^{\pi/2}\cos^{2u-1}\theta\sin^{2v-1}\theta\,d\theta = \Gamma(u+v)\beta(u,v). \quad\blacksquare$$

Figure 6.11 shows the qualitative form of the beta function. To evaluate $\beta(u,v)$, we rely on the relation $\beta(u,v) = \Gamma(u)\Gamma(v)/\Gamma(u+v)$ and tabulated gamma function (Table 6.1). The following dynamics problem illustrates the technique.

6.55 Example: A ball bearing is released from the lip of a hemispherical bowl of radius 1 foot. How long does it take to reach the bottom?

The diagram to the right locates the ball bearing as a function of θ, the angle swept out as it rolls toward the bottom of the bowl. Assuming a pivot at the center top of the bowl, the clockwise torque is $mgr\cos\theta$, where m is the mass of the ball bearing and $g = 32.2$ is the acceleration imparted by gravity. About the same pivot, the inertial moment is mr^2. The angular acceleration induced by the torque is $\alpha = d\omega/dt$, where $\omega = d\theta/dt$ is the angular velocity.

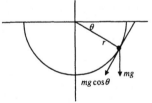

The rotational form of Newton's law equates torque with the product of the moment of inertia and the angular acceleration. Therefore,

$$mgr\cos\theta = mr^2\alpha$$

$$\frac{d\omega}{dt} = \frac{g\cos\theta}{r}.$$

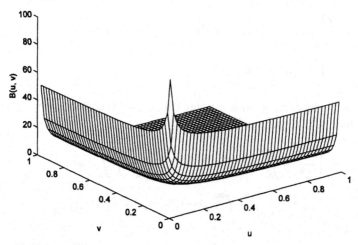

Figure 6.11. The beta function $\beta(u, v)$

Since $d\omega/dt = (d\omega/d\theta)(d\theta/dt) = \omega \cdot d\omega/d\theta$, we can separate variables as follows.

$$\int \omega \, d\omega = \int \frac{g \cos \theta}{r} \, d\theta$$

$$\frac{\omega^2}{2} = \frac{g \sin \theta}{r}$$

The last equation follows because the integration constants are zero on both sides: the particle starts from rest at location $\theta = 0$. Consequently,

$$\frac{d\theta}{dt} = \omega = \left(\frac{2g \sin \theta}{r} \right)^{1/2}$$

$$\int_0^{\pi/2} \frac{d\theta}{\sqrt{\sin \theta}} = \left(\frac{2g}{r} \right)^{1/2} \int_0^T dt = \left(\frac{2g}{r} \right)^{1/2} T,$$

where T is the time for the ball bearing to achieve a rotation of $\pi/2$ about the pivot. We can rewrite the θ integral to reveal a β function.

$$T = \left(\frac{r}{2g} \right)^{1/2} \cdot \frac{1}{2} \cdot 2 \cdot \int_0^{\pi/2} \sin^{2(1/4)-1} \theta \cos^{2(1/2)-1} \theta \, d\theta = \left(\frac{r}{8g} \right)^{1/2} \beta(1/4, 1/2)$$

$$= \left(\frac{r}{8g} \right)^{1/2} \cdot \frac{\Gamma(1/4)\Gamma(1/2)}{\Gamma(3/4)} = \left(\frac{r\pi}{8g} \right)^{1/2} \frac{\Gamma(1/4)}{\Gamma(3/4)} = \left(\frac{r\pi}{8g} \right)^{1/2} \frac{4\Gamma(5/4)}{\Gamma(3/4)}$$

We have $r = 1$, and $\Gamma(5/4) = 0.9064$, $\Gamma(3/4) = 1.2254$ are in the range of Table 6.1. Therefore,

$$T = \left(\frac{\pi}{8(32.2)} \right)^{1/2} \left(\frac{4(0.9064)}{1.2254} \right) = 0.327 \text{ seconds.} \quad \square$$

The following theorem establishes the descriptive features of the beta random variable.

6.56 Theorem: The mean and variance of a beta random variable with parameters (u, v) are

$$\mu = \frac{u}{u + v} \qquad\qquad \sigma^2 = \frac{uv}{(u + v)^2(u + v + 1)}.$$

PROOF: The computations are straightforward.

$$\mu = \frac{1}{\beta(u, v)} \int_0^1 x x^{u-1}(1 - x)^{v-1} \, dx = \frac{\beta(u + 1, v)}{\beta(u, v)}$$

$$= \frac{\Gamma(u + 1)\Gamma(v)}{\Gamma(u + v + 1)} \cdot \frac{\Gamma(u + v)}{\Gamma(u)\Gamma(v)} = \frac{u\Gamma(u)\Gamma(v)}{(u + v)\Gamma(u + v)} \cdot \frac{\Gamma(u + v)}{\Gamma(u)\Gamma(v)} = \frac{u}{u + v}$$

Letting X be the random variable in question, we have

$$E[X^2] = \frac{1}{\beta(u, v)} \int_0^1 x^2 x^{u-1}(1 - x)^{v-1} \, dx = \frac{\beta(u + 2, v)}{\beta(u, v)}$$

$$= \frac{\Gamma(u + 2)\Gamma(v)}{\Gamma(u + v + 2)} \cdot \frac{\Gamma(u + v)}{\Gamma(u)\Gamma(v)}$$

$$= \frac{u(u + 1)\Gamma(u)\Gamma(v)}{(u + v)(u + v + 1)\Gamma(u + v)} \cdot \frac{\Gamma(u + v)}{\Gamma(u)\Gamma(v)} = \frac{u(u + 1)}{(u + v)(u + v + 1)}.$$

Consequently,

$$\sigma^2 = \frac{u(u + 1)}{(u + v)(u + v + 1)} - \frac{u^2}{(u + v)^2} = \frac{uv}{(u + v)^2(u + v + 1)}. \blacksquare$$

Exercises

6.35 Prove that $\Gamma(t) = (1/2^{t-1}) \int_0^\infty x^{2t-1} e^{-x^2/2} \, dx$, and thereby complete the proof of Theorem 6.44.

6.36 Show that $\Gamma(t) = \int_0^1 \left[\ln\left(\frac{1}{x}\right)\right]^{t-1} dx$.

6.37 Evaluate $A = \int_0^1 \left(\frac{\ln(1/x)}{x}\right)^{1/2} dx$.

6.38 Suppose that Z has a standard normal distribution. Show that the even moments are

$$E[Z^{2k}] = \frac{2^k}{\sqrt{\pi}}\Gamma\left(k + \frac{1}{2}\right).$$

6.39 Let X be a gamma random variable with parameters (α, λ). Verify that

$$\Psi'_X(0) = E[X] = \mu_X$$
$$\Psi''_X(0) = E[X^2] = \sigma_X^2 + \mu_X^2.$$

6.40 A client simultaneously sends the same database request to three servers. The response time of each server is an Erlang random variable with parameters $(5, 2.5)$. To ensure accuracy, the client waits for all three responses. What is the average waiting time?

*6.41 Let Z be standard normal, and let $X = Z^2$. Derive the density function for X. What are μ_X and σ_X^2?

6.42 Let X have a beta distribution with parameters (m, n), where m, n are positive integers. Show that the kth moment of X is

$$E[X^k] = \prod_{i=0}^{n-1} \left(\frac{m + i}{m + k + i} \right).$$

6.43 Evaluate the integral $\int_0^{\pi/2} \sqrt{\tan \theta} \, d\theta$.

*6.44 Starting from rest at the point $x_0 > 0$, a particle moves in one dimension under the influence of an attractive force that pulls it toward the origin. At a distance x from the origin, the magnitude of the force is k/x. Derive an expression for the time required for the particle to first reach the origin.

6.4   The χ^2 and related distributions

The next chapter deals with parameter estimation, a decision problem in which samples are taken from a distribution with unknown parameters. The optimal action is the announcement of estimates for these parameters. The distributions of this section will prove useful in that context. They are all combinations of standard normal random variables. If Z is standard normal, we first show that Z^2 has a gamma distribution.

$$F_{Z^2}(x) = \Pr(Z^2 \le x) = \Pr(-\sqrt{x} \le Z \le \sqrt{x}) = \int_{-\sqrt{x}}^{\sqrt{x}} \frac{1}{\sqrt{2\pi}} e^{-t^2/2} \, dt$$

$$= \frac{2}{\sqrt{2\pi}} \int_0^{\sqrt{x}} e^{-t^2/2} \, dt$$

$$f_{Z^2}(x) = F'_{Z^2}(x) = \frac{2}{\sqrt{2\pi}} e^{-x/2} \left(\frac{1}{2\sqrt{x}} \right) = \frac{1}{\Gamma(1/2)} \left(\frac{1}{2} \right)^{1/2} x^{(1/2)-1} e^{-(1/2)x}$$

This is a gamma distribution with parameters $(\alpha, \lambda) = (1/2, 1/2)$.

6.57 Definition: Let $Z_1, Z_2, \ldots, Z_n$ be independent standard normal random variables. Let $X = \sum_{k=1}^n Z_k^2$. Then X is a χ^2 random variable with n degrees of freedom. ∎

6.58 Theorem: Let X be a χ^2 random variable with n degrees of freedom. Then X has a gamma distribution with parameters $(\alpha = n/2, \lambda = 1/2)$.

PROOF: Let $X = Z_1^2 + Z_2^2 + \ldots + Z_n^2$, where each Z_k is standard normal. The moment generating function for X is the nth power of the generating function for Z_k^2. Because the latter has a gamma distribution with parameters $(\alpha, \lambda) = (1/2, 1/2)$, we obtain its generating function from Theorem 6.48. Consequently,

$$\Psi_X(t) = \prod_{k=1}^{n} \Psi_{Z_k^2}(t) = \left[\left(\frac{1/2}{(1/2) - t}\right)^{1/2}\right]^n = \left(\frac{1/2}{(1/2) - t}\right)^{n/2},$$

which is the moment generating function of a gamma distribution with parameters $(\alpha = n/2, \lambda = 1/2)$. ∎

The features of the χ^2 distribution then follow from the known expressions for the gamma distribution.

6.59 Theorem: Let X be a χ^2 random variable with n degrees of freedom. Then

$$\mu_X = n$$
$$\sigma_X^2 = 2n$$
$$\Psi_X(t) = \left(\frac{1}{1 - 2t}\right)^{n/2}$$
$$f_X(x) = \frac{1}{\Gamma(n/2)}(1/2)^{n/2}x^{(n/2)-1}e^{-(1/2)x}, \text{ for } x \geq 0.$$

PROOF: These properties follow from Theorem 6.48. ∎

We use the generic notation $\chi^2_{(n)}$ for a random variable that has a χ^2 distribution with n degrees of freedom. Figure 6.12 illustrates the behavior of the density curves with increasing degrees of freedom. Evaluating specific probabilities involves a difficult integral. For example, to obtain $\Pr(\chi^2_{(5)} \leq 10)$, we must evaluate

$$\Pr(\chi^2_{(5)} \leq 10) = \frac{(1/2)^{5/2}}{\Gamma(5/2)} \int_0^{10} t^{(5/2)-1}e^{-t/2}\,dt.$$

As for the standard normal distribution, tables are available for this purpose. Traditionally, the tables give selected percentiles. That is, for selected probabilities $p_1 < p_2 < \ldots < p_k$, the tabulation contains *percentiles* $x_1 < x_2 < \ldots < x_k$ such that $\Pr(\chi^2_{(n)} \leq x_i) = p_i$. We use the notation $\chi^2_{(n)p_i}$ to denote a percentile x_i. For example, $\chi^2_{(5)0.025} = 0.8312$, which means that $\Pr(\chi^2_{(5)} \leq 0.8312) = 0.025$.

As illustrated in the next example, the tables are most useful for p_i near zero or 1, and therefore more entries appear at these extremes. Table 6.2 lists the lower percentiles: $\chi^2_{(n)0.005}$ through $\chi^2_{(n)0.4}$; Table 6.3 continues with $\chi^2_{(n)0.6}$ through $\chi^2_{(n)0.995}$. More extensive tables appear in Appendix B.

6.60 Example: A full adder is a circuit that constructs the sum of two binary numbers. When the inputs change, the carries must propagate from the least significant bit to the most significant bit, and this process takes some time to stabilize. The time depends on the inputs and on capacitances in the circuit. The inputs are variable, of course, and across many adders the capacitances are also slightly

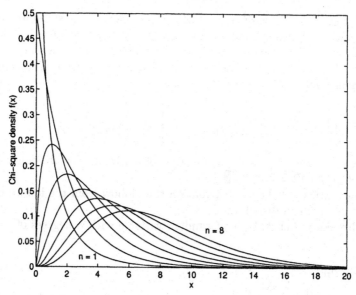

Figure 6.12. χ^2 densities with degrees of freedom ranging from 1 through 8

different. The time to develop the sum is therefore a random variable. Let T be this time. Suppose we know, perhaps unrealistically, that T is normally distributed with mean 50 nanoseconds. Four measurements, taken with randomly chosen circuits and inputs, yield the samples: $48.2, 51.6, 45.0, 47.3$ nanoseconds. What can you say about the variance σ_T^2?

This problem involves parameter estimation, which is treated more systematically in the next chapter. For the moment, we can apply a simple transformation to involve the χ^2 distribution. Suppose that $T_1, T_2, \ldots, T_n$ are independent samples from a normal distribution with known mean μ and unknown variance σ^2. Then each $(T_i - \mu)/\sigma$ is standard normal (Theorem 6.4), and $\sum_{i=1}^{n}(T_i - \mu)^2/\sigma^2$ has the χ^2 distribution with n degrees of freedom. We apply this observation to the case at hand, in which $n = 4$, $\mu = 50.0$, and the T_i are the four measurements. We have access to an observation of the random variable $S = \sum_{i=1}^{4}(T_i - \mu)^2$, and we know that S/σ^2 is $\chi_{(4)}^2$. Ninety-five percent of this χ^2 distribution lies between $\chi_{(4)0.025}^2$ and $\chi_{(4)0.975}^2$, and Table 6.2 shows these percentiles to be 0.4844 and 11.1433. Consequently,

$$0.95 = \Pr\left(0.4844 \le \frac{S}{\sigma^2} \le 11.1433\right) = \Pr\left(\frac{S}{11.1433} \le \sigma^2 \le \frac{S}{0.4844}\right).$$

Of course, S is a random variable and varies from sample to sample. Nevertheless, we expect that 95 samples out of every 100 will produce an S value such that $S/11.1433 \le \sigma^2 \le S/0.4844$. In this case, we have observed the value $S = s$, where

$$s = \sum_{i=1}^{4}(T_i - \mu)^2 = (48.2 - 50)^2 + (51.6 - 50)^2 + (45.0 - 50)^2 + (47.3 - 50)^2 = 38.09.$$

	Percentiles							
n	$\chi^2_{.005}$	$\chi^2_{.01}$	$\chi^2_{.02}$	$\chi^2_{.025}$	$\chi^2_{.05}$	$\chi^2_{.1}$	$\chi^2_{.2}$	$\chi^2_{.4}$
1	0.0000	0.0002	0.0006	0.0010	0.0039	0.0158	0.0642	0.2750
2	0.0100	0.0201	0.0404	0.0506	0.1026	0.2107	0.4463	1.0217
3	0.0717	0.1148	0.1848	0.2158	0.3518	0.5844	1.0052	1.8692
4	0.2070	0.2971	0.4294	0.4844	0.7107	1.0636	1.6488	2.7528
5	0.4117	0.5543	0.7519	0.8312	1.1455	1.6103	2.3425	3.6555
6	0.6757	0.8721	1.1344	1.2373	1.6354	2.2041	3.0701	4.5702
7	0.9893	1.2390	1.5643	1.6899	2.1673	2.8331	3.8223	5.4932
8	1.3444	1.6465	2.0325	2.1797	2.7326	3.4895	4.5936	6.4226
9	1.7349	2.0879	2.5324	2.7004	3.3251	4.1682	5.3801	7.3570
10	2.1559	2.5582	3.0591	3.2470	3.9403	4.8652	6.1791	8.2955
11	2.6032	3.0535	3.6087	3.8157	4.5748	5.5778	6.9887	9.2373
12	3.0738	3.5706	4.1783	4.4038	5.2260	6.3038	7.8073	10.1820
13	3.5650	4.1069	4.7654	5.0088	5.8919	7.0415	8.6339	11.1291
14	4.0747	4.6604	5.3682	5.6287	6.5706	7.7895	9.4673	12.0785
15	4.6009	5.2293	5.9849	6.2621	7.2609	8.5468	10.3070	13.0297
16	5.1422	5.8122	6.6142	6.9077	7.9616	9.3122	11.1521	13.9827
17	5.6972	6.4078	7.2550	7.5642	8.6718	10.0852	12.0023	14.9373
18	6.2648	7.0149	7.9062	8.2307	9.3905	10.8649	12.8570	15.8932
19	6.8440	7.6327	8.5670	8.9065	10.1170	11.6509	13.7158	16.8504
20	7.4338	8.2604	9.2367	9.5908	10.8508	12.4426	14.5784	17.8088
21	8.0337	8.8972	9.9146	10.2829	11.5913	13.2396	15.4446	18.7683
22	8.6427	9.5425	10.6000	10.9823	12.3380	14.0415	16.3140	19.7288
23	9.2604	10.1957	11.2926	11.6886	13.0905	14.8480	17.1865	20.6902
24	9.8862	10.8564	11.9918	12.4012	13.8484	15.6587	18.0618	21.6525
25	10.5197	11.5240	12.6973	13.1197	14.6114	16.4734	18.9398	22.6156

TABLE 6.2. Lower percentiles for χ^2 distributions with n degrees of freedom

Unless this sample constitutes one of the exceptions, we can conclude

$$3.42 = \frac{38.09}{11.1433} \le \sigma^2 \le \frac{38.09}{0.4844} = 78.63.$$

If the sample is one of the exceptions, this conclusion is wrong. Traditionally, we announce the result with a qualifying phrase. There is a 95% chance that the unknown variance lies in the interval $[3.42, 78.63]$. Equivalently, the 95% confidence interval for σ^2 is $[3.42, 78.63]$. This statement must be viewed with caution. The unknown variance is a fixed quantity, not a random one. It either lies in the confidence interval or it does not. Hence the probability that $\sigma^2 \in [3.42, 78.63]$ is either zero or 1—not 0.95. What does vary randomly is the interval itself. If we take four new measurements, we obtain different endpoints for the 95% confidence interval. If we repeat the process many times, we obtain a sequence of intervals, 95% of which will contain the fixed point σ^2. Because the repetitions are independent, any one of them has a 95% chance of capturing the variance. The proper interpretation of a confidence interval computed in this manner is that it is a sample from a long sequence of such intervals, 95% of which contain the parameter in question.

The interval $[3.42, 78.63]$ is rather wide, and we suspect that a shorter interval corresponds to a lower confidence level. That is, if we are satisfied with a 80%

				Percentiles				
n	$\chi^2_{.6}$	$\chi^2_{.8}$	$\chi^2_{.9}$	$\chi^2_{.95}$	$\chi^2_{.975}$	$\chi^2_{.98}$	$\chi^2_{.99}$	$\chi^2_{.995}$
1	0.7083	1.6424	2.7055	3.8415	5.0239	5.4119	6.6349	7.8795
2	1.8326	3.2189	4.6052	5.9915	7.3778	7.8240	9.2103	10.5966
3	2.9462	4.6416	6.2514	7.8147	9.3484	9.8374	11.3450	12.8384
4	4.0446	5.9886	7.7794	9.4877	11.1433	11.6678	13.2767	14.8603
5	5.1318	7.2889	9.2352	11.0699	12.8309	13.3882	15.0863	16.7496
6	6.2108	8.5581	10.6446	12.5916	14.4494	15.0332	16.8119	18.5475
7	7.2832	9.8033	12.0171	14.0675	16.0138	16.6240	18.4799	20.2786
8	8.3505	11.0301	13.3616	15.5073	17.5346	18.1682	20.0903	21.9553
9	9.4136	12.2421	14.6836	16.9189	19.0225	19.6786	21.6645	23.5848
10	10.4732	13.4420	15.9872	18.3070	20.4831	21.1607	23.2088	25.1882
11	11.5298	14.6314	17.2750	19.6752	21.9202	22.6182	24.7257	26.7569
12	12.5838	15.8120	18.5494	21.0261	23.3368	24.0542	26.2178	28.3025
13	13.6356	16.9848	19.8119	22.3620	24.7356	25.4716	27.6886	29.8210
14	14.6853	18.1508	21.0641	23.6848	26.1189	26.8726	29.1412	31.3193
15	15.7332	19.3107	22.3071	24.9958	27.4883	28.2593	30.5779	32.8013
16	16.7795	20.4651	23.5418	26.2962	28.8452	29.6330	31.9999	34.2672
17	17.8244	21.6146	24.7690	27.5871	30.1910	30.9950	33.4082	35.7185
18	18.8679	22.7595	25.9894	28.8693	31.5264	32.3462	34.8052	37.1557
19	19.9102	23.9004	27.2036	30.1436	32.8524	33.6875	36.1911	38.5826
20	20.9514	25.0375	28.4120	31.4105	34.1696	35.0196	37.5662	39.9980
21	21.9915	26.1711	29.6151	32.6706	35.4789	36.3434	38.9322	41.4011
22	23.0307	27.3015	30.8133	33.9245	36.7807	37.6595	40.2894	42.7957
23	24.0689	28.4288	32.0069	35.1725	38.0756	38.9683	41.6384	44.1813
24	25.1063	29.5533	33.1962	36.4150	39.3641	40.2704	42.9798	45.5585
25	26.1430	30.6752	34.3816	37.6525	40.6464	41.5661	44.3141	46.9279

TABLE 6.3. Upper percentiles for χ^2 distributions with n degrees of freedom

probability that the interval contains σ^2, we use $\chi^2_{(4)0.1} = 1.0636$ and $\chi^2_{(4)0.9} = 7.7794$. This gives

$$0.8 = \Pr\left(\frac{38.09}{7.7794} \le \sigma^2 \le \frac{38.09}{1.0636}\right) = \Pr(4.90 \le \sigma^2 \le 35.81).$$

[4.90, 35.81] is indeed a tighter interval than [3.42, 78.63]. We now have σ^2 confined to a smaller interval but with less confidence.

Suppose that the sample still produced $s = \sum_{i=1}^{25}(T_i - 50)^2 = 38.09$, but with 25 measurements. What does this entail for the confidence intervals? Because $\chi^2_{(25)0.025} = 13.1197$ and $\chi^2_{(25)0.975} = 40.6464$, we have

$$0.95 = \Pr\left(\frac{38.09}{40.6464} \le \sigma^2 \le \frac{38.09}{13.1197}\right) = \Pr(0.94 \le \sigma^2 \le 2.90).$$

[0.94, 2.90] is a much more compact 95% confidence interval, but this result comports with intuition. If 25 measurements manage to confine the squared deviation from the mean to 38.09, the variability must be much smaller than the earlier case, which accumulates 38.09 in only 4 measurements. $\square$

The F distribution has two parameters, also called degrees of freedom, and is the weighted ratio of two χ^2 random variables.

6.61 Definition: Let X, Y be independent random variables such that X is χ^2 with n_1 degrees of freedom and Y is χ^2 with n_2 degrees of freedom. Let $W = (X/n_1)/(Y/n_2) = (n_2 X)/(n_1 Y)$. The distribution of W is called an F distribution with (n_1, n_2) degrees of freedom. ∎

To derive the density function of the F distribution, we start with the joint density of X and Y, two independent χ^2 random variables with n_1 and n_2 degrees of freedom, respectively. The density is zero when $x < 0$ or $y < 0$. For $x \geq 0, y \geq 0$, it is

$$f_{XY}(x, y) = f_X(x) f_Y(y) = \left(\frac{x^{(n_1/2)-1} e^{-x/2}}{2^{n_1/2} \Gamma(n_1/2)} \right) \left(\frac{y^{(n_2/2)-1} e^{-y/2}}{2^{n_2/2} \Gamma(n_2/2)} \right)$$

$$f_{XY}(x, y) = \left(\frac{1}{2^{(n_1+n_2)/2} \Gamma(n_1/2) \Gamma(n_2/2)} \right) x^{(n_1/2)-1} y^{(n_2/2)-1} e^{-(x+y)/2}.$$

Let $W = (n_2 X)/(n_1 Y)$. Obviously, W is nonnegative, and we have, for a fixed $w > 0$,

$$F_W(w) = \Pr(W \leq w) = \Pr\left(\frac{n_2 X}{n_1 Y} \leq w \right) = \Pr\left(X \leq \frac{w n_1 Y}{n_2} \right).$$

This is the volume under the joint density surface in the first quadrant region defined by the line $y \geq [n_2/(n_1 w)] x$.

$$F_W(w) = \int_{y=0}^{\infty} \int_{x=0}^{n_1 w y/n_2} f_{XY}(x, y) \, dx \, dy$$

$$= \frac{1}{2^{(n_1+n_2)/2} \Gamma(n_1/2) \Gamma(n_2/2)}$$

$$\cdot \int_{y=0}^{\infty} y^{(n_2/2)-1} e^{-y/2} \left(\int_{x=0}^{n_1 w y/n_2} x^{(n_1/2)-1} e^{-x/2} \, dx \right) dy$$

We differentiate to obtain the density.

$$f(w) = \frac{1}{2^{(n_1+n_2)/2} \Gamma(n_1/2) \Gamma(n_2/2)}$$

$$\cdot \int_{y=0}^{\infty} y^{(n_2/2)-1} e^{-y/2} \left[\left(\frac{n_1 w y}{n_2} \right)^{(n_1/2)-1} \exp\left(\frac{-n_1 w y}{2 n_2} \right) \right] \left(\frac{n_1 y}{n_2} \right) dy$$

$$= \frac{(n_1/n_2)^{n_1/2}}{2^{(n_1+n_2)/2} \Gamma(n_1/2) \Gamma(n_2/2)} w^{(n_1/2)-1}$$

$$\cdot \int_{y=0}^{\infty} y^{(n_1+n_2)/2-1} \exp\left(-\frac{y}{2} \left(1 + \frac{n_1 w}{n_2} \right) \right) dy$$

We manipulate the integral to reveal a gamma function.

$$f(w) = \frac{(n_1/n_2)^{n_1/2} w^{(n_1/2)-1}}{2^{(n_1+n_2)/2} \Gamma(n_1/2) \Gamma(n_2/2)} \left[\frac{1}{2}\left(1 + \frac{n_1 w}{n_2}\right)\right]^{-(n_1+n_2)/2} \cdot I,$$

where the integral I is

$$I = \int_{y=0}^{\infty} \left[\frac{1}{2}\left(1 + \frac{n_1 w}{n_2}\right)\right]^{(n_1+n_2)/2} y^{[(n_1+n_2)/2]-1} \exp\left(-\frac{1}{2}\left(1 + \frac{n_1 w}{n_2}\right) y\right) dy$$

$$= \Gamma((n_1 + n_2)/2).$$

Consequently,

$$f(w) = \frac{\Gamma[(n_1 + n_2)/2](n_1/n_2)^{n_1/2} w^{(n_1/2)-1} 2^{(n_1+n_2)/2}}{2^{(n_1+n_2)/2} \Gamma(n_1/2) \Gamma(n_2/2)} \left(1 + \frac{n_1 w}{n_2}\right)^{-(n_1+n_2)/2}$$

$$= \frac{\Gamma((n_1 + n_2)/2)}{\Gamma(n_1/2)\Gamma(n_2/2)} \left(\frac{n_1}{n_2}\right)^{n_1/2} w^{(n_1/2)-1} \left(1 + \frac{n_1 w}{n_2}\right)^{-(n_1+n_2)/2}.$$

Although this density is somewhat complicated, the mean and variance calculations are straightforward because the defining integrals also reduce to gamma functions.

$$\mu_W = E[W] = \int_0^{\infty} w f_W(w)\, dw$$

$$= \frac{\Gamma((n_1 + n_2)/2)}{\Gamma(n_1/2)\Gamma(n_2/2)} \left(\frac{n_1}{n_2}\right)^{n_1/2} \int_0^{\infty} w^{n_1/2} \left(1 + \frac{n_1 w}{n_2}\right)^{-(n_1+n_2)/2} dw$$

$$= \frac{\Gamma((n_1 + n_2)/2)}{\Gamma(n_1/2)\Gamma(n_2/2)} \int_0^{\infty} \left(\frac{n_1 w/n_2}{1 + n_1 w/n_2}\right)^{n_1/2} (1 + n_1 w/n_2)^{-n_2/2}\, dw$$

We find the following integral transformation useful at this point.

$$t = \frac{n_1 w/n_2}{1 + n_1 w/n_2} = \left(\frac{n_1}{n_2}\right)\left(\frac{w}{1 + n_1 w/n_2}\right)$$

$$dt = \left(\frac{n_1}{n_2}\right)\left(\frac{(1 + n_1 w/n_2) - w(n_1/n_2)}{(1 + n_1 w/n_2)^2}\right) dw = \left(\frac{n_1/n_2}{(1 + n_1 w/n_2)^2}\right) dw$$

These equations imply that $w = (n_2/n_1)[t/(1-t)]$ and $1 + n_1 w/n_2 = (1-t)^{-1}$. The substitution gives

$$\mu_W = \frac{\Gamma((n_1 + n_2)/2)}{\Gamma(n_1/2)\Gamma(n_2/2)} \left(\frac{n_2}{n_1}\right)$$

$$\cdot \int_0^{\infty} \left(\frac{n_1 w/n_2}{1 + n_1 w/n_2}\right)^{n_1/2} (1 + n_1 w/n_2)^{-n_2/2+2} \left(\frac{n_1/n_2}{(1 + n_1 w/n_2)^2}\right) dw$$

$$= \frac{n_2 \Gamma((n_1 + n_2)/2)}{n_1 \Gamma(n_1/2)\Gamma(n_2/2)} \int_0^{1} t^{n_1/2}(1 - t)^{n_2/2-2}\, dt,$$

in which we recognize an integral form of a beta function. Continuing,

$$
\begin{aligned}
\mu_W &= \left(\frac{n_2\Gamma((n_1+n_2)/2)}{n_1\Gamma(n_1/2)\Gamma(n_2/2)}\right)\beta(n_1/2+1, n_2/2-1)\\
&= \left(\frac{n_2\Gamma((n_1+n_2)/2)}{n_1\Gamma(n_1/2)\Gamma(n_2/2)}\right)\left(\frac{\Gamma(n_1/2+1)\Gamma(n_2/2-1)}{\Gamma((n_1+n_2)/2)}\right)\\
&= \frac{n_2(n_1/2)\Gamma(n_1/2)\Gamma(n_2/2-1)}{n_1\Gamma(n_1/2)(n_2/2-1)\Gamma(n_2/2-1)} = \frac{n_2 n_1/2}{n_1(n_2/2-1)} = \frac{n_2}{n_2-2}.
\end{aligned}
$$

The beta function is defined only for positive arguments, which requires $n_2/2-1 > 0$ or $n_2 > 2$. One of the exercises asks you to show that the mean does not exist if $n_2 \le 2$. We now pursue a similar integral for $E[W^2]$:

$$
E[W^2] = \int_0^\infty w^2 f_W(w)\,dw
$$

$$
\begin{aligned}
E[W^2] &= \frac{\Gamma((n_1+n_2)/2)}{\Gamma(n_1/2)\Gamma(n_2/2)}\left(\frac{n_1}{n_2}\right)^{n_1/2}\int_0^\infty w^{(n_1/2)+1}\left(1+\frac{n_1 w}{n_2}\right)^{-(n_1+n_2)/2}dw\\
&= \frac{\Gamma((n_1+n_2)/2)}{\Gamma(n_1/2)\Gamma(n_2/2)}\left(\frac{n_2}{n_1}\right)\int_0^\infty\left(\frac{n_1 w/n_2}{1+n_1 w/n_2}\right)^{1+n_1/2}(1+n_1 w/n_2)^{1-n_2/2}\,dw\\
&= \frac{\Gamma((n_1+n_2)/2)}{\Gamma(n_1/2)\Gamma(n_2/2)}\left(\frac{n_2}{n_1}\right)^2\int_0^\infty\left(\frac{n_1 w/n_2}{1+n_1 w/n_2}\right)^{(n_1/2)+1}\\
&\qquad\cdot(1+n_1 w/n_2)^{-n_2/2+3}(n_1/n_2)(1+n_1 w/n_2)^{-2}\,dw.
\end{aligned}
$$

At this point, we apply the same integral transformation that we used to reduce the expression for μ_W.

$$
\begin{aligned}
E[W^2] &= \frac{\Gamma((n_1+n_2)/2)}{\Gamma(n_1/2)\Gamma(n_2/2)}\left(\frac{n_2}{n_1}\right)^2\int_0^1 t^{(n_1/2)+1}(1-t)^{(n_2/2-3)}\,dt\\
&= \frac{\Gamma((n_1+n_2)/2)}{\Gamma(n_1/2)\Gamma(n_2/2)}\left(\frac{n_2}{n_1}\right)^2\beta(n_1/2+2, n_2/2-2)\\
&= \frac{\Gamma((n_1+n_2)/2)}{\Gamma(n_1/2)\Gamma(n_2/2)}\left(\frac{n_2}{n_1}\right)^2\frac{\Gamma(n_1/2+2)\Gamma(n_2/2-2)}{\Gamma((n_1+n_2)/2)}\\
&= \left(\frac{n_2}{n_1}\right)^2\cdot\frac{(n_1/2+1)(n_1/2)}{(n_2/2-1)(n_2/2-2)} = \frac{n_2^2(n_1+2)}{n_1(n_2-2)(n_2-4)},
\end{aligned}
$$

provided that $n_2 > 4$. Consequently,

$$
\sigma_W^2 = \frac{n_2^2(n_1+2)}{n_1(n_2-2)(n_2-4)} - \left(\frac{n_2}{n_2-2}\right)^2 = \frac{2n_2^2(n_1+n_2-2)}{n_1(n_2-2)^2(n_2-4)}.
$$

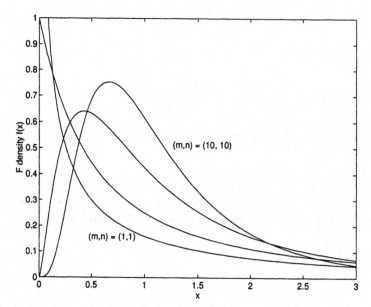

Figure 6.13. F densities with $(1,1), (2,2), (5,5)$, and $(10,10)$ degrees of freedom

---

6.62 Theorem: Let W have an F distribution with (n_1, n_2) degrees of freedom. Then

$$f_W(w) = \frac{\Gamma((n_1 + n_2)/2)}{\Gamma(n_1/2)\Gamma(n_2/2)} \left(\frac{n_1}{n_2}\right)^{n_1/2} w^{(n_1/2)-1} \left(1 + \frac{n_1 w}{n_2}\right)^{-(n_1+n_2)/2}$$

$$\mu_W = \frac{n_2}{n_2 - 2}$$

$$\sigma_W^2 = \frac{2n_2^2(n_1 + n_2 - 2)}{n_1(n_2 - 2)^2(n_2 - 4)}.$$

The expressions for the mean and variance are valid for $n_2 > 2$ and $n_2 > 4$ respectively. For other values of n_2, the defining integrals diverge.

PROOF: See discussion above. ∎

Several F densities appear in Figure 6.13. These densities are also diffi-cult to integrate, so tables are available for selected useful probabilities. We denote by $F_{(m,n)p}$ the pth percentile of the F distribution with (m, n) degrees of freedom. For example, if X is a random variable having an F distribu-tion with $(3, 2)$ degrees of freedom, then $\Pr(X \le F_{(3,2)0.9}) = 0.9$. Percentile tabulation is more cumbersome because the two degrees of freedom use both tabular dimensions. This means a separate table for each percentile. That is, a tabulation of $F_{(m,n)0.9}$ runs m horizontally and n vertically, and requires a third dimension, or separate tables, for p values other than 0.9. Table 6.4 circumvents this difficulty by placing five entries in each cell. From top to bottom, they are $F_{(m,n)0.8}, F_{(m,n)0.9}, F_{(m,n)0.95}, F_{(m,n)0.975}$, and $F_{(m,n)0.99}$.

There is a simplification; we need not list the lower percentiles because we can compute them from an available upper percentile. That is, $F_{(m,n)0.1}$ is related to $F_{(n,m)0.9}$. Specifically, if X and Y are independent χ^2 random variables with m and n degrees of freedom, respectively, then $((X/m)/(Y/n) \le t)$ and $((Y/n)/(X/m) \ge 1/t)$ are the same event. Consequently,

$$p = \Pr\left(\frac{X/m}{Y/n} \le F_{(m,n)p}\right) = \Pr\left(\frac{Y/n}{X/m} \ge \frac{1}{F_{(m,n)p}}\right)$$

$$= 1 - \Pr\left(\frac{Y/n}{X/m} \le \frac{1}{F_{(m,n)p}}\right).$$

$$1 - p = \Pr\left(\frac{Y/n}{X/m} \le \frac{1}{F_{(m,n)p}}\right)$$

$$F_{(n,m)(1-p)} = \frac{1}{F_{(m,n)p}}$$

This relationship allows us to compute, say, $F_{(3,2)0.1} = 1/F_{(2,3)0.9}$. With this extension, Table 6.4 also provides $F_{(m,n)0.01}, F_{(m,n)0.025}, F_{(m,n)0.05}, F_{(m,n)0.1}$, and $F_{(m,n)0.2}$. More extensive tables appear in Appendix B.

6.63 Example: We return to the context of Example 6.60, in which T is the time required for a full adder to develop its result. We again assume that T has a normal distribution, but now we know neither the mean nor the variance of that distribution. Two measurements, taken with randomly chosen circuits and inputs, yield the samples: 48.2 and 51.6 nanoseconds. What can you infer about the unknown mean μ and variance σ^2?

Let the sample be (T_1, T_2). These are independent and normally distributed with common mean μ and variance σ^2. Consequently, they are bivariate normal. Consider the following quantities, the first of which is the sample mean.

$$\overline{T} = \frac{T_1 + T_2}{2}$$

$$S = (T_1 - \overline{T})^2 + (T_2 - \overline{T})^2 = \left(T_1 - \frac{T_1 + T_2}{2}\right)^2 + \left(T_2 - \frac{T_1 + T_2}{2}\right)^2$$

$$= \left(\frac{T_1 - T_2}{2}\right)^2 + \left(\frac{T_2 - T_1}{2}\right)^2 = \left(\frac{T_1 - T_2}{\sqrt{2}}\right)^2$$

Being a linear combination of normal random variables, $X = (T_1 - T_2)/(\sigma\sqrt{2})$ is normal. Consequently,

$$\mu_X = \left(\frac{1}{\sigma\sqrt{2}}\right)\mu_{T_1} - \left(\frac{1}{\sigma\sqrt{2}}\right)\mu_{T_2} = \left(\frac{1}{\sigma\sqrt{2}}\right)(\mu - \mu) = 0$$

$$\sigma_X^2 = \left(\frac{1}{\sigma\sqrt{2}}\right)^2 \sigma_{T_1}^2 + 2\left(\frac{1}{\sigma\sqrt{2}}\right)\left(\frac{1}{\sigma\sqrt{2}}\right)\mathrm{cov}(T_1, T_2) + \left(\frac{1}{\sigma\sqrt{2}}\right)^2 \sigma_{T_2}^2$$

$$= \left(\frac{1}{2\sigma^2}\right)(\sigma^2 + \sigma^2) = 1.$$

n \ m	1	2	3	4	5	6	7	8	9	10
1	9.47	12.00	13.06	13.65	14.01	14.26	14.44	14.58	14.68	14.77
	39.88	49.49	53.58	55.82	57.28	58.20	58.88	59.40	59.84	60.20
	161.4	199.4	216.0	224.7	229.9	234.0	236.7	238.8	240.3	241.7
	649.3	801.0	865.1	900.1	921.1	937.3	948.3	953.9	965.3	971.0
	4063	4985	5414	5650	5731	5901	5901	5988	5988	6078
2	3.56	4.00	4.16	4.24	4.28	4.32	4.34	4.36	4.37	4.38
	8.53	9.00	9.16	9.24	9.29	9.33	9.35	9.37	9.38	9.39
	18.51	19.00	19.17	19.25	19.30	19.33	19.35	19.37	19.38	19.40
	38.51	39.00	39.20	39.25	39.30	39.33	39.36	39.37	39.39	39.40
	98.50	99.00	99.95	99.25	99.29	99.33	99.36	99.37	99.39	99.40
3	2.68	2.89	2.94	2.96	2.97	2.97	2.97	2.98	2.98	2.98
	5.54	5.46	5.39	5.34	5.31	5.28	5.27	5.25	5.24	5.23
	10.13	9.55	9.28	9.12	9.01	8.94	8.89	8.85	8.81	8.79
	17.44	16.04	15.44	15.10	14.88	14.73	14.62	14.54	14.47	14.42
	34.12	30.82	29.48	28.71	28.24	27.91	27.67	27.49	27.35	27.23
4	2.35	2.47	2.48	2.48	2.48	2.47	2.47	2.47	2.46	2.46
	4.54	4.32	4.19	4.11	4.05	4.01	3.98	3.95	3.94	3.92
	7.71	6.94	6.59	6.39	6.26	6.16	6.09	6.04	6.00	5.96
	12.22	10.65	9.98	9.60	9.36	9.20	9.07	8.98	8.90	8.84
	21.20	18.00	16.70	15.98	15.52	15.21	14.98	14.80	14.66	14.55
5	2.18	2.26	2.25	2.24	2.23	2.22	2.21	2.20	2.20	2.19
	4.06	3.78	3.62	3.52	3.45	3.40	3.37	3.34	3.32	3.30
	6.61	5.79	5.41	5.19	5.05	4.95	4.88	4.82	4.77	4.74
	10.01	8.43	7.76	7.39	7.15	6.98	6.85	6.76	6.68	6.62
	16.26	13.27	12.06	11.39	10.97	10.67	10.46	10.29	10.16	10.05
6	2.07	2.13	2.11	2.09	2.08	2.06	2.05	2.04	2.03	2.03
	3.78	3.46	3.29	3.18	3.11	3.05	3.02	2.98	2.96	2.94
	5.99	5.14	4.76	4.53	4.39	4.28	4.21	4.15	4.10	4.06
	8.81	7.26	6.60	6.23	5.99	5.82	5.70	5.60	5.52	5.46
	13.75	10.92	9.78	9.15	8.75	8.47	8.26	8.10	7.98	7.87
7	2.00	2.04	2.02	1.99	1.97	1.96	1.94	1.93	1.93	1.92
	3.59	3.26	3.07	2.96	2.88	2.83	2.78	2.75	2.72	2.70
	5.59	4.74	4.35	4.12	3.97	3.87	3.79	3.73	3.68	3.64
	8.07	6.54	5.89	5.52	5.29	5.12	4.99	4.90	4.82	4.76
	12.25	9.55	8.45	7.85	7.46	7.19	6.99	6.84	6.72	6.62

TABLE 6.4. 80th, 90th, 95th, 97.5th and 99th percentiles for F distributions with (m, n) degrees of freedom

X is standard normal, and therefore $S/\sigma^2 = [(T_1 - T_2)/(\sigma\sqrt{2})]^2 = X^2$ has a χ^2 distribution with one degree of freedom. We construct an 80% confidence interval from the percentiles in Table 6.2.

$$0.8 = \Pr\left(\chi^2_{(1)0.1} \leq \frac{S}{\sigma^2} \leq \chi^2_{(1)0.9}\right) = \Pr\left(0.016 \leq \frac{S}{\sigma^2} \leq 2.71\right)$$

$$= \Pr\left(\frac{S}{2.71} \leq \sigma^2 \leq \frac{S}{0.016}\right)$$

For the case at hand, $\overline{T} = (48.2 + 51.6)/2 = 49.9$ and $S = (48.2 - 49.9)^2 + (51.6 - 49.9)^2 = 5.78$. This means that σ^2 lies in the interval $[5.78/2.71, 5.78/0.016] = [2.13, 361.25]$ with probability 0.8. This large interval is hardly better than not knowing σ^2 at all. Moreover, we must carefully attend the interpretation of such a confidence interval. Remember that σ^2 is unknown, but fixed. It is not a random variable. Rather, the interval endpoints vary as the sample (T_1, T_2) varies. Eighty percent of such intervals capture σ^2. In this context, we say that our particular computed interval captures σ^2 with probability 0.8. The uncomfortable interval width arises from the small sample size. We will investigate variance estimation with larger samples in the next chapter.

Can we construct a confidence interval for the unknown mean? Note that

$$\mu_{\overline{T}} = \left(\frac{1}{2}\right)\mu_{T_1} + \left(\frac{1}{2}\right)\mu_{T_2} = \frac{1}{2}(\mu + \mu) = \mu$$

$$\sigma_{\overline{T}}^2 = \left(\frac{1}{2}\right)^2 \sigma_{T_1}^2 + \left(\frac{1}{2}\right)\left(\frac{1}{2}\right)\text{cov}(T_1, T_2) + \left(\frac{1}{2}\right)^2 \sigma_{T_2}^2$$

$$= \left(\frac{1}{4}\right)(\sigma^2 + \sigma^2) = \frac{\sigma^2}{2}.$$

Because it is a linear combination of T_1 and T_2, $\overline{T}$ is normal. Consequently, $Y = (\overline{T} - \mu)/(\sigma/\sqrt{2})$ is standard normal. Perhaps surprisingly, this quantity is independent of $X = (T_1 - T_2)/(\sigma\sqrt{2})$, whose square is S/σ^2. Because both X and Y are normal, we can show independence by showing zero covariance.

$$\text{cov}(X, Y) = E[XY] - (0)(0) = E\left[\frac{T_1 - T_2}{\sigma\sqrt{2}} \cdot \frac{\sqrt{2}(\overline{T} - \mu)}{\sigma}\right]$$

$$= \frac{1}{\sigma^2}E\left[(T_1 - T_2)\left(\frac{T_1 + T_2}{2} - \mu\right)\right] = \frac{1}{2\sigma^2}E[T_1^2 - T_2^2] - \frac{\mu}{\sigma^2}E[T_1 - T_2]$$

$$= \frac{1}{2\sigma^2}[\sigma^2 + \mu^2 - (\sigma^2 + \mu^2)] - \frac{\mu}{\sigma^2}(\mu - \mu) = 0$$

Consequently, $Y^2/X^2 = Y^2/(S/\sigma^2)$ has an F distribution with $(1, 1)$ degrees of freedom. But,

$$\frac{Y^2}{S/\sigma^2} = \frac{2(\overline{T} - \mu)^2/\sigma^2}{S/\sigma^2} = \frac{2(\overline{T} - \mu)^2}{S}.$$

So, for a 95% confidence interval,

$$0.95 = \Pr\left(\frac{2(\overline{T} - \mu)^2}{S} \leq F_{(1,1)0.95}\right) = \Pr\left(|\overline{T} - \mu| \leq \sqrt{\frac{F_{(1,1)0.95}S}{2}}\right)$$

$$= \Pr\left(\overline{T} - \sqrt{\frac{F_{(1,1)0.95}S}{2}} \leq \mu \leq \overline{T} + \sqrt{\frac{F_{(1,1)0.95}S}{2}}\right).$$

We have $\overline{T} = 49.9$ and $S = 5.78$. Table 6.4 gives $F_{(1,1)0.95} = 161.35$, and we then compute the 95% confidence interval for μ to be 49.9 ± 21.59 or $[28.3, 71.5]$. □

6.64 Definition: A continuous random variable X has a t distribution with n degrees of freedom if its density is symmetric about zero and X^2 has an F distribution with $(1, n)$ degrees of freedom. ∎

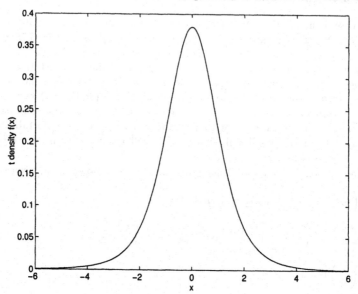

Figure 6.14. t density with 5 degrees of freedom

Figure 6.14 illustrates a typical t density. If X has a t density, the symmetry of the graph shows that $\Pr(X \leq -x) = 1 - \Pr(X \leq x)$. Following the pattern established with the χ^2 and F distributions, we use $t_{(n)}$ to indicate a random variable having a t distribution with n degrees of freedom, and we use $t_{(n)p}$ to denote the corresponding percentiles. That is, for example, $\Pr(t_{(5)} \leq t_{(5)0.95}) = 0.95$. Consequently, for $p \geq 0.5$, we have

$$\Pr(t_{(n)} \leq -t_{(n)p}) = 1 - \Pr(t_{(n)} \leq t_{(n)p}) = 1 - p,$$

which implies that $t_{(n)1-p} = -t_{(n)p}$. Of course, this forces $t_{(n)0.5} = 0$.

We now calculate the density for $t_{(n)}$. Suppose that $x \geq 0$. Then, because $t_{(n)}^2$ has an F distribution with $(1, n)$ degrees of freedom, the symmetric slice $S = (-x \leq t_{(n)} \leq x)$ has probability

$$
\begin{aligned}
\Pr(S) &= \Pr(-x \leq t_{(n)} \leq x) = \Pr(t_{(n)}^2 \leq x^2) \\
&= \int_0^{x^2} \frac{\Gamma((n+1)/2)}{\Gamma(n/2)\Gamma(1/2)} \left(\frac{1}{n}\right)^{1/2} w^{(1/2)-1} \left(1 + \frac{w}{n}\right)^{-(n+1)/2} dw \\
&= \frac{\Gamma((n+1)/2)}{\sqrt{n\pi}\,\Gamma(n/2)} \int_0^{x^2} w^{-1/2} \left(1 + \frac{w}{n}\right)^{-(n+1)/2} dw.
\end{aligned}
$$

From Figure 6.14, we see that $\Pr(t_{(n)} \leq x)$ is the probability of this central slice $(-x \leq t_{(n)} \leq x)$ plus the probability of the left tail $(t_{(n)} < -x)$. However, the probabilities of the left and right tails are equal, and together they complement the central slice. Therefore,

$$\Pr(t_{(n)} \le x) = \Pr(-x \le t_{(n)} \le x) + \frac{1}{2}[1 - \Pr(-x \le t_{(n)} \le x)]$$

$$= \frac{1 + \Pr(-x \le t_{(n)} \le x)}{2}$$

$$= \frac{1}{2}\left[1 + \frac{\Gamma((n+1)/2)}{\sqrt{n\pi}\Gamma(n/2)} \int_0^{x^2} w^{-1/2}\left(1 + \frac{w}{n}\right)^{-(n+1)/2} dw\right].$$

The desired density is the derivative of this expression with respect to x.

$$f_{t_{(n)}}(x) = \frac{1}{2}\left(\frac{\Gamma((n+1)/2)}{\sqrt{n\pi}\Gamma(n/2)}(x^2)^{-1/2}\left(1 + \frac{x^2}{n}\right)^{-(n+1)/2}\right) \cdot (2x)$$

$$= \frac{\Gamma((n+1)/2)}{\sqrt{n\pi}\Gamma(n/2)}\left(1 + \frac{x^2}{n}\right)^{-(n+1)/2}$$

If $x < 0$, then $\Pr(t_{(n)} \le x)$ is the probability of a left tail, which is one-half of the complement of the central section. That is,

$$\Pr(t_{(n)} \le x) = \frac{1 - \Pr(x \le t_{(n)} \le -x)}{2}$$

$$\Pr(t_{(n)}) = \frac{1}{2}\left[1 - \frac{\Gamma((n+1)/2)}{\sqrt{n\pi}\Gamma(n/2)} \int_0^{x^2} w^{-1/2}\left(1 + \frac{w}{n}\right)^{-(n+1)/2} dw\right].$$

Differentiating again yields

$$f_{t_{(n)}}(x) = \frac{-1}{2}\left(\frac{\Gamma((n+1)/2)}{\sqrt{n\pi}\Gamma(n/2)}(x^2)^{-1/2}\left(1 + \frac{x^2}{n}\right)^{-(n+1)/2}\right) \cdot (2x).$$

Because x is negative, $(x^2)^{-1/2} = 1/|x|$, and $x/|x| = -1$, and this case reduces to the same expression as was obtained for $x \ge 0$.

6.65 Theorem: $t_{(n)}$ has mean $\mu_{t_{(n)}} = 0$, variance $\sigma^2_{t_{(n)}} = \dfrac{n}{n-2}$, and density

$$f_{t_{(n)}}(x) = \frac{\Gamma((n+1)/2)}{\sqrt{n\pi}\Gamma(n/2)}\left(1 + \frac{x^2}{n}\right)^{-(n+1)/2}.$$

For the mean, the defining integral is absolutely convergent for $n > 1$; for the variance, $n > 2$ is necessary.

PROOF: The discussion above established the density, which we will call simply $f(x)$ for the duration of this proof. Because the density is symmetrically distributed about zero, we have $\mu_{t_{(n)}} = 0$, provided that the defining integral exists. Consider the integral $\int_{-\infty}^{\infty} |xf(x)|\, dx$. For $n > 1$, the following calculation demonstrates convergence. Assume that $a < 0 < b$ and transform the integral with $y = x^2/n$, $dy = 2x\, dx/n$.

$$A = A(a, b) = \int_a^b |xf(x)|\, dx$$

$$= \frac{\Gamma((n+1)/2)}{\sqrt{n\pi}\,\Gamma(n/2)} \cdot$$

$$\left[\int_a^0 (-x)\left(1 + \frac{x^2}{n}\right)^{-(n+1)/2} dx + \int_0^b x \left(1 + \frac{x^2}{n}\right)^{-(n+1)/2} dx \right]$$

$$A = \frac{n\Gamma((n+1)/2)}{2\sqrt{n\pi}\,\Gamma(n/2)} \left[-\int_{a^2/n}^0 (1+y)^{-(n+1)/2}\, dy + \int_0^{b^2/n} (1+y)^{-(n+1)/2}\, dy \right]$$

$$= \frac{n\Gamma((n+1)/2)}{2\sqrt{n\pi}\,\Gamma(n/2)} \left[\int_0^{a^2/n} (1+y)^{-(n+1)/2}\, dy + \int_0^{b^2/n} (1+y)^{-(n+1)/2}\, dy \right]$$

$$= \frac{-n\Gamma((n+1)/2)}{(n-1)\sqrt{n\pi}\,\Gamma(n/2)} \left[\left(\frac{1}{1+a^2/n}\right)^{(n-1)/2} - 1 + \left(\frac{1}{1+b^2/n}\right)^{(n-1)/2} - 1 \right]$$

Consequently,

$$\int_{-\infty}^{\infty} |xf(x)|\, dx = \lim_{a \to -\infty, b \to \infty} A(a, b) = \frac{2n\Gamma((n+1)/2)}{(n-1)\sqrt{n\pi}\,\Gamma(n/2)}.$$

For $n = 1$, however, the same process yields

$$\int_{-\infty}^{\infty} |xf(x)|\, dx = \lim_{a, b \to \infty} \frac{\Gamma(1)}{2\sqrt{\pi}\sqrt{\pi}} \left[\int_0^{a^2} (1+y)^{-1}\, dy + \int_0^{b^2} (1+y)^{-1}\, dy \right]$$

$$= \frac{1}{2\pi\sqrt{n}} \lim_{a, b \to \infty} \left[\ln(1+a^2) + \ln(1+b^2) \right] = \infty.$$

This suggests that the defining integral for the mean does not exist for $n = 1$. Indeed, the integral splits into two parts, one tending to $-\infty$ and the other toward $+\infty$.

When the mean exists, it is zero. Therefore, the variance is $E[t_{(n)}^2]$, which is the mean of an F distribution with $(1, n)$ degrees of freedom. From Theorem 6.62, this value is $n/(n-2)$, provided that $n > 2$. ∎

Table 6.5 lists the upper percentiles for t distributions with varying degrees of freedom. More extensive tables appear in Appendix B. As noted earlier, the lower percentiles are available from the relation $t_{(n)1-p} = -t_{(n)p}$. Moreover, the percentiles are related to the those from the F distribution. In particular, $F_{(1,n)p} = t_{(n)q}^2$, when $q = (1+p)/2$. For example, the squares of

n	$t_{.55}$	$t_{.60}$	$t_{.65}$	$t_{.70}$	$t_{.75}$	$t_{.80}$	$t_{.85}$	$t_{.90}$	$t_{.95}$	$t_{.975}$	$t_{.99}$	$t_{.995}$	$t_{.9995}$
1	0.158	0.325	0.510	0.727	1.000	1.376	1.963	3.078	6.314	12.706	31.821	63.657	636.619
2	0.142	0.289	0.445	0.617	0.816	1.061	1.386	1.886	2.920	4.303	6.965	9.925	31.598
3	0.137	0.277	0.424	0.584	0.765	0.979	1.250	1.638	2.353	3.182	4.541	5.841	12.925
4	0.134	0.271	0.414	0.569	0.741	0.941	1.190	1.533	2.132	2.776	3.747	4.604	8.610
5	0.132	0.267	0.408	0.559	0.727	0.920	1.156	1.476	2.015	2.571	3.365	4.032	6.868
6	0.131	0.265	0.404	0.553	0.718	0.906	1.134	1.440	1.943	2.447	3.143	3.707	5.959
7	0.130	0.263	0.402	0.549	0.711	0.896	1.119	1.415	1.895	2.365	2.998	3.499	5.408
8	0.130	0.262	0.399	0.546	0.706	0.889	1.108	1.397	1.860	2.306	2.896	3.355	5.041
9	0.129	0.261	0.398	0.543	0.703	0.883	1.100	1.383	1.833	2.262	2.821	3.250	4.781
10	0.129	0.260	0.397	0.542	0.700	0.879	1.093	1.372	1.813	2.228	2.764	3.169	4.587

TABLE 6.5. Percentiles for the t distribution with n degrees of freedom

column $t_{(n)0.975}$ give the 95th percentile entries in the first column of Table 6.4. We can easily derive this relationship as follows.

Let $x = F_{(1,n)p}$. Then

$$
\begin{aligned}
p &= \Pr(F_{(1,n)} \leq x) = \Pr(-\sqrt{x} \leq t_{(n)} \leq \sqrt{x}) \\
&= \Pr(t_{(n)} \leq \sqrt{x}) - \Pr(t_{(n)} \leq -\sqrt{x}) \\
&= \Pr(t_{(n)} \leq \sqrt{x}) - [1 - \Pr(t_{(n)} \leq \sqrt{x})] = 2\Pr(t_{(n)} \leq \sqrt{x}) - 1
\end{aligned}
$$

$$
\frac{1+p}{2} = \Pr(t_{(n)} \leq \sqrt{x}).
$$

This last equation implies that $\sqrt{x} = t_{(n)q}$, where $q = (1+p)/2$. The required relation then appears upon squaring both sides of this result. It is also true that $F_{(n,1)p} = t_{(n)q}^{-2}$, when $q = (2-p)/2$. We leave this proof for the exercises. Given the close relationship between the t and F distribution, it is not surprising that we can investigated the circuit of Example 6.63 with a t distribution. The next example provides the details.

6.66 Example: Recall the context of Examples 6.60 and 6.63. T is the time required for a full adder to develop its result, and we assume that T has a normal distribution with unknown parameters μ and σ^2. Two measurements, taken with randomly chosen circuits and inputs, yield the samples: 48.2 and 51.6 nanoseconds. The goal of this exercise is to arrive at a 95% confidence interval for σ^2 through a t distribution.

As before, let the sample be (T_1, T_2). These are independent and normally distributed with common mean μ and variance σ^2. Example 6.63 constructed the quantities $X = (T_1 - T_2)/(\sigma\sqrt{2})$ and $Y = (\overline{T} - \mu)/(\sigma/\sqrt{2})$, where $\overline{T}$ is the sample mean. Moreover, the example found X and Y to be independent and standard normal, and it pursued the consequences associated with Y^2/X^2 having an F distribution with $(1,1)$ degrees of freedom. However, we note the the square root of this quantity must have a t distribution with one degree of freedom. That is,

$$
\frac{Y}{X} = \frac{(\overline{T} - \mu)\sqrt{2}}{\sigma} \cdot \frac{\sigma\sqrt{2}}{T_1 - T_2} = \frac{2(\overline{T} - \mu)}{T_1 - T_2}
$$

has this t distribution. From Table 6.5, $t_{(1)0.975} = 12.706$. Consequently,

$$0.95 = \Pr\left(t_{(1)0.025} \le \frac{2(\overline{T} - \mu)}{T_1 - T_2} \le t_{(1)0.975}\right) = \Pr\left(-12.706 \le \frac{2(\overline{T} - \mu)}{T_1 - T_2} \le 12.706\right).$$

We manipulate the interval to obtain left and right bounds on μ:

$$0.95 = \Pr\left(\overline{T} - \frac{12.706|T_1 - T_2|}{2} \le \mu \le \overline{T} + \frac{12.706|T_1 - T_2|}{2}\right) = \Pr(28.3 \le \mu \le 70.5),$$

which is the same interval as that obtained in Example 6.63. $\square$

The χ^2, F, and t distributions all derive from squares of independent standard normal random variables. Each has a corresponding noncentral version that derives from independent normal random variables with nonzero means and common variance 1. Two of these will find use in the next chapter, so we derive their densities.

6.67 Definition: Suppose that $Z_1, Z_2, \ldots, Z_n$ are independent normal random variables with common variance $\sigma^2 = 1$. Their means are $\mu_1, \mu_2, \ldots, \mu_n$, respectively. Then $Y = \sum_{i=1}^{n} Z_i^2$ has a *noncentral χ^2 distribution* with n degrees of freedom and noncentrality parameter $\lambda = (1/2) \sum_{i=1}^{n} \mu_i^2$.

Let U be a noncentral χ^2 random variable with n degrees of freedom and noncentrality parameter λ. Let V be a central χ^2 random variable with m degrees of freedom. Then $W = (U/n)/(V/m)$ has a *noncentral F distribution* with (n, m) degrees of freedom and noncentrality parameter λ. When context does not make the matter clear, the original χ^2 and F distributions will be called *central distributions*. ∎

We first derive the density of Y, a noncentral χ^2 random variable with n degrees of freedom and noncentrality parameter λ. We have $Y = \sum_{i=1}^{n} Z_i^2$, where Z_i is normal with parameters $(\mu_i, 1)$. Consequently, the moment generating function of Z_i^2 is

$$\Psi_{Z_i^2}(t) = E[e^{tZ_i^2}] = \frac{1}{\sqrt{2\pi}} \int_{-\infty}^{\infty} e^{tz^2} e^{-(z-\mu_i)^2/2} \, dz$$

$$= \frac{1}{\sqrt{2\pi}} \int_{-\infty}^{\infty} \exp\left(\left(t - \frac{1}{2}\right)z^2 + \mu_i z - \frac{\mu_i^2}{2}\right) dz$$

$$= \frac{1}{\sqrt{2\pi}} \int_{-\infty}^{\infty} \exp\left[-\left(\frac{1 - 2t}{2}\right)\left[\left(z - \frac{\mu_i}{1 - 2t}\right)^2 - \frac{2t\mu_i^2}{(1 - 2t)^2}\right]\right] dz$$

$$= \frac{1}{\sqrt{2\pi}} e^{t\mu_i^2/(1-2t)} \int_{-\infty}^{\infty} \exp\left(-\frac{1}{2}\left[\sqrt{1 - 2t}\left(z - \frac{\mu_i}{1 - 2t}\right)\right]^2\right) dz$$

$$= \frac{1}{\sqrt{2\pi}} e^{t\mu_i^2/(1-2t)} \int_{-\infty}^{\infty} e^{-y^2/2} \frac{1}{\sqrt{1 - 2t}} \, dy = \frac{e^{t\mu_i^2/(1-2t)}}{\sqrt{1 - 2t}}.$$

Because the Z_i^2 are independent, we have, for $t < 1/2$,

$$\Psi_Y(t) = \prod_{i=1}^{n} \Psi_{Z_i^2}(t) = (1 - 2t)^{-n/2} \exp\left(\frac{t(\mu_1^2 + \mu_2^2 + \ldots + \mu_n^2)}{1 - 2t}\right)$$

$$= = (1 - 2t)^{-n/2} e^{2\lambda t/(1-2t)}.$$

Recall that the moment generating function for the central χ^2 distribution with n degrees of freedom is $(1 - 2t)^{-n/2}$. Therefore $\Psi_Y(t)$ reduces to the proper form when the noncentrality parameter $\lambda = 0$. Also,

$$\frac{2\lambda t}{1 - 2t} = -\lambda + \frac{\lambda}{1 - 2t},$$

which allows the reduction

$$\Psi_Y(t) = (1 - 2t)^{-n/2} e^{-\lambda + \lambda/(1-2t)} = (1 - 2t)^{-n/2} e^{-\lambda} \sum_{k=0}^{\infty} \frac{1}{k!} \left(\frac{\lambda}{1 - 2t}\right)^k$$

$$\Psi_Y(t) = \sum_{k=0}^{\infty} \frac{e^{-\lambda}\lambda^k}{k!}(1 - 2t)^{-(n+2k)/2}.$$

Note that the operation of creating a moment generating function from a density is linear in the following sense. Suppose that $f_1(\cdot), f_2(\cdot), \ldots$ are densities for random variables $X_1, X_2, \ldots$ and $a_1, a_2, \ldots$ are constants that sum to 1. Define $g(x) = \sum_{i=1}^{\infty} a_i f_i(x)$. Then $g(x)$ is a density because it is nonnegative with total integral 1. Moreover, if we let W be the random variable associated with density $g(\cdot)$, we have

$$\Psi_W(t) = E[e^{tW}] = \int_{-\infty}^{\infty} e^{tw} g(w)\, dw = \sum_{i=1}^{\infty} a_i \int_{-\infty}^{\infty} e^{tw} f_i(w)\, dw$$

$$= \sum_{i=1}^{\infty} a_i E[e^{tX_i}] = \sum_{i=1}^{\infty} a_i \Psi_{X_i}(t).$$

Reviewing the equation for $\Psi_Y(t)$, we note that the terms $(1 - 2t)^{-(n+2k)/2}$ are the moment generating functions of central χ^2 distributions with degrees of freedom $n + 2k$. Moreover, the coefficients satisfy

$$\sum_{k=0}^{\infty} \frac{e^{-\lambda}\lambda^k}{k!} = e^{-\lambda} \sum_{k=0}^{\infty} \frac{\lambda^k}{k!} = e^{-\lambda} e^{\lambda} = 1.$$

Consequently,

$$\Psi_Y(t) = \sum_{k=0}^{\infty} \frac{e^{-\lambda}\lambda^k}{k!} \Psi_{\chi^2(n+2k)}(t)$$

$$f_Y(t) = \sum_{k=0}^{\infty} \frac{e^{-\lambda}\lambda^k}{k!} f_{\chi^2(n+2k)}(t)$$

$$= \sum_{k=0}^{\infty} \left(\frac{e^{-\lambda}\lambda^k}{k!2^{(n+2k)/2}\Gamma((n + 2k)/2)}\right) t^{(n+2k)/2-1} e^{-t/2}.$$

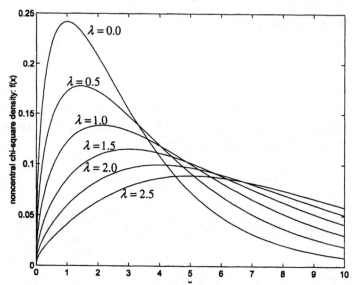

Figure 6.15. Noncentral χ^2 distribution with 3 degrees of freedom for increasing noncentrality parameter λ

This is the noncentral χ^2 density with n degrees of freedom and noncentrality parameter λ. Figure 6.15 displays a noncentral χ^2 distribution. Note that the graph flattens and shifts to the right with increasing noncentrality parameter λ. For a given x_0, therefore, $\Pr(Y > x_0)$ increases with increasing λ.

We now verify that the mean of Y is $n + 2\lambda$, which confirms the right shift in the graph. Recall that $Y = \sum_{i=1}^{n} Z_i^2$, where Z_i is normal with parameters $(\mu_i, 1)$. Consequently, because $\sum_{i=1}^{n}(Z_i - \mu_i)^2$ is χ^2 with n degrees of freedom,

$$E[Y] = \sum_{i=1}^{n} E[Z_i^2] = \sum_{i=1}^{n} E[(Z_i - \mu_i + \mu_i)^2]$$

$$= \sum_{i=1}^{n} E[(Z_i - \mu_i)^2] + 2\sum_{i=1}^{n} \mu_i E[Z_i - \mu_i] + \sum_{i=1}^{n} \mu_i^2 = n + 2\lambda.$$

Turning now to the noncentral F distribution, we imitate the derivation in Theorem 6.62 to calculate its density. The process involves a number of complicated integrals, which are very similar to those in the theorem cited. Therefore, we present an abbreviated derivation. Suppose that X and Y are independent and $W = (X/n_1)(Y/n_2)$, where X is noncentral χ^2 with n_1 degrees of freedom and noncentrality parameter λ and Y has a central χ^2 distribution with n_2 degrees of freedom. The joint density of X and Y is then the product of the noncentral χ^2 derived above and the simpler central χ^2 of Theorem 6.59.

$$f_{XY}(x,y) = \left(\sum_{k=0}^{\infty} \frac{e^{-\lambda}\lambda^k x^{(n_1+2k)/2-1}e^{-x/2}}{k!2^{(n_1+2k)/2}\Gamma((n_1+2k)/2)} \right) \left(\frac{y^{n_2/2-1}e^{-y/2}}{2^{n_2/2}\Gamma(n_2/2)} \right)$$

The cumulative distribution of W is an integral of this density over the appropriate region.

$$F_W(w) = \Pr(W \le w) = \Pr\left(X \le \frac{n_1 w Y}{n_2}\right) = \int_{y=0}^{\infty} \int_{x=0}^{n_1 w y / n_2} f_{XY}(x, y) \, dx \, dy$$

Substituting for $f_{XY}(\cdot)$, we have

$$F_W(w) = \sum_{k=0}^{\infty} e^{-\lambda} \lambda^k \int_{y=0}^{\infty} \frac{y^{n_2/2-1} e^{-y/2}}{2^{n_2/2} \Gamma(n_2/2)}$$

$$\cdot \int_{x=0}^{n_1 w y / n_2} \frac{x^{(n_1+2k)/2-1} e^{-x/2}}{k! 2^{(n_1+2k)/2} \Gamma((n_1+2k)/2)} \, dx \, dy$$

The density follows through differentiation.

$$f_W(w) = \sum_{k=0}^{\infty} e^{-\lambda} \lambda^k$$

$$\cdot \int_{y=0}^{\infty} \frac{y^{n_2/2-1} e^{-y/2}}{2^{n_2/2} \Gamma(n_2/2)} \left[\frac{(n_1 w y / n_2)^{(n_1+2k)/2-1} e^{-n_1 w y / 2 n_2}}{k! 2^{(n_1+2k)/2} \Gamma((n_1+2k)/2)} \right] \left(\frac{n_1 y}{n_2} \right) dy$$

$$= \sum_{k=0}^{\infty} \frac{e^{-\lambda} \lambda^k (n_1/n_2)^{(n_1+2k)/2}}{k! 2^{(n_1+n_2+2k)/2} \Gamma(n_2/2) \Gamma((n_1+2k)/2)} w^{(n_1+2k)/2-1}$$

$$\cdot \int_{y=0}^{\infty} y^{(n_1+n_2+2k)/2-1} e^{-(1+n_1 w / n_2) y / 2} \, dy \qquad (6.4)$$

Letting $\tau = (1/2)[1 + n_1 w / n_2]$ and $t = (n_1 + n_2 + 2k)/2$, we have

$$I = \int_{y=0}^{\infty} y^{(n_1+n_2+2k)/2-1} e^{-(1+n_1 w / n_2) y / 2} \, dy = \tau^{-t} \tau^t \int_{y=0}^{\infty} y^{t-1} e^{-\tau y} \, dy$$

$$= \tau^{-t} \Gamma(t) = \left[\frac{1}{2} \left(1 + \frac{n_1 w}{n_2} \right) \right]^{-(n_1+n_2+2k)/2} \Gamma((n_1 + n_2 + 2k)/2). \qquad (6.5)$$

6.68 Theorem: If X has a noncentral χ^2 distribution with n degrees of freedom and noncentrality parameter λ, then its mean is $\mu_X = n + 2\lambda$ and its density is

$$f_X(t) = \sum_{k=0}^{\infty} \left(\frac{e^{-\lambda} \lambda^k}{k! 2^{(n+2k)/2} \Gamma((n+2k)/2)} \right) t^{(n+2k)/2-1} e^{-t/2}.$$

If Y has a noncentral F distribution with (n_1, n_2) degrees of freedom and noncentrality parameter λ, then its mean is $[n_2/(n_2 - 2)][1 + 2\lambda/n_1]$ and its density is

$$f_Y(t) = \sum_{k=0}^{\infty} \left(\frac{e^{-\lambda} \lambda^k}{k!} \right) \left(\frac{\Gamma((n_1 + n_2 + 2k)/2)}{\Gamma((n_1 + 2k)/2) \Gamma(n_2/2)} \right) \left(\frac{n_1}{n_2} \right)^{(n_1+2k)/2}$$

$$\cdot w^{(n_1+2k)/2-1} \left(1 + \frac{n_1 w}{n_2} \right)^{-(n_1+n_2+2k)/2}.$$

PROOF: The preceding discussion establishes the mean and density of the noncentral χ^2 distribution. The density of the noncentral F distribution follows by substituting the integral expression I (Equation 6.5) into the $f_W(\cdot)$ as given in Equation 6.4. There remains only the mean of the noncentral F density, which we calculate as follows.

$$
\begin{aligned}
\mu_Y &= \int_0^\infty t f_Y(t)\, dt \\
&= \sum_{k=0}^\infty \frac{e^{-\lambda}\lambda^k \Gamma((n_1 + n_2 + 2k)/2)}{k!\,\Gamma((n_1 + 2k)/2)\Gamma(n_2/2)} \\
&\quad \cdot \int_0^\infty \left(\frac{n_1}{n_2}\right)^{(n_1+2k)/2} t^{(n_1+2k)/2}\left(1 + \frac{n_1 t}{n_2}\right)^{-(n_1+n_2+2k)/2} dt
\end{aligned}
$$

We use the substitution

$$
\begin{aligned}
\tau &= \frac{n_1 t/n_2}{1 + n_1 t/n_2} \\
d\tau &= \frac{(n_1/n_2)\, dt}{(1 + n_1 t/n_2)^2} \\
t &= \left(\frac{n_2}{n_1}\right)\frac{\tau}{1-\tau}
\end{aligned}
$$

$$
1 + \frac{n_1 t}{n_2} = \frac{1}{1-\tau}
$$

on the integrand above, which we designate as I.

$$
\begin{aligned}
I &= \frac{n_2}{n_1}\int_0^1 \tau^{(n_1+2k)/2}\left(\frac{1}{1-\tau}\right)^{-n_2/2+2} d\tau \\
&= \frac{n_2}{n_1}\int_0^1 \tau^{(n_1+2k)/2}(1-\tau)^{n_2/2-2}\, d\tau = \frac{n_2}{n_1}\beta((n_1+2k)/2+1, n_2/2-1) \\
&= \frac{n_2\Gamma((n_1+2k)/2+1)\Gamma(n_2/2-1)}{n_1\Gamma((n_1+n_2+2k)/2)}.
\end{aligned}
$$

Substituting in the expression above for μ_Y, we have

$$
\begin{aligned}
\mu_Y &= \sum_{k=0}^\infty \frac{e^{-\lambda}\lambda^k \Gamma((n_1+n_2+2k)/2)}{k!\,\Gamma((n_1+2k)/2)\Gamma(n_2/2)} \cdot \frac{n_2\Gamma((n_1+2k)/2+1)\Gamma(n_2/2-1)}{n_1\Gamma((n_1+n_2+2k)/2)} \\
&= \sum_{k=0}^\infty \frac{e^{-\lambda}\lambda^k}{k!} \cdot \frac{n_2(n_1+2k)/2}{n_1(n_2/2-1)} = \sum_{k=0}^\infty \frac{e^{-\lambda}\lambda^k}{k!} \cdot \frac{n_2(n_1+2k)}{n_1(n_2-2)} \\
&= \frac{n_2 e^{-\lambda}}{n_2-2}\sum_{k=0}^\infty \left(1 + \frac{2k}{n_1}\right)\frac{\lambda^k}{k!} = \frac{n_2 e^{-\lambda}}{n_2-2}\left[\sum_{k=0}^\infty \frac{\lambda^k}{k!} + \frac{2}{n_1}\sum_{k=0}^\infty \frac{k\lambda^k}{k!}\right]
\end{aligned}
$$

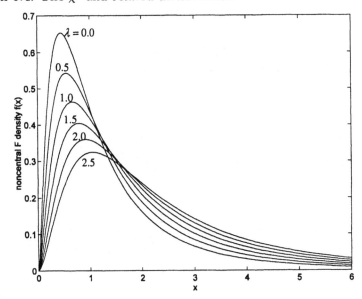

Figure 6.16. Noncentral F distribution with $(5, 6)$ degrees of freedom and increasing noncentrality parameter λ

$$\mu_Y = \frac{n_2 e^{-\lambda}}{n_2 - 2}\left[e^\lambda + \frac{2}{n_1}\sum_{k=1}^{\infty}\frac{\lambda^k}{(k-1)!}\right] = \frac{n_2 e^{-\lambda}}{n_2 - 2}\left[e^\lambda + \frac{2\lambda}{n_1}\sum_{k=0}^{\infty}\frac{\lambda^k}{k!}\right]$$

$$= \frac{n_2}{n_2 - 2}\left(1 + \frac{2\lambda}{n_1}\right). \blacksquare$$

Figure 6.16 shows a collection of F densities. All have $(5, 6)$ degrees of freedom, but they have different noncentrality parameters, varying from 0.0 through 2.5. Notice that the density flattens and shifts to the right with increasing noncentrality parameter. As with the noncentral χ^2 distribution, this means that for a given x_0, the probability mass to the right of x_0 increases with increasing λ. This observation will be important in the statistical estimation techniques of the next chapter. Tables are available for the percentiles of the noncentral distributions, but a separate tabulation is necessary for each noncentrality value. When we need specific percentiles or probabilities, we will, with computer assistance of course, use numerical integration on the densities.

Exercises

6.45 Let X be a χ^2 random variable with n degrees of freedom. Derive an expression for the kth moment $E[X^k]$.

6.46 When operated at the specified power levels, a memory chip's temperature stabilizes at T. Across many chips, this temperature is a random variable, but it is known to be normally distributed with mean 80 degrees Celsius. The variance is not known, but six random measurements yield temperatures 79.8, 80.2, 78.9, 81.1, 81.2, and 78.9. Construct a 95% confidence interval for the variance.

6.47 Construct a 90% confidence level for the variance of the temperature distribution in the preceding exercise.

6.48 For the F distribution with (n_1, n_2) degrees of freedom, show that the total allocated probability is 1. That is, verify that the following integral evaluates to 1.

$$\int_{w=0}^{\infty} \frac{\Gamma((n_1 + n_2)/2)}{\Gamma(n_1/2)\Gamma(n_2/2)} \left(\frac{n_1}{n_2}\right)^{n_1/2} w^{(n_1/2)-1} \left(1 + \frac{n_1 w}{n_2}\right)^{-(n_1+n_2)/2} dw$$

*6.49 Let W have an F distribution with (n_1, n_2) degrees of freedom. Suppose that $n_2 = 1$ or $n_2 = 2$. Show that the defining integral for $E[W]$ does not converge.

6.50 Derive an expression for $E[W^k]$, where W has an F distribution with (n_1, n_2) degrees of freedom.

*6.51 Let $f_{(m,n)}(\cdot)$ be the density function of an F random variable with (m, n) degrees of freedom. Show that

$$D = \int_0^t f_{(m,n)}(w)\, dw - \frac{m}{m+2} \int_0^t f_{(m+2,n)}\left(\frac{mw}{m+2}\right) dw$$

$$= \frac{2(mt/n)^{m/2}}{m\beta(m/2, n/2)} \left(1 + \frac{mt}{n}\right)^{-(m+n)/2}.$$

*6.52 Let (X_1, X_2, X_3) be independent observations of the random variable X, which is normally distributed with mean μ and variance σ^2. Let $\overline{X}$ be the sample mean, and let $S = \sum_{i=1}^{3}(X_i - \overline{X})^2$. By considering the quantities

$$V_1 = \frac{1}{\sigma}\left[\left(\frac{1}{2}\sqrt{2}\right) X_1 - \left(\frac{1}{2}\sqrt{2}\right) X_3\right]$$

$$V_2 = \frac{1}{\sigma}\left[-\left(\frac{1}{6}\sqrt{6}\right) X_1 + \left(\frac{1}{3}\sqrt{6}\right) X_2 - \left(\frac{1}{6}\sqrt{6}\right) X_3\right],$$

show that S/σ^2 has a χ^2 distribution with two degrees of freedom.

*6.53 As in Example 6.63, let T be the settling time for a full adder circuit. Assume that T is normal with unknown mean μ and variance σ^2. Extending the example, suppose that we now have three measurements, taken with randomly chosen circuits and inputs. They are 48.2, 49.9, and 51.6 nanoseconds. Compute an 80% confidence interval for σ^2.

*6.54 In the context of the preceding exercise, compute an 80% confidence interval for μ.

6.55 Show that $F_{(n,1)p} = t_{(n)q}^{-2}$, when $q = (2-p)/2$.

*6.56 Let Y have a noncentral χ^2 distribution with n degrees of freedom and noncentrality parameter λ. Verify that $E[Y] = n + 2\lambda$ by directly integrating the density function. That is, show that $\int_0^\infty t f_Y(t)\, dt = n + 2\lambda$, where

$$f_Y(t) = \sum_{k=0}^\infty \left(\frac{e^{-\lambda}\lambda^k}{k!2^{(n+2k)/2}\Gamma((n+2k)/2)} t^{(n+2k)/2-1} \right) e^{-t/2}.$$

*6.57 Compute the variance of Y, a noncentral χ^2 random variable with n degrees of freedom and noncentrality parameter λ.

*6.58 Verify that the noncentral F distribution allocates total probability 1. That is, verify that

$$\int_0^\infty \sum_{k=0}^\infty \left(\frac{e^{-\lambda}\lambda^k}{k!} \right) \left(\frac{\Gamma((n_1+n_2+2k)/2)}{\Gamma((n_1+2k)/2)\Gamma(n_2/2)} \right) \left(\frac{n_1}{n_2} \right)^{(n_1+2k)/2}$$
$$\cdot w^{(n_1+2k)/2-1} \left(1 + \frac{n_1 w}{n_2} \right)^{-(n_1+n_2+2k)/2} dw = 1.$$

*6.59 Compute the variance of X, which has a noncentral F distribution with (n_1, n_2) degrees of freedom and noncentrality parameter λ.

6.5   Computer simulations

Chapter 3 presented algorithms for generating samples from selected discrete distributions. This section extends those ideas to continuous distributions. A general technique introduced in Chapter 3 is the inverse transform. Specifically, if U is a continuous uniform random variable with range $[0,1)$ and X is a discrete random variable with cumulative distribution $F_X(\cdot)$, then $Y = F_X^{-1}(U)$ has the same distribution as X. We use the system utility rand() to approximate U and compute Y samples from U samples. The notation requires some explanation because F_X is not invertible for discrete X. In any case, Algorithm 3.5 provides the details, and Theorem 3.6 proves the algorithm's correctness for discrete distributions. The proof extends easily to continuous distributions.

6.69 Theorem: Let X be a continuous random variable with cumulative distribution $F_X(\cdot)$. Then, assuming that F_X is monotone increasing and therefore invertible, $Y = F_X^{-1}(U)$ has the same distribution as X, given that U has a continuous uniform distribution on $[0,1)$.

```
function erlang(n, lambda)    // exponential (lambda) if n = 1
    x = 1.0;                  // ; Erlang (n, lambda) if n > 1
    for (i = 0; i < n; i++)
        x = x *(1 - rand());
    return -ln(x)/lambda;
```

Figure 6.17. Algorithm for random observations from an Erlang or exponential distribution

PROOF: Recall that $F_U(x) = x$ for $0 \leq x \leq 1$. Consequently, since $0 \leq F_X(x) \leq 1$,

$$F_Y(x) = \Pr(Y \leq x) = \Pr(F_X^{-1}(U) \leq x) = \Pr(U \leq F_X(x))$$
$$= F_U(F_X(x)) = F_X(x). \blacksquare$$

The exponential distribution function is particularly easy to invert. If X is exponential with parameter λ, then for $x \geq 0$,

$$F_X(x) = 1 - e^{-\lambda x}$$
$$x = \frac{-\ln(1 - F_X(x))}{\lambda}$$
$$F_X^{-1}(y) = \frac{-\ln(1 - y)}{\lambda}.$$

For uniform U on $[0, 1)$, therefore, $-[\ln(1 - U)]/\lambda$ is exponentially distributed with parameter λ. Indeed, we derived this result, without mention of its inverse transform origins, as Equation 3.4, which found use in generating Poisson interarrival times for client-server simulations. Because an Erlang random variable is simply a sum of independent exponentials, we can extend the analysis to cover this case as well. Specifically,

$$Y = \frac{-\ln(1 - U_1)}{\lambda} + \frac{-\ln(1 - U_2)}{\lambda} + \ldots + \frac{-\ln(1 - U_n)}{\lambda}$$
$$= \frac{-1}{\lambda} \ln \left(\prod_{i=1}^{n} (1 - U_i) \right)$$

is Erlang with parameters (n, λ), provided that $U_1, U_2, \ldots, U_n$ are independent and uniform on $[0, 1)$. Algorithm 6.17 exploits this relationship to generate either exponential or Erlang samples.

Certain distribution functions are difficult to invert. Consider the standard normal distribution $\Phi(x) = (1/\sqrt{2\pi}) \int_{-\infty}^{x} e^{-y^2/2} \, dy$. The inversion must find, for a given $p \in [0, 1]$, a value x such that $\Phi(x) = p$. In other words, it must locate a particular percentile. We can approach such problems with a rejection filter, a concept also introduced for discrete random variables in Chapter 3. Only a slight modification is needed. Recall that the method generates samples from a related distribution but selectively discards some

```
function simulateY ( )
   while (true)
      x = simulateX();
      u = rand();
      if (u < f_Y (x) / (K * f_X (x)))
         return x;
```

Figure 6.18. Rejection filter algorithm adapted for continuous distributions

of them. Suppose that we can generate X samples, but we need Y samples. Algorithm 6.18 delivers the Y samples required, provided that the constant K is such that $f_Y(x) \leq K f_X(x)$ for all x. It is essentially Algorithm 3.12.

To prove the algorithm correct, let U_i be the uniform random variables returned by rand() on sequential passes through the while-loop. These are independent of the X_i as returned by simulateX(). For each (X_i, U_i) pair, the joint density is $f_U(u) \cdot f_X(x) = f_X(x)$ when $0 \leq u \leq 1$ and zero elsewhere. Let T_i be the event where iteration i satisfies the exit criterion. Then $\Pr(T_i)$ is the integral of this density over the (x, u) such that $f_X(x) > 0$ and $u < f_Y(x)/(K f_X(x))$. That is, if $D_X = \{x \mid f_X(x) > 0\}$, we have

$$\begin{aligned}\Pr(T_i) &= \int_{x \in D_X} \int_{u=0}^{f_Y(x)/(K f_X(x))} f_X(x)\,du\,dx \\ &= \int_{x \in D_X} \frac{f_Y(x)}{K f_X(x)} \cdot f_X(x)\,dx = \frac{1}{K} \int_{x \in D_X} f_Y(x)\,dx.\end{aligned}$$

The condition $f_Y(x) \leq K f_X(x)$ forces $f_Y(x) = 0$ for $x \notin D_X$. Consequently,

$$\Pr(T_i) = \frac{1}{K} \int_{-\infty}^{\infty} f_Y(x)\,dx = \frac{1}{K}.$$

By a similar calculation, the intersection event $(X_i \leq t) \cap T_i$ has probability

$$\begin{aligned}\Pr((X_i \leq t) \cap T_i) &= \int_{x \in D_X, x \leq t} \int_{u=0}^{f_Y(x)/(K f_X(x))} f_X(x)\,du\,dx \\ &= \frac{1}{K} \int_{-\infty}^{t} f_Y(x)\,dx = \frac{F_Y(t)}{K}.\end{aligned}$$

If R is the algorithm return, the probability that $(R \leq t)$ is the sum of the probabilities of the disjoint events $[(X_i \leq t) \cap T_i] \cap \overline{T_{i-1}} \cap \overline{T_{i-2}} \cap \ldots \cap \overline{T_1}$, for $i = 1, 2, \ldots$. The distribution of R is then

$$\Pr(R \leq t) = \sum_{i=1}^{\infty} \left(1 - \frac{1}{K}\right)^{i-1} \frac{F_Y(t)}{K} = \frac{F_Y(t)}{K} \sum_{i=0}^{\infty} \left(1 - \frac{1}{K}\right)^{i} = F_Y(t).$$

6.70 Example: Use a rejection filter to generate sample from a gamma distribution with parameters $(\alpha = 5/2, \lambda = 2)$.

This gamma distribution has density $f_Y(y) = 2^{5/2}y^{3/2}e^{-2y}/\Gamma(5/2)$, for $y \geq 0$. We can implement a rejection filter over the exponential $f_X(y) = ke^{-ky}$, provided that the ratio $f_Y(y)/f_X(y)$ is bounded. This ratio is

$$\frac{f_Y(y)}{f_X(y)} = \frac{2^{5/2}y^{3/2}e^{-2y}}{\Gamma(5/2)} \cdot \frac{1}{ke^{-ky}} = \left(\frac{2^{5/2}}{k\Gamma(5/2)}\right)y^{3/2}e^{-(2-k)y},$$

which is bounded when $k < 2$. For convenience, we choose $k = 1$, and solve for the maximum ratio value.

$$r(y) = \frac{f_Y(y)}{f_X(y)} = \left(\frac{2^{5/2}}{\Gamma(5/2)}\right)y^{3/2}e^{-y}$$

$$r'(y) = \frac{2^{5/2}}{\Gamma(5/2)}\left[\frac{3}{2}y^{1/2}e^{-y} + y^{3/2}(-e^{-y})\right] = \left(\frac{2^{5/2}}{\Gamma(5/2)}\right)y^{1/2}e^{-y}\left[\frac{3}{2} - y\right] = 0$$

The maximum ratio is therefore $r_0 = [2^{5/2}/\Gamma(5/2)](3/2)^{3/2}e^{-3/2} = 1.75$, and it occurs at $y_0 = 3/2$. The required generator is then Algorithm 6.18 with $K = 1.75$. Figure 6.19 illustrates a test of this generator. The continuous curve is the density $f_Y(\cdot)$. The points marked with the small circles show the distribution of returns from Algorithm 6.18. The following calculations produced the circle locations. Let the random variable R denote the algorithm's returns. The graph's horizontal span, running from $x = 0$ through $x = 6$, contains 25 cells of equal width $\Delta x = 6/25 = 0.24$, giving the partition $0 = x_0 < x_1 < x_2 < \ldots < x_{25} = 6$. The fraction of the returns f_i falling in a particular cell $(x_i, x_{i+1}]$ approximates $\Pr(x_i < Y \leq x_{i+1})$. Of course, $f_Y(x_i)\Delta x$ also approximates this probability. Therefore, an accurate generator should have $f_Y(x_i) \approx f_i/\Delta x$. The circles represent the quantities $f_i/\Delta x$, and they track the continuous curve rather closely. $\square$

The example provides a template for generating any gamma distribution with parameters $(\alpha \geq 1, \lambda > 0)$, because the corresponding densities $f_X(x)$ approach zero as $x \to 0^+$ or $x \to \infty$. If $0 < \alpha < 1$, however, the density is unbounded as $x \to 0^+$ and cannot be dominated by any multiple of an exponential density. In these cases, it is possible to use a mixture of the inverse transform and rejection filter methods, as illustrated in the next example.

6.71 Example: Compose an algorithm to generate samples from a gamma distribution with parameters $(\alpha = 1/2, \lambda = 2)$.

For positive x, the gamma density is

$$f_Y(x) = \left(\frac{2^{1/2}}{\Gamma(1/2)}\right)x^{-1/2}e^{-2x} = \left(\sqrt{\frac{2}{\pi}}\right)x^{-1/2}e^{-2x},$$

which is unbounded as $x \to 0^+$. To use a rejection filter, we must find a constant K and a reference density $f_X(x)$ such that $f_Y(x) \leq Kf_X(x)$. A simple exponential, as used in the preceding example, will not suffice in this case. However, we can modify the exponential by substituting the form $C_1x^{-1/2}$ on the interval $(0, 1]$. That is, we choose the reference $f_X(x)$ to be

$$f_X(x) = \begin{cases} C_1x^{-1/2}, & 0 < x \leq 1 \\ C_2e^{-x}, & x > 1. \end{cases}$$

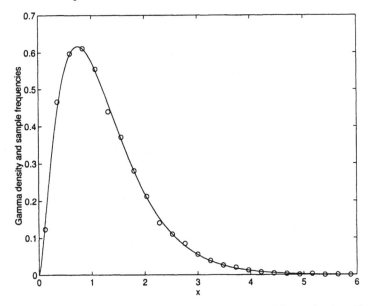

Figure 6.19. Performance of a gamma generator with a rejection filter (see Example 6.70)

We must choose C_1, C_2 such that the total probability is 1. For convenience, we assign half of the probability to each component. Then

$$\frac{1}{2} = C_1 \int_0^1 x^{-1/2}\, dx = 2C_1 x^{1/2}\Big|_0^1 = 2C_1$$

$$\frac{1}{2} = C_2 \int_1^\infty e^{-x}\, dx = -C_2 e^{-x}\Big|_1^\infty = C_2 e^{-1}.$$

Therefore, $C_1 = 1/4$ and $C_2 = e/2$. Continuing, we find that $f_X(x)$ and the ratio of interest are

$$f_X(x) = \begin{cases} \dfrac{1}{4\sqrt{x}}, & 0 < x \le 1 \\[2mm] \dfrac{e^{-(x-1)}}{2}, & x > 1 \end{cases}$$

$$r(x) = \frac{f_Y(x)}{f_X(x)} = \begin{cases} \dfrac{\sqrt{2/\pi}\,x^{-1/2}e^{-2x}}{1/(4\sqrt{x})} = \left(\sqrt{\dfrac{32}{\pi}}\right) e^{-2x}, & 0 < x \le 1 \\[3mm] \dfrac{\sqrt{2/\pi}\,x^{-1/2}e^{-2x}}{e^{-(x-1)}/2} = \left(\sqrt{\dfrac{8}{\pi}}\right) x^{-1/2}e^{-(x+1)}, & x > 1. \end{cases}$$

In the span $0 < x \le 1$, the ratio $r(x)$ achieves its maximum as $x \to 0^+$. That maximum is $\sqrt{32/\pi} = 3.2$. On the span $1 < x < \infty$, a maximum of 0.22 occurs as $x \to 1^-$. Therefore, we take $K = 3.2$ in the rejection filter algorithm. We must still obtain an inverse transform generator for the reference density $f_X(x)$. Integrating

```
function modexp
    y = rand();
    if (y ≤ 1/2)
        return 4 * y * y;
    else
        return 1 - ln (2 - 2 * y);
```

Figure 6.20. Generation algorithm for a mixed density, which combines $e^{-(x-1)}/2$ and $1/(4\sqrt{x})$ (see Example 6.71)

the density, we find that

$$F_X(x) = \begin{cases} \frac{1}{4}\int_0^x t^{-1/2}\,dt = \frac{x^{1/2}}{2}, & 0 < x \le 1 \\ \frac{1}{2} + \frac{1}{2}\int_1^x e^{-(t-1)}\,dt = 1 - \frac{e^{-(x-1)}}{2}, & x > 1. \end{cases}$$

Consequently,

$$F_X^{-1}(y) = \begin{cases} 4y^2, & 0 < y \le \frac{1}{.2} \\ 1 - \ln(2 - 2y), & \frac{1}{2} < y \le 1. \end{cases}$$

Algorithm 6.20 returns samples from the reference distribution. Figure 6.21 tests this generator. It plots the density $f_X(x)$ and indicates the distribution of the generator returns with the small circles. As in the preceding example, these circles locate the fraction of the returns that fall in a small cell centered horizontally on the circle. The fractions are multiplied by the cell width to be comparable with the density function.

Finally, the rejection filter of Algorithm 6.18, with $K = 3.2$, now correctly produces samples from the gamma distribution with parameters $(1/2, 2)$. Figure 6.22 illustrates how the returns cluster near the desired density. $\square$

The density $x^{u-1}(1 - x)^{v-1}/\beta(u, v)$ for $0 < x < 1$ characterizes the beta random variable with parameters (u, v). This density is bounded when $u, v \ge 1$. Consequently, we can implement a rejection filter generator using the continuous uniform distribution on $[0, 1)$ as a reference. When $0 < u < 1$ or $0 < v < 1$, or both, the density is unbounded near one end or the other of the $(0, 1)$ interval. In this case, we need an unbounded reference density such that the ratio remains bounded. To generate samples from $f_Y(x) = x^{-1/2}(1 - x)^{1/2}/\beta(1/2, 3/2)$ for $0 < x < 1$, for example, we can use the reference $f_X(x) = 1/(2\sqrt{x})$ on $0 < x < 1$. The ratio of interest on the interval $(0, 1)$ is then

$$r(x) = \frac{f_Y(x)}{f_X(x)} = \frac{(1 - x)^{1/2}}{x^{1/2}\beta(1/2, 3/2)} \cdot 2x^{1/2} = \frac{2(1 - x)^{1/2}}{\beta(1/2, 3/2)},$$

which achieves a maximum of $2/\beta(1/2, 3/2) = 4/\pi = 1.273$ as $x \to 0^+$. The exercises ask that you complete the rejection filter algorithms for beta distributions.

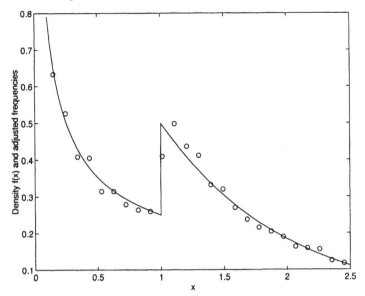

Figure 6.21. Performance of the mixed density generator (see Example 6.71)

The χ^2 random variable with n degrees of freedom has a gamma density with parameters $(\alpha = n/2, \lambda = 1/2)$, and we can generate samples with the rejection filters illustrated in the last two examples. Since $\chi^2_{(1)}$ is the square of a standard normal, the square roots of the $\chi^2_{(1)}$ returns, evenly divided between plus and minus square roots, will have a standard normal distribution. We prove this statement as follows. Suppose that R and U are independent random variables such that R is $\chi^2_{(1)}$ and U is uniform on $[0,1)$. Let

$$Y = \begin{cases} \sqrt{R}, & 0 \leq U < 1/2 \\ -\sqrt{R}, & 1/2 \leq U \leq 1. \end{cases}$$

We claim that Y is standard normal. The joint density of R and U is

$$f_{R,U}(r,u) = \frac{1}{\Gamma(1/2)} \left(\frac{1}{2}\right)^{1/2} r^{-1/2} e^{-r/2}, \text{ for } r > 0 \text{ and } 0 \leq u < 1.$$

The density is zero outside this region. Consequently, for $t \geq 0$, we have

$$F_Y(t) = \Pr(Y \leq t) = \Pr\left(U \geq \frac{1}{2}\right) + \Pr\left(\left(U < \frac{1}{2}\right) \cap \left(\sqrt{R} \leq t\right)\right)$$

$$= \frac{1}{2} + \frac{1}{2}\Pr(R \leq t^2) = \frac{1}{2} + \frac{1}{2}\int_0^{t^2} \frac{1}{\Gamma(1/2)} \left(\frac{1}{2}\right)^{1/2} r^{-1/2} e^{-r/2} \, dr$$

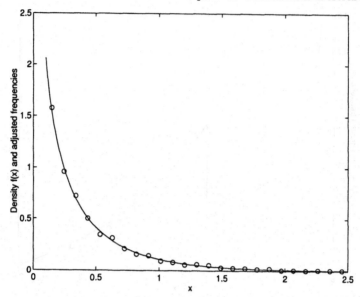

Figure 6.22. Rejection filter performance for the unbounded gamma density $(1/2, 2)$ (see Example 6.71)

$$F_Y(t) = \frac{1}{2} + \frac{1}{2\sqrt{2\pi}} \int_0^{t^2} r^{-1/2} e^{-r/2}\, dr$$

$$f_Y(t) = F_Y'(t) = \frac{1}{2\sqrt{2\pi}}[(t^2)^{-1/2} e^{-t^2/2}](2t) = \frac{1}{\sqrt{2\pi}} e^{-t^2/2}. \tag{6.6}$$

The calculation for $t < 0$ is similar, and we conclude that Y has a standard normal density. Algorithm 6.23 assembles these observations into a functional generator. The reference density

$$f_X(x) = \begin{cases} \dfrac{1}{4\sqrt{x}}, & 0 < x \le 1 \\[2mm] \dfrac{e^{-(x-1)/2}}{4}, & x > 1 \\[2mm] 0, & \text{elsewhere} \end{cases}$$

provides a bounded ratio with the $\chi^2_{(1)}$ density.

$$r(x) = \frac{f_{\chi^2_{(1)}}(x)}{f_X(x)} = \begin{cases} \dfrac{1}{\sqrt{2\pi}} x^{-1/2} e^{-x/2} \cdot 4x^{1/2} = \dfrac{4e^{-x/2}}{\sqrt{2\pi}}, & 0 < x \le 1 \\[3mm] \dfrac{1}{\sqrt{2\pi}} x^{-1/2} e^{-x/2} \cdot \dfrac{4}{e^{-(x-1)/2}} = \dfrac{4}{\sqrt{2\pi e x}}, & x > 1 \end{cases}$$

```
function stdnormal()
    searching = true;
    while (searching)
        u = rand();
        if (u < 0.5)
            t = 4.0 * u * u;
        else
            t = 1.0 - 2.0 * ln (2.0 - 2.0 * u);
        u = rand();
        if ((t <= 1.0) && (u < exp (-t/2)))
            searching = false;
        else if ((t > 1.0) && (u < 1/sqrt(e * t)))
            searching = false;
    u = rand();
    if (u < 0.5)
        return sqrt(t);
    else
        return - sqrt(t);
```

Figure 6.23. Algorithm for random observations from a standard normal distribution

---

We have $r(x) \leq 4/\sqrt{2\pi}$ for all $x > 0$. The exit bound for the rejection filter is then

$$\frac{f_{\chi^2_{(1)}}(x)}{(4/\sqrt{2\pi})f_X(x)} = \begin{cases} e^{-x/2}, & 0 < x \leq 1 \\ \dfrac{1}{\sqrt{ex}}, & x > 1. \end{cases}$$

Integrating $f_X(x)$, we have

$$F_X(x) = \begin{cases} \int_0^x [1/(4\sqrt{t})]\, dt = (1/2)\sqrt{x}, & 0 < x \leq 1 \\ 1/2 + \int_1^x (1/4)e^{-(t-1)/2}\, dt = 1 - (1/2)e^{-(x-1)/2}, & x > 1 \end{cases}$$

$$F_X^{-1}(y) = \begin{cases} 4y^2, & 0 \leq y \leq 1/2 \\ 1 - 2\ln(2 - 2y), & 1/2 < y \leq 1. \end{cases}$$

Inside the searching loop, Algorithm 6.23 first uses this $F_X^{-1}(\cdot)$ to convert the uniform random variable u to the reference random variable t. It then generates a new uniform random u, and checks it against the exit bound noted above. A successful comparison terminates the searching loop and releases the t value as a sample from the $\chi^2_{(1)}$ distribution. A final independent call to rand() provides a symmetric decision to return $\sqrt{t}$ or $-\sqrt{t}$.

Figure 6.24 tests the standard normal generator in the manner established with the earlier examples. The standard normal density appears as a solid curve, while the small circles indicate the fraction of the generator's returns that fall in a small cell centered horizontally on the circle. These fractions are adjusted for the cell width so as to be comparable with the density. The generator performs as expected.

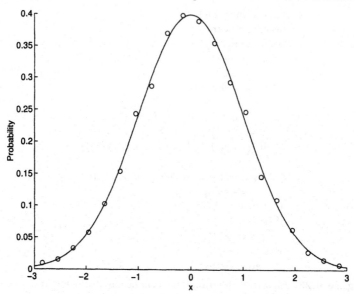

Figure 6.24. Evaluating a rejection filter generator for a standard normal density

It is possible to generate a standard normal sample with a direct computation over a pair of independent uniform observations. If Z_1 and Z_2 are independent and standard normal, we interpret (Z_1, Z_2) as a point in the plane. In polar coordinates (R, Θ), the $\chi^2_{(2)}$ random variable $R^2 = Z_1^2 + Z_2^2$ has a gamma distribution with parameters $(2/2, 1/2) = (1, 1/2)$. Therefore,

$$f_{R^2}(r) = \frac{(1/2)^1 r^{1-1} e^{-r/2}}{\Gamma(1)} = \frac{1}{2} e^{-r/2}$$

$$f_{Z_1, Z_2}(x, y) = \frac{1}{2\pi} e^{-(x^2+y^2)/2}$$

$$f_{R, \Theta}(r, \theta) = (e^{-r^2/2})/(2\pi)$$
$$F_{R^2, \Theta}(r, \theta) = \Pr((R^2 \leq r) \cap (\Theta \leq \theta)) = \Pr((R \leq \sqrt{r}) \cap (\Theta \leq \theta))$$
$$= \int_{\rho=0}^{\sqrt{r}} \int_{\phi=0}^{\theta} \frac{1}{2\pi} e^{-\rho^2/2} \, d\phi \rho \, d\rho = \left(\frac{\theta}{2\pi}\right) \left(1 - e^{-r/2}\right).$$

A double differentiation now gives the joint density of R^2 and θ.

$$f_{R^2, \Theta}(r, \theta) = \partial_{12} F_{R^2, \Theta}(r, \theta) = \left(\frac{1}{2\pi}\right) \left(\frac{1}{2} e^{-r/2}\right)$$

We now integrate out the radial component to obtain the density of θ.

$$f_{\Theta}(\theta) = \int_0^{\infty} f_{R^2, \Theta}(r, \theta) \, dr = \frac{1}{2\pi} \int_0^{\infty} \frac{1}{2} e^{-r/2} \, dr = \frac{1}{2\pi}$$
$$f_{R^2, \Theta}(r, \theta) = f_{R^2}(r) \cdot f_{\Theta}(\theta)$$

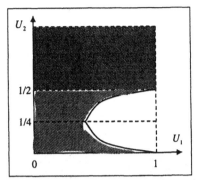

Figure 6.25. Sketch of area contributing to event $(Z \leq t)$ (see Theorem 6.72)

So R^2 and Θ are independent. Moreover, R^2 has an exponential distribution with parameter $\lambda = 1/2$, and Θ is uniformly distributed over $[0, 2\pi]$. Now, since $Z_1 = R\cos\Theta = \sqrt{R^2}\cos\Theta$, we can simulate a standard normal as the square root of an exponential random variable times the cosine of a uniform one. Because we know that U uniform on $[0, 1)$ gives $-2\ln(1 - U)$ exponential with parameter $\lambda = 1/2$, we conjecture the following theorem.

6.72 Theorem: Let U_1, U_2 be independent uniform random variables on $[0, 1)$. Then $Z = \sqrt{-2\ln(1 - U_1)}\sin(2\pi U_2)$ is standard normal.

PROOF: We compute $\Pr(Z \leq t)$ for $t \geq 0$. If $1/2 \leq U_2 < 1$, $\sin(2\pi U_2) \leq 0$, which implies that $Z \leq 0 \leq t$. This event occurs with probability $1/2$. Within $0 \leq U_2 < 1/2$, we must have

$$\sqrt{-2\ln(1 - U_1)}\sin(2\pi U_2) \leq t$$

$$\sqrt{-2\ln(1 - U_1)} \leq \frac{t}{\sin(2\pi U_2)} = t\csc(2\pi U_2).$$

Solving for the bounds on U_1, we obtain

$$\ln(1 - U_1) \geq -\frac{1}{2}(t\csc(2\pi U_2))^2$$

$$U_1 \leq 1 - e^{-(t\csc(2\pi U_2))^2/2}.$$

The shaded area of Figure 6.25 indicates the (U_1, U_2) points that contribute to the event $(Z \leq t)$. The line $U_1 = 1 - \exp(-[t\csc(2\pi U_2)]^2/2)$ forms the curved boundary of the region. Consequently,

$$F_Z(t) = \Pr(Z \leq t) = \frac{1}{2} + 2\int_{y=0}^{1/4}\int_{x=0}^{1-e^{-(t\csc(2\pi y))^2/2}} \quad (1)\ dx\,dy$$

$$= \frac{1}{2} + 2\int_{y=0}^{1/4}\left[1 - e^{-(t\csc(2\pi y))^2/2}\right]dy$$

$$= 1 - 2\int_0^{1/4} e^{-(t\csc(2\pi y))^2/2}\,dy.$$

```
function directnormal( )
    x = -2.0 * ln (1.0 - rand());      // x is exponential with parameter 1/2
    y = 2 * π * rand();       // y is uniform on (0, 2 π)
    return sqrt(x) * sin (y);    // return is standard normal
```

Figure 6.26. Direct computation of a standard normal random variable

---

We let
$$v = \cot(2\pi y)$$
$$1 + v^2 = 1 + \cot^2(2\pi y) = \csc^2(2\pi y)$$
$$dv = -\csc^2(2\pi y) \cdot (2\pi)\, dy = -2\pi(1 + v^2)\, dy,$$
which transforms the integral as follows.

$$F_Z(t) = 1 - 2\int_\infty^0 e^{-t^2(1+v^2)/2}\left(\frac{-dv}{2\pi(1+v^2)}\right) = 1 - \frac{1}{\pi}\int_0^\infty \frac{e^{-t^2(1+v^2)/2}}{(1+v^2)}\, dv$$

$$f_Z(t) = F_Z'(t) = \frac{-1}{\pi}\int_0^\infty \frac{-t(1+v^2)e^{-t^2(1+v^2)/2}}{(1+v^2)}\, dv = \frac{e^{-t^2/2}}{\pi}\int_0^\infty te^{-t^2v^2/2}\, dv$$

A final transformation, $w = tv$, $dw = t\, dv$, gives

$$f_Z(z) = \frac{e^{-t^2/2}}{\pi}\int_0^\infty e^{-w^2/2}\, dw = \frac{e^{-t^2/2}}{\pi}\cdot\frac{\sqrt{2\pi}}{2} = \frac{e^{-t^2/2}}{\sqrt{2\pi}}.$$

A similar computation holds if $t < 0$ and yields the same result. We conclude that Z is standard normal. ∎

Algorithm 6.26 exploits the theorem to return samples from a standard normal distribution. It merely performs the required calculations on a pair of independent observations from the uniform distribution on $[0, 1)$. A test generated 10000 samples, sorted them into 20 cells, and plotted the fraction in each cell, adjusted for comparability with the standard normal density function. The plot was similar to Figure 6.24. One of the exercises asks you to carry out this test.

Exercises

6.60 Continuous random variable X has the density function $f(x) = x$ for $0 \le x \le \sqrt{2}$ and zero elsewhere. Write an algorithm to generate samples from this distribution.

6.61 Continuous random variable Y has density

$$f_Y(y) = \begin{cases} 4y, & 0 \le y < 1/2 \\ 12x^2/7, & 1/2 \le y \le 1 \\ 0, & \text{elsewhere.} \end{cases}$$

Write an algorithm to generate samples from this distribution.

6.62 Let N be the number of while-loop iterations in Algorithm 6.18. In terms of the bounding constant K, compute μ_N and σ_N^2.

6.63 Write a rejection filter algorithm to generate samples from a gamma distribution with parameters $(\alpha = 7/3, \lambda = 1)$. Test the generator by constructing a plot similar to Figure 6.22.

6.64 Write a rejection filter algorithm to generate samples from a beta distribution with parameters $(3/2, 5/2)$. Use the continuous uniform density on $[0, 1)$ for a reference density.

6.65 Write a rejection filter algorithm to generate samples from a beta distribution with parameters $(1/2, 3/2)$. Use the reference density $f_X(x) = 1/(2\sqrt{x})$ for $0 < x < 1$.

6.66 Write a rejection filter algorithm to generate samples from a beta distribution with parameters $(1/2, 1/2)$.

6.67 Write a algorithm to generate samples from a normal distribution with parameters $(\mu = 1.0, \sigma^2 = 2)$.

6.68 The discussion culminating in Equation 6.6 shows, for $t \geq 0$, that Algorithm 6.23 correctly generates samples from a standard normal distribution. Complete the proof for the case $t < 0$.

6.69 Show that $f_Z(t) = (1/\sqrt{2\pi})e^{-t^2/2}$ for $t < 0$ and thereby complete the proof of Theorem 6.72.

6.70 Let U_1, U_2 be independent uniform random variables on $[0, 1)$. Show that $Z = \sqrt{-2\ln(1 - U_1)} \cos(2\pi U_2)$ is standard normal.

6.71 Write a program to produce a graph of the standard normal density, annotated with points that display the aggregate behavior of points generated by Algorithm 6.26. Divide the span $[-3.0, 3.0]$ into 20 cells of equal size, and compute the fraction of 10000 samples that fall in each cell. This fraction, divided by the cell width, is an approximation of the density at the cell center. For each cell, plot a small circle with horizontal location at the cell center and vertical location at the adjusted fraction. Compare the plot with Figure 6.24.

*6.72 Let (X, Y) be the random variables returned by the program to the right. Show that they have a uniform distribution on the circle $x^2 + y^2 \leq 1$. Also show that Z, defined as follows is standard normal.

$$Z = X\sqrt{\frac{-2\ln(X^2 + Y^2)}{(X^2 + Y^2)}}$$

```
function unifcircle( )
    searching = 1;
    while (searching)
        x = 2.0 * rand() - 1.0;
        y = 2.0 * rand() - 1.0;
        if (x * x + y * y ≤ 1.0)
            searching = 0;
    return (x, y);
```

6.6   Summary

The univariate normal distribution with parameters (μ, σ^2) has density $f(x) = e^{(x-\mu)^2/2\sigma^2}/(\sigma\sqrt{2\pi})$. Its mean and variance are μ and σ^2. If $\mu = 0$ and $\sigma^2 = 1$, the distribution is termed standard normal, and its percentiles are available in tables. Because an arbitrary univariate normal distribution X, with parameters (μ, σ^2), scales via $Z = (X - \mu)/\sigma$ to standard normal, the standard percentile tables apply in the general case. For the general univariate normal X, the moment generating function is $\Psi_X(t) = \exp(\mu t + \sigma^2 t^2/2)$.

A pair of random variables $\mathbf{X} = (X_1, X_2)$ has a bivariate normal distribution with parameters $(\boldsymbol{\mu}, \boldsymbol{\Sigma})$ when the joint density is

$$f(\mathbf{x}) = f(x_1, x_2) = \frac{1}{2\pi\sqrt{\det \boldsymbol{\Sigma}}} \exp\left(-\frac{1}{2}(\mathbf{x} - \boldsymbol{\mu})'\boldsymbol{\Sigma}^{-1}(\mathbf{x} - \boldsymbol{\mu})\right).$$

The vector $\boldsymbol{\mu}$ contains the means of X_1 and X_2, while the covariance matrix $\boldsymbol{\Sigma}$ contains the individual variances on the diagonal and the covariances in the off-diagonal entries. It is necessarily a symmetric matrix.

The marginal and conditional densities are also normal. The parameters of the marginals are the individual means and variances extracted from $\boldsymbol{\mu}$ and $\boldsymbol{\Sigma}$. The mean of a conditional distribution varies with the value of the conditioning variable. For example, $\mu_{1|2}$, the mean of X_1, given $X_2 = x_2$, is $\mu_1 + \sigma_{12}(x_2 - \mu_2)/\sigma_{22}$, where σ_{ij} denotes a component of $\boldsymbol{\Sigma}$. The variance of a conditional distribution, however, is *not* dependent on the value of the conditioning variable. For example, $\sigma_{11|2}$, the variance of X_1, given $X_2 = x_2$, is $\sigma_{11} - \sigma_{12}^2/\sigma_{22}$.

Independent random variables are always uncorrelated. In the case of normal random variables, the exponential density forces uncorrelated to be equivalent to independent. If $\mathbf{Y} = \mathbf{TX}$, for a 2×2 matrix $\mathbf{T}$ with nonzero determinant, then $\mathbf{Y}$ is bivariate normal with parameters $(\mathbf{T}\boldsymbol{\mu}, \mathbf{T}\boldsymbol{\Sigma}\mathbf{T}')$.

The bivariate normal concept extends to multivariate normal, which applies jointly to a vector of n random variables. The parameters are $\boldsymbol{\mu}$, a vector of the individual means, and $\boldsymbol{\Sigma}$, an $n \times n$ symmetric, positive definite matrix with the individual variances on the diagonal and the covariances in the off-diagonal entries. The matrix formulation of the density then remains as given in the bivariate case above. For a $n \times n$ matrix $\mathbf{T}$ with nonzero determinant, $\mathbf{TX}$ remains multivariate normal with parameters $(\mathbf{T}\boldsymbol{\mu}, \mathbf{T}\boldsymbol{\Sigma}\mathbf{T}')$. For any selection of $\mathbf{X}$ components, the corresponding marginal density is multivariate normal with parameters obtained by selecting the same components from $\boldsymbol{\mu}$ and $\boldsymbol{\Sigma}$.

Convergence for a sequence of random variables has several definitions. Since random variables are actually functions that map a probability space into the real numbers, the various notions correspond to different ways in which a sequence of functions can converge. We say that X_n converges strongly to Y if $\Pr(\{\omega| \lim X_n(\omega) = Y(\omega)\}) = 1$. On the other hand, we say that the convergence is weak if $\lim \Pr(|X_n - Y| \geq \epsilon) = 0$ for every

$\epsilon > 0$. Strong convergence implies weak convergence, but the converse is not necessarily true. When X_n converges strongly, we say that it converges almost surely or almost everywhere. Weak convergence is also called convergence in probability. Finally, we say that X_n converges to Y is quadratic mean if $\lim E[(X_n - Y)^2] = 0$. Convergence in quadratic mean also implies weak convergence. The weak law of large numbers states that the sequence $Y_n = (1/n) \sum_{i=1}^{n} X_i$ converges weakly to the constant μ, where the X_i have identical independent distributions with mean μ and finite variance σ^2.

The sine and cosine moment generating functions $\Psi_s(t)$ and $\Psi_c(t)$ have expansions that are closely related to the ordinary moment generating function. Except for alternating signs, the sine variation picks out the odd terms, and the cosine version selects the even terms. Consequently, when it exists as a convergent power series expansion, $\Psi(t)$ determines $(\Psi_s(t), \Psi_c(t))$, and vice versa. An inversion formula permits recovery of the cumulative distribution function from the (Ψ_s, Ψ_c) pair, which means that a moment generating function corresponds to a unique cumulative distribution function. This property is useful because it allows an unambiguous identification of a distribution from its moment generating function. This technique was used to prove the central limit theorem, which states that the sum of independent identically distributed random variables approaches a normal distribution.

The gamma and beta functions, for positive arguments, are

$$\Gamma(t) = \int_0^\infty x^{t-1} e^{-x} \, dx$$

$$\beta(u, v) = \int_0^1 x^{u-1} (1 - x)^{v-1} \, dx = \frac{\Gamma(u)\Gamma(v)}{\Gamma(u + v)}.$$

Each gives rise to a probability distribution of the same name. A gamma distribution with positive parameters (α, λ) has density $\lambda^\alpha x^{\alpha-1} e^{-\lambda x} / \Gamma(\alpha)$ on the positive real axis. If $\alpha = n$ is an integer, the distribution is called an Erlang distribution. If $n = 1$, it is called an exponential distribution. The Erlang random variable is the sum of n independent exponential random variables, each with parameter λ. The general gamma random variable with parameters (α, λ) has mean α/λ and variance α/λ^2. The beta distribution with positive parameters (u, v) has density $x^{u-1} (1 - x)^{v-1} / \beta(u, v)$ on the interval $[0, 1]$. It has mean $u/(u + v)$ and variance $uv/[(u + v)^2 (u + v + 1)]$.

The χ^2 random variable with n degrees of freedom is the sum of the squares of n independent standard normal random variables. It is a special form of a gamma random variable with parameters $(\alpha = n/2, \lambda = 1/2)$. Consequently, it has mean n and variance $2n$. An F random variable is the ratio of two independent χ^2 random variables, each divided by its degrees of freedom. That is, if X is χ^2 with n_1 degrees of freedom and Y is independently χ^2 with n_2 degrees of freedom, then $(X/n_1)/(Y/n_2)$ has an F distribution with (n_1, n_2) degrees of freedom. Its mean is $n_2/(n_2 - 2)$. If Y has a symmetric distribution about zero and Y^2 has an F distribution with $(1, n)$ degrees of freedom, then we say that Y is a t random variable. As functions of the degrees of freedom, percentiles for the χ^2, F, and t distributions appear in tables, which are useful in statistical inference problems. All three distributions have noncentral equivalents. The noncentral χ^2 with n degrees of freedom and noncentrality parameter $\lambda = (\mu_1^2 + \mu_2^2 + \ldots + \mu_n^2)/2$ is the sum of the squares of n independent normal random variables with common variance 1 and means $\mu_1, \mu_2, \ldots, \mu_n$. The noncentral F distribution with (n_1, n_2) degrees of freedom and noncentrality parameter λ is the ratio $(X/n_1)/(Y/n_2)$, where X is noncentral χ^2 with n_1 degrees of freedom and noncentrality parameter λ and Y is central χ^2 with n_2 degrees of freedom. The noncentral t random variable is the symmetric square root of the corresponding noncentral F. These noncentral distributions are used to judge the merit of certain statistical tests and are also available in tables, although the tabulation becomes increasingly difficult due to the large number of parameters.

We can generate observations from a continuous population X by exploiting the fact that $Y = F_X^{-1}(U)$ has the same distribution as X. Here $F_X(\cdot)$ is the cumulative distribution function for X and U has a continuous uniform distribution on $[0, 1)$. Using a system utility, typically called rand(), to approximate samples from U, we transform them into samples from X. This is a generalization of the inverse transform technique that was developed in an earlier chapter for discrete random variables. Moreover, we can adapt the rejection filter algorithm, also developed for discrete distributions, to the continuous case. The underlying concept is unchanged. If we want samples from population X but find it more convenient to generate samples from a related population Y, we filter the Y returns by discarding some of them. In particular, if there exists a constant K, such that $f_Y(x) \leq K f_X(x)$ for all x, we discard the Y return y when $u \geq f_Y(y)/(K f_X(y))$, where u is chosen from an independent random population, uniform on $[0, 1)$. To generate observations from a standard normal distribution, we can exploit the fact that $Z = \sin(2\pi U_2)\sqrt{-2\ln(1 - U_1)}$ is standard normal when (U_1, U_2) are independent with uniform distributions on $[0, 1)$.

Historical Notes

Porter[66] states that the development of the normal distribution was "practically coextensive with the history of statistical mathematics during the nineteenth century." Actually, Abraham De Moivre, in the first third of the previous century, produced the normal distribution as the limiting form of binomial distributions, and shortly thereafter both Lagrange and Laplace noted that the limit enabled a readily computable bound on the probability of a given error in the statistical mean. Around 1800, astronomers began using the method of least squares, which determines the optimal curve of a specified type matching a set of observations. This technique, introduced by Legendre, proved empirically suitable, and eventually, Gauss proved its optimality properties under the assumption that the observational errors follow a normal distribution. Many statistical observations are such that an underlying deterministic magnitude is indeed obscured by normally distributed errors, and consequently, the normal distribution is frequently called the error curve or the Gaussian error curve.

Computer science students may be interested to know that Alan Turing, one of the discipline's founders, proved the limiting tendency of a variety of distributions toward the normal, bell-shaped curve. In this endeavor, he rediscovered the central limit theorem, the first proof of which goes back to DeMoivre. Nevertheless, the proof demonstrated the capabilities of the young Turing, and he was made a Cambridge don. Davis [17] provides this anecdote while describing Turing's part in the evolution of the digital computer.

Early applications emphasized the error aspect of nondeterministic variation. In the early nineteenth century, Adolphe Quetelet introduced the average or "normal" man, a creature characterized by moderate features in all dimensions: physical, mental, and moral. Actual human beings differed from this ideal only through error, and because the error followed the celebrated Gaussian distribution, that distribution acquired the name "normal" distribution. Several of the famous names of nineteenth century statistics, such as Francis Galton and Karl Pearson, were proponents of eugenics, which considers policies that might increase the small fraction of humanity that exhibits exemplary physical, mental, and moral attributes. They felt that society would improve if the segment to the far right of the normal distribution were to procreate more profusely. Of course, the high fertility of the segment to the far left was condemned as injurious to the future of humankind. This debate continues today in, for example, the controversy surrounding statistical correlations between intelligence and race; see Herrnstein[35] or Fraser[22].

William Gosset, a chemist at the Guiness brewery in Dublin, invented the t-test in connection with quality control experiments for brewing beer. He published the work under the name Student, with the result that the t distribution is frequently called Student's t.

The gamma and beta functions originated with Euler and are sometimes known as Eulerian integrals.

Further Reading

The normal distribution forms a central exhibit in every probability text. Consequently, there are many sources for alternative views concerning the theoretical development and practical application of this distribution and the related χ^2, F, and t forms. For example, Lindgren[51] and Games and Klare[24] provide theoretical and applications-oriented treatments, respectively. Anderson[2] describes in great detail the matrix operations involved in the analysis of multivariate normal distributions. See Ross[73] for simulation techniques for normal and other random variables.

The sine and cosine variations of the moment generating function are convenient for deriving an inversion formula because they permit an analysis without complex numbers. Because complex numbers are not necessary in any other part of the text, this approach is appropriate here. However, most derivations in the literature use complex numbers and base the inversion formula on the expansion of $E[e^{itX}]$ rather than $E[e^{tX}]$. Indeed, there are many inversion formulas, and all are equivalent. Lindgren[51] and Parzen[61] provide variations. The inversion formula is essentially an inverse Fourier transform and is treated in many texts on Fourier analysis.

Gamma and beta functions are standard fare for any advanced calculus text. Hildebrand[37] and Taylor[86], for example, cover the material.

Chapter 7

Parameter Estimation

In the parameter estimation problem, we know the general form of a distribution but not the detailed parameters. For example, we may know that the response time of a database server is normally distributed, but we do not know the defining parameters (μ, σ^2). Our goal is to develop statistical techniques, based on sample observations, that provide estimates of the unknown parameters.

These techniques extend the decision theory of Chapter 4 to continuous distributions. As discussed in that chapter, decision theory allows us to select an optimal action when faced with an unknown state of nature. Here, we regard the states of nature as the spectrum of possible parameter values. The unknown parameters of a normal distribution, for example, form a two-dimensional continuum, in contrast with the states in Chapter 4, which were discrete and finite. The optimal action is the announcement of a good estimate for the unknown parameters, which means that the space of available actions is identical with the states of nature. As in the discrete case, we judge the estimate's merit with an expected loss calculation.

The first section below introduces the measures of bias, consistency, and efficiency, which assess the merit of an estimator and facilitate comparisons with competitors. An estimator, as we shall see, is a random variable, and the expected value of a good estimator should equal the parameter that it approximates. An unbiased estimator satisfies this intuitively plausible criterion. Consistency and efficiency then relate to the variance of an unbiased estimator. Precise definitions of these quantities appear later.

The remaining sections deal with normal populations, an important subcase because the central limit theorem frequently allows us to assume a normal distribution. For a normal population, we first develop effective estimators for the mean and variance. A mathematical interlude then investigates techniques for decomposing the sum of squares associated with the sample variance. In this manner, the total variance is seen as a sum of components from different sources. Building on this decomposition, the final sections treat analysis of variance and linear regression. These procedures uncover relationships among the means of normal subpopulations and construct estimators for the relationship parameters.

7.1 Bias, consistency, and efficiency

Suppose that we have a random variable X characterized by a distribution from a given family. The family might be exponential, normal, gamma, or any of the other continuous distributions that appeared in the preceding chapter. However, we do not know the parameter values that distinguish the actual distribution from others in the same family. Suppose that the parameters are $\theta = (\theta_1, \ldots, \theta_m)$. To emphasize the dependence on θ, we write the density as $f_{X|\theta}(\cdot)$, which we read as the density of X given θ. Given the sample $\mathbf{X} = (X_1, X_2, \ldots, X_n)$, our task is to establish a vector $\mathbf{S} = (S_1, \ldots, S_m)$, in which each component $S_i(X_1, X_2, \ldots, X_n)$ is a statistic that estimates the corresponding θ_i. For example, suppose that the time T between client requests arriving at a database server is known to be exponential but with a single unknown parameter. That is, T has a density in the family $f_{T|\lambda}(t) = \lambda e^{-\lambda t}$, which is indexed by λ. Using approaches developed in this chapter, we can sample the arrival stream and apply an appropriate statistic to estimate the unknown λ.

In general, a function $f_{X|\theta}(\cdot)$ characterizes a random variable X, and our analysis goal is to determine a good estimate for θ. In estimating the parameters, we are giving our best judgment as to which density actually prevails at the time of the sample. As with discrete decision theory, we need a loss function for the cost of announcing the estimates $S_1(\mathbf{x}), \ldots, S_m(\mathbf{x})$, calculated from the sample values $\mathbf{x} = (x_1, \ldots, x_n)$, when the state of nature is actually $\theta = (\theta_1, \ldots, \theta_m)$. Although this function could vary with the application at hand, we will adopt a single function for all cases. This simplification facilitates the subsequent analysis, and moreover, the chosen function is widely applicable. It is the quadratic loss function defined below. This function assigns zero loss when the estimate is exactly correct, and it assigns increasing losses with increasing error between the estimate and the actual prevailing parameter. Unfortunately, unless the permissible parameter values are confined to some finite interval, this loss function is unbounded. For each state of nature, the maximum loss is infinite, which means that a minimax strategy is not appropriate. However, the function works well for other approaches, which seek to minimize the expected loss. Continuing the terminology of Chapter 4, we use the term *risk* for expected loss.

7.1 Definition: Suppose that $f_{X|\theta}(\cdot)$ is a family of density functions indexed by $\theta = (\theta_1, \ldots, \theta_m)$. Each family member is a potential density for the random variable X. For a sample $\mathbf{X} = X_1, X_2, \ldots, X_n$, suppose that statistics $\mathbf{S}(\mathbf{X}) = (S_1(\mathbf{X}), \ldots, S_m(\mathbf{X}))$ provide estimates of θ. The *quadratic loss function* associated with these parameters and their estimators is $L_{\mathbf{S}}(\mathbf{x}; \theta) = \sum_{i=1}^{m} (S_i(\mathbf{x}) - \theta_i)^2$. This expression gives the loss that occurs when state θ prevails and the sample value is $\mathbf{x}$. The corresponding *quadratic risk* is the expected value of this function when the parameter vector θ actually governs the distribution of $\mathbf{S(X)}$. That is, the quadratic risk is $\hat{L}_{\mathbf{S}}(\theta) = E_\theta[L_{\mathbf{S}}(\cdot; \theta)]$. The subscript on the expectation operator indicates the density for the calculation. ∎

Suppose that statistic S_i estimates θ_i. We can calculate this statistic's contribution to the quadratic risk as follows. If we write $\mu_{S_i|\theta} = E_\theta[S_i(\cdot)]$ and $\sigma^2_{S_i|\theta} = E_\theta[(S_i(\cdot) - \mu_{S_i|\theta})^2]$ for the conditional mean and variance of S_i, we have

$$
\begin{aligned}
E_\theta[(S_i(\cdot) - \theta_i)^2] &= E_\theta\left[[(S_i(\cdot) - \mu_{S_i|\theta}) - (\theta_i - \mu_{S_i|\theta})]^2\right] \\
&= E_\theta[(S_i(\cdot) - \mu_{S_i|\theta})^2] - 2(\theta_i - \mu_{S_i|\theta})E_\theta[S_i(\cdot) - \mu_{S_i|\theta}] \\
&\quad + (\mu_{S_i|\theta} - \theta_i)^2 \\
&= \sigma^2_{S_i|\theta} + (\mu_{S_i|\theta} - \theta_i)^2.
\end{aligned}
$$

To minimize the risk, we need to minimize the conditional variance of each estimator, and we need to minimize the additional term, which the following definition establishes as the square of the estimator's bias.

7.2 Definition: Let $\mathbf{S} = (S_1(X_1, \ldots, X_n), \ldots, S_m(X_1, \ldots, X_n))$ be a statistical estimator of $\theta = (\theta_1, \ldots, \theta_m)$, based on a sample from a population with density $f_{X|\theta}(\cdot)$. The *bias* of estimator component S_i is

$$
b_{S_i}(\theta) = \mu_{S_i|\theta} - \theta_i = E_\theta[S_i(\cdot)] - \theta_i.
$$

An estimator with zero bias is called an *unbiased estimator*. ∎

Thus the risk of estimators $S_1, \ldots, S_m$ relates to their variances and biases: $\hat{L}_{\mathbf{S}}(\theta) = \sum_{i=1}^m [\sigma^2_{S_i|\theta} + b^2_{S_i}(\theta)]$. To minimize the risk, we search among the class of unbiased estimators for members with the smallest variance. For the balance of this section, we consider only cases with one parameter. The next section, which takes up normal distributions, uses the more general definitions established above.

7.3 Example: Based on a sample $\mathbf{X} = X_1, X_2, \ldots, X_n$, we construct some unbiased estimators for the parameter τ in the exponential class $f_{X|\tau}(x) = (1/\tau)e^{-x/\tau}$. First, consider the sample mean $\overline{X} = (1/n)\sum_{i=1}^n X_i$. We have

$$
\mu_{\overline{X}|\tau} = \mu_{X|\tau} = \int_0^\infty \frac{xe^{-x/\tau}\,dx}{\tau} = \tau,
$$

which implies that $\overline{X}$ is unbiased. Its variance is

$$
\sigma^2_{\overline{X}|\tau} = \frac{\sigma^2_{X|\tau}}{n} = \frac{1}{n}\int_0^\infty \frac{(x - \tau)^2 e^{-x/\tau}\,dx}{\tau} = \frac{\tau^2}{n}.
$$

Consequently, $\hat{L}_{\overline{X}}(\tau) = \tau^2/n$, which approaches zero with increasing n.

Now consider the estimator X_1. This estimator simply ignores most of the sample. Intuitively, we suspect that this is a poor estimator, but it is unbiased:

$\mu_{X_1|\tau} = \mu_{X|\tau} = \tau$. Its variance, however, is the variance of the underlying population, $\sigma^2_{X|\tau} = \tau^2$. Therefore, for $n > 1$, the risk is higher than that of the sample mean: $\hat{L}_{X_1}(\tau) = \tau^2 > \tau^2/n = \hat{L}_{\overline{X}}(\tau)$.

Finally, consider the estimate $Y = X_{(n)}$, the sample maximum. For any t, we have $Y \leq t$ if and only if $X_i \leq t$ for all i. Consequently,

$$F_{Y|\tau}(t) = \Pr(Y \leq t) = [F_{X|\tau}(t)]^n = \left(\int_0^t f_{X|\tau}(x)\, dx \right)^n$$

$$= \left(\int_0^t \frac{1}{\tau} e^{-x/\tau}\, dx \right)^n = \left(1 - e^{-t/\tau} \right)^n$$

$$f_{Y|\tau}(t) = F'_{Y|\tau}(t) = \frac{ne^{-t/\tau}}{\tau} \left(1 - e^{-t/\tau} \right)^{n-1}.$$

To check for bias, we examine the mean.

$$\mu_{Y|\tau} = \int_0^\infty t f_{Y|\tau}(t)\, dt = \frac{n}{\tau} \int_0^\infty t e^{-t/\tau} \left(1 - e^{-t/\tau} \right)^{n-1} dt$$

$$= \frac{n}{\tau} \int_0^\infty t e^{-t/\tau} \left(\sum_{i=0}^{n-1} C_{n-1,i}(-1)^{n-1-i} e^{-(n-1-i)t/\tau} \right) dt$$

$$= \frac{n}{\tau} \sum_{i=0}^{n-1} C_{n-1,i}(-1)^{n-1-i} \int_0^\infty t e^{-(n-i)t/\tau}\, dt$$

$$= \frac{n}{\tau} \sum_{i=0}^{n-1} C_{n-1,i}(-1)^{n-1-i} \frac{\tau^2}{(n-i)^2} = n\tau \sum_{i=0}^{n-1} C_{n-1,i} \frac{(-1)^{n-1-i}}{(n-i)^2}$$

It is convenient to sum the terms in the reverse direction. Letting $j = n-1-i$, we have

$$\mu_{Y|\tau} = n\tau \sum_{j=0}^{n-1} C_{n-1,j}(-1)^j \frac{1}{(j+1)^2}.$$

Chapter 1 covered techniques for reducing such sums. You may want to review the derivation of Equation 1.6, which illustrates how to sum terms of the form $iC_{n,i}$. That approach uses derivatives of $(x+1)^n$. Here, it is convenient to use integrals of $(1-x)^{n-1}$.

$$(1-x)^{n-1} = \sum_{j=0}^{n-1} C_{n-1,j}(-1)^j x^j$$

$$\int_0^y (1-x)^{n-1}\, dx = \frac{-(1-x)^n}{n} \Big|_0^y = \frac{1-(1-y)^n}{n}$$

$$= \sum_{j=0}^{n-1} C_{n-1,j}(-1)^j \int_0^y x^j\, dx = \sum_{j=0}^{n-1} C_{n-1,j}(-1)^j \frac{y^{j+1}}{j+1}$$

$$\frac{1-(1-y)^n}{ny} = \sum_{j=0}^{n-1} C_{n-1,j}(-1)^j \frac{y^j}{j+1}$$

We iterate this process to establish a second $(j+1)$ factor in the denominator of the summand.

$$\frac{1}{n}\int_0^1 \frac{1-(1-y)^n}{y}\, dy = \sum_{j=0}^{n-1} C_{n-1,j}(-1)^j \int_0^1 \frac{y^j}{j+1}\, dy = \sum_{j=0}^{n-1} C_{n-1,j}(-1)^j \frac{1}{(j+1)^2}$$

We now have the desired sum as an integral, which we proceed to evaluate using the transformation $t = 1 - y$, $dt = -\, dy$.

$$\mu_{Y|\tau} = \tau \int_0^1 \frac{1-(1-y)^n}{y}\, dy = \tau \int_0^1 \frac{1-t^n}{1-t}\, dt = \tau \int_0^1 \left(\sum_{i=0}^{n-1} t^i \right) dt$$

$$= \tau \sum_{i=0}^{n-1} \left[\frac{t^{i+1}}{i+1} \right]_0^1 = \tau \left[1 + \frac{1}{2} + \frac{1}{3} + \ldots + \frac{1}{n} \right]$$

Except for the case $n = 1$, the estimator $Y = X_{(n)}$ is biased. Moreover, the bias increases with increasing sample size. A similar analysis shows that $\sigma^2_{Y|\tau} = \tau^2[1 + 1/2^2 + 1/3^2 + \ldots + 1/n^2]$. We conclude that both components of the risk $\hat{L}_Y(\tau)$ increase with increasing sample size. $\square$

Because the risk contains two components, the estimator's bias and variance, we seek to minimize both. One approach first ascertains the class of unbiased estimators and then chooses the member with the smallest variance. This leads to the comparison of competing estimators, for which further criteria are appropriate.

7.4 Definition: For each $i = 1, 2, \ldots$, let S_i be an estimator of the parameter θ that appears in the population density $f_{X|\theta}(\cdot)$. The sequence has *zero asymptotic risk* if S_i converges in quadratic mean to the constant θ. That is, for each θ, we have $\lim_{n\to\infty} E_\theta[(S_n - \theta)^2] = \lim_{n\to\infty} \hat{L}_{S_n}(\theta) = 0$. The $E_\theta(\cdot)$ notation indicates that we compute the expected value using the density $f_{X|\theta}(\cdot)$. ∎

7.5 Definition: For each $i = 1, 2, \ldots$, let S_i be an estimator of the parameter θ that appears in the population density $f_{X|\theta}(\cdot)$. The sequence is *consistent* if S_i converges weakly to the constant θ. That is, for each θ, we have $\lim_{n\to\infty} \mathrm{Pr}_\theta(|S_n - \theta| \geq \epsilon) = 0$, for arbitrary $\epsilon > 0$. Again, the $\mathrm{Pr}_\theta(\cdot)$ notation indicates that we compute the probability of the event $(|S_n - \theta| \geq \epsilon)$ with the density $f_{X|\theta}(\cdot)$. Note that the weak convergence must hold for all θ. ∎

7.6 Definition: Let S_1 and S_2 be estimators of the parameter θ that appears in the population density $f_{X|\theta}(\cdot)$. The *relative efficiency* of S_1 with respect to S_2 is $\mathrm{eff}(S_1, S_2; \theta) = \hat{L}_{S_2}(\theta)/\hat{L}_{S_1}(\theta)$. For unbiased estimators, the definition reduces to a comparison of variances. The estimator with the smaller variance is the more efficient. ∎

7.7 Example: Continuing the context of Example 7.3, let X have a density from the family $f_{X|\tau}(t) = e^{-t/\tau}/\tau$. For a sample $X_1, X_2, \ldots, X_n$, let $\overline{X}_n = (1/n)\sum_{i=1}^n X_i$, $Y_n = X_{(n)}$, and $Z_n = X_1$ be three estimators of τ. These are the sample mean, the sample maximum, and the first sample point. With increasing sample size, each defines a sequence of random variables, for which we discuss the asymptotic risk, the consistency, and the relative efficiency.

From the preceding example, we have

$$\hat{L}_{\overline{X}_n}(\tau) = \frac{\tau^2}{n}$$

$$\hat{L}_{Y_n}(\tau) = \tau^2 \left[\left(\sum_{i=1}^n \frac{1}{i^2} \right) + \left(\sum_{i=2}^n \frac{1}{i} \right)^2 \right]$$

$$\hat{L}_{Z_n}(\tau) = \tau^2.$$

Consequently, $\overline{X}_n$ has zero asymptotic risk, while Y_n and Z_n do not. Since convergence in quadratic mean implies weak convergence, the sequence $\overline{X}_n$ is consistent. As for Y_n, we excerpt the density from the preceding example:

$$f_{Y_n|\tau}(t) = \frac{ne^{-t/\tau}}{\tau} \left(1 - e^{-t/\tau} \right)^{n-1}.$$

For an arbitrary ϵ in the range $0 < \epsilon \leq \tau$, we compute as follows, using the transformation $y = 1 - e^{-t/\tau}$.

$$\Pr(|Y_n - \tau| \geq \epsilon) = 1 - \Pr(|Y_n - \tau| < \epsilon) = 1 - \Pr(\tau - \epsilon < Y_n < \tau + \epsilon)$$

$$= 1 - \int_{\tau-\epsilon}^{\tau+\epsilon} \frac{ne^{-t/\tau}}{\tau}(1 - e^{-t/\tau})^{n-1}\, dt = 1 - \int_{1-e^{-(1-\epsilon/\tau)}}^{1-e^{-(1+\epsilon/\tau)}} ny^{n-1}\, dy$$

$$= 1 - \left[\left(1 - e^{-(1+\epsilon/\tau)} \right)^n - \left(1 - e^{-(1-\epsilon/\tau)} \right)^n \right] = 1 - [r_1^n - r_2^n],$$

where r_1 and r_2 are fractions in the range $(0,1)$. So, $\lim_{n\to\infty} \Pr(|Y_n - \tau| \geq \epsilon) = 1$, and the sequence Y_n is not consistent. For Z_n, we compute, for $0 < \epsilon \leq \tau$,

$$\Pr(|Z_n - \tau| \geq \epsilon) = 1 - \Pr(|Z_n - \tau| < \epsilon) = 1 - \Pr(\tau - \epsilon) < Z_n < \tau + \epsilon)$$

$$= 1 - \int_{\tau-\epsilon}^{\tau+\epsilon} \frac{1}{\tau} e^{-t/\tau}\, dt = 1 - \left(e^{-(1-\epsilon/\tau)} - e^{-(1+\epsilon/\tau)} \right),$$

which is a positive quantity, independent of n. Therefore, the sequence Z_n is not consistent.

For a given n, the efficiencies of $\overline{X}_n$ with respect to Y_n and Z_n are

$$\text{eff}(\overline{X}_n, Y_n, \tau) = \frac{\hat{L}_{Y_n}(\tau)}{\hat{L}_{\overline{X}_n}(\tau)} = n \left[\left(\sum_{i=1}^n \frac{1}{i^2} \right) + \left(\sum_{i=2}^n \frac{1}{i} \right)^2 \right] > n$$

$$\text{eff}(\overline{X}_n, Z_n, \tau) = \frac{\hat{L}_{Z_n}(\tau)}{\hat{L}_{\overline{X}_n}(\tau)} = n.$$

$\overline{X}_n$ is more efficient that either Y_n or Z_n, and the relative efficiency increases with the sample size. Among these three estimators, the sample mean is the best. It is unbiased and consistent; it has zero asymptotic risk; and it is increasingly more efficient than either of its competitors. In a contest between Y_n and Z_n, we see that Y_n is biased, while Z_n is unbiased. Neither is consistent, and neither has zero asymptotic risk. As for efficiency,

$$\text{eff}(Z_n, Y_n, \tau) = \frac{\hat{L}_{Y_n}(\tau)}{\hat{L}_{Z_n}(\tau)} = \left[\left(\sum_{i=1}^n \frac{1}{i^2} \right) + \left(\sum_{i=2}^n \frac{1}{i} \right)^2 \right] > 1.$$

Consequently, Z_n, which ignores all but the initial observation in the sample, is more efficient than Y_n, the sample maximum. Moreover, the efficiency advantage increases with sample size. $\square$

As noted in the definition, relative efficiency for unbiased estimators is simply the ratios of their variances. A smaller variance translates to a greater efficiency and a smaller risk. While this criterion provides a rational comparison between two competing estimators, it gives little guidance in a search for the most efficient estimator. A lower bound on the variance across all unbiased estimators would be very helpful. If an estimator achieves this lower bound, then we know that it cannot be surpassed in efficiency. For a sample $\mathbf{X} = (X_1, X_2, \dots, X_n)$, the joint density is $f_{\mathbf{X}|\theta}(x_1, \dots, x_n) = \prod_{i=1}^{n} f_{X|\theta}(x_i)$. Suppose, for each fixed $\mathbf{x} = (x_1, x_2, \dots, x_n)$, that this joint density is a differentiable function of θ. That is, $f_{\mathbf{X}|\theta}(x_1, \dots, x_n) = g(\theta; x_1, x_2, \dots, x_n)$, and we postulate that $\partial_1 g$ exists. In this case, we can sometimes manipulate the random variable

$$W = \frac{\partial}{\partial \theta} \ln g(\theta; X_1, X_2, \dots, X_n)$$

to find the desired lower bound. The next example illustrates the details for the density family $(1/\tau)e^{-t/\tau}$.

7.8 Example: The parameter τ indexes the density family $f_{X|\tau}(t) = e^{-t/\tau}/\tau$. In the preceding two examples, we found the estimator $\overline{X}_n$, the sample mean, to be unbiased and more efficient than its competitors X_1, the first observation, and $X_{(n)}$, the sample maximum. As the sample size increases, the sequence $\overline{X}_n$ is consistent, has zero asymptotic risk, and each sequence element is increasingly more efficient than the corresponding element of $X_{(n)}$. We show now that $\overline{X}_n$ cannot be surpassed in efficiency by any unbiased estimator operating on a sample of size n.

Let $\mathbf{X} = (X_1, X_2, \dots, X_n)$ be a sample, and define

$$W = \frac{\partial}{\partial \tau} \ln \left(\prod_{i=1}^{n} f_{X|\tau}(X_i) \right) = \frac{\partial}{\partial \tau} \ln \left[\frac{1}{\tau^n} \exp \left(\frac{-\sum_{i=1}^{n} X_i}{\tau} \right) \right]$$

$$= \frac{\partial}{\partial \tau} \left(-n \ln \tau - \frac{\sum_{i=1}^{n} X_i}{\tau} \right) = \frac{-n}{\tau} + \frac{\sum_{i=1}^{n} X_i}{\tau^2}.$$

We have

$$\mu_{W|\tau} = E_\tau[W] = \frac{-n}{\tau} + \frac{1}{\tau^2} \sum_{i=1}^{n} E[X_i] = \frac{-n}{\tau} + \frac{1}{\tau^2}(n\tau) = 0$$

$$\sigma^2_{W|\tau} = E_\tau[W^2] - (E_\tau[W])^2 = E_\tau[W^2] = \sum_{i=1}^{n} \frac{1}{\tau^4} \mathrm{var}(X_i) = \frac{n\tau^2}{\tau^4} = \frac{n}{\tau^2}.$$

Now, let $S = g(X_1, X_2, \dots, X_n)$ be any unbiased estimator of τ, which means that $E_\tau[S] = \tau$. We compute

$$1 = \frac{\partial}{\partial \tau}(\tau) = \frac{\partial}{\partial \tau} E_\tau[S]$$

$$= \frac{\partial}{\partial \tau} \int_0^\infty \dots \int_0^\infty g(x_1, \dots, x_n) \frac{1}{\tau^n} \exp \left(\frac{-\sum_{i=1}^{n} x_i}{\tau} \right) dx_1 \dots dx_n.$$

We now manipulate the integrand to reveal an expression for W:

$$1 = \int_0^\infty \cdots \int_0^\infty g(x_1,\ldots,x_n) \left[\frac{-n}{\tau^{n+1}} + \frac{1}{\tau^n} \cdot \frac{\sum_{i=1}^n x_i}{\tau^2} \right] \exp\left(\frac{-\sum_{i=1}^n x_i}{\tau} \right) dx_1 \ldots dx_n$$

$$= \int_0^\infty \cdots \int_0^\infty g(x_1,\ldots,x_n) \left[\frac{-n}{\tau} + \frac{\sum_{i=1}^n x_i}{\tau^2} \right] \left(\frac{1}{\tau^n} \right) \exp\left(\frac{-\sum_{i=1}^n x_i}{\tau} \right) dx_1 \ldots dx_n$$

$$= E_\tau[WS].$$

Since $E_\tau[W] = 0$, we conclude that $\mathrm{cov}_\tau(W,S) = E_\tau[WS] - E_\tau[W]E_\tau[S] = 1$. Because the correlation of W and S must be in the range $[-1, 1]$, we can continue to obtain the desired lower bound on the variance of S.

$$\rho^2(W,S) = \frac{\mathrm{cov}^2(W,S|\tau)}{\sigma_{W|\tau}^2 \sigma_{S|\tau}^2} = \frac{1}{(n/\tau^2)\sigma_{S|\tau}^2} \le 1$$

We conclude that $\sigma_{S|\tau}^2 \ge \tau^2/n$. Thus any unbiased estimator must have variance no smaller than τ^2/n, which is precisely the variance of the estimator $\overline{X}_n$. It follows that no unbiased estimator utilizing a sample of size n can surpass $\overline{X}_n$ in efficiency. $\square$

The technique illustrated in the example works in the general case, provided that the logarithm of the sample joint density presents a differentiable function of the parameter and the mathematical manipulations are justified. Suppose that the samples come from a population X with parameterized density $f_{X|\theta}(\cdot)$. For a sample $\mathbf{X} = (X_1, X_2, \ldots, X_n)$, we write the joint density as $f(x_1, x_2, \ldots, x_n, \theta) = \prod_{i=1}^n f_{X|\theta}(x_i)$ to emphasize the required differentiability with respect to θ. Let

$$W = \frac{\partial}{\partial \theta} \ln f(X_1, \ldots, X_n, \theta) = \left(\frac{1}{f(X_1, \ldots, X_n, \theta)} \right) \frac{\partial}{\partial \theta} f(X_1, \ldots, X_n, \theta).$$

As in the example, $E_\theta[W] = 0$:

$$E_\theta[W] = \int_{-\infty}^\infty \cdots \int_{-\infty}^\infty \frac{1}{f(\mathbf{x}, \theta)} \cdot \frac{\partial f(\mathbf{x}, \theta)}{\partial \theta} \cdot f(\mathbf{x}, \theta) \, dx_1 \ldots dx_n$$

$$= \frac{\partial}{\partial \theta} \int_{-\infty}^\infty \cdots \int_{-\infty}^\infty f(\mathbf{x}, \theta) \, dx_1 \ldots dx_n = \frac{\partial}{\partial \theta}(1) = 0.$$

For any unbiased estimator $S = g(X_1, \ldots, X_n)$, the covariance of S and W is then the expected value of the product. Assuming that the interchange of differentiation and integration is valid, we have

$$\mathrm{cov}(W,S) = \int_{-\infty}^\infty \cdots \int_{-\infty}^\infty \frac{g(\mathbf{x})}{f\mathbf{x}, \theta)} \cdot \frac{\partial f(\mathbf{x}, \theta)}{\partial \theta} \cdot f(\mathbf{x}, \theta) \, dx_1 \ldots dx_n$$

$$= \frac{\partial}{\partial \theta} \int_{-\infty}^\infty \cdots \int_{-\infty}^\infty g(\mathbf{x})f(\mathbf{x}, \theta) \, dx_1 \ldots dx_n = \frac{\partial}{\partial \theta} E_\theta[S] = \frac{\partial}{\partial \theta}(\theta) = 1.$$

As in the example, we conclude that

$$\rho^2(W,S) = \frac{\mathrm{cov}^2(W,S|\theta)}{\sigma_{W|\theta}^2 \sigma_{S|\theta}^2} = \frac{1}{\sigma_{W|\theta}^2 \sigma_{S|\theta}^2} \le 1.$$

Consequently, $\sigma_{S|\theta}^2 \geq 1/\sigma_{W|\theta}^2$.

7.9 Theorem: Let $f(x_1, x_2, \ldots, x_n, \theta) = \prod_{i=1}^n f_{X|\theta}(x_i)$ be the joint density of the sample $\mathbf{X} = (X_1, X_2, \ldots, X_n)$ from a population X with parameterized density $f_{X|\theta}(\cdot)$. If this joint density is a differentiable function of θ and if the interchange of differentiation and integration in the discussion above is valid, then for any unbiased estimator S of θ, we have the bound $\sigma_{S|\theta}^2 \geq 1/I(\theta)$, where

$$I(\theta) = E_\theta\left[\left(\frac{\partial}{\partial\theta}\ln f(\mathbf{X}, \theta)\right)^2\right].$$

PROOF: See discussion above. ∎

7.10 Definition: Let $\mathbf{X} = (X_1, \ldots, X_n)$ be a sample over a population X with parameterized density $f_{X|\theta}(\cdot)$. Denote the joint density of the sample by $f(\mathbf{x}, \theta) = \prod_{i=1}^n f_{X|\theta}(x_i)$. The *information* in the sample is the quantity

$$I(\theta) = E_\theta\left[\left(\frac{\partial}{\partial\theta}\ln f(\mathbf{X}, \theta)\right)^2\right].$$

If $S = g(X_1, \ldots, X_n)$ is an unbiased estimator of θ, we say that S is *efficient* if $\sigma_{S|\theta}^2 = 1/I(\theta)$. ∎

In this terminology, we have shown that the sample mean is an efficient estimator of the parameter τ for the family $f_{X|\tau}(t) = (1/\tau)e^{-t/\tau}$. At this point, we can determine if a given unbiased estimator is efficient, but we still need a technique for discovering that estimator. The lower bound derivation uses the fact that the correlation of two random variables lies in the range $[-1, 1]$. A further observation is helpful here. The lower bound is achieved precisely when the correlation is ± 1, which occurs when a linear relationship exists between S and W. That is, if S is an unbiased estimator for τ, with respect to the family $f_{X|\tau}(t) = (1/\tau)e^{-t/\tau}$, and if $\sigma_{S|\tau}^2$ achieves the lower bound τ^2/n, then S and W have the relationship $S = k_1(\tau)W + k_2(\tau)$. The correlation must be ± 1 for all values of τ, but the coefficients in the linear relationship need not be the same for different τ values. Rather, they can be functions of τ, as indicated above. Because S is unbiased, we have

$$\tau = E_\tau[S] = k_1(\tau)E_\tau[W] + k_2(\tau) = k_2(\tau).$$

This forces S to assume the form

$$S = k_1(\tau)W + \tau = k_1(\tau)\left(\frac{-n}{\tau} + \frac{\sum_{i=1}^n X_i}{\tau^2}\right) + \tau$$

$$= \frac{k_1(\tau)}{\tau^2}\sum_{i=1}^n X_i + \left(\tau - \frac{nk_1(\tau)}{\tau}\right).$$

As an estimator of τ, S cannot depend on τ. Rather, it must be a function of the $X_1, X_2, \ldots, X_n$ alone. This forces $\tau - nk_1(\tau)/\tau = 0$. Solving for k_1, we have $k_1(\tau) = \tau^2/n$. That is, an unbiased estimator S linearly related to

W must be $S = (1/n) \sum_{i=1}^{n} X_i = \overline{X}_n$. This analysis leads us to the efficient estimator $\overline{X}_n$.

If we know the distribution of an estimator, we can calculate a confidence interval. The exercise, however, presents two difficulties, one conceptual and one technical. The conceptual problem lies in the interpretation of the confidence interval, while the technical obstacle is often the complex distribution of the estimator. The first example below considers an overly simple estimator, which presents a simple distribution. In this context, we discuss the interpretation of an estimator. To illustrate the technical difficulties, we follow with a second example that uses a more complicated, but more realistic, estimator.

7.11 Example: Let $\mathbf{X} = (X_1, X_2, \ldots, X_n)$ be a sample from a population X having parameterized density $f_{X|\tau}(t) = e^{-t/\tau}/\tau$, for $t \geq 0$. Consider the estimator X_1, the initial observation. As an estimator of τ, it is unbiased, and its risk and variance are both equal to τ^2. It has the same distribution as the population X, so we can compute the probability that the estimator lies in any given interval. For a given confidence level, say 0.9 or 90%, there are many intervals $[a, b]$ such that $\Pr(X_1 \in [a, b]) = 0.9$. We can remove the ambiguity by stipulating that the remaining 10% probability must divide equally between the events $(X_1 < a)$ and $(X_1 > b)$. These conditions suffice to determine $[a, b]$ as follows.

$$0.05 = \Pr_\tau(X_1 < a) = \int_0^a \frac{e^{-t/\tau}}{\tau}\, dt = \left. -e^{-t/\tau} \right|_0^a = 1 - e^{-a/\tau}$$

$$a = -\tau \ln(0.95) = 0.0513\tau$$

$$0.05 = \Pr_\tau(X_1 > b) = \int_b^\infty \frac{e^{-t/\tau}}{\tau}\, dt = \left. e^{-t/\tau} \right|_b^\infty = e^{-b/\tau}$$

$$b = -\tau \ln(0.05) = 2.996\tau$$

The event $(0.0513\tau \leq X_1 \leq 2.996\tau)$ then has probability 0.9. The two inequalities defining this event are equivalent to $(0.3338X_1 \leq \tau \leq 19.50X_1)$. If we observe the value $X_1 = 0.2$, for example, we report the interval $[0.3338(0.2), 19.50(0.2)] = [0.0668, 3.9]$ as the 90% confidence interval for τ. This adds some perspective to the estimated value 0.2. In particular, it says that the estimate is not very credible. To achieve a 90% confidence band around this estimate, we must extend the left extreme almost to zero and the right extreme to some twenty times the estimate. This is no surprise; an estimator that ignores most of the sample is hardly promising.

More important is the interpretation of the confidence interval. It does not mean that τ falls in $[0.0668, 3.9]$ nine out of every ten samples. Recall that τ is a fixed but unknown parameter. It is either in the interval, or it is not. It does not shift its residence from sample to sample. What shifts from sample to sample are the interval endpoints because they are calculated from the statistic X_1, which assumes a new value with each sample. Over a long sequence of samples, 90% of the corresponding intervals, all conceivably different, contain the fixed τ. The reported interval $[0.0668, 3.9]$ is simply a random excerpt from this sequence. The chances are then nine out of ten that it is one of those that do contain τ. $\square$

7.12 Example: Again let $\mathbf{X} = (X_1, X_2, \ldots, X_n)$ be a sample from a population X having parameterized density $f_{X|\tau}(t) = e^{-t/\tau}/\tau$, for $t \geq 0$. Now consider the estimator $\overline{X}$, the sample mean. As an estimator of τ, it is unbiased and efficient. Its risk and variance are both equal to τ^2/n. Its distribution, however, is not that of the underlying population. X has an exponential distribution with parameter $1/\tau$. Consequently, $S = X_1 + X_2 + \ldots + X_n$ has an Erlang distribution with parameters $(n, 1/\tau)$. That is,

$$f_{S|\tau}(t) = \frac{(1/\tau)^n t^{n-1} e^{-t/\tau}}{(n-1)!}$$

$$F_{\overline{X}|\tau}(x) = \mathrm{Pr}_\tau(\overline{X} \le x) = \mathrm{Pr}_\tau(S \le nx) = \int_0^{nx} \frac{t^{n-1} e^{-t/\tau}}{\tau^n (n-1)!} \, dt$$

$$f_{\overline{X}|\tau}(x) = F'_{\overline{X}|\tau}(x) = \frac{n(nx)^{n-1} e^{-nx/\tau}}{\tau^n (n-1)!}.$$

We now attempt to construct a 90% confidence level by computing points a and b such that the events $(\overline{X} < a)$ and $(\overline{X} > b)$ each has probability 0.05. The relevant equations are

$$0.05 = \mathrm{Pr}_\tau(\overline{X} < a) = \int_0^a \frac{n(nx)^{n-1} e^{-nx/\tau}}{\tau^n (n-1)!} \, dx$$

$$0.05 = \mathrm{Pr}_\tau(\overline{X} > b) = \int_b^\infty \frac{n(nx)^{n-1} e^{-nx/\tau}}{\tau^n (n-1)!} \, dx.$$

While it is possible to evaluate the integrals, the resulting expressions are very difficult to solve for a and b. We illustrate with the equation for a, where we first employ the transformation $y = nx$.

$$0.05 = \int_0^a \frac{n(nx)^{n-1} e^{-nx/\tau}}{\tau^n (n-1)!} \, dx = \int_0^{na} \frac{y^{n-1} e^{-y/\tau}}{\tau^n (n-1)!} \, dy$$

We have already had occasion to integrate the form $x^n e^{-\lambda x}$. The result, whose derivation appears following Equation 6.3 is

$$\int_0^t x^n e^{-\lambda x} \, dx = \frac{n!}{\lambda^{n+1}} - e^{-\lambda t} \sum_{i=0}^n \frac{n! t^i}{i! \lambda^{n+1-i}}.$$

Exploiting this expression, we continue the computation for a.

$$0.05 = \frac{1}{\tau^n (n-1)!} \left[\frac{(n-1)!}{(1/\tau)^n} - e^{-na/\tau} \sum_{i=0}^{n-1} \frac{(n-1)!(na)^i}{i!(1/\tau)^{n-i}} \right]$$

$$= 1 - e^{-na/\tau} \sum_{i=1}^{n-1} \frac{(na/\tau)^i}{i!}$$

From here, we must proceed with numerical methods, and we must specify n, the sample size. This latter constraint is a major irritant because the freedom to increase the sample size provides an avenue for shrinking an uncomfortably large confidence interval. Hence, we prefer to conduct the analysis with a general n as far as possible. Letting $\beta = na/\tau$, we have $e^{-\beta} \sum_{i=0}^{n-1} \beta^i/i! = 0.95$. With some computerized assistance, we discover that $n = 10$ yields the solution $\beta = 5.428$, which in turn corresponds to $a = \beta\tau/n = 0.5428\tau$. A similar analysis of the b

equation produces $e^{-\gamma} \sum_{i=0}^{n-1} \gamma^i/i! = 0.05$, where $\gamma = nb/\tau$. For $n = 10$, a numerical solution gives $\gamma = 15.712$, which corresponds to $b = 1.5712\tau$. We conclude that the event $(0.5428\tau \leq \overline{X} \leq 1.5712\tau)$ has probability 0.9. An equivalent event is $(0.6365\overline{X} \leq \tau \leq 1.842\overline{X})$.

For an observed $\overline{X}$ value of 0.2, we report $[0.1273, 0.3684]$ as the 90% confidence interval. The interpretation, of course, is the same as in the preceding example. The much narrower interval reflects the superiority of the $\overline{X}$ estimator in comparison with X_1. We expect yet narrower confidence intervals for larger sample sizes. Unfortunately, we must resort to the numerical solution for each n. For an observed $\overline{X} = 0.2$, the confidence intervals I_n are as follows for various sample sizes n.

n	10	30	50	70	90
I_n	[0.1273, 0.3684]	[0.1517, 0.2781]	[0.1609, 0.2566]	[0.1659, 0.2461]	[0.1695, 0.2400]

The central limit theorem provides a shortcut around these burdensome computations. We know that $S = \sum_{i=1}^{n} X_i$ has mean $n\tau$ and variance $n\tau^2$. For large n, it has an approximate normal distribution. Consequently, $(S - \tau)/(\tau\sqrt{n})$ is nearly standard normal. From Table 2.1, we note the 95th and 5th percentiles of the standard normal distribution to be ± 1.645. Therefore,

$$0.9 = \Pr\left(-1.645 \leq \frac{S - n\tau}{\tau\sqrt{n}} \leq 1.645\right) = \Pr(\tau(n - 1.645\sqrt{n}) \leq S \leq \tau(n + 1.645\sqrt{n}))$$

$$= \Pr\left(\tau\left(1 - \frac{1.645}{\sqrt{n}}\right) \leq \overline{X} \leq \tau\left(1 + \frac{1.645}{\sqrt{n}}\right)\right)$$

$$= \Pr\left(\frac{\overline{X}\sqrt{n}}{\sqrt{n} + 1.645} \leq \tau \leq \frac{\overline{X}\sqrt{n}}{\sqrt{n} - 1.645}\right).$$

For an observed value of $\overline{X} = 0.2$, these approximation, J_n, for the confidence intervals produces the following values.

n	10	30	50	70	90
J_n	[0.1316, 0.4168]	[0.1538, 0.2859]	[0.1623, 0.2606]	[0.1671, 0.2489]	[0.1704, 0.2420]

These computations compare favorably with those computed above, and the correspondence grows closer as the sample size increases. ☐

Exercises

*7.1 As in Example 7.3, let $X_1, X_2, \ldots, X_n$ be a sample from population X, for which the parameterized density is $f_{X|\tau}(t) = e^{-t/\tau}/\tau$ for $t \geq 0$. As an estimator of τ, consider the sample maximum $Y = X_{(n)}$. Show that the variance of Y is $\sigma^2_{Y|\tau} = \tau^2[1 + 1/2^2 + 1/3^2 + \ldots + 1/n^2]$, as asserted in Example 7.3. Show also that $\sigma^2_{Y|\tau}$ approaches a finite limit as $n \to \infty$.

*7.2 For $\theta > 0$, consider the parameterized density family $f_{X|\theta}(x) = 1/\theta$ for $0 \leq x \leq \theta$ and zero elsewhere. Let $\overline{X}_n$ be the mean of a sample of size n. As an estimator of θ, compute the bias of $\overline{X}_n$. Is the sequence $\overline{X}_n$ consistent? Does it have zero asymptotic risk?

*7.3 For the family of the preceding exercise, consider the estimator $X_{(n)}$, the sample maximum. Compute its bias. Is the sequence $X_{(n)}$ consistent? Does it have zero asymptotic risk?

7.4 Consider a family of normal densities, each with unit variance. That is, $f_{X|\mu}(x) = (1/\sqrt{2\pi})e^{-(x-\mu)^2/2}$. The mean μ indexes the family. Given a sample $\mathbf{X} = (X_1, \ldots, X_n)$ and an unbiased estimator $S = g(X_1, \ldots, X_n)$ of μ, determine a lower bound on the variance of S. Show that the estimator $\overline{X} = (\sum X_i)/n$ achieves this lower bound.

7.5 Consider the densities $f_{X|\tau}(t) = e^{-t/(\tau-2)}/(\tau - 2)$ for $t \geq 0$ and zero elsewhere. Find an efficient estimator for τ, assuming that $\tau > 2$.

7.6 Consider again the family of normal densities in Exercise 7.4. For a sample $\mathbf{X} = (X_1, X_2, \ldots, X_n)$, we use $\overline{X}_n = (1/n)\sum_{i=1}^n X_i$ as an unbiased estimator of μ. If the observed $\overline{X}_{50}$ is 1.6, determine the 90% confidence interval. Approximately what size sample is necessary to reduce the width of the 90% confidence interval to 0.2μ?

7.7 Consider the density family $f_{X|\tau}(x) = x^2 e^{-x/\tau}/(2\tau^3)$ for $x \geq 0$ and zero elsewhere. Determine an efficient estimator for τ. Compute a 90% confidence interval for τ as a function of this estimator.

*7.8 Show that the information in a sample is additive in the following sense. Let population X have density $f_{X|\theta}(\cdot)$. Consider the sample $X_1, \ldots, X_n$ followed by $Y_1, Y_2, \ldots, Y_m$. Let $I_1(\theta)$ be the information in the subsample $X_1, X_2, \ldots, X_n$; let $I_2(\theta)$ be that in the subsample $Y_1, Y_2, \ldots, Y_m$. Show that $I(\theta) = I_1(\theta) + I_2(\theta)$, where $I(\theta)$ is the information associated with the full sample.

7.2 Normal inference

Two parameters, μ and σ^2, index the normal family

$$f_{X|\mu,\sigma^2}(x) = \frac{1}{\sigma\sqrt{2\pi}}\exp\left(\frac{-(x-\mu)^2}{2\sigma^2}\right). \qquad (7.1)$$

If we know either of these two parameters, we can easily construct an efficient estimator for the other. When both parameters are unknown, the construction is more difficult. We first develop the single-parameter cases.

Suppose that $\mathbf{X} = X_1, X_2, \ldots, X_n$ is a sample from a normal distribution with known variance σ^2. We wish to establish an efficient estimator for the mean μ and analyze the estimator's distribution to compute confidence intervals. We consider σ^2 constant in Equation 7.1, and we regard μ as an index. Because the sample mean $\overline{X} = (1/n)\sum_{i=1}^n X_i$ is a linear combination of normal random variables, its distribution is also normal. Consequently, we can use the percentile tables for the standard normal distribution to compute confidence intervals. We first verify that the sample mean is an efficient estimator. We have

$$E_\mu[\overline{X}] = \frac{1}{n}\sum_{i=1}^{n} E_\mu[X_i] = \mu$$

$$E_\mu[\sigma_{\overline{X}}^2] = \left(\frac{1}{n^2}\right) E_\mu[\sigma_{X_i}^2] = \frac{\sigma^2}{n}.$$

This shows that the sample mean is unbiased, but we must still verify that its variance achieves the lower bound $1/I(\mu)$. As a function of μ, the sample distribution is $f(\mu) = (2\pi\sigma^2)^{-n/2}\exp\left(-[1/(2\sigma^2)]\sum_{i=1}^{n}(x_i - \mu)^2\right)$. We calculate $I(\mu)$, the information in the sample.

$$I(\mu) = E_\mu\left[\left(\frac{\partial}{\partial\mu}\ln f_{X|\mu}(\cdot)\right)^2\right]$$

$$= E_\mu\left[\left(\frac{\partial}{\partial\mu}[-(n/2)\ln(2\pi\sigma^2) - (1/2\sigma^2)\sum_{i=1}^{n}(X_i - \mu)^2]\right)^2\right]$$

$$= E_\mu\left[\left(\frac{\sum_{i=1}^{n}(X_i - \mu)}{\sigma^2}\right)^2\right]$$

$$I(\mu) = \frac{1}{\sigma^4}\left[\sum_{i=1}^{n} E[(X_i - \mu)^2] + \sum_{i\neq j} E[(X_i - \mu)(X_j - \mu)]\right] = \frac{n\sigma^2}{\sigma^4} = \frac{n}{\sigma^2}$$

We conclude that the variance of the sample mean is precisely $1/I(\mu)$. The sample mean is an efficient estimator.

Because $\overline{X}$ is normal with parameters $(\mu, \sigma^2/n)$, the transformed quantity $Z = (\overline{X}-\mu)/(\sigma/\sqrt{n}) = \sqrt{n}(\overline{X}-\mu)/\sigma$ is standard normal. If p is a fraction in the range $(0,1)$, we denote by z_p the $100p$th percentile of the standard normal distribution. That is, $\Phi(z_p) = \Pr(Z \leq z_p) = (1/\sqrt{2\pi})\int_{-\infty}^{z_p} e^{-x^2/2}\,dx = p$. From the symmetry of the normal density, we know $z_{1-p} = -z_p$. Table 2.1 lists these percentiles. For a $100p\%$ confidence interval, we initially seek two values a and b, for which $\Pr_\mu(\overline{X} \leq a) = (1-p)/2$ and $\Pr_\mu(\overline{X} \geq b) = (1-p)/2$. We analyze the first equation to determine a.

$$\frac{1-p}{2} = \Pr(\overline{X} \leq a) = \Pr_\mu\left(\frac{\overline{X} - \mu}{\sigma/\sqrt{n}} \leq \frac{a - \mu}{\sigma/\sqrt{n}}\right)$$

$$= \Pr\left(Z \leq \frac{(a - \mu)\sqrt{n}}{\sigma}\right)$$

$$\frac{(a - \mu)\sqrt{n}}{\sigma} = z_{(1-p)/2}$$

$$a = \mu + \frac{\sigma z_{(1-p)/2}}{\sqrt{n}} = \mu - \frac{\sigma z_{1-(1-p)/2}}{\sqrt{n}} = \mu - \frac{\sigma z_{(1+p)/2}}{\sqrt{n}}$$

A similar analysis yields $b = \mu + \sigma z_{(1+p)/2}/\sqrt{n}$. We conclude that

$$\Pr\left(\mu - \frac{\sigma z_{(1+p)/2}}{\sqrt{n}} \leq \overline{X} \leq \mu + \frac{\sigma z_{(1+p)/2}}{\sqrt{n}}\right) = p,$$

or equivalently,

$$\Pr\left(\overline{X} - \frac{\sigma z_{(1+p)/2}}{\sqrt{n}} \leq \mu \leq \overline{X} + \frac{\sigma z_{(1+p)/2}}{\sqrt{n}}\right) = p.$$

This last formula gives the desired confidence intervals.

7.13 Example: Over a representative mix of client requests, a database server exhibits random response times. The response time T, measured in seconds, is known to have a normal distribution with variance 0.004. Using a sample of size 10, we wish to estimate the mean of the distribution and to report a 90% confidence interval. Suppose that the sample contains the following observations.

i	1	2	3	4	5	6	7	8	9	10
T_i	1.72	1.88	1.84	1.81	1.75	1.92	1.80	1.68	1.80	1.81

As an estimate of μ, the distribution mean, we report the sample mean $\overline{T} = (1/10)\sum_{i=1}^{10} T_i = 1.80$. For a 90% confidence interval, we first obtain the percentile $z_{(1+0.9)/2} = z_{0.95} = 1.645$ from Table 2.1 and then compute the interval as discussed above.

$$\left[\overline{T} - z_{0.95}\sqrt{\frac{0.004}{10}}, \overline{T} + z_{0.95}\sqrt{\frac{0.004}{10}}\right] = [1.767, 1.833] \;\square$$

Consider now the situation where we know the mean μ of a normal distribution, and we want to estimate the variance σ^2. In Equation 7.1, μ is now constant and σ^2 indexes the family. For a sample $\mathbf{X} = X_1, \ldots, X_n$, the quantity $S = (1/n)\sum_{i=1}^{n}(X_i - \mu)^2$ has expected value $E_{\sigma^2}[S] = \sigma^2$. Consequently, S is an unbiased estimator of σ^2. To compute the variance of S, we note that $Z_i = (X_i - \mu)/\sigma$ is standard normal. Theorem 6.6 then gives the following moments.

$$E[Z_i^n] = \begin{cases} (2k-1)(2k-3)\ldots(3)(1), & n = 2k, k = 1, 2, 3, \ldots \\ 0, & n = 2k-1, k = 1, 2, 3, \ldots \end{cases}$$

This observation is useful for the calculation

$$\text{var}_{\sigma^2}(S) = E_{\sigma^2}[S^2] - (E_{\sigma^2}[S]))^2 = E\left[\left(\frac{1}{n}\sum_{i=1}^{n}(X_i - \mu)^2\right)^2\right] - \sigma^4$$

$$= \frac{1}{n^2}E\left[\sum_{i=1}^{n}(X_i - \mu)^4 + \sum_{i \neq j}(X_i - \mu)^2(X_j - \mu)^2\right] - \sigma^4$$

$$= \frac{\sigma^4}{n^2}E\left[\sum_{i=1}^{n}\left(\frac{X_i - \mu}{\sigma}\right)^4\right] + \frac{(n^2 - n)\sigma^4}{n^2} - \sigma^4.$$

Using the aforementioned moments, this expression simplifies to

$$\text{var}_{\sigma^2}(S) = \frac{\sigma^4}{n^2}\sum_{i=1}^{n}E[Z_i^4] - \frac{\sigma^4}{n} = \frac{\sigma^4}{n^2}\sum_{i=1}^{n}(3)(1) - \frac{\sigma^4}{n} = \sigma^4\left(\frac{3}{n} - \frac{1}{n}\right) = \frac{2\sigma^4}{n}.$$

Is this the minimum possible variance for an unbiased estimator of σ^2? We investigate the information in the sample.

$$I(\sigma^2) = E_{\sigma^2}\left[\left(\frac{\partial}{\partial\sigma^2}\ln\left((2\pi\sigma^2)^{-n/2}\exp\left(\frac{-1}{2\sigma^2}\sum_{i=1}^{n}(X_i - \mu)^2\right)\right)\right)^2\right]$$

$$= E\left[\left(\frac{\partial}{\partial\sigma^2}\left[\frac{-n}{2}\ln(2\pi\sigma^2) - \frac{1}{2\sigma^2}\sum_{i=1}^{n}(X_i - \mu)^2\right]\right)^2\right]$$

$$= E\left[\left(\frac{-n}{2}\cdot\frac{1}{2\pi\sigma^2}\cdot(2\pi) + \frac{1}{2}\cdot\frac{1}{\sigma^4}\sum_{i=1}^{n}(X_i - \mu)^2\right)^2\right]$$

$$= E\left[\left(\frac{1}{2\sigma^4}\left[-n\sigma^2 + \sum_{i=1}^{n}(X_i - \mu)^2\right]\right)^2\right]$$

$$= \frac{1}{4\sigma^8}E\left[n^2\sigma^4 - 2n\sigma^2\sum_{i=1}^{n}(X_i - \mu)^2 + \left(\sum_{i=1}^{n}(X_i - \mu)^2\right)^2\right]$$

$$= \frac{1}{4\sigma^8}\left[n^2\sigma^4 - 2n\sigma^2(n\sigma^2) + \sum_{i=1}^{n}E[(X_i - \mu)^4]\right.$$

$$\left. + \sum_{i\neq j}E[(X_i - \mu)^2(X_j - \mu)^2]\right]$$

$$= \frac{1}{4\sigma^8}\left[-n^2\sigma^4 + \sigma^4\sum_{i=1}^{n}E\left[\left(\frac{X_i - \mu}{\sigma}\right)^4\right] + (n^2 - n)\sigma^4\right]$$

$$= \frac{1}{4\sigma^8}\left[\sigma^4\sum_{i=1}^{n}(3)(1) - n\sigma^4\right] = \frac{n}{2\sigma^4}.$$

The variance of the estimator S is precisely $1/I(\sigma^2)$, and we conclude that it is an efficient estimator. To compute confidence intervals, we again exploit the fact that $(X_i - \mu)/\sigma$ is standard normal. Then

$$\frac{nS}{\sigma^2} = \sum_{i=1}^{n}\left(\frac{X_i - \mu}{\sigma}\right)^2$$

has a χ^2 distribution with n degrees of freedom. Therefore, for a fraction p in the range $(0, 1)$,

$$\Pr\left(\chi^2_{(n),(1-p)/2} \leq \frac{nS}{\sigma^2} \leq \chi^2_{(n),1-(1-p)/2}\right) = p.$$

Equivalently,

$$\Pr\left(\frac{nS}{\chi^2_{(n),(1+p)/2}} \le \sigma^2 \le \frac{nS}{\chi^2_{(n),(1-p)/2}}\right) = p.$$

Given a value for the estimator S, the last equation provides a $100p\%$ confidence interval.

7.14 Example: We use the same context and data as in the preceding example, but we now assume a known mean. Specifically, the response time T of a database server, measured in seconds, is known to have a normal distribution with mean 1.80. Using a sample of size 10, we wish to estimate the variance and to report a 90% confidence interval. Suppose that the sample contains the following observations.

i	1	2	3	4	5	6	7	8	9	10
T_i	1.72	1.88	1.84	1.81	1.75	1.92	1.80	1.68	1.80	1.81

We report $S = (1/10)\sum_{i=1}^{10}(T_i - 1.80)^2 = 0.00459$ as an estimate of σ^2, the distribution variance. For a 90% confidence interval, we consult Tables 6.3 and 6.2 to obtain the percentiles

$$\chi^2_{(10),(1+0.9)/2} = \chi^2_{10,0.95} = 18.307$$

$$\chi^2_{(10),(1-0.9)/2} = \chi^2_{(10),0.05} = 3.9403.$$

We then compute the interval as discussed above.

$$\left[\frac{nS}{\chi^2_{(10),0.95}}, \frac{nS}{\chi^2_{(10),0.05}}\right] = \left[\frac{(10)(0.00459)}{18.307}, \frac{(10)(0.00459)}{3.9403}\right] = [0.0025, 0.0116] \;\square$$

7.15 Theorem: Based on a sample $X_1, X_2, \dots, X_n$ from a normal population X with known variance σ^2, the sample mean $\overline{X} = (1/n)\sum_{i=1}^{n} X_i$ is an unbiased and efficient estimator of the population mean μ. For $p \in (0,1)$, the $100p$th confidence interval is

$$I_p = \left[\overline{X} - \frac{\sigma z_{(1+p)/2}}{\sqrt{n}}, \overline{X} + \frac{\sigma z_{(1+p)/2}}{\sqrt{n}}\right],$$

where z_q is the $100q$th percentile of the standard normal distribution.

Based on a sample $X_1, X_2, \dots, X_n$ from a normal population X with known mean μ, the quantity $S = (1/n)\sum_{i=1}^{n}(X_i - \mu)^2$ is an unbiased and efficient estimator of the population variance σ^2. For $p \in (0,1)$, the $100p$th confidence interval is

$$J_p = \left[\frac{nS}{\chi^2_{(1+p)/2,(n)}}, \frac{nS}{\chi^2_{(1-p)/2,(n)}}\right],$$

where $\chi^2_{q,(n)}$ is the $100q$th percentile of the χ^2 distribution with n degrees of freedom.

PROOF: See development above. ∎

Theorem 7.15 is seldom useful in practice because, in the typical situation, both the mean and variance are unknown. That is, the density family of Equation 7.1 is indexed by the parameter pair (μ, σ^2). The sample mean and sample variance suggest themselves as estimators of μ and σ^2, respectively. For a sample $X_1, \ldots, X_n$, these are $\overline{X} = (1/n) \sum_{i=1}^{n} X_i$ and $s_X^2 = (1/n) \sum_{i=1}^{n} (X_i - \overline{X})^2$. Of course, the sample mean remains an unbiased estimator of μ in the two-parameter situation. We use $E_{\mu,\sigma^2}[\cdot]$ to indicate which density function to use for the calculation, but we drop the subscript when it is clear from context.

$$E_{\mu,\sigma^2}[\overline{X}] = \frac{1}{n} \sum_{i=1}^{n} E[X_i] = \mu$$

$$E_{\mu,\sigma^2}[s_X^2] = \frac{1}{n} E\left[\sum_{i=1}^{n} (X_i - \overline{X})^2 \right] = \frac{1}{n} E\left[\sum_{i=1}^{n} ((X_i - \mu) - (\overline{X} - \mu))^2 \right]$$

$$= \frac{1}{n} \left\{ \sum_{i=1}^{n} E[(X_i - \mu)^2] + nE[(\overline{X} - \mu)^2] - 2E\left[(\overline{X} - \mu) \sum_{i=1}^{n} (X_i - \mu) \right] \right\}$$

$$= \frac{1}{n} \left\{ n\sigma^2 + n \cdot \frac{\sigma^2}{n} - 2E\left[(\overline{X} - \mu) \left(\left(\sum_{i=1}^{n} X_i \right) - n\mu \right) \right] \right\}$$

$$= \frac{1}{n} \left\{ n\sigma^2 + \sigma^2 - 2E[(\overline{X} - \mu)(n\overline{X} - n\mu)] \right\}$$

$$= \frac{1}{n} \left\{ n\sigma^2 + \sigma^2 - 2nE[(\overline{X} - \mu)^2] \right\}$$

$$= \frac{1}{n} \left[n\sigma^2 + \sigma^2 - 2n \cdot \frac{\sigma^2}{n} \right] = \frac{(n-1)\sigma^2}{n}$$

We see that s_X^2 is biased, although the bias grows less with increasing sample size. Consequently, we use a related estimator, the unbiased sample variance.

7.16 Definition: For a sample $\mathbf{X} = X_1, X_2, \ldots, X_n$, the quantity

$$\hat{s}_X^2 = \frac{1}{n-1} \sum_{i=1}^{n} (X_i - \overline{X})^2$$

is called the *unbiased sample variance*. Here, of course, $\overline{X}$ is the sample mean. ∎

The unbiased sample variance is, as the name suggests, an unbiased estimator of the population variance. It is simply a constant multiple of the sample variance.

$$E[\hat{s}_X^2] = E\left[\frac{1}{n-1} \sum_{i=1}^{n} (X_i - \overline{X})^2 \right] = E\left[\left(\frac{n}{n-1} \right) \left(\frac{1}{n} \right) \sum_{i=1}^{n} (X_i - \overline{X})^2 \right]$$

$$= \frac{n}{n-1} E[s_X^2] = \sigma^2$$

We can now define two information measures for a sample:

$$I_1(\mu, \sigma^2) = E_{\mu, \sigma^2}\left[\left(\frac{\partial}{\partial \mu}\ln\left(\prod_{i=1}^{n}(2\pi\sigma^2)^{-1/2}e^{(X_i-\mu)^2/(2\sigma^2)}\right)\right)^2\right]$$

$$I_2(\mu, \sigma^2) = E_{\mu, \sigma^2}\left[\left(\frac{\partial}{\partial \sigma^2}\ln\left(\prod_{i=1}^{n}(2\pi\sigma^2)^{-1/2}e^{(X_i-\mu)^2/(2\sigma^2)}\right)\right)^2\right].$$

Because the earlier single-parameter calculations remain valid with the substitution of E_{μ,σ^2} for E_μ and E_{σ^2}, we have $I_1(\mu,\sigma^2) = n/\sigma^2$ and $I_2(\mu,\sigma^2) = n/(2\sigma^4)$. Because $\overline{X}$ and $\hat{s}_X^2$ are unbiased estimators of μ and σ^2, we conclude that

$$\mathrm{var}_{\mu,\sigma^2}(\overline{X}) \geq \frac{1}{I_1(\mu,\sigma^2)} = \frac{\sigma^2}{n}$$

$$\mathrm{var}_{\mu,\sigma^2}(\hat{s}_X^2) \geq \frac{1}{I_2(\mu,\sigma^2)} = \frac{2\sigma^4}{n}.$$

Because the variance of $\overline{X}$ is precisely σ^2/n, the sample mean achieves this lower bound and cannot be surpassed by any unbiased estimator for μ. What about the unbiased sample variance? From Theorem 4.13, we obtain the variance of the sample variance, the biased version:

$$\mathrm{var}_{\mu,\sigma^2}(s_X^2) = \frac{\delta_4 - \sigma^4}{n} - \frac{2(\delta_4 - 2\sigma^4)}{n^2} + \frac{\delta_4 - 3\sigma^4}{n^3},$$

where δ_4 is the fourth central moment $E_{\mu,\sigma^2}[(X-\mu)^4]$. Earlier we noted that $E[Z^4] = 3$, when Z is a standard normal random variable. Consequently,

$$\delta_4 = E_{\mu,\sigma^2}[(X-\mu)^4] = \sigma^4 E\left[\left(\frac{X-\mu}{\sigma}\right)^4\right] = \sigma^4 E[Z^4] = 3\sigma^4$$

$$\mathrm{var}_{\mu,\sigma^2}(s_X^2) = \frac{2\sigma^4}{n} - \frac{2\sigma^4}{n^2} = \frac{2(n-1)\sigma^4}{n^2}.$$

Because the unbiased sample variance is a constant multiple of its biased counterpart, we have

$$\mathrm{var}_{\mu,\sigma^2}(\hat{s}_X^2) = \mathrm{var}\left(\frac{n}{n-1}\cdot s_X^2\right) = \frac{n^2}{(n-1)^2}\cdot\frac{2(n-1)\sigma^4}{n^2}$$

$$= \frac{2\sigma^4}{n-1} = \left(\frac{n}{n-1}\right)\frac{2\sigma^4}{n} = \left(\frac{n}{n-1}\right)\frac{1}{I_2(\mu,\sigma^2)}.$$

We conclude that variance of the unbiased sample variance does not quite achieve the lower bound, but the difference grows less with increasing sample size. In this case, we say that the unbiased sample variance is *asymptotically efficient*.

At this point, we have effective estimators for the population mean and variance, but we cannot calculate confidence intervals unless we know the distribution of the estimators. While it is true that $(\overline{X} - \mu)/\sigma$ is standard normal, we cannot convert an interval between two standard normal percentiles into corresponding limits on μ. The difficulty is that σ is unknown. The computation runs as follows, but it is ultimately unsuccessful because the resulting confidence interval has endpoints that are functions of σ. We use a 90% confidence interval to illustrate the point.

$$0.9 = \Pr\left(z_{0.05} \le \frac{\sqrt{n}(\overline{X} - \mu)}{\sigma} \le z_{0.95}\right)$$

$$= \Pr\left(\overline{X} - \frac{\sigma z_{0.95}}{\sqrt{n}} \le \mu \le \overline{X} - \frac{\sigma z_{0.05}}{\sqrt{n}}\right)$$

$$= \Pr\left(\overline{X} - \frac{\sigma z_{0.95}}{\sqrt{n}} \le \mu \le \overline{X} + \frac{\sigma z_{0.95}}{\sqrt{n}}\right)$$

The solution to this dilemma lies with the fact that $n(\overline{X} - \mu)^2/\sigma^2$ and $(n-1)\hat{s}_X^2/\sigma^2$ are *independent* χ^2 random variables with degrees of freedom 1 and $n - 1$, respectively. The proof, however, requires a digression into matrix algebra, and we defer the details to the upcoming mathematical interlude. The result appears as Theorem 7.24 in the next section. Assuming this result for the moment, the statistic

$$T = \sqrt{\frac{n(\overline{X} - \mu)/[(1)\sigma^2]}{(n - 1)\hat{s}_X^2/[(n - 1)\sigma^2]}} = \frac{\sqrt{n}(\overline{X} - \mu)}{\sqrt{\hat{s}_X^2}}$$

has a t distribution with $n - 1$ degrees of freedom and is independent of σ^2. If $t_{p,(n-1)}$ indicates the $100p$th percentile of this distribution, we can construct a $100p$ confidence level as follows. Recall that the t distribution is symmetric, and therefore $t_{p,(n-1)} = -t_{1-p,(n-1)}$.

$$p = \Pr(t_{(1-p)/2,(n-1)} \le T \le t_{1-(1-p)/2,(n-1)})$$

$$= \Pr\left(t_{(1-p)/2,(n-1)} \le \frac{\sqrt{n}(\overline{X} - \mu)}{\sqrt{\hat{s}_X^2}} \le t_{(1+p)/2,(n-1)}\right)$$

$$= \Pr\left(\frac{-t_{(1+p)/2,(n-1)}\sqrt{\hat{s}_X^2}}{\sqrt{n}} \le \overline{X} - \mu \le \frac{t_{(1+p)/2,(n-1)}\sqrt{\hat{s}_X^2}}{\sqrt{n}}\right)$$

$$= \Pr\left(\overline{X} - t_{(1+p)/2,(n-1)}\sqrt{\frac{\hat{s}_X^2}{n}} \le \mu \le \overline{X} + t_{(1+p)/2,(n-1)}\sqrt{\frac{\hat{s}_X^2}{n}}\right).$$

Confidence intervals for the variance are simpler. Assuming again that $(n-1)\hat{s}_X^2/\sigma^2$ has a χ^2 distribution with $n - 1$ degrees of freedom, the computation proceeds as in the case of a known mean, except that we use $\hat{s}_X^2$ instead of $S = \left(\sum(X_i - \mu)^2\right)/n$ and we reduce the degrees of freedom by 1. That is, for the $100p$th confidence level, we have

$$p = \Pr\left(\chi^2_{(1-p)/2,(n-1)} \leq \frac{(n-1)\hat{s}_X^2}{\sigma^2} \leq \chi^2_{1-(1-p)/2,(n-1)}\right)$$

$$= \Pr\left(\frac{(n-1)\hat{s}_X^2}{\chi^2_{(1+p)/2,(n-1)}} \leq \sigma^2 \leq \frac{(n-1)\hat{s}_X^2}{\chi^2_{(1-p)/2,(n-1)}}\right).$$

7.17 Theorem: Let $X_1, X_2, \ldots, X_n$ be a sample from a normal population X with unknown mean μ and unknown variance σ^2. The sample mean and unbiased sample variance are $\overline{X} = (1/n)\sum_{i=1}^{n} X_i$ and $\hat{s}_X^2 = [1/(n-1)]\sum_{i=1}^{n}(X_i - \overline{X})^2$, respectively. Then $\overline{X}$ is an unbiased and efficient estimator for μ, and $\hat{s}_X^2$ is an unbiased and asymptotically efficient estimator for σ^2. For $0 < p < 1$, the $100p$th confidence intervals are

$$\overline{X} - t_{(1+p)/2,(n-1)}\sqrt{\frac{\hat{s}_X^2}{n}} \leq \mu \leq \overline{X} + t_{(1+p)/2,(n-1)}\sqrt{\frac{\hat{s}_X^2}{n}}$$

$$\frac{(n-1)\hat{s}_X^2}{\chi^2_{(1+p)/2,(n-1)}} \leq \sigma^2 \leq \frac{(n-1)\hat{s}_X^2}{\chi^2_{(1-p)/2,(n-1)}}.$$

PROOF: See discussion above. ∎

7.18 Example: Continuing with the database example, assume now that neither the mean nor the variance is known. Specifically, the response time X of a database server, measured in seconds, is known to have a normal distribution with unknown mean μ and unknown variance σ^2. Using a sample of size 10, we wish to estimate the mean and variance and to report a 90% confidence interval for each estimate. The sample contains the following observations.

i	1	2	3	4	5	6	7	8	9	10
X_i	1.72	1.88	1.84	1.81	1.75	1.92	1.80	1.68	1.80	1.81

The estimate for the mean remains $\overline{X} = 1.80$. We calculate

$$\hat{s}_X^2 = \frac{1}{n-1}\sum_{i=1}^{n}(X_i - \overline{X})^2 = \frac{1}{9}\sum_{i=1}^{10}(X_i - 1.80)^2 = 0.0051.$$

Table 6.5 lists t percentiles, and we note that $p = 0.90$ for a 90% confidence interval. The table gives $t_{(1+p)/2,(n-1)} = t_{0.95,(9)} = 1.833$, and the confidence interval is then

$$\left[\overline{X} - t_{0.95,(9)}\sqrt{\frac{\hat{s}_X^2}{n}}, \overline{X} + t_{0.95,(9)}\sqrt{\frac{\hat{s}_X^2}{n}}\right] = 1.80 \mp 1.833\sqrt{0.00051} = [1.76, 1.84].$$

The estimate for the variance is $\hat{s}_X^2 = 0.0051$ and its 90% confidence interval is

$$\left[\frac{(n-1)\hat{s}_X^2}{\chi^2_{(1+p)/2,(n-1)}}, \frac{(n-1)\hat{s}_X^2}{\chi^2_{(1-p)/2,(n-1)}}\right] = [(9)(0.0051)/16.9189, (9)(0.0051)/3.3251]$$

$$= [0.0027, 0.0138].$$

We obtained the χ^2 percentiles from Table 6.2. □

When we must estimate both the mean and variance of a normal random variable, it is perhaps more reasonable to specify a confidence *region* in the (μ, σ^2) plane that captures the actual mean and variance with the desired probability. You might think that the Cartesian product of the two separate confidence intervals would suffice. That is, if the 90% confidence interval for μ is $[a, b]$ and the 90% confidence interval for σ^2 is $[c, d]$, you might argue that the points $\{(x, y) \mid a \le x \le b, c \le y \le d\}$ should constitute an 81% confidence region for (μ, σ^2). This approach does not quite work out because the t statistic that gives the mean confidence interval is not independent of the χ^2 statistic that gives the variance confidence interval. However, we are assuming that $n(\overline{X} - \mu)^2/\sigma^2$ and $(n - 1)\hat{s}_X^2/\sigma^2$ are independent χ^2 random variables with 1 and $n - 1$ degrees of freedom, respectively. This implies that $\sqrt{n}(\overline{X} - \mu)/\sigma$ is standard normal and still independent of the quantity $(n - 1)\hat{s}_X^2/\sigma^2$. These observations permit the following calculation.

Suppose that p is a fraction in the range $0 < p < 1$, and we wish a $100p\%$ two-dimensional confidence region in the (μ, σ^2) plane. Assume that we want to extract the region symmetrically, in the sense that it is centered at $(\overline{X}, \hat{s}_X^2)$ and each of the two independent expressions contributes equally. That is, we extract the center of the $(\overline{X} - \mu)\sqrt{n}/\sigma$ distribution, from z_t to z_{1-t}, and the $(n - 1)\hat{s}_X^2/\sigma^2$ distribution from $\chi^2_{t,(n-1)}$ to $\chi^2_{1-t,(n-1)}$. To determine an appropriate value for t, we calculate as follows.

$$p = \Pr\left(z_t \le \frac{\sqrt{n}(\overline{X} - \mu)}{\sigma} \le z_{1-t} \text{ and } \chi^2_{t,(n-1)} \le \frac{(n - 1)\hat{s}_X^2}{\sigma^2} \le \chi^2_{1-t,(n-1)}\right)$$

$$= \Pr\left(z_t \le \frac{\sqrt{n}(\overline{X} - \mu)}{\sigma} \le z_{1-t}\right) \cdot \Pr\left(\chi^2_{t,(n-1)} \le \frac{(n - 1)\hat{s}_X^2}{\sigma^2} \le \chi^2_{1-t,(n-1)}\right)$$

$$= (1 - 2t)^2$$

Solving, we obtain $t = (1 - \sqrt{p})/2$. With this t value, the equations

$$z_t \le \frac{\sqrt{n}(\overline{X} - \mu)}{\sigma} \le z_{1-t}$$

$$\chi^2_{t,(n-1)} \le \frac{(n - 1)\hat{s}_X^2}{\sigma^2} \le \chi^2_{1-t,(n-1)}$$

define the desired $100p\%$ confidence region in the (μ, σ^2) plane. The second equation describes the horizontal slice

$$\frac{(n - 1)\hat{s}_X^2}{\chi^2_{1-t,(n-1)}} \le \sigma^2 \le \frac{(n - 1)\hat{s}_X^2}{\chi^2_{t,(n-1)}},$$

which is just the width of the $100(1 - 2t)\%$ confidence interval for the variance. The first equation, however, is equivalent to

$$-z_{1-t} \le \frac{\sqrt{n}(\overline{X} - \mu)}{\sigma} \le z_{1-t}$$

$$\frac{n(\overline{X} - \mu)^2}{\sigma^2} \le z^2_{1-t}.$$

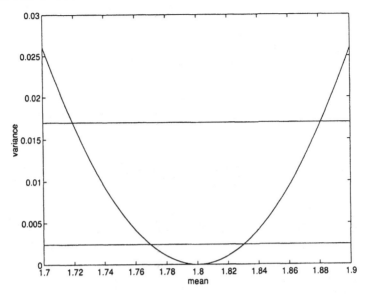

Figure 7.1. Two-dimensional confidence region for mean and variance estimators (see Example 7.19)

That is, $\sigma^2 \geq n(\mu - \overline{X})^2/z_{1-t}^2$, which describes a parabola, opening upward, with vertex at $(\mu = \overline{X}, \sigma^2 = 0)$. The region above this parabola, yet within the horizontal slice, is the desired $100p\%$ confidence region.

7.19 Example: Calculate a 90% confidence region for the data of the preceding example.

From the calculations in the cited example, we have $\overline{X} = 1.80$ and $\hat{s}_X^2 = 0.0051$. The sample size is $n = 10$. For $p = 0.9$, we determine the parameter t and the related standard normal and χ^2 percentiles.

$$t = \frac{1 - \sqrt{p}}{2} = 0.025$$

$$z_{1-t} = z_{0.975} = 1.96$$
$$\chi_{t,(n-1)}^2 = \chi_{0.025,(9)}^2 = 2.7004$$
$$\chi_{1-t,(n-1)}^2 = \chi_{0.975,(9)}^2 = 19.0025$$

The parabola $\sigma^2 \geq n(\mu - \overline{X})^2/z_{1-t}^2 = 2.603(\mu - 1.80)^2$ and the horizontal slice $0.0024 = (n-1)\hat{s}_X^2/\chi_{1-t,(n-1)}^2 \leq \sigma^2 \leq (n-1)\hat{s}_X^2/\chi_{t,(n-1)}^2 = 0.0170$ define the desired 90% confidence region. Figure 7.1 shows this region. $\square$

Exercises

7.9 The random variable X has a normal distribution with a known variance $\sigma_X^2 = 7.32$. A sample of size 10 provides the estimate $\overline{X} = 13.1$ for the unknown mean. Compute the 95% confidence interval for this estimate.

7.10 The random variable Y has a normal distribution with a known variance $\sigma_Y^2 = 16.0$. We want to estimate the unknown mean with a 95% confidence interval of width 0.1. How large should the sample be?

7.11 The random variable X has a normal distribution with a known mean $\mu = 13.1$. A sample of size 20 yields $S = (1/20) \sum_{i=1}^{20} (X_i - \mu)^2 = 3.84$. Taking this value as an estimate of the unknown population variance σ^2, compute the 95% confidence interval.

7.12 The random variable Y has a normal distribution with a known mean μ. We wish to estimate the unknown variance σ^2. Find the sample size n, which will deliver a 95% confidence interval with width equal to one-tenth the estimate $S = (1/n) \sum_{i=1}^{n} (Y_i - \mu)^2$.

7.13 Population X has a normal distribution with unknown mean and unknown variance. A sample gives the following data points.

i	1	2	3	4	5	6	7	8
X_i	21.8	17.9	18.2	19.9	22.0	21.3	20.4	22.1

i	9	10	11	12	13	14	15
X_i	17.3	24.0	18.8	19.3	19.5	18.4	21.0

Estimate the mean and variance and provide a 95% confidence interval for each.

7.14 Sketch the two-dimensional 95% confidence region for the mean and variance estimates from the preceding exercise.

7.15 A normal population X yields the following sample.

i	1	2	3	4	5	6	7	8
X_i	76.79	76.79	79.35	73.83	75.19	78.31	70.35	70.53

i	9	10	11	12	13	14	15
X_i	75.30	76.71	70.08	73.83	70.67	74.17	76.87

If the mean and variance are both unknown, then $\hat{s}_X^2 = 9.16$ estimates the variance with a 90% confidence interval of $[5.41, 19.52]$. However, if we know the mean to be $\mu = 70.0$, then $S = 29.57$ estimates the variance with 90% confidence interval of $[17.74, 61.1]$. Verify these calculations. Explain why the second estimate for the variance is so much higher and why its confidence interval does not even include the first estimate.

7.3 Sums of squares

In estimation problems, we frequently encounter expressions involving sums of squares. If the terms involve independent normal random variables, we can easily manipulate the expression into a χ^2 random variable. The problem of estimating the variance σ^2 of a normal population with a known mean μ is an example. For a sample $\mathbf{X} = (X_1, X_2, \ldots, X_n)$, the expression $S = (1/n) \sum_{i=1}^{n} (X_i - \mu)^2$ involves a sum of squares, each associated with an independent normal random variable. Multiplying by n/σ^2, we derive $nS/\sigma^2 = \sum_{i=1}^{n}[(X_i - \mu)/\sigma]^2$, which we recognize as a χ^2 random variable with n degrees of freedom. We used this result to obtain confidence intervals for the population variance estimate. A difficulty arises, however, when the terms do not involve independent random variables. The sample variance $s_X^2 = (1/n)\sum_{i=1}^{n}(X_i - \overline{X})^2 = (1/n)\sum_{i=1}^{n}[X_i - (1/n)\sum_{j=1}^{n} X_j]^2$ is a case in point. The squares in the summation have the following form.

$$[(1 - 1/n)X_1 + (-1/n)X_2 + \ldots + (-1/n)X_n]^2$$
$$[(-1/n)X_1 + (1 - 1/n)X_2 + \ldots + (-1/n)X_n]^2$$
$$\vdots \quad \vdots \quad \quad \vdots \quad \vdots \quad \vdots \quad \quad \quad \vdots$$
$$[(-1/n)X_1 + (-1/n)X_2 + \ldots + (1 - 1/n)X_n]^2$$

The terms are squares of linear combinations of independent normal random variables. However, there may be algebraic relationships among these linear combinations, and they may not be independent. Consequently, a manipulation into χ^2 form is not immediately apparent. This section provides the necessary techniques for such a transformation.

If we have a $m \times n$ matrix $\mathbf{A} = [a_{ij}]$, we denote its row vectors by $\mathbf{a}_{i\cdot}$, for $1 \leq i \leq m$, and its column vectors by $\mathbf{a}_{\cdot j}$, for $1 \leq j \leq n$. Consider a collection of m linear combinations, which we write in matrix form as follows.

$$\mathbf{Y} = \begin{bmatrix} Y_1 \\ \vdots \\ Y_m \end{bmatrix} = \mathbf{AX} = \begin{bmatrix} a_{11} & a_{12} & \ldots & a_{1n} \\ \vdots & \vdots & & \vdots \\ a_{m1} & a_{m2} & \ldots & a_{mn} \end{bmatrix} \begin{bmatrix} X_1 \\ \vdots \\ X_n \end{bmatrix}$$

$\mathbf{A}$ is an $m \times n$ matrix, which we can regard as a collection of m row vectors: $\mathbf{a}_{1\cdot}, \mathbf{a}_{2\cdot}, \ldots, \mathbf{a}_{m\cdot}$. It is possible that all m rows form an independent set. In the general case, however, the size of the largest independent set is r, the rank of the matrix, and $r \leq \min(m, n)$.

For example, the linear combinations that appear in ns_X^2 above are

$$\begin{bmatrix} Y_1 \\ Y_2 \\ \vdots \\ Y_n \end{bmatrix} = \begin{bmatrix} 1 - 1/n & -1/n & \ldots & -1/n \\ -1/n & 1 - 1/n & \ldots & -1/n \\ \vdots & \vdots & & \vdots \\ -1/n & -1/n & \ldots & 1 - 1/n \end{bmatrix} \begin{bmatrix} X_1 \\ X_2 \\ \vdots \\ X_n \end{bmatrix}.$$

The row vectors satisfy linear relationship $\mathbf{a}_{1\cdot} + \mathbf{a}_{2\cdot} + \ldots + \mathbf{a}_{n\cdot} = \mathbf{0}$, in which the coefficients are all 1. Evidently, the full set of n rows is not an independent set. The rank of this matrix is then some value between 1 and $(n - 1)$. At this point, it is not clear that ns_X^2 reduces to a χ^2 form. To show that such a reduction is indeed possible, we digress into an algebraic analysis of the sum of squares $Q = \sum_{i=1}^{m} y_i^2$, where each y_i is a linear combination of $(x_1, x_2, \ldots, x_n)$. That is, $\mathbf{y} = \mathbf{Ax}$:

$$\begin{bmatrix} y_1 \\ \vdots \\ y_m \end{bmatrix} = \begin{bmatrix} a_{11} & a_{12} & \ldots & a_{1n} \\ \vdots & \vdots & & \vdots \\ a_{m1} & a_{m2} & \ldots & a_{mn} \end{bmatrix} \begin{bmatrix} x_1 \\ \vdots \\ x_n \end{bmatrix}.$$

In this notation $Q(\mathbf{x}) = \mathbf{y}'\mathbf{y} = (\mathbf{A}\mathbf{x})'(\mathbf{A}\mathbf{x}) = \mathbf{x}'\mathbf{A}'\mathbf{A}\mathbf{x}$, which shows that Q is ultimately a function of $\mathbf{x}$. Suppose that the rank of $\mathbf{A}$ is r. Relabeling the y_i if necessary, we assume that the first r row vectors of $\mathbf{A}$ form an independent set. The submatrix consisting of just these first r rows has rank r and therefore contains a square $r \times r$ submatrix with nonzero determinant (Theorem A.75). Relabeling the x_i if necessary, we assume that this submatrix is the first r columns. For k in the range $r < k \leq m$, consider a row vector $\mathbf{a}_{k\cdot}$. Because $\mathbf{a}_{1\cdot}, \mathbf{a}_{2\cdot}, \ldots, \mathbf{a}_{r\cdot}, \mathbf{a}_{k\cdot}$ is a dependent set, there exist coefficients $c_1, c_2, \ldots, c_r, c_k$, not all zero, such that $c_k \mathbf{a}_{k\cdot} + \sum_{i=1}^{r} c_i \mathbf{a}_{i\cdot} = \mathbf{0}$. If $c_k = 0$, the equation reduces to a linear combination of the first r rows, and the independence of these rows forces the remaining $c_i = 0$. This contradicts the fact that nonzero entries must appear among the c_i, so we conclude that $c_k \neq 0$. We can then solve for $\mathbf{a}_{k\cdot}$ in terms of the independent set: $\mathbf{a}_{k\cdot} = (-1/c_k) \sum_{i=1}^{r} c_i \mathbf{a}_{i\cdot}$. This dependence transfers to y_k.

$$
y_k = \sum_{j=1}^{n} a_{kj} x_j = \sum_{j=1}^{n} \left(\frac{-1}{c_k} \sum_{i=1}^{r} c_i a_{ij} \right) x_j = \sum_{i=1}^{r} \left(\frac{-c_i}{c_k} \right) \sum_{j=1}^{n} a_{ij} x_j
$$

$$
= \sum_{i=1}^{r} \left(\frac{-c_i}{c_k} \right) y_i.
$$

We now have, for $r+1 \leq k \leq m$, that y_k is a linear combination of $y_1, \ldots, y_r$. In particular, setting $b_{ki} = -c_i/c_k$ for $r+1 \leq k \leq m$ and $1 \leq i \leq r$, we have

$$
\begin{bmatrix} y_{r+1} \\ y_{r+2} \\ \vdots \\ y_m \end{bmatrix} = \begin{bmatrix} b_{r+1,1} & b_{r+1,2} & \cdots & b_{r+1,r} \\ b_{r+2,1} & b_{r+2,2} & \cdots & b_{r+2,r} \\ \vdots & \vdots & & \vdots \\ b_{m,1} & b_{m,2} & \cdots & b_{m,r} \end{bmatrix} \begin{bmatrix} y_1 \\ y_2 \\ \vdots \\ y_r \end{bmatrix}.
$$

These linear combinations substitute for the terms $y_{r+1}^2, y_{r+2}^2, \ldots, y_m^2$ and simplify the sum of squares expression.

$$
Q = \sum_{i=1}^{r} y_i^2 + \sum_{k=r+1}^{m} \left(\sum_{i=1}^{r} b_{ki} y_i \right)^2 = \sum_{i=1}^{r} y_i^2 + \sum_{k=r+1}^{m} \sum_{i=1}^{r} \sum_{j=1}^{r} b_{ki} b_{kj} y_i y_j
$$

$$
= \sum_{i=1}^{r} \left(1 + \sum_{k=r+1}^{m} b_{ki}^2 \right) y_i^2 + \sum_{i=1}^{r} \sum_{j \neq i} \left(\sum_{k=r+1}^{m} b_{ki} b_{kj} \right) y_i y_j
$$

Now define the symmetric $r \times r$ matrix $\mathbf{D} = [d_{ij}]$ by

$$
d_{ij} = \begin{cases} 1 + \sum_{k=r+1}^{m} b_{ki}^2, & i = j \\ \sum_{k=r+1}^{m} b_{ki} b_{kj}, & i \neq j. \end{cases}
$$

We then have $Q = \breve{\mathbf{y}}' \mathbf{D} \breve{\mathbf{y}}$, where $\breve{\mathbf{y}}$ is the truncated $\mathbf{y}$ vector containing the first r entries of $\mathbf{y}$. We now show that $\mathbf{D}$ is positive definite. Recall that we have relabeled the x_i and y_i if necessary to ensure that the submatrix

$$\overline{\mathbf{A}} = \begin{bmatrix} a_{11} & a_{12} & \cdots & a_{1r} \\ a_{21} & a_{22} & \cdots & a_{2r} \\ \vdots & \vdots & & \vdots \\ a_{r1} & a_{r2} & \cdots & a_{rr} \end{bmatrix}$$

has nonzero determinant. Therefore, $\overline{\mathbf{A}}$ is invertible. Consequently, for any r-vector $\check{\mathbf{y}}$, we can find an r-vector $\check{\mathbf{x}}$ such that $\check{\mathbf{y}} = \overline{\mathbf{A}}\check{\mathbf{x}}$. We extend this $\check{\mathbf{x}}$ vector to an n-vector by appending zeros for components $r+1, r+2, \ldots, n$. Call this n-vector $\mathbf{x}$. Applying $\mathbf{A}$ to $\mathbf{x}$, we obtain $\mathbf{y} = \mathbf{A}\mathbf{x}$, in which the first r components of $\mathbf{y}$ form $\check{\mathbf{y}}$. We then have $0 \leq \mathbf{y}'\mathbf{y} = Q(\mathbf{x}) = \check{\mathbf{y}}'\mathbf{D}\check{\mathbf{y}}$, for an arbitrary r-vector $\check{\mathbf{y}}$. Also, if $\check{\mathbf{y}} \neq \mathbf{0}$, then neither is the extension $\mathbf{y}$, which implies that $\check{\mathbf{y}}'\mathbf{D}\check{\mathbf{y}} = Q = \mathbf{y}'\mathbf{y} > 0$. We conclude that D is positive definite.

7.20 Theorem: Let $\mathbf{A}$ be a $m \times n$ matrix of rank r. Then there exists a symmetric positive definite $r \times r$ matrix $\mathbf{D}$ such that, for all n-vectors $\mathbf{x}$, we have $\mathbf{x}'\mathbf{A}'\mathbf{A}\mathbf{x} = \mathbf{x}'\mathbf{A}_1'\mathbf{D}\mathbf{A}_1\mathbf{x}$, where $\mathbf{A}_1$ is an $r \times n$ matrix containing r independent rows from $\mathbf{A}$.

PROOF: In the discussion above, let $\mathbf{y} = \mathbf{A}\mathbf{x}$. $\mathbf{A}_1$ is then the r independent rows from $\mathbf{A}$ that identify the truncated vector $\check{\mathbf{y}}$. The development above assumed that these rows were the uppermost in $\mathbf{A}$, but the argument remains valid, with more complicated notation, if the r independent rows appear at arbitrary positions in the matrix. ∎

We can refine Theorem 7.20 to show that the sum of the squares of m linear combinations reduces to the sum of the squares of r different linear combinations, where r is the rank of the matrix that elaborates the original linear combinations. Specifically, we have the following theorem.

7.21 Theorem: Let $\mathbf{A}$ be a $m \times n$ matrix of rank r. Then there exists an $r \times n$ matrix $\mathbf{B}$, also of rank r, such that, for all n-vectors $\mathbf{x}$, we have $\mathbf{x}'\mathbf{A}'\mathbf{A}\mathbf{x} = \mathbf{x}'\mathbf{B}'\mathbf{B}\mathbf{x}$.

PROOF: From Theorem 7.20, we have $\mathbf{x}'\mathbf{A}'\mathbf{A}\mathbf{x} = \mathbf{x}'\mathbf{A}_1'\mathbf{D}\mathbf{A}_1\mathbf{x}$, where D is an $r \times r$ symmetric positive definite matrix and $\mathbf{A}_1$ consists of r independent rows from $\mathbf{A}$. Invoking Theorem 6.14, we obtain an invertible $\mathbf{G}$ such that $\mathbf{G}'\mathbf{D}\mathbf{G} = \mathbf{I}$, the $r \times r$ identity matrix. Let $\mathbf{B} = \mathbf{G}^{-1}\mathbf{A}_1$. If follows that

$$\mathbf{D} = (\mathbf{G}')^{-1}\mathbf{I}\mathbf{G}^{-1} = (\mathbf{G}^{-1})'\mathbf{G}^{-1}$$

$$\mathbf{A}_1'\mathbf{D}\mathbf{A}_1 = \mathbf{A}_1'(\mathbf{G}^{-1})'\mathbf{G}^{-1}\mathbf{A}_1 = (\mathbf{G}^{-1}\mathbf{A}_1)'(\mathbf{G}^{-1}\mathbf{A}_1) = \mathbf{B}'\mathbf{B}.$$

Consequently, $\mathbf{x}'\mathbf{A}'\mathbf{A}\mathbf{x} = \mathbf{x}'\mathbf{B}'\mathbf{B}\mathbf{x}$ for all n-vectors $\mathbf{x}$. It remains to show that the rank of $\mathbf{B}$ is r. We note that $\mathbf{A}_1$ has rank r and therefore has r independent columns. We assume that the first r columns are independent. (Otherwise, we interchange columns to bring the independent ones to the front. This process interchanges the corresponding columns in $\mathbf{B} = \mathbf{G}^{-1}\mathbf{A}_1$, which does not change its rank.) We can then write

$$\mathbf{B} = \mathbf{G}^{-1}[\mathbf{H}|\mathbf{K}] = [\mathbf{G}^{-1}\mathbf{H}|\mathbf{G}^{-1}\mathbf{K}],$$

where $\mathbf{H}$ contains the first r columns of $\mathbf{A}_1$. Since $\mathbf{G}^{-1}$ is invertible, it has nonzero determinant. Because $\mathbf{H}$ is an $r \times r$ submatrix with r independent columns, it also has nonzero determinant. Consequently, $\det \mathbf{G}^{-1}\mathbf{H} \neq 0$. Then $\mathbf{B}$ is an $r \times n$ matrix containing an $r \times r$ submatrix with nonzero determinant. We conclude that the rank of $\mathbf{B}$ is r. $\blacksquare$

Now suppose that we have n independent normal random variables with common variance 1. That is, $\mathbf{X}$ is multivariate normal with parameters $(\boldsymbol{\mu}, \mathbf{I})$, where $\mathbf{I}$ is the $n \times n$ identity matrix. Suppose further that we have the sum of squares of the $\mathbf{X}$ components as the sum of k quadratic forms in those same variables. That is,

$$Q = \sum_{i=1}^{n} X_i^2 = \mathbf{X}'\mathbf{X} = Q_1 + Q_2 + \ldots + Q_k,$$

where $Q_j = (\mathbf{A}_j\mathbf{X})'(\mathbf{A}_j\mathbf{X})$ with the rank of $\mathbf{A}_j$ equal to r_j. Each $\mathbf{A}_j$ is an $m_j \times n$ matrix, which necessitates $r_j \leq n$. Invoking Theorem 7.21, we have

$$Q_j = \mathbf{X}'\mathbf{B}_j'\mathbf{B}_j\mathbf{X}$$

$$Q = \mathbf{X}'\mathbf{I}\mathbf{X} = \sum_{j=1}^{k} \mathbf{X}'\mathbf{A}_j'\mathbf{A}_j\mathbf{X} = \sum_{j=1}^{k} \mathbf{X}'\mathbf{B}_j'\mathbf{B}_j\mathbf{X},$$

where each $\mathbf{B}_j$ is an $r_j \times n$ matrix of full rank r_j. Let $\mathbf{C}$ be the matrix obtained by stacking the $\mathbf{B}_j$. That is, $\mathbf{C}' = [\,\mathbf{B}_1' \ \ \mathbf{B}_2' \ \ \cdots \ \ \mathbf{B}_k'\,]$. Note that $\mathbf{C}$ has $\sum_{j=1}^{k} r_j$ rows and n columns, which implies that $\mathbf{C}'\mathbf{C}$ is a square $n \times n$ matrix. Writing $\mathbf{D} = \mathbf{C}'\mathbf{C}$, we have $Q = \mathbf{X}'\mathbf{X} = \mathbf{X}'\mathbf{C}'\mathbf{C}\mathbf{X} = \mathbf{X}'\mathbf{D}\mathbf{X}$. This relation holds for all $\mathbf{X}$ values, so we can use particular values to probe for the $\mathbf{D}$ components, which we denote by d_{ij}. Of course, $\mathbf{D} = \mathbf{C}'\mathbf{C}$ is symmetric, so $d_{ij} = d_{ji}$. If $X_i = 1$ and $X_j = 0$ for $j \neq i$, we have $1 = \mathbf{X}'\mathbf{X} = \mathbf{X}'\mathbf{D}\mathbf{X} = d_{ii}$. We now know that $\mathbf{D}$ has 1s on the diagonal. Continuing the probe, fix two distinct positions, $i \neq j$, and set $X_i = X_j = 1$ and $X_k = 0$ for $k \neq i, j$. This produces $2 = \mathbf{X}'\mathbf{X} = \mathbf{X}'\mathbf{D}\mathbf{X} = d_{ii} + d_{ji} + d_{ij} + d_{jj} = 2 + 2d_{ij}$, which implies that $d_{ij} = d_{ji} = 0$. We conclude that $\mathbf{D} = \mathbf{C}'\mathbf{C}$ is the $n \times n$ identity matrix. So the rank of $\mathbf{C}'\mathbf{C}$ is n.

What can we say about the rank of $\mathbf{C}$ itself? If $\mathbf{c}_{\cdot i}$ is the ith column of $\mathbf{C}$, then $d_{ij} = \mathbf{c}_{\cdot i}'\mathbf{c}_{\cdot j}$. Therefore,

$$\mathbf{c}_{\cdot i}'\mathbf{c}_{\cdot j} = \begin{cases} 1, & i = j \\ 0, & i \neq j. \end{cases}$$

Suppose that we have constants $t_1, t_2, \ldots, t_n$ such that $\sum_{i=0}^{n} t_i \mathbf{c}_{\cdot i} = \mathbf{0}$. Then, for any j,

$$0 = \mathbf{c}_{\cdot j}'\mathbf{0} = \mathbf{c}_{\cdot j}' \sum_{i=0}^{n} t_i \mathbf{c}_{\cdot i} = \sum_{i=1}^{n} t_i \mathbf{c}_{\cdot j}'\mathbf{c}_{\cdot i} = t_j.$$

Thus the n columns of $\mathbf{C}$ are independent, which means that the rank of $\mathbf{C}$ is n. This forces $\mathbf{C}$ to have n independent rows, which means that $r_1 + r_2 + \ldots + r_k \geq n$.

Moreover, if $r_1 + r_2 + \ldots + r_k$ is precisely equal to n, then $\mathbf{C}$ is a square $n \times n$ matrix. Then $\mathbf{C}'\mathbf{C} = \mathbf{I}$ means that $\mathbf{C}' = \mathbf{C}^{-1}$. We infer that $\mathbf{Y} = \mathbf{C}\mathbf{X}$ is multivariate normal with parameters $(C\mu, \mathbf{C}'\mathbf{I}\mathbf{C}) = (C\mu, \mathbf{I})$. That is, the components of $\mathbf{Y} = \mathbf{C}\mathbf{X}$ are independent normal random variables, all with variance 1. Recall that $\mathbf{C}$ is the stacked $\mathbf{B}_i$ matrices. Therefore, $\mathbf{Y}$ has $r_1 + r_2 + \ldots + r_k = n$ components, of which the first r_1 are $\mathbf{B}_1\mathbf{X}$, the second r_2 are $\mathbf{B}_2\mathbf{X}$, and so forth through the last r_k, which are $\mathbf{B}_k\mathbf{X}$. For the jth segment, consisting of r_j components from Y, the means ν_j are $\nu_j = \mathbf{B}_j\mu$. Consequently, the $Q_j = (\mathbf{B}_j\mathbf{X})'(\mathbf{B}_j\mathbf{X})$ are independent noncentral χ^2 random variables with r_j degrees of freedom and noncentrality parameters

$$\lambda_j = \frac{1}{2}\nu_j'\nu_j = \frac{1}{2}(\mathbf{B}_j\mu)'(\mathbf{B}_j\mu) = \frac{1}{2}(\mathbf{A}_j\mu)'(\mathbf{A}_j\mu).$$

We record this conclusion as the following theorem.

7.22 Theorem: Let $\mathbf{X}$ be an n-vector of independent random variables, multivariate normal with parameters $(\mu, \mathbf{I})$. Suppose further that $\mathbf{X}'\mathbf{X}$ is the sum of k quadratic forms as follows.

$$\mathbf{X}'\mathbf{X} = \sum_{i=1}^{n} X_i^2 = \sum_{i=1}^{k} Q_i$$

$$Q_i = (\mathbf{A}_i\mathbf{X})'(\mathbf{A}_i\mathbf{X})$$

$$\mathrm{rank}(\mathbf{A}_i) = r_i$$

Then $\sum_{i=1}^{k} r_i \geq n$. Moreover, if $\sum_{i=1}^{k} r_i = n$, then the Q_i are independent noncentral χ^2 random variables, with degrees of freedom r_i and noncentrality parameters $\lambda_i = (1/2)(\mathbf{A}_i\mu)'(\mathbf{A}_i\mu)$.

PROOF: See discussion above. ∎

Theorem 7.22 is a strengthened version of Cochran's theorem, which we state below.

7.23 Theorem: (Cochran) Suppose that $\mathbf{X}$ contains n independent standard normal random variables and $Q = \mathbf{X}'\mathbf{X} = \sum_{i=1}^{k} Q_j$, where $Q_j = (\mathbf{A}_j\mathbf{X})'(\mathbf{A}_j\mathbf{X})$ and A_j has rank r_j. Then $r_1 + r_2 + \ldots + r_k \geq n$. Furthermore, if $r_1 + r_2 + \ldots + r_k = n$, then the Q_j are have independent central χ^2 distributions with degrees of freedom r_j.

PROOF: The parameters of $\mathbf{X}$ are $(\mu = \mathbf{0}, \mathbf{\Sigma} = \mathbf{I})$. From the preceding theorem, the Q_j have independent noncentral χ^2 distributions with degrees of freedom r_j and noncentrality parameters

$$\lambda_j = \frac{1}{2}(\mathbf{A}_j\mu)'(\mathbf{A}_j\mu) = \frac{1}{2}(\mathbf{A}_j\mathbf{0})'(\mathbf{A}_j\mathbf{0}) = \frac{1}{2}\mathbf{0}'\mathbf{0} = 0.$$

The Q_j are therefore independent central χ^2 random variables with degrees of freedom r_j. ∎

This result is crucial for justifying the estimation argument of the preceding section, in which an unknown variance complicates the estimate for the mean of a normal population. The context for the argument is as follows. $\mathbf{X} = (X_1, X_2, \ldots, X_n)$ is a random sample from a population with unknown mean μ and unknown variance σ^2. We wish to use the sample mean, $\overline{X} = (1/n) \sum X_i$, to estimate μ. We write $\sum_{i=1}^{n}[(X_i - \mu)/\sigma]^2 = \sum_{i=1}^{n}[(X_i - \overline{X} + \overline{X} - \mu)/\sigma]^2$ and expand the expression to obtain two quadratic forms.

$$\sum_{i=1}^{n}\left(\frac{X_i - \mu}{\sigma}\right)^2 = \sum_{i=1}^{n}\left(\frac{X_i - \overline{X}}{\sigma}\right)^2 + n\left(\frac{\overline{X} - \mu}{\sigma}\right)^2 + 2\left(\frac{\overline{X} - \mu}{\sigma^2}\right)\sum_{i=1}^{n}(X_i - \overline{X})$$

$$= \sum_{i=1}^{n}\left(\frac{X_i - \overline{X}}{\sigma}\right)^2 + n\left(\frac{\overline{X} - \mu}{\sigma}\right)^2$$

Let $\mathbf{Z}$ be the vector with components $(X_i - \mu)/\sigma$. Then $\mathbf{Z}$ is a vector of independent standard normal random variables. That is, $\mathbf{Z}$ is multivariate normal with parameters $(\mathbf{0}, \mathbf{I})$. The left side of the equation above is then $\mathbf{Z}'\mathbf{Z}$. We rewrite the right side as quadratic forms in the $\mathbf{Z}$ components.

$$\frac{X_i - \overline{X}}{\sigma} = \frac{1}{\sigma}[(X_i - \mu) - (\overline{X} - \mu)] = \frac{1}{\sigma}\left[(X_i - \mu) - \frac{1}{n}\sum_{j=1}^{n}(X_j - \mu)\right]$$

$$= \left(\frac{-1}{n}\right)\frac{X_1 - \mu}{\sigma} + \ldots + \left(\frac{-1}{n}\right)\frac{X_{i-1} - \mu}{\sigma} + \left(1 - \frac{1}{n}\right)\frac{X_i - \mu}{\sigma}$$

$$+ \left(\frac{-1}{n}\right)\frac{X_{i+1} - \mu}{\sigma} + \ldots + \left(\frac{-1}{n}\right)\frac{X_n - \mu}{\sigma}$$

In matrix form, we have

$$\sum_{i=1}^{n}\left(\frac{X_i - \overline{X}}{\sigma}\right)^2 = (\mathbf{A}_1\mathbf{Z})'(\mathbf{A}_1\mathbf{Z}) = \mathbf{Z}'\mathbf{A}_1'\mathbf{A}_1\mathbf{Z} = Q_1,$$

where $\mathbf{A}_1 = \begin{bmatrix} 1 - 1/n & -1/n & -1/n & \ldots & -1/n \\ -1/n & 1 - 1/n & -1/n & \ldots & -1/n \\ \vdots & \vdots & \vdots & & \vdots \\ -1/n & -1/n & -1/n & \ldots & 1 - 1/n \end{bmatrix}.$

Also, $(\overline{X} - \mu)/\sigma = [1/(n\sigma)]\sum_{i=1}^{n}(X_i - \mu) = (1/n)\sum_{i=1}^{n}(X_i - \mu)/\sigma$, which implies that

$$n\left(\frac{\overline{X} - \mu}{\sigma}\right)^2 = \left(\sum_{i=1}^{n}\frac{1}{\sqrt{n}}\cdot\frac{X_i - \mu}{\sigma}\right)^2 = (\mathbf{A}_2\mathbf{Z})'(\mathbf{A}_2\mathbf{Z}) = \mathbf{Z}'\mathbf{A}_2'\mathbf{A}_2\mathbf{Z} = Q_2$$

where $\mathbf{A}_2 = \begin{bmatrix} 1/\sqrt{n} & 1/\sqrt{n} & 1/\sqrt{n} & \ldots & 1/\sqrt{n} \end{bmatrix}$. Because the sum of its rows is zero, $\mathbf{A}_1$ has a dependent row set. If q is the rank of $\mathbf{A}_1$, we have $q \leq n - 1$. Because $\mathbf{A}_2$ has a single row, its rank is evidently 1. We have $\mathbf{Z}'\mathbf{Z} = \mathbf{Z}'\mathbf{A}_1'\mathbf{A}\mathbf{Z} + \mathbf{Z}'\mathbf{A}_2'\mathbf{A}_2\mathbf{Z}$, and Cochran's theorem asserts that $q + 1 \geq n$, which implies that $q \geq n - 1$. Because we already have $q \leq n - 1$, we conclude that

$q = n - 1$. The sum of the ranks of A_1 and A_2 is then precisely n, so the second half of Cochran's theorem also applies. That is,

$$Q_1 = \frac{(n-1)\hat{s}_X^2}{\sigma^2} = \sum_{i=1}^{n}\left(\frac{X_i - \overline{X}}{\sigma}\right)^2 \qquad Q_2 = \frac{n(\overline{X} - \mu)^2}{\sigma^2}$$

are independent χ^2 random variables with $n - 1$ and 1 degrees of freedom, respectively. Consequently, statistic T below has a t distribution with $n - 1$ degrees of freedom.

$$T = \sqrt{\frac{n(\overline{X} - \mu)^2/\sigma^2}{\hat{s}_X^2/\sigma^2}} = \frac{\sqrt{n}(\overline{X} - \mu)}{\sqrt{\hat{s}_X^2}}$$

7.24 Theorem: Let $\mathbf{X} = (X_1, \ldots, X_n)$ be a sample from a normal population with unknown mean μ and unknown variance σ^2. Let $\overline{X}$ and $\hat{s}_X^2$ denote the sample mean and the unbiased sample variance. Then $Q_1 = (n-1)\hat{s}_X^2/\sigma^2$ and $Q_2 = n(\overline{X} - \mu)^2/\sigma^2$ are independent χ^2 random variables with $n - 1$ and 1 degrees of freedom, respectively. Moreover, $T = \sqrt{n}(\overline{X} - \mu)/\sqrt{\hat{s}_X^2}$ has a t distribution with $n - 1$ degrees of freedom.

PROOF: See discussion above. ∎

Exercises

7.16 Suppose that $\mathbf{A}$ is a 2×2 matrix having rank 1. Show that there exist constants a and b such that $\mathbf{x}'\mathbf{A}'\mathbf{A}\mathbf{x} = (ax_1 + bx_2)^2$ for all $\mathbf{x} \in \mathcal{R}^2$.

7.17 Verify that the matrix $\mathbf{A}$ to the right has rank 2. Find a 2×3 matrix $\mathbf{B}$ such that $\mathbf{x}'\mathbf{B}'\mathbf{B}\mathbf{x} = \mathbf{x}'\mathbf{A}'\mathbf{A}\mathbf{x}$ for all 3-vectors $\mathbf{x}$.
$$\begin{bmatrix} 1 & -4 & 2 \\ 0 & 1 & 3 \\ 2 & -7 & 7 \end{bmatrix}$$

*7.18 Suppose that X_{ij}, for $1 \le i \le m$ and $1 \le j \le n$, are independent random variables, each normally distributed with parameters (μ, σ^2). For $1 \le i \le m$, define $\overline{X}_i = (1/n)\sum_{j=1}^{n} X_{ij}$ and $\overline{X} = [1/(mn)]\sum_{i=1}^{m}\sum_{j=1}^{n} X_{ij}$. Show that

$$Q_1 = \sum_{i=1}^{m}\sum_{j=1}^{n}\left(\frac{X_{ij} - \overline{X}_i}{\sigma}\right)^2 \qquad Q_3 = mn\left(\frac{\overline{X} - \mu}{\sigma}\right)^2$$

$$Q_2 = n\sum_{i=1}^{m}\left(\frac{\overline{X}_i - \overline{X}}{\sigma}\right)^2$$

are independent χ^2 random variables with $m(n-1), m-1$, and 1 degrees of freedom, respectively.

7.4 Analysis of variance

Using the confidence interval approach of Section 7.2, we can test a null hypothesis concerning the means of two populations. Specifically, suppose that

X and Y are normally distributed with means μ_1 and μ_2, respectively. We consider a null hypothesis that the means are equal, and we endeavor to construct a test that maintains the probability of an erroneous rejection of the null hypothesis at 0.05. The difference $D = X - Y$ is normally distributed with mean $\mu_1 - \mu_2$, which is zero if the null hypothesis prevails. We take a sample of n pairs and calculate the pairwise differences to obtain a sample $D_1, D_2, \ldots, D_n$ from D. We let σ^2 be the variance of D, and we denote the sample mean and unbiased sample variance by $\overline{D}$ and $\hat{s}_D^2$ in the traditional manner. From Theorem 7.17,

$$Q_1 = \frac{n[\overline{D} - (\mu_1 - \mu_2)]^2}{\sigma^2} \qquad\qquad Q_2 = \frac{(n-1)\hat{s}_D^2}{\sigma^2}$$

are independent χ^2 random variables with degrees of freedom 1 and $n - 1$, respectively. If the null hypothesis is true, then $Q_1/[Q_2/(n-1)]$ has an F distribution with $(1, n-1)$ degrees of freedom. Consequently,

$$T = \sqrt{\frac{Q_1}{Q_2/(n-1)}} = \overline{D}\sqrt{\frac{n}{\hat{s}_D^2}}$$

has a t distribution with $n - 1$ degrees of freedom. There is then only a 5% chance that T lies outside the percentile interval $[t_{0.025;(n-1)}, t_{0.975;(n-1)}]$. If we reject the null hypothesis when T, as computed from the sample, lies outside this interval, we suffer an error probability of 0.05. Since $t_{0.025;(n-1)} = -t_{0.975;(n-1)}$, a T-value *inside* the interval is equivalent to

$$-t_{0.975;(n-1)} \leq \overline{D}\sqrt{n/\hat{s}_D^2} \leq t_{0.975;(n-1)}$$

$$-t_{0.975;(n-1)}\sqrt{\hat{s}_D^2/n} \leq -\overline{D} \leq t_{0.975;(n-1)}\sqrt{\hat{s}_D^2/n}.$$

Adding D across this inequality, we obtain an interval about zero:

$$\overline{D} - t_{0.975;(n-1)}\sqrt{\hat{s}_D^2/n} \leq 0 \leq \overline{D} + t_{0.975;(n-1)}\sqrt{\hat{s}_D^2/n}.$$

The last interval is the 95% confidence interval associated with $\overline{D}$ as an estimate of the mean $\mu_1 - \mu_2$. Our test, then, rejects the null hypothesis if the 95% confidence interval for the D mean excludes zero. Although we have maintained the probability of an erroneous rejection at 0.05, we have not calculated the probability of an erroneous acceptance. We defer this analysis for the moment because it follows from the more general results in this section.

   The general situation is that we have m normal populations, and we seek to test the null hypothesis that all their means are equal. Suppose, for example, that each of m database servers purports to answer queries with a low mean response time. Before subscribing to a particular service, we undertake to determine if there is indeed any difference in their mean response times. Of course, if we do conclude that differences exist, further testing may be necessary to identify the best service. Nevertheless, the first concern is to ascertain if there is sufficient difference to warrant a more serious study.

This situation occurs frequently in attempts to determine the efficacy of experimental drugs, and the terminology reflects this particular application. The typical test administers the drug to a homogeneous population of patients, all suffering from the same malady. A patient's dose contains some amount of the drug, say $0, 2, 4$, or 6 milligrams. The drug itself is called a *factor* because it claims to influence the patient's recovery. The dose amount is called a *level*. In this case, we have one factor with four levels. Four subpopulations receive treatments, each at a different level. Our initial goal is to ascertain if there is any difference in the mean recovery time across these subpopulations. If there is, then further testing is necessary to determine the optimal level. If there is no difference, the drug is dismissed as inappropriate for the particular malady in question.

To develop a general approach to this sort of problem, we suppose that there are m levels of some factor that influences the mean of a random population. We take data at each level and obtain an array of random variables. $X_{ij}, 1 \leq j \leq n_i$ are the n_i observations at level i. We assume that the X_{ij} are independent normally distributed random variables with parameters (μ_i, σ^2). That is, the mean may vary with the level, but the variance remains constant. This constant variance assumption is called a *homoskedastic constraint*. Let H_0 denote the null hypothesis, which states that the means μ_i are all equal. We want to test the null hypothesis against the alternative H_1, which asserts that there is a significant difference among the means. Let

$$n = \sum_{i=1}^{m} n_i \qquad\qquad \overline{X}_i = (1/n_i) \sum_{j=1}^{n_i} X_{ij}$$

$$\overline{\mu} = (1/n) \sum_{i=1}^{m} n_i \mu_i \qquad\qquad \overline{X} = (1/n) \sum_{i=1}^{m} \sum_{j=1}^{n_i} X_{ij}.$$

$\overline{\mu}$ is a weighted average of the μ_i. Under the null hypothesis, $\mu_i = \overline{\mu}$ for all i. For $1 \leq i \leq m$, $\overline{X}_i$ is the sample mean within level i. $\overline{X}$ is the sample mean computed across all n sample values. Now consider the following decomposition of a sum of squares of normal random variables: $Q = \sum_{i=1}^{m} \sum_{j=1}^{n_i} (X_{ij}/\sigma)^2$.

$$Q = \frac{1}{\sigma^2} \sum_{i=1}^{m} \sum_{j=1}^{n_i} (X_{ij} - \overline{X}_i + \overline{X}_i - \overline{X} + \overline{X})^2$$

$$= \frac{1}{\sigma^2} \left[\sum_{i=1}^{m} \sum_{j=1}^{n_i} (X_{ij} - \overline{X}_i)^2 + \sum_{i=1}^{m} \sum_{j=1}^{n_i} (\overline{X}_i - \overline{X})^2 + \sum_{i=1}^{m} \sum_{j=1}^{n_i} \overline{X}^2 \right.$$

$$+ 2 \sum_{i=1}^{m} \sum_{j=1}^{n_i} (X_{ij} - \overline{X}_i)(\overline{X}_i - \overline{X}) + 2 \sum_{i=1}^{m} \sum_{j=1}^{n_i} (X_{ij} - \overline{X}_i)\overline{X}$$

$$\left. + 2 \sum_{i=1}^{m} \sum_{j=1}^{n_i} (\overline{X}_i - \overline{X})\overline{X} \right] \qquad\qquad (7.2)$$

The cross-product terms all sum to zero. For example,

$$
C = \sum_{i=1}^{m} \sum_{j=1}^{n_i} (X_{ij} - \overline{X}_i)(\overline{X}_i - \overline{X}) = \sum_{i=1}^{m} (\overline{X}_i - \overline{X}) \sum_{j=1}^{n_i} (X_{ij} - \overline{X}_i)
$$

$$
= \sum_{i=1}^{m} (\overline{X}_i - \overline{X}) \left[\left(\sum_{j=1}^{n_i} X_{ij} \right) - n_i \overline{X}_i \right] = \sum_{i=1}^{m} (\overline{X}_i - \overline{X}) \left[n_i \overline{X}_i - n_i \overline{X}_i \right] = 0,
$$

and the other two cross-products admit similar analyses. This leaves the three sums of squares. In the second, the summand is independent of index j, while in the third, it is independent of both i and j. We then rewrite the expression as

$$
\sum_{i=1}^{m} \sum_{j=1}^{n_i} \left(\frac{X_{ij}}{\sigma} \right)^2 = \frac{1}{\sigma^2} \left[\sum_{i=1}^{m} \sum_{j=1}^{n_i} (X_{ij} - \overline{X}_i)^2 + \sum_{i=1}^{m} n_i (\overline{X}_i - \overline{X})^2 + n \overline{X}^2 \right]
$$

$$
= Q_1 + Q_2 + Q_3.
$$

We want to manipulate this expression into three quadratic forms with ranks summing precisely to n. This will establish the three sums on the right as independent χ^2 random variables and allow us to quantify an improbable sampling event that advocates rejecting the null hypothesis. Define $Z_{ij} = X_{ij}/\sigma$ and $\mathbf{Z}_i' = [Z_{i,1} \ \ Z_{i,2} \ \cdots \ Z_{i,n_i}]$. Each Z_{ij} now has mean μ_i/σ and variance 1. To obtain the proper form for Q_1, we rewrite $(X_{ij} - \overline{X}_i)/\sigma$ as follows.

$$
\frac{X_{ij} - \overline{X}_i}{\sigma} = \frac{X_{ij}}{\sigma} - \frac{1}{n_i} \sum_{k=1}^{n_i} \frac{X_{ik}}{\sigma} = Z_{ij} - \frac{1}{n_i} \sum_{k=1}^{n_i} Z_{ik}
$$

For each i, we then have $(1/\sigma^2) \sum_{j=1}^{n_i} (X_{ij} - \overline{X}_i)^2 = (\mathbf{A}_i \mathbf{Z}_i)'(\mathbf{A}_i \mathbf{Z}_i)$, where

$$
\mathbf{A}_i = \begin{bmatrix} 1 - 1/n_i & -1/n_i & \cdots & -1/n_i \\ -1/n_i & 1 - 1/n_i & \cdots & -1/n_i \\ \vdots & \vdots & & \vdots \\ -1/n_i & -1/n_i & \cdots & 1 - 1/n_i \end{bmatrix}
$$

is a $n_i \times n_i$ matrix. The row vectors of $\mathbf{A}_i$ sum to zero, which implies that $\mathbf{A}_i$ contains at most $n_i - 1$ independent rows. Now define $\mathbf{Z}$, an n-vector, and $\mathbf{B}$, a $n \times n$ matrix, by stacking the $\mathbf{Z}_i$ and $\mathbf{A}_i$ as follows.

$$
\mathbf{Z}' = [\mathbf{Z}_1' \ \mathbf{Z}_2' \ \cdots \ \mathbf{Z}_m'] \qquad \mathbf{B} = \begin{bmatrix} \mathbf{A}_1 & 0 & \cdots & 0 \\ 0 & \mathbf{A}_2 & \cdots & 0 \\ \vdots & \vdots & & \vdots \\ 0 & 0 & \cdots & \mathbf{A}_m \end{bmatrix}
$$

$\mathbf{Z}$ is multivariate normal with parameters $(\boldsymbol{\mu}, \mathbf{I})$, where $\boldsymbol{\mu}$ is an n-vector comprised of m segments. The first segment contains n_1 entries, all μ_1/σ. The second segment contains n_2 entries, all μ_2/σ, and so forth. On its diagonal, $\mathbf{B}$ contains square boxes of varying sizes, corresponding to the various n_i. The $\mathbf{0}$ entries are not necessarily square, but they indicate zero entries for all matrix elements outside the diagonal boxes. Because $\mathbf{B}$ contains at most $(n_1 - 1) + (n_2 - 1) + \ldots + (n_m - 1) = n - m$ independent rows, we conclude that $\text{rank}(\mathbf{B}) \leq n - m$. Moreover,

$$Q_1 = \frac{1}{\sigma^2} \sum_{i=1}^{m} \sum_{j=1}^{n_i} (X_{ij} - \overline{X}_i)^2 = \sum_{i=1}^{m} (\mathbf{A}_i \mathbf{Z})' (\mathbf{A}_i \mathbf{Z}_i) = (\mathbf{BZ})'(\mathbf{BZ}) = \mathbf{Z}'\mathbf{B}'\mathbf{BZ}.$$

We treat Q_2 in a similar manner.

$$Q_2 = \sum_{i=1}^{m} \left(\frac{\sqrt{n_i}(\overline{X}_i - \overline{X})}{\sigma} \right)^2$$

$$\frac{\sqrt{n_i}(\overline{X}_i - \overline{X})}{\sigma} = \frac{\sqrt{n_i}}{\sigma} \left(\frac{1}{n_i} \sum_{j=1}^{n_i} X_{ij} - \frac{1}{n} \sum_{k=1}^{m} \sum_{j=1}^{n_k} X_{kj} \right)$$

$$= \sqrt{n_i} \left[\sum_{j=1}^{n_i} \left(\frac{1}{n_i} - \frac{1}{n} \right) Z_{ij} - \sum_{k \neq i} \sum_{j=1}^{n_k} \left(\frac{1}{n} \right) Z_{kj} \right] = \mathbf{c}_i' \mathbf{Z},$$

where $\mathbf{c}_i'$ is a partitioned row vector of total length n. It is the concatenation of m smaller row vectors d_{ik}', $1 \leq k \leq m$. Each d_{ik}' is of length n_i and contains elements

$$d_{ik}' = \begin{cases} [-\sqrt{n_i}/n \quad -\sqrt{n_i}/n \quad \cdots \quad -\sqrt{n_i}/n], k \neq i \\ [\sqrt{n_i}(1/n_i - 1/n) \, \sqrt{n_i}(1/n_i - 1/n) \ldots \sqrt{n_i}(1/n_i - 1/n)], k = i. \end{cases}$$

That is, $\mathbf{c}_i' = [\, d_{i1}' \,|\, \cdots \,|\, d_{i,i-1}' \,|\, d_{ii}' \,|\, d_{i,i+1}' \,|\, \cdots \,|\, d_{i,m}' \,]$. The entries of the d_{ii}' components are all $\sqrt{n_i}[(1/n_i) - (1/n)]$; all the remaining entries are $(-\sqrt{n_i})/n$. We speak of the first n_1 components as the first *aggregated* component of $\mathbf{c}_i$. The next n_2 components are the second aggregated component, and so forth. We stack the $\mathbf{c}_i'$ rows to obtain the $m \times n$ matrix $\mathbf{C}$. That is, $\mathbf{C}' = [\, \mathbf{c}_1 \, \mathbf{c}_2 \, \cdots \, \mathbf{c}_m \,]$. Construct a matrix $\mathbf{D}$ by multiplying the ith row of $\mathbf{C}$ by $\sqrt{n_i}$. Let s_k denote a column sum in the kth aggregated component $\mathbf{D}$. We have

$$s_k = \sqrt{n_1} \left(\frac{-\sqrt{n_1}}{n} \right) + \ldots + \sqrt{n_{k-1}} \left(\frac{-\sqrt{n_{k-1}}}{n} \right) + \sqrt{n_k} \left[\sqrt{n_k} \left(\frac{1}{n_k} - \frac{1}{n} \right) \right]$$

$$+ \sqrt{n_{k+1}} \left(\frac{-\sqrt{n_{k+1}}}{n} \right) + \ldots + \sqrt{n_m} \left(\frac{-\sqrt{n_m}}{n} \right),$$

which evaluates to $1 - (1/n) \sum_{i=1}^{m} n_i = 0$. That is, we have a linear combination of the m rows of C equal to zero. This shows that the rank of C cannot exceed $m - 1$. Moreover,

$$Q_2 = \sum_{i=1}^{m} (\mathbf{c}_i' \mathbf{Z}_i)^2 = (\mathbf{CZ})'(\mathbf{CZ}) = \mathbf{Z}'\mathbf{C}'\mathbf{CZ}.$$

Finally, we address Q_3.

$$Q_3 = \frac{n\overline{X}^2}{\sigma^2} = n\left(\frac{1}{n}\sum_{i=1}^{m}\sum_{j=1}^{n_i}\frac{X_{ij}}{\sigma}\right)^2 = \left(\sum_{i=1}^{m}\sum_{j=1}^{n_i}\frac{1}{\sqrt{n}}Z_{ij}\right)^2 = \mathbf{Z'E'EZ},$$

where E is a $1 \times n$ matrix (row vector) with $1/\sqrt{n}$ in each component. Clearly, the rank of E is 1. We now have

$$\mathbf{Z'Z} = Q_1 + Q_2 + Q_3 = \mathbf{Z'B'BZ} + \mathbf{Z'C'CZ} + \mathbf{Z'E'EZ}.$$

Theorem 7.22 asserts that the sum of the ranks of $\mathbf{B}$, $\mathbf{C}$, and $\mathbf{E}$ must exceed n. However, we know that the sum of the ranks is at most $(n-m)+(m-1)+1 = n$. We conclude that the sum of the ranks is exactly n and invoke the second part of the theorem to obtain Q_1, Q_2, and Q_3 as independent noncentral χ^2 random variables with degrees of freedom $n - m, m - 1$, and 1, respectively. Let $\boldsymbol{\mu}_i$ be an n_i-vector containing μ_i/σ in each component. Note that

$$\mathbf{A}_i\boldsymbol{\mu}_i = \begin{bmatrix} 1-1/n_i & -1/n_i & \cdots & -1/n_i \\ -1/n_i & 1-1/n_i & \cdots & -1/n_i \\ \vdots & \vdots & & \vdots \\ -1/n_i & -1/n_i & \cdots & 1-1/n_i \end{bmatrix}\begin{bmatrix} \mu_i/\sigma \\ \mu_i/\sigma \\ \vdots \\ \mu_i/\sigma \end{bmatrix} = \frac{\mu_i}{\sigma}\begin{bmatrix} 0 \\ 0 \\ \vdots \\ 0 \end{bmatrix} = \begin{bmatrix} 0 \\ 0 \\ \vdots \\ 0 \end{bmatrix}.$$

The noncentrality parameters λ_i of the Q_i are then as follows.

$$\lambda_1 = \frac{1}{2}(\mathbf{B}\boldsymbol{\mu})'(\mathbf{B}\boldsymbol{\mu}) = \frac{1}{2}\sum_{i=1}^{m}(\mathbf{A}_i\boldsymbol{\mu}_i)'(\mathbf{A}_i\boldsymbol{\mu}_i) = \frac{1}{2}\sum_{i=1}^{m}\mathbf{0'0} = 0$$

$$\lambda_2 = \frac{1}{2}(\mathbf{C}\boldsymbol{\mu})'(\mathbf{C}\boldsymbol{\mu}) = \sum_{i=1}^{m}(\mathbf{c}_i'\boldsymbol{\mu}_i)^2$$

$$= \frac{1}{2}\sum_{i=1}^{m}\left[n_1\left(\frac{-\sqrt{n_i}}{n}\right)\left(\frac{\mu_1}{\sigma}\right) + n_2\left(\frac{-\sqrt{n_i}}{n}\right)\left(\frac{\mu_2}{\sigma}\right)\right.$$

$$+ \ldots + n_i\sqrt{n_i}\left(\frac{1}{n_i} - \frac{1}{n}\right)\left(\frac{\mu_i}{\sigma}\right) + n_{i+1}\left(\frac{-\sqrt{n_i}}{n}\right)\left(\frac{\mu_{i+1}}{\sigma}\right)$$

$$+ \ldots + n_m\left.\left(\frac{-\sqrt{n_i}}{n}\right)\left(\frac{\mu_m}{\sigma}\right)\right]^2$$

$$= \frac{1}{2}\sum_{i=1}^{m}\left[\sqrt{n_i}\frac{\mu_i}{\sigma} - \frac{\sqrt{n_i}}{n\sigma}\sum_{k=1}^{m}n_k\mu_k\right]^2 = \frac{1}{2\sigma^2}\sum_{i=1}^{m}n_i(\mu_i - \overline{\mu})^2$$

$$\lambda_3 = \frac{1}{2}(\mathbf{E}\boldsymbol{\mu})'(\mathbf{E}\boldsymbol{\mu}) = \frac{1}{2}\left(\frac{1}{\sqrt{n}}\sum_{i=1}^{m}\frac{n_i\mu_i}{\sigma}\right)^2 = \frac{1}{2n\sigma^2}\left(\sum_{i=1}^{m}n_i\mu_i\right)^2 = \frac{n\overline{\mu}^2}{2\sigma^2}$$

Under either H_0 or H_1, Q_1 has a central χ^2 distribution with $n - m$ degrees of freedom. On the other hand, Q_2 has a central χ^2 distribution under H_0

but a noncentral distribution under H_1. In either case, the degrees of freedom are $m - 1$. The following theorem summarizes the discussion to this point.

7.25 Theorem: Let X_{ij}, for $1 \leq i \leq m, 1 \leq j \leq n_i$, be independent normal random variables with common variance σ^2. Within the subset having a common i index, the means are equal. That is, mean$(X_{ij}) = \mu_i$ for all j. Define

$$n = \sum_{i=1}^{m} n_i \qquad\qquad \overline{X} = (1/n)\sum_{i=1}^{m}\sum_{j=1}^{n_i} X_{ij}$$
$$\overline{X}_i = (1/n_i)\sum_{j=1}^{n_i} X_{ij} \qquad\qquad \overline{\mu} = (1/n)\sum_{i=1}^{m} n_i\mu_i.$$

Then the quantity

$$R = \frac{\dfrac{1}{m-1}\sum_{i=1}^{m} n_i(\overline{X}_i - \overline{X})^2}{\dfrac{1}{n-m}\sum_{i=1}^{m}\sum_{j=1}^{n_i}(X_{ij} - \overline{X}_i)^2}$$

has a noncentral F distribution with $(m - 1, n - m)$ degrees of freedom and noncentrality parameter $\lambda = [1/(2\sigma^2)]\sum_{i=1}^{m} n_i(\mu_i - \overline{\mu})^2$.

PROOF: Setting $R = [Q_2/(m-1)]/[Q_1/(n-m)]$, the result follows from the discussion above. ∎

This theorem proves very useful for certain statistical questions. Before applying the theorem, however, we review some terminology from Section 4.4, established in the context of discrete hypothesis testing. Traditionally, H_0 and H_1 denote the null hypothesis and its alternative. A type I error is the incorrect rejection of the null hypothesis. For a given test, α denotes the probability of a type I error and is called the significance level of the test. A type II error is the incorrect acceptance of the null hypothesis. For a given test, β denotes the probability of a type II error. Each test uses some statistic S, in the sense that the S value computed from a sample determines acceptance or rejection of the null hypothesis. Consequently, we identify a test with the critical region of its statistic, which contains those S values that reject the null hypothesis. If the critical region is C_S, then $\alpha = \Pr(C_S|H_0)$ and $\beta = 1 - \Pr(C_S|H_1)$. In general, composite H_0 and H_1 each contain several states of nature θ, which means that α and β can vary with the particular state of nature that prevails within the hypothesis. The graph of $\Pr(C_S|\theta)$ versus θ is the power curve of the test. For a good test, of course, the power curve is low for $\theta \in H_0$ and high for $\theta \in H_1$. The next example illustrates these concepts.

7.26 Definition: An *analysis of variance* is a statistical test for a significant difference among the means of several subpopulations. It rests on a decomposition of the sums of squares that form the sample variance within subpopulations and across the entire population. A ratio statistic involving these sums of squares provides an accessible distribution, a noncentral F distribution, that enables calculation of the type I and type II error probabilities. ∎

7.27 Example: Suppose that three database servers compete for our business. Each purports to have the smallest mean response time, averaged over a query mix particular to our activity. We wish to construct a statistical test for the null hypothesis H_0, which states that there is no significant difference among the services. We assume that the response times of all services are normally distributed with the same variance σ^2. The means μ_i may or may not differ, and the test is constructed to answer that question. We use a sample of size 30, with 10 observations from each service. Let $X_{ij}, 1 \le i \le 3, 1 \le j \le 10$, be the corresponding random variables. X_{ij} is the response time of the jth observation from service i. We denote the actual observed values by x_{ij}. We set the significance level α, the probability of a type I error, at 0.05. In the notation of Theorem 7.25, we have

$$m = 3 \qquad\qquad \overline{X}_i = (1/n_i) \sum_{j=1}^{n_i} X_{ij} = (1/10) \sum_{j=1}^{10} X_{ij}$$
$$n_1 = n_2 = n_3 = 10 \qquad \overline{X} = (1/n) \sum_{i=1}^{m} \sum_{j=1}^{n_i} X_{ij} = (1/30) \sum_{i=1}^{3} \sum_{j=1}^{10} X_{ij}$$
$$n = \sum_{i=1}^{m} n_i = 30 \qquad \overline{\mu} = (1/n) \sum_{i=1}^{m} n_i \mu_i = (1/3) \sum_{i=1}^{3} \mu_i.$$

The quantity

$$R = \frac{\dfrac{1}{m-1} \sum_{i=1}^{m} n_i (\overline{X}_i - \overline{X})^2}{\dfrac{1}{n-m} \sum_{i=1}^{m} \sum_{j=1}^{n_i} (X_{ij} - \overline{X}_i)^2} = \frac{135 \sum_{i=1}^{3} (\overline{X}_i - \overline{X})^2}{\sum_{i=1}^{3} \sum_{j=1}^{10} (X_{ij} - \overline{X})^2}$$

then has a noncentral F distribution with $(m-1, n-m) = (2, 27)$ degrees of freedom and noncentrality parameter

$$\lambda = \frac{1}{2\sigma^2} \sum_{i=1}^{m} n_i (\mu_i - \overline{\mu})^2 = \frac{5}{\sigma^2} \sum_{i=1}^{3} (\mu_i - \overline{\mu})^2.$$

From Theorem 6.68, we compute the mean of R:

$$\mu_R = \frac{n-m}{n-m-2} \left(1 + \frac{2\lambda}{m-1}\right) = 1.08(1 + \lambda) = 1.08 \left(1 + \frac{5}{\sigma^2} \sum_{i=1}^{3} (\mu_i - \overline{\mu})^2 \right).$$

Because this mean shifts to the right with increasing differences among the μ_i, we associate larger R values with increasing evidence for rejecting the null hypothesis. If the null hypothesis is true, then $\lambda = 0$ and R has a central F distribution with $(2, 27)$ degrees of freedom. To achieve the desired $\alpha = 0.05$, we choose the region $(R > F_{0.95;(2,27)})$ as the critical region of our test. That is, we reject the null hypothesis if R is greater than the 95th percentile of the central F distribution with $(2, 27)$ degrees of freedom. From the table in Appendix B, we find that $F_{0.95;(2,27)} = 3.354$. We now take the data and obtain the following table.

Response times (milliseconds): x_{ij}									
$j = 1$	2	3	4	5	6	7	8	9	10
$i = 1$ 732.99	625.37	515.02	570.32	360.70	839.23	511.81	859.41	552.81	674.33
$i = 2$ 10.77	159.77	549.20	172.20	415.47	227.99	272.88	30.13	45.91	496.91
$i = 3$ 341.02	190.21	196.97	522.35	338.75	452.70	429.34	501.24	430.10	81.72

We compute the various approximations of the means and the corresponding R value.

$$\overline{x}_1 = (1/10) \sum_{j=1}^{10} x_{1j} = 624.20 \qquad\qquad \overline{x}_3 = (1/10) \sum_{j=1}^{10} x_{3j} = 348.44$$
$$\overline{x}_2 = (1/10) \sum_{j=1}^{10} x_{2j} = 238.12 \qquad\qquad \overline{x} = (1/30) \sum_{i=1}^{3} \sum_{j=1}^{10} x_{ij} = 403.59$$

$$R = \frac{135 \sum_{i=1}^{3} (\overline{x}_i - \overline{x})^2}{\sum_{i=1}^{3} \sum_{j=1}^{10} (x_{ij} - \overline{x})^2} = 14.17$$

Observing an R value greater than 3.354, we reject the null hypothesis and conclude that the services do have different mean response times. Further testing is necessary to identify the best service. Because we reject the null hypothesis, a type II error is not possible, and the chance of a type I error is 0.05 or less.

In truth, a normal random number generator constructed the data. For parameters, it used $\sigma = 200$, $\mu_1 = 500$, $\mu_2 = 300$, and $\mu_3 = 350$. The analysis of variance test reached the correct conclusion.

Now, however, consider new data, again constructed with a normal generator, but with parameters $\sigma = 200$, $\mu_1 = 500$, $\mu_2 = 540$, and $\mu_3 = 460$. The subpopulation means are now closer together. Will the analysis conclude that they are different? The new data are as follows.

Response times (milliseconds): x_{ij}									
$j = 1$	2	3	4	5	6	7	8	9	10
$i = 1$ 575.01	725.03	645.73	24.51	445.24	435.41	563.60	397.77	499.59	821.30
$i = 2$ 709.53	593.62	355.30	525.90	569.58	428.58	472.66	623.05	851.56	51.14
$i = 3$ 240.36	684.53	576.33	405.73	542.84	264.44	255.71	523.54	763.22	609.89

We again compute the various approximations of the means and the corresponding R value.

$$\bar{x}_1 = (1/10)\sum_{j=1}^{10} x_{1j} = 513.32 \qquad \bar{x}_3 = (1/10)\sum_{j=1}^{10} x_{3j} = 486.66$$
$$\bar{x}_2 = (1/10)\sum_{j=1}^{10} x_{2j} = 518.09 \qquad \bar{x} = (1/30)\sum_{i=1}^{3}\sum_{j=1}^{10} x_{ij} = 506.02$$

$$R = \frac{135\sum_{i=1}^{3}(\bar{x}_i - \bar{x})^2}{\sum_{i=1}^{3}\sum_{j=1}^{10}(x_{ij} - \bar{x})^2} = 0.07$$

The critical region remains ($R > 3.354$). Because the calculated R value is well outside the critical region, we do not reject the null hypothesis at the predetermined 5% significance level. Because we secretly concocted the data from known normal distributions, we know that there is indeed a difference in the means. Therefore, the analysis of variance has reached an erroneous conclusion; it suffers a type II error.

In general, we do not know the parameters of the competing distributions, so we cannot compute the probability of a type II error. In this artificially constructed case, however, we do know the parameters, and we can calculate the noncentrality factor for the R distribution.

$$\bar{\mu} = \frac{500 + 540 + 460}{3} = 500$$

$$\lambda = \frac{5}{\sigma^2}\sum_{i=1}^{3}(\mu_i - \bar{\mu})^2 = \frac{5}{(200)^2}[(500 - 500)^2 + (540 - 500)^2 + (460 - 500)^2] = 0.4$$

The noncentral F density appears as an ungainly expression in Theorem 6.68. We use a numerical integration routine to compute $\Pr(R \leq 3.354|\lambda = 0.4) = 0.8921$. Consequently, β, the probability of a type II error, is 0.8921 when $\lambda = 0.4$. This is a high probability of erroneously accepting the null hypothesis. Therefore, we cannot be overly critical of the analysis of variance technique because it reached an erroneous conclusion in this case. The small difference among the means resulted in a high probability of an erroneous acceptance of the null hypothesis. Indeed, we can plot the power curve of the test by calculating $\Pr(R > 3.354|\lambda)$ as a function

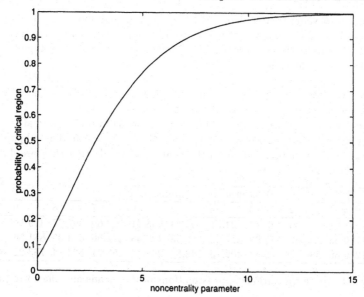

Figure 7.2. Power curve for the F test (see Example 7.27)

of λ. Figure 7.2 shows the result. At $\lambda = 0$, the curve is at height 0.05, which is α, the probability of a type I error. For positive values of λ, the difference between the curve height and 1 is β, the probability of a type II error. The noncentrality parameter must approach 10 before β becomes comparable to the 0.05 value that we stipulated for α.

The composite alternative H_1 is composed of all the nonzero noncentrality parameters. In general, when H_1 prevails, we do not know the exact noncentrality parameter. Consequently, in a test of H_0, equal means, versus H_1, unequal means, we draw no conclusion when the sample statistic falls outside the critical region. We do not reject the null hypothesis, and because we cannot quantify the error probability, we do not accept it either. We simply state that the test is inconclusive or that the data are not statistically significant. $\square$

The example shows how the probability of a type II error varies with the noncentrality parameter. As the noncentrality parameter increases, the probability of a type II error from a fixed sample decreases. Unfortunately, the noncentrality parameter depends on the unknown common variance of the subpopulations. However, this is not a serious obstacle because a significant difference among the means must be associated with a reference. If we adopt an expression such as $[\sum n_i(\mu_i - \overline{\mu})^2]/n$ as a measure of how much variation exists among the means, then a value of 10, say, may represent a significant difference, if the variance is small, but it may be insignificant if the variance is large. The quantity $\sqrt{\sum (n_i/n)(\mu_i - \overline{\mu})^2}$ is a weighted average of the squared deviations $(\mu_i - \overline{\mu})^2$, and it has the same units as the sample observations. In the example above, these units are milliseconds. When this quantity is comparable to σ, we conclude that the difference among the means is significant.

Consequently, we index the possibilities within H_1 by the quantity

$$\delta = \frac{1}{\sigma}\sqrt{\frac{1}{n}\sum_{i=1}^{m}n_i(\mu_i - \overline{\mu})^2}.$$

Clearly, $\lambda = n\delta^2/2$, so this approach is equivalent to indexing the possibilities with λ. However, we can interpret δ as the difference among the means, expressed in σ units. If we have $\delta = 1$, for example, we say that the difference among the means is comparable to one standard deviation. Typically, such a difference is significant, whereas $\delta = 0.1$ is not. In a particular application, we specify a δ value that we consider significant and adjust the sample size to ensure that the probability of a type II error is small when the actual difference among the means corresponds to that δ. We continue the preceding example to illustrate the point.

7.28 Example: We continue with the three database servers of the preceding example. Again, the null hypothesis is H_0, which states that all three services yield the same mean response time. We again assume that the response times are normally distributed with common variance σ^2, and we again set the significance level α, the probability of a type I error, at 0.05. Now, however, we are free to adjust the sample size to gain some control over β, the probability of a type II error. Specifically, if we do not reject the null hypothesis, we want a mean difference of $\delta = 0.5$ to have a 5% or less probability of occurrence. Larger differences among the means will, of course, have even smaller probabilities of occurrence. Smaller differences among the means will have larger probabilities of occurrence, which means larger probabilities of type II errors when we accept the null hypothesis. In effect, we are establishing a transition zone, $0 < \delta < 0.5$, in which we acknowledge larger probabilities of erroneously accepting the null hypothesis. Within that zone, however, we have decided that differences among the means are not large, and therefore we do not object to an erroneous acceptance.

For convenience, we decide to take equal-sized samples from the subpopulations. That is, $n_1 = n_2 = n_3$. For each prospective total sample size $n = n_1+n_2+n_3$, we use $\alpha = 0.05$ to determine the critical region. It is the region $(R > F_{0.95;(2,n-3)})$ to the right of the 95th percentile of the central F distribution with $(2, n-3)$ degrees of freedom. When $\delta = 0.5$, we have $\lambda = n(0.5)^2/2 = 0.125n$. Under this condition, the probability of a type II error is the probability that $G \leq F_{0.95;(2,n-3)}$, where G has a noncentral F distribution with $(2, n-3)$ degrees of freedom and noncentrality parameter $0.125n$. We integrate the noncentral density for various n values to obtain the following table. The percentiles $F_{0.95;(2,n-3)}$ are interpolations in Table 4 of Appendix B.

n	6	12	18	24	30	36	42	48	54	60	66
$F_{0.95;(2,n-3)}$	9.55	4.26	3.68	3.47	3.35	3.29	3.24	3.21	3.18	3.16	3.14
$\beta = \Pr(G \leq F_{0.95;(2,n-3)})$	0.90	0.76	0.61	0.48	0.36	0.27	0.20	0.14	0.10	0.07	0.05

We see that β achieves the required 0.05 value when $n = 66$. Consequently, we set the total sample size at 66, and we take 22 observations from each of the three services. The critical region is $(R > F_{0.95;(2,63)}) = (R > 3.14)$. If we reject the null hypothesis, our probability of a type I error will be 0.05 or less. If we accept the null hypothesis *and* there is indeed a difference among the means sufficient to produce $\delta = 0.5$ or more, then our probability of a type II error will be 0.05 or less.

Suppose that we now observe the following data.

i	$x_{ij}, 1 \le j \le 22$									
1	398.46	677.06	450.38	354.75	410.99	377.42	458.17	612.43	287.22	570.32	726.60
	530.00	640.63	489.52	903.70	684.83	137.18	506.99	138.43	705.64	578.92	627.88
2	796.84	972.48	557.99	594.52	745.15	817.58	398.93	512.00	629.98	125.43	853.73
	416.74	852.70	464.71	748.96	786.08	586.79	734.49	596.51	732.83	402.53	475.74
3	658.95	253.96	425.43	60.63	297.70	223.86	325.46	573.30	573.56	612.00	409.86
	477.90	166.92	287.85	632.08	557.74	465.74	128.53	442.93	456.01	296.97	436.46

We calculate the subpopulation means and the overall sample mean to obtain $\overline{x}_1 = 512.16$, $\overline{x}_2 = 627.40$, $\overline{x}_3 = 398.36$, and $\overline{x} = 512.64$, which combine to produce $R = 8.33$. Consequently, we reject the null hypothesis.

In this example, we performed numerical integrations of the noncentral F densities to obtain a trial-and-error appreciation of the relationship between sample size and β. This relationship is also available in tables. Table 7.1 gives the power function of the F test for $\alpha = 0.05, m = 3$. The table's horizontal dimension corresponds to various δ values; the vertical dimension corresponds to the number of observations from each subpopulation. Under the column $\delta = 0.5$, we scan down until the power exceeds 0.95. We arrive at line 22, which corresponds to a total sample size of $3 \times 22 = 66$. This agrees with the result obtained by direct integration of the noncentral densities. More extensive power tables appear in Appendix B.

For the situation analyzed above, a normal random number generator produced the data using parameters $\sigma = 200$, $\mu_1 = 500$, $\mu_2 = 623$, and $\mu_3 = 377$. These means give $\overline{\mu} = 500$ and combine to give

$$\delta = \frac{1}{200}\sqrt{\frac{1}{3}[(500 - 500)^2 + (623 - 500)^2 + (377 - 500)^2]} = 0.5,$$

so we are satisfied with the rejection. Suppose, however, that we construct data corresponding to $\delta = 0.1$. It is then quite likely that we will erroneously accept the null hypothesis. The normal generator, now using $\sigma = 200$, $\mu_1 = 500$, $\mu_2 = 500 + 24$, and $\mu_3 = 500 - 24$, produced the following table. The parameters combine to give a small $\delta = 0.1$, which reflects the much smaller variation among the means.

i	$x_{ij}, 1 \le j \le 22$									
1	732.99	625.37	515.02	570.32	360.70	839.23	511.81	859.41	552.81	674.33	210.77
	359.77	749.20	372.20	615.47	427.99	472.88	230.13	245.91	696.91	491.02	340.21
2	370.97	696.35	512.75	626.70	603.34	675.24	604.10	255.72	599.01	749.03	669.73
	48.51	469.24	459.41	587.60	421.77	523.59	845.30	693.53	577.62	339.30	509.90
3	505.58	364.58	408.66	559.05	787.56	12.86	256.36	700.53	592.33	421.73	558.84
	280.44	271.71	539.54	779.22	625.89	374.46	653.06	426.38	330.75	386.99	353.42

From these data, we calculate $R = 0.98$, which is less than the threshold 3.14. Consequently, we accept the null hypothesis and suffer a type II error. □

Figure 7.3 states the general algorithm for a one-factor analysis of variance. The algorithm's inputs are the error probabilities, α and β, and the δ value associated with a significant difference among the means.

n	0.100	0.200	0.300	0.400	0.500	0.600	0.700	0.800	0.900	1.000
2	0.052	0.058	0.068	0.082	0.100	0.122	0.149	0.180	0.214	0.253
3	0.054	0.067	0.090	0.124	0.168	0.225	0.292	0.368	0.450	0.534
4	0.057	0.077	0.114	0.168	0.241	0.331	0.434	0.542	0.648	0.743
5	0.059	0.087	0.138	0.214	0.314	0.433	0.561	0.683	0.788	0.870
6	0.061	0.097	0.163	0.260	0.386	0.528	0.668	0.788	0.879	0.938
7	0.064	0.108	0.188	0.307	0.455	0.613	0.754	0.863	0.933	0.972
8	0.066	0.118	0.214	0.353	0.520	0.686	0.822	0.913	0.964	0.988
9	0.068	0.129	0.240	0.398	0.580	0.748	0.873	0.947	0.982	0.995
10	0.071	0.139	0.266	0.442	0.635	0.800	0.911	0.968	0.991	0.998
11	0.073	0.150	0.292	0.485	0.685	0.843	0.938	0.981	0.995	0.999
12	0.076	0.161	0.318	0.525	0.729	0.878	0.957	0.989	0.998	1.000
13	0.078	0.172	0.343	0.564	0.769	0.906	0.971	0.993	0.999	1.000
14	0.080	0.184	0.369	0.601	0.803	0.928	0.981	0.996	0.999	1.000
15	0.083	0.195	0.394	0.635	0.834	0.945	0.987	0.998	1.000	1.000
16	0.085	0.206	0.419	0.667	0.860	0.958	0.991	0.999	1.000	1.000
17	0.088	0.218	0.444	0.698	0.882	0.969	0.994	0.999	1.000	1.000
18	0.090	0.229	0.468	0.726	0.902	0.977	0.996	1.000	1.000	1.000
19	0.093	0.241	0.491	0.752	0.918	0.983	0.998	1.000	1.000	1.000
20	0.095	0.252	0.514	0.776	0.932	0.987	0.999	1.000	1.000	1.000
21	0.098	0.264	0.536	0.798	0.944	0.990	0.999	1.000	1.000	1.000
22	0.100	0.275	0.558	0.818	0.953	0.993	0.999	1.000	1.000	1.000
23	0.103	0.287	0.579	0.837	0.962	0.995	1.000	1.000	1.000	1.000
24	0.106	0.298	0.599	0.853	0.969	0.996	1.000	1.000	1.000	1.000
25	0.108	0.310	0.619	0.869	0.974	0.997	1.000	1.000	1.000	1.000
26	0.111	0.321	0.638	0.883	0.979	0.998	1.000	1.000	1.000	1.000
27	0.113	0.333	0.656	0.895	0.983	0.999	1.000	1.000	1.000	1.000
28	0.116	0.344	0.674	0.907	0.986	0.999	1.000	1.000	1.000	1.000
29	0.119	0.356	0.691	0.917	0.989	0.999	1.000	1.000	1.000	1.000
30	0.121	0.367	0.707	0.926	0.991	0.999	1.000	1.000	1.000	1.000

TABLE 7.1. Power function of one-factor analysis of variance for $\alpha = 0.05, m = 3$ (see Example 7.28)

We can extend the one-factor analysis of variance to accommodate two or more factors. The technique involves a more extensive decomposition of the sum of squares appearing in the sample variance. We illustrate the approach with two factors. First, note that the one-factor model assumes that each observation is a constant, depending on the subpopulation, plus a random effect. That is, if i indexes the subpopulations and j indexes the repeated observations within each subpopulation, then

$$X_{ij} = \overline{\mu} + (\mu_i - \overline{\mu}) + (X_{ij} - \mu_i) = \overline{\mu} + \alpha_i + N_{ij},$$

where the N_{ij} are independent and normally distributed with zero mean and with common variance σ^2. The null hypothesis then states that the α_i are all zero.

With two factors, the observations are X_{ijk}, where $1 \leq i \leq p$ indexes

1. Assume there are m normal subpopulations with identical variances and potentially different means. The null hypothesis H_0 is that all the means are equal. The alternative H_1 is that the means are not all equal.

2. The samples from the subpopulations will be equal size. That is, $n_1 = n_2 = \ldots = n_m = n$. The total sample size is then mn.

3. From a table of F percentiles, determine the threshold $t = F_{1-\alpha;(m-1,mn-m)}$, the $100(1-\alpha)$th percentile of the central F distribution with $(m-1, mn-m)$ degrees of freedom.

4. Tabulate, for various n, the probability of $(G_n \leq t)$, where G_n has a noncentral F distribution with $(m-1, mn-m)$ degrees of freedom and noncentrality parameter $n\delta^2/2$. These probabilities can be obtained by numerically integrating the appropriate noncentral F density function, which appears in Theorem 6.68. Curves for approximating this calculation are available in the literature. Choose n such that $\Pr(G \leq t) < \beta$.

5. Take the sample. Let $x_{ij}, 1 \leq j \leq n$, be the observations at level i.

6. Compute $\bar{x}_i = (1/n) \sum_{j=1}^{n} x_{ij}$
$$\bar{x} = [1/(mn)] \sum_{i=1}^{m} \sum_{j=1}^{n} x_{ij}$$
$$R = \frac{[n/(m-1)] \sum_{i=1}^{m} (\bar{x}_i - \bar{x})^2}{[1/(mn-m)] \sum_{i=1}^{m} \sum_{j=1}^{n} (x_{ij} - \bar{x}_i)^2}.$$

7. The critical region is $(R > t)$. If $R > t$, reject the null hypothesis. The probability of a type I error is α or less. If $R \leq t$, accept the null hypothesis. In the latter case, the probability of a type II error is β or less, provided that the actual means are such that

$$\sqrt{\frac{1}{m} \sum_{i=1}^{m} (\mu_i - \bar{\mu})^2} \geq \delta\sigma.$$

Figure 7.3. Algorithm for one-factor analysis of variance

---

the various levels of factor A and $1 \leq j \leq q$ indexes the levels of factor B. For each fixed (i, j) pair, k indexes the observations taken at levels i and j of the two factors. In general, the number of observations may be different for each (i, j) pair, and therefore $1 \leq k \leq n_{ij}$ denotes the span of k. However, we will consider only the case in which each (i, j) pair has the same number of observations. That is, $n_{ij} = m$ for all i, j. The X_{ijk} are independent and normally distributed with a common variance σ^2. The means, μ_{ij}, are potentially different. In the spirit of the one-factor analysis, we make the following definitions.

$$n_{i\cdot} = \sum_{j=1}^{q} n_{ij} = mq \qquad \mu_{i\cdot} = (1/n_{i\cdot}) \sum_{j=1}^{q} n_{ij}\mu_{ij} = (1/q) \sum_{j=1}^{q} \mu_{ij}$$
$$n_{\cdot j} = \sum_{i=1}^{p} n_{ij} = mp \qquad \mu_{\cdot j} = (1/n_{\cdot j}) \sum_{i=1}^{p} n_{ij}\mu_{ij} = (1/p) \sum_{i=1}^{p} \mu_{ij}$$

The n-expressions are the total observations at level i of factor A and the total observations at level j of factor B. Under the current assumption that $n_{ij} = m$ for all i, j, these expressions are independent of i and j. The μ-expressions are the weighted average of the means at level i of factor A and the weighted average of the means at level j of factor B. In the interests of simpler notation, we have dropped the traditional overline that typically denotes an average. Instead, the dot indicates a sum or an average over the index that would appear in the dot's location.

We also have expressions for the total number of observations in the entire experiment and the overall mean.

$$n = \sum_{i=1}^{p}\sum_{j=1}^{q} n_{ij} = \sum_{i=1}^{p} n_{i\cdot} = \sum_{j=1}^{q} n_{\cdot j} = mpq$$

$$\mu = \frac{1}{n}\sum_{i=1}^{p}\sum_{j=1}^{q} n_{ij}\mu_{ij} = \frac{1}{pq}\sum_{i=1}^{p}\sum_{j=1}^{q}\mu_{ij} = \frac{1}{q}\sum_{j=1}^{q}\mu_{\cdot j} = \frac{1}{p}\sum_{i=1}^{p}\mu_{i\cdot}.$$

In these terms, the general model is

$$
\begin{aligned}
X_{ijk} &= \mu + (\mu_{i\cdot} - \mu) + (\mu_{\cdot j} - \mu) + (\mu_{ij} - \mu_{i\cdot} - \mu_{\cdot j} + \mu) + (X_{ijk} - \mu_{ij})\\
&= \mu + \alpha_i + \beta_j + \gamma_{ij} + N_{ijk},
\end{aligned}
\tag{7.3}
$$

where the N_{ijk} are independent and normally distributed with parameters $(0, \sigma^2)$. We have

$$\sum_{i=1}^{p}\alpha_i = \sum_{i=1}^{p}(\mu_{i\cdot} - \mu) = \left(\sum_{i=1}^{p}\frac{1}{q}\sum_{j=1}^{q}\mu_{ij}\right) - \mu p = \mu p - \mu p = 0$$

$$\sum_{j=1}^{q}\beta_j = \sum_{j=1}^{q}(\mu_{\cdot j} - \mu) = \left(\sum_{j=1}^{q}\frac{1}{p}\sum_{i=1}^{p}\mu_{ij}\right) - \mu q = \mu q - \mu q = 0$$

$$
\begin{aligned}
\sum_{i=1}^{p}\gamma_{ij} &= \sum_{i=1}^{p}(\mu_{ij} - \mu_{i\cdot} - \mu_{\cdot j} + \mu) = \left(\sum_{i=1}^{p}\mu_{ij}\right) - \left(\sum_{i=1}^{p}\frac{1}{q}\sum_{j=1}^{q}\mu_{ij}\right) - p\mu_{\cdot j} + p\mu\\
&= p\mu_{\cdot j} - p\mu - p\mu_{\cdot j} + p\mu = 0
\end{aligned}
$$

$$
\begin{aligned}
\sum_{j=1}^{q}\gamma_{ij} &= \sum_{j=1}^{q}(\mu_{ij} - \mu_{i\cdot} - \mu_{\cdot j} + \mu) = \left(\sum_{j=1}^{q}\mu_{ij}\right) - \left(\sum_{j=1}^{q}\frac{1}{p}\sum_{i=1}^{p}\mu_{ij}\right) - q\mu_{i\cdot} + q\mu\\
&= q\mu_{i\cdot} - q\mu - q\mu_{i\cdot} + q\mu = 0.
\end{aligned}
$$

We distinguish four sample means over selected subpopulations. For each (i, j) pair, we have a sample mean over the m observations X_{ijk}. For each level i of factor A, we have a mean over the larger group X_{ijk}, where $1 \leq j \leq q$ and $1 \leq k \leq m$. Similarly, we have a sample mean for each level j of factor B. Finally, we have an overall sample mean across all observations. The notation is as follows.

$$X_{ij\cdot} = \frac{1}{n_{ij}}\sum_{k=1}^{n_{ij}} X_{ijk} = \frac{1}{m}\sum_{k=1}^{m} X_{ijk}$$

$$X_{i..} = \frac{1}{n_{i.}} \sum_{j=1}^{q} \sum_{k=1}^{n_{ij}} X_{ijk} = \frac{1}{mq} \sum_{j=1}^{q} \sum_{k=1}^{m} X_{ijk}$$

$$X_{.j.} = \frac{1}{n_{.j}} \sum_{i=1}^{p} \sum_{k=1}^{n_{ij}} X_{ijk} = \frac{1}{mp} \sum_{i=1}^{p} \sum_{k=1}^{m} X_{ijk}$$

$$X_{...} = \frac{1}{n} \sum_{i=1}^{p} \sum_{j=1}^{q} \sum_{k=1}^{n_{ij}} X_{ijk} = \frac{1}{mpq} \sum_{i=1}^{p} \sum_{j=1}^{q} \sum_{k=1}^{m} X_{ijk} = \frac{1}{q} \sum_{j=1}^{q} X_{.j.} = \frac{1}{p} \sum_{i=1}^{p} X_{i..}$$

Note that

$$E[X_{ij.}] = \frac{1}{m} \sum_{k=1}^{m} E[X_{ijk}] = \frac{1}{m} \sum_{k=1}^{m} \mu_{ij} = \mu_{ij}$$

$$E[X_{i..}] = \frac{1}{mq} \sum_{j=1}^{q} \sum_{k=1}^{m} E[X_{ijk}] = \frac{1}{mq} \sum_{j=1}^{q} m\mu_{ij} = \mu_{i.}$$

$$E[X_{.j.}] = \frac{1}{mp} \sum_{i=1}^{p} \sum_{k=1}^{m} E[X_{ijk}] = \frac{1}{mp} \sum_{i=1}^{p} m\mu_{ij} = \mu_{.j}$$

$$E[X_{...}] = \frac{1}{mpq} \sum_{i=1}^{p} \sum_{j=1}^{q} \sum_{k=1}^{m} E[X_{ijk}] = \frac{1}{mpq} \sum_{i=1}^{p} \sum_{j=1}^{q} m\mu_{ij} = \mu.$$

Because it introduces linear combinations with means $(\mu_{i.} - \mu)$, $(\mu_{.j} - \mu)$, and $(\mu_{ij} - \mu_{i.} - \mu_{.j} + \mu)$, Equation 7.3 then suggests the following decomposition of the total sum of squares.

$$\sum_{i=1}^{p} \sum_{j=1}^{q} \sum_{k=1}^{m} X_{ijk}^2 = \sum_{i=1}^{p} \sum_{j=1}^{q} \sum_{k=1}^{m} \Big[(X_{ijk} - X_{ij.}) + (X_{i..} - X_{...}) + (X_{.j.} - X_{...})$$

$$+ (X_{ij.} - X_{i..} - X_{.j.} + X_{...}) + X_{...} \Big]^2$$

Expanding the square, we obtain the squares of the five constituents plus ten cross-product terms. As in the one-factor case, the cross-product terms are all zero. We illustrate with one of them.

$$\sum_{i=1}^{p} \sum_{j=1}^{q} \sum_{k=1}^{m} (X_{ijk} - X_{ij.})(X_{i..} - X_{...}) = \sum_{i=1}^{p} (X_{i..} - X_{...}) \sum_{j=1}^{q} \left(-mX_{ij.} + \sum_{k=1}^{m} X_{ijk} \right)$$

$$= \sum_{i=1}^{p} (X_{i..} - X_{...}) \sum_{j=1}^{q} (-mX_{ij.} + mX_{ij.}) = 0.$$

The remaining squares constitute the decomposition

$$\sum_{i=1}^{p}\sum_{j=1}^{q}\sum_{k=1}^{m} X_{ijk}^2 = \sum_{i=1}^{p}\sum_{j=1}^{q}\sum_{k=1}^{m}(X_{ijk} - X_{ij\cdot})^2 + mq\sum_{i=1}^{p}(X_{i\cdot\cdot} - X_{\cdots})^2$$

$$+ mp\sum_{j=1}^{q}(X_{\cdot j\cdot} - X_{\cdots})^2 \tag{7.4}$$

$$+ m\sum_{i=1}^{p}\sum_{j=1}^{q}(X_{ij\cdot} - X_{i\cdot\cdot} - X_{\cdot j\cdot} + X_{\cdots})^2 + mpqX_{\cdots}^2$$

$$= Q_1 + Q_2 + Q_3 + Q_4 + Q_5, \tag{7.5}$$

which is a sum of five quadratic forms. The terms are all squares of linear combinations of the X_{ijk}. We can express the decomposition in matrix terms by creating an mpq-vector from the X_{ijk}. The first m components are $X_{1,1,1}$ through $X_{1,1,m}$. These are followed by $X_{1,2,1}$ through $X_{1,2,m}$. We continue to stack the observations at levels $(1, j)$ through $j = q$ before beginning the second level of factor A. This technique linearizes the three-dimensional observation matrix, and the notation is as follows.

$$\mathbf{X}_{ij} = \begin{bmatrix} X_{ij,1} \\ \vdots \\ X_{ij,m} \end{bmatrix} \qquad \mathbf{X}_i = \begin{bmatrix} X_{i,1} \\ \vdots \\ X_{i,q} \end{bmatrix} \qquad \mathbf{X} = \begin{bmatrix} X_1 \\ \vdots \\ X_p \end{bmatrix}$$

The decomposition now reads $\mathbf{X}'\mathbf{X} = \sum_{t=1}^{5} Q_t = \sum_{t=1}^{5}(\mathbf{A}_t\mathbf{X})'(\mathbf{A}_t\mathbf{X})$, where the $\mathbf{A}_t$ reflect the individual Q_t constructions. In particular, we need $\mathbf{A}_1$ such that

$$Q_1 = (\mathbf{A}_1\mathbf{X})'(\mathbf{A}_1\mathbf{X}) = \sum_{i=1}^{p}\sum_{j=1}^{q}\sum_{k=1}^{m}(X_{ijk} - X_{ij\cdot})^2.$$

To this end, for each (i, j) pair, construct $\mathbf{B}_{ij}$ as the $m \times m$ matrix

$$\mathbf{B}_{ij} = \begin{bmatrix} 1 - 1/m & -1/m & \cdots & -1/m \\ -1/m & 1 - 1/m & \cdots & -1/m \\ \vdots & \vdots & & \vdots \\ -1/m & -1/m & \cdots & 1 - 1/m \end{bmatrix},$$

then arrange these along the main diagonal of a larger matrix $\mathbf{B}_i$:

$$\mathbf{B}_i = \begin{bmatrix} \mathbf{B}_{i,1} & 0 & \cdots & 0 \\ 0 & \mathbf{B}_{i,2} & \cdots & 0 \\ \vdots & \vdots & & \vdots \\ 0 & 0 & \cdots & \mathbf{B}_{i,q} \end{bmatrix}.$$

$\mathbf{B}_i$ is size $mq \times mq$. Finally, we form $\mathbf{A}_1$ by aligning the $\mathbf{B}_i$ along the diagonal of a yet larger $mpq \times mpq$ matrix:

$$\mathbf{A}_1 = \begin{bmatrix} \mathbf{B}_1 & \mathbf{0} & \cdots & \mathbf{0} \\ \mathbf{0} & \mathbf{B}_2 & \cdots & \mathbf{0} \\ \vdots & \vdots & & \vdots \\ \mathbf{0} & \mathbf{0} & \cdots & \mathbf{B}_p \end{bmatrix}.$$

By construction, $(\mathbf{A}_1\mathbf{X})'(\mathbf{A}_1\mathbf{X}) = \sum_{i=1}^{p}\sum_{j=1}^{q}\sum_{k=1}^{m}(X_{ijk} - X_{ij\cdot})^2 = Q_1$. Because the rows of each $\mathbf{B}_{ij}$ sum to the zero vector, each has at most $m-1$ independent rows. $\mathbf{B}_i$ then has at most $\sum_{j=1}^{q}(m-1) = mq - q$ independent rows, and $\mathbf{A}_1$ has at most $\sum_{i=1}^{p}(mq-q) = mpq - pq$ independent rows.

Next, we need $\mathbf{A}_2$ such that

$$Q_2 = (\mathbf{A}_2\mathbf{X})'(\mathbf{A}_2\mathbf{X}) = mq\sum_{i=1}^{p}(X_{i\cdot\cdot} - X_{\cdots})^2$$

$$= \sum_{i=1}^{p}mq\left(\frac{1}{mq}\sum_{j=1}^{q}\sum_{k=1}^{m}X_{ijk} - \frac{1}{mpq}\sum_{t=1}^{p}\sum_{j=1}^{q}\sum_{k=1}^{m}X_{tjk}\right)^2$$

$$= \sum_{i=1}^{p}\left[\sum_{j=1}^{q}\sum_{k=1}^{m}\sqrt{mq}\left(\frac{1}{mq} - \frac{1}{mpq}\right)X_{ijk} + \sum_{t\neq i}\sum_{j=1}^{q}\sum_{k=1}^{m}\left(\frac{-\sqrt{mq}}{mpq}\right)X_{tjk}\right]^2$$

$$= \sum_{i=1}^{p}\left[\sum_{j=1}^{q}\sum_{k=1}^{m}\frac{1}{\sqrt{mq}}\left(1 - \frac{1}{p}\right)X_{ijk} + \sum_{t\neq i}\sum_{j=1}^{q}\sum_{k=1}^{m}\left(\frac{-1}{p\sqrt{mq}}\right)X_{tjk}\right]^2.$$

Consequently, $\mathbf{A}_2$ is a $p \times mpq$ matrix, in which each row has p segments of length mq. Row i contains the constant $-1/(p\sqrt{mq})$ in all segments except the ith, where the entries are $(1/\sqrt{mq})(1 - 1/p)$. So, $\mathbf{A}_2$ takes the following form.

$$\begin{bmatrix} \dfrac{1-1/p}{\sqrt{mq}} & \cdots & \dfrac{1-1/p}{\sqrt{mq}} & \dfrac{-1/p}{\sqrt{mq}} & \cdots & \dfrac{-1/p}{\sqrt{mq}} & \cdots & \dfrac{-1/p}{\sqrt{mq}} & \cdots & \dfrac{-1/p}{\sqrt{mq}} \\ \dfrac{-1/p}{\sqrt{mq}} & \cdots & \dfrac{-1/p}{\sqrt{mq}} & \dfrac{1-1/p}{\sqrt{mq}} & \cdots & \dfrac{1-1/p}{\sqrt{mq}} & \cdots & \dfrac{-1/p}{\sqrt{mq}} & \cdots & \dfrac{-1/p}{\sqrt{mq}} \\ \vdots & \vdots & \vdots & \vdots & \vdots & \vdots & & \vdots & \vdots & \vdots \\ \dfrac{-1/p}{\sqrt{mq}} & \cdots & \dfrac{-1/p}{\sqrt{mq}} & \dfrac{-1/p}{\sqrt{mq}} & \cdots & \dfrac{-1/p}{\sqrt{mq}} & \cdots & \dfrac{1-1/p}{\sqrt{mq}} & \cdots & \dfrac{1-1/p}{\sqrt{mq}} \end{bmatrix}$$

Each column of $\mathbf{A}_2$ sums to zero, which means that its rank is at most $p-1$. Turning to $\mathbf{A}_3$, we find a construction parallel to that for $\mathbf{A}_2$. In particular, we want

$$Q_3 = (\mathbf{A}_3\mathbf{X})'(\mathbf{A}_3\mathbf{X}) = mp\sum_{j=1}^{q}(X_{\cdot j\cdot} - X_{\cdots})^2.$$

Expanding the square, we have

$$Q_3 = \sum_{j=1}^{q} mp \left(\frac{1}{mp} \sum_{i=1}^{p} \sum_{k=1}^{m} X_{ijk} - \frac{1}{mpq} \sum_{i=1}^{p} \sum_{t=1}^{q} \sum_{k=1}^{m} X_{itk} \right)^2$$

$$= \sum_{j=1}^{q} \left[\sum_{i=1}^{p} \sum_{k=1}^{m} \sqrt{mp} \left(\frac{1}{mp} - \frac{1}{mpq} \right) X_{ijk} + \sum_{i=1}^{p} \sum_{t\neq j} \sum_{k=1}^{m} \left(\frac{-\sqrt{mp}}{mpq} \right) X_{itk} \right]^2$$

$$= \sum_{j=1}^{q} \left[\sum_{i=1}^{p} \sum_{k=1}^{m} \frac{1}{\sqrt{mp}} \left(1 - \frac{1}{q} \right) X_{ijk} + \sum_{i=1}^{p} \sum_{t\neq j} \sum_{k=1}^{m} \left(\frac{-1}{q\sqrt{mp}} \right) X_{itk} \right]^2.$$

Consequently, $\mathbf{A}_3$ is a $q \times mpq$ matrix. Each row has p segments of length mq; each segment has q subsegments of length m. In the jth row, a segment has entries $-1/(q\sqrt{mp})$ in all subsegments except the jth, where it has $(1/\sqrt{mp})(1 - 1/q)$. The matrix below depicts a vertical slice through the an arbitrary segment. All p segments have the same appearance. The slice contains q rows, and the vertical bars delimit subsegments of m columns each.

$$\left[\begin{array}{cccc|ccc|c} \dfrac{1-1/q}{\sqrt{mp}} & \dfrac{1-1/q}{\sqrt{mp}} & \cdots & \dfrac{1-1/q}{\sqrt{mp}} & \dfrac{-1/q}{\sqrt{mp}} & \dfrac{-1/q}{\sqrt{mp}} & \cdots & \dfrac{-1/q}{\sqrt{mp}} & \cdots \\ \dfrac{-1/q}{\sqrt{mp}} & \dfrac{-1/q}{\sqrt{mp}} & \cdots & \dfrac{-1/q}{\sqrt{mp}} & \dfrac{1-1/q}{\sqrt{mp}} & \dfrac{1-1/q}{\sqrt{mp}} & \cdots & \dfrac{1-1/q}{\sqrt{mp}} & \cdots \\ \vdots & \vdots & & \vdots & \vdots & \vdots & & \vdots & \\ \dfrac{-1/q}{\sqrt{mp}} & \dfrac{-1/q}{\sqrt{mp}} & \cdots & \dfrac{-1/q}{\sqrt{mp}} & \dfrac{-1/q}{\sqrt{mp}} & \dfrac{-1/q}{\sqrt{mp}} & \cdots & \dfrac{-1/q}{\sqrt{mp}} & \cdots \end{array} \right]$$

The sum of any column is evidently zero, which means that $\mathbf{A}_3$ has at most $q - 1$ independent rows.

For $\mathbf{A}_4$, we need

$$Q_4 = (\mathbf{A}_4 \mathbf{X})'(\mathbf{A}_4 \mathbf{X}) = m \sum_{i=1}^{p} \sum_{j=1}^{q} (X_{ij.} - X_{i..} - X_{.j.} + X_{...})^2$$

$$= \sum_{i=1}^{p} \sum_{j=1}^{q} \left[\sum_{k=1}^{m} \frac{\sqrt{m}}{m} X_{ijk} - \sum_{t=1}^{q} \sum_{k=1}^{m} \frac{\sqrt{m}}{mq} X_{itk} \right.$$

$$\left. - \sum_{s=1}^{p} \sum_{k=1}^{m} \frac{\sqrt{m}}{mp} X_{sjk} + \sum_{s=1}^{p} \sum_{t=1}^{q} \sum_{k=1}^{m} \frac{\sqrt{m}}{mpq} X_{stk} \right]^2.$$

Consequently, $\mathbf{A}_4$ is of size $pq \times mpq$. We envision it as p groups of q rows each. As before, each row has p segments, each containing q subsegments of length m. The matrix, therefore, exhibits a checkerboard pattern, in which each tile is located vertically as row j in group i and horizontally as subsegment t in segment s. In segment s, subsegment t, the jth row in group i contains the constant c_{ijst} repeated m times. That constant is

$$
c_{ijst} = \begin{cases}
\sqrt{m}\left(\dfrac{1}{m} - \dfrac{1}{mq} - \dfrac{1}{mp} + \dfrac{1}{mpq}\right), & \text{if } s = i \text{ and } t = j \\[2mm]
\sqrt{m}\left(\dfrac{-1}{mq} + \dfrac{1}{mpq}\right), & \text{if } s = i \text{ and } t \neq j \\[2mm]
\sqrt{m}\left(\dfrac{-1}{mp} + \dfrac{1}{mpq}\right), & \text{if } s \neq i \text{ and } t = j \\[2mm]
\sqrt{m}\left(\dfrac{1}{mpq}\right), & \text{if } s \neq i \text{ and } t \neq j.
\end{cases}
$$

For a j in the range $1 \leq j \leq q$, consider the row subset obtained by choosing the jth row from each of the p groups. We claim that every column sum in this subset is zero. Consider a column in segment s, subsegment t. The sum is

$$
\sum_{i=1}^{p} c_{ijst} = \begin{cases}
\sqrt{m}\left(\dfrac{1}{m} - \dfrac{1}{mq} - \dfrac{1}{mp} + \dfrac{1}{mpq} - \dfrac{p-1}{mp} + \dfrac{p-1}{mpq}\right) = 0, & \text{if } t = j \\[2mm]
\sqrt{m}\left(\dfrac{-1}{mq} + \dfrac{1}{mpq} + \dfrac{p-1}{mpq}\right) = 0, & \text{if } t \neq j.
\end{cases}
$$

Each row in the last group is therefore a linear combination of similarly positioned rows in the first $p - 1$ groups. We can thus restrict our search for independent rows to the first $p - 1$ groups. But, within any such group, each column sum is again zero. In group i, for example, consider again a column in segment s, subsegment t. The sum is

$$
\sum_{j=1}^{q} c_{ijst} = \begin{cases}
\sqrt{m}\left(\dfrac{1}{m} - \dfrac{1}{mq} - \dfrac{1}{mp} + \dfrac{1}{mpq} - \dfrac{q-1}{mq} + \dfrac{q-1}{mpq}\right) = 0, & \text{if } s = i \\[2mm]
\sqrt{m}\left(\dfrac{-1}{mp} + \dfrac{1}{mpq} + \dfrac{q-1}{mpq}\right) = 0, & \text{if } s \neq i.
\end{cases}
$$

Consequently, each of the first $p-1$ groups contains at most $q-1$ independent rows. The rank of $\mathbf{A}_4$ is then at most $(p-1)(q-1)$.

Finally, we need $\mathbf{A}_5$ such that

$$
Q_5 = (\mathbf{A}_5\mathbf{X})'(\mathbf{A}_5\mathbf{X}) = mpqX_{...}^2 = \left(\sum_{i=1}^{p}\sum_{j=1}^{q}\sum_{k=1}^{m}\frac{\sqrt{mpq}}{mpq}X_{ijk}\right)^2.
$$

$\mathbf{A}_5$ is a $1 \times mpq$ matrix with $1/\sqrt{mpq}$ is each entry. Its rank is 1. Combining the results, we have

$$
\sum_{i=1}^{5} \operatorname{rank}(Q_i/\sigma^2) \leq (mpq - pq) + (p - 1) + (q - 1) + (p-1)(q-1) + 1 = mpq.
$$

Invoking Theorem 7.22, we conclude that Q_1/σ^2 through Q_5/σ^2 have independent χ^2 distributions with respective degrees of freedom $mpq - pq$, $p - 1$, $q - 1$, $(p - 1)(q - 1)$, and 1. The noncentrality parameter λ_i of Q_i/σ^2 is

$\mu' A_i \mu / 2$, where μ is an mpq vector containing m copies of μ_{ij}/σ is segment i, subsegment j. Therefore,

$$\lambda_1 = \frac{1}{2\sigma^2} \sum_{i=1}^{p} \sum_{j=1}^{q} \sum_{k=1}^{m} \left(\mu_{ij} - \frac{1}{m} \sum_{u=1}^{m} \mu_{ij} \right)^2 = 0$$

$$\lambda_2 = \frac{1}{2\sigma^2} \sum_{i=1}^{p} \sum_{j=1}^{q} \sum_{k=1}^{m} \left(\frac{1}{mq} \sum_{t=1}^{q} \sum_{u=1}^{m} \mu_{it} - \frac{1}{mpq} \sum_{s=1}^{p} \sum_{t=1}^{q} \sum_{u=1}^{m} \mu_{st} \right)^2$$

$$= \frac{mq}{2\sigma^2} \sum_{i=1}^{p} (\mu_{\cdot i} - \mu)^2 = \frac{mq}{2} \sum_{i=1}^{p} \left(\frac{\alpha_i}{\sigma} \right)^2$$

$$\lambda_3 = \frac{1}{2\sigma^2} \sum_{i=1}^{p} \sum_{j=1}^{q} \sum_{k=1}^{m} \left(\frac{1}{mp} \sum_{s=1}^{p} \sum_{u=1}^{m} \mu_{sj} - \frac{1}{mpq} \sum_{s=1}^{p} \sum_{t=1}^{q} \sum_{u=1}^{m} \mu_{st} \right)^2$$

$$= \frac{mp}{\sigma^2} \sum_{j=1}^{q} (\mu_{\cdot j} - \mu)^2 = \frac{mp}{2} \sum_{j=1}^{q} \left(\frac{\beta_j}{\sigma} \right)^2$$

$$\lambda_4 = \frac{1}{2\sigma^2} \sum_{i=1}^{p} \sum_{j=1}^{q} \sum_{k=1}^{m} \left(\frac{1}{m} \sum_{u=1}^{m} \mu_{ij} - \frac{1}{mq} \sum_{t=1}^{q} \sum_{u=1}^{m} \mu_{it} - \frac{1}{mp} \sum_{s=1}^{p} \sum_{u=1}^{m} \mu_{sj} \right.$$
$$\left. + \frac{1}{mpq} \sum_{s=1}^{p} \sum_{t=1}^{q} \sum_{u=1}^{m} \mu_{st} \right)^2$$

$$= \frac{m}{2\sigma^2} \sum_{i=1}^{p} \sum_{j=1}^{q} (\mu_{ij} - \mu_{i\cdot} - \mu_{\cdot j} + \mu)^2 = \frac{m}{2} \sum_{i=1}^{p} \sum_{j=1}^{q} \left(\frac{\gamma_{ij}}{\sigma} \right)^2$$

$$\lambda_5 = \frac{1}{2\sigma^2} \sum_{i=1}^{p} \sum_{j=1}^{q} \sum_{k=1}^{m} \left(\frac{1}{mpq} \sum_{s=1}^{p} \sum_{t=1}^{q} \sum_{u=1}^{m} \mu_{ij} \right)^2 = \frac{mpq}{2} \left(\frac{\mu}{\sigma} \right)^2.$$

Certain weighted ratios of the Q_i/σ^2 then have F distributions that provide error probabilities in testing hypothesis about the α_i, β_j, and γ_{ij}. In particular, we have the following theorem.

7.29 Theorem: For $1 \le i \le p, 1 \le j \le q, 1 \le k \le m$, let X_{ijk} be independent, normally distributed, random variables with means μ_{ij} and common variance σ^2. Assume that p, q, m are all greater than 1. Note that μ_{ij} is the common mean for random variables X_{ijk} for $1 \le k \le m$. Let

$$\mu_{i\cdot} = (1/q) \sum_{j=1}^{q} \mu_{ij} \qquad\qquad X_{ij\cdot} = (1/m) \sum_{k=1}^{m} X_{ijk}$$
$$\mu_{\cdot j} = (1/p) \sum_{i=1}^{p} \mu_{ij} \qquad\qquad X_{i\cdot\cdot} = [1/(mq)] \sum_{j=1}^{q} \sum_{k=1}^{m} X_{ijk}$$
$$\mu = [1/(pq)] \sum_{i=1}^{p} \sum_{j=1}^{q} \mu_{ij} \qquad\qquad X_{\cdot j\cdot} = [1/(mp)] \sum_{i=1}^{p} \sum_{k=1}^{m} X_{ijk}$$
$$X_{\cdots} = [1(mpq)] \sum_{i=1}^{p} \sum_{j=1}^{q} \sum_{k=1}^{m} X_{ijk}.$$

Then the sums of squares in Table 7.2 have independent χ^2 distributions with the properties indicated. Their ratios have F distributions, also with the properties listed.

Quantity	Degrees of freedom	Noncentrality parameter
$\frac{1}{\sigma^2}\sum_{i=1}^{p}\sum_{j=1}^{q}\sum_{k=1}^{m}(X_{ijk}-X_{ij\cdot})^2$	$pq(m-1)$	0
$\frac{mq}{\sigma^2}\sum_{i=1}^{p}(X_{i\cdot\cdot}-X_{\cdots})^2$	$p-1$	$\frac{mq}{2}\sum_{i=1}^{p}\left(\frac{\alpha_i}{\sigma}\right)^2$
$\frac{mp}{\sigma^2}\sum_{j=1}^{q}(X_{\cdot j\cdot}-X_{\cdots})^2$	$q-1$	$\frac{mp}{2}\sum_{j=1}^{q}\left(\frac{\beta_j}{\sigma}\right)^2$
$\frac{m}{\sigma^2}\sum_{i=1}^{p}\sum_{j=1}^{q}(X_{ij\cdot}-X_{i\cdot\cdot}-X_{\cdot j\cdot}+X_{\cdots})^2$	$(p-1)(q-1)$	$\frac{m}{2}\sum_{i=1}^{p}\sum_{j=1}^{q}\left(\frac{\gamma_{ij}}{\sigma}\right)^2$
$\left(\frac{mpq}{\sigma^2}\right)X_{\cdots}^2$	1	$\frac{mpq}{2}\left(\frac{\mu}{\sigma}\right)^2$
$\frac{(mpq-pq)mq\sum_{i=1}^{p}(X_{i\cdot\cdot}-X_{\cdots})^2}{(p-1)\sum_{i=1}^{p}\sum_{j=1}^{q}\sum_{k=1}^{m}(X_{ijk}-X_{ij\cdot})^2}$	$\frac{p-1}{mpq-pq}$	$\frac{mq}{2}\sum_{i=1}^{p}\left(\frac{\alpha_i}{\sigma}\right)^2$
$\frac{(mpq-pq)mp\sum_{j=1}^{q}(X_{\cdot j\cdot}-X_{\cdots})^2}{(q-1)\sum_{i=1}^{p}\sum_{j=1}^{q}\sum_{k=1}^{m}(X_{ijk}-X_{ij\cdot})^2}$	$\frac{q-1}{mpq-pq}$	$\frac{mp}{2}\sum_{j=1}^{q}\left(\frac{\beta_j}{\sigma}\right)^2$
$\frac{(mpq-pq)m\sum_{i=1}^{p}\sum_{j=1}^{q}(X_{ij\cdot}-X_{i\cdot\cdot}-X_{\cdot j\cdot}+X_{\cdots})^2}{(p-1)(q-1)\sum_{i=1}^{p}\sum_{j=1}^{q}\sum_{k=1}^{m}(X_{ijk}-X_{ij\cdot})^2}$	$\frac{(p-1)(q-1)}{mpq-pq}$	$\frac{m}{2}\sum_{i=1}^{p}\sum_{j=1}^{q}\left(\frac{\gamma_{ij}}{\sigma}\right)^2$

TABLE 7.2. Components of a sum of squares decomposition for a two-way analysis of variance (multiple samples per cell)

These quantities typically arise in sampling pq subpopulations in connection with a two-way analysis of variance. Multiple observations ($m > 1$) come from each subpopulation, which is identified with an (i,j) pair. We commonly refer to the various i values as levels of factor A, while the j values are levels of factor B. The hypotheses of interest are (H_A) the $\alpha_i = (\mu_{i\cdot} - \mu)$ are all zero, (H_B) the $\beta_j = (\mu_{\cdot j} - \mu)$ are all zero, and (H_{AB}) the $\gamma_{ij} = (\mu_{ij} - \mu_{i\cdot} - \mu_{\cdot j} + \mu)$ are all zero.

PROOF: The discussion above establishes the χ^2 distributions. Appropriate ratios then produce the F distributions. ∎

7.30 Example: Suppose that database services are available from vendors in three remote cities. In each city, a vendor can run any of four software packages on the server. We let the city be factor A with levels $i = 1, 2, 3$ corresponding to the three choices. We let the software package be factor B with levels $j = 1, 2, 3, 4$ corresponding to the four possibilities. The total time required to respond to a client request is a random quantity, depending on the round-trip network delay to the destination city and the execution time on the server. Accordingly, if X_{ij} is the response time for city i using software package j, our model is $X_{ij} = \mu + \alpha_i + \beta_j + \gamma_{ij} + N(0, \sigma^2)$, where α_i is due only to the city chosen, β_j is due only to the software package chosen, and γ_{ij} is an interaction time attributable to both. μ is a constant offset, which represents the overall average response time over all cities and software packages. $N(0, \sigma^2)$ is a normally distributed random component.

For each city i and software package j, we take two observations of the total time required to answer a fixed client request. The observations are X_{ijk} with $1 \le k \le 2$. In the notation of Theorem 7.29, $p = 3$, $q = 4$, and $m = 2$. The data tabulation appears to the right.

Response time X_{ijk} in milliseconds			
$j = 1$	$j = 2$	$j = 3$	$j = 4$
$i = 1$ 766.47	750.84	667.62	632.03
758.86	775.41	640.96	663.92
$i = 2$ 651.06	653.73	558.16	549.37
654.97	662.33	544.91	538.70
$i = 3$ 540.15	529.55	448.08	439.18
573.99	540.08	430.92	462.19

The sample means across the relevant subpopulations are

$$[X_{ij\cdot}] = \begin{bmatrix} 762.67 & 763.12 & 654.29 & 647.98 \\ 653.02 & 658.03 & 551.54 & 544.03 \\ 557.07 & 534.82 & 439.50 & 450.68 \end{bmatrix}$$

$$[X_{i\cdot\cdot}] = \begin{bmatrix} 707.02 \\ 601.65 \\ 495.52 \end{bmatrix}$$

$$[X_{\cdot j\cdot}] = \begin{bmatrix} 657.59 & 651.99 & 548.44 & 547.56 \end{bmatrix}$$

$$X_{\cdots} = 601.40.$$

The relevant sums of squares and F-ratios are

$$Q_1 = \sum_{i=1}^{3}\sum_{j=1}^{4}\sum_{k=1}^{2}(X_{ijk} - X_{ij\cdot})^2 = 2424$$

$$Q_2 = (2)(4)\sum_{i=1}^{3}(X_{i\cdot\cdot} - X_{\cdots})^2 = 178929$$

$$Q_3 = (2)(3)\sum_{j=1}^{4}(X_{\cdot j\cdot} - X_{\cdots})^2 = 68513$$

$$Q_4 = (2)\sum_{i=1}^{3}\sum_{j=1}^{4}(X_{ij\cdot} - X_{i\cdot\cdot} - X_{\cdot j\cdot} + X_{\cdots})^2 = 767.3$$

$$F_1 = \frac{pq(m-1)Q_4}{(p-1)(q-1)Q_1} = 0.63$$

$$F_2 = \frac{pq(m-1)Q_2}{(p-1)Q_1} = 442.9$$

$$F_3 = \frac{pq(m-1)Q_3}{(q-1)Q_1} = 113.1.$$

We first test, at the 5% significance level, the null hypothesis that the γ_{ij} are all zero. Under this hypothesis, F_1 has a central F distribution with $((p-1)(q-1), pq(m-1)) = (6, 12)$ degrees of freedom. The 95th percentile of this distribution is 2.996. Because $0.63 < 2.996$, we accept the hypothesis that all the γ_{ij} are zero.

Next, we test, again at the 5% significance level, the null hypothesis that the α_i are all zero. Under this hypothesis F_2 has a central F distribution with $(p-1, pq(m-1)) = (2, 12)$ degrees of freedom. The 95th percentile of this distribution is 3.885. Because $442.9 > 3.885$, we reject this hypothesis. Finally, the null hypothesis that the β_j are all zero imposes on F_3 a central F distribution with $(3, 12)$ degrees of freedom. Because $113.1 > 3.490$, the 95th percentile of that distribution, we reject all-zero status for the β_j. We conclude that both city and software package are influential in the total server response time. The displacement of the mean response time, however, is the sum of an offset due to the city and another due to the software package. There is no need for a third term for displacement particular to a city-software combination.

Taken as a whole, the three decisions incur some probability of error that is difficult to evaluate. We can, however, compute the error probability inherent in

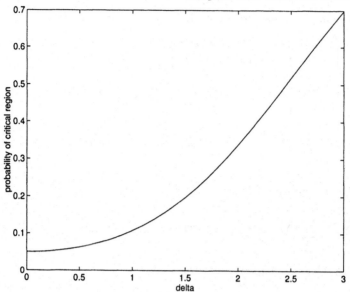

Figure 7.4. Power function of test evaluating all $\gamma_{ij} = 0$ (see Example 7.30)

each separate decision. In rejecting the hypothesis that the α_i are all zero, we risk a type I error with probability 0.05 because, if the parameters are indeed all zero, this is the probability that F_2 exceeds $F_{0.95;(2,12)} = 3.885$. Similarly, the probability of a type I error in rejecting the hypothesis that the β_j are all zero is likewise 0.05. In accepting the hypothesis that the γ_{ij} are all zero, we risk a type II error, the probability of which varies with the true γ_{ij} values. As in the one-factor case, an overall measure of the γ_{ij} sizes is

$$\delta = \frac{m}{2} \sqrt{\sum_{i=1}^{p} \sum_{j=1}^{q} \left(\frac{\gamma_{ij}}{\sigma} \right)^2 }.$$

For a given $\delta > 0$, the probability of a type II error is $\Pr(F_3 \le F_{0.95;(6,12)}) = \Pr(F_3 \le 2.996)$, where F_3 has a noncentral F distribution with $(6, 12)$ degrees of freedom and noncentrality parameter $\lambda = m\delta^2/2$. This error probability is 1 minus the power function of the test at the given δ. Figure 7.4 graphs this power function. It shows a rather unpleasant picture. While the error is understandably large for small δ, it remains above 0.5 even for δ approaching 2.5. The remedy lies in increasing the number of samples per cell. If we increase m from 2 to 20, the relevant F distribution then has $((p-1)(q-1), pq(m-1)) = (6, 228)$ degrees of freedom, and the critical region threshold is $F_{0.95;(6,228)} = 2.139$. The power function of the improved test appears as Figure 7.5, where we see the type II error probability falling off more rapidly with increasing δ. $\square$

All $\gamma_{ij} = 0$ in the example is intuitively appealing because there is no reason for a particular software package to function better or worse when moved to a different city. When all $\gamma_{ij} = 0$, we say that the effects of factors A and B are *additive*. The rationale for this terminology is as follows. Suppose that i_1 and i_2 are two levels of factor A and j_1, j_2 are two levels of factor B.

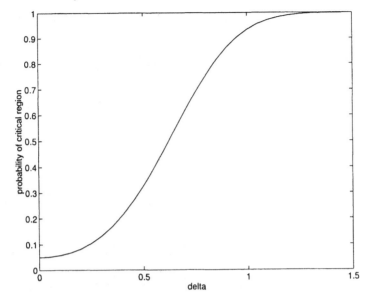

Figure 7.5. Power function of revised test with 20 samples per cell (see Example 7.30)

If all γ_{ij} are zero, we have, for any j,

$$\mu_{i_1,j} - \mu_{i_2,j} = (\mu_{i_1,j} - \mu_{i_1\cdot} - \mu_{\cdot j} + \mu) - (\mu_{i_2,j} - \mu_{i_2\cdot} - \mu_{\cdot j} + \mu) + \mu_{i_1\cdot} - \mu_{i_2\cdot}$$
$$= \gamma_{i_1,j} - \gamma_{i_2,j} + \mu_{i_1\cdot} - \mu_{i_2\cdot} = \mu_{i_1\cdot} - \mu_{i_2\cdot}.$$

That is, comparing two cells in the same column, the difference in means is independent of the column. If factor A does affect the outcomes, it does so uniformly across the levels of factor B. Similarly,

$$\mu_{i,j_1} - \mu_{i,j_2} = (\mu_{i,j_1} - \mu_{i\cdot} - \mu_{\cdot j_1} + \mu) - (\mu_{i,j_2} - \mu_{i\cdot} - \mu_{\cdot j_2} + \mu) + \mu_{\cdot j_1} - \mu_{\cdot j_2}$$
$$= \gamma_{i,j_1} - \gamma_{i,j_2} + \mu_{\cdot j_1} - \mu_{\cdot j_2} = \mu_{\cdot j_1} - \mu_{\cdot j_2}.$$

Comparing two cells in the same row, the difference in means is independent of the row. If factor B does affect the outcomes, it does so uniformly across the levels of factor A.

In the last example, we suspected that the factors were additive, and we are pleased that the data confirm this conclusion. In such circumstances, we can assume the simpler model $X_{ij} = \mu + \alpha_i + \beta_j + N_{ij}(0, \sigma^2)$, where the $N_{ij}(0, \sigma^2)$ are independent normally distributed random variables with parameters $(0, \sigma^2)$. This model allows an analysis even when there is a single observation per cell. Recall that $m > 1$ is necessary to apply Theorem 7.29 in order that the denominator of the F expression have nonzero degrees of freedom. We now sketch the applicable decomposition of the sums of squares for this special case. Letting X_{ij} denote the single observation in cell (i, j), for $1 \leq i \leq p, 1 \leq j \leq q$, we define simpler approximations of the means.

$X_{i\cdot} = (1/q) \sum_{j=1}^{q} X_{ij} \quad X_{\cdot j} = (1/p) \sum_{i=1}^{p} X_{ij} \quad X_{\cdot\cdot} = [1/(pq)] \sum_{i=1}^{p} \sum_{j=1}^{q} X_{ij}$

We decompose the sum of squares of the observations as follows.

$$\sum_{i=1}^{p} \sum_{j=1}^{q} X_{ij}^2 = \sum_{i=1}^{p} \sum_{j=1}^{q} [(X_{i\cdot} - X_{\cdot\cdot}) + (X_{\cdot j} - X_{\cdot\cdot}) + (X_{ij} - X_{i\cdot} - X_{\cdot j} + X_{\cdot\cdot}) + X_{\cdot\cdot}]^2$$

$$= q \sum_{i=1}^{p} (X_{i\cdot} - X_{\cdot\cdot})^2 + p \sum_{j=1}^{q} (X_{\cdot j} - X_{\cdot\cdot})^2$$

$$+ \sum_{i=1}^{p} \sum_{j=1}^{q} (X_{ij} - X_{i\cdot} - X_{\cdot j} + X_{\cdot\cdot})^2 + pq X_{\cdot\cdot}^2$$

$$= Q_1 + Q_2 + Q_3 + Q_4 \tag{7.6}$$

The cross-product terms all sum to zero, and after some effort identifying the matrices associated with the quadratic forms, Theorem 7.22 again implies that the Q_i/σ^2 have independent χ^2 distributions with respective degrees of freedom $p-1, q-1, (p-1)(q-1)$, and 1. These computations are essentially the same as those employed in the earlier analysis of $m > 1$ observations per cell. The exercises provide an opportunity to fill in the details. We summarize the results in the following theorem and then consider an image-processing example.

7.31 Theorem: For $1 \leq i \leq p, 1 \leq j \leq q$, let X_{ij} be independent, normally distributed, random variables with means μ_{ij} and common variance σ^2. Assume that p, q are greater than 1. Across the subpopulations of interest, define the the true means and the corresponding sample means as follows.

$$\mu_{i\cdot} = (1/q) \sum_{j=1}^{q} \mu_{ij} \qquad\qquad X_{i\cdot} = (1/q) \sum_{j=1}^{q} X_{ij}$$

$$\mu_{\cdot j} = (1/p) \sum_{i=1}^{p} \mu_{ij} \qquad\qquad X_{\cdot j} = (1/p) \sum_{i=1}^{p} X_{ij}$$

$$\mu = [1/(pq)] \sum_{i=1}^{p} \sum_{j=1}^{q} \mu_{ij} \qquad\qquad X_{\cdot\cdot} = [1/(pq)] \sum_{i=1}^{p} \sum_{j=1}^{q} X_{ij}$$

Let $\alpha_i = \mu_{i\cdot} - \mu$ and $\beta_j = \mu_{\cdot j} - \mu$ and assume that the μ_{ij} satisfy $\mu_{ij} = \mu + (\mu_{i\cdot} - \mu) + (\mu_{\cdot j} - \mu) = \mu + \alpha_i + \beta_j$.

Then the sums of squares in Table 7.3 have independent χ^2 distributions with the properties indicated. Their ratios have F distributions, also with the properties listed. These quantities arise in sampling pq subpopulations in connection with a two-way analysis of variance. A single observation X_{ij} comes from each subpopulation. We refer to the various i values as levels of factor A, while the j values are levels of factor B. The assumption that $\mu_{ij} = \mu + \alpha_i + \beta_j$ constitutes an assertion that the two factors are additive. The hypotheses of interest are (H_A) the α_i are all zero and (H_B) the β_j are

Quantity	Form	Degrees of freedom	Noncentrality parameter
$\frac{q}{\sigma^2}\sum_{i=1}^{p}(X_{i\cdot}-X_{\cdot\cdot})^2$	χ^2	$p-1$	$\frac{q}{2}\sum_{i=1}^{p}\left(\frac{\alpha_i}{\sigma}\right)^2$
$\frac{p}{\sigma^2}\sum_{j=1}^{q}(X_{\cdot j}-X_{\cdot\cdot})^2$	χ^2	$q-1$	$\frac{p}{2}\sum_{j=1}^{q}\left(\frac{\beta_j}{\sigma}\right)^2$
$\frac{1}{\sigma^2}\sum_{i=1}^{p}\sum_{j=1}^{q}(X_{ij}-X_{i\cdot}-X_{\cdot j}+X_{\cdot\cdot})^2$	χ^2	$(p-1)(q-1)$	0
$\frac{pq}{\sigma^2}X_{\cdot\cdot}^2$	χ^2	1	$\frac{pq}{2}\left(\frac{\mu}{\sigma}\right)^2$
$\dfrac{q(q-1)\sum_{i=1}^{p}(X_{i\cdot}-X_{\cdot\cdot})^2}{\sum_{i=1}^{p}\sum_{j=1}^{q}(X_{ij}-X_{i\cdot}-X_{\cdot j}+X_{\cdot\cdot})^2}$	F	$\dfrac{p-1}{(p-1)(q-1)}$	$\frac{q}{2}\sum_{i=1}^{p}\left(\frac{\alpha_i}{\sigma}\right)^2$
$\dfrac{p(p-1)\sum_{j=1}^{q}(X_{\cdot j}-X_{\cdot\cdot})^2}{\sum_{i=1}^{p}\sum_{j=1}^{q}(X_{ij}-X_{i\cdot}-X_{\cdot j}+X_{\cdot\cdot})^2}$	F	$\dfrac{q-1}{(p-1)(q-1)}$	$\frac{p}{2}\sum_{j=1}^{q}\left(\frac{\beta_j}{\sigma}\right)^2$

TABLE 7.3. Components of a sum of squares decomposition for a two-way analysis of variance (one observation per cell)

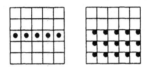

Figure 7.6. Scanning window encounters a horizontal line (see Example 7.32)

---

all zero.

PROOF: Matrix algebra computations, similar to those used to establish Theorem 7.29, show that the decomposition of Equation 7.6 leads to independent χ^2 distributions with the appropriate properties. Selected ratios then produce the desired F distributions. ∎

7.32 Example: In a rectangular image of $p \times q$ pixels, each pixel has a brightness level in the integer range 0 (black) to 8 (white). In this context, we wish to detect horizontal and vertical lines. Our detector is a sliding window, $m \times m$ pixels in size, that we scan across the image. We insist that m be an odd integer, so that the window has a well-defined center pixel. When the window intersects a horizontal line in the image, the general situation is one of those depicted in Figure 7.6, depending on whether the regions above and below the line return to the background brightness or continue at the (low) level of the line. The former indicates a true horizontal line, while the latter indicates the edge of an object. In either case, we wish to detect this situation. The cells with dark circles represent pixels with a lowered brightness level, while the unmarked cells indicate brightness that varies randomly. A human, we conjecture, detects a line by noting the pattern of darkened pixels as it stands out against the random background. We wish to design our line detector in a similar manner.

The appearance of a horizontal line or edge in the window significantly decreases the mean brightness of the occluded rows, while a window with no line or edge exhibits the same mean across all its rows. We model the brightness of the pixel in row i, column j, by $B_{ij} = \mu + \alpha_i + \beta_j + N_{ij}(0, \sigma^2)$. The α_i and β_j are offsets from the overall average brightness that correspond to row and columns effects, respectively. We assume the N_{ij} to have independent normal distributions with parameters $(0, \sigma^2)$. The null hypothesis that the window does not intersect a horizontal line corresponds to all $\alpha_i = 0$. Applying Theorem 7.31, we construct, at the 5% significance level, the test

$$F_h = \frac{m(m-1)\sum_{i=1}^{m}(B_{i\cdot} - B_{\cdot\cdot})^2}{\sum_{i=1}^{m}\sum_{j=1}^{m}(B_{ij} - B_{i\cdot} - B_{\cdot j} + B_{\cdot\cdot})^2} > F_{0.95;((m-1),(m-1)^2)}.$$

Using this test, we construct a *new* pixel array as follows. We first initialize all cells in the new array to maximum white. We then address the center cell of the window as it scans across the original array. If $F_h > F_{0.95;((m-1),(m-1)^2)}$, we change the pixel under the window center, in the new image, to maximum dark. Otherwise, we leave it alone, and it remains maximum white as initialized. We then slide the window one pixel to the right and repeat the process. Upon reaching the right side of the image, we start over at the left, one pixel lower. Upon reaching the lower right corner of the image, we conclude $(p - 2\lfloor m/2 \rfloor)(q - 2\lfloor m/2 \rfloor)$ such tests, thereby producing a new image just slightly smaller than the original. In the new image, all pixels are at brightness levels 0 (white) or 8 (black). The black pixels are the centers of windows for which the null hypothesis has been rejected. Because the window should reject the null hypothesis when a horizontal line or edge appears anywhere in the window, a horizontal line one pixel wide will trigger m consecutive rejections as the window proceeds downward over it in the vertical direction. Consequently, the line detected will be widened, which may or may not be significant for further processing.

A similar test for vertical lines uses

$$F_v = \frac{m(m-1)\sum_{j=1}^{m}(B_{\cdot j} - B_{\cdot\cdot})^2}{\sum_{i=1}^{m}\sum_{j=1}^{m}(B_{ij} - B_{i\cdot} - B_{\cdot j} + B_{\cdot\cdot})^2} > F_{0.95;((m-1),(m-1)^2)}.$$

If we mark the center pixel when either F_h or F_v exceeds its threshold, we detect both horizontal and vertical lines. The algorithm appears as Figure 7.7. It accepts pix, a matrix of p × q pixels, the window size m, and the F test threshold. It returns a new image pixa, in which the pixels are black or white, depending on whether the associated window rejects the horizontal or vertical null hypothesis.

Figure 7.8 plots the brightness levels of pixels in a 100 × 100 grid. A normal random number generator produced these levels with a mean of 4 and a variance of 1. Superimposed on this random background is a cross, centered in the frame. In the horizontal (x) direction, its extent is about 20% of the frame width; in the vertical direction (y), it extends to about 80% of the frame height. The horizontal and vertical members are each one pixel wide. Their pixel values are three greater than the background, provided that the sum does not exceed eight. In Figure 7.8, the cross is barely visible to the naked eye.

In any case, Figure 7.9 shows the new pixel matrix generated by the algorithm described above. It uses a scanning window of size 5 × 5 pixels. The cross is clearly delineated from the background, although several type I and type II errors are apparent. The type I errors are pixels on the upper plane. They represent rejection of the null hypothesis when it was actually true. The type II errors occur where a pixel is missing from the cross. In this case, the null hypothesis was false, but it was accepted. Notice that the one-pixel members are thickened as expected. A smaller scanning window reduces this distortion, but at the expense of more type I and type II errors. Figure 7.10 illustrates the results obtained with a 3 × 3 scanning window. □

```
function linedetect (pix, p, q, m, threshold)
    // inputs are the p × q pixel matrix pix, window size m, and F-test threshold
    border = floor(m/2);
    pp = p - 2 * border; qq = q - 2 * border;    // new pix is of size pp × qq
    for (i = 1; i <= pp; i++)
        for (j = 1; j <= qq; j++)
            pixa[i,j] = 0.0;        // new pix initially all white
    for (i = 1; i <= pp; i++)
        for (j = 1; j <= qq; j++)
            for (s = 1; s <= m; s++)
                for (t = 1; t <= m; t++)
                    x[s,t] = pix[i + s - 1, j + t - 1];
                        // fill window x with data from surrounding pix
            for (s = 1; s <= m; s++)        // compute row means
                xid[s] = 0.0;
                for (t = 1; t <= m; t++)
                    xid[s] = xid[s] + x[s,t];
                xid[s] = xid[s]/m;
            for (t = 1; t <= m; t++)        // compute column means
                xdj[t] = 0.0;
                for (s = 1; s <= m; s++)
                    xdj[t] = xdj[t] + x[s,t];
                xdj[t] = xdj[t]/m;
            xdd = 0.0;        // compute overall mean
            for (t = 1; t <= m; t++)
                xdd = xdd + xdj[t];
            xdd = xdd/m;
            den = 0.0        // compute common denominator for F statistic
            for (s = 1; s <= m; s++)
                for (t = 1; t <= m; t++)
                    den = den + (x[s,t] - xid[s] - xdj[t] + xdd)
                            * (x[s,t] - xid[s] - xdj[t] + xdd);
            Fh = 0.0        // compute F statistic for horizontal line detection
            for (s = 1; s <= m; s++)
                Fh = Fh + (xid[s] - xdd) * (xid[s] - xdd);
            Fh = m * (m - 1) * Fh / den;
            if (Fh > threshold)
                // if horizontal line detected, mark pixa at window center
                pixa[i, j] = 8.0;
            else        // else compute F statistic for vertical lines
                Fv = 0.0;
                for (t = 1; t <= m; t++)
                    Fv = Fv + (xdj[t] - xdd) * (xdj[t] - xdd);
                Fv = m * (m - 1) * Fv / den;
                if (Fv > threshold)        // vertical line detected, mark pixa
                    pixa[i,j] = 8.0;
    return pixa;
```

Figure 7.7. Algorithm for horizontal and vertical line detection (see Example 7.32)

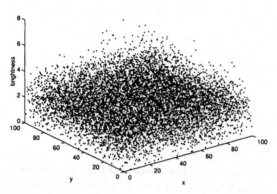

Figure 7.8. A cross hidden in a random background (see Example 7.32)

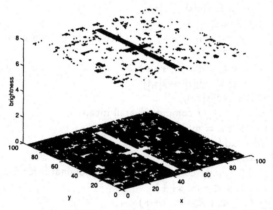

Figure 7.9. 5×5 pixel window discovers the concealed cross (see Example 7.32)

Exercises

7.19 Verify that all three cross-product terms in Equation 7.2 are zero.

7.20 Suppose that a one-factor analysis of variance has m subpopulations. The data are $x_{ij}, 1 \leq i \leq m, 1 \leq j \leq n_i$. Considering $[x_{ij}]$ as a matrix with rows of possibly unequal length, we let T_i be the total of the ith row, T the sum of all the entries, and $n = \sum_{i=1}^{m} n_i$. Show that the following formulas provide alternative computations for the sums of squares required in the analysis.

$$Q_1 = \sum_{i=1}^{m} \sum_{j=1}^{n_i} (x_{ij} - \overline{x})^2 = \sum_{i=1}^{m} \sum_{j=1}^{n_i} x_{ij}^2 - \frac{T^2}{n}$$

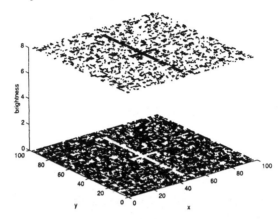

Figure 7.10. 3 × 3 pixel window finds a more accurate cross but with
more errors (see Example 7.32)

$$Q_2 = \sum_{i=1}^{m} n_i(\overline{x}_i - \overline{x})^2 = \left(\sum_{i=1}^{m} \frac{T_{i\cdot}^2}{n_i}\right) - \frac{T^2}{n}$$

$$Q_3 = \sum_{i=1}^{m}\sum_{j=1}^{n_i}(x_{ij} - \overline{x}_i)^2 = \sum_{i=1}^{m}\sum_{j=1}^{n_i} x_{ij}^2 - \sum_{i=1}^{m} \frac{T_{i\cdot}^2}{n_i}$$

7.21 A professor wishes to know if student hair color makes a difference in
exam scores. He gives the same test to all students and obtains the
following data.

blonde	81.65	76.27	70.75	73.52	63.03	86.96	70.59	87.97	72.64	78.72
black	56.54	63.99	83.46	64.61	76.77	67.40	69.64			
brown	68.55	61.01	61.35	77.62	68.44	74.13	72.97	76.56	73.00	

At the 5% significance level, test the null hypothesis that there is no
difference in the mean exam scores across the three groups.

7.22 Wheat is grown on one-acre plots, which are identical in terms of soil
conditions and climate. All plots are cultivated in the same manner,
except that they are subjected to varying dosages, in gallons per acre,
of an experimental insecticide. The resulting yields are as follows.

Dosage	Yield (bushels per acre)									
0	415.0	445.0	429.2	304.9	389.0	387.1	412.7	379.6	399.9	464.3
20	473.9	450.7	403.1	437.2	445.9	417.7	426.5	456.6	502.3	342.2
40	316.1	404.9	383.3	349.1	376.6	320.9	319.1	372.7	420.6	390.0

At the 5% significance level, test the null hypothesis that the mean yield
is independent of the insecticide dosage. What is β, the probability of

a type II error if $\mu_2 = \mu_3 = \mu_1 + \sigma$, where σ^2 is the common variance of the subpopulations. Assuming that each dosage is applied to the same number of plots, what size sample is necessary to reduce this β to 0.05?

7.23 A two-processor system contains a single common memory. The memory access time for processor A is not constant because of random contention from processor B. The time may be longer or shorter depending on the activity in processor B. Access time samples for different kinds of processor B activity are as follows.

B activity	Memory access time (nanoseconds) for processor A
Text editing	379.7 435.4 390.1 371.0 382.2 375.5 391.6 422.5 357.4 414.1
Net browsing	405.3 366.0 388.1 357.9 440.7 397.0 287.4 361.4 287.7 401.1
Computation	335.8 345.6 355.0 390.1 307.2 314.5 344.6 359.1 275.4 298.0

At the 5% significance level, test the null hypothesis that the mean access time is independent of processor B activity. What is β, the probability of a type II error if there is sufficient variation among the means to give $\delta = 0.6$? Assuming an experimental design that takes the same number of observations for each processor B activity, what size sample is necessary to reduce this β to 0.05?

7.24 Patients in three age brackets take a drug in four dosages: 0 (placebo), 10, 20, and 30 milligrams per day. The drug claims to reduce the recovery time for the common cold. Three patients appear in each age-dose combination. To study the results with a two-factor analysis of variance, we let the three age brackets constitute factor A, with levels $i = 1, 2, 3$ corresponding to young, middle-aged, and old. Factor B is the dosage, with levels $j = 1, 2, 3, 4$ corres-

Recovery time (hours) for age (i) and dosage (j)			
$j = 1$	2	3	4
$i = 1$	48.30	67.43	74.59	140.12
	37.54	21.08	119.69	133.01
	26.50	35.98	99.10	98.17
$i = 2$	32.03	74.92	84.02	132.50
	11.07	37.22	84.70	147.50
	58.92	61.55	117.23	139.57
$i = 3$	26.18	42.80	98.88	77.45
	60.94	47.29	110.27	119.52
	30.28	23.01	107.93	118.54

ponding to 0, 10, 20, 30 milligrams per day. We model the recovery time as a random variable R_{ij} that satisfies $R_{ij} = \mu + \alpha_i + \beta_j + \gamma_{ij} + N(0, \sigma^2)$. On the data to the right, test at the 5% significance level the hypotheses: (a) the γ_{ij} are all zero, (b) the α_i are all zero, and (c) the β_j are all zero.

*7.25 For $1 \le i \le 3$, let μ_i be the mean of subpopulation i. Given the following data, test the null hypothesis that $\mu_2 = \mu_1 + 10$ and $\mu_3 = \mu_1 + 20$. Set the significance level of the test at 0.05.

i	x_{ij}									
1	100.8	50.3	123.2	79.5	123.1	84.3	112.7	116.4	96.5	111.2
2	107.5	121.1	88.1	95.4	138.1	97.6	114.7	78.3	102.0	94.6
3	114.7	139.5	139.6	143.4	123.2	130.0	98.9	111.0	145.4	138.0

7.26 In a two-factor analysis of variance, factor A has levels $1, 2, 3$ corresponding to three bank tellers. Factor B has levels $j = 1, 2, 3, 4$ corresponding to the time of day: morning, noon, afternoon, and evening. The variable under study is customer transaction time, and the assumed model is $T_{ij} = \mu + \alpha_i + \beta_j + \gamma_{ij} + N(0, \sigma^2)$. A sample of two transactions

Transaction time (seconds) for teller (i) and time-of-day (j)			
$j = 1$	2	3	4
$i = 1$ \ 66.36	66.53	166.16	168.28
49.78	83.59	86.11	140.44
$i = 2$ \ 59.96	87.96	113.04	139.57
92.13	73.86	157.45	166.35
$i = 3$ \ 76.95	78.27	146.63	190.32
65.36	93.30	129.57	174.99

in each (i, j) cell produces the data to the right above. At the 5% significance level, test the hypotheses: (a) the γ_{ij} are all zero, (b) the α_i are all zero, and (c) the β_j are all zero.

7.27 Show that the cross-product terms from the following expansion are all zero.

$$\sum_{i=1}^{p}\sum_{j=1}^{q}\sum_{k=1}^{m} X_{ijk}^2 = \sum_{i=1}^{p}\sum_{j=1}^{q}\sum_{k=1}^{m} \Big[(X_{ijk} - X_{ij.}) + (X_{i..} - X_{...})$$
$$+ (X_{.j.} - X_{...}) + (X_{ij.} - X_{i..} - X_{.j.} + X_{...}) + X_{...} \Big]^2$$

7.28 The sum of squares

$$\sum_{i=1}^{p}\sum_{j=1}^{q} X_{ij}^2 = \sum_{i=1}^{p}\sum_{j=1}^{q} \Big[(X_{i.} - X_{..}) + (X_{.j} - X_{..})$$
$$+ (X_{ij} - X_{i.} - X_{.j} + X_{..}) + X_{..} \Big]^2$$

appears in the analysis of a two-factor analysis of variance with one observation per cell. Verify that the expression reduces to the following by showing that the cross-product terms sum to zero.

$$\sum_{i=1}^{p}\sum_{j=1}^{q} X_{ij}^2 = q \sum_{i=1}^{p}(X_{i.} - X_{..})^2 + p \sum_{j=1}^{q}(X_{.j} - X_{..})^2$$
$$+ \sum_{i=1}^{p}\sum_{j=1}^{q}(X_{ij} - X_{i.} - X_{.j} + X_{..})^2 + pq X_{..}^2$$

*7.29 Continuing with the decomposition of the preceding problem, let

$$Q_1 = q\sum_{i=1}^{p}(X_{i.} - X_{..})^2 \qquad Q_3 = \sum_{i=1}^{p}\sum_{j=1}^{q}(X_{ij} - X_{i.} - X_{.j} + X_{..})^2$$
$$Q_2 = p\sum_{j=1}^{q}(X_{.j} - X_{..})^2 \qquad Q_4 = pq X_{..}^2.$$

Stack the observations X_{ij} as follows to obtain a pq-vector $\mathbf{X}$. That is, $\mathbf{X}_i' = \begin{bmatrix} X_{i1} & X_{i2} & \dots & X_{iq} \end{bmatrix}$ and $\mathbf{X}' = \begin{bmatrix} \mathbf{X}_1 & \mathbf{X}_2 & \dots & \mathbf{X}_p \end{bmatrix}$. Find matrices $\mathbf{A}_k, 1 \leq k \leq 4$, such that $Q_k = (\mathbf{A}_k\mathbf{X})'(\mathbf{A}_k\mathbf{X})$ and

$$\text{rank}(\mathbf{A}_k) = \begin{cases} p-1, & k=1 \\ q-1, & k=2 \\ (p-1)(q-1), & k=3 \\ 1, & k=4. \end{cases}$$

*7.30 Implement the algorithm of Figure 7.7 and show that it is blind to diagonal lines.

*7.31 Modify the algorithm of Figure 7.7 so that it detects only horizontal lines. Implement the code and test it with a cross containing a horizontal and a vertical member. The image processed should contain only the horizontal member, plus sporadic type I and type II errors.

*7.32 Modify the algorithm of Figure 7.7 so that it detects only intersections of horizontal and vertical lines. Test it with the cross pattern.

7.5   Linear regression

Consider again the one-factor analysis of variance problem. We have m sub-populations, and $X_{ij}, 1 \leq j \leq n_i$, are observations of the ith such subpopulation. We model the variation by $X_{ij} = \mu + \alpha_i + N_{ij}(0, \sigma^2)$. The overall mean is μ, while α_i is an offset associated with the ith subpopulation. Of course, N_{ij} adds the random components, which we assume to be independent and normally distributed. The first hypothesis of interest asserts that the α_i are all zero. That is, the subpopulations all have the same mean.

As a variation on this problem, suppose that we want to test the hypothesis that the mean increases by 1.5 as we move from one subpopulation to the next in the sequence $i = 1, 2, \dots, m$. This hypothesis is true if and only if the hypothesis of equal means is true for the related problem: $Y_{ij} = X_{ij} - 1.5(i-1)$. The model $Y_{ij} = \mu + \alpha_i + N_{ij}(0, \sigma^2)$ becomes $X_{ij} - 1.5(i-1) = \mu + \alpha_i + N_{ij}(0, \sigma^2)$. The hypothesis that the α_i are all zero corresponds to $X_{ij} = \mu + 1.5(i-1) + N_{ij}(0, \sigma^2)$. That is, a linear relationship exists among the means of the subpopulations when they are aligned in the order $i = 1, 2, \dots, m$. It is convenient to replace i with t_i because, in the general case, the index does not assume equally spaced integral values. We retain the notation $X_{ij}, 1 \leq i \leq m, 1 \leq j \leq n_i$ for the sample, but X_{ij} is now the jth observation of subpopulation X_{t_i}.

7.33 Definition: For real t, let the family X_t have independent normal distributions with a common variance σ^2 but with potentially different means μ_t. A *linear regression* is a test of the linear hypothesis $\mu_t = A + Bt$. The regression also estimates the unknown constants A and B. ∎

With X_t as in the definition, consider a sample that selects n_i observations from the subpopulation X_{t_i}, for $1 \leq i \leq m$. Denote the observations by $X_{ij}, 1 \leq i \leq m, 1 \leq j \leq n_i$. Define $n = \sum_{i=1}^{m} n_i$ and $\bar{t} = (1/n) \sum_{i=1}^{m} n_i t_i$. If the subpopulation means are indeed linearly related, then

$$X_{ij} = \mu_{t_i} - (X_{ij} - \mu_{t_i}) = (A + Bt_i) + (X_{ij} - \mu_{t_i}) = A + Bt_i + N_{ij}(0, \sigma^2),$$

where the N_{ij} are independent and normally distributed with the parameters indicated. As will become apparent in the upcoming algebra, it is convenient to express the model as follows.

$$X_{ij} = A + B\bar{t} + B(t_i - \bar{t}) + N_{ij} = \alpha + \beta(t_i - \bar{t}) + N_{ij},$$

where $\alpha = A + B\bar{t}$ and $\beta = B$. Henceforth, we deal with α and β as the unknown parameters in the linear relationship. If we should need the A, B mentioned in the definition, we can retrieve them from $B = \beta$ and $A = \alpha - \beta\bar{t}$.

The *least squares estimators* for α and β are the quantities $\hat{\alpha}$ and $\hat{\beta}$ that minimize the expression

$$E(\alpha, \beta) = \sum_{i=1}^{m} \sum_{j=1}^{n_i} [X_{ij} - (\alpha + \beta(t_i - \bar{t}))]^2.$$

We locate the minimum where both partial derivatives vanish.

$$\partial_1 E(\alpha, \beta) = \sum_{i=1}^{m} \sum_{j=1}^{n_i} 2[X_{ij} - \alpha - \beta(t_i - \bar{t})](-1) = 0$$

$$\partial_2 E(\alpha, \beta) = \sum_{i=1}^{m} \sum_{j=1}^{n_i} 2[X_{ij} - \alpha - \beta(t_i - \bar{t})][-(t_i - \bar{t})] = 0$$

These reduce to the simultaneous equations

$$n\alpha + \left(\sum_{i=1}^{m} n_i(t_i - \bar{t}) \right) \beta = \sum_{i=1}^{m} \sum_{j=1}^{n_i} X_{ij}$$

$$\left(\sum_{i=1}^{m} n_i(t_i - \bar{t}) \right) \alpha + \left(\sum_{i=1}^{m} n_i(t_i - \bar{t})^2 \right) \beta = \sum_{i=1}^{m} \sum_{j=1}^{n_i} X_{ij}(t_i - \bar{t}),$$

and the reason for using (α, β), rather than the original (A, B), becomes apparent. Because

$$\sum_{i=1}^{m} n_i(t_i - \bar{t}) = \left(\sum_{i=1}^{m} n_i t_i \right) - \bar{t} \left(\sum_{i=1}^{m} n_i \right) = n\bar{t} - n\bar{t} = 0,$$

one term drops out of each equation. The minimizing values are then

$$\hat{\alpha} = \frac{1}{n} \sum_{i=1}^{m} \sum_{j=1}^{n_i} X_{ij} \qquad \hat{\beta} = \frac{\sum_{i=1}^{m} \sum_{j=1}^{n_i} X_{ij}(t_i - \bar{t})}{\sum_{i=1}^{m} n_i(t_i - \bar{t})^2}.$$

Continuing with the assumption that the linear relationship does hold, we have

$$E[X_{ij}] = E[\alpha + \beta(t_i - \bar{t}) + N_{ij}] = \alpha + \beta(t_i - \bar{t})$$

$$E[\hat{\alpha}] = \frac{1}{n}\sum_{i=1}^{m}\sum_{j=1}^{n_i} E[X_{ij}] = \frac{1}{n}\sum_{i=1}^{m} n_i[\alpha + \beta(t_i - \bar{t})]$$

$$= \frac{1}{n}\left(\alpha\sum_{i=1}^{m} n_i + \beta\sum_{i=1}^{m} n_i(t_i - \bar{t})\right) = \alpha$$

$$E[\hat{\beta}] = \frac{\sum_{i=1}^{m}\sum_{j=1}^{n_i}(t_i - \bar{t})E[X_{ij}]}{\sum_{i=1}^{m}\sum_{j=1}^{n_i} n_i(t_i - \bar{t})^2} = \frac{\sum_{i=1}^{m} n_i(t_i - \bar{t})[\alpha + \beta(t_i - \bar{t})]}{\sum_{i=1}^{m}\sum_{j=1}^{n_i} n_i(t_i - \bar{t})^2}$$

$$= \frac{\alpha\sum_{i=1}^{m} n_i(t_i - \bar{t}) + \beta\sum_{i=1}^{m} n_i(t_i - \bar{t})^2}{\sum_{i=1}^{m}\sum_{j=1}^{n_i} n_i(t_i - \bar{t})^2} = \beta.$$

Therefore, $\hat{\alpha}$ and $\hat{\beta}$ are unbiased estimates of α and β. It follows that $\hat{\alpha} + \hat{\beta}(t-\bar{t})$ is an unbiased estimate of $\alpha + \beta(t-\bar{t})$ for any value of t. Indeed, in the class of unbiased linear combinations of the observations, $\hat{\alpha} + \hat{\beta}(t-\bar{t})$ has the smallest variance and is therefore the most efficient. To see how this happens, fix a t and suppose that $T = \sum_{i=1}^{m}\sum_{j=1}^{n_i} c_{ij}X_{ij}$ is an unbiased estimate of $\alpha + \beta(t-\bar{t})$. Because it is unbiased, we must have

$$E[T] = \alpha + \beta(t-\bar{t}) = \sum_{i=1}^{m}\sum_{j=1}^{n_i} c_{ij}E[X_{ij}] = \sum_{i=1}^{m}\sum_{j=1}^{n_i} c_{ij}E[\alpha + \beta(t_i - \bar{t}) + N_{ij}]$$

$$= \sum_{i=1}^{m}\sum_{j=1}^{n_i} c_{ij}[\alpha + \beta(t_i - \bar{t})] = \alpha\left(\sum_{i=1}^{m}\sum_{j=1}^{n_i} c_{ij}\right) + \beta\left(\sum_{i=1}^{m}\sum_{j=1}^{n_i} c_{ij}(t_i - \bar{t})\right).$$

This relation must hold regardless of α and β, so we infer the following constraints on the c_{ij}.

$$\sum_{i=1}^{m}\sum_{j=1}^{n_i} c_{ij} = 1 \qquad\qquad \sum_{i=1}^{m}\sum_{j=1}^{n_i} c_{ij}(t_i - \bar{t}) = (t - \bar{t}) \qquad (7.7)$$

Because T is a linear combination of the independent X_{ij}, we have

$$\text{var}(T) = \sum_{i=1}^{m}\sum_{j=1}^{n_i} c_{ij}^2\text{var}(X_{ij}) = \sigma^2\sum_{i=1}^{m}\sum_{j=1}^{n_i} c_{ij}^2.$$

We now ask what c_{ij} coefficients minimize var(T), subject to the constraints noted above. Consider the related quantity

$$v = \text{var}(T) + \lambda_1\left(1 - \sum_{i=1}^{m}\sum_{j=1}^{n_i} c_{ij}\right) + \lambda_2\left((t-\bar{t}) - \sum_{i=1}^{m}\sum_{j=1}^{n_i} c_{ij}(t_i - \bar{t})\right)$$

$$= \sigma^2\sum_{i=1}^{m}\sum_{j=1}^{n_i} c_{ij}^2 + \lambda_1\left(1 - \sum_{i=1}^{m}\sum_{j=1}^{n_i} c_{ij}\right) + \lambda_2\left((t-\bar{t}) - \sum_{i=1}^{m}\sum_{j=1}^{n_i} c_{ij}(t_i - \bar{t})\right),$$

a function of the c_{ij} and the two additional parameters λ_1 and λ_2. Considering v unconstrained in this enlarged independent variable space, we seek its minimum in the traditional manner.

$$\frac{\partial v}{\partial c_{ij}} = 2c_{ij}\sigma^2 - \lambda_1 - \lambda_2(t_i - \bar{t}) = 0$$

$$\frac{\partial v}{\partial \lambda_1} = 1 - \sum_{i=1}^{m}\sum_{j=1}^{n_i} c_{ij} = 0$$

$$\frac{\partial v}{\partial \lambda_2} = (t - \bar{t}) - \sum_{i=1}^{m}\sum_{j=1}^{n_i} c_{ij}(t_i - \bar{t}) = 0$$

Solving, we have

$$c_{ij} = \frac{1}{2\sigma^2}[\lambda_1 + \lambda_2(t_i - \bar{t})]$$

$$0 = 1 - \sum_{i=1}^{m}\sum_{j=1}^{n_i} c_{ij} = 1 - \frac{1}{2\sigma^2}\sum_{i=1}^{m}\sum_{j=1}^{n_i}[\lambda_1 + \lambda_2(t_i - \bar{t})] = 1 - \frac{n\lambda_1}{2\sigma^2}$$

$$\lambda_1 = \frac{2\sigma^2}{n}$$

$$0 = (t - \bar{t}) - \sum_{i=1}^{m}\sum_{j=1}^{n_i}\frac{(t_i - \bar{t})}{2\sigma^2}\left[\frac{2\sigma^2}{n} + \lambda_2(t_i - \bar{t})\right] = (t - \bar{t}) - \frac{\lambda_2}{2\sigma^2}\sum_{i=1}^{m} n_i(t_i - \bar{t})^2$$

$$\lambda_2 = \frac{2\sigma^2(t - \bar{t})}{\sum_{i=1}^{m} n_i(t_i - \bar{t})^2}$$

$$c_{ij} = \frac{1}{2\sigma^2}\left(\frac{2\sigma^2}{n} + \frac{2\sigma^2(t - \bar{t})(t_i - \bar{t})}{\sum_{i=1}^{m} n_i(t_i - \bar{t})^2}\right) = \frac{1}{n} + \frac{(t - \bar{t})(t_i - \bar{t})}{\sum_{i=1}^{m} n_i(t_i - \bar{t})^2}.$$

This $c_{ij}, \lambda_1, \lambda_2$ point minimizes the unconstrained quantity v, and the point satisfies the constraints of Equation 7.7. Consequently, v does not decrease when we move in *any* direction from this minimum point. In particular, v does not decrease if we move in a direction where the constraints remain satisfied. But along such constrained paths, $v = \text{var}(T)$. We conclude that the minimum point of v is a constrained minimum for $\text{var}(T)$. Therefore, T as an unbiased linear combination of the observations has a minimum when

$$c_{ij} = \frac{1}{n} + \frac{(t - \bar{t})(t_i - \bar{t})}{\sum_{i=1}^{m} n_i(t_i - \bar{t})^2}.$$

In this case, however,

$$T = \sum_{i=1}^{m}\sum_{j=1}^{n_i} c_{ij}X_{ij} = \frac{1}{n}\sum_{i=1}^{m}\sum_{j=1}^{n_i} X_{ij} + (t - \bar{t})\frac{\sum_{i=1}^{m}\sum_{j=1}^{n_i} X_{ij}(t_i - \bar{t})}{\sum_{i=1}^{m} n_i(t_i - \bar{t})^2}$$

$$= \hat{\alpha} + \hat{\beta}(t - \bar{t}).$$

For any t, therefore, $\hat{\alpha} + \hat{\beta}(t - \bar{t})$ is the minimum variance estimate of $\alpha + \beta(t - \bar{t})$ over all unbiased linear combinations of the observations.

We now consider the distributions of $\hat{\alpha}$ and $\hat{\beta}$, regardless of any linearity relationship among the means. Both are linear combinations of independent normal random variables. The means and variances are then

$$\text{mean}(\hat{\alpha}) = \frac{1}{n} \sum_{i=1}^{m} \sum_{j=1}^{n_i} \text{mean}(X_{ij}) = \frac{1}{n} \sum_{i=1}^{m} n_i \mu_{t_i}$$

$$\text{mean}(\hat{\beta}) = \frac{\sum_{i=1}^{m} \sum_{j=1}^{n_i} (t_i - \bar{t}) \text{mean}(X_{ij})}{\sum_{i=1}^{m} n_i (t_i - \bar{t})^2} = \frac{\sum_{i=1}^{m} n_i \mu_{t_i} (t_i - \bar{t})}{\sum_{i=1}^{m} n_i (t_i - \bar{t})^2}$$

$$\text{var}(\hat{\alpha}) = \frac{1}{n^2} \sum_{i=1}^{m} \sum_{j=1}^{n_i} \text{var}(X_{ij}) = \frac{n\sigma^2}{n^2} = \frac{\sigma^2}{n}$$

$$\text{var}(\hat{\beta}) = \frac{\sum_{i=1}^{m} \sum_{j=1}^{n_i} (t_i - \bar{t})^2 \text{var}(X_{ij})}{\left(\sum_{i=1}^{m} n_i (t_i - \bar{t})^2\right)^2} = \frac{\sigma^2 \sum_{i=1}^{m} n_i (t_i - \bar{t})^2}{\left(\sum_{i=1}^{m} n_i (t_i - \bar{t})^2\right)^2}$$

$$= \frac{\sigma^2}{\sum_{i=1}^{m} n_i (t_i - \bar{t})^2}.$$

The covariance $\hat{\alpha}$ and $\hat{\beta}$ is

$$\text{cov}(\hat{\alpha}, \hat{\beta}) = E\left[\left(\hat{\alpha} - \frac{1}{n} \sum_{i=1}^{m} n_i \mu_{t_i}\right)\left(\hat{\beta} - \frac{\sum_{i=1}^{m} n_i \mu_{t_i} (t_i - \bar{t})}{\sum_{i=1}^{m} n_i (t_i - \bar{t})^2}\right)\right]$$

$$= E\left[\left(\frac{1}{n} \sum_{i=1}^{m} \sum_{j=1}^{n_i} (X_{ij} - \mu_{t_i})\right)\right.$$

$$\left. \cdot \left(\frac{1}{\sum_{i=1}^{m} n_i (t_i - \bar{t})^2} \sum_{i=1}^{m} \sum_{j=1}^{n_i} (t_i - \bar{t})(X_{ij} - \mu_{t_i})\right)\right]$$

$$= \frac{1}{n \sum_{i=1}^{m} n_i (t_i - \bar{t})^2} \sum_{i=1}^{m} \sum_{j=1}^{n_i} \sum_{k=1}^{m} \sum_{l=1}^{n_i} (t_i - \bar{t}) E[(X_{ij} - \mu_{t_i})(X_{kl} - \mu_{t_k})].$$

Now, $E[(X_{ij} - \mu_{t_i})(X_{kl} - \mu_{t_k})] = \text{cov}(X_{ij}, X_{kl}) = \sigma^2$, if $i = k$ and $j = l$. It is zero otherwise. The quadruple sum above then reduces to

$$\text{cov}(\hat{\alpha}, \hat{\beta}) = \frac{1}{n \sum_{i=1}^{m} n_i (t_i - \bar{t})^2} \sum_{i=1}^{m} \sum_{j=1}^{n_i} \sigma^2 (t_i - \bar{t})$$

$$= \frac{\sigma^2}{n \sum_{i=1}^{m} n_i (t_i - \bar{t})^2} \sum_{i=1}^{m} n_i (t_i - \bar{t}) = 0.$$

Thus $\hat{\alpha}$ and $\hat{\beta}$ are uncorrelated, and being normal, they are then independent. We summarize the results to this point with the following theorem.

7.34 Theorem: Let X_t be a family of independent normal random variables with means μ_t and common variance σ^2. For a fixed set $t_1 < t_2 < \ldots < t_m$, let $X_{ij}, 1 \le j \le n_i$ be a sample from X_{t_i}. If

$$n = \sum_{i=1}^{m} n_i \qquad \hat{\alpha} = \frac{1}{n} \sum_{i=1}^{m} \sum_{j=1}^{n_i} X_{ij}$$

$$\bar{t} = \frac{1}{n} \sum_{i=1}^{m} n_i t_i \qquad \hat{\beta} = \frac{\sum_{i=1}^{m} \sum_{j=1}^{n_i} X_{ij}(t_i - \bar{t})}{\sum_{i=1}^{m} n_i (t_i - \bar{t})^2},$$

then $(\hat{\alpha}, \hat{\beta})$ has a bivariate normal distribution with mean vector and covariance matrix

$$\mu = \begin{bmatrix} \frac{1}{n} \sum_{i=1}^{m} n_i \mu_{t_i} \\ \dfrac{\sum_{i=1}^{m} n_i \mu_{t_i}(t_i - \bar{t})}{\sum_{i=1}^{m} n_i (t_i - \bar{t})^2} \end{bmatrix} \qquad \Sigma = \begin{bmatrix} \dfrac{\sigma^2}{n} & 0 \\ 0 & \dfrac{\sigma^2}{\sum_{i=1}^{m} n_i (t_i - \bar{t})^2} \end{bmatrix}.$$

Moreover, if the means are linearly related by $\mu_t = \alpha + \beta(t - \bar{t})$, the mean vector is $\mu = [\,\alpha\ \beta\,]'$, in which case $\hat{\alpha}$ and $\hat{\beta}$ are unbiased consistent estimators for α and β.

PROOF: The discussion above establishes all the assertions. Recall that an unbiased estimator is consistent if its variance tends to zero with increasing sample size. The variance of $\hat{\beta}$ depends on the spacing of the t_i. Therefore, its variance can remain large, despite an increasing sample size, if the t_i spacing is reduced sufficiently. For example, after choosing n observations and establishing the corresponding $\bar{t}$, we can choose all further observations from $X_{\bar{t}}$. The $\hat{\beta}$ variance then remains unchanged as $n \to \infty$. Therefore, we should interpret the consistency assertion as applying for a fixed t_i set. Increasing sample size then means increasing n_i, and the variance does indeed approach zero in this circumstance. ∎

We wish to consider two kinds of problems. In the first kind, we know that a linear relationship exists among the means μ_t, and we want estimates of the slope and intercept parameters. In the second kind, we want to test the hypothesis that the linear relationship holds. We first study the estimation problem in the context of a known linear relationship.

Suppose that $\mu_t = A + Bt$, for unknown constants A, B. We fix a set $t_i, 1 \le i \le m$, and take a sample $X_{ij}, 1 \le j \le n_i$. As in the discussion above, the observations X_{ij} come from the subpopulation X_{t_i}, $n = \sum_{i=1}^{m} n_i$, and $\bar{t} = (1/n) \sum_{i=1}^{m} n_i t_i$. As before, we write

$$\mu_t = A + Bt = A + \bar{t} + B(t - \bar{t}) = \alpha + \beta(t - \bar{t}),$$

and proceed to obtain estimates for α and β. The point estimates $\hat{\alpha}$ and $\hat{\beta}$ are unbiased, and noting the distributions in Theorem 7.34, we can strive for small variances by selecting a large sample and by choosing the t_i to be as spread out as possible, thereby producing a large $\sum_{i=1}^{m} n_i (t_i - \bar{t})^2$ in the denominator of $\text{var}(\hat{\beta})$. Because we know that $\hat{\alpha}$ has a normal distribution with parameters $(\alpha, \sigma^2/n)$, we might attempt to construct a confidence interval by reducing it to standard normal. A 90% confidence interval involves the following computation, where Z is standard normal and z_p is the 100pth percentile.

$$0.9 = \Pr(z_{0.05} < Z < z_{0.95}) = \Pr\left(-1.645 < \frac{\hat{\alpha} - \alpha}{\sigma/\sqrt{n}} < 1.645\right)$$

$$= \Pr\left(\hat{\alpha} - \frac{1.645\sigma}{\sqrt{n}} < \alpha < \hat{\alpha} + \frac{1.645\sigma}{\sqrt{n}}\right)$$

We are frustrated at this point. We cannot actually calculate this confidence interval because we do not know σ. The solution lies with a ratio of quantities, so constructed that the unknown variance cancels and leaves an F distribution. Consider the following decomposition of a sum of squares in the original observations.

$$\sum_{i=1}^{m}\sum_{j=1}^{n_i} X_{ij}^2 = \sum_{i=1}^{m}\sum_{j=1}^{n_i} [(X_{ij} - \hat{\alpha} - \hat{\beta}(t_i - \bar{t})) + \hat{\alpha} + \hat{\beta}(t_i - \bar{t})]^2$$

$$= \sum_{i=1}^{m}\sum_{j=1}^{n_i}[X_{ij} - \hat{\alpha} - \hat{\beta}(t_i - \bar{t})]^2 + \sum_{i=1}^{m}\sum_{j=1}^{n_i}\hat{\alpha}^2 + \sum_{i=1}^{m}\sum_{j=1}^{n_i}\hat{\beta}^2(t_i - \bar{t})^2$$

$$+ 2\sum_{i=1}^{m}\sum_{j=1}^{n_i}\hat{\alpha}[X_{ij} - \hat{\alpha} - \hat{\beta}(t_i - \bar{t})]$$

$$+ 2\sum_{i=1}^{m}\sum_{j=1}^{n_i}\hat{\beta}(t_i - \bar{t})[X_{ij} - \hat{\alpha} - \hat{\beta}(t_i - \bar{t}] + 2\sum_{i=1}^{m}\sum_{j=1}^{n_i}\hat{\alpha}\hat{\beta}(t_i - \bar{t}).$$

The cross-product terms all sum to zero. We illustrate with the first of them.

$$2\sum_{i=1}^{m}\sum_{j=1}^{n_i}\hat{\alpha}[X_{ij} - \hat{\alpha} - \hat{\beta}(t_i - \bar{t})] = 2\left[\hat{\alpha}\sum_{i=1}^{m}\sum_{j=1}^{n_i}X_{ij} - n\hat{\alpha}^2 - \hat{\alpha}\hat{\beta}\sum_{i=1}^{m}n_i(t_i - \bar{t})\right]$$

$$= 2\left(n\hat{\alpha}^2 - n\hat{\alpha}^2\right) = 0.$$

There remain three quadratic forms,

$$\sum_{i=1}^{m}\sum_{j=1}^{n_i} X_{ij}^2 = \sum_{i=1}^{m}\sum_{j=1}^{n_i}[X_{ij} - \hat{\alpha} - \hat{\beta}(t_i - \bar{t})]^2 + n\hat{\alpha}^2 + \hat{\beta}^2\sum_{i=1}^{m}n_i(t_i - \bar{t})^2$$

$$= Q_1 + Q_2 + Q_3,$$

for which we must discover the matrix equivalents. In the usual manner, we obtain the n-vector $\mathbf{X}$ by first stacking the n_i observations for each i and then stacking these minivectors. That is, $\mathbf{X}_i' = [\, X_{i,1} \ \ X_{i,2} \ \ \dots \ \ X_{i,n_i} \,]$ and $\mathbf{X}' = [\, \mathbf{X}_1' \ \ \mathbf{X}_2' \ \ \dots \ \ \mathbf{X}_m' \,]$. The vector contains m segments, of length $n_1, n_2, \dots, n_m$ in order from first to last. The matrices for the Q_k will exhibit similar structures. It is convenient to have a concise symbol for the denominator of the $\hat{\beta}$ formula, so let $D = \sum_{i=1}^{m} n_i(t_i - \bar{t})^2$. Starting with Q_1, we write

$$Q_1 = \sum_{i=1}^{m}\sum_{j=1}^{n_i}\left[X_{ij} - \frac{1}{n}\sum_{k=1}^{m}\sum_{l=1}^{n_k}X_{kl} - \frac{t_i - \bar{t}}{D}\sum_{k=1}^{m}\sum_{l=1}^{n_k}X_{kl}(t_k - \bar{t})\right]^2 .$$

We construct the $n \times n$ matrix $\mathbf{A}_1$ with m groups containing $n_1, n_2, \ldots, n_m$ rows each as we proceed from top to bottom. Each row has the same structure: m segments of $n_1, n_2, \ldots, n_m$ entries proceeding from left to right. The jth row of group i has coefficients determined by the (i, j) summand in the equation just above. That is, in segment k, position l, the entry is

$$a = \begin{cases} 1 - \dfrac{1}{n} - \dfrac{(t_k - \bar{t})^2}{D}, & \text{if } k = i \text{ and } l = j \\[2mm] -\dfrac{1}{n} - \dfrac{(t_k - \bar{t})^2}{D}, & \text{if } k = i \text{ and } l \neq j \\[2mm] -\dfrac{1}{n} - \dfrac{(t_k - \bar{t})(t_i - \bar{t})}{D}, & \text{if } k \neq i. \end{cases}$$

By construction, we have $Q_1 = (\mathbf{A}_1\mathbf{X})'(\mathbf{A}_1\mathbf{X})$. Let the rows of $\mathbf{A}_1$ be $\mathbf{a}_{ij}$, where i denotes the row group and j the individual row within the group. Of course, $1 \leq i \leq m$ and $1 \leq j \leq n_i$. We claim that the following linear combinations sum to zero: $\sum_{i=1}^{m}\sum_{j=1}^{n_i}\mathbf{a}_{ij} = \mathbf{0}$ and $\sum_{i=1}^{m}\sum_{j=1}^{n_i}(t_i - \bar{t})\mathbf{a}_{ij} = \mathbf{0}$. To see why these linear constraints hold, consider column l in segment k of the first linear combination. Scanning down the column, we encounter one i group for which $i = k$ and $m - 1$ groups for which $i \neq k$. Within the group where $i = k$, we find one row with $j = l$ and $n_k - 1$ rows where $j \neq l$. The sum of this column is then

$$\text{sum} = 1 - \frac{1}{n} - \frac{(t_k - \bar{t})^2}{D} + (n_k - 1)\left(-\frac{1}{n} - \frac{(t_k - \bar{t})^2}{D}\right)$$
$$+ \sum_{i \neq k} n_i \left(-\frac{1}{n} - \frac{(t_k - \bar{t})(t_i - \bar{t})}{D}\right)$$
$$= 1 - \frac{1}{n}\sum_{i=1}^{m} n_i - \frac{(t_k - \bar{t})}{D}\sum_{i=1}^{m} n_i(t_i - \bar{t}) = 1 - 1 - 0 = 0.$$

Similarly, column l in segment k in the second linear combination is

$$\text{sum} = (t_k - \bar{t})\left(1 - \frac{1}{n} - \frac{(t_k - \bar{t})^2}{D}\right) + (n_k - 1)(t_k - \bar{t})\left(-\frac{1}{n} - \frac{(t_k - \bar{t})^2}{D}\right)$$
$$+ \sum_{i \neq k} n_i(t_i - \bar{t})\left(-\frac{1}{n} - \frac{(t_k - \bar{t})(t_i - \bar{t})}{D}\right)$$
$$= (t_k - \bar{t}) - \frac{1}{n}\sum_{i=1}^{m} n_i(t_i - \bar{t}) - (t_k - \bar{t})\sum_{i=1}^{m} n_i \frac{(t_i - \bar{t})^2}{D}$$
$$= (t_k - \bar{t}) - 0 - (t_k - \bar{t})\frac{D}{D} = 0.$$

Thus, for either of the vectors

$$\mathbf{z}' = \begin{bmatrix} t_1 - \bar{t} & \cdots & t_1 - \bar{t} & t_2 - \bar{t} & \cdots & t_2 - \bar{t} & \cdots & t_m - \bar{t} & \cdots & t_m - \bar{t} \end{bmatrix}$$
$$\mathbf{z}' = \begin{bmatrix} 1 & 1 & \cdots & 1 \end{bmatrix},$$

we have $\mathbf{A}_1 \mathbf{z} = \mathbf{0}$. Because these two vectors are linearly independent, the null space of $\mathbf{A}_1$ is of dimension 2 or more. Theorem A.79 then asserts that the rank of $\mathbf{A}_1$ is at most $n - 2$.

For Q_2, we have

$$Q_2 = n\hat{\alpha}^2 = n \left(\frac{1}{n} \sum_{i=1}^{m} \sum_{j=1}^{n_i} X_{ij} \right)^2 = (\mathbf{A}_2 \mathbf{X})'(\mathbf{A}_2 \mathbf{X}),$$

where $\mathbf{A}_2$ is a $1 \times n$ matrix that contains $1/\sqrt{n}$ in each position. Its rank is clearly 1. For Q_3, we have a similar expansion.

$$Q_3 = \hat{\beta}^2 D = D \left(\frac{1}{D} \sum_{i=1}^{m} \sum_{j=1}^{n_i} X_{ij}(t_i - \bar{t}) \right)^2 = (\mathbf{A}_3 \mathbf{X})'(\mathbf{A}_3 \mathbf{X}),$$

where $\mathbf{A}_3$ is a $1 \times n$ matrix containing m segments of length $n_1, n_2, \ldots, n_m$. Segment i contains n_i entries, all equal to $(t_i - \bar{t})/\sqrt{D}$. As a $1 \times n$ matrix, $\mathbf{A}_3$ has rank at most 1. We now invoke Theorem 7.22 to conclude that the Q_k/σ^2 have independent χ^2 distributions with $n - 2, 1, 1$ degrees of freedom, respectively. We compute the noncentrality factors by $\lambda_k = (\mathbf{A}_k \mu)'(\mathbf{A}_k \mu)/2$, for $k = 1, 2, 3$. Here, μ is $E[\mathbf{X}]/\sigma$, an n-vector with segments of length $n_1, n_2, \ldots, n_m$. Segment i contains n_i entries, all equal to μ_{t_i}/σ. We can therefore compute λ_k by replacing X_{ij} with μ_{t_i}/σ in Q_k. Accordingly, because $\mu_{t_i} = \alpha + \beta(t_i - \bar{t})$, we have

$$\lambda_1 = \frac{1}{2\sigma^2} \sum_{i=1}^{m} \sum_{j=1}^{n_i} \left[\mu_{t_i} - \frac{1}{n} \sum_{k=1}^{m} \sum_{l=1}^{n_k} \mu_{t_k} - \frac{(t_i - \bar{t})}{D} \sum_{k=1}^{m} \sum_{l=1}^{n_k} \mu_{t_k}(t_k - \bar{t}) \right]^2$$

$$= \frac{1}{2\sigma^2} \sum_{i=1}^{m} \sum_{j=1}^{n_i} \left[\alpha + \beta(t_i - \bar{t}) - \frac{1}{n} \sum_{k=1}^{m} n_k(\alpha - \beta(t_k - \bar{t})) \right.$$

$$\left. - \frac{(t_i - \bar{t})}{D} \sum_{k=1}^{m} n_k(\alpha + \beta(t_k - \bar{t}))(t_k - \bar{t}) \right]^2$$

$$= \frac{1}{2\sigma^2} \sum_{i=1}^{m} n_i[\alpha + \beta(t_i - \bar{t}) - \alpha - \beta(t_i - \bar{t})]^2 = 0$$

$$\lambda_2 = \frac{n}{2\sigma^2} \left[\frac{1}{n} \sum_{i=1}^{m} \sum_{j=1}^{n_i} \mu_{t_i} \right]^2 = \frac{1}{2n\sigma^2} \left[\sum_{i=1}^{m} n_i(\alpha + \beta(t_i - \bar{t})) \right]^2 = \frac{n}{2} \left(\frac{\alpha}{\sigma} \right)^2$$

$$\lambda_3 = \frac{D}{2\sigma^2} \left[\frac{1}{D} \sum_{i=1}^{m} \sum_{j=1}^{n_i} \mu_{t_i}(t_i - \bar{t}) \right]^2 = \frac{1}{2D\sigma^2} \left[\sum_{i=1}^{m} n_i(t_i - \bar{t})(\alpha + \beta(t_i - \bar{t})) \right]^2$$

$$= \frac{D}{2} \left(\frac{\beta}{\sigma} \right)^2.$$

Because Q_2/σ^2 has one degree of freedom, $\sqrt{Q_2}/\sigma = \hat{\alpha}\sqrt{n}/\sigma$ is normal with parameters $(\sqrt{2\lambda_2}, 1)$. Consequently, $(\hat{\alpha} - \alpha)\sqrt{n}/\sigma$ is standard normal and $Q_2'/\sigma^2 = n(\hat{\alpha} - \alpha)^2/\sigma^2$ has a central χ^2 distribution with one degree of freedom. Q_2' remains independent of Q_1 and Q_3 because it is a function of Q_2 only:

$$Q_2' = n(\hat{\alpha} - \alpha)^2 = n \left(\sqrt{\frac{Q_2}{n}} - \alpha \right)^2.$$

Similarly, $Q_3'/\sigma^2 = D(\hat{\beta} - \beta)^2/\sigma^2$ has a central χ^2 distribution with one degree of freedom, independent of Q_1 and Q_2'. We summarize with the following theorem.

7.35 Theorem: Let X_t be a family of independent normal random variables with common variance σ^2 and means μ_t. The means are linearly related but the slope and intercept parameters are unknown. For the fixed set of t-values, $t_1 < t_2 < \dots < t_m$, let $X_{ij}, 1 \leq j \leq n_i$, be a sample from subpopulation X_{t_i}. Define

$$n = \sum_{i=1}^{m} n_i \qquad \hat{\alpha} = \frac{1}{n} \sum_{i=1}^{m} \sum_{j=1}^{n_i} X_{ij}$$

$$\bar{t} = \frac{1}{n} \sum_{i=1}^{m} n_i t_i \qquad \hat{\beta} = \frac{1}{D} \sum_{i=1}^{m} \sum_{j=1}^{n_i} X_{ij}(t_i - \bar{t})$$

$$D = \sum_{i=1}^{m} n_i(t_i - \bar{t})^2 \qquad \hat{\sigma}^2 = \frac{1}{n} \sum_{i=1}^{m} \sum_{j=1}^{n_i} [X_{ij} - \hat{\alpha} - \hat{\beta}(t_i - \bar{t})]^2.$$

For unknown α and β, let $\alpha + \beta(t - \bar{t})$ be the linear relation among the means. Then $\hat{\alpha}$ and $\hat{\beta}$ are unbiased estimators for α and β, respectively, and the ratios $R_1 = (n-2)(\hat{\alpha} - \alpha)^2/\hat{\sigma}^2$ and $R_2 = (n-2)D(\hat{\beta} - \beta)^2/(n\hat{\sigma}^2)$ have central F distributions with $(1, n-2)$ degrees of freedom.

PROOF: The F distributions follow from ratios of the quadratic forms discussed above. ∎

7.36 Example: Consider the process that reads data from a disk storage unit into a memory buffer. There is an initial delay, called a latency delay, while the data's starting sector rotates under the read head. The disk unit then transmits data as the disk continues to rotate into the data's storage sectors. Because the data may not occupy contiguous sectors, additional latency delays may occur when the disk must reacquire a new continuation point. The latency delays are, of course, random variables. When needed, the target sector could be under the read head, or it could be an entire revolution distant. The number of such delays is also random. If the disk is empty when the data are recorded, it can write the information on contiguous sectors. In this case, a read operation incurs only the initial latency delay. However, if the disk has few free sectors, it may have to record the data in discontinuous segments, which forces a subsequent read operation to incur many delays. In any case, it is reasonable to model the read time as a linear function of the data length. There is an initial offset, of random duration, followed by a transmission that is a clear function of the data length. The need for random restart points introduces a further random element in the transmission, but over many such operations, we expect the mean values to be linear in the data length.

Let T_t be the time to read data of length t. We measure t in kilobytes and T_t in milliseconds. Each T_t has mean μ_t, and we assume that all T_t share the common variance σ^2. We assume further that a linear relationship exists among the means, and we want to estimate the parameters of that line. Suppose that we take the following data points. We have five observations from each of the ten subpopulations. We denote the unknown relationship among the means by $\mu_t = \alpha + \beta(t - \bar{t})$, for unknown α and β. In the notation of Theorem 7.35, we have $m = 10$ and $n_i = 5$ for $1 \le i \le m$.

T_{ij} (milliseconds)									
t_i (kilobytes)									
j	100	200	300	400	500	600	700	800	900	1000
1	58.19	98.11	90.14	137.31	174.41	214.69	245.89	269.79	318.05	332.74
2	51.17	76.77	99.86	140.23	164.58	213.17	255.67	278.15	310.50	335.61
3	43.98	99.43	125.25	124.41	165.02	217.86	250.50	267.34	294.96	345.41
4	47.58	79.44	100.67	125.44	186.24	213.22	210.00	273.97	306.08	360.31
5	33.92	87.37	116.53	154.84	174.27	190.51	237.43	294.95	308.93	308.13

We compute

$$n = \sum_{i=1}^{m} n_i = 50$$

$$\bar{t} = (1/n) \sum_{i=1}^{m} n_i t_i = 550$$

$$D = \sum_{i=1}^{m} n_i (t_i - \bar{t})^2 = 825000$$

$$\hat{\alpha} = (1/n) \sum_{i=1}^{m} \sum_{j=1}^{n_i} T_{ij} = 192.18$$

$$\hat{\beta} = (1/D) \sum_{i=1}^{m} \sum_{j=1}^{n_i} T_{ij}(t_i - \bar{t}) = 0.324$$

$$\hat{\sigma}^2 = (1/n) \sum_{i=1}^{m} \sum_{j=1}^{n_i} [T_{ij} - \hat{\alpha} - \hat{\beta}(t_i - \bar{t})]^2 = 149.37.$$

Our estimate of the linear relationship is then $\mu_t = 192.18 + 0.324(t - 550)$. Using the F ratios asserted by the theorem, we can compute confidence intervals for $\hat{\alpha}$ and $\hat{\beta}$. Suppose that we want 95% confidence intervals. Since $(n-2)(\hat{\alpha} - \alpha)^2/\hat{\sigma}^2$ has a central F distribution with $(1, n-2)$ degrees of freedom, the square root $(\alpha - \hat{\alpha})\sqrt{(n-2)/\hat{\sigma}^2}$ has a t distribution with $(n-2)$ degrees of freedom. Letting $t_{p;(n-2)}$ denote the $100p$th percentile of this distribution, we obtain $t_{0.975;(48)} =$

$2.011, t_{0.025;(48)} = -2.011$ from Table 5 in Appendix B. We then compute

$$0.95 = \Pr\left(t_{0.025;(n-2)} \le (\alpha - \hat{\alpha})\sqrt{\frac{n-2}{\hat{\sigma}^2}} \le t_{0.975;(n-2)} \right)$$

$$= \Pr\left(-2.011\sqrt{\frac{\hat{\sigma}^2}{n-2}} \le \alpha - \hat{\alpha} \le 2.011\sqrt{\frac{\hat{\sigma}^2}{n-2}} \right)$$

$$= \Pr\left(\alpha \in \hat{\alpha} \mp 2.011\sqrt{\frac{\hat{\sigma}^2}{n-2}} \right) = \Pr(188.63 \le \alpha \le 195.73).$$

This gives the 95% confidence interval for $\hat{\alpha}$ as an estimator for α. For $\hat{\beta}$, we proceed in a similar manner.

$$0.95 = \Pr\left(t_{0.025;(n-2)} \le (\beta - \hat{\beta})\sqrt{\frac{(n-2)D}{n\hat{\sigma}^2}} \le t_{0.975;(n-2)} \right)$$

$$= \Pr\left(\beta \in \hat{\beta} \mp 2.011\sqrt{\frac{n\hat{\sigma}^2}{(n-2)D}} \right) = \Pr(0.311 \le \beta \le 0.336).$$

We now have separate 95% confidence intervals for each of $\hat{\alpha}$ and $\hat{\beta}$. Suppose that we want a 95% confidence *region* for the joint $(\hat{\alpha}, \hat{\beta})$ estimate. Because the F ratios above both use $\hat{\sigma}^2$ in the denominator, they are not independent. Therefore, we cannot simply compute the 97.5% confidence levels for $\hat{\alpha}$ and $\hat{\beta}$ and assert that (α, β) lies within their Cartesian product with probability $0.975^2 = 0.95$. Instead, we note that

$$\frac{\left[\dfrac{n(\hat{\alpha} - \alpha)^2}{\sigma^2} + \dfrac{D(\hat{\beta} - \beta)^2}{\sigma^2} \right]\left(\dfrac{1}{2}\right)}{\left[\dfrac{n\hat{\sigma}^2}{\sigma^2} \right]\left(\dfrac{1}{n-2}\right)} = \frac{(n-2)[n(\hat{\alpha} - \alpha)^2 + D(\hat{\beta} - \beta)^2]}{2n\hat{\sigma}^2}$$

has a central F distribution with $(2, n-2)$ degrees of freedom. This follows from the discussion preceding Theorem 7.35, where it was noted that $Q_2'/\sigma^2 = n(\hat{\alpha} - \alpha)^2/\sigma^2$ and $Q_3'/\sigma^2 = D(\hat{\beta} - \beta)^2/\sigma^2$ have central χ^2 distributions with one degree of freedom, independent of each other and of Q_1/σ^2. Consequently,

$$0.95 = \Pr\left(\frac{(n-2)[n(\hat{\alpha} - \alpha)^2 + D(\hat{\beta} - \beta)^2]}{2n\hat{\sigma}^2} \le F_{0.95;(2,n-2)} \right)$$

$$= \Pr\left(\frac{(\hat{\alpha} - \alpha)^2}{2\hat{\sigma}^2 F_{0.95;(2,n-2)}/(n-2)} + \frac{(\hat{\beta} - \beta)^2}{2n\hat{\sigma}^2 F_{0.95;(2,n-2)}/((n-2)D)} \le 1 \right).$$

In the α-β plane, this equation describes the interior of an ellipse, centered at $(\hat{\alpha}, \hat{\beta})$ with semimajor and semiminor axes of length

$$a = \sqrt{\frac{2\hat{\sigma}^2 F_{0.95;(2,n-2)}}{(n-2)}} \qquad b = \sqrt{\frac{2n\hat{\sigma}^2 F_{0.95;(2,n-2)}}{(n-2)D}}$$

parallel to the α and β axes, respectively. In the case at hand, $F_{0.95;(2,48)} = 3.193$ and the corresponding confidence region is the interior of the ellipse $(\alpha - 192.18)/(4.46)^2 + (\beta - 0.324)^2/(0.0347)^2 \le 1$. Figure 7.11 graphs this ellipse and also shows the rectangular region formed by the 97.5% confidence intervals for $\hat{\alpha}$ and $\hat{\beta}$ separately. $\square$

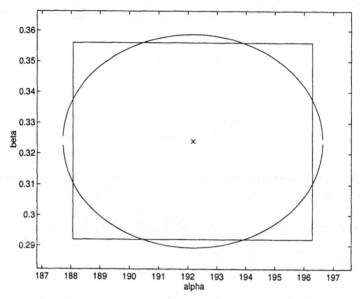

Figure 7.11. Elliptical joint $(\hat{\alpha}, \hat{\beta})$ confidence region versus rectangular confidence intervals (see Example 7.36)

Figure 7.12 generalizes the example to a regression algorithm for estimating the parameters of a linear relationship among the means of a family of independent normal populations. The final items deal with prediction intervals, which we now discuss.

Suppose that we have in hand the results of a linear regression analysis on the sample $X_{ij}, 1 \leq i \leq m, 1 \leq j \leq n_i$, and we now plan to take an observation from X_t, where t may or may not be among the t_i used in the analysis. Can we predict this observation? It is clear that we cannot do so deterministically because X_t is a random variable with mean $\alpha + \beta(t - \bar{t})$. Even if we know α and β exactly, there will still be random variation from the prediction $\alpha + \beta(t - \bar{t})$ due to the unknown variance σ^2. Having only the imperfect estimates $\hat{\alpha}$ and $\hat{\beta}$, we expect even greater variation from the prediction $\hat{\alpha} + \hat{\beta}(t - \bar{t})$. We derive now a prediction interval associated with a particular confidence value, say 95%. Before delving into the computation, we note the interpretation of this interval. To obtain a frequency-of-occurrence interpretation, we hold fixed the points $t_1 < t_2 < \ldots < t_m$ and the subsample sizes $n_1, n_2, \ldots, n_m$. We then execute the following process repeatedly.

- Take a sample $X_{ij}, 1 \leq i \leq m, 1 \leq j \leq n_i$.

- Compute the estimators $\hat{\alpha}$ and $\hat{\beta}$, with the associated intermediate values $\bar{t}, D$, and $\hat{\sigma}^2$.

- Predict value $\hat{\alpha} + \hat{\beta}(t - \bar{t})$ for X_t.

- Compute the 95% prediction interval.

1. The linear regression context is a family of independent normal random variables X_t with common variance σ^2. The means are a linear function of t. The algorithm input is a sample with n_i observations from X_{t_i}, for $t_1 < t_2 < \ldots < t_m$. Let X_{ij} denote the jth sample from X_{t_i}.

2. Define $n = \sum_{i=1}^m n_i$, $\bar{t} = (1/n)\sum_{i=1}^n n_i t_i$ and write the linear relationship among the means as $\mu_t = \alpha + \beta(t - \bar{t})$.

3. $\hat{\alpha}$ and $\hat{\beta}$ are unbiased estimators for α and β, where

$$D = \sum_{i=1}^m n_i(t_i - \bar{t})^2 \qquad\qquad \hat{\beta} = (1/D)\sum_{i=1}^m \sum_{j=1}^{n_i} X_{ij}(t_i - \bar{t})$$

$$\hat{\alpha} = (1/n)\sum_{i=1}^m \sum_{j=1}^{n_i} X_{ij} \qquad\qquad \hat{\sigma}^2 = (1/n\sum_{i=1}^m \sum_{j=1}^{n_i}[X_{ij} - \hat{\alpha} - \hat{\beta}(t_i - \bar{t})]^2.$$

4. Let $I_{p,\hat{\alpha}}$ and $J_{p,\hat{\beta}}$ be the $100p\%$ confidence intervals for α and β respectively. Then

$$I_{p,\hat{\alpha}} = \left[\hat{\alpha} - t_{(1+p)/2;(n-2)}\sqrt{\frac{\hat{\sigma}^2}{n-2}}, \hat{\alpha} + t_{(1+p)/2;(n-2)}\sqrt{\frac{\hat{\sigma}^2}{n-2}}\right]$$

$$J_{p,\hat{\beta}} = \left[\hat{\beta} - t_{(1+p)/2;(n-2)}\sqrt{\frac{n\hat{\sigma}^2}{(n-2)D}}, \hat{\beta} + t_{(1+p)/2;(n-2)}\sqrt{\frac{n\hat{\sigma}^2}{(n-2)D}}\right],$$

where $t_{(1+p)/2;(n-2)}$ is the $100(1+p)/2$ percentile of the central t distribution with $(n-2)$ degrees of freedom.

5. The $100p\%$ confidence region for (α, β) is the interior of the ellipse

$$\frac{(\alpha - \hat{\alpha})^2}{\dfrac{2\hat{\sigma}^2 F_{p;(2,n-2)}}{(n-2)}} + \frac{(\beta - \hat{\beta})^2}{\dfrac{2n\hat{\sigma}^2 F_{p;(2,n-2)}}{(n-2)D}} \leq 1,$$

where $F_{p;(2,n-2)}$ is the $100p$th percentile of the central F distribution with $(2, n-2)$ degrees of freedom.

6. For general t, not necessarily among the t_i, define the prediction interval

$$K_{p,t,\hat{\alpha},\hat{\beta}} \in \hat{\alpha} + \hat{\beta}(t - \bar{t}) \mp t_{(1+p)/2;(n-2)}\sqrt{\frac{n\hat{\sigma}^2(1 + 1/n + (t - \bar{t})^2/D)}{(n-2)}}.$$

This is the $100p\%$ prediction interval for a new observation from X_t. The point estimate is $\hat{\alpha} + \hat{\beta}(t - \bar{t})$.

7. Given a new set of observations $Y_1, Y_2, \ldots, Y_k$ all taken from subpopulation X_t, for an unknown t, define $\bar{Y} = (1/k)\sum_{i=1}^k Y_i$. The maximum likelihood estimate for t is $\hat{t} = \bar{t} + (\bar{Y} - \hat{\alpha})/\hat{\beta}$, provided that $\hat{\beta} \neq 0$. The $100p\%$ confidence interval is the set of t'-values that satisfy

$$\frac{(n-2)[\bar{Y} - \hat{\alpha} - \hat{\beta}(t' - \bar{t})]^2}{n\hat{\sigma}^2[1/k + 1/n + (t' - \bar{t})^2/D]} < F_{p;(1,n-2)}.$$

Figure 7.12. Regression algorithm for estimating the parameters of a linear relationship among subpopulation means

- Finally, take the actual observation X_t.

We will find that the actual observation X_t lies in the corresponding prediction interval for 95% of the repetitions.

Now to the actual calculation. We first note that the difference $Y = X_t - [\hat{\alpha} + \hat{\beta}(t - \bar{t})]$ is normal, being a linear combination of normal random variables, and it has parameters

$$\text{mean}(Y) = E[X_t] - E[\hat{\alpha}] - (t - \bar{t})E[\hat{\beta}] = \alpha + \beta(t - \bar{t}) - \alpha - (t - \bar{t})\beta = 0$$

$$\text{var}(Y) = \text{var}(X_t) + \text{var}(\hat{\alpha}) + (t - \bar{t})^2 \text{var}(\hat{\beta}) = \sigma^2 \left[1 + \frac{1}{n} + \frac{(t - \bar{t})^2}{D}\right].$$

The variance expression follows because X_t, a new observation, is independent of the X_{ij} that determine $\hat{\alpha}$ and $\hat{\beta}$. Of course, $\hat{\alpha}$ and $\hat{\beta}$ are independent of each other as well. The variance exhibits a first term σ^2, which is the variation expected, even if the coefficients α and β are known exactly. The remaining terms arise from the imperfect approximations $\hat{\alpha}$ and $\hat{\beta}$.

It follows that $Y/[\sigma\sqrt{1 + 1/n + (t - \bar{t})^2/D}]$ is standard normal and its square is χ^2 with one degree of freedom. Moreover, this square is independent of the $\hat{\sigma}^2$ derived in the regression analysis. To see why this follows, recall the proof of Theorem 7.35, which derives the three independent χ^2 quantities: $Q_1/\sigma^2 = (1/\sigma^2) \sum_{i=1}^{m} \sum_{j=1}^{n_i} [X_{ij} - \hat{\alpha} - \hat{\beta}(t_i - \bar{t})]^2 = n\hat{\sigma}^2/\sigma^2$, $Q_2/\sigma^2 = n\hat{\alpha}^2/\sigma^2$, and $Q_3/\sigma^2 = D\hat{\beta}^2/\sigma^2$. Of course, X_t is independent of all these because X_t is not among the X_{t_i}. Now

$$Y = X_t - \hat{\alpha} - \hat{\beta}(t - \bar{t}) = X_t - \sqrt{\frac{\sigma^2}{n}\left(\frac{Q_2}{\sigma^2}\right)} - (t - \bar{t})\sqrt{\frac{\sigma^2}{D}\left(\frac{Q_3}{\sigma^2}\right)}$$

$$= g(X_t, Q_2, Q_3),$$

a function of X_t, Q_2, and Q_3, all of which are independent of Q_1/σ^2. Recalling that $Q_1/\sigma^2 = n\hat{\sigma}^2/\sigma^2$ has a central χ^2 distribution with $n - 2$ degrees of freedom, we conclude that

$$R = \frac{\dfrac{Y^2}{\sigma^2(1 + 1/n + (t - \bar{t})^2/D)}}{\dfrac{n\hat{\sigma}^2}{(n - 2)\sigma^2}} = \frac{(n - 2)[X_t - \hat{\alpha} - \hat{\beta}(t - \bar{t})]^2}{n\hat{\sigma}^2(1 + 1/n + (t - \bar{t})^2/D)}$$

has a central F distribution with $(1, n - 2)$ degrees of freedom. Its square root then has a t distribution with $(n - 2)$ degrees of freedom, and we can construct our $100p\%$ prediction interval as follows.

$$p = \Pr\left(-t_{(1+p)/2;(n-2)} \le E \le t_{(1+p)/2;(n-2)}\right),$$

where

$$E = [X_t - \hat{\alpha} - \hat{\beta}(t - \bar{t})]\sqrt{\frac{(n - 2)}{n\hat{\sigma}^2(1 + 1/n + (t - \bar{t})^2/D)}}.$$

Rearranging, we find the equivalent expression:

$$p = \Pr\left(X_t \in \hat{\alpha} + \hat{\beta}(t - \bar{t}) \pm t_{(1+p)/2;(n-2)}\sqrt{\frac{n\hat{\sigma}^2(1 + 1/n + (t - \bar{t})^2/D)}{(n - 2)}}\right).$$

As anticipated, the prediction interval widens with increasing $(t - \bar{t})$, which corresponds to less certain prediction when t is far removed from the t_i used in the regression analysis.

7.37 Example: Continuing with the context of the preceding example, we have a regression analysis relating the random disk access time to data length. The 10 data lengths used in the analysis were $100, 200, \ldots, 1000$ kilobytes. The sample involved five observations for each subpopulation. Thus $m = 10$ and $n_i = 5$ for $1 \leq i \leq m$. The analysis provided the following intermediate results.

$$n = \sum_{i=1}^{m} n_i = 50$$

$$\bar{t} = (1/n)\sum_{i=1}^{m} n_i t_i = 550$$

$$D = \sum_{i=1}^{m} n_i(t_i - \bar{t})^2 = 825000$$

$$\hat{\alpha} = (1/n)\sum_{i=1}^{m}\sum_{j=1}^{n_i} T_{ij} = 192.18$$

$$\hat{\beta} = (1/D)\sum_{i=1}^{m}\sum_{j=1}^{n_i} T_{ij}(t_i - \bar{t}) = 0.324$$

$$\hat{\sigma}^2 = (1/n)\sum_{i=1}^{m}\sum_{j=1}^{n_i}[T_{ij} - \hat{\alpha} - \hat{\beta}(t_i - \bar{t})]^2 = 149.37$$

We now prepare to read a $t = 842$-kilobyte data block from the disk, and we wish a 95% prediction interval for the access time T_t. Following the discussion above, we ascertain $t_{(1+0.95)/2;(48)} = t_{0.975;(48)} = 2.011$ and compute as follows.

$$0.95 = \Pr\left(\hat{\alpha} + \hat{\beta}(t - \bar{t}) - 2.011\sqrt{\frac{n\hat{\sigma}^2(1 + 1/n + (t - \bar{t})^2/D)}{(n - 2)}} \leq T_t\right.$$

$$\left. \leq \hat{\alpha} + \hat{\beta}(t - \bar{t}) + 2.011\sqrt{\frac{n\hat{\sigma}^2(1 + 1/n + (t - \bar{t})^2/D)}{(n - 2)}}\right)$$

$$= \Pr(260.21 \leq T_t \leq 313.37)$$

We announce an estimate of $\hat{\alpha} + \hat{\beta}(t - \bar{t}) = 286.79$ milliseconds for T_t with a 95% prediction interval of 260.21 to 313.37 milliseconds. We append an interpretation. In a sequence of trials, each of which carries out a regression with the fixed points t_i and the fixed subpopulation sample sizes, followed by the computation of a 95% prediction interval, and then followed by an observation at $t = 842$, the final observation will fall inside the computed prediction interval for 95% of the sequence. Of

course, both the observed T_t and the endpoints of the prediction interval will vary
from trial to trial in the sequence. $\square$

Finally, suppose that, at the conclusion of the regression analysis, we
encounter k new observations, all taken from a single subpopulation $X_{t'}$. The
identifying t', however, is unknown. We wish to provide an estimate and a
corresponding confidence interval for t'. We denote the new observations by
$Y_i, 1 \le i \le k$. They have normal distributions with means $\alpha + \beta(t' - \bar{t})$ and
common variance σ^2. They are independent of each other and of the X_{ij}
used in the regression analysis. Consider the joint density of the original n
observations plus the new Y_i:

$$L(\alpha, \beta, t') = f(X_{ij}, Y_i | \alpha, \beta, t)$$

$$= \prod_{i=1}^{m} \prod_{j=1}^{n_i} \frac{1}{\sigma\sqrt{2\pi}} \exp\left(\frac{1}{2\sigma^2}[X_{ij} - \alpha - \beta(t_i - \bar{t})]^2\right)$$

$$\cdot \prod_{i=1}^{k} \frac{1}{\sigma\sqrt{2\pi}} \exp\left(\frac{1}{2\sigma^2}[Y_i - \alpha - \beta(t' - \bar{t})]^2\right).$$

Considered as a function of the unknown parameters α, β, t', for fixed sample
values X_{ij} and Y_i, this is a likelihood function. The maximum likelihood
estimates for the parameters are those that maximize this function. The
calculus is easier if we work with the natural logarithm.

$$\ln L = \frac{(n+k)\ln(2\pi\sigma^2)}{2} - \frac{1}{2\sigma^2}\sum_{i=1}^{m}\sum_{j=1}^{n_i}[X_{ij} - \alpha - \beta(t_i - \bar{t})]^2$$

$$- \frac{1}{2\sigma^2}\sum_{i=1}^{k}[Y_i - \alpha - \beta(t' - \bar{t})]^2$$

Because $\ln L$ is monotone in L, both achieve their maxima at the same point.
We proceed by calculating the partial derivatives of $\ln L$.

$$\frac{\partial \ln L}{\partial \alpha} = \frac{-1}{2\sigma^2}\sum_{i=1}^{m}\sum_{j=1}^{n_i} 2[X_{ij} - \alpha - \beta(t_i - \bar{t})](-1)$$

$$+ \frac{-1}{2\sigma^2}\sum_{i=1}^{k} 2[Y_i - \alpha - \beta(t' - \bar{t})](-1)$$

$$= 0$$

$$\frac{\partial \ln L}{\partial \beta} = \frac{-1}{2\sigma^2}\sum_{i=1}^{m}\sum_{j=1}^{n_i} 2[X_{ij} - \alpha - \beta(t_i - \bar{t})][-(t_i - \bar{t})]$$

$$+ \frac{-1}{2\sigma^2}\sum_{i=1}^{k} 2[Y_i - \alpha - \beta(t' - \bar{t})][-(t' - \bar{t})]$$

$$= 0$$

$$\frac{\partial \ln L}{\partial t'} = \frac{-1}{2\sigma^2}\sum_{i=1}^{k} 2[Y_i - \alpha - \beta(t' - \bar{t})](-\beta) = 0$$

We use the notation $\tilde{\alpha}, \tilde{\beta}$, and $\hat{t}$ for the minimizing values. In the equations above, we replace the α, β, and t' occurrences with $\tilde{\alpha}, \tilde{\beta}$, and $\hat{t}$, respectively. We then define $\overline{Y} = (1/k)\sum_{i=1}^{k} Y_i$ and recall that $\sum_{i=1}^{m}\sum_{j=1}^{n_i} X_{ij} = n\hat{\alpha}$ and $\sum_{i=1}^{m}\sum_{j=1}^{n_i} X_{ij}(t_i - \bar{t}) = D\hat{\beta}$. The partial derivative equations then appear as

$$(n+k)\tilde{\alpha} + k(\hat{t}-\bar{t})\tilde{\beta} = n\hat{\alpha} + k\overline{Y}$$
$$k(\hat{t}-\bar{t})\tilde{\alpha} + [D + k(\hat{t}-\bar{t})^2]\tilde{\beta} = D\hat{\beta} + k(\hat{t}-\bar{t})\overline{Y}$$
$$k\tilde{\alpha} + k(\hat{t}-\bar{t})\tilde{\beta} = k\overline{Y}.$$

Substituting $k\overline{Y}$ from the third equation into the first yields $\tilde{\alpha} = \hat{\alpha}$. Similarly, substituting $(\hat{t}-\bar{t})$ times $k\overline{Y}$ from the third equation into the second gives $\tilde{\beta} = \hat{\beta}$. Back substituting these values into the first equation gives $\hat{t} = \bar{t} + (\overline{Y}-\hat{\alpha})/\hat{\beta}$, provided that $\hat{\beta} \neq 0$. If $\hat{\beta} = 0$, the regression line is horizontal and provides an estimate of the means that does not vary with location t across the family X_t. In this case, the analysis cannot extract any information about the location t' of the new observations. If $\hat{\beta} \neq 0$, however, $\hat{t}$ is the maximum likelihood estimate for t'. Moreover, the quantity

$$\hat{\beta}(\hat{t}-t') = \hat{\beta}(\hat{t}-\bar{t}) - \hat{\beta}(t'-\bar{t}) = \overline{Y} - \hat{\alpha} - \hat{\beta}(t'-\bar{t})$$

has mean and variance

$$\text{mean}(\hat{\beta}(\hat{t}-t')) = E[\overline{Y}] - E[\hat{\alpha}] - (t'-\bar{t})E[\hat{\beta}] = \alpha + \beta(t'-\bar{t}) - \alpha - \beta(t'-\bar{t}) = 0$$
$$\text{var}(\hat{\beta}(\hat{t}-t')) = \text{var}(Y) + \text{var}(\hat{\alpha}) + (t'-\bar{t})^2\text{var}(\hat{\beta}) = \frac{\sigma^2}{k} + \frac{\sigma^2}{n} + \frac{\sigma^2(t'-\bar{t})^2}{D}.$$

The variance expression follows because $\overline{Y}$ depends only on the new observations and is therefore independent of $\hat{\alpha}$ and $\hat{\beta}$, which derive from the original X_{ij}. Then

$$\left(\frac{\hat{\beta}(\hat{t}-t')}{\sigma\sqrt{1/k + 1/n + (t'-\bar{t})^2/D}}\right)^2 = \frac{[\overline{Y} - \hat{\alpha} - \hat{\beta}(t'-\bar{t})]^2}{\sigma^2[1/k + 1/n + (t'-\bar{t})^2/D]}$$

is χ^2 with one degree of freedom. Being a function of the new observations, $\hat{\alpha}$, and $\hat{\beta}$, it remains independent of $\hat{\sigma}^2$. Therefore,

$$\frac{\dfrac{\hat{\beta}^2(\hat{t}-t')^2}{\sigma^2[1/k + 1/n + (t'-\bar{t})^2/D]}}{\dfrac{n\hat{\sigma}^2}{(n-2)\sigma^2}} = \frac{(n-2)[\overline{Y} - \hat{\alpha} - \hat{\beta}(t'-\bar{t})]^2}{n\hat{\sigma}^2[1/k + 1/n + (t'-\bar{t})^2/D]}$$

has a central F distribution with $(1, n-2)$ degrees of freedom. We obtain the $100p\%$ confidence interval for t' by solving for the set of t' values for which this expression is less than $F_{p;(1,n-2)}$. Note that this interval always contains $\hat{t}$ because the numerator square is zero when $t' = \hat{t}$. The next example illustrates the technique.

7.38 Example: We continue with the context of Examples 7.36 and 7.37. We have a linear regression that relates disk access time to the length of the data transferred. The model is $T = \hat{\alpha} + \hat{\beta}(t - \bar{t})$, where T is in milliseconds, t is in kilobytes, and $\hat{\alpha} = 192.18$, $\hat{\beta} = 0.324$, and $\bar{t} = 550$. The previous analysis also provides $n = 50$, $D = 825000$, and $\hat{\sigma}^2 = (1/n)\sum_{i=1}^{m}\sum_{j=1}^{n_i}[T_{ij} - \hat{\alpha} - \hat{\beta}(t_i - \bar{t})]^2 = 149.37$. The read times associated with six new data transfers, all of the same unknown length, are

i	1	2	3	4	5	6
Y_i	309.28	302.82	296.20	299.52	286.94	315.65

from which we ascertain $k = 6$ and $\overline{Y} = 301.74$. Our estimate for the unknown data length is then

$$\hat{t} = \bar{t} + \frac{\overline{Y} - \hat{\alpha}}{\hat{\beta}} = 550 + \frac{301.74 - 192.18}{0.324} = 888.14.$$

To compute a 95% confidence interval, we first look up the percentile, $F_{0.95;(1,48)} = 4.044$. We then solve for the t values that satisfy

$$\frac{(n-2)[\overline{Y} - \hat{\alpha} - \hat{\beta}(t - \bar{t})]^2}{n\hat{\sigma}^2[1/k + 1/n + (t - \bar{t})^2/D]} < 4.044.$$

In this case, the equation becomes $(t - 550)^2 - 681.27(t - 550) + 114060 < 0$, which yields $846.23 < t < 935.05$. Note that the confidence interval contains $\hat{t} = 888.14$. $\square$

The regression analyses studied to this point have all assumed that a linear relationship exists among the subpopulation means. They have endeavored to estimate the linearity parameters and to predict the value or location of a new observation. We now turn to the case where the linear relationship is in doubt. In particular, we establish a null hypothesis that the linear relationship exists and develop a test for a prespecified significance level. We maintain the notation established above. If the null hypothesis is true, the offset random variables $X_{ij} - \alpha - \beta(t_i - \bar{t})$ all have the same mean, which is zero. If α and β were known, the test would be a straightforward analysis of variance, which would test the validity of a *particular* linear relationship. Unfortunately, α and β are unknown, and we want to know if *any* linear relationship holds among the means. To this end, we compute $\hat{\alpha}, \hat{\beta}, \hat{\sigma}^2$ exactly as in the preceding discussion. In proving Theorem 7.35, we established the decomposition

$$\sum_{i=1}^{m}\sum_{j=1}^{n_i} X_{ij}^2 = \sum_{i=1}^{m}\sum_{j=1}^{n_i}[X_{ij} - \hat{\alpha} - \hat{\beta}(t_i - \bar{t})]^2 + n\hat{\alpha}^2 + \hat{\beta}^2\sum_{i=1}^{m} n_i(t_i - \bar{t})^2$$

$$= Q_1 + Q_2 + Q_3$$

and showed that the Q_k/σ^2 have independent χ^2 distributions with $n - 2, 1, 1$ degrees of freedom, respectively. We now propose a further decomposition of Q_1. Accordingly, define $X_{i\cdot} = (1/n_i)\sum_{j=1}^{n_i} X_{ij}$. Then, because the cross-product terms again sum to zero, we have

$$\sum_{i=1}^{m}\sum_{j=1}^{n_i} X_{ij}^2 = \sum_{i=1}^{m}\sum_{j=1}^{n_i}[(X_{ij} - X_{i\cdot}) + (X_{i\cdot} - \hat{\alpha} - \hat{\beta}(t_i - \bar{t})) + \hat{\alpha} + \hat{\beta}(t_i - \bar{t})]^2$$

$$= \sum_{i=1}^{m}\sum_{j=1}^{n_i}(X_{ij} - X_{i\cdot})^2 + \sum_{i=1}^{m} n_i[X_{i\cdot} - \hat{\alpha} - \hat{\beta}(t_i - \bar{t})]^2 + n\hat{\alpha}^2 + D\hat{\beta}^2$$

$$= Q_{1a} + Q_{1b} + Q_2 + Q_3.$$

We continue to denote by $\mathbf{X}$ the vector that stacks the X_{ij} for each i and then stacks the resulting m segments. From the proof of Theorem 7.35, we extract the matrices $\mathbf{A}_2$ and $\mathbf{A}_3$, each of rank 1, such that $Q_2 = (\mathbf{A}_2\mathbf{X})'(\mathbf{A}_2\mathbf{X})$ and $Q_3 = (\mathbf{A}_3\mathbf{X})'(\mathbf{A}_3\mathbf{X})$.

For Q_{1a}, we seek a matrix $\mathbf{A}_{1a}$ such that $Q_{1a} = (\mathbf{A}_{1a}\mathbf{X})'(\mathbf{A}_{1a}\mathbf{X})$. Accordingly, we write

$$Q_{1a} = \sum_{i=1}^{m}\sum_{j=1}^{n_i}\left[X_{ij} - \frac{1}{n_i}\sum_{l=1}^{n_i} X_{il}\right]^2,$$

from which we can read the structure of $\mathbf{A}_{1a}$. It is an $n \times n$ matrix, consisting of m row groups. Group i contains n_i rows. Each row contains n entries, which we identify in m segments. The kth segment contains n_k entries. Row j of group i contains

$$\text{entry } l \text{ of segment } k = \begin{cases} 0, & k \neq i \\ 1 - 1/n_i, & k = i, l = j \\ -1/n_i, & k = i, l \neq j. \end{cases}$$

Summing the entries in column l of segment k across the rows within group i, we encounter one entry $(1 - 1/n_i)$ and $(n_i - 1)$ entries $(-1/n_i)$. The sum is zero. Consequently, there are at most $(n_i - 1)$ independent rows in each group. The rank of the matrix is at most $\sum_{i=1}^{m}(n_i - 1) = n - m$.

For Q_{1b}, we seek $\mathbf{A}_{1b}$ such that $Q_{1b} = (\mathbf{A}_{1b}\mathbf{X})'(\mathbf{A}_{1b}\mathbf{X})$.

$$Q_{1b} = \sum_{i=1}^{m} n_i\left[\frac{1}{n_i}\sum_{l=1}^{n_i} X_{il} - \frac{1}{n}\sum_{k=1}^{m}\sum_{l=1}^{n_k} X_{kl} - \frac{(t_i - \bar{t})}{D}\sum_{k=1}^{m}\sum_{l=1}^{n_k} X_{kl}(t_k - \bar{t})\right]^2.$$

$\mathbf{A}_{1b}$ is an $m \times n$ matrix, for which row i has

$$\text{entry } l \text{ of segment } k = \begin{cases} \sqrt{n_k}\left(\dfrac{1}{n_k} - \dfrac{1}{n} - \dfrac{(t_k - \bar{t})^2}{D}\right), & k = i \\ \sqrt{n_i}\left(-\dfrac{1}{n} - \dfrac{(t_i - \bar{t})(t_k - \bar{t})}{D}\right), & k \neq i. \end{cases}$$

We multiply row i by $\sqrt{n_i}$ and sum. In column l of segment k, we have

$$\text{sum} = n_k\left(\frac{1}{n_k} - \frac{1}{n} - \frac{(t_k - \bar{t})^2}{D}\right) + \sum_{i \neq k} n_i\left(-\frac{1}{n} - \frac{(t_i - \bar{t})(t_k - \bar{t})}{D}\right)$$

$$= 1 - \frac{1}{n}\sum_{i=1}^{m} n_i - \frac{(t_k - \bar{t})}{D}\sum_{i=1}^{m} n_i(t_i - \bar{t}) = 0.$$

Now we multiply row i by $\sqrt{n_i}(t_i - \bar{t})$ and sum. In column l of segment k, we have

$$\text{sum} = n_k(t_k - \bar{t})\left(\frac{1}{n_k} - \frac{1}{n} - \frac{(t_k - \bar{t})^2}{D}\right) + \sum_{i \neq k} n_i(t_i - \bar{t})\left(-\frac{1}{n} - \frac{(t_i - \bar{t})(t_k - \bar{t})}{D}\right)$$

$$= (t_k - \bar{t}) - \frac{1}{n}\sum_{i=1}^{m} n_i(t_i - \bar{t}) - \frac{(t_k - \bar{t})}{D}\sum_{i=1}^{m} n_i(t_i - \bar{t})^2 = 0.$$

The null space of $\mathbf{A}_{1b}$ is therefore of dimension at least 2, which means that the rank is at most $m - 2$. We invoke Theorem 7.22 to conclude that $Q_{1a}/\sigma^2, Q_{1b}/\sigma^2, Q_2/\sigma^2$, and Q_3/σ^2 have independent χ^2 distributions with $n - m, m - 2, 1$, and 1 degrees of freedom, respectively. We worked out the noncentrality parameters of Q_2/σ^2 and Q_3/σ^2 earlier. Of interest here are the parameters for Q_{1a}/σ^2 and Q_{1b}/σ^2. As before, we obtain them by substituting μ_{t_i}/σ for X_{ij}/σ in the quadratic form. Writing μ_i for μ_{t_i}, we obtain

$$\lambda_{1a} = \frac{1}{2\sigma^2}\sum_{i=1}^{m}\sum_{j=1}^{n_i}\left(\mu_i - \frac{1}{n_i}\sum_{l=1}^{n_i}\mu_i\right)^2 = 0$$

$$\lambda_{1b} = \frac{1}{2\sigma^2}\sum_{i=1}^{m} n_i\left(\frac{1}{n_i}\sum_{l=1}^{n_i}\mu_i - \frac{1}{n}\sum_{k=1}^{m}\sum_{l=1}^{n_k}\mu_k - \frac{(t_i - \bar{t})}{D}\sum_{k=1}^{m}\sum_{l=1}^{n_k}\mu_k(t_k - \bar{t})\right)^2$$

$$= \frac{1}{2\sigma^2}\sum_{i=1}^{m} n_i\left(\mu_i - \frac{1}{n}\sum_{k=1}^{m} n_k\mu_k - \frac{(t_i - \bar{t})}{D}\sum_{k=1}^{m} n_k\mu_k(t_k - \bar{t})\right)^2.$$

If $\lambda_{1b} = 0$, then for each μ_i we have

$$\mu_i = \frac{1}{n}\sum_{k=1}^{m} n_k\mu_k + \left(\frac{\sum_{k=1}^{m} n_k\mu_k(t_k - \bar{t})}{D}\right)(t_i - \bar{t}) = \alpha + \beta(t_i - \bar{t}),$$

a linear relationship among the μ_i. Conversely, if $\mu_i = \alpha + \beta(t_i - \bar{t})$, we can verify that $\lambda_{1b} = 0$. We conclude that

$$\frac{Q_{1b}/[(m-2)\sigma^2]}{Q_{1a}/[(n-m)\sigma^2]} = \frac{(n-m)\sum_{i=1}^{m} n_i[X_{i\cdot} - \hat{\alpha} - \hat{\beta}(t_i - \bar{t})]^2}{(m-2)\sum_{i=1}^{m}\sum_{j=1}^{n_i}(X_{ij} - X_{i\cdot})^2}$$

has an F distribution with $(m - 2, n - m)$ degrees of freedom. If the null hypothesis is true, it is a central distribution. Otherwise, it is noncentral with noncentrality parameter λ_{1b}.

7.39 Theorem: Let X_t be a family of independent normal random variables with common variance σ^2 and means μ_t. For the fixed set of t-values, $t_1 < t_2 < \ldots < t_m$, let $X_{ij}, 1 \leq j \leq n_i$, be a sample from subpopulation X_{t_i}. Write μ_i for μ_{t_i}. Define

$$n = \sum_{i=1}^{m} n_i \qquad \hat{\alpha} = (1/n)\sum_{i=1}^{m}\sum_{j=1}^{n_i} X_{ij}$$
$$\bar{t} = (1/n)\sum_{i=1}^{m} n_i t_i \qquad \hat{\beta} = (1/D)\sum_{i=1}^{m}\sum_{j=1}^{n_i} X_{ij}(t_i - \bar{t})$$
$$D = \sum_{i=1}^{m} n_i(t_i - \bar{t})^2 \qquad X_{i\cdot} = (1/n_i)\sum_{j=1}^{n_i} X_{ij}.$$

Then

$$L = \frac{(n-m)\sum_{i=1}^{m} n_i [X_{i\cdot} - \hat{\alpha} - \hat{\beta}(t_i - \bar{t})]^2}{(m-2)\sum_{i=1}^{m}\sum_{j=1}^{n_i}(X_{ij} - X_{i\cdot})^2}$$

has an F distribution with $(m-2, n-m)$ degrees of freedom. The noncentrality parameter is

$$\lambda = \frac{1}{2\sigma^2}\sum_{i=1}^{m} n_i \left(\mu_i - \frac{1}{n}\sum_{k=1}^{m} n_k \mu_k - \frac{(t_i - \bar{t})}{D}\sum_{k=1}^{m} n_k \mu_k (t_k - \bar{t})\right)^2.$$

$\lambda = 0$ if and only if $\mu_i = \alpha + \beta(t_i - \bar{t})$ for some α, β. For a null hypothesis asserting a linear relationship among the means, the critical region for a test at significance level $1 - p$ is $L > F_{p;(m-2,n-m)}$.

PROOF: The assertions all follow from the preceding discussion. ∎

The statistic L of Theorem 7.39 provides a test for linearity across the subpopulation means. When constructed at significance level $1-p$, the chance of a type I error in rejecting linearity is $1 - p$ or less. However, the chance of a type II error in accepting linearity depends on the degree of nonlinearity actually present. Because the noncentrality parameter λ is zero if and only if linearity holds, we can use λ, or some function of it, as a measure of the nonlinearity. As was the case with the analysis of variance tests, this nonlinearity is most conveniently expressed in σ-units. That is, we define

$$\delta = \frac{1}{\sigma}\sqrt{\frac{1}{n}\sum_{i=1}^{m} n_i \left(\mu_i - \frac{1}{n}\sum_{k=1}^{m} n_k \mu_k - \frac{(t_i - \bar{t})}{D}\sum_{k=1}^{m} n_k \mu_k (t_k - \bar{t})\right)^2},$$

and we express the power of the test as a function of δ. Clearly, $\lambda = n\delta^2/2$. Therefore, the test's power is

$$\mathcal{P}(\delta) = \Pr(L > F_{1-p;(m-2,n-m)}|\delta),$$

which is the probability that a noncentral F random variable with $(m-2, n-m)$ degrees of freedom and noncentrality parameter $n\delta^2/2$ exceeds the percentile threshold $F_{1-p;(m-2,n-m)}$. Because the type II error probability is $1 - \mathcal{P}(\delta)$ and because $\mathcal{P}$ is a continuous function of δ, the type II error probability is unavoidably large when δ is small. This difficulty is not serious because accepting linearity when only a slight degree of nonlinearity is present may be harmless. A good test will, however, exhibit a power function that rises rapidly with δ. This reduces the type II error probability when a significant amount of nonlinearity is present. The next example blends a controlled amount of nonlinearity into a sample and investigates the consequences.

7.40 Example: The following data were generated with a normal random number generator to follow the pattern

$$X_{ij} = 50.0 + t_i + 0.05t_i^2 + N_{ij}(0, 400),$$

where N_{ij} is normal with mean zero and variance $\sigma^2 = 400$. The quadratic term causes the means to deviate slightly from linear.

				t_i				
j	10	20	30	40	50	60	70	80
1	78.34	79.72	150.79	165.44	206.92	258.26	364.05	442.18
2	63.64	127.93	114.39	177.54	238.97	271.58	352.71	422.36
3	30.29	84.94	111.14	194.43	234.65	277.71	369.81	471.53
4	81.13	86.51	107.80	191.97	241.23	283.31	367.50	475.39
5	46.70	109.58	135.58	152.94	231.56	291.61	360.53	459.75

We compute

$$m = 8 \qquad\qquad D = \sum_{i=1}^{m} n_i(t_i - \bar{t})^2 = 21000$$

$$n_i = 5, \text{ for all } i \qquad \hat{\alpha} = (1/n)\sum_{i=1}^{m}\sum_{j=1}^{n_i} X_{ij} = 222.81$$

$$n = \sum_{i=1}^{m} n_i = 40 \qquad \hat{\beta} = (1/D)\sum_{i=1}^{m}\sum_{j=1}^{n_i} X_{ij}(t_i - \bar{t}) = 5.47.$$

$$\bar{t} = (1/n)\sum_{i=1}^{m} n_i t_i = 45.0$$

Also,

$$X_{i\cdot} = \left(\frac{1}{n_1}\sum_{j=1}^{n_1} X_{1j}, \ldots, \frac{1}{n_m}\sum_{j=1}^{n_m} X_{mj} \right)$$

$$= (60.02, 97.74, 123.94, 176.46, 230.67, 276.49, 362.92, 454.24)$$

$$L = \frac{(n-m)\sum_{i=1}^{m} n_i[X_{i\cdot} - \hat{\alpha} - \hat{\beta}(t_i - \bar{t})]^2}{(m-2)\sum_{i=1}^{m}\sum_{j=1}^{n_i}(X_{ij} - X_{i\cdot})^2} = 12.153.$$

Because we know the true means, we can compute

$$\mu_i = (65, 90, 125, 170, 225, 290, 365, 450)$$

$$\lambda = \frac{1}{2\sigma^2}\sum_{i=1}^{m} n_i \left(\mu_i - \frac{1}{n}\sum_{k=1}^{m}(5)\mu_k - \frac{(t_i - \bar{t})}{D}\sum_{k=1}^{m} n_k\mu_k(t_k - \bar{t}) \right)^2 = 26.25,$$

which gives $\delta = \sqrt{2\lambda/n} = 1.15$. Recall that δ measures the nonlinearity in σ units. Here, we have nonlinearity equivalent in size to one standard deviation, and accepting the linearity hypothesis is a serious error. To construct a test at the 5% significance level, we establish the critical region as

$$L > F_{0.95;(m-2,n-m)} = F_{0.95;(6,32)} = 2.404.$$

The chance erroneously accepting linearity is $\Pr(L < 2.404|\delta = 1.15)$, which is the probability that a noncentral F random variable with $(6,32)$ degrees of freedom and noncentrality parameter $\lambda = n\delta^2/2 = 26.25$ is less than 2.404. We can obtain this value by integrating the noncentral F density of Theorem 6.68. A numerical integration gives 0.0002, a comforting result. Indeed, since the observed L is 12.153, which is much larger than the threshold 2.404, we reject the linearity hypothesis.

By repeating the numerical integration of the noncentral F distribution for varying δ, we obtain Figure 7.13, which graphs the power function of the test ($L > 2.404$). The plot shows that the probability of erroneously accepting the linearity hypothesis is small, say less than 0.1, when δ exceeds 0.75.

Suppose that we now decrease the quadratic effect by generating $Y_{ij} = 50 + t_i + 0.005t_i^2 + N_{ij}(0, 400)$. The new data appear below.

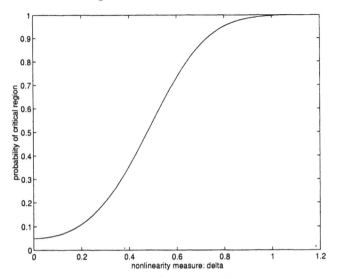

Figure 7.13. Power function of the hypothesis test with sample size 40
(see Example 7.40)

				t_i				
j	10	20	30	40	50	60	70	80
1	41.37	92.85	64.00	106.39	100.27	117.88	102.05	148.53
2	69.54	96.18	63.36	83.15	114.51	108.34	127.49	181.80
3	35.42	87.62	142.25	109.46	90.24	119.26	136.02	150.45
4	65.63	48.40	79.15	95.13	131.64	160.81	123.91	136.43
5	68.92	92.03	74.73	94.86	100.94	138.61	137.52	167.63

Some computations remain unchanged, such as $\bar{t}$ and D, while others do not. The
full set of new computations is as follows.

$$m = 8 \qquad\qquad D = \sum_{i=1}^{m} n_i(t_i - \bar{t})^2 = 21000$$

$$n_i = 5, \text{ for all } i \qquad \hat{\alpha} = (1/n)\sum_{i=1}^{m}\sum_{j=1}^{n_i} X_{ij} = 105.12$$

$$n = \sum_{i=1}^{m} n_i = 40 \qquad \hat{\beta} = (1/D)\sum_{i=1}^{m}\sum_{j=1}^{n_i} X_{ij}(t_i - \bar{t}) = 1.26$$

$$\bar{t} = (1/n)\sum_{i=1}^{m} n_i t_i = 45.0$$

$$X_{i\cdot} = \left(\frac{1}{n_1}\sum_{j=1}^{n_1} X_{1j} \dots \frac{1}{n_m}\sum_{j=1}^{n_m} X_{mj}\right)$$

$$= (56.18, 83.42, 84.70, 97.80, 107.52, 128.98, 125.40, 156.97)$$

$$L = \frac{(n-m)\sum_{i=1}^{m} n_i[X_{i\cdot} - \hat{\alpha} - \hat{\beta}(t_i - \bar{t})]^2}{(m-2)\sum_{i=1}^{m}\sum_{j=1}^{n_i}(X_{ij} - X_{i\cdot})^2} = 0.758$$

Because the data are artificially constructed, we know the true means: $50 + t_i + 0.005t_i^2$. Therefore,

$$\mu_i = 60.5, 72.0, 84.5, 98.0, 112.5, 128.0, 144.5, 162.0)$$

$$\lambda = \frac{1}{2\sigma^2}\sum_{i=1}^{m} n_i\left(\mu_i - \frac{1}{n}\sum_{k=1}^{m}(5)\mu_k - \frac{(t_i - \bar{t})}{D}\sum_{k=1}^{m} n_i\mu_k(t_k - \bar{t})\right)^2 = 0.262,$$

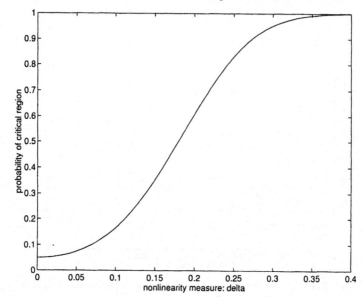

Figure 7.14. Power function of the hypothesis test with sample size 240
(see Example 7.40)

which gives the new $\delta = \sqrt{2\lambda/n} = 0.11$. Consulting Figure 7.13, we see that the test's power function is about 0.07 at $\delta = 0.11$, which means that the probability of erroneously accepting the linearity hypothesis is about $1 - 0.07 = 0.93$. So, we should not be surprised if the test accepts linearity. Indeed, because $0.758 < 2.404 = F_{0.95;(6,32)}$, we do accept linearity. This does not indicate a bad test because the data are very nearly linear.

We can carry out the computations for the type II error probabilities here only because the contrived data let us compute the nonlinearity measure δ. In general, however, this is not possible. We can, nevertheless, present a plot of the test's power function. If we are unhappy with the power curve, we can make it rise faster with increasing δ by taking a larger sample. For example, if we enlarge the sample size to 240 by taking 30 observations at each t_i, the threshold of the critical region decreases to $F_{0.95;(6,232)} = 2.138$. The power function then assumes the shape shown in Figure 7.14. At $\delta = 0.11$, the power function is now about 0.21, which gives 0.79 as the probability of a type II error. This is still quite high, and if the data give the same $L = 0.758$, linearity is still accepted. $\square$

Exercises

7.33 Let $X_{ij}, 1 \le i \le m, 1 \le j \le n_i$, be observations from subpopulations X_{t_i} as in the definition of a linear regression. Define

$$X_{i\cdot} = \frac{1}{n_i} \sum_{j=1}^{n_i} X_{ij} \qquad X_{\cdot\cdot} = \frac{1}{n} \sum_{i=1}^{m} \sum_{j=1}^{n_i} X_{ij}.$$

Show that the minimum of $E(\alpha, \beta) = \sum_{i=1}^{m} \sum_{j=1}^{n_i} [X_{ij} - (\alpha + \beta(t_i - \bar{t}))]^2$

occurs at $\alpha = X.., \beta = \left(\sum_{i=1}^{m} n_i(t_i - \bar{t})(X_i. - X..)\right) / \left(\sum_{i=1}^{m} n_i(t_i - \bar{t})^2\right)$.

7.34 In the context of Theorem 7.35, justify the notation

$$\hat{\sigma}^2 = \frac{1}{n} \sum_{i=1}^{m} \sum_{j=1}^{n_i} [X_{ij} - \hat{\alpha} - \hat{\beta}(t_i - \bar{t})]^2$$

by showing that $E[\hat{\sigma}^2] = \sigma^2$, the common variance of the X_{ij}.

7.35 Show that the cross-product terms are all zero in the expansion

$$\sum_{i=1}^{m} \sum_{j=1}^{n_i} X_{ij}^2 = \sum_{i=1}^{m} \sum_{j=1}^{n_i} [(X_{ij} - \hat{\alpha} - \hat{\beta}(t_i - \bar{t})) + \hat{\alpha} + \hat{\beta}(t_i - \bar{t})]^2$$

$$= \sum_{i=1}^{m} \sum_{j=1}^{n_i} [X_{ij} - \hat{\alpha} - \hat{\beta}(t_i - \bar{t})]^2 + \sum_{i=1}^{m} \sum_{j=1}^{n_i} \hat{\alpha}^2$$

$$+ \sum_{i=1}^{m} \sum_{j=1}^{n_i} \hat{\beta}^2 (t_i - \bar{t})^2 + 2 \sum_{i=1}^{m} \sum_{j=1}^{n_i} \hat{\alpha}[X_{ij} - \hat{\alpha} - \hat{\beta}(t_i - \bar{t})]$$

$$+ 2 \sum_{i=1}^{m} \sum_{j=1}^{n_i} \hat{\beta}(t_i - \bar{t})[X_{ij} - \hat{\alpha} - \hat{\beta}(t_i - \bar{t}]$$

$$+ 2 \sum_{i=1}^{m} \sum_{j=1}^{n_i} \hat{\alpha}\hat{\beta}(t_i - \bar{t}).$$

*7.36 In the context of Theorem 7.35, show that $\hat{\alpha}$ and $\hat{\beta}$ are the maximum likelihood estimates for α and β. Recall that the maximum likelihood estimates are those that maximize the likelihood function. The likelihood function, given a particular sample $X_{ij}, 1 \le i \le m, 1 \le j \le n_i$, is the probability density of the sample, considered as a function of the parameters α and β. In this case,

$$L(\alpha, \beta) = \prod_{i=1}^{m} \prod_{j=1}^{n_i} \frac{1}{\sqrt{2\pi\sigma^2}} \exp\left(\frac{[X_{ij} - (\alpha + \beta(t - \bar{t})]^2}{2\sigma^2}\right).$$

7.37 Subpopulations X_t have independent normal distributions with a common variance and a linear relationship among the means. Conduct a linear regression on the data to the right. Estimate the linear relationship $\mu_t = \alpha + \beta(t - \bar{t})$. Find the unbiased estimates $\hat{\alpha}$ and $\hat{\beta}$, their individual 95% confidence intervals. In addition, find their joint

			X_{ij}		
			t_i		
j	84	93	110	128	150
1	191.62	330.46	354.56	394.06	345.10
2	316.25	221.04	282.07	347.70	478.34
3	294.31	345.13	315.15	371.05	347.58
4	258.78	376.87	249.65	339.88	421.14
5	277.94	256.78	371.54	337.58	357.78
6	347.52	235.47	282.01	413.39	384.42
7	269.61	237.98	363.71	358.07	419.89

95% confidence region. Finally, determine the 95% prediction interval for an observation from X_{175}.

7.38 Subpopulations X_t have independent normal distributions with a common variance and a linear relationship among the means. Conduct a linear regression on the following data to estimate the linear relationship $\mu_t = \alpha + \beta(t - \bar{t})$. Find the unbiased estimates $\hat{\alpha}$ and $\hat{\beta}$, their individual 95% confidence intervals, and their joint 95% confidence region.

				X_{ij}			
				t_i			
j	10	15	25	32	46	58	63
1	88.44	121.51	130.61	173.93	152.88	147.94	100.83
2	118.54	151.45	136.90	151.95	111.67	152.03	182.26
3	168.20	178.25	133.14	164.32	175.09	168.87	133.40
4	156.11	161.11	142.17	147.43	160.92	176.97	182.14
5	137.03	133.00	159.42	193.73	166.40	130.16	138.76
6	152.36	164.15	123.06	169.26	171.65	142.80	170.54
7	121.24	138.80	154.71	108.04	191.28	155.99	174.69

7.39 Given the data below, construct a linear regression to obtain $\mu_t = \hat{\alpha} + \hat{\beta}(t - \bar{t})$. For an unknown t, we obtain the following new observations from the subpopulation X_t: $71.53, 88.59, 92.96$. Obtain the maximum likelihood estimate for t and construct a 95% confidence interval.

				X_{ij}			
				t_i			
j	50	63	75	102	141	155	168
1	83.30	74.18	109.92	86.59	149.88	172.50	184.36
2	72.54	108.94	72.22	131.69	161.27	187.50	167.78
3	61.50	78.28	96.55	111.10	158.93	179.57	177.96
4	67.03	90.43	77.80	96.02	166.12	117.45	210.13
5	46.07	44.08	82.29	96.70	159.01	159.52	194.95
6	93.92	58.98	58.01	129.23	124.17	158.54	183.36

7.40 A water control board has the following inflation-adjusted costs for well installations in its county. The well depth in in feet and the cost is in dollars.

Depth	54	83	85	91	97	110
Cost	1517.45	2323.65	2747.61	3462.91	1605.85	2590.90

	115	135	153
	3821.32	4030.83	4036.32

The board is aware that a landowner installed a new well at a cost of $2450. The board wishes to estimate the well depth without making a site visit. Perform a regression analysis to obtain the estimate and provide a 70% confidence interval. What assumptions must you make?

7.41 The following sum of squares appeared in the proof of Theorem 7.39. Show that the cross-product terms sum to zero.

$$\sum_{i=1}^{m}\sum_{j=1}^{n_i} X_{ij}^2 = \sum_{i=1}^{m}\sum_{j=1}^{n_i} \Big[(X_{ij} - X_{i\cdot}) + (X_{i\cdot} - \hat{\alpha} - \hat{\beta}(t_i - \bar{t}))$$

$$+ \hat{\alpha} + \hat{\beta}(t_i - \bar{t})\Big]^2$$

7.42 Test the data below for linearity at the 5% significance level. If you accept linearity, compute $\hat{\alpha}$ and $\hat{\beta}$ and their 95% confidence region.

| | | | X_{ij} | | |
| | | | t_i | | |
j	200	300	400	500	600
1	103.30	41.08	59.10	57.50	56.95
2	92.54	55.98	44.02	72.50	45.36
3	81.50	94.92	44.70	64.57	21.53
4	87.03	57.22	77.23	2.45	38.59
5	66.07	81.55	58.88	44.52	42.96
6	113.92	62.80	70.27	43.54	28.86
7	81.18	67.29	67.93	56.36	33.27
8	115.94	43.01	75.12	39.78	48.30
9	85.28	44.59	68.01	49.96	71.16
10	97.43	89.69	33.17	82.13	8.89

7.43 Test the data below for linearity at the 5% significance level. If you accept linearity, compute $\hat{\alpha}$ and $\hat{\beta}$ and their 95% confidence region.

| | | | | X_{ij} | | | |
| | | | | t_i | | | |
j	26	34	48	62	93	104	110
1	60.80	75.13	116.08	129.16	269.86	270.05	293.32
2	105.21	101.91	106.52	157.47	247.97	274.95	300.56
3	94.39	125.88	112.14	173.10	193.21	279.64	258.69
4	77.33	110.55	108.78	153.44	230.19	297.21	270.00
5	91.04	85.41	116.86	164.50	193.33	255.76	281.80
6	63.20	113.27	132.28	149.39	250.05	259.41	231.34

7.6 Summary

An estimator of an unknown parameter is a statistic, that is, a function of a sample, that claims to approximate the parameter. If the estimator's expected value is precisely equal to the parameter, the estimator is unbiased. For a sample $\mathbf{X}$, the estimate $S(\mathbf{X})$ is an unbiased estimator of θ if $E_\theta[S(\mathbf{X})] = \theta$.

In general, the bias is the excess of the expected value over the parameter. If we judge merit with a quadratic risk function $\hat{L}_S(\theta) = E_\theta[(S(\mathbf{X}) - \theta)^2]$, we find that the risk associated with a particular estimator is the sum of its variance and the square of its bias. Therefore, an estimator is the more meritorious when it exhibits the smaller bias and the smaller variance. A sequence of estimators S_n has zero asymptotic risk if the corresponding quadratic risks converge to zero. A slightly weaker condition is consistency, which applies if $\Pr_\theta(|S_n - \theta| \geq \epsilon) \to 0$ for every $\epsilon > 0$. The relative efficiency of two estimators is the ratio of their quadratic risks; the estimator with the smaller risk has the greater efficiency. For unbiased estimators, the relative efficiency is the ratio of their variances. If the density of the sample, $f(\mathbf{X}|\theta)$ is a differentiable function of θ, we can derive a lower bound on the variance of any unbiased estimator S, as a function of the information in the sample $I(\theta)$:

$$I(\theta) = E_\theta\left[\left(\frac{\partial}{\partial\theta}\ln f(\mathbf{X},\theta)\right)^2\right].$$

The bound is $\sigma^2_{S|\theta} \geq 1/I(\theta)$, and an efficient estimator is one that achieves this lower bound.

If we know the distribution of an estimator for θ, we can compute a $100p\%$ confidence interval $[a, b]$. A loose interpretation says that $\theta \in [a, b]$ with probability p. This is somewhat misleading. In truth, θ is a constant, while the computed $[a, b]$ is a function of the sample values. Each new sample gives a different confidence interval. The more accurate interpretation refers to a sequence of confidence intervals, computed from a sequence of independent samples. In such a sequence, $100p\%$ of the intervals contain θ. Because the reported $[a, b]$ is one member of such a sequence, there is a $100p\%$ chance that it contains θ. In any case, knowing the symmetric distribution of estimator $S(\mathbf{X})$, we can write

$$p = \Pr_\theta(\theta + x_{(1-p)/2} \leq S \leq \theta + x_{(1+p)/2})$$
$$= \Pr(S - x_{(1+p)/2} \leq \theta \leq S - x_{(1-p)/2}),$$

which provides the confidence interval. Here, $x_{(1-p)/2}$ and $x_{(1+p)/2}$ are percentiles of the estimator's distribution.

For a normal population with a known variance σ^2, the sample mean $\overline{X} = (1/n)\sum_{i=1}^{n} X_i$ is an unbiased efficient estimator for the population mean μ. Its $100p\%$ confidence interval is

$$[a, b] = \left[\overline{X} - z_{(1+p)/2}\left(\frac{\sigma}{\sqrt{n}}\right), \overline{X} + z_{(1+p)/2}\left(\frac{\sigma}{\sqrt{n}}\right)\right],$$

where z_q is the $100q$th percentile of the standard normal distribution. For a normal population with a known mean μ, the quantity $S = (1/n)\sum_{i=1}^{n}(X_i - \mu)^2$ is an unbiased efficient estimator for the variance σ^2. Its $100p\%$ confidence interval is

$$[a, b] = \left[\frac{nS}{\chi^2_{(1+p)/2;(n)}}, \frac{nS}{\chi^2_{(1-p)/2;(n)}}\right],$$

where $\chi^2_{q;(n)}$ is the $100q$th percentile of the χ^2 distribution with n degrees of freedom. If both μ and σ^2 are unknown, then the sample mean and unbiased sample variance $\hat{s}^2_X = [1/(n-1)]\sum_{i=1}^n (X_i - \overline{X})^2$ are joint estimates for μ and σ^2. Their $100p\%$ confidence intervals are, respectively,

$$[a,b]_{\overline{X}} = \left[\overline{X} - t_{(1+p)/2;(n-1)}\sqrt{\frac{\hat{s}^2_X}{n}}, \overline{X} + t_{(1+p)/2;(n-1)}\sqrt{\frac{\hat{s}^2_X}{n}} \right]$$

$$[a,b]_{\hat{s}^2_X} = \left[\frac{(n-1)\hat{s}^2_X}{\chi^2_{(1+p)/2;(n-1)}}, \frac{(n-1)\hat{s}^2_X}{\chi^2_{(1-p)/2;(n-1)}} \right],$$

where $t_{q;(n)}$ is the $100q$th percentile of the t distribution with n degrees of freedom. Both estimators are unbiased. The sample mean is efficient, while the unbiased sample variance is efficient in the limit of large sample sizes. The joint $100p\%$ confidence region in the μ-σ^2 plane is the area within the parabola given by the first equation below and between the horizontal lines given by the remaining equations. Here, $t = (1 - \sqrt{p})/2$.

$$\sigma^2 = n(\mu - \overline{X})^2/z^2_{1-t}$$
$$\sigma^2 = (n-1)\hat{s}^2_X/\chi^2_{1-t;(n-1)}$$
$$\sigma^2 = (n-1)\hat{s}^2_X/\chi^2_{t;(n-1)}$$

If, for an n-vector $\mathbf{X}$ of independent normal random variables with parameters $(\boldsymbol{\mu}, \mathbf{I})$, we can write $\mathbf{X}'\mathbf{X} = \sum_{i=1}^k Q_i$, where each Q_i is a quadratic form $(\mathbf{A}_i\mathbf{X})'(\mathbf{A}_i\mathbf{X})$ with rank$(\mathbf{A}_i) = r_i$, then we must have $\sum_{i=1}^k r_i \geq n$. Moreover, if $\sum_{i=1}^k r_i = n$, then the Q_i are independent χ^2 random variables with noncentrality parameters $\lambda_i = (1/2)(\mathbf{A}_i\boldsymbol{\mu})'(\mathbf{A}_i\boldsymbol{\mu})$ and r_i degrees of freedom. If $\boldsymbol{\mu} = \mathbf{0}$, the Q_i have independent central χ^2 distributions. Analysis of variance and linear regression applications make extensive use of these decompositions.

An analysis of variance tests a finite collection of subpopulations for a common mean. It assumes they all have a common variance σ^2. Suppose that we have m subpopulations, X_i, with potentially different means μ_i. Let $X_{ij}, 1 \leq j \leq n_i$, be a sample from subpopulation i. Define the total sample size, the weighted average mean, the overall sample mean, and the sample means within subpopulations by

$$n = \sum_{i=1}^m n_i \qquad\qquad \overline{X} = (1/n)\sum_{i=1}^m \sum_{j=1}^{n_i} X_{ij}$$
$$\overline{\mu} = (1/n)\sum_{i=1}^m \mu_i \qquad\qquad \overline{X}_i = (1/n_i)\sum_{j=1}^{n_i} X_{ij}.$$

The ratio

$$R = \frac{\dfrac{1}{m-1}\sum_{i=1}^m n_i(\overline{X}_i - \overline{X})^2}{\dfrac{1}{n-m}\sum_{i=1}^m \sum_{j=1}^{n_i}(X_{ij} - \overline{X}_i)^2}$$

then has a noncentral F distribution with $(m-1, n-m)$ degrees of freedom and noncentrality parameter $\lambda = 1/(2\sigma^2)\sum_{i=1}^m (\mu_i - \overline{\mu})^2$. The null hypothesis asserts that there is no difference among the subpopulation means. With $F_{p;(s,t)}$ denoting the $100p$th percentile of the F distribution with (s,t) degrees of freedom, the test with critical region $R > F_{p;(m-1,n-m)}$ has significance level p, which is the probability of erroneously rejecting the hypothesis. The probability of erroneously accepting the hypothesis depends on the inequality that actually exists among the subpopulation means. The quantity

$$\delta = \frac{1}{\sigma}\sqrt{\frac{1}{n}\sum_{i=1}^{m} n_i(\mu_i - \overline{\mu})^2}$$

increases as the inequality among the means increases. Therefore, it provides a measure of the inequality. The power of the test is then $\Pr(R > F_{p;(m-1,n-m)})$, computed with respect to a noncentral F distribution with $(m-1, n-m)$ degrees of freedom and noncentrality parameter $\lambda = n\delta^2/2$.

A linear regression tests a continuum of subpopulations for a linear relationship among the means. That is, if X_t is a family of subpopulations with means μ_t and common variance σ^2, the regression tests for a relationship of the form $\mu_t = \alpha + \beta(t - t_0)$ and estimates parameters that define the line. Specifically, if $X_{ij}, 1 \le j \le n_i$, is a sample from subpopulation X_{t_i} for chosen index points $t_1 < t_2 < \ldots < t_m$, we calculate

$$\hat{\alpha} = \frac{1}{n}\sum_{i=1}^{m}\sum_{j=1}^{n_i} X_{ij} \quad \hat{\beta} = \frac{1}{D}\sum_{i=1}^{m}\sum_{j=1}^{n_i} X_{ij}(t_i - \overline{t}) \quad \hat{\sigma}^2 = \frac{1}{n}[X_{ij} - \hat{\alpha} - \hat{\beta}(t_i - \overline{t})]^2,$$

where $\overline{t} = (1/n)\sum_{i=1}^{m} n_i t_i$ is the average t value and $D = \sum_{i=1}^{m} n_i(t_i - \overline{t})^2$ measures the dispersion of the t values. If $X_{i\cdot} = (1/n_i)\sum_{j=1}^{n_i} X_{ij}$ are the subpopulation sample means, the ratio

$$L = \frac{(n-m)\sum_{i=1}^{m} n_i[X_{i\cdot} - \hat{\alpha} - \hat{\beta}(t_i - \overline{t})]^2}{(m-2)\sum_{i=1}^{m}\sum_{j=1}^{n_i}(X_{ij} - X_{i\cdot})^2}$$

has a noncentral F distribution with $(m-2, n-m)$ degrees of freedom. The noncentrality parameter is

$$\lambda = \frac{1}{2\sigma^2}\sum_{i=1}^{m} n_i \left(\mu_i - \frac{1}{n}\sum_{k=1}^{m} n_k\mu_k - \frac{(t_i - \overline{t})}{D}\sum_{k=1}^{m} n_k\mu_k(t_k - \overline{t})\right)^2.$$

A linear relationship holds among the means if and only if $\lambda = 0$. Hence we take $\delta = \sqrt{2\lambda/n}$, a measure of nonlinearity that is commensurate with σ. In this situation, the null hypothesis asserts a linear relationship. The test with critical region $L > F_{p;(m-2,n-m)}$ has significance level p, which is the probability of erroneously rejecting the null hypothesis. The probability of erroneously accepting the hypothesis depends on the actual deviation from linearity, as measured by δ. The power of the test is then $\Pr(L > F_{p;(m-2,n-m)})$, computed with respect to a noncentral F distribution with $(m-2, n-m)$ degrees of freedom and noncentrality parameter $\lambda = n\delta^2/2$.

If the linear hypothesis is accepted, or if we know from other considerations that a linear relationship holds among the means, we estimate the regression line by $\hat{\alpha} + \hat{\beta}(t - \bar{t})$. Supposing the actual relationship to be $\alpha + \beta(t - \bar{t})$, we discover that $\hat{\alpha}$ and $\hat{\beta}$ are unbiased estimators for α and β. The $100p\%$ confidence intervals for α and β are, respectively,

$$I_{p,\hat{\alpha}} = \left[\hat{\alpha} - t_{(1+p)/2;(n-2)} \sqrt{\frac{\hat{\sigma}^2}{n-2}} \leq \alpha \leq \hat{\alpha} + t_{(1+p)/2;(n-2)} \sqrt{\frac{\hat{\sigma}^2}{n-2}} \right]$$

$$J_{p,\hat{\beta}} = \left[\hat{\beta} - t_{(1+p)/2;(n-2)} \sqrt{\frac{n\hat{\sigma}^2}{(n-2)D}} \leq \beta \leq \hat{\beta} + t_{(1+p)/2;(n-2)} \sqrt{\frac{n\hat{\sigma}^2}{(n-2)D}} \right].$$

The $100p\%$ confidence region for the joint estimate $(\hat{\alpha}, \hat{\beta})$ is the interior of the ellipse $(\alpha - \hat{\alpha})^2/a^2 + (\beta - \hat{\beta})^2/b^2 = 1$, with semimajor and semiminor axes

$$a = \sqrt{\frac{2\hat{\sigma}^2 F_{p;(2,n-2)}}{(n-2)}} \qquad b = \sqrt{\frac{2n\hat{\sigma}^2 F_{p;(2,n-2)}}{(n-2)D}}.$$

We can use the estimated regression line to predict a new observation from X_t, where t is not necessarily among the t_i used in the analysis. The prediction is $\hat{\alpha} + \hat{\beta}(t - \bar{t})$ and its $100p\%$ confidence interval is

$$K_{p,t,\hat{\alpha},\hat{\beta}} = \left[\hat{\alpha} + \hat{\beta}(t - \bar{t}) - t_{(1+p)/2;(n-2)} \sqrt{\frac{n\hat{\sigma}^2(1 + 1/n + (t - \bar{t})^2/D)}{(n-2)}}, \right.$$

$$\left. \hat{\alpha} + \hat{\beta}(t - \bar{t}) + t_{(1+p)/2;(n-2)} \sqrt{\frac{n\hat{\sigma}^2(1 + 1/n + (t - \bar{t})^2/D)}{(n-2)}} \right].$$

Finally, we may wish to infer the value of the control variable t associated with a new set of observations $Y_i, 1 \leq i \leq k$. Defining the mean of the new observations by $\bar{Y} = (1/k)\sum_{i=1}^{k} Y_i$, we discover the maximum likelihood estimate for t to be $\hat{t} = \bar{t} + (\bar{Y} - \hat{\alpha})/\hat{\beta}$, provided that $\hat{\beta} \neq 0$. The $100p\%$ confidence interval is the set of t'-values that satisfy

$$\frac{(n-2)[\bar{Y} - \hat{\alpha} - \hat{\beta}(t' - \bar{t})]^2}{n\hat{\sigma}^2[1/k + 1/n + (t' - \bar{t})^2/D]} < F_{p;(1,n-2)},$$

an interval that must contain $\hat{t}$.

Historical Notes

Around 1900, Karl Pearson published a sequence of papers that took a statistical approach to biological problems, including heredity and evolution. He contributed to the theory of regression analysis and introduced the χ^2 test for statistical significance. He was apparently the first to use the term *standard deviation*. The British statistician Ronald A. Fisher invented analysis of variance as a technique for studying plant genetics. Working in the early twentieth century, he calculated the distributions of statistics appropriate for small samples and introduced the maximum likelihood estimator.

These contributions rank him among the founders of modern statistics. His plant-breeding experiments led to conjectures about fitness as a genetic trait, and he was instrumental in establishing the Cambridge University Eugenics Society. In 1933, he followed Karl Pearson as Galton Professor of Eugenics at University College, London. Fisher reportedly expressed the view that "natural selection is a mechanism for generating an exceedingly high degree of improbability."

Although he never advocated state controlled-marriage, Fisher did feel that eugenics should have some influence on government policy. He cited differential birth rates between classes as a reason for the fall of civilizations in the past, and he hinted that Western civilization may come to the same end. Rather than increasing with need, he felt that government subsidies should increase with family income. As a family earned more, and therefore demonstrated its survival advantage, it should receive compensation to offset the cost of rearing children. Even in Fisher's day, the middle twentieth century, this was a controversial argument that never saw implementation. Imagine the reception it would find today. See Box[7] for further details on Fisher's
• views. George Snedecor named the F distribution in Fisher's honor.

Further Reading

The analysis of variance commands a vast literature, to which this chapter's selections constitute only a brief introduction. Indeed, the examples here seem to raise many more questions than they answer. When we reject the null hypothesis, we conclude that a factor has a significant influence on the observed data. At this point, however, we are hardly finished. The crucial question is then which levels have the optimal effects. There are also questions of the form: Is the average of the first three row means equal to the average of the remaining rows? Studies of this sort involve hypotheses formulated as linear combinations of the row or column means. In any case, the classical reference in the matter is Scheffé[77]. Guenther[29] provides a more accessible introduction to scientific applications, although all distributions are presented without proof. This chapter's image processing example was adapted from Kurz[49], where enhancements of the basic model may be found.

Statistical hypothesis testing, based on the techniques established in this chapter, appear in a wide variety of contexts. One seeks to determine, for example, if a manufacturing process has drifted from specifications, if an experimental drug has an effect against a disease, if a collection of sick patients implies that an epidemic is possible, or if the response time from a new database server does indeed represent an improvement over the current server. Czitrom and Spagon[15] provide informative case studies from industrial processes, while Grant and Ewens[28] discuss statistical approaches in bioinformatics. Hastie et al.[34] investigate statistical methods associated with data mining in large databases.

Appendix A

Analytical Tools

The text assumes a basic knowledge of differential and integral calculus and at least a first course in matrix algebra. This appendix reviews selected topics in these areas and extends them as needed for arguments that appear in the text's main body. In particular, it reviews sets and functions, operations with limits, convergence properties of the real numbers, the basic features of integrals, and the properties of permutations and determinants. Some of this material is immediately applicable in the first chapter, which begins the treatment of probability theory. Many of the concepts, however, find use only in later chapters, and so it is most appropriate to use this appendix as a reference. The main chapters note when to review certain appendix sections, but the review needed depends on the reader's background. All readers will benefit from a quick overview of the topics, which introduce notation and demonstrate the spirit of the proofs pursued in the text.

The conceptual definitions include an explanation of any unusual symbols. However, we assume a familiarity with traditional abbreviations, such as the quantifiers $\exists x$ for "there exists an x such that ..." and $\forall x$ for "for all x it is true that ...."

A.1 Sets and functions

A set is simply a collection of objects, and the notation $x \in A$ asserts that x is a member of set A. We define a set by listing all its members, as in $A = \{1, 2, 8, 14\}$, by inferring its membership with a pattern, as in $A = \{2, 4, 6, 8, \ldots\}$, or by stating a condition for membership, as in $A = \{x | (\exists y)(x = 2y)\}$. The following definitions establish further notation for sets.

A.1 Definition: Given two sets, X and Y, we say that X is a *subset* of Y, or that X is contained in Y, if for every $x \in X$, we also have $x \in Y$. We write $X \subset Y$ for this relationship. We denote the empty set by ϕ. For sets contained in a well-defined universal set Ω, we define the following operations and relationships.

- $X \cap Y$, the *intersection* of X and Y, contains just those elements that are members of both X and Y; for an indexed collection $\{X_i\}$, we denote by $\cap_i X_i$ the set that contains elements common to all sets in the collection.

- $X \cup Y$, the *union* of X and Y, contains those elements that are members of either X or Y or both; for an indexed collection, we denote by $\cup_i X_i$ the set that contains elements from one or more sets in the collection.

- $X - Y$, the *difference* of X over Y, contains those elements of X that are not members of Y.

- $\overline{X}$, the *complement* of X in Ω, is $\Omega - X$.

- $X \Delta Y$, the *symmetric difference* between X and Y is $(X - Y) \cup (Y - X)$.

- X and Y are called *disjoint* sets or *mutually exclusive* sets if $X \cap Y = \phi$.

- A finite or infinite collection of mutually exclusive sets $X_1, X_2, \ldots$ is a *partition* of a set X if $X = \cup_i X_i$. ∎

Note that $X = Y$ if and only if $X \subset Y$ and $Y \subset X$. That is, every element of one set is a member of the other, and vice versa. Also, if X, Y are both subsets of Ω, then $X - Y = X \cap \overline{Y}$. The reductions summarized in the following theorem are frequently useful in working with sets.

A.2 Theorem: Union and intersection are commutative and associative operations, and they distribute across each other. That is, the following hold. We assume that X, Y, Z are all subsets of a larger set Ω, so that complements are well-defined.

- (Commutative rules) $X \cup Y = Y \cup X$ and $X \cap Y = Y \cap X$.

- (Associative rules) $(X \cup Y) \cup Z = X \cup (Y \cup Z)$ and $(X \cap Y) \cap Z = X \cap (Y \cap Z)$. Consequently, we normally omit parentheses in such expressions.

- (Distributive rules) $X \cup (Y \cap Z) = (X \cup Y) \cap (X \cup Z)$ and $X \cap (Y \cup Z) = (X \cap Y) \cup (X \cap Z)$.

A complement propagates into a union and becomes an intersection; a complement propagates into an intersection and becomes a union. These rules, known as De Morgan's laws, appear symbolically as follows:

- $\overline{X \cup Y} = \overline{X} \cap \overline{Y}$, and

- $\overline{X \cap Y} = \overline{X} \cup \overline{Y}$.

PROOF: The commutative, associative, and distributive properties follow directly from the definitions of union and intersection by the usual method. That is, the proof shows that the left side of the purported equation is a subset of the right side, and vice versa. Consider, for example, the first of the distributive properties. If x is a member of the left-side set, then $x \in X \cup (Y \cap Z)$, which means that $x \in X$ or $x \in (Y \cap Z)$. If indeed $x \in X$, then $x \in (X \cup Y)$ and $x \in (X \cup Z)$, which implies that x is a member of their intersection, the right-side set. On the other hand, if $x \in (Y \cap Z)$, then x belongs to both Y and Z, which implies that $x \in (X \cup Y)$ and $x \in (X \cup Z)$. Again x is a member of their intersection, which is the right side of the equation. The remaining assertions admit similar proofs.

For the first of De Morgan's laws, note that $x \in \overline{X \cup Y}$ if and only if $x \notin X$ and $x \notin Y$. But this holds if and only if $x \in \overline{X}$ and $x \in \overline{Y}$, which holds if and only if $x \in \overline{X} \cap \overline{Y}$. This establishes the first law, and a similar proof gives the second law. ∎

Functions are mappings between sets in the sense that a function establishes an unambiguous partner in its destination set for each element in its source set. The following definitions clarify this association and establish the traditional vocabulary for discussing functions.

A.3 Definition:

- A function, f, associates each element of a source set, the domain D_1, with an element of a destination or target set, the codomain D_2. To emphasize the domain and codomain, we write $f : D_1 \to D_2$. For $x \in D_1$, we write $f(x)$ for the corresponding element in D_2.

- In general, we say that f maps D_1 *into* D_2. In the special case where every element $y \in D_2$ is the target of at least one $x \in D_1$, we say that f maps D_1 *onto* R_2, and we refer to f as a *surjective function*.

- In another special case where f maps distinct elements of D_1 to distinct elements of D_2, we say that f is *one-to-one*, and we refer to f as an *injective* function. Thus, f is injective when $f(x) = f(y)$ implies that $x = y$.

- If f is both surjective and injective, we say that f establishes a *one-to-one correspondence* between D_1 and D_2, and we refer to f as a *bijective function*.

- For $A \subset D_1$, the *image* of A is the set $f(A) = \{f(x) | x \in A\}$. $f(D_1)$ is the *range* of f.

- For $B \subset D_2$, the *inverse image* of B is the set $f^{-1}(B) = \{x \in D_1 | f(x) \in B\}$. For the inverse image of a single point, we write $f^{-1}(y)$ instead of $f^{-1}(\{y\})$.

- When we refer to the function itself, we use the notation f or $f(\cdot)$, as opposed to $f(x)$, which designates the value of the function at the point x. ∎

We use one-to-one correspondence to categorize sets as countable or uncountable, according to the following definition.

A.4 Definition: A set is *countably infinite* if it is in one-to-one correspondence with the positive integers. A set is *countable* if it is finite or if it is countably infinite. For a finite set A, we use the notation $|A|$ to indicate the number of elements in A. A set is *uncountable* if it is not countable. ∎

The following example illustrates some aspects of functions, using the ceiling function $\lceil x \rceil$, which means the smallest integer greater than or equal to x. There is also a floor function $\lfloor x \rfloor$, which means the largest integer less than or equal to x. We use $\mathcal{N}^+$ for the natural numbers $\{1, 2, 3, \ldots\}$.

A.5 Example: $f : \mathcal{N}^+ \to \mathcal{N}^+$ via $f(x) = \lceil x/2 \rceil$ is onto but not one-to-one. Every codomain element is the target of at least one domain value. For example, 7 comes from both 13 and 14 because $\lceil 13/2 \rceil = \lceil 14/2 \rceil = 7$. However, it is not one-to-one; the distinct elements 13 and 14 both map to 7.

$g : \mathcal{N}^+ \to \mathcal{N}^+$ via $g(x) = 2x$ is one-to-one but not onto. The codomain element 3, for example, is not the target of any domain element. Moreover, if $g(x_1) = g(x_2)$, then $2x_1 = 2x_2$ and consequently, $x_1 = x_2$. So distinct domain elements must map to distinct range elements.

If $\mathcal{M}^+$ is the set of even natural numbers, then the same $g : \mathcal{N}^+ \to \mathcal{M}^+$ via $g(x) = 2x$ is now both one-to-one and onto. This function establishes a one-to-one correspondence between the natural numbers and the even natural numbers. It also demonstrates that the set of even natural numbers is countable. □

A.6 Theorem: Inverse images preserve intersections, unions, differences, complements, and symmetric differences. That is, if $f : D_1 \to D_2$, then

- $f^{-1}(X \cap Y) = f^{-1}(X) \cap f^{-1}(Y)$,

- $f^{-1}(X \cup Y) = f^{-1}(X) \cup f^{-1}(Y)$,

- $f^{-1}(X - Y) = f^{-1}(X) - f^{-1}(Y)$,

- $f^{-1}(\overline{X}) = \overline{f^{-1}(X)}$,

- $f^{-1}(X \Delta Y) = f^{-1}(X) \Delta f^{-1}(Y)$.

PROOF: It is understood, of course, that the set operations on the left sides occur in D_2, while those on the right sides occur in D_1. In particular, a complement within D_2, such as $\overline{X}$, means $D_2 - X$. Similarly, a complement in D_1, such as $\overline{f^{-1}(X)}$, means $D_1 - f^{-1}(X)$.

For the first assertion, we reason as follows. If $x \in f^{-1}(X \cap Y)$, then by the definition of an inverse image, $f(x) \in X \cap Y$. So $f(x) \in X$, which places $x \in f^{-1}(X)$, and $f(x) \in Y$, which places $x \in f^{-1}(Y)$. Therefore, $x \in f^{-1}(X) \cap f^{-1}(Y)$, which proves that $f^{-1}(X \cap Y) \subset f^{-1}(X) \cap f^{-1}(Y)$. Reversing the argument, we can show that if x starts out in $f^{-1}(X) \cap f^{-1}(Y)$, then it must also belong to $f^{-1}(X \cap Y)$. So the subset relation holds in both directions, and we conclude that the sets in the first assertion are indeed equal.

The second assertion follows in the same manner. Using the notation $\Longleftrightarrow$ to mean "if and only if," we proceed as follows. $x \in f^{-1}(X \cup Y) \Longleftrightarrow$

$f(x) \in X \cup Y \iff f(x) \in X$ or $f(x) \in Y$. The first alternative holds if and only if $x \in f^{-1}(X)$, and the second holds if and only if $x \in f^{-1}(Y)$. So $x \in f^{-1}(X \cup Y) \iff x \in f^{-1}(X) \cup f^{-1}(Y)$. The left and right sides of the assertions are then subsets of each other, and so must be equal.

Also, $x \in f^{-1}(X - Y) \iff f(x) \in X - Y \iff f(x) \in X$ and $f(x) \notin Y \iff x \in f^{-1}(X)$ and $x \notin f^{-1}(Y) \iff x \in f^{-1}(X) - f^{-1}(Y)$, which proves the third assertion.

Noting that $f^{-1}(D_2) = D_1$, we have

$$f^{-1}(\overline{X}) = f^{-1}(D_2 - X) = f^{-1}(D_2) - f^{-1}(X) = D_1 - f^{-1}(X) = \overline{f^{-1}(X)}$$
$$f^{-1}(X \Delta Y) = f^{-1}((X - Y) \cup (Y - X)) = f^{-1}(X - Y) \cup f^{-1}(Y - X)$$
$$= [f^{-1}(X) - f^{-1}(Y)] \cup [f^{-1}(Y) - f^{-1}(X)] = f^{-1}(X) \Delta f^{-1}(Y),$$

which prove the last two parts of the theorem. ∎

In this text, the most frequently occurring domains and codomains are the nonnegative integers $\mathcal{N}$, the positive integers $\mathcal{N}^+$, and the real numbers $\mathcal{R}$. Although we often describe a function with a formula, such a compact expression is not necessary. A computer program or a verbal description can suffice, provided that the description uses a finite number of symbols and the process described transforms each domain element unambiguously into its function value in a finite number of steps. Suppose, for example, that we want to show that a countable collection of countable sets is countable. If we denote the countable sets by $A_1, A_2, \ldots$, we need a one-to-one function from $\mathcal{N}^+$ onto $\cup_{n=1}^{\infty} A_n$. We describe the function $f(\cdot)$ by arranging the values $f(1), f(2), f(3), \ldots$ in a sequence. If the values are all distinct, this arrangement suffices to define the function. Given an argument n, the function returns the nth value in the sequence. Because each A_n is countable, there is a one-to-one function $f_n : \mathcal{N}^+ \to A_n$. We enumerate A_1 by listing the elements $f_1(1), f_1(2), f_1(3), \ldots$, and we follow a similar process for the remaining sets.

Denoting $f_i(j)$ by a_{ij}, we have the display to the right. The entries in a given line are all different, although there may be some duplication between lines. We start the descriptive sequence for $f(\cdot)$ as follows. The sequence first lists the elements with subscripts summing to 1, then those with subscripts summing to 2, and so forth. That is, the new sequence starts

$$A_1 = a_{11}, a_{12}, a_{13}, \ldots$$
$$A_2 = a_{21}, a_{22}, a_{23}, \ldots$$
$$\vdots \qquad \vdots$$
$$A_n = a_{n1}, a_{n2}, a_{n3}, \ldots$$
$$\vdots \qquad \vdots$$

$a_{11}, a_{12}, a_{21}, a_{13}, a_{22}, a_{31}, a_{14}, a_{23}, a_{32}, a_{41}, a_{15}, \ldots$. Although this is an infinite sequence, it reaches any specified element in a finite number of steps. We refine the sequence by deleting elements that appear again after their initial appearance. This defines a new sequence, which implicitly defines $f(\cdot)$. That is, $f(n)$ is the nth entry in the refined list. Note that the process does not proceed by first crossing out the duplicate entries and then locating the nth element. This approach does not define a function because the deletion process may not terminate, in which case the value $f(n)$ does not appear after finitely many steps. Instead, the functional process constructs the refined list while it watches simultaneously for the nth persistent element. When the process installs an entry because the subscript sum so demands, it immediately checks for an earlier duplicate. This is a finite search because it involves only the list generated to the current point. If an earlier duplicate occurs, the new entry is deleted. If an entry survives this initial challenge, it remains forever in the list. Consequently, when evaluating $f(n)$, the process stops at the nth stable entry. This list prefix is sufficient to determine $f(n)$.

This informal description provides complete information on the one-to-one function mapping $\mathcal{N}^+$ to $\cup_{n=1}^{\infty} A_n$ and thereby establishes the countability of the union. In truth, the description assumes that each A_n is countably infinite, whereas countable actually means finite or countably infinite. It is a simple task to modify the functional description to handle a mixture of finite and countably infinite sets. If one or more of the A_n are finite, the process merely deletes each nonexistent a_{ij} when the subscript sum first commands its appearance. That is, an element survives its initial check if its subscripts correspond to an existing element and there is no prior duplicate. With this additional rule, the function correctly maps any countable union of countable sets. Although the constructed function does not admit description with a formula, it nevertheless defines a function, and the existence of this function proves the following theorem.

A.7 Theorem: A countable collection of countable sets is countable.

PROOF: See discussion above. ∎

Exercises

A.1 Let X, Y be subsets of Ω. Show that $X \cup Y = (X \cap \overline{Y}) \cup (Y \cap \overline{X}) \cup (X \cap Y)$. Show also that the three components, $(X \cap \overline{Y}), (Y \cap \overline{X})$, and $(X \cap Y)$, are mutually disjoint.

A.2 Let X, Y, Z be subsets of Ω, and consider the U_i defined to the right. Show that $X \cup Y \cup Z = \cup_{i=1}^{7} U_i$. Show also that $U_i \cap U_j = \phi$, when $i \neq j$.

$$U_1 = X \cap \overline{Y \cup Z} \qquad U_5 = X \cap Z \cap \overline{Y}$$
$$U_2 = Y \cap \overline{X \cup Z} \qquad U_6 = Y \cap Z \cap \overline{X}$$
$$U_3 = Z \cap \overline{X \cup Y} \qquad U_7 = X \cap Y \cap Z$$
$$U_4 = X \cap Y \cap \overline{Z}$$

A.3 Complete the proof of De Morgan's laws by showing that $\overline{X \cap Y} = \overline{X} \cup \overline{Y}$.

A.4 (Principle of inclusion and exclusion). Prove the following statement for finite sets A_i. How does the formula extend to the union of n components?

$$|A_1 \cup A_2 \cup A_3| = |A_1| + |A_2| + |A_3| - |A_1 \cap A_2| - |A_1 \cap A_3|$$
$$- |A_2 \cap A_3| + |A_1 \cap A_2 \cap A_3|.$$

A.5 Suppose that $f : D_1 \to D_2$. Show that $f(X \cup Y) = f(X) \cup f(Y)$ but that, in general, $f(X \cap Y) \subset f(X) \cap f(Y)$. What further condition on f forces equality in the latter case?

A.6 Show that $y_1 \neq y_2$ forces $f^{-1}(y_1)$ and $f^{-1}(y_2)$ to be mutually exclusive.

A.7 Suppose that $|A| = m, |B| = n$ with $m \leq n$. How many functions map A into B? How many injective functions map A into B?

*A.8 Show that the positive rational numbers are countable.

A.2 Limits

We use the standard notation for intervals of real numbers. For example, (a, b), $[a, b)$, $(a, b]$, and $[a, b]$ all represent intervals. A parenthesis means that the endpoint is not included, while a square bracket means that the endpoint is included. For example, $[0, 1] = \{x \in \mathcal{R} \mid 0 \leq x \leq 1\}$. Plus or minus infinity may appear as an excluded endpoint, as in $[10, \infty)$.

An infinite sequence of real numbers, $a_1, a_2, \ldots$, may or may not have a limit. If it does, the limit may be a real number, or it may be $\pm\infty$. The definition and subsequent analysis differ accordingly and often entail a case-by-case exposition. We can unify the concepts somewhat with the notion of a neighborhood, which is an interval associated with a point or with $\pm\infty$.

A.8 Definition: Let v be a real number. For each real $\epsilon > 0$, a *neighborhood* of v is $\mathcal{I} = \mathcal{I}(v, \epsilon) = \{x \mid |x - v| < \epsilon\}$. A *deleted neighborhood* of v is a neighborhood with the point v deleted: $\overline{\mathcal{I}} = \overline{\mathcal{I}}(v, \epsilon) = \{x \mid x \neq v \text{ and } |x - v| < \epsilon\}$. For each real R, define a *neighborhood of* ∞ by $\mathcal{I} = \mathcal{I}(\infty, R) = \{x \mid x > R\}$ and a *neighborhood of* $-\infty$ by $\mathcal{I} = \mathcal{I}(-\infty, R) = \{x \mid x < R\}$. ∎

We use the symbol $\mathcal{I}$ for neighborhood because neighborhoods are intervals. A neighborhood of a real number v has the form $(v - \epsilon, v + \epsilon)$. A neighborhood of ∞ has the form (R, ∞); a neighborhood of $-\infty$ appears as $(-\infty, R)$. When v is a real number, a v-neighborhood contains v itself and some range of numbers that are close to v. If ϵ is small, the neighborhood is correspondingly small because it contains only numbers that differ from v by less than ϵ. An ∞-neighborhood contains all numbers beyond some specified lower bound R. Such neighborhoods are always very large because there are infinitely many numbers beyond any specified point. Nevertheless the neighborhood imposes some constraint on membership: all members must be larger than R. Similar comments apply to a $(-\infty)$-neighborhood. We define the limit of a sequence of real numbers in terms of neighborhoods. The intuitive idea is that the sequence has a limit if we can tame its tail. That is, it has a limit if every neighborhood of that limit captures a tail of the sequence.

A.9 Definition: Let A be a real number or $\pm\infty$. $\lim_{n \to \infty} a_n = A$ means that for every A-neighborhood $\mathcal{I}$ there exists an integer N such that $n \geq N$ implies that $a_n \in \mathcal{I}$. In this case, we say that a_n approaches A, and we use the shorthand $a_n \to A$. When a sequence has a real limit, we say that it *converges*. If the limit is $\pm\infty$, we say that it *diverges to* $\pm\infty$. The remaining possibility is that the sequence has no limit. In this case, it oscillates in such a manner that every potential limit, finite or infinite, has some neighborhood that cannot capture a sequence tail. We summarize this behavior by saying that the sequence *diverges*. ∎

Frequently, it is obvious whether or not a sequence stabilizes to a limit, but occasionally we must actually invoke the definition to prove it. The next example illustrates such situations and shows how to construct a proof if necessary.

A.10 Example: Let $a_n = 1/n$. Then $a_n \to 0$. In particular, given $\epsilon > 0$, choose $N > 1/\epsilon$. Then $n \geq N$ implies that $|a_n - 0| = |a_n| = 1/n \leq 1/N < 1/(1/\epsilon) = \epsilon$. That is, the neighborhood $\mathcal{I}(0, \epsilon)$ captures the sequence tail $a_N, a_{N+1}, \ldots$.

Now, let $a_n = (n+1)/(2n+6)$. Then $a_n \to 1/2$. Indeed,

$$\left| \frac{n+1}{2n+6} - \frac{1}{2} \right| = \left| \frac{(n+1) - (n+3)}{2(n+3)} \right| = \left| \frac{1}{n+3} \right| < \frac{1}{n} < \epsilon,$$

when $n > 1/\epsilon$.

Next, let $a_n = (n+1)^2/(2n+6)$. Then $a_n \to \infty$. Noting that $2n + 6 < 2n + n = 3n$ for $n > 6$, we reason as follows for $n > 6$.

$$\frac{(n+1)^2}{2n+6} > \frac{n^2 + 2n + 1}{3n} = \frac{n}{3} + \frac{2}{3} + \frac{1}{n} > \frac{n}{3} > R,$$

when $n > 3R$. Hence, given R, we choose $N > \max(6, 3R)$. Then $n \geq N$ implies that $a_n > R$.

Let $a_n = (-1)^n$. This sequence has no limit. Because it alternates forever between plus and minus 1, it can never reside permanently in any neighborhood of the form $(a, 1/4)$ when a is a real number. Such neighborhoods have width $1/2$, and because the sequence jumps by 2 in moving from one term to the next, if one term occupies the neighborhood, the next term does not. It also escapes the neighborhoods $(\infty, 2)$ and $(-\infty, -2)$ by the simple expedient of never entering them. In conclusion, if A is a real number or $\pm\infty$, some neighborhood of A fails to capture a sequence tail. The sequence therefore diverges.

Let $a_n = 1 + (-1)^n/n$. Then $a_n \to 1$. There is oscillation on either side of 1, but these variations become smaller and smaller with increasing n. A neighborhood of the form $(1 - \epsilon, 1 + \epsilon)$ therefore contains all terms at and beyond some fixed N. If $\epsilon = 0.001$, for example, then $N = 1001$ suffices. □

Given two sequences, we can form a new sequence through term-by-term arithmetic. For example, the sequences $\{a_n\}, \{b_n\}$ give rise to sequences $\{a_n + b_n\}, \{a_n - b_n\}, \{a_n b_n\}$, and $\{a_n/b_n\}$. The last is well-defined only if the denominators are never zero. In general, limits commute with arithmetic operations. That is, if $a_n \to a, b_n \to b$, we can expect that $a_n + b_n \to a + b$, and so forth. Although this observation greatly facilitates many estimation arguments involving limits, we must be cautious when the limits involve infinity or a zero in the denominator. In particular, we use Table A.1 to ascertain the proper limit. When the arithmetic applied to the sequence limits yields an indeterminate form, one of the entries in the last column of the table, we must analyze the specific situation to determine the limit, if it exists at all.

The next theorem proves, for selected cases, the validity of such arithmetic with limits. It uses a convenient property of real numbers, called the

Forms yielding ∞	Forms yielding $-\infty$	Forms yielding 0	Indeterminate forms
$\infty + \infty$	$-\infty + (-\infty)$	r/∞	$\infty + (-\infty)$
$\infty - (-\infty)$	$-\infty - \infty$	$r/(-\infty)$	$-\infty + \infty$
$\infty + r$	$-\infty + r$		$\infty - \infty$
$r + \infty$	$r + (-\infty)$		$-\infty - (-\infty)$
$\infty - r$	$-\infty - r$		$0 \cdot \infty$
$\infty \cdot \infty$	$-\infty \cdot \infty$		$\infty \cdot 0$
$-\infty \cdot -\infty$	$\infty \cdot (-\infty)$		$0 \cdot (-\infty)$
$\infty \cdot p$	$-\infty \cdot p$		$(-\infty) \cdot 0$
$p \cdot \infty$	$p \cdot (-\infty)$		∞/∞
∞/p	$-\infty/p$		$\infty/(-\infty)$
$-\infty/n$	∞/n		$(-\infty)/\infty$
			$(-\infty)/(-\infty)$
			$0/0$

TABLE A.1. Arithmetic involving infinities ($p > 0, n < 0, r$ are real numbers)

---

triangle inequality. This inequality asserts that

$$|a| - |b| \leq \big||a| - |b|\big| \leq |a \pm b| \leq |a| + |b|,$$

for any real numbers a and b. You can verify it by exhaustively checking all combinations of positive and negative a, b in the case where a has the greater magnitude and again in the case where b is larger.

A.11 Theorem: Suppose that $a_n \to A$ and $b_n \to B$, where A, B are real numbers or $\pm\infty$. Then the following related limits hold.

- $a_n + b_n \to A + B$.

- $a_n - b_n \to A - B$.

- $Ca_n \to CA$, for any constant C.

- $a_n b_n \to AB$.

- If all $b_n \neq 0$, so that a_n/b_n is well-defined, and if $B \neq 0$, then $a_n/b_n \to A/B$.

Table A.1 specifies the results of any arithmetic involving infinities or zero denominators. Such arithmetic must not yield an indeterminate form.

PROOF: There are many cases to consider as A, B can be any combination of real numbers, ∞, and $-\infty$. We select a few cases to illustrate the technique. Consider $a_n + b_n \to A + B$, where A, B are both real numbers. Given $\epsilon > 0$, choose N_1 such that $n \geq N_1$ implies that $|a_n - A| < \epsilon/2$, and choose N_2 such that $n \geq N_2$ implies that $|b_n - B| < \epsilon/2$. These choices are possible because $a_n \to A$ and $b_n \to B$ mean that a tail of $\{a_n\}$ lies in $\mathcal{I}(A, \epsilon/2)$ and a tail of $\{b_n\}$ lies in $\mathcal{I}(B, \epsilon/2)$. Then $n \geq N = \max(N_1, N_2)$ implies that $|a_n + b_n - (A+B)| \leq |a_n - A| + |b_n - B| < \epsilon/2 + \epsilon/2 = \epsilon$. So $a_n + b_n \in \mathcal{I}(A+B, \epsilon)$ when $n \geq N$, and $a_n + b_n \to A + B$.

Consider $a_n/b_n \to A/B$, where A, B are both real numbers and $B \neq 0$. $|B|/2 > 0$ provides an ϵ-value to use with the definition of $b_n \to B$. In particular, there exists N_1 such that $n \geq N_1$ implies that $|b_n - B| < |B|/2$. The triangle inequality then gives $|B| - |b_n| \leq |b_n - B| < |B|/2$ or $|b_n| > |B|/2$ when $n \geq N_1$. For $n \geq N_1$, we can then reason as follows.

$$\left| \frac{a_n}{b_n} - \frac{A}{B} \right| = \frac{|Ba_n - Ab_n|}{|Bb_n|} < \frac{|Ba_n - Ab_n|}{B^2/2} = \left(\frac{2}{B^2} \right) |a_n B - b_n A|$$

$$= \left(\frac{2}{B^2} \right) |a_n B - AB + AB - b_n A|$$

$$\leq \left(\frac{2}{B^2} \right) |a_n B - AB| + \left(\frac{2}{B^2} \right) |b_n A - AB|$$

$$= \frac{2|a_n - A|}{|B|} + \frac{2|A| \cdot |b_n - B|}{B^2}.$$

Now $a_n \to A$, so there exists N_2 such that $n \geq N_2$ implies that $|a_n - A| < \epsilon |B|/4$. Now $n \geq \max(N_1, N_2)$ implies that

$$\left| \frac{a_n}{b_n} - \frac{A}{B} \right| < \frac{\epsilon}{2} + \frac{2|A| \cdot |b_n - B|}{B^2}.$$

If $A = 0$, this last value is less than ϵ and we can take $N = \max(N_1, N_2)$. If $A \neq 0$, $b_n \to B$ means there exists N_3 such that $n \geq N_3$ implies that $|b_n - B| < \epsilon B^2/(4|A|)$. Then, for $N = \max(N_1, N_2, N_3)$, $n \geq N$ implies that

$$\left| \frac{a_n}{b_n} - \frac{A}{B} \right| < (\epsilon/2) + (\epsilon/2) = \epsilon.$$

Hence, whether $A = 0$ or not, we have $a_n/b_n \in \mathcal{I}(A/B, \epsilon)$ when $n \geq N$.

Consider the case $a_n + b_n \to A + B$, where $A = \infty$ and B is a real number. We must show that $a_n + b_n \to \infty$. Given R defining a neighborhood of ∞, choose N_1 such that $n \geq N_1$ implies that $a_n > R - (B - 1)$. Choose N_2 such that $n \geq N_2$ implies that $|b_n - B| < 1$. Then $n \geq N_2$ implies that $B - 1 < b_n < B + 1$. Let $N = \max(N_1, N_2)$. Then $n \geq N$ implies that $a_n + b_n > R - (B - 1) + (B - 1) = R$. Hence $a_n + b_n \to \infty$.

Similar arguments establish the remaining cases. ∎

Given a sequence $\{a_n\}$, we build a subsequence by choosing terms from ever increasing locations in $\{a_n\}$. We denote a subsequence by $a_{i_1}, a_{i_2}, \ldots$, where $i_1, i_2, \ldots$ is itself an increasing sequence of positive integers. For example, the choice $a_{14}, a_{32}, a_{56}, a_{204}, \ldots$ is a subsequence. It is clear that $a_n \to A$ if and only if $a_{i_j} \to A$ for every subsequence. Indeed, suppose that every subsequence converges to A, a real number. Then the sequence itself, being a special case of a subsequence, converges to A. Conversely, if $a_n \to A$, then given $\epsilon > 0$ there exists N such that $n \geq N$ implies that $|a_n - A| < \epsilon$. For any subsequence $a_{i_1}, a_{i_2}, \ldots$, the subscripts must eventually exceed N. That is, there exists M such that $j \geq M$ implies that $i_j \geq N$, which in turn implies that $|a_{i_j} - A| < \epsilon$. Hence the subsequence converges to A. Similar reasoning establishes the result when A is $\pm\infty$.

Another useful fact relates the limits of convergent sequences for which the individual terms satisfy an order relationship. That is, if $a_n \to A, b_n \to B$ are two convergent sequences for which $n \geq N$ implies that $a_n \leq b_n$, then $A \leq B$. To see why this is so, suppose to the contrary that $A > B$. Then $(A - B)/4 > 0$, and there exists N_1, N_2 such that $n \geq N_1$ implies that $|a_n - A| < (A - B)/4$ and $n \geq N_2$ implies that $|b_n - B| < (A - B)/4$. For $n \geq \max(N_1, N_2)$, we have

$$A - (A - B)/4 < a_n < A + (A - B)/4$$
$$B - (A - B)/4 < b_n < B + (A - B)/4$$
$$a_n - b_n > A - (A - B)/4 - [B + (A - B)/4] = (A - B)/2 > 0$$

This gives $a_n > b_n$, a contradiction because $a_n \leq b_n$ for sufficiently large n. Hence we must have $A \leq B$. By taking the $b_n = k$, a constant, we obtain the related result: if $a_n \to A$ and $a_n \leq k$ for $n \geq N$, then $A \leq k$.

A.12 Theorem: The following are useful facts about limits.

- $a_n \to A$ if and only if every subsequence also converges to A.

- $a_n \to \pm\infty$ if and only if every subsequence also diverges to $\pm\infty$.

- If $a_n \to A, b_n \to B$ are convergent sequences and if there exists N such that $n \geq N$ implies that $a_n \leq b_n$, then $A \leq B$.

- If $a_n \to A$ and there exists N such that $n \geq N$ implies that $a_n \leq k$, a constant, then $A \leq k$.

- If $a_n \to \infty$ and there exists N such that $n \geq N$ implies that $a_n \leq b_n$, then $b_n \to \infty$.

- For $j \in \{1, 2, 3, \ldots, K\}$, a finite set, let $\lim_{n \to \infty} a_{jn} = A_j$, a real number. Then given $\epsilon > 0$, there exists M such that $n \geq M$ implies that $|a_{jn} - A_j| < \epsilon$ simultaneously for $j = 1, 2, \ldots, K$.

PROOF: The discussion above establishes all of the theorem except the last two items. For the first of these, if R is given, choose M such that $n \geq M$ implies that $a_n > R$. Then for $n \geq \max(N, M)$, we have $b_n \geq a_n > R$, which shows that $b_n \to \infty$. For the last item, given $\epsilon > 0$, choose N_j such that $n \geq N_j$ implies that $|a_{jn} - A_j| < \epsilon$. Now choose $M = \max(N_1, N_2, \ldots, N_K)$. For $n \geq M$, we have $n > N_j$ for all j. Therefore, $n \geq M$ implies that $|a_{jn} - A_j| < \epsilon$ for all j. ∎

Estimation arguments with limits make frequent use of the theorem above but usually provide no further justification beyond setting up the required conditions. A situation that invokes the theorem's last result, for example, typically uses the phrase: "Because there are only a finite number of sequences in question, we can choose N such that $n > N$ implies that all of them are simultaneously within ϵ of their respective limits."

The limit concept applies to series as well as sequences. The infinite sum $\sum_{i=0}^{\infty} a_i$ is an alternative notation for $\lim_{n \to \infty} \sum_{i=0}^{n} a_i$. Moreover, we can extend the concept to functions that map the real numbers $\mathcal{R}$ to the real numbers.

A.13 Definition: Let $f : D \to R$, where both D and R are subsets of the real numbers. $\lim_{x \to x_0} f(x) = A$ means that for every A-neighborhood $\mathcal{J}$ there exists a deleted x_0-neighborhood $\overline{\mathcal{I}}$ such that $x \in \overline{\mathcal{I}} \cap D$ implies that $f(x) \in \mathcal{J}$. We abbreviate the concept by saying $f(x) \to A$ as $x \to x_0$. Note that x_0 and/or A may be $\pm\infty$. ∎

It is convenient to have a one-sided limit concept. The limit from above is the limit of the functional values as x approaches x_0 through domain values greater than x_0. The limit from below has a similar definition.

A.14 Definition: Let $f : D \to R$, where D and R are subsets of the real numbers. $\lim_{x \to x_0^+} f(x) = A$ means that for every A-neighborhood $\mathcal{J}$ there exists a deleted x_0-neighborhood $\overline{\mathcal{I}}$ such that $x \in \overline{\mathcal{I}} \cap D$ and $x > x_0$ implies that $f(x) \in \mathcal{J}$. We speak of A as the limit of $f(x)$ as x approaches x_0 from above.

$\lim_{x \to x_0^-} f(x) = A$ means that for every A-neighborhood $\mathcal{J}$ there exists a deleted x_0-neighborhood $\overline{\mathcal{I}}$ such that $x \in \overline{\mathcal{I}} \cap D$ and $x < x_0$ implies that $f(x) \in \mathcal{J}$. We speak of A as the limit of $f(x)$ as x approaches x_0 from below. ∎

A.15 Example: Let

$$f(x) = \begin{cases} 0, & x < 1 \\ 1, & x \geq 1. \end{cases}$$

Then $\lim_{x \to 1+} f(x) = 1$, while $\lim_{x \to 1-} f(x) = 0$. □

A.16 Definition: Let $f : D \to R$, where both D and R are subsets of the real numbers. If $f(x) \to f(x_0)$ as $x \to x_0$, then we say that $f(\cdot)$ is *continuous* at x_0. If $\lim_{x \to x_0^+} f(x) = f(x_0)$, we say that $f(\cdot)$ is *continuous from the right* or *continuous from above* at x_0. If $\lim_{x \to x_0^-} f(x) = f(x_0)$, we say that $f(\cdot)$ is *continuous from the left* or *continuous from below* at x_0.

If $f(\cdot)$ is continuous at every point in its domain, we say that $f(\cdot)$ is a *continuous function*. If the size of the deleted x-neighborhood that responds to an $f(x)$-neighborhood $\mathcal{J}(f(x), \epsilon)$ is independent of $x \in D$, we say that $f(\cdot)$ is *uniformly continuous* on D. That is, $f(\cdot)$ is uniformly continuous on D if, given $\epsilon > 0$, there exists $\delta > 0$ such that, for all $x \in D$, $y \in \overline{\mathcal{I}}(x, \delta)$ implies that $f(y) \in \mathcal{J}(f(x), \epsilon)$. ∎

A.17 Example: Suppose that $f : \mathcal{R} \to [0, \infty)$ via

$$f(x) = \begin{cases} x^2, & x \neq 2 \\ 0, & x = 2. \end{cases}$$

$\lim_{x \to 2} f(x) = 4$ since $x \neq 2, x \in (\sqrt{4 - \epsilon}, \sqrt{4 + \epsilon})$ implies that $x^2 \in (4 - \epsilon, 4 + \epsilon)$, provided $\epsilon < 4$. If $\epsilon \geq 4$, use $x \neq 2, x \in (3/2, 5/2)$ implies that $x^2 \in (9/4, 25/4) \subset$

$(0, 8) \subset (4 - \epsilon, 4 + \epsilon)$. $f(\cdot)$ is not continuous at 2 because this limit is not equal to $f(2)$. It is, however, continuous at all other points.

The function $g(x) = x^2$ is a continuous function because it is continuous at each point in its domain. However, it is not uniformly continuous. To see why this is so, suppose that $\overline{\mathcal{I}}(x, \delta)$ is the fixed-size domain neighborhood that purports to map into the range neighborhood $\mathcal{J}(f(x), \epsilon)$ for all $x \in \mathcal{R}$. For $x > \delta$, $f(\overline{\mathcal{I}}(x, \delta)) = ((x - \delta)^2, (x + \delta)^2) - \{x\}$. The width of this image is $4x\delta$, which grows arbitrarily large with increasing x. In particular, for $x > \epsilon/(2\delta)$ it is larger than 2ϵ, which is the width of $\mathcal{J}(f(x), \epsilon)$. Hence, it cannot map into $\mathcal{J}(f(x), \epsilon)$ for these large values of x. Reaching this contradiction, we conclude that $g(\cdot)$ is not uniformly continuous.

Suppose that $h : [1, 2] \to \mathcal{R}$ via $h(x) = x^2$. Because its restricted domain does not allow arbitrarily large values, $h(\cdot)$ is uniformly continuous. In particular,

$$f([1, 2] \cap (x - \delta, x + \delta) - \{x\}) \subset ((x - \delta)^2, (x + \delta)^2) = (x^2 - \gamma_1, x^2 + \gamma_2),$$

where $\gamma_1 = 2x\delta - \delta^2$ and $\gamma_2 = 2x\delta + \delta^2$. Given $\epsilon > 0$, choose $\delta = \min(1, \epsilon/5)$. Then

$$\gamma_1 = 2x\delta - \delta^2 \leq 4\delta - \delta^2 = \delta(4 - \delta) \leq 4(\epsilon/5) < 4(\epsilon/4) = \epsilon$$
$$\gamma_2 = 2x\delta + \delta^2 \leq 4\delta + \delta^2 = \delta(4 + \delta) \leq (\epsilon/5)(5) = \epsilon.$$

Hence $y \in [1, 2] \cap \overline{\mathcal{I}}(x, \delta)$ implies that $h(y) \in \mathcal{J}(h(y), \epsilon)$. As the choice of δ did not depend on $x \in [1, 2]$, $h(\cdot)$ is uniformly continuous on that domain. $\square$

A.18 Definition: Let D, a subset of the real numbers, be the common domain for a sequence of functions $f_n : D \to \mathcal{R}, n = 1, 2, \ldots$ and $f : D \to \mathcal{R}$. We say that $\{f_n(\cdot)\}$ *converges pointwise* to $f(\cdot)$ if $f_n(x) \to f(x)$ for each $x \in D$. If the N that responds to a $f(x)$-neighborhood $\mathcal{I}(f(x), \epsilon)$ is independent of $x \in D$, we say that $f_n(\cdot)$ *converges uniformly* to $f(\cdot)$ on D. That is, $f_n(\cdot)$ converges uniformly to $f(\cdot)$ on D if, given $\epsilon > 0$, there exists N such that $n \geq N$ implies that $|f_n(x) - f(x)| < \epsilon$ simultaneously for all $x \in D$. ∎

A.19 Example: Let $f(x) = 0$ and define the sequence $\{f_n(\cdot)\}$ as follows.

$$f_n(x) = \begin{cases} 0, & x \leq n \\ 1, & x > n \end{cases}$$

Then $f_n(x) \to f(x)$ at each point x. Given $\epsilon > 0$, choose $N = \lceil x \rceil$. Then $n \geq N$ implies that $n \geq \lceil x \rceil \geq x$. So $n \geq N$ implies that $f_n(x) = 0$, and therefore, $|f_n(x) - f(x)| = |f_n(x) - 0| = 0 < \epsilon$. However, the convergence is not uniform. If $\epsilon = 1/4$, then each N admits x-values such that $n \geq N$ and $f_n(x) \notin \mathcal{I}(0, 1/4)$. For instance, $n = N + 1$ and $x = N + 2$ suffice, since $f_n(x) = 1$. This means that no N has the property that $n \geq N$ implies that $f_n(x) \in \mathcal{I}(0, 1/4)$ simultaneously for all x. $\square$

A.20 Theorem: Let $f_n(\cdot)$ be a sequence of continuous functions that converge uniformly to $f(\cdot)$ on domain D. Then $f(\cdot)$ is continuous on D.

PROOF: We will show that $f(\cdot)$ is continuous at an arbitrary point $x \in D$. For any $y \in D$ and any n, we have

$$|f(y) - f(x)| \leq |f(y) - f_n(y)| + |f_n(y) - f_n(x)| + |f_n(x) - f(x)|.$$

Given $\epsilon > 0$, choose N such that $n \geq N$ implies that $|f_n(z) - f(z)| < \epsilon/3$ for all $z \in D$. Fix $n \geq N$ and choose $\delta > 0$ such that $0 < |y - x| < \delta$ implies that $|f_n(y) - f_n(x)| < \epsilon/3$. Then $0 < |y - x| < \delta$ forces $|f(y) - f(x)| < 3(\epsilon/3) = \epsilon$. ∎

The limit processes discussed above also apply in $\mathcal{R}^n$, the space of n-vectors with components in $\mathcal{R}$. We denote the components of $\mathbf{x} \in \mathcal{R}^n$ by $(x_1, x_2, \ldots, x_n)$. We write $\|\mathbf{x}\|$ for the largest absolute component of $\mathbf{x}$, and we denote by $\mathbf{x} \pm \mathbf{y}$ the component-wise addition or subtraction of n-vectors. We write $\mathbf{x} \leq \mathbf{y}$ when each component of $\mathbf{x}$ is less than or equal to the corresponding component of $\mathbf{y}$. $\mathbf{x} < \mathbf{y}$ means that each $\mathbf{x}$ component is strictly less than the corresponding $\mathbf{y}$ component. Boldface $\mathbf{0}$ is the n-vector with zero in each component; ∞ is the n-vector with ∞ in each component. Symbolically, we have

$$\|\mathbf{x}\| \;=\; \max_{1 \leq k \leq n} |x_k|$$

$$\mathbf{x} + \mathbf{y} \;=\; (x_1 + y_1, x_2 + y_2, \ldots, x_n + y_n)$$

$$\mathbf{x} - \mathbf{y} \;=\; (x_1 - y_1, x_2 - y_2, \ldots, x_n - y_n)$$

$$\mathbf{x} \leq \mathbf{y} \;\Longleftrightarrow\; x_1 \leq y_1, x_2 \leq y_2, \ldots, x_n \leq y_n$$

$$\mathbf{x} < \mathbf{y} \;\Longleftrightarrow\; x_1 < y_1, x_2 < y_2, \ldots, x_n < y_n.$$

For example, $\|(-4, 2)\| = 4, \|(-4, 2) + (3, -5)\| = \|(-1, -3)\| = 3$, and $(0, 2) < (1, 3)$. With this notation, the required definitions are straightforward extensions of the corresponding one-dimensional concepts.

A.21 Definition: Given $\epsilon > 0$, a *neighborhood* $\mathcal{I}(\mathbf{x}; \epsilon)$ of a point $\mathbf{x} \in \mathcal{R}^n$ is $\{\mathbf{y} \mid \|\mathbf{y} - \mathbf{x}\| < \epsilon\}$. A *deleted neighborhood* is $\overline{\mathcal{I}}(\mathbf{x}; \epsilon) = \{\mathbf{y} \mid \mathbf{y} \neq \mathbf{x}$ and $\|\mathbf{y} - \mathbf{x}\| < \epsilon\}$. For a real number R, we call $\{\mathbf{y} \mid y_1 > R, y_2 > R, \ldots, y_n > R\}$ a neighborhood of ∞ and denote it by $\mathcal{I}(\infty; R)$. Similarly, $\mathcal{I}(-\infty, R) = \{\mathbf{y} \mid y_1 < R, y_2 < R, \ldots, y_n < R\}$ is a neighborhood of $-\infty$. Deleted neighborhoods of $\pm\infty$ are the same as the corresponding neighborhoods. That is, $\overline{\mathcal{I}}(\infty, R) = \mathcal{I}(\infty, R)$ and similarly for $-\infty$. ∎

When a discussion involves many points in $\mathcal{R}^n$, we use upper subscripts enclosed in parentheses to distinguish the points. Lower subscripts denote components. For example $x_j^{(i)}$ is the jth component of point $\mathbf{x}^{(i)}$.

A.22 Definition: Let $\mathbf{x}^{(1)}, \mathbf{x}^{(2)}, \ldots$ be a sequence of points in $\mathcal{R}^n$. We say that $\mathbf{x}^{(i)}$ converges to $\mathbf{a}$, where $\mathbf{a}$ is a point in $\mathcal{R}^n$ or $\pm\infty$, if for every neighborhood $\mathcal{I}$ of $\mathbf{a}$, there exists an index N such that $i \geq N$ implies that $\mathbf{x}^{(i)} \in \mathcal{I}$. We abbreviate this feature as $\mathbf{x}^{(i)} \to \mathbf{a}$. ∎

As in the one-dimensional case, the neighborhood concept allows a unified definition for limits, whether they be points in $\mathcal{R}^n$ or $\pm\infty$. However, the analysis that builds on the definition depends on the particular form. If $\mathbf{x}^{(i)} \to \mathbf{a} \in \mathcal{R}^n$, for example, then for each $\epsilon > 0$, there exists N such that $i \geq N$ implies that $\|\mathbf{x}^{(i)} - \mathbf{a}\| < \epsilon$. If, however, $\mathbf{x}^{(i)} \to \infty$, then for each $R > 0$, there exists N such that $i \geq N$ implies that $x_j^{(i)} > R$ for $j = 1, 2, \ldots, n$.

A.23 Definition: Let $f : \mathcal{R}^n \to \mathcal{R}$. We write $\lim_{\mathbf{x} \to \mathbf{a}} f(\mathbf{x}) = L$ if for every neighborhood $\mathcal{I}$ of L, there exists a deleted neighborhood $\overline{\mathcal{J}}$ of $\mathbf{a}$ such that $\mathbf{x} \in \overline{\mathcal{J}}$ implies that $f(\mathbf{x}) \in \mathcal{I}$. The approach point $\mathbf{a}$ can be a point in $\mathcal{R}^n$, or it can be $\pm\infty$. The limit L can be a real number, or it can be $\pm\infty$. ∎

The neighborhood concept again facilitates a unified definition, but subsequent analysis depends on whether $\mathbf{a}$ and/or L are real points or $\pm\infty$. For example, $\lim_{\mathbf{x}\to\infty} f(\mathbf{x}) = L \in \mathcal{R}$ means that given $\epsilon > 0$, there exists $R > 0$ such that $x_j^{(i)} > R$ for $1 \le j \le n$ implies that $|f(\mathbf{x}) - L| < \epsilon$. Continuity and one-sided limits also extend to $\mathcal{R}^n$.

A.24 Definition: $f : \mathcal{R}^n \to \mathcal{R}$ is *continuous* at a point $\mathbf{x}$ if $\lim_{\mathbf{y}\to\mathbf{x}} f(\mathbf{y}) = f(\mathbf{x})$. A continuous function is one that is continuous at every point. ∎

A.25 Definition: For $\mathbf{a} \in \mathcal{R}^n$, we write $\lim_{\mathbf{x}\to\mathbf{a}+} f(\mathbf{x}) = L$ if for every neighborhood $\mathcal{I}$ of L, there exists a deleted neighborhood $\overline{\mathcal{J}}$ of $\mathbf{a}$ such that $\mathbf{x} \in \overline{\mathcal{J}}, \mathbf{x} \ge \mathbf{a}$ implies that $f(\mathbf{x}) \in \mathcal{I}$. Similarly, we write $\lim_{\mathbf{x}\to\mathbf{a}-} f(\mathbf{x}) = L$ if for every neighborhood $\mathcal{I}$ of L, there exists a deleted neighborhood $\overline{\mathcal{J}}$ of $\mathbf{a}$ such that $\mathbf{x} \in \overline{\mathcal{J}}, \mathbf{x} \le \mathbf{a}$ implies that $f(\mathbf{x}) \in \mathcal{I}$. ∎

A.26 Example: Suppose that $f(x, y) = (x + 2)/y$. Substituting the limits for the arguments, we suspect that $\lim_{(x,y)\to(1,2)} f(x, y) = f(1, 2) = 3/2$. Rigorously, we proceed as follows. Given a neighborhood $\mathcal{I}(3/2; \epsilon)$, we investigate

$$\left| \frac{x+2}{y} - \frac{3}{2} \right| = \left| \frac{2(x-1) - 3(y-2)}{2y} \right| \le \frac{1}{2}(2|x-1| + 3|y-2|),$$

where the final inequality holds for $y > 1$. Consequently, we respond with the deleted neighborhood $\overline{\mathcal{J}}((1, 2); \min(1, \epsilon/2))$. Then $(x, y) \in \overline{\mathcal{J}}$ implies that

$$\|(x, y) - (1, 2)\| < \min(1, \epsilon/2)$$
$$|x - 1| < \epsilon/2$$
$$|y - 2| < \epsilon/2 < 2\epsilon/3$$
$$|y - 2| < 1$$
$$y > 1$$
$$\left| f(x, y) - \frac{3}{2} \right| = \left| \frac{x+2}{y} - \frac{3}{2} \right| \le \frac{1}{2}(2|x-1| + 3|y-2|) < |x-1| + \frac{3}{2}|y-2|$$
$$< \frac{\epsilon}{2} + \left(\frac{3}{2} \right) \left(\frac{2\epsilon}{3} \right) = \epsilon$$
$$f(x, y) \in \mathcal{I}.$$

Hence, $\lim_{(x,y)\to(1,2)} f(x, y) = 3/2 = f(1, 2)$, and f is continuous at $(1, 2)$.

As $(x, y) \to (0, 0)$, however, $f(x, y)$ does not approach a limit. Every deleted neighborhood of $(0, 0)$ contains points that map to arbitrarily large values, both positive and negative. Consider the deleted neighborhood $\overline{\mathcal{J}}((0, 0); \epsilon)$. It contains the points $(0, \pm\epsilon/2), (0, \pm\epsilon/4), (0, \pm\epsilon/8), \ldots$, which map to $\pm 4, \pm 8, \pm 16, \ldots$. For any proposed limit L, whether finite or infinite, these values escape all neighborhoods of L except $(-\infty, \infty)$. Consequently, $\lim_{(x,y)\to(0,0)} f(x, y)$ does not exist. □

As in the one-dimension case, we can characterize functional convergence through the convergence of related sequences.

A.27 Theorem: Let $f : \mathcal{R}^n \to \mathcal{R}$. Then $\lim_{\mathbf{y}\to\mathbf{x}} f(\mathbf{x}) = L$ if and only if $\lim_{i\to\infty} f(\mathbf{y}^{(i)}) = L$ for every sequence $\mathbf{y}^{(i)} \to \mathbf{x}$ with all $\mathbf{y}^{(i)} \ne \mathbf{x}$. L may be $\pm\infty$, and $\mathbf{x}$ may be $\pm\infty$.

PROOF: Suppose that $\lim_{\mathbf{y}\to\mathbf{x}} f(\mathbf{x}) = L$ and $\mathbf{y}^{(i)} \to \mathbf{x}, y^{(i)} \ne x$. Given a neighborhood $\mathcal{I}$ of L, there exists a deleted neighborhood $\overline{\mathcal{J}}$ of $\mathbf{x}$ such that $\mathbf{y} \in \overline{\mathcal{J}}$ implies that $f(\mathbf{y}) \in \mathcal{I}$. Also, there exists an index N such that $i \geq N$ implies that $\mathbf{y}^{(i)} \in \mathcal{J}$. Since $\mathbf{y}^{(i)} \neq \mathbf{x}$, we have $\mathbf{y}^{(i)} \in \overline{\mathcal{J}}$ for $i \geq N$. Consequently, $i \geq N$ implies that $f(\mathbf{y}^{(i)}) \in \mathcal{I}$, which shows that $\lim_{i\to\infty} f(\mathbf{y}^{(i)}) = L$.

Conversely, suppose that $\lim_{i\to\infty} f(\mathbf{y}^{(i)}) = L$ for every sequence $\mathbf{y}^{(i)} \to \mathbf{x}$ with $\mathbf{y}^{(i)} \neq \mathbf{x}$. For purposes of deriving a contradiction, suppose that $f(\mathbf{y})$ does not approach L as $\mathbf{y} \to \mathbf{x}$. Then there must exist a neighborhood $\mathcal{I}$ of L such that every deleted neighborhood of $\mathbf{x}$ contains infinitely many points that map outside of $\mathcal{I}$. Define a sequence of steadily shrinking deleted neighborhoods of $\mathbf{x}$. If $\mathbf{x} \in \mathcal{R}^n$, these neighborhoods are $\overline{\mathcal{J}}(\mathbf{x}, 1/2), \overline{\mathcal{J}}(\mathbf{x}, 1/3), \overline{\mathcal{J}}(\mathbf{x}, 1/4)$, and so forth. If $\mathbf{x} = \infty$, they are $\mathcal{J}(\infty, 1), \mathcal{J}(\infty, 2), \mathcal{J}(\infty, 3), \ldots$. If $\mathbf{x} = -\infty$, they are $\mathcal{J}(-\infty, -1)$, $\mathcal{J}(-\infty, -2)$, $\mathcal{J}(-\infty, -3)$, and so forth. Choose $\mathbf{y}^{(1)}$ from the first neighborhood such that $f(\mathbf{y}^{(1)}) \notin \mathcal{I}$. Then choose $\mathbf{y}^{(2)} \neq \mathbf{y}^{(1)}$ in the second neighborhood such that $f(\mathbf{y}^{(2)}) \notin \mathcal{I}$, and so forth. Each $\mathbf{y}^{(i)}$ differs from all points previously chosen, resides in a smaller deleted neighborhood of $\mathbf{x}$, and maps outside $\mathcal{I}$. Consequently, $\mathbf{y}^{(i)} \to \mathbf{x}$ and $\mathbf{y}^{(i)} \neq \mathbf{x}$, but $f(\mathbf{y}^{(i)})$ does not converge to L—a contradiction. ∎

A.28 Theorem: Suppose that $f : \mathcal{R}^n \to \mathcal{R}$. Let L be a real number or $\pm\infty$. If $\mathbf{x} \in \mathcal{R}^n$ or $\mathbf{x} = \infty$, the following are equivalent.

(a) $\lim_{\mathbf{y}\to\mathbf{x}^-} f(\mathbf{y}) = L$

(b) $\lim_{i\to\infty} f(\mathbf{y}^{(i)}) = L$ for every sequence $\mathbf{y}^{(i)} \to \mathbf{x}$ with all $\mathbf{y}^{(i)} \neq \mathbf{x}$ and all $\mathbf{y}^{(i)} \leq \mathbf{x}$

(c) $\lim_{i\to\infty} f(\mathbf{y}^{(i)}) = L$ for every *increasing* sequence $\mathbf{y}^{(i)} \to \mathbf{x}$ with all $\mathbf{y} \neq \mathbf{x}$ and $\mathbf{y}^{(1)} \leq \mathbf{y}^{(2)} \leq \mathbf{y}^{(3)} \leq \ldots \leq \mathbf{x}$

If $\mathbf{x} \in \mathcal{R}^n$ or $\mathbf{x} = -\infty$, the following are equivalent.

(a) $\lim_{\mathbf{y}\to\mathbf{x}^+} f(\mathbf{y}) = L$

(b) $\lim_{i\to\infty} f(\mathbf{y}^{(i)}) = L$ for every sequence $\mathbf{y}^{(i)} \to \mathbf{x}$ with all $\mathbf{y}^{(i)} \neq \mathbf{x}$ and all $\mathbf{y}^{(i)} \geq \mathbf{x}$

(c) $\lim_{i\to\infty} f(\mathbf{y}^{(i)}) = L$ for every *decreasing* sequence $\mathbf{y}^{(i)} \to \mathbf{x}$ with all $\mathbf{y}^{(i)} \neq \mathbf{x}$ and $\mathbf{y}^{(1)} \geq \mathbf{y}^{(2)} \geq \mathbf{y}^{(3)} \geq \ldots \geq \mathbf{x}$

PROOF: The theorem involves many cases. The limit L can be a real number or $\pm\infty$. The approach point x can also be real or infinite. The approach can progress from below or from above. We prove the specific case where L and x are real numbers and the approach is from below. The other cases admit similar proofs.

To show that (a) implies (b), suppose that $\lim_{\mathbf{y}\to\mathbf{x}^-} f(\mathbf{y}) = L$, and let $\mathbf{y}^{(i)} \to \mathbf{x}$ with all $\mathbf{y}^{(i)} \neq \mathbf{x}$ and all $\mathbf{y}^{(i)} \leq \mathbf{x}$. We need to show that $\lim_{i\to\infty} f(\mathbf{y}^{(i)}) = L$. Challenged with an $\epsilon > 0$, we appeal to the functional limit to obtain a $\delta > 0$ such that $\mathbf{y} \neq \mathbf{x}$, $||\mathbf{x}-\mathbf{y}|| < \delta$ implies that $|f(\mathbf{y})-L| < \epsilon$. Now we appeal to the sequential limit $\mathbf{y}^{(i)} \to \mathbf{x}$ to obtain a position N such that $i \geq N$ implies that $||\mathbf{x} - \mathbf{y}^{(i)}|| < \delta$. Consequently, $i \geq N$ implies that $|f(\mathbf{y}^{(i)}) - L| < \epsilon$. We conclude that $\lim_{i \to \infty} f(\mathbf{y}^{(i)}) = L$.

It is clear that (b) implies (c), because an increasing sequence approaching $\mathbf{x}$ from the left is still a sequence approaching $\mathbf{x}$ from the left. Finally, we use a proof by contradiction to establish that (c) implies (a). Specifically, suppose that $f(\mathbf{y})$ does not approach L as $\mathbf{y} \to \mathbf{x}^-$. We obtain the contradiction by explicitly constructing an increasing sequence $\mathbf{y}^{(i)} \to x$ for which $f(\mathbf{y}^{(i)})$ does not approach L. The notation $D(\mathbf{a}, \mathbf{b})$ is the smallest absolute coordinate difference between the two points. That is, $D(\mathbf{a}, \mathbf{b}) = \min_{1 \leq i \leq n} |a_i - b_i|$. Because $\lim_{\mathbf{y} \to \mathbf{x}^-} f(\mathbf{y})$ is not L, there exists an $\epsilon > 0$ such that every $\delta > 0$ allows $|f(\mathbf{z}) - L| > \epsilon$ for some $\mathbf{z} \neq \mathbf{x}$ such that $\mathbf{z} \leq \mathbf{x}$ and $||\mathbf{z} - \mathbf{x}|| < \delta$. Consequently, we construct the sequence $\mathbf{z}^{(1)} \leq \mathbf{z}^{(2)} \leq \ldots \leq \mathbf{x}$ as follows. Choose $\mathbf{z}^{(1)}$ such that $\mathbf{z}^{(1)} \neq \mathbf{x}, \mathbf{z}^{(1)} \leq \mathbf{x}, ||\mathbf{z}^{(1)} - \mathbf{x}|| < 1$, and $|f(\mathbf{z}^{(1)}) - L| > \epsilon$. Then choose $\mathbf{z}^{(2)}$ such that $\mathbf{z}^{(2)} \neq \mathbf{x}, \mathbf{z}^{(2)} \leq \mathbf{x}, ||\mathbf{z}^{(2)} - \mathbf{x}|| < \min(D(\mathbf{x}, \mathbf{z}^{(1)}), 1/2)$, and $|f(\mathbf{z}^{(2)}) - L| > \epsilon$. This forces $\mathbf{z}^{(1)} \leq \mathbf{z}^{(2)} \leq \mathbf{x}$. Continue with $\mathbf{z}^{(3)} \leq \mathbf{x}$ such that $\mathbf{z}^{(3)} \neq \mathbf{x}$ and $||\mathbf{z}^{(3)} - \mathbf{x}|| < \min(D(\mathbf{x}, \mathbf{z}^{(2)}), 1/3)$ and $|f(\mathbf{z}^{(3)}) - L| > \epsilon$. This forces $\mathbf{z}^{(1)} \leq \mathbf{z}^{(2)} \leq \mathbf{z}^{(3)} \leq \mathbf{x}$. Having $\mathbf{z}^{(1)} \leq \mathbf{z}^{(2)} \leq \cdots \leq \mathbf{z}^{(i)} \leq \mathbf{x}$, and none equal to $\mathbf{x}$, we choose $\mathbf{z}^{(i+1)} \leq \mathbf{x}$ such that $\mathbf{z}^{(i+1)} \neq \mathbf{x}, ||\mathbf{z}^{(i+1)}, \mathbf{x}|| < \min(D(\mathbf{x}, \mathbf{z}^{(i)}), 1/(i+1))$, and $|f(\mathbf{z}^{(i+1)}) - L| > \epsilon$. The construction defines an increasing sequence $\mathbf{z}^{(1)} \leq \mathbf{z}^{(2)} \leq \ldots \leq \mathbf{x}$ with $||\mathbf{z}^{(i)} - \mathbf{x}|| < 1/i$ for all i. Moreover, for all i, we have $\mathbf{z}^{(i)} \neq \mathbf{x}$. Therefore, $\mathbf{z}^{(i)} \to \mathbf{x}^-$. Moreover, $|f(\mathbf{z}^{(i)}) - L| > \epsilon$ for all i. Consequently, it is not possible that $\lim_{i \to \infty} f(\mathbf{z}^{(i)})$ is L. ∎

We say that $f : \mathcal{R}^n \to \mathcal{R}$ is *monotone nondecreasing* if $\mathbf{x} \leq \mathbf{y}$ implies that $f(\mathbf{x}) \leq f(\mathbf{y})$.

A.29 Example: Let

$$f(x, y) = \begin{cases} -1, & x + y < 0 \\ x + y + 1, & x + y \geq 0. \end{cases}$$

Show that $f(\cdot, \cdot)$ is monotone nondecreasing and calculate the one-sided limits as (x, y) approaches $(0, 0)^+$ and $(0, 0)^-$.

Consulting the figure to the right below, we see that $f(x, y) = -1$ in area A_1 and $f(x, y) = x + y + 1$ in area A_2, which includes the boundary line $x + y = 0$. Consider two points (x_1, y_1) and (x_2, y_2) such that $(x_1, y_1) \leq (x_2, y_2)$. Then (x_2, y_2) must lie to the northeast of (x_1, y_1), possibly directly above or directly to the right.

We discern three cases.

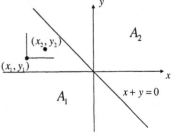

(a) If both points are in A_1, $f(x_1, y_1) = -1 = f(x_2, y_2)$.

(b) If they are both in A_2, $f(x_1, y_1) = x_1 + y_1 + 1 \leq x_2 + y_2 + 1 = f(x_2, y_2)$.

(c) If one is in A_1 and the other in A_2, it must be (x_2, y_2) that lies in A_2. Then $f(x_1, y_1) = -1 < 0 \leq x_2 + y_2 < x_2 + y_2 + 1 = f(x_2, y_2)$.

Therefore, $f(\cdot, \cdot)$ is monotone nondecreasing.

To evaluate the one-sided limits, we can choose convenient monotone sequences approaching $(0, 0)$ from above and below. Approaching from above, we choose points $(x_i, y_i) = (1/i, 1/i)$ for $i = 1, 2, \ldots$, which gives

$$\lim_{(x,y) \to (0,0)+} f(x, y) = \lim_{i \to \infty} f(1/i, 1/i) = \lim_{i \to \infty} (1/i + 1/i + 1) = 1.$$

Approaching from below, we choose $(x_i, y_i) = (-1/i, -1/i)$ for $i = 1, 2, \ldots$, which gives

$$\lim_{(x,y) \to (0,0)-} f(x, y) = \lim_{i \to \infty} f(-1/i, -1/i) = -1. \; \square$$

Exercises

A.9 The sequence $a_n = \cos n\pi$ is obviously divergent because it oscillates forever between plus and minus 1. However, as an exercise in exploring the negation of the limit definition, show that the limit does not exist by finding, for every purported limit A, an A-neighborhood that cannot capture a sequence tail.

A.10 Verify the triangle inequality: $|a \pm b| \le |a| + |b|$.

A.11 Theorem A.11 involves many more cases than those elaborated in the abbreviated proof. Prove the following additional cases.

   – If $a_n \to \infty, b_n \to \infty$, then $a_n b_n \to \infty$.
   – If A, B are real numbers and $a_n \to A, b_n \to B$, then $a_n b_n \to AB$.
   – If A is a real number such that $a_n \to A, b_n \to \infty$, then $a_n/b_n \to 0$.

A.12 Give an example in which $a_n \to \infty, b_n \to 0$, but a_n/b_n has no limit, finite or infinite.

A.13 Suppose $\{a_n\}$ is such that the limit of every subsequence is infinity. Show that $a_n \to \infty$.

A.14 Among other consequences, Theorem A.12 asserts that $a_n < k$ for all n implies that $\lim_{n \to \infty} a_n \le k$. Show that this result cannot be sharpened to a strict inequality by exhibiting a constant k and sequence $\{a_n\}$ such that $a_n < k$ for all n but $\lim_{n \to \infty} a_n = k$.

A.15 Show that $f : \mathcal{R} \to \mathcal{R}$ is continuous at x_0 if and only if it is both continuous from the right and continuous from the left at x_0.

A.16 Let $f : \mathcal{R} \to \mathcal{R}$. Show that $f(\cdot)$ is continuous at x_0 if and only if $f(x_n) \to f(x_0)$ for every sequence $\{x_n\}$ such that $x_n \to x_0$.

*A.17 Let $f : \mathcal{R} \to \mathcal{R}$. For real numbers x and L, show that $\lim_{y \to x-} f(y) = L$ if and only if $\lim_{n \to \infty} f(y_n) = L$ for every increasing sequence $y_1 \le y_2 \le y_3 \le \ldots < x$ such that $y_n \to x$.

A.18 Show that $f : (0,1) \to \mathcal{R}$ via $f(x) = 1/x$ is not uniformly continuous on its domain.

A.19 Let $f_n(x) = x^n$, for $0 \le x \le 1/2$. Show that $f_n(\cdot) \to 0$ uniformly on $[0, 1/2]$.

A.20 Let $f_n(x) = x^n$, for $0 \le x \le 1$. What is the limiting function? Is the convergence uniform on $[0, 1]$?

A.21 In terms of the defining ϵ for a neighborhood of a point $\mathbf{a}$ and the R for a neighborhood of $\pm\infty$, translate the meaning of the following limits.

(a) $\lim_{\mathbf{x}\to\mathbf{a}} f(\mathbf{x}) = L \in \mathcal{R}$ (d) $\lim_{\mathbf{x}\to\infty} f(\mathbf{x}) = \infty$

(b) $\lim_{\mathbf{x}\to\infty} f(\mathbf{x}) = L \in \mathcal{R}$ (e) $\lim_{\mathbf{x}\to-\infty} f(\mathbf{x}) = L \in \mathcal{R}$

(c) $\lim_{\mathbf{x}\to\mathbf{a}} f(\mathbf{x}) = \infty$

*A.22 Let $f : \mathcal{R}^n \to \mathcal{R}$. Although it omits the detailed proof, Theorem A.28 asserts the equivalence of the following three statements.

(a) $\lim_{\mathbf{y}\to\mathbf{x}^-} f(\mathbf{y}) = L$

(b) $\lim_{i\to\infty} f(\mathbf{y}^{(i)}) = L$ for every sequence $\mathbf{y}^{(i)} \to \mathbf{x}$ with $\mathbf{y}^{(i)} < \mathbf{x}$ for all i

(c) $\lim_{i\to\infty} f(\mathbf{y}^{(i)}) = L$ for every increasing sequence $\mathbf{y}^{(i)} \to \mathbf{x}$ with $\mathbf{y}^{(1)} \le \mathbf{y}^{(2)} \le \mathbf{y}^{(3)} \le \ldots < \mathbf{x}$

Prove these results for the case where $\mathbf{x} \in \mathcal{R}^n$ and L is a real number.

*A.23 Prove the equivalence in the preceding exercise for the case where $\mathbf{x} = \infty$ and L is a real number.

A.3   Structure of the real numbers

The real numbers possess some special properties that facilitate analysis. For example, we can argue under certain circumstances that a limit must exist even though we cannot easily identify it. This section reviews those properties that will prove useful in justifying certain probability concepts in the text. The development also provides practice with the limiting arguments of the last section. All sets here are sets of real numbers.

A sequence of digits, with an embedded decimal point, uniquely determines a real number. For example, $d_1 d_2 d_3 . d_4 d_5 \ldots$ is a real number. We accept this proposition as a given fact, but in truth, the exact nature of a real number requires substantial argument. The historical notes at the appendix's conclusion provide a brief discussion. In any case, this observation holds the key to inferring the existence of a limit without actually calculating its value.

A.30 Definition: A sequence $\{a_i\}$ is a *Cauchy sequence* if $\lim_{m,n\to\infty} |a_n - a_m| = 0$. That is, $\{a_i\}$ is Cauchy if, given $\epsilon > 0$, there exists integer N such that $|a_n - a_m| < \epsilon$ for $m, n \ge N$. ∎

It turns out that Cauchy sequence and convergent sequence are one and the same concept. The next theorem establishes this result.

A.31 Theorem: The sequence $\{a_i\}$ converges if and only if it is a Cauchy sequence.

PROOF: Suppose that $\{a_i\}$ converges to a. Let $\epsilon > 0$ be given. There exists N such that $|a_n - a| < \epsilon/2$ for $n \geq N$. Then $n, m \geq N$ implies that $|a_n - a_m| = |(a_n - a) - (a_m - a)| \leq |a_n - a| + |a_m - a| < \epsilon/2 + \epsilon/2 = \epsilon$. Thus $\lim_{n,m\to\infty} |a_n - a_m| = 0$, and the sequence is Cauchy.

Conversely, suppose that the sequence is Cauchy. We will construct the decimal expansion of its limit, one digit at a time, and thereby show that it does converge. Because the sequence is Cauchy, there is an N_1 such that $|a_{N_1} - a_n| < 0.25$ for $n \geq N_1$. That is, all but finitely many of the entries are within 0.25 of a_{N_1}. This implies that there is a rightmost interval $[k_0, k_0 + 1)$, with integer k_0, that contains infinitely many sequence entries. If a_{N_1} falls near the center of such an interval $[k_0, k_0 + 1)$, then the span $a_{N_1} \pm 0.25$ does not intersect any other interval, and $[k_0, k_0 + 1)$ is the only interval containing infinitely many sequence entries. If a_{N_1} falls near an end of $[k_0, k_0 + 1)$, then the span $a_{N_1} \pm 0.25$ can intersect the preceding or subsequent interval, but not both. In these cases, both $[k_0, k_0 + 1)$ and one of its neighbors can contain infinitely many sequence entries. Still, one of them is rightmost, and relabeling if necessary, let this rightmost one be $[k_0, k_0 + 1)$. The integer part of the desired limit is k_0.

Now, divide $[k_0, k_0 + 1)$ into ten disjoint, equal parts: $[k_0 + i/10, k_0 + (i + 1)/10)$, for $0 \leq i \leq 9$. One of these subintervals must contain infinitely many sequence entries. Identify k_1, the next digit on the limit expansion, from the left end of the rightmost such subinterval. That is, $[k_0 + k_1/10, k_0 + (k_1 + 1)/10)$ contains infinitely many sequence entries. We then divide this subinterval into ten parts and identify k_2 by the fact that $[k_0 + k_1/10 + k_2/100, k_0 + k_1/10 + (k_2 + 1)/100)$ contains infinitely many sequence entries. Continuing this process, we obtain the expansion $a = k_0.k_1 k_2 k_3 \ldots$. The process also defines a sequence of nested intervals: $I_0 \supset I_1 \supset I_2 \ldots$, with the length of $I_j = |I_j| = 10^{-j}$ and $a \in I_j$ for all j. Each I_j contains infinitely many sequence entries.

Given $\epsilon > 0$, the Cauchy nature of the sequence allows us to establish an N such that $|a_n - a_m| < \epsilon/2$ for $m, n \geq N$. Choose j such that $|I_j| = 10^{-j} < \epsilon/2$. Since I_j contains infinitely many entries from the sequence, it must contain at least one, call it a_M, with $M > N$. Then $|a_M - a| < 10^{-j} < \epsilon/2$, and for any $n \geq M$, $|a_n - a_M| < \epsilon/2$. Hence $n \geq M$ implies that $|a_n - a| \leq |a_n - a_M| + |a_M - a| < \epsilon/2 + \epsilon/2 = \epsilon$. Therefore, $\lim_{n\to\infty} a_n = a$. ∎

Theorem A.31 is very useful because it allows us to work with convergent sequences without knowing the actual limit. The theorems below note several results that follow immediately.

A.32 Definition: We say that the series $\sum_{n=0}^{\infty} a_n$ *converges absolutely* if $\sum_{n=1}^{\infty} |a_n|$ converges. ∎

A.33 Theorem: If $\sum_n a_n$ converges absolutely, then it converges in the ordinary sense.

PROOF: By hypothesis, $\sum_n |a_n|$ converges, which means, according to the preceding theorem, that its partial sums form a Cauchy sequence. Let S_n denote the partial sum $\sum_{i=1}^n |a_i|$. Given $\epsilon > 0$, there exists an N such that $|S_m - S_n| = \sum_{i=n+1}^m |a_i| < \epsilon$ for $N \le n < m$. Repeated use of the triangle inequality gives $|\sum_{i=n+1}^m a_i| \le \sum_{i=n+1}^m |a_i| < \epsilon$ for $N \le n < m$. This says that the partial sums from the original series form a Cauchy sequence. Applying Theorem A.31 once again, we can assert that the partial sums from the original series are convergent. Therefore, the original series converges. ∎

A.34 Theorem: If $\sum_n a_n$ converges, then $\lim_{n \to \infty} a_n = 0$.

PROOF: The partial sums from the series form a Cauchy sequence. Given $\epsilon > 0$, there exists N such that $|\sum_{i=n+1}^m a_i| < \epsilon$ when $N \le n < m$. Letting $m = n+1$, we have that $|a_{n+1}| < \epsilon$ for $n \ge N$. That is, $|a_n| < \epsilon$ for $n \ge N+1$. Therefore, $\lim_{n \to \infty} a_n = 0$. ∎

A.35 Theorem: (dominated convergence) If $|a_i| \le |b_i|$ for $i \ge N$ and if $\sum_i |b_i|$ converges, then $\sum_i a_i$ converges absolutely.

PROOF: $\sum_i |b_i|$ convergent implies that the partial sums form a Cauchy sequence. Given $\epsilon > 0$, there exists an integer M such that $\sum_{i=n+1}^m |b_i| < \epsilon$ for $M \le n < m$. But then

$$\left| \sum_{i=n+1}^m a_i \right| \le \sum_{i=n+1}^m |a_i| \le \sum_{i=n+1}^m |b_i| < \epsilon,$$

for $M \le n < m$. So the partial sums from both $\sum_i a_i$ and $\sum_i |a_i|$ are Cauchy sequences, and $\sum_i a_i$ converges absolutely. ∎

The tools above find repeated use in estimation arguments, which appear throughout the text. As an initial example, we show how absolute convergence justifies reordering the summands in an infinite summation. To see that this is not always possible, consider the following series. $\lfloor x \rfloor$ means the largest integer that does not exceed x. For example, $\lfloor 5/2 \rfloor = 2$, $\lfloor 3 \rfloor = 3$, $\lfloor -5/2 \rfloor = -3$. Let

$$a_i = \frac{(-1)^{i+1}}{\lfloor (i+1)/2 \rfloor}.$$

The sum then takes the form

$$\sum_{i=1}^\infty a_i = 1 - 1 + \frac{1}{2} - \frac{1}{2} + \frac{1}{3} - \frac{1}{3} + \dots .$$

Evidently, the partial sums are

$$A_n = \sum_{i=1}^n a_i = \begin{cases} \dfrac{2}{n+1}, & n \text{ odd} \\ 0, & n \text{ even}. \end{cases}$$

Therefore, $A_n \to 0$, and $\sum_{n=1}^\infty a_n = 0$. We can, however, describe a rearrangement of the summands that does not converge at all. Starting from any fixed m_1, the sum

$$\frac{1}{m_1} + \frac{1}{m_1 + 1} + \frac{1}{m_1 + 2} + \dots$$

grows without bound as we include more and more terms. As shown in Figure A.1,

$$\sum_{n=m_1}^{m_2} \frac{1}{n} > \int_{m_1}^{m_2+1} \frac{\mathrm{d}x}{x} = \ln \frac{m_2 + 1}{m_1},$$

which approaches infinity as $m_2 \to \infty$. The desired rearrangement operates as follows. First choose enough of the positive terms to drive the partial sum to a value larger than 1. This is possible because the sum of the positive terms approaches infinity. Suppose that this phase uses up the positive terms through $1/n_1$. Now choose negative terms which, when added to the positive terms already chosen, drive the partial sum to a value smaller than -1. This is possible because the sum of the negative terms approaches negative infinity. Suppose that this process uses up the negative terms through $-1/n_2$. Now, starting with the leftmost unused positive term $1/(n_1 + 1)$, choose enough additional positive terms to drive the accumulating partial sum back to a value greater than 1. Again, this is possible because the sum of the positive terms, starting with $1/(n_1 + 1)$, still diverges to infinity. Suppose that this process uses up the positive terms through $1/n_3$. Now, starting with the negative term $-1/(n_2 + 1)$, choose enough negative terms to drive the partial sum back to a value less than -1. This is possible because the sum of the negative terms, starting at $-1/(n_2 + 1)$, still diverges to negative infinity. Suppose that this operation uses up the negative terms through $-1/n_4$. The pattern should be clear. We alternate each sequence of positive summands with a sequence of negative summands, continually driving the accumulating partial sum back and forth across the span $[-1, 1]$. Because the partial sum oscillates forever in this manner, it cannot converge. We therefore have a rearrangement of the series that does not converge to the same value as the original series. As a matter of fact, the rearrangement does not converge at all. The next theorem asserts that absolute convergence precludes this disturbing behavior.

A.36 Theorem: If the series $\sum_{n=1}^{\infty} a_n$ converges absolutely, then all rearrangements converge to the same limit.

PROOF: Absolute convergence implies ordinary convergence, so $\sum_{n=1}^{\infty} a_n = a$, for some real number a. Suppose that $g(\cdot)$ is a one-to-one function that maps the positive integers onto the positive integers. That is, $\{g(1), g(2), g(3), \ldots\}$ is a rearrangement of $\{1, 2, 3, \ldots\}$. We must show that $\sum_{k=1}^{\infty} a_{g(k)} = a$. Because the original series of absolute values converges, the partial sums $\sum_{i=1}^{n} |a_i|$ form a Cauchy sequence. Therefore, for a given $\epsilon > 0$, there exists N such that $N \le n < m$ implies that $\sum_{i=n+1}^{m} |a_i| < \epsilon$. Consider the positive integers k such that $g(k) \in \{1, 2, 3, \ldots, N\}$. Because $g(\cdot)$ is injective, there are N such values. Suppose that K is the largest. Let $K + 1 \le i < j$, and consider the indices $\{g(i), g(i + 1), g(i + 2), \ldots, g(j)\}$. These values are all greater than N. If n_1, n_2 are the smallest and largest among them, then $N < n_1 < n_2$, and $\sum_{k=i}^{j} |a_{g(k)}| \le \sum_{k=n_1}^{n_2} |a_k| < \epsilon$.

    This shows that the partial sums $\sum_{k=1}^{n} |a_{g(k)}|$ form a Cauchy sequence

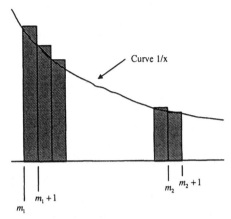

Figure A.1. $\sum_{n=m_1}^{m_2} 1/n$ dominates the integral $\int_{m_1}^{m_2+1} dx/x$

and therefore converge. This means that the series $\sum_{k=1}^{\infty} a_{g(k)}$ converges absolutely. Absolute convergence implies ordinary convergence, but it remains to show that $\sum_{k=1}^{\infty} a_{g(k)} = a$.

Let $\epsilon > 0$ be given. Working with the convergence and absolute convergence of $\sum a_i$, choose N such that $N \leq n < m$ implies both the following constraints: $|\sum_{i=1}^{n} a_i - a| < \epsilon/2$ and $\sum_{i=n+1}^{m} |a_i| < \epsilon/2$. Again let K be the largest k-value such that $g(k) \in \{1, 2, \ldots, N\}$. Then $k \geq K+1$ implies that

$$\left| \sum_{i=1}^{k} a_{g(i)} - a \right| = \left| \sum_{i=1}^{N} a_i + \sum_{i \in I} a_{g(i)} - a \right|,$$

where $I = \{i | 1 \leq i \leq k, g(i) > N\}$. Let $M = \max\{g(i) | i \in I\}$. Then $M \geq N+1$ and

$$\left| \sum_{i=1}^{k} a_{g(i)} - a \right| \leq \left| \sum_{i=1}^{N} a_i - a \right| + \sum_{i \in I} |a_{g(i)}| < \frac{\epsilon}{2} + \sum_{i=N+1}^{M} |a_i| < \frac{\epsilon}{2} + \frac{\epsilon}{2} = \epsilon.$$

This proves that $\sum_{k=1}^{\infty} a_{g(k)} = a$. ∎

We look now to special features associated with bounded sets of real numbers.

A.37 Definition: Let A be a set of real numbers. A point x is a *limit point* of A if there exists a sequence $x_1, x_2, \ldots$ such that $x_n \in A$ for $n = 1, 2, \ldots$ and $x_n \to x$. ∎

Note that a set need not contain its limit points. For example, the interval $(0, 1)$ contains the sequence $x_n = 1/n$, and $x_n \to 0$. So 0 is a limit point of $(0, 1)$, but $0 \notin (0, 1)$. Some sets, however, do contain all their limit points, and we distinguish such sets with a special name.

A.38 Definition: We say that a set is *closed* if it contains all its limit points. We say that a set is *open* if each of its points possesses a neighborhood that is completely within the set. ∎

From the discussion above, we see that $(0,1)$ is not closed. It omits zero, which is one of its limit points. However, $(0,1)$ is open. If $x \in (0,1)$, then $0 < x < 1$. Let $\epsilon = (1/2) \cdot \min(1-x, x)$. In other words, ϵ is one-half the distance to the closest endpoint. Then the neighborhood $(x-\epsilon, x+\epsilon) \subset (0,1)$. This shows that $(0,1)$ contains a neighborhood around each of its points and is therefore an open set. The interval $(0,1]$ is not closed because it omits zero, which is still a limit point. Moreover, it is not open because every neighborhood of 1, which must be of the form $(1-\epsilon, 1+\epsilon)$ for some $\epsilon > 0$, extends beyond 1 and must therefore include some numbers greater than 1. That is, $(1-\epsilon, 1+\epsilon) \not\subset (0,1]$ for any positive ϵ.

We argue by contradiction that intervals of the form $[a,b]$ are closed. Suppose that $[a,b]$ is not closed. That is, suppose that it has a limit point $x \notin [a,b]$. Then $x < a$ or $x > b$. Consider first the case $x < a$. Let $x_1, x_2, \ldots$ be a sequence such that $x_i \in [a,b]$ and $x_i \to x$. Choose N such that $i \geq N$ implies that $|x_i - x| < (a-x)/2$. Then

$$x_i - x \leq |x_i - x| < \frac{a-x}{2}$$

$$x_i < \frac{a+x}{2} = a - \frac{a-x}{2} < a,$$

which asserts that $x \notin [a,b]$, a contradiction. The case $x > b$ admits a similar contradiction. We conclude that $[a,b]$ contains all its limit points and is therefore closed. Analogous reasoning shows that intervals of the form (a,b) are open. In keeping with these observations, we refer to intervals of the form (a,b) as open intervals, while those of the form $[a,b]$ are closed intervals. The hybrids, $(a,b]$ and $[a,b)$, are neither open nor closed. We call them half-open intervals.

A.39 Definition: A real number a is an *upper bound* for a set A if $x \leq a$ for every $x \in A$. An upper bound u is a *least upper bound* for A if $u \leq a$ for every upper bound a of A. Similarly, a real number b is a *lower bound* for a set A if $b \leq x$ for every $x \in A$. A lower bound l is a *greatest lower bound* for A if $b \leq l$ for every lower bound b of A. The least upper bound is also called the *supremum* over A, and we abbreviate it as lub A or sup A. The greatest lower bound is also known as the *infimum* over A, and we abbreviate it as glb A or inf A. ∎

Some sets do not have least upper bounds. The positive integers, for example, have no upper bound and therefore no least upper bound. However, the completeness theorem for the real numbers asserts that a set bounded from above always has a least upper bound.

A.40 Theorem: If a set of real numbers A has an upper bound, then it has a unique least upper bound. If it has a lower bound, then it has a unique greatest lower bound.

PROOF: We prove the least upper bound statement. The proof for the greatest lower bound is similar.

Let I_0 be the rightmost interval with integer endpoints $(i-1, i]$ containing one or more members of A. Because A has an upper bound, these intervals must eventually leave A behind as they proceed to the right, toward $+\infty$. Therefore, a rightmost interval intersecting A must exist. That is I_0. Suppose that $I_0 = (i_0 - 1, i_0]$.

Consider the two subintervals $(i_0 - 1, i_0 - 1/2]$ and $(i_0 - 1/2, i_0]$. Let I_1 be the rightmost that contains points in A. We continue this construction. At each stage, we divide the current interval in half and designate the next interval as the rightmost of the two components that contains an element of A. The process gives a sequence of nested intervals, $I_0 \supset I_1 \supset I_2 \supset \ldots$, and the width of the jth is $|I_j| = 1/2^j$. Define the sequence $\{x_j\}$ by letting x_j equal the right endpoint of interval I_j.

By construction, $a \leq x_j$ for each x_j and for all $a \in A$. That is, each x_j is an upper bound for A. Moreover, for any two indices i, j, either $x_i, x_j \in I_i$ or $x_i, x_j \in I_j$, depending on which interval has the smaller index. Therefore, $|x_i - x_j| \leq (1/2)^{\min(i,j)}$. This means that $|x_i - x_j| \to 0$ as $i, j \to \infty$, that $\{x_i\}$ is a Cauchy sequence, and finally, that $x_i \to x$, a real number. For any $a \in A$, we have $a \leq x_i$ for all i, which forces $a \leq x$. Therefore, x is an upper bound for A.

If u is an upper bound and $u < x$, then there exists integer J such that $j \geq J$ implies that $|x - x_j| < (x - u)/2$, or equivalently, $x_j > u + (x - u)/2$. Now, fix $j > J$ such that $1/2^j < (x - u)/2$. Then the entire interval I_j lies to the right of u. Since I_j contains points of A, we have points of A greater than u, which contradicts the upper bound status of u. We conclude that $u \geq x$, which shows that x is a least upper bound. If y is another least upper bound, then the definition forces both $x \leq y$ and $y \leq x$. That is, we must have $y = x$. Hence x is the unique least upper bound for A. ∎

This result is also useful in proving the convergence of sequences when the actual limit is unknown. In particular, if a sequence is monotone, either nondecreasing or nonincreasing, and does not diverge to infinity, then it must converge. Moreover, a monotone nondecreasing sequence converges to the least upper bound of its bounded range, and a monotone nonincreasing sequence converges to the greatest lower bound of its bounded range. The following theorem provides an exact statement.

A.41 Theorem: Suppose that the sequence $\{x_n\}$ satisfies $x_n \leq x_{n+1}$ for all n and there exists a real number a such that $x_n \leq a$ for all n. Then $\{x_n\}$ converges to $x = \text{lub}\,\{x_n\}$. Also, if $y_n \geq y_{n+1}$ and there exists a real number b such that $y_n \geq b$ for all n, then $\{y_n\}$ converges to $y = \text{glb}\,\{y_n\}$.

PROOF: The set $\{x_1, x_2, \ldots\}$ is bounded above by a. Therefore, it has a least upper bound x. Given $\epsilon > 0$, the number $x - \epsilon$ cannot be an upper bound. So, there is an element x_N such that $x - \epsilon < x_N \leq x$. Because the sequence is nondecreasing, we have $x - \epsilon < x_N \leq x_n \leq x$ whenever $n \geq N$. Stated differently, $n \geq N$ implies that $|x - x_n| < \epsilon$. So $x_n \to x$. The proof for a nonincreasing sequence bounded from below is similar. ∎

A useful fact concerning the supremum is that it can only increase when the field of competitors is enlarged. Similarly, the infimum can only decrease. Moreover, we can factor a constant from the argument of a supremum or infimum, but the resulting expression varies depending on the sign of the constant. The next theorem records these properties.

A.42 Theorem: For real-valued functions $f(\cdot)$ and $g(\cdot)$,

$$\sup\{f(x) + g(x)|x \in [a, b]\} \leq \sup\{f(x)|x \in [a, b]\} + \sup\{g(x)|x \in [a, b]\}$$
$$\inf\{f(x) + g(x)|x \in [a, b]\} \geq \inf\{f(x)|x \in [a, b]\} + \inf\{g(x)|x \in [a, b]\}.$$

For any set $X \subset \mathcal{R}$ and any constant k, let $kX = \{kx|x \in X\}$. Then

$$\sup(kX) = \begin{cases} k \sup X, & k > 0 \\ k \inf X, & k < 0 \end{cases}$$

$$\inf(kX) = \begin{cases} k \inf X, & k > 0 \\ k \sup X, & k < 0. \end{cases}$$

PROOF: In the first relation, the right side represents a search for the largest $f(x) + g(y)$, where x and y can vary independently over $[a, b]$. The left side considers only the candidates $f(x)+g(x)$, which is a subset of those considered on the right. A similar comment applies to the second relation. Rigorously, we argue as follows. Let

$$A = \sup\{f(x) + g(x)|x \in [a, b]\}$$
$$A_f = \sup\{f(x)|x \in [a, b]\}$$
$$A_g = \sup\{g(x)|x \in [a, b]\}.$$

Given $\epsilon > 0$, we know that $A-\epsilon$ cannot be an upper bound of $\{f(x)+g(x)|x \in [a, b]\}$. Consequently, there exists $x_0 \in [a, b]$ such that $f(x_0) + g(x_0) > A - \epsilon$. Also, $f(x_0) \leq A_f$ and $g(x_0) \leq A_g$, from which it follows that

$$A - \epsilon < f(x_0) + g(x_0) \leq A_f + A_g$$
$$A < A_f + A_g + \epsilon.$$

Because $\epsilon > 0$ is arbitrary, we conclude that $A \leq A_f + A_g$. A similar argument establishes the corresponding property for the infimum.

Turning now to a final pair of properties, suppose first that $k > 0$. In this case, let $A = \sup(kX)$ and $B = \sup X$. Then, for all $x \in X$, we have both the following arguments.

$$\begin{array}{cc} kx \leq A & x \leq B \\ x \leq A/k & kx \leq kB \\ B = \sup X \leq A/k & A = \sup(kX) \leq kB \\ A \geq kB \end{array}$$

We conclude that $A = kB$. The argument is basically the same for $k < 0$, except we let $B = \inf X$:

$$\begin{array}{cc} kx \leq A & x \geq B \\ x \geq A/k & kx \leq kB \\ B = \inf X \geq A/k & A = \sup(kX) \leq kB. \\ A \geq kB \end{array}$$

Consequently, A is again equal to kB. Combining the cases, we have

$$\sup(kX) = \begin{cases} k \sup X, & k > 0 \\ k \inf X, & k < 0. \end{cases}$$

The argument for $\inf(kX)$ is similar. ∎

The following point may seem overly technical, but it has many ramifications among the properties of real numbers. Suppose that a collection of open intervals, possibly uncountable, is such that the union of all its members contains a closed bounded interval $[a, b]$. Then most of the open intervals are superfluous for the cover. The following theorem states the property precisely. We use the term *cover* of A for a collection of sets that contains A in the union of its members. If a subset of the collection also contains A in the union of its members, we refer to the subcollection as a *subcover*.

A.43 Theorem: Let $\mathcal{G}$ be a collection of open intervals such that $[a, b] \subset \cup_{G \in \mathcal{G}} G$. Then, there exists a finite subcollection $(a_1, b_1), \ldots, (a_n, b_n)$ such that $(a_i, b_i) \in \mathcal{G}$ for $i = 1, 2, \ldots, n$ and $[a, b] \subset \cup_{i=1}^{n}(a_i, b_i)$. This property is usually stated as: Every open cover of a closed bounded interval admits a finite subcover.

PROOF: We establish the result by contradiction. Suppose that no finite cover exists. Let $I_0 = [a, b]$, and let $c = (a + b)/2$. If both $[a, c]$ and $[c, b]$ admit finite subcovers from $\mathcal{G}$, so does $[a, b]$. Therefore, one of $[a, c], [c, b]$ admits no finite subcover. Call this interval I_1. We form I_2 by dividing I_1 in half and choosing a half that does not admit a finite subcover. We continue the process to generate a sequence of nested intervals $I_0 \supset I_1 \supset I_2 \supset \ldots$, none of which admits a finite subcover. Let $a = a_0 \le a_1 \le \ldots$ be the left endpoints of the intervals; let $b = b_0 \ge b_1 \ge \ldots$ be the corresponding right endpoints. The a_i are all less than b and therefore constitute an increasing sequence bounded from above. By Theorem A.41, we have $a_i \to x = \text{lub} \{a_1, a_2, \ldots\}$. Because all $a_i \in [a, b]$, x is a limit point of $[a, b]$. Because $[a, b]$ is closed, $x \in [a, b]$. Indeed, $a_i < b_j$ for every i, j, whence $x \le b_j$ for every j. So $x \in [a_j, b_j]$ for all j. Since $x \in [a, b]$, there exists G such that $x \in G \in \mathcal{G}$. Because G is open, there exists $\epsilon > 0$ such that $(x - \epsilon, x + \epsilon) \subset G$. By construction, $b_i - a_i = (b - a)/2^i$. Choose N such that $n \ge N$ implies that $(b - a)/2^n < \epsilon$. For $n \ge N$, we have $I_n \subset (x - \epsilon, x + \epsilon) \subset G$. That is, I_n admits a finite subcover for $n \ge N$. Indeed, it admits a subcover of size one because the single element G contains I_n. This contradiction establishes the theorem. ∎

A.44 Theorem: A continuous function on a closed bounded interval is uniformly continuous and bounded.

PROOF: Let $f(\cdot)$ be continuous on $[a, b]$. Given $\epsilon > 0$, we must exhibit a $\delta > 0$ such that, for all $x \in [a, b]$, $|x - y| < \delta$ implies that $|f(x) - f(y)| < \epsilon$. Accordingly, we accept $\epsilon > 0$ and work with the continuity at each point $x \in [a, b]$ to obtain $\delta_x > 0$ such that $|x - y| < \delta_x$ and $y \in [a, b]$ imply that $|f(x) - f(y)| < \epsilon/2$. The uncountable collection

$$\mathcal{G} = \{(x - \delta_x/2, x + \delta_x/2) | x \in [a, b]\}$$

covers $[a, b]$. Invoking the preceding theorem, there exists a finite subcover

$$\mathcal{G}' = \{(x_1 - \delta_{x_1}/2, x_1 + \delta_{x_1}/2), \ldots, (x_n - \delta_{x_n}/2, x_n + \delta_{x_n}/2)\}$$

that also covers $[a, b]$. Let $\delta = (1/2)\min\{\delta_{x_1}, \delta_{x_2}, \ldots, \delta_{x_n}\} > 0$. Now, suppose that $x, y \in [a, b]$ and $|x - y| < \delta$. For some i, we must have $x \in (x_i - \delta_{x_i}/2, x + \delta_{x_i}/2)$, which means that $|x - x_i| < \delta_{x_i}/2$. Consequently,

$$|y - x_i| = |y - x + x - x_i| \leq |y - x| + |x - x_i| < \delta + \delta_{x_i}/2$$
$$< \delta_{x_i}/2 + \delta_{x_i}/2 = \delta_{x_i}$$
$$|f(x) - f(y)| \leq |f(x) - f(x_i)| + |f(x_i) - f(y)| < \epsilon/2 + \epsilon/2 = \epsilon,$$

and f is uniformly continuous.

Now, repeat the construction above for $\epsilon = 1$ to obtain a finite subcover

$$\mathcal{H}' = \{(x_1 - \delta_{x_1}, x_1 + \delta_{x_1}), (x_2 - \delta_{x_2}, x_2 + \delta_{x_2}), \ldots, (x_n - \delta_{x_n}, x_n + \delta_{x_n})\}$$

such that $y \in [a, b]$ and $|y - x_i| < \delta_{x_i}$ imply that $|f(y) - f(x_i)| < 1$. Let $M = 1 + \max_{i=1}^{n} |f(x_i)|$. For any $y \in [a, b]$, we must have $y \in (x_i - \delta_{x_i}, x_i + \delta_{x_i})$ for some i. The following inequality then shows that $f(\cdot)$ is bounded on $[a, b]$.

$$|f(y)| = |f(y) - f(x_i)| + |f(x_i)| < 1 + |f(x_i)| < 1 + \max_{1 \leq j \leq 1} |f(x_j)| = M \blacksquare$$

We conclude with a final useful feature of absolute convergence. It permits summation order interchange in a double infinite sum. To motivate the discussion, we first note that it is possible to have a grid of numbers, a_{ij}, such that $\sum_i \sum_j a_{ij}$ is not the same as $\sum_j \sum_i a_{ij}$. Indeed, one expression may converge, while the other diverges. Table A.2 illustrates the point. It lays out the initial terms and the general pattern for $\sum_i \sum_j a_{ij}$, where

$$a_{ij} = \begin{cases} 1, & j = 1 \\ \dfrac{-(2^i - 1)}{2^{i+j-1}}, & j > 1. \end{cases}$$

For any i, we can manipulate the row sum to reveal a convergent geometric series. That is,

$$\sum_{j=1}^{\infty} a_{ij} = 1 + \sum_{j=2}^{\infty} \frac{-(2^i - 1)}{2^{i+j-1}} = 1 - \frac{2^i - 1}{2^i} \sum_{j=1}^{\infty} \frac{1}{2^j}$$

$$= 1 - \frac{2^i - 1}{2^{i+1}} \sum_{j=0}^{\infty} \frac{1}{2^j} = 1 - \frac{2^i - 1}{2^i} = \frac{1}{2^i}$$

$$\sum_{i=1}^{\infty} \sum_{j=1}^{\infty} a_{ij} = \sum_{i=1}^{\infty} \frac{1}{2^i} = \frac{1}{2} \sum_{i=0}^{\infty} \frac{1}{2^i} = \frac{1}{2}(2) = 1.$$

However, the reverse summation $\sum_{j=1}^{\infty} \sum_{i=1}^{\infty} a_{ij}$ is divergent because the sum over the first column diverges to infinity. Although this fact alone is

$\vdots$	$\vdots$	$\vdots$	$\vdots$	$\vdots$	$\cdots$	$\vdots$	$\cdots$
i	1	$-(2^i-1)/2^{i+1}$	$-(2^i-1)/2^{i+2}$	$-(2^i-1)/2^{i+3}$	$\cdots$	$-(2^i-1)/2^{i+j-1}$	$\cdots$
$\vdots$	$\vdots$	$\vdots$	$\vdots$	$\vdots$	$\cdots$	$\vdots$	$\cdots$
5	1	$-31/2^6$	$-31/2^7$	$-31/2^8$	$\cdots$	$-31/2^{j+4}$	$\cdots$
4	1	$-15/2^5$	$-15/2^6$	$-15/2^7$	$\cdots$	$-15/2^{j+3}$	$\cdots$
3	1	$-7/2^4$	$-7/2^5$	$-7/2^6$	$\cdots$	$-7/2^{j+2}$	$\cdots$
2	1	$-3/2^3$	$-3/2^4$	$-3/2^5$	$\cdots$	$-3/2^{j+1}$	$\cdots$
1	1	$-1/2^2$	$-1/2^3$	$-1/2^4$	$\cdots$	$-1/2^j$	$\cdots$
	1	2	3	4	$\cdots$	j	$\cdots$

TABLE A.2. A double sum, $\sum_i \sum_j a_{ij}$, that does not allow the interchange of the summation order.

sufficient to preclude an initial summation by columns, it turns out that the remaining columns also exhibit divergent sums. One of the exercises asks you to prove this observation. We will see that absolute convergence, which is not a property of this example, precludes these pathological cases and allows us to interchange summations whenever it suits our convenience.

First, consider the case where $\sum_{i=1}^{\infty} \sum_{j=1}^{\infty} a_{ij} = a$, a real number, and all the a_{ij} are nonnegative. In this case, we can use upper bound arguments to show that the reversed summation order produces the same limit. Let a_{i*} denote $\sum_{j=1}^{\infty} a_{ij}$. Then the partial sums $\sum_{i=1}^{m} a_{i*} \to a$. We have $\sum_{i=1}^{m} \sum_{j=1}^{n} a_{ij} \le \sum_{i=1}^{m} a_{i*} \le a$ for any (m,n) pair. So, a is an upper bound of the set $S = \left\{ \sum_{i=1}^{m} \sum_{j=1}^{n} a_{ij} \,\middle|\, m,n = 1,2,3,\ldots \right\}$. We argue that $a-\epsilon$ cannot be an upper bound for any $\epsilon > 0$. Indeed, given such an ϵ, we choose M such that $a - \sum_{i=1}^{m} a_{i*} < \epsilon/2$ for all $m \ge M$. Then choose N such that $a_{i*} - \sum_{j=1}^{n} a_{ij} < \epsilon/(2M)$ for all $n \ge N$ and all i in the finite span $1,2,\ldots,M$. Then

$$a - \sum_{i=1}^{M} \sum_{j=1}^{N} a_{ij} = a - \sum_{i=1}^{M} a_{i*} + \sum_{i=1}^{M} \left(a_{i*} - \sum_{j=1}^{N} a_{ij} \right) < \epsilon/2 + M[\epsilon/(2M)] = \epsilon.$$

This implies, of course, that $\sum_{i=1}^{M} \sum_{j=1}^{N} a_{ij} > a - \epsilon$, which shows that $a - \epsilon$ cannot be an upper bound. We conclude that a is the least upper bound of S.

Now, consider the reversed summation. The partial sums $\sum_{j=1}^{n} \sum_{i=1}^{m} a_{ij}$ are finite and therefore the summation order is irrelevant. The set S then contains all the partial sums of the reversed summation. Then, for any j, we have $\sum_{i=1}^{m} a_{ij} \le \sum_{k=1}^{j} \sum_{i=1}^{m} a_{ij} \le a$, which implies convergence of the inner sums: $\sum_{i=1}^{\infty} a_{ij} = a_{*j}$. Also, we must have $\sum_{j=1}^{n} a_{*j} \le a$ for all n. If one such sum were larger than a, we could produce a finite sum $\sum_{j=1}^{N} \sum_{i=1}^{N} a_{ij}$ that is also larger than a and thereby contradict the least upper bound status of a. Consequently, we must also have $\sum_{j=1}^{\infty} a_{*j}$ convergent. Knowing now that the reversed summation converges, we can repeat the above argument to show that it converges to the least upper bound of S, which is a.

So, we can reverse the summation order for double convergent sums of nonnegative summands. It turns out that nonnegative summands is not necessary, provided that the convergence is absolute. The following theorem states the condition precisely.

A.45 Theorem: Suppose that $\sum_{i=1}^{\infty} \sum_{j=1}^{\infty} a_{ij}$ converges absolutely. That is, for each i, $\sum_{j=1}^{\infty} |a_{ij}| = A_i$, a real number, and $\sum_{i=1}^{\infty} A_i = A$, also a real number. Then $\sum_{i=1}^{\infty} \sum_{j=1}^{\infty} a_{ij}$ and $\sum_{j=1}^{\infty} \sum_{i=1}^{\infty} a_{ij}$ both converge to the same real value.

PROOF: Because, for each i, $\sum_j a_{ij}$ converges absolutely, it must converge in the ordinary sense. Let $\sum_{j=1}^{\infty} a_{ij} = a_{i*}$. Moreover, since $|\sum_{j=1}^{n} a_{ij}| \leq \sum_{j=1}^{n} |a_{ij}| \leq A_i$ for all n, we have $|a_{i*}| \leq A_i$. Dominated convergence (Theorem A.35) then asserts that $\sum_i a_{i*}$ converges. Let $\sum_{i=1}^{\infty} a_{i*} = a$. Substituting the definition of a_{i*}, we obtain $\sum_i \sum_j a_{ij} = a$.

Because the summands are nonnegative, we can sum $\sum_{i=1}^{\infty} \sum_{j=1}^{\infty} |a_{ij}|$ in either order to obtain the same result. Consequently, $\sum_{j=1}^{\infty} \sum_{i=1}^{\infty} a_{ij}$ also converges absolutely. We repeat the argument above to conclude that $\sum_{i=1}^{\infty} a_{ij} = a_{*j}$ for $j = 1, 2, \ldots$ and $\sum_{j=1}^{\infty} a_{*j} = \sum_{j=1}^{\infty} \sum_{i=1}^{\infty} a_{ij} = b$, where the a_{*j} and b are real numbers. It remains to show that $b = a$, or equivalently, that $|a - b| < \epsilon$ for an arbitrary $\epsilon > 0$.

We use the triangle inequality to write $|a - b| \leq \sum_{k=1}^{5} T_k$ and then show that each $T_k < \epsilon/5$. Let S be the collection of partial absolute sums $\sum_{i=1}^{m} \sum_{j=1}^{n} |a_{ij}|$, and let A be its least upper bound. Choose M, N such that $A - \epsilon/5 < \sum_{i=1}^{m} \sum_{j=1}^{n} |a_{ij}| \leq A$ for all (m, n) pairs such that $m \geq M$ and $n \geq N$. For any choice (m, n, m', n') such that $m > m' > M$ and $n > n' > N$, we have

$$
\begin{aligned}
T_3 &= \left| \sum_{i=1}^{m} \sum_{j=1}^{n} a_{ij} - \sum_{i=1}^{m'} \sum_{j=1}^{n'} a_{ij} \right| \\
&= \left| \sum_{i=1}^{m'} \sum_{j=n'+1}^{n} a_{ij} + \sum_{i=m'+1}^{m} \sum_{j=1}^{n'} a_{ij} + \sum_{i=m'+1}^{m} \sum_{j=n'+1}^{n} a_{ij} \right| \\
&\leq \sum_{i=1}^{m'} \sum_{j=n'+1}^{n} |a_{ij}| + \sum_{i=m'+1}^{m} \sum_{j=1}^{n'} |a_{ij}| + \sum_{i=m'+1}^{m} \sum_{j=n'+1}^{n} |a_{ij}| \\
&= \sum_{i=1}^{m} \sum_{j=1}^{n} |a_{ij}| - \sum_{i=1}^{m'} \sum_{j=1}^{n'} |a_{ij}| < \frac{\epsilon}{5}.
\end{aligned}
$$

The last inequality follows because both double sums are members of S that lie in the interval $(A - \epsilon/5, A]$. Now, we exercise felicitous choices for m, n, m', and n' to achieve the same bound on the remaining T_i.

First, we exploit $\sum_{j=1}^{n'} a_{*j} \to b$ to fix $n' > N$ such that the T_5 inequality below holds. Next, we use $\sum_{i=1}^{m'} a_{ij} \to a_{*j}$ to choose $m' > M$ such that

$\left|\sum_{i=1}^{m'} a_{ij} - a_{*j}\right| < \epsilon/(5n')$, simultaneously for $j = 1, 2, \ldots, n'$. This forces the T_4 inequality below to hold. Now, noting that $\sum_{i=1}^{m} a_{i*} \to a$, choose $m > m' > M$ to achieve the T_1 inequality. Finally, because $\sum_{j=1}^{n} a_{ij} \to a_{i*}$, we can choose $n > n' > N$ such that, simultaneously for $i = 1, 2, \ldots, m$, we have $\left|\sum_{j=1}^{n} a_{ij} - a_{i*}\right| < \epsilon/(5m)$. This forces the T_2 inequality.

$$T_5 = \left|\sum_{j=1}^{n'} a_{*j} - b\right| < \frac{\epsilon}{5}$$

$$T_4 = \sum_{j=1}^{n'}\left|\sum_{i=1}^{m'} a_{ij} - a_{*j}\right| < n' \cdot \frac{\epsilon}{5n'} = \frac{\epsilon}{5}$$

$$T_1 = \left|a - \sum_{i=1}^{m} a_{i*}\right| < \frac{\epsilon}{5}$$

$$T_2 = \sum_{i=1}^{m}\left|a_{i*} - \sum_{j=1}^{n} a_{ij}\right| < m \cdot \frac{\epsilon}{5m} = \frac{\epsilon}{5}$$

With these choices of m, n, m', and n', we have

$$|a - b| \leq T_1 + T_2 + T_3 + T_4 + T_5 < 5(\epsilon/5) = \epsilon. \blacksquare$$

Exercises

A.24 Show that $\sum_i 1/i$ diverges.

A.25 Show that $\sum_i 1/i^2$ converges.

*A.26 Show that $\sum_i (-1)^i/i$ converges, but not absolutely.

*A.27 Suppose that the sequence $x_1, x_2, \ldots$ converges to x. Define a new sequence by averaging leading segments of the x_i. Specifically, let $y_n = (1/n)\sum_{i=1}^{n} x_i$. Show that the sequence $y_1, y_2, \ldots$ also converges to x.

A.28 Table A.2 exhibits a grid of values a_{ij} for which $\sum_i \sum_j a_{ij}$ is not the same as $\sum_j \sum_i a_{ij}$. Show that all the column sums diverge.

*A.29 A set is *bounded* if there exists numbers M, N such that $S \subset (M, N)$. Show that a bounded infinite set of real numbers must have at least one limit point.

A.30 Suppose that $\sum_{i=0}^{\infty} a_i$ converges. Show that $\lim_{i\to\infty} |a_i| = 0$.

*A.31 Given a sequence $\{a_n\}$, suppose that $0 < \lim_{n\to\infty} |a_{n+1}/a_n| = a < 1$. Show that the series $\sum_{n=0}^{\infty} a_n$ converges.

A.32 Show that the complement of an open set must be closed.

A.33 Show that the interval $[x, \infty)$ is closed and that (x, ∞) is open. Construct a subset that is both open and closed.

A.34 Show that a set of real numbers bounded from below has a greatest lower bound.

A.35 Show that a nonincreasing sequence that is bounded from below must converge.

A.36 Complete the proof of Theorem A.42 by showing that

$$\inf\{f(x) + g(x) | x \in [a, b]\} \geq \inf\{f(x) | x \in [a, b]\} + \inf\{g(x) | x \in [a, b]\}.$$

A.37 Suppose that $x_n \to x$, a real number. Show that the set $\{x_i\}$ has a least upper bound.

A.4   Riemann-Stieltjes integrals

The familiar integral from elementary calculus is more accurately called the Riemann integral to distinguish it from extended forms, such as the Lebesgue integral. Although the latter is more appropriate for advanced probability calculations, the Riemann integral suffices for this text. Of course, the Riemann integral is most closely associated with the area under a curve, obtained as a limit of a sum of rectangular approximations. The Riemann-Stieltjes integral is a generalization in which the width of each rectangular component receives more or less weight, depending on the component position. This section reviews the definitions and major results needed to use Riemann integrals. Of course, it is not possible to recapitulate all of calculus here, so we concentrate on the integral as the limit of approximating sums, particularly the Riemann-Stieltjes form, which finds useful application in integrations with respect to a probability cumulative distribution function.

A.46 Definition: Let $[a, b]$ be a closed interval of real numbers. A *partition* P of $[a, b]$ is a finite sequence $a = x_0 \leq x_1 \leq x_2 \leq \ldots \leq x_n = b$. If the sequence extends through x_n, we say that P contains n subintervals, $[x_{i-1}, x_i]$ for $i = 1, 2, \ldots, n$. The elements $x_0, x_1, \ldots, x_n$ are the *boundary points* of the partition. The *mesh* of the partition is $\text{mesh}(P) = \max_{1 \leq i \leq n}(x_i - x_{i-1})$. If P_1 and P_2 are two partitions of $[a, b]$, we say that P_2 *refines* P_1, if each boundary point of P_1 is also a boundary point of P_2. When P_2 refines P_1, we also say that P_1 *coarsens* P_2. ∎

The mesh of a partition is the width of its largest subinterval. When P_2 refines P_1, P_2 contains all the boundary points of P_1, plus some additional boundary points. Therefore, each subinterval of the coarser partition P_1 is exactly the union of some number of adjacent subintervals from the finer partition P_2. This forces $\text{mesh}(P_2) \leq \text{mesh}(P_1)$.

A.47 Definition: A *singularity* of a function $f(\cdot)$ is a point x_0 such that the set $\{f(x) \mid x \in \mathcal{I}(x_0, \epsilon)\}$ is unbounded for every neighborhood $\mathcal{I}(x_0, \epsilon)$. We say that $f(\cdot)$ has *isolated singularities* on $[a, b]$ if there exists a partition P such that the boundary points of P, except possibly one or both endpoints, are the singularities of $f(\cdot)$. We say that partition P marks the singularities of $f(\cdot)$ on $[a, b]$. ▮

Consider, for example, the function $f(x) = 1/x$ for $x > 0$ and zero elsewhere. It has a singularity at $x = 0$. The partition $x_0 = -1, x_1 = 0, x_2 = 1$ marks the singularities and suffices to show that $f(\cdot)$ has isolated singularities on $[-1, 1]$. The partition $x_0 = 0, x_1 = 4$ serves a similar purpose on $[0, 4]$. Integrals in this text will involve only functions with isolated singularities. We note that $f(\cdot)$ is bounded on any subspan $[c, d]$ that includes no singularities. This is fairly easy to prove, given Theorem A.43, which asserts that an open cover of a closed bounded interval must admit a finite subcover. For each $x \in [c, d]$, there is a neighborhood $(x - \delta_x, x + \delta_x)$ and a bound M_x such that $|f(x)| < M_x$ for x in the neighborhood. These neighborhoods form an uncountable cover of $[c, d]$. Suppose that a finite subcover involves neighborhoods centered on $x_1, x_2, \ldots, x_n$. Let $M = \max\{M_{x_1}, M_{x_2}, \ldots, M_{x_n}\}$. Then, for $y \in [c, d]$, we must have $y \in (x_i - \delta_{x_i}, x_i + \delta_{x_i})$ for some i, which implies that $|f(y)| < M_{x_i} \leq M$.

A.48 Theorem: Suppose that $f(\cdot)$ has no singularities on the closed finite interval $[a, b]$. Then there exists M such that $|f(x)| < M$ for all $x \in [a, b]$.

PROOF: See preceding discussion. ▮

A.49 Definition: We define the integral in steps, each generalizing the previous. If $f(\cdot)$ has no singularities on $[a, b]$ and if $[a, b]$ is a finite interval, we say that the integral, if it exists, is a *proper* integral. Otherwise, it is an *improper* integral. We first define proper integrals and then extend the definition to improper integral through limit processes.

- Suppose first that $f(\cdot)$ has no singularities on the finite interval $[a, b]$. From the preceding theorem, $f(\cdot)$ must be bounded on $[a, b]$. That is, $|f(x)| \leq M$ for all $x \in [a, b]$. Let $a = x_0 \leq x_1 \leq \ldots \leq x_n = b$ be a partition of $[a, b]$. The *upper sum* and *lower sum* of $f(\cdot)$ on partition P are, respectively,

$$S_f(P) = \sum_{i=1}^{n} (x_i - x_{i-1}) \cdot \sup\{f(x) \mid x_{i-1} \leq x \leq x_i\}$$

$$s_f(P) = \sum_{i=1}^{n} (x_i - x_{i-1}) \cdot \inf\{f(x) \mid x_{i-1} \leq x \leq x_i\}.$$

Because $\sum_{i=1}^{n} (x_i - x_{i-1}) = b - a$, we have $S_f(P) \geq -M(b - a)$ and $s_f(P) \leq M(b-a)$. These bounds are independent of the particular partition P. Let $\mathcal{P}$ be the collection of all partitions of $[a, b]$. The bounds then imply that $U_f = \inf\{S_f(P) \mid P \in \mathcal{P}\}$ and $L_f = \sup\{s_f(P) \mid P \in \mathcal{P}\}$ exist. If $U_f = L_f$, we say that $f(\cdot)$ is *integrable* on $[a, b]$ and write

$\int_a^b f = U_f = L_f$. This is the Riemann integral of $f(\cdot)$. The alternative notation $\int_a^b f(x)\,dx$ is frequently employed, although the dummy variable x in no way influences the value of the integral. For $a > b$, we define $\int_a^b f = -\int_b^a f$, if the latter exists.

- Suppose further that $F(\cdot)$ is a monotone nondecreasing function on $[a, b]$. We modify the upper and lower sums slightly to

$$S_{f,F}(P) = \sum_{i=1}^n [F(x_i) - F(x_{i-1})] \cdot \sup\{f(x) \mid x_{i-1} \le x \le x_i\}$$
$$\le M[F(b) - F(a)]$$
$$s_{f,F}(P) = \sum_{i=1}^n [F(x_i) - F(x_{i-1})] \cdot \inf\{f(x) \mid x_{i-1} \le x \le x_i\}$$
$$\ge -M[F(b) - F(a)].$$

As the sums remain bounded, $U_{f,F} = \inf\{S_{f,F}(P) \mid P \in \mathcal{P}\}$ and $L_{f,F} = \sup\{s_{f,F}(P) \mid P \in \mathcal{P}\}$ exist. If $U_{f,F} = L_{f,F}$, we say that $f(\cdot)$ is *integrable with respect to F* on $[a, b]$ and write $\int_a^b f\,dF = U_{f,F} = L_{f,F}$. This is the Riemann-Stieltjes integral of $f(\cdot)$. The alternative notation $\int_a^b f(x)\,dF(x)$ also appears. If $F(x) = x$, the Riemann-Stieltjes integral reverts to the ordinary Riemann integral. If $a > b$, we define $\int_a^b f\,dF = -\int_b^a f\,dF$, assuming that the latter exists. Hereafter, whenever we refer to the Riemann-Stieltjes form $\int f\,dF$ or use the phrase integrable with respect to F, we implicitly assume that F is a monotone nondecreasing function.

- If $f(\cdot)$ has a singularity at a or at b or at both, we say that $f(\cdot)$ is integrable on $[a, b]$ and write $\int_a^b f = L$ if the appropriate limit exists. Depending on the location of the singularity, the limit is $L = \lim_{c \to a+} \int_c^b f$, or $L = \lim_{d \to b-} \int_a^d f$, or $L = \lim_{c \to a+, d \to b-} \int_c^d f$.

  Similarly, we say that $f(\cdot)$ is integrable with respect to F and write $\int_a^b f\,dF = L$ if the appropriate limit exists. The limiting form again depends on the location of the singularity: $L = \lim_{c \to a+} \int_c^b f\,dF$, or $L = \lim_{d \to b-} \int_a^d f\,dF$, or $L = \lim_{c \to a+, d \to b-} \int_c^d f\,dF$.

- If $f(\cdot)$ has isolated singularities on $[a, b]$, let P, composed of $(a = x_0 < x_1 < \ldots < x_n = b)$, mark the singularities. We say that $f(\cdot)$ is integrable on $[a, b]$ if $f(\cdot)$ is integrable on all components of P. In this case, we write $\int_a^b f = \sum_{i=1}^n \int_{x_{i-1}}^{x_i} f$. If $f(\cdot)$ is integrable with respect to F on all components of P, we say that $f(\cdot)$ is integrable with respect to F on $[a, b]$ and write $\int_a^b f\,dF = \sum_{i=1}^n \int_{x_{i-1}}^{x_i} f\,dF$.

- Finally, we say that $f(\cdot)$ is integrable, or integrable with respect to F, on $(-\infty, b], [a, \infty)$, or $(-\infty, \infty)$ if the appropriate limit exists. The notations are

$$\int_{-\infty}^{b} f = \lim_{a \to -\infty} \int_{a}^{b} f \qquad \qquad \int_{-\infty}^{b} f \, dF = \lim_{a \to -\infty} \int_{a}^{b} f \, dF$$
$$\int_{a}^{\infty} f = \lim_{b \to \infty} \int_{a}^{b} f \qquad \qquad \int_{a}^{\infty} f \, dF = \lim_{b \to \infty} \int_{a}^{b} f \, dF$$
$$\int_{-\infty}^{\infty} f = \lim_{\substack{a \to -\infty \\ b \to \infty}} \int_{a}^{b} f \qquad \qquad \int_{-\infty}^{\infty} f \, dF = \lim_{\substack{a \to -\infty \\ b \to \infty}} \int_{a}^{b} f \, dF. \quad \blacksquare$$

The traditional example of a nonintegrable function is $f(x) = 1$ for rational x and $f(x) = -1$ for irrational x. On all partitions P of $[0, 1]$, we have $S_f(P) = 1$ and $s_f(P) = -1$. Consequently, $L_f = -1 < U_f = 1$, and the function is not integrable on $[0, 1]$. Of course, this function is very volatile; it is discontinuous at every point. The proof is beyond the text's scope, but integrable functions are precisely those with a very sparse set of discontinuities.

Of course, $s_f(P) \leq S_f(P)$ for every partition P, but it is also true that $s_f(P) \leq S_f(Q)$ for any pair of partitions P and Q. This follows by considering their common refinement R, which includes all points of both P and Q. A component $[x_{i-1}, x_i]$ of Q consists of one or more R components, say $[y_{i0}, y_{i1}], \dots, [y_{i,k_i-1}, y_{i,k_i}]$. Also, $y_{i0} = x_{i-1}$ and $y_{i,k_i} = x_i$ because the R components must fit exactly in the parent Q component. Because the supremum of f over an R component can be no larger than that over its parent Q component, we can bound $S_f(Q)$ as follows.

$$S_f(Q) = \sum_{i=1}^{n} (x_i - x_{i-1}) \sup\{f(x) \mid x \in [x_{i-1}, x_i]\}$$
$$= \sum_{i=1}^{n} \left[\sum_{j=1}^{k_i} (y_{i,j} - y_{i,j-1}) \right] \sup\{f(x) \mid x \in [x_{i-1}, x_i]\}$$

Applying the aforementioned bound on the supremum, we have

$$S_f(Q) \geq \sum_{i=1}^{n} \sum_{j=1}^{k_i} (y_{i,j} - y_{i,j-1}) \sup\{f(x) \mid x \in [y_{i,j-1}, y_{i,j}]\} = S_f(R).$$

A similar argument shows that $s_f(P) \leq s_f(R)$. Refining a partition raises the lower sum and lowers the upper sum. Combining these results, we then have $s_f(P) \leq s_f(R) \leq S_f(R) \leq S_f(Q)$, as expected.

Now, fix a partition P. We have $S_f(Q) \geq s_f(P)$ and $s_f(Q) \leq S_f(P)$ for all $Q \in \mathcal{P}$, which implies that

$$L_f = \sup\{s_f(Q) \mid Q \in \mathcal{P}\} \leq s_f(P) \leq S_f(P) \leq \inf\{S_f(Q) \mid Q \in \mathcal{P}\} = U_f.$$

We see that $L_f \leq U_f$, whether f is integrable or not. If they are equal, it is integrable; if $L_f < U_f$, it is not. The same argument applies if we integrate with respect to a monotone nondecreasing F. That is, $L_{f,F} \leq U_{f,F}$ always, whether f is integrable with respect to F or not.

In the case where a limit appears on both ends of the integration span, we insist that the limits be independent. That is, $\lim_{c \to a+, d \to b-} \int_c^d f = L$ means for every $\epsilon > 0$, there exist $\delta_1, \delta_2 > 0$ such that $a < c < a + \delta_1$ and $b - \delta_2 < d < b$ imply that $|\int_c^d f - L| < \epsilon$. Example A.54 illustrates this point. First, however, we need methods for evaluating the base integrals to which we apply limiting arguments. The following theorem provides a useful technical device, which facilitates the proofs of other properties of integrals.

A.50 Theorem: Suppose that f has no singularities on the finite interval $[a, b]$. Then the following are equivalent.

(a) f is integrable with respect to F on $[a, b]$.

(b) For every $\epsilon > 0$, there exist partitions P and Q of $[a, b]$ such that $S_{f,F}(P) - s_{f,F}(Q) < \epsilon$.

(c) For every $\epsilon > 0$, there exists a partition P such that $S_{f,F}(P) - s_{f,F}(P) < \epsilon$.

PROOF: Let $\mathcal{P}$ denote the collection of all partitions of $[a, b]$. First, suppose that f is integrable with respect to F on $[a, b]$. Let $A = \int_a^b f \, dF$. Given $\epsilon > 0$, $A + \epsilon/2$ cannot be a lower bound of $\{S_{f,F}(P) | P \in \mathcal{P}\}$, nor can $A - \epsilon/2$ be an upper bound of $\{s_{f,F}(P) | P \in \mathcal{P}\}$. Consequently, there exist partitions P and Q such that $S_{f,F}(P) < A + \epsilon/2$ and $s_{f,F}(Q) > A - \epsilon/2$. Then, however,

$$0 \le S_{f,F}(P) - s_{f,F}(Q) \le A + \epsilon/2 - (A - \epsilon/2) = \epsilon.$$

This shows that (a) implies (b). Now, assume (b). Given $\epsilon > 0$, (b) asserts the existence of partitions P and Q such that $S_{f,F}(P) - s_{f,F}(Q) < \epsilon$. Let R be the common refinement of P and Q. Moving to R from either P or Q raises the lower sum and lowers the upper sum. Therefore,

$$0 \le S_{f,F}(R) - s_{f,F}(R) \le S_{f,F}(P) - s_{f,F}(Q) < \epsilon,$$

which shows that (b) implies (c). Now, assume (c). For arbitrary $\epsilon > 0$, we have a partition P such that $S_{f,F}(P) - s_{f,F}(P) < \epsilon$. Then $U_{f,F} - L_{f,F} \le S_{f,F}(P) - s_{f,F}(P) < \epsilon$, and we conclude that $U_{f,F} = L_{f,F}$. ∎

The discussion below establishes further properties of the basic integral without singularities on a finite interval $[a, b]$. For any partition $a = x_0 \le x_1 \le x_2 \ldots \le x_n = b$, we have

$$\sum_{i=1}^{n} [F(x_i) - F(x_{i-1})] = F(x_1) - F(x_0) + \ldots + F(x_n) - F(x_{n-1})$$

$$= F(b) - F(a).$$

Therefore, for a constant k, all partitions give the same upper and lower sum:

$$S_{k,F}(P) = \sum_{i=1}^{n} k[F(x_i) - F(x_{i-1})] = k[F(b) - F(a)]$$

$$s_{k,F}(P) = \sum_{i=1}^{n} k[F(x_i) - F(x_{i-1})] = k[F(b) - F(a)]$$

$$U_{k,F} = \inf\{S_{k,F}(P)\} = k[F(b) - F(a)]$$

$$L_{k,F} = \sup\{s_{k,F}(P)\} = k[F(b) - F(a)].$$

Consequently, k is integrable with respect to F on $[a, b]$, and $\int_a^b k \, dF = k[F(b) - F(a)]$.

We next show that the integral is linear. That is, if f and g are integrable with respect to F on $[a, b]$, so is $Af + Bg$ for constants A and B. Moreover, $\int_a^b (Af + Bg) \, dF = A \int_a^b f \, dF + B \int_a^b g \, dF$. Linearity follows from properties of the supremum and infimum operations as asserted in Theorem A.42. There are, however, a number of cases to consider. We first show that $\int_a^b kf = k \int_a^b f$ for constant k. For $k = 0$, we have $\int_a^b 0 = 0 \cdot [F(b) - F(a)] = 0$, so the property is certainly true if $k = 0$. If $k > 0$, it factors from infimum and supremum operations:

$$S_{kf,F}(P) = \sum_{i=1}^{n} [F(x_i) - F(x_{i-1})] \sup_{x_{i-1} \le x \le x_i} kf(x) = kS_{f,F}(P)$$

$$U_{kf,F} = \inf_P S_{kf,F}(P) = \inf_P [kS_{f,F}(P)] = k \inf_P S_{f,F} = kU_{f,F} = k \int_a^b f,$$

and similarly for the lower sums. Consequently, $\int_a^b kf \, dF = k \int_a^b f \, dF$. The case for $k < 0$ follows in an analogous fashion, except that factoring k from a supremum translates it to an infimum and vice versa.

Now consider partition P, as given above, applied to $\int_a^b (f + g)$. Again using Theorem A.42, we have

$$S_{f+g,F}(P) = \sum_{i=1}^{n} [F(x_i) - F(x_{i-1})] \sup_{x_{i-1} \le x \le x_i} [f(x) + g(x)]$$

$$\le \sum_{i=1}^{n} [F(x_i) - F(x_{i-1})] \sup_{x_{i-1} \le x \le x_i} f(x)$$

$$+ \sum_{i=1}^{n} [F(x_i) - F(x_{i-1})] \sup_{x_{i-1} \le x \le x_i} g(x) = S_{f,F}(P) + S_{g,F}(P).$$

Similarly, we can show that $s_{f+g,F}(P) \ge s_{f,F}(P) + s_{g,F}(P)$, which gives the two-sided bound

$$s_{f,F}(P) + s_{g,F}(P) \le s_{f+g,F}(P) \le S_{f+g,F}(P) \le S_{f,F}(P) + S_{g,F}(P).$$

Taking the supremum on the left and the infimum on the right, we have

$$\int_a^b f\,\mathrm{d}F + \int_a^b g\,\mathrm{d}F \;\le\; L_{f+g,F} \;\le\; U_{f+g,F} \;\le\; \int_a^b f\,\mathrm{d}F + \int_a^b g\,\mathrm{d}F.$$

Consequently, $f + g$ is integrable with respect to F, and its integral is the sum of the component integrals. Combining this result with $\int_a^b kf\,\mathrm{d}F = k\int_a^b f\,\mathrm{d}F$ established above yields the desired linearity: $\int_a^b (Af + Bg)\,\mathrm{d}F = \int_a^b Af\,\mathrm{d}F + \int_a^b Bg\,\mathrm{d}F = A\int_a^b f\,\mathrm{d}F + B\int_a^b g\,\mathrm{d}F$.

Continuing with another property of the basic Riemann-Stieltjes integral, suppose that f is integrable with respect to F on $[a, b]$ and $f(x) \ge 0$ on $[a, b]$. Then all approximating sums $S_{f,F}(P)$ and $s_{f,F}(P)$ are nonnegative. It follows that $L_{f,F}$ and $U_{f,F}$ are nonnegative, which proves that $\int_a^b f\,\mathrm{d}F \ge 0$. If $f(x) \ge g(x)$ on $[a, b]$ and both are integrable with respect to F, then the linearity proved above permits the following expansion: $0 \le \int_a^b (f - g)\,\mathrm{d}F = \int_a^b f\,\mathrm{d}F - \int_a^b g\,\mathrm{d}F$, which proves that $\int_a^b f\,\mathrm{d}F \ge \int_a^b g\,\mathrm{d}F$.

We now wish to show that the integrability of f implies the integrability of $|f|$. Suppose that $x, y \in [x_{i-1}, x_i]$, a partition component and the supremum and infimum operations below refer to this component. We have

$$|f(x)| - |f(y)| \;\le\; |f(x) - f(y)| \;\le\; \sup f(t) - \inf f(t).$$

Because this bound holds for arbitrary x, y in the interval, it remains true when we take the supremum as x varies and then the infimum as y varies. That is, $\sup |f(x)| - \inf |f(y)| \le \sup f(t) - \inf f(t)$. Consequently, letting the difference $D = S_{|f|,F}(P) - s_{|f|,F}(P)$, we have

$$
\begin{aligned}
D &= \sum_{i=1}^{n}[F(x_i) - F(x_{i-1})]\left(\sup_{x_{i-1}\le x\le x_i} |f(x)| - \inf_{x_{i-1}\le x\le x_i} |f(x)| \right) \\
&\le \sum_{i=1}^{n}[F(x_i) - F(x_{i-1})]\left(\sup_{x_{i-1}\le t\le x_i} f(t) - \inf_{x_{i-1}\le t\le x_i} f(t) \right) \\
&= S_{f,F}(P) - s_{f,F}(P).
\end{aligned}
$$

Because f is integrable, we can make the right side arbitrarily small. It follows that we can make the left side arbitrarily small, and then Theorem A.50 asserts that $|f|$ is integrable with respect to F on $[a, b]$.

Since $|f|$ is nonnegative, $\int_a^b |f|\,\mathrm{d}F \ge 0$. Moreover, $-|f(x)| \le f(x) \le |f(x)|$. It follows that

$$-\int_a^b |f|\,\mathrm{d}F \;\le\; \int_a^b f\,\mathrm{d}F \;\le\; \int_a^b |f|\,\mathrm{d}F$$

$$\left| \int_a^b f\,\mathrm{d}F \right| \;\le\; \int_a^b |f|\,\mathrm{d}F.$$

For integrable f and g, we then have

$$f(x) + g(x) \leq |f(x) + g(x)| \leq |f(x)| + |g(x)|$$
$$-[f(x) + g(x)] \leq |f(x) + g(x)| \leq |f(x)| + |g(x)|$$

$$\left| \int_a^b (f + g) \, dF \right| \leq \int_a^b |f + g| \, dF \leq \int_a^b |f| \, dF + \int_a^b |g| \, dF,$$

which establishes the triangle inequality for integrals.

We now show that f integrable on $[a, b]$ implies that it is integrable on subintervals contained in $[a, b]$. First, consider the degenerate interval $[c, c]$. Over $[c, c]$, all partitions are of the form $c = x_1 = x_2 = \ldots = x_n = c$. Therefore, $s_{f,F}(P) = S_{f,F}(P) = 0$ for all partitions. It follows that $L_{f,F} = U_{f,F} = 0$, $\int_c^c f \, dF$ exists, and it is equal to zero. Now, suppose that c is any point in $[a, b]$. We want to prove that f integrable with respect to F on $[a, b]$ implies integrability on $[a, c]$ and $[c, b]$. Moreover, $\int_a^b f \, dF = \int_a^c f \, dF + \int_c^b f \, dF$.

The construction is as follows. Let $\mathcal{P}$ be the collection of all partitions of $[a, b]$. Let $\mathcal{P}'$ be the subcollection of partitions that include point c as a boundary point. Because f is integrable on $[a, b]$, we can choose, given arbitrary $\epsilon > 0$, partition $P \in \mathcal{P}$ such that $0 \leq S_{f,F}(P) - s_{f,F}(P) < \epsilon$. Construct $R \in \mathcal{P}'$ by adding the point c to P, if it is not already there as a boundary point. As noted earlier, moving to the refinement R lowers the upper sum and raises the lower sum. It follows that

$$0 \leq S_{f,F}(R) - s_{f,F}(R) \leq S_{f,F}(P) - s_{f,F}(P) < \epsilon.$$

This shows that there are partitions containing boundary point c, for which the upper and lower sums are arbitrarily close. It follows that $U'_{f,F} = L'_{f,F}$, where

$$U'_{f,F} = \inf\{S_{f,F}(R) \mid R \in \mathcal{P}'\} \geq \inf\{S_{f,F}(R) \mid R \in \mathcal{P}\} = \int_a^b f \, dF$$

$$L'_{f,F} = \sup\{s_{f,F}(R) \mid R \in \mathcal{P}'\} \leq \sup\{s_{f,F}(R) \mid R \in \mathcal{P}\} = \int_a^b f \, dF.$$

The far right inequality holds because the supremum increases and the infimum decreases when the field of candidates enlarges. We now have $L'_{f,F} \leq \int_a^b f \, dF \leq U'_{f,F} = L'_{f,F}$. That is, we obtain the integral by considering only partitions that include boundary point c. On such partitions, the upper and lower sums break into two parts. One part sums partition components lying in $[a, c]$; the other sums components in $[c, b]$. Now, given $\epsilon > 0$, choose $R \in \mathcal{P}'$ such that $0 \leq S_{f,F}(R) - s_{f,F}(R) \leq \epsilon$. Denote by R^- and R^+ the partition components in $[a, c]$ and $[c, b]$, respectively. Then

$$[S_{f,F}(R^-) - s_{f,F}(R^-)] + [S_{f,F}(R^+) - s_{f,F}(R^+)] = S_{f,F}(R) - s_{f,F}(R) < \epsilon,$$

which forces $S_{f,F}(R^-) - s_{f,F}(R^-) < \epsilon$ and renders f integrable with respect to F on $[a, c]$. Similarly, it is integrable with respect to F on $[c, b]$. Finally,

$$s_{f_F}(R) = s_{f,F}(R^-) + s_{f,F}(R^+) \leq \left\{ \begin{array}{c} \int_a^b f \, dF \\ \int_a^c f \, dF + \int_c^b f \, dF \end{array} \right\}$$

$$\leq S_{f,F}(R^-) + S_{f,F}(R^+) = S_{f,F}(R) < s_{f,F}(R) + \epsilon,$$

from which we conclude that $\int_a^b f \, dF = \int_a^c f \, dF + \int_c^b f \, dF$.

We can repeat the process to break the interval into more than two parts. That is, if $a = x_0 \leq x_1 \leq \ldots \leq x_n = b$, we have $\int_a^b f \, dF = \sum_{i=1}^n \int_{x_{i-1}}^{x_i} f \, dF$. Finally, if the x_i are scattered arbitrarily in $[a, b]$, some of the subintervals $[x_{i-1}, x_i]$ may have $x_{i-1} > x_i$. This reverses the sense of the integral over such subintervals, but the integrals over all subintervals continue to sum to $\int_a^b f$. For example, suppose that $a < c_1 < c_2 < b$, and consider

$$\int_a^{c_2} f \, dF + \int_{c_2}^{c_1} f \, dF + \int_{c_1}^b f \, dF = \int_a^{c_1} f \, dF + \int_{c_1}^{c_2} f \, dF - \int_{c_1}^{c_2} f \, dF + \int_{c_1}^b f \, dF$$

$$= \int_a^b f \, dF.$$

We summarize the properties developed to this point in the next theorem, which also adds two further entries.

A.51 Theorem: Assume that $f(\cdot)$ and $g(\cdot)$ are integrable with respect to nondecreasing F over $[a, b]$. Unless otherwise noted, the interval may be infinite or semi-infinite. Then the following properties hold. Each is stated for the integral with respect to F, but, setting $F(x) = x$, we have them for the ordinary Riemann integral also.

- For constant K and finite interval $[a, b]$, we have $\int_a^b K \, dF = K[F(b) - F(a)]$.

- The integral is linear. That is, for real constants A and B, $\int_a^b (Af + Bg) \, dF = A \int_a^b f \, dF + B \int_a^b g \, dF$.

- If $f(x) \geq 0$ for $x \in [a, b]$, then $\int_a^b f \, dF \geq 0$. Consequently, if $f(x) \geq g(x)$ on $[a, b]$, then $\int_a^b f \, dF \geq \int_a^b g \, dF$.

- $|f|$ is integrable with respect to F on $[a, b]$ and $\left| \int_a^b f \, dF \right| \leq \int_a^b |f| \, dF$. Also, $\left| \int_a^b (f + g) \, dF \right| \leq \int_a^b |f + g| \, dF \leq \int_a^b |f| \, dF + \int_a^b |g| \, dF$.

- $\int_c^c f = 0$ for any $c \in [a, b]$.

- The integral decomposes into a sum of integrals on the subintervals of a partition. That is, if $\{x_0, x_1, \ldots, x_n\}$ are points in $[a, b]$, then f is integrable with respect to F on the partition components, and $\int_a^b f \, dF = \sum_{i=1}^n \int_{x_{i-1}}^{x_i} f \, dF$.

- If F is continuous at x, then $G(x) = \int_a^x f \, dF$ is continuous at x. The continuity is one-sided if $x = a$ or $x = b$.

- If f is continuous at x and F is differentiable at x, then G is differentiable at x and $G'(x) = f(x)F'(x)$.

PROOF: We establish the properties for functions with no singularities on a finite interval $[a, b]$. The more general versions follow by taking the appropriate limits. Having no singularities on $[a, b]$, f is bounded there. Suppose that $|f(x)| < M$ for $x \in [a, b]$.

The first six items follow from the preceding discussion. We need to establish the final two properties, which concern the related function $G(x) = \int_a^x f \, dF$. This integral is now known to exist because it spans a subinterval of $[a, b]$. We first address the continuity of G at x, assuming that F is continuous there. For any $y \in [a, b]$,

$$|G(x) - G(y)| = \left| \int_a^x f \, dF - \int_a^y f \, dF \right| = \left| \int_y^x f \, dF \right| \leq \left| \int_y^x |f| \, dF \right|$$
$$\leq M|F(x) - F(y)|.$$

Given $\epsilon > 0$, we invoke the continuity of F at x to choose δ such that $0 < |x - y| < \delta$ implies that $|F(x) - F(y)| < \epsilon/M$, which then implies that $|G(x) - G(y)| < \epsilon$. This shows that G is continuous at x.

Finally, suppose that f is continuous at x and F is differentiable there. For positive Δx such that $[x, x + \Delta x] \subset [a, b]$, we investigate the behavior of the absolute difference $Q = |[G(x + \Delta x) - G(x)]/\Delta x - f(x)F'(x)|$ as Δx approaches zero.

$$Q = \left| \frac{\int_x^{x+\Delta x} f \, dF}{\Delta x} - f(x)F'(x) \right|$$
$$\leq \left| \frac{\int_x^{x+\Delta x} f \, dF}{\Delta x} - f(x)\frac{F(x + \Delta x) - F(x)}{\Delta x} \right|$$
$$+ \left| f(x)\frac{F(x + \Delta x) - F(x)}{\Delta x} - f(x)F'(x) \right|$$
$$= \left| \frac{\int_x^{x+\Delta x}(f - f(x)) \, dF}{\Delta x} \right| + |f(x)| \left| \frac{F(x + \Delta x) - F(x)}{\Delta x} - F'(x) \right|$$
$$Q \leq \frac{\int_x^{x+\Delta x} |f - f(x)| \, dF}{\Delta x} + |f(x)| \left| \frac{F(x + \Delta x) - F(x)}{\Delta x} - F'(x) \right|$$

The proof now diverges into a number of cases, the most general being where both $f(x) \neq 0$ and $F'(x) \neq 0$. Of course, because F is monotone nondecreasing, $F'(x) \neq 0$ implies that $F'(x) > 0$. In this case, given $\epsilon > 0$, choose positive Δx such that $|f(t) - f(x)| < \epsilon/[2(F'(x) + 1)]$ for all $t \in [x, x + \Delta x]$

and $|[F(x + \Delta x) - F(x)]/\Delta x - F'(x)| < \min\{1, \epsilon/(2|f(x)|)\}$. Then

$$Q \leq \left(\frac{\epsilon}{2(F'(x) + 1)}\right) \left(\frac{F(x + \Delta x) - F(x)}{\Delta x}\right) + |f(x)| \left(\frac{\epsilon}{2|f(x)|}\right)$$

$$\leq \left(\frac{\epsilon}{2(F'(x) + 1)}\right) (F'(x) + 1) + \frac{\epsilon}{2} = \epsilon.$$

The other cases are similar, as is the case for $\Delta x < 0$. We conclude that $G'(x) = f(x)F'(x)$. ∎

It is now fairly easy to show than continuous functions are integrable on closed finite intervals; we simply use Theorem A.44, which asserts the uniform continuity of such functions.

A.52 Theorem: If $f(\cdot)$ is continuous on $[a, b]$, then it is integrable with respect to F on $[a, b]$. Moreover,

$$\int_a^b f \, dF = \begin{cases} \lim_{\text{mesh}(P) \to 0} S_{f,F}(P) \\ \lim_{\text{mesh}(P) \to 0} s_{f,F}(P) \\ \lim_{\text{mesh}(P) \to 0} T_{f,F}(P), \end{cases}$$

where $T_{f,F}(P) = \sum_{i=1}^n f(x_i)[F(x_i) - F(x_{i-1})]$ for partition $P : a = x_0 \leq x_1 \leq \ldots \leq x_n = b$.

PROOF: We refer to $T_{f,F}(P)$ as a right sum because it differs from $S_{f,F}(P)$ in its use of $f(x_i)$ at the right end of a partition component instead of $\sup\{f(x)|x \in [x_{i-1}, x_i]\}$.

First, if $F(b) = F(a)$, then all upper sums, lower sums, and right sums are zero, regardless of the partition. In this case, therefore, f is integrable and indeed,

$$0 = \int_a^b f \, dF = \lim_{\text{mesh}(P) \to 0} S_{f,F}(P) = \lim_{\text{mesh}(P) \to 0} s_{f,F}(P) = \lim_{\text{mesh}(P) \to 0} T_{f,F}(P).$$

So, we need only consider the case where $A = F(b) - F(a) > 0$. Accordingly, given $\epsilon > 0$, we exploit the uniform continuity of f to obtain $\delta > 0$ such that $x, y \in [a, b], |x - y| < \delta$ implies that $|f(x) - f(y)| < \epsilon/(2A)$. Let P be a partition with boundary points $a = x_1 \leq x_2 \leq \ldots \leq x_n = b$ and with $\text{mesh}(P) < \delta$. For any $x, y \in [x_{i-1}, x_i]$, we have $|f(x) - f(y)| < \epsilon/(2A)$. Consequently, $\sup f(x) - \inf f(y) \leq \epsilon/(2A)$ over the partition component. As this inequality holds for each partition component, we have

$$S_{f,P} - s_{f,P} = \sum_{i=1}^n [F(x_i) - F(x_{i-1})] \left(\sup_{x_{i-1} \leq x \leq x_i} f(x) - \inf_{x_{i-1} \leq x \leq x_i} f(x)\right)$$

$$\leq \epsilon/(2A) \sum_{i=1}^n [F(x_i) - F(x_{i-1})] = \epsilon/(2A) \cdot A = \epsilon/2 < \epsilon.$$

This proves that f is integrable with respect to F on $[a, b]$. The argument also shows that, for any partition P having $\operatorname{mesh}(P) < \delta$,

$$\left| S_{f,F}(P) - \int_a^b f \, dF \right| = S_{f,F}(P) - \int_a^b f \, dF \le S_{f,F}(P) - s_{f,F}(P) < \epsilon.$$

This shows that $\int_a^b f \, dF = \lim_{\operatorname{mesh}(P) \to 0} S_{f,F}(P)$. A similar argument applies to the limit of the lower sums.

For the right sums, we simply note that $[F(x_i) - F(x_{i-1})] \inf f(x) \le [F(x_i) - F(x_{i-1})] f(x_i) \le [F(x_i) - F(x_{i-1})] \sup f(x)$ for any partition component, which implies that $s_{f,F}(P) \le T_{f,F}(P) \le S_{f,F}(P)$ for all partitions P. Consequently, for $\operatorname{mesh}(P) < \delta$, we have both $\int_a^b f \, dF$ and $T_{f,F}(P)$ in the interval $[s_{f,F}(P), S_{f,F}(P)]$ of width less than ϵ. Then, $\lim_{\operatorname{mesh}(P) \to 0} T_{f,F}(P) = \int_a^b f \, dF$. ∎

Although some manipulation of the endpoint contributions may be necessary, we can extend the theorem to piecewise continuous functions by breaking the integration interval into consecutive spans where the integrand is continuous. This extension covers the familiar polynomial, trigonometric, exponential, and logarithmic functions as well as their compositions. All integrals in this text fall into these categories. On infinite or semi-infinite intervals, the integrability depends on the limit processes involved.

We turn now to the practical matter of evaluating integrals. The last item of Theorem A.51 leads to the most common method for evaluating the integral of a continuous function. We determine the collection of functions $G(x)$ such that $G'(x) = f(x)$. One of these must satisfy $G(b) = \int_a^b f$. The boundary condition, $G(a) = 0$, determines the particular function. Algorithmically, the procedure is as follows. We first determine the collection $\mathcal{G} = \{ G \mid G' = f \}$. Any two members of $\mathcal{G}$ differ by an additive constant. So, we choose an arbitrary member, $G_0(x)$, knowing that the desired member is $G_0(x) + C$ for some constant C. Consequently, $\int_a^x f = G_0(x) + C$ for any $x \in [a, b]$. Setting $x = a$, we see that $C = -G_0(a)$. The usual notation for this operation is

$$\int_a^b f = \int_a^b f(t) \, dt = G_0(t) \big|_a^b = G_0(b) - G_0(a).$$

This is the familiar antiderivative method for integral evaluation. We can, for example, compute

$$\int_{-\pi/2}^{\pi/2} \cos x \, dx = \sin x \big|_{-\pi/2}^{\pi/2} = \sin \pi/2 - \sin(-\pi/2) = 1 - (-1) = 2.$$

If F is differentiable and f is continuous, we apply the same reasoning to conclude $\int_a^b f \, dF = H_0(b) - H_0(a)$, where H_0 is any function satisfying $H_0'(y) = f(y) F'(y)$ on $[a, b]$. For example, $\sin x$ is monotone increasing on

$[-\pi/2, \pi/2]$, so we can integrate with respect to $\sin x$ on this interval. Consequently,

$$\int_{-\pi/2}^{\pi/2} (\cos x)\, d(\sin x) = \int_{-\pi/2}^{\pi/2} (\cos^2 x)\, dx = \int_{-\pi/2}^{\pi/2} \frac{1 + \cos 2x}{2}\, dx$$

$$= \left[\frac{x}{2} + \frac{\sin 2x}{4} \right]_{-\pi/2}^{\pi/2}$$

$$= \left(\frac{\pi}{2} + \frac{\sin \pi}{4} \right) - \left(\frac{-\pi}{2} + \frac{\sin(-\pi)}{4} \right) = \pi.$$

Continuing this theme, we write $G = f \circ g$ when $G(x) = f(g(x))$, and we call G the *composition* of f and g. Suppose that we have such a composition G and both f and g are differentiable. Then

$$G'(x) = f'(g(x)) \cdot g'(x)$$

$$\int_a^b (f' \circ g) \cdot g' = G(b) - G(a) = f(g(b)) - f(g(a)) = \int_{g(a)}^{g(b)} f'.$$

Replacing f' with h, we have the formula $\int_a^b (h \circ g) \cdot g' = \int_{g(a)}^{g(b)} h$. This transformation is called *substitution of variables*. The traditional incantation is that we substitute $y = g(x)$, $dy = g'(x)\, dx$ to transform the integral: $\int_a^b f(g(x))g'(x)\, dx = \int_{g(a)}^{g(b)} f(y)\, dy$. Using the substitution $y = \sin x$, $dy = (\cos x)\, dx$, for example, we evaluate the following integral.

$$\int_{\pi/4}^{3\pi/4} \frac{\cos x}{\sin x}\, dx = \int_{\sqrt{2}/2}^{\sqrt{2}/2} \frac{dy}{y} = 0$$

We can exploit differentiation formulas to obtain further integration techniques. We illustrate one more, which arises from the formula $(uv)' = u'v + uv'$. Hence,

$$\int_a^b uv' = \int_a^b (uv)' - \int_a^b u'v = uv\big|_a^b - \int_a^b u'v.$$

This process is commonly called *integration by parts*. For example, we let $u = \sin x, v' = \cos x, v = \sin x, u' = \cos x$ to compute

$$\int_0^{\pi/2} \sin x \cos x\, dx = \sin^2 x\big|_0^{\pi/2} - \int_0^{\pi/2} \cos x \sin x\, dx$$

$$2 \int_0^{\pi/2} \sin x \cos x\, dx = \sin^2(\pi/2) - \sin^2(0) = 1$$

$$\int_0^{\pi/2} \sin x \cos x\, dx = \frac{1}{2}.$$

Except perhaps for the extensions to the Riemann-Stieltjes form, these formulas should be familiar from calculus. We summarize them in the following theorem. They are sufficient for all integrals encountered in this text.

A.53 Theorem: The following are practical integration techniques.

- If f is piecewise continuous,

$$\int_a^b f \, dF = \lim_{\text{mesh}(P) \to 0} S_{f,F}(P) = \lim_{\text{mesh}(P) \to 0} s_{f,F}(P).$$

- If f is piecewise continuous and F is differentiable, $\int_a^b f \, dF = \int_a^b f \cdot F'$.

- $\int_a^b g' = g(b) - g(a)$.

- (Transformation of variables) $\int_a^b (f \circ g) \cdot g' = \int_{g(a)}^{g(b)} f$.

- (Integration by parts) $\int_a^b (uv') = uv \big|_a^b - \int_a^b vu'$.

PROOF: See discussion above. ∎

A.54 Example: Figure A.2 illustrates a function that becomes unbounded as x approaches ± 1. Specifically, $f(x) = x/(1 - |x|)$, for $-1 < x < 1$. We evaluate the integral as follows. Let a, b be such that $-1 < a < 0$ and $0 < b < 1$. We have

$$\int_a^b f(x) \, dx = \int_a^0 \frac{x \, dx}{1 + x} + \int_0^b \frac{x \, dx}{1 - x}$$

$$= \int_a^0 \left(1 - \frac{1}{1 + x}\right) dx + \int_0^b \left(-1 + \frac{1}{1 - x}\right) dx$$

$$= [x - \ln(1 + x)]\big|_a^0 + [-x - \ln(1 - x)]\big|_0^b = -(a + b) + \ln\left(\frac{1 + a}{1 - b}\right)$$

Then $\int_{-1}^{1} f(x) \, dx = \lim[-(a + b) + \ln[(1 + a)/(1 - b)]]$, where the limit involves $a \to -1^+$ and $b \to 1^-$. The quantity $(a + b) \to 0$, but the second term has no limit. As a and b approach -1^+ and 1^-, respectively, the expression assumes all values in the range $[0, \infty)$. This is true no matter how small the neighborhoods around -1 and 1 are chosen. To see how this happens, suppose that $\mathcal{I}(-1, \delta_1), \mathcal{I}(1, \delta_2)$ are arbitrary neighborhoods of -1 and 1, and let r be any nonnegative real number. Let $\delta = \min(\delta_1, \delta_2, 1/2) > 0$. Select an a such that $-1 < a < -1 + \delta$, and set $b = 1 - (1 + a)e^{-r} < 1$. Then

$$a = (1 - b)e^r - 1$$

$$1 - \delta < -a = 1 - (1 - b)e^r < 1 - (1 - b) = b < 1$$

$$\ln\left(\frac{1 + a}{1 - b}\right) = \ln\left(\frac{(1 - b)e^r}{(1 - b)}\right) = r.$$

Thus, $a \in \mathcal{I}(-1, \delta_1)$, $b \in \mathcal{I}(1, \delta_2)$, and $\ln((1 + a)/(1 - b)) = r$. Consequently, there is no limit. The result is manifest in the limiting form of $\ln((1 + a)/(1 - b))$, which is $\ln(0^+/0^+)$, an indeterminate form. Even though the graph suggests that the positive and negative areas should cancel, they do so only when the limits are symmetric. That is, $\int_{-a}^{a} f(x) \, dx = 0$ for all a as $a \to 1$. The integral is defined, however, for *independent* limits. We must have $a \to -1^+$ and $b \to 1^-$ without any constraining relationship between them. Without some constraint, the limit does not exist.

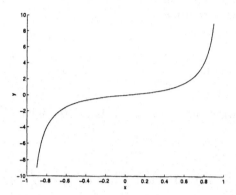

Figure A.2. An unbounded integrand: $f(x) = x/(1 - |x|)$ on $(-1, 1)$
(see Example A.54)

If we change the function slightly to $f(x) = x/\sqrt{1 - |x|}$, the graph retains the qualitative appearance of Figure A.2, but the integral now exists and assumes the value zero as suggested by the figure.

$$\int_{-1}^{1} f(x)\,dx = \int_{-1}^{0} x(1 + x)^{-1/2}\,dx + \int_{0}^{1} x(1 - x)^{-1/2}\,dx$$

In the integral over $(-1, 0]$, we use the substitution $y = (1 + x)^{1/2}$, $dy = [(1 + x)^{-1/2}/2]\,dx$; in the other part, we use $y = (1 - x)^{1/2}$, $dy = [-(1 - x)^{1/2}/2]\,dx$.

$$\int_{-1}^{1} f(x)\,dx = \int_{0}^{1} (y^2 - 1) \cdot 2\,dy + \int_{1}^{0} (1 - y^2) \cdot (-2\,dy)$$

$$= 2\left[\frac{y^3}{3} - y\right]_{0}^{1} + 2\left[y - \frac{y^3}{3}\right]_{0}^{1} = 2\left(\frac{1}{3} - 1\right) + 2\left(1 - \frac{1}{3}\right) = 0 \;\square$$

A final theorem provides technical support for certain arguments involving the interchange of summations and integrals. These appear in connection with the moment generating function of a general random variable, as discussed in Chapter 5.

A.55 Theorem: Suppose that $\sum_{n=1}^{\infty} f_n(x)$ converges uniformly to $f(x)$ for $x \in [a, b]$. Suppose further that each f_n and f are integrable with respect to nondecreasing F on $[a, b]$. Then $\int_{a}^{b} \left(\sum_{n=1}^{\infty} f_n\right)\,dF = \sum_{n=1}^{\infty} \int_{a}^{b} f_n\,dF$.
PROOF: If $F(b) - F(a) = 0$, all integrals across $[a, b]$ with respect to F are zero, and the theorem is trivially true. Accordingly, we assume $F(b) > F(a)$. For arbitrary $\epsilon > 0$, there exists N such that $m \geq N$ implies simultaneously for all $x \in [a, b]$ that $|f(x) - \sum_{n=1}^{m} f_n(x)| < \epsilon/[F(b) - F(a)]$. Equivalently,

$$f(x) - \frac{\epsilon}{F(b) - F(a)} < \sum_{n=1}^{m} f_n(x) < f(x) + \frac{\epsilon}{F(b) - F(a)}.$$

Theorem A.51 and the linearity of the integral then give

$$\int_a^b f \, dF - \epsilon < \int_a^b \left(\sum_{n=1}^m f_n \right) dF = \sum_{n=1}^m \int_a^b f_n \, dF < \int_a^b f \, dF + \epsilon,$$

for all $m \geq N$. Letting $m \to \infty$, we obtain

$$\int_a^b f \, dF - \epsilon \leq \sum_{n=1}^\infty \int_a^b f_n \, dF < \int_a^b f \, dF + \epsilon.$$

Because $\epsilon > 0$ is arbitrary, we conclude that $\int_a^b \left(\sum_{n=1}^\infty f_n \right) dF = \int_a^b f \, dF = \sum_{n=1}^\infty \int_a^b f_n \, dF$. ∎

Exercises

A.38 Evaluate $\int_{-1}^1 \frac{dx}{x^{1/3}}$.

A.39 Integrate by parts: (a) $\int_0^\infty t e^{-t} \, dt$, (b) $\int_0^\infty t^2 e^{-t} \, dt$, and (c) $\int_0^\infty t^n e^{-t} \, dt$.

A.40 For the following F, compute (a) $\int_0^\infty x \, dF$, (b) $\int_0^{10} (x^2 + 2) \, dF$, and (c) $\int_0^3 x^2 \, dF$.

$$F(x) = \begin{cases} 0, & x < 1 \\ 0.2, & 1 \leq x < 3 \\ 0.7, & 3 \leq x < 5 \\ 1.0, & x \geq 5. \end{cases}$$

A.41 For the following F, compute (a) $\int_0^\infty x \, dF$, (b) $\int_0^\infty x(x+1) \, dF$, and (c) $\int_0^3 x^2 \, dF$.

$$F(x) = \begin{cases} 0, & x < 0 \\ 0.2 + 0.8(1 - e^{-x}), & x \geq 0 \end{cases}$$

*A.42 Suppose that f, f^2, g, g^2, and fg are all integrable with respect to F on $[a, b]$. Show that

$$\left[\int_a^b fg \, dF \right]^2 \leq \int_a^b f^2 \, dF \cdot \int_a^b g^2 \, dF.$$

*A.43 Theorem A.51 asserts, among other properties, that f integrable on finite $[a, b]$ implies that $|f|$ integrable. Provide a counterexample for the converse. That is, construct a function f on a finite $[a, b]$ such that $|f|$ is integrable but f is not.

A.44 Suppose that f is continuous on the finite interval $[a, b]$. Let F be monotone nondecreasing on the interval. For partition P with boundary markers $a = x_0 \leq x_1 \leq \ldots \leq x_n = b$, define

$$V_{f,F}(P; y_1, \ldots, y_n) = \begin{cases} \sum_{i=1}^n f(y_i)[F(x_i) - F(x_{i-1})], & y_i \in [x_{i-1}, x_i] \\ S_{f,F}(P), & \text{otherwise,} \end{cases}$$

where $S_{f,F}(P)$ is the traditional upper sum. Show that

$$\lim_{\text{mesh}(P) \to 0} V_{f,F}(P; y_1, y_2, \ldots) = \int_a^b f \, dF.$$

That is, given $\epsilon > 0$, there exists $\delta > 0$ such that $\text{mesh}(P) < \delta$ implies that $\left| \int_a^b f \, dF - V_{f,F}(P; y_1, y_2, \ldots) \right| < \epsilon$, independent of the choices for $y_1, y_2, \ldots$.

A.5   Permutations and determinants

Operating on a finite set $S_n = \{1, 2, \ldots, n\}$ of integers, a permutation is a bijective function $\sigma : S_n \to S_n$. Consequently, we can describe each of the $n!$ possible permutations with a table that lists the domain-range pairs of the corresponding function. The following tables, for example, give several permutations of S_4.

x	1 2 3 4
$\sigma_1(x)$	2 4 3 1

x	1 2 3 4
$\sigma_2(x)$	4 3 2 1

x	1 2 3 4
$\iota(x)$	1 2 3 4

It is convenient to suppress the domain row and describe a permutation on S_n as an n-vector. In this notation, σ_1 above appears as $[2, 4, 3, 1]$. We use brackets for this notation to free parentheses for a related concept to be introduced shortly. We compose permutations as functions. That is, we compute $\sigma_3 = \sigma_1 \circ \sigma_2$ by $\sigma_3(x) = \sigma_1(\sigma_2(x))$. The composition of two bijections is itself a bijection, so σ_3 is also a permutation on S_n. With σ_1 and σ_2 as given above, we have $\sigma_1 \circ \sigma_2(1) = \sigma_1(\sigma_2(1)) = \sigma_1(4) = 1$, and so forth for the remaining values. The final result is or $\sigma_3 = \sigma_1 \circ \sigma_2 = [1, 3, 4, 2]$.

The permutation ι, defined by $\iota(x) = x$ for all $x \in S_n$, is the identity permutation. For any permutation σ, we have $\iota \circ \sigma = \sigma \circ \iota = \sigma$. Each σ possesses an inverse σ^{-1} such that $\sigma \circ \sigma^{-1} = \sigma^{-1} \circ \sigma = \iota$. The inverse permutation is easily computed; it is simply the inverse function. For example, if $\sigma = [2, 4, 3, 1]$, then $\sigma^{-1} = [4, 1, 3, 2]$. We have shown that the permutations on S_n, together with the binary operation of composition, form a mathematical object called a group.

A.56 Definition: Let $\circ$ be a binary operator on the set G. The pair $(G, \circ)$ is a *group* if

- $x, y \in G$ implies that $x \circ y \in G$;

- $x \circ (y \circ z) = (x \circ y) \circ z$;

- there exists an identity element $e \in G$ such that $e \circ x = x \circ e = x$ for each $x \in G$; and

- for each $x \in G$, there exists an inverse $x^{-1} \in G$ such that $x \circ x^{-1} = x^{-1} \circ x = e$. ∎

A.57 Theorem: The permutations on S_n under binary composition form a group, denoted perm(S_n).

PROOF: See discussion above. ∎

Groups have been studied extensively, and many properties of permutations derive from their group status. In this text, permutations arise in certain probability problems, such as those encountered in the introductory sections of Chapter 1, and they also figure prominently in computations with determinants. We continue to derive some properties that are useful in computing determinants. The first such feature is parity. Each permutation on S_n is either even or odd, depending on the number of inversions that it contains.

A.58 Definition: Suppose that $\sigma = [x_1, x_2, \ldots, x_n]$ is a permutation on S_n. Exploiting the ordering inherent in this notation, we say that x_i *precedes* x_j in σ if $i < j$. That is, x_i precedes x_j if x_i appears to the left of x_j in the n-vector description of σ. Let m_j be the number of elements preceding x_j that are greater than x_j. Then N_σ, the number of *inversions* in σ, is $N_\sigma = \sum_{j=1}^{n} m_j$. If N_σ is even, we say that σ is an *even* permutation; otherwise, it is an *odd* permutation. The *sign* of σ is defined as sign$(\sigma) = 1$ if σ is an even permutation and sign$(\sigma) = -1$ if σ is odd. ∎

Consider $\sigma = [2, 4, 3, 1]$ on S_4. There are no elements preceding $i_1 = 2$ and therefore $m_1 = 0$. Obviously, $m_1 = 0$ for any permutation. As for $i_2 = 4$, it has a single predecessor, which is smaller. Therefore, $m_2 = 0$ also. Continuing with $i_3 = 3$, we find two predecessors, one of which is larger than 3. This gives $m_3 = 1$. Finally, $i_4 = 1$, and it has three predecessors, all larger, which gives $m_4 = 3$. The total number of inversions is then $0 + 0 + 1 + 3 = 4$. The permutation is even.

We investigate the change in inversions associated with a transposition of two elements. Continuing with $[2, 4, 3, 1]$ as the original permutation, we transpose the 2 and 3 to obtain $[3, 4, 2, 1]$, which now has 5 inversions. The transposition changed the parity of the permutation. This property is a general one, which we state as the next theorem.

A.59 Theorem: Let $\sigma \in$ perm(S_n), and form σ' by transposing two elements of σ. Then σ' and σ have opposite parities.

PROOF: Let $\sigma = [i_1, i_2, \ldots, i_n]$, and suppose that the interchange of i_j and i_k forms σ'. In calculating N_σ, we compare each of the $C_{n,2}$ element pairs precisely once. If the larger precedes the smaller in the permutation order, the comparison contributes one inversion to the accumulating count. When i_j and i_k are compared in σ', the result is opposite from that encountered in the σ comparison. This changes the parity of the permutation, provided that comparisons involving just one of i_j or i_k remain balanced.

Suppose that t is an element such that $t \neq i_j$ and $t \neq i_k$. There are a number of cases, distinguished by the size of t and its location. If t is to the left of both i_j and i_k in σ, it remains to their left in σ'. The comparison of i_j with t then gives the same result in both permutations, and so does the comparison of i_k with t. If t lies to the right of i_j and i_k in σ, the same observation applies. There remains only the case where t appears between i_j and i_k in σ. If t is less than both i_j and i_k in σ, one of the comparisons reveals an inversion and the other does not. The same situation prevails after the transposition in σ'. If t is greater than both i_j and i_k, the same remark applies.

So, the only possibility for changing the number of inversions, other than that associated with comparing i_j with i_k, happens when t appears between i_j and i_k in σ and t is between i_j and i_k in size. One arrangement in this scenario is $i_j < t < i_k$. If the σ-order is $(\ldots, i_j, \ldots, t, \ldots, i_k, \ldots)$, the (i_j, t) and (i_k, t) comparisons yield zero inversions in σ, but they give two inversions in σ'. If the σ-order is $(\ldots, i_k, \ldots, t, \ldots, i_j, \ldots)$, these comparisons add two inversions in σ, but they contribute zero inversions in σ'. A similar analysis holds if $i_k < t < i_j$.

Having exhausted all cases, we conclude that the transposition changes the number of inversions by one through the direct comparison of i_j with i_k, and it changes the number of inversions by zero or two through comparisons involving just one of i_j or i_k. The net change is odd, which swaps the permutation parity. ∎

For a given σ, we write $\sigma^0(x) = x$, $\sigma^1(x) = \sigma(x)$, $\sigma^2(x) = \sigma(\sigma(x))$, and so forth. We also use σ^{-k} for $(\sigma^{-1})^k$. Consider the sequence obtained by repeatedly applying σ to an initial element x: $\sigma^0(x), \sigma^1(x), \sigma^2(x), \ldots$. Because there are but finitely many elements in S_n, the sequence must repeat. Suppose that $\sigma^j(x) = \sigma^i(x)$, for $j > i$, is the first repetition. Then, applying σ^{-i} to both sides, we have $x = \sigma^{j-i}(x)$, another repetition at index $j - i$. Also, $0 < j - i \leq j$. Because j is the first repetition, we must have $i = 0$. That is, $\sigma^j(x) = x$ is the first repetition. Consequently, the sequence repeats the segment $x, \sigma^1(x), \sigma^2(x), \ldots, \sigma^{j-1}(x)$ over and over. Moreover, all j elements in the segment are distinct, and σ moves each to its neighbor on the right. The rightmost wraps around to the beginning. Consider, for example, $\sigma = [4, 5, 1, 6, 7, 8, 2, 3]$ on S_8. Starting with 6, we generate the sequence $6, 8, 3, 1, 4, 6, 8, 3, 1, 4, \ldots$. The repeating segment is $(6, 8, 3, 1, 4)$. Starting instead with 3, we obtain $3, 1, 4, 6, 8, 3, 1, 4, 6, 8, \ldots$ with repeating segment $(3, 1, 4, 6, 8)$. This is just a rotation of the preceding segment.

A.60 Definition: Let σ be a permutation on S_n, a *cycle* in σ is a finite sequence of distinct elements $(i_1, i_2, \ldots, i_k)$, such that each $i_j \in S_n$ and

$$\sigma(i_j) = \begin{cases} i_{j+1}, & 1 \leq j < k \\ i_1, & j = k. \end{cases}$$

With each cycle $c = (i_1, i_2, \ldots, i_k)$ in σ, we associate the permutation σ_c, defined by

$$\sigma_c(x) = \begin{cases} \sigma(x), & x \in c \\ x, & x \notin c. \end{cases}$$

Evidently, σ_c leaves unchanged all entries external to the cycle. An entry in the cycle maps to its right neighbor, with the last entry mapping to the first. We use the parenthesized k-vector notation $(i_1, i_2, \dots, i_k)$ interchangeably for a cycle and its associated permutation. The empty cycle () then denotes the identity permutation. A *transposition* is a cycle of length 2. ∎

We do not distinguish between $(i_1, i_2, \dots, i_k)$ and any rotation of this pattern, such as $(i_{k-1}, i_k, i_1, i_2, \dots, i_{k-2})$, because all rotations represent the same associated permutation. Consequently, a cycle can have many representations, all rotations of one another. With this qualification, distinct cycles in σ cannot have any members in common. To see why this is so, suppose that two distinct cycles have a common member. Then, continued application of σ brings all of one cycle into the span of the other, and vice versa. They are then the same cycle, up to rotation, which contradicts their distinctness.

Now, any cycle, $c = (i_1, i_2, \dots, i_k)$, considered as a permutation, is the composition of 2-cycles, also considered as permutations. Specifically, $c = (i_1, i_2, \dots, i_k) = (i_1, i_k) \circ (i_1, i_{k-1}) \circ \dots \circ (i_1, i_2)$. We can verify this equality directly. If $x \notin c$, then

$$(i_1, i_2, \dots, i_k)(x) = x$$
$$(i_1, i_k) \circ (i_1, i_{k-1}) \circ \dots \circ (i_1, i_2)(x) = x,$$

because all the permutations involved leave x unchanged. If $x = i_1 \in c$, we have

$$(i_1, i_2, \dots, i_k)(i_1) = i_2$$
$$(i_1, i_k) \circ (i_1, i_{k-1}) \circ \dots \circ (i_1, i_2)(i_1) = (i_1, i_k) \circ (i_1, i_{k-1}) \circ \dots \circ (i_1, i_3)(i_2) = i_2,$$

because none of the remaining transpositions mentions i_2. If $x = i_j$ for $1 < j < k$, then x passes unchanged through the rightmost transpositions until it encounters (i_1, i_j). That is,

$$(i_1, i_k) \circ (i_1, i_{k-1}) \circ \dots \circ (i_1, i_2)(i_j) = (i_1, i_k) \circ (i_1, i_{k-1}) \circ \dots \circ (i_1, i_j)(i_j)$$
$$= (i_1, i_k) \circ (i_1, i_{k-1}) \circ \dots \circ (i_1, i_{j+1})(i_1)$$
$$= i_{j+1}.$$

Finally, if $x = i_k$, it passes unchanged through all transpositions except the last. That is,

$$(i_1, i_k) \circ (i_1, i_{k-1}) \circ \dots \circ (i_1, i_2)(i_k) = (i_1, i_k)(i_k) = i_1.$$

In all cases, therefore, the cycle has the same effect as the composition of the transpositions.

We can write σ as the composition of its cycles, each considered as a permutation, in any order. The order is immaterial because a cycle leaves unchanged those elements not mentioned in its k-vector. Consequently, because each cycle is a composition of transpositions, so is σ. While the number of transpositions in such a representation is not unique, it is always even or odd in step with the parity of σ. This happens because we can construct σ from the identity permutation by applying the transpositions successively. According to Theorem A.59, each transposition changes the parity of the accumulating construction. Because the identity permutation is even, the number of applied transpositions must be even if σ is even and odd if σ is odd.

We can use this result to relate the sign of a composition to that of its factors. Specifically, suppose that σ_1 and σ_2 are two permutations on S_n that exhibit the following representations as transposition sequences: $\sigma_1 = \mu_1 \circ \mu_2 \circ \ldots \circ \mu_s$ and $\sigma_2 = \nu_1 \circ \nu_2 \circ \ldots \circ \nu_t$. Then $\sigma_1 \circ \sigma_2 = \mu_1 \circ \mu_2 \circ \ldots \circ \mu_s \circ \nu_1 \circ \nu_2 \circ \ldots \circ \nu_t$. Consequently,

$$\text{sign}(\sigma_1 \circ \sigma_2) = \begin{cases} 1, & s+t \text{ even} \\ -1, & s+t \text{ odd} \end{cases} = \begin{cases} 1, & s \text{ and } t \text{ have same parity} \\ -1, & s \text{ and } t \text{ have opposite parity} \end{cases}$$
$$= \text{sign}(\sigma_1) \cdot \text{sign}(\sigma_2).$$

A.61 Theorem: The following properties hold for $\sigma, \tau \in \text{perm}(S_n)$.

(a) If we express $\sigma = \mu_1 \circ \mu_2 \circ \ldots \circ \mu_s$ as a composition of 2-cycles (transpositions), then s is even if and only if σ is even.

(b) $\text{sign}(\sigma \circ \tau) = \text{sign}(\sigma) \cdot \text{sign}(\tau)$.

(c) $\text{sign}(\sigma) = \text{sign}(\sigma^{-1})$.

PROOF: The discussion above establishes the first two properties. For the last, we note that $\sigma \circ \sigma^{-1} = \iota$, the identity permutation, which is even. We then have $\text{sign}(\sigma) \cdot \text{sign}(\sigma^{-1}) = \text{sign}(\iota) = 1$, which implies that σ and σ^{-1} have the same sign. ∎

A.62 Example: Operating on S_5, let $\sigma = [2, 1, 5, 4, 3]$ and $\tau = [1, 2, 4, 5, 3]$. Verify the parity relationship of Theorem A.61.

N_σ, the number of inversions in σ is $0+1+0+1+2 = 4$. $N_\tau = 0+0+0+0+2 = 2$. Hence $\text{sign}(\sigma) = \text{sign}(\tau) = 1$. We verify that $\text{sign}(\sigma \circ \tau) = 1 \cdot 1 = 1$. We first compute $\sigma \circ \tau = [2, 1, 4, 3, 5]$, for which the number of inversions is $0+1+0+1+0 = 2$. The parity is even, which means that $\text{sign}(\sigma \circ \tau) = 1$, as expected. We also compute $\sigma^{-1} = [2, 1, 5, 4, 3]$, for which the number of inversions is $0+1+0+1+2 = 4$. Again as expected, σ and σ^{-1} have the same parity. □

Associated with a square matrix, a determinant is a single number that captures many important aspects of the array. Its definition uses permutations to direct a calculation on the matrix elements.

A.63 Definition: Let

$$\mathbf{A} = \begin{bmatrix} a_{11} & a_{12} & \cdots & a_{1n} \\ \vdots & \vdots & \vdots & \vdots \\ a_{n1} & a_{n2} & \cdots & a_{nn} \end{bmatrix}$$

be a square matrix of real numbers. The *determinant* of $\mathbf{A}$ is

$$\det \mathbf{A} = \sum_{\sigma \in \text{perm}(S_n)} \left(\text{sign}(\sigma) \prod_{i=1}^{n} a_{i,\sigma(i)} \right) = \sum_{\sigma \in \text{perm}(S_n)} \text{sign}(\sigma) a_{1,\sigma(1)} \cdots a_{n,\sigma(n)}. \blacksquare$$

This definition envisions the determinant as a sum of $n!$ terms, each a product of n factors. The factors come from separate rows, and the entry in each row is selected according to the permutation governing the term. Almost all properties of determinants have a dual nature. One part expresses some feature associated with the matrix rows; the other expresses the same feature as applied to the columns. For example, we can rephrase the definition to emphasize the source of the factors in separate columns, rather than rows, as follows. For a given σ, we have $\prod_{i=1}^{n} a_{i,\sigma(i)} = \prod_{k=1}^{n} a_{\sigma^{-1}(k),k}$ because as i runs through $1, 2, \ldots, n$, so does $\sigma(i)$, although in a potentially different order. Therefore,

$$\det \mathbf{A} = \sum_{\sigma \in \text{perm}(S_n)} \text{sign}(\sigma^{-1}) \prod_{k=1}^{n} a_{\sigma^{-1}(k),k} = \sum_{\sigma^{-1} \in \text{perm}(S_n)} \text{sign}(\sigma^{-1}) \prod_{k=1}^{n} a_{\sigma^{-1}(k),k}.$$

The last equality follows because a term in the first summation arising from σ arises from σ^{-1} in the second. Therefore,

$$\det \mathbf{A} = \sum_{\sigma \in \text{perm}(S_n)} \text{sign}(\sigma) a_{\sigma(1),1} a_{\sigma(2),2} \cdots a_{\sigma(n),n}.$$

We regard this reformulated expression as part of the definition.

A.64 Example: Compute the determinant of

$$\mathbf{A} = \begin{bmatrix} 4 & 2 & -1 \\ 0 & 1 & -2 \\ 3 & -4 & 1 \end{bmatrix}.$$

Let the elements of $\mathbf{A}$ be a_{ij}. The six permutations of S_3, together with their signs and the corresponding determinant terms, are as follows.

Permutation	As cycles	Sign	Determinant term
$[1, 2, 3]$	$()$	$+$	$+a_{11}a_{22}a_{33} = +(4)(1)(1) = 4$
$[1, 3, 2]$	$(2, 3)$	$-$	$-a_{11}a_{23}a_{32} = -(4)(-2)(-4) = -32$
$[2, 1, 3]$	$(1, 2)$	$-$	$-a_{12}a_{21}a_{33} = -(2)(0)(1) = 0$
$[2, 3, 1]$	$(1, 3) \circ (1, 2)$	$+$	$+a_{12}a_{23}a_{31} = +(2)(-2)(3) = -12$
$[3, 1, 2]$	$(1, 2) \circ (1, 3)$	$+$	$+a_{13}a_{21}a_{32} = +(-1)(0)(-4) = 0$
$[3, 2, 1]$	$(1, 3)$	$-$	$-a_{13}a_{22}a_{31} = -(-1)(1)(3) = 3$
det $\mathbf{A}$			-37

The determinant is -37, the sum of the contributions from the six permutations. □

The following two theorems establish the basic properties of determinants.

A.65 Theorem: Let $\mathbf{A}$ be a square $n \times n$ matrix, with a_{ij} denoting the entry in row i column j. Then,

(a) If $\mathbf{A}$ contains a row or a column of zeros, then $\det \mathbf{A} = 0$.

(b) If we form $\mathbf{B}$ by interchanging two rows, or two columns, of $\mathbf{A}$, then $\det \mathbf{B} = - \det \mathbf{A}$.

(c) If $\mathbf{A}$ has two identical rows, or two identical columns, then $\det \mathbf{A} = 0$.

(d) If we form $\mathbf{B}$ by multiplying a row, or a column, by a constant k, then $\det \mathbf{B} = k \det \mathbf{A}$.

(e) If we form $\mathbf{B}$ by adding a multiple of row i to a different row j, then $\det \mathbf{B} = \det \mathbf{A}$. The determinant is also unchanged if we add a multiple of a column to a different column.

(f) If $\mathbf{B}$ is an arbitrary $n \times n$ matrix, then $\det \mathbf{AB} = \det \mathbf{A} \cdot \det \mathbf{B}$.

(g) If $\mathbf{A}$ possesses an inverse, then $\det \mathbf{A} \det \mathbf{A}^{-1} = 1$.

(h) The determinant of $\mathbf{A}$ equals the determinant of its transpose. That is, $\det \mathbf{A} = \det \mathbf{A}'$.

PROOF: We prove the properties for rows; the proofs for columns are similar. Item (a) is apparent from the definition because a zero row forces a zero factor into each term. For the second item, suppose that we interchange rows i and j to form $\mathbf{B}$. As σ indexes through $\text{perm}(S_n)$, so does $\sigma \circ (i, j)$. Consequently,

$$\det \mathbf{A} = \sum_{\sigma \in \text{perm}(S_n)} \text{sign}(\sigma \circ (i, j)) a_{1, \sigma \circ (i,j)(1)} a_{2, \sigma \circ (i,j)(2)} \cdots a_{n, \sigma \circ (i,j)(n)}.$$

Now, composing with a transposition changes the sign of a permutation, so $\text{sign}(\sigma \circ (i, j)) = -\text{sign}(\sigma)$. Also, $\sigma \circ (i, j)(k) = \sigma(k)$, when $k \neq i$ and $k \neq j$. For the two exceptions, we have $\sigma \circ (i, j)(i) = \sigma(j)$ and $\sigma \circ (i, j)(j) = \sigma(i)$. With these substitutions, we have

$$\det \mathbf{A} = - \sum_{\sigma \in \text{perm}(S_n)} \text{sign}(\sigma) a_{i, \sigma(j)} a_{j, \sigma(i)} \prod_{k \neq i, j} a_{k, \sigma(k)}$$

$$= - \sum_{\sigma \in \text{perm}(S_n)} \text{sign}(\sigma) b_{j, \sigma(j)} b_{i, \sigma(i)} \prod_{k \neq i, j} b_{k, \sigma(k)} = - \det \mathbf{B},$$

where $\mathbf{B}$ has components b_{ij}. For item (c), if rows i and j are identical, then interchanging them leaves the matrix unaltered. Consequently, $\det \mathbf{A} = -\det \mathbf{A}$, which forces $\det \mathbf{A} = 0$. Item (d) follows directly from the definition because each term acquires the factor k precisely once. For item (e), suppose that we form $\mathbf{B}$ by multiplying row i of $\mathbf{A}$ by c and then adding it to row j. Then

$$\det \mathbf{B} = \sum_{\sigma \in \mathrm{perm}(S_n)} \mathrm{sign}(\sigma) a_{i,\sigma(i)} (a_{j,\sigma(j)} + c a_{i,\sigma(j)}) \prod_{k \neq i,j} a_{k,\sigma(k)}$$

$$= \sum_{\sigma \in \mathrm{perm}(S_n)} \mathrm{sign}(\sigma) \prod_{k=1}^{n} a_{k,\sigma(k)} + c \sum_{\sigma \in \mathrm{perm}(S_n)} \mathrm{sign}(\sigma) a_{i,\sigma(i)} a_{i,\sigma(j)} \prod_{k \neq i,j} a_{k,\sigma(k)}.$$

The second sum is the determinant of a matrix with identical rows i and j. Consequently, it is zero. The first sum is simply $\det \mathbf{A}$.

For item (f), we first note that the ith row, jth column, of $\mathbf{AB}$ contains $\sum_{k=1}^{n} a_{ik} b_{kj}$. From the definition of the determinant, we have

$$\det \mathbf{AB} = \sum_{\sigma \in \mathrm{perm}(S_n)} \mathrm{sign}(\sigma) \left(\sum_{k_1=1}^{n} a_{1,k_1} b_{k_1,\sigma(1)} \right) \cdots \left(\sum_{k_n=1}^{n} a_{n,k_n} b_{k_n,\sigma(n)} \right)$$

$$= \sum_{k_1=1}^{n} \cdots \sum_{k_n=1}^{n} a_{1,k_1} \cdots a_{n,k_n} \sum_{\sigma \in \mathrm{perm}(S_n)} \mathrm{sign}(\sigma) b_{k_1,\sigma(1)} \cdots b_{k_n,\sigma(n)}.$$

The interior sum over the permutations on S_n is the determinant of a matrix whose rows 1 to n are rows $k_1, k_2, \ldots, k_n$ of $\mathbf{B}$. If there are one or more repetitions among $k_1, k_2, \ldots, k_n$, that determinant is zero. Therefore, the n^n terms generated by the initial n summations produce only $n!$ nonzero values. These occur when $k_1, k_2, \ldots, k_n$ are all distinct. In that case, however, they are a permutation on S_n. The interior sum is then the determinant of a matrix whose rows are the rearranged rows of $\mathbf{B}$. This means the interior sum is $\pm \det \mathbf{B}$, depending on the parity of the permutation $(k_1, k_2, \ldots, k_n)$. If the parity is even, the interior sum is $\det \mathbf{B}$, because it is the determinant of a matrix that arises from $\mathbf{B}$ through an even number of transposition of rows. If the parity is odd, the interior sum is $- \det \mathbf{B}$. Therefore, writing τ for the generic permutation $(k_1, k_2, \ldots, k_n)$, we have

$$\det \mathbf{AB} = \sum_{\tau \in \mathrm{perm}(S_n)} a_{1,\tau(1)} a_{2,\tau(2)} \cdots a_{n,\tau(n)} \mathrm{sign}(\tau) \det \mathbf{B} = \det \mathbf{A} \cdot \det \mathbf{B}.$$

Item (g) then follows immediately because $\mathbf{AA}^{-1} = \mathbf{I}$, which is the $n \times n$ identity matrix. The determinant of $\mathbf{I}$ is 1 because only the term associated with the identity permutation takes all its factors from the main diagonal. All others contain a zero factor. Item (h) follows immediately from the alternative definition of a determinant, which emphasizes factors chosen from the columns. In particular, let $\mathbf{B} = [b_{ij}] = \mathbf{A}'$.

$$\det \mathbf{A} = \sum_{\sigma \in \mathrm{perm}(S_n)} \mathrm{sign}(\sigma) a_{\sigma(1),1} a_{\sigma(2),2} \cdots a_{\sigma(n),n}$$

$$= \sum_{\sigma \in \mathrm{perm}(S_n)} \mathrm{sign}(\sigma) b_{1,\sigma(1)} b_{2,\sigma(2)} \cdots b_{n,\sigma(n)} = \det \mathbf{B} \ \blacksquare$$

A.66 Example: In the preceding example, we computed

$$\det \mathbf{A} = \det \begin{bmatrix} 4 & 2 & -1 \\ 0 & 1 & -2 \\ 3 & -4 & 1 \end{bmatrix} = -37.$$

We form $\mathbf{B}$ by interchanging rows 1 and 3 in $\mathbf{A}$; we construct $\mathbf{C}$ from $\mathbf{A}$ by adding twice the first column to the third column. We expect $\det \mathbf{B} = +37$ and $\det \mathbf{C} = -37$.

$$\mathbf{B} = \begin{bmatrix} 3 & -4 & 1 \\ 0 & 1 & -2 \\ 4 & 2 & -1 \end{bmatrix} \qquad\qquad \mathbf{C} = \begin{bmatrix} 4 & 2 & 7 \\ 0 & 1 & -2 \\ 3 & -4 & 7 \end{bmatrix}$$

Denoting the elements of $\mathbf{B}$ and $\mathbf{C}$ by b_{ij} and c_{ij}, respectively, we organize the permutation calculation as before.

σ	$[1,2,3]$	$[1,3,2]$	$[2,1,3]$	$[2,3,1]$	$[3,1,2]$	$[3,2,1]$	Total
$\mathrm{sign}(\sigma)$	$+$	$-$	$-$	$+$	$+$	$-$	
B elements	$b_{11}b_{22}b_{33}$	$b_{11}b_{23}b_{32}$	$b_{12}b_{21}b_{33}$	$b_{12}b_{23}b_{31}$	$b_{13}b_{21}b_{32}$	$b_{13}b_{22}b_{31}$	
Signed product	-3	12	0	32	0	-4	37
C elements	$c_{11}c_{22}c_{33}$	$c_{11}c_{23}c_{32}$	$c_{12}c_{21}c_{33}$	$c_{12}c_{23}c_{31}$	$c_{13}c_{21}c_{32}$	$c_{13}c_{22}c_{31}$	
Signed product	28	-32	0	-12	0	-21	-37

Finally, define $\mathbf{D}$ as follows and compute $\mathbf{AD}$.

$$\mathbf{D} = [d_{ij}] = \begin{bmatrix} 1 & -2 & 1 \\ 2 & -1 & 1 \\ 3 & 0 & 0 \end{bmatrix} \qquad\qquad \mathbf{AD} = [x_{ij}] = \begin{bmatrix} 5 & -10 & 6 \\ -4 & -1 & 1 \\ -2 & -2 & -1 \end{bmatrix}$$

The computations below show that $\det \mathbf{D} = -3$ and also verify that $\det \mathbf{AD} = (-37)(-3) = 111$.

σ	$[1,2,3]$	$[1,3,2]$	$[2,1,3]$	$[2,3,1]$	$[3,1,2]$	$[3,2,1]$	Total
$\mathrm{sign}(\sigma)$	$+$	$-$	$-$	$+$	$+$	$-$	
D elements	$d_{11}d_{22}d_{33}$	$d_{11}d_{23}d_{32}$	$d_{12}d_{21}d_{33}$	$d_{12}d_{23}d_{31}$	$d_{13}d_{21}d_{32}$	$d_{13}d_{22}d_{31}$	
Signed product	0	0	0	-6	0	3	-3
AD elements	$x_{11}x_{22}x_{33}$	$x_{11}x_{23}x_{32}$	$x_{12}x_{21}x_{33}$	$x_{12}x_{23}x_{31}$	$x_{13}x_{21}x_{32}$	$x_{13}x_{22}x_{31}$	
Signed product	5	10	40	20	48	-12	111

The determinants are as expected. $\square$

The example illustrates determinant computations with a tabular arrangement. While suitable for small matrices, the table becomes uncomfortably large with increasing matrix size. The table includes a row for each permutation on S_n, and there are $n!$ such permutations. An alternative approach uses a recursive reduction, which expresses the determinant of an $n \times n$ matrix in terms of several determinants of $(n-1) \times (n-1)$ matrices. These operational formulas depend on the concepts of minors and cofactors.

A.67 Definition: Suppose that $\mathbf{A}$ is an $n \times n$ matrix with entries a_{ij}. The *minor* of a_{ij} is the determinant of the $(n-1) \times (n-1)$ matrix obtained from

A by deleting row i and column j. We denote the minor of a_{ij} by M_{ij}. The *cofactor* of a_{ij} is $(-1)^{i+j}M_{ij}$. ∎

A.68 Theorem: Let **A** be a square $n \times n$ matrix with entries a_{ij}. Then,

(a) If $n = 1$, then $\det \mathbf{A} = a_{11}$. If $n = 2$, then $\det \mathbf{A} = a_{11}a_{22} - a_{12}a_{21}$.

(b) For any row i or column j, the determinant of **A** has an expansion involving cofactors. That is,

$$\det \mathbf{A} = \sum_{j=1}^{n} a_{ij}(-1)^{i+j}M_{ij} = \sum_{i=1}^{n} a_{ij}(-1)^{i+j}M_{ij},$$

for arbitrary row i in the first sum and arbitrary column j in the second.

(c) The expansion via row i, using the minors from a different row, produces zero. That is, $\sum_{j=1}^{n} a_{ij}(-1)^{i+j}M_{kj} = 0$ when $k \neq i$. Similarly, for column j, we have $\sum_{i=1}^{n} a_{ij}(-1)^{i+j}M_{ik} = 0$ when $k \neq j$.

(d) If $\det \mathbf{A} \neq 0$, then **A** is invertible. If b_{ij} denotes the row i column j entry of $\mathbf{A}^{-1}$, then $b_{ij} = (-1)^{i+j}M_{ji}/[\det \mathbf{A}]$.

PROOF: Item (a) follows immediately from the definition. The only permutation on S_1 is the identity. On S_2, the only permutations are $[1,2]$ and $[2,1]$; the first is even and the second is odd. Items (b) and (c) have dual expressions. We prove the expressions for rows; the expressions for columns admit similar proofs. For item (b), we consider first the special case when $i = 1$. We decompose the sum in the definition into n groups, each distinguished by the value of $\sigma(1)$.

$$\det \mathbf{A} = \sum_{\sigma \in \mathrm{perm}(S_n)} \mathrm{sign}(\sigma)a_{1,\sigma(1)}a_{2,\sigma(2)} \cdots a_{n,\sigma(n)}$$

$$= \sum_{\sigma(1)=1} \mathrm{sign}(\sigma)a_{11}a_{2,\sigma(2)} \cdots a_{n,\sigma(n)} + \sum_{\sigma(1)=2} \mathrm{sign}(\sigma)a_{12}a_{2,\sigma(2)} \cdots a_{n,\sigma(n)}$$

$$+ \ldots + \sum_{\sigma(1)=n} \mathrm{sign}(\sigma)a_{1n}a_{2,\sigma(2)} \cdots a_{n,\sigma(n)}$$

All permutations in the first group have the form $[1, x, x, \ldots, x]$, where the x symbols designate some arrangement of the integers $\{2, 3, \ldots, n\}$. There is precisely one such arrangement for each permutation in S_{n-1}, in the sense that $[i_1, i_2, \ldots, i_{n-1}] \in \mathrm{perm}(S_{n-1})$ corresponds to $[1, i_1 + 1, i_2 + 1, \ldots, i_{n-1} + 1]$. If $n = 4$, for example, then $[2, 1, 3] \in \mathrm{perm}(S_3)$ corresponds to $[1, 3, 2, 4]$ in the subset of $\mathrm{perm}(S_4)$ having $\sigma(1) = 1$. Moreover, the sign of $[i_1, i_2, \ldots, i_{n-1}] \in \mathrm{perm}(S_{n-1})$ is the same as the sign of $[1, i_1+1, i_2+1, \ldots, i_{n-1}+1] \in \mathrm{perm}(S_n)$ because increasing all components by 1 and inserting a 1 to the extreme left leaves the number of inversions unchanged. The first group then constitutes the subtotal

$$\sum_{\sigma(1)=1} \text{sign}(\sigma)a_{11}\cdots a_{n,\sigma(n)} = a_{11} \sum_{\tau\in\text{perm}(S_{n-1})} \text{sign}(\tau)a_{2,\tau(1)+1}\cdots a_{n,\tau(n-1)+1}$$

$$= a_{11} \sum_{\tau\in\text{perm}(S_{n-1})} \text{sign}(\tau)a'_{1,\tau(1)}\cdots a'_{n-1,\tau(n-1)}$$

$$= a_{11}(-1)^{1+1}M_{11},$$

where the $a'_{kl} = a_{k+1,l+1}, 1 \le k \le n-1, 1 \le l \le n-1$, constitute the matrix obtained by striking the first row and first column of $\mathbf{A}$. Consider now permutations in the jth group, which all have the form $[j,x,x,x,\ldots,x]$. Now the x symbols designate a rearrangement of the integers $\{1,2,\ldots,n\} - \{j\}$. We again associate each such rearrangement with a permutation on S_{n-1} by

$$[i_1,i_2,\ldots,i_{n-1}] \in \text{perm}(S_{n-1}) \leftrightarrow [j,j_1,j_2,\ldots,j_{n-1}] \in \text{perm}(S_n),$$

where

$$j_k = \begin{cases} i_k, & i_k < j \\ i_k+1, & i_k \ge j. \end{cases}$$

If $j = 3$, for example, we associate $[3,1,2] \in S_3$ with $[3,4,1,2]$. Consider the number of inversions in $[i_1,i_2,\ldots,i_{n-1}] \in S_{n-1}$ in comparison with $[j,j_1,j_2,\ldots,j_{n-1}]$. In comparing j_k with j_l, where $k < l$, we have the following cases.

- $i_k < j$ and $i_l < j$. This means that $j_k = i_k$ and $j_l = i_l$. The comparison (j_k,j_l) then yields an inversion precisely when the comparison (i_k,i_l) does so.

- $i_k \ge j$ and $i_l \ge j$. This case implies that $j_k = i_k + 1$ and $j_l = i_l + 1$. Again, the comparison (j_k,j_l) shows an inversion just when (i_k,i_l) does.

- $i_k < j$ and $i_l \ge j$. Now, we have $j_k = i_k, j_l = i_l + 1$, and $i_k < j \le i_l$. The (i_k,i_l) comparison does not yield an inversion, and neither does (j_k,j_l) because $j_k = i_k < i_l < i_l + 1 = j_l$.

- $i_k \ge j$ and $i_l < j$. In this final case, we have $j_k = i_k + 1, j_l = i_l$, and $i_l < j \le i_k$. Consequently, the (i_k,i_l) comparison shows an inversion, but so does (j_k,j_l) because $j_l = i_l < i_l + 1 = j_k$.

So, the number of inversions in $[i_1,\ldots,i_{n-1}]$ equals the number in its associate $[j,j_1,\ldots,j_{n-1}]$ that are generated by comparisons not involving j. To get the total number of inversions in $[j,j_1,j_2,\ldots,j_{n-1}]$, we must add inversions generated by comparisons of the form (j,j_k). Among the j_k are the entries $1,2,\ldots,j-1$, and j is greater than all of them. This produces $j-1$ additional inversions. Also among the j_k are the entries $j+1,j+2,\ldots,n$, but j is less than all of them and produces no further inversions. It follows that

$$\text{sign}([i_1,i_2,\ldots,i_{n-1}]) = (-1)^{j-1}\text{sign}([j,j_1,j_2,\ldots,j_{n-1}]).$$

For example, if $n = 5$ and $j = 3$, we have $[2, 4, 1, 3] \in \text{perm}(S_4) \leftrightarrow [3, 2, 5, 1, 4]$.
$[2, 4, 1, 3]$ contains 3 pairs in which the larger precedes the smaller, and the pattern $[\bullet, 2, 5, 1, 4]$ also contains 3 such pairs. When we insert the initial 3, we obtain $[3, 2, 5, 1, 4]$, which has $j - 1 = 2$ additional inversions. The jth group then constitutes the subtotal

$$\sum_{\sigma(1)=j} \text{sign}(\sigma) a_{1j} \cdots a_{n,\sigma(n)} = a_{1j}(-1)^{j-1} \sum_{\tau \in \text{perm}(S_{n-1})} a_{2,\tau'(1)} a_{3,\tau'(2)} \cdots a_{n,\tau'(n-1)},$$

where

$$\tau'(k) = \begin{cases} \tau(k), & \tau(k) < j \\ \tau(k) + 1, & \tau(k) \geq j. \end{cases}$$

If a'_{kl} is the matrix obtained from $\mathbf{A}$ by striking the first row and the jth column, we have

$$a'_{kl} = \begin{cases} a_{k+1,l}, & l < j \\ a_{k+1,l+1}, & l \geq j. \end{cases}$$

If $\tau(l) < j$, we have $a_{k+1,\tau'(l)} = a_{k+1,\tau(l)} = a'_{k,\tau(l)}$. If $\tau(l) \geq j$, we likewise have $a_{k+1,\tau'(l)} = a_{k+1,\tau(l)+1} = a'_{k,\tau(l)}$. Because $(-1)^{j-1} = (-1)^{j+1}$, the jth group then takes the form

$$\sum_{\sigma(1)=j} \text{sign}(\sigma) a_{1j} \cdots a_{n,\sigma(n)} = a_{1j}(-1)^{j+1} \sum_{\tau \in \text{perm}(S_{n-1})} a'_{1,\tau(1)} \cdots a'_{n-1,\tau(n-1)}$$

$$= a_{1j}(-1)^{j+1} M_{1j}.$$

Assembling these components, we have the desired expression for the determinant as an expansion in the cofactors of the first row. That is, $\det \mathbf{A} = \sum_{j=1}^{n} a_{1j}(-1)^{1+j} M_{1j}$. If $i > 1$, we first bubble the ith row to the top by successively interchanging it with the row immediately above it. This involves $i - 1$ transpositions and produces a new matrix $\overline{\mathbf{A}}$. The submatrix obtained by striking row 1 column j from $\overline{\mathbf{A}}$ is the same as that obtained by striking row i column j from $\mathbf{A}$. The determinant via first-row minors of $\overline{\mathbf{A}}$ is then $(-1)^{i-1} \det \mathbf{A}$. That is,

$$\det \mathbf{A} = (-1)^{i-1} \sum_{j=1}^{n} a_{ij}(-1)^{1+j} M_{ij} = \sum_{j=1}^{n} a_{ij}(-1)^{i+j} M_{ij}.$$

For item (c), we construct a new matrix $\mathbf{B} = [b_{ij}]$ from $\mathbf{A}$ by replacing row k with row i. If $k \neq i$, $\mathbf{B}$ now has two identical rows, and therefore $\det \mathbf{B} = 0$. The submatrix formed by deleting row k column j from $\mathbf{B}$ is the same as that formed by the same operations on $\mathbf{A}$. Denoting the determinants of these submatrices by M_{kj}, we have

$$0 = \det \mathbf{B} = \sum_{j=1}^{n} b_{kj}(-1)^{i+j} M_{kj} = \sum_{j=1}^{n} a_{ij}(-1)^{i+j} M_{kj}.$$

For item (d), we need only verify that the purported A^{-1} satisfies $AA^{-1} = A^{-1}A = I$, the $n \times n$ identity matrix. Let c_{ij} be the row i column j entry of AA^{-1}. We have

$$c_{ij} = \sum_{k=1}^{n} a_{ik} b_{kj} = \sum_{k=1}^{n} a_{ik} \cdot \frac{(-1)^{k+j} M_{jk}}{\det A} = \begin{cases} 1, & i = j \\ 0, & i \neq j. \end{cases}$$

The verification for $A^{-1}A = I$ is similar. ∎

A.69 Example: Continuing with the matrix A from earlier examples, we know that

$$\det A = \det \begin{bmatrix} 4 & 2 & -1 \\ 0 & 1 & -2 \\ 3 & -4 & 1 \end{bmatrix} = -37.$$

We verify an equivalent computation via minors along row 2.

$$\det A = (0) \cdot (-1)^{2+1} \det \begin{bmatrix} 2 & -1 \\ -4 & 1 \end{bmatrix} + (1)(-1)^{2+2} \det \begin{bmatrix} 4 & -1 \\ 3 & 1 \end{bmatrix}$$

$$+ (-2)(-1)^{2+3} \det \begin{bmatrix} 4 & 2 \\ 3 & -4 \end{bmatrix}$$

$$= (1)(4+3) - (-2)(-16-6) = -37.$$

Via column 3, we have

$$\det A = (-1)(-1)^{1+3} \det \begin{bmatrix} 0 & 1 \\ 3 & -4 \end{bmatrix} + (-2)(-1)^{2+3} \det \begin{bmatrix} 4 & 2 \\ 3 & -4 \end{bmatrix}$$

$$+ (1)(-1)^{3+3} \det \begin{bmatrix} 4 & 2 \\ 0 & 1 \end{bmatrix}$$

$$= (-1)(0-3) - (-2)(-16-6) + (1)(4-0) = -37.$$

The expansion along row 1 using minor from row 3 yields the expected zero.

$$\sum_{j=1}^{3} a_{1j}(-1)^{3+j} M_{3j} = (4)(-1)^{3+1} \det \begin{bmatrix} 2 & -1 \\ 1 & -2 \end{bmatrix} + (2)(-1)^{3+2} \det \begin{bmatrix} 4 & -1 \\ 0 & -2 \end{bmatrix}$$

$$+ (-1)(-1)^{3+3} \det \begin{bmatrix} 4 & 2 \\ 0 & 1 \end{bmatrix}$$

$$= (4)(-4+1) - (2)(-8+0) + (-1)(4-0) = 0$$

To invert A, we assemble the cofactors and arrange them in transpose order. That is, the entry in row i, column j, is $(-1)^{i+j} M_{ji} / \det A$.

$$(-1)^{1+1} M_{11} = \det \begin{bmatrix} 1 & -2 \\ -4 & 1 \end{bmatrix} = 1 - 8 = -7$$

$(-1)^{1+2} M_{12} = -\det \begin{bmatrix} 0 & -2 \\ 3 & 1 \end{bmatrix} = -(0+6) = -6$

$(-1)^{1+3} M_{13} = \det \begin{bmatrix} 0 & 1 \\ 3 & -4 \end{bmatrix} = 0 - 3 = -3$

$(-1)^{2+1} M_{21} = -\det \begin{bmatrix} 2 & -1 \\ -4 & 1 \end{bmatrix} = -(2-4) = 2$

$(-1)^{2+2} M_{22} = \det \begin{bmatrix} 4 & -1 \\ 3 & 1 \end{bmatrix} = 4 + 3 = 7$

$(-1)^{2+3} M_{23} = -\det \begin{bmatrix} 4 & 2 \\ 3 & -4 \end{bmatrix} = -(-16-6) = 22$

$(-1)^{3+1} M_{31} = \det \begin{bmatrix} 2 & -1 \\ 1 & -2 \end{bmatrix} = -4 + 1 = -3$

$(-1)^{3+2} M_{32} = -\det \begin{bmatrix} 4 & -1 \\ 0 & -2 \end{bmatrix} = -(-8+0) = 8$

$(-1)^{3+3} M_{33} = \det \begin{bmatrix} 4 & 2 \\ 0 & 1 \end{bmatrix} = 4 - 0 = 4$

We then have

$$\mathbf{A}^{-1} = \begin{bmatrix} 7/37 & -2/37 & 3/37 \\ 6/37 & -7/37 & -8/37 \\ 3/37 & -22/37 & -4/37 \end{bmatrix}.$$

Multiplication then verifies that $\mathbf{A}\mathbf{A}^{-1} = \mathbf{A}^{-1}\mathbf{A} = \mathbf{I}$. $\square$

At our convenience, we consider a matrix as a collection of row vectors or column vectors. In either case, there is a fundamental relationship between the determinant and these vector collections.

A.70 Definition: Let $\mathbf{v}_1, \mathbf{v}_2, \dots, \mathbf{v}_n$ be a collection of row or column vectors. If there exist constants $c_1, c_2, \dots, c_n$, not all zero, for which $\sum_{i=1}^n c_i \mathbf{v}_i = \mathbf{0}$, then the vectors are *dependent*. If $\sum_{i=1}^n c_i \mathbf{v}_i = \mathbf{0}$ implies that $c_1 = c_2 = \dots = c_n = 0$, then the vectors are *independent*. ∎

If $\mathbf{A}$ is a matrix, not necessarily square, row independence is not affected by transposing any two rows or any two columns. This fact follows from a short reflection on the nature of the independence condition. Let the rows be

$\mathbf{a}_{1\cdot} = (a_{11}, a_{12}, \dots, a_{1n})$

$\vdots \qquad\qquad \vdots$

$\mathbf{a}_{m\cdot} = (a_{m1}, a_{m2}, \dots, a_{mn}).$

We interchange rows k and l to obtain $\overline{\mathbf{A}}$. Then a nonzero set $(c_1, c_2, \dots, c_m)$ such that $\sum_{i=1}^m c_i \mathbf{a}_{i\cdot} = \mathbf{0}$ shows the dependency of the $\mathbf{A}$ rows and suffices to demonstrate also the $\overline{\mathbf{A}}$ row dependency. We simply swap c_k and c_l when applying the coefficients to the $\overline{\mathbf{A}}$ rows.

Similar reasoning holds if we form $\overline{\mathbf{A}}$ by swapping two columns of $\mathbf{A}$. The condition $\sum_{i=1}^m c_i \mathbf{a}_{i\cdot} = \mathbf{0}$ means that $\sum_{i=1}^m c_i a_{ij} = 0$ for every column j. Interchanging two columns does not change this fact. So, the same linear combination that shows row dependence in $\mathbf{A}$ also shows row dependence in $\overline{\mathbf{A}}$.

A.71 Theorem: If the rows of a matrix are independent, they remain so after the interchange of any two rows or any two columns. If the rows are dependent, they remain so after such interchanges. Similarly, the dependency status of the the columns is unchanged by swapping any two rows or any two columns.

PROOF: See discussion above. ∎

A.72 Theorem: $\det \mathbf{A} = 0$ if and only if the row vectors form a dependent set. Similarly, $\det \mathbf{A} = 0$ if and only if the column vectors form a dependent set.

PROOF: We prove the result associated with the row vectors. A similar proof establishes the parallel claim for the column vectors. Suppose that the row vectors $\mathbf{u}_1 = (a_{11}, a_{12}, \dots, a_{1n}), \dots, \mathbf{u}_n = (a_{n1}, a_{n2}, \dots, a_{nn})$ are dependent. Then there exist constants $c_1, c_2, \dots, c_n$, not all zero, such that $\sum_{i=1}^{n} c_i \mathbf{u}_i = \mathbf{0}$.

Suppose that $c_i \neq 0$. Then $-\mathbf{u}_i = \sum_{j \neq i} \frac{c_j \mathbf{u}_j}{c_i}$. For each $j \neq i$, we add c_j/c_i times row j to row i. This forms a new matrix $\mathbf{B}$ in which the ith row is zero. Consequently, $\det \mathbf{B} = 0$. The operations that form $\mathbf{B}$, however, leave the determinant unchanged. Therefore, $\det \mathbf{A} = 0$.

To prove the converse, we attack its contrapositive. That is, we show that independent rows implies a nonzero determinant. To this end, suppose that the rows of $\mathbf{A}$ are independent. It follows that row 1 cannot contain all zeros. Otherwise, $1 \cdot \mathbf{u}_1 + 0 \cdot \mathbf{u}_2 + \dots + 0 \cdot \mathbf{u}_n = \mathbf{0}$, which contradicts independence. We move a nonzero entry in the first row to the leftmost position by transposing columns. This operation multiplies the determinant by $(-1)^{t_1}$, where t_1 is zero, if a_{11} is already nonzero, or 1, if a transposition is necessary. The rows are now

$$\mathbf{x}_{1\cdot} = (x_{11}, x_{12}, \dots, x_{1n})$$
$$\vdots \qquad \vdots$$
$$\mathbf{x}_{n\cdot} = (x_{n1}, x_{n2}, \dots, x_{nn})$$

By the preceding theorem, the rows remain independent. Moreover, $x_{11} \neq 0$. We multiply row 1 by $(-x_{j1}/x_{11})$ and add it to row j for $j = 2, 3, \dots, n$. The determinant remains $(-1)^{t_1} \det \mathbf{A}$, and the matrix has the form

$$\mathbf{y}_{1\cdot} = (y_{11}, y_{12}, \dots, y_{1n}) = \mathbf{x}_{1\cdot}$$
$$\mathbf{y}_{i\cdot} = (0, y_{i2}, \dots, y_{in}) = \mathbf{x}_{i\cdot} - x_{i1}\mathbf{x}_{1\cdot}/x_{11}, \qquad \text{for } 2 \leq i \leq n.$$

We claim that the new rows remain independent. If $\sum_{i=1}^{n} c_i \mathbf{y}_{i\cdot} = \mathbf{0}$, it follows that

$$0 = c_1 \mathbf{x}_{1\cdot} + \sum_{i=2}^{n} c_i \left(\mathbf{x}_{i\cdot} - \frac{x_{i1}\mathbf{x}_{1\cdot}}{x_{11}} \right) = \left(c_1 - \sum_{i=2}^{n} \frac{c_i x_{i1}}{x_{11}} \right) \mathbf{x}_{1\cdot} + \sum_{i=2}^{n} c_i \mathbf{x}_{i\cdot}.$$

The independence of the x_i, then forces $c_2 = c_3 = \ldots = c_n = 0$ and then $c_1 - \sum_{i=2}^{n} c_i x_{i1}/x_{11} = c_1 = 0$. We conclude that the y_i are independent.

We now repeat this exercise with the second row. It cannot contain all zeros, so we permute a nonzero entry to the column 2 position. Assuming t_2 transpositions, we now have the determinant equal to $(-1)^{t_1+t_2} \det \mathbf{A}$. We multiply the second row by appropriate constants and add to the other rows to install zeros in the second column, except in row 2. We now have a matrix of row vectors

$$\mathbf{z}_1. = (z_{11}, 0, z_{13}, \ldots, z_{1n})$$

$$\vdots \qquad \qquad \vdots$$

$$\mathbf{z}_n. = (0, 0, z_{n3}, \ldots, z_{nn}),$$

in which z_{11} and z_{22} are the only nonzero entries in columns 1 and 2. The determinant remains $(-1)^{t_1+t_2} \det \mathbf{A}$ and the rows remain independent. We continue in this fashion to obtain a matrix $\mathbf{B}$ of the form

$$\mathbf{B} = \begin{bmatrix} b_{11} & 0 & 0 & \ldots & 0 \\ \vdots & \vdots & \vdots & & \vdots \\ 0 & 0 & 0 & \ldots & b_{nn} \end{bmatrix},$$

in which the b_{ii} are nonzero and the remaining entries are zero. We have

$$\det \mathbf{A} = (-1)^{t_1+t_2+\ldots+t_n} \det \mathbf{B} = (-1)^{t_1+t_2+\ldots+t_n} b_{11} b_{22} \cdots b_{nn} \neq 0. \; \blacksquare$$

Suppose that we have a collection of n-vectors, which we arrange as the rows of matrix $\mathbf{A}$.

$$\mathbf{A} = \begin{bmatrix} a_{11} & a_{12} & \ldots & a_{1n} \\ \vdots & \vdots & & \vdots \\ a_{m1} & a_{m2} & \ldots & a_{mn} \end{bmatrix} = \begin{bmatrix} \mathbf{a}_1. \\ \vdots \\ \mathbf{a}_m. \end{bmatrix}$$

If this collection is independent, we claim that it can contain at most n vectors. That is, $m \leq n$. Suppose, to the contrary, that $m > n$. We divide the vectors into two groups. The first contains n vectors; the second contains $m - n$. Let $\mathbf{x} = (x_1, \ldots, x_n)$ and $\mathbf{y} = (y_1, \ldots, y_{m-n})$ be two row vectors for which we will determine specific values as the argument progresses. Specifically, we want to find values, not all zero, such that

$$\sum_{i=1}^{n} x_i \mathbf{a}_i. + \sum_{i=n+1}^{m} y_{i-n} \mathbf{a}_i. = \mathbf{0}. \tag{A.1}$$

This is equivalent to the matrix equation

$$\begin{bmatrix} x_1 & \ldots & x_n \end{bmatrix} \begin{bmatrix} a_{11} & \ldots & a_{1n} \\ \vdots & & \vdots \\ a_{n1} & \ldots & a_{nn} \end{bmatrix} = -\begin{bmatrix} y_1 & \ldots & y_{m-n} \end{bmatrix} \begin{bmatrix} a_{n+1,1} & \ldots & a_{n+1,n} \\ \vdots & & \vdots \\ a_{m1} & \ldots & a_{mn} \end{bmatrix}$$

or $\mathbf{x}\mathbf{A}_1 = -\mathbf{y}\mathbf{A}_2$, where the rows of $\mathbf{A}_1$ are the first group of n vectors and the rows of $\mathbf{A}_2$ are the second group. The rows of $\mathbf{A}_1$, being a subset of the original rows, remain independent. This is because any nonzero linear combination of the subset extends to a nonzero linear combination of the full set by assigning a zero coefficient to vectors outside the subset. Consequently, $\det \mathbf{A}_1 \neq 0$ and $\mathbf{A}_1$ is invertible. Choose any *nonzero* values for the $\mathbf{y}$ components and calculate $\mathbf{x}$ values as follows.

$$\mathbf{x}\mathbf{A}_1 = -\mathbf{y}\mathbf{A}_2$$
$$\mathbf{x} = -\mathbf{y}\mathbf{A}_2\mathbf{A}_1^{-1}.$$

The $\mathbf{x}$ and $\mathbf{y}$ values together then constitute a set of constants satisfying Equation A.1, which shows the full set of vectors to be dependent. This contradiction means that we must have $m \leq n$.

A.73 Theorem: A collection of $n + 1$ or more n-vectors is a dependent set.

PROOF: See preceding discussion. ∎

A.74 Definition: The *rank* of a matrix, not necessarily square, is the largest number of independent rows. ∎

A.75 Theorem: The rank of a matrix, not necessarily square, is the largest number of independent columns. The rank is also the size of the largest square submatrix with a nonzero determinant. A submatrix is an array formed by deleting selected rows and columns from the original matrix.

PROOF: Call the matrix $\mathbf{A} = [a_{ij}]$, where $1 \leq i \leq m, 1 \leq j \leq n$. Let the row vectors be $\mathbf{a}_{1.}, \mathbf{a}_{2.}, \ldots, \mathbf{a}_{m.}$ and the column vectors be $\mathbf{a}_{.1}, \mathbf{a}_{.2}, \ldots, \mathbf{a}_{.n}$. Let r, s, t be, respectively, the largest number of independent rows, the largest number of independent columns, and the size of the largest square submatrix with nonzero determinant. The rows are n-vectors, and we know that a collection of independent n-vectors can contain at most n members. Consequently, $r \leq n$. A group of dependent rows remains dependent when we remove one or more columns. Indeed, suppose that rows $i_1, i_2, \ldots, i_m$ are dependent. Then there exists constants, $c_1, c_2, \ldots, c_m$, not all zero, such that $\sum_{k=1}^{m} c_k \mathbf{a}_{i_k.} = \mathbf{0}$. This is equivalent to $\sum_{k=1}^{m} c_k a_{i_k,j} = 0$ for $j = 1, 2, \ldots, n$. If we shorten the rows to $\overline{\mathbf{a}}_{i_k.}$ by removing some columns, the equation above remains valid for the remaining columns j. Consequently, $\sum_{k=1}^{m} c_k \overline{\mathbf{a}}_{i_k.} = \mathbf{0}$, and the shortened rows remain dependent. Now, since any group of $r + 1$ or more rows must be dependent, any square submatrix with $r + 1$ or more rows must contain a dependent row set and therefore have determinant zero. It follows that $t \leq r \leq n$. We now demonstrate that $t = r$ by constructing a submatrix of size $r \times r$ with nonzero determinant. The argument is the same as that used in Theorem A.72, so we abbreviate the presentation.

Let rows $i_1, i_2, \ldots, i_r$ be independent. Row i_1 cannot contain all zeros, so we select a column j_1 with a nonzero entry. We multiply row i_1 by appropriate constants and add to rows $i_2, i_3, \ldots, i_r$ to produce zeros in these rows in column j_1. This operation leaves unchanged the determinant of any $r \times r$ submatrix selected from this row set. The modified rows remain independent, and consequently, row i_2 must have a nonzero entry. We choose column j_2 with a nonzero entry. Note $j_2 \neq j_1$ because row i_2 now has a zero in column j_1. We multiply row i_2 by appropriate constants and add to rows $i_1, i_3, \ldots, i_r$ to produce zeros in these rows in column j_2. We continue in this fashion until we have r columns, $\{j_1, j_2, \ldots, j_r\}$, each containing all zeros except in cells $(i_1, j_1), (i_2, j_2), \ldots, (i_r, j_r)$. We extract the square submatrix containing rows $i_1, i_2, \ldots, i_r$ and columns $j_1, j_2, \ldots, j_r$. This submatrix has the same determinant as the corresponding submatrix selected from $\mathbf{A}$. The columns $j_1, j_2, \ldots, j_r$ may not be in order, but we can permute them to obtain $j_1 < j_2 < \ldots < j_r$ with only a sign change in the determinant. Doing so produces a matrix with nonzero entries on the main diagonal and zeros everywhere else. Because the determinant is then the product of the diagonal entries, it is nonzero. The corresponding $r \times r$ submatrix from $\mathbf{A}$ has this same determinant, except possibly for a change of sign. Consequently, we have an $r \times r$ submatrix with nonzero determinant, which shows that $t = r \leq n$.

We can repeat the argument using columns instead of rows to obtain $t = s \leq m$. We conclude that $t = r = s \leq \min(m, n)$. ∎

A.76 Example: Find the rank of matrix $\mathbf{A}$ to the right below.

We can perform row operations, such as swapping two rows or adding to a row multiples of other rows, without changing the absolute value of the determinant of a submatrix containing those rows. We can perform the corresponding column operations with similar impunity. Consequently, we multiply the first row by $1/2$ and add to the second row to create a column of zeros under the row 1 column 1 entry. We then transpose columns to get a nonzero entry in row 2 column 2 and clear the rest of the column. The progression appears as follows.

$$\begin{bmatrix} 4 & 1 & -2 & 3 \\ -2 & 0 & 1 & 2 \\ 0 & -3 & 2 & 2 \\ 0 & -2 & 2 & 9 \end{bmatrix}$$

$$\begin{bmatrix} 4 & 1 & -2 & 3 \\ -2 & 0 & 1 & 2 \\ 0 & -3 & 2 & 2 \\ 0 & -2 & 2 & 9 \end{bmatrix} \rightarrow \begin{bmatrix} 4 & 1 & -2 & 3 \\ 0 & 1/2 & 0 & 7/2 \\ 0 & -3 & 2 & 2 \\ 0 & -2 & 2 & 9 \end{bmatrix} \rightarrow \begin{bmatrix} 4 & 0 & -2 & -4 \\ 0 & 1/2 & 0 & 7/2 \\ 0 & 0 & 2 & 23 \\ 0 & 0 & 2 & 23 \end{bmatrix} \rightarrow \begin{bmatrix} 4 & 0 & 0 & 19 \\ 0 & 1/2 & 0 & 7/2 \\ 0 & 0 & 2 & 23 \\ 0 & 0 & 0 & 0 \end{bmatrix}$$

The 3×3 submatrix in the upper left corner has nonzero determinant, which establishes the rank as 3. □

We can easily extend a collection of independent n-vectors to its maximum size. Suppose that the collection contains m vectors and $m < n$. Arranged as row vectors, the collection forms a matrix of rank m. Hence it must have m independent columns. Let j be one of the dependent columns and add a new row vector with zeros in all locations except column j, which contains a 1. Now consider the determinant of the square submatrix containing the original m independent columns plus column j. This submatrix $\mathbf{M}$ has the following form, where we write column j on the far right. Permuting column j to the far right does not change the magnitude of the determinant.

$$\mathbf{M} = \begin{bmatrix} x_{11} & x_{12} & \cdots & x_{1m} & y_1 \\ \vdots & \vdots & & \vdots & \vdots \\ x_{m1} & x_{m2} & \cdots & x_{mm} & y_m \\ 0 & 0 & \cdots & 0 & 1 \end{bmatrix}.$$

Expanding by minors along row $m+1$, we see that det $\mathbf{M}$ is the determinant of the upper left $m \times m$ corner, which is nonzero because it contains the original m independent columns. So $\mathbf{M}$ is a square $(m+1) \times (m+1)$ submatrix with nonzero determinant, which implies that the rank of the augmented matrix is $m+1$ and all $m+1$ rows are independent. We can continue to add rows in this manner as long as a dependent column remains. Since dependent columns exist while the number of rows is less than n, the construction stops with n independent rows, of which the first m are the original n-vectors.

A.77 Theorem: We can augment a collection of independent n-vectors of size $m < n$ to a full collection of n independent n-vectors.

PROOF: See argument above. ∎

A.78 Definition: Let $\mathbf{A}$ be an $n \times n$ matrix and consider the collection of n-vectors.

$$\mathcal{N} = \{\mathbf{x} \mid \mathbf{Ax} = \mathbf{0}\}$$

The *nullity* of $\mathbf{A}$ is the size of the largest subset of independent vectors from $\mathcal{N}$. ∎

By Theorem A.73, we can have at most n independent n-vectors. Consequently, nullity$(\mathbf{A}) \leq n$. The next theorem provides a more exact result.

A.79 Theorem: Let $\mathbf{A}$ be an $n \times n$ matrix. Then nullity$(\mathbf{A}) = n - \text{rank}(\mathbf{A})$.

PROOF: If rank$(\mathbf{A}) = n$, then det $\mathbf{A} \neq 0$ and $\mathbf{A}$ is invertible. Consequently, $\mathbf{Ax} = \mathbf{0}$ implies that $\mathbf{x} = \mathbf{A}^{-1}\mathbf{0} = \mathbf{0}$. The zero vector alone is not an independent collection because any multiple of it is still zero. This means that no independent vectors exist in $\mathcal{N}$, and nullity$(A) = 0 = n - \text{rank}(A)$.

If rank$(\mathbf{A}) = 0$, then $\mathbf{A}$ contains all zero entries. Therefore, $\mathbf{Ax} = \mathbf{0}$ for all n-vectors $\mathbf{x}$. By Theorem A.77, we can grow an initially empty collection to contain n independent n-vectors, which means that nullity$(A) = n = n - \text{rank}(\mathbf{A})$.

There remain the intermediate cases, where rank$(A) = r$ for some $0 < r < n$. In this case, there are r independent columns in $\mathbf{A}$, and any larger group is a dependent subset. Denote the columns of $\mathbf{A}$ by $\mathbf{a}_{.k}$, and suppose that the independent columns are $1, 2, \ldots, r$. Define an $n \times r$ matrix $B = [b_{ij}]$ by

$$b_{ij} = \begin{cases} 1, & i = j \\ 0, & \text{otherwise.} \end{cases}$$

Let $\mathbf{b}_{.j}$ denote column j from $\mathbf{B}$. Each $\mathbf{b}_{.j}$ contains a single 1 in component j. If, for example, $n = 5$ and $r = 3$,

$$\mathbf{B} = \begin{bmatrix} 1 & 0 & 0 \\ 0 & 1 & 0 \\ 0 & 0 & 1 \\ 0 & 0 & 0 \\ 0 & 0 & 0 \end{bmatrix}.$$

In $\mathbf{B}$, the ones appear in different locations across the r column vectors, which forces them to constitute an independent set. Also, $\mathbf{Ab}_{.j} = \mathbf{a}_{.j}$ for

$1 \leq j \leq r$. Suppose that we can choose m independent vectors $x_1, x_2, \ldots, x_m$ such that $Ax_j = 0$ for $j = 1, 2, \ldots, m$. Consider a linear combination $0 = \sum_{j=1}^{r} c_j b_{.j} + \sum_{j=1}^{m} d_j x_j$. Premultiplying both sides by A gives

$$0 = \sum_{j=1}^{r} c_j A b_{.j} + \sum_{j=1}^{m} d_j A x_j \cdot = \sum_{j=1}^{r} c_j a_{.j} + \sum_{j=1}^{m} d_j 0 = \sum_{j=1}^{r} c_j a_{.j},$$

which forces $c_1 = c_2 = \ldots = c_r = 0$. We now have $\sum_{j=1}^{m} d_j x_j = 0$, and the independence of the x_j forces $d_1 = d_2 = \ldots = d_m = 0$. We conclude that the $b_{.j}$ and x_j combine to form an independent set of n-vectors. Consequently, $r + m \leq n$. Equivalently, $m \leq n - r$, and the nullity of A cannot exceed $n - r$. To show that it is exactly $n - r$, we simply expand the collection $\{b_{.j}\}$ to an independent collection of size n by adding $n - r$ new n-vectors. Recall that columns $r + 1, r + 2, \ldots, n$ of A are dependent on the first r independent columns. That is, there are linear combinations such that $a_{.(r+j)} = \sum_{i=1}^{r} p_{ji} a_{.i}$, for $1 \leq j \leq n - r$. Construct an $n \times (n - r)$ matrix $Y = [y_{ij}]$ via

$$y_{ij} = \begin{cases} -1, & i = r + j \\ p_{ji}, & i \leq r \\ 0, & \text{otherwise.} \end{cases}$$

Let y_j denote the jth column of Y. Then $Ay_j = -a_{.(r+j)} + \sum_{i=1}^{r} p_{ji} a_{.i} = -a_{.(r+j)} + a_{.(r+j)} = 0$. Consider the set $\{b_{.j} \mid 1 \leq j \leq r\} \cup \{y_j \mid 1 \leq j \leq n - r\}$. For the example noted above, where $n = 5$ and $r = 3$, the augmented collection of columns is

$$[b_{.1} b_{.2} b_{.3} y_1 y_2] = \begin{bmatrix} 1 & 0 & 0 & p_{11} & p_{21} \\ 0 & 1 & 0 & p_{12} & p_{22} \\ 0 & 0 & 1 & p_{13} & p_{23} \\ 0 & 0 & 0 & -1 & 0 \\ 0 & 0 & 0 & 0 & -1 \end{bmatrix}.$$

Each column vector of Y contains a (-1) in a row where other Y-vectors have a zero. This forces the set to be independent. These columns constitute $n - r$ independent vectors y_j such that $Ay_j = 0$. Therefore, the nullity of A at least $n - r$. Combined with the previous bound, this forces the nullity of A to be exactly $n - r$. ∎

There are a number of shortcuts for calculating determinants and inverses for matrices with a special format. We list several that are useful in the text's concluding chapters.

A.80 Theorem: The following techniques apply to partitioned matrices.

(a) If A has zero entries above the main diagonal, then $\det A$ is the product of the main diagonal entries.

(b) If an $n \times n$ matrix takes the form

$$A = \left[\begin{array}{c|c} B & C \\ \hline 0 & D \end{array} \right],$$

where $\mathbf{B}$ and $\mathbf{D}$ are square but not necessarily of equal size, then $\det \mathbf{A} = \det \mathbf{B} \cdot \det \mathbf{D}$.

(c) If an $n \times n$ matrix takes the form

$$\mathbf{A} = \left[\begin{array}{c|c} \mathbf{B} & 0 \\ \hline 0 & \mathbf{D} \end{array} \right],$$

where the submatrices $\mathbf{B}$ and $\mathbf{D}$ are square but not necessarily of equal size, then

$$\mathbf{A}^{-1} = \left[\begin{array}{c|c} \mathbf{B}^{-1} & 0 \\ \hline 0 & \mathbf{D}^{-1} \end{array} \right].$$

PROOF: For item (a), consider the expansion

$$\det \mathbf{A} = \sum_{\sigma \in \mathrm{perm}(S_n)} \mathrm{sign}(\sigma) a_{1,\sigma(1)} a_{2,\sigma(2)} \cdots a_{n,\sigma(n)}.$$

To be nonzero, a term must have $\sigma(1) \leq 1$, which implies that $\sigma(1) = 1$. It must also have $\sigma(2) \leq 2$. Since the permutation is a bijective mapping from S_n to S_n, and since $\sigma(1) = 1$, we must have $\sigma(2) = 2$. We continue in this fashion to conclude that $\sigma(i) = i$ for all i. So the determinant expansion has only one nonzero term, corresponding to $\sigma = [1, 2, 3, \ldots, n]$. This permutation has no inversions and is therefore even. It gives the determinant as the product of the main diagonal entries.

For item (b), suppose that $\mathbf{B}$ is $k \times k$ and $\mathbf{D}$ is $(n - k) \times (n - k)$. We consider a similar determinant expansion and note that the nonzero terms must involve permutations that do not mix the groups $\{1, 2, \ldots, k\}$ and $\{k + 1, k + 2, \ldots, n\}$. If i is in the first group and $\sigma(i)$ is in the second, then the factor $a_{i,\sigma(i)} = 0$, which means that the term involving that permutation is zero. If i is in the second group and $\sigma(i)$ is in the first, then there remain only $k - 1$ elements in the first group as targets of $\{1, 2, \ldots, k\}$ under σ. This is not enough targets to accommodate all the elements of the first group, so there must exist a j in the first with $\sigma(j)$ in the second, which again produces a zero factor. Each σ then corresponds to a pair of permutations, one on S_k and one on S_{n-k}. Moreover, the number of inversions in σ is the sum of the inversions in the pair. This follows because $\sigma(1), \ldots, \sigma(k)$ all lie to the left of position $k + 1$ and therefore add no transpositions to the counts for locations $k + 1, \ldots, n$. The sign of σ is then the product of the signs of the subset pair. The correspondence is one-to-one because each pair of permutations on the subsets combines to form a single permutation on S_n that yields a nonzero determinant term. Notationally, we write ν for the bijective mapping from $\{k + 1, k + 2, \ldots, n\}$ to itself and ν' for the corresponding mapping from $\{1, 2, \ldots, n - k\}$ to itself. That is, $\nu'(i) = \nu(i + k) - k$. Discarding the zero terms, we then have

$$\det \mathbf{A} = \left(\sum_{\tau \in \mathrm{perm}(S_k)} \mathrm{sign}(\tau) a_{1,\tau(1)} a_{2,\tau(2)} \cdots a_{k,\tau(k)} \right)$$

$$\cdot \left(\sum_{\nu \in \mathrm{perm}(\{k+1,k+2,\dots,n\})} \mathrm{sign}(\nu) a_{k+1,\nu(k+1)} a_{k+2,\nu(k+2)} \cdots a_{n,\nu(n)} \right)$$

$$= \left(\sum_{\tau \in \mathrm{perm}(S_k)} \mathrm{sign}(\tau) b_{1,\tau(1)} b_{2,\tau(2)} \cdots b_{k,\tau(k)} \right)$$

$$\cdot \left(\sum_{\nu' \in \mathrm{perm}(S_{n-k})} \mathrm{sign}(\nu') d_{1,\nu'(1)} d_{2,\nu'(2)} \cdots d_{n-k,\nu'(n-k)} \right)$$

$$= \det \mathbf{B} \det \mathbf{D}.$$

Item (c) is a straightforward matrix multiplication. ∎

Exercises

A.45 Compute the number of inversions in $\sigma = [2,5,7,1,3,4,8,6]$. Compute the number of inversions in σ^{-1} to verify that it has the same parity as σ.

A.46 Suppose that $\mathbf{A}, \mathbf{B}$, and $\mathbf{C}$ are identical matrices, except for the ith row. The ith row of $\mathbf{A}$ is $u_1, u_2, \dots, u_n$; the ith row of $\mathbf{B}$ is $v_1, v_2, \dots, v_n$. The ith row of $\mathbf{C}$ is $u_1 + v_1, u_2 + v_2, \dots, u_n + v_n$. Show that $\det \mathbf{C} = \det \mathbf{A} + \det \mathbf{B}$.

A.47 For the matrix $\mathbf{A}$ below, compute $\det \mathbf{A}$ directly from the definition.

$$\mathbf{A} = \begin{bmatrix} a_{11} & a_{12} & a_{13} \\ a_{21} & a_{22} & a_{23} \\ a_{31} & a_{32} & a_{33} \end{bmatrix}$$

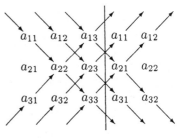

Verify the shortcut operation illustrated to the right, which proceeds as follows. First, copy columns 1 and 2 to the right of column 3. Form the determinant by adding the products on the full southeast diagonals and subtracting those on the full northeast diagonals.

A.48 With matrix $\mathbf{A}$ shown to the right, compute $\det \mathbf{A}$ directly from the definition. Verify the calculation with an expansion involving the minors of the second row.

$$\begin{bmatrix} 3 & -1 & -1 & -1 \\ -1 & 3 & -1 & -1 \\ -1 & -1 & 3 & -1 \\ -1 & -1 & -1 & 3 \end{bmatrix}$$

A.49 For $\mathbf{A}$ below, where $ad - bc \neq 0$, compute $\mathbf{A}^{-1}$.

$$\mathbf{A} = \begin{bmatrix} a & b \\ c & d \end{bmatrix}$$

*A.50 Suppose that $\mathbf{A}$ is an $n \times n$ matrix with $\det \mathbf{A} \neq 0$. Construct the matrices $\mathbf{B}_i$, for $1 \leq i \leq n$, by substituting the column vector $\mathbf{b}$ for column i in $\mathbf{A}$. Show that the solution to $\mathbf{Ax} = \mathbf{b}$ is $x_i = \det \mathbf{B}_i / \det \mathbf{A}$, where x_i is the ith component of column vector $\mathbf{x}$.

A.51 Show that the dependency status of a collection of m vectors is unchanged if a multiple of one vector is added to a different vector.

A.52 Compute the rank of

$$\mathbf{A} = \begin{bmatrix} 1 & -1 & 2 & 0 \\ 1 & 3 & -3 & -1 \\ 3 & 1 & 1 & -1 \\ 7 & 1 & 4 & -2 \end{bmatrix}.$$

A.53 For the $n \times n$ matrix $\mathbf{A} = [a_{ij}]$, prove that the following expansion in the cofactors of column j produces $\det \mathbf{A}$: $\det \mathbf{A} = \sum_{i=1}^{n} (-1)^{i+j} a_{ij} M_{ij}$.

A.54 For the $n \times n$ matrix $\mathbf{A}$, prove that the expansion via column j, using the minors from a different column, produces zero. That is, verify that the sum $\sum_{i=1}^{n} a_{ij} (-1)^{i+j} M_{ik} = 0$, when $k \neq j$.

A.55 Show that the rank of a matrix is equal to the rank of its transpose.

A.56 An $n \times n$ matrix $\mathbf{A}$ is called positive definite if (1) $\mathbf{x}' \mathbf{Ax} \geq 0$ for all n-vectors $\mathbf{x}$ and (2) $\mathbf{xAx} = 0$ implies that $\mathbf{x} = \mathbf{0}$. Show that a positive definite matrix has a nonzero determinant.

Historical Notes

The precise manipulation of limit processes troubled mathematicians for over 2000 years. At the very dawn of serious mathematical inquiry, the early Greeks struggled to calculate areas enclosed by curved lines. Eudoxus, who lived from about 408 to 355 B.C., invented a technique called the method of exhaustion. Our knowledge of the method, however, derives from the presentation in Book XII of Euclid's *Elements*. Although it does not explicitly mention limits, the method computes an area as a sequence of rectilinear approximations. As such, it foreshadows the integral calculus by about 2000 years. Kline[45] provides the details.

The ϵ-δ definition of a functional limit is due to Weierstrass and Heine in the late nineteenth century. Before this precise, verifiable criterion, mathematicians spoke with more or less ambiguity of differences that can be made as small as convenient, the freedom to discard infinitely small quantities, and the merits of "indefinite approach." Indeed, the greatest mathematicians of the seventeenth and eighteenth centuries—Euler, Leibniz, Newton, Lagrange, Wallis—used infinite series with scarcely any concern about convergence. Whenever a discrepancy arose, they chose to regard it as a paradox.

Kline[45] discusses the controversy that arose over the expansion $1/(1 + x) = 1 - x + x^2 - x^3 + \ldots$, when $x = 1$. The left side gives $1/2$, but the right side is rather open to interpretation. Grouping the terms in pairs suggests that the sum should be zero; grouping in pairs after the initial 1 suggests that the sum should be 1. Leibniz opined that the sum is $1/2$ because the alternative sums, zero and one, are equally plausible and therefore should be combined in an arithmetic mean.

Augustin-Louis Cauchy, whose name figures prominently in the convergence results of this chapter, had the misfortune to be born just before the French Revolution. The years of subsequent unrest threatened the development of his considerable mathematical talents. He was discovered, however, by Laplace and Lagrange, who provided much needed encouragement. He excelled at his studies and found his first career in preparing port facilities for Napoleon's planned English invasion. His early mathematical research constituted an overload above his bureaucratic duties.

In Cauchy's time, the idea of a rigorous proof was not well-defined, and many purported proofs were actually unorganized evidence in support of an intuitive or empirical concept. Cauchy established higher standards for proofs, particularly concerning the convergence of infinite sequences.

Cauchy was assertively religious, and some historians suggest that his intolerance in this area led to poor relations with other scientists. For whatever reason, his life was never free of political and religious battles, some of which cost him honors that he should have won on mathematical talent alone. In addition to Cauchy sequences, there are the Cauchy integral theorem and the Cauchy existence criterion for the solution of certain partial differential equations. Boyer[8] states, however, that the famous Cauchy convergence criterion appeared earlier in an obscure work by the Cambridge scholar Edward Waring.

When we accepted a sequence of decimal digits as a real number, we skipped lightly over another mathematical subtlety that stirred controversy for over 2000 years. While the integers and the rational numbers are abstractions of real-world concepts (e.g., 4 apples or 1/3 bushel of grain), the irrational entries at first seem associated only with geometric figures. The early Greeks knew that the hypotenuse of a right isosceles triangle with unit legs cannot be expressed as the ratio of integers. These troublesome lengths were termed *incommensurate*, which means that they cannot be expressed as ratios of rational lengths. The Greeks tried hard to ignore incommensurate numbers, typically by restricting their use to geometric problems. More drastic action was sometimes required. Legend has it, for example, that one Hippasus, a member of the famed Pythagorean school, was tossed into the sea for uncovering an incommensurate value.

Unfortunately, irrational numbers are very common; they constitute an uncountably dense subset of the reals. Despite many unsatisfactory explanations, the issue remained intractable until the nineteenth century. Around that time, the common approach identified irrational numbers with convergent sequences of rationals, despite the obvious logical problem that the sequence converged to no known number. A flurry of activity on the subject occurred in the decade 1865–1875, during which time Charles Méray, Georg Cantor, Eduard Heine, and Richard Dedekind all introduced essentially the same solution. In Dedekind's terms, a real number is a cut, which divides the rationals into two classes such that a member of the left class is always less than a member of the right. Strictly within the rationals, the left class may have a least upper bound or the right class may have a greatest lower bound, in which case the cut is a rational number. But, if neither extremum exists, it is an irrational number. Of course, there remain the details that show how to calculate with cuts. In particular, it is necessary to prove that a sequence of decimal digits defines a unique cut. Kline[45] discusses the history of irrational numbers.

Further Reading

Mathematical analysis, including arguments with limits and the structure of the real numbers, is well presented in Rudin's[75] classic book. Another text of particular interest, although first published in 1908, is Hardy's[33] introduction to analysis. Hardy develops limit processes with great care and is especially guarded in using the term *infinity* outside a well-defined context. This attitude contrasts with more recent texts, which treat infinity as a reasonably controllable quantity. This change of perspective reflects an additional century's digestion of the concept, which continues to produce surprises. Davis[17] presents an overview of the changing appreciation of the infinite over the last several hundred years.

Numerous texts treats matrix algebra. One of the clearest, in my opinion, is Noble[57], which provides complete operational details. Herstein[36] gives a more theoretical treatment. For discrete mathematics in general, Johnsonbaugh[39], Skvarcius and Robinson[81], and Stanat and McAllister[83] provide introductory texts. Knuth[47] introduces more advanced combinatorics associated with numerical algorithms.

The tools discussed in this appendix, together with the mathematical interludes that appear in the chapters, provide the background that might be expected of an undergraduate attempting a rigorous study of probability and statistics. This is, of course, the book's intended audience. More advanced mathematical tools and their applications to probability theory may be found in, for example, Fristedt and Gray[23] and Feller's[20] second volume.

Appendix B

Statistical Tables

1. $\Phi(x)$, the cumulative distribution function of a standard normal random variable

2. Percentiles of the standard normal distribution

3. Percentiles of the χ^2 distribution

4. Ninety-fifth percentiles of the F distribution

5. Percentiles of the t distribution

6. Power function for one-factor analysis of variance F test

Table 1

$\Phi(x)$, the cumulative distribution function of a standard normal random variable

$$\Phi(x) = \frac{1}{\sqrt{2\pi}} \int_{-\infty}^{x} e^{-y^2/2} \, dy$$

$$\Phi(-x) = 1 - \Phi(x)$$

x	0	1	2	3	4	5	6	7	8	9
0.00	.5000	.5040	.5080	.5120	.5160	.5199	.5239	.5279	.5319	.5359
0.10	.5398	.5438	.5478	.5517	.5557	.5596	.5636	.5675	.5714	.5753
0.20	.5793	.5832	.5871	.5910	.5948	.5987	.6026	.6064	.6103	.6141
0.30	.6179	.6217	.6255	.6293	.6331	.6368	.6406	.6443	.6480	.6517
0.40	.6554	.6591	.6628	.6664	.6700	.6736	.6772	.6808	.6844	.6879
0.50	.6915	.6950	.6985	.7019	.7054	.7088	.7123	.7157	.7190	.7224
0.60	.7257	.7291	.7324	.7357	.7389	.7422	.7454	.7486	.7517	.7549
0.70	.7580	.7611	.7642	.7673	.7704	.7734	.7764	.7794	.7823	.7852
0.80	.7881	.7910	.7939	.7967	.7995	.8023	.8051	.8078	.8106	.8133
0.90	.8159	.8186	.8212	.8238	.8264	.8289	.8315	.8340	.8365	.8389
1.00	.8413	.8438	.8461	.8485	.8508	.8531	.8554	.8577	.8599	.8621
1.10	.8643	.8665	.8686	.8708	.8729	.8749	.8770	.8790	.8810	.8830
1.20	.8849	.8869	.8888	.8907	.8925	.8944	.8962	.8980	.8997	.9015
1.30	.9032	.9049	.9066	.9082	.9099	.9115	.9131	.9147	.9162	.9177
1.40	.9192	.9207	.9222	.9236	.9251	.9265	.9279	.9292	.9306	.9319
1.50	.9332	.9345	.9357	.9370	.9382	.9394	.9406	.9418	.9429	.9441
1.60	.9452	.9463	.9474	.9484	.9495	.9505	.9515	.9525	.9535	.9545
1.70	.9554	.9564	.9573	.9582	.9591	.9599	.9608	.9616	.9625	.9633
1.80	.9641	.9649	.9656	.9664	.9671	.9678	.9686	.9693	.9699	.9706
1.90	.9713	.9719	.9726	.9732	.9738	.9744	.9750	.9756	.9761	.9767
2.00	.9772	.9778	.9783	.9788	.9793	.9798	.9803	.9808	.9812	.9817
2.10	.9821	.9826	.9830	.9834	.9838	.9842	.9846	.9850	.9854	.9857
2.20	.9861	.9864	.9868	.9871	.9875	.9878	.9881	.9884	.9887	.9890
2.30	.9893	.9896	.9898	.9901	.9904	.9906	.9909	.9911	.9913	.9916
2.40	.9918	.9920	.9922	.9925	.9927	.9929	.9931	.9932	.9934	.9936
2.50	.9938	.9940	.9941	.9943	.9945	.9946	.9948	.9949	.9951	.9952
2.60	.9953	.9955	.9956	.9957	.9959	.9960	.9961	.9962	.9963	.9964
2.70	.9965	.9966	.9967	.9968	.9969	.9970	.9971	.9972	.9973	.9974
2.80	.9974	.9975	.9976	.9977	.9977	.9978	.9979	.9979	.9980	.9981
2.90	.9981	.9982	.9982	.9983	.9984	.9984	.9985	.9985	.9986	.9986

Table 2

Percentiles z_p of the standard normal distribution

$$\Phi(z_p) = \frac{1}{\sqrt{2\pi}} \int_{-\infty}^{z_p} e^{-y^2/2}\, dy = p$$

p	z_p
0.001	-3.090
0.005	-2.576
0.010	-2.326
0.020	-2.054
0.025	-1.960
0.050	-1.645
0.100	-1.282
0.200	-0.842
0.300	-0.524
0.400	-0.253
0.500	0.000
0.600	0.253
0.700	0.524
0.800	0.842
0.900	1.282
0.950	1.645
0.975	1.960
0.980	2.054
0.990	2.326
0.995	2.576
0.999	3.090

Table 3

Percentiles $\chi^2_{p(n)}$ of the χ^2_n, the χ^2 random variable with n degrees of freedom

$$\Pr(\chi^2_n \leq \chi^2_{p(n)}) = p$$

n	p							
	0.005	0.01	0.02	0.025	0.05	0.1	0.2	0.4
1	0.0000	0.0002	0.0006	0.0010	0.0039	0.0158	0.0642	0.2750
2	0.0100	0.0201	0.0404	0.0506	0.1026	0.2107	0.4463	1.0217
3	0.0717	0.1148	0.1848	0.2158	0.3518	0.5844	1.0052	1.8692
4	0.2070	0.2971	0.4294	0.4844	0.7107	1.0636	1.6488	2.7528
5	0.4117	0.5543	0.7519	0.8312	1.1455	1.6103	2.3425	3.6555
6	0.6757	0.8721	1.1344	1.2373	1.6354	2.2041	3.0701	4.5702
7	0.9893	1.2390	1.5643	1.6899	2.1673	2.8331	3.8223	5.4932
8	1.3444	1.6465	2.0325	2.1797	2.7326	3.4895	4.5936	6.4226
9	1.7349	2.0879	2.5324	2.7004	3.3251	4.1682	5.3801	7.3570
10	2.1559	2.5582	3.0591	3.2470	3.9403	4.8652	6.1791	8.2955
11	2.6032	3.0535	3.6087	3.8157	4.5748	5.5778	6.9887	9.2373
12	3.0738	3.5706	4.1783	4.4038	5.2260	6.3038	7.8073	10.182
13	3.5650	4.1069	4.7654	5.0088	5.8919	7.0415	8.6339	11.129
14	4.0747	4.6604	5.3682	5.6287	6.5706	7.7895	9.4673	12.079
15	4.6009	5.2293	5.9849	6.2621	7.2609	8.5468	10.307	13.030
16	5.1422	5.8122	6.6142	6.9077	7.9616	9.3122	11.152	13.983
17	5.6972	6.4078	7.2550	7.5642	8.6718	10.085	12.002	14.937
18	6.2648	7.0149	7.9062	8.2307	9.3905	10.865	12.857	15.893
19	6.8440	7.6327	8.5670	8.9065	10.117	11.651	13.716	16.850
20	7.4338	8.2604	9.2367	9.5908	10.851	12.443	14.578	17.809
21	8.0337	8.8972	9.9146	10.283	11.591	13.240	15.445	18.768
22	8.6427	9.5425	10.600	10.982	12.338	14.042	16.314	19.729
23	9.2604	10.196	11.293	11.689	13.091	14.848	17.187	20.690
24	9.8862	10.856	11.992	12.401	13.848	15.659	18.062	21.653
25	10.520	11.524	12.697	13.120	14.611	16.473	18.940	22.616
26	11.160	12.198	13.409	13.844	15.379	17.292	19.820	23.579
27	11.808	12.879	14.125	14.573	16.151	18.114	20.703	24.544
28	12.461	13.565	14.848	15.308	16.928	18.939	21.588	25.509
29	13.121	14.257	15.575	16.047	17.708	19.768	22.475	26.475
30	13.787	14.953	16.306	16.791	18.493	20.599	23.364	27.442
40	20.707	22.164	23.838	24.433	26.509	29.051	32.345	37.134
50	27.991	29.707	31.664	32.357	34.764	37.689	41.449	46.864
60	35.535	37.485	39.699	40.482	43.188	46.459	50.641	56.620
70	43.275	45.442	47.893	48.758	51.739	55.329	59.898	66.396
80	51.172	53.540	56.213	57.153	60.391	64.278	69.207	76.188
90	59.196	61.754	64.635	65.647	69.126	73.291	78.558	85.993
100	67.328	70.065	73.142	74.222	77.929	82.358	87.945	95.808
120	83.852	86.923	90.367	91.573	95.705	100.62	106.81	115.47
140	100.66	104.03	107.82	109.14	113.66	119.03	125.76	135.15
160	117.68	121.35	125.44	126.87	131.76	137.55	144.78	154.86
180	134.88	138.82	143.21	144.74	149.97	156.15	163.87	174.58
200	152.24	156.43	161.10	162.73	168.28	174.84	183.00	194.32

Table 3 (continued)

Percentiles $\chi^2_{p(n)}$ of χ^2_n, the χ^2 random variable with n degrees of freedom

$$\Pr(\chi^2_n \leq \chi^2_{p(n)}) = p$$

| n | \multicolumn{8}{c}{p} |
	0.6	0.8	0.9	0.95	0.975	.98	0.99	0.995
1	0.7083	1.6424	2.7055	3.8415	5.0239	5.4119	6.6349	7.8795
2	1.8326	3.2189	4.6052	5.9915	7.3778	7.8240	9.2103	10.597
3	2.9462	4.6416	6.2514	7.8147	9.3484	9.8374	11.345	12.838
4	4.0446	5.9886	7.7794	9.4877	11.143	11.668	13.277	14.860
5	5.1318	7.2889	9.2352	11.070	12.831	13.388	15.086	16.750
6	6.2108	8.5581	10.645	12.592	14.449	15.033	16.812	18.548
7	7.2832	9.8033	12.017	14.068	16.014	16.624	18.480	20.279
8	8.3505	11.030	13.362	15.507	17.535	18.168	20.090	21.955
9	9.4136	12.242	14.684	16.919	19.023	19.679	21.665	23.585
10	10.473	13.442	15.987	18.307	20.483	21.161	23.209	25.188
11	11.530	14.631	17.275	19.675	21.920	22.618	24.726	26.757
12	12.584	15.812	18.549	21.026	23.337	24.054	26.218	28.303
13	13.636	16.985	19.812	22.362	24.736	25.472	27.689	29.821
14	14.685	18.151	21.064	23.685	26.119	26.873	29.141	31.319
15	15.733	19.311	22.307	24.996	27.488	28.259	30.578	32.801
16	16.780	20.465	23.542	26.296	28.845	29.633	32.000	34.267
17	17.824	21.615	24.769	27.587	30.191	30.995	33.408	35.719
18	18.868	22.760	25.989	28.869	31.526	32.346	34.805	37.156
19	19.910	23.900	27.204	30.144	32.852	33.688	36.191	38.583
20	20.951	25.038	28.412	31.411	34.170	35.020	37.566	39.998
21	21.992	26.171	29.615	32.671	35.479	36.343	38.932	41.401
22	23.031	27.302	30.813	33.925	36.781	37.660	40.289	42.796
23	24.069	28.429	32.007	35.173	38.076	38.968	41.638	44.181
24	25.106	29.553	33.196	36.415	39.364	40.270	42.980	45.559
25	26.143	30.675	34.382	37.653	40.646	41.566	44.314	46.928
26	27.179	31.795	35.563	38.885	41.923	42.856	45.642	48.290
27	28.214	32.912	36.741	40.113	43.195	44.140	46.963	49.645
28	29.249	34.027	37.916	41.337	44.461	45.419	48.278	50.993
29	30.283	35.139	39.088	42.557	45.722	46.693	49.588	52.335
30	31.316	36.250	40.256	43.773	46.979	47.962	50.892	53.671
40	41.622	47.269	51.805	55.758	59.342	60.436	63.691	66.766
50	51.892	58.164	63.167	67.505	71.420	72.613	76.154	79.490
60	62.135	68.972	74.397	79.082	83.298	84.580	88.380	91.952
70	72.358	79.715	85.527	90.531	95.023	96.388	100.43	104.22
80	82.566	90.405	96.578	101.88	106.63	108.07	112.33	116.32
90	92.761	101.05	107.57	113.15	118.14	119.65	124.12	128.30
100	102.95	111.67	118.50	124.34	129.56	131.14	135.81	140.17
120	123.29	132.81	140.23	146.57	152.21	153.92	158.95	163.64
140	143.60	153.85	161.83	168.61	174.65	176.47	181.84	186.85
160	163.90	174.83	183.31	190.52	196.92	198.85	204.53	209.82
180	184.17	195.74	204.70	212.30	219.04	221.08	227.06	232.62
200	204.43	216.61	226.02	233.99	241.06	243.19	249.45	255.26

Table 4

Ninety-fifth percentiles of the F distribution with m/n degrees of freedom

n	1	2	3	4	5	6	7	8
1	161.4	199.4	216.0	224.7	229.9	234.0	236.7	238.8
2	18.51	19.00	19.17	19.25	19.30	19.33	19.35	19.37
3	10.13	9.552	9.277	9.117	9.013	8.941	8.887	8.845
4	7.709	6.944	6.591	6.388	6.256	6.163	6.094	6.041
5	6.608	5.786	5.410	5.192	5.050	4.950	4.876	4.818
6	5.987	5.143	4.757	4.534	4.387	4.284	4.207	4.147
7	5.591	4.737	4.347	4.120	3.972	3.866	3.787	3.726
8	5.318	4.459	4.066	3.838	3.687	3.581	3.500	3.438
9	5.117	4.257	3.863	3.633	3.482	3.374	3.293	3.230
10	4.965	4.103	3.708	3.478	3.326	3.217	3.135	3.072
11	4.845	3.982	3.587	3.357	3.204	3.095	3.012	2.948
12	4.747	3.885	3.490	3.259	3.106	2.996	2.913	2.849
13	4.667	3.806	3.411	3.179	3.025	2.915	2.832	2.767
14	4.600	3.739	3.344	3.112	2.958	2.848	2.764	2.699
15	4.543	3.682	3.287	3.056	2.901	2.790	2.707	2.641
16	4.494	3.634	3.239	3.007	2.852	2.741	2.657	2.591
17	4.452	3.592	3.197	2.965	2.810	2.699	2.614	2.548
18	4.414	3.555	3.160	2.928	2.773	2.661	2.577	2.510
19	4.381	3.522	3.127	2.895	2.740	2.628	2.544	2.477
20	4.351	3.493	3.098	2.866	2.711	2.599	2.514	2.447
21	4.325	3.467	3.072	2.840	2.685	2.573	2.488	2.420
22	4.301	3.443	3.049	2.817	2.661	2.549	2.464	2.397
23	4.279	3.422	3.028	2.796	2.640	2.528	2.442	2.375
24	4.260	3.403	3.009	2.776	2.621	2.508	2.423	2.355
25	4.247	3.385	2.991	2.759	2.603	2.490	2.405	2.337
26	4.2251	3.369	2.975	2.743	2.587	2.474	2.388	2.321
27	4.210	3.354	2.960	2.728	2.572	2.459	2.373	2.305
28	4.196	3.340	2.947	2.714	2.558	2.445	2.359	2.291
29	4.183	3.328	2.934	2.701	2.545	2.432	2.346	2.278
30	4.171	3.316	2.922	2.690	2.534	2.421	2.334	2.266
40	4.085	3.232	2.839	2.606	2.449	2.336	2.249	2.180
50	4.034	3.183	2.790	2.557	2.400	2.286	2.199	2.130
60	4.001	3.150	2.758	2.525	2.368	2.254	2.167	2.097
70	3.978	3.128	2.736	2.503	2.346	2.231	2.143	2.074
80	3.960	3.111	2.719	2.486	2.329	2.214	2.126	2.056
90	3.947	3.098	2.706	2.473	2.316	2.201	2.113	2.043
100	3.936	3.087	2.696	2.463	2.305	2.191	2.103	2.032
120	3.920	3.072	2.680	2.447	2.290	2.175	2.087	2.016
140	3.909	3.061	2.669	2.436	2.279	2.164	2.076	2.005
160	3.900	3.053	2.661	2.428	2.271	2.156	2.067	1.997
180	3.894	3.046	2.655	2.422	2.264	2.149	2.061	1.990
200	3.888	3.041	2.650	2.417	2.259	2.144	2.056	1.985
220	3.884	3.037	2.646	2.413	2.255	2.140	2.051	1.981
240	3.880	3.033	2.642	2.409	2.252	2.137	2.048	1.977
260	3.877	3.030	2.639	2.406	2.249	2.134	2.045	1.974
280	3.875	3.028	2.637	2.404	2.246	2.131	2.042	1.972
300	3.873	3.026	2.635	2.402	2.244	2.129	2.040	1.969
320	3.871	3.024	2.633	2.400	2.242	2.127	2.038	1.967

Table 4 (continued)

Ninety-fifth percentiles of the F distribution with m/n degrees of freedom

					m				
n	9	10	11	12	13	14	15	16
1	240.3	241.7	243.1	243.8	244.6	245.3	246.0	246.8
2	19.39	19.40	19.41	19.41	19.42	19.42	19.43	19.43
3	8.812	8.786	8.763	8.745	8.729	8.715	8.703	8.692
4	5.999	5.964	5.936	5.912	5.891	5.873	5.858	5.844
5	4.772	4.735	4.704	4.678	4.655	4.636	4.619	4.604
6	4.099	4.060	4.027	3.999	3.976	3.956	3.938	3.922
7	3.677	3.637	3.603	3.575	3.550	3.529	3.511	3.494
8	3.388	3.347	3.313	3.284	3.259	3.237	3.218	3.202
9	3.179	3.137	3.103	3.073	3.048	3.025	3.006	2.989
10	3.020	2.978	2.943	2.913	2.887	2.865	2.845	2.828
11	2.896	2.854	2.818	2.788	2.761	2.739	2.719	2.701
12	2.796	2.753	2.717	2.687	2.660	2.637	2.617	2.599
13	2.714	2.671	2.635	2.604	2.577	2.554	2.533	2.515
14	2.646	2.602	2.566	2.534	2.507	2.484	2.463	2.445
15	2.588	2.544	2.507	2.475	2.448	2.424	2.403	2.385
16	2.538	2.493	2.456	2.425	2.397	2.373	2.352	2.333
17	2.494	2.450	2.413	2.381	2.353	2.329	2.308	2.289
18	2.456	2.412	2.374	2.342	2.314	2.290	2.269	2.250
19	2.423	2.378	2.340	2.308	2.280	2.256	2.234	2.215
20	2.393	2.348	2.310	2.278	2.250	2.225	2.203	2.184
21	2.366	2.321	2.283	2.250	2.222	2.197	2.176	2.156
22	2.342	2.297	2.259	2.226	2.197	2.173	2.151	2.131
23	2.320	2.275	2.236	2.204	2.175	2.150	2.128	2.109
24	2.300	2.255	2.216	2.183	2.155	2.130	2.108	2.088
25	2.282	2.236	2.198	2.165	2.136	2.111	2.089	2.069
26	2.265	2.220	2.181	2.148	2.119	2.094	2.072	2.052
27	2.250	2.204	2.166	2.132	2.103	2.078	2.056	2.036
28	2.236	2.190	2.151	2.118	2.089	2.064	2.041	2.021
29	2.223	2.177	2.138	2.105	2.075	2.050	2.027	2.007
30	2.211	2.165	2.126	2.092	2.063	2.037	2.015	1.995
40	2.124	2.077	2.038	2.003	1.974	1.948	1.924	1.904
50	2.073	2.026	1.986	1.952	1.921	1.895	1.871	1.850
60	2.040	1.993	1.952	1.917	1.887	1.860	1.836	1.815
70	2.017	1.969	1.928	1.893	1.863	1.836	1.812	1.790
80	1.999	1.951	1.910	1.875	1.845	1.817	1.793	1.772
90	1.986	1.938	1.897	1.861	1.830	1.803	1.779	1.757
100	1.975	1.927	1.886	1.850	1.819	1.792	1.768	1.746
120	1.959	1.910	1.869	1.834	1.803	1.775	1.751	1.728
140	1.947	1.899	1.858	1.822	1.791	1.763	1.738	1.716
160	1.939	1.890	1.849	1.813	1.782	1.754	1.729	1.707
180	1.932	1.884	1.842	1.806	1.775	1.747	1.722	1.700
200	1.927	1.878	1.837	1.801	1.769	1.741	1.717	1.694
220	1.923	1.874	1.832	1.796	1.765	1.737	1.712	1.690
240	1.919	1.870	1.829	1.793	1.761	1.733	1.708	1.686
260	1.916	1.867	1.826	1.790	1.758	1.730	1.705	1.683
280	1.913	1.865	1.823	1.787	1.755	1.727	1.702	1.680
300	1.911	1.862	1.821	1.785	1.753	1.725	1.700	1.677
320	1.909	1.860	1.819	1.783	1.751	1.723	1.698	1.675

Table 4 (continued)

Ninety-fifth percentiles of the F distribution with m/n degrees of freedom

n	m							
	17	18	19	20	21	22	23	24
1	246.8	247.6	247.6	248.3	248.3	248.3	249.0	249.0
2	19.44	19.44	19.44	19.45	19.45	19.45	19.45	19.45
3	8.683	8.675	8.667	8.660	8.654	8.648	8.643	8.639
4	5.832	5.821	5.811	5.803	5.795	5.787	5.781	5.774
5	4.590	4.579	4.568	4.558	4.549	4.541	4.534	4.527
6	3.908	3.896	3.884	3.874	3.865	3.856	3.849	3.841
7	3.480	3.467	3.455	3.445	3.435	3.426	3.418	3.410
8	3.187	3.173	3.161	3.150	3.140	3.131	3.123	3.115
9	2.974	2.960	2.948	2.936	2.926	2.917	2.908	2.900
10	2.812	2.798	2.785	2.774	2.764	2.754	2.745	2.737
11	2.685	2.671	2.658	2.646	2.636	2.626	2.617	2.609
12	2.583	2.568	2.555	2.544	2.533	2.523	2.514	2.505
13	2.499	2.484	2.471	2.459	2.448	2.438	2.429	2.420
14	2.428	2.413	2.400	2.388	2.377	2.367	2.357	2.349
15	2.368	2.353	2.340	2.328	2.316	2.306	2.297	2.288
16	2.317	2.302	2.288	2.276	2.264	2.254	2.244	2.235
17	2.272	2.257	2.243	2.230	2.219	2.208	2.199	2.190
18	2.233	2.217	2.203	2.191	2.179	2.168	2.159	2.150
19	2.198	2.182	2.168	2.155	2.144	2.133	2.123	2.114
20	2.167	2.151	2.137	2.124	2.112	2.102	2.092	2.082
21	2.139	2.123	2.109	2.096	2.084	2.073	2.063	2.054
22	2.114	2.098	2.084	2.071	2.059	2.048	2.038	2.028
23	2.091	2.075	2.061	2.048	2.036	2.025	2.014	2.005
24	2.070	2.054	2.040	2.027	2.015	2.004	1.993	1.984
25	2.051	2.035	2.021	2.007	1.995	1.984	1.974	1.964
26	2.034	2.018	2.003	1.990	1.978	1.966	1.956	1.946
27	2.018	2.002	1.987	1.974	1.961	1.950	1.940	1.930
28	2.003	1.987	1.972	1.959	1.946	1.935	1.924	1.915
29	1.989	1.973	1.958	1.945	1.932	1.921	1.910	1.901
30	1.976	1.960	1.945	1.932	1.919	1.908	1.897	1.887
40	1.885	1.868	1.853	1.839	1.826	1.814	1.803	1.793
50	1.831	1.814	1.798	1.784	1.771	1.759	1.748	1.737
60	1.796	1.778	1.763	1.748	1.735	1.722	1.711	1.700
70	1.771	1.753	1.737	1.722	1.709	1.696	1.685	1.674
80	1.752	1.734	1.718	1.703	1.689	1.677	1.665	1.654
90	1.737	1.720	1.703	1.688	1.675	1.662	1.650	1.639
100	1.726	1.708	1.691	1.676	1.663	1.650	1.638	1.627
120	1.709	1.690	1.674	1.659	1.645	1.632	1.620	1.608
140	1.696	1.678	1.661	1.646	1.632	1.619	1.607	1.595
160	1.687	1.669	1.652	1.637	1.622	1.609	1.597	1.586
180	1.680	1.661	1.645	1.629	1.615	1.602	1.589	1.578
200	1.674	1.656	1.639	1.623	1.609	1.596	1.583	1.572
220	1.669	1.651	1.634	1.618	1.604	1.591	1.579	1.567
240	1.665	1.647	1.630	1.614	1.600	1.587	1.574	1.563
260	1.662	1.644	1.627	1.611	1.597	1.583	1.571	1.559
280	1.659	1.641	1.624	1.608	1.594	1.580	1.568	1.556
300	1.657	1.638	1.621	1.606	1.591	1.578	1.565	1.554
320	1.655	1.636	1.619	1.603	1.589	1.576	1.563	1.551

Table 4 (continued)

Ninety-fifth percentiles of the F distribution with m/n degrees of freedom

n	25	26	27	28	29	30	40	50
				m				
1	249.0	249.8	249.8	249.8	249.8	249.8	251.3	252.0
2	19.46	19.46	19.46	19.46	19.46	19.46	19.47	19.48
3	8.634	8.630	8.626	8.623	8.620	8.617	8.594	8.581
4	5.769	5.763	5.759	5.754	5.750	5.746	5.717	5.699
5	4.521	4.515	4.510	4.505	4.500	4.496	4.464	4.444
6	3.835	3.829	3.823	3.818	3.813	3.808	3.774	3.754
7	3.404	3.397	3.391	3.386	3.381	3.376	3.340	3.319
8	3.108	3.102	3.095	3.090	3.084	3.079	3.043	3.020
9	2.893	2.886	2.880	2.874	2.869	2.864	2.826	2.803
10	2.730	2.723	2.716	2.710	2.705	2.700	2.661	2.637
11	2.601	2.594	2.588	2.582	2.576	2.570	2.531	2.507
12	2.498	2.491	2.484	2.478	2.472	2.466	2.426	2.401
13	2.412	2.405	2.398	2.392	2.386	2.380	2.339	2.314
14	2.341	2.333	2.326	2.320	2.314	2.308	2.266	2.241
15	2.280	2.272	2.265	2.259	2.253	2.247	2.204	2.178
16	2.227	2.220	2.212	2.206	2.200	2.194	2.151	2.124
17	2.181	2.174	2.167	2.160	2.154	2.148	2.104	2.077
18	2.141	2.134	2.126	2.119	2.113	2.107	2.063	2.035
19	2.106	2.098	2.090	2.084	2.077	2.071	2.026	1.999
20	2.074	2.066	2.059	2.052	2.045	2.039	1.994	1.966
21	2.045	2.037	2.030	2.023	2.016	2.010	1.965	1.936
22	2.020	2.012	2.004	1.997	1.990	1.984	1.938	1.909
23	1.996	1.988	1.981	1.973	1.967	1.961	1.914	1.885
24	1.975	1.967	1.959	1.952	1.945	1.939	1.892	1.863
25	1.955	1.947	1.939	1.932	1.926	1.919	1.872	1.842
26	1.938	1.929	1.921	1.914	1.907	1.901	1.853	1.823
27	1.921	1.913	1.905	1.898	1.891	1.884	1.836	1.806
28	1.906	1.897	1.889	1.882	1.875	1.869	1.820	1.790
29	1.891	1.883	1.875	1.868	1.861	1.854	1.806	1.775
30	1.878	1.870	1.862	1.854	1.847	1.841	1.792	1.761
40	1.783	1.775	1.766	1.759	1.751	1.744	1.693	1.660
50	1.727	1.718	1.710	1.702	1.694	1.687	1.634	1.600
60	1.690	1.681	1.672	1.664	1.656	1.649	1.594	1.559
70	1.664	1.654	1.646	1.637	1.629	1.622	1.566	1.530
80	1.644	1.634	1.626	1.617	1.609	1.602	1.545	1.508
90	1.629	1.619	1.610	1.601	1.593	1.586	1.528	1.491
100	1.616	1.607	1.598	1.589	1.581	1.573	1.515	1.477
120	1.598	1.588	1.579	1.570	1.562	1.554	1.495	1.457
140	1.585	1.575	1.566	1.557	1.549	1.541	1.481	1.442
160	1.575	1.565	1.556	1.547	1.538	1.531	1.470	1.430
180	1.567	1.557	1.548	1.539	1.531	1.523	1.462	1.422
200	1.561	1.551	1.542	1.533	1.524	1.516	1.455	1.415
220	1.556	1.546	1.537	1.528	1.519	1.511	1.450	1.409
240	1.552	1.542	1.532	1.523	1.515	1.507	1.445	1.404
260	1.548	1.538	1.529	1.520	1.511	1.503	1.441	1.400
280	1.546	1.535	1.526	1.517	1.508	1.500	1.438	1.396
300	1.543	1.533	1.523	1.514	1.505	1.497	1.435	1.393
320	1.541	1.530	1.521	1.512	1.503	1.495	1.432	1.391

Table 4 (continued)

Ninety-fifth percentiles of the F distribution with m/n degrees of freedom

				m				
n	60	70	80	90	100	120	140	160
1	252.0	252.8	252.8	252.8	252.8	253.6	253.6	253.6
2	19.48	19.48	19.48	19.49	19.49	19.49	19.49	19.49
3	8.572	8.566	8.561	8.557	8.554	8.549	8.546	8.544
4	5.688	5.679	5.673	5.668	5.664	5.658	5.654	5.651
5	4.431	4.422	4.415	4.410	4.405	4.398	4.394	4.390
6	3.740	3.730	3.722	3.717	3.712	3.705	3.700	3.696
7	3.304	3.294	3.286	3.280	3.275	3.267	3.262	3.258
8	3.005	2.994	2.986	2.980	2.975	2.967	2.961	2.957
9	2.787	2.776	2.768	2.761	2.756	2.748	2.742	2.737
10	2.621	2.610	2.601	2.594	2.588	2.580	2.574	2.570
11	2.490	2.478	2.469	2.462	2.457	2.448	2.442	2.437
12	2.384	2.372	2.363	2.356	2.350	2.341	2.335	2.330
13	2.297	2.284	2.275	2.267	2.261	2.252	2.246	2.241
14	2.223	2.210	2.201	2.193	2.187	2.178	2.171	2.166
15	2.160	2.147	2.137	2.130	2.123	2.114	2.107	2.102
16	2.106	2.093	2.083	2.075	2.068	2.059	2.052	2.047
17	2.058	2.045	2.035	2.027	2.020	2.011	2.004	1.998
18	2.017	2.003	1.993	1.985	1.978	1.968	1.961	1.956
19	1.980	1.966	1.955	1.947	1.940	1.930	1.923	1.917
20	1.946	1.932	1.922	1.913	1.907	1.896	1.889	1.883
21	1.916	1.902	1.892	1.883	1.876	1.866	1.858	1.853
22	1.889	1.875	1.864	1.855	1.849	1.838	1.830	1.825
23	1.865	1.850	1.839	1.830	1.823	1.813	1.805	1.799
24	1.842	1.828	1.816	1.808	1.800	1.790	1.782	1.776
25	1.822	1.807	1.796	1.787	1.779	1.768	1.760	1.754
26	1.803	1.788	1.776	1.767	1.760	1.749	1.741	1.735
27	1.785	1.770	1.758	1.749	1.742	1.731	1.723	1.716
28	1.769	1.754	1.742	1.733	1.725	1.714	1.706	1.699
29	1.754	1.738	1.726	1.717	1.710	1.698	1.690	1.684
30	1.740	1.724	1.712	1.703	1.695	1.683	1.675	1.669
40	1.637	1.621	1.608	1.598	1.589	1.577	1.567	1.560
50	1.576	1.558	1.544	1.534	1.525	1.511	1.502	1.494
60	1.534	1.516	1.502	1.491	1.481	1.467	1.457	1.449
70	1.505	1.486	1.471	1.459	1.450	1.435	1.424	1.416
80	1.482	1.463	1.448	1.436	1.426	1.411	1.400	1.391
90	1.465	1.445	1.429	1.417	1.407	1.391	1.380	1.371
100	1.450	1.430	1.415	1.402	1.392	1.376	1.364	1.355
120	1.429	1.408	1.392	1.379	1.368	1.352	1.340	1.330
140	1.414	1.392	1.376	1.363	1.352	1.335	1.322	1.312
160	1.402	1.381	1.364	1.350	1.339	1.321	1.308	1.298
180	1.393	1.371	1.354	1.340	1.329	1.311	1.298	1.287
200	1.386	1.364	1.346	1.332	1.321	1.302	1.289	1.278
220	1.380	1.357	1.340	1.326	1.314	1.296	1.282	1.271
240	1.375	1.352	1.335	1.320	1.308	1.290	1.276	1.265
260	1.370	1.348	1.330	1.316	1.303	1.285	1.270	1.259
280	1.367	1.344	1.326	1.311	1.299	1.280	1.266	1.255
300	1.363	1.341	1.323	1.308	1.296	1.277	1.262	1.251
320	1.361	1.338	1.320	1.305	1.293	1.273	1.259	1.247

Table 4 (continued)

Ninety-fifth percentiles of the F distribution with m/n degrees of freedom

					m			
n	180	200	220	240	260	280	300	320
1	253.6	253.6	253.6	253.6	253.6	253.6	253.6	253.6
2	19.49	12.00	6.000	3.000	2.000	2.000	2.000	2.000
3	8.542	8.540	12.000	6.000	3.000	2.000	2.000	2.000
4	5.648	5.646	5.645	6.000	3.500	3.000	2.000	2.000
5	4.387	4.385	4.383	4.382	6.000	3.000	2.000	2.000
6	3.693	3.690	3.689	3.687	3.685	3.500	3.000	2.000
7	3.255	3.252	3.250	3.249	3.247	3.246	3.000	2.000
8	2.954	2.951	2.949	2.947	2.946	2.945	3.000	2.000
9	2.734	2.731	2.729	2.727	2.726	2.724	3.000	3.000
10	2.566	2.563	2.561	2.559	2.558	2.556	2.555	3.000
11	2.434	2.431	2.428	2.426	2.425	2.423	2.422	3.000
12	2.326	2.323	2.321	2.319	2.317	2.316	2.314	3.000
13	2.237	2.234	2.232	2.230	2.228	2.226	2.225	3.000
14	2.162	2.159	2.157	2.155	2.153	2.151	2.150	2.149
15	2.098	2.095	2.092	2.090	2.088	2.087	2.085	2.084
16	2.043	2.040	2.037	2.035	2.033	2.031	2.030	2.028
17	1.994	1.991	1.988	1.986	1.984	1.982	1.981	1.980
18	1.951	1.948	1.945	1.943	1.941	1.939	1.938	1.936
19	1.913	1.910	1.907	1.905	1.902	1.901	1.899	1.898
20	1.879	1.875	1.873	1.870	1.868	1.866	1.865	1.863
21	1.848	1.845	1.842	1.839	1.837	1.835	1.834	1.832
22	1.820	1.816	1.814	1.811	1.809	1.807	1.806	1.804
23	1.795	1.791	1.788	1.785	1.783	1.781	1.780	1.778
24	1.771	1.768	1.764	1.762	1.760	1.758	1.756	1.755
25	1.750	1.746	1.743	1.740	1.738	1.736	1.735	1.733
26	1.730	1.726	1.723	1.720	1.718	1.716	1.714	1.713
27	1.712	1.708	1.704	1.702	1.700	1.698	1.696	1.694
28	1.694	1.691	1.687	1.685	1.682	1.680	1.679	1.677
29	1.679	1.675	1.671	1.669	1.666	1.664	1.663	1.661
30	1.664	1.660	1.656	1.654	1.651	1.649	1.647	1.646
40	1.555	1.551	1.547	1.544	1.541	1.539	1.537	1.535
50	1.488	1.483	1.480	1.476	1.473	1.471	1.469	1.467
60	1.443	1.438	1.434	1.430	1.427	1.424	1.422	1.420
70	1.410	1.404	1.400	1.396	1.393	1.390	1.388	1.386
80	1.384	1.379	1.374	1.370	1.367	1.364	1.361	1.359
90	1.364	1.358	1.353	1.349	1.346	1.343	1.340	1.338
100	1.348	1.342	1.337	1.333	1.329	1.326	1.323	1.321
120	1.322	1.316	1.311	1.307	1.303	1.300	1.297	1.294
140	1.304	1.298	1.292	1.288	1.284	1.280	1.277	1.275
160	1.290	1.283	1.278	1.273	1.269	1.265	1.262	1.259
180	1.279	1.272	1.266	1.261	1.257	1.253	1.250	1.247
200	1.270	1.263	1.257	1.252	1.247	1.244	1.240	1.237
220	1.262	1.255	1.249	1.244	1.239	1.236	1.232	1.229
240	1.256	1.248	1.242	1.237	1.233	1.229	1.225	1.222
260	1.250	1.243	1.237	1.231	1.227	1.223	1.219	1.216
280	1.246	1.238	1.232	1.226	1.222	1.218	1.214	1.211
300	1.241	1.234	1.228	1.222	1.217	1.213	1.210	1.206
320	1.238	1.230	1.224	1.218	1.213	1.209	1.206	1.202

Table 5

Percentiles, $t_{p;(n)}$ of the t distribution with n degrees of freedom

Note: $t_{1-p;(n)} = -t_{p;(n)}$

n				p				
	0.6000	0.7000	0.8000	0.9000	0.9500	0.9750	0.9800	0.9900
1	0.325	0.727	1.376	3.078	6.314	12.706	15.895	31.820
2	0.289	0.617	1.061	1.886	2.920	4.303	4.849	6.965
3	0.277	0.584	0.979	1.638	2.353	3.182	3.482	4.541
4	0.271	0.569	0.941	1.533	2.132	2.776	2.999	3.747
5	0.267	0.559	0.920	1.476	2.015	2.571	2.757	3.365
6	0.265	0.553	0.906	1.440	1.943	2.447	2.612	3.143
7	0.263	0.549	0.896	1.415	1.895	2.365	2.517	2.998
8	0.262	0.546	0.889	1.397	1.860	2.306	2.449	2.896
9	0.261	0.543	0.883	1.383	1.833	2.262	2.398	2.821
10	0.260	0.541	0.879	1.372	1.812	2.228	2.359	2.764
11	0.260	0.540	0.876	1.363	1.796	2.201	2.328	2.718
12	0.259	0.539	0.873	1.356	1.782	2.179	2.303	2.681
13	0.259	0.538	0.870	1.350	1.771	2.160	2.282	2.650
14	0.258	0.537	0.868	1.345	1.761	2.145	2.264	2.624
15	0.258	0.536	0.866	1.341	1.753	2.131	2.249	2.603
16	0.258	0.535	0.865	1.337	1.746	2.120	2.235	2.584
17	0.257	0.534	0.863	1.333	1.740	2.110	2.224	2.567
18	0.257	0.534	0.862	1.330	1.734	2.101	2.214	2.552
19	0.257	0.533	0.861	1.328	1.729	2.093	2.205	2.539
20	0.257	0.533	0.860	1.325	1.725	2.086	2.197	2.528
21	0.257	0.532	0.859	1.323	1.721	2.080	2.189	2.518
22	0.256	0.532	0.858	1.321	1.717	2.074	2.183	2.508
23	0.256	0.532	0.858	1.319	1.714	2.069	2.177	2.500
24	0.256	0.531	0.857	1.318	1.711	2.064	2.172	2.492
25	0.256	0.531	0.856	1.316	1.708	2.060	2.167	2.485
26	0.256	0.531	0.856	1.315	1.706	2.056	2.162	2.479
27	0.256	0.531	0.855	1.314	1.703	2.052	2.158	2.473
28	0.256	0.530	0.855	1.313	1.701	2.048	2.154	2.467
29	0.256	0.530	0.854	1.311	1.699	2.045	2.150	2.462
30	0.256	0.530	0.854	1.310	1.697	2.042	2.147	2.457
40	0.255	0.529	0.851	1.303	1.684	2.021	2.123	2.423
50	0.255	0.528	0.849	1.299	1.676	2.009	2.109	2.403
60	0.254	0.527	0.848	1.296	1.671	2.000	2.099	2.390
70	0.254	0.527	0.847	1.294	1.667	1.994	2.093	2.381
80	0.254	0.526	0.846	1.292	1.664	1.990	2.088	2.374
90	0.254	0.526	0.846	1.291	1.662	1.987	2.084	2.369
100	0.254	0.526	0.845	1.290	1.660	1.984	2.081	2.364
120	0.254	0.526	0.845	1.289	1.658	1.980	2.076	2.358
140	0.254	0.526	0.844	1.288	1.656	1.977	2.073	2.353
160	0.254	0.525	0.844	1.287	1.654	1.975	2.071	2.350
180	0.254	0.525	0.844	1.286	1.653	1.973	2.069	2.347
200	0.254	0.525	0.843	1.286	1.652	1.972	2.067	2.345
220	0.254	0.525	0.843	1.285	1.652	1.971	2.066	2.343
240	0.254	0.525	0.843	1.285	1.651	1.970	2.065	2.342
260	0.254	0.525	0.843	1.285	1.651	1.969	2.064	2.341
280	0.254	0.525	0.843	1.285	1.650	1.968	2.063	2.340
300	0.254	0.525	0.843	1.284	1.650	1.968	2.063	2.339
320	0.254	0.525	0.843	1.284	1.650	1.967	2.062	2.338

Table 6

Power function for one-factor analysis of variance F test $(m = 2)$

$\Pr(G > F_{0.05;(m-1,mn-m)})$, where G is noncentral $F_{m-1,mn-m}$ with $\lambda = mn\delta^2/2$

	δ													
n	0.100	0.200	0.300	0.400	0.500	0.600	0.700	0.800	0.900	1.000	1.500	2.000	2.500	3.000
2	0.052	0.057	0.067	0.079	0.095	0.114	0.137	0.161	0.189	0.218	0.387	0.565	0.719	0.836
3	0.054	0.067	0.089	0.119	0.159	0.207	0.263	0.325	0.392	0.463	0.783	0.948	0.993	0.999
4	0.057	0.077	0.111	0.160	0.223	0.299	0.385	0.476	0.568	0.657	0.939	0.996	1.000	1.000
5	0.059	0.087	0.134	0.201	0.286	0.386	0.495	0.603	0.704	0.791	0.985	1.000	1.000	1.000
6	0.061	0.096	0.156	0.241	0.347	0.468	0.591	0.705	0.802	0.876	0.996	1.000	1.000	1.000
7	0.064	0.106	0.179	0.281	0.406	0.541	0.672	0.785	0.870	0.929	0.999	1.000	1.000	1.000
8	0.066	0.116	0.201	0.320	0.461	0.608	0.740	0.845	0.917	0.960	1.000	1.000	1.000	1.000
9	0.068	0.126	0.224	0.358	0.513	0.667	0.796	0.890	0.947	0.978	1.000	1.000	1.000	1.000
10	0.071	0.135	0.246	0.395	0.562	0.718	0.841	0.922	0.967	0.988	1.000	1.000	1.000	1.000
11	0.073	0.145	0.268	0.431	0.607	0.763	0.877	0.946	0.980	0.994	1.000	1.000	1.000	1.000
12	0.076	0.155	0.290	0.466	0.649	0.802	0.906	0.963	0.988	0.997	1.000	1.000	1.000	1.000
13	0.078	0.165	0.312	0.499	0.687	0.835	0.928	0.974	0.993	0.998	1.000	1.000	1.000	1.000
14	0.080	0.175	0.333	0.531	0.721	0.863	0.946	0.983	0.996	0.999	1.000	1.000	1.000	1.000
15	0.083	0.185	0.355	0.562	0.753	0.887	0.959	0.988	0.997	1.000	1.000	1.000	1.000	1.000
16	0.085	0.195	0.376	0.591	0.781	0.907	0.969	0.992	0.998	1.000	1.000	1.000	1.000	1.000
17	0.087	0.205	0.396	0.619	0.807	0.924	0.977	0.995	0.999	1.000	1.000	1.000	1.000	1.000
18	0.090	0.215	0.417	0.645	0.830	0.938	0.983	0.997	0.999	1.000	1.000	1.000	1.000	1.000
19	0.092	0.224	0.437	0.670	0.851	0.949	0.987	0.998	1.000	1.000	1.000	1.000	1.000	1.000
20	0.095	0.234	0.456	0.693	0.869	0.959	0.991	0.999	1.000	1.000	1.000	1.000	1.000	1.000
21	0.097	0.244	0.475	0.716	0.885	0.967	0.993	0.999	1.000	1.000	1.000	1.000	1.000	1.000
22	0.099	0.254	0.494	0.736	0.900	0.973	0.995	0.999	1.000	1.000	1.000	1.000	1.000	1.000
23	0.102	0.264	0.512	0.756	0.913	0.978	0.996	1.000	1.000	1.000	1.000	1.000	1.000	1.000
24	0.104	0.274	0.530	0.774	0.924	0.983	0.997	1.000	1.000	1.000	1.000	1.000	1.000	1.000
25	0.107	0.283	0.547	0.791	0.934	0.986	0.998	1.000	1.000	1.000	1.000	1.000	1.000	1.000
26	0.109	0.293	0.564	0.807	0.942	0.989	0.999	1.000	1.000	1.000	1.000	1.000	1.000	1.000
27	0.111	0.303	0.581	0.822	0.950	0.991	0.999	1.000	1.000	1.000	1.000	1.000	1.000	1.000
28	0.114	0.312	0.597	0.836	0.957	0.993	0.999	1.000	1.000	1.000	1.000	1.000	1.000	1.000
29	0.116	0.322	0.612	0.849	0.963	0.994	0.999	1.000	1.000	1.000	1.000	1.000	1.000	1.000
30	0.119	0.332	0.627	0.861	0.968	0.995	1.000	1.000	1.000	1.000	1.000	1.000	1.000	1.000
31	0.121	0.341	0.642	0.873	0.972	0.996	1.000	1.000	1.000	1.000	1.000	1.000	1.000	1.000
32	0.124	0.350	0.656	0.883	0.976	0.997	1.000	1.000	1.000	1.000	1.000	1.000	1.000	1.000
33	0.126	0.360	0.670	0.893	0.979	0.998	1.000	1.000	1.000	1.000	1.000	1.000	1.000	1.000
34	0.128	0.369	0.684	0.902	0.982	0.998	1.000	1.000	1.000	1.000	1.000	1.000	1.000	1.000
35	0.131	0.378	0.697	0.910	0.985	0.999	1.000	1.000	1.000	1.000	1.000	1.000	1.000	1.000
36	0.133	0.388	0.709	0.917	0.987	0.999	1.000	1.000	1.000	1.000	1.000	1.000	1.000	1.000
37	0.136	0.397	0.721	0.924	0.989	0.999	1.000	1.000	1.000	1.000	1.000	1.000	1.000	1.000
38	0.138	0.406	0.733	0.931	0.990	0.999	1.000	1.000	1.000	1.000	1.000	1.000	1.000	1.000
39	0.141	0.415	0.744	0.937	0.992	0.999	1.000	1.000	1.000	1.000	1.000	1.000	1.000	1.000
40	0.143	0.424	0.755	0.942	0.993	1.000	1.000	1.000	1.000	1.000	1.000	1.000	1.000	1.000
41	0.146	0.432	0.765	0.947	0.994	1.000	1.000	1.000	1.000	1.000	1.000	1.000	1.000	1.000
42	0.148	0.441	0.776	0.952	0.995	1.000	1.000	1.000	1.000	1.000	1.000	1.000	1.000	1.000
43	0.150	0.450	0.785	0.956	0.996	1.000	1.000	1.000	1.000	1.000	1.000	1.000	1.000	1.000
44	0.153	0.458	0.795	0.960	0.996	1.000	1.000	1.000	1.000	1.000	1.000	1.000	1.000	1.000
45	0.155	0.467	0.804	0.964	0.997	1.000	1.000	1.000	1.000	1.000	1.000	1.000	1.000	1.000
46	0.158	0.475	0.812	0.967	0.997	1.000	1.000	1.000	1.000	1.000	1.000	1.000	1.000	1.000
47	0.160	0.484	0.821	0.970	0.998	1.000	1.000	1.000	1.000	1.000	1.000	1.000	1.000	1.000
48	0.163	0.492	0.829	0.973	0.998	1.000	1.000	1.000	1.000	1.000	1.000	1.000	1.000	1.000
49	0.165	0.500	0.836	0.975	0.998	1.000	1.000	1.000	1.000	1.000	1.000	1.000	1.000	1.000
50	0.168	0.508	0.844	0.977	0.999	1.000	1.000	1.000	1.000	1.000	1.000	1.000	1.000	1.000

Table 6 (continued)
Power function for one-factor analysis of variance F test ($m = 3$)
$\Pr(G > F_{0.05;(m-1,mn-m)})$, where G is noncentral $F_{m-1,mn-m}$ with $\lambda = mn\delta^2/2$

	δ													
n	0.100	0.200	0.300	0.400	0.500	0.600	0.700	0.800	0.900	1.000	1.500	2.000	2.500	3.000
2	0.052	0.058	0.068	0.082	0.100	0.122	0.149	0.180	0.214	0.253	0.479	0.703	0.863	0.949
3	0.054	0.067	0.090	0.124	0.168	0.225	0.292	0.368	0.450	0.534	0.873	0.985	0.999	1.000
4	0.057	0.077	0.114	0.168	0.241	0.331	0.434	0.542	0.648	0.743	0.979	1.000	1.000	1.000
5	0.059	0.087	0.138	0.214	0.314	0.433	0.561	0.683	0.788	0.870	0.997	1.000	1.000	1.000
6	0.061	0.097	0.163	0.260	0.386	0.528	0.668	0.788	0.879	0.938	1.000	1.000	1.000	1.000
7	0.064	0.108	0.188	0.307	0.455	0.613	0.754	0.863	0.933	0.972	1.000	1.000	1.000	1.000
8	0.066	0.118	0.214	0.353	0.520	0.686	0.822	0.913	0.964	0.988	1.000	1.000	1.000	1.000
9	0.068	0.129	0.240	0.398	0.580	0.748	0.873	0.947	0.982	0.995	1.000	1.000	1.000	1.000
10	0.071	0.139	0.266	0.442	0.635	0.800	0.911	0.968	0.991	0.998	1.000	1.001	1.000	1.000
11	0.073	0.150	0.292	0.485	0.685	0.843	0.938	0.981	0.995	0.999	1.000	1.000	1.000	1.000
12	0.076	0.161	0.318	0.525	0.729	0.878	0.957	0.989	0.998	1.000	1.000	1.000	1.000	1.000
13	0.078	0.172	0.343	0.564	0.769	0.906	0.971	0.993	0.999	1.000	1.000	1.000	1.000	1.000
14	0.080	0.184	0.369	0.601	0.803	0.928	0.981	0.996	0.999	1.000	1.000	1.000	1.000	1.000
15	0.083	0.195	0.394	0.635	0.834	0.945	0.987	0.998	1.000	1.000	1.000	1.000	1.000	1.000
16	0.085	0.206	0.419	0.667	0.860	0.958	0.991	0.999	1.000	1.000	1.000	1.000	1.000	1.000
17	0.088	0.218	0.444	0.698	0.882	0.969	0.994	0.999	1.000	1.000	1.000	1.000	1.000	1.000
18	0.090	0.229	0.468	0.726	0.902	0.977	0.996	1.000	1.000	1.000	1.000	1.000	1.000	1.000
19	0.093	0.241	0.491	0.752	0.918	0.983	0.998	1.000	1.000	1.000	1.000	1.000	1.000	1.000
20	0.095	0.252	0.514	0.776	0.932	0.987	0.999	1.000	1.000	1.000	1.000	1.000	1.000	1.000
21	0.098	0.264	0.536	0.798	0.944	0.990	0.999	1.000	1.000	1.000	1.000	1.000	1.000	1.000
22	0.100	0.275	0.558	0.818	0.953	0.993	0.999	1.000	1.000	1.000	1.000	1.000	1.000	1.000
23	0.103	0.287	0.579	0.837	0.962	0.995	1.000	1.000	1.000	1.000	1.000	1.000	1.000	1.000
24	0.106	0.298	0.599	0.853	0.969	0.996	1.000	1.000	1.000	1.000	1.000	1.000	1.000	1.000
25	0.108	0.310	0.619	0.869	0.974	0.997	1.000	1.000	1.000	1.000	1.000	1.000	1.000	1.000
26	0.111	0.321	0.638	0.883	0.979	0.998	1.000	1.000	1.000	1.000	1.000	1.000	1.000	1.000
27	0.113	0.333	0.656	0.895	0.983	0.999	1.000	1.000	1.000	1.000	1.000	1.000	1.000	1.000
28	0.116	0.344	0.674	0.907	0.986	0.999	1.000	1.000	1.000	1.000	1.000	1.000	1.000	1.000
29	0.119	0.356	0.691	0.917	0.989	0.999	1.000	1.000	1.000	1.000	1.000	1.000	1.000	1.000
30	0.121	0.367	0.707	0.926	0.991	0.999	1.000	1.000	1.000	1.000	1.000	1.000	1.000	1.000
31	0.124	0.378	0.723	0.935	0.993	1.000	1.000	1.000	1.000	1.000	1.000	1.000	1.000	1.000
32	0.126	0.389	0.738	0.942	0.994	1.000	1.000	1.000	1.000	1.000	1.000	1.000	1.000	1.000
33	0.129	0.401	0.752	0.949	0.995	1.000	1.000	1.000	1.000	1.000	1.000	1.000	1.000	1.000
34	0.132	0.412	0.766	0.955	0.996	1.000	1.000	1.000	1.000	1.000	1.000	1.000	1.000	1.000
35	0.134	0.423	0.779	0.960	0.997	1.000	1.000	1.000	1.000	1.000	1.000	1.000	1.000	1.000
36	0.137	0.433	0.792	0.965	0.998	1.000	1.000	1.000	1.000	1.000	1.000	1.000	1.000	1.000
37	0.140	0.444	0.803	0.969	0.998	1.000	1.000	1.000	1.000	1.000	1.000	1.000	1.000	1.000
38	0.143	0.455	0.815	0.973	0.998	1.000	1.000	1.000	1.000	1.000	1.000	1.000	1.000	1.000
39	0.145	0.466	0.826	0.976	0.999	1.000	1.000	1.000	1.000	1.000	1.000	1.000	1.000	1.000
40	0.148	0.476	0.836	0.979	0.999	1.000	1.000	1.000	1.000	1.000	1.000	1.000	1.000	1.000
41	0.151	0.486	0.846	0.982	0.999	1.000	1.000	1.000	1.000	1.000	1.000	1.000	1.000	1.000
42	0.153	0.497	0.855	0.984	0.999	1.000	1.000	1.000	1.000	1.000	1.000	1.000	1.000	1.000
43	0.156	0.507	0.864	0.986	1.000	1.000	1.000	1.000	1.000	1.000	1.000	1.000	1.000	1.000
44	0.159	0.517	0.872	0.988	1.000	1.000	1.000	1.000	1.000	1.000	1.000	1.000	1.000	1.000
45	0.162	0.527	0.880	0.989	1.000	1.000	1.000	1.000	1.000	1.000	1.000	1.000	1.000	1.000
46	0.164	0.537	0.887	0.991	1.000	1.000	1.000	1.000	1.000	1.000	1.000	1.000	1.000	1.000
47	0.167	0.546	0.894	0.992	1.000	1.000	1.000	1.000	1.000	1.000	1.000	1.000	1.000	1.000
48	0.170	0.556	0.901	0.993	1.000	1.000	1.000	1.000	1.000	1.000	1.000	1.000	1.000	1.000
49	0.173	0.565	0.907	0.994	1.000	1.000	1.000	1.000	1.000	1.000	1.000	1.000	1.000	1.000
50	0.176	0.575	0.913	0.995	1.000	1.000	1.000	1.000	1.000	1.000	1.000	1.000	1.000	1.000

Table 6 (continued)

Power function for one-factor analysis of variance F test $(m = 4)$

$\Pr(G > F_{0.05;(m-1,mn-m)})$, where G is noncentral $F_{m-1,mn-m}$ with $\lambda = mn\delta^2/2$

							δ							
n	0.100	0.200	0.300	0.400	0.500	0.600	0.700	0.800	0.900	1.000	1.500	2.000	2.500	3.000
2	0.052	0.059	0.070	0.086	0.107	0.134	0.167	0.206	0.250	0.299	0.581	0.821	0.946	0.989
3	0.055	0.069	0.094	0.132	0.185	0.252	0.333	0.424	0.520	0.616	0.936	0.997	1.000	1.000
4	0.057	0.079	0.120	0.183	0.269	0.375	0.495	0.617	0.729	0.822	0.994	1.000	1.000	1.000
5	0.059	0.090	0.147	0.235	0.354	0.492	0.635	0.762	0.860	0.927	1.000	1.000	1.000	1.000
6	0.062	0.101	0.175	0.289	0.436	0.597	0.745	0.859	0.933	0.973	1.000	1.000	1.000	1.000
7	0.064	0.113	0.204	0.343	0.515	0.687	0.828	0.920	0.969	0.990	1.000	1.000	1.000	1.000
8	0.067	0.124	0.234	0.397	0.587	0.762	0.887	0.957	0.987	0.997	1.000	1.000	1.000	1.000
9	0.069	0.136	0.264	0.449	0.652	0.821	0.927	0.977	0.994	0.999	1.000	1.000	1.000	1.000
10	0.072	0.148	0.294	0.499	0.710	0.868	0.954	0.988	0.998	1.000	1.000	1.000	1.000	1.000
11	0.075	0.161	0.325	0.547	0.760	0.904	0.972	0.994	0.999	1.000	1.000	1.000	1.000	1.000
12	0.077	0.173	0.355	0.592	0.803	0.931	0.983	0.997	1.000	1.000	1.000	1.000	1.000	1.000
13	0.080	0.186	0.385	0.634	0.840	0.951	0.990	0.999	1.000	1.000	1.000	1.000	1.000	1.000
14	0.082	0.199	0.414	0.673	0.870	0.965	0.994	0.999	1.000	1.000	1.000	1.000	1.000	1.000
15	0.085	0.212	0.444	0.709	0.896	0.976	0.997	1.000	1.000	1.000	1.000	1.000	1.000	1.000
16	0.088	0.225	0.472	0.742	0.917	0.983	0.998	1.000	1.000	1.000	1.000	1.000	1.000	1.000
17	0.090	0.238	0.500	0.772	0.934	0.989	0.999	1.000	1.000	1.000	1.000	1.000	1.000	1.000
18	0.093	0.251	0.527	0.799	0.948	0.992	0.999	1.000	1.000	1.000	1.000	1.000	1.000	1.000
19	0.096	0.265	0.553	0.823	0.959	0.995	1.000	1.000	1.000	1.000	1.000	1.000	1.000	1.000
20	0.099	0.278	0.579	0.845	0.968	0.996	1.000	1.000	1.000	1.000	1.000	1.000	1.000	1.000
21	0.101	0.291	0.603	0.865	0.975	0.998	1.000	1.000	1.000	1.000	1.000	1.000	1.000	1.000
22	0.104	0.305	0.627	0.882	0.981	0.998	1.000	1.000	1.000	1.000	1.000	1.000	1.000	1.000
23	0.107	0.318	0.649	0.898	0.985	0.999	1.000	1.000	1.000	1.000	1.000	1.000	1.000	1.000
24	0.110	0.332	0.671	0.912	0.988	0.999	1.000	1.000	1.000	1.000	1.000	1.000	1.000	1.000
25	0.113	0.345	0.692	0.924	0.991	1.000	1.000	1.000	1.000	1.000	1.000	1.000	1.000	1.000
26	0.116	0.359	0.711	0.934	0.993	1.000	1.000	1.000	1.000	1.000	1.000	1.000	1.000	1.000
27	0.119	0.372	0.730	0.943	0.995	1.000	1.000	1.000	1.000	1.000	1.000	1.000	1.000	1.000
28	0.121	0.385	0.748	0.951	0.996	1.000	1.000	1.000	1.000	1.000	1.000	1.000	1.000	1.000
29	0.124	0.398	0.765	0.958	0.997	1.000	1.000	1.000	1.000	1.000	1.000	1.000	1.000	1.000
30	0.127	0.412	0.781	0.964	0.998	1.000	1.000	1.000	1.000	1.000	1.000	1.000	1.000	1.000
31	0.130	0.425	0.796	0.970	0.998	1.000	1.000	1.000	1.000	1.000	1.000	1.000	1.000	1.000
32	0.133	0.437	0.810	0.974	0.999	1.000	1.000	1.000	1.000	1.000	1.000	1.000	1.000	1.000
33	0.136	0.450	0.824	0.978	0.999	1.000	1.000	1.000	1.000	1.000	1.000	1.000	1.000	1.000
34	0.139	0.463	0.836	0.981	0.999	1.000	1.000	1.000	1.000	1.000	1.000	1.000	1.000	1.000
35	0.142	0.475	0.848	0.984	0.999	1.000	1.000	1.000	1.000	1.000	1.000	1.000	1.000	1.000
36	0.145	0.488	0.859	0.987	1.000	1.000	1.000	1.000	1.000	1.000	1.000	1.000	1.000	1.000
37	0.148	0.500	0.870	0.989	1.000	1.000	1.000	1.000	1.000	1.000	1.000	1.000	1.000	1.000
38	0.151	0.512	0.879	0.990	1.000	1.000	1.000	1.000	1.000	1.000	1.000	1.000	1.000	1.000
39	0.154	0.524	0.889	0.992	1.000	1.000	1.000	1.000	1.000	1.000	1.000	1.000	1.000	1.000
40	0.158	0.536	0.897	0.993	1.000	1.000	1.000	1.000	1.000	1.000	1.000	1.000	1.000	1.000
41	0.161	0.548	0.905	0.994	1.000	1.000	1.000	1.000	1.000	1.000	1.000	1.000	1.000	1.000
42	0.164	0.559	0.912	0.995	1.000	1.000	1.000	1.000	1.000	1.000	1.000	1.000	1.000	1.000
43	0.167	0.570	0.919	0.996	1.000	1.000	1.000	1.000	1.000	1.000	1.000	1.000	1.000	1.000
44	0.170	0.582	0.926	0.997	1.000	1.000	1.000	1.000	1.000	1.000	1.000	1.000	1.000	1.000
45	0.173	0.592	0.932	0.997	1.000	1.000	1.000	1.000	1.000	1.000	1.000	1.000	1.000	1.000
46	0.176	0.603	0.937	0.998	1.000	1.000	1.000	1.000	1.000	1.000	1.000	1.000	1.000	1.000
47	0.180	0.614	0.942	0.998	1.000	1.000	1.000	1.000	1.000	1.000	1.000	1.000	1.000	1.000
48	0.183	0.624	0.947	0.998	1.000	1.000	1.000	1.000	1.000	1.000	1.000	1.000	1.000	1.000
49	0.186	0.634	0.951	0.999	1.000	1.000	1.000	1.000	1.000	1.000	1.000	1.000	1.000	1.000
50	0.189	0.644	0.955	0.999	1.000	1.000	1.000	1.000	1.000	1.000	1.000	1.000	1.000	1.000

Table 6 (continued)

Power function for one-factor analysis of variance F test $(m = 5)$

$\Pr(G > F_{0.05;(m-1,mn-m)})$, where G is noncentral $F_{m-1,mn-m}$ with $\lambda = mn\delta^2/2$

							δ							
n	0.100	0.200	0.300	0.400	0.500	0.600	0.700	0.800	0.900	1.000	1.500	2.000	2.500	3.000
2	0.052	0.060	0.072	0.091	0.115	0.147	0.187	0.234	0.287	0.347	0.673	0.899	0.982	0.998
3	0.055	0.070	0.099	0.142	0.203	0.281	0.376	0.480	0.587	0.689	0.970	0.999	1.000	1.000
4	0.058	0.082	0.127	0.199	0.298	0.421	0.555	0.685	0.797	0.881	0.998	1.000	1.000	1.000
5	0.060	0.093	0.158	0.259	0.394	0.550	0.701	0.825	0.911	0.961	1.000	1.000	1.000	1.000
6	0.063	0.106	0.189	0.320	0.487	0.661	0.809	0.910	0.965	0.989	1.000	1.000	1.000	1.000
7	0.065	0.118	0.222	0.381	0.573	0.752	0.883	0.956	0.987	0.997	1.000	1.000	1.000	1.000
8	0.068	0.131	0.256	0.441	0.649	0.823	0.931	0.979	0.995	0.999	1.000	1.000	1.000	1.000
9	0.071	0.144	0.290	0.499	0.716	0.877	0.960	0.991	0.998	1.000	1.000	1.000	1.000	1.000
10	0.074	0.158	0.325	0.554	0.773	0.915	0.978	0.996	1.000	1.000	1.000	1.000	1.000	1.000
11	0.076	0.172	0.359	0.605	0.821	0.943	0.988	0.998	1.000	1.000	1.000	1.000	1.000	1.000
12	0.079	0.186	0.393	0.653	0.860	0.962	0.994	0.999	1.000	1.000	1.000	1.000	1.000	1.000
13	0.082	0.200	0.427	0.696	0.891	0.975	0.997	1.000	1.000	1.000	1.000	1.000	1.000	1.000
14	0.085	0.215	0.460	0.736	0.917	0.984	0.998	1.000	1.000	1.000	1.000	1.000	1.000	1.000
15	0.088	0.230	0.492	0.771	0.937	0.990	0.999	1.000	1.000	1.000	1.000	1.000	1.000	1.000
16	0.091	0.245	0.524	0.803	0.952	0.994	1.000	1.000	1.000	1.000	1.000	1.000	1.000	1.000
17	0.094	0.260	0.554	0.831	0.964	0.996	1.000	1.000	1.000	1.000	1.000	1.000	1.000	1.000
18	0.097	0.275	0.584	0.856	0.973	0.998	1.000	1.000	1.000	1.000	1.000	1.000	1.000	1.000
19	0.100	0.290	0.612	0.877	0.980	0.999	1.000	1.000	1.000	1.000	1.000	1.000	1.000	1.000
20	0.103	0.305	0.639	0.896	0.986	0.999	1.000	1.000	1.000	1.000	1.000	1.000	1.000	1.000
21	0.106	0.321	0.664	0.912	0.989	0.999	1.000	1.000	1.000	1.000	1.000	1.000	1.000	1.000
22	0.109	0.336	0.689	0.926	0.992	1.000	1.000	1.000	1.000	1.000	1.000	1.000	1.000	1.000
23	0.112	0.351	0.712	0.938	0.994	1.000	1.000	1.000	1.000	1.000	1.000	1.000	1.000	1.000
24	0.115	0.366	0.734	0.948	0.996	1.000	1.000	1.000	1.000	1.000	1.000	1.000	1.000	1.000
25	0.118	0.382	0.754	0.957	0.997	1.000	1.000	1.000	1.000	1.000	1.000	1.000	1.000	1.000
26	0.121	0.397	0.773	0.964	0.998	1.000	1.000	1.000	1.000	1.000	1.000	1.000	1.000	1.000
27	0.124	0.412	0.792	0.970	0.999	1.000	1.000	1.000	1.000	1.000	1.000	1.000	1.000	1.000
28	0.128	0.427	0.809	0.976	0.999	1.000	1.000	1.000	1.000	1.000	1.000	1.000	1.000	1.000
29	0.131	0.442	0.824	0.980	0.999	1.000	1.000	1.000	1.000	1.000	1.000	1.000	1.000	1.000
30	0.134	0.456	0.839	0.983	0.999	1.000	1.000	1.000	1.000	1.000	1.000	1.000	1.000	1.000
31	0.138	0.471	0.853	0.986	1.000	1.000	1.000	1.000	1.000	1.000	1.000	1.000	1.000	1.000
32	0.141	0.485	0.865	0.989	1.000	1.000	1.000	1.000	1.000	1.000	1.000	1.000	1.000	1.000
33	0.144	0.499	0.877	0.991	1.000	1.000	1.000	1.000	1.000	1.000	1.000	1.000	1.000	1.000
34	0.148	0.513	0.888	0.993	1.000	1.000	1.000	1.000	1.000	1.000	1.000	1.000	1.000	1.000
35	0.151	0.527	0.898	0.994	1.000	1.000	1.000	1.000	1.000	1.000	1.000	1.000	1.000	1.000
36	0.154	0.541	0.907	0.995	1.000	1.000	1.000	1.000	1.000	1.000	1.000	1.000	1.000	1.000
37	0.158	0.554	0.916	0.996	1.000	1.000	1.000	1.000	1.000	1.000	1.000	1.000	1.000	1.000
38	0.161	0.567	0.924	0.997	1.000	1.000	1.000	1.000	1.000	1.000	1.000	1.000	1.000	1.000
39	0.165	0.580	0.931	0.997	1.000	1.000	1.000	1.000	1.000	1.000	1.000	1.000	1.000	1.000
40	0.168	0.593	0.937	0.998	1.000	1.000	1.000	1.000	1.000	1.000	1.000	1.000	1.000	1.000
41	0.172	0.605	0.943	0.998	1.000	1.000	1.000	1.000	1.000	1.000	1.000	1.000	1.000	1.000
42	0.175	0.617	0.949	0.999	1.000	1.000	1.000	1.000	1.000	1.000	1.000	1.000	1.000	1.000
43	0.179	0.629	0.954	0.999	1.000	1.000	1.000	1.000	1.000	1.000	1.000	1.000	1.000	1.000
44	0.182	0.641	0.958	0.999	1.000	1.000	1.000	1.000	1.000	1.000	1.000	1.000	1.000	1.000
45	0.186	0.653	0.962	0.999	1.000	1.000	1.000	1.000	1.000	1.000	1.000	1.000	1.000	1.000
46	0.189	0.664	0.966	0.999	1.000	1.000	1.000	1.000	1.000	1.000	1.000	1.000	1.000	1.000
47	0.193	0.675	0.970	1.000	1.000	1.000	1.000	1.000	1.000	1.000	1.000	1.000	1.000	1.000
48	0.196	0.686	0.973	1.000	1.000	1.000	1.000	1.000	1.000	1.000	1.000	1.000	1.000	1.000
49	0.200	0.696	0.975	1.000	1.000	1.000	1.000	1.000	1.000	1.000	1.000	1.000	1.000	1.000
50	0.204	0.706	0.978	1.000	1.000	1.000	1.000	1.000	1.000	1.000	1.000	1.000	1.000	1.000

Table 6 (continued)

Power function for one-factor analysis of variance F test ($m = 6$)

$\Pr(G > F_{0.05;(m-1,mn-m)})$, where G is noncentral $F_{m-1,mn-m}$ with $\lambda = mn\delta^2/2$

	δ													
n	0.100	0.200	0.300	0.400	0.500	0.600	0.700	0.800	0.900	1.000	1.500	2.000	2.500	3.000
2	0.053	0.061	0.074	0.095	0.124	0.161	0.207	0.262	0.325	0.394	0.750	0.946	0.994	1.000
3	0.055	0.072	0.103	0.152	0.221	0.311	0.418	0.534	0.649	0.752	0.986	1.000	1.000	1.000
4	0.058	0.084	0.135	0.215	0.327	0.465	0.610	0.744	0.850	0.923	1.000	1.000	1.000	1.000
5	0.061	0.097	0.168	0.282	0.434	0.603	0.758	0.874	0.945	0.980	1.000	1.000	1.000	1.000
6	0.064	0.110	0.203	0.350	0.535	0.718	0.859	0.943	0.982	0.995	1.000	1.000	1.000	1.000
7	0.066	0.124	0.240	0.418	0.626	0.806	0.922	0.976	0.995	0.999	1.000	1.000	1.000	1.000
8	0.069	0.138	0.278	0.484	0.704	0.871	0.958	0.990	0.998	1.000	1.000	1.000	1.000	1.000
9	0.072	0.153	0.316	0.547	0.771	0.916	0.979	0.996	1.000	1.000	1.000	1.000	1.000	1.000
10	0.075	0.168	0.355	0.605	0.825	0.947	0.989	0.999	1.000	1.000	1.000	1.000	1.000	1.000
11	0.078	0.183	0.393	0.659	0.868	0.967	0.995	1.000	1.000	1.000	1.000	1.000	1.000	1.000
12	0.081	0.199	0.431	0.707	0.902	0.980	0.998	1.000	1.000	1.000	1.000	1.000	1.000	1.000
13	0.084	0.215	0.468	0.751	0.928	0.988	0.999	1.000	1.000	1.000	1.000	1.000	1.000	1.000
14	0.087	0.231	0.504	0.789	0.948	0.993	1.000	1.000	1.000	1.000	1.000	1.000	1.000	1.000
15	0.090	0.248	0.539	0.823	0.962	0.996	1.000	1.000	1.000	1.000	1.000	1.000	1.000	1.000
16	0.093	0.265	0.572	0.852	0.973	0.998	1.000	1.000	1.000	1.000	1.000	1.000	1.000	1.000
17	0.097	0.282	0.605	0.877	0.981	0.999	1.000	1.000	1.000	1.000	1.000	1.000	1.000	1.000
18	0.100	0.298	0.635	0.898	0.987	0.999	1.000	1.000	1.000	1.000	1.000	1.000	1.000	1.000
19	0.103	0.315	0.665	0.916	0.991	1.000	1.000	1.000	1.000	1.000	1.000	1.000	1.000	1.000
20	0.106	0.333	0.692	0.931	0.994	1.000	1.000	1.000	1.000	1.000	1.000	1.000	1.000	1.000
21	0.110	0.350	0.718	0.944	0.996	1.000	1.000	1.000	1.000	1.000	1.000	1.000	1.000	1.000
22	0.113	0.367	0.742	0.955	0.997	1.000	1.000	1.000	1.000	1.000	1.000	1.000	1.000	1.000
23	0.117	0.384	0.765	0.963	0.998	1.000	1.000	1.000	1.000	1.000	1.000	1.000	1.000	1.000
24	0.120	0.400	0.786	0.970	0.999	1.000	1.000	1.000	1.000	1.000	1.000	1.000	1.000	1.000
25	0.123	0.417	0.806	0.976	0.999	1.000	1.000	1.000	1.000	1.000	1.000	1.000	1.000	1.000
26	0.127	0.434	0.824	0.981	0.999	1.000	1.000	1.000	1.000	1.000	1.000	1.000	1.000	1.000
27	0.131	0.450	0.841	0.985	1.000	1.000	1.000	1.000	1.000	1.000	1.000	1.000	1.000	1.000
28	0.134	0.467	0.856	0.988	1.000	1.000	1.000	1.000	1.000	1.000	1.000	1.000	1.000	1.000
29	0.138	0.483	0.870	0.991	1.000	1.000	1.000	1.000	1.000	1.000	1.000	1.000	1.000	1.000
30	0.141	0.499	0.883	0.993	1.000	1.000	1.000	1.000	1.000	1.000	1.000	1.000	1.000	1.000
31	0.145	0.515	0.895	0.994	1.000	1.000	1.000	1.000	1.000	1.000	1.000	1.000	1.000	1.000
32	0.149	0.530	0.906	0.995	1.000	1.000	1.000	1.000	1.000	1.000	1.000	1.000	1.000	1.000
33	0.152	0.546	0.916	0.996	1.000	1.000	1.000	1.000	1.000	1.000	1.000	1.000	1.000	1.000
34	0.156	0.561	0.925	0.997	1.000	1.000	1.000	1.000	1.000	1.000	1.000	1.000	1.000	1.000
35	0.160	0.575	0.933	0.998	1.000	1.000	1.000	1.000	1.000	1.000	1.000	1.000	1.000	1.000
36	0.164	0.590	0.940	0.998	1.000	1.000	1.000	1.000	1.000	1.000	1.000	1.000	1.000	1.000
37	0.167	0.604	0.947	0.999	1.000	1.000	1.000	1.000	1.000	1.000	1.000	1.000	1.000	1.000
38	0.171	0.618	0.952	0.999	1.000	1.000	1.000	1.000	1.000	1.000	1.000	1.000	1.000	1.000
39	0.175	0.631	0.958	0.999	1.000	1.000	1.000	1.000	1.000	1.000	1.000	1.000	1.000	1.000
40	0.179	0.645	0.963	0.999	1.000	1.000	1.000	1.000	1.000	1.000	1.000	1.000	1.000	1.000
41	0.183	0.657	0.967	1.000	1.000	1.000	1.000	1.000	1.000	1.000	1.000	1.000	1.000	1.000
42	0.187	0.670	0.971	1.000	1.000	1.000	1.000	1.000	1.000	1.000	1.000	1.000	1.000	1.000
43	0.191	0.682	0.974	1.000	1.000	1.000	1.000	1.000	1.000	1.000	1.000	1.000	1.000	1.000
44	0.195	0.694	0.977	1.000	1.000	1.000	1.000	1.000	1.000	1.000	1.000	1.000	1.000	1.000
45	0.198	0.706	0.980	1.000	1.000	1.000	1.000	1.000	1.000	1.000	1.000	1.000	1.000	1.000
46	0.202	0.717	0.982	1.000	1.000	1.000	1.000	1.000	1.000	1.000	1.000	1.000	1.000	1.000
47	0.206	0.728	0.984	1.000	1.000	1.000	1.000	1.000	1.000	1.000	1.000	1.000	1.000	1.000
48	0.210	0.739	0.986	1.000	1.000	1.000	1.000	1.000	1.000	1.000	1.000	1.000	1.000	1.000
49	0.215	0.749	0.988	1.000	1.000	1.000	1.000	1.000	1.000	1.000	1.000	1.000	1.000	1.000
50	0.219	0.760	0.989	1.000	1.000	1.000	1.000	1.000	1.000	1.000	1.000	1.000	1.000	1.000

Table 6 (continued)
Power function for one-factor analysis of variance F test ($m = 7$)
$\Pr(G > F_{0.05;(m-1,mn-m)})$, where G is noncentral $F_{m-1,mn-m}$ with $\lambda = mn\delta^2/2$

n	δ 0.100	0.200	0.300	0.400	0.500	0.600	0.700	0.800	0.900	1.000	1.500	2.000	2.500	3.000
2	0.053	0.061	0.077	0.100	0.132	0.174	0.226	0.290	0.362	0.440	0.812	0.973	0.998	1.000
3	0.056	0.074	0.108	0.162	0.239	0.340	0.458	0.584	0.703	0.805	0.994	1.000	1.000	1.000
4	0.058	0.087	0.142	0.231	0.356	0.507	0.661	0.794	0.891	0.950	1.000	1.000	1.000	1.000
5	0.061	0.100	0.179	0.305	0.473	0.652	0.806	0.910	0.966	0.990	1.000	1.000	1.000	1.000
6	0.064	0.115	0.218	0.380	0.580	0.766	0.897	0.965	0.991	0.998	1.000	1.000	1.000	1.000
7	0.067	0.130	0.258	0.454	0.674	0.849	0.948	0.987	0.998	1.000	1.000	1.000	1.000	1.000
8	0.070	0.145	0.300	0.525	0.752	0.906	0.976	0.996	0.999	1.000	1.000	1.000	1.000	1.000
9	0.073	0.161	0.342	0.591	0.816	0.944	0.989	0.999	1.000	1.000	1.000	1.000	1.000	1.000
10	0.077	0.178	0.384	0.652	0.866	0.967	0.995	1.000	1.000	1.000	1.000	1.000	1.000	1.000
11	0.080	0.195	0.426	0.706	0.904	0.981	0.998	1.000	1.000	1.000	1.000	1.000	1.000	1.000
12	0.083	0.212	0.467	0.754	0.932	0.990	0.999	1.000	1.000	1.000	1.000	1.000	1.000	1.000
13	0.086	0.230	0.507	0.796	0.953	0.994	1.000	1.000	1.000	1.000	1.000	1.000	1.000	1.000
14	0.090	0.248	0.545	0.833	0.967	0.997	1.000	1.000	1.000	1.000	1.000	1.000	1.000	1.000
15	0.093	0.266	0.582	0.863	0.978	0.998	1.000	1.000	1.000	1.000	1.000	1.000	1.000	1.000
16	0.096	0.285	0.617	0.889	0.985	0.999	1.000	1.000	1.000	1.000	1.000	1.000	1.000	1.000
17	0.100	0.303	0.651	0.911	0.990	1.000	1.000	1.000	1.000	1.000	1.000	1.000	1.000	1.000
18	0.103	0.322	0.682	0.929	0.994	1.000	1.000	1.000	1.000	1.000	1.000	1.000	1.000	1.000
19	0.107	0.340	0.712	0.943	0.996	1.000	1.000	1.000	1.000	1.000	1.000	1.000	1.000	1.000
20	0.110	0.359	0.739	0.955	0.997	1.000	1.000	1.000	1.000	1.000	1.000	1.000	1.000	1.000
21	0.114	0.378	0.764	0.965	0.998	1.000	1.000	1.000	1.000	1.000	1.000	1.000	1.000	1.000
22	0.118	0.397	0.788	0.972	0.999	1.000	1.000	1.000	1.000	1.000	1.000	1.000	1.000	1.000
23	0.121	0.415	0.810	0.978	0.999	1.000	1.000	1.000	1.000	1.000	1.000	1.000	1.000	1.000
24	0.125	0.433	0.830	0.983	1.000	1.000	1.000	1.000	1.000	1.000	1.000	1.000	1.000	1.000
25	0.129	0.452	0.848	0.987	1.000	1.000	1.000	1.000	1.000	1.000	1.000	1.000	1.000	1.000
26	0.133	0.470	0.864	0.990	1.000	1.000	1.000	1.000	1.000	1.000	1.000	1.000	1.000	1.000
27	0.137	0.488	0.879	0.992	1.000	1.000	1.000	1.000	1.000	1.000	1.000	1.000	1.000	1.000
28	0.140	0.505	0.893	0.994	1.000	1.000	1.000	1.000	1.000	1.000	1.000	1.000	1.000	1.000
29	0.144	0.522	0.905	0.996	1.000	1.000	1.000	1.000	1.000	1.000	1.000	1.000	1.000	1.000
30	0.148	0.539	0.916	0.997	1.000	1.000	1.000	1.000	1.000	1.000	1.000	1.000	1.000	1.000
31	0.152	0.556	0.926	0.998	1.000	1.000	1.000	1.000	1.000	1.000	1.000	1.000	1.000	1.000
32	0.156	0.573	0.935	0.998	1.000	1.000	1.000	1.000	1.000	1.000	1.000	1.000	1.000	1.000
33	0.160	0.589	0.943	0.999	1.000	1.000	1.000	1.000	1.000	1.000	1.000	1.000	1.000	1.000
34	0.164	0.604	0.950	0.999	1.000	1.000	1.000	1.000	1.000	1.000	1.000	1.000	1.000	1.000
35	0.169	0.620	0.956	0.999	1.000	1.000	1.000	1.000	1.000	1.000	1.000	1.000	1.000	1.000
36	0.173	0.635	0.962	0.999	1.000	1.000	1.000	1.000	1.000	1.000	1.000	1.000	1.000	1.000
37	0.177	0.649	0.967	1.000	1.000	1.000	1.000	1.000	1.000	1.000	1.000	1.000	1.000	1.000
38	0.181	0.664	0.971	1.000	1.000	1.000	1.000	1.000	1.000	1.000	1.000	1.000	1.000	1.000
39	0.185	0.677	0.975	1.000	1.000	1.000	1.000	1.000	1.000	1.000	1.000	1.000	1.000	1.000
40	0.190	0.691	0.978	1.000	1.000	1.000	1.000	1.000	1.000	1.000	1.000	1.000	1.000	1.000
41	0.194	0.704	0.981	1.000	1.000	1.000	1.000	1.000	1.000	1.000	1.000	1.000	1.000	1.000
42	0.198	0.717	0.983	1.000	1.000	1.000	1.000	1.000	1.000	1.000	1.000	1.000	1.000	1.000
43	0.202	0.729	0.986	1.000	1.000	1.000	1.000	1.000	1.000	1.000	1.000	1.000	1.000	1.000
44	0.207	0.741	0.988	1.000	1.000	1.000	1.000	1.000	1.000	1.000	1.000	1.000	1.000	1.000
45	0.211	0.752	0.989	1.000	1.000	1.000	1.000	1.000	1.000	1.000	1.000	1.000	1.000	1.000
46	0.216	0.763	0.991	1.000	1.000	1.000	1.000	1.000	1.000	1.000	1.000	1.000	1.000	1.000
47	0.220	0.774	0.992	1.000	1.000	1.000	1.000	1.000	1.000	1.000	1.000	1.000	1.000	1.000
48	0.224	0.785	0.993	1.000	1.000	1.000	1.000	1.000	1.000	1.000	1.000	1.000	1.000	1.000
49	0.229	0.795	0.994	1.000	1.000	1.000	1.000	1.000	1.000	1.000	1.000	1.000	1.000	1.000

Table 6 (continued)

Power function for one-factor analysis of variance F test $(m = 8)$

$\Pr(G > F_{0.05;(m-1,mn-m)})$, where G is noncentral $F_{m-1,mn-m}$ with $\lambda = mn\delta^2/2$

						δ								
n	0.100	0.200	0.300	0.400	0.500	0.600	0.700	0.800	0.900	1.000	1.500	2.000	2.500	3.000
2	0.053	0.062	0.079	0.104	0.139	0.187	0.246	0.317	0.398	0.484	0.861	0.986	1.000	1.000
3	0.056	0.075	0.112	0.171	0.257	0.368	0.497	0.630	0.750	0.847	0.997	1.000	1.000	1.000
4	0.059	0.089	0.149	0.247	0.385	0.547	0.706	0.835	0.922	0.969	1.000	1.000	1.000	1.000
5	0.062	0.104	0.189	0.328	0.509	0.696	0.845	0.937	0.980	0.995	1.000	1.000	1.000	1.000
6	0.065	0.119	0.232	0.409	0.621	0.807	0.925	0.979	0.996	0.999	1.000	1.000	1.000	1.000
7	0.068	0.135	0.276	0.489	0.717	0.884	0.966	0.993	0.999	1.000	1.000	1.000	1.000	1.000
8	0.071	0.152	0.321	0.564	0.793	0.933	0.986	0.998	1.000	1.000	1.000	1.000	1.000	1.000
9	0.075	0.170	0.367	0.632	0.853	0.963	0.994	0.999	1.000	1.000	1.000	1.000	1.000	1.000
10	0.078	0.188	0.413	0.694	0.898	0.980	0.998	1.000	1.000	1.000	1.000	1.000	1.000	1.000
11	0.081	0.206	0.457	0.748	0.930	0.989	0.999	1.000	1.000	1.000	1.000	1.000	1.000	1.000
12	0.085	0.225	0.501	0.795	0.953	0.995	1.000	1.000	1.000	1.000	1.000	1.000	1.000	1.000
13	0.088	0.244	0.543	0.835	0.969	0.997	1.000	1.000	1.000	1.000	1.000	1.000	1.000	1.000
14	0.092	0.264	0.584	0.868	0.980	0.999	1.000	1.000	1.000	1.000	1.000	1.000	1.000	1.000
15	0.095	0.284	0.622	0.895	0.987	0.999	1.000	1.000	1.000	1.000	1.000	1.000	1.000	1.000
16	0.099	0.304	0.658	0.918	0.992	1.000	1.000	1.000	1.000	1.000	1.000	1.000	1.000	1.000
17	0.103	0.324	0.692	0.936	0.995	1.000	1.000	1.000	1.000	1.000	1.000	1.000	1.000	1.000
18	0.107	0.345	0.724	0.950	0.997	1.000	1.000	1.000	1.000	1.000	1.000	1.000	1.000	1.000
19	0.110	0.365	0.753	0.962	0.998	1.000	1.000	1.000	1.000	1.000	1.000	1.000	1.000	1.000
20	0.114	0.385	0.780	0.971	0.999	1.000	1.000	1.000	1.000	1.000	1.000	1.000	1.000	1.000
21	0.118	0.405	0.804	0.978	0.999	1.000	1.000	1.000	1.000	1.000	1.000	1.000	1.000	1.000
22	0.122	0.425	0.826	0.983	1.000	1.000	1.000	1.000	1.000	1.000	1.000	1.000	1.000	1.000
23	0.126	0.445	0.847	0.988	1.000	1.000	1.000	1.000	1.000	1.000	1.000	1.000	1.000	1.000
24	0.130	0.465	0.865	0.991	1.000	1.000	1.000	1.000	1.000	1.000	1.000	1.000	1.000	1.000
25	0.134	0.485	0.881	0.993	1.000	1.000	1.000	1.000	1.000	1.000	1.000	1.000	1.000	1.000
26	0.138	0.504	0.896	0.995	1.000	1.000	1.000	1.000	1.000	1.000	1.000	1.000	1.000	1.000
27	0.142	0.523	0.909	0.996	1.000	1.000	1.000	1.000	1.000	1.000	1.000	1.000	1.000	1.000
28	0.147	0.541	0.921	0.997	1.000	1.000	1.000	1.000	1.000	1.000	1.000	1.000	1.000	1.000
29	0.151	0.560	0.931	0.998	1.000	1.000	1.000	1.000	1.000	1.000	1.000	1.000	1.000	1.000
30	0.155	0.577	0.940	0.999	1.000	1.000	1.000	1.000	1.000	1.000	1.000	1.000	1.000	1.000
31	0.160	0.595	0.948	0.999	1.000	1.000	1.000	1.000	1.000	1.000	1.000	1.000	1.000	1.000
32	0.164	0.612	0.955	0.999	1.000	1.000	1.000	1.000	1.000	1.000	1.000	1.000	1.000	1.000
33	0.168	0.628	0.962	0.999	1.000	1.000	1.000	1.000	1.000	1.000	1.000	1.000	1.000	1.000
34	0.173	0.645	0.967	1.000	1.000	1.000	1.000	1.000	1.000	1.000	1.000	1.000	1.000	1.000
35	0.177	0.660	0.972	1.000	1.000	1.000	1.000	1.000	1.000	1.000	1.000	1.000	1.000	1.000
36	0.182	0.676	0.976	1.000	1.000	1.000	1.000	1.000	1.000	1.000	1.000	1.000	1.000	1.000
37	0.186	0.690	0.979	1.000	1.000	1.000	1.000	1.000	1.000	1.000	1.000	1.000	1.000	1.000
38	0.191	0.705	0.982	1.000	1.000	1.000	1.000	1.000	1.000	1.000	1.000	1.000	1.000	1.000
39	0.195	0.719	0.985	1.000	1.000	1.000	1.000	1.000	1.000	1.000	1.000	1.000	1.000	1.000
40	0.200	0.732	0.987	1.000	1.000	1.000	1.000	1.000	1.000	1.000	1.000	1.000	1.000	1.000
41	0.205	0.745	0.989	1.000	1.000	1.000	1.000	1.000	1.000	1.000	1.000	1.000	1.000	1.000
42	0.209	0.757	0.991	1.000	1.000	1.000	1.000	1.000	1.000	1.000	1.000	1.000	1.000	1.000
43	0.214	0.769	0.992	1.000	1.000	1.000	1.000	1.000	1.000	1.000	1.000	1.000	1.000	1.000

Table 6 (continued)
Power function for one-factor analysis of variance F test ($m = 9$)
$\Pr(G > F_{0.05;(m-1,mn-m)})$, where G is noncentral $F_{m-1,mn-m}$ with $\lambda = mn\delta^2/2$

n	δ 0.100	0.200	0.300	0.400	0.500	0.600	0.700	0.800	0.900	1.000	1.500	2.000	2.500	3.00
2	0.053	0.063	0.081	0.108	0.147	0.199	0.265	0.344	0.433	0.526	0.898	0.994	1.000	1.00
3	0.056	0.077	0.116	0.181	0.275	0.396	0.534	0.672	0.791	0.881	0.999	1.000	1.000	1.00
4	0.059	0.091	0.156	0.263	0.412	0.584	0.746	0.869	0.944	0.980	1.000	1.000	1.000	1.00
5	0.063	0.107	0.199	0.350	0.544	0.735	0.877	0.956	0.988	0.998	1.000	1.000	1.000	1.00
6	0.066	0.123	0.245	0.438	0.660	0.842	0.946	0.987	0.998	1.000	1.000	1.000	1.000	1.00
7	0.069	0.141	0.293	0.522	0.755	0.911	0.978	0.997	1.000	1.000	1.000	1.000	1.000	1.00
8	0.072	0.159	0.342	0.600	0.828	0.952	0.992	0.999	1.000	1.000	1.000	1.000	1.000	1.00
9	0.076	0.178	0.391	0.670	0.883	0.975	0.997	1.000	1.000	1.000	1.000	1.000	1.000	1.00
10	0.079	0.197	0.440	0.732	0.923	0.988	0.999	1.000	1.000	1.000	1.000	1.000	1.000	1.00
11	0.083	0.217	0.488	0.785	0.950	0.994	1.000	1.000	1.000	1.000	1.000	1.000	1.000	1.00
12	0.087	0.238	0.534	0.829	0.968	0.997	1.000	1.000	1.000	1.000	1.000	1.000	1.000	1.00
13	0.090	0.259	0.578	0.866	0.980	0.999	1.000	1.000	1.000	1.000	1.000	1.000	1.000	1.00
14	0.094	0.280	0.620	0.896	0.988	0.999	1.000	1.000	1.000	1.000	1.000	1.000	1.000	1.00
15	0.098	0.302	0.659	0.920	0.993	1.000	1.000	1.000	1.000	1.000	1.000	1.000	1.000	1.00
16	0.102	0.323	0.696	0.939	0.996	1.000	1.000	1.000	1.000	1.000	1.000	1.000	1.000	1.00
17	0.106	0.345	0.729	0.954	0.997	1.000	1.000	1.000	1.000	1.000	1.000	1.000	1.000	1.00
18	0.110	0.367	0.760	0.966	0.999	1.000	1.000	1.000	1.000	1.000	1.000	1.000	1.000	1.00
19	0.114	0.389	0.789	0.975	0.999	1.000	1.000	1.000	1.000	1.000	1.000	1.000	1.000	1.00
20	0.118	0.410	0.814	0.981	1.000	1.000	1.000	1.000	1.000	1.000	1.000	1.000	1.000	1.00
21	0.122	0.432	0.838	0.986	1.000	1.000	1.000	1.000	1.000	1.000	1.000	1.000	1.000	1.00
22	0.126	0.453	0.858	0.990	1.000	1.000	1.000	1.000	1.000	1.000	1.000	1.000	1.000	1.00
23	0.131	0.475	0.877	0.993	1.000	1.000	1.000	1.000	1.000	1.000	1.000	1.000	1.000	1.000
24	0.135	0.495	0.893	0.995	1.000	1.000	1.000	1.000	1.000	1.000	1.000	1.000	1.000	1.000
25	0.139	0.516	0.908	0.996	1.000	1.000	1.000	1.000	1.000	1.000	1.000	1.000	1.000	1.000
26	0.144	0.536	0.921	0.997	1.000	1.000	1.000	1.000	1.000	1.000	1.000	1.000	1.000	1.000
27	0.148	0.556	0.932	0.998	1.000	1.000	1.000	1.000	1.000	1.000	1.000	1.000	1.000	1.000
28	0.153	0.575	0.942	0.999	1.000	1.000	1.000	1.000	1.000	1.000	1.000	1.000	1.000	1.000
29	0.157	0.594	0.950	0.999	1.000	1.000	1.000	1.000	1.000	1.000	1.000	1.000	1.000	1.000
30	0.162	0.613	0.958	0.999	1.000	1.000	1.000	1.000	1.000	1.000	1.000	1.000	1.000	1.000
31	0.167	0.631	0.964	1.000	1.000	1.000	1.000	1.000	1.000	1.000	1.000	1.000	1.000	1.000
32	0.171	0.648	0.970	1.000	1.000	1.000	1.000	1.000	1.000	1.000	1.000	1.000	1.000	1.000
33	0.176	0.665	0.974	1.000	1.000	1.000	1.000	1.000	1.000	1.000	1.000	1.000	1.000	1.000
34	0.181	0.681	0.978	1.000	1.000	1.000	1.000	1.000	1.000	1.000	1.000	1.000	1.000	1.000
35	0.186	0.697	0.982	1.000	1.000	1.000	1.000	1.000	1.000	1.000	1.000	1.000	1.000	1.000
36	0.191	0.713	0.985	1.000	1.000	1.000	1.000	1.000	1.000	1.000	1.000	1.000	1.000	1.000
37	0.196	0.727	0.987	1.000	1.000	1.000	1.000	1.000	1.000	1.000	1.000	1.000	1.000	1.000
38	0.201	0.742	0.989	1.000	1.000	1.000	1.000	1.000	1.000	1.000	1.000	1.000	1.000	1.000

Bibliography

[1] Allen, Arnold O. *Probability, Statistics, and Queueing Theory*, 2nd Ed., Academic Press, San Diego, CA, 1990.

[2] Anderson, T. W. *An Introduction to Multivariate Statistics*, John Wiley & Sons, New York, 1958.

[3] Applebaum, David. *Probability and Information*, Cambridge University Press, Cambridge, 1996.

[4] Banks, Jerry; Carson, John S. *Discrete-Event System Simulation*, Prentice Hall, Upper Saddle River, NJ, 1984.

[5] Bell, E. T. *Men of Mathematics*, Simon & Schuster, New York, 1937.

[6] Best, Michael J.; Ritter, Klaus. *Linear Programming*, Prentice Hall, Upper Saddle River, NJ, 1985.

[7] Box, Joan Fisher. *R. A. Fisher: The Life of a Scientist*, John Wiley & Sons, New York, 1978.

[8] Boyer, Carl B. *A History of Mathematics*, John Wiley & Sons, New York, 1968.

[9] Brassard, Gilles; Bratley, Paul. *Fundamentals of Algorithmics*, Prentice Hall, Upper Saddle River, NJ, 1996.

[10] Carlson, William; Thorne, Betty. *Applied Statistical Methods for Business, Economics, and the Social Sciences*, Prentice Hall, Upper Saddle River, NJ, 1997.

[11] Cohen, Daniel I. A. *Basic Techniques of Combinatorial Theory*, John Wiley & Sons, New York, 1978.

[12] Constantine, Gregory M. *Combinatorial Theory and Statistical Design*, John Wiley & Sons, New York, 1987.

[13] Corman, Thomas H.; Leiserson, Charles E.; Rivest, Ronald L. *Introduction to Algorithms*, MIT Press and McGraw-Hill, New York, 1990.

[14] Cox, D. R.; Miller, H. D. *The Theory of Stochastic Processes*, John Wiley & Sons, New York, 1965.

[15] Czitrom, Veronica; Spagon, Patrick D. *Statistical Case Studies for Industrial Process Improvement*, American Statistical Association and the Society for Industrial and Applied Mathematics, Philadelphia, 1997.

[16] David, F. N. *Games, Gods, and Gambling*, Charles Griffin & Company, London, 1962.

[17] Davis, Martin. *The Universal Computer*, W.W. Norton & Company, New York, 2000.

[18] Earman, John. *Bayes or Bust! A Critical Examination of Bayesian Confirmation Theory*, MIT Press, Cambridge, MA, 1992.

[19] Feller, William. *An Introduction to Probability Theory and Its Applications*, Volume I, 3rd Ed., revised printing, John Wiley & Sons, New York, 1968.

[20] Feller, William. *An Introduction to Probability Theory and Its Applications*, Volume II, 2nd Ed., John Wiley & Sons, New York, 1971.

[21] Fraser, G. A. S. *Probability and Statistics: Theory and Applications*, Wadsworth Publishing Co., Belmont, CA, 1976.

[22] Fraser, Steven (editor). *The Bell Curve Wars: Race, Intelligence, and the Future of America*, Basic Books, New York, 1995.

[23] Fristedt, Bert; Gray, Lawrence. *A Modern Approach to Probability Theory*, Birkhäuser, Boston, 1997.

[24] Games, Paul A.; Klare, George R. *Elementary Statistics*, McGraw-Hill, New York, 1967.

[25] Gentle, James E. *Random Number Generation and Monte Carlo Methods*, Springer-Verlag, New York, 1998.

[26] Gillies, D. A. *An Objective Theory of Probability*, Methuen & Company, London, 1973.

[27] Graham, Ronald L.; Knuth, Donald E.; Patashnik, Oren. *Concrete Mathematics*, Addison-Wesley, Reading, MA, 1989.

[28] Grant, Gregory R.; Ewens, Warren J. *Statistical Methods in Bioinformatics: An Introduction*, Springer-Verlag, New York, 2001.

[29] Guenther, William C. *Analysis of Variance*, Prentice Hall, Upper Saddle River, NJ, 1964.

[30] Hacking, Ian. *The Emergence of Probability*, Cambridge University Press, Cambridge, 1975.

[31] Hacking, Ian. *The Taming of Chance*, Cambridge University Press, Cambridge, 1990.

[32] Halmos, P. R. *Measure Theory*, Van Nostrand, New York, 1950.

[33] Hardy, G. H. *A Course in Pure Mathematics*, Cambridge University Press, Cambridge, 1908.

[34] Hastie, Trevor; Tibshirani, Robert; Friedman, Jerome. *The Elements of Statistical Learning*, Springer-Verlag, New York, 2001.

[35] Herrnstein, Richard J. *The Bell Curve: Intelligence and Class Structure in American Life*, Free Press, New York, 1994.

[36] Herstein, I. *Topics in Algebra*, Blaisdell Publishing Company, Waltham, MA, 1964.

[37] Hildebrand, F. B. *Advanced Calculus for Applications*, 2nd Ed., Prentice Hall, Upper Saddle River, NJ, 1976.

[38] Johnson, James L. *Database: Models, Languages, Design*, Oxford University Press, New York, 1997.

[39] Johnsonbaugh, Richard. *Discrete Mathematics*, Addison-Wesley, Reading, MA, 1996.

[40] Kadane, Joseph B.; Schum, David A. *A Probabilistic Analysis of the Sacco and Vanzetti Evidence*, John Wiley & Sons, New York, 1996.

[41] Karian, Zaven A.; Dudewicz, Edward J. *Modern Statistical Systems and GPSS Simulation*, 2nd Ed., CRC Press, Boca Raton, FL, 1999.

[42] Karlin, Samuel; Taylor, Howard M. *A First Course in Stochastic Processes*, Academic Press, San Diego, CA, 1975.

[43] Katz, Victor J. *A History of Mathematics*, Harper Collins, New York, 1993.

[44] Kelton, W. David; Sadowski, Randall P.; Sadowski, Deborah A. *Simulation with Arena*, McGraw-Hill, New York, 1998.

[45] Kline, Morris. *Mathematical Thought from Ancient to Modern Times*, Oxford University Press, Oxford, 1972.

[46] Knuth, Donald E. *The Art of Computer Programming*, Volume I: *Fundamental Algorithms*, 2nd Ed., Addison-Wesley, Reading, MA, 1973.

[47] Knuth, Donald E. *The Art of Computer Programming*, Volume II: *Seminumerical Algorithms*, 2nd Ed., Addison-Wesley, Reading, MA, 1981.

[48] Kolmogorov, A. N. *Grundbegriffe der Wahrscheinlichkeitsrechnung*, 1933; English translation by N. Morrison, *Foundations of the Theory of Probability*, Chelsea Publishing Company, New York, 1956.

[49] Kurz, Ludwik; Benteftifa, M. Hafed. *Analysis of Variance in Statistical Image Processing*, Cambridge University Press, Cambridge, 1997.

[50] Lehmer, D. H. Mathematical methods in large-scale computing units, *Proceedings of the Second Symposium on Large Scale Digital Computing Machinery*, Harvard University Press, Cambridge, MA, 1951, pp. 141–146.

[51] Lindgren, B. W. *Statistical Theory*, 2nd Ed., Macmillan, New York, 1962.

[52] Maistrov, L. E. *Probability Theory: A Historical Sketch*, 1967; English translation by Samuel Kotz, Academic Press, New York, 1974.

[53] Mosteller, Frederick. *Fifty Challenging Problems in Probability with Solutions*, Addison-Wesley, Reading, MA, 1965; republished by Dover Publications, New York, 1987.

[54] Mott, Joe L.; Kandel, Abraham; Baker, Theodore P. *Discrete Mathematics for Computer Scientists*, Reston, Reston, VA, 1983.

[55] Nance, Richard E. A history of discrete event simulation programming languages, *ACM SIGPLAN Notices*, Vol. 28, No. 3, March 1993, pp. 149–169.

[56] Nelson, Randolph. *Probability, Stochastic Processes, and Queuing Theory: The Mathematics of Computer Performance Modeling*, Springer-Verlag, New York, 1995.

[57] Noble, Ben. *Applied Linear Algebra*, Prentice Hall, Upper Saddle River, NJ, 1969.

[58] Ore, Oystein. *Cardano: The Gambling Scholar*, Princeton University Press, Princeton, NJ, 1953.

[59] Papoulis, Athanasios. *Probability, Random Variables, and Stochastic Processes*, McGraw-Hill, New York, 1965.

[60] Park, Stephen K.; Miller, Keith W. Random number generators: Good ones are hard to find, *Communications of the ACM*, Vol. 31, 1988, pp. 1192–1201.

[61] Parzen, Emanuel. *Modern Probability Theory and Its Applications*, John Wiley & Sons, New York, 1960.

[62] Parzen, Emanuel. *Stochastic Processes*, Holden-Day, San Francisco, 1962.

[63] Paulos, John Allen. *Once upon a Number: The Hidden Mathematical Logic of Stories*, Basic Books, New York, 1998.

[64] Pearson, E. S.; Kendall, M. G. (editors). *Studies in the History of Statistics and Probability*, Charles Griffin & Co., London, 1970.

[65] Pegden, C. Dennis; Shannon, Robert E.; Sadowski, Randall P. *Introduction to Simulation using SIMAN*, 2nd Ed., McGraw-Hill, New York, 1995.

[66] Porter, Theodore M. *The Rise of Statistical Thinking: 1820–1900*, Princeton University Press, Princeton, NJ, 1986.

[67] Pratt, John W.; Raiffa, Howard; Schlaifer, Robert. *Introduction to Statistical Decision Theory*, MIT Press, Cambridge, MA, 1995.

[68] Rawlings, Gregory J. E. *Compared to What? An Introduction to the Analysis of Algorithms*, Computer Science Press, New York, 1992.

[69] Riordan, John. *An Introduction to Combinatorial Analysis*, John Wiley & Sons, New York, 1958.

[70] Roberts, J. M. *History of the World*, Oxford University Press, Oxford, 1993.

[71] Rockafellar, R. Tyrrell. *Convex Analysis*, Princeton University Press, Princeton, NJ, 1970.

[72] Ross, Sheldon M. *Introduction to Probability Models*, 5th Ed., Academic Press, San Diego, CA, 1993.

[73] Ross, Sheldon M. *Simulation*, 2nd Ed., Academic Press, San Diego, CA, 1997.

[74] Royden, H. L. *Real Analysis*, 3rd Ed., Prentice Hall, Upper Saddle River, NJ, 1988.

[75] Rudin, Walter. *Principles of Mathematical Analysis*, McGraw-Hill, New York, 1964.

[76] Rudin, Walter, *Real and Complex Analysis*, McGraw-Hill, New York, 1966.

[77] Scheffé, Henry. *The Analysis of Variance*, John Wiley & Sons, New York, 1959.

[78] Schriber, Thomas J. *An Introduction to Simulation using GPSS/H*, John Wiley & Sons, New York, 1991.

[79] Schrijver, Alexander. *Theory of Linear and Integer Programming*, John Wiley & Sons, New York, 1986.

[80] Shapiro, Barbara J. *Probability and Uncertainty in Seventeenth Century England*, Princeton University Press, Princeton, NJ, 1983.

[81] Skvarcius, R.; Robinson, W. B. *Discrete Mathematics with Computer Science Applications*, Benjamin-Cummings, San Francisco, 1986.

[82] Sposito, V. A. *Linear and Nonlinear Programming*, Iowa State University Press, Ames, IA, 1975.

[83] Stanat, Donald F.; McAllister, David F. *Discrete Mathematics in Computer Science*, Prentice Hall, Upper Saddle River, NJ, 1977.

[84] Stigler, Stephen M. *The History of Statistics: The Measurement of Uncertainty before 1900*, Harvard University Press, Cambridge, MA, 1986.

[85] Taylor, Angus E.; Mann, W. Robert. *Advanced Calculus*, 2nd Ed., John Wiley & Sons, New York, 1972.

[86] Taylor, Howard M.; Karlin, Samuel. *An Introduction to Stochastic Modeling*, 3rd Ed., Academic Press, San Diego, CA, 1998.

[87] Todhunter, *A History of the Mathematical Theory of Probability from the time of Pascal to that of Laplace*, Chelsea Publishing Company, Cambridge, 1865; New York, 1965.

[88] Vilenkin, Naum Ya. *Combinatorics*, Academic Press, San Diego, CA, 1971 (English translation of the Russian text *Kombinatorika*).

[89] von Plato, Jan. *Creating Modern Probability*, Cambridge University Press, Cambridge, 1994.

[90] Weatherford, Roy. *Philosophical Foundations of Probability Theory*, Routledge & Kegan Paul, New York, 1982.

[91] Weiss, Lionel. *Statistical Decision Theory*, McGraw-Hill, New York, 1961.

[92] Wigmore, J. *The Science of Proof: As Given by Logic, Psychology, and General Experience and Illustrated in Judicial Trials*, 3rd Ed., Little, Brown, Boston, 1937.

[93] Yandell, Benjamin. *The Honors Class: Hilbert's Problems and their Solvers*, A. K. Peters, Natick, MA, 2002.

[94] Zehna, Peter W. *Probability Distributions and Statistics*. Allyn & Bacon, Needham Heights, MA, 1970.

Index